Student's Solutions Manual to Accompany

Algebra and Trigonometry
Fourth Edition

David Cohen
Department of Mathematics
University of California
Los Angeles, California

Prepared by
Ross Rueger
Department of Mathematics
College of the Sequoias
Visalia, California

West Publishing Company
Minneapolis/St. Paul   New York   Los Angeles   San Francisco

WEST'S COMMITMENT TO THE ENVIRONMENT

In 1906, West Publishing Company began recycling materials left over from the production of books. This began a tradition of efficient and responsible use of resources. Today, 100% of our legal bound volumes are printed on acid-free, recycled paper consisting of 50% new fibers. West recycles nearly 27,700,000 pounds of scrap paper annually—the equivalent of 229,300 trees. Since the 1960s, West has devised ways to capture and recycle waste inks, solvents, oils, and vapors created in the printing process. We also recycle plastics of all kinds, wood, glass, corrugated cardboard, and batteries, and have eliminated the use of polystyrene book packaging. We at West are proud of the longevity and the scope of our commitment to the environment.

West pocket parts and advance sheets are printed on recyclable paper and can be collected and recycled with newspapers. Staples do not have to be removed. Bound volumes can be recycled after removing the cover.

Production, Prepress, Printing and Binding by West Publishing Company.

 *TEXT IS PRINTED ON 10% POST CONSUMER RECYCLED PAPER*

Contents

Appendix

Preface

This <u>Student's Solutions Manual</u> contains complete solutions to all odd-numbered regular and graphing utility exercises of <u>Algebra and Trigonometry</u> by David Cohen. It also contains complete solutions to odd and even exercises for each chapter test. I have attempted to format solutions for readability and accuracy, and apologize to you for any errors that you may encounter. If you have any comments, suggestions, error corrections, or alternative solutions please feel free to drop me a note.

Please use this manual with some degree of caution. Be sure that you have attempted a solution, and re-attempted it, before you look it up in this manual. Mathematics can only be learned by **doing**, and not by observing! As you use this manual, do not just read the solution but work it along with the manual, using my solution to check your work. If you use this manual in that fashion then it should be helpful to you in your studying.

I would like to thank a number of people for their assistance in preparing this manual. Thanks go to Peter Marshall, Jane Bass, and Chris Hurney at West Educational Publishing for their valuable assistance and support. Special thanks go to Charles Heuer of Concordia College for his meticulous error-checking of my solutions, as well as numerous suggestions for improvement of solutions.

I wish to express my deepest appreciation to David Cohen for continuing his tradition of excellence with this textbook. His addition of graphing utility exercises in this fourth edition will help you in conceptualizing the notion of a function. That notion is the heart of calculus and mathematics in general.

for their support: my father and mother
for her patience: carolyn

Ross Rueger
College of the Sequoias
915 South Mooney Boulevard
Visalia, CA 93277

August, 1996

Chapter One
Fundamental Concepts

1.1 Sets of Real Numbers

1. (a) natural number, integer, rational number
    (b) integer, rational number

3. (a) rational number
    (b) irrational number

5. (a) natural number, integer, rational number
    (b) rational number

7. (a) rational number
    (b) rational number

9. irrational number

11. irrational number

13. Since $\frac{11}{4} \approx 2.75$, we have the following graph:

15.   Since $1 + \sqrt{2} \approx 2.4$, we have the following graph:

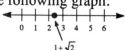

17.   Since $\sqrt{2} - 1 \approx 0.4$, we have the following graph:

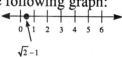

19.   Since $\sqrt{2} + \sqrt{3} \approx 3.1$, we have the following graph:

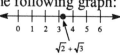

21.   Since $\dfrac{1 + \sqrt{2}}{2} \approx \dfrac{2.4}{2} \approx 1.2$, we have the following graph:

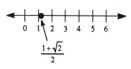

23.   Draw the graph:

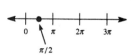

25.   Draw the graph:

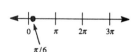

27.   Draw the graph:

29.   Draw the graph:

31.   Since $1 \approx \frac{\pi}{3}$, draw the following graph:

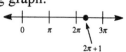

33. Since $\dfrac{\sqrt{139}-5}{3} \approx 2.26$, draw the following graph:

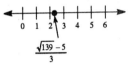

35. False

37. True (since $-2 = -2$, it is also true that $-2 \le -2$)

39. False

41. False (since $2\pi \approx 6.2$)

43. True (since $2\sqrt{2} \approx 2.8$)

45. Graph the interval $(2, 5)$:

47. Graph the interval $[1, 4]$:

49. Graph the interval $[0, 3)$:

51. Graph the interval $(-3, \infty)$:

53. Graph the interval $[-1, \infty)$:

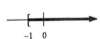

55. Graph the interval $(-\infty, 1)$:

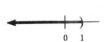

57. Graph the interval $(-\infty, \pi]$:

59. (a) The common value to six decimal places is 3.863703.
    (b) The common value (unrounded) to six decimal places is 3.162277.
    (c) The common value to six decimal places is 1.847759.

61. (a) Since $(4/3)^4 \approx 3.16$, it agrees with π to one decimal place.

(b) Since $\frac{22}{7} \approx 3.142$, it agrees with π to two decimal places.

(c) Since $\frac{355}{113} \approx 3.1415929$, it agrees with π to six decimal places.

(d) Since $\frac{63}{25}\left(\frac{17+15\sqrt{5}}{7+15\sqrt{5}}\right) \approx 3.1415926538$, it agrees with π to nine decimal places.

63. (a) We need to find two irrational numbers a and b such that their product is rational. If we choose $a = \sqrt{2}$ and $b = \sqrt{8}$, then $ab = \sqrt{2} \cdot \sqrt{8} = \sqrt{16} = 4$, which is rational.

(b) We need to find two irrational numbers a and b such that their product is irrational. If we choose $a = \sqrt{2}$ and $b = \sqrt{3}$, then $ab = \sqrt{2} \cdot \sqrt{3} = \sqrt{6}$, which is irrational.

65. (a) Raising 2 to the 1/2 power results in $2^{1/2} = \sqrt{2}$, which is irrational.

(b) Raising $\sqrt{2}$ to the 2 power results in $\left(\sqrt{2}\right)^2 = 2$, which is rational.

1.2 Absolute Value

1. $|3| = 3$

3. $|-6| = 6$

5. $|-1+3| = |2| = 2$

7. Simplify the expression:
$$\left|-\frac{4}{5}\right| - \frac{4}{5} = \frac{4}{5} - \frac{4}{5} = 0$$

9. Simplify the expression:
$$|-6+2| - |4| = |-4| - |4| = 4 - 4 = 0$$

11. Simplify the expression:
$$\left||-8| + |-9|\right| = |8+9| = |17| = 17$$

13. Simplify the expression:
$$\left|\frac{27-5}{5-27}\right| = \left|\frac{22}{-22}\right| = |-1| = 1$$

15. Simplify the expression:
$$|7(-8)| - |7||-8| = |-56| - 7(8) = 56 - 56 = 0$$

17. Substituting $a = -2$ and $b = 3$:
$$|a-b|^2 = |-2-3|^2 = |-5|^2 = (5)^2 = 25$$

19. Substituting $a = -2$, $b = 3$, and $c = -4$:
$$|c| - |b| - |a| = |-4| - |3| - |-2| = 4 - 3 - 2 = -1$$

21. Substituting $a = -2$, $b = 3$, and $c = -4$:
$$|a+b|^2 - |b+c|^2 = |-2+3|^2 - |3+(-4)|^2 = |1|^2 - |-1|^2 = (1)^2 - (1)^2 = 1 - 1 = 0$$

23. Substituting $a = -2$ and $b = 3$:
$$\frac{a+b+|a-b|}{2} = \frac{-2+3+|-2-3|}{2} = \frac{1+|-5|}{2} = \frac{1+5}{2} = \frac{6}{2} = 3$$

25. Since $\sqrt{2} - 1 > 0$:
$$|\sqrt{2}-1| - 1 = (\sqrt{2}-1) - 1 = \sqrt{2} - 2$$

27. Since $x \geq 3$, $x - 3 \geq 0$ and thus:
$$|x-3| = x - 3$$

29. Since $t^2 + 1 > 0$:
$$|t^2+1| = t^2 + 1$$

31. Since $-\sqrt{3} - 4 < 0$:
$$|-\sqrt{3}-4| = -(-\sqrt{3}-4) = \sqrt{3} + 4$$

33. Since $x < 3$, $x - 3 < 0$ and $x - 4 < 0$, and thus:
$$|x-3| + |x-4| = -(x-3) + [-(x-4)] = -x + 3 - x + 4 = -2x + 7$$

35. Since $3 < x < 4$, $x - 3 > 0$ and $x - 4 < 0$, and thus:
$$|x-3| + |x-4| = (x-3) + [-(x-4)] = x - 3 - x + 4 = 1$$

37. Since $-\frac{5}{2} < x < -\frac{3}{2}$, $x + 1 < 0$ and $x + 3 > 0$, and thus:
$$|x+1| + 4|x+3| = [-(x+1)] + 4(x+3) = -x - 1 + 4x + 12 = 3x + 11$$

39. The absolute value equality can be written as $|x - 4| = 8$.

41. The absolute value equality can be written as $|x - 1| = \frac{1}{2}$.

43. The absolute value inequality can be written as $|x - 1| \geq \frac{1}{2}$.

45. The absolute value inequality can be written as $|y - (-4)| < 1$, or $|y + 4| < 1$.

47. The absolute value inequality can be written as $|y - 0| < 3$, or $|y| < 3$.

49. The absolute value inequality can be written as $|x^2 - a^2| < M$.

51. Graph the interval $|x| < 4$:

53. Graph the interval $|x| > 1$:

55. Graph the interval $|x - 5| < 3$:

57. Graph the interval $|x - 3| \le 4$:

59. Graph the interval $\left| x + \frac{1}{3} \right| < \frac{3}{2}$:

61. Graph the interval $|x - 5| \ge 2$:

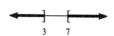

63. (a) Graph the interval $|x - 2| < 1$:

(b) Graph the interval $0 < |x - 2| < 1$, noting that $x = 2$ is excluded:

(c) The interval in (b) does not include 2.

65. (a) Clearly min $(6, 1) = 1$, now substituting $a = 6$ and $b = 1$:
$$\min(6,1) = \frac{6 + 1 - |6 - 1|}{2} = \frac{7 - 5}{2} = \frac{2}{2} = 1$$
Thus the equation is verified.

(b)   Clearly min $(1, -6) = -6$, now substituting $a = 1$ and $b = -6$:

$$\min(1, -6) = \frac{1 + (-6) - |1 - (-6)|}{2} = \frac{-5 - |1 + 6|}{2} = \frac{-5 - 7}{2} = \frac{-12}{2} = -6$$

Thus the equation is verified.

(c)   Clearly min $(-6, -6) = -6$, now substituting $a = -6$ and $b = -6$:

$$\min(-6, -6) = \frac{-6 + (-6) - |-6 - (-6)|}{2} = \frac{-12 - |-6 + 6|}{2} = \frac{-12}{2} = -6$$

Thus the equation is verified.

67.   Using the triangle inequality twice:

$$|a + b + c| = |a + (b + c)| \leq |a| + |b + c| \leq |a| + |b| + |c|$$

69.   Consider three cases: $a = b$, $a > b$, and $a < b$.

case 1:   If $a = b$, then max $(a, b) = a$, now verifying:

$$\frac{a + b + |a - b|}{2} = \frac{a + a + |a - a|}{2} = \frac{2a}{2} = a$$

Thus the equation is verified.

case 2:   If $a > b$, then max $(a, b) = a$, and since $a - b > 0$, we have:

$$\frac{a + b + |a - b|}{2} = \frac{a + b + a - b}{2} = \frac{2a}{2} = a$$

Thus the equation is verified.

case 3:   If $a < b$, then max $(a, b) = b$, and since $a - b < 0$, we have:

$$\frac{a + b + |a - b|}{2} = \frac{a + b + -(a - b)}{2} = \frac{a + b - a + b}{2} = \frac{2b}{2} = b$$

Thus the equation is verified.

71.   (a)   Property 1(b).

(b) $a + b \leq |a| + |b|$

(c)   Since $(-a) + (-b) \leq |a| + |b|$, $-(a + b) \leq |a| + |b|$.

(d)   Since $a + b \leq |a| + |b|$ and $-(a + b) \leq |a| + |b|$, $|a + b| \leq |a| + |b|$, since $|a + b|$ is either $a + b$ or $-(a + b)$.

1.3 Integer Exponents. Scientific Notation

1. Substituting $x = -2$:
$$2(-2)^3 - (-2) + 4 = 2(-8) + 2 + 4 = -10$$

3. Substituting $x = -\frac{1}{2}$:
$$\frac{1 - 2\left(-\frac{1}{2}\right)^2}{1 + 2\left(-\frac{1}{2}\right)^3} = \frac{1 - 2\left(\frac{1}{4}\right)}{1 + 2\left(-\frac{1}{8}\right)} = \frac{1 - \frac{1}{2}}{1 - \frac{1}{4}} = \frac{\frac{1}{2}}{\frac{3}{4}} = \frac{2}{3}$$

5. Substituting $x = 2$:
$$\frac{2^2 + 2^3 - 2^2}{2^2 + 3^2 - 3^2} = \frac{8}{4} = 2$$

7. The expression can be written as $(x + y)^2$.

9. The expression can be written as $(x + y)^2 + 3$.

11. The expression can be written as $\left[\frac{1}{2}\left(x^2 - 2y^3\right)\right]^2$.

13. The expression can be written as $\left|x - 1\right|^3$.

15. (a) $a^3 a^{12} = a^{3+12} = a^{15}$
   (b) $(a + 1)^3 (a + 1)^{12} = (a + 1)^{15}$
   (c) $(a + 1)^{12} (a + 1)^3 = (a + 1)^{15}$

17. (a) $yy^2 y^8 = y^{1+2+8} = y^{11}$
   (b) $(y + 1)(y + 1)^2 (y + 1)^8 = (y + 1)^{11}$
   (c) $\left[(y + 1)(y + 1)^8\right]^2 = \left[(y + 1)^9\right]^2 = (y + 1)^{18}$

19. (a) Using properties of exponents:
$$\frac{\left(x^2 + 3\right)^{10}}{\left(x^2 + 3\right)^9} = \left(x^2 + 3\right)^{10-9} = x^2 + 3$$

   (b) Using properties of exponents:
$$\frac{\left(x^2 + 3\right)^9}{\left(x^2 + 3\right)^{10}} = \left(x^2 + 3\right)^{9-10} = \left(x^2 + 3\right)^{-1} = \frac{1}{x^2 + 3}$$

(c)   Using properties of exponents:

$$\frac{12^{10}}{12^9} = 12^{10-9} = 12$$

21.   (a)   Using properties of exponents:

$$\frac{t^{15}}{t^9} = t^{15-9} = t^6$$

(b)   Using properties of exponents:

$$\frac{t^9}{t^{15}} = t^{9-15} = t^{-6} = \frac{1}{t^6}$$

(c)   Using properties of exponents:

$$\frac{(t^2+3)^{15}}{(t^2+3)^9} = (t^2+3)^{15-9} = (t^2+3)^6$$

23.   (a)   Using properties of exponents:

$$\frac{x^6 y^{15}}{x^2 y^{20}} = \frac{x^{6-2}}{y^{20-15}} = \frac{x^4}{y^5}$$

(b)   Using properties of exponents:

$$\frac{x^2 y^{20}}{x^6 y^{15}} = \frac{y^{20-15}}{x^{6-2}} = \frac{y^5}{x^4}$$

(c)   Using properties of exponents:

$$\left(\frac{x^2 y^{20}}{x^6 y^{15}}\right)^2 = \left(\frac{y^{20-15}}{x^{6-2}}\right)^2 = \left(\frac{y^5}{x^4}\right)^2 = \frac{y^{10}}{x^8}$$

25.   (a) $4(x^3)^2 = 4x^6$

(b) $(4x^3)^2 = 4^2 x^6 = 16x^6$

(c)   Using properties of exponents:

$$\frac{(4x^2)^3}{(4x^3)^2} = \frac{4^3 x^6}{4^2 x^6} = 4^{3-2} = 4$$

27.   (a)   Since $x^0 = 1$ for any value of $x \neq 0$, $64^0 = 1$.

(b)   Since $x^0 = 1$ for any value of $x \neq 0$, $(64^3)^0 = 1$.

(c)   Since $x^0 = 1$ for any value of $x \neq 0$, $(64^0)^3 = (1)^3 = 1$.

29. (a) Evaluating negative exponents:
$$10^{-1} + 10^{-2} = \tfrac{1}{10} + \tfrac{1}{100} = \tfrac{10}{100} + \tfrac{1}{100} = \tfrac{11}{100}$$

(b) Using the result from part (a):
$$\left(10^{-1} + 10^{-2}\right)^{-1} = \left(\tfrac{11}{100}\right)^{-1} = \tfrac{100}{11}$$

(c) Evaluating negative exponents:
$$\left[\left(10^{-1}\right)\left(10^{-2}\right)\right]^{-1} = \left[\tfrac{1}{10} \cdot \tfrac{1}{100}\right]^{-1} = \left(\tfrac{1}{1000}\right)^{-1} = 1000$$

31. Evaluating negative exponents:
$$\left(\tfrac{1}{3}\right)^{-1} + \left(\tfrac{1}{4}\right)^{-1} = 3 + 4 = 7$$

33. By first adding fractions:
$$\left(\tfrac{2}{3} + \tfrac{3}{2}\right)^{-1} = \left(\tfrac{4}{6} + \tfrac{9}{6}\right)^{-1} = \left(\tfrac{13}{6}\right)^{-1} = \tfrac{6}{13}$$

35. (a) Evaluating negative exponents:
$$5^{-2} + 10^{-2} = \tfrac{1}{25} + \tfrac{1}{100} = \tfrac{4}{100} + \tfrac{1}{100} = \tfrac{5}{100} = \tfrac{1}{20}$$

(b) Evaluating negative exponents:
$$(5 + 10)^{-2} = (15)^{-2} = \tfrac{1}{225}$$

37. Using properties of exponents:
$$\left(a^2 b c^0\right)^{-3} = \frac{1}{\left(a^2 b\right)^3} = \frac{1}{a^6 b^3}$$

39. Using properties of exponents:
$$\left(a^{-2} b^{-1} c^3\right)^{-2} = a^4 b^2 c^{-6} = \frac{a^4 b^2}{c^6}$$

41. Using properties of exponents:
$$\left(\frac{x^3 y^{-2} z}{x y^2 z^{-3}}\right)^{-3} = \frac{x^{-9} y^6 z^{-3}}{x^{-3} y^{-6} z^9} = \frac{y^{6-(-6)}}{x^{-3+9} z^{9+3}} = \frac{y^{12}}{x^6 z^{12}}$$

43. Using properties of exponents:
$$\left(\frac{x^4 y^{-8} z^2}{x y^2 z^{-6}}\right)^{-2} = \frac{x^{-8} y^{16} z^{-4}}{x^{-2} y^{-4} z^{12}} = \frac{y^{16+4}}{x^{-2+8} z^{12+4}} = \frac{y^{20}}{x^6 z^{16}}$$

45. Using properties of exponents:
$$\left(\frac{a^{-2} b^{-3} c^{-4}}{a^2 b^3 c^4}\right)^2 = \frac{a^{-4} b^{-6} c^{-8}}{a^4 b^6 c^8} = \frac{1}{a^{4+4} b^{6+6} c^{8+8}} = \frac{1}{a^8 b^{12} c^{16}}$$

47. Using properties of exponents:
$$\frac{x^2}{y^{-3}} \div \frac{x^2}{y^3} = \frac{x^2}{y^{-3}} \cdot \frac{y^3}{x^2} = \frac{x^2 y^3}{x^2 y^{-3}} = y^6$$

49. Subtracting exponents:
$$\frac{b^{p+1}}{b^p} = b^{p+1-p} = b$$

51. Adding exponents:
$$(x^p)(x^p) = x^{p+p} = x^{2p}$$

53. Using properties of exponents:
$$(a^2 q b^{-p})^3 = a^6 q b^{-3p} = \frac{a^6 q}{b^{3p}}$$

55. Factoring $9 = 3^2$ and $12 = 2^2 \cdot 3$:
$$\frac{2^8 \cdot 3^{15}}{9 \cdot 3^{10} \cdot 12} = \frac{2^8 \cdot 3^{15}}{3^2 \cdot 3^{10} \cdot 2^2 \cdot 3} = \frac{2^8 \cdot 3^{15}}{2^2 \cdot 3^{13}} = 2^6 \cdot 3^2 = 576$$

57. Factoring $24 = 2^3 \cdot 3$, $32 = 2^5$, and $12 = 2^2 \cdot 3$:
$$\frac{24^5}{32 \cdot 12^4} = \frac{(2^3 \cdot 3)^5}{2^5 \cdot (2^2 \cdot 3)^4} = \frac{2^{15} \cdot 3^5}{2^5 \cdot 2^8 \cdot 3^4} = \frac{2^{15} \cdot 3^5}{2^{13} \cdot 3^4} = 2^2 \cdot 3 = 12$$

59. 9.29×10^7 miles

61. 6.68×10^4 mph

63. 2.5×10^{19} miles

65. (a) 8.6688×10^1 days
    (b) 1.0604772×10^4 days
    (c) 8.9424×10^4 days

67. (a) 1.0×10^{-9} seconds
    (b) 1.0×10^{-18} seconds
    (c) 1.0×10^{-24} seconds

69. Using the properties of exponents:
$$\frac{a^{3x+y}}{a^{2x} a^{x+y}} = \frac{a^{3x+y}}{a^{3x+y}} = 1$$

71. Using properties of exponents:

$$\frac{\left(x^{5n+1}\right)^n}{\left(x^n\right)^{5n}}\cdot\frac{1}{x^{n-2}}=\frac{x^{5n^2+n}}{x^{5n^2}}\cdot\frac{1}{x^{n-2}}=\frac{x^{5n^2+n}}{x^{5n^2+n-2}}=x^{5n^2+n-5n^2-n+2}=x^2$$

73. Converting units:

$$\frac{1\,\text{sec}}{2.9979\times10^{10}\,\text{cm}}\cdot\frac{1\,\text{min}}{60\,\text{sec}}\cdot\frac{160930\,\text{cm}}{1\,\text{mi}}\cdot\left(92.9\times10^6\,\text{mi}\right)\approx8\,\text{min}$$

75. Raising both sides of the equation $p=b^x$ to the power of y yields $p^y=b^{xy}$. Similarly, raising both sides of $q=b^y$ to the power of x yields $q^x=b^{xy}$. Now we use these results to substitute for p^y and q^x in the equation $b^2=\left(p^y q^x\right)^z$ to obtain:

$$b^2=\left(b^{xy}b^{xy}\right)^z \text{ or } b^2=b^{2xyz}$$

From this last equation we conclude (by equating exponents) that:
$$2=2xyz, \text{ or } xyz=1$$

1.4 Nth Roots

1. false

3. true

5. true

7. true

9. true

11. (a) $\sqrt[3]{-64}=-4$
  (b) $\sqrt[4]{-64}$ is undefined

13. (a) $\sqrt[3]{\frac{8}{125}}=\frac{2}{5}$
  (b) $\sqrt[3]{-\frac{8}{125}}=-\frac{2}{5}$

15. (a) $\sqrt{-16}$ is undefined
  (b) $\sqrt[4]{-16}$ is undefined

17. (a) $\sqrt[4]{\frac{256}{81}}=\frac{4}{3}$
  (b) $\sqrt[3]{-\frac{27}{125}}=-\frac{3}{5}$

19. (a) $\sqrt[5]{-32} = -2$

   (b) $-\sqrt[5]{-32} = -(-2) = 2$

21. (a) Factoring $18 = 9 \cdot 2$:
   $$\sqrt{18} = \sqrt{9}\sqrt{2} = 3\sqrt{2}$$

   (b) Factoring $54 = 27 \cdot 2$:
   $$\sqrt[3]{54} = \sqrt[3]{27}\,\sqrt[3]{2} = 3\sqrt[3]{2}$$

23. (a) Factoring $98 = 49 \cdot 2$:
   $$\sqrt{98} = \sqrt{49}\sqrt{2} = 7\sqrt{2}$$

   (b) Factoring $-64 = -32 \cdot 2$, we have:
   $$\sqrt[5]{-64} = \sqrt[5]{-32}\,\sqrt[5]{2} = -2\sqrt[5]{2}$$

25. (a) $\sqrt{\frac{25}{4}} = \frac{5}{2}$

   (b) $\sqrt[4]{\frac{16}{625}} = \frac{2}{5}$

27. (a) Factoring $8 = 4 \cdot 2$:
   $$\sqrt{2} + \sqrt{8} = \sqrt{2} + \sqrt{4}\sqrt{2} = \sqrt{2} + 2\sqrt{2} = 3\sqrt{2}$$

   (b) Factoring $16 = 8 \cdot 2$:
   $$\sqrt[3]{2} + \sqrt[3]{16} = \sqrt[3]{2} + \sqrt[3]{8}\,\sqrt[3]{2} = \sqrt[3]{2} + 2\sqrt[3]{2} = 3\sqrt[3]{2}$$

29. (a) Factoring $50 = 25 \cdot 2$ and $128 = 64 \cdot 2$:
   $$4\sqrt{50} - 3\sqrt{128} = 4\sqrt{25}\sqrt{2} - 3\sqrt{64}\sqrt{2} = 20\sqrt{2} - 24\sqrt{2} = -4\sqrt{2}$$

   (b) Factoring $32 = 16 \cdot 2$ and $162 = 81 \cdot 2$:
   $$\sqrt[4]{32} + \sqrt[4]{162} = \sqrt[4]{16}\,\sqrt[4]{2} + \sqrt[4]{81}\,\sqrt[4]{2} = 2\sqrt[4]{2} + 3\sqrt[4]{2} = 5\sqrt[4]{2}$$

31. (a) 0.3 [because $(0.3)^2 = 0.09$]

   (b) 0.2 [because $(0.2)^3 = 0.008$]

33. Simplifying each radical yields:
   $$4\sqrt{24} - 8\sqrt{54} + 2\sqrt{6} = 4\sqrt{4}\sqrt{6} - 8\sqrt{9}\sqrt{6} + 2\sqrt{6} = 8\sqrt{6} - 24\sqrt{6} + 2\sqrt{6} = -14\sqrt{6}$$

35. $\sqrt{\sqrt{64}} = \sqrt{8} = \sqrt{4}\sqrt{2} = 2\sqrt{2}$

37. (a) $\sqrt{36x^2} = 6x$, since $x > 0$

   (b) $\sqrt{36y^2} = -6y$, since $y < 0$

39. (a) Multiplying radicals yields:
$$\sqrt{ab^2}\sqrt{a^2b} = \sqrt{a^3b^3} = ab\sqrt{ab}$$

(b) Multiplying radicals yields:
$$\sqrt{ab^3}\sqrt{a^3b} = \sqrt{a^4b^4} = a^2b^2$$

41. Simplifying the radical yields:
$$\sqrt{72a^3b^4c^5} = \sqrt{36a^2b^4c^4}\sqrt{2ac} = 6ab^2c^2\sqrt{2ac}$$

43. Simplifying the radical yields:
$$\sqrt[4]{16a^4b^5} = \sqrt[4]{16a^4b^4}\cdot\sqrt[4]{b} = 2ab\sqrt[4]{b}$$

45. Simplifying the radical yields:
$$\sqrt{18a^3b^2} = \sqrt{9a^2b^2}\sqrt{2a} = 3ab\sqrt{2a}$$

47. Simplifying the radical yields:
$$\sqrt[3]{\frac{16a^{12}b^2}{c^9}} = \frac{\sqrt[3]{8a^{12}}\sqrt[3]{2b^2}}{\sqrt[3]{c^9}} = \frac{2a^4\sqrt[3]{2b^2}}{c^3}$$

49. Simplifying the radical yields:
$$\sqrt[6]{\frac{5a^7}{a^{-5}b^6}} = \sqrt[6]{\frac{5a^{12}}{b^6}} = \frac{\sqrt[6]{5}\sqrt[6]{a^{12}}}{\sqrt[6]{b^6}} = \frac{a^2\sqrt[6]{5}}{b}$$

51. Rationalizing the denominator:
$$\frac{4}{\sqrt{7}}\cdot\frac{\sqrt{7}}{\sqrt{7}} = \frac{4\sqrt{7}}{7}$$

53. Rationalizing the denominator:
$$\frac{1}{\sqrt{8}}\cdot\frac{\sqrt{2}}{\sqrt{2}} = \frac{\sqrt{2}}{\sqrt{16}} = \frac{\sqrt{2}}{4}$$

55. Rationalizing the denominator:
$$\frac{1}{1+\sqrt{5}}\cdot\frac{1-\sqrt{5}}{1-\sqrt{5}} = \frac{1-\sqrt{5}}{1-5} = \frac{1-\sqrt{5}}{-4} = \frac{\sqrt{5}-1}{4}$$

57. Rationalizing the denominator:
$$\frac{1+\sqrt{3}}{1-\sqrt{3}}\cdot\frac{1+\sqrt{3}}{1+\sqrt{3}} = \frac{1+\sqrt{3}+\sqrt{3}+3}{1-3} = \frac{4+2\sqrt{3}}{-2} = -2-\sqrt{3}$$

59. Rationalizing the denominator and simplifying:
$$\frac{1}{\sqrt{5}}+4\sqrt{45} = \frac{1}{\sqrt{5}}\cdot\frac{\sqrt{5}}{\sqrt{5}}+4\sqrt{9}\sqrt{5} = \frac{\sqrt{5}}{5}+12\sqrt{5} = \frac{\sqrt{5}}{5}+\frac{60\sqrt{5}}{5} = \frac{61\sqrt{5}}{5}$$

61. Rationalizing the denominator:

$$\frac{1}{\sqrt[3]{25}} \cdot \frac{\sqrt[3]{5}}{\sqrt[3]{5}} = \frac{\sqrt[3]{5}}{\sqrt[3]{125}} = \frac{\sqrt[3]{5}}{5}$$

63. Rationalizing the denominator:

$$\frac{3}{\sqrt[4]{3}} \cdot \frac{\sqrt[4]{3^3}}{\sqrt[4]{3^3}} = \frac{3\sqrt[4]{27}}{3} = \sqrt[4]{27}$$

65. Rationalizing the denominator:

$$\frac{1}{\sqrt[4]{2ab^5}} \cdot \frac{\sqrt[4]{8a^3b^3}}{\sqrt[4]{8a^3b^3}} = \frac{\sqrt[4]{8a^3b^3}}{\sqrt[4]{16a^4b^8}} = \frac{\sqrt[4]{8a^3b^3}}{2ab^2}$$

67. Rationalizing the denominator:

$$\frac{3}{\sqrt[5]{16a^4b^9}} \cdot \frac{\sqrt[5]{2ab}}{\sqrt[5]{2ab}} = \frac{3\sqrt[5]{2ab}}{\sqrt[5]{32a^5b^{10}}} = \frac{3\sqrt[5]{2ab}}{2ab^2}$$

69. (a) Rationalizing the denominator:

$$\frac{2}{\sqrt{3}+1} \cdot \frac{\sqrt{3}-1}{\sqrt{3}-1} = \frac{2(\sqrt{3}-1)}{3-1} = \frac{2(\sqrt{3}-1)}{2} = \sqrt{3}-1$$

(b) Rationalizing the denominator:

$$\frac{2}{\sqrt{3}-1} \cdot \frac{\sqrt{3}+1}{\sqrt{3}+1} = \frac{2(\sqrt{3}+1)}{3-1} = \frac{2(\sqrt{3}+1)}{2} = \sqrt{3}+1$$

71. (a) Rationalizing the denominator:

$$\frac{x}{\sqrt{x}-2} \cdot \frac{\sqrt{x}+2}{\sqrt{x}+2} = \frac{x(\sqrt{x}+2)}{x-4}$$

(b) Rationalizing the denominator:

$$\frac{x}{\sqrt{x}-y} \cdot \frac{\sqrt{x}+y}{\sqrt{x}+y} = \frac{x(\sqrt{x}+y)}{x-y^2}$$

73. (a) Rationalizing the denominator:

$$\frac{\sqrt{x}+\sqrt{2}}{\sqrt{x}-\sqrt{2}} \cdot \frac{\sqrt{x}+\sqrt{2}}{\sqrt{x}+\sqrt{2}} = \frac{x+2\sqrt{2x}+2}{x-2}$$

(b) Rationalizing the denominator:

$$\frac{\sqrt{x}+\sqrt{a}}{\sqrt{x}-\sqrt{a}} \cdot \frac{\sqrt{x}+\sqrt{a}}{\sqrt{x}+\sqrt{a}} = \frac{x+2\sqrt{ax}+a}{x-a}$$

75. Rationalizing the denominator:
$$\frac{1}{\sqrt{x+1}-\sqrt{x}}\cdot\frac{\sqrt{x+1}+\sqrt{x}}{\sqrt{x+1}+\sqrt{x}}=\frac{\sqrt{x+1}+\sqrt{x}}{x+1-x}=\sqrt{x+1}+\sqrt{x}$$

77. Rationalizing the denominator:
$$\frac{-2}{\sqrt{x+h}-\sqrt{x}}\cdot\frac{\sqrt{x+h}+\sqrt{x}}{\sqrt{x+h}+\sqrt{x}}=\frac{-2\left(\sqrt{x+h}+\sqrt{x}\right)}{x+h-x}=\frac{-2\left(\sqrt{x+h}+\sqrt{x}\right)}{h}$$

79. Rationalizing the numerator:
$$\frac{\sqrt{x}-\sqrt{5}}{x-5}\cdot\frac{\sqrt{x}+\sqrt{5}}{\sqrt{x}+\sqrt{5}}=\frac{x-5}{(x-5)\left(\sqrt{x}+\sqrt{5}\right)}=\frac{1}{\sqrt{x}+\sqrt{5}}$$

81. Rationalizing the numerator:
$$\frac{\sqrt{x}+\sqrt{a}}{x-a}\cdot\frac{\sqrt{x}-\sqrt{a}}{\sqrt{x}-\sqrt{a}}=\frac{x-a}{(x-a)\left(\sqrt{x}-\sqrt{a}\right)}=\frac{1}{\sqrt{x}-\sqrt{a}}$$

83. Rationalizing the numerator:
$$\frac{\sqrt{2+h}+\sqrt{2}}{h}\cdot\frac{\sqrt{2+h}-\sqrt{2}}{\sqrt{2+h}-\sqrt{2}}=\frac{2+h-2}{h\left(\sqrt{2+h}-\sqrt{2}\right)}=\frac{h}{h\left(\sqrt{2+h}-\sqrt{2}\right)}=\frac{1}{\sqrt{2+h}-\sqrt{2}}$$

85. Rationalizing the numerator:
$$\frac{\sqrt{x+h}-\sqrt{x}}{h}\cdot\frac{\sqrt{x+h}+\sqrt{x}}{\sqrt{x+h}+\sqrt{x}}=\frac{x+h-x}{h\left(\sqrt{x+h}+\sqrt{x}\right)}=\frac{h}{h\left(\sqrt{x+h}+\sqrt{x}\right)}=\frac{1}{\sqrt{x+h}+\sqrt{x}}$$

87. Let $a=9$ and $b=16$, then:
$$\sqrt{a+b}=\sqrt{9+16}=\sqrt{25}=5$$
$$\sqrt{a}+\sqrt{b}=\sqrt{9}+\sqrt{16}=3+4=7$$
So the formula is not valid.

89. Let $u=1$ and $v=8$, then:
$$\sqrt[3]{u+v}=\sqrt[3]{1+8}=\sqrt[3]{9}\approx2.08$$
$$\sqrt[3]{u}+\sqrt[3]{v}=\sqrt[3]{1}+\sqrt[3]{8}=1+2=3$$
So the formula is not valid.

91. (a) To six decimal places, both values are 1.645751.

    (b) Following the hint, square each side:
$$\left(\sqrt{7}-1\right)^2=7-2\sqrt{7}+1=8-2\sqrt{7}$$

93. First rationalize the denominator in each fraction:

$$\frac{\sqrt{a}}{\sqrt{a}+\sqrt{b}} = \frac{\sqrt{a}}{\sqrt{a}+\sqrt{b}} \cdot \frac{\sqrt{a}-\sqrt{b}}{\sqrt{a}-\sqrt{b}} = \frac{a-\sqrt{ab}}{a-b}$$

$$\frac{\sqrt{b}}{\sqrt{a}-\sqrt{b}} = \frac{\sqrt{b}}{\sqrt{a}-\sqrt{b}} \cdot \frac{\sqrt{a}+\sqrt{b}}{\sqrt{a}+\sqrt{b}} = \frac{\sqrt{ab}+b}{a-b}$$

Now adding the two fractions:

$$\frac{\sqrt{a}}{\sqrt{a}+\sqrt{b}} + \frac{\sqrt{b}}{\sqrt{a}-\sqrt{b}} = \frac{\sqrt{a}-\sqrt{ab}}{a-b} + \frac{\sqrt{ab}+b}{a-b} = \frac{a-\sqrt{ab}+\sqrt{ab}+b}{a-b} = \frac{a+b}{a-b}$$

95. Following the hint:

$$\frac{1}{1+\sqrt{2}+\sqrt{3}} \cdot \frac{\left(1+\sqrt{2}\right)-\sqrt{3}}{\left(1+\sqrt{2}\right)-\sqrt{3}} = \frac{\left(1+\sqrt{2}\right)-\sqrt{3}}{\left(1+\sqrt{2}\right)^2 - \left(\sqrt{3}\right)^2}$$

$$= \frac{1+\sqrt{2}-\sqrt{3}}{1+2\sqrt{2}+2-3}$$

$$= \frac{1+\sqrt{2}-\sqrt{3}}{2\sqrt{2}} \cdot \frac{\sqrt{2}}{\sqrt{2}}$$

$$= \frac{\sqrt{2}+2-\sqrt{6}}{4}$$

1.5 Rational Exponents

1. $a^{3/5} = \sqrt[5]{a^3} = \left(\sqrt[5]{a}\right)^3$

3. $5^{2/3} = \sqrt[3]{5^2} = \left(\sqrt[3]{5}\right)^2$

5. $\left(x^2+1\right)^{3/4} = \sqrt[4]{\left(x^2+1\right)^3} = \left(\sqrt[4]{x^2+1}\right)^3$

7. $2^{xy/3} = \sqrt[3]{2^{xy}} = \left(\sqrt[3]{2}\right)^{xy}$

9. $\sqrt[3]{p^2} = p^{2/3}$

11. $\sqrt[7]{(1+u)^4} = (1+u)^{4/7}$

13. $\sqrt[p]{\left(a^2+b^2\right)^3} = \left(a^2+b^2\right)^{3/p}$

15. $16^{1/2} = \sqrt{16} = 4$

17. $\left(\frac{1}{36}\right)^{1/2} = \sqrt{\frac{1}{36}} = \frac{1}{6}$

19. $(-16)^{1/2} = \sqrt{-16}$, which is undefined

21. $625^{1/4} = \sqrt[4]{625} = 5$

23. $8^{1/3} = \sqrt[3]{8} = 2$

25. $8^{2/3} = \left(\sqrt[3]{8}\right)^2 = (2)^2 = 4$

27. $(-32)^{1/5} = \sqrt[5]{-32} = -2$

29. $(-1000)^{1/3} = \sqrt[3]{-1000} = -10$

31. $49^{-1/2} = \left(\sqrt{49}\right)^{-1} = 7^{-1} = \frac{1}{7}$

33. $(-49)^{-1/2} = \left(\sqrt{-49}\right)^{-1}$, which is undefined

35. $(36)^{-3/2} = \left(\sqrt{36}\right)^{-3} = 6^{-3} = \dfrac{1}{6^3} = \frac{1}{216}$

37. $125^{2/3} = \left(\sqrt[3]{125}\right)^2 = (5)^2 = 25$

39. $(-1)^{3/5} = \left(\sqrt[5]{-1}\right)^3 = (-1)^3 = -1$

41. Evaluating the rational exponents:
$$32^{4/5} - 32^{-4/5} = \left(\sqrt[5]{32}\right)^4 - \left(\sqrt[5]{32}\right)^{-4} = 2^4 - 2^{-4} = 16 - \frac{1}{16} = \frac{255}{16}$$

43. Evaluating the rational exponents:
$$\begin{aligned}
\left(\frac{9}{16}\right)^{-5/2} - \left(\frac{1000}{27}\right)^{4/3} &= \left(\frac{16}{9}\right)^{5/2} - \left(\frac{1000}{27}\right)^{4/3} \\
&= \left(\sqrt{\frac{16}{9}}\right)^5 - \left(\sqrt[3]{\frac{1000}{27}}\right)^4 \\
&= \left(\frac{4}{3}\right)^5 - \left(\frac{10}{3}\right)^4 \\
&= \frac{1024}{243} - \frac{10000}{81} \\
&= -\frac{28976}{243}
\end{aligned}$$

45. $\left(2a^{1/3}\right)\left(3a^{1/4}\right) = 6a^{7/12}$, since $\frac{1}{3} + \frac{1}{4} = \frac{4}{12} + \frac{3}{12} = \frac{7}{12}$

47. Writing with rational exponents, then simplifying:
$$\sqrt[4]{\dfrac{64a^{2/3}}{a^{1/3}}} = \sqrt[4]{64a^{1/3}} = \left(2^6\,a^{1/3}\right)^{1/4} = 2^{3/2}\,a^{1/12}$$

49. Subtracting exponents yields:
$$\dfrac{\left(x^2+1\right)^{3/4}}{\left(x^2+1\right)^{-1/4}} = \left(x^2+1\right)^{4/4} = x^2+1$$

51. (a) $\sqrt{3}\,\sqrt[3]{6} = 3^{1/2}\,6^{1/3} = 3^{1/2}2^{1/3}3^{1/3} = 2^{1/3}\,3^{5/6}$
    (b) $3^{1/2}\,6^{1/3} = 3^{3/6}\,6^{2/6} = \sqrt[6]{3^3\,6^2} = \sqrt[6]{972}$

53. (a) $\sqrt[3]{6}\,\sqrt[4]{2} = 6^{1/3}\,2^{1/4} = 2^{1/3}3^{1/3}2^{1/4} = 2^{7/12}\,3^{1/3}$
    (b) $6^{1/3}\,2^{1/4} = 6^{4/12}\,2^{3/12} = \sqrt[12]{6^4\,2^3} = \sqrt[12]{10368}$

55. (a) $\sqrt[3]{x^2}\,\sqrt[5]{y^4} = x^{2/3}\,y^{4/5}$
    (b) $x^{2/3}\,y^{4/5} = x^{10/15}\,y^{12/15} = \sqrt[15]{x^{10}\,y^{12}}$

57. (a) Using rational exponents:
    $$\sqrt[4]{x^a}\,\sqrt[3]{x^b}\,\sqrt{x^{a/6}} = x^{a/4}\,x^{b/3}\,x^{a/12} = x^{3a/12}\,x^{4b/12}\,x^{a/12} = x^{(4a+4b)/12} = x^{(a+b)/3}$$

    (b) $x^{(a+b)/3} = \sqrt[3]{x^{a+b}}$

59. $\sqrt[3]{(x+1)^2} = (x+1)^{2/3}$

61. $\left(\sqrt[5]{x+y}\right)^2 = (x+y)^{2/5}$

63. Using rational exponents:
    $$\sqrt[3]{\sqrt{x}} + \sqrt{\sqrt[3]{x}} = \left(x^{1/2}\right)^{1/3} + \left(x^{1/3}\right)^{1/2} = x^{1/6} + x^{1/6} = 2x^{1/6}$$

65. Using rational exponents:
    $$\sqrt{\sqrt[3]{x}\,\sqrt[4]{y}} = \left(x^{1/3}\,y^{1/4}\right)^{1/2} = x^{1/6}\,y^{1/8}$$

67. Compute $9^{10/9} \approx 11.5$ and $10^{9/10} \approx 7.9$. Thus $9^{10/9}$ is larger.

69. Using $a = 9$ and $b = 16$:
    $$(a+b)^{1/2} = (9+16)^{1/2} = 25^{1/2} = 5$$
    $$a^{1/2} + b^{1/2} = 9^{1/2} + 16^{1/2} = 3+4 = 7$$
    So the formula is not valid.

71. Using $u = 1$ and $v = 8$:

$$(u+v)^{1/3} = (1+8)^{1/3} = 9^{1/3} \approx 2.08$$
$$u^{1/3} + v^{1/3} = 1^{1/3} + 8^{1/3} = 1 + 2 = 3$$

So the formula is not valid.

73. Using $x = 4$ and $m = 2$:

$$x^{1/m} = 4^{1/2} = 2$$
$$\frac{1}{x^m} = \frac{1}{4^2} = \frac{1}{16}$$

So the formula is not valid.

75. Complete the table:

n	2	5	10	100	10^3	10^4	10^5	10^6
$n^{1/n}$	1.4142	1.3797	1.2589	1.0471	1.0069	1.0009	1.0001	1.0000

77. (a) Note that $2^{2/3}$ is less than 2 and $2^{3/2}$ is greater than 2, thus $2^{3/2}$ is the larger number.

(b) Note that $5^{1/2} = \sqrt{5}$ and $5^{-2} = \frac{1}{25}$, thus $5^{1/2}$ is the larger number.

(c) $2^{1/2}$ is larger. (One way to see this is to raise both numbers to the sixth power.)

(d) $\left(\frac{1}{2}\right)^{1/3}$ is larger. (One way to see this is to raise both numbers to the sixth power.)

(e) Note that $10^{1/10}$ is larger than 1, but $\left(\frac{1}{10}\right)^{10}$ is much less than 1, thus $10^{1/10}$ is the larger number.

79. Without a calculator, $(0.5)^{1/3} = (0.5)^{4/12}$ while $(0.5)^{1/4} = (0.5)^{3/12}$, and thus $(0.5)^{1/3} = (0.5)^{1/12}(0.5)^{1/4}$, and thus, since $(0.5)^{1/12} < 1$, $(0.5)^{1/3}$ is closer to zero. Using a calculator, we compute $(0.5)^{1/3} \approx 0.7937$ while $(0.5)^{1/4} \approx 0.8409$, so $(0.5)^{1/3}$ is closer to zero.

81. Using rational exponents:

$$\frac{a-b}{a+b}\sqrt{\frac{a+b}{a-b}} = \frac{a-b}{a+b} \cdot \frac{(a+b)^{1/2}}{(a-b)^{1/2}} = \frac{a-b}{(a-b)^{1/2}} \cdot \frac{(a+b)^{1/2}}{a+b} = \frac{(a-b)^{1/2}}{(a+b)^{1/2}} = \left(\frac{a-b}{a+b}\right)^{1/2}$$

1.6 Polynomials

1. (a) The domain is the set of all real numbers.
(b) The domain is the set of all non-negative real numbers.

3. (a) The domain is the set of all real numbers.
(b) The domain is the set of all real numbers.

5. (a) The domain is the set of all non-negative real numbers.
   (b) The domain is the set of all positive real numbers except 1.

7. (a) The degree is 3 and the coefficients are 4, –2, – 6 and –1.
   (b) The degree is 2 and the coefficients are a, b and c.

9. (a) The degree is 1 and the coefficient is 6.
   (b) The degree is 3 and the coefficients are –1, 1 and 6.

11. $\left(12x^2 - 4x + 2\right) + \left(8x^2 + 6x - 1\right) = 20x^2 + 2x + 1$

13. $\left(x^2 - x - 1\right) - \left(x^2 + x + 1\right) = x^2 - x - 1 - x^2 - x - 1 = -2x - 2$

15. Subtracting and combining like terms:
$$\left(2x^2 - 4x - 4\right) - \left(6x^2 + 5\right) - (8x - 1) = 2x^2 - 4x - 4 - 6x^2 - 5 - 8x + 1$$
$$= -4x^2 - 12x - 8$$

17. Carrying out the operations and combining like terms:
$$\left(2x^2 - 6x - 1\right) + \left(5x^2 - 5x - 1\right) - \left(3x^2 + 8x + 12\right)$$
$$= 2x^2 - 6x - 1 + 5x^2 - 5x - 1 - 3x^2 - 8x - 12$$
$$= 4x^2 - 19x - 14$$

19. Subtracting and combining like terms:
$$\left(ax^2 + bx + c\right) - \left(2ax^2 - 3bx + c\right) = ax^2 + bx + c - 2ax^2 + 3bx - c = -ax^2 + 4bx$$

21. $2x\left(x^2 - 4x - 5\right) = 2x^3 - 8x^2 - 10x$

23. $(x - 1)(x - 2) = x^2 - 3x + 2$

25. $(2x + 4)(x + 1) = 2x^2 + 6x + 4$

27. $\left(x^2 + 3x\right)\left(x^2 + x\right) = x^4 + 4x^3 + 3x^2$

29. $(2xy - 3)(2xy - 1) = 4x^2y^2 - 8xy + 3$

31. Carrying out the multiplication:
$$\left(x + \frac{1}{x}\right)\left(x + \frac{1}{x}\right) = x^2 + 1 + 1 + \frac{1}{x^2} = x^2 + 2 + \frac{1}{x^2}$$

33. Carrying out the multiplication:
$$\left(\sqrt{a+b} + 1\right)\left(\sqrt{a+b} + 3\right) = \left(\sqrt{a+b}\right)^2 + 4\sqrt{a+b} + 3 = a + b + 4\sqrt{a+b} + 3$$

35. Carrying out the multiplication:
$$\left(x^{1/2} - 1\right)\left(x^{1/2} - 2\right) = \left(x^{1/2}\right)^2 - 3x^{1/2} + 2 = x - 3x^{1/2} + 2$$

37. Carrying out the multiplication:
$$\begin{aligned}
(y + 2)(y^2 - 3y - 5) &= y(y^2 - 3y - 5) + 2(y^2 - 3y - 5)\\
&= y^3 - 3y^2 - 5y + 2y^2 - 6y - 10\\
&= y^3 - y^2 - 11y - 10
\end{aligned}$$

39. Carrying out the multiplication:
$$\begin{aligned}
(x - 2y + z)(x + 2y - z) &= (x - (2y - z))(x + (2y - z))\\
&= x^2 - (2y - z)^2\\
&= x^2 - (4y^2 - 4yz + z^2)\\
&= x^2 - 4y^2 + 4yz - z^2
\end{aligned}$$

41.　(a)　Using the special product for $(A - B)(A + B)$:
$$(x - y)(x + y) = x^2 - y^2$$

　　(b)　Using the special product for $(A - B)(A + B)$:
$$(x^2 - 5)(x^2 + 5) = (x^2)^2 - (5)^2 = x^4 - 25$$

43.　(a)　Using the special product for $(A - B)(A + B)$:
$$(A - 4)(A + 4) = A^2 - 16$$

　　(b)　Using the special product for $(A - B)(A + B)$:
$$[(a + b) - 4][(a + b) + 4] = (a + b)^2 - 16 = a^2 + 2ab + b^2 - 16$$

45.　(a)　Using the special product for $(A - B)^2$:
$$(x - 8)^2 = x^2 - 2(x)(8) + 64 = x^2 - 16x + 64$$

　　(b)　Using the special product for $(A - B)^2$:
$$(2x^2 - 5)^2 = (2x^2)^2 - 2(2x^2)(5) + 25 = 4x^4 - 20x^2 + 25$$

47.　(a)　Using the special product for $(A + B)^2$:
$$\left(\sqrt{x} + \sqrt{y}\right)^2 = \left(\sqrt{x}\right)^2 + 2\left(\sqrt{x}\right)\left(\sqrt{y}\right) + \left(\sqrt{y}\right)^2 = x + 2\sqrt{xy} + y$$

　　(b)　Using the special product for $(A - B)^2$:
$$\begin{aligned}
\left(\sqrt{x + y} - \sqrt{x}\right)^2 &= \left(\sqrt{x + y}\right)^2 - 2\left(\sqrt{x + y}\right)\left(\sqrt{x}\right) + \left(\sqrt{x}\right)^2\\
&= x + y - 2\sqrt{x^2 + xy} + x\\
&= 2x + y - 2\sqrt{x^2 + xy}
\end{aligned}$$

49. (a) Using the special product for $(A+B)^3$:
$$(a+1)^3 = a^3 + 3(a^2)(1) + 3(a)(1)^2 + (1)^3 = a^3 + 3a^2 + 3a + 1$$

(b) Using the special product for $(A-B)^3$:
$$(3x^2 - 2a^2)^3 = (3x^2)^3 - 3(3x^2)^2(2a^2) + 3(3x^2)(2a^2)^2 - (2a^2)^3$$
$$= 27x^6 - 54x^4a^2 + 36x^2a^4 - 8a^6$$

51. (a) Using the special product for $(A+B)(A^2 - AB + B^2)$:
$$(x+1)(x^2 - x + 1) = x^3 + 1$$

(b) Using the special product for $(A+B)(A^2 - AB + B^2)$:
$$(x^2 + 1)(x^4 - x^2 + 1) = (x^2)^3 + (1)^3 = x^6 + 1$$

53. Using long division:

$$
\begin{array}{r}
x - 5 \\
x - 3 \,\overline{\big)\, x^2 - 8x + 4} \\
\underline{x^2 - 3x} \\
-5x + 4 \\
\underline{-5x + 15} \\
-11
\end{array}
$$

The quotient is $x - 5$ and the remainder is -11. Write the equation:
$$x^2 - 8x + 4 = (x - 3)(x - 5) - 11$$

55. Using long division:

$$
\begin{array}{r}
x - 11 \\
x + 5 \,\overline{\big)\, x^2 - 6x - 2} \\
\underline{x^2 + 5x} \\
-11x - 2 \\
\underline{-11x - 55} \\
53
\end{array}
$$

The quotient is $x - 11$ and the remainder is 53. Write the equation:
$$x^2 - 6x - 2 = (x + 5)(x - 11) + 53$$

57. Using long division:

$$\begin{array}{r}
3x^2 - \frac{3}{2}x - \frac{1}{4} \\
2x+1\overline{\smash{\big)}\,6x^3 + 0x^2 - 2x + 3} \\
\underline{6x^3 + 3x^2} \\
-3x^2 - 2x \\
\underline{-3x^2 - \frac{3}{2}x} \\
-\frac{1}{2}x + 3 \\
\underline{-\frac{1}{2}x - \frac{1}{4}} \\
\frac{13}{4}
\end{array}$$

The quotient is $3x^2 - \frac{3}{2}x - \frac{1}{4}$ and the remainder is $\frac{13}{4}$. Write the equation:

$$6x^3 - 2x + 3 = (2x+1)\left(3x^2 - \frac{3}{2}x - \frac{1}{4}\right) + \frac{13}{4}$$

59. Using long division:

$$\begin{array}{r}
t^4 - 3t^3 + 9t^2 - 27t + 81 \\
t+3\overline{\smash{\big)}\,t^5 + 0t^4 + 0t^3 + 0t^2 + 0t + 2} \\
\underline{t^5 + 3t^4} \\
-3t^4 + 0t^3 \\
\underline{-3t^4 - 9t^3} \\
9t^3 + 0t^2 \\
\underline{9t^3 + 27t^2} \\
-27t^2 + 0t \\
\underline{-27t^2 - 81t} \\
81t + 2 \\
\underline{81t + 243} \\
-241
\end{array}$$

The quotient is $t^4 - 3t^3 + 9t^2 - 27t + 81$ and the remainder is –241. Write the equation:

$$t^5 + 2 = (t+3)\left(t^4 - 3t^3 + 9t^2 - 27t + 81\right) - 241$$

61. Using long division:

$$\begin{array}{r}
u^5 + 2u^4 + 4u^3 + 8u^2 + 16u + 32 \\
u - 2\ \overline{\smash{\big)}\ u^6 + 0u^5 + 0u^4 + 0u^3 + 0u^2 + 0u - 64} \\
\underline{u^6 - 2u^5} \\
2u^5 + 0u^4 \\
\underline{2u^5 - 4u^4} \\
4u^4 + 0u^3 \\
\underline{4u^4 - 8u^3} \\
8u^3 + 0u^2 \\
\underline{8u^3 - 16u^2} \\
16u^2 + 0u \\
\underline{16u^2 - 32u} \\
32u - 64 \\
\underline{32u - 64} \\
0
\end{array}$$

The quotient is $u^5 + 2u^4 + 4u^3 + 8u^2 + 16u + 32$ and the remainder is 0. Write the equation:

$$u^6 - 64 = (u - 2)\left(u^5 + 2u^4 + 4u^3 + 8u^2 + 16u + 32\right) + 0$$

63. Using long division:

$$\begin{array}{r}
5x^2 + 15x + 17 \\
x^2 - 3x + 5\ \overline{\smash{\big)}\ 5x^4 + 0x^3 - 3x^2 + 0x + 2} \\
\underline{5x^4 - 15x^3 + 25x^2} \\
15x^3 - 28x^2 + 0x \\
\underline{15x^3 - 45x^2 + 75x} \\
17x^2 - 75x + 2 \\
\underline{17x^2 - 51x + 85} \\
-24x - 83
\end{array}$$

The quotient is $5x^2 + 15x + 17$ and the remainder is $-24x - 83$. We write the equation:

$$5x^4 - 3x^2 + 2 = \left(x^2 - 3x + 5\right)\left(5x^2 + 15x + 17\right) + (-24x - 83)$$

65. Using long division:

$$\begin{array}{r}
t^2 - 2t - 4 \\
t^2 - 2t + 4\ \overline{\smash{\big)}\ t^4 - 4t^3 + 4t^2 + 0t - 16} \\
\underline{t^4 - 2t^3 + 4t^2} \\
-2t^3 + 0t^2 + 0t \\
\underline{-2t^3 + 4t^2 - 8t} \\
-4t^2 + 8t - 16 \\
\underline{-4t^2 + 8t - 16} \\
0
\end{array}$$

The quotient is $t^2 - 2t - 4$ and the remainder is 0. Write the equation:

$$t^4 - 4t^3 + 4t^2 - 16 = \left(t^2 - 2t + 4\right)\left(t^2 - 2t - 4\right) + 0$$

67. Using long division:

$$
\begin{array}{r}
z^4 + z^3 + z^2 + z + 1 \\
z-1 \overline{\smash{\big)}\, z^5 + 0z^4 + 0z^3 + 0z^2 + 0z - 1} \\
\underline{z^5 - 1z^4} \\
z^4 + 0z^3 \\
\underline{z^4 - 1z^3} \\
z^3 + 0z^2 \\
\underline{z^3 - 1z^2} \\
z^2 + 0z \\
\underline{z^2 - 1z} \\
z - 1 \\
\underline{z - 1} \\
0
\end{array}
$$

The quotient is $z^4 + z^3 + z^2 + z + 1$ and the remainder is 0. Write the equation:

$$z^5 - 1 = (z-1)\left(z^4 + z^3 + z^2 + z + 1\right) + 0$$

69. Using the special product for $(A - B)(A + B)$:

$$(a + b - 3)(a + b + 3) = (a+b)^2 - (3)^2 = a^2 + 2ab + b^2 - 9$$

71. Since $(a + b - x - y) = (a + b) - (x + y)$, use the special product for $(A - B)(A + B)$ to obtain:

$$(a + b + x + y)(a + b - x - y) = (a+b)^2 - (x+y)^2 = a^2 + 2ab + b^2 - x^2 - 2xy - y^2$$

73. Using the special product for $(A + B)^2$:

$$\left(\sqrt{a} + \sqrt{b}\right)^2 = \left(\sqrt{a}\right)^2 + 2\left(\sqrt{a}\right)\left(\sqrt{b}\right) + \left(\sqrt{b}\right)^2 = a + 2\sqrt{ab} + b$$

75. Using the special product for $(A - B)\left(A^2 + AB + B^2\right)$:

$$\left(x^{1/3} - y^{1/3}\right)\left(x^{2/3} + x^{1/3}y^{1/3} + y^{2/3}\right) = \left(x^{1/3}\right)^3 - \left(y^{1/3}\right)^3 = x - y$$

77. (a) Completing the tables:

x	$\dfrac{x^2 - 16}{x - 4}$
3.9	7.9
3.99	7.99
3.999	7.999
3.9999	7.9999
3.99999	7.99999

x	$\dfrac{x^2 - 16}{x - 4}$
4.1	8.1
4.01	8.01
4.001	8.001
4.0001	8.0001
4.00001	8.00001

(b) As x approaches 4, the value of the expression $\dfrac{x^2 - 16}{x - 4}$ approaches 8.

(c) Since $x^2 - 16 = (x+4)(x-4)$:
$$\frac{x^2-16}{x-4} = \frac{(x+4)(x-4)}{x-4} = x+4, \text{ if } x \neq 4$$
Thus as x approaches 4, we would expect $x+4$ to approach 8.

79. First compute the three products:
$$(b-c)(b+c-a) = b^2 + bc - ab - bc - c^2 + ac = b^2 - c^2 - ab + ac$$
$$(c-a)(c+a-b) = c^2 + ac - bc - ac - a^2 + ab = c^2 - a^2 - bc + ab$$
$$(a-b)(a+b-c) = a^2 + ab - ac - ab - b^2 + bc = a^2 - b^2 - ac + bc$$
By adding the three expressions we've obtained and combining like terms, we obtain the sum of 0, as required.

81. Using the special product for $(A-B)(A+B)$:
$$\left(x^2+8-4x\right)\left(x^2+8+4x\right) = \left(x^2+8\right)^2 - (4x)^2 = x^4 + 16x^2 + 64 - 16x^2 = x^4 + 64$$

83. (a) Using long division:
$$\begin{array}{r} ax + (b+ar) \\ x-r \ \overline{)\ ax^2 + bx + c} \\ \underline{ax^2 - arx} \\ (b+ar)x + c \\ \underline{(b+ar)x - r(b+ar)} \\ c + r(b+ar) \end{array}$$
The remainder is $c + r(b+ar) = ar^2 + br + c$.

(b) Using long division:
$$\begin{array}{r} ax^2 + (b+ar)x + (c+br+ar^2) \\ x-r \ \overline{)\ ax^3 + bx^2 + cx + d} \\ \underline{ax^3 - arx^2} \\ (b+ar)x^2 + cx \\ \underline{(b+ar)x^2 - r(b+ar)x} \\ (c+br+ar^2)x + d \\ \underline{(c+br+ar^2)x - r(c+br+ar^2)} \\ d + r(c+br+ar^2) \end{array}$$
The remainder is $d + r(c+br+ar^2) = ar^3 + br^2 + cr + d$.

(c) A reasonable guess would be $ar^4 + br^3 + cr^2 + dr + e$, which can be verified.

85. (a) Completing the table:

n	1	2	3	4	5	6
$\dfrac{n^2}{n+1}$	$\dfrac{1}{2}$	$\dfrac{4}{3}$	$\dfrac{9}{4}$	$\dfrac{16}{5}$	$\dfrac{25}{6}$	$\dfrac{36}{7}$

(b)   Using long division:

$$\begin{array}{r} n-1 \\ n+1 \overline{\smash{\big)}\, n^2 + 0n + 0} \\ \underline{n^2 + 1n} \\ -1n + 0 \\ \underline{-1n - 1} \\ 1 \end{array}$$

Since the remainder is always 1, $\dfrac{n^2}{n+1}$ can never be a natural number.

1.7 Factoring

1. (a) Using a difference of squares:
$$x^2 - 64 = (x+8)(x-8)$$

(b) Using a common factor of $7x^2$:
$$7x^4 + 14x^2 = 7x^2(x^2 + 2)$$

(c) Using a common factor of z then a difference of squares:
$$121z - z^3 = z(121 - z^2) = z(11 + z)(11 - z)$$

(d) Using a difference of squares:
$$a^2b^2 - c^2 = (ab)^2 - (c)^2 = (ab + c)(ab - c)$$

3. (a) Factoring by trial and error:
$$x^2 + 2x - 3 = (x+3)(x-1)$$

(b) Factoring by trial and error:
$$x^2 - 2x - 3 = (x-3)(x+1)$$

(c) Factoring by trial and error, we find that $x^2 - 2x + 3$ is irreducible.

(d) Factoring by trial and error:
$$-x^2 + 2x + 3 = (-x+3)(x+1)$$
Note that we could also use a common factor of -1, then use trial and error:
$$-x^2 + 2x + 3 = -1(x^2 - 2x - 3) = -(x-3)(x+1)$$

5. (a) By the sum of cubes:
$$x^3 + 1 = (x+1)(x^2 - x + 1)$$

(b) Since $6^3 = 216$, use the sum of cubes to obtain:
$$x^3 + 216 = x^3 + 6^3 = (x+6)(x^2 - 6x + 36)$$

(c) Using a common factor of 8 then a difference of cubes:
$$1000 - 8x^6 = 8(125 - x^6) = 8(5^3 - x^6) = 8(5 - x^2)(25 + 5x^2 + x^4)$$

(d) By the difference of cubes:
$$64a^3x^3 - 125 = (4ax)^3 - 5^3 = (4ax - 5)(16a^2x^2 + 20ax + 25)$$

7. (a) Using a difference of squares:
$$144 - x^2 = (12 + x)(12 - x)$$

(b) Since this is a sum of squares, the expression $144 + x^2$ is irreducible.

(c) Using a difference of squares:
$$144 - (y - 3)^2 = (12 + y - 3)(12 - y + 3) = (9 + y)(15 - y)$$

9. (a) Using a common factor of h^3 then a difference of squares:
$$h^3 - h^5 = h^3(1 - h^2) = h^3(1 + h)(1 - h)$$

(b) Using a common factor of h^3 then a difference of squares:
$$100h^3 - h^5 = h^3(100 - h^2) = h^3(10 + h)(10 - h)$$

(c) Using a common factor of $(h + 1)^3$ then a difference of squares:
$$100(h + 1)^3 - (h + 1)^5 = (h + 1)^3[100 - (h + 1)^2]$$
$$= (h + 1)^3(10 + h + 1)(10 - h - 1)$$
$$= (h + 1)^3(11 + h)(9 - h)$$

11. (a) Factoring by trial and error:
$$x^2 - 13x + 40 = (x - 8)(x - 5)$$

(b) Factoring by trial and error, we find that $x^2 - 13x - 40$ is irreducible.

13. (a) Factoring by trial and error:
$$x^2 + 5x - 36 = (x + 9)(x - 4)$$

(b) Factoring by trial and error:
$$x^2 - 13x + 36 = (x - 9)(x - 4)$$

15. (a) Factoring by trial and error:
$$3x^2 - 22x - 16 = (3x + 2)(x - 8)$$

(b) Factoring by trial and error, we find that $3x^2 - x - 16$ is irreducible.

17. **(a)** Factoring by trial and error:
$$6x^2 + 13x - 5 = (3x - 1)(2x + 5)$$

(b) Factoring by trial and error:
$$6x^2 - x - 5 = (6x + 5)(x - 1)$$

19. **(a)** Using the special product for $(A + B)^2$:
$$t^4 + 2t^2 + 1 = (t^2 + 1)^2$$

(b) Using the special product for $(A - B)^2$ and the difference of squares:
$$t^4 - 2t^2 + 1 = (t^2 - 1)^2 = (t + 1)^2(t - 1)^2$$

(c) By trial and error, we find that $t^4 - 2t^2 - 1$ is irreducible.

21. **(a)** Using the common factor x:
$$4x^3 - 20x^2 - 25x = x(4x^2 - 20x - 25)$$

(b) Using the common factor x and the special product for $(A - B)^2$::
$$4x^3 - 20x^2 + 25x = x(4x^2 - 20x + 25) = x(2x - 5)^2$$

23. **(a)** Using grouping:
$$ab - bc + a^2 - ac = b(a - c) + a(a - c) = (a - c)(b + a)$$

(b) Using grouping:
$$(u + v)x - xy + (u + v)^2 - (u + v)y = x(u + v - y) + (u + v)(u + v - y)$$
$$= (u + v - y)(x + u + v)$$

25. Using grouping:
$$x^2z^2 + xzt + xyz + yt = xz(xz + t) + y(xz + t) = (xz + t)(xz + y)$$

27. Using the special product for $(A - B)^2$:
$$a^4 - 4a^2b^2c^2 + 4b^4c^4 = (a^2 - 2b^2c^2)^2$$

29. Since this is a sum of squares, $A^2 + B^2$ is irreducible.

31. Since $4^3 = 64$, using the sum of cubes factoring yields:
$$x^3 + 64 = (x + 4)(x^2 - 4x + 16)$$

33. Using the difference of cubes factoring:
$$(x+y)^3 - y^3 = (x+y-y)\left[(x+y)^2 + (x+y)(y) + y^2\right]$$
$$= x\left(x^2 + 2xy + y^2 + xy + y^2 + y^2\right)$$
$$= x\left(x^2 + 3xy + 3y^2\right)$$

35. Using the difference of cubes factoring, then grouping:
$$x^3 - y^3 + x - y = (x-y)(x^2 + xy + y^2) + (x-y)(1) = (x-y)(x^2 + xy + y^2 + 1)$$

37. (a) Using the difference of squares factoring (twice):
$$p^4 - 1 = (p^2+1)(p^2-1) = (p^2+1)(p+1)(p-1)$$

(b) Using the difference of squares factoring (three times):
$$p^8 - 1 = (p^4+1)(p^4-1)$$
$$= (p^4+1)(p^2+1)(p^2-1)$$
$$= (p^4+1)(p^2+1)(p+1)(p-1)$$

39. Using the special product for $(A+B)^3$:
$$x^3 + 3x^2 + 3x + 1 = (x+1)^3$$
Note: If you do not recognize the special product, you can also use grouping:
$$x^3 + 3x^2 + 3x + 1 = (x^3+1) + (3x^2 + 3x)$$
$$= (x+1)(x^2 - x + 1) + 3x(x+1)$$
$$= (x+1)(x^2 - x + 1 + 3x)$$
$$= (x+1)(x^2 + 2x + 1)$$
$$= (x+1)(x+1)^2$$
$$= (x+1)^3$$

41. Since this is a sum of squares, $x^2 + 16y^2$ is irreducible.

43. Using the difference of squares factoring:
$$\tfrac{25}{16} - c^2 = \left(\tfrac{5}{4} + c\right)\left(\tfrac{5}{4} - c\right)$$

45. Using the difference of squares factoring (twice):
$$z^4 - \tfrac{81}{16} = \left(z^2 + \tfrac{9}{4}\right)\left(z^2 - \tfrac{9}{4}\right) = \left(z^2 + \tfrac{9}{4}\right)\left(z + \tfrac{3}{2}\right)\left(z - \tfrac{3}{2}\right)$$

47. Using the difference of cubes factoring:
$$\frac{125}{m^3 n^3} - 1 = \left(\frac{5}{mn} - 1\right)\left(\frac{25}{m^2 n^2} + \frac{5}{mn} + 1\right)$$

49. Using the special product for $(A+B)^2$:
$$\tfrac{1}{4}x^2 + xy + y^2 = \left(\tfrac{1}{2}x + y\right)^2$$

51. Using the common factor $x-a$, then a difference of squares:
$$64(x-a)^3 - x + a = (x-a)\left[64(x-a)^2 - 1\right]$$
$$= (x-a)[8(x-a)+1][8(x-a)-1]$$
$$= (x-a)(8x-8a+1)(8x-8a-1)$$

53. Using grouping:
$$x^2 - a^2 + y^2 - 2xy = \left(x^2 - 2xy + y^2\right) - a^2$$
$$= (x-y)^2 - a^2$$
$$= (x-y-a)(x-y+a)$$

55. Using the common factor x, then trial and error:
$$21x^3 + 82x^2 - 39x = x\left(21x^2 + 82x - 39\right) = x(7x-3)(3x+13)$$

57. Using grouping, the special product for $(A-B)^2$, and the difference of squares factoring:
$$12xy + 25 - 4x^2 - 9y^2 = 5^2 - \left(4x^2 - 12xy + 9y^2\right)$$
$$= 5^2 - (2x-3y)^2$$
$$= [5 - (2x-3y)][5 + (2x-3y)]$$
$$= (5 - 2x + 3y)(5 + 2x - 3y)$$

59. Factoring by trial and error:
$$ax^2 + (a+b)x + b = (ax+b)(x+1)$$

61. Using the common factor $(x+1)^{1/2}$:
$$(x+1)^{1/2} - (x+1)^{3/2} = (x+1)^{1/2}\left[1 - (x+1)^{2/2}\right]$$
$$= (x+1)^{1/2}(1 - x - 1)$$
$$= (x+1)^{1/2}(-x)$$
$$= -x(x+1)^{1/2}$$

63. Using the common factor $(x+1)^{-3/2}$:
$$(x+1)^{-1/2} - (x+1)^{-3/2} = (x+1)^{-3/2}\left[(x+1)^1 - 1\right]$$
$$= (x+1)^{-3/2}(x)$$
$$= x(x+1)^{-3/2}$$

65. Using the common factor $\frac{1}{3}(2x+3)^{1/2}$:

$$(2x+3)^{1/2} - \frac{1}{3}(2x+3)^{3/2} = \frac{1}{3}(2x+3)^{1/2}\left[3 - (2x+3)^1\right]$$
$$= \frac{1}{3}(2x+3)^{1/2}(3 - 2x - 3)$$
$$= \frac{1}{3}(2x+3)^{1/2}(-2x)$$
$$= -\frac{2}{3}x(2x+3)^{1/2}$$

67. (a) Using the difference of squares factoring:
$$100^2 - 99^2 = (100+99)(100-99) = (199)(1) = 199$$

(b) Using the difference of cubes factoring:
$$8^3 - 6^3 = (8-6)(8^2 + 8\cdot 6 + 6^2)$$
$$= 2(64 + 48 + 36)$$
$$= 2(148)$$
$$= 296$$

(c) Using the difference of squares factoring:
$$1000^2 - 999^2 = (1000+999)(1000-999) = (1999)(1) = 1999$$

69. Factoring the sum of cubes and then grouping:
$$A^3 + B^3 + 3AB(A+B) = (A+B)(A^2 - AB + B^2) + (A+B)(3AB)$$
$$= (A+B)(A^2 - AB + B^2 + 3AB)$$
$$= (A+B)(A^2 + 2AB + B^2)$$
$$= (A+B)(A+B)^2$$
$$= (A+B)^3$$

71. Using the common factor $x(a^2 + x^2)^{-3/2}$:

$$2x(a^2 + x^2)^{-1/2} - x^3(a^2 + x^2)^{-3/2} = x(a^2 + x^2)^{-3/2}\left[2(a^2 + x^2) - x^2\right]$$
$$= x(a^2 + x^2)^{-3/2}(2a^2 + 2x^2 - x^2)$$
$$= x(a^2 + x^2)^{-3/2}(2a^2 + x^2)$$

73. Using grouping (first with fourth term, second with third term):
$$y^4 - p^2q^2 - (p+q)y^3 + pq(p+q)y = (y^2 + pq)(y^2 - pq) - (p+q)(y)(y^2 - pq)$$
$$= (y^2 - pq)(y^2 + pq - py - qy)$$
$$= (y^2 - pq)(y^2 - (p+q)y + pq)$$
$$= (y^2 - pq)(y - p)(y - q)$$

75. First multiply out some terms, then factor by grouping:

$$(1+t)^2(1+u^2)-(1+u)^2(1+t^2)$$
$$=\left[(1+t)^2-(1+u)^2\right]+\left[u^2(1+t)^2-t^2(1+u)^2\right]$$
$$=(1+t+1+u)(1+t-1-u)+(u+ut+t+ut)(u+ut-t-ut)$$
$$=(-2-u-t)(u-t)+(u+t+2ut)(u-t)$$
$$=(u-t)(-2-u-t+u+t+2ut)$$
$$=(u-t)(2ut-2)$$
$$=2(u-t)(ut-1)$$

77. Adding and subtracting $16x^2$:

$$x^4+64=x^4+16x^2+64-16x^2$$
$$=\left(x^2+8\right)^2-16x^2$$
$$=\left(x^2+8+4x\right)\left(x^2+8-4x\right)$$
$$=\left(x^2+4x+8\right)\left(x^2-4x+8\right)$$

79. Factor as follows:

$$(x+y)^2+(x+z)^2-(z+t)^2-(y+t)^2$$
$$=\left[(x+y)^2-(y+t)^2\right]+\left[(x+z)^2-(z+t)^2\right]$$
$$=\left[x+y-(y+t)\right]\left[x+y+(y+t)\right]+\left[x+z-(z+t)\right]\left[x+z+(z+t)\right]$$
$$=(x-t)(x+2y+t)+(x-t)(x+2z+t)$$
$$=(x-t)(x+2y+t+x+2z+t)$$
$$=(x-t)(2x+2y+2z+2t)$$
$$=2(x-t)(x+y+z+t)$$

81. Factor as follows:

$$(b-c)^3+(c-a)^3+(a-b)^3$$
$$=b^3-3b^2c+3bc^2-c^3+c^3-3c^2a+3ca^2-a^3+a^3-3a^2b+3ab^2-b^3$$
$$=-3b^2c+3bc^2-3c^2a+3ca^2-3a^2b+3ab^2$$
$$=\left(3bc^2-3c^2a\right)+\left(3ca^2-3b^2c\right)+\left(3ab^2-3a^2b\right)$$
$$=3c^2(b-a)+3c\left(a^2-b^2\right)+3ab(b-a)$$
$$=3c^2(b-a)-3c\left(b^2-a^2\right)+3ab(b-a)$$
$$=3(b-a)\left[c^2-c(b+a)+ab\right]$$
$$=3(b-a)\left[c^2-cb-ca+ab\right]$$
$$=3(b-a)\left[c(c-b)-a(c-b)\right]$$
$$=3(b-a)\left[(c-b)(c-a)\right]$$
$$=3(b-a)(c-a)(c-b)$$

1.8 Fractional Expressions

1. Factor and reduce:

$$\frac{x^2-9}{x+3}=\frac{(x-3)(x+3)}{x+3}=x-3$$

3. Factor and reduce:

$$\frac{x+2}{x^4-16}=\frac{x+2}{\left(x^2-4\right)\left(x^2+4\right)}=\frac{x+2}{(x-2)(x+2)\left(x^2+4\right)}=\frac{1}{(x-2)\left(x^2+4\right)}$$

5. Factor and reduce:

$$\frac{x^2+2x+4}{x^3-8}=\frac{x^2+2x+4}{(x-2)\left(x^2+2x+4\right)}=\frac{1}{x-2}$$

7. Factor and reduce:

$$\frac{9ab-12b^2}{6a^2-8ab}=\frac{3b(3a-4b)}{2a(3a-4b)}=\frac{3b}{2a}$$

9. Factor and reduce:

$$\frac{a^3+a^2+a+1}{a^2-1}=\frac{a^2(a+1)+(a+1)}{(a-1)(a+1)}=\frac{(a+1)\left(a^2+1\right)}{(a-1)(a+1)}=\frac{a^2+1}{a-1}$$

11. Factor and reduce:

$$\frac{x^3-y^3}{(x-y)^3}=\frac{(x-y)\left(x^2+xy+y^2\right)}{(x-y)^3}=\frac{x^2+xy+y^2}{(x-y)^2}$$

13. Multiplying and factoring:

$$\frac{2}{x-2}\cdot\frac{x^2-4}{x+2}=\frac{2(x-2)(x+2)}{(x-2)(x+2)}=2$$

15. Multiplying and factoring:

$$\frac{x^2-x-2}{x^2+x-12}\cdot\frac{x^2-3x}{x^2-4x+4}=\frac{(x-2)(x+1)}{(x+4)(x-3)}\cdot\frac{x(x-3)}{(x-2)^2}=\frac{x(x+1)}{(x+4)(x-2)}=\frac{x^2+x}{(x+4)(x-2)}$$

17. Dividing and factoring:

$$\frac{x^3+y^3}{x^2-4xy+3y^2}\div\frac{(x+y)^3}{x^2-2xy-3y^2}=\frac{(x+y)\left(x^2-xy+y^2\right)}{(x-3y)(x-y)}\cdot\frac{(x+y)(x-3y)}{(x+y)^3}$$

$$=\frac{x^2-xy+y^2}{(x-y)(x+y)}$$

19. Multiplying and factoring:

$$\frac{x^2 + xy - 2y^2}{x^2 - 5xy + 4y^2} \cdot \frac{x^2 - 7xy + 12y^2}{x^2 + 5xy + 6y^2} = \frac{(x+2y)(x-y)}{(x-4y)(x-y)} \cdot \frac{(x-4y)(x-3y)}{(x+3y)(x+2y)} = \frac{x-3y}{x+3y}$$

21. Obtaining common denominators and subtracting yields:

$$\frac{4}{x} - \frac{2}{x^2} = \frac{4x}{x^2} - \frac{2}{x^2} = \frac{4x-2}{x^2}$$

23. Obtaining common denominators and subtracting yields:

$$\frac{6}{a} - \frac{a}{6} = \frac{36}{6a} - \frac{a^2}{6a} = \frac{36 - a^2}{6a}$$

25. Obtaining common denominators and adding yields:

$$\frac{1}{x+3} + \frac{3}{x+2} = \frac{x+2}{(x+3)(x+2)} + \frac{3(x+3)}{(x+3)(x+2)} = \frac{x+2+3(x+3)}{(x+3)(x+2)} = \frac{4x+11}{(x+3)(x+2)}$$

27. Obtaining common denominators and subtracting yields:

$$\frac{3x}{x-2} - \frac{6}{x^2-4} = \frac{3x}{x-2} - \frac{6}{(x-2)(x+2)} = \frac{3x(x+2)-6}{(x-2)(x+2)} = \frac{3x^2+6x-6}{(x-2)(x+2)}$$

29. Obtaining common denominators and adding yields:

$$\frac{a}{x-1} + \frac{2ax}{(x-1)^2} + \frac{3ax^2}{(x-1)^3} = \frac{a(x-1)^2 + 2ax(x-1) + 3ax^2}{(x-1)^3}$$

$$= \frac{ax^2 - 2ax + a + 2ax^2 - 2ax + 3ax^2}{(x-1)^3}$$

$$= \frac{6ax^2 - 4ax + a}{(x-1)^3}$$

31. Obtaining common denominators and adding yields:

$$\frac{x}{x^2-9} + \frac{x-1}{x^2-5x+6} = \frac{x}{(x-3)(x+3)} + \frac{x-1}{(x-3)(x-2)}$$

$$= \frac{x(x-2) + (x-1)(x+3)}{(x-3)(x+3)(x-2)}$$

$$= \frac{x^2 - 2x + x^2 + 2x - 3}{(x-3)(x+3)(x-2)}$$

$$= \frac{2x^2 - 3}{(x-3)(x+3)(x-2)}$$

33. Factoring out −1 yields:

$$\frac{4}{x-5} - \frac{4}{5-x} = \frac{4}{x-5} + \frac{4}{x-5} = \frac{8}{x-5}$$

35. Obtaining common denominators and combining yields:

$$\frac{a^2+b^2}{a^2-b^2}+\frac{a}{a+b}+\frac{b}{b-a}=\frac{a^2+b^2}{(a-b)(a+b)}+\frac{a}{a+b}-\frac{b}{a-b}$$

$$=\frac{a^2+b^2+a(a-b)-b(a+b)}{(a-b)(a+b)}$$

$$=\frac{a^2+b^2+a^2-ab-ab-b^2}{(a-b)(a+b)}$$

$$=\frac{2a^2-2ab}{(a-b)(a+b)}$$

$$=\frac{2a(a-b)}{(a-b)(a+b)}$$

$$=\frac{2a}{a+b}$$

37. Obtaining common denominators and subtracting yields:

$$\frac{1}{x^2+x-20}-\frac{1}{x^2-8x+16}=\frac{1}{(x+5)(x-4)}-\frac{1}{(x-4)^2}$$

$$=\frac{x-4-(x+5)}{(x+5)(x-4)^2}$$

$$=\frac{-9}{(x+5)(x-4)^2}$$

39. Obtaining common denominators and subtracting yields:

$$\frac{2q+p}{2p^2-9pq-5q^2}-\frac{p+q}{p^2-5pq}=\frac{2q+p}{(2p+q)(p-5q)}-\frac{p+q}{p(p-5q)}$$

$$=\frac{(2q+p)p-(p+q)(2p+q)}{p(2p+q)(p-5q)}$$

$$=\frac{2pq+p^2-2p^2-3pq-q^2}{p(2p+q)(p-5q)}$$

$$=\frac{-p^2-pq-q^2}{p(2p+q)(p-5q)}$$

41. Obtaining common denominators and combining yields:

$$\frac{x}{(x-y)(x-z)}+\frac{y}{(y-z)(y-x)}+\frac{z}{(z-x)(z-y)}$$

$$=\frac{x}{(x-y)(x-z)}-\frac{y}{(y-z)(x-y)}+\frac{z}{(x-z)(y-z)}$$

$$=\frac{x(y-z)-y(x-z)+z(x-y)}{(x-y)(x-z)(y-z)}$$

$$=\frac{xy-xz-xy+yz+xz-yz}{(x-y)(x-z)(y-z)}$$

$$=\frac{0}{(x-y)(x-z)(y-z)}$$

$$=0$$

43. Obtaining common denominators and combining yields:

$$\frac{y+z}{x^2-xy-xz+yz}-\frac{x+z}{xy-xz-y^2+yz}+\frac{x+y}{xy-yz-xz+z^2}$$

$$=\frac{y+z}{x(x-y)-z(x-y)}-\frac{x+z}{x(y-z)-y(y-z)}+\frac{x+y}{y(x-z)-z(x-z)}$$

$$=\frac{y+z}{(x-y)(x-z)}-\frac{x+z}{(y-z)(x-y)}+\frac{x+y}{(x-z)(y-z)}$$

$$=\frac{(y+z)(y-z)-(x+z)(x-z)+(x+y)(x-y)}{(x-y)(x-z)(y-z)}$$

$$=\frac{y^2-z^2-x^2+z^2+x^2-y^2}{(x-y)(x-z)(y-z)}$$

$$=\frac{0}{(x-y)(x-z)(y-z)}$$

$$=0$$

45. Obtaining common denominators and combining yields:

$$\frac{x^2+x-a-1}{a^2-x^2}+\frac{x+1}{x+a}+\frac{x-a}{x-1}=\frac{x^2+x-a-1}{a^2-x^2}+\frac{x+1}{x+a}\cdot\frac{x-1}{x-a}$$

$$=\frac{x^2+x-a-1}{-\left(x^2-a^2\right)}+\frac{x^2-1}{x^2-a^2}$$

$$=\frac{-x^2-x+a+1+x^2-1}{x^2-a^2}$$

$$=\frac{-x+a}{(x+a)(x-a)}$$

$$=\frac{-(x-a)}{(x+a)(x-a)}$$

$$=\frac{-1}{x+a}$$

47. Multiplying by $\dfrac{x}{x}$ yields:

$$\frac{\frac{1}{x}+1}{\frac{1}{x}-1}\cdot\frac{x}{x}=\frac{1+x}{1-x}$$

49. Multiplying by $\dfrac{xa}{xa}$ yields:

$$\frac{\frac{1}{x}-\frac{1}{a}}{x-a}\cdot\frac{xa}{xa}=\frac{a-x}{xa(x-a)}=-\frac{1}{ax}$$

51. Multiplying by $\dfrac{x}{x}$ yields:

$$\frac{1+\frac{4}{x}}{\frac{3}{x}-2}\cdot\frac{x}{x}=\frac{x+4}{3-2x}$$

53. Multiplying by $\dfrac{a}{a}$ yields:

$$\frac{a-\frac{1}{a}}{1+\frac{1}{a}}\cdot\frac{a}{a}=\frac{a^2-1}{a+1}=\frac{(a+1)(a-1)}{a+1}=a-1$$

55. Multiplying by $\dfrac{2(2+h)}{2(2+h)}$ yields:

$$\frac{\frac{1}{2+h}-\frac{1}{2}}{h}\cdot\frac{2(2+h)}{2(2+h)}=\frac{2-2-h}{2h(2+h)}=\frac{-h}{2h(2+h)}=-\frac{1}{4+2h}$$

57. Multiplying by $\dfrac{a^2x^2}{a^2x^2}$ yields:

$$\frac{\frac{a}{x^2}+\frac{x}{a^2}}{a^2-ax+x^2}\cdot\frac{a^2x^2}{a^2x^2}=\frac{a^3+x^3}{a^2x^2(a^2-ax+x^2)}=\frac{(a+x)(a^2-ax+x^2)}{a^2x^2(a^2-ax+x^2)}=\frac{a+x}{a^2x^2}$$

59. Rewriting the expression without negative exponents:

$$\left(x^{-1}+2\right)^{-1}=\frac{1}{x^{-1}+2}=\frac{1}{\frac{1}{x}+2}\cdot\frac{x}{x}=\frac{x}{1+2x}$$

61. Rewriting the expression without negative exponents:

$$\left(\frac{1}{a^{-1}}+\frac{1}{a^{-2}}\right)^{-1}=\left(a+a^2\right)^{-1}=\frac{1}{a+a^2}$$

63.   Rewriting the expression without negative exponents:

$$x(x+y)^{-1} + y(x-y)^{-1} = \frac{x}{x+y} + \frac{y}{x-y} = \frac{x(x-y)+y(x+y)}{x^2-y^2} = \frac{x^2+y^2}{x^2-y^2}$$

65.   Rewriting the expression without negative exponents:

$$\begin{aligned}
\left(t^2 + t^{-2} + 2\right)\left(t + t^{-1}\right)^{-1} &= \frac{t^2 + t^{-2} + 2}{t + t^{-1}} \\
&= \frac{t^2 + \frac{1}{t^2} + 2}{t + \frac{1}{t}} \cdot \frac{t^2}{t^2} \\
&= \frac{t^4 + 1 + 2t^2}{t^3 + t} \\
&= \frac{\left(t^2 + 1\right)^2}{t\left(t^2 + 1\right)} \\
&= \frac{t^2 + 1}{t}
\end{aligned}$$

67.   Combining fractions and simplifying yields:

$$\frac{x+y}{2(x^2+y^2)} - \frac{1}{2(x+y)} + \frac{x-y}{x^2-y^2} - \frac{x^3-y^3}{x^4-y^4}$$

$$= \frac{x+y}{2(x^2+y^2)} - \frac{1}{2(x+y)} + \frac{1}{x+y} - \frac{(x-y)(x^2+xy+y^2)}{(x-y)(x+y)(x^2+y^2)}$$

$$= \frac{x+y}{2(x^2+y^2)} - \frac{1}{2(x+y)} + \frac{1}{x+y} - \frac{x^2+xy+y^2}{(x+y)(x^2+y^2)}$$

$$= \frac{(x+y)(x+y) - (x^2+y^2) + 2(x^2+y^2) - 2(x^2+xy+y^2)}{2(x^2+y^2)(x+y)}$$

$$= \frac{x^2 + 2xy + y^2 - x^2 - y^2 + 2x^2 + 2y^2 - 2x^2 - 2xy - 2y^2}{2(x^2+y^2)(x+y)}$$

$$= \frac{0}{2(x^2+y^2)(x+y)}$$

$$= 0$$

69. Multiplying by $\dfrac{(a-b)(a+b)}{(a-b)(a+b)}$ yields:

$$\frac{(a-b)(a+b)}{(a-b)(a+b)} \cdot \frac{\frac{a+b}{a-b}+\frac{a-b}{a+b}}{\frac{a-b}{a+b}-\frac{a+b}{a-b}} \cdot \frac{ab(b^2-a^2)}{a^2+b^2} = \frac{(a+b)^2+(a-b)^2}{(a-b)^2-(a+b)^2} \cdot \frac{ab(b-a)(b+a)}{a^2+b^2}$$

$$= \frac{2a^2+2b^2}{-4ab} \cdot \frac{ab(b-a)(b+a)}{a^2+b^2}$$

$$= \frac{2(a^2+b^2)}{-4} \cdot \frac{(b-a)(b+a)}{a^2+b^2}$$

$$= \frac{(b-a)(b+a)}{-2}$$

$$= \frac{b^2-a^2}{-2}$$

$$= \frac{a^2-b^2}{2}$$

71. Rewriting the expression without negative exponents:

$$\left[x-\left(x+x^{-1}\right)^{-1}\right]^{-1} - \left[x+\left(x-x^{-1}\right)^{-1}\right]^{-1}$$

$$= \frac{1}{x-\left(x+x^{-1}\right)^{-1}} - \frac{1}{x+\left(x-x^{-1}\right)^{-1}}$$

$$= \frac{1}{x-\frac{1}{x+x^{-1}}} - \frac{1}{x+\frac{1}{x-x^{-1}}}$$

$$= \frac{1}{x-\frac{1}{x+x^{-1}}} \cdot \frac{x+x^{-1}}{x+x^{-1}} - \frac{1}{x+\frac{1}{x-x^{-1}}} \cdot \frac{x-x^{-1}}{x-x^{-1}}$$

$$= \frac{x+x^{-1}}{x^2+1-1} - \frac{x-x^{-1}}{x^2-1+1}$$

$$= \frac{x+\frac{1}{x}}{x^2} - \frac{x-\frac{1}{x}}{x^2}$$

$$= \frac{x^2+1}{x^3} - \frac{x^2-1}{x^3}$$

$$= \frac{x^2+1-x^2+1}{x^3}$$

$$= \frac{2}{x^3}$$

73. Combining fractions and simplifying yields:

$$\frac{ap+q}{ax-bx-a^2+ab}+\frac{bp+q}{bx-ax-b^2+ab}$$

$$=\frac{ap+q}{x(a-b)-a(a-b)}+\frac{bp+q}{x(b-a)-b(b-a)}$$

$$=\frac{ap+q}{(a-b)(x-a)}+\frac{bp+q}{(b-a)(x-b)}$$

$$=\frac{ap+q}{(a-b)(x-a)}-\frac{bp+q}{(a-b)(x-b)}$$

$$=\frac{(ap+q)(x-b)-(bp+q)(x-a)}{(a-b)(x-a)(x-b)}$$

$$=\frac{apx+qx-abp-bq-bpx-qx+apb+aq}{(a-b)(x-a)(x-b)}$$

$$=\frac{apx-bq-bpx+aq}{(a-b)(x-a)(x-b)}$$

$$=\frac{a(px+q)-b(px+q)}{(a-b)(x-a)(x-b)}$$

$$=\frac{(px+q)(a-b)}{(a-b)(x-a)(x-b)}$$

$$=\frac{px+q}{(x-a)(x-b)}$$

75. Simplifying the complex fractions:

$$\frac{\frac{1}{a}-\frac{a-x}{a^2+x^2}}{\frac{1}{x}-\frac{x-a}{x^2+a^2}}+\frac{\frac{1}{a}-\frac{a+x}{a^2+x^2}}{\frac{1}{x}-\frac{x+a}{x^2+a^2}}=\frac{ax(a^2+x^2)}{ax(a^2+x^2)}\bullet\frac{\frac{1}{a}-\frac{a-x}{a^2+x^2}}{\frac{1}{x}-\frac{x-a}{x^2+a^2}}+\frac{ax(a^2+x^2)}{ax(a^2+x^2)}\bullet\frac{\frac{1}{a}-\frac{a+x}{a^2+x^2}}{\frac{1}{x}-\frac{x+a}{x^2+a^2}}$$

$$=\frac{x(a^2+x^2)-(a-x)ax}{a(a^2+x^2)-(x-a)ax}+\frac{x(a^2+x^2)-(a+x)ax}{a(a^2+x^2)-(x+a)ax}$$

$$=\frac{a^2x+x^3-a^2x+ax^2}{a^3+ax^2-ax^2+a^2x}+\frac{a^2x+x^3-a^2x-ax^2}{a^3+ax^2-ax^2-a^2x}$$

$$=\frac{x^3+ax^2}{a^3+a^2x}+\frac{x^3-ax^2}{a^3-a^2x}$$

$$=\frac{x^2(x+a)}{a^2(a+x)}+\frac{x^2(x-a)}{a^2(a-x)}$$

$$=\frac{x^2}{a^2}+\frac{x^2}{a^2}(-1)$$

$$=0$$

77. Combining fractions yields:

$$\frac{x^2 - qr}{(p-q)(p-r)} + \frac{x^2 - rp}{(q-r)(q-p)} + \frac{x^2 - pq}{(r-p)(r-q)}$$

$$= \frac{x^2 - qr}{(p-q)(p-r)} - \frac{x^2 - rp}{(q-r)(p-q)} + \frac{x^2 - pq}{(p-r)(q-r)}$$

$$= \frac{(x^2 - qr)(q-r) - (x^2 - rp)(p-r) + (x^2 - pq)(p-q)}{(p-q)(p-r)(q-r)}$$

$$= \frac{qx^2 - q^2r - rx^2 + qr^2 - px^2 + p^2r + rx^2 - pr^2 + px^2 - p^2q - qx^2 + pq^2}{(p-q)(p-r)(q-r)}$$

$$= \frac{-q^2r + qr^2 + p^2r - pr^2 - p^2q + pq^2}{(p-q)(p-r)(q-r)}$$

$$= \frac{(p^2r - q^2r) - (pr^2 - qr^2) - (p^2q - pq^2)}{(p-q)(p-r)(q-r)}$$

$$= \frac{r(p-q)(p+q) - r^2(p-q) - pq(p-q)}{(p-q)(p-r)(q-r)}$$

$$= \frac{(p-q)[r(p+q) - r^2 - pq]}{(p-q)(p-r)(q-r)}$$

$$= \frac{rp + rq - r^2 - pq}{(p-r)(q-r)}$$

$$= \frac{p(r-q) + r(q-r)}{(p-r)(q-r)}$$

$$= \frac{-p(q-r) + r(q-r)}{(p-r)(q-r)}$$

$$= \frac{(q-r)(-p+r)}{(p-r)(q-r)}$$

$$= -1$$

79. (a)   The sum and the product both equal $\frac{1}{42}$.

(b)   After adding the three fractions and combining the many like terms appearing in the resulting numerator, we obtain:

$$\frac{b-c}{1+bc}+\frac{c-a}{1+ca}+\frac{a-b}{1+ab}=\frac{ab^2-ac^2+bc^2-a^2b+a^2c-b^2c}{(1+bc)(1+ca)(1+ab)}$$

$$=\frac{\left(ab^2-b^2c\right)+\left(a^2c-ac^2\right)+\left(bc^2-a^2b\right)}{(1+bc)(1+ca)(1+ab)}$$

$$=\frac{b^2(a-c)+ac(a-c)+b\left(c^2-a^2\right)}{(1+bc)(1+ca)(1+ab)}$$

$$=\frac{b^2(a-c)+ac(a-c)-b(a-c)(a+c)}{(1+bc)(1+ca)(1+ab)}$$

$$=\frac{(a-c)\left(b^2+ac-b(a+c)\right)}{(1+bc)(1+ca)(1+ab)}$$

$$=\frac{(a-c)\left(b^2+ac-ab-bc\right)}{(1+bc)(1+ca)(1+ab)}$$

$$=\frac{(a-c)(c(a-b)-b(a-b))}{(1+bc)(1+ca)(1+ab)}$$

$$=\frac{(a-c)((a-b)(c-b))}{(1+bc)(1+ca)(1+ab)}$$

$$=\frac{(c-a)(a-b)(b-c)}{(1+bc)(1+ca)(1+ab)}$$

$$=\frac{b-c}{1+bc}\bullet\frac{c-a}{1+ca}\bullet\frac{a-b}{1+ab}$$

81.   For convenience, let $z=x-y$. Then $-z=y-x$, therefore:

$$\left(1+a^z\right)^{-1}+\left(1+a^{-z}\right)^{-1}=\frac{1}{1+a^z}+\frac{1}{1+\frac{1}{a^z}}$$

$$=\frac{1}{1+a^z}+\frac{a^z}{a^z}\bullet\frac{1}{1+\frac{1}{a^z}}$$

$$=\frac{1}{1+a^z}+\frac{a^z}{a^z+1}$$

$$=\frac{1+a^z}{1+a^z}$$

$$=1$$

1.9 Solving Equations (Review and Preview)

1. Substituting $x = -2$ into the equation:
$$4x - 5 = 4(-2) - 5 = -8 - 5 = -13$$
So $x = -2$ is a solution to the equation.

3. Substituting $y = -3$ into each side of the equation:
$$\frac{2}{y-1} - \frac{3}{y} = \frac{2}{-3-1} - \frac{3}{-3} = -\frac{1}{2} + 1 = \frac{1}{2}$$
$$\frac{7}{y^2 - y} = \frac{7}{(-3)^2 - (-3)} = \frac{7}{9+3} = \frac{7}{12}$$
So $y = -3$ is not a solution to the equation.

5. Substituting $m = \frac{1}{4}$ into the equation:
$$m^2 + m - \frac{5}{16} = \left(\frac{1}{4}\right)^2 + \left(\frac{1}{4}\right) - \left(\frac{5}{16}\right) = \frac{1}{16} + \frac{1}{4} - \frac{5}{16} = \frac{5}{16} - \frac{5}{16} = 0$$
So $m = \frac{1}{4}$ is a solution to the equation.

7. Solving for x:
$$2x - 3 = -5$$
$$2x = -2$$
$$x = -1$$

9. Solving for x:
$$-6x + 1 = 49$$
$$-6x = 48$$
$$x = -8$$

11. Solving for m:
$$2m - 1 + 3m + 5 = 6m - 8$$
$$5m + 4 = 6m - 8$$
$$-m = -12$$
$$m = 12$$

13. Solving for x:
$$(x + 2)(x + 1) = x^2 + 11$$
$$x^2 + 3x + 2 = x^2 + 11$$
$$3x + 2 = 11$$
$$3x = 9$$
$$x = 3$$

15.   Multiplying by 15 and then solving for x:

$$\frac{x}{3}+\frac{2x}{5}=-\tfrac{11}{5}$$

$$15\left(\frac{x}{3}\right)+15\left(\frac{2x}{5}\right)=15\left(-\tfrac{11}{5}\right)$$

$$5x+6x=-33$$

$$11x=-33$$

$$x=-3$$

17.   Multiplying by 3 and then solving for x:

$$1-\frac{y}{3}=6$$

$$3(1)-3\left(\frac{y}{3}\right)=3(6)$$

$$3-y=18$$

$$-y=15$$

$$y=-15$$

19.   Multiplying by 4 and then solving for x:

$$\frac{x-1}{4}+\frac{2x+3}{-1}=0$$

$$4\left(\frac{x-1}{4}+\frac{2x+3}{-1}\right)=4(0)$$

$$x-1-4(2x+3)=0$$

$$x-1-8x-12=0$$

$$-7x=13$$

$$x=-\tfrac{13}{7}$$

21.   Multiplying by $x^2-9=(x+3)(x-3)$ and then solving for x:

$$\frac{1}{x-3}-\frac{2}{x+3}=\frac{1}{x^2-9}$$

$$\frac{(x-3)(x+3)}{x-3}-\frac{2(x-3)(x+3)}{x+3}=\frac{(x-3)(x+3)}{(x-3)(x+3)}$$

$$x+3-2(x-3)=1$$

$$x+3-2x+6=1$$

$$-x=-8$$

$$x=8$$

Check: Replacing x by 8 in the original equation yields:

$$\tfrac{1}{5}-\tfrac{2}{11}=\tfrac{1}{55}$$

$$\tfrac{11}{55}-\tfrac{10}{55}=\tfrac{1}{55}, \text{ which is true}$$

23. Multiplying by $x^2 - 25 = (x + 5)(x - 5)$ and then solving for x:

$$\frac{1}{x-5} + \frac{1}{x+5} = \frac{2x+1}{x^2-25}$$

$$\frac{(x+5)(x-5)}{x-5} + \frac{(x+5)(x-5)}{x+5} = \frac{(2x+1)(x^2-25)}{x^2-25}$$

$$x + 5 + x - 5 = 2x + 1$$

$$2x = 2x + 1$$

$$0 = 1$$

This last equation is false regardless of the value assigned to x. Consequently the original equation has no solution.

25. Multiplying by $2x^2 + 3x + 1 = (2x + 1)(x + 1)$ and then solving for x:

$$\frac{3}{2x+1} - \frac{4}{x+1} = \frac{2}{2x^2+3x+1}$$

$$\frac{3(2x+1)(x+1)}{2x+1} - \frac{4(2x+1)(x+1)}{x+1} = \frac{2(2x+1)(x+1)}{(2x+1)(x+1)}$$

$$3(x+1) - 4(2x+1) = 2$$

$$3x + 3 - 8x - 4 = 2$$

$$-5x - 1 = 2$$

$$-5x = 3$$

$$x = -\frac{3}{5}$$

Check: Replacing x by $-\frac{3}{5}$ in the original equation yields:

$$\frac{3}{-\frac{6}{5}+1} - \frac{4}{-\frac{3}{5}+1} = \frac{2}{2\left(\frac{9}{25}\right)-\frac{9}{5}+1}$$

$$\frac{15}{-1} - \frac{20}{-3+5} = \frac{50}{18-45+25}$$

$$-15 - 10 = -25$$

$$-25 = -25, \text{ which is true}$$

27. Multiplying by $x - 1$ and then solving for x:

$$\frac{2(x+1)}{x-1} - 3 = \frac{5x-1}{x-1}$$

$$2(x+1) - 3(x-1) = 5x - 1$$

$$2x + 2 - 3x + 3 = 5x - 1$$

$$-x + 5 = 5x - 1$$

$$-6x = -6$$

$$x = 1$$

This shows that if the original equation has a solution, it must be $x = 1$. However, the value $x = 1$ does not satisfy the original equation, since it yields zeros in the denominators. Consequently the original equation has no solution.

29. Multiplying by x and then solving for x:

$$\frac{1}{x} = \frac{4}{x} - 1$$
$$1 = 4 - x$$
$$x = 3$$

Check: Replacing x by 3 in the original equation yields:

$$\tfrac{1}{3} = \tfrac{4}{3} - 1$$
$$\tfrac{1}{3} = \tfrac{4}{3} - \tfrac{3}{3}, \text{ which is true}$$

31. (a) Multiplying by $3x$ and then solving for x:

$$3x \cdot \frac{2}{3x} = 3x \cdot \frac{3}{x}$$
$$2 = 9$$

Since there are no values of x which can make this last statement true, there is no solution.

(b) Multiplying by $3x(x + 1)$ and then solving for x:

$$3x(x+1) \cdot \frac{2}{3x} = 3x(x+1) \cdot \frac{3}{x+1}$$
$$2x + 2 = 9x$$
$$2 = 7x$$
$$x = \tfrac{2}{7}$$

Check: Replacing x by $\tfrac{2}{7}$ in the original equation yields:

$$\frac{2}{\frac{6}{7}} = \frac{3}{\frac{2}{7}+1}$$
$$\tfrac{7}{3} = \tfrac{7}{3}, \text{ which is true}$$

(c) Multiplying by $3x$ and then solving for x:

$$3x \cdot \frac{2}{3x} = 3x\left(\frac{3}{x}+1\right)$$
$$2 = 9 + 3x$$
$$-7 = 3x$$
$$x = -\tfrac{7}{3}$$

Check: Replacing x by $-\tfrac{7}{3}$ in the original equation yields:

$$\frac{2}{-7} = \frac{3}{-\frac{7}{3}}+1$$
$$-\tfrac{2}{7} = -\tfrac{9}{7}+1, \text{ which is true}$$

33. Factoring:
$$x^2 - 5x - 6 = 0$$
$$(x - 6)(x + 1) = 0$$
$$x = 6 \text{ or } x = -1$$

35. Factoring:
$$10z^2 - 13z - 3 = 0$$
$$(5z + 1)(2z - 3) = 0$$
$$z = -\tfrac{1}{5} \text{ or } z = \tfrac{3}{2}$$

37. Factoring:
$$(x + 1)^2 - 4 = 0$$
$$(x + 1)^2 - 2^2 = 0$$
$$(x + 1 - 2)(x + 1 + 2) = 0$$
$$(x - 1)(x + 3) = 0$$
$$x = 1 \text{ or } x = -3$$

39. Factoring:
$$x(2x - 13) = -6$$
$$2x^2 - 13x + 6 = 0$$
$$(2x - 1)(x - 6) = 0$$
$$x = \tfrac{1}{2} \text{ or } x = 6$$

41. Factoring:
$$x(x + 1) = 156$$
$$x^2 + x - 156 = 0$$
$$(x + 13)(x - 12) = 0$$
$$x = -13 \text{ or } x = 12$$

43. Using the quadratic formula with $a = 1$, $b = -1$, and $c = -5$:
$$x = \frac{-(-1) \pm \sqrt{(-1)^2 - 4(1)(-5)}}{2(1)} = \frac{1 \pm \sqrt{21}}{2} \approx -1.79, 2.79$$

45. Using the quadratic formula with $a = 2$, $b = 3$, and $c = -4$:
$$x = \frac{-3 \pm \sqrt{(3)^2 - 4(2)(-4)}}{2(2)} = \frac{-3 \pm \sqrt{41}}{4} \approx -2.35, 0.85$$

47. The equation is equivalent to $x^2 + 6x = -2$, or $x^2 + 6x + 2 = 0$. Using the quadratic formula with $a = 1$, $b = 6$, and $c = 2$:
$$x = \frac{-6 \pm \sqrt{(6)^2 - 4(1)(2)}}{2(1)} = \frac{-6 \pm \sqrt{28}}{2} = \frac{-6 \pm 2\sqrt{7}}{2} = -3 \pm \sqrt{7} \approx -5.65, -0.35$$

49. The equation is equivalent to $2x^2 + \sqrt{2}\,x - 10 = 0$. Using the quadratic formula with $a = 2$, $b = \sqrt{2}$, and $c = -10$:

$$x = \frac{-\sqrt{2} \pm \sqrt{(\sqrt{2})^2 - 4(2)(-10)}}{2(2)} = \frac{-\sqrt{2} \pm \sqrt{82}}{4} \approx -2.62, 1.91$$

51. The equation is equivalent to $12x^2 - 25x + 12 = 0$. Using the quadratic formula with $a = 12$, $b = -25$, and $c = 12$:

$$x = \frac{-(-25) \pm \sqrt{(-25)^2 - 4(12)(12)}}{2(12)} = \frac{25 \pm \sqrt{49}}{24} = \frac{25 \pm 7}{24} = \frac{3}{4}, \frac{4}{3}$$

53. Taking square roots yields:

$$x^2 = 24$$
$$x = \pm\sqrt{24}$$
$$x = \pm 2\sqrt{6}$$

55. Taking square roots yields:

$$\tfrac{1}{8} - t^2 = 0$$
$$t^2 = \tfrac{1}{8}$$
$$t = \pm\sqrt{\tfrac{1}{8}}$$
$$t = \pm\frac{1}{2\sqrt{2}} \cdot \frac{\sqrt{2}}{\sqrt{2}}$$
$$t = \pm\frac{\sqrt{2}}{4}$$

57. (a) Multiplying terms and then factoring:

$$u(u + 18) = -81$$
$$u^2 + 18u + 81 = 0$$
$$(u + 9)^2 = 0$$
$$u = -9$$

(b) Multiplying terms and converting to standard form:

$$u(u + 18) = 81$$
$$u^2 + 18u - 81 = 0$$

This equation does not factor. Applying the quadratic formula:

$$u = \frac{-18 \pm \sqrt{18^2 - 4(1)(-81)}}{2(1)} = \frac{-18 \pm \sqrt{648}}{2} = \frac{-18 \pm 18\sqrt{2}}{2} = -9 \pm 9\sqrt{2}$$

59. Solving for x:
$$3ax - 2b = b + 3$$
$$3ax = 3b + 3$$
$$ax = b + 1$$
$$x = \frac{b+1}{a}$$

61. Solving for x:
$$ax + b = bx + a$$
$$ax - bx = a - b$$
$$x(a - b) = a - b$$
$$x = \frac{a-b}{a-b}$$
$$x = 1$$
Note that the last step required $a \neq b$.

63. Multiplying by x and solving for x yields:
$$\frac{1}{x} = a + b$$
$$1 = x(a + b)$$
$$x = \frac{1}{a+b}$$

65. Multiplying by abx and solving for x yields:
$$\frac{1}{a} - \frac{1}{x} = \frac{1}{x} - \frac{1}{b}$$
$$bx - ab = ab - ax$$
$$bx + ax = ab + ab$$
$$x(b + a) = 2ab$$
$$x = \frac{2ab}{b+a}$$

67. Solving for x by factoring:
$$2y^2x^2 - 3yx + 1 = 0$$
$$(2yx - 1)(yx - 1) = 0$$
$$x = \frac{1}{2y} \text{ or } x = \frac{1}{y}$$

69. Solving for x by factoring:

$$(x-p)^2 + (x-q)^2 = p^2 + q^2$$
$$x^2 - 2px + p^2 + x^2 - 2qx + q^2 = p^2 + q^2$$
$$2x^2 - 2px - 2qx = 0$$
$$x^2 - px - qx = 0$$
$$x(x-p-q) = 0$$
$$x = 0 \text{ or } x = p+q$$

71. Solving for x:

$$a^2(a-x) = b^2(b+x) - 2abx$$
$$a^3 - a^2x = b^3 + b^2x - 2abx$$
$$a^3 - b^3 = a^2x + b^2x - 2abx$$
$$a^3 - b^3 = x(a^2 + b^2 - 2ab)$$
$$a^3 - b^3 = x(a-b)^2$$
$$x = \frac{(a-b)(a^2 + ab + b^2)}{(a-b)^2}$$
$$x = \frac{a^2 + ab + b^2}{a-b}$$

73. Multiplying by $(a-b)(b-c)$ and solving for x:

$$\frac{a-x}{a-b} - 2 = \frac{c-x}{b-c}$$
$$(a-x)(b-c) - 2(a-b)(b-c) = (c-x)(a-b)$$
$$ab - bx - ac + cx - 2ab + 2b^2 + 2ac - 2bc = ac - ax - bc + bx$$
$$-ab - bx + ac + cx + 2b^2 - 2bc = ac - ax - bc + bx$$
$$ax - 2bx + cx = ab - 2b^2 + bc$$
$$x(a - 2b + c) = b(a - 2b + c)$$
$$x = b \qquad \text{(assuming } a - 2b + c \neq 0)$$

75. Multiplying by $(x - a)(x - b)$ and solving for x:
$$\frac{x - a}{x - b} = \frac{b - x}{a - x} = \frac{x - b}{x - a}$$
$$(x - a)^2 = (x - b)^2$$
$$x^2 - 2ax + a^2 = x^2 - 2bx + b^2$$
$$-2ax + a^2 = -2bx + b^2$$
$$2bx - 2ax = b^2 - a^2$$
$$2x(b - a) = b^2 - a^2$$
$$x = \frac{b^2 - a^2}{2(b - a)}$$
$$x = \frac{(b - a)(b + a)}{2(b - a)}$$
$$x = \frac{a + b}{2} \qquad \text{(assuming } a \neq b\text{)}$$

77. Multiplying by $1 + rt$ and solving for r:
$$d = \frac{r}{1 + rt}$$
$$d(1 + rt) = r$$
$$d + drt = r$$
$$drt - r = -d$$
$$r(dt - 1) = -d$$
$$r = \frac{-d}{dt - 1}$$
$$r = \frac{d}{1 - dt}$$

79. Solving for h:
$$S = 2\pi r^2 + 2\pi rh$$
$$S - 2\pi r^2 = 2\pi rh$$
$$h = \frac{S - 2\pi r^2}{2\pi r}$$

81. Multiplying through by the least common denominator $x(x + 5)$ yields:
$$3x + 4(x + 5) = 2x(x + 5)$$
$$7x + 20 = 2x^2 + 10x$$
$$-2x^2 - 3x + 20 = 0$$
$$2x^2 + 3x - 20 = 0$$
$$(2x - 5)(x + 4) = 0$$
$$x = \tfrac{5}{2} \text{ or } x = -4$$
Both of these values check in the original equation.

83. Multiplying through by the least common denominator $6x + 1$ yields:
$$-6x^2 + 5x - 1 = 0$$
$$6x^2 - 5x + 1 = 0$$
$$(3x - 1)(2x - 1) = 0$$
$$x = \tfrac{1}{3} \text{ or } x = \tfrac{1}{2}$$
Both values check in the original equation.

85. Multiplying through by the least common denominator $(x + 2)(x - 2)$ yields:
$$x(x + 2) + x(x - 2) = 8$$
$$2x^2 - 8 = 0$$
$$x^2 - 4 = 0$$
$$(x - 2)(x + 2) = 0$$
$$x = 2 \text{ or } x = -2$$
However, neither value satisfies the original equation. Thus the original equation has no solution.

87. (a) Solve by factoring:
$$x^3 - 13x^2 + 42x = 0$$
$$x(x^2 - 13x + 42) = 0$$
$$x(x - 7)(x - 6) = 0$$
$$x = 0, 6, 7$$

(b) First factor the equation as $x(x^2 - 6x + 1) = 0$. So $x = 0$ is one solution. Apply the quadratic formula to find the other two solutions:
$$x = \frac{-(-6) \pm \sqrt{(-6)^2 - 4(1)(1)}}{2(1)} = \frac{6 \pm \sqrt{32}}{2} = \frac{6 \pm 4\sqrt{2}}{2} = 3 \pm 2\sqrt{2}$$

89. Multiplying by $(x + p)(x + q)(x - 2pq)$ and solving for x:
$$q(x + p)(x - 2pq) - p(x + q)(x - 2pq) = (q - p)(x + q)(x + p)$$
After expanding and simplifying, this last equation becomes:
$$qx + px + 2pqx = -pq$$
$$x(q + p + 2pq) = -pq$$
$$x = \frac{-pq}{q + p + 2qp}$$

91. Multiplying by ab and solving for x:
$$a(x-a)+b(x-b)+x-1=2b+2a+2ab$$
$$ax-a^2+bx-b^2+x-1=2b+2a+2ab$$
$$x(a+b+1)=a^2+b^2+2b+2a+2ab+1$$
$$x(a+b+1)=\left(a^2+2ab+b^2\right)+2a+2b+1$$
$$x(a+b+1)=(a+b)^2+2(a+b)+1$$
$$x(a+b+1)=(a+b+1)^2$$
$$x=a+b+1$$

<u>Chapter One Review Exercises</u>

1. Factoring as a difference of squares:
$$a^2-16b^2=a^2-(4b)^2=(a-4b)(a+4b)$$

3. Factoring as a difference of cubes:
$$8-(a+1)^3=\left[2-(a+1)\right]\left[4+2(a+1)+(a+1)^2\right]$$
$$=(1-a)\left(4+2a+2+a^2+2a+1\right)$$
$$=(1-a)\left(7+4a+a^2\right)$$

5. Factoring the common factor x then using the special product for $(A+B)^2$:
$$a^2x^3+2ax^2b+b^2x=x\left(a^2x^2+2abx+b^2\right)=x(ax+b)^2$$

7. Factoring by trial and error:
$$8x^2+6x+1=(4x+1)(2x+1)$$

9. Factoring the common factor a^4x^4 then as a difference of squares (twice):
$$a^4x^4-x^8a^8=a^4x^4\left(1-x^4a^4\right)$$
$$=a^4x^4\left(1-x^2a^2\right)\left(1+x^2a^2\right)$$
$$=a^4x^4(1-xa)(1+xa)\left(1+x^2a^2\right)$$

11. Factoring by grouping:
$$8+12a+6a^2+a^3=\left(8+a^3\right)+\left(12a+6a^2\right)$$
$$=(2+a)\left(4-2a+a^2\right)+6a(2+a)$$
$$=(2+a)\left(4-2a+a^2+6a\right)$$
$$=(2+a)\left(4+4a+a^2\right)$$
$$=(2+a)(2+a)^2$$
$$=(2+a)^3$$

13. Factoring the common factor z^3 then by trial and error:
$$4x^2y^2z^3 - 3xyz^3 - z^3 = z^3(4x^2y^2 - 3xy - 1) = z^3(4xy + 1)(xy - 1)$$

15. Factoring as a difference of squares then as a sum and difference of cubes:
$$1 - x^6 = \left(1^3\right)^2 - \left(x^3\right)^2$$
$$= \left(1 - x^3\right)\left(1 + x^3\right)$$
$$= (1 - x)\left(1 + x + x^2\right)(1 + x)\left(1 - x + x^2\right)$$

17. Factoring by grouping the first three terms:
$$a^2x^2 + 2abx + b^2 - 4a^2b^2x^2 = (ax + b)^2 - (2abx)^2 = (ax + b - 2abx)(ax + b + 2abx)$$

19. Factoring by grouping pairs of terms:
$$a^2 - b^2 + ac - bc + a^2b - b^2a = (a - b)(a + b) + c(a - b) + ab(a - b)$$
$$= (a - b)(a + b + c + ab)$$

21. $4^{3/2} = \left(\sqrt{4}\right)^3 = 2^3 = 8$

23. $\left((3025)^{1/2}\right)^0 = 1$ since $x^0 = 1$ for any real number x

25. Evaluating the rational exponent:
$$8^{-4/3} = \left(\sqrt[3]{8}\right)^{-4} = 2^{-4} = \frac{1}{2^4} = \frac{1}{16}$$

27. Evaluating the rational exponent:
$$(-243)^{-2/5} = \left(\sqrt[5]{-243}\right)^{-2} = (-3)^{-2} = \frac{1}{(-3)^2} = \frac{1}{9}$$

29. Using properties of exponents:
$$\left(a^2b^6c^8\right)^{1/2} = a^{2/2}b^{6/2}c^{8/2} = ab^3c^4$$

31. Multiplying radicals, we have:
$$\sqrt{a^3b^5}\sqrt{4ab^3} = \sqrt{4a^4b^8} = 2a^2b^4$$

33. Factoring $16 = 8 \cdot 2$ and $-54 = -27 \cdot 2$, we have:
$$\sqrt[3]{16} - \sqrt[3]{-54} = \sqrt[3]{8}\sqrt[3]{2} - \sqrt[3]{-27}\sqrt[3]{2} = 2\sqrt[3]{2} - (-3)\sqrt[3]{2} = 5\sqrt[3]{2}$$

35. Simplifying radicals, we have:
$$\sqrt{24a^2b^3} + ba\sqrt{54b} = \sqrt{4a^2b^2}\sqrt{6b} + ba\sqrt{9}\sqrt{6b} = 2ab\sqrt{6b} + 3ab\sqrt{6b} = 5ab\sqrt{6b}$$

37. $\sqrt{t^2} = |t|$

39. $\sqrt[4]{16x^4} = |2x| = 2|x|$

41. (a) Using rational exponents:
$$\sqrt[3]{x}\,\sqrt[4]{x^3} = x^{1/3}x^{3/4} = x^{4/12}x^{9/12} = x^{13/12}$$

(b) Writing as a radical:
$$x^{13/12} = \sqrt[12]{x^{13}} = x\sqrt[12]{x}$$

43. (a) Using rational exponents:
$$\sqrt{\sqrt[3]{t}\,\sqrt[5]{t^4}} = \left(t^{1/3}t^{4/5}\right)^{1/2} = \left(t^{5/15}t^{12/15}\right)^{1/2} = \left(t^{17/15}\right)^{1/2} = t^{17/30}$$

(b) Writing as a radical:
$$t^{17/30} = \sqrt[30]{t^{17}}$$

45. Rationalizing denominators:
$$\frac{6}{\sqrt{3}} \cdot \frac{\sqrt{3}}{\sqrt{3}} = \frac{6\sqrt{3}}{3} = 2\sqrt{3}$$

47. Rationalizing denominators:
$$\frac{1}{\sqrt{6}-\sqrt{3}} \cdot \frac{\sqrt{6}+\sqrt{3}}{\sqrt{6}+\sqrt{3}} = \frac{\sqrt{6}+\sqrt{3}}{6-3} = \frac{\sqrt{6}+\sqrt{3}}{3}$$

49. Rationalizing denominators:
$$\frac{\sqrt{a^2+x^2}+\sqrt{a^2-x^2}}{\sqrt{a^2+x^2}-\sqrt{a^2-x^2}} \cdot \frac{\sqrt{a^2+x^2}+\sqrt{a^2-x^2}}{\sqrt{a^2+x^2}+\sqrt{a^2-x^2}} = \frac{\left(a^2+x^2\right)+2\sqrt{a^4-x^4}+\left(a^2-x^2\right)}{\left(a^2+x^2\right)-\left(a^2-x^2\right)}$$
$$= \frac{2a^2+2\sqrt{a^4-x^4}}{2x^2}$$
$$= \frac{a^2+\sqrt{a^4-x^4}}{x^2}$$

51. Rationalizing denominators:

$$\frac{1}{\sqrt{2}+\sqrt{3}+\sqrt{5}}\cdot\frac{\sqrt{2}+\sqrt{3}-\sqrt{5}}{\sqrt{2}+\sqrt{3}-\sqrt{5}}=\frac{\sqrt{2}+\sqrt{3}-\sqrt{5}}{\left(\sqrt{2}+\sqrt{3}\right)^2-5}$$

$$=\frac{\sqrt{2}+\sqrt{3}-\sqrt{5}}{2+2\sqrt{6}+3-5}$$

$$=\frac{\sqrt{2}+\sqrt{3}-\sqrt{5}}{2\sqrt{6}}\cdot\frac{\sqrt{6}}{\sqrt{6}}$$

$$=\frac{\sqrt{12}+\sqrt{18}-\sqrt{30}}{2(6)}$$

$$=\frac{2\sqrt{3}+3\sqrt{2}-\sqrt{30}}{12}$$

53. Rationalizing numerators:

$$\frac{\sqrt{t}-a}{t-a^2}\cdot\frac{\sqrt{t}+a}{\sqrt{t}+a}=\frac{t-a^2}{\left(t-a^2\right)\left(\sqrt{t}+a\right)}=\frac{1}{\sqrt{t}+a}$$

55. Factoring the denominator:

$$\frac{\sqrt[3]{x}-2}{x-8}=\frac{x^{1/3}-2}{\left(x^{1/3}\right)^3-2^3}=\frac{x^{1/3}-2}{\left(x^{1/3}-2\right)\left(x^{2/3}+2x^{1/3}+4\right)}=\frac{1}{\sqrt[3]{x^2}+2\sqrt[3]{x}+4}$$

57. Using the special product for $(A+B)^2$:
$$(3x+1)^2=(3x)^2+2(3x)(1)+(1)^2=9x^2+6x+1$$

59. Using the special product for $(A+B)^2$:
$$\left(3x^2+y^2\right)^2=\left(3x^2\right)^2+2\left(3x^2\right)\left(y^2\right)+\left(y^2\right)^2=9x^4+6x^2y^2+y^4$$

61. Using the special product for $(A+B)^3$:
$$(2a+3)^3=(2a)^3+3(2a)^2(3)+3(2a)(3)^2+3^3=8a^3+36a^2+54a+27$$

63. Using the special product for $(A-B)\left(A^2+AB+B^2\right)$:
$$(1-3a)\left(1+3a+9a^2\right)=1^3-(3a)^3=1-27a^3$$

65. Using the special product for $(A-B)(A+B)$:
$$\left(x^{1/2}-y^{1/2}\right)\left(x^{1/2}+y^{1/2}\right)=\left(x^{1/2}\right)^2-\left(y^{1/2}\right)^2=x-y$$

67. Using the special product for $(A+B)^3$:
$$\left(x^{1/3}+y^{1/3}\right)^3=\left(x^{1/3}\right)^3+3\left(x^{1/3}\right)^2\left(y^{1/3}\right)+3\left(x^{1/3}\right)\left(y^{1/3}\right)^2+\left(y^{1/3}\right)^3$$
$$=x+3x^{2/3}y^{1/3}+3x^{1/3}y^{2/3}+y$$

69. Using the special product for $(A - B)^2$:

$$\begin{aligned}
\left(x^2 - 3x + 1\right)^2 &= \left[x^2 - (3x - 1)\right]^2 \\
&= x^4 - 2x^2(3x - 1) + (3x - 1)^2 \\
&= x^4 - 6x^3 + 2x^2 + 9x^2 - 6x + 1 \\
&= x^4 - 6x^3 + 11x^2 - 6x + 1
\end{aligned}$$

71. Solving the equation:

$$\begin{aligned}
5 - 9x &= 2 \\
-9x &= -3 \\
x &= \tfrac{1}{3}
\end{aligned}$$

73. Solving the equation:

$$\begin{aligned}
(t - 4)(t + 3) &= (t + 5)^2 \\
t^2 - t - 12 &= t^2 + 10t + 25 \\
-11t &= 37 \\
t &= -\tfrac{37}{11}
\end{aligned}$$

75. Solving the equation:

$$\begin{aligned}
\frac{2t - 1}{t + 2} &= 5 \\
2t - 1 &= 5t + 10 \\
-3t &= 11 \\
t &= -\tfrac{11}{3}
\end{aligned}$$

77. Solving the equation:

$$\begin{aligned}
\frac{2y - 5}{4y + 1} &= \frac{y - 1}{2y + 5} \\
4y^2 - 25 &= 4y^2 - 3y - 1 \\
3y &= 24 \\
y &= 8
\end{aligned}$$

79. Solving the equation:

$$\begin{aligned}
12x^2 + 2x - 2 &= 0 \\
6x^2 + x - 1 &= 0 \\
(3x - 1)(2x + 1) &= 0 \\
x &= \tfrac{1}{3}, -\tfrac{1}{2}
\end{aligned}$$

81. Multiplying by 2, solve the equation:

$$\begin{aligned}
\tfrac{1}{2}x^2 + x - 12 &= 0 \\
x^2 + 2x - 24 &= 0 \\
(x + 6)(x - 4) &= 0 \\
x &= -6, 4
\end{aligned}$$

83. Solving the equation:
$$\frac{x}{5-x} = \frac{-2}{11-x}$$
$$11x - x^2 = -10 + 2x$$
$$-x^2 + 9x + 10 = 0$$
$$x^2 - 9x - 10 = 0$$
$$(x - 10)(x + 1) = 0$$
$$x = 10, -1$$

85. Solving the equation:
$$\frac{1}{3x-7} - \frac{2}{5x-5} - \frac{3}{3x+1} = 0$$
$$(5x - 5)(3x + 1) - 2(3x - 7)(3x + 1) - 3(3x - 7)(5x - 5) = 0$$
$$15x^2 - 10x - 5 - 18x^2 + 36x + 14 - 45x^2 + 150x - 105 = 0$$
$$-48x^2 + 176x - 96 = 0$$
$$3x^2 - 11x + 6 = 0$$
$$(3x - 2)(x - 3) = 0$$
$$x = \tfrac{2}{3}, 3$$

87. Multiply by 2 to obtain $2t^2 + 2t - 1 = 0$. Now using the quadratic formula:
$$t = \frac{-2 \pm \sqrt{4 - 4(2)(-1)}}{2(2)} = \frac{-2 \pm \sqrt{12}}{4} = \frac{-2 \pm 2\sqrt{3}}{4} = \frac{-1 \pm \sqrt{3}}{2}$$

89. Factoring the equation yields $x^2(x^2 - x + 1) = 0$, so $x = 0$ is one solution. Using the quadratic formula:
$$x = \frac{-(-1) \pm \sqrt{(-1)^2 - 4(1)(1)}}{2(1)} = \frac{1 \pm \sqrt{-3}}{2}, \text{ which is not a real number}$$
So $x = 0$ is the only real solution.

91. Using absolute value, the inequality is $|x - a| < \tfrac{1}{2}$.

93. Using absolute value, the equality is $|x - (-1)| = 5$, or $|x + 1| = 5$.

95. If $|x - 5| = 0$, then $x - 5 = 0$, so $x = 5$.

97. Since $2 - \sqrt{6} < 0$:
$$|2 - \sqrt{6}| = -(2 - \sqrt{6}) = \sqrt{6} - 2$$

99. (a) Since $x < 3$, $x - 3 < 0$ and thus:
$$|x - 3| = -(x - 3) = 3 - x$$

(b)   Since $x > 3$, $x - 3 > 0$ and thus:
$$|x - 3| = x - 3$$

101. (a)   If $x < -2$, then $x + 2 < 0$ and $x - 1 < 0$, so:
$$|x + 2| + |x - 1| = -(x + 2) - (x - 1) = -x - 2 - x + 1 = -2x - 1$$

(b)   If $-2 < x < 1$, then $x + 2 > 0$ and $x - 1 < 0$, so:
$$|x + 2| + |x - 1| = x + 2 - (x - 1) = x + 2 - x + 1 = 3$$

(c)   If $x > 1$, then $x + 2 > 0$ and $x - 1 > 0$, so:
$$|x + 2| + |x - 1| = x + 2 + x - 1 = 2x + 1$$

103. In scientific notation the number is 1.4×10^{-3}.

105. In scientific notation the number is 1.2001×10^{1}.

107. In scientific notation the number is:
$$\left(5.24 \times 10^{1}\right)\left(10^{8}\right) = 5.24 \times 10^{9}$$

109. Simplifying parentheses, we obtain:
$$1 - 2[3 - 4(1 - 5)] = 1 - 2[3 - 4(-4)] = 1 - 2(19) = -37$$

111. Multiplying out parentheses:
$$1 + 3\left(x^2 - 5x - 4\right) - \left[1 - \left(15x - 3x^2\right)\right] = 1 + 3x^2 - 15x - 12 - \left(1 - 15x + 3x^2\right)$$
$$= 3x^2 - 15x - 11 - 1 + 15x - 3x^2$$
$$= -12$$

113. Multiplying out parentheses:
$$(x - 1)(x + 1)(x + 2) = \left(x^2 - 1\right)(x + 2) = x^3 - x + 2x^2 - 2 = x^3 + 2x^2 - x - 2$$

115. Multiplying out parentheses:
$$x^2 + 4 - (x - 1)(x - 2) = x^2 + 4 - \left(x^2 - 3x + 2\right) = x^2 + 4 - x^2 + 3x - 2 = 3x + 2$$

117. Sketch the interval $(3, 9)$:

119. Sketch the intervals $(-\infty, -2]$ and $[0, \infty)$:

121. (a)   Sketch the interval $(-1, 9)$, excluding the point $x = 4$:

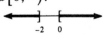

(b)　Sketch the interval $(-1, 9)$:

123.　Complete the table:

x	$1+\frac{x}{2}$	$\sqrt{1+x}$
0.1	1.05	1.048808
0.01	1.005	1.004987
0.001	1.0005	1.000499

125.　(a)　To six decimal places, its value is 0.447214.

(b)　To six decimal places, its value is 0.447214.

(c)　Rationalize the denominator:

$$\frac{\sqrt{3-\sqrt{5}}}{\sqrt{2}+\sqrt{7-3\sqrt{5}}} \cdot \frac{\sqrt{2}-\sqrt{7-3\sqrt{5}}}{\sqrt{2}-\sqrt{7-3\sqrt{5}}} = \frac{\sqrt{6-2\sqrt{5}}-\sqrt{\left(3-\sqrt{5}\right)\left(7-3\sqrt{5}\right)}}{2-\left(7-3\sqrt{5}\right)}$$

$$= \frac{\sqrt{6-2\sqrt{5}}-\sqrt{21-16\sqrt{5}+15}}{-5+3\sqrt{5}}$$

$$= \frac{\sqrt{6-2\sqrt{5}}-\sqrt{36-16\sqrt{5}}}{-5+3\sqrt{5}}$$

$$= \frac{\sqrt{6-2\sqrt{5}}-2\sqrt{9-4\sqrt{5}}}{-5+3\sqrt{5}}$$

(d)　Squaring each side:

$$\left(\sqrt{5}-1\right)^2 = 5-2\sqrt{5}+1 = 6-2\sqrt{5}, \text{ so } \sqrt{5}-1 = \sqrt{6-2\sqrt{5}}$$

$$\left(\sqrt{5}-2\right)^2 = 5-4\sqrt{5}+4 = 9-4\sqrt{5}, \text{ so } \sqrt{5}-2 = \sqrt{9-4\sqrt{5}}$$

(e)　The expression from (c) becomes:

$$\frac{\sqrt{5}-1-2\left(\sqrt{5}-2\right)}{-5+3\sqrt{5}} = \frac{\sqrt{5}-1-2\sqrt{5}+4}{-5+3\sqrt{5}}$$

$$= \frac{3-\sqrt{5}}{-5+3\sqrt{5}} \cdot \frac{-5-3\sqrt{5}}{-5-3\sqrt{5}}$$

$$= \frac{-15-4\sqrt{5}+15}{25-45}$$

$$= \frac{-4\sqrt{5}}{-20}$$

$$= \frac{\sqrt{5}}{5}$$

127. First compute x^3:

$$x^3 = \left(a + \sqrt{a^2 + b^3}\right) + 3\left(a + \sqrt{a^2 + b^3}\right)^{2/3}\left(a - \sqrt{a^2 + b^3}\right)^{1/3}$$

$$+ 3\left(a + \sqrt{a^2 + b^3}\right)^{1/3}\left(a - \sqrt{a^2 + b^3}\right)^{2/3} + \left(a - \sqrt{a^2 + b^3}\right)$$

$$= 2a + 3\left(a + \sqrt{a^2 + b^3}\right)^{1/3}\left(-b^3\right)^{1/3} + 3\left(a - \sqrt{a^2 + b^3}\right)^{1/3}\left(-b^3\right)^{1/3}$$

$$= 2a - 3b\left(a + \sqrt{a^2 + b^3}\right)^{1/3} - 3b\left(a - \sqrt{a^2 + b^3}\right)^{1/3}$$

Now simplify $x^3 + 3bx - 2a$ as:

$$2a - 3b\left(a + \sqrt{a^2 + b^3}\right)^{1/3} - 3b\left(a - \sqrt{a^2 + b^3}\right)^{1/3}$$

$$+ 3b\left(a + \sqrt{a^2 + b^3}\right)^{1/3} + 3b\left(a - \sqrt{a^2 + b^3}\right)^{1/3} - 2a = 0$$

Chapter One Test

1. (a) Factoring by difference of cubes:

$$64 - (x-2)^3 = \left[4 - (x-2)\right]\left[16 + 4(x-2) + (x-2)^2\right]$$

$$= (4 - x + 2)(16 + 4x - 8 + x^2 - 4x + 4)$$

$$= (6 - x)(12 + x^2)$$

(b) Since this is a sum of squares, $w^2y^2 + 16$ is irreducible.

2. (a) Using trial and error, $2x^2 + 3x - 8$ is irreducible.

(b) Factoring the common factor z and using trial and error:

$$6x^2y^2z + xyz - z = z\left(6x^2y^2 + xy - 1\right) = z(3xy - 1)(2xy + 1)$$

3. Factoring the common factor $3x\left(1 - x^2\right)^{-3/2}$:

$$6x\left(1 - x^2\right)^{-1/2} - 3x^2\left(1 - x^2\right)^{-3/2} = 3x\left(1 - x^2\right)^{-3/2}\left[2\left(1 - x^2\right) - x\right]$$

$$= 3x\left(1 - x^2\right)^{-3/2}\left(2 - 2x^2 - x\right)$$

$$= -3x\left(1 - x^2\right)^{-3/2}\left(2x^2 + x - 2\right)$$

4. Factoring using grouping:

$$9ax + 6bx - 6ay - 4by = 3x(3a + 2b) - 2y(3a + 2b) = (3a + 2b)(3x - 2y)$$

5. Factoring $54 = 27 \cdot 2$ and $-2000 = -1000 \cdot 2$:

$$\sqrt[3]{54} - \sqrt[3]{-2000} = \sqrt[3]{27} \cdot \sqrt[3]{2} - \sqrt[3]{-1000} \cdot \sqrt[3]{2}$$

$$= 3 \cdot \sqrt[3]{2} + 10 \cdot \sqrt[3]{2}$$

$$= 13\sqrt[3]{2}$$

6.   (a)   Using rational exponents:
$$\sqrt[5]{x^2}\,\sqrt[7]{x^3} = x^{2/5}x^{3/7} = x^{14/35+15/35} = x^{29/35}$$

  (b)   Writing in radical form:
$$x^{29/35} = \sqrt[35]{x^{29}}$$

7.   Rationalizing the denominator:
$$\frac{\sqrt{x^2+1}+\sqrt{x^2-1}}{\sqrt{x^2+1}-\sqrt{x^2-1}}\cdot\frac{\sqrt{x^2+1}+\sqrt{x^2-1}}{\sqrt{x^2+1}+\sqrt{x^2-1}} = \frac{\left(x^2+1\right)+2\sqrt{x^4-1}+\left(x^2-1\right)}{\left(x^2+1\right)-\left(x^2-1\right)}$$
$$= \frac{2x^2+2\sqrt{x^4-1}}{2}$$
$$= x^2 + \sqrt{x^4-1}$$

8.   Using the special product for $(A+B)^3$:
$$\left(2a+3b^2\right)^3 = (2a)^3 + 3(2a)^2\left(3b^2\right) + 3(2a)\left(3b^2\right)^2 + \left(3b^2\right)^3$$
$$= 8a^3 + 36a^2b^2 + 54ab^4 + 27b^6$$

9.   If $6 < x < 7$, then $x-6 > 0$ and $x-7 < 0$, so:
$$|x-6| + |x-7| = (x-6) - (x-7) = x-6-x+7 = 1$$

10.   (a)   This is 3.96×10^9 light-years in scientific notation.
  (b)   This is 6.750×10^{-7} meters in scientific notation.

11.   (a)   Since the distance from x to 4 is less than $\frac{1}{10}$, we have:
$$4 - \tfrac{1}{10} < x < 4 + \tfrac{1}{10}$$
$$\tfrac{39}{10} < x < \tfrac{41}{10}$$
The interval is $\left(\frac{39}{10}, \frac{41}{10}\right)$.

  (b)   Since $x \geq -2$, the interval is $[-2, \infty)$.

12.   Multiplying by $(x+4)(x-4)$:
$$\frac{2}{x+4} - \frac{1}{x-4} = \frac{-7}{x^2-16}$$
$$2(x-4) - (x+4) = -7$$
$$2x - 8 - x - 4 = -7$$
$$x = 5$$

13. Multiplying by $6(1-x)(2-x)$:

$$\frac{1}{1-x} + \frac{4}{2-x} = \frac{11}{6}$$
$$6(2-x) + 4(6)(1-x) = 11(1-x)(2-x)$$
$$12 - 6x + 24 - 24x = 22 - 33x + 11x^2$$
$$-11x^2 + 3x + 14 = 0$$
$$11x^2 - 3x - 14 = 0$$
$$(11x - 14)(x + 1) = 0$$
$$x = \tfrac{14}{11}, -1$$

14. (a) Solve by factoring:

$$x^2 + 4x = 5$$
$$x^2 + 4x - 5 = 0$$
$$(x + 5)(x - 1) = 0$$
$$x = -5, 1$$

(b) Solve by writing the equation as $x^2 + 4x - 1 = 0$, then using the quadratic formula:

$$x = \frac{-4 \pm \sqrt{(4)^2 - 4(1)(-1)}}{2(1)} = \frac{-4 \pm \sqrt{20}}{2} = \frac{-4 \pm 2\sqrt{5}}{2} = -2 \pm \sqrt{5}$$

15. Multiplying by $cx + d$:

$$\frac{ax + b}{cx + d} = e$$
$$ax + b = cex + de$$
$$ax - cex = de - b$$
$$x(a - ce) = de - b$$
$$x = \frac{de - b}{a - ce}$$

16. Solving for x:

$$ax = bx + a^2 - b^2$$
$$ax - bx = a^2 - b^2$$
$$x(a - b) = (a + b)(a - b)$$
$$x = \frac{(a + b)(a - b)}{a - b} = a + b$$

17.   Using long division:

$$
\begin{array}{r}
x - 11 \\
x + 2 \enclose{longdiv}{x^2 - 9x + 5} \\
\underline{x^2 + 2x} \\
-11x + 5 \\
\underline{-11x - 22} \\
27
\end{array}
$$

The quotient is $x - 11$ and the remainder is 27.

18.   Using long division:

$$
\begin{array}{r}
2x^2 - \frac{1}{2} \\
2x^2 - 1 \enclose{longdiv}{4x^4 - 3x^2 + 2} \\
\underline{4x^4 - 2x^2} \\
-x^2 + 2 \\
\underline{-x^2 + \frac{1}{2}} \\
\frac{3}{2}
\end{array}
$$

The quotient is $2x^2 - \frac{1}{2}$ and the remainder is $\frac{3}{2}$.

Chapter Two
Coordinates and Graphs

2.1 Rectangular Coordinates

1. Plotting the points:

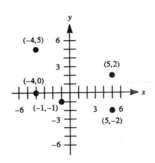

3. (a) Draw the right triangle *PQR*:

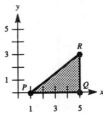

(b) Since the base is $b = 5 - 1 = 4$ and the height is $h = 3 - 0 = 3$, the area is given by:

5. (a) Here $(x_1, y_1) = (0, 0)$ and $(x_2, y_2) = (-3, 4)$, so by the distance formula:

$$d = \sqrt{(-3-0)^2 + (4-0)^2} = \sqrt{9+16} = \sqrt{25} = 5$$

   (b) Here $(x_1, y_1) = (2, 1)$ and $(x_2, y_2) = (7, 13)$, so:

$$d = \sqrt{(7-2)^2 + (13-1)^2} = \sqrt{25+144} = \sqrt{169} = 13$$

7. (a) Here $(x_1, y_1) = (-5, 0)$ and $(x_2, y_2) = (5, 0)$, so:

$$d = \sqrt{[5-(-5)]^2 + (0-0)^2} = \sqrt{100+0} = \sqrt{100} = 10$$

   (b) Here $(x_1, y_1) = (0, -8)$ and $(x_2, y_2) = (0, 1)$, so:

$$d = \sqrt{(0-0)^2 + [1-(-8)]^2} = \sqrt{0+81} = \sqrt{81} = 9$$

Note that we really don't need to use the distance formula for either (a) or (b), since in each case one of the coordinates (either x or y) is the same. Draw quick graphs and you can find the distance by inspection:

   (a) This graph indicates the distance is 10:

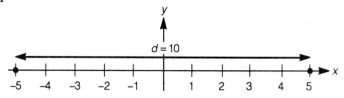

   (b) This graph indicates the distance is 9:

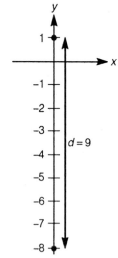

9. Here $(x_1, y_1) = (1, \sqrt{3})$ and $(x_2, y_2) = (-1, -\sqrt{3})$, so:

$$d = \sqrt{(-1-1)^2 + (-\sqrt{3}-\sqrt{3})^2} = \sqrt{(-2)^2 + (-2\sqrt{3})^2} = \sqrt{4+12} = 4$$

11. (a) Calculate the distance of each point from the origin:

$$(3,-2): \quad d = \sqrt{(3-0)^2 + (-2-0)^2} = \sqrt{9+4} = \sqrt{13}$$

$$\left(4,\tfrac{1}{2}\right): \quad d = \sqrt{(4-0)^2 + \left(\tfrac{1}{2}-0\right)^2} = \sqrt{16+\tfrac{1}{4}} = \sqrt{16.25}$$

So $\left(4,\tfrac{1}{2}\right)$ is farther from the origin.

(b) Calculate the distance of each point from the origin:

$$(-6,7): \quad d = \sqrt{(-6-0)^2 + (7-0)^2} = \sqrt{36+49} = \sqrt{85}$$

$$(9,0): \quad d = \sqrt{(9-0)^2 + (0-0)^2} = \sqrt{81+0} = \sqrt{81}$$

So $(-6,7)$ is farther from the origin.

13. We will graph each triangle and then determine (using the converse of the Pythagorean theorem) whether $a^2 + b^2 = c^2$.

(a) Graph the points:

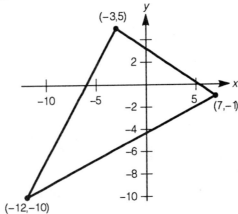

Now calculate the distances:

$$a = \sqrt{(-3-7)^2 + [5-(-1)]^2} = \sqrt{100+36} = \sqrt{136}$$

$$b = \sqrt{[-12-(-3)]^2 + (-10-5)^2} = \sqrt{81+225} = \sqrt{306}$$

$$c = \sqrt{(-12-7)^2 + [-10-(-1)]^2} = \sqrt{361+81} = \sqrt{442}$$

Now check the converse of the Pythagorean theorem:

$$a^2 + b^2 = 136 + 306 = 442 = c^2$$

So the triangle is a right triangle.

(b)   Graph the points:

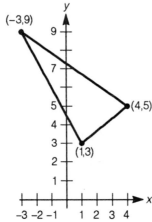

Now calculate the distances:

$$a = \sqrt{(-3-1)^2 + (9-3)^2} = \sqrt{16+36} = \sqrt{52}$$

$$b = \sqrt{(4-1)^2 + (5-3)^2} = \sqrt{9+4} = \sqrt{13}$$

$$c = \sqrt{(-3-4)^2 + (9-5)^2} = \sqrt{49+16} = \sqrt{65}$$

Now check the converse of the Pythagorean theorem:

$$a^2 + b^2 = 52 + 13 = 65 = c^2$$

So the triangle is a right triangle.

(c)   Graph the points:

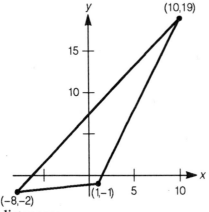

Now calculate the distances:

$$a = \sqrt{(-8-1)^2 + [-2-(-1)]^2} = \sqrt{81+1} = \sqrt{82}$$

$$b = \sqrt{(10-1)^2 + [19-(-1)]^2} = \sqrt{81+400} = \sqrt{481}$$

$$c = \sqrt{[10-(-8)]^2 + [19-(-2)]^2} = \sqrt{324+441} = \sqrt{765}$$

Now check the converse of the Pythagorean theorem:

$$a^2 + b^2 = 82 + 481 = 563 \neq 765 = c^2$$

So the triangle is not a right triangle.

15. Let $(x_1, y_1) = (1, -4)$, $(x_2, y_2) = (5, 3)$, and $(x_3, y_3) = (13, 17)$, so using the formula from Exercise 14(b) we have:

$$\text{Area} = \left(\tfrac{1}{2}\right)\left|1(3) - 5(-4) + 5(17) - 13(3) + 13(-4) - 1(17)\right|$$
$$= \left(\tfrac{1}{2}\right)\left|3 + 20 + 85 - 39 - 52 - 17\right|$$
$$= \left(\tfrac{1}{2}\right)\left|0\right|$$
$$= 0$$

For the area of the triangle to be 0 it must be that these three points do not form a triangle. The only way that could occur is if the three points are collinear; that is, they all lie on the same line.

17. The circle is in standard form, so its center is (3, 1) and its radius is $\sqrt{25} = 5$. To determine if the point (–1, 4) lies on the circle, substitute into the equation:

$$(x - 3)^2 + (y - 1)^2 = (-1 - 3)^2 + (4 - 1)^2 = (-4)^2 + (3)^2 = 16 + 9 = 25$$

So (–1, 4) lies on the circle.

19. The circle is in standard form, so its center is (–4, –2) and its radius is $\sqrt{20} = 2\sqrt{5}$. To determine if the point (0, 1) lies on the circle, substitute into the equation:

$$(x + 4)^2 + (y + 2)^2 = (0 + 4)^2 + (1 + 2)^2 = (4)^2 + (3)^2 = 16 + 9 = 25 \neq 20$$

So (0, 1) does not lie on the circle.

21. The circle is in standard form, so its center is (0, 0) and its radius is $\sqrt{1} = 1$. To determine if the point $\left(\tfrac{1}{2}, \tfrac{\sqrt{3}}{2}\right)$ lies on the circle, substitute into the equation:

$$x^2 + y^2 = \left(\tfrac{1}{2}\right)^2 + \left(\tfrac{\sqrt{3}}{2}\right)^2 = \tfrac{1}{4} + \tfrac{3}{4} = 1$$

So $\left(\tfrac{1}{2}, \tfrac{\sqrt{3}}{2}\right)$ lies on the circle.

23. (a) Here $(x_1, y_1) = (3, 2)$ and $(x_2, y_2) = (9, 8)$, so by the midpoint formula:

$$M = \left(\frac{3+9}{2}, \frac{2+8}{2}\right) = \left(\tfrac{12}{2}, \tfrac{10}{2}\right) = (6, 5)$$

(b) Here $(x_1, y_1) = (-4, 0)$ and $(x_2, y_2) = (5, -3)$, so by the midpoint formula:

$$M = \left(\frac{-4+5}{2}, \frac{0-3}{2}\right) = \left(\tfrac{1}{2}, -\tfrac{3}{2}\right)$$

(c) Here $(x_1, y_1) = (3, -6)$ and $(x_2, y_2) = (-1, -2)$, so by the midpoint formula:

$$M = \left(\frac{3-1}{2}, \frac{-6-2}{2}\right) = \left(\tfrac{2}{2}, -\tfrac{8}{2}\right) = (1, -4)$$

25. (a) Since $\overline{PQ}$ is the diameter of the circle, its midpoint must be the center of the circle. Using the midpoint formula:
$$\text{center} = \left(\tfrac{-4+6}{2}, \tfrac{-2+4}{2}\right) = \left(\tfrac{2}{2}, \tfrac{2}{2}\right) = (1,1)$$

(b) The radius of the circle is the distance from this center $(1, 1)$ to point Q (or point P), so using the distance formula:
$$r = \sqrt{(6-1)^2 + (4-1)^2} = \sqrt{25+9} = \sqrt{34}$$

(c) Using the standard form $(x-h)^2 + (y-k)^2 = r^2$ for a circle, the equation is $(x-1)^2 + (y-1)^2 = 34$.

27. Solving the equation for y:
$$(x+5)^2 + (y+3)^2 = 4$$
$$(y+3)^2 = 4 - (x+5)^2$$
$$y+3 = \pm\sqrt{4-(x+5)^2}$$
$$y = -3 \pm \sqrt{4-(x+5)^2}$$

(a) First graph $y = -3 + \sqrt{4-(x+5)^2}$:

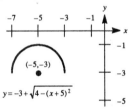

(b) Now graph $y = -3 - \sqrt{4-(x+5)^2}$:

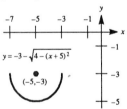

(c) Putting the two graphs together results in the circle $(x+5)^2 + (y+3)^2 = 4$:

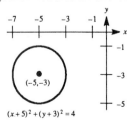

29. (a) Sketch the parallelogram $ABCD$:

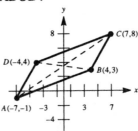

(b) For AC, let $(x_1, y_1) = (-7, -1)$ and $(x_2, y_2) = (7, 8)$, so the midpoint is:
$$M_1 = \left(\tfrac{-7+7}{2}, \tfrac{-1+8}{2}\right) = \left(0, \tfrac{7}{2}\right)$$
For BD, let $(x_1, y_1) = (4, 3)$ and $(x_2, y_2) = (-4, 4)$, so the midpoint is:
$$M_2 = \left(\tfrac{4-4}{2}, \tfrac{3+4}{2}\right) = \left(0, \tfrac{7}{2}\right)$$

(c) It appears that the midpoints of the two diagonals of a parallelogram are the same. Stated more concisely, the diagonals of a parallelogram bisect each other.

31. (a) Since $(AB)^2 = 68$, $(BC)^2 = 40$, and $(AC)^2 = 20$:
$$(AB)^2 + (BC)^2 + (AC)^2 = 128$$

(b) Let the midpoints be $X, Y,$ and Z, so:
$$X = M_{AB} = (5, 2)$$
$$Y = M_{BC} = (6, 4)$$
$$Z = M_{AC} = (2, 3)$$
Then the medians are:
$$AY = \sqrt{(6-1)^2 + (4-1)^2} = \sqrt{25+9} = \sqrt{34}$$
$$BZ = \sqrt{(2-9)^2 + (3-3)^2} = \sqrt{49+0} = \sqrt{49} = 7$$
$$CX = \sqrt{(5-3)^2 + (2-5)^2} = \sqrt{4+9} = \sqrt{13}$$
So the required value is:
$$(AY)^2 + (BZ)^2 + (CX)^2 = 34 + 49 + 13 = 96$$

(c) This ratio is $\tfrac{128}{96} = \tfrac{4}{3}$.

33. Since the center of the circle is $(3, 2)$, we know the equation will take the form $(x-3)^2 + (y-2)^2 = r^2$, where r is the radius of the circle. We can find r since the point $(-2, -10)$ must satisfy the equation for the circle:
$$(-2-3)^2 + (-10-2)^2 = r^2$$
$$25 + 144 = r^2$$
$$169 = r^2$$
The equation of the circle is $(x-3)^2 + (y-2)^2 = 169$.

35.   Since the radius is 3, the equation of the circle is $(x-3)^2 + (y-5)^2 = 9$.

37.   Using the Pythagorean theorem:

$$a^2 = 1^2 + 1^2 = 2, \text{ so } a = \sqrt{2}$$

$$b^2 = 1^2 + \left(\sqrt{2}\right)^2 = 1 + 2 = 3, \text{ so } b = \sqrt{3}$$

$$c^2 = 1^2 + \left(\sqrt{3}\right)^2 = 1 + 3 = 4, \text{ so } c = 2$$

$$d^2 = 1^2 + 2^2 = 1 + 4 = 5, \text{ so } d = \sqrt{5}$$

$$e^2 = 1^2 + \left(\sqrt{5}\right)^2 = 1 + 5 = 6, \text{ so } e = \sqrt{6}$$

$$f^2 = 1^2 + \left(\sqrt{6}\right)^2 = 1 + 6 = 7, \text{ so } f = \sqrt{7}$$

$$g^2 = 1^2 + \left(\sqrt{7}\right)^2 = 1 + 7 = 8, \text{ so } g = \sqrt{8} = 2\sqrt{2}$$

39.   Using the hint and squaring:

$$13 = \sqrt{(12-0)^2 + (t-2)^2}$$
$$169 = 144 + (t-2)^2$$
$$25 = (t-2)^2$$

So either $t - 2 = 5$ and thus $t = 7$, or $t - 2 = -5$ and thus $t = -3$. The values of t are 7 and -3.

41.   **(a)**   Call (x, y) the coordinates of B. Clearly $y = c$, since the top and bottom of the parallelogram must be parallel, and thus since the bottom is horizontal the top must be also. Now compute the lengths OC and OB using the distance formula:

$$OC = \sqrt{(b-0)^2 + (c-0)^2} = \sqrt{b^2 + c^2}$$
$$AB = \sqrt{(x-a)^2 + (c-0)^2} = \sqrt{(x-a)^2 + c^2}$$

Since $OC = AB$, we have:

$$\sqrt{b^2 + c^2} = \sqrt{(x-a)^2 + c^2}$$
$$b^2 + c^2 = (x-a)^2 + c^2$$
$$b^2 = (x-a)^2$$

Taking roots, we have $x - a = \pm b$, or $x = a \pm b$. But clearly $x = a - b$ doesn't make sense, as $x \geq a$ from the figure, so $x = a + b$.

(b)   Find the midpoints of the two diagonals $\overline{OB}$ and $\overline{AC}$:

For $O(0,0)$ and $B(a+b,c)$, $M_{\overline{OB}} = \left(\frac{a+b}{2}, \frac{c}{2}\right)$

For $A(a,c)$ and $C(b,c)$, $M_{\overline{AC}} = \left(\frac{a+b}{2}, \frac{c}{2}\right)$

(c)   Clearly our two midpoints from (b) are equal. Since the two midpoints on the diagonals are equal they must bisect each other.

43. Using the hint and the lengths of our diagonals from Exercise 42, and squaring each side:

$$\sqrt{(a+b)^2 + c^2} = \sqrt{(b-a)^2 + c^2}$$
$$(a+b)^2 + c^2 = b^2 - 2ab + a^2 + c^2$$
$$4ab = 0$$
$$a = 0 \text{ or } b = 0$$

Now $a = 0$ is impossible, since our figure would no longer be a parallelogram. So $b = 0$. But if $b = 0$, then we must have a rectangle (look at the figure). This proves the desired result.

45. First compute the left-hand side of the equality:

$$AB = \sqrt{(2-0)^2 + (0-0)^2} = \sqrt{4} = 2$$
$$AC = \sqrt{(2a-0)^2 + (2b-0)^2} = \sqrt{4a^2 + 4b^2}$$

So $AB^2 + AC^2 = 4 + 4a^2 + 4b^2$. For the right-hand side, first find:

$$M = \left(\frac{2a+a}{2}, \frac{2b+0}{2}\right) = (a+1, b)$$
$$BM = \sqrt{(a+1-2)^2 + (b-0)^2} = \sqrt{(a-1)^2 + b^2}$$
$$AM = \sqrt{(a+1-0)^2 + (b-0)^2} = \sqrt{(a+1)^2 + b^2}$$

Therefore:

$$2(BM^2 + AM^2) = 2\left[(a-1)^2 + b^2 + (a+1)^2 + b^2\right]$$
$$= 2(a^2 - 2a + 1 + b^2 + a^2 + 2a + 1 + b^2)$$
$$= 2(2a^2 + 2b^2 + 2)$$
$$= 4a^2 + 4b^2 + 4$$

So $AB^2 + AC^2 = 2(BM^2 + AM^2)$.

47. (a) Using the distance formula:

$$PM = \sqrt{\left(\frac{x_1 + x_2}{2} - x_1\right)^2 + \left(\frac{y_1 + y_2}{2} - y_1\right)^2}$$
$$= \sqrt{\left(\frac{x_2 - x_1}{2}\right)^2 + \left(\frac{y_2 - y_1}{2}\right)^2}$$
$$= \frac{\sqrt{(x_2 - x_1)^2 + (y_2 - y_1)^2}}{2}$$

$$MQ = \sqrt{\left(x_2 - \frac{x_1 + x_2}{2}\right)^2 + \left(y_2 - \frac{y_1 + y_2}{2}\right)^2}$$
$$= \sqrt{\left(\frac{x_2 - x_1}{2}\right)^2 + \left(\frac{y_2 - y_1}{2}\right)^2}$$
$$= \frac{\sqrt{(x_2 - x_1)^2 + (y_2 - y_1)^2}}{2}$$

Thus $PM = MQ$.

(b)   Adding the two distances together:
$$PM + MQ = 2(PM) = \sqrt{(x_2 - x_1)^2 + (y_2 - y_1)^2} = PQ$$

49. (a)   Each side of the outermost quadrilateral is the hypotenuse of the triangle in Figure A, so each side has a length of c. Also, since each angle of the outermost quadrilateral is the sum of the two acute angles from the triangle, the measure of each angle is $90°$. Since the quadrilateral has angles of $90°$ and sides of equal length, it must be a square.

(b)   Each side of the innermost quadrilateral is the difference of the two legs of the triangle in Figure A, so the length is $b - a$.

(c)   The area of each triangle is $\frac{1}{2}ab$, and the area of the innermost square is $(b-a)^2$. Since there are four triangles, the figure area is:
$$4\left(\tfrac{1}{2}ab\right) + (b-a)^2 = 2ab + b^2 - 2ab + a^2 = a^2 + b^2$$
Since the figure has an area of c^2, we have the result $a^2 + b^2 = c^2$.

51. (a)   Using the three-step technique to convert the equation to standard form:
$$x^2 - 10x + y^2 + 2y + 17 = 0$$
$$x^2 - 10x + y^2 + 2y = -17$$
$$\left(x^2 - 10x + 25\right) + y^2 + 2y = -17 + 25$$
$$(x - 5)^2 + y^2 + 2y = 8$$
$$(x - 5)^2 + \left(y^2 + 2y + 1\right) = 8 + 1$$
$$(x - 5)^2 + (y + 1)^2 = 9$$
The equation is now in standard form. The center of the circle is $(5, -1)$ and the radius is 3.

(b)   Using the three-step technique to convert the equation to standard form:
$$x^2 + y^2 + 8x - 6y = -24$$
$$x^2 + 8x + y^2 - 6y = -24$$
$$\left(x^2 + 8x + 16\right) + y^2 - 6y = -24 + 16$$
$$(x + 4)^2 + y^2 - 6y = -8$$
$$(x + 4)^2 + \left(y^2 - 6y + 9\right) = -8 + 9$$
$$(x + 4)^2 + (y - 3)^2 = 1$$
The equation is now in standard form. The center of the circle is $(-4, 3)$ and the radius is 1.

(c) Using the three-step technique to convert the equation to standard form:
$$4x^2 - 4x + 4y^2 - 63 = 0$$
$$4x^2 - 4x + 4y^2 = 63$$
$$x^2 - x + y^2 = \frac{63}{4}$$
$$\left(x^2 - x + \frac{1}{4}\right) + y^2 = \frac{63}{4} + \frac{1}{4}$$
$$\left(x - \frac{1}{2}\right)^2 + y^2 = 16$$

The equation is now in standard form. The center of the circle is $\left(\frac{1}{2}, 0\right)$ and the radius is 4.

(d) Using the three-step technique to convert the equation to standard form:
$$9x^2 + 54x + 9y^2 - 6y + 64 = 0$$
$$9x^2 + 54x + 9y^2 - 6y = -64$$
$$x^2 + 6x + y^2 - \frac{2}{3}y = -\frac{64}{9}$$
$$\left(x^2 + 6x + 9\right) + y^2 - \frac{2}{3}y = -\frac{64}{9} + 9$$
$$(x+3)^2 + y^2 - \frac{2}{3}y = \frac{17}{9}$$
$$(x+3)^2 + \left(y^2 - \frac{2}{3}y + \frac{1}{9}\right) = \frac{17}{9} + \frac{1}{9}$$
$$(x+3)^2 + \left(y - \frac{1}{3}\right)^2 = 2$$

The equation is now in standard form. The center of the circle is $\left(-3, \frac{1}{3}\right)$ and the radius is $\sqrt{2}$.

53. (a) Compute the areas:

      Rectangle: (length)(width) $= (5)(4) = 20$

      Triangles: $A_1 = \frac{1}{2}(\text{base})(\text{height}) = \frac{1}{2}(4)(4) = 8$

                     $A_2 = \frac{1}{2}(\text{base})(\text{height}) = \frac{1}{2}(5)(1) = \frac{5}{2}$

                     $A_3 = \frac{1}{2}(\text{base})(\text{height}) = \frac{1}{2}(3)(1) = \frac{3}{2}$

Therefore:
$$\text{Area}_{\triangle ABC} = 20 - 8 - \frac{5}{2} - \frac{3}{2} = 8 \text{ sq. units}$$

(b) Compute the areas:

      Rectangle: (length)(width) $= (9)(3) = 27$

      Triangles: $A_1 = \frac{1}{2}(\text{base})(\text{height}) = \frac{1}{2}(9)(1) = \frac{9}{2}$

                     $A_2 = \frac{1}{2}(\text{base})(\text{height}) = \frac{1}{2}(2)(3) = 3$

                     $A_3 = \frac{1}{2}(\text{base})(\text{height}) = \frac{1}{2}(6)(3) = 9$

Therefore:
$$\text{Area}_{\triangle ABC} = 27 - \frac{9}{2} - 3 - 9 = \frac{21}{2} \text{ sq. units}$$

(c)   Compute the areas:

Rectangle:

$$(\text{length})(\text{width}) = (x_2 - x_1)(y_3 - y_1) = x_2 y_3 - x_1 y_3 - x_2 y_1 + x_1 y_1$$

Triangles:

$$A_1 = \tfrac{1}{2}(\text{base})(\text{height}) = \tfrac{1}{2}(y_3 - y_1)(x_3 - x_1) = \tfrac{1}{2}(x_3 y_3 - x_3 y_1 - x_1 y_3 + x_1 y_1)$$

$$A_2 = \tfrac{1}{2}(\text{base})(\text{height}) = \tfrac{1}{2}(y_3 - y_2)(x_2 - x_3) = \tfrac{1}{2}(x_2 y_3 - x_2 y_2 - x_3 y_3 + x_3 y_2)$$

$$A_3 = \tfrac{1}{2}(\text{base})(\text{height}) = \tfrac{1}{2}(x_2 - x_1)(y_2 - y_1) = \tfrac{1}{2}(x_2 y_2 - x_2 y_1 - x_1 y_2 + x_1 y_1)$$

Therefore:

$$\text{Area}_{\triangle ABC} = \text{Rectangle} - \text{Triangles}$$
$$= \tfrac{1}{2}(2x_2 y_3 - 2x_1 y_3 - 2x_2 y_1 + 2x_1 y_1 - x_3 y_3 + x_3 y_1 + x_1 y_3 - x_1 y_1 - x_2 y_3$$
$$+ x_2 y_2 + x_3 y_3 - x_3 y_2 - x_2 y_2 + x_2 y_1 - x_1 y_2 - x_1 y_1)$$
$$= \tfrac{1}{2}(x_2 y_3 - x_1 y_3 - x_2 y_1 + x_3 y_1 - x_3 y_2 + x_1 y_2)$$

This is the desired result.

Graphing Utility Exercises for Section 2.1

1.   Solving for y yields $y = \pm\sqrt{36 - x^2}$. Graphing each function:

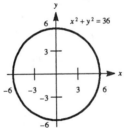

3.   Solving for y yields $y = \pm\sqrt{2 - x^2}$. Graphing each function:

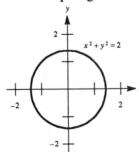

5.   Solving the equation for y:

$$(x+2)^2 + (y-4)^2 = 9$$
$$(y-4)^2 = 9 - (x+2)^2$$
$$y - 4 = \pm\sqrt{9 - (x+2)^2}$$
$$y = 4 \pm \sqrt{9 - (x+2)^2}$$

Graphing each function:

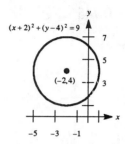

7.   Solving the equation for y:

$$(x-12)^2 + (y+10)^2 = 90$$
$$(y+10)^2 = 90 - (x-12)^2$$
$$y + 10 = \pm\sqrt{90 - (x-12)^2}$$
$$y = -10 \pm \sqrt{90 - (x-12)^2}$$

Graphing each function:

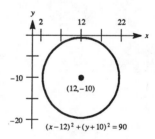

9.   (a)   Graphing the circle:

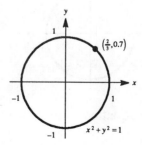

(b)   Substituting $x = \frac{2}{3}$ into the equation:

$$\left(\frac{2}{3}\right)^2 + y^2 = 1$$
$$\frac{4}{9} + y^2 = 1$$
$$y^2 = \frac{5}{9}$$
$$y = \sqrt{\frac{5}{9}} = \frac{\sqrt{5}}{3} \approx 0.745$$

11.   (a)   Graphing the circle:

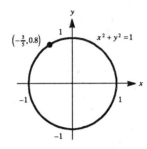

(b)   Substituting $x = -\frac{3}{5}$ into the equation:

$$\left(-\frac{3}{5}\right)^2 + y^2 = 1$$
$$\frac{9}{25} + y^2 = 1$$
$$y^2 = \frac{16}{25}$$
$$y = \sqrt{\frac{16}{25}} = \frac{4}{5} = 0.8$$

13.   (a)   Graphing the circle:

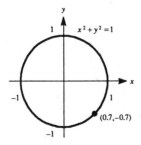

(b)   Substituting $x = 0.7$ into the equation:

$$(0.7)^2 + y^2 = 1$$
$$0.49 + y^2 = 1$$
$$y^2 = 0.51$$
$$y = -\sqrt{0.51} \approx -0.714$$

15. (a)   Graphing the ellipse:

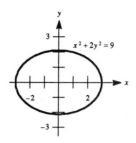

(b)   The curves appear to intersect at $(-3, 0)$ and $(3, 0)$:

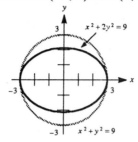

(c)   Substituting the points $(\pm 3, 0)$ into the two equations:
$$(\pm 3)^2 + (0)^2 = 9$$
$$(\pm 3)^2 + 2(0)^2 = 9$$
Thus the intersection points are indeed $(-3, 0)$ and $(3, 0)$.

17. (a)   Graphing the hyperbola:

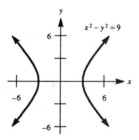

(b)   Graphing the two lines $y = x$ and $y = -x$:

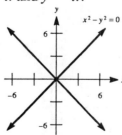

(c)   Putting the two graphs together:

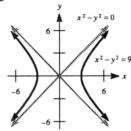

19. Solving for y yields $y = \pm\sqrt{x^3 - 9}$. Graphing the curve:

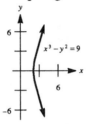

21. (a)   Solving for y yields $y = \pm\sqrt[4]{9 - x^4}$.

(b)   Solving for y yields $y = \pm\sqrt[6]{9 - x^6}$. Graphing the two curves:

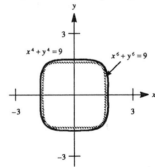

(c)   Solving for y yields $y = \pm\sqrt[8]{9 - x^8}$. Graphing the curve:

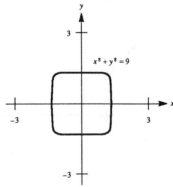

23. Using the zero-product property, we must have $x^2 - y^2 - 1 = 0$ or $x^2 - y^2 = 0$. So $x^2 - y^2 = 1$ or $x^2 = y^2$. Graphing the hyperbola and the two lines:

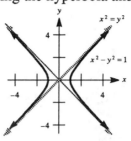

2.2 Graphs and Equations, A Second Look

1. Substituting $x = 8$:
$$y = \tfrac{1}{2}(8) + 3 = 4 + 3 = 7 \neq 6$$
The point $(8, 6)$ does not lie on the graph.

3. Substituting $(x, y) = (4, 3)$:
$$3(4)^2 + (3)^2 = 3(16) + 9 = 48 + 9 = 57 \neq 52$$
The point $(4, 3)$ does not lie on the graph.

5. Substituting $x = a$:
$$y = 4a$$
The point $(a, 4a)$ lies on the graph.

7. Since $(2, -3)$ lies on the graph, it must satisfy the equation:
$$y = ax + 6$$
$$-3 = a(2) + 6$$
$$-3 = 2a + 6$$
$$-9 = 2a$$
$$a = -\tfrac{9}{2}$$

9. Since $(-5, 2)$ lies on the graph, it must satisfy the equation:
$$y = a(x - 6)^2 - 2$$
$$2 = a(-5 - 6)^2 - 2$$
$$2 = 121a - 2$$
$$4 = 121a$$
$$a = \tfrac{4}{121}$$

11. Since $(1, -3)$ lies on the graph, it must satisfy the equation:
$$x^2 + ay^2 = 28$$
$$(1)^2 + a(-3)^2 = 28$$
$$1 + 9a = 28$$
$$9a = 27$$
$$a = 3$$

13. Since the denominator cannot equal 0, we must have $x \neq 4$. To assist us in graphing the curve, simplify when $x \neq 4$:
$$y = \frac{x^2 + x - 20}{x - 4} = \frac{(x+5)(x-4)}{x-4} = x + 5$$
Now graph the line $y = x + 5$, leaving an open circle at the point $(4, 9)$:

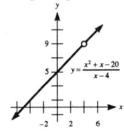

15. Since the denominator cannot equal 0, we must have $x \neq 3$. To assist us in graphing the curve, simplify when $x \neq 3$:
$$y = \frac{x^2 - 9}{x - 3} = \frac{(x+3)(x-3)}{x-3} = x + 3$$
Now graph the line $y = x + 3$, leaving an open circle at the point $(3, 6)$:

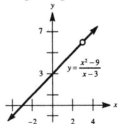

17. Since the denominator cannot equal 0, we must have $x^2 - 1 \neq 0$, so $x \neq \pm 1$. To simplify the fraction factor the numerator by grouping:
$$-x^3 + 4x^2 + x - 4 = -x^2(x - 4) + 1(x - 4) = (x - 4)\left(1 - x^2\right)$$
Now simplify the fraction when $x \neq \pm 1$:
$$y = \frac{-x^3 + 4x^2 + x - 4}{x^2 - 1} = \frac{(x-4)\left(1 - x^2\right)}{x^2 - 1} = -1(x - 4) = -x + 4$$

Now graph the line $y = -x + 4$, leaving open circles at the points $(1, 3)$ and $(-1, 5)$:

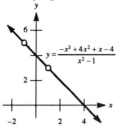

19. To find the x-intercept, substitute $y = 0$ into the equation $3x + 4y = 12$:
$$3x + 4(0) = 12$$
$$3x = 12$$
$$x = 4$$
To find the y-intercept, substitute $x = 0$ into the equation $3x + 4y = 12$:
$$3(0) + 4y = 12$$
$$4y = 12$$
$$y = 3$$
Now graph the line:

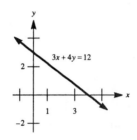

21. To find the x-intercept, substitute $y = 0$ into the equation $y = 2x - 4$:
$$0 = 2x - 4$$
$$-2x = -4$$
$$x = 2$$
To find the y-intercept, substitute $x = 0$ into the equation $y = 2x - 4$:
$$y = 2(0) - 4$$
$$y = -4$$
Now graph the line:

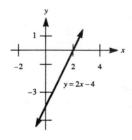

23. To find the x-intercept, substitute $y = 0$ into the equation $x + y = 1$:
$$x + 0 = 1$$
$$x = 1$$
To find the y-intercept, substitute $x = 0$ into the equation $x + y = 1$:
$$0 + y = 1$$
$$y = 1$$
Now graph the line:

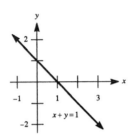

25. (a) To find the y-intercepts, let $x = 0$:
$$y = (0)^2 + 3(0) + 2 = 0 + 0 + 2 = 2$$
To find the x-intercepts, let $y = 0$:
$$x^2 + 3x + 2 = 0$$
$$(x + 2)(x + 1) = 0$$
$$x = -1, -2$$

(b) To find the y-intercepts, let $x = 0$:
$$y = (0)^2 + 2(0) + 3 = 0 + 0 + 3 = 3$$
To find the x-intercepts, we must solve the equation $x^2 + 2x + 3 = 0$. Using the quadratic formula:
$$x = \frac{-2 \pm \sqrt{(2)^2 - 4(1)(3)}}{2(1)} = \frac{-2 \pm \sqrt{4 - 12}}{2} = \frac{-2 \pm \sqrt{-8}}{2}$$
Since this equation has no real solutions, there are no x-intercepts.

27. (a) To find the y-intercepts, let $x = 0$:
$$y = (0)^2 + 0 - 1 = 0 + 0 - 1 = -1$$
To find the x-intercepts, we must solve the equation $x^2 + x - 1 = 0$. Using the quadratric formula:
$$x = \frac{-1 \pm \sqrt{(1)^2 - 4(1)(-1)}}{2(1)} = \frac{-1 \pm \sqrt{1 + 4}}{2} = \frac{-1 \pm \sqrt{5}}{2}$$

(b) To find the y-intercepts, let $x = 0$:
$$y = (0)^2 + 0 + 1 = 0 + 0 + 1 = 1$$
To find the x-intercepts, we must solve the equation $x^2 + x + 1 = 0$. Using the quadratic formula:
$$x = \frac{-1 \pm \sqrt{(1)^2 - 4(1)(1)}}{2(1)} = \frac{-1 \pm \sqrt{1 - 4}}{2} = \frac{-1 \pm \sqrt{-3}}{2}$$
Since this equation has no real solutions, there are no x-intercepts.

29. To find the y-intercepts, let $x = 0$:
$$y = 11(0) - 2(0)^2 - (0)^3 = 0 - 0 - 0 = 0$$
To find the x-intercepts, let $y = 0$:
$$0 = 11x - 2x^2 - x^3$$
$$0 = x\left(11 - 2x - x^2\right)$$
So $x = 0$ is clearly one x-intercept. Find the other two using the quadratic formula:
$$x = \frac{2 \pm \sqrt{(-2)^2 - 4(11)(-1)}}{2(-1)} = \frac{2 \pm \sqrt{48}}{-2} = -1 \pm 2\sqrt{3}$$
The other two x-intercepts are $-1 - 2\sqrt{3} \approx -4.46$ and $-1 + 2\sqrt{3} \approx 2.46$.

31. To find the y-intercepts, let $x = 0$:
$$y^2 - 4y - 8 = 0$$
$$(y - 6)(y + 2) = 0$$
$$y = -2, 6$$
To find the x-intercepts, let $y = 0$:
$$3x = (0)^2 - 4(0) - 8$$
$$3x = -8$$
$$x = -\frac{8}{3}$$

33. To find the y-intercepts, let $x = 0$:
$$(0 - 2)^2 + (y - 1)^2 = 9$$
$$4 + (y - 1)^2 = 9$$
$$(y - 1)^2 = 5$$
$$y - 1 = \pm\sqrt{5}$$
$$y = 1 \pm \sqrt{5}$$
The two y-intercepts are $1 - \sqrt{5} \approx -1.24$ and $1 + \sqrt{5} \approx 3.24$.
To find the x-intercepts, let $y = 0$:
$$(x - 2)^2 + (0 - 1)^2 = 9$$
$$(x - 2)^2 + 1 = 9$$
$$(x - 2)^2 = 8$$
$$x - 2 = \pm\sqrt{8} = \pm 2\sqrt{2}$$
$$x = 2 \pm 2\sqrt{2} \approx -0.83, 4.83$$
The two x-intercepts are $2 - 2\sqrt{2} \approx -0.83$ and $2 + 2\sqrt{2} \approx 4.83$.

35. (a)   Set up a table of values:

x	-3	-2	-1	0	1	2	3
$y = \lvert x \rvert$	3	2	1	0	1	2	3

Now graph the equation:

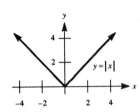

(b)   Set up a table of values:

x	-3	-2	-1	0	1	2	3
$y = x^2$	9	4	1	0	1	4	9

Now graph the equation:

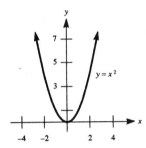

(c)   Set up a table of values:

x	-2	-1	0	1	2
$y = x^3$	-8	-1	0	1	8

Now graph the equation:

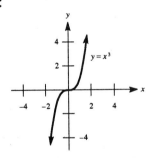

37.   We "piece together" the curves $y = \lvert x \rvert$ when $x \leq 0$ and $y = x^2$ when $x > 0$:

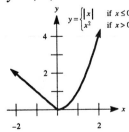

39. (a) We "piece together" the curves $y = \sqrt{x}$ when $0 \le x \le 1$ and $y = \dfrac{1}{x}$ when $1 < x < 2$:

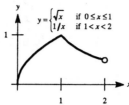

(b) This graph is identical to that from part (a), except the point (1, 1) is excluded:

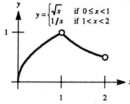

41. We "piece together" the curves $y = x^3$ when $-1 \le x < 0$, $\;y = \sqrt{x}$ when $0 \le x < 1$, and $y = \dfrac{1}{x}$ when $1 \le x \le 3$:

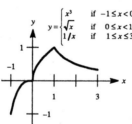

43. (a) From the graph it appears that $0°$ F corresponds to $-18°$ C, which is the (approximate) C-intercept of the graph.

(b) Substituting $F = 0°$ into the equation and solving for C:

$$\tfrac{9}{5}C + 32 = 0$$
$$\tfrac{9}{5}C = -32$$
$$C = -\tfrac{160}{9} \approx -17.8°$$

This value is consistent with our approximation from part (a).

45. (a) Tracing up to the curve from $x = 2$, $y = \sqrt{2} \approx 1.4$.

(b) Tracing up to the curve from $x = 3$, $y = \sqrt{3} \approx 1.7$.

(c) Since $\sqrt{ab} = \sqrt{a} \cdot \sqrt{b}$, $\sqrt{6} = \sqrt{2 \cdot 3} = \sqrt{2} \cdot \sqrt{3} \approx (1.4)(1.7) \approx 2.4$.

47. (a) Tracing up to the curve from $t = 0$, $N = 500$ bacteria.

(b) Since $(0, 500)$ lies on the curve, find where $(t, 1000)$ would be on the curve, as the population would now be double. Tracing down from $N = 1000$, $t = 1.5$ hours. So the population will double in 1.5 hours.

(c) As in (b), find where $(t, 2500)$ would be on the curve. Tracing down from $N = 2500$, $t = 3.5$ hours.

(d) Between $t = 0$ and $t = 1$, the population has grown from $N = 500$ to $N = 800$, so it has increased at an average rate of 300 bacteria per hour. Between $t = 3$ and $t = 4$, the population has grown from $N = 2000$ to $N = 3000$, so it has increased at an average rate of 1000 bacteria per hour. So the population has increased more rapidly between $t = 3$ and $t = 4$.

49. Since the y-coordinate for point A is $0.4 = \frac{2}{5}$, substitute to find x:

$$\frac{1}{2}x^2 = \frac{2}{5}$$
$$x^2 = \frac{4}{5}$$
$$x = \sqrt{\frac{4}{5}} = \frac{2}{5}\sqrt{5} \approx 0.894$$

Since the y-coordinate for point B is $0.6 = \frac{3}{5}$, substitute to find x:

$$\frac{1}{2}x^2 = \frac{3}{5}$$
$$x^2 = \frac{6}{5}$$
$$x = \sqrt{\frac{6}{5}} = \frac{1}{5}\sqrt{30} \approx 1.095$$

Graphing Utility Exercises for Section 2.2

1. The graph verifies that $y = -\frac{3}{4}(x - 2)^2 + 3$ passes through the origin:

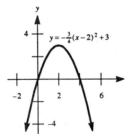

3. Simplifying the fraction:

$$y = \frac{2x^2 + 9x + 10}{x+2} = \frac{(2x+5)(x+2)}{x+2} = 2x+5, \text{ if } x \neq -2$$

Graphing the line with the point $(-2, 1)$ excluded:

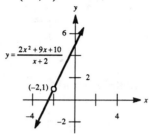

5. Since $x^2 - 6x + 8 = (x-4)(x-2)$, the missing points occur at $x = 2$ and $x = 4$. Using long division:

$$\frac{2x^3 - 11x^2 + 10x + 8}{x^2 - 6x + 8} = 2x + 1, \text{ if } x \neq 2, 4$$

Graphing the line with the points $(2, 5)$ and $(4, 9)$ excluded:

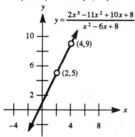

7. (a) Graphing the equation:

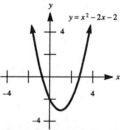

The x-intercepts are approximately -0.73 and 2.73.

(b) Using the quadratic formula:

$$x = \frac{2 \pm \sqrt{(-2)^2 - 4(1)(-2)}}{2} = \frac{2 \pm \sqrt{4+8}}{2} = \frac{2 \pm 2\sqrt{3}}{2} = 1 \pm \sqrt{3} \approx -0.732, 2.732$$

These values check our results from part (a).

9. (a) Graphing the equation:

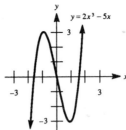

The x-intercepts are approximately 0 and ± 1.58.

(b) The equation factors as $y = x\left(2x^2 - 5\right)$, so $x = 0$ is one x-intercept. Solving for the other two x-intercepts:

$$2x^2 - 5 = 0$$
$$2x^2 = 5$$
$$x^2 = \tfrac{5}{2}$$
$$x = \pm\sqrt{\tfrac{5}{2}} = \pm\frac{\sqrt{10}}{2} \approx \pm 1.581$$

These values check our results from part (a).

11. (a) Graphing the equation:

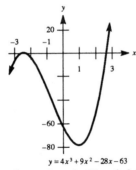

$$y = 4x^3 + 9x^2 - 28x - 63$$

The x-intercepts are approximately -2.65, -2.25, and 2.65.

(b) Factoring by grouping:

$$y = 4x^3 + 9x^2 - 28x - 63 = x^2(4x + 9) - 7(4x + 9) = (4x + 9)\left(x^2 - 7\right)$$

So the x-intercepts are $-\tfrac{9}{4} = -2.25$ and $\pm\sqrt{7} \approx \pm 2.646$, which check our results from part (a).

13. (a) Graphing the equation:

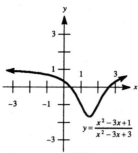

$$y = \frac{x^2 - 3x + 1}{x^2 - 3x + 3}$$

The x-intercepts are approximately 0.38 and 2.62.

(b) Using the quadratic formula (for the numerator):

$$x = \frac{3 \pm \sqrt{(-3)^2 - 4(1)(1)}}{2} = \frac{3 \pm \sqrt{9 - 4}}{2} = \frac{3 \pm \sqrt{5}}{2} \approx 0.382, 2.618$$

These values check our results from part (a).

15. Graphing the equation:

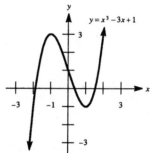

$$y = x^3 - 3x + 1$$

The x-intercepts are approximately -1.879, 0.347, and 1.532.

17. Graphing the equation:

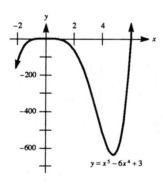

$$y = x^5 - 6x^4 + 3$$

Zooming in near the origin:

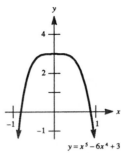

$$y = x^5 - 6x^4 + 3$$

The x-intercepts are approximately –0.815, 0.875, and 5.998.

19. Graphing $y = |x|$:

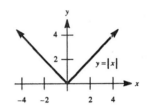

Graphing $y = x^2$:

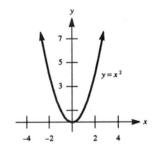

Graphing $y = x^3$:

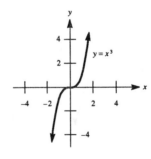

21. (a) Graphing $y = x^2$ using the specified viewing rectangle:

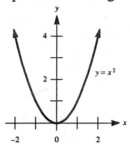

(b) Graphing $y = x^3$ using the specified viewing rectangle:

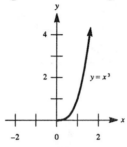

(c) Graphing the two curves together:

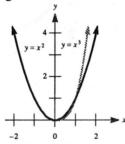

The two curves intersect at the points $(0, 0)$ and $(1, 1)$. Note that $x^2 > x^3$ on the interval $(0, 1)$.

23. Graphing the curve:

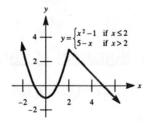

25.   Graphing the curve:

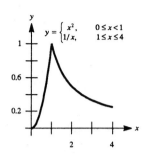

27.   Graphing the curve:

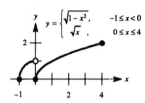

29.   Graphing the curve:

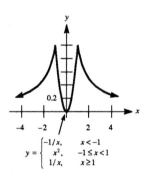

31.   Graphing the curve:

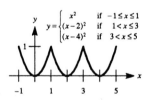

2.3 Equations of Lines

1.   (a)   Here $(x_1, y_1) = (-3, 2)$ and $(x_2, y_2) = (1, -6)$, so:

$$\text{slope} = \frac{-6-2}{1-(-3)} = -\frac{8}{4} = -2$$

(b)   Here $(x_1, y_1) = (2, -5)$ and $(x_2, y_2) = (4, 1)$, so:

$$\text{slope} = \frac{1-(-5)}{4-2} = \frac{6}{2} = 3$$

(c) Here $(x_1, y_1) = (-2, 7)$ and $(x_2, y_2) = (1, 0)$, so:

$$\text{slope} = \frac{0-7}{1-(-2)} = -\frac{7}{3}$$

(d) Here $(x_1, y_1) = (4, 5)$ and $(x_2, y_2) = (5, 8)$, so:

$$\text{slope} = \frac{8-5}{5-4} = \frac{3}{1} = 3$$

3. (a) Here $(x_1, y_1) = (1, 1)$ and $(x_2, y_2) = (-1, -1)$, so:

$$\text{slope} = \frac{-1-1}{-1-1} = \frac{-2}{-2} = 1$$

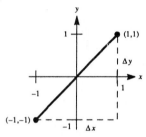

(b) Here $(x_1, y_1) = (0, 5)$ and $(x_2, y_2) = (-8, 5)$, so:

$$\text{slope} = \frac{5-5}{-8-0} = \frac{0}{-8} = 0$$

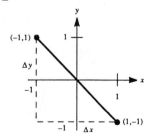

(c) Here $(x_1, y_1) = (-1, 1)$ and $(x_2, y_2) = (1, -1)$, so:

$$\text{slope} = \frac{-1-1}{1-(-1)} = \frac{-2}{2} = -1$$

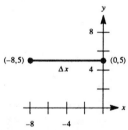

(d)   Here $(x_1, y_1) = (a, b)$ and $(x_2, y_2) = (b, a)$, so:

$$\text{slope} = \frac{a-b}{b-a} = -1$$

5. m_3 is smallest (since it is negative)
m_2 is next (it appears to be near zero)
m_4 is next (it is positive but not as steep as m_1)
m_1 is largest (it is steeper than m_4)

7.   Compute the slopes:

$$\text{slope}_{AB} = \frac{\frac{1}{2} - (-2)}{2 - (-8)} = \frac{\frac{5}{2}}{10} = \frac{1}{4}$$

$$\text{slope}_{BC} = \frac{-1 - \frac{1}{2}}{11 - 2} = \frac{-\frac{3}{2}}{9} = -\frac{1}{6}$$

Since these slopes are different the three points cannot be collinear.

9.   Compute the slopes:

$$\text{slope}_{AB} = \frac{4 - (-5)}{3 - 0} = \frac{9}{3} = 3$$

$$\text{slope}_{BC} = \frac{-8 - 4}{-1 - 3} = \frac{-12}{-4} = 3$$

Since these slopes are equal the three points are collinear.

11.   (a)   Here $(x_1, y_1) = (-2, 1)$ and $m = -5$, so by the point-slope formula:

$$y - 1 = -5[x - (-2)]$$
$$y - 1 = -5(x + 2)$$
$$y - 1 = -5x - 10$$
$$y = -5x - 9$$

(b)   Here $(x_1, y_1) = (4, -4)$ and $m = 4$, so by the point-slope formula:

$$y - (-4) = 4(x - 4)$$
$$y + 4 = 4x - 16$$
$$y = 4x - 20$$

(c) Here $(x_1, y_1) = \left(-6, -\frac{2}{3}\right)$ and $m = \frac{1}{3}$, so by the point-slope formula:

$$y - \left(-\tfrac{2}{3}\right) = \left(\tfrac{1}{3}\right)[x - (-6)]$$
$$y + \tfrac{2}{3} = \left(\tfrac{1}{3}\right)(x + 6)$$
$$y + \tfrac{2}{3} = \tfrac{1}{3}x + 2$$
$$y = \tfrac{1}{3}x + \tfrac{4}{3}$$

13. (a) First find the slope:
$$m = \frac{-6 - 8}{-3 - 4} = \frac{-14}{-7} = 2$$
Using $(x_1, y_1) = (4, 8)$ in the point-slope formula:
$$y - 8 = 2(x - 4)$$
$$y - 8 = 2x - 8$$
$$y = 2x$$

(b) First find the slope:
$$m = \frac{-10 - 0}{3 - (-2)} = \frac{-10}{5} = -2$$
Using $(x_1, y_1) = (-2, 0)$ in the point-slope formula:
$$y - 0 = -2[x - (-2)]$$
$$y = -2(x + 2)$$
$$y = -2x - 4$$

(c) First find the slope:
$$m = \frac{-1 - (-2)}{4 - (-3)} = \tfrac{1}{7}$$
Using $(x_1, y_1) = (4, -1)$ in the point-slope formula:
$$y - (-1) = \tfrac{1}{7}(x - 4)$$
$$y + 1 = \tfrac{1}{7}x - \tfrac{4}{7}$$
$$y = \tfrac{1}{7}x - \tfrac{11}{7}$$

15. Since vertical lines have the form $x = $ constant, and $(-3, 4)$ is on the line, the equation is $x = -3$.

17. Since horizontal lines have the form $y = $ constant, and $(-3, 4)$ is on the line, the equation is $y = 4$.

19. The y-axis is vertical, so its equation must have the form $x = $ constant. Since $(0, 0)$ is on the y-axis, the equation is $x = 0$.

21. (a) Using the slope-intercept formula with $m = -4$ and $b = 7$, we have $y = -4x + 7$.
    (b) Using the slope-intercept formula with $m = 2$ and $b = \frac{3}{2}$, we have $y = 2x + \frac{3}{2}$.
    (c) Using the slope-intercept formula with $m = -\frac{4}{3}$ and $b = 14$, we have $y = -\frac{4}{3}x + 14$.

23. (a)   Use the point-slope formula:
$$y - (-1) = 4[x - (-3)]$$
$$y + 1 = 4(x + 3)$$
$$y + 1 = 4x + 12$$
$$y = 4x + 11$$
Now draw the graph:

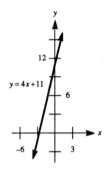

(b)   Use the point-slope formula:
$$y - 0 = \tfrac{1}{2}\left(x - \tfrac{5}{2}\right)$$
$$y = \tfrac{1}{2}x - \tfrac{5}{4}$$
Now draw the graph:

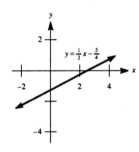

(c)   Find the slope between the points $(6, 0)$ and $(0, 5)$:
$$m = \frac{5 - 0}{0 - 6} = -\tfrac{5}{6}$$
Since 5 is the y-intercept, by the slope-intercept formula:
$$y = -\tfrac{5}{6}x + 5$$
Now draw the graph:

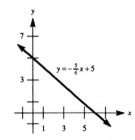

(d) Use the point-slope formula with the point $(-2, 0)$:
$$y - 0 = \tfrac{3}{4}[x - (-2)]$$
$$y = \tfrac{3}{4}(x + 2)$$
$$y = \tfrac{3}{4}x + \tfrac{3}{2}$$
Now draw the graph:

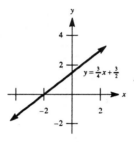

(e) First find the slope:
$$m = \frac{6 - 2}{2 - 1} = \tfrac{4}{1} = 4$$
Using the point $(1, 2)$ in the point-slope formula:
$$y - 2 = 4(x - 1)$$
$$y - 2 = 4x - 4$$
$$y = 4x - 2$$
Now draw the graph:

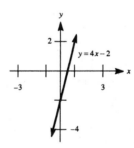

25. First find the x- and y-intercepts of the circle by letting $x = 0$ and $y = 0$:

$$y = 0 \qquad\qquad x = 0$$
$$x^2 + 4x + 4 = 0 \qquad\qquad y^2 - 4y + 4 = 0$$
$$(x + 2)^2 = 0 \qquad\qquad (y - 2)^2 = 0$$
$$x = -2 \qquad\qquad y = 2$$

So the line passes through the points $(-2, 0)$ and $(0, 2)$. Finding its slope:
$$m = \frac{2 - 0}{0 - (-2)} = \tfrac{2}{2} = 1$$

So $y = x + 2$ is the equation (in slope-intercept form) of the line. Draw the graph:

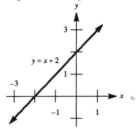

27.   If the line is parallel to the x-axis, then its slope is 0. Since this line is of the form $y = $ constant, for $(-3, 4)$ to lie on the line the equation is $y = 4$, or $y - 4 = 0$ in the desired form. Draw the graph:

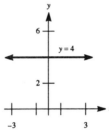

29.   (a)   If $x = 0$, then $y = 3$ and if $y = 0$, then $x = 5$. So the x-intercept is 5 and the y-intercept is 3. Thus:
$$\text{area} = \tfrac{1}{2}(5)(3) = \tfrac{15}{2}$$
$$\text{perimeter} = 5 + 3 + \sqrt{5^2 + 3^2} = 8 + \sqrt{34}$$

   (b)   If $x = 0$, then $y = -3$ and if $y = 0$, then $x = 5$. So the x-intercept is 5 and the y-intercept is -3. Thus:
$$\text{area} = \tfrac{1}{2}(5)(3) = \tfrac{15}{2}$$
$$\text{perimeter} = 5 + 3 + \sqrt{5^2 + 3^2} = 8 + \sqrt{34}$$

31.   (a)   Find the slopes of each:

$$3x - 4y = 12 \qquad\qquad 4x - 3y = 12$$
$$-4y = -3x + 12 \qquad\qquad -3y = -4x + 12$$
$$y = \tfrac{3}{4}x - 3 \qquad\qquad y = \tfrac{4}{3}x - 4$$
$$m = \tfrac{3}{4} \qquad\qquad m = \tfrac{4}{3}$$

The lines are not parallel (slopes aren't the same), the lines are not perpendicular $\left(\tfrac{3}{4} \cdot \tfrac{4}{3} = 1, \text{ not } -1\right)$, so they are neither.

   (b)   Find the slopes of each:

$$y = 5x - 16 \qquad\qquad y = 5x + 2$$
$$m = 5 \qquad\qquad m = 5$$

Since these slopes are the same, the lines are parallel.

(c) Find the slopes of each:

$$5x - 6y = 25$$
$$-6y = -5x + 25$$
$$y = \tfrac{5}{6}x - \tfrac{25}{6}$$
$$m = \tfrac{5}{6}$$

$$6x + 5y = 0$$
$$5y = -6x$$
$$y = -\tfrac{6}{5}x$$
$$m = -\tfrac{6}{5}$$

Since $\left(\tfrac{5}{6}\right)\left(-\tfrac{6}{5}\right) = -1$, the lines are perpendicular.

(d) Find the slopes of each:

$$y = -\tfrac{2}{3}x - 1$$
$$m = -\tfrac{2}{3}$$

$$y = \tfrac{3}{2}x - 1$$
$$m = \tfrac{3}{2}$$

Since $\left(-\tfrac{2}{3}\right)\left(\tfrac{3}{2}\right) = -1$, the lines are perpendicular.

(e) Find the slopes of each:

$$-2x - 5y = 1$$
$$-5y = 2x + 1$$
$$y = -\tfrac{2}{5}x - \tfrac{1}{5}$$
$$m = -\tfrac{2}{5}$$

$$y - \tfrac{2}{5}x - 4 = 0$$
$$y = \tfrac{2}{5}x + 4$$
$$m = \tfrac{2}{5}$$

The lines are not parallel (slopes aren't the same), the lines are not perpendicular $\left(-\tfrac{2}{5} \cdot \tfrac{2}{5} = -\tfrac{4}{25}, \text{ not } -1\right)$, so they are neither.

(f) Find the slopes of each:

$$x = 8y + 3$$
$$8y = x - 3$$
$$y = \tfrac{1}{8}x - \tfrac{3}{8}$$
$$m = \tfrac{1}{8}$$

$$4y - \tfrac{1}{2}x = 32$$
$$4y = \tfrac{1}{2}x + 32$$
$$y = \tfrac{1}{8}x + 8$$
$$m = \tfrac{1}{8}$$

Since these slopes are the same, the lines are parallel.

33. First find the slope:

$$2x - 5y = 10$$
$$-5y = -2x + 10$$
$$y = \tfrac{2}{5}x - 2$$

So the slope is $\tfrac{2}{5}$. Using the point $(-1, 2)$ in the point-slope formula:

$$y - 2 = \tfrac{2}{5}[x - (-1)]$$
$$y - 2 = \tfrac{2}{5}(x + 1)$$
$$y - 2 = \tfrac{2}{5}x + \tfrac{2}{5}$$
$$y = \tfrac{2}{5}x + \tfrac{12}{5} \qquad [\text{form } y = mx + b]$$
$$5y = 2x + 12$$
$$2x - 5y + 12 = 0 \qquad [\text{form } Ax + By + C = 0]$$

35.   First find the slope of $4y - 3x = 1$:
$$4y = 3x + 1$$
$$y = \tfrac{3}{4}x + \tfrac{1}{4}$$
Since this slope is $\tfrac{3}{4}$, the perpendicular line slope is $-\tfrac{4}{3}$. Now use the point $(4, 0)$ in the point-slope formula:
$$y - 0 = -\tfrac{4}{3}(x - 4)$$
$$y = -\tfrac{4}{3}x + \tfrac{16}{3} \qquad [\text{form } y = mx + b]$$
$$3y = -4x + 16$$
$$4x + 3y - 16 = 0 \qquad [\text{form } Ax + By + C = 0]$$

37.   First find the slope of $3x - 5y = 25$:
$$-5y = -3x + 25$$
$$y = \tfrac{3}{5}x - 5$$
So the slope is $\tfrac{3}{5}$. Now find the y-intercept of $6x - y + 11 = 0$:
$$y = 6x + 11$$
So the y-intercept is 11. Writing the equation:
$$y = \tfrac{3}{5}x + 11 \qquad [\text{form } y = mx + b]$$
$$5y = 3x + 55$$
$$3x - 5y + 55 = 0 \qquad [\text{form } Ax + By + C = 0]$$

39.   (a)   The center is $(0, 0)$ and the radius is 5. Drawing the graph:

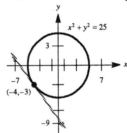

   (b)   The tangent line will be perpendicular to the radius drawn from the center $(0, 0)$ and the point $(-4, -3)$. Find the slope:
$$m = \frac{-3 - 0}{-4 - 0} = \frac{-3}{-4} = \tfrac{3}{4}$$
So the perpendicular line will have slope $= -\tfrac{4}{3}$. Use the point $(-4, -3)$ in the point-slope formula:
$$y - (-3) = -\tfrac{4}{3}[x - (-4)]$$
$$y + 3 = -\tfrac{4}{3}(x + 4)$$
$$y + 3 = -\tfrac{4}{3}x - \tfrac{16}{3}$$
$$y = -\tfrac{4}{3}x - \tfrac{25}{3}$$

41. Let $(x_1, y_1) = (3, 9)$ and $(x_2, y_2) = [3 + h, (3 + h)^2]$, so:

$$m = \frac{(3+h)^2 - 9}{3 + h - 3} = \frac{9 + 6h + h^2 - 9}{h} = \frac{6h + h^2}{h} = 6 + h$$

43. Let $(x_1, y_1) = (x, x^3)$ and $(x_2, y_2) = [x + h, (x + h)^3]$. Therefore the slope is given by:

$$\frac{(x+h)^3 - x^3}{x + h - x} = \frac{x^3 + 3x^3h + 3xh^2 + h^3 - x^3}{h}$$

$$= \frac{3x^2h + 3xh^2 + h^3}{h}$$

$$= \frac{h(3x^2 + 3xh + h^2)}{h}$$

$$= 3x^2 + 3xh + h^2$$

45. Let $(x_1, y_1) = (x, \frac{1}{x})$ and $(x_2, y_2) = (x + h, \frac{1}{x+h})$, so:

$$\text{slope} = \frac{\frac{1}{x+h} - \frac{1}{x}}{x + h - x} = \frac{\frac{1}{x+h} - \frac{1}{x}}{h}$$

Multiply the numerator and denominator by $x(x + h)$ to obtain:

$$\text{slope} = \frac{x - x - h}{hx(x + h)} = \frac{-h}{hx(x + h)} = \frac{-1}{x(x + h)}$$

47. Draw the graph:

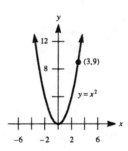

The completed table is:

x	2.5	2.9	2.99	2.999	2.9999
y	6.25	8.41	8.9401	8.994001	8.99940001
Δx	0.5	0.1	0.01	0.001	0.0001
Δy	2.75	0.59	0.0599	0.005999	0.00059999
m	5.5	5.9	5.99	5.999	5.9999

It would appear that as x approaches 3, the slope of these lines (called secant lines) approaches 6. So we would estimate that the slope of the tangent line to the curve $y = x^2$ at $T(3, 9)$ is 6.

49. Let the coordinates of P be (x, x^3). Since the slope of the line passing through P and $(1, 1)$ is $\frac{3}{4}$, we have:

$$\frac{x^3 - 1}{x - 1} = \frac{3}{4}$$

$$\frac{(x-1)(x^2 + x + 1)}{x - 1} = \frac{3}{4}$$

$$x^2 + x + 1 = \frac{3}{4}$$

$$4x^2 + 4x + 4 = 3$$

$$4x^2 + 4x + 1 = 0$$

$$(2x + 1)^2 = 0$$

$$2x + 1 = 0$$

$$x = -\frac{1}{2}$$

$$y = \left(-\frac{1}{2}\right)^3 = -\frac{1}{8}$$

Thus the point P is $\left(-\frac{1}{2}, -\frac{1}{8}\right)$.

51. Let the coordinates of P be $\left(x, \frac{1}{x}\right)$. Since the slope of the line through P and $\left(2, \frac{1}{2}\right)$ is $-\frac{1}{16}$:

$$\frac{\frac{1}{x} - \frac{1}{2}}{x - 2} = -\frac{1}{16}$$

$$\frac{2 - x}{2x(x - 2)} = -\frac{1}{16}$$

$$\frac{-1}{2x} = -\frac{1}{16}$$

$$2x = 16$$

$$x = 8$$

So the point P is $\left(8, \frac{1}{8}\right)$.

53. Use the point-slope formula to find the equation of the line:

$$y - 6 = -5(x - 3)$$
$$y - 6 = -5x + 15$$
$$y = -5x + 21$$

To find the x-intercept, let $y = 0$:

$$-5x + 21 = 0$$
$$5x = 21$$
$$x = \frac{21}{5}$$

To find the y-intercept, let $x = 0$, so $y = 21$. The area of the triangle is given by:

$$\text{Area} = \frac{1}{2}(\text{base})(\text{height}) = \frac{1}{2} \cdot \frac{21}{5} \cdot 21 = 44.1 \text{ square units}$$

55. (a) Draw the graph:

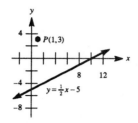

(b) Since the slope must be –2, use (1, 3) in the point-slope formula:
$$y - 3 = -2(x - 1)$$
$$y - 3 = -2x + 2$$
$$y = -2x + 5$$

(c) Set the two equations equal:
$$-2x + 5 = \tfrac{1}{2}x - 5$$
$$-4x + 10 = x - 10$$
$$-5x = -20$$
$$x = 4$$
$$y = -2(4) + 5 = -3$$
The intersection point is (4, –3).

(d) Find the distance from (1, 3) to (4, –3):
$$d = \sqrt{(4 - 1)^2 + (-3 - 3)^2} = \sqrt{9 + 36} = \sqrt{45} = 3\sqrt{5}$$

57. (a) Since the slope is m and the point (2, 1) lies on the line, use the point-slope formula to obtain $y - 1 = m(x - 2)$.

(b) For the x-intercept, let $y = 0$:
$$-1 = m(x - 2)$$
$$-1 = mx - 2m$$
$$mx = 2m - 1$$
$$x = \frac{2m - 1}{m}$$
For the y-intercept, let $x = 0$:
$$y - 1 = m(-2)$$
$$y - 1 = -2m$$
$$y = -2m + 1$$

(c) The area is given by:
$$\text{Area} = \tfrac{1}{2}(\text{base})(\text{height}) = \frac{1}{2}\left(\frac{2m - 1}{m}\right)(-2m + 1)$$

(d)   Since the area is 4:

$$\frac{1}{2}\left(\frac{2m-1}{m}\right)(-2m+1) = 4$$
$$(2m-1)(-2m+1) = 8m$$
$$-4m^2 + 4m - 1 = 8m$$
$$4m^2 + 4m + 1 = 0$$

(e)   Solve for m:

$$(2m+1)^2 = 0$$
$$2m+1 = 0$$
$$2m = -1$$
$$m = -\tfrac{1}{2}$$

59. (a)   Since $\dfrac{\Delta x}{\Delta P} = -\tfrac{2}{3}$, we have the equation:

$$x = 280 - \tfrac{2}{3}(P - 195) = 410 - \tfrac{2}{3}P$$

(b)   If $P = 270$:

$$x = 410 - \tfrac{2}{3}(270) = 410 - 180 = 230$$

So 230 units can be sold in a month.

(c)   If $x = 205$:

$$205 = 410 - \tfrac{2}{3}P$$
$$\tfrac{2}{3}P = 205$$
$$P = \tfrac{3}{2}(205) = 307.5$$

The price would be $307.50 per unit.

61. (a)   Since A and C lie on the line $y = m_1 x + b_1$, and the x-coordinates of A and C are 0 and 1, respectively, the y-coordinates will be b_1 and $m_1 + b_1$, so $A = (0, b_1)$ and $C = (1, m_1 + b_1)$. Similarly, the points B and D are $B = (0, b_2)$ and $D = (1, m_2 + b_2)$.

(b)   Verify the results:

$$AB = \sqrt{(0-0)^2 + (b_1 - b_2)^2} = \sqrt{(b_1 - b_2)^2} = |b_1 - b_2|,$$
$$\text{which is } b_1 - b_2 \text{ since } b_1 > b_2$$
$$CD = \sqrt{(1-1)^2 + [(m_1 + b_1) - (m_2 + b_2)]^2}$$
$$= \sqrt{[(m_1 + b_1) - (m_2 + b_2)]^2}$$
$$= |(m_1 + b_1) - (m_2 + b_2)|$$
$$= m_1 + b_1 - (m_2 + b_2) \text{ since } m_1 + b_1 > m_2 + b_2$$

(c) Since $AB = CD$:
$$b_1 - b_2 = m_1 + b_1 - m_2 - b_2$$
$$0 = m_1 - m_2$$
$$m_1 = m_2$$

63. Multiplying both numerator and denominator by -1:
$$\frac{y_2 - y_1}{x_2 - x_1} \cdot \frac{-1}{-1} = \frac{y_1 - y_2}{x_1 - x_2}$$
This identity tells us that, in calculating slope, either point can be chosen as the starting point.

65. (a) See the figure:

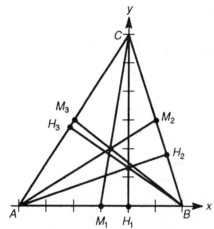

Find the midpoints:
$$M_1 = (-1, 0), \; M_2 = (1, 3), \; M_3 = (-2, 3)$$
Substitute in the equation $(x - h)^2 + (y - k)^2 = r^2$, and simplify to obtain the equations:
$$1 + 2h + h^2 + k^2 = r^2$$
$$10 - 2h + h^2 - 6k + k^2 = r^2$$
$$13 + 4h + h^2 - 6k + k^2 = r^2$$
Subtracting the first equation from the other two yields:
$$4h + 6k = 9$$
$$2h - 6k = -12$$
Adding these yields $h = -\frac{1}{2}$, and substituting into the first equation yields $k = \frac{11}{6}$, and finally we find that $r^2 = \frac{65}{18}$. So the circle is $\left(x + \frac{1}{2}\right)^2 + \left(y - \frac{11}{6}\right)^2 = \frac{65}{18}$. This can be written in the form $3x^2 + 3y^2 + 3x - 11y = 0$.

(b)   Now find H_1, H_2 and H_3. $H_1 = (0, 0)$ by inspection. Since $\overline{BC}$ has a slope $= -3$, $\overline{AH_2}$ has a slope $= \frac{1}{3}$, thus $y = \frac{1}{3}x + \frac{4}{3}$. $\overline{BC}$ has an equation of $y = -3x + 6$, so we find the intersection point to be $\left(\frac{7}{5}, \frac{9}{5}\right)$. So $H_2 = \left(\frac{7}{5}, \frac{9}{5}\right)$. Similarly $\overline{AC}$ has a slope $= \frac{3}{2}$, so $\overline{BH_3}$ has a slope $= -\frac{2}{3}$, thus $y = -\frac{2}{3}x + \frac{4}{3}$. $\overline{AC}$ has an equation of $y = \frac{3}{2}x + 6$, so we find the intersection point to be $\left(-\frac{28}{13}, \frac{36}{13}\right)$. So $H_3 = \left(-\frac{28}{13}, \frac{36}{13}\right)$. We find the equation of the circle passing through these three points and get $\left(x + \frac{1}{2}\right)^2 + \left(y - \frac{11}{6}\right)^2 = \frac{130}{36}$, which simplifies to $3x^2 + 3y^2 + 3x - 11y = 0$.

(c)   The equations of the three altitudes are $x = 0$, $y = \frac{1}{3}x + \frac{4}{3}$, and $y = -\frac{2}{3}x + \frac{4}{3}$. Since $x = 0$, we get $y = \frac{4}{3}$, so they all intersect at the point $P\left(0, \frac{4}{3}\right)$.

(d)   Compute $N_1 = \left(-2, \frac{2}{3}\right)$, $N_2 = \left(1, \frac{2}{3}\right)$, and $N_3 = \left(0, \frac{11}{3}\right)$. We check that N_1, N_2, and N_3 lie on $3x^2 + 3y^2 + 3x - 11y = 0$, which they all do.

(e)   Rather than using the hint, substitute in $(x - h)^2 + (y - k)^2 = r^2$ directly and simplify, yielding:
$$36 + h^2 - 12k + k^2 = r^2$$
$$16 + h^2 + 8h + k^2 = r^2$$
$$4 - 4h + h^2 + k^2 = r^2$$
Subtracting the first equation from the other two yields:
$$2h + 3k = 5$$
$$-h + 3k = 8$$
This has a solution of $h = -1$ and $k = \frac{7}{3}$. So the center of the circle is $\left(-1, \frac{7}{3}\right)$.

(f)   Find the radius of the circumcircle using the point $(0, 6)$ and the center $\left(-1, \frac{7}{3}\right)$:
$$r^2 = (0+1)^2 + \left(6 - \frac{7}{3}\right)^2 = \frac{130}{9}, \text{ so } r = \frac{\sqrt{130}}{3}$$
From (a) we have the radius of the nine-point circle is $\frac{\sqrt{130}}{6}$, which is one-half of the radius of the circumcircle.

(g)   The midpoint of $\overline{QP}$ is $\left(-\frac{1}{2}, \frac{11}{6}\right)$. Note that this is the center of the nine-point circle from (a).

(h)   First find the medians:
$$CM_1: \quad m = 6, \text{ so } y = 6x + 6$$
$$BM_3: \quad m = -\frac{3}{4}, \text{ so } y = -\frac{3}{4}x + \frac{3}{2}$$
$$AM_2: \quad m = \frac{3}{5}, \text{ so } y = \frac{3}{5}x + \frac{12}{5}$$

Now find the intersections:
$$6x + 6 = -\tfrac{3}{4}x + \tfrac{3}{2}, \text{ so } x = -\tfrac{2}{3} \text{ and } y = 2$$
$$6x + 6 = \tfrac{3}{5}x + \tfrac{12}{5}, \text{ so } x = -\tfrac{2}{3} \text{ and } y = 2$$
$$-\tfrac{3}{4}x + \tfrac{3}{2} = \tfrac{3}{5}x + \tfrac{12}{5}, \text{ so } x = -\tfrac{2}{3} \text{ and } y = 2$$
So the intersection point is $\left(-\tfrac{2}{3}, 2\right)$.

(i)   Find the equation of $\overline{QP}$ using $m = -1$, so $y = -x + \tfrac{4}{3}$. By the distance formula:
$$PG = \sqrt{\left(-\tfrac{2}{3} - 0\right)^2 + \left(2 - \tfrac{4}{3}\right)^2} = \tfrac{2\sqrt{2}}{3}$$
$$GQ = \sqrt{\left(-1 + \tfrac{2}{3}\right)^2 + \left(\tfrac{7}{3} - 2\right)^2} = \tfrac{\sqrt{2}}{3}$$
So $PG = 2 \cdot GQ$.

Graphing Utility Exercises for Section 2.3

1.   (a)   Since the lines pass through the origin, the y-intercepts are 0. So the equations of the four lines are $y = x$, $y = 2x$, $y = 3x$, and $y = 10x$. Sketch the graphs:

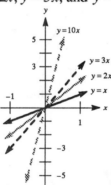

As the positive slopes increase, the lines become steeper.

(b)   Since the lines pass through the origin, the y-intercepts are 0. So the equations of the four lines are $y = -x$, $y = -2x$, $y = -3x$, and $y = -10x$. Sketch the graphs:

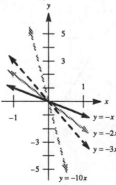

As the negative slopes decrease, the lines become steeper.

3.   Graph the three parallel lines:

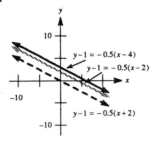

5.   (a)   Solving for y:

$$\frac{x}{2} + \frac{y}{3} = 1$$
$$3x + 2y = 6$$
$$2y = -3x + 6$$
$$y = -\frac{3}{2}x + 3$$

Now graph the line:

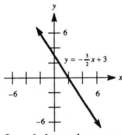

Note that the x-intercept is 2 and the y-intercept is 3.

(b)   Solving for y:

$$\frac{x}{-2} + \frac{y}{-3} = 1$$
$$3x + 2y = -6$$
$$2y = -3x - 6$$
$$y = -\frac{3}{2}x - 3$$

Now graph the line:

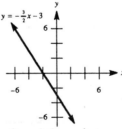

Note that the x-intercept is –2 and the y-intercept is –3.

(c)   Solving for y:

$$\frac{x}{6} + \frac{y}{5} = 1$$
$$5x + 6y = 30$$
$$6y = -5x + 30$$
$$y = -\frac{5}{6}x + 5$$

Now graph the line:

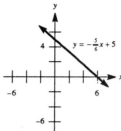

Note that the x-intercept is 6 and the y-intercept is 5.

(d)   Solving for y:

$$\frac{x}{-6} + \frac{y}{-5} = 1$$
$$5x + 6y = -30$$
$$6y = -5x - 30$$
$$y = -\frac{5}{6}x - 5$$

Now graph the line:

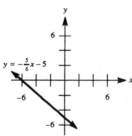

Note that the x-intercept is – 6 and the y-intercept is –5.

(e)   It appears that $\frac{x}{a} + \frac{y}{b} = 1$ is a line with an x-intercept of a and y-intercept of b.

7.   The center of the circle is $(0, 0)$, so the slope between $(0, 0)$ and $(-2, 4)$ is –2. The perpendicular slope must be $\frac{1}{2}$. Now using the point-slope formula:

$$y - 4 = \frac{1}{2}(x + 2)$$
$$y - 4 = \frac{1}{2}x + 1$$
$$y = \frac{1}{2}x + 5$$

Graphing the circle and the tangent line:

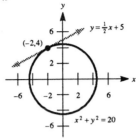

9.   The center of the circle is $(10, -10)$, so the slope between $(10, -10)$ and $(18, 5)$ is:

$$\frac{5-(-10)}{18-10} = \frac{15}{8}$$

The perpendicular slope must be $-\frac{8}{15}$. Now using the point-slope formula:

$$y - 5 = -\frac{8}{15}(x - 18)$$
$$y - 5 = -\frac{8}{15}x + \frac{48}{5}$$
$$y = -\frac{8}{15}x + \frac{73}{5}$$

Graphing the circle and the tangent line:

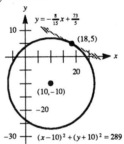

11.   First find the slope of the given line:

$$3x + 4y = 12$$
$$4y = -3x + 12$$
$$y = -\frac{3}{4}x + 3$$

So the line passing through the origin which is parallel to this line is $y = -\frac{3}{4}x$. Graphing each line:

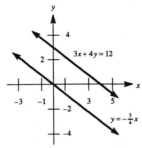

13. **(a)** Draw the triangle with vertices $A(-4, 0)$, $B(2, 0)$, and $C(0, 6)$:

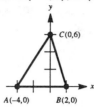

(b) Using the midpoint formula:

$$M_{AB} = \left(\tfrac{-4+2}{2}, \tfrac{0+0}{2}\right) = \left(-\tfrac{2}{2}, \tfrac{0}{2}\right) = (-1, 0)$$

$$M_{BC} = \left(\tfrac{2+0}{2}, \tfrac{0+6}{2}\right) = \left(\tfrac{2}{2}, \tfrac{6}{2}\right) = (1, 3)$$

$$M_{AC} = \left(\tfrac{-4+0}{2}, \tfrac{0+6}{2}\right) = \left(-\tfrac{4}{2}, \tfrac{6}{2}\right) = (-2, 3)$$

Now add the medians to the graph of the triangle:

The medians appear to intersect at the point $(-1, 2)$.

(c) For the median from A to side BC, the line passes through the points $(-4, 0)$ and $(1, 3)$. Finding the slope:

$$m = \frac{3-0}{1+4} = \frac{3}{5}$$

Using the point-slope formula with the point $(1, 3)$:

$$y - 3 = \tfrac{3}{5}(x - 1)$$

$$y - 3 = \tfrac{3}{5}x - \tfrac{3}{5}$$

$$y = \tfrac{3}{5}x + \tfrac{12}{5}$$

For the median from B to side AC, the line passes through the points $(2, 0)$ and $(-2, 3)$. Finding the slope:

$$m = \frac{3-0}{-2-2} = -\tfrac{3}{4}$$

Using the pont-slope formula with the point $(2, 0)$:

$$y - 0 = -\tfrac{3}{4}(x - 2)$$

$$y = -\tfrac{3}{4}x + \tfrac{3}{2}$$

For the median from C to side AB, the line passes through the points $(0, 6)$ and $(-1, 0)$. Finding the slope:

$$m = \frac{0-6}{-1-0} = 6$$

Since the y-intercept is 6, its equation is $y = 6x + 6$.

Solving any pair of simultaneous equations will yield the coordinates of the centroid. Using $y = -\frac{3}{4}x + \frac{3}{2}$ and $y = 6x + 6$, solve the system:

$$-\frac{3}{4}x + \frac{3}{2} = 6x + 6$$
$$-3x + 6 = 24x + 24$$
$$-27x = 18$$
$$x = -\frac{2}{3}$$
$$y = 6\left(-\frac{2}{3}\right) + 6 = 2$$

The exact coordinates of the centroid are $\left(-\frac{2}{3}, 2\right)$, which agrees with our estimate from part (b), to the nearest integer.

2.4 Symmetry and Graphs

1.   (a)   The reflection of $\overline{AB}$ about the x-axis is given by the graph:

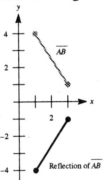

(b)   The reflection of $\overline{AB}$ about the y-axis is given by the graph:

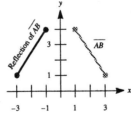

(c)   The reflection of $\overline{AB}$ about the origin is given by the graph:

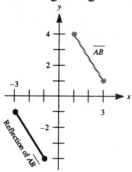

(d)   The reflection of $\overline{AB}$ about the line $y = x$ is given by the graph:

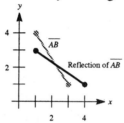

3.   (a)   The reflection of $\overline{AB}$ about the x-axis is given by the graph:

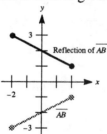

(b)   The reflection of $\overline{AB}$ about the y-axis is given by the graph:

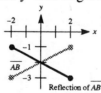

(c)   The reflection of $\overline{AB}$ about the origin is given by the graph:

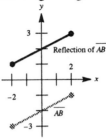

(d)   The reflection of $\overline{AB}$ about the line $y = x$ is given by the graph:

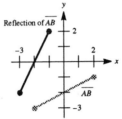

5.   (a)   The reflection of $\overline{AB}$ about the x-axis is given by the graph:

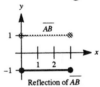

(b)   The reflection of $\overline{AB}$ about the y-axis is given by the graph:

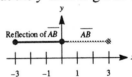

(c)   The reflection of $\overline{AB}$ about the origin is given by the graph:

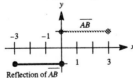

(d)   The reflection of $\overline{AB}$ about the line $y = x$ is given by the graph:

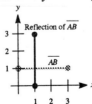

7. Complete the table:

	Symmetric about the x - axis	Symmetric about the y - axis	Symmetric about the origin
$y = x^2$	no	yes	no
$y = x^3$	no	no	yes
$y = \sqrt{x}$	no	no	no

9. The x-intercepts are 2 and -2, and the y-intercept is 4. The graph is symmetric about the y-axis.

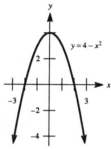

11. There are no x- or y-intercepts. The graph is symmetric about the origin.

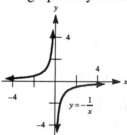

13. The x- and y-intercepts are both 0. The graph is symmetric about the y-axis.

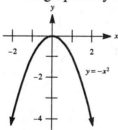

15. There are no x- or y-intercepts. The graph is symmetric about the origin.

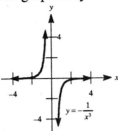

17. The x- and y-intercepts are both 0. The graph is symmetric about the y-axis.

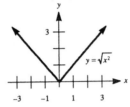

19. The x- and y-intercepts are both 1. The graph does not possess any of the three types of symmetry.

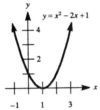

21. The x-intercept is 2 and there is no y-intercept. The graph is symmetric about the x-axis.

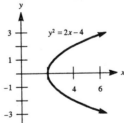

23. To find the x-intercepts, use the quadratic formula:
$$x = \frac{-1 \pm \sqrt{(1)^2 - 4(2)(-4)}}{2(2)} = \frac{-1 \pm \sqrt{33}}{4}$$

The y-intercept is -4. The graph does not possess any of the three types of symmetry.

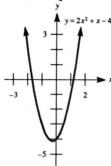

25. (a) The x- and y-intercepts are both 2. The graph does not possess any of the three types of symmetry.

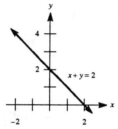

(b) The x-intercepts are ± 2 and the y-intercept is 2. The graph is symmetric about the y-axis.

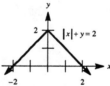

27. (a) The x-intercept for each graph is the same. Setting $y = 0$, we obtain:
$$\frac{3}{4}x - 2 = 0$$
$$\frac{3}{4}x = 2$$
$$x = \frac{8}{3}$$

The y-intercept for $y = \frac{3}{4}x - 2$ is -2, so the y-intercept for $y = \left|\frac{3}{4}x - 2\right|$ is $|-2| = 2$.

(b) The two graphs are identical on the interval $\left[\frac{8}{3}, \infty\right)$.

(c)   The graph of $y = \left|\frac{3}{4}x - 2\right|$ can be obtained by reflecting $y = \frac{3}{4}x - 2$ about the x-axis for the interval $\left(-\infty, \frac{8}{3}\right)$. For the interval $\left[\frac{8}{3}, \infty\right)$, no reflection is necessary.

29.   Let the points $P = (7, -1)$ and $Q = (-1, 7)$. We must first show that the line segment $\overline{PQ}$ is perpendicular to the line $y = x$. Line segment $\overline{PQ}$ has a slope of:

$$m = \frac{7 - (-1)}{-1 - 7} = \frac{8}{-8} = -1$$

Since $y = x$ has a slope of 1, and $1(-1) = -1$, the two lines are perpendicular. Next, we must show that P and Q are equidistant from $y = x$. Find the midpoint of line segment $\overline{PQ}$:

$$M = \left(\frac{7-1}{2}, \frac{-1+7}{2}\right) = (3, 3)$$

Since $(3, 3)$ lies on the line $y = x$, and since $\overline{PQ}$ is perpendicular to $y = x$, and $PM = QM$, by the definition of symmetry P and Q are symmetric about the line $y = x$.

31.   Let the points $P = (a, b)$ and $Q = (b, a)$. We must first show that the line segment $\overline{PQ}$ is perpendicular to the line $y = x$. Line segment $\overline{PQ}$ has a slope of:

$$m = \frac{a - b}{b - a} = -1$$

Since $y = x$ has a slope of 1, and $1(-1) = -1$, the two lines are perpendicular. Next, show that P and Q are equidistant from $y = x$. We find the midpoint of line segment $\overline{PQ}$:

$$M = \left(\frac{a+b}{2}, \frac{b+a}{2}\right) = \left(\frac{a+b}{2}, \frac{a+b}{2}\right)$$

Since $\left(\frac{a+b}{2}, \frac{a+b}{2}\right)$ lies on the line $y = x$, and since $\overline{PQ}$ is perpendicular to $y = x$, and $PM = QM$, by the definition of symmetry P and Q are symmetric about the line $y = x$.

33.   Let's pick the points $P(0, 3)$, $Q(2, 4)$, $R(4, 5)$. Then the reflected points will be $P'(3, 0)$, $Q'(4, 2)$, and $R'(5, 4)$. We will show that P', Q', and R' are collinear by computing the slopes:

$$m_{P'Q'} = \frac{2 - 0}{4 - 3} = \frac{2}{1} = 2$$

$$m_{Q'R'} = \frac{4 - 2}{5 - 4} = \frac{2}{1} = 2$$

So P', Q', and R' all are collinear (lie on the same line). There are two ways to find the equation of the line.

1st way: Use the point $(3, 0)$ and $m = 2$ in the point-slope formula:

$$y - 0 = 2(x - 3)$$
$$y = 2x - 6$$

2nd way: Realize that this line must be the inverse function of $y = \frac{1}{2}x + 3$.

Exchange x and y and solve for y:
$$x = \frac{1}{2}y + 3$$
$$x - 3 = \frac{1}{2}y$$
$$y = 2x - 6$$

In either case, we obtain the same equation.

35. The x-intercepts are found by setting $y = 0$:
$$x^3 - 27x = 0$$
$$x(x - 3\sqrt{3})(x + 3\sqrt{3}) = 0$$
$$x = 0, \pm 3\sqrt{3} \approx 0, \pm 5.2$$

When $x = 0$, $y = (0)^3 - 27(0) = 0$, so the y-intercept is 0. Using the point $(-x, -y)$, we find the graph is symmetric with respect to the origin:
$$-y = (-x)^3 - 27(-x)$$
$$-y = -x^3 + 27x$$
$$y = x^3 - 27x$$

Set up a table of values:

x	0	0.5	1	1.5	2	2.5	3	3.5	4	4.5	5	5.5	6
y	0	−13.4	−26	−37.1	−46	−51.9	−54	−51.6	−44	−30.4	−10	17.9	54

Now graph the curve:

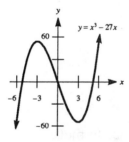

37. Let $P = (8, 2)$ and $Q = (4, 8)$. We first find the slope m_1 of the line segment $\overline{PQ}$ perpendicular to the line $y = mx + b$:
$$m_1 = \frac{8 - 2}{4 - 8} = -\frac{6}{4} = -\frac{3}{2}$$

So $-\frac{3}{2}m = -1$, and thus $m = \frac{2}{3}$. So $y = \frac{2}{3}x + b$.

Next, we must find the value of b so that P and Q are equidistant from $y = \frac{2}{3}x + b$. Call the point $A\left(a, \frac{2}{3}a + b\right)$ on $y = \frac{2}{3}x + b$. Use the distance formula:
$$PA = \sqrt{(a - 8)^2 + \left(\frac{2}{3}a + b - 2\right)^2}$$
$$QA = \sqrt{(a - 4)^2 + \left(\frac{2}{3}a + b - 8\right)^2}$$

Now, since $PA = QA$:

$$\sqrt{(a-8)^2 + \left(\tfrac{2}{3}a+b-2\right)^2} = \sqrt{(a-4)^2 + \left(\tfrac{2}{3}a+b-8\right)^2}$$

$$(a-8)^2 + \left(\tfrac{2}{3}a+b-2\right)^2 = (a-4)^2 + \left(\tfrac{2}{3}a+b-8\right)^2$$

$$a^2 - 16a + 64 + \left(\tfrac{2}{3}a+b\right)^2 - 4\left(\tfrac{2}{3}a+b\right) + 4 = a^2 - 8a + 16 + \left(\tfrac{2}{3}a+b\right)^2 - 16\left(\tfrac{2}{3}a+b\right) + 64$$

$$-16a - 4\left(\tfrac{2}{3}a+b\right) + 4 = -8a + 16 - 16\left(\tfrac{2}{3}a+b\right)$$

$$12\left(\tfrac{2}{3}a+b\right) = 8a + 12$$

$$8a + 12b = 8a + 12$$

$$12b = 12$$

$$b = 1$$

So $y = \tfrac{2}{3}x + 1$ is the line of symmetry, and so $m = \tfrac{2}{3}$ and $b = 1$.

39. **(a)** Let $P = (a, b)$ and $Q = (x, y)$. Q is the point we are trying to find. Since $\overline{PQ}$ is perpendicular to $y = 3x$, which has a slope of 3:

$$\frac{y-b}{x-a} = -\tfrac{1}{3}$$

$$3y - 3b = -x + a$$

$$x + 3y = a + 3b$$

Now call $C(c, 3c)$ a point on the line $y = 3x$:

$$\frac{3c-b}{c-a} = -\tfrac{1}{3}$$

$$9c - 3b = -c + a$$

$$10c = a + 3b$$

$$c = \frac{a+3b}{10}$$

Since $PC = QC$, by the distance formula:

$$\sqrt{(a-c)^2 + (b-3c)^2} = \sqrt{(x-c)^2 + (y-3c)^2}$$

$$\left(a - \tfrac{a+3b}{10}\right)^2 + \left(b - \tfrac{3a+9b}{10}\right)^2 = \left(x - \tfrac{a+3b}{10}\right)^2 + \left(y - \tfrac{3a+9b}{10}\right)^2$$

Recall that $x + 3y = a + 3b$, and thus $3x + 9y = 3a + 9b$. Substituting these values into the right-hand side, we obtain:

$$\left(a - \tfrac{a+3b}{10}\right)^2 + \left(b - \tfrac{3a+9b}{10}\right)^2 = \left(x - \tfrac{x+3y}{10}\right)^2 + \left(y - \tfrac{3x+9y}{10}\right)^2$$

$$\left(\tfrac{9a-3b}{10}\right)^2 + \left(\tfrac{b-3a}{10}\right)^2 = \left(\tfrac{9x-3y}{10}\right)^2 + \left(\tfrac{y-3x}{10}\right)^2$$

$$81a^2 - 54ab + 9b^2 + b^2 - 6ab + 9a^2 = 81x^2 - 54xy + 9y^2 + y^2 - 6xy + 9x^2$$

$$90a^2 - 60ab + 10b^2 = 90x^2 - 60xy + 10y^2$$

$$9a^2 - 6ab + b^2 = 9x^2 - 6xy + y^2$$

$$(3a - b)^2 = (3x - y)^2$$

$$|3a - b| = |3x - y|$$

So, either $3x - y = 3a - b$ or $3x - y = -(3a - b) = -3a + b$.

Case 1:

$x + 3y = a + 3b$, so $x = a + 3b - 3y$

$3x - y = 3a - b$

Substituting:

$$3(a + 3b - 3y) - y = 3a - b$$

$$3a + 9b - 9y - y = 3a - b$$

$$-10y = -10b$$

$$y = b$$

$$x = a$$

But if $(x, y) = (a, b)$, then the point lies **on** the line $y = 3x$.

Note: An alternate approach is that if $3x - y = 3a - b$, then $\dfrac{y - b}{x - a} = 3$. But this is impossible, since $\dfrac{y - b}{x - a} = -\tfrac{1}{3}$.

Case 2:

$x + 3y = a + 3b$, so $x = a + 3b - 3y$

$3x - y = -3a + b$

Substituting:

$$3(a + 3b - 3y) - y = -3a + b$$

$$3a + 9b - 9y - y = -3a + b$$

$$-10y = -6a - 8b$$

$$y = \frac{3a + 4b}{5}$$

Since $x = a + 3b - 3y$, we have:

$$x = a + 3b - 3\left(\tfrac{3a+4b}{5}\right)$$

$$= a + 3b - \frac{9a + 12b}{5}$$

$$= \frac{5a + 15b - 9a - 12b}{5}$$

$$= \frac{-4a + 3b}{5}$$

So the reflected point (x, y) is $\left(\tfrac{3b-4a}{5}, \tfrac{3a+4b}{5}\right)$.

(b) Calling P and Q as in (a):

$$\frac{y - b}{x - a} = -\frac{1}{m}$$

$$my - mb = -x + a$$

$$x + my = a + mb$$

Let $C(c, mc)$ be a point on the line $y = mx$:

$$\frac{mc - b}{c - a} = -\frac{1}{m}$$

$$m^2 c - mb = a - c$$

$$\left(m^2 + 1\right)c = a + mb$$

$$c = \frac{a + mb}{m^2 + 1}$$

Now since $PC = QC$:

$$\sqrt{(a - c)^2 + (b - mc)^2} = \sqrt{(x - c)^2 + (y - mc)^2}$$

$$\left(a - \frac{a + mb}{m^2 + 1}\right)^2 + \left(b - \frac{am + m^2 b}{m^2 + 1}\right)^2 = \left(x - \frac{a + mb}{m^2 + 1}\right)^2 + \left(y - \frac{am + m^2 b}{m^2 + 1}\right)^2$$

Recall that $x + my = a + mb$, and thus $mx + m^2 y = am + m^2 b$. Substituting these values into the right-hand side:

$$\left(a - \frac{a + mb}{m^2 + 1}\right)^2 + \left(b - \frac{am + m^2 b}{m^2 + 1}\right)^2 = \left(x - \frac{x + my}{m^2 + 1}\right)^2 + \left(y - \frac{mx + m^2 y}{m^2 + 1}\right)^2$$

$$\left(\frac{m^2 a - mb}{m^2 + 1}\right)^2 + \left(\frac{b - am}{m^2 + 1}\right)^2 = \left(\frac{m^2 x - my}{m^2 + 1}\right)^2 + \left(\frac{y - mx}{m^2 + 1}\right)^2$$

$$m^2 a^2 - 2abm + b^2 = m^2 x^2 - 2mxy + y^2$$

$$(ma - b)^2 = (mx - y)^2$$

$$|ma - b| = |mx - y|$$

So, either $mx - y = ma - b$ or $mx - y = -(ma + b) = -ma + b$.

Case 1:

$x + my = a + mb$, so $x = a + mb - my$

$mx - y = ma - b$

Substituting:

$$m(a + mb - my) - y = ma - b$$

$$ma + m^2b - m^2y - y = ma - b$$

$$(m^2 + 1)b = (m^2 + 1)y$$

$$b = y$$

$$a = x$$

But, if $(x, y) = (a, b)$, then the point lies **on** the line $y = mx$.

Note: An alternate approach is that if $mx - y = ma - b$, then $\dfrac{y - b}{x - a} = m$.

But this is impossible, since $\dfrac{y - b}{x - a} = -\dfrac{1}{m}$.

Case 2:

$x + my = a + mb$, so $x = a + mb - my$

$mx - y = -ma + b$

Substituting:

$$m(a + mb - my) - y = -ma + b$$

$$ma + m^2b - m^2y - y = -ma + b$$

$$-(m^2 + 1)y = -2ma + b(1 - m^2)$$

$$y = \frac{2ma + (m^2 + 1)b}{m^2 + 1}$$

Since $x = a + mb - my$, we have:

$$x = a + mb - \frac{2m^2a + m(m^2 - 1)b}{m^2 + 1}$$

$$= \frac{am^2 + a + m^3b + mb - 2m^2a - m^3b + mb}{m^2 + 1}$$

$$= \frac{2mb - (m^2 - 1)a}{m^2 + 1}$$

So the reflected point (x, y) is:

$$\left(\frac{2mb - (m^2 - 1)a}{m^2 + 1}, \frac{2ma + (m^2 - 1)b}{m^2 + 1} \right)$$

Note: Using $m = 3$ and our answer from (a) verifies this formula.

Graphing Utility Exercises for Section 2.4

1.   Graphing $y = x^2 - 3x$ in the standard viewing rectangle:

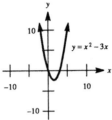

Now graphing $y = x^2 - 3x$ using the suggested settings:

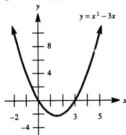

The graph does not possess any of the three types of symmetry.

3.   Graphing $y = 2^x$ in the standard viewing rectangle:

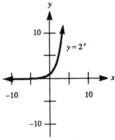

Now graphing $y = 2^x$ using the suggested settings:

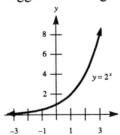

The graph does not possess any of the three types of symmetry.

5. Graphing $y = \dfrac{1}{x^2 - x}$ in the standard viewing rectangle:

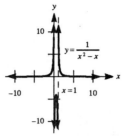

Now graphing $y = \dfrac{1}{x^2 - x}$ using the suggested settings:

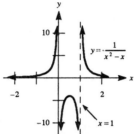

The graph does not possess any of the three types of symmetry.

7. Graphing $y = x^2 - 0.2x - 15$ in the standard viewing rectangle:

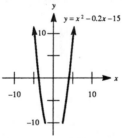

Now graphing $y = x^2 - 0.2x - 15$ using other settings:

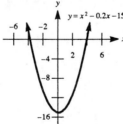

The graph does not possess any of the three types of symmetry.

9. Graphing $y = \sqrt{|x|}$ in the standard viewing rectangle:

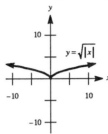

Now graphing $y = \sqrt{|x|}$ using other settings:

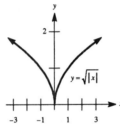

The graph appears to possess y-axis symmetry. We replace (x, y) with $(-x, y)$:

$$y = \sqrt{|-x|} = \sqrt{|x|}$$

So the graph does possess y-axis symmetry.

11. Graphing $y = 2x - x^3 - x^5 + x^7$ in the standard viewing rectangle:

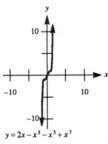

Now graphing $y = 2x - x^3 - x^5 + x^7$ using other settings:

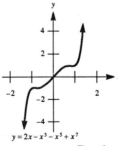

The graph appears to have origin symmetry. Replace (x, y) with $(-x, -y)$:

$$-y = 2(-x) - (-x)^3 - (-x)^5 + (-x)^7$$
$$-y = -2x + x^3 + x^5 - x^7$$
$$y = 2x - x^3 - x^5 + x^7$$

So the graph does possess origin symmetry.

13. Graphing $y = x^4 - 10x^2 + \frac{1}{4}x$ in the standard viewing rectangle:

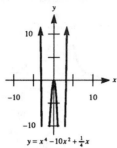

Now graphing $y = x^4 - 10x^2 + \frac{1}{4}x$ using other settings:

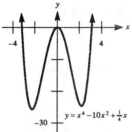

The graph does not possess any of the three types of symmetry.

15. Graphing $y = -\dfrac{4}{x}$ for the given viewing rectangle:

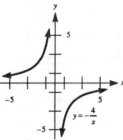

Now expanding the viewing rectangle:

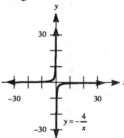

Note how closely the axes approximate the curve.

17. (a)   Graphing $y = \dfrac{3x-6}{x+1}$ using the standard viewing rectangle:

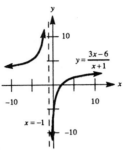

(b)   Now expanding the viewing rectangle:

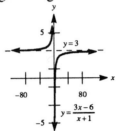

The line $y = 3$ appears to be a horizontal asymptote for this curve.

(c)   Completing the tables:

x	10	100	1000	10,000	100,000
$\dfrac{3x-6}{x+1}$	2.18182	2.91089	2.99101	2.99910	2.99991

x	-10	-100	-1000	$-10,000$	$-100,000$
$\dfrac{3x-6}{x+1}$	4	3.09091	3.00901	3.00090	3.00009

(d)   Graphing $y = \dfrac{3x-6}{x+1}$ using the given viewing rectangle:

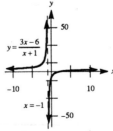

Note that the curve appears to have $x = -1$ as a vertical asymptote.

Chapter Two Review Exercises

1. Find the slope:
    $$m = \frac{6-2}{-6-(-4)} = \frac{4}{-2} = -2$$
    Using the point $(-4, 2)$ in the point-slope formula:
    $$y - 2 = -2[x - (-4)]$$
    $$y - 2 = -2(x + 4)$$
    $$y - 2 = -2x - 8$$
    $$y = -2x - 6$$

3. Using the point-slope formula:
    $$y - (-3) = \tfrac{1}{4}[x - (-2)]$$
    $$y + 3 = \tfrac{1}{4}(x + 2)$$
    $$y + 3 = \tfrac{1}{4}x + \tfrac{1}{2}$$
    $$y = \tfrac{1}{4}x - \tfrac{5}{2}$$

5. Find the slope between the points $(-4, 0)$ and $(0, 8)$:
    $$m = \frac{8-0}{0-(-4)} = \tfrac{8}{4} = 2$$
    Using the slope-intercept formula, we have $y = 2x + 8$.

7. Since the line is parallel to the x-axis, its equation will be of the form $y = $ constant (horizontal line). Since $(0, -2)$ is on the line, its equation is $y = -2$.

9. Find the slope of $x + y + 1 = 0$ by writing it in slope-intercept form, and obtain $y = -x - 1$. So its slope is -1, and thus the perpendicular slope would be 1. Use the point $(1, 2)$ in the point-slope formula:
    $$y - 2 = 1(x - 1)$$
    $$y - 2 = x - 1$$
    $$y = x + 1$$

11. The centers of the circles are $(-2, -1)$ and $(2, 8)$. Find the slope between $(-2, -1)$ and $(2, 8)$:
    $$m = \frac{8-(-1)}{2-(-2)} = \tfrac{9}{4}$$
    Use the point $(2, 8)$ in the point-slope formula:
    $$y - 8 = \tfrac{9}{4}(x - 2)$$
    $$4y - 32 = 9x - 18$$
    $$-9x + 4y - 14 = 0$$
    $$9x - 4y + 14 = 0$$

13. Find the midpoint of the line segment joining $(-2, -3)$ and $(6, -5)$:
$$M = \left(\frac{-2+6}{2}, \frac{-3-5}{2}\right) = \left(\frac{4}{2}, -\frac{8}{2}\right) = (2, -4)$$
Now find the slope between $(0, 0)$ and $(2, -4)$:
$$m = \frac{-4-0}{2-0} = \frac{-4}{2} = -2$$
Using the slope-intercept formula, we obtain $y = -2x$, or $2x + y = 0$.

15. The center is $(3, -4)$. Now find the slope of the radius drawn from $(3, -4)$ to $(0, 0)$:
$$m = \frac{0-(-4)}{0-3} = \frac{4}{-3} = -\frac{4}{3}$$
So the slope of the perpendicular tangent line is $\frac{3}{4}$. Using the slope-intercept formula, we obtain $y = \frac{3}{4}x$, or $3x - 4y = 0$.

17. Call the x-intercept a. Then the y-intercept is $2 - a$, and we find the slope from each to $(2, -1)$:
$$m = \frac{-1-0}{2-a} = \frac{-1}{2-a}$$
$$m = \frac{-1-(2-a)}{2-0} = \frac{-1-2+a}{2} = \frac{a-3}{2}$$
Since these two slopes must be equal:
$$\frac{-1}{2-a} = \frac{a-3}{2}$$
$$-2 = (2-a)(a-3)$$
$$-2 = -a^2 + 5a - 6$$
$$0 = a^2 - 5a + 4$$
$$0 = (a-4)(a-1)$$
$$a = 1 \text{ or } a = 4$$
When $a = 1$, we have $m = \frac{1-3}{2} = \frac{-2}{2} = -1$. When $a = 4$, we have $m = \frac{4-3}{2} = \frac{1}{2}$. Using the point $(2, -1)$ and each of these slopes in the point-slope formula:
$$y-(-1) = -1(x-2) \qquad\qquad y-(-1) = \tfrac{1}{2}(x-2)$$
$$y+1 = -x+2 \qquad\qquad\qquad 2y+2 = x-2$$
$$x+y-1 = 0 \qquad\qquad\qquad\quad x-2y-4 = 0$$
Both of these lines satisfy the given conditions.

19. Using the distance formula with $(x_1, y_1) = (-1, 2)$ and $(x_2, y_2) = (4, -10)$:
$$d = \sqrt{[4-(-1)]^2 + (-10-2)^2} = \sqrt{25+144} = \sqrt{169} = 13$$

21. Find the distance of each from the origin:
$$d_1 = \sqrt{(15-0)^2 + (6-0)^2} = \sqrt{225+36} = \sqrt{261}$$
$$d_2 = \sqrt{(16-0)^2 + (2-0)^2} = \sqrt{256+4} = \sqrt{260}$$
The point $(15, 6)$ is farther from the origin.

23. For x-axis symmetry, replace (x, y) with $(x, -y)$:
$$-y = x^4 - 2x^2$$
$$y = -x^4 + 2x^2$$
The graph is not symmetric about the x-axis.
For y-axis symmetry, replace (x, y) with $(-x, y)$:
$$y = (-x)^4 - 2(-x)^2 = x^4 - 2x^2$$
The graph is symmetric about the y-axis.
For origin symmetry, replace (x, y) with $(-x, -y)$:
$$-y = (-x)^4 - 2(-x)^2$$
$$-y = x^4 - 2x^2$$
$$y = -x^4 + 2x^2$$
The graph is not symmetric about the origin.

25. For x-axis symmetry, replace (x, y) with $(x, -y)$:
$$-y = x^3 + 5x$$
$$y = -x^3 - 5x$$
The graph is not symmetric about the x-axis.
For y-axis symmetry, replace (x, y) with $(-x, y)$:
$$y = (-x)^3 + 5(-x) = -x^3 - 5x$$
The graph is not symmetric about the y-axis.
For origin symmetry, replace (x, y) with $(-x, -y)$:
$$-y = (-x)^3 + 5(-x)$$
$$-y = -x^3 - 5x$$
$$y = x^3 + 5x$$
The graph is symmetric about the origin.

27. For x-axis symmetry, replace (x, y) with $(x, -y)$:
$$(-y)^2 = [x + (-y)]^4$$
$$y^2 = (x - y)^4$$
The graph is not symmetric about the x-axis.
For y-axis symmetry, replace (x, y) with $(-x, y)$:
$$y^2 = (-x + y)^4$$
The graph is not symmetric about the y-axis.
For origin symmetry, replace (x, y) with $(-x, -y)$:
$$(-y)^2 = [-x + (-y)]^4$$
$$y^2 = (-x - y)^4$$
$$y^2 = (x + y)^4$$
The graph is symmetric about the origin.

29. For x-axis symmetry, replace (x, y) with $(x, -y)$:
$$-y = 3x - \frac{1}{x}$$
$$y = -3x + \frac{1}{x}$$
The graph is not symmetric about the x-axis.

For y-axis symmetry, replace (x, y) with $(-x, y)$:
$$y = 3(-x) - \frac{1}{-x} = -3x + \frac{1}{x}$$
The graph is not symmetric about the y-axis.
For origin symmetry, replace (x, y) with $(-x, -y)$:
$$-y = 3(-x) - \frac{1}{-x}$$
$$-y = -3x + \frac{1}{x}$$
$$y = 3x - \frac{1}{x}$$
The graph is symmetric about the origin.

31. For x-axis symmetry, replace (x, y) with $(x, -y)$:
$$-y = 2^x - 2^{-x}$$
$$y = -2^x + 2^{-x}$$
The graph is not symmetric about the x-axis.
For y-axis symmetry, replace (x, y) with $(-x, y)$:
$$y = 2^{-x} - 2^{-(-x)} = 2^{-x} - 2^x$$
The graph is not symmetric about the y-axis.
For origin symmetry, replace (x, y) with $(-x, -y)$:
$$-y = 2^{-x} - 2^{-(-x)}$$
$$-y = 2^{-x} - 2^x$$
$$y = 2^x - 2^{-x}$$
The graph is symmetric about the origin.

33. The graph is symmetric about the x-axis only.

35. The graph is symmetric about the x-axis, the y-axis, the origin, and the line $y = x$.

37. The graph is symmetric about the x-axis only.

39. The graph is symmetric about the x-axis only.

41. Using the distance formula and substituting $y = -\frac{2}{3}x + 1$:
$$\sqrt{x^2 + \left(-\frac{2}{3}x + 1\right)^2} = \sqrt{10}$$
$$x^2 + \frac{4}{9}x^2 - \frac{4}{3}x + 1 = 10$$
$$\frac{13}{9}x^2 - \frac{4}{3}x - 9 = 0$$
$$13x^2 - 12x - 81 = 0$$
$$(13x + 27)(x - 3) = 0$$
$$x = -\frac{27}{13}, 3$$
$$y = \frac{31}{13}, -1$$
The two points are $\left(-\frac{27}{13}, \frac{31}{13}\right)$ and $(3, -1)$.

43. Since the sum of the x- and y-coordinates is 2 and $y = x^2$:

$$x + x^2 = 2$$
$$x^2 + x - 2 = 0$$
$$(x + 2)(x - 1) = 0$$
$$x = -2, 1$$
$$y = 4, 1$$

The two points are $(-2, 4)$ and $(1, 1)$.

45. Draw the graph:

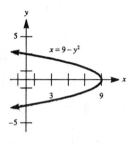

47. This is a circle with center $= (0, 0)$ and radius $= 1$. Draw the graph:

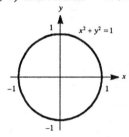

49. Draw the graph:

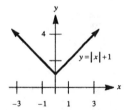

51. Draw the graph:

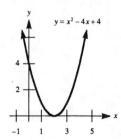

53. Draw the graph:

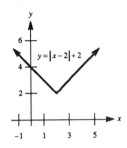

55. Write the line in slope-intercept form:
$$3x + y = 0$$
$$y = -3x$$
Now draw the graph:

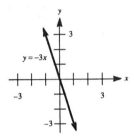

57. This is a circle with center = $(2, -3)$ and radius = $\sqrt{13}$.

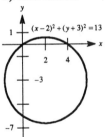

59. As long as $x \neq 4$, we can simplify:
$$y = \frac{x^2 - 16}{x - 4} = \frac{(x + 4)(x - 4)}{(x - 4)} = x + 4$$
This is a line in slope-intercept form. Now draw the graph:

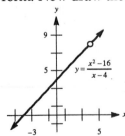

61. Setting each factor equal to 0, we have $4x - y + 4 = 0$ and thus $y = 4x + 4$, or $4x + y - 4 = 0$ and thus $y = -4x + 4$. So the graph consists of two lines:

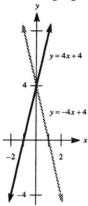

63. Set up the equation involving slope:

$$\frac{t-1}{5-2} = 6$$
$$\frac{t-1}{3} = 6$$
$$t - 1 = 18$$
$$t = 19$$

65. The slope between the points $(-2, -8)$ and (x, x^3) is given by:

$$m = \frac{x^3 - (-8)}{x - (-2)} = \frac{x^3 + 8}{x + 2} = \frac{(x+2)(x^2 - 2x + 4)}{x + 2} = x^2 - 2x + 4$$

67. The midpoint of the hypotenuse BC would be:

$$M = \left(\frac{0 + 2c}{2}, \frac{2b + 0}{2}\right) = \left(\frac{2c}{2}, \frac{2b}{2}\right) = (c, b)$$

Compute the distances:

$$MA = \sqrt{(c-0)^2 + (b-0)^2} = \sqrt{c^2 + b^2}$$
$$MB = \sqrt{(c-0)^2 + (b-2b)^2} = \sqrt{c^2 + b^2}$$
$$MC = \sqrt{(c-2c)^2 + (b-0)^2} = \sqrt{c^2 + b^2}$$

These distances are all the same.

69. (a) Compute the distance:

$$d = \sqrt{(5-2)^2 + (-6-5)^2} = \sqrt{9 + 121} = \sqrt{130}$$

(b) Compute the slope:

$$m = \frac{-6-5}{5-2} = -\frac{11}{3}$$

(c)   Compute the midpoint:
$$M = \left(\frac{2+5}{2}, \frac{5-6}{2} \right) = \left(\tfrac{7}{2}, -\tfrac{1}{2} \right)$$

71. (a)   Compute the distance:
$$d = \sqrt{ \left(-\tfrac{\sqrt{3}}{2} - \tfrac{\sqrt{3}}{2} \right)^2 + \left(-\tfrac{1}{2} - \tfrac{1}{2} \right)^2 } = \sqrt{ \left(-\sqrt{3} \right)^2 + (-1)^2 } = \sqrt{3+1} = \sqrt{4} = 2$$

(b)   Compute the slope:
$$m = \frac{ -\tfrac{1}{2} - \tfrac{1}{2} }{ -\tfrac{\sqrt{3}}{2} - \tfrac{\sqrt{3}}{2} } = \frac{-1}{-\sqrt{3}} = \frac{1}{\sqrt{3}} = \frac{\sqrt{3}}{3}$$

(c)   Compute the midpoint:
$$M = \left(\frac{ \tfrac{\sqrt{3}}{2} - \tfrac{\sqrt{3}}{2} }{2}, \frac{ \tfrac{1}{2} - \tfrac{1}{2} }{2} \right) = (0,0)$$

73.   Find the slope:
$$m = \frac{1-2}{4-1} = -\tfrac{1}{3}$$
Using $(1, 2)$ in the point-slope formula:
$$y - 2 = -\tfrac{1}{3}(x-1)$$
$$y - 2 = -\tfrac{1}{3}x + \tfrac{1}{3}$$
$$y = -\tfrac{1}{3}x + \tfrac{7}{3}$$
So the x- and y-intercepts are $\tfrac{7}{3}$ (if $x = 0$) and 7 (if $y = 0$). Then the area is given by:
$$\text{Area} = \tfrac{1}{2}(\text{base})(\text{height}) = \tfrac{1}{2} \cdot 7 \cdot \tfrac{7}{3} = \tfrac{49}{6}$$

75. (a)   Find each midpoint:
$$M_1 = \left(\frac{-5+7}{2}, \frac{3+7}{2} \right) = (1,5)$$
$$M_2 = \left(\frac{3+7}{2}, \frac{1+7}{2} \right) = (5,4)$$
$$M_3 = \left(\frac{-5+3}{2}, \frac{3+1}{2} \right) = (-1,2)$$
Now use the result from Exercise 74 to find the desired points. Let P_1, P_2, and P_3 be the desired points on medians CM_1, AM_2, and BM_3. Then:
$$P_1 = \left(\tfrac{1}{3}(3) + \tfrac{2}{3}(1), \tfrac{1}{3}(1) + \tfrac{2}{3}(5) \right) = \left(1 + \tfrac{2}{3}, \tfrac{1}{3} + \tfrac{10}{3} \right) = \left(\tfrac{5}{3}, \tfrac{11}{3} \right)$$
$$P_2 = \left(\tfrac{1}{3}(-5) + \tfrac{2}{3}(5), \tfrac{1}{3}(3) + \tfrac{2}{3}(4) \right) = \left(-\tfrac{5}{3} + \tfrac{10}{3}, 1 + \tfrac{8}{3} \right) = \left(\tfrac{5}{3}, \tfrac{11}{3} \right)$$
$$P_3 = \left(\tfrac{1}{3}(7) + \tfrac{2}{3}(-1), \tfrac{1}{3}(7) + \tfrac{2}{3}(2) \right) = \left(\tfrac{7}{3} - \tfrac{2}{3}, \tfrac{7}{3} + \tfrac{4}{3} \right) = \left(\tfrac{5}{3}, \tfrac{11}{3} \right)$$
They all intersect at the same point.

(b)   Find each midpoint:

$$M_1 = \left(\frac{0+2b}{2},\frac{0+2c}{2}\right) = (b,c)$$

$$M_1 = \left(\frac{2b+2a}{2},\frac{2c+0}{2}\right) = (a+b,c)$$

$$M_3 = \left(\frac{0+2a}{2},\frac{0+0}{2}\right) = (a,0)$$

Now use the result from Exercise 74 to find the desired points. Let P_1, P_2, and P_3 be the desired points on medians CM_3, BM_1, and AM_2. Then:

$$P_1 = \left(\tfrac{1}{3}(2b)+\tfrac{2}{3}(a),\tfrac{1}{3}(2c)+\tfrac{2}{3}(0)\right) = \left(\frac{2b}{3}+\frac{2a}{3},\frac{2c}{3}\right)$$

$$P_2 = \left(\tfrac{1}{3}(2a)+\tfrac{2}{3}(b),\tfrac{1}{3}(0)+\tfrac{2}{3}(c)\right) = \left(\frac{2a}{3}+\frac{2b}{3},\frac{2c}{3}\right)$$

$$P_3 = \left(\tfrac{1}{3}(0)+\tfrac{2}{3}(a+b),\tfrac{1}{3}(0)+\tfrac{2}{3}(c)\right) = \left(\frac{2a}{3}+\frac{2b}{3},\frac{2c}{3}\right)$$

Again all these intersect at the same point. All medians of a triangle intersect at a point that is $\frac{2}{3}$ of the distance from each vertex to the midpoint of the opposite side.

77.   The center is $(2, -1)$ and the radius is $\sqrt{20} = 2\sqrt{5}$. Since $(1, -2)$ is the midpoint of the chord, the chord must be perpendicular to the radius line drawn to that point. Find the distance a of the line segment joining $(2, -1)$ and $(1, -2)$:

$$a = \sqrt{(1-2)^2+[-2-(-1)]^2} = \sqrt{(-1)^2+(-1)^2} = \sqrt{1+1} = \sqrt{2}$$

We can find d (half of the length of the chord) by the Pythagorean theorem:

$$d^2 + \left(\sqrt{2}\right)^2 = \left(\sqrt{20}\right)^2$$
$$d^2 + 2 = 20$$
$$d^2 = 18$$
$$d = \sqrt{18} = 3\sqrt{2}$$

The length of the chord is $2d = 2(3\sqrt{2}) = 6\sqrt{2}$.

79.   Using the hint:

$$m^2 = (a+c-0)^2 + (b-0)^2$$
$$m^2 = a^2 + 2ac + c^2 + b^2$$
$$s^2 = (2a-0)^2 + (2b-0)^2 = 4a^2 + 4b^2$$
$$t^2 = (2c)^2 = 4c^2$$
$$u^2 = (2a-2c)^2 + (2b-0)^2 = 4a^2 - 8ac + 4c^2 + 4b^2$$

Working from the right-hand side:

$$\tfrac{1}{2}\left(s^2+t^2\right) - \tfrac{1}{4}u^2 = \tfrac{1}{2}\left(4a^2+4b^2+4c^2\right) - \tfrac{1}{4}\left(4a^2-8ac+4c^2+4b^2\right)$$
$$= 2a^2 + 2b^2 + 2c^2 - a^2 + 2ac - c^2 - b^2$$
$$= a^2 + b^2 + c^2 + 2ac$$
$$= m^2$$

81. (a) The reflection of $\overline{AB}$ about the line $y = x$ is given by the graph:

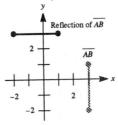

(b) The reflection of $\overline{AB}$ about the x-axis is given by the graph:

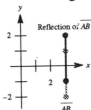

(c) The reflection of $\overline{AB}$ about the y-axis is given by the graph:

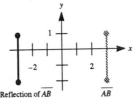

(d) The reflection of $\overline{AB}$ about the origin is given by the graph:

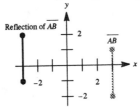

83. Here $(x_0, y_0) = (1, 2)$, $m = \frac{1}{2}$, and $b = -5$:

$$d = \frac{2 - \frac{1}{2}(1) - (-5)}{\sqrt{1 + \left(\frac{1}{2}\right)^2}} = \frac{2 - \frac{1}{2} + 5}{\sqrt{1 + \frac{1}{4}}} = \frac{\frac{13}{2}}{\sqrt{\frac{5}{4}}} = \frac{13\sqrt{5}}{5}$$

85. Here $(x_0, y_0) = (-1, -3)$, $A = 2$, $B = 3$ and $C = -6$:

$$d = \frac{|2(-1) + 3(-3) + (-6)|}{\sqrt{2^2 + 3^2}} = \frac{|-2 - 9 - 6|}{\sqrt{13}} = \frac{17}{\sqrt{13}} = \frac{17\sqrt{13}}{13}$$

87. **(a)** If the center is (h, k) and the circle is tangent to the coordinate axes, then the points of tangency are $(h, 0)$ and $(0, k)$. Since the distances from the center to the coordinate axes are the same, $h = k = r$ (the radius) and the center must be (r, r). Write the line in slope-intercept form:

$$3x + 4y = 12$$
$$4y = -3x + 12$$
$$y = -\tfrac{3}{4}x + 3$$

Find the distance from (r, r) to this line, using the fact that this distance is also r:

$$r = \frac{\left| r + \tfrac{3}{4}r - 3 \right|}{\sqrt{1 + (3/4)^2}} = \frac{\left| \tfrac{7}{4}r - 3 \right|}{5/4}$$

So we have the equation:

$$\tfrac{5}{4}r = \left| \tfrac{7}{4}r - 3 \right|$$

Thus we have:

$$\tfrac{5}{4}r = \tfrac{7}{4}r - 3 \qquad \text{or} \qquad -\tfrac{5}{4}r = \tfrac{7}{4}r - 3$$
$$5r = 7r - 12 \qquad\qquad\qquad -5r = 7r - 12$$
$$-2r = -12 \qquad\qquad\qquad -12r = -12$$
$$r = 6 \qquad\qquad\qquad\qquad r = 1$$

Clearly $r = 6$ corresponds to a circle above the line, so $r = 1$. Thus the equation of the circle is $(x - 1)^2 + (y - 1)^2 = 1$.

(b) We have the coordinates $S(1, 0)$ and $U(0, 1)$. We must find the coordinates of T. We know the slope of the center to $3x + 4y = 12$ is $4/3$, so use the point $(1, 1)$ in the point-slope formula:

$$y - 1 = \tfrac{4}{3}(x - 1)$$
$$y - 1 = \tfrac{4}{3}x - \tfrac{4}{3}$$
$$y = \tfrac{4}{3}x - \tfrac{1}{3}$$

Now find the intersection point of this line with our given line:

$$\tfrac{4}{3}x - \tfrac{1}{3} = -\tfrac{3}{4}x + 3$$
$$16x - 4 = -9x + 36$$
$$25x = 40$$
$$x = \tfrac{8}{5}$$

Substituting, we find $y = 9/5$, so we have the point $T(8/5, 9/5)$. We now can find the required equations. For $\overline{AT}$ the slope is $m = \dfrac{9/5 - 0}{8/5 - 0} = \tfrac{9}{8}$, so using the point $(0, 0)$ in the point-slope equation yields:

$$y - 0 = \tfrac{9}{8}(x - 0)$$
$$y = \tfrac{9}{8}x$$

For $\overline{BU}$ the slope is $m = \dfrac{0-1}{4-0} = -\dfrac{1}{4}$, so using the point $(0, 1)$ in the point-slope equation yields:

$$y - 1 = -\tfrac{1}{4}(x - 0)$$
$$y = -\tfrac{1}{4}x + 1$$

For $\overline{CS}$ the slope is $m = \dfrac{3-0}{0-1} = -3$, so using the point $(1, 0)$ in the point-slope equation yields:

$$y - 0 = -3(x - 1)$$
$$y = -3x + 3$$

(c) To find the required intersection points, set the equations equal. For $\overline{AT}$ and $\overline{CS}$ we have:

$$\tfrac{9}{8}x = -3x + 3$$
$$\tfrac{33}{8}x = 3$$
$$x = \tfrac{24}{33} = \tfrac{8}{11}$$
$$y = \tfrac{9}{11}$$

The intersection point is $(8/11, 9/11)$. For $\overline{AT}$ and $\overline{BU}$ we have:

$$\tfrac{9}{8}x = -\tfrac{1}{4}x + 1$$
$$9x = -2x + 8$$
$$11x = 8$$
$$x = \tfrac{8}{11}$$
$$y = \tfrac{9}{11}$$

The intersection point is $(8/11, 9/11)$. Observe that the point of intersection is the same in both cases.

Chapter Two Test

1. First find the slope:

$$m = \frac{8 - (-2)}{3 - 1} = \tfrac{10}{2} = 5$$

Now using the point $(3, 8)$ in the point-slope formula, we have:

$$y - 8 = 5(x - 3)$$
$$y - 8 = 5x - 15$$
$$y = 5x - 7$$

2.   First write $5x + 6y = 30$ in slope-intercept form:
$$5x + 6y = 30$$
$$6y = -5x + 30$$
$$y = -\tfrac{5}{6}x + 5$$

So the perpendicular slope is $\tfrac{6}{5}$. Now using the point $(2, -1)$ in the point-slope formula:
$$y - (-1) = \tfrac{6}{5}(x - 2)$$
$$y + 1 = \tfrac{6}{5}x - \tfrac{12}{5}$$
$$y = \tfrac{6}{5}x - \tfrac{17}{5}$$

3.   First find the midpoint of the line segment joining the points $(-5, -2)$ and $(3, 8)$:
$$M = \left(\frac{-5 + 3}{2}, \frac{-2 + 8}{2}\right) = \left(-\tfrac{2}{2}, \tfrac{6}{2}\right) = (-1, 3)$$

Since the line passes through the origin, its slope is given by:
$$m = \frac{3 - 0}{-1 - 0} = \frac{3}{-1} = -3$$

So the equation is $y = -3x$.

4.   The center of the circle is $(4, -5)$. The slope from this center to the origin is therefore:
$$m = \frac{-5 - 0}{4 - 0} = -\tfrac{5}{4}$$

So the tangent line slope, which is perpendicular to this radius slope, is $\tfrac{4}{5}$. So the equation of the tangent line is $y = \tfrac{4}{5}x$.

5.   Compute the distance from each point to the origin:
$$d_1 = \sqrt{(3 - 0)^2 + (9 - 0)^2} = \sqrt{9 + 81} = \sqrt{90}$$
$$d_2 = \sqrt{(5 - 0)^2 + (8 - 0)^2} = \sqrt{25 + 64} = \sqrt{89}$$

So $(3, 9)$ is farther from the origin.

6.   (a)   To test for symmetry about the x-axis, replace (x, y) with $(x, -y)$:
$$-y = x^3 + 5x$$
$$y = -x^3 - 5x$$

Since the equation is changed, there is no x-axis symmetry. To test for symmetry about the y-axis, replace (x, y) with $(-x, y)$:
$$y = (-x)^3 + 5(-x) = -x^3 - 5x$$

Since the equation is changed, there is no y-axis symmetry. To test for symmetry about the origin, replace (x, y) with $(-x, -y)$:
$$-y = (-x)^3 + 5(-x)$$
$$-y = -x^3 - 5x$$
$$y = x^3 + 5x$$

Since the equation is unchanged, there is origin symmetry.

(b)   To test for symmetry about the x-axis, replace (x, y) with $(x, -y)$:

$$-y = 3^x + 3^{-x}$$
$$y = -3^x - 3^{-x}$$

Since the equation is changed, there is no x-axis symmetry. To test for symmetry about the y-axis, replace (x, y) with $(-x, y)$:

$$y = 3^{-x} + 3^{-(-x)} = 3^{-x} + 3^x$$

Since the equation is unchanged, there is y-axis symmetry. To test for symmetry about the origin, replace (x, y) with $(-x, -y)$:

$$-y = 3^{-x} + 3^{-(-x)}$$
$$-y = 3^{-x} + 3^x$$
$$y = -3^{-x} - 3^x$$

Since the equation is changed, there is no origin symmetry.

(c)   To test for symmetry about the x-axis, replace (x, y) with $(x, -y)$:

$$(-y)^2 = 5x^2 + x$$
$$y^2 = 5x^2 + x$$

Since the equation is unchanged, there is x-axis symmetry. To test for symmetry about the y-axis, replace (x, y) with $(-x, y)$:

$$y^2 = 5(-x)^2 + (-x) = 5x^2 - x$$

Since the equation is changed, there is no y-axis symmetry. To test for symmetry about the origin, replace (x, y) with $(-x, -y)$:

$$(-y)^2 = 5(-x)^2 + (-x)$$
$$y^2 = 5x^2 - x$$

Since the equation is changed, there is no origin symmetry.

7.   To find the x-intercept, let $y = 0$:

$$3x = 15$$
$$x = 5$$

To find the y-intercept, let $x = 0$:

$$-5y = 15$$
$$y = -3$$

Now draw the graph:

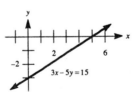

8.   To find the x-intercepts, let $y = 0$:

$$(x - 1)^2 + (0 + 2)^2 = 9$$
$$(x - 1)^2 + 4 = 9$$
$$(x - 1)^2 = 5$$
$$x - 1 = \pm\sqrt{5}$$
$$x = 1 \pm \sqrt{5}$$

To find the y-intercepts, let $x = 0$:

$$(0-1)^2 + (y+2)^2 = 9$$
$$1 + (y+2)^2 = 9$$
$$(y+2)^2 = 8$$
$$y + 2 = \pm\sqrt{8} = \pm 2\sqrt{2}$$
$$y = -2 \pm 2\sqrt{2}$$

Now draw the graph:

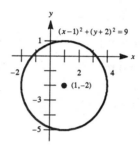

9.   To find the x-intercepts, let $y = 0$:

$$0 = \frac{2}{x^2}$$
$$0 = 2$$

Since this is impossible, there are no x-intercepts. Since $x \neq 0$, there are no y-intercepts. Now draw the graph:

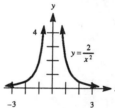

10.   The slope of the line passing through $(-4, 3)$ and $(t, 13)$ is given by:

$$m = \frac{13-3}{t-(-4)} = \frac{10}{t+4}$$

Setting the slope $m = 2$:

$$\frac{10}{t+4} = 2$$
$$10 = 2t + 8$$
$$2 = 2t$$
$$1 = t$$

11. Substituting the point $(4, 5)$ into the equation:

$$y = k(x-2)^2 + 7$$
$$5 = k(4-2)^2 + 7$$
$$5 = 4k + 7$$
$$-2 = 4k$$
$$k = -\tfrac{1}{2}$$

12. Substituting $x = -\tfrac{1}{2}$ into $y = 4x^2 - 8x$:

$$y = 4\left(-\tfrac{1}{2}\right)^2 - 8\left(-\tfrac{1}{2}\right) = 4\left(\tfrac{1}{4}\right) + 4 = 1 + 4 = 5$$

So $\left(-\tfrac{1}{2}, 5\right)$ lies on the graph of $y = 4x^2 - 8x$.

13. (a)   Sketch the graph, noting that $x \neq -1/2$:

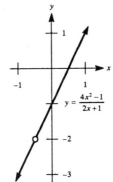

(b)   Sketch the graph:

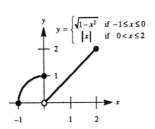

(c)   Sketch the graph:

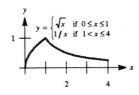

14. (a)   The reflection of $\overline{AB}$ about the x-axis is given by the graph:

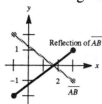

(b)   The reflection of $\overline{AB}$ about the y-axis is given by the graph:

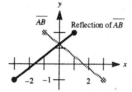

(c)   The reflection of $\overline{AB}$ about the origin is given by the graph:

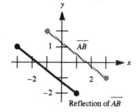

(d)   The reflection of $\overline{AB}$ about the line $y = x$ is given by the graph:

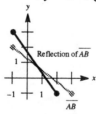

15. First find the slope of the line:

$$m = \frac{4-2}{1-6} = -\frac{2}{5}$$

Using the point-slope formula with the point $(1, 4)$:

$$y - 4 = -\tfrac{2}{5}(x-1)$$
$$5y - 20 = -2x + 2$$
$$2x + 5y = 22$$

Now find the x- and y-intercepts:

  let $x = 0$ let $y = 0$

  $5y = 22$ $2x = 22$

  $y = \frac{22}{5}$ $x = 11$

So the area is given by:

$$\text{Area} = \tfrac{1}{2}(\text{base})(\text{height}) = \tfrac{1}{2} \cdot 11 \cdot \tfrac{22}{5} = \tfrac{121}{5} \text{ sq. units}$$

16. Since the denominator cannot equal zero, the domain is $x \neq 3$. Factoring the numerator:

$$y = \frac{x^2 + x - 6}{x + 3} = \frac{(x+3)(x-2)}{x+3} = x - 2, \text{ if } x \neq -3$$

Thus the graph will consist of a line with a hole at the point $(-3, -5)$. Drawing the graph:

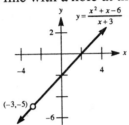

Chapter Three
Equations and Inequalities

3.1 Quadratic Equations: Theory and Examples

1. Solve by completing the square:

$$x^2 + 8x - 2 = 0$$
$$x^2 + 8x = 2$$
$$x^2 + 8x + 16 = 2 + 16$$
$$(x+4)^2 = 18$$
$$x + 4 = \pm\sqrt{18} = \pm 3\sqrt{2}$$
$$x = -4 \pm 3\sqrt{2} \approx -8.24, 0.24$$

3. Solve by completing the square:

$$x^2 + 4x + 1 = 0$$
$$x^2 + 4x = -1$$
$$x^2 + 4x + 4 = -1 + 4$$
$$(x+2)^2 = 3$$
$$x + 2 = \pm\sqrt{3}$$
$$x = -2 \pm \sqrt{3} \approx -3.73, -0.27$$

5.   Solve by completing the square:

$$2y^2 - 5y - 2 = 0$$
$$y^2 - \tfrac{5}{2}y - 1 = 0$$
$$y^2 - \tfrac{5}{2}y = 1$$
$$y^2 - \tfrac{5}{2}y + \tfrac{25}{16} = 1 + \tfrac{25}{16}$$
$$\left(y - \tfrac{5}{4}\right)^2 = \tfrac{41}{16}$$
$$y - \tfrac{5}{4} = \pm\sqrt{\tfrac{41}{16}} = \pm\frac{\sqrt{41}}{4}$$
$$y = \frac{5 \pm \sqrt{41}}{4} \approx -0.35,\ 2.85$$

7.   Solve by completing the square:

$$4y^2 + 8y + 5 = 0$$
$$y^2 + 2y + \tfrac{5}{4} = 0$$
$$y^2 + 2y = -\tfrac{5}{4}$$
$$y^2 + 2y + 1 = -\tfrac{5}{4} + 1$$
$$(y + 1)^2 = -\tfrac{1}{4}$$

The equation has no real roots.

9.   Solve by completing the square:

$$4s^2 - 20s + 25 = 0$$
$$s^2 - 5s + \tfrac{25}{4} = 0$$
$$s^2 - 5s = -\tfrac{25}{4}$$
$$s^2 - 5s + \tfrac{25}{4} = -\tfrac{25}{4} + \tfrac{25}{4}$$
$$\left(s - \tfrac{5}{2}\right)^2 = 0$$
$$s - \tfrac{5}{2} = 0$$
$$s = \tfrac{5}{2}$$

11. Using the quadratic formula with $a = 1$, $b = -8$, and $c = 6$:

$$x = \frac{-(-8) \pm \sqrt{(-8)^2 - 4(1)(6)}}{2(1)}$$

$$= \frac{8 \pm \sqrt{64 - 24}}{2}$$

$$= \frac{8 \pm \sqrt{40}}{2}$$

$$= \frac{8 \pm 2\sqrt{10}}{2}$$

$$= 4 \pm \sqrt{10}$$

$$\approx 0.84, 7.16$$

13. Using the quadratic formula with $a = -3$, $b = 1$, and $c = 3$:

$$x = \frac{-1 \pm \sqrt{(1)^2 - 4(-3)(3)}}{2(-3)} = \frac{-1 \pm \sqrt{1 + 36}}{-6} = \frac{-1 \pm \sqrt{37}}{-6} = \frac{1 \pm \sqrt{37}}{6} \approx -0.85, 1.18$$

15. Rewrite the equation as $y^2 + 8y + 1 = 0$. Now using the quadratic formula with $a = 1$, $b = 8$, and $c = 1$:

$$y = \frac{-8 \pm \sqrt{(8)^2 - 4(1)(1)}}{2(1)}$$

$$= \frac{-8 \pm \sqrt{64 - 4}}{2}$$

$$= \frac{-8 \pm \sqrt{60}}{2}$$

$$= \frac{-8 \pm 2\sqrt{15}}{2}$$

$$= -4 \pm \sqrt{15}$$

$$\approx -7.87, -0.13$$

17. Rewrite the equation as $t^2 + 3t + 4 = 0$. Now using the quadratic formula with $a = 1$, $b = 3$, and $c = 4$:

$$t = \frac{-3 \pm \sqrt{(3)^2 - 4(1)(4)}}{2(1)} = \frac{-3 \pm \sqrt{9 - 16}}{2} = \frac{-3 \pm \sqrt{-7}}{2}$$

The equation has no real roots.

19. Using the quadratic formula with $a = 252$, $b = 188$, and $c = -225$:

$$x = \frac{-188 \pm \sqrt{(188)^2 - 4(252)(-225)}}{2(252)}$$

$$= \frac{-188 \pm \sqrt{35344 + 226800}}{504}$$

$$= \frac{-188 \pm \sqrt{262144}}{504}$$

$$= \frac{-188 \pm 512}{504}$$

$$= -\frac{25}{18}, \frac{9}{14}$$

21. Finding the sum and the product of the roots:

$$r_1 + r_2 = -b = -8$$
$$r_1 \cdot r_2 = c = -20$$

23. First divide by 4 to write the equation as $y^2 - 7y + \frac{9}{4} = 0$. Now finding the sum and the product of the roots:

$$r_1 + r_2 = -b = -(-7) = 7$$
$$r_1 \cdot r_2 = c = \frac{9}{4}$$

25. If $a = 1$, we can find c and b:

$$c = r_1 \cdot r_2 = 3 \cdot 11 = 33$$
$$b = -\left(r_1 + r_2\right) = -(3 + 11) = -14$$

Thus the equation is $x^2 - 14x + 33 = 0$.

27. If $a = 1$, we can find c and b:

$$c = r_1 \cdot r_2 = \left(1 - \sqrt{2}\right)\left(1 + \sqrt{2}\right) = 1 - 2 = -1$$
$$b = -\left(r_1 + r_2\right) = -\left(1 - \sqrt{2} + 1 + \sqrt{2}\right) = -2$$

Thus the equation is $x^2 - 2x - 1 = 0$.

29. If $a = 1$, we can find c and b:

$$c = r_1 \cdot r_2 = \left(3 + \sqrt{3}\right)\left(3 - \sqrt{3}\right) = 9 - 3 = 6$$
$$b = -\left(r_1 + r_2\right) = -\left(3 + \sqrt{3} + 3 - \sqrt{3}\right) = -6$$

Thus the equation is $x^2 - 6x + 6 = 0$.

31. If $a = 1$, we can find c and b:
$$c = r_1 \cdot r_2 = \tfrac{1}{2}\left(2 + \sqrt{5}\right) \cdot \tfrac{1}{2}\left(2 - \sqrt{5}\right) = \tfrac{1}{4}(4 - 5) = -\tfrac{1}{4}$$
$$b = -\left(r_1 + r_2\right) = -\left(1 + \tfrac{1}{2}\sqrt{5} + 1 - \tfrac{1}{2}\sqrt{5}\right) = -2$$

Thus the equation is $x^2 - 2x - \tfrac{1}{4} = 0$. Multiplying by 4 (to produce integer coefficients) results in the equation $4x^2 - 8x - 1 = 0$.

33. If $r_1 = \sqrt{2}$ and $r_2 = \sqrt{5}$, note that:
$$c = r_1 r_2 = \sqrt{2} \cdot \sqrt{5} = \sqrt{10}$$
$$b = -\left(r_1 + r_2\right) = -\left(\sqrt{2} + \sqrt{5}\right)$$

So the roots are $\sqrt{2}$ and $\sqrt{5}$.

35. Using the quadratic formula:
$$x = \frac{-3 \pm \sqrt{9 - 4(1)(2.249)}}{2(1)} \approx -1.47, \, -1.53$$

37. Solving for r using the quadratic formula:
$$2\pi r^2 + 2\pi rh = 20\pi$$
$$2\pi r^2 + 2\pi rh - 20\pi = 0$$
$$2\pi\left(r^2 + rh - 10\right) = 0$$
$$r^2 + rh - 10 = 0$$
$$r = \frac{-h \pm \sqrt{h^2 - 4(-10)}}{2(1)} = \frac{-h \pm \sqrt{h^2 + 40}}{2}$$

39. Solving for t by factoring:
$$-16t^2 + v_0 t = 0$$
$$t\left(-16t + v_0\right) = 0$$
$$t = 0 \text{ or } t = \frac{v_0}{16}$$

41. Solving for x by factoring:
$$x^3 + bx^2 - 2b^2 x = 0$$
$$x\left(x^2 + bx - 2b^2\right) = 0$$
$$x(x + 2b)(x - b) = 0$$
$$x = 0, \, -2b, \, b$$

43. Given $a = 1$, $b = -12$, $c = 16$, compute the discriminant:
$$b^2 - 4ac = 144 - 4(1)(16) = 80 > 0$$
The equation has two real roots.

45. Given $a = 4$, $b = -5$, $c = -\frac{1}{2}$, compute the discriminant:
$$b^2 - 4ac = 25 - 4(4)\left(-\tfrac{1}{2}\right) = 33 > 0$$
The equation has two real roots.

47. Given $a = 1$, $b = \sqrt{3}$, $c = \frac{3}{4}$, compute the discriminant:
$$b^2 - 4ac = 3 - 4(1)\left(\tfrac{3}{4}\right) = 0$$
The equation has one real root.

49. Given $a = 1$, $b = -\sqrt{5}$, $c = 1$, compute the discriminant:
$$b^2 - 4ac = 5 - 4(1)(1) = 1 > 0$$
The equation has two real roots.

51. Set the discriminant equal to 0:
$$b^2 - 4ac = 0$$
$$144 - 4(1)(k) = 0$$
$$4k = 144$$
$$k = 36$$

53. Set the discriminant equal to 0:
$$b^2 - 4ac = 0$$
$$k^2 - 4(1)(5) = 0$$
$$k = \pm\sqrt{20} = \pm 2\sqrt{5}$$

55. Multiplying through by the least common denominator $x(x + 5)$ gives us:
$$3x + 4(x + 5) = 2x(x + 5)$$
$$7x + 20 = 2x^2 + 10x$$
$$-2x^2 - 3x + 20 = 0$$
$$2x^2 + 3x - 20 = 0$$
$$(2x - 5)(x + 4) = 0$$
$$x = \tfrac{5}{2} \text{ or } x = -4$$
Both of these values check in the original equation.

57. Multiplying through by the least common denominator $6x + 1$ gives us:
$$-6x^2 + 5x - 1 = 0$$
$$6x^2 - 5x + 1 = 0$$
$$(3x - 1)(2x - 1) = 0$$
$$x = \tfrac{1}{3} \text{ or } x = \tfrac{1}{2}$$
Both values check in the original equation.

59. Multiplying through by the least common denominator $(x + 1)(x - 2)(x - 3)$ gives us:
$$3x^2 - 6x - 3 + (5 - 2x)(x + 1) = 0$$
$$3x^2 - 6x - 3 - 2x^2 + 3x + 5 = 0$$
$$x^2 - 3x + 2 = 0$$
$$(x - 2)(x - 1) = 0$$
$$x = 2 \text{ or } x = 1$$
Of these two values, only $x = 1$ checks in the original equation.

61. Multiplying through by the least common denominator $(x - 1)(x + 1)(x - 3)$ gives us:
$$2x(x + 3) - (x - 1)(x + 1) = 0$$
$$2x^2 + 6x - x^2 + 1 = 0$$
$$x^2 + 6x + 1 = 0$$
$$x = \frac{-6 \pm \sqrt{36 - 4}}{2} = \frac{-6 \pm 4\sqrt{2}}{2} = -3 \pm 2\sqrt{2}$$
Both values check in the original equation.

63. Since $x^2 + 4x + 3 = (x + 1)(x + 3)$, multiply through by the least common denominator $(x + 1)(x + 3)$ to yield:
$$(x - 1)(x + 3) - (x + 1)(x + 1) + 4 = 0$$
$$\left(x^2 + 2x - 3\right) - \left(x^2 + 2x + 1\right) + 4 = 0$$
$$-4 + 4 = 0$$
Since this last equation is true for any real number, any value of x (except $x = -1$ or $x = -3$, which yield 0 denominators) is a solution to the equation.

65. (a) When the ball lands $h = 0$. Therefore:
$$-16t^2 + 96t = 0$$
$$-16t(t - 6) = 0$$
$$t = 0 \text{ or } t = 6$$
The value $t = 0$ gives the time when the ball is first thrown. Consequently the ball lands after 6 seconds.

(b) With $h = 80$, the equation becomes:
$$80 = -16t^2 + 96t$$
$$16t^2 - 96t + 80 = 0$$
$$t^2 - 6t + 5 = 0$$
$$(t - 5)(t - 1) = 0$$
$$t = 5 \text{ or } t = 1$$
At $t = 1$ second the ball is 80 ft. high and rising. At $t = 5$ seconds the ball is 80 ft high and falling.

67. (a) Using the quadratic formula with $a = 1$, $b = 3$, and $c = 1$:
$$x = \frac{-3 \pm \sqrt{(3)^2 - 4(1)(1)}}{2(1)} = \frac{-3 \pm \sqrt{9 - 4}}{2} = \frac{-3 \pm \sqrt{5}}{2} = \tfrac{1}{2}\left(-3 \pm \sqrt{5}\right)$$

(b) Rationalizing the denominator:

$$\frac{1}{\frac{1}{2}\left(-3-\sqrt{5}\right)} \cdot \frac{-3+\sqrt{5}}{-3+\sqrt{5}} = \frac{-3+\sqrt{5}}{\frac{1}{2}(9-5)} = \frac{-3+\sqrt{5}}{2} = \frac{1}{2}\left(-3+\sqrt{5}\right)$$

(c) Yes. Since $r_1 r_2 = c$, and $c = 1$, then $r_1 r_2 = 1$ and thus $r_1 = \dfrac{1}{r_2}$. Thus the roots of the

equation $x^2 + 3x + 1 = 0$ are reciprocals.

69. (a) Using the hint:

$$r_1^2 + r_2^2 = \left(r_1 + r_2\right)^2 - 2r_1 r_2 = (-b)^2 - 2c = b^2 - 2c$$

(b) Using the hint:

$$\frac{1}{r_1^2} + \frac{1}{r_2^2} = \frac{r_1^2 + r_2^2}{r_1^2 r_2^2} = \frac{b^2 - 2c}{c^2}$$

71. Using the quadratic formula:

$$x = \frac{-b \pm \sqrt{b^2 - 4(a)(-a)}}{2a} = \frac{-b \pm \sqrt{b^2 + 4a^2}}{2a}$$

Since $a \neq 0$, the quantity $b^2 + 4a^2 > 0$, thus the equation has two real roots.

73. Since r_1 and r_2 are roots:

$$r_1^2 + br_1 + c = 0$$
$$r_2^2 + br_2 + c = 0$$

Let $r_1 = 2r_2$, so:

$$\left(2r_2\right)^2 + b\left(2r_2\right) + c = 0$$
$$4r_2^2 + 2br_2 + c = 0$$

Solving these equations (involving r_2) for c:

$$c = -r_2^2 - br_2$$
$$c = -4r_2^2 - 2br_2$$

Setting these two expressions equal:

$$-r_2^2 - br_2 = -4r_2^2 - 2br_2$$
$$3r_2^2 + br_2 = 0$$
$$r_2\left(3r_2 + b\right) = 0$$
$$r_2 = 0 \ \text{ or } \ r_2 = -\frac{b}{3}$$

If $r_2 = 0$, then $r_1 = 0$ and thus $b = c = 0$, so $2b^2 - 9c = 0$. If $r_2 = -\dfrac{b}{3}$, we can substitute:

$$r_2^2 + br_2 + c = 0$$

$$\left(-\frac{b}{3}\right)^2 + b\left(-\frac{b}{3}\right) + c = 0$$

$$\frac{b^2}{9} - \frac{b^2}{3} + c = 0$$

$$b^2 - 3b^2 + 9c = 0$$

$$-2b^2 + 9c = 0$$

$$2b^2 - 9c = 0$$

An alternate (and easier) solution is, since $r_1 = 2r_2$:

$$b = -\left(r_1 + r_2\right) = -3r_2$$

$$c = r_1 r_2 = 2r_2^2$$

Therefore:

$$2b^2 - 9c = 2\left(-3r_2\right)^2 - 9\left(2r_2^2\right) = 18r_2^2 - 18r_2^2 = 0$$

75. Let r_1 and r_2 be the roots such that $r_1 = 3r_2$. Therefore:

$$r_1^2 + br_1 + c = 0$$

$$r_2^2 + br_2 + c = 0$$

Substitute $r_1 = 3r_2$, so:

$$\left(3r_2\right)^2 + b\left(3r_2\right) + c = 0$$

$$9r_2^2 + 3br_2 + c = 0$$

Solving these equations (involving r_2) for c:

$$c = -r_2^2 - br_2$$

$$c = -9r_2^2 - 3br_2$$

Setting these two expressions equal:

$$-r_2^2 - br_2 = -9r_2^2 - 3br_2$$

$$8r_2^2 + 2br_2 = 0$$

$$2r_2\left(4r_2 + b\right) = 0$$

$$r_2 = 0 \quad \text{or} \quad r_2 = -\frac{b}{4}$$

If $r_2 = 0$, then $r_1 = 0$ and thus $b = c = 0$, so $3b^2 = 16c$. If $r_2 = -\dfrac{b}{4}$, we can substitute:

$$r_2^2 + br_2 + c = 0$$

$$\left(-\frac{b}{4}\right)^2 + b\left(-\frac{b}{4}\right) + c = 0$$

$$\frac{b^2}{16} - \frac{b^2}{4} + c = 0$$

$$b^2 - 4b^2 + 16c = 0$$

$$-3b^2 + 16c = 0$$

$$3b^2 = 16c$$

Again, an easier solution is since $r_1 = 3r_2$:

$$b = -(r_1 + r_2) = -4r_2$$

$$c = r_1 r_2 = 3r_2^2$$

Therefore:

$$3b^2 = 3(-4r_2)^2 = 48r_2^2 = 16(3r_2^2) = 16c$$

77. First rewrite the equation as $x^2 - 2(3k + 1)x + 7(2k + 3) = 0$. Since the roots are equal, set the discriminant equal to 0:

$$[-2(3k+1)]^2 - 4(1)[7(2k+3)] = 0$$

$$4(9k^2 + 6k + 1) - 28(2k + 3) = 0$$

$$36k^2 + 24k + 4 - 56k - 84 = 0$$

$$36k^2 - 32k - 80 = 0$$

$$9k^2 - 8k - 20 = 0$$

$$(9k + 10)(k - 2) = 0$$

$$k = -\frac{10}{9} \text{ or } k = 2$$

79. If the roots are A and B, then:

$$x^2 + Ax + B = (x - A)(x - B)$$

$$x^2 + Ax + B = x^2 - (A + B)x + AB$$

Setting the coefficients equal:

$$-(A + B) = A \qquad \text{and} \qquad AB = B$$

$$-A - B = A \qquad\qquad B(A - 1) = 0$$

$$2A + B = 0 \qquad\qquad B = 0 \text{ or } A = 1$$

If $B = 0$, then $A = 0$, and if $A = 1$, then $B = -2$. So the values of A and B are:

$$A = 0 \text{ and } B = 0$$

$$A = 1 \text{ and } B = -2$$

81. (a) Making the desired substitution $x = y + k$:
$$(y + k)^2 + (y + k) - 1 = 0$$
$$y^2 + 2ky + k^2 + y + k - 1 = 0$$
$$y^2 + (2k + 1)y = 1 - k - k^2$$

(b) When $k = -\frac{1}{2}$:
$$y^2 + (-1 + 1)y = 1 + \tfrac{1}{2} - \tfrac{1}{4}$$
$$y^2 = \tfrac{5}{4}$$

(c) Solving for y, $y = \pm \frac{\sqrt{5}}{2}$. Since $x = y + k$, $x = -\frac{1}{2} \pm \frac{\sqrt{5}}{2}$.

83. Factoring and using the sum and product of the roots:
$$a^2 b + ab^2 = ab(a + b) = (q)(-p) = -pq$$

85. First compute the sum and product of the roots:
$$r_1 + r_2 = \frac{\sqrt{a}}{\sqrt{a} + \sqrt{a - b}} + \frac{\sqrt{a}}{\sqrt{a} - \sqrt{a - b}}$$
$$= \frac{\left(\sqrt{a}\right)\left(\sqrt{a} - \sqrt{a - b}\right) + \left(\sqrt{a}\right)\left(\sqrt{a} + \sqrt{a - b}\right)}{\left(\sqrt{a} + \sqrt{a - b}\right)\left(\sqrt{a} - \sqrt{a - b}\right)}$$
$$= \frac{a - \sqrt{a^2 - ab} + a + \sqrt{a^2 - ab}}{a - (a - b)}$$
$$= \frac{2a}{b}$$

$$r_1 r_2 = \frac{\sqrt{a}}{\sqrt{a} + \sqrt{a - b}} \cdot \frac{\sqrt{a}}{\sqrt{a} - \sqrt{a - b}} = \frac{a}{a - (a - b)} = \frac{a}{b}$$

Thus $c = r_1 r_2 = \frac{a}{b}$ and $b = -\left(r_1 + r_2\right) = -\frac{2a}{b}$. So the equation can be written as

$$x^2 - \frac{2a}{b}x + \frac{a}{b} = 0, \text{ or } bx^2 - 2ax + a = 0.$$

87. The equation can be solved by means of the quadratic formula. Alternately, we have the following solution by factoring:
$$[abx - (a + b)][(a + b)x + 2] = 0$$
$$abx = a + b \quad \text{or} \quad (a + b)x + 2 = 0$$
$$x = \frac{a + b}{ab} \quad \text{or} \quad x = \frac{-2}{a + b}$$

The two solutions are $\dfrac{a + b}{ab}$ and $\dfrac{-2}{a + b}$.

Graphing Utility Exercises for Section 3.1

1.   (a)   Graphing the equation:

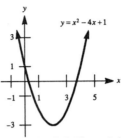

The x-intercepts are approximately 0.268 and 3.732.

(b)   Using the quadratic formula:

$$x = \frac{-(-4) \pm \sqrt{(-4)^2 - 4(1)(1)}}{2(1)}$$

$$= \frac{4 \pm \sqrt{16 - 4}}{2}$$

$$= \frac{4 \pm \sqrt{12}}{2}$$

$$= \frac{4 \pm 2\sqrt{3}}{2}$$

$$= 2 \pm \sqrt{3}$$

$$\approx 0.2679, \, 3.7321$$

These intercepts are consistent with those found in (a).

3.   (a)   Graphing the equation:

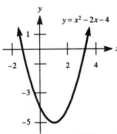

The x-intercepts are approximately −1.236 and 3.236.

(b) Using the quadratic formula:

$$x = \frac{-(-2) \pm \sqrt{(-2)^2 - 4(1)(-4)}}{2(1)}$$

$$= \frac{2 \pm \sqrt{4 + 16}}{2}$$

$$= \frac{2 \pm \sqrt{20}}{2}$$

$$= \frac{2 \pm 2\sqrt{5}}{2}$$

$$= 1 \pm \sqrt{5}$$

$$\approx -1.2361, 3.2361$$

These intercepts are consistent with those found in (a).

5. (a) Graphing the equation:

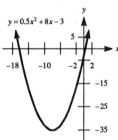

The x-intercepts are approximately -16.367 and 0.367.

(b) Using the quadratic formula:

$$x = \frac{-8 \pm \sqrt{(8)^2 - 4(0.5)(-3)}}{2(0.5)} = \frac{-8 \pm \sqrt{64 + 6}}{1} = -8 \pm \sqrt{70} \approx -16.3666, 0.3666$$

These intercepts are consistent with those found in (a).

7. (a) Graphing the equation:

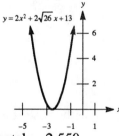

The x-intercept is approximately -2.550.

(b)   Using the quadratic formula:

$$x = \frac{-2\sqrt{26} \pm \sqrt{\left(2\sqrt{26}\right)^2 - 4(2)(13)}}{2(2)}$$

$$= \frac{-2\sqrt{26} \pm \sqrt{104 - 104}}{4}$$

$$= \frac{-2\sqrt{26}}{4}$$

$$= \frac{-\sqrt{26}}{2}$$

$$\approx -2.5495$$

This intercept is consistent with that found in (a).

9.   (a)   Graphing the equation:

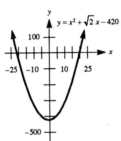

The x-intercepts are approximately -21.213 and 19.799.

(b)   Using the quadratic formula:

$$x = \frac{-\sqrt{2} \pm \sqrt{\left(\sqrt{2}\right)^2 - 4(1)(-420)}}{2(1)}$$

$$= \frac{-\sqrt{2} \pm \sqrt{2 + 1680}}{2}$$

$$= \frac{-\sqrt{2} \pm \sqrt{1682}}{2}$$

$$= \frac{-\sqrt{2} \pm 29\sqrt{2}}{2}$$

$$= -15\sqrt{2}, 14\sqrt{2}$$

$$\approx -21.2132, 19.7990$$

These intercepts are consistent with those found in (a).

11. (a)   Graphing the two equations:

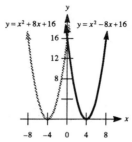

(b)   The root of $x^2 + 8x + 16 = 0$ appears to be –4, while the root of $x^2 - 8x + 16 = 0$ appears to be 4. The roots are opposites.

(c)   Solving each equation:
$$x^2 + 8x + 16 = 0$$
$$(x + 4)^2 = 0$$
$$x = -4$$
$$x^2 - 8x + 16 = 0$$
$$(x - 4)^2 = 0$$
$$x = 4$$
The results support the response for part (b).

13. (a)   Graphing the desired equations:

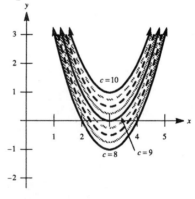

(b)   It appears when $c = 9$ the two x-intercepts merge into one.

(c)   Solving the equation when $c = 9$:
$$x^2 - 6x + 9 = 0$$
$$(x - 3)^2 = 0$$
$$x = 3$$
This confirms our observation from part (b).

(d)   When $c > 9$ the graphs appear to have no x-intercepts. Picking $c = 10$, we have:

$$x = \frac{-(-6) \pm \sqrt{(-6)^2 - 4(1)(10)}}{2(1)} = \frac{6 \pm \sqrt{36 - 40}}{2} = \frac{6 \pm \sqrt{-4}}{2}$$

Note that this is not a real number.

3.2 Other Types of Equations

1.   If $x - 5 \geq 0$, the equation becomes:
$$x - 5 = 1$$
$$x = 6$$
If $x - 5 < 0$, the equation becomes:
$$-(x - 5) = 1$$
$$-x + 5 = 1$$
$$-x = -4$$
$$x = 4$$
The solutions of the original equation are 6 and 4.

3.   If $x + 6 \geq 0$, the equation becomes:
$$x + 6 = \tfrac{1}{2}$$
$$x = -\tfrac{11}{2}$$
If $x + 6 < 0$, the equation becomes:
$$-(x + 6) = \tfrac{1}{2}$$
$$-x - 6 = \tfrac{1}{2}$$
$$-x = \tfrac{13}{2}$$
$$x = -\tfrac{13}{2}$$
The solutions of the original equation are $-\tfrac{11}{2}$ and $-\tfrac{13}{2}$.

5.   If $6x - 5 \geq 0$, the equation becomes:
$$6x - 5 = 25$$
$$6x = 30$$
$$x = 5$$
If $6x - 5 < 0$, the equation becomes:
$$-(6x - 5) = 25$$
$$-6x + 5 = 25$$
$$-6x = 20$$
$$x = -\tfrac{10}{3}$$
The solutions of the original equation are 5 and $-\tfrac{10}{3}$.

7. Squaring each side of the equation:
$$|x+3| = 2x-2$$
$$(x+3)^2 = (2x-2)^2$$
$$x^2 + 6x + 9 = 4x^2 - 8x + 4$$
$$0 = 3x^2 - 14x - 5$$
$$0 = (3x+1)(x-5)$$
$$x = -\tfrac{1}{3}, 5$$
After checking in the original equation, note that $x = 5$ is the only solution.

9. Squaring each side of the equation:
$$|2x-1| = 1 - \tfrac{1}{2}x$$
$$(2x-1)^2 = \left(1 - \tfrac{1}{2}x\right)^2$$
$$4x^2 - 4x + 1 = 1 - x + \tfrac{1}{4}x^2$$
$$16x^2 - 16x + 4 = 4 - 4x + x^2$$
$$15x^2 - 12x = 0$$
$$3x(5x-4) = 0$$
$$x = 0, \tfrac{4}{5}$$
Both values check in the original equation.

11. Factoring:
$$3x^2 - 48x = 0$$
$$3x(x-16) = 0$$
$$x = 0, 16$$
The solutions are 0 and 16.

13. Factoring:
$$t^3 - 125 = 0$$
$$(t-5)\left(t^2 + 5t + 25\right) = 0$$
Setting the first factor equal to zero yields $t = 5$. Setting the second factor equal to zero yields a quadratic equation with no real roots. The only solution is 5.

15. Factoring:
$$7x^4 - 28x^2 = 0$$
$$7x^2\left(x^2 - 4\right) = 0$$
$$7x^2(x-2)(x+2) = 0$$
$$x = 0, 2, -2$$
The solutions are 0, 2, and –2.

17. Factoring:

$$225(x-1)-x^2(x-1)=0$$
$$(x-1)(225-x^2)=0$$
$$(x-1)(15-x)(15+x)=0$$
$$x=1,15,-15$$

The solutions are 1, 15, and –15.

19. Factoring:

$$4y^3-20y^2+25y=0$$
$$y(4y^2-20y+25)=0$$
$$y(2y-5)^2=0$$
$$y=0,\tfrac{5}{2}$$

The solutions are 0 and $\tfrac{5}{2}$.

21. Factoring:

$$t^4+2t^3-3t^2=0$$
$$t^2(t^2+2t-3)=0$$
$$t^2(t+3)(t-1)=0$$
$$t=0,-3,1$$

The solutions are 0, –3, and 1.

23. Factoring:

$$6x=23x^2+4x^3$$
$$4x^3+23x^2-6x=0$$
$$x(4x^2+23x-6)=0$$
$$x(4x-1)(x+6)=0$$
$$x=0,\tfrac{1}{4},-6$$

The solutions are 0, $\tfrac{1}{4}$, and -6.

25. Factoring:

$$x^4-x^2=6$$
$$x^4-x^2-6=0$$
$$(x^2-3)(x^2+2)=0$$
$$x=\pm\sqrt{3}$$

Note that $x^2+2\neq0$. The solutions are $-\sqrt{3}$ and $\sqrt{3}$.

27. Factoring:
$$4y^2 = 5 - y^4$$
$$y^4 + 4y^2 - 5 = 0$$
$$(y^2 + 5)(y^2 - 1) = 0$$
$$(y^2 + 5)(y + 1)(y - 1) = 0$$
$$y = -1, 1$$
Note that $y^2 + 5 \neq 0$. The solutions are -1 and 1.

29. Factoring:
$$3t^2 + 2 = 9t^4$$
$$9t^4 - 3t^2 - 2 = 0$$
$$(3t^2 + 1)(3t^2 - 2) = 0$$
$$3t^2 = 2$$
$$t = \pm\sqrt{\tfrac{2}{3}} = \pm\tfrac{\sqrt{6}}{3}$$
Note that $3t^2 + 1 \neq 0$. The solutions are $\pm\tfrac{\sqrt{6}}{3}$.

31. Taking roots:
$$(x - 2)^3 - 5 = 0$$
$$(x - 2)^3 = 5$$
$$x - 2 = \sqrt[3]{5}$$
$$x = 2 + \sqrt[3]{5} \approx 3.71$$
The only solution is $2 + \sqrt[3]{5} \approx 3.71$.

33. Taking roots:
$$(x + 4)^5 + 16 = 0$$
$$(x + 4)^5 = -16$$
$$x + 4 = -\sqrt[5]{16}$$
$$x = -4 - \sqrt[5]{16} \approx -5.74$$
The only solution is $-4 - \sqrt[5]{16} \approx -5.74$.

35. (a) Taking roots:
$$(x - 3)^4 - 30 = 0$$
$$(x - 3)^4 = 30$$
$$x - 3 = \pm\sqrt[4]{30}$$
$$x = 3 \pm \sqrt[4]{30} \approx 5.34, 0.66$$
The solutions are $3 + \sqrt[4]{30} \approx 5.34$ and $3 - \sqrt[4]{30} \approx 0.66$.

(b)   Solving for x:
$$(x-3)^4 + 30 = 0$$
$$(x-3)^4 = -30$$
Since $(x-3)^4 \geq 0$ for real values of x, there are no solutions to this equation.

37.   Factoring:
$$y^4 + 4y^2 - 5 = 0$$
$$(y^2 - 1)(y^2 + 5) = 0$$
$$(y-1)(y+1)(y^2 + 5) = 0$$
$$y = 1, -1$$
Note that $y^2 + 5 \neq 0$. The solutions are ± 1.

39.   Factoring:
$$9t^4 - 3t^2 - 2 = 0$$
$$(3t^2 - 2)(3t^2 + 1) = 0$$
$$3t^2 - 2 = 0$$
$$t^2 = \tfrac{2}{3}$$
$$t = \pm\sqrt{\tfrac{2}{3}} = \pm\tfrac{\sqrt{6}}{3}$$
Note that $3t^2 + 1 \neq 0$. The solutions are $\pm\tfrac{\sqrt{6}}{3}$.

41.   Factoring:
$$x^6 - 10x^4 + 24x^2 = 0$$
$$x^2(x^4 - 10x^2 + 24) = 0$$
$$x^2(x^2 - 6)(x^2 - 4) = 0$$
$$x^2(x^2 - 6)(x-2)(x+2) = 0$$
$$x = 0, \pm\sqrt{6}, \pm 2$$
The solutions are 0, $\pm\sqrt{6}$, and ± 2.

43.   Letting $x^2 = t$ and $x^4 = t^2$ yields:
$$t^2 + t - 1 = 0$$
$$t = \frac{-1 \pm \sqrt{1 - 4(-1)}}{2} = \frac{-1 \pm \sqrt{5}}{2}$$
Since t must be non-negative (because $t = x^2$) choose the value:
$$t = \frac{-1 + \sqrt{5}}{2}$$
Therefore:
$$x^2 = \frac{-1 + \sqrt{5}}{2}$$
$$x = \pm\sqrt{\frac{-1 + \sqrt{5}}{2}}$$
The solutions are $\pm\sqrt{\frac{-1+\sqrt{5}}{2}}$.

45. Letting $x^2 = t$ and $x^4 = t^2$ yields:

$$t^2 + 3t - 2 = 0$$

$$t = \frac{-3 \pm \sqrt{9 - 4(-2)}}{2} = \frac{-3 \pm \sqrt{17}}{2}$$

Since t must be non-negative (because $t = x^2$) choose the value:

$$t = \frac{-3 + \sqrt{17}}{2}$$

Therefore:

$$x^2 = \frac{-3 + \sqrt{17}}{2}$$

$$x = \pm\sqrt{\frac{-3 + \sqrt{17}}{2}}$$

The solutions are $\pm\sqrt{\frac{-3+\sqrt{17}}{2}}$.

47. Letting $x^3 = t$ and $x^6 = t^2$, the equation becomes:

$$t^2 + 7t - 8 = 0$$

$$(t + 8)(t - 1) = 0$$

$$t = -8, 1$$

So $x^3 = -8$ or $x^3 = 1$, thus $x = -2$ or $x = 1$. The solutions are -2 and 1.

49. Let $t^{-1} = x$ and $t^{-2} = x^2$. Then the equation becomes:

$$x^2 - 7x + 12 = 0$$

$$(x - 3)(x - 4) = 0$$

$$x = 3, 4$$

So $t^{-1} = 3$ or $t^{-1} = 4$, thus $t = \frac{1}{3}$ or $t = \frac{1}{4}$. The solutions are $\frac{1}{3}$ and $\frac{1}{4}$.

51. Letting $y^{-1} = t$ and $y^{-2} = t^2$, we have the equation:

$$12t^2 - 23t + 5 = 0$$

$$(3t - 5)(4t - 1) = 0$$

$$t = \tfrac{5}{3}, \tfrac{1}{4}$$

So $y^{-1} = \frac{5}{3}$ or $y^{-1} = \frac{1}{4}$, thus $y = \frac{3}{5}$ or $y = 4$. The solutions are $\frac{3}{5}$ and 4.

53. Letting $x^{-2} = t$ and $x^{-4} = t^2$, we have the equation:

$$4t^2 - 33t - 27 = 0$$

$$(4t + 3)(t - 9) = 0$$

$$t = -\tfrac{3}{4}, 9$$

So $x^{-2} = -\frac{3}{4}$ or $x^{-2} = 9$. If $x^{-2} = -\frac{3}{4}$ then $x^2 = -\frac{4}{3}$, which is impossible for real values of x since $x^2 \geq 0$. If $x^{-2} = 9$ then $x^2 = \frac{1}{9}$ and thus $x = \pm\frac{1}{3}$. The solutions are $\pm\frac{1}{3}$.

55. Raising both sides of the given equation to the power $\frac{3}{2}$ yields:

$$t = \pm 9^{3/2} = \pm\left(\sqrt{9}\right)^3 = \pm 27$$

Since $t = \pm 27$ both satisfy the original equation, the solutions are ± 27.

57. Taking fourth roots, we have $x = \pm\sqrt[4]{81} = \pm 3$. Since $x = \pm 3$ both satisfy the original equation, the solutions are ± 3.

59. Taking cube roots:

$$(y-1)^3 = 7$$
$$y - 1 = \sqrt[3]{7}$$
$$y = 1 + \sqrt[3]{7}$$

The solution is $1 + \sqrt[3]{7}$.

61. Taking fifth roots:

$$(t+1)^5 = -243$$
$$t + 1 = \sqrt[5]{-243}$$
$$t + 1 = -3$$
$$t = -4$$

The solution is -4.

63. Letting $x^{2/3} = t$ and $x^{4/3} = t^2$, the equation becomes:

$$9t^2 - 10t + 1 = 0$$
$$(t-1)(9t-1) = 0$$
$$t = 1, \tfrac{1}{9}$$

So $x^{2/3} = 1$ or $x^{2/3} = \frac{1}{9}$. Thus we have either of the two equations:

$$x^{2/3} = 1$$
$$\left(\sqrt[3]{x}\right)^2 = 1$$
$$\sqrt[3]{x} = \pm 1$$
$$x = \pm 1$$

$$x^{2/3} = \tfrac{1}{9}$$
$$\left(\sqrt[3]{x}\right)^2 = \tfrac{1}{9}$$
$$\sqrt[3]{x} = \pm\tfrac{1}{3}$$
$$x = \pm\tfrac{1}{27}$$

The solutions are ± 1 and $\pm\frac{1}{27}$.

65. Set $y = 0$ to produce the equation $x^4 - 2x^2 - 1 = 0$. Using the quadratic formula:

$$x^2 = \frac{-(-2) \pm \sqrt{(-2)^2 - 4(1)(-1)}}{2(1)} = \frac{2 \pm \sqrt{4+4}}{2} = \frac{2 \pm \sqrt{8}}{2} = \frac{2 \pm 2\sqrt{2}}{2} = 1 \pm \sqrt{2}$$

Since $1 - \sqrt{2} < 0$, $x^2 = 1 - \sqrt{2}$ is impossible. Thus:

$$x^2 = 1 + \sqrt{2}$$
$$x = \pm\sqrt{1 + \sqrt{2}}$$
$$x \approx \pm 1.55$$

Note that these two intercepts are consistent with the graph.

67. Squaring both sides yields:
$$\sqrt{1-3x} = 2$$
$$1 - 3x = 4$$
$$-3x = 3$$
$$x = -1$$
Upon checking, -1 is the solution.

69. Isolating the radical and squaring yields:
$$\sqrt{x} + 6 = x$$
$$\sqrt{x} = x - 6$$
$$x = x^2 - 12x + 36$$
$$0 = x^2 - 13x + 36$$
$$0 = (x - 9)(x - 4)$$
$$x = 4, 9$$
Upon checking, 9 is the solution ($x = 4$ does not check).

71. Isolating the radical and squaring yields:
$$x - \sqrt{3 - x} = -3$$
$$x + 3 = \sqrt{3 - x}$$
$$x^2 + 6x + 9 = 3 - x$$
$$x^2 + 7x + 6 = 0$$
$$(x + 6)(x + 1) = 0$$
$$x = -1, -6$$
Upon checking, -1 is the solution ($x = -6$ does not check).

73. Isolating the radical and squaring yields:
$$4x + \sqrt{2x + 5} = 0$$
$$\sqrt{2x + 5} = -4x$$
$$2x + 5 = 16x^2$$
$$0 = 16x^2 - 2x - 5$$
$$0 = (8x - 5)(2x + 1)$$
$$x = -\tfrac{1}{2}, \tfrac{5}{8}$$
Upon checking, $-\tfrac{1}{2}$ is the solution ($x = \tfrac{5}{8}$ does not check).

75. Letting $x^2 = t$ and $x^4 = t^2$:
$$\sqrt{t^2 - 13t + 37} = 1$$
$$t^2 - 13t + 37 = 1$$
$$t^2 - 13t + 36 = 0$$
$$(t - 9)(t - 4) = 0$$
$$t = 4, 9$$
Since $x^2 = t, x^2 = 4$ or $x^2 = 9$, thus $x = \pm 2$ or $x = \pm 3$. Upon checking, ± 2 and ± 3 are solutions.

77. Isolating $\sqrt{1-2x}$ and squaring yields:

$$\sqrt{1-2x} = 4 - \sqrt{x+5}$$
$$1 - 2x = 16 - 8\sqrt{x+5} + x + 5$$
$$8\sqrt{x+5} = 3x + 20$$
$$64(x+5) = 9x^2 + 120x + 400$$
$$0 = 9x^2 + 56x + 80$$
$$0 = (x+4)(9x+20)$$
$$x = -4, -\tfrac{20}{9}$$

Upon checking, -4 and $-\tfrac{20}{9}$ are solutions.

79. Isolating $\sqrt{3+2t}$ and squaring yields:

$$\sqrt{3+2t} = 1 - \sqrt{-1+4t}$$
$$3 + 2t = 1 - 2\sqrt{-1+4t} - 1 + 4t$$
$$2\sqrt{-1+4t} = 2t - 3$$
$$4(-1+4t) = 4t^2 - 12t + 9$$
$$0 = 4t^2 - 28t + 13$$
$$0 = (2t-1)(2t-13)$$
$$t = \tfrac{1}{2}, \tfrac{13}{2}$$

Upon checking, neither of these values are solutions.

81. Isolating $\sqrt{3y+3}$ and squaring yields:

$$\sqrt{2y-3} + \sqrt{3y-2} = \sqrt{3y+3}$$
$$2y - 3 + 2\sqrt{2y-3}\sqrt{3y-2} + 3y - 2 = 3y + 3$$
$$2\sqrt{6y^2 - 13y + 6} = -2y + 8$$
$$\sqrt{6y^2 - 13y + 6} = -y + 4$$
$$6y^2 - 13y + 6 = y^2 - 8y + 16$$
$$5y^2 - 5y - 10 = 0$$
$$y^2 - y - 2 = 0$$
$$(y-2)(y+1) = 0$$
$$y = -1, 2$$

Upon checking, 2 is the solution ($y = -1$ does not check).

83. Letting $x^2 = t$ and $x^4 = t^2$:

$$t^2 - 2t - 4 = 0$$
$$t = \frac{2 \pm \sqrt{20}}{2} = \frac{2 \pm 2\sqrt{5}}{2} = 1 \pm \sqrt{5}$$

We want the positive value for t here, because $t = x^2$. Thus:

$$x^2 = 1 + \sqrt{5}$$
$$x = \pm\sqrt{1 + \sqrt{5}}$$

Using a calculator and rounding to two decimal places, we find these values of x to be ± 1.80.

85. Squaring both sides yields:

$$\sqrt{\sqrt{x}+\sqrt{a}}+\sqrt{\sqrt{x}-\sqrt{a}}=\sqrt{2\sqrt{x}+2\sqrt{b}}$$

$$\left(\sqrt{\sqrt{x}+\sqrt{a}}+\sqrt{\sqrt{x}-\sqrt{a}}\right)^2=\left(\sqrt{2\sqrt{x}+2\sqrt{b}}\right)^2$$

$$\sqrt{x}+\sqrt{a}+2\sqrt{\sqrt{x}+\sqrt{a}}\sqrt{\sqrt{x}-\sqrt{a}}+\sqrt{x}-\sqrt{a}=2\sqrt{x}+2\sqrt{b}$$

$$\sqrt{\sqrt{x}+\sqrt{a}}\sqrt{\sqrt{x}-\sqrt{a}}=\sqrt{b}$$

$$\left(\sqrt{x}+\sqrt{a}\right)\left(\sqrt{x}-\sqrt{a}\right)=b$$

$$x-a=b$$

$$x=a+b$$

Upon checking, we find that this value of x satisfies the original equation. *Note*: The following fact is useful in carrying out the check. Two non-negative quantities are equal if and only if their squares are equal.

87. Let $t=\sqrt{x}$. Then the given equation can be written as:

$$\frac{t-a}{t}=\frac{t+a}{t-b}$$

$$t^2+at=t^2-at-bt+ab$$

$$2at+bt=ab$$

$$t(2a+b)=ab$$

$$t=\frac{ab}{2a+b}$$

Thus $x=t^2=\dfrac{a^2b^2}{(2a+b)^2}$. This value of x checks in the original equation.

89. Let $t=x^2-x-1$. Then the given equation becomes:

$$\sqrt{t}-\frac{2}{\sqrt{t}}=1$$

$$t-2=\sqrt{t}$$

$$t^2-4t+4=t$$

$$t^2-5t+4=0$$

$$(t-4)(t-1)=0$$

$$t=4 \ \text{ or } \ t=1$$

Upon checking, $t=4$ satisfies the original equation involving t, but $t=1$ does not. With $t=4$ we have:

$$x^2-x-1=4$$

$$x^2-x-5=0$$

Now applying the quadratic formula:

$$x=\frac{1\pm\sqrt{1-4(-5)}}{2}=\frac{1\pm\sqrt{21}}{2}$$

91. With $t = \dfrac{x-a}{x}$ the given equation becomes:

$$\sqrt{t} + \frac{4}{\sqrt{t}} = 5$$
$$t + 4 = 5\sqrt{t}$$
$$t^2 + 8t + 16 = 25t$$
$$t^2 - 17t + 16 = 0$$
$$(t-1)(t-16) = 0$$
$$t = 1, 16$$

Note that both $t = 1$ and $t = 16$ satisfy the original equation involving t.
If $t = 1$:

$$1 = \frac{x-a}{x}$$
$$x = x - a$$
$$0 = -a$$

Since $a \neq 0$, we discard this case. If $t = 16$:

$$16 = \frac{x-a}{x}$$
$$16x = x - a$$
$$15x = -a$$
$$x = -\frac{a}{15}$$

93. (a)　We'll show the verfication for $x = \sqrt{3 + \sqrt{5}}$. (The other three cases are similar.)
With $x = \sqrt{3 + \sqrt{5}}$ the equation becomes:

$$\left(\sqrt{3+\sqrt{5}}\right)^4 - 6\left(\sqrt{3+\sqrt{5}}\right)^2 + 4 = 0$$
$$\left(3+\sqrt{5}\right)^2 - 6\left(3+\sqrt{5}\right) + 4 = 0$$
$$9 + 6\sqrt{5} + 5 - 18 - 6\sqrt{5} + 4 = 0$$
$$(9 + 5 - 18 + 4) + 6\sqrt{5} - 6\sqrt{5} = 0$$
$$0 = 0$$

(b)　With $x = \frac{1}{8}$:

$$4\left(\tfrac{1}{8}\right)^{4/3} + 15\left(\tfrac{1}{8}\right)^{2/3} - 4 = 0$$
$$4\left(\sqrt[3]{\tfrac{1}{8}}\right)^4 + 15\left(\sqrt[3]{\tfrac{1}{8}}\right)^2 - 4 = 0$$
$$4\left(\tfrac{1}{2}\right)^4 + 15\left(\tfrac{1}{2}\right)^2 - 4 = 0$$
$$\tfrac{4}{16} + \tfrac{15}{4} - 4 = 0$$
$$4 - 4 = 0$$

The verification for $x = -\frac{1}{8}$ is similar.

95. (a) When $x = -5$, the curve $y = \sqrt{8 + 2x}$ has no value while the curve
$y = \sqrt{15 + 3x} - \sqrt{5 + x}$ is 0. Thus the left-hand curve must be
$y = \sqrt{15 + 3x} - \sqrt{5 + x}$ and the right-hand curve must be $y = \sqrt{8 + 2x}$.

(b) The root of the given equation will be the x-coordinate of the intersection point of
these two curves. This appears to be approximately -3.6.

(c) Solve for x by squaring each side of the equation:
$$\sqrt{8 + 2x} = \sqrt{15 + 3x} - \sqrt{5 + x}$$
$$8 + 2x = 15 + 3x - 2\sqrt{(15 + 3x)(5 + x)} + 5 + x$$
$$8 + 2x = 20 + 4x - 2\sqrt{3(5 + x)^2}$$
$$-12 - 2x = -2(5 + x)\sqrt{3}$$
$$6 + x = 5\sqrt{3} + x\sqrt{3}$$
$$6 - 5\sqrt{3} = \left(\sqrt{3} - 1\right)x$$
$$x = \frac{6 - 5\sqrt{3}}{\sqrt{3} - 1} \cdot \frac{\sqrt{3} + 1}{\sqrt{3} + 1} = \frac{\sqrt{3} - 9}{2}$$
$$x \approx -3.634$$

97. Letting $x^2 + x = t$, the given equation becomes:
$$\sqrt{t} + \frac{1}{\sqrt{t}} = \frac{5}{2}$$
$$2t + 2 = 5\sqrt{t}$$
$$4t^2 + 8t + 4 = 25t$$
$$4t^2 - 17t + 4 = 0$$
$$(4t - 1)(t - 4) = 0$$
$$t = \tfrac{1}{4} \text{ or } t = 4$$

Both of these values for t check in the equation $\sqrt{t} + \dfrac{1}{\sqrt{t}} = \dfrac{5}{2}$.

If $t = \tfrac{1}{4}$:
$$x^2 + x = \tfrac{1}{4}$$
$$4x^2 + 4x - 1 = 0$$
Using the quadratic formula yields:
$$x = \frac{-4 \pm \sqrt{16 - 4(4)(-1)}}{2(4)} = \frac{-4 \pm \sqrt{32}}{8} = \frac{-4 \pm 4\sqrt{2}}{8} = \frac{-1 \pm \sqrt{2}}{2} \approx 0.21, -1.21$$

If $t = 4$:
$$x^2 + x = 4$$
$$x^2 + x - 4 = 0$$
Using the quadratic formula yields:
$$x = \frac{-1 \pm \sqrt{1 - 4(1)(-4)}}{2(1)} = \frac{-1 \pm \sqrt{17}}{2} \approx 1.56, -2.56$$
The solutions are 0.21, -1.21, 1.56, and -2.56.

99. If we rewrite using rational exponents:

$$(a+x)^{2/m} + 2(a-x)^{2/m} = 3(a+x)^{1/m}(a-x)^{1/m}$$

$$(a+x)^{2/m} - 3(a+x)^{1/m}(a-x)^{1/m} + 2(a-x)^{2/m} = 0$$

$$\left[(a+x)^{1/m} - 2(a-x)^{1/m}\right]\left[(a+x)^{1/m} - (a-x)^{1/m}\right] = 0$$

Setting the first factor equal to 0 yields:

$$(a+x)^{1/m} = 2(a-x)^{1/m}$$

$$a+x = 2^m(a-x)$$

$$a+x = 2^m a - 2^m x$$

$$(2^m+1)x = a(2^m-1)$$

$$x = \frac{a(2^m-1)}{2^m+1}$$

Setting the second factor equal to 0 yields:

$$(a+x)^{1/m} = (a-x)^{1/m}$$

$$a+x = a-x$$

$$2x = 0$$

$$x = 0$$

The solutions are $x = 0$ and $x = \dfrac{a(2^m-1)}{2^m+1}$.

<u>Graphing Utility Exercises for Section 3.2</u>

1.   (a)   Graphing the two equations:

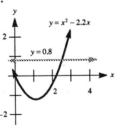

(b)   Zooming in with a new viewing rectangle:

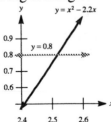

To three decimal places the root is 2.518.

(c) Using the quadratic formula to find the positive root for the equation
$x^2 - 2.2x - 0.8 = 0$:

$$x = \frac{-(-2.2) \pm \sqrt{(-2.2)^2 - 4(1)(-0.8)}}{2(1)}$$

$$= \frac{2.2 \pm \sqrt{4.84 + 3.2}}{2}$$

$$= \frac{2.2 \pm \sqrt{8.04}}{2}$$

$$\approx 2.5177$$

3.   (a) Graphing the equation:

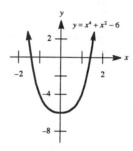

(b) There are two roots at ± 1.414. These values agree with those found in Example 3(b).

5.   (a) Graphing the curve $y = (x-1)^4 - 15$:

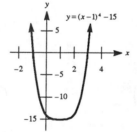

There are two roots r_1 and r_2 where $-1 < r_1 < 0$ and $2 < r_2 < 3$.

(b) To three decimal places the roots are -0.968 and 2.968.

(c) The values are $1 - \sqrt[4]{15} \approx -0.9680$ and $1 + \sqrt[4]{15} \approx 2.9680$. These values are consistent with those found in part (b).

7.   (a)   Graphing the curve $y = x^2 - 5x + 3$:

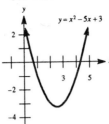

The roots are approximately 0.697 and 4.303.

(b)   Using the quadratic formula:
$$x = \frac{-(-5) \pm \sqrt{(-5)^2 - 4(1)(3)}}{2(1)} = \frac{5 \pm \sqrt{25 - 12}}{2} = \frac{5 \pm \sqrt{13}}{2} \approx 0.6972, 4.3028$$
These roots are consistent with those found in part (a).

9.   (a)   Graphing the curve $y = x - 5\sqrt{x} + 3$:

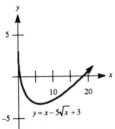

The roots are approximately 0.486 and 18.514.

(b)   Solving the equation:
$$x - 5\sqrt{x} = -3$$
$$x + 3 = 5\sqrt{x}$$
$$(x + 3)^2 = \left(5\sqrt{x}\right)^2$$
$$x^2 + 6x + 9 = 25x$$
$$x^2 - 19x + 9 = 0$$
Using the quadratic formula:
$$x = \frac{-(-19) \pm \sqrt{(-19)^2 - 4(1)(9)}}{2(1)}$$
$$= \frac{19 \pm \sqrt{361 - 36}}{2}$$
$$= \frac{19 \pm \sqrt{325}}{2}$$
$$= \frac{19 \pm 5\sqrt{13}}{2}$$
$$\approx 0.4861, 18.5139$$
These roots are consistent with those found in part (a).

11. (a) Graphing the curve $y = |x-3| - 7x$:

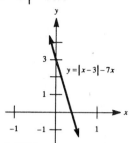

The root is approximately 0.375.

(b) Solving the equation:
$$|x-3| - 7x = 0$$
$$|x-3| = 7x$$
$$(x-3)^2 = (7x)^2$$
$$x^2 - 6x + 9 = 49x^2$$
$$0 = 48x^2 + 6x - 9$$
$$0 = 3\left(16x^2 + 2x - 3\right)$$
$$0 = 3(8x - 3)(2x + 1)$$
$$x = -\tfrac{1}{2}, \tfrac{3}{8}$$

Note that $x = -\tfrac{1}{2}$ is an extraneous solution. The root $x = \tfrac{3}{8}$ is consistent with that found in part (a).

13. (a) Graphing the curve $y = 2x - \sqrt{12x - 5}$:

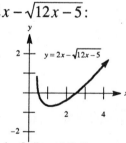

The roots are approximately 0.5 and 2.5.

(b) Solving the equation:
$$2x - \sqrt{12x - 5} = 0$$
$$2x = \sqrt{12x - 5}$$
$$4x^2 = 12x - 5$$
$$4x^2 - 12x + 5 = 0$$
$$(2x - 5)(2x - 1) = 0$$
$$x = \tfrac{1}{2}, \tfrac{5}{2}$$

These roots are consistent with those found in part (a).

15.   (a) $(x-2)^3 - 4 = 0$

   (a)   Graphing the curve $y = (x-2)^3 - 4$:

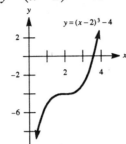

   The root is approximately 3.587.

   (b)   Solving the equation:
   $$(x-2)^3 - 4 = 0$$
   $$(x-2)^3 = 4$$
   $$x - 2 = \sqrt[3]{4}$$
   $$x = 2 + \sqrt[3]{4} \approx 3.5874$$
   This root is consistent with that found in part (a).

   (b) $(x-2)^4 - 4 = 0$

   (a)   Graphing the curve $y = (x-2)^4 - 4$:

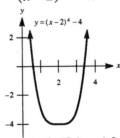

   The roots are approximately 0.586 and 3.414.

   (b)   Solving the equation:
   $$(x-2)^4 - 4 = 0$$
   $$(x-2)^4 = 4$$
   $$x - 2 = \pm\sqrt[4]{4}$$
   $$x - 2 = \pm\sqrt{2}$$
   $$x = 2 \pm \sqrt{2} \approx 0.5858,\ 3.4142$$
   These roots are consistent with those found in part (a).

17.   (a)   Graphing the curve $y = \sqrt{2x-1} - \sqrt{x-2} - 2$:

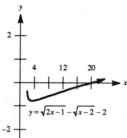

The root is approximately 19.944.

(b)   Solving the equation:

$$\sqrt{2x-1} - \sqrt{x-2} = 2$$
$$\sqrt{2x-1} = 2 + \sqrt{x-2}$$
$$2x-1 = 4 + 4\sqrt{x-2} + x - 2$$
$$2x-1 = x + 2 + 4\sqrt{x-2}$$
$$x - 3 = 4\sqrt{x-2}$$
$$(x-3)^2 = 16(x-2)$$
$$x^2 - 6x + 9 = 16x - 32$$
$$x^2 - 22x + 41 = 0$$

Using the quadratic formula:

$$x = \frac{-(-22) \pm \sqrt{(-22)^2 - 4(1)(41)}}{2(1)}$$
$$= \frac{22 \pm \sqrt{484 - 164}}{2}$$
$$= \frac{22 \pm \sqrt{320}}{2}$$
$$= \frac{22 \pm 8\sqrt{5}}{2}$$
$$= 11 \pm 4\sqrt{5}$$
$$\approx 2.0557, 19.9443$$

Since $x \approx 2.0557$ is extraneous, the root is $x \approx 19.9443$. This root is consistent with that found in part (a).

19. (a) Graphing the curve $y = \dfrac{\sqrt{x}-4}{\sqrt{x}+3} + \dfrac{1}{2}$:

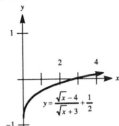

The root is approximately 2.778.

(b) Solving the equation:

$$\frac{\sqrt{x}-4}{\sqrt{x}+3} + \frac{1}{2} = 0$$

$$\frac{\sqrt{x}-4}{\sqrt{x}+3} = -\frac{1}{2}$$

$$2\sqrt{x} - 8 = -\sqrt{x} - 3$$

$$3\sqrt{x} = 5$$

$$9x = 25$$

$$x = \tfrac{25}{9}$$

This root is consistent with that found in part (a).

21. (a) Graphing the curve $y = x^{2/3} - x^{1/3} - 1$:

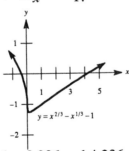

The roots are approximately –0.236 and 4.236.

(b) Solving the equation $x^{2/3} - x^{1/3} - 1 = 0$, use the quadratic formula:

$$x^{1/3} = \frac{-(-1) \pm \sqrt{(-1)^2 - 4(1)(-1)}}{2(1)} = \frac{1 \pm \sqrt{1+4}}{2} = \frac{1 \pm \sqrt{5}}{2}$$

Cubing each side, we have two roots:

$$x = \left(\frac{1 - \sqrt{5}}{2} \right)^3 \approx -0.2361$$

$$x = \left(\frac{1 + \sqrt{5}}{2} \right)^3 \approx 4.2361$$

These roots are consistent with those found in part (a).

3.3 Applied Problems (A Warmup)

1. (a) The problem asks us to find two numbers. Let x and y denote the two numbers.

(b) Complete the table:

In English	In algebraic notation
Find two numbers.	x and y
Their sum is -9.	$x + y = -9$
The sum of their squares is 153.	$x^2 + y^2 = 153$

(c) From part (b) we have the equations:
$$x + y = -9$$
$$x^2 + y^2 = 153$$
From the first equation, $y = -9 - x$, and therefore:
$$x^2 + (-9 - x)^2 = 153$$
$$x^2 + 81 + 18x + x^2 = 153$$
$$2x^2 + 18x - 72 = 0$$
$$x^2 + 9x - 36 = 0$$

(d) Solving the equation:
$$x^2 + 9x - 36 = 0$$
$$(x + 12)(x - 3) = 0$$
$$x = -12, 3$$

(e) The two numbers are -12 and 3. Note that $-12 + 3 = 9$ and $(-12)^2 + 3^2 = 153$.

3. (a) The problem asks us to find two numbers. Let x and y denote the two numbers.

(b) Complete the table:

In English	In algebraic notation
Find two numbers.	x and y
Their sum is 17.	$x + y = 17$
Their product is 52.	$xy = 52$

(c)   From part (b) we have the equations:
$$x + y = 17$$
$$xy = 52$$
From the first equation, $y = 17 - x$, and therefore:
$$x(17 - x) = 52$$
$$17x - x^2 = 52$$
$$-x^2 + 17x - 52 = 0$$
$$x^2 - 17x + 52 = 0$$

(d)   Solving the equation:
$$x^2 - 17x + 52 = 0$$
$$(x - 13)(x - 4) = 0$$
$$x = 4, 13$$

(e)   The two numbers are 4 and 13. Note that $4 + 13 = 17$ and $4 \cdot 13 = 52$.

5.   (a)   The problem asks us to find two positive consecutive integers. Let x and $x + 1$ denote these consecutive integers.

(b)   Complete the table:

In English	In algebraic notation
Find two consecutive integers.	x and $x + 1$
The sum of their squares is 365.	$x^2 + (x + 1)^2 = 365$

(c)   Working from the equation $x^2 + (x + 1)^2 = 365$ obtained in part (b):
$$x^2 + x^2 + 2x + 1 = 365$$
$$2x^2 + 2x - 364 = 0$$
$$x^2 + x - 182 = 0$$

(d)   Factoring:
$$x^2 + x - 182 = 0$$
$$(x + 14)(x - 13) = 0$$
$$x = -14, 13$$
Since $x > 0$, $x = 13$.

(e)   The two numbers are 13 and 14. Note that $13^2 + 14^2 = 365$.

7.   (a)   The problem asks us to find two numbers. Let x and y denote the two numbers.

(b)   Complete the table:

In English	In algebraic notation
Find two numbers.	x and y
Their average is 4.	$\frac{1}{2}(x+y)=4$
The average of their squares is 80.	$\frac{1}{2}(x^2+y^2)=80$

(c)   From part (b) we have the equations
$$\frac{1}{2}(x+y)=4$$
$$\frac{1}{2}(x^2+y^2)=80$$
From the first equation $x+y=8$, so $y=8-x$, and therefore:
$$\frac{1}{2}\left[x^2+(8-x)^2\right]=80$$
$$x^2+64-16x+x^2=160$$
$$2x^2-16x-96=0$$
$$x^2-8x-48=0$$

(d)   Solving the equation:
$$x^2-8x-48=0$$
$$(x-12)(x+4)=0$$
$$x=-4,12$$

(e)   The two numbers are –4 and 12. Note that their average is 4 and the average of their squares is 80.

9.   (a)   The problem is asking for the dimensions of the rectangle. Label the length l and the width w, and draw the sketch:

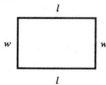

(b)   Complete the table:

In English	In algebraic notation
Find the width and length of a rectangle.	w and l
The perimeter is 40 cm.	$2w+2l=40$
The area is 64 cm^2.	$lw=64$

(c)   From part (b) we have the equations:
$$2w + 2l = 40$$
$$lw = 64$$
From the first equation we have $w + l = 20$, so $l = 20 - w$, and therefore:
$$(20 - w)w = 64$$
$$20w - w^2 = 64$$
$$-w^2 + 20w - 64 = 0$$
$$w^2 - 20w + 64 = 0$$

(d)   Solving the equation:
$$w^2 - 20w + 64 = 0$$
$$(w - 4)(w - 16) = 0$$
$$w = 4, 16$$

(e)   The width is 4 cm and the length is 16 cm. Note that the perimeter is 40 cm and the area is 64 cm^2.

11.   (a)   The problem asks for the number of years required until the total interest earned on all three accounts is \$8,760. Let t denote the required number of years.

(b)   Apply the formula $I = Prt$ to each account. After t years, the interest on the first account is $(3000)(0.07)t$, the interest on the second account is $(5000)(0.08)t$, and the interest on the third account is $(10000)(0.085)t$. Since we want the sum of these three amounts to be \$8,760, the required equation is:
$$(3000)(0.07)t + (5000)(0.08)t + (10000)(0.085)t = 8760$$

(c)   Solving the equation:
$$(3000)(0.07)t + (5000)(0.08)t + (10000)(0.085)t = 8760$$
$$210t + 400t + 850t = 8760$$
$$1460t = 8760$$
$$t = 6$$

(d)   After 6 years the total interest will amount to \$8,760. Note that this time satisfies the original problem.

13.   (a)   Let r represent the interest rate for the smaller account (as a percent), so $r + 1$ will be the rate for the larger account. Applying the formula $I = Prt$, the first account will generate $(7400)\left(\frac{r}{100}\right)(10)$ in interest, so the total amount in the account (including the initial principal) is $7400 + (7400)\left(\frac{r}{100}\right)(10) = 7400\left(1 + \frac{r}{10}\right)$. Similarly, the second account will generate $(12600)\left(\frac{1+r}{100}\right)(10)$ in interest, so the total amount in the account is $12600 + (12600)\left(\frac{1+r}{100}\right)(10) = 12600\left(1 + \frac{1+r}{10}\right)$. Since we want the sum of these amounts to be \$31,260, the required equation is:
$$7400\left(1 + \frac{r}{10}\right) + 12600\left(1 + \frac{1+r}{10}\right) = 31260$$

(b) Solving the equation:

$$7400\left(1+\tfrac{r}{10}\right)+12600\left(1+\tfrac{1+r}{10}\right)=31260$$
$$7400+740r+12600+1260+1260r=31260$$
$$2000r+21260=31260$$
$$2000r=10000$$
$$r=5$$

(c) The interest rate is 5%. Note that this rate satisfies the original problem.

15. (a) The problem is asking you to find a and b.

(b) Completing the table:

In English	In algebraic notation
Find two numbers.	a and b
(a,b) is in the first quadrant.	$a>0, \quad b>0$
(a,b) is on the line $y=2x+3$.	$b=2a+3$
The distance from (a,b) to $(0,0)$ is 5.	$\sqrt{a^2+b^2}=5$

(c) Substituting for b:

$$\sqrt{a^2+b^2}=5$$
$$\sqrt{a^2+(2a+3)^2}=5$$
$$\sqrt{a^2+4a^2+12a+9}=5$$
$$\sqrt{5a^2+12a+9}=5$$

(d) Solving the equation:

$$\sqrt{5a^2+12a+9}=5$$
$$5a^2+12a+9=25$$
$$5a^2+12a-16=0$$

Using the quadratic formula:

$$a=\frac{-12\pm\sqrt{(12)^2-4(5)(-16)}}{2(5)}$$
$$=\frac{-12\pm\sqrt{144+320}}{10}$$
$$=\frac{-12\pm\sqrt{464}}{10}$$
$$=\frac{-12\pm4\sqrt{29}}{10}$$
$$=\frac{-6\pm2\sqrt{29}}{5}$$

Since $a>0$, we have $a=\dfrac{-6+2\sqrt{29}}{5}\approx0.954$.

(e) Finding b:

$$b = 2a + 3 = 2\left(\frac{-6 + 2\sqrt{29}}{5}\right) + 3 = \frac{-12 + 4\sqrt{29}}{5} + \frac{15}{5} = \frac{3 + 4\sqrt{29}}{5} \approx 4.908$$

The point (a, b) is $\left(\dfrac{-6 + 2\sqrt{29}}{5}, \dfrac{3 + 4\sqrt{29}}{5}\right) \approx (0.954, 4.908)$.

17. (a) The problem is asking you to find a and b.

(b) Completing the table:

In English	In algebraic notation
Find two numbers.	a and b
(a,b) is in the third quadrant.	$a < 0, \quad b < 0$
(a,b) is on the graph of $y = \dfrac{1}{x}$.	$b = \dfrac{1}{a}$
The distance from (a,b) to $(0,0)$ is $\dfrac{\sqrt{257}}{4}$.	$\sqrt{a^2 + b^2} = \dfrac{\sqrt{257}}{4}$

(c) Substituting for b:

$$\sqrt{a^2 + b^2} = \frac{\sqrt{257}}{4}$$

$$\sqrt{a^2 + \frac{1}{a^2}} = \frac{\sqrt{257}}{4}$$

(d) Solving the equation:

$$\sqrt{a^2 + \frac{1}{a^2}} = \frac{\sqrt{257}}{4}$$

$$a^2 + \frac{1}{a^2} = \frac{257}{16}$$

$$16a^4 + 16 = 257a^2$$

$$16a^4 - 257a^2 + 16 = 0$$

$$\left(16a^2 - 1\right)\left(a^2 - 16\right) = 0$$

$$(4a + 1)(4a - 1)(a + 4)(a - 4) = 0$$

$$a = -\tfrac{1}{4}, \tfrac{1}{4}, -4, 4$$

Since $a < 0$, we have $a = -\tfrac{1}{4}$ or $a = -4$.

(e) Finding b for each value of a:

$$a = -\tfrac{1}{4}: \quad b = \frac{1}{-1/4} = -4$$

$$a = -4: \quad b = \frac{1}{-4} = -\tfrac{1}{4}$$

The point (a, b) is either $\left(-\tfrac{1}{4}, -4\right)$ or $\left(-4, -\tfrac{1}{4}\right)$.

19. (a) Since $A = \tfrac{1}{2}bh$, and $b + h = 7$, solve to get $h = 7 - b$ and substitute:

$$A = \tfrac{1}{2}bh = \tfrac{1}{2}b(7 - b) = \tfrac{7}{2}b - \tfrac{1}{2}b^2$$

(b) Solving for b:

$$\tfrac{7}{2}b - \tfrac{1}{2}b^2 = A$$

$$7b - b^2 = 2A$$

$$0 = b^2 - 7b + 2A$$

Using the quadratic formula:

$$b = \frac{-(-7) \pm \sqrt{(-7)^2 - 4(1)(2A)}}{2(1)} = \frac{7 \pm \sqrt{49 - 8A}}{2}$$

(c) Completing the table:

b	1	2	3	4	5	6
A	3	5	6	6	5	3

(d) The values $b = 3$ and $b = 4$ yield the largest area.

(e) When $b = 3.5$ cm, we have $A = 6.125$ cm^2.

(f) No. Using the expression obtained in part (b), note that:

$$49 - 8A \geq 0$$

$$49 \geq 8A$$

$$A \leq \tfrac{49}{8} = 6.125$$

Thus $A > 6.5$ would be impossible.

21. (a) Since the height is four times the radius, we can write $h = 4r$ where h represents the height and r represents the radius.

(b) The general formula for the volume is $V = \pi r^2 h$. Replacing h with $4r$, we obtain:

$$V = \pi r^2(4r) = 4\pi r^3$$

(c) Since $h = 4r$, $r = \tfrac{h}{4}$. Replacing r with $\tfrac{h}{4}$:

$$V = \pi\left(\tfrac{h}{4}\right)^2 \cdot h = \tfrac{1}{16}\pi h^3$$

23.   (a)   Since the sum of the radius r and the height h is 4 cm, $r + h = 4$.

(b)   Since $r + h = 4$, $h = 4 - r$. Using the general formula for the volume, $V = \pi r^2 h$, and replacing h with $4 - r$:
$$V = \pi r^2 (4 - r)$$

(c)   Complete the table:

r	1.00	1.50	2.00	2.50	3.00	3.50
V	9.42	17.67	25.13	29.45	28.27	19.24

(d)   From the table, $r = 2.5$ cm yields the largest volume.

(e)   When $r = 2.6$ cm, $V \approx 29.73$ cm^3. When $r = 2.7$ cm, $V \approx 29.77$ cm^3.

(f)   No. The largest value for V is approximately 29.77 cm^3, occurring when $r = 2.7$ cm.

25.   (a)   Since the volume is given by $V = \frac{1}{3}\pi r^2 h$, we have $\pi = \frac{1}{3}\pi r^2 h$, so $r^2 h = 3$.

(b)   Solving the equation from part (a) for h, $h = \frac{3}{r^2}$. Using the formula for the lateral surface area S:
$$S = \pi r \sqrt{r^2 + h^2} = \pi r \sqrt{r^2 + \left(\frac{3}{r^2}\right)^2} = \pi r \sqrt{r^2 + \frac{9}{r^4}} = \frac{\pi r \sqrt{r^6 + 9}}{r^2} = \frac{\pi}{r}\sqrt{r^6 + 9}$$

(c)   Complete the table:

r	0.250	0.500	0.750	1.000	1.250	1.500	1.750	2.000
S	37.700	18.866	12.690	9.935	8.997	9.457	11.026	13.421

(d)   The least surface area occurs when $r = 1.25$ cm.

(e)   When $r = 1.30$ cm, we have $S \approx 8.986$ cm^2.

(f)   Yes. When $r = 1.29$ cm, we have $S \approx 8.984$ cm^2.

27.   (a)   Solving by factoring:
$$x^2 - 8x - 84 = 0$$
$$(x - 14)(x + 6) = 0$$
$$x = -6, 14$$

(b) Using the quadratic formula:

$$x = \frac{-(-8) \pm \sqrt{(-8)^2 - 4(1)(-84)}}{2(1)}$$

$$= \frac{8 \pm \sqrt{64 + 336}}{2}$$

$$= \frac{8 \pm \sqrt{400}}{2}$$

$$= \frac{8 \pm 20}{2}$$

$$= -6, 14$$

Our solutions agree with those in part (a).

29. (a) $(-12, -5)$ is in the third quadrant.

(b) Substituting $x = -12$:

$$y = \tfrac{1}{2}(-12) + 1 = -6 + 1 = -5$$

So $(-12, -5)$ lies on the line $y = \tfrac{1}{2}x + 1$.

(c) Using the distance formula:

$$\sqrt{(-12 - 0)^2 + (-5 - 0)^2} = \sqrt{144 + 25} = \sqrt{169} = 13$$

Graphing Utility Exercises for Section 3.3

1. (a) Graphing the two curves:

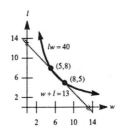

(b) The intersection points are $(5, 8)$ and $(8, 5)$, which agree with Example 2 in the text.

(c) Graphing the equation:

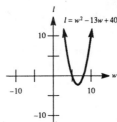

The w-intercepts are 5 and 8. They are the same as the w-coordinates of the intersection points.

3.   Graphing the line $y = 3x - 1$ and the circle $x^2 + y^2 = 9^2$:

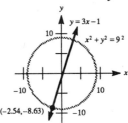

The intersection point is approximately $(-2.54, -8.63)$. Substituting $y = 3x - 1$ into the equation of the circle:

$$x^2 + y^2 = 81$$
$$x^2 + (3x - 1)^2 = 81$$
$$x^2 + 9x^2 - 6x + 1 = 81$$
$$10x^2 - 6x - 80 = 0$$
$$5x^2 - 3x - 40 = 0$$

Using the quadratic formula:

$$x = \frac{-(-3) \pm \sqrt{(-3)^2 - 4(5)(-40)}}{2(5)} = \frac{3 \pm \sqrt{9 + 800}}{10} = \frac{3 \pm \sqrt{809}}{10} \approx -2.544, \, 3.144$$

Since $x < 0$ in the third quadrant, $x = \dfrac{3 - \sqrt{809}}{10} \approx -2.544$. Finding y:

$$y = 3\left(\frac{3 - \sqrt{809}}{10}\right) - 1 = \frac{9 - 3\sqrt{809}}{10} - 1 = \frac{-1 - 3\sqrt{809}}{10} \approx -8.633$$

The intersection point is $\left(\dfrac{3 - \sqrt{809}}{10}, \dfrac{-1 - 3\sqrt{809}}{10}\right) \approx (-2.544, -8.633)$. This agrees with the intersection point found graphically.

5.   Graphing the curve $y = \dfrac{1}{x}$ and the circle $x^2 + y^2 = 6^2$:

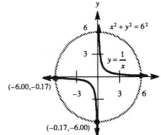

The intersection points are approximately $(-0.17, -6.00)$ and $(-6.00, -0.17)$.

Substituting $y = \dfrac{1}{x}$ into the equation of the circle:

$$x^2 + y^2 = 36$$

$$x^2 + \frac{1}{x^2} = 36$$

$$x^4 + 1 = 36x^2$$

$$x^4 - 36x^2 + 1 = 0$$

Using the quadratic formula:

$$x^2 = \frac{-(-36) \pm \sqrt{(-36)^2 - 4(1)(1)}}{2(1)}$$

$$= \frac{36 \pm \sqrt{1296 - 4}}{2}$$

$$= \frac{36 \pm \sqrt{1292}}{2}$$

$$= \frac{36 \pm 2\sqrt{323}}{2}$$

$$= 18 \pm \sqrt{323}$$

Since $x < 0$ in the third quadrant:

$$x = -\sqrt{18 \pm \sqrt{323}} \approx -0.167, \, -5.998$$

Finding y:

$$y = \frac{1}{-\sqrt{18 - \sqrt{323}}} \approx -5.998$$

$$y = \frac{1}{-\sqrt{18 + \sqrt{323}}} \approx -0.167$$

The intersection points are $\left(-\sqrt{18 - \sqrt{323}}, \, \dfrac{1}{-\sqrt{18 - \sqrt{323}}} \right) \approx (-0.167, -5.998)$ and

$\left(-\sqrt{18 + \sqrt{323}}, \, \dfrac{1}{-\sqrt{18 + \sqrt{323}}} \right) \approx (-5.998, -0.167)$. These agree with the intersection points found graphically.

7.   Graphing the curve $y = \dfrac{5}{x}$ and the circle $x^2 + y^2 = 3^2$:

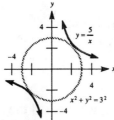

Note that the two curves do not intersect.

9.   (a)   Graphing the curve:

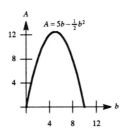

$$A = 5b - \tfrac{1}{2}b^2$$

(b)   Completing the table:

b	0	1	2	3	4	5	6	7	8	9	10
A	0	4.5	8	10.5	12	12.5	12	10.5	8	4.5	0

(c)   When $b = 5$ the area is the largest. This agrees with the graph, which has a maximum value at the point (5, 12.5).

3.4 Applications

1.   Denote the two numbers by x and $7 - x$. Since the difference is 17:
$$x - (7 - x) = 17$$
$$2x = 24$$
$$x = 12$$
Then $7 - x = 7 - 12 = -5$. The two numbers are 12 and -5.

3.   Denoting the two numbers by x and $17 - x$:
$$x(17 - x) = 52$$
$$x^2 - 17x + 52 = 0$$
$$(x - 13)(x - 4) = 0$$
$$x = 13 \text{ or } x = 4$$
If $x = 13$, then $17 - x = 4$. Similarly, if $x = 4$, then $17 - x = 13$. So in either case, the two numbers are 13 and 4.

5.   Let x denote the number. Then:
$$x + 2x = 63$$
$$3x = 63$$
$$x = 21$$
The number is 21.

7.   Let x denote the positive number. Then:
$$x(x + 4) = 96$$
$$x^2 + 4x - 96 = 0$$
$$(x + 12)(x - 8) = 0$$
$$x = -12 \text{ or } x = 8$$
Since x is supposed to be positive, we discard the first solution. The required positive number is 8.

9. Letting x denote the required number:
$$\frac{3+x}{7+x} = \frac{2}{3}$$
$$9 + 3x = 14 + 2x$$
$$x = 14 - 9$$
$$x = 5$$
The required number is 5.

11. Call the numbers x and y. Then we have $x - y = \frac{1}{3}(x + y)$. Solving this for x in terms of y yields $x = 2y$. Since the sum of the squares is 180:
$$(2y)^2 + y^2 = 180$$
$$5y^2 = 180$$
$$y^2 = 36$$
$$y = \pm 6$$
Since y is supposed to be positive, $y = 6$ and $x = 2(6) = 12$. The two numbers are 12 and 6.

13. Letting x denote the required score:
$$\frac{70 + 77 + 75 + x}{4} = 75$$
$$222 + x = 300$$
$$x = 300 - 222$$
$$x = 78$$
The required score is 78.

15. Letting x denote the fourth number:
$$\frac{1}{4}(80 + 22 + 62 + x) = 52$$
$$164 + x = 208$$
$$x = 208 - 164$$
$$x = 44$$
The fourth number is 44.

17. Let x denote the length of the hypotenuse. Then the lengths of the other two sides are $x - 8$ and $x - 1$, so:
$$(x-1)^2 + (x-8)^2 = x^2$$
$$x^2 - 18x + 65 = 0$$
$$(x-13)(x-5) = 0$$
$$x = 13 \text{ or } x = 5$$
If $x = 13$, the other two sides are $13 - 8 = 5$, and $13 - 1 = 12$. The second solution $x = 5$ must be discarded, for in that case $x - 8$ is negative. The required lengths are 5 cm, 12 cm, and 13 cm.

19.   Letting x denote the length of a side in the original square:
$$(x+2)^2 = x^2 + 14$$
$$4x = 10$$
$$x = \tfrac{5}{2}$$
The length of a side in the original square is 2.5 cm.

21.   If x and y are the dimensions of the rectangle:
$$2x + 2y = 104$$
$$x + y = 52$$
$$y = 52 - x$$
Since the area is 640 cm^2:
$$x(52 - x) = 640$$
$$-x^2 + 52x = 640$$
$$x^2 - 52x + 640 = 0$$
$$(x - 20)(x - 32) = 0$$
$$x = 20 \ \text{ or } \ x = 32$$
Now if $x = 20$, then $y = 52 - 20 = 32$. Similarly, if $x = 32$, then $y = 52 - 32 = 20$. So in either case the shorter side is 20 cm.

23.   If the shortest side is 2 in., there are 10 in. left for the other two sides. Call those sides x and $10 - x$. Then:
$$2^2 + x^2 = (10 - x)^2$$
$$4 + x^2 = 100 - 20x + x^2$$
$$20x = 96$$
$$x = \tfrac{96}{20} = 4.8$$
The other two sides of the triangle are 4.8 in. and $10 - 4.8 = 5.2$ in.

25.   Let x denote the length of the base. Then the legs each have length $3x$, so:
$$x + 3x + 3x = 70$$
$$7x = 70$$
$$x = 10$$
The three sides are 10 cm, 30 cm, and 30 cm.

27.   Letting t represent the number of years until the total interest is $5840:
$$(4000)(0.07)t + (5000)(0.09)t = 5840$$
$$280t + 450t = 5840$$
$$730t = 5840$$
$$t = \tfrac{5840}{730} = 8$$
Thus it will take 8 years until the interest is $5840.

29. Let t be the number of years until the balance in the account reaches $21,000. Then:

$$10,000 + (10,000)(0.11)t = 21,000$$
$$10,000 + 1100t = 21,000$$
$$100 + 11t = 210$$
$$11t = 110$$
$$t = 10$$

It will take 10 years for the balance to reach $21,000.

31. Let i be the overall interest rate on the entire investment of $11,500. Then:

$$(11,500)(i)(1) = (7500)(0.075)(1) + (4000)(0.05)(1)$$
$$11500i = 762.5$$
$$i = \frac{762.5}{11500} \approx 0.06630$$

To convert this value to a percent, multiply by 100. Rounding off to the nearest tenth of one percent, we obtain $i \approx 6.6\%$.

33. Let the interest rate for the $8000 account be $r\%$. Then the rate for the $9000 account is $(r + 1)\%$. Since the $9000 account earns $165 more than the $8000 account each year:

$$(8000)\left(\tfrac{r}{100}\right) + 165 = (9000)\left(\tfrac{r+1}{100}\right)$$
$$80r + 165 = 90r + 90$$
$$-10r = -75$$
$$r = 7.5$$

The interest rate on the $8000 account is 7.5%, and the rate on the $9000 account is 8.5%.

35. Let x be the number of cc of the 10% solution, and let $200 - x$ be the number of cc of the 35% solution. Note the following table:

type of solution	cc	% acid	total acid (cc)
10% solution	x	10%	$0.10(x)$
35% solution	$200 - x$	35%	$0.35(200 - x)$
mixture	200	25%	$0.25(200)$

Since the total acid in the mixture comes from that in the two given solutions:

$$0.10(x) + 0.35(200 - x) = 0.25(200)$$
$$10x + 35(200 - x) = 25(200)$$
$$10x + 7000 - 35x = 5000$$
$$-25x = -2000$$
$$x = 80$$

Then $200 - x = 200 - 80 = 120$. The student should use 80 cm^3 of the 10% solution and 120 cm^3 of the 35% solution.

37. Let x denote the required number of pounds of the first type of coffee. Note the following table:

type of coffee	pounds	cost / pound	total cost
first type	x	$5.20	$5.20(x)$
second type	5	$5.80	$5.80(5)$
blend	$x+5$	$5.35	$5.35(x+5)$

Looking at the total cost column in our table, we obtain the equation:
$$5.20(x) + 5.80(5) = 5.35(x+5)$$
$$520x + 580(5) = 535(x+5)$$
$$520x + 2900 = 535x + 2675$$
$$-15x = -225$$
$$x = 15$$
Thus 15 pounds of the first type of coffee are required.

39. Let x denote the required number of tons of the second ore. Note the following table:

type of ore	tons	% iron	total iron (tons)
first type	16	20%	$0.20(16)$
second type	x	35%	$0.35(x)$
blend	$16+x$	25%	$0.25(16+x)$

By considering the total iron:
$$0.20(16) + 0.35(x) = 0.25(16+x)$$
$$20(16) + 35x = 25(16+x)$$
$$320 + 35x = 400 + 25x$$
$$10x = 80$$
$$x = 8$$
Thus 8 tons of the second ore are required.

41. Denoting the lengths of the two shorter sides by x and $x-7$:
$$\tfrac{1}{2}x(x-7) = 60$$
$$x^2 - 7x - 120 = 0$$
$$(x-15)(x+8) = 0$$
$$x = 15 \text{ or } x = -8$$
Discard the negative root since x represents a length. Thus $x = 15$ and $x - 7 = 8$.
Therefore the hypotenuse is:
$$\sqrt{15^2 + 8^2} = \sqrt{289} = 17 \text{ cm}$$

43. The area of the square and the rectangle are $(1-x)^2$ and $(1)(x)$, respectively. Therefore:
$$(1-x)^2 = x$$
$$x^2 - 3x + 1 = 0$$
$$x = \frac{3 \pm \sqrt{9-4}}{2} = \frac{3 \pm \sqrt{5}}{2}$$
Choose the negative root here, since the quantity $1-x$ is supposed to be positive.
Therefore $x = \frac{3-\sqrt{5}}{2}$.

45. Label point P as $P(3, y)$. Since the distance from P to $(1, 0)$ is 4 units, use the distance formula:

$$\sqrt{(3-1)^2 + (y-0)^2} = 4$$
$$\sqrt{4+y^2} = 4$$
$$4 + y^2 = 16$$
$$y^2 = 12$$
$$y = \pm\sqrt{12} = \pm 2\sqrt{3}$$

Since P is in the fourth quadrant, $y < 0$ and so we choose $y = -2\sqrt{3}$. Now find the distance from $P\left(3, -2\sqrt{3}\right)$ to the origin:

$$d = \sqrt{(3-0)^2 + \left(-2\sqrt{3}-0\right)^2} = \sqrt{9+12} = \sqrt{21}$$

47. First draw the sketch and label appropriate sides:

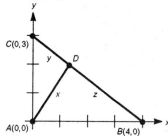

Since $BC = 5$, we have $y + z = 5$, so $y = 5 - z$. Also $x^2 + y^2 = 9$ and $x^2 + z^2 = 16$, so $x^2 = 9 - y^2$ and $x^2 = 16 - z^2$, thus $9 - y^2 = 16 - z^2$, or $z^2 - y^2 = 7$. Substituting for y:

$$z^2 - y^2 = 7$$
$$z^2 - (5-z)^2 = 7$$
$$z^2 - 25 + 10z - z^2 = 7$$
$$10z - 25 = 7$$
$$10z = 32$$
$$z = 3.2$$

So $y = 5 - 3.2 = 1.8$. Now substitute to find x:

$$x = \sqrt{9 - y^2} = \sqrt{9 - (1.8)^2} = \sqrt{5.76} = 2.4$$

So $AD = 2.4$, $BD = 3.2$, and $CD = 1.8$.
An alternate solution can be found geometrically. The area of the triangle is $\frac{1}{2}(4)(3) = 6$. But the area can also be found (using AD as the height of the triangle and BC as the base) by:

$$\tfrac{1}{2}(x)(y+z) = A$$
$$\tfrac{1}{2}(x)(5) = 6$$
$$5x = 12$$
$$x = 2.4$$

We can then find y and z:
$$y = \sqrt{9 - x^2} = \sqrt{9 - (2.4)^2} = \sqrt{3.24} = 1.8$$
$$z = 5 - y = 5 - 1.8 = 3.2$$

49. (a) When the ball lands $h = 0$. Therefore:
$$-16t^2 + 96t = 0$$
$$-16t(t - 6) = 0$$
$$t = 0 \text{ or } t = 6$$
The value $t = 0$ gives the time when the ball is first thrown. Consequently the ball lands after 6 seconds.

(b) With $h = 80$, the equation becomes:
$$80 = -16t^2 + 96t$$
$$16t^2 - 96t + 80 = 0$$
$$t^2 - 6t + 5 = 0$$
$$(t - 5)(t - 1) = 0$$
$$t = 5 \text{ or } t = 1$$
At $t = 1$ second the ball is 80 ft. high and rising. At $t = 5$ seconds the ball is 80 ft high and falling.

51. After t minutes, the distance covered by Emily and Ernie are $\frac{t}{9}$ and $\frac{t}{11}$, respectively. Emily will lap Ernie when her distance exceeds Ernie's by $\frac{1}{4}$ mile. In that case we have:
$$\frac{t}{9} = \frac{t}{11} + \frac{1}{4}$$
$$44t = 36t + 99$$
$$8t = 99$$
$$t = \frac{99}{8}$$

It will take $12\frac{3}{8}$ minutes for Emily to lap Ernie on the track. This is about 12 minutes and 22 seconds.

53. Let r and $r - 5$ denote the speeds of the eastbound and westbound trains, respectively. When $t = 6$ hours the total distance traveled by the two trains is 630 mi, so:
$$r(6) + (r - 5)6 = 630$$
$$6r + 6r - 30 = 630$$
$$12r = 660$$
$$r = 55$$
The speed of the eastbound train is 55 mph.

55. Let r and $r + 50$ represent the speeds of the slower and faster planes, respectively.
When $t = 2$ the total distance covered is 1500, so:

$$r(2) + (r + 50)2 = 1500$$
$$4r = 1400$$
$$r = 350$$

The speed of the slower plane is 350 mph.

57. Let D be the distance that the boat can travel out to sea, and let t be the time for that
part of the trip. Then the time for the return trip is $2 - t$. For the trip out to sea, we have
$D = 20t$, and for the trip back $D = 12(2 - t)$. Solving the first equation for t yields
$t = \frac{D}{20}$, and then substituting in the second equation:

$$D = 12\left(2 - \frac{D}{20}\right)$$
$$D = 24 - \frac{3D}{5}$$
$$5D = 120 - 3D$$
$$8D = 120$$
$$D = 15$$

The boat can travel 15 miles out to sea before heading back.

59. Let x represent the additional amount that must be invested. Then the total of the two
deposits and one investment (in dollars) is $2650 + 3300 + x$, or $5950 + x$. Since the
overall return on this investment is to be 10%:

$$0.1(5950 + x) = 0.081(2650) + 0.092(3300) + 0.115x$$
$$595 + 0.1x = 214.65 + 303.60 + 0.115x$$
$$-0.015x = -76.75$$
$$x \approx 5116.67$$

The additional investment at 11.5% should be $5117.

61. Let Linda's initial rate be r. When Larry has covered 2 miles, the elapsed time is
$\frac{2}{1/15} = 30$ minutes. In that 30 minutes Linda has walked a distance of $r(30)$. At 1:15
(75 minutes past noon), the distances covered by Larry and Linda are equal, and
consequently:

Larry's total distance	=	Linda's distance in first 30 min.	+	Linda's distance in next 45 min.
$\left(\frac{1}{15}\right)(75)$	=	$30r$	+	$(2r)(45)$

Therefore:

$$5 = 30r + 90r$$
$$5 = 120r$$
$$r = \frac{5}{120} = \frac{1}{24}$$

Linda's initial rate was $\frac{1}{24}$ mile/minute.

63.   (a)   Using the hint, $AP = x$, $PB = y$, and $AB = x + y$. Therefore:

$$\frac{AB}{AP} = \frac{AP}{PB}$$

$$\frac{x+y}{x} = \frac{x}{y}$$

$$xy + y^2 = x^2$$

$$y^2 + xy - x^2 = 0$$

Using the quadratic formula to solve for y:

$$y = \frac{-x \pm \sqrt{x^2 - 4(1)\left(-x^2\right)}}{2(1)}$$

$$= \frac{-x \pm \sqrt{x^2 + 4x^2}}{2}$$

$$= \frac{-x \pm \sqrt{5x^2}}{2}$$

$$= \frac{-x \pm x\sqrt{5}}{2}$$

$$= \frac{\left(-1 \pm \sqrt{5}\right)x}{2}$$

Since $x > 0$ and $y > 0$, choose $y = \dfrac{\left(-1 + \sqrt{5}\right)x}{2}$. Therefore the ratio is:

$$\frac{AP}{PB} = \frac{x}{y}$$

$$= \frac{x}{\frac{\left(-1+\sqrt{5}\right)x}{2}}$$

$$= \frac{2}{-1+\sqrt{5}} \cdot \frac{-1-\sqrt{5}}{-1-\sqrt{5}}$$

$$= \frac{-2\left(1+\sqrt{5}\right)}{1-5}$$

$$= \frac{-2\left(1+\sqrt{5}\right)}{-4}$$

$$= \frac{1+\sqrt{5}}{2}$$

(b) (i)   Substitute $\phi = \dfrac{1+\sqrt{5}}{2}$ into the equation $x^2 - x - 1 = 0$:

$$\left(\frac{1+\sqrt{5}}{2}\right)^2 - \frac{1+\sqrt{5}}{2} - 1 = \frac{1+2\sqrt{5}+5}{4} - \frac{1+\sqrt{5}}{2} - 1$$

$$= \frac{6+2\sqrt{5}}{4} + \frac{-1-\sqrt{5}}{2} - 1$$

$$= \frac{3+\sqrt{5}}{2} + \frac{-1-\sqrt{5}}{2} - \frac{2}{2}$$

$$= \frac{3+\sqrt{5}-1-\sqrt{5}-2}{2}$$

$$= 0$$

Thus ϕ is a positive root of the equation $x^2 - x - 1 = 0$.

(ii)   Substitute $\phi = \dfrac{1+\sqrt{5}}{2}$ into the equation $\phi^2 = \phi + 1$:

$$\phi^2 = \left(\frac{1+\sqrt{5}}{2}\right)^2 = \frac{1+2\sqrt{5}+5}{4} = \frac{6+2\sqrt{5}}{4} = \frac{3+\sqrt{5}}{2}$$

$$\phi + 1 = \frac{1+\sqrt{5}}{2} + 1 = \frac{1+\sqrt{5}}{2} + \frac{2}{2} = \frac{3+\sqrt{5}}{2}$$

Therefore $\phi^2 = \phi + 1$. Alternatively, since ϕ satisfies $x^2 - x - 1 = 0$ from part (i), we have $\phi^2 - \phi - 1 = 0$, thus $\phi^2 = \phi + 1$.

(iii) Substitute $\phi = \dfrac{1+\sqrt{5}}{2}$ into the equation $\phi^3 = 2\phi + 1$:

$$\phi^3 = \left(\frac{1+\sqrt{5}}{2}\right)^3 = \frac{1+3\sqrt{5}+15+5\sqrt{5}}{8} = \frac{16+8\sqrt{5}}{8} = 2+\sqrt{5}$$

$$2\phi + 1 = 2\left(\frac{1+\sqrt{5}}{2}\right) + 1 = 1 + \sqrt{5} + 1 = 2 + \sqrt{5}$$

Therefore $\phi^3 = 2\phi + 1$. Alternatively, since $\phi^2 = \phi + 1$ from part (ii), we have:
$$\phi^3 = \phi \cdot \phi^2 = \phi(\phi+1) = \phi^2 + \phi = (\phi+1) + \phi = 2\phi + 1$$

(iv)   Substitute $\phi = \dfrac{1+\sqrt{5}}{2}$ into the equation $\phi^{-1} = \phi - 1$:

$$\phi^{-1} = \left(\frac{1+\sqrt{5}}{2}\right)^{-1} = \frac{2}{1+\sqrt{5}} \cdot \frac{1-\sqrt{5}}{1-\sqrt{5}} = \frac{2\left(1-\sqrt{5}\right)}{1-5} = \frac{2\left(1-\sqrt{5}\right)}{-4} = \frac{-1+\sqrt{5}}{2}$$

$$\phi - 1 = \frac{1+\sqrt{5}}{2} - 1 = \frac{1+\sqrt{5}}{2} - \frac{2}{2} = \frac{-1+\sqrt{5}}{2}$$

Therefore $\phi^{-1} = \phi - 1$. Alternatively, since ϕ satisfies $x^2 - x - 1 = 0$ from part (i), we have $\phi^2 - \phi - 1 = 0$. Dividing by ϕ:

$$\phi^2 - \phi - 1 = 0$$

$$\phi - 1 - \frac{1}{\phi} = 0$$

$$\phi - 1 = \frac{1}{\phi}$$

(v)   Substitute $\phi = \dfrac{1+\sqrt{5}}{2}$ into the equation $\phi^{-2} = -\phi + 2$:

$$\phi^{-2} = \left(\frac{1+\sqrt{5}}{2}\right)^{-2}$$

$$= \left(\frac{2}{1+\sqrt{5}}\right)^{2}$$

$$= \frac{4}{1 + 2\sqrt{5} + 5}$$

$$= \frac{4}{6 + 2\sqrt{5}}$$

$$= \frac{2}{3 + \sqrt{5}} \cdot \frac{3 - \sqrt{5}}{3 - \sqrt{5}}$$

$$= \frac{2\left(3 - \sqrt{5}\right)}{9 - 5}$$

$$= \frac{2\left(3 - \sqrt{5}\right)}{4}$$

$$= \frac{3 - \sqrt{5}}{2}$$

$$-\phi + 2 = -\frac{1+\sqrt{5}}{2} + 2 = \frac{-1-\sqrt{5}}{2} + \frac{4}{2} = \frac{3 - \sqrt{5}}{2}$$

Therefore $\phi^{-2} = -\phi + 2$. Alternatively, since $\phi^{-1} = \phi - 1$ (part iv) and $\phi^2 = \phi + 1$ (part ii), we have:

$$\phi^{-2} = \left(\phi^{-1}\right)^2 = (\phi - 1)^2 = \phi^2 - 2\phi + 1 = (\phi + 1) - 2\phi + 1 = -\phi + 2$$

65. Let D denote the distance from town A to town B. Then the time to go from town A to town B is $\frac{D}{v_1}$, and the time for the return trip is $\frac{D}{v_2}$. Thus:

$$\text{average speed} = \frac{\text{total distance}}{\text{total time}} = \frac{2D}{\frac{D}{v_1} + \frac{D}{v_2}} = \frac{2}{\frac{1}{v_1} + \frac{1}{v_2}} = \frac{2v_1v_2}{v_1 + v_2} \text{ mph}$$

67. Let t denote the required number of years. Then:

$$d\left(\tfrac{r}{100}\right)t + D\left(\tfrac{R}{100}\right)t = d + D$$
$$drt + DRt = 100(d + D)$$
$$t(dr + DR) = 100(d + D)$$
$$t = \frac{100(d + D)}{dr + DR} \text{ years}$$

69. Let x denote the length of the hypotenuse. Since the perimeter is a and the length of the shortest side is b, the length of the third side is $a - b - x$. By the Pythagorean theorem:

$$x^2 = (a - b - x)^2 + b^2$$
$$x^2 = (a - b)^2 - 2(a - b)x + x^2 + b^2$$
$$2(a - b)x = a^2 - 2ab + 2b^2$$
$$x = \frac{a^2 - 2ab + 2b^2}{2(a - b)} \text{ inches}$$

71. Denoting the length of a side of the original square by x:

$$(ax)^2 = x^2 + b^2$$
$$a^2x^2 - x^2 = b^2$$
$$x^2(a^2 - 1) = b^2$$
$$x^2 = \frac{b^2}{a^2 - 1}$$
$$x = \frac{b}{\sqrt{a^2 - 1}}$$

The length of each side in the original square is $\dfrac{b}{\sqrt{a^2 - 1}}$ units. This quantity can also be written as $\dfrac{b\sqrt{a^2 - 1}}{a^2 - 1}$ units.

73. Let x and $16 - x$ denote the lengths of the two pieces. So the two circumferences are x and $16 - x$, respectively. We can find the corresponding radii by writing the formula $C = 2\pi r$ as $r = \dfrac{C}{2\pi}$. Thus the corresponding radii are $\dfrac{x}{2\pi}$ and $\dfrac{16 - x}{2\pi}$, respectively. Now if the sum of the two areas is 12 cm^2:

$$\pi\left(\frac{x}{2\pi}\right)^2 + \pi\left(\frac{16 - x}{2\pi}\right)^2 = 12$$

$$\frac{x^2}{4\pi} + \frac{256 - 32x + x^2}{4\pi} = 12$$

$$2x^2 - 32x + 256 = 48\pi$$

$$x^2 - 16x + 128 = 24\pi$$

$$x^2 - 16x + (128 - 24\pi) = 0$$

Applying the quadratic formula with $a = 1$, $b = -16$, and $c = 128 - 24\pi$ yields:

$$x = \frac{16 \pm \sqrt{256 - 4(128 - 24\pi)}}{2}$$

$$= \frac{16 \pm \sqrt{96\pi - 256}}{2}$$

$$= \frac{16 \pm 4\sqrt{6\pi - 16}}{2}$$

$$= 8 \pm 2\sqrt{6\pi - 16}$$

Both roots here yield positive values for x. First consider the positive sign. Then $x = 8 + 2\sqrt{6\pi - 16}$ and consequently $16 - x = 8 - 2\sqrt{6\pi - 16}$ (which is the same as the value obtained for x using the negative root). Similarly, if we begin with $x = 8 - 2\sqrt{6\pi - 16}$, we find that $16 - x$ is $8 + 2\sqrt{6\pi - 16}$. So in either case the lengths are:

$$8 + 2\sqrt{6\pi - 16} \quad \text{and} \quad 8 - 2\sqrt{6\pi - 16}$$

The smaller of these two numbers is $8 - 2\sqrt{6\pi - 16}$. Using a calculator and rounding to two decimal places, we find the required length to be 4.62 cm.

75. Call w and l the width and length, respectively. The area is represented by wl. Thus we have $wl + (l - w) = 183$ and $w + l = 27$. Solving the second equation for l yields $l = 27 - w$, now substitute:

$$wl + l - w = 183$$

$$w(27 - w) + (27 - w) - w = 183$$

$$27w - w^2 + 27 - w - w = 183$$

$$-w^2 + 25w + 27 = 183$$

$$0 = w^2 - 25w + 156$$

$$0 = (w - 12)(w - 13)$$

$$w = 12, 13$$

If $w = 12$, $l = 15$ and $A = 180$. If $w = 13$, $l = 14$ and $A = 182$. There are two possible rectangles (width = 12, length = 15, area = 180 or width = 13, length = 14, area = 182) for this situation.

77. Call the side x, thus we have the following figure:

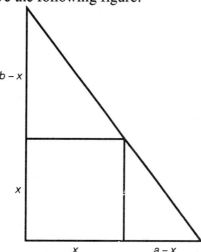

By similar triangles:

$$\frac{x}{a-x} = \frac{b-x}{x}$$

$$x^2 = ab - bx - ax + x^2$$

$$(a+b)x = ab$$

$$x = \frac{ab}{a+b}$$

79. Let $PV = a$, $PU = b$, $QU = c$, and $QT = d$. Note that $ST = a + b$ and $VS = c + d$.
Setting areas equal:

Area of $\triangle VPS$ = Area of $\triangle SQT$

$$\tfrac{1}{2}(a)(c+d) = \tfrac{1}{2}(d)(a+b)$$

$$ac + ad = ad + bd$$

$$ac = bd$$

$$b = \frac{ac}{d}$$

Similarly:

Area of $\triangle UPQ$ = Area of $\triangle SQT$

$$\tfrac{1}{2}(bc) = \tfrac{1}{2}(d)(a+b)$$

$$bc = ad + bd$$

Substituting for b:

$$bc = ad + bd$$

$$\frac{ac}{d} \cdot c = ad + \frac{ac}{d} \cdot d$$

$$\frac{ac^2}{d} = ad + ac$$

$$\frac{c^2}{d} = d + c$$

Dividing through by d yields:

$$\frac{c^2}{d^2} = 1 + \frac{c}{d}$$

$$\frac{c^2}{d^2} - \frac{c}{d} - 1 = 0$$

$$\left(\frac{c}{d}\right)^2 - \left(\frac{c}{d}\right) - 1 = 0$$

Thus $\frac{c}{d}$ is the positive root of the equation $x^2 - x - 1 = 0$. By Exercise 63, ϕ is also this

solution. Thus $\frac{c}{d} = \phi$, so $\frac{QU}{QT} = \phi$. Finally, since $ac = bd$, we have $\frac{c}{d} = \frac{b}{a}$, and thus

$\frac{b}{a} = \phi$, so $\frac{PU}{PV} = \phi$.

Special thanks go to David Cohen, who wrote this rather creative solution.

3.5 Inequalities

1. Solving the inequality:
    $$x + 5 < 4$$
    $$x < -1$$
    The solution set is $(-\infty, -1)$.

3. Solving the inequality:
    $$1 - 3x \leq 0$$
    $$-3x \leq -1$$
    $$x \geq \tfrac{1}{3}$$
    The solution set is $\left[\tfrac{1}{3}, \infty\right)$.

5. Solving the inequality:
    $$4x + 6 < 3(x - 1) - x$$
    $$4x + 6 < 2x - 3$$
    $$2x < -9$$
    $$x < -\tfrac{9}{2}$$
    The solution set is $\left(-\infty, -\tfrac{9}{2}\right)$.

7. Solving the inequality:
    $$1 - 2(t + 3) - t \leq 1 - 2t$$
    $$1 - 2t - 6 - t \leq 1 - 2t$$
    $$-6 - t \leq 0$$
    $$-t \leq 6$$
    $$t \geq -6$$
    The solution set is $[-6, \infty)$.

9. Multiplying by 15 then solving the inequality:

$$\frac{3x}{5} - \frac{x-1}{3} < 1$$
$$9x - 5(x-1) < 15$$
$$4x + 5 < 15$$
$$4x < 10$$
$$x < \tfrac{5}{2}$$

The solution set is $\left(-\infty, \tfrac{5}{2}\right)$.

11. Multiplying by 20 then solving the inequality:

$$\frac{x-1}{4} - \frac{2x+3}{5} \leq x$$
$$5(x-1) - 4(2x+3) \leq 20x$$
$$5x - 5 - 8x - 12 \leq 20x$$
$$-23x \leq 17$$
$$x \geq -\tfrac{17}{23}$$

The solution set is $\left[-\tfrac{17}{23}, \infty\right)$.

13. Solving the double inequality:

$$-2 \leq x - 6 \leq 0$$
$$4 \leq x \leq 6$$

The solution set is [4, 6].

15. Multiplying by 3 then solving the double inequality:

$$-1 \leq \frac{1-4t}{3} \leq 1$$
$$-3 \leq 1 - 4t \leq 3$$
$$-4 \leq -4t \leq 2$$
$$1 \geq t \geq -\tfrac{1}{2}$$
$$-\tfrac{1}{2} \leq t \leq 1$$

The solution set is $\left[-\tfrac{1}{2}, 1\right]$.

17. Multiplying by 2 then solving the double inequality:

$$0.99 < \frac{x}{2} - 1 < 0.999$$
$$1.98 < x - 2 < 1.998$$
$$3.98 < x < 3.998$$

The solution set is (3.98, 3.998).

19. (a) The inequality is equivalent to $-\tfrac{1}{2} \leq x \leq \tfrac{1}{2}$, so the solution set is $\left[-\tfrac{1}{2}, \tfrac{1}{2}\right]$.

  (b) The inequality is equivalent to $x > \tfrac{1}{2}$ or $x < -\tfrac{1}{2}$, so the solution set is

$$\left(-\infty, -\tfrac{1}{2}\right) \cup \left(\tfrac{1}{2}, \infty\right).$$

21. (a) The inequality is equivalent to $x > 0$ or $x < 0$, so the solution set is $(-\infty, 0) \cup (0, \infty)$.

(b) Since $|x| \geq 0$ for all x, there is no solution to this inequality.

23. (a) Solving the inequality:
$$x - 2 < 1$$
$$x < 3$$
The solution set is $(-\infty, 3)$.

(b) Solving the inequality:
$$|x - 2| < 1$$
$$-1 < x - 2 < 1$$
$$1 < x < 3$$
The solution set is $(1, 3)$.

(c) Solving the inequality:
$$|x - 2| > 1$$
$$x - 2 > 1 \quad \text{or} \quad x - 2 < -1$$
$$x > 3 \quad \text{or} \quad x < 1$$
The solution set is $(-\infty, 1) \cup (3, \infty)$.

25. (a) Solving the inequality:
$$1 - x \leq 5$$
$$-x \leq 4$$
$$x \geq -4$$
The solution set is $[-4, \infty)$.

(b) Solving the inequality:
$$|1 - x| \leq 5$$
$$-5 \leq 1 - x \leq 5$$
$$-6 \leq -x \leq 4$$
$$6 \geq x \geq -4$$
The solution set is $[-4, 6]$.

(c) Solving the inequality:
$$|1 - x| > 5$$
$$1 - x > 5 \quad \text{or} \quad 1 - x < -5$$
$$-x > 4 \quad \text{or} \quad -x < -6$$
$$x < -4 \quad \text{or} \quad x > 6$$
The solution set is $(-\infty, -4) \cup (6, \infty)$.

27. (a) Solving the inequality:
$$a - x < c$$
$$-x < c - a$$
$$x > a - c$$
The solution set is $(a - c, \infty)$.

(b) Solving the inequality:
$$|a - x| < c$$
$$-c < a - x < c$$
$$-c - a < -x < c - a$$
$$a + c > x > a - c$$
The solution set is $(a - c, a + c)$.

(c) Solving the inequality:
$$|a - x| \ge c$$
$$a - x \ge c \quad \text{or} \quad a - x \le -c$$
$$-x \ge c - a \quad \text{or} \quad -x \le -c - a$$
$$x \le a - c \quad \text{or} \quad x \ge a + c$$
The solution set is $(-\infty, a - c] \cup [a + c, \infty)$.

29. Solving the inequality:
$$\left| \frac{x - 2}{3} \right| < 4$$
$$-4 < \frac{x - 2}{3} < 4$$
$$-12 < x - 2 < 12$$
$$-10 < x < 14$$
The solution set is $(-10, 14)$.

31. Solving the inequality:
$$\left| \frac{x + 1}{2} - \frac{x - 1}{3} \right| < 1$$
$$-1 < \frac{x + 1}{2} - \frac{x - 1}{3} < 1$$
$$-6 < 3(x + 1) - 2(x - 1) < 6$$
$$-6 < x + 5 < 6$$
$$-11 < x < 1$$
The solution set is $(-11, 1)$.

33. (a) Solving the inequality:
$$|(x + h)^2 - x^2| < 3h^2$$
$$|x^2 + 2xh + h^2 - x^2| < 3h^2$$
$$|2xh + h^2| < 3h^2$$
$$-3h^2 < 2xh + h^2 < 3h^2$$
$$-4h^2 < 2xh < 2h^2$$
Since $h > 0$, we can divide by $2h$ to obtain $-2h < x < h$. The solution set is $(-2h, h)$.

(b) Since $h < 0$, when we divide the inequality in (a) by $2h$, we reverse the inequalities to obtain $-2h > x > h$. The solution set is $(h, -2h)$.

35. (a) The solution set will be of the form $[a, \infty)$.

(b) The value of a is approximately 0.3.

(c) Solving the inequality:
$$7x - 2 \geq 0$$
$$7x \geq 2$$
$$x \geq \frac{2}{7}$$
The solution set is $\left[\frac{2}{7}, \infty\right)$, where $a = \frac{2}{7}$.

37. (a) The solution set will be of the form $[a, b]$.

(b) The values are $a \approx 0.1$ and $b \approx 0.6$.

(c) Solving the inequality:
$$|8x - 3| - 2 \leq 0$$
$$|8x - 3| \leq 2$$
$$-2 \leq 8x - 3 \leq 2$$
$$1 \leq 8x \leq 5$$
$$\frac{1}{8} \leq x \leq \frac{5}{8}$$
The solution set is $\left[\frac{1}{8}, \frac{5}{8}\right]$, where $a = \frac{1}{8}$ and $b = \frac{5}{8}$.

39. Since $F = \frac{9}{5}C + 32$ is the conversion from Fahrenheit to Celsius temperature, we substitute and solve the resulting inequality:
$$-280° \leq F \leq 260°$$
$$-280° \leq \frac{9}{5}C + 32 \leq 260°$$
$$-312° \leq \frac{9}{5}C \leq 228°$$
$$-173.3° \leq C \leq 126.7°$$
To the nearest 5° C, the corresponding interval is $-175° \leq C \leq 125°$.

41. First solve the equation $F = \frac{9}{5}C + 32$ for C. This yields:
$$C = \frac{5F - 160}{9}$$
The following inequalities are then equivalent:
$$-25 \leq C \leq 475$$
$$-25 \leq \frac{5F - 160}{9} \leq 475$$
$$-225 \leq 5F - 160 \leq 4275$$
$$-65 \leq 5F \leq 4435$$
$$-13° \leq F \leq 887°$$
The corresponding range on the Fahrenheit scale is $-13° \leq F \leq 887°$.

43. Using the expression for C found in Exercise 41:

$$5100 \le C \le 6500$$

$$5100 \le \frac{5F - 160}{9} \le 6500$$

$$45900 \le 5F - 160 \le 58500$$

$$46060 \le 5F \le 58660$$

$$9212 \le F \le 11732$$

Rounding to the nearest $100°$, the corresponding range is $9200° \le F \le 11700°$.

45. (a) Complete the table:

a	b	$\sqrt{ab}$ (G.M.)	$\dfrac{a+b}{2}$ (A.M.)	Which is larger, G.M. or A.M.?
1	2	1.4142	1.5	A.M.
1	3	1.7320	2.0	A.M.
1	4	2.0000	2.5	A.M.
2	3	2.4495	2.5	A.M.
3	4	3.4641	3.5	A.M.
5	10	7.0711	7.5	A.M.
9	10	9.4868	9.5	A.M.
99	100	99.4987	99.5	A.M.
999	1000	999.4999	999.5	A.M.

(b) Using the hint and squaring each side, proving the inequality is equivalent to proving the following inequality:

$$ab \le \left(\frac{a+b}{2}\right)^2$$

Note that the inequality $(a - b)^2 \ge 0$ is true for all a and b, thus:

$$(a - b)^2 \ge 0$$

$$a^2 - 2ab + b^2 \ge 0$$

$$a^2 + b^2 \ge 2ab$$

Working from the right-hand side of the original inequality:

$$\left(\frac{a+b}{2}\right)^2 = \frac{a^2 + 2ab + b^2}{4} = \frac{a^2 + b^2}{4} + \frac{ab}{2} \ge \frac{2ab}{4} + \frac{ab}{2} = \frac{ab}{2} + \frac{ab}{2} = ab$$

This proves the required inequality.

47.   Consider three cases.

case 1:   If $x < 1$, then $x - 1 < 0$ and $x - 2 < 0$. Therefore:
$$-(x-1)-(x-2)<3$$
$$-x+1-x+2<3$$
$$-2x+3<3$$
$$-2x<0$$
$$x>0$$
Since $x < 1$, this is the interval $(0, 1)$.

case 2:   If $1 \le x < 2$, then $x - 1 \ge 0$ and $x - 2 < 0$. Therefore:
$$(x-1)-(x-2)<3$$
$$x-1-x+2<3$$
$$1<3$$
Thus the interval $[1, 2)$ is a solution.

case 3:   If $x \ge 2$, then $x - 1 \ge 0$ and $x - 2 \ge 0$. Therefore:
$$(x-1)+(x-2)<3$$
$$2x-3<3$$
$$2x<6$$
$$x<3$$
Since $x \ge 2$, this is the interval $[2, 3)$.
Putting our solutions together, the solution is $(0, 3)$.

49.   (a)   Complete the table:

x	y	$\dfrac{x}{y}+\dfrac{y}{x}$	True or False $\dfrac{x}{y}+\dfrac{y}{x} \ge 2$
1	1	2.0000	true
2	3	2.1667	true
3	5	2.2667	true
4	7	2.3214	true
5	9	2.3556	true
9	10	2.0111	true
49	50	2.0004	true
99	100	2.0001	true

(b)   Using the inequality in Exercise 45(b) and taking $a = \frac{x}{y}$ and $b = \frac{y}{x}$:

$$\sqrt{ab} \le \frac{a+b}{2}$$

$$\sqrt{\frac{x}{y} \cdot \frac{y}{x}} \le \frac{\frac{x}{y}+\frac{y}{x}}{2}$$

$$1 \le \frac{\frac{x}{y}+\frac{y}{x}}{2}$$

$$2 \le \frac{x}{y}+\frac{y}{x}$$

This proves the required inequality.

51.   (a)   Since $a + b$ is the diameter of the circle, $CE = EF = \frac{a+b}{2}$, as they are radii of the circle.

(b)   Consider the triangle:

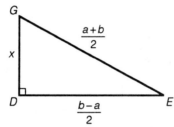

Note that $DE = b - \dfrac{a+b}{2} = \dfrac{b-a}{2}$. Therefore:

$$\left(\frac{b-a}{2}\right)^2 + x^2 = \left(\frac{a+b}{2}\right)^2$$

$$\frac{b^2 - 2ab + a^2}{4} + x^2 = \frac{a^2 + 2ab + b^2}{4}$$

$$-2ab + 4x^2 = 2ab$$

$$4x^2 = 4ab$$

$$x^2 = ab$$

$$x = \sqrt{ab}$$

Thus $DG = \sqrt{ab}$. Now consider the triangle:

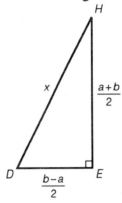

Therefore:

$$\left(\frac{b-a}{2}\right)^2 + \left(\frac{a+b}{2}\right)^2 = x^2$$

$$\frac{b^2 - 2ab + a^2}{4} + \frac{a^2 + 2ab + b^2}{4} = x^2$$

$$\frac{2a^2 + 2b^2}{4} = x^2$$

$$\frac{a^2 + b^2}{2} = x^2$$

$$\sqrt{\frac{a^2 + b^2}{2}} = x$$

Thus $DH = \sqrt{\dfrac{a^2 + b^2}{2}}$.

(c) Note that $DG \le EH$ follows from the figure, and $EH \le DH$ since DH is the hypotenuse of the above triangle. Therefore:

$$\sqrt{ab} \le \frac{a+b}{2} \le \sqrt{\frac{a^2 + b^2}{2}}$$

53. (a) Let l be the length, so $2x + 2l = 30$ and thus $l = 15 - x$. Thus the area is given by $A = x(15 - x)$.

(b) Using the inequality from Exercise 45(b):

$$x(15 - x) \le \left(\frac{x + 15 - x}{2}\right)^2 = \left(\frac{15}{2}\right)^2 = \frac{225}{4}$$

(c) Set $A = \frac{225}{4}$ and solve for x:

$$x(15 - x) = \tfrac{225}{4}$$
$$15x - x^2 = \tfrac{225}{4}$$
$$60x - 4x^2 = 225$$
$$4x^2 - 60x + 225 = 0$$
$$(2x - 15)^2 = 0$$
$$x = \tfrac{15}{2}$$

Since the length is also $\frac{15}{2}$, the rectangle is a square of dimensions $\frac{15}{2}$ ft by $\frac{15}{2}$ ft.

55. Using the hint:

$$\sqrt[4]{stuv} = \sqrt{\sqrt{st}\,\sqrt{uv}} \le \frac{\sqrt{st} + \sqrt{uv}}{2} \le \frac{\frac{s+t}{2} + \frac{u+v}{2}}{2} = \frac{s+t+u+v}{4}$$

3.6 More on Inequalities

1. (a) Using the graph, the solution is $(-\infty, -1] \cup [4, \infty)$.

(b) Using the graph, the solution is $[-1, 4]$.

3. (a) From the graph, there is no solution.

(b) From the graph, the solution is all real numbers, or $(-\infty, \infty)$.

5. (a) Using the graph, the solution is $[-1, 1] \cup [3, \infty)$.

(b) Using the graph, the solution is $(-\infty, -1) \cup (1, 3)$.

7. First factor $x^2 + x - 6 = (x + 3)(x - 2)$, so the key numbers are -3 and 2. Now draw the sign chart:

Interval	Test Number	$x + 3$	$x - 2$	$(x+3)(x-2)$
$(-\infty, -3)$	-4	neg.	neg.	pos.
$(-3, 2)$	0	pos.	neg.	neg.
$(2, \infty)$	3	pos.	pos.	pos.

So $x^2 + x - 6 < 0$ on the interval $(-3, 2)$.

9. First factor $x^2 - 11x + 18 = (x - 2)(x - 9)$, so the key numbers are 2 and 9. Now draw the sign chart:

Interval	Test Number	$x - 2$	$x - 9$	$(x-2)(x-9)$
$(-\infty, 2)$	0	neg.	neg.	pos.
$(2, 9)$	3	pos.	neg.	neg.
$(9, \infty)$	10	pos.	pos.	pos.

So $x^2 - 11x + 18 > 0$ on the intervals $(-\infty, 2) \cup (9, \infty)$.

11. First simplify and factor the inequality:

$$9x - x^2 \leq 20$$
$$-x^2 + 9x - 20 \leq 0$$
$$x^2 - 9x + 20 \geq 0$$
$$(x - 4)(x - 5) \geq 0$$

So the key numbers are 4 and 5. Now draw the sign chart:

Interval	Test Number	$x - 4$	$x - 5$	$(x - 4)(x - 5)$
$(-\infty, 4)$	0	neg.	neg.	pos.
$(4, 5)$	$\frac{9}{2}$	pos.	neg.	neg.
$(5, \infty)$	6	pos.	pos.	pos.

So $x^2 - 9x + 20 \geq 0$ (and consequently the original inequality is satisfied) on the intervals $(-\infty, 4] \cup [5, \infty)$.

13. First factor $x^2 - 16 = (x + 4)(x - 4)$, so the key numbers are -4 and 4. Now draw the sign chart:

Interval	Test Number	$x - 4$	$x + 4$	$(x - 4)(x + 4)$
$(-\infty, -4)$	-5	neg.	neg.	pos.
$(-4, 4)$	0	neg.	pos.	neg.
$(4, \infty)$	5	pos.	pos.	pos.

So $x^2 - 16 \geq 0$ on the intervals $(-\infty, -4] \cup [4, \infty)$.

15. First simplify and factor the inequality:

$$16x^2 + 24x < -9$$
$$16x^2 + 24x + 9 < 0$$
$$(4x + 3)^2 < 0$$

Since $(4x + 3)^2 \geq 0$ for all real values of x, this inequality has no solution.

17. First factor $x^3 + 13x^2 + 42x = x(x^2 + 13x + 42) = x(x + 7)(x + 6)$, so the key numbers are $0, -7$ and -6. Now draw the sign chart:

Interval	Test Number	x	$x + 7$	$x + 6$	$x(x + 7)(x + 6)$
$(-\infty, -7)$	-8	neg.	neg.	neg.	neg.
$(-7, -6)$	$-\frac{13}{2}$	neg.	pos.	neg.	pos.
$(-6, 0)$	-1	neg.	pos.	pos.	neg.
$(0, \infty)$	1	pos.	pos.	pos.	pos.

So $x^3 + 13x^2 + 42x > 0$ on the intervals $(-7, -6) \cup (0, \infty)$.

19. First simplify and factor the inequality:
$$225x \le x^3$$
$$225x - x^3 \le 0$$
$$x(225 - x^2) \le 0$$
$$x(15 + x)(15 - x) \le 0$$

So the key numbers are 0, –15, and 15. Now draw the sign chart:

Interval	Test Number	x	$15 - x$	$15 + x$	$x(15 - x)(15 + x)$
$(-\infty, -15)$	-16	neg.	pos.	neg.	pos.
$(-15, 0)$	-1	neg.	pos.	pos.	neg.
$(0, 15)$	1	pos.	pos.	pos.	pos.
$(15, \infty)$	16	pos.	neg.	pos.	neg.

So $225 - x^3 \le 0$ (and consequently the original inequality is satisfied) on the intervals $[-15, 0] \cup [15, \infty)$.

21. Since $2x^2 + 1 \ge 1$ for all real numbers x, the inequality is satisfied for the interval $(-\infty, \infty)$.

23. First factor $12x^3 + 17x^2 + 6x = x(12x^2 + 17x + 6) = x(4x + 3)(3x + 2)$, so the key numbers are 0, $-\frac{3}{4}$ and $-\frac{2}{3}$. Now draw the sign chart:

Interval	Test Number	x	$4x + 3$	$3x + 2$	$x(4x + 3)(3x + 2)$
$\left(-\infty, -\frac{3}{4}\right)$	-1	neg.	neg.	neg.	neg.
$\left(-\frac{3}{4}, -\frac{2}{3}\right)$	$-\frac{17}{24}$	neg.	pos.	neg.	pos.
$\left(-\frac{2}{3}, 0\right)$	$-\frac{1}{3}$	neg.	pos.	pos.	neg.
$(0, \infty)$	1	pos.	pos.	pos.	pos.

So $12x^3 + 17x^2 + 6x < 0$ on the intervals $\left(-\infty, -\frac{3}{4}\right) \cup \left(-\frac{2}{3}, 0\right)$.

25. The key numbers are found by solving the equation $x^2 + x - 1 = 0$. Using the quadratic formula we have:
$$x = \frac{-1 \pm \sqrt{1 - 4(-1)}}{2} = \frac{-1 \pm \sqrt{5}}{2}$$

For purposes of picking appropriate test numbers, note that:
$$\frac{-1 + \sqrt{5}}{2} \approx 0.6 \quad \text{and} \quad \frac{-1 - \sqrt{5}}{2} \approx -1.6$$

Draw the sign chart:

Interval	Test Number	$x + 1.6$	$x - 0.6$	$x^2 + x - 1$
$\left(-\infty, \frac{-1-\sqrt{5}}{2}\right)$	-2	neg.	neg.	pos.
$\left(\frac{-1-\sqrt{5}}{2}, \frac{-1+\sqrt{5}}{2}\right)$	0	pos.	neg.	neg.
$\left(\frac{-1+\sqrt{5}}{2}, \infty\right)$	1	pos.	pos.	pos.

So $x^2 + x - 1 > 0$ on the intervals $\left(-\infty, \frac{-1-\sqrt{5}}{2}\right) \cup \left(\frac{-1+\sqrt{5}}{2}, \infty\right)$.

27. The key numbers are found by using the quadratic formula to solve $x^2 - 8x + 2 = 0$. We have:

$$x = \frac{8 \pm \sqrt{64 - 4(2)}}{2} = \frac{8 \pm \sqrt{56}}{2} = \frac{8 \pm 2\sqrt{14}}{2} = 4 \pm \sqrt{14}$$

For purposes of picking appropriate test numbers, note that $4 + \sqrt{14} \approx 7.7$ and $4 - \sqrt{14} \approx 0.3$. Draw the sign chart:

Interval	Test Number	$x - 7.7$	$x - 0.3$	$x^2 - 8x + 2$
$(-\infty, 4 - \sqrt{14})$	0	neg.	neg.	pos.
$(4 - \sqrt{14}, 4 + \sqrt{14})$	4	neg.	pos.	neg.
$(4 + \sqrt{14}, \infty)$	8	pos.	pos.	pos.

So $x^2 - 8x + 2 \leq 0$ on the interval $\left[4 - \sqrt{14}, 4 + \sqrt{14}\right]$.

29. The key numbers are -4, -3, and 1. Draw the sign chart:

Interval	Test Number	$x - 1$	$x + 3$	$x + 4$	$(x - 1)(x + 3)(x + 4)$
$(-\infty, -4)$	-5	neg.	neg.	neg.	neg.
$(-4, -3)$	$-\frac{7}{2}$	neg.	neg.	pos.	pos.
$(-3, 1)$	0	neg.	pos.	pos.	neg.
$(1, \infty)$	2	pos.	pos.	pos.	pos.

So $(x - 1)(x + 3)(x + 4) \geq 0$ on the intervals $[-4, -3] \cup [1, \infty)$.

31. The key numbers are -4, -5, and -6. Draw the sign chart:

Interval	Test Number	$x + 4$	$x + 5$	$x + 6$	$(x + 4)(x + 5)(x + 6)$
$(-\infty, -6)$	-7	neg.	neg.	neg.	neg.
$(-6, -5)$	$-\frac{11}{2}$	neg.	neg.	pos.	pos.
$(-5, -4)$	$-\frac{9}{2}$	neg.	pos.	pos.	neg.
$(-4, \infty)$	0	pos.	pos.	pos.	pos.

So $(x + 4)(x + 5)(x + 6) < 0$ on the intervals $(-\infty, -6) \cup (-5, -4)$.

33. The key numbers are $-\frac{1}{3}, \frac{1}{3}$ and 2. Draw the sign chart:

Interval	Test Number	$(x-2)^2$	$(3x+1)^3$	$3x-1$	product
$\left(-\infty,-\frac{1}{3}\right)$	-1	pos.	neg.	neg.	pos.
$\left(-\frac{1}{3},\frac{1}{3}\right)$	0	pos.	pos.	neg.	neg.
$\left(\frac{1}{3},2\right)$	1	pos.	pos.	pos.	pos.
$(2,\infty)$	3	pos.	pos.	pos.	pos.

So $(x-2)^2(3x+1)^3(3x-1) > 0$ on the intervals $\left(-\infty,-\frac{1}{3}\right) \cup \left(\frac{1}{3},2\right) \cup (2,\infty)$.

35. The key numbers are $3, -1, -\frac{1}{2}$ and $-\frac{2}{3}$. Draw the sign chart:

Interval	Test Number	$(x-3)^2$	$(x+1)^4$	$(2x+1)^4$	$3x+2$	product
$(-\infty,-1)$	-2	pos.	pos.	pos.	neg.	neg.
$\left(-1,-\frac{2}{3}\right)$	$-\frac{3}{4}$	pos.	pos.	pos.	neg.	neg.
$\left(-\frac{2}{3},-\frac{1}{2}\right)$	$-\frac{7}{12}$	pos.	pos.	pos.	pos.	pos.
$\left(-\frac{1}{2},3\right)$	0	pos.	pos.	pos.	pos.	pos.
$(3,\infty)$	4	pos.	pos.	pos.	pos.	pos.

So $(x-3)^2(x+1)^4(2x+1)^4(3x+2) \le 0$ on the intervals $\left(-\infty,-1\right] \cup \left[-1,-\frac{2}{3}\right]$, which is equivalent to the interval $\left(-\infty,-\frac{2}{3}\right]$. It is also zero at 3 and $-1/2$, so the inequality is satisfied for the values $\left(-\infty,-\frac{2}{3}\right] \cup \left\{3,-\frac{1}{2}\right\}$.

37. First simplify and factor the inequality:
$$20 \ge x^2\left(9-x^2\right)$$
$$20 \ge 9x^2 - x^4$$
$$x^4 - 9x^2 + 20 \ge 0$$
$$\left(x^2-4\right)\left(x^2-5\right) \ge 0$$
$$(x-2)(x+2)\left(x-\sqrt{5}\right)\left(x+\sqrt{5}\right) \ge 0$$

The key numbers are $-2, 2, -\sqrt{5}$, and $\sqrt{5}$. Draw the sign chart:

Interval	Test Number	$x-\sqrt{5}$	$x-2$	$x+2$	$x+\sqrt{5}$	x^4-9x^2+20
$\left(-\infty,-\sqrt{5}\right)$	-3	neg.	neg.	neg.	neg.	pos.
$\left(-\sqrt{5},-2\right)$	-2.1	neg.	neg.	neg.	pos.	neg.
$(-2,2)$	0	neg.	neg.	pos.	pos.	pos.
$\left(2,\sqrt{5}\right)$	2.1	neg.	pos.	pos.	pos.	neg.
$\left(\sqrt{5},\infty\right)$	3	pos.	pos.	pos.	pos.	pos.

So $x^4 - 9x^2 + 20 \ge 0$ (and consequently the original inequality is satisfied) on the intervals $\left(-\infty,-\sqrt{5}\right] \cup [-2,2] \cup \left[\sqrt{5},\infty\right)$.

39. First simplify and factor the inequality:
$$9(x-4) - x^2(x-4) < 0$$
$$(x-4)(9-x^2) < 0$$
$$(x-4)(3+x)(3-x) < 0$$
The key numbers are 4, –3 and 3. Draw the sign chart:

Interval	Test Number	$x-4$	$3+x$	$3-x$	$(x-4)(3+x)(3-x)$
$(-\infty, -3)$	-4	neg.	neg.	pos.	pos.
$(-3, 3)$	0	neg.	pos.	pos.	neg.
$(3, 4)$	3.5	neg.	pos.	neg.	pos.
$(4, \infty)$	6	pos.	pos.	neg.	neg.

So $(x+4)(3+x)(3-x) < 0$ (and consequently the original inequality is satisfied) on the intervals $(-3, 3) \cup (4, \infty)$.

41. First simplify and factor the inequality:
$$4(x^2-9) - (x^2-9)^2 > -5$$
$$4x^2 - 36 - x^4 + 18x^2 - 81 > -5$$
$$-x^4 + 22x^2 - 112 > 0$$
$$x^4 - 22x^2 + 112 < 0$$
$$(x^2-8)(x^2-14) < 0$$
$$(x-2\sqrt{2})(x+2\sqrt{2})(x-\sqrt{14})(x+\sqrt{14}) < 0$$
The key numbers are $2\sqrt{2}, -2\sqrt{2}, \sqrt{14},$ and $-\sqrt{14}$. Draw the sign chart:

Interval	Test Number	x^2-8	x^2-14	x^4-22x^2+112
$(-\infty, -\sqrt{14})$	-5	pos.	pos.	pos.
$(-\sqrt{14}, -2\sqrt{2})$	-3	pos.	neg.	neg.
$(-2\sqrt{2}, 2\sqrt{2})$	0	neg.	neg.	pos.
$(2\sqrt{2}, \sqrt{14})$	3	pos.	neg.	neg.
$(\sqrt{14}, \infty)$	5	pos.	pos.	pos.

So $x^4 - 22x^2 + 112 < 0$ (and consequently the original inequality is satisfied) on the intervals $(-\sqrt{14}, -2\sqrt{2}) \cup (2\sqrt{2}, \sqrt{14})$.

43. One of the key numbers is 4. The others are found by solving the equation $2x^2 - 6x - 1 = 0$. With the quadratic formula:
$$x = \frac{6 \pm \sqrt{36 - 4(2)(-1)}}{2(2)} = \frac{6 \pm \sqrt{44}}{4} = \frac{6 \pm 2\sqrt{11}}{4} = \frac{3 \pm \sqrt{11}}{2}$$

Thus, the key numbers are 4 and $\frac{3\pm\sqrt{11}}{2}$. For purposes of picking appropriate test numbers, note that $\frac{3+\sqrt{11}}{2} \approx 3.2$ and that $\frac{3-\sqrt{11}}{2} \approx -0.2$. Draw the sign chart:

Interval	Test Number	$x-4$	$x-3.2$	$x+0.2$	$(x-4)(2x^2-6x-1)$
$(-\infty,-0.2)$	-1	neg.	neg.	neg.	neg.
$(-0.2,3.2)$	0	neg.	neg.	pos.	pos.
$(3.2,4)$	3.5	neg.	pos.	pos.	neg.
$(4,\infty)$	5	pos.	pos.	pos.	pos.

So $(x-4)(2x^2-6x-1)<0$ on the intervals $\left(-\infty,\frac{3-\sqrt{11}}{2}\right)\cup\left(\frac{3+\sqrt{11}}{2},4\right)$.

45. First factor the inequality:
$$x^3+2x^2-x-2>0$$
$$x^2(x+2)-(x+2)>0$$
$$(x+2)(x^2-1)>0$$
$$(x+2)(x+1)(x-1)>0$$
The key numbers are -2, -1 and 1. Draw the sign chart:

Interval	Test Number	$x-1$	$x+1$	$x+2$	$(x-1)(x+1)(x+2)$
$(-\infty,-2)$	-3	neg.	neg.	neg.	neg.
$(-2,-1)$	-1.5	neg.	neg.	pos.	pos.
$(-1,1)$	0	neg.	pos.	pos.	neg.
$(1,\infty)$	2	pos.	pos.	pos.	pos.

So $x^3+2x^2-x-2>0$ on the intervals $(-2,-1)\cup(1,\infty)$.

47. The key numbers are -1 and 1. Draw the sign chart:

Interval	Test Number	$x-1$	$x+1$	$\frac{x-1}{x+1}$
$(-\infty,-1)$	-2	neg.	neg.	pos.
$(-1,1)$	0	neg.	pos.	neg.
$(1,\infty)$	2	pos.	pos.	pos.

Thus the quotient is negative on $(-1, 1)$ and it is zero when $x = 1$. So $\dfrac{x-1}{x+1} \le 0$ on the interval $(-1, 1]$.

49. The key numbers are 2 and $\frac{3}{2}$. Draw the sign chart:

Interval	Test Number	$2-x$	$3-2x$	$\frac{2-x}{3-2x}$
$(-\infty,\frac{3}{2})$	0	pos.	pos.	pos.
$(\frac{3}{2},2)$	$\frac{7}{4}$	pos.	neg.	neg.
$(2,\infty)$	3	neg.	neg.	pos.

Thus the quotient is positive on $\left(-\infty,\frac{3}{2}\right)\cup(2,\infty)$ and it is zero when $x=2$. So $\dfrac{2-x}{3-2x}\geq 0$ on the intervals $\left(-\infty,\frac{3}{2}\right)\cup[2,\infty)$.

51. Factoring, we have $\dfrac{(x+1)(x-9)}{x}<0$. So the key numbers are -1, 0, and 9. Draw the sign chart:

Interval	Test Number	$x-9$	x	$x+1$	$\frac{(x-9)(x+1)}{x}$
$(-\infty,-1)$	-2	neg.	neg.	neg.	neg.
$(-1,0)$	$-\frac{1}{2}$	neg.	neg.	pos.	pos.
$(0,9)$	1	neg.	pos.	pos.	neg.
$(9,\infty)$	10	pos.	pos.	pos.	pos.

Thus the quotient is negative on the intervals $(-\infty,-1)\cup(0,9)$.

53. Factoring the inequality yields:
$$\frac{2x^3+5x^2-7x}{3x^2+7x+4}>0$$
$$\frac{x(2x^2+5x-7)}{(3x+4)(x+1)}>0$$
$$\frac{x(2x+7)(x-1)}{(3x+4)(x+1)}>0$$

So the key numbers are 0, $-\frac{7}{2}$, 1, $-\frac{4}{3}$ and -1. Draw the sign chart:

Interval	Test Number	$x-1$	x	$x+1$	$3x+4$	$2x+7$	$\frac{x(2x+7)(x-1)}{(3x+4)(x+1)}$
$\left(-\infty,-\frac{7}{2}\right)$	-4	neg.	neg.	neg.	neg.	neg.	neg.
$\left(-\frac{7}{2},-\frac{4}{3}\right)$	-2	neg.	neg.	neg.	neg.	pos.	pos.
$\left(-\frac{4}{3},-1\right)$	$-\frac{7}{6}$	neg.	neg.	neg.	pos.	pos.	neg.
$(-1,0)$	$-\frac{1}{2}$	neg.	neg.	pos.	pos.	pos.	pos.
$(0,1)$	$\frac{1}{2}$	neg.	pos.	pos.	pos.	pos.	neg.
$(1,\infty)$	2	pos.	pos.	pos.	pos.	pos.	pos.

Thus the quotient is positive on the intervals $\left(-\frac{7}{2},-\frac{4}{3}\right)\cup(-1,0)\cup(1,\infty)$.

55. First simplify the inequality:
$$\frac{x}{x+1}-1>0$$
$$\frac{x-x-1}{x+1}>0$$
$$\frac{-1}{x+1}>0$$

This is true whenever $x+1<0$, so $x<-1$. So the solution is $(-\infty,-1)$.

57. First simplify the inequality:

$$\frac{1}{x} - \frac{1}{x+1} \leq 0$$

$$\frac{x+1-x}{x(x+1)} \leq 0$$

$$\frac{1}{x(x+1)} \leq 0$$

So the key numbers are 0 and –1. Draw the sign chart:

Interval	Test Number	x	$x+1$	$\frac{1}{x(x+1)}$
$(-\infty, -1)$	-2	neg.	neg.	pos.
$(-1, 0)$	-0.5	neg.	pos.	neg.
$(0, \infty)$	1	pos.	pos.	pos.

So $\dfrac{1}{x(x+1)} \leq 0$ (and consequently the original inequality is satisfied) on the interval $(-1, 0)$.

59. First simplify the inequality:

$$\frac{x+2}{x+5} - 1 \leq 0$$

$$\frac{x+2-x-5}{x+5} \leq 0$$

$$\frac{-3}{x+5} \leq 0$$

This is true whenever $x + 5 > 0$, so $x > -5$. So the solution is $(-5, \infty)$.

61. First simplify the inequality:

$$\frac{1}{x-2} - \frac{1}{x-1} - \frac{1}{6} \geq 0$$

$$\frac{6(x-1) - 6(x-2) - (x-2)(x-1)}{6(x-2)(x-1)} \geq 0$$

$$\frac{-x^2 + 3x + 4}{6(x-2)(x-1)} \geq 0$$

$$\frac{x^2 - 3x - 4}{6(x-2)(x-1)} \leq 0$$

$$\frac{(x-4)(x+1)}{6(x-2)(x-1)} \leq 0$$

So the key numbers are –1, 1, 2, and 4. Draw the sign chart:

Interval	Test Number	$x-4$	$x-2$	$x-1$	$x+1$	$\frac{(x-4)(x+1)}{6(x-2)(x-1)}$
$(-\infty,-1)$	-2	neg.	neg.	neg.	neg.	pos.
$(-1,1)$	0	neg.	neg.	neg.	pos.	neg.
$(1,2)$	$\frac{3}{2}$	neg.	neg.	pos.	pos.	pos.
$(2,4)$	3	neg.	pos.	pos.	pos.	neg.
$(4,\infty)$	5	pos.	pos.	pos.	pos.	pos.

So the quotient is negative on the intervals $(-1,1)\cup(2,4)$ and zero at $x=4$ and $x=-1$, so our original inequality is satisfied on the intervals $[-1,1)\cup(2,4]$.

63.   First simplify the inequality:
$$\frac{1+x}{1-x}-\frac{1-x}{1+x}+1<0$$
$$\frac{(1+x)^2-(1-x)^2+(1-x^2)}{(1-x)(1+x)}<0$$
$$\frac{-x^2+4x+1}{(1-x)(1+x)}<0$$
$$\frac{x^2-4x-1}{(x-1)(x+1)}<0$$

The denominator is zero when $x=\pm 1$. Using the quadratic formula, we find the numerator is zero when $x=2\pm\sqrt{5}$. Thus, the key numbers are ± 1 and $2\pm\sqrt{5}$. For purposes of picking appropriate test numbers, note that $2+\sqrt{5}\approx 4.2$ and that $2-\sqrt{5}\approx -0.2$. Draw the sign chart:

Interval	Test Number	$x-4.2$	$x-1$	$x+0.2$	$x+1$	$\frac{x^2-4x-1}{(x-1)(x+1)}$
$(-\infty,-1)$	-2	neg.	neg.	neg.	neg.	pos.
$(-1,-0.2)$	-0.5	neg.	neg.	neg.	pos.	neg.
$(-0.2,1)$	0	neg.	neg.	pos.	pos.	pos.
$(1,4.2)$	2	neg.	pos.	pos.	pos.	neg.
$(4.2,\infty)$	5	pos.	pos.	pos.	pos.	pos.

So $\dfrac{x^2-4x-1}{(x-1)(x+1)}<0$ (and consequently the original inequality is satisfied) on the intervals $\left(-1,2-\sqrt{5}\right)\cup\left(1,2+\sqrt{5}\right)$.

65. First simplify the inequality:

$$\frac{3-2x}{3+2x} - \frac{1}{x} > 0$$

$$\frac{x(3-2x) - (3+2x)}{x(3+2x)} > 0$$

$$\frac{x - 2x^2 - 3}{x(3+2x)} > 0$$

$$\frac{2x^2 - x + 3}{x(3+2x)} < 0$$

The denominator is zero when $x = 0$ and $x = -\frac{3}{2}$. By using the quadratic formula, we find that there are no real numbers for which the numerator is zero. Also note that $2x^2 - x + 3$ is positive for all x. Thus 0 and $-\frac{3}{2}$ are the only key numbers. Draw the sign chart:

Interval	Test Number	$2x+3$	x	$\frac{2x^2-x+3}{x(2x+3)}$
$\left(-\infty, -\frac{3}{2}\right)$	-2	neg.	neg.	pos.
$\left(-\frac{3}{2}, 0\right)$	-1	pos.	neg.	neg.
$(0, \infty)$	1	pos.	pos.	pos.

So $\dfrac{2x^2 - x + 3}{x(2x+3)} < 0$ (and consequently the original inequality is satisfied) on the interval $\left(-\frac{3}{2}, 0\right)$.

67. First simplify the inequality:

$$1 + \frac{1}{x} - \frac{1}{1+x} \geq 0$$

$$\frac{x(1+x) + (1+x) - x}{x(1+x)} \geq 0$$

$$\frac{x^2 + x + 1}{x(1+x)} \geq 0$$

The only key numbers are 0 and -1. By using the quadratic formula, we find that there are no real numbers for which the numerator $x^2 + x + 1$ is zero. Note that $x^2 + x + 1$ is positive for all values of x. Draw the sign chart:

Interval	Test Number	$x+1$	x	$\frac{x^2+x+1}{x(x+1)}$
$(-\infty, -1)$	-2	neg.	neg.	pos.
$(-1, 0)$	$-\frac{1}{2}$	pos.	neg.	neg.
$(0, \infty)$	1	pos.	pos.	pos.

So $\dfrac{x^2 + x + 1}{x(1+x)} \geq 0$ (and consequently the original inequality is satisfied) on the intervals $(-\infty, -1) \cup (0, \infty)$.

69. (a) Solve the inequality:
$$x^2 - 4x - 5 \geq 0$$
$$(x - 5)(x + 1) \geq 0$$
The key numbers are 5 and –1. Draw the sign chart:

Interval	Test Number	$x - 5$	$x + 1$	$(x - 5)(x + 1)$
$(-\infty, -1)$	-2	neg.	neg.	pos.
$(-1, 5)$	0	neg.	pos.	neg.
$(5, \infty)$	6	pos.	pos.	pos.

The solution set for $(x - 5)(x + 1) \geq 0$ is $(-\infty, -1] \cup [5, \infty)$.

(b) Solve the inequality $\dfrac{1}{x^2 - 4x - 5} \geq 0$. This will have the same solutions as in (a), except the endpoints $x = -1$ and $x = 5$ are not included. The solution set for this is $(-\infty, -1) \cup (5, \infty)$.

71. The solutions will be real provided the discriminant $b^2 - 4ac$ is non-negative. Thus:
$$b^2 - 4 \geq 0$$
$$(b - 2)(b + 2) \geq 0$$
The key numbers are ± 2. Draw the sign chart:

Interval	Test Number	$b - 2$	$b + 2$	$(b - 2)(b + 2)$
$(-\infty, -2)$	-3	neg.	neg.	pos.
$(-2, 2)$	0	neg.	pos.	neg.
$(2, \infty)$	3	pos.	pos.	pos.

The solution set consists of the two intervals $(-\infty, -2] \cup [2, \infty)$. For values of b in either of these two intervals, the equation $x^2 + bx + 1 = 0$ will have real solutions.

73. If $x = 1$ is a solution of $\dfrac{2a + x}{x - 2a} < 1$, then:
$$\frac{2a + 1}{1 - 2a} < 1$$
$$\frac{2a + 1}{1 - 2a} - 1 < 0$$
$$\frac{2a + 1 - (1 - 2a)}{1 - 2a} < 0$$
$$\frac{4a}{1 - 2a} < 0$$

The key numbers are 0 and $\frac{1}{2}$. Draw the sign chart:

Interval	Test Number	$4a$	$1-2a$	$\frac{4a}{1-2a}$
$(-\infty,0)$	-1	neg.	pos.	neg.
$\left(0,\frac{1}{2}\right)$	$\frac{1}{4}$	pos.	pos.	pos.
$\left(\frac{1}{2},\infty\right)$	1	pos.	neg.	neg.

The allowable values of a are those numbers in either of the intervals $(-\infty,0)\cup\left(\frac{1}{2},\infty\right)$.

75. Using the Pythagorean theorem, we find the hypotenuse is $\sqrt{x^2+(1-x)^2}$, or $\sqrt{2x^2-2x+1}$. If this is less than $\frac{\sqrt{17}}{5}$, then:

$$\sqrt{2x^2-2x+1}<\frac{\sqrt{17}}{5}$$
$$\left(5\sqrt{2x^2-2x+1}\right)^2<\left(\sqrt{17}\right)^2$$
$$50x^2-50x+25<17$$
$$50x^2-50x+8<0$$
$$25x^2-25x+4<0$$
$$(5x-1)(5x-4)<0$$

The key numbers here are $\frac{1}{5}$ and $\frac{4}{5}$. Draw the sign chart:

Interval	Test Number	$5x-1$	$5x-4$	$(5x-1)(5x-4)$
$(-\infty,\frac{1}{5})$	0	neg.	neg.	pos.
$\left(\frac{1}{5},\frac{4}{5}\right)$	$\frac{2}{5}$	pos.	neg.	neg.
$\left(\frac{4}{5},\infty\right)$	1	pos.	pos.	pos.

The solution set is the interval $\left(\frac{1}{5},\frac{4}{5}\right)$.

77. For the cylinder:
$$\frac{V}{S}=\frac{\pi r^2 h}{2\pi r^2+2\pi rh}=\frac{\pi r^2}{2\pi r^2+2\pi r}=\frac{r}{2r+2}$$

The condition $\dfrac{V}{S}<\dfrac{1}{3}$ then becomes:

$$\frac{r}{2r+2}<\frac{1}{3}$$

Since r is positive here, we can multiply through by the positive quantity $3(2r+2)$. This yields:

$$3r<2r+2$$
$$r<2$$

Since $r>0$, the possible values of r are in the interval $(0,2)$.

79. (a) Complete the table:

x	1	2	5	10	100	200	10,000
$\dfrac{x^2+1000}{2x^2+x}$	333.67	100.40	18.64	5.24	0.55	0.51	0.50

(b) Simplify the inequality:

$$\frac{x^2+1000}{2x^2+x} < \frac{1}{2}$$

$$\frac{x^2+1000}{x(2x+1)} - \frac{1}{2} < 0$$

$$\frac{2(x^2+1000)-x(2x+1)}{2x(2x+1)} < 0$$

$$\frac{-x+2000}{2x(2x+1)} < 0$$

Thus the key numbers are 2000, 0, and $-\frac{1}{2}$. Draw the sign chart:

Interval	Test Number	$2x+1$	x	$-x+2000$	$\frac{-x+2000}{2x(2x+1)}$
$\left(-\infty,-\frac{1}{2}\right)$	-1	neg.	neg.	pos.	pos.
$\left(-\frac{1}{2},0\right)$	$-\frac{1}{4}$	pos.	neg.	pos.	neg.
$(0,2000)$	1	pos.	pos.	pos.	pos.
$(2000,\infty)$	2001	pos.	pos.	neg.	neg.

The solution set for $\dfrac{x^2+1000}{2x^2+x} < \frac{1}{2}$ is $\left(-\frac{1}{2},0\right)\cup(2000,\infty)$. So the positive real numbers which satisfy the inequality are those in the interval $(2000, \infty)$.

(c) The inequality holds when $n > 2000$, and the smallest natural number fulfilling this condition is $n = 2001$.

81. Multiply this out to obtain:

$$x^2-2ax+a^2-x^2+2bx-b^2 > \frac{(a-b)^2}{4}$$

$$(2b-2a)x+(a^2-b^2) > \frac{(a-b)^2}{4}$$

$$8(b-a)x+4(a+b)(a-b) > (a-b)^2$$

Since $a > b$, $a - b > 0$, so we can divide by $a - b$ and preserve the inequality. Thus:

$$-8x+4(a+b) > a-b$$
$$-8x+4a+4b > a-b$$
$$-8x > -3a-5b$$
$$x < \frac{3a+5b}{8}$$

So the solution is $\left(-\infty,\dfrac{3a+5b}{8}\right)$.

Graphing Utility Exercises for Section 3.6

1.   Graphing the equation:

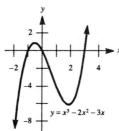

Note that $y > 0$ when $-1 < x < 0$ and when $x > 3$. Thus the curve is positive on the set $(-1,0) \cup (3,\infty)$, which represents the solution set for the inequality.

3.   (a)   Graphing the equation:

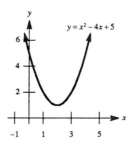

(b)   Since $y > 0$ for all values of x, the solution set is all real numbers, or $(-\infty,\infty)$.

(c)   There are no real numbers for which $y < 0$, so the solution set is empty.

5.   Graphing the curve $y = x^2 - 5x + 3$:

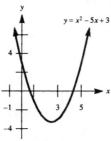

Thus $y \leq 0$ on the interval $[0.697, 4.303]$. Finding the x-intercepts algebraically, use the quadratic formula:

$$x = \frac{-(-5) \pm \sqrt{(-5)^2 - 4(1)(3)}}{2(1)} = \frac{5 \pm \sqrt{25 - 12}}{2} = \frac{5 \pm \sqrt{13}}{2} \approx 0.697,\ 4.303$$

These values agree with those found graphically.

7. Graphing the curve $y = 2x^2 - 5x - 1$:

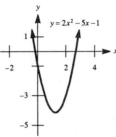

Thus $y \geq 0$ on the set $(-\infty, -0.186] \cup [2.686, \infty)$. Finding the x-intercepts algebraically, use the quadratic formula:

$$x = \frac{-(-5) \pm \sqrt{(-5)^2 - 4(2)(-1)}}{2(2)} = \frac{5 \pm \sqrt{25+8}}{4} = \frac{5 \pm \sqrt{33}}{4} \approx -0.186, 2.686$$

These values agree with those found graphically.

9. Graphing the curve $y = 0.25x^2 - 6x - 2$:

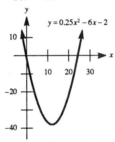

Thus $y < 0$ on the interval $(-0.329, 24.329)$. Finding the x-intercepts algebraically, use the quadratic formula:

$$x = \frac{-(-6) \pm \sqrt{(-6)^2 - 4(0.25)(-2)}}{2(0.25)}$$

$$= \frac{6 \pm \sqrt{36+2}}{0.5}$$

$$= \frac{6 \pm \sqrt{38}}{0.5}$$

$$= 12 \pm 2\sqrt{38}$$

$$\approx -0.329, 24.329$$

These values agree with those found graphically.

11. Graphing the curve $y = x^4 - 2x^2 - 1$:

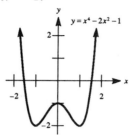

Thus $y > 0$ on the set $(-\infty, -1.554) \cup (1.554, \infty)$. Finding the x-intercepts algebraically, use the quadratic formula:

$$x^2 = \frac{-(-2) \pm \sqrt{(-2)^2 - 4(1)(-1)}}{2(1)} = \frac{2 \pm \sqrt{4+4}}{2} = \frac{2 \pm 2\sqrt{2}}{2} = 1 \pm \sqrt{2}$$

Since $1 - \sqrt{2} < 0$, we have $x^2 = 1 + \sqrt{2}$, thus:

$$x = \pm\sqrt{1 + \sqrt{2}} \approx \pm 1.554$$

These values agree with those found graphically.

13. Graphing the curve $y = \dfrac{x^2 - 5}{x^2 + 1}$:

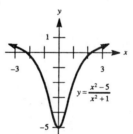

Thus $y \leq 0$ on the interval $[-2.236, 2.236]$. To find the x-intercepts algebraically, set the numerator equal to 0:

$$x^2 - 5 = 0$$
$$x^2 = 5$$
$$x = \pm\sqrt{5} \approx \pm 2.236$$

Note that $x^2 + 1 \neq 0$. These values agree with those found graphically.

15. Graphing the curve $y = \dfrac{x^2 + 1}{x^2 - 5}$:

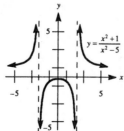

Thus $y \le 0$ on the interval $(-2.236, 2.236)$. To find the asymptotes algebraically, set the denominator equal to 0:

$$x^2 - 5 = 0$$
$$x^2 = 5$$
$$x = \pm\sqrt{5} \approx \pm2.236$$

Note that $x^2 + 1 \ne 0$. These values agree with those found graphically.

17. Graphing the curve $y = x^3 + 2x + 1$:

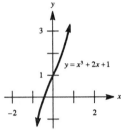

Thus $y \ge 0$ on the interval $[-0.453, \infty)$.

19. Graphing the curve $y = x^4 - 2x + 1$:

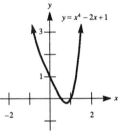

Thus $y > 0$ on the set $(-\infty, 0.544) \cup (1, \infty)$.

21. Graphing the curve $y = x - \dfrac{x^3}{3!} + \dfrac{x^5}{5!} - \dfrac{x^7}{7!}$:

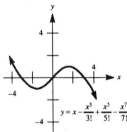

Thus $y < 0$ on the set $(-3.079, 0) \cup (3.079, \infty)$.

23. Graphing the curve $y = \dfrac{x^2 - 5x - 5}{x^2 + 5x + 5}$:

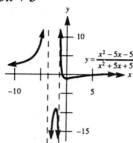

Thus $y \geq 0$ on the set $(-\infty, -3.618) \cup (-1.382, -0.854] \cup [5.854, \infty)$.

Chapter Three Review Exercises

1. True

3. False (For instance: $x = 5$)

5. False (For instance: $x = 1$)

7. True

9. False (For instance: $x = 0$, $y = -1$)

11. False (For instance: $a = -5$, $b = 1$)

13. Solve the equation:
$$12x^2 + 2x - 2 = 0$$
$$6x^2 + x - 1 = 0$$
$$(3x - 1)(2x + 1) = 0$$
$$x = \tfrac{1}{3}, -\tfrac{1}{2}$$

15. Multiplying by 2, solve the equation:
$$\tfrac{1}{2}x^2 + x - 12 = 0$$
$$x^2 + 2x - 24 = 0$$
$$(x + 6)(x - 4) = 0$$
$$x = -6, 4$$

17.   Solve the equation:
$$\frac{x}{5-x} = \frac{-2}{11-x}$$
$$11x - x^2 = -10 + 2x$$
$$-x^2 + 9x + 10 = 0$$
$$x^2 - 9x - 10 = 0$$
$$(x - 10)(x + 1) = 0$$
$$x = 10, -1$$

19.   Solve the equation:
$$\frac{1}{3x-7} - \frac{2}{5x-5} - \frac{3}{3x+1} = 0$$
$$(5x - 5)(3x + 1) - 2(3x - 7)(3x + 1) - 3(3x - 7)(5x - 5) = 0$$
$$15x^2 - 10x - 5 - 18x^2 + 36x + 14 - 45x^2 + 150x - 105 = 0$$
$$-48x^2 + 176x - 96 = 0$$
$$3x^2 - 11x + 6 = 0$$
$$(3x - 2)(x - 3) = 0$$
$$x = \tfrac{2}{3}, 3$$

21.   Multiply by 2 to obtain $2t^2 + 2t - 1 = 0$. Now using the quadratic formula:
$$t = \frac{-2 \pm \sqrt{4 - 4(2)(-1)}}{2(2)} = \frac{-2 \pm \sqrt{12}}{4} = \frac{-2 \pm 2\sqrt{3}}{4} = \frac{-1 \pm \sqrt{3}}{2}$$

23.   Since $y^4 = 9$, $y^2 = 3$ or $y^2 = -3$ (which has no real solutions). For $y^2 = 3$, we have $y = \pm\sqrt{3}$.

25.   Factoring yields $x(x^4 - 2x^2 - 2) = 0$, so $x = 0$ is one solution. Using the quadratic formula:
$$x^2 = \frac{2 \pm \sqrt{4 - 4(-2)}}{2} = \frac{2 \pm \sqrt{12}}{2} = \frac{2 \pm 2\sqrt{3}}{2} = 1 \pm \sqrt{3}$$
Note that $1 - \sqrt{3}$ is negative, so no real solutions are obtained from $x^2 = 1 - \sqrt{3}$. Consequently, $x^2 = 1 + \sqrt{3}$ and $x = \pm\sqrt{1 + \sqrt{3}}$. So the solutions are $\pm\sqrt{1 + \sqrt{3}}$ and 0.

27.   Multiplying by x^2:
$$x^2 + 14x + 48 = 0$$
$$(x + 6)(x + 8) = 0$$
$$x = -6, -8$$

29.   Let $t = x^{1/4}$ and $t^2 = x^{1/2}$. Then:
$$t^2 - 13t + 36 = 0$$
$$(t - 9)(t - 4) = 0$$
$$t = 9 \text{ or } t = 4$$
If $t = 9$, then $x^{1/4} = 9$ and $x = 9^4 = 6561$. If $t = 4$, then $x^{1/4} = 4$ and $x = 4^4 = 256$. So the solutions are 6561 and 256.

31. Squaring:
$$\sqrt{4-3x} = 5$$
$$4 - 3x = 25$$
$$-3x = 21$$
$$x = -7$$
This value checks in the original equation.

33. Squaring:
$$\sqrt{4x+3} = \sqrt{11-8x} - 1$$
$$4x + 3 = 11 - 8x - 2\sqrt{11-8x} + 1$$
$$12x - 9 = -2\sqrt{11-8x}$$
$$144x^2 - 216x + 81 = 44 - 32x$$
$$144x^2 - 184x + 37 = 0$$
$$(4x - 1)(36x - 37) = 0$$
$$x = \tfrac{1}{4} \text{ or } x = \tfrac{37}{36}$$
Upon checking, we find only $x = \tfrac{1}{4}$ checks in the original equation.

35. Isolating the radical and squaring:
$$\sqrt{x+48} - \sqrt{x} = 4$$
$$\sqrt{x+48} = \sqrt{x} + 4$$
$$x + 48 = x + 8\sqrt{x} + 16$$
$$32 = 8\sqrt{x}$$
$$4 = \sqrt{x}$$
$$16 = x$$
This value checks in the original equation.

37. Multiplying by $\sqrt{x+6}\sqrt{x-6}$:
$$\frac{2}{\sqrt{x^2-36}} + \frac{1}{\sqrt{x+6}} - \frac{1}{\sqrt{x-6}} = 0$$
$$2 + \sqrt{x-6} - \sqrt{x+6} = 0$$
$$2 + \sqrt{x-6} = \sqrt{x+6}$$
$$4 + 4\sqrt{x-6} + x - 6 = x + 6$$
$$4\sqrt{x-6} = 8$$
$$\sqrt{x-6} = 2$$
$$x - 6 = 4$$
$$x = 10$$
This value checks in the original equation.

39. Squaring:

$$\sqrt{x+7} - \sqrt{x+2} = \sqrt{x-1} - \sqrt{x-2}$$
$$x+7 - 2\sqrt{x+7}\sqrt{x+2} + x+2 = x-1 - 2\sqrt{x-1}\sqrt{x-2} + x-2$$
$$-2\sqrt{x+7}\sqrt{x+2} + 9 = -2\sqrt{x-1}\sqrt{x-2} - 3$$
$$-2\sqrt{x+7}\sqrt{x+2} = -2\sqrt{x-1}\sqrt{x-2} - 12$$
$$\sqrt{x+7}\sqrt{x+2} = \sqrt{x-1}\sqrt{x-2} + 6$$
$$(x+7)(x+2) = (x-1)(x-2) + 12\sqrt{x-1}\sqrt{x-2} + 36$$
$$x^2 + 9x + 14 = x^2 - 3x + 2 + 12\sqrt{x-1}\sqrt{x-2} + 36$$
$$12x - 24 = 12\sqrt{x-1}\sqrt{x-2}$$
$$x - 2 = \sqrt{x-1}\sqrt{x-2}$$
$$x^2 - 4x + 4 = x^2 - 3x + 2$$
$$-x = -2$$
$$x = 2$$

This value checks in the original equation.

41. Solve the equation for x:

$$4x^2y^2 - 4xy = -1$$
$$4x^2y^2 - 4xy + 1 = 0$$
$$(2xy - 1)^2 = 0$$
$$2xy - 1 = 0$$
$$x = \frac{1}{2y}$$

43. Multiplying through by a^2bx yields:

$$a^2bx^2 + abx - a^2x = 2b + 2a$$
$$a^2bx^2 + (ab - a^2)x - 2(b+a) = 0$$

Applying the quadratic formula:

$$x = \frac{-(ab - a^2) \pm \sqrt{(ab - a^2)^2 - 4(a^2b)[-2(b+a)]}}{2(a^2b)}$$
$$= \frac{-ab + a^2 \pm \sqrt{a^4 + 6a^3b + 9a^2b^2}}{2a^2b}$$
$$= \frac{-ab + a^2 \pm \sqrt{(a^2 + 3ab)^2}}{2a^2b}$$
$$= \frac{-ab + a^2 \pm (a^2 + 3ab)}{2a^2b}$$

So the two solutions are:

$$x = \frac{-ab + a^2 + a^2 + 3ab}{2a^2b} = \frac{2a^2 + 2ab}{2a^2b} = \frac{2a(a+b)}{2a^2b} = \frac{a+b}{ab}$$
$$x = \frac{-ab + a^2 - a^2 - 3ab}{2a^2b} = \frac{-4ab}{2a^2b} = -\frac{2}{a}$$

Both of these values satisfy the original equation.

45. Multiplying by $abx(x + a + b)$ yields:
$$\frac{1}{x + a + b} = \frac{1}{x} + \frac{1}{a} + \frac{1}{b}$$
$$abx = ab(x + a + b) + bx(x + a + b) + ax(x + a + b)$$
$$abx = abx + a^2b + ab^2 + bx^2 + abx + b^2x + ax^2 + a^2x + abx$$
$$0 = (a + b)x^2 + \left(a^2 + 2ab + b^2\right)x + \left(a^2b + ab^2\right)$$
$$0 = (a + b)x^2 + (a + b)^2 x + ab(a + b)$$
$$0 = x^2 + (a + b)x + ab \qquad \text{(since } a + b \neq 0\text{)}$$
$$0 = (x + a)(x + b)$$
$$x = -a \text{ or } x = -b$$
Both of these values satisfy the original equation.

47. Letting $t = \frac{1}{\sqrt{x}}$ and $t^2 = \frac{1}{x}$:
$$\left(a^2 - b^2\right)t^2 - 2\left(a^2 + b^2\right)t + \left(a^2 - b^2\right) = 0$$
$$[(a - b)t - (a + b)][(a + b)t - (a - b)] = 0$$
Setting each factor equal to zero:
$$(a - b)t - (a + b) = 0 \qquad\qquad (a + b)t - (a - b) = 0$$

$$t = \frac{a + b}{a - b} \qquad\qquad\qquad t = \frac{a - b}{a + b}$$

$$\frac{1}{\sqrt{x}} = \frac{a + b}{a - b} \qquad\qquad\qquad \frac{1}{\sqrt{x}} = \frac{a - b}{a + b}$$

$$\sqrt{x} = \frac{a - b}{a + b} \qquad\qquad\qquad \sqrt{x} = \frac{a + b}{a - b}$$

$$x = \left(\frac{a - b}{a + b}\right)^2 \qquad\qquad x = \left(\frac{a + b}{a - b}\right)^2$$

Both of these values check in the original equation.

49. Solve the inequality:
$$-1 < \frac{1 - 2(1 + x)}{3} < 1$$
$$-3 < 1 - 2 - 2x < 3$$
$$-2 < -2x < 4$$
$$1 > x > -2$$
The solution set is $(-2, 1)$.

51. The inequality is equivalent to $-\frac{1}{2} \leq x \leq \frac{1}{2}$, so the solution set is $\left[-\frac{1}{2}, \frac{1}{2}\right]$.

53. Solve the inequality:
$$|x + 4| < \frac{1}{10}$$
$$-\frac{1}{10} < x + 4 < \frac{1}{10}$$
$$-\frac{41}{10} < x < -\frac{39}{10}$$
The solution set is $\left(-\frac{41}{10}, -\frac{39}{10}\right)$.

55. Solve the inequality $|2x-1| \geq 5$:

$$2x - 1 \geq 5 \quad \text{or} \quad -(2x-1) \geq 5$$
$$2x \geq 6 \qquad\qquad 2x - 1 \leq -5$$
$$x \geq 3 \qquad\qquad\quad 2x \leq -4$$
$$\qquad\qquad\qquad\qquad x \leq -2$$

The solution set is $(-\infty, -2] \cup [3, \infty)$.

57. Factor $x^2 + 3x - 40 = (x-5)(x+8)$, so the key numbers are 5 and -8. Now draw the sign chart:

Interval	Test Number	$x+8$	$x-5$	$(x+8)(x-5)$
$(-\infty, -8)$	-9	neg.	neg.	pos.
$(-8, 5)$	0	pos.	neg.	neg.
$(5, \infty)$	6	pos.	pos.	pos.

So the product is negative on the interval $(-8, 5)$.

59. The key numbers are found by using the quadratic formula to solve $x^2 - 6x - 1 = 0$. The result is $x = 3 \pm \sqrt{10}$. For purposes of picking appropriate test numbers, note that $3 + \sqrt{10} \approx 6.2$ and that $3 - \sqrt{10} \approx -0.2$. Now draw the sign chart:

Interval	Test Number	$x-6.2$	$x+0.2$	x^2-6x-1
$\left(-\infty, 3-\sqrt{10}\right)$	-1	neg.	neg.	pos.
$\left(3-\sqrt{10}, 3+\sqrt{10}\right)$	0	neg.	pos.	neg.
$\left(3+\sqrt{10}, \infty\right)$	7	pos.	pos.	pos.

So the product is negative on the interval $\left(3-\sqrt{10}, 3+\sqrt{10}\right)$.

61. First factor the inequality:

$$x^4 - 34x^2 + 225 < 0$$
$$\left(x^2 - 9\right)\left(x^2 - 25\right) < 0$$
$$(x-3)(x+3)(x-5)(x+5) < 0$$

The key numbers are -3, 3, -5, and 5. Now draw the sign chart:

Interval	Test Number	$x-5$	$x-3$	$x+3$	$x+5$	$\left(x^2-9\right)\left(x^2-25\right)$
$(-\infty, -5)$	-6	neg.	neg.	neg.	neg.	pos.
$(-5, -3)$	-4	neg.	neg.	neg.	pos.	neg.
$(-3, 3)$	0	neg.	neg.	pos.	pos.	pos.
$(3, 5)$	4	neg.	pos.	pos.	pos.	neg.
$(5, \infty)$	6	pos.	pos.	pos.	pos.	pos.

So the product is negative on the intervals $(-5, -3) \cup (3, 5)$.

63. The key numbers are 7 and –2. Now draw the sign chart:

Interval	Test Number	$(x+2)^3$	$(x-7)^2$	$\frac{(x-7)^2}{(x+2)^3}$
$(-\infty,-2)$	–3	neg.	pos.	neg.
$(-2,7)$	0	pos.	pos.	pos.
$(7,\infty)$	8	pos.	pos.	pos.

The quotient is positive on the intervals $(-2,7)\cup(7,\infty)$, and zero at $x=7$. The single interval $(-2,\infty)$ satisfies the inequality.

65. Factor the inequality as $\dfrac{(x-1)(x-9)}{(x+1)(x^2-x+1)}\le 0$. The key numbers are 1, 9, and –1. Draw the sign chart:

Interval	Test Number	$x-9$	$x-1$	$x+1$	$\frac{(x-1)(x-9)}{x^3+1}$
$(-\infty,-1)$	–2	neg.	neg.	neg.	neg.
$(-1,1)$	0	neg.	neg.	pos.	pos.
$(1,9)$	2	neg.	pos.	pos.	neg.
$(9,\infty)$	10	pos.	pos.	pos.	pos.

The quotient is negative on the intervals $(-\infty,-1)\cup(1,9)$, and is zero at $x=1$ and $x=9$. The solution set is $(-\infty,-1)\cup[1,9]$.

67. First simplify and factor the inequality:
$$\frac{1-2x}{1+2x}-\frac{1}{2}\le 0$$
$$\frac{2(1-2x)-(1+2x)}{2(1+2x)}\le 0$$
$$\frac{-6x+1}{2(1+2x)}\le 0$$

The key numbers are $\frac{1}{6}$ and $-\frac{1}{2}$. Now draw the sign chart:

Interval	Test Number	$2x+1$	$-6x+1$	$\frac{-6x+1}{2(2x+1)}$
$\left(-\infty,-\frac{1}{2}\right)$	–1	neg.	pos.	neg.
$\left(-\frac{1}{2},\frac{1}{6}\right)$	0	pos.	pos.	pos.
$\left(\frac{1}{6},\infty\right)$	1	pos.	neg.	neg.

The quotient is negative on the intervals $\left(-\infty,-\frac{1}{2}\right)\cup\left(\frac{1}{6},\infty\right)$, and zero at $\frac{1}{6}$. The solution set is $\left(-\infty,-\frac{1}{2}\right)\cup\left[\frac{1}{6},\infty\right)$.

69.   First simplify the inequality:

$$\sqrt{x} - \frac{5}{\sqrt{x}} - 4 \le 0$$

$$\frac{x - 5 - 4\sqrt{x}}{\sqrt{x}} \le 0$$

Looking at the denominator, we see that one key number is $x = 0$. The others, if any, are found by solving the equation:

$$x - 5 - 4\sqrt{x} = 0$$

Writing this as $x - 5 = 4\sqrt{x}$, and then squaring both sides:

$$x^2 - 10x + 25 = 16x$$
$$x^2 - 26x + 25 = 0$$
$$(x - 25)(x - 1) = 0$$
$$x = 25 \text{ or } x = 1$$

The value $x = 25$ satisfies the equation $x - 5 - 4\sqrt{x} = 0$, but $x = 1$ is an extraneous root. Thus the key numbers are 0 and 25. Now draw the sign chart:

Interval	Test Number	$\sqrt{x}$	$x - 5 - 4\sqrt{x}$	$\frac{x-5-4\sqrt{x}}{\sqrt{x}}$
$(-\infty, 0)$	-1		undefined	
$(0, 25)$	1	pos.	neg.	neg.
$(25, \infty)$	36	pos.	pos.	pos.

The quotient is negative on the interval $(0, 25)$ and zero at $x = 25$. So the solution set is $(0, 25]$.

71.   The discriminant is $b^2 - 4ac = 1 - 4k^2$. Solve for when this discriminant is non-negative:

$$1 - 4k^2 \ge 0$$
$$(1 - 2k)(1 + 2k) \ge 0$$

The key numbers here are $k = \pm\frac{1}{2}$. Draw the sign chart:

Interval	Test Number	$1 + 2k$	$1 - 2k$	$1 - 4k^2$
$\left(-\infty, -\frac{1}{2}\right)$	-1	neg.	pos.	neg.
$\left(-\frac{1}{2}, \frac{1}{2}\right)$	0	pos.	pos.	pos.
$\left(\frac{1}{2}, \infty\right)$	1	pos.	neg.	neg.

The product is positive on $\left(-\frac{1}{2}, \frac{1}{2}\right)$, and zero at $x = -\frac{1}{2}$ and $x = \frac{1}{2}$. So the values of k are chosen from the interval $\left[-\frac{1}{2}, \frac{1}{2}\right]$.

73.   The discriminant is:

$$b^2 - 4ac = (k + 1)^2 - 4(2k) = k^2 + 2k + 1 - 8k = k^2 - 6k + 1$$

To find the key numbers for $k^2 - 6k + 1 \ge 0$, use the quadratic formula:

$$k = \frac{6 \pm \sqrt{36 - 4}}{2} = \frac{6 \pm \sqrt{32}}{2} = \frac{6 \pm 4\sqrt{2}}{2} = 3 \pm 2\sqrt{2}$$

For purposes of picking appropriate test numbers, note that $3+2\sqrt{2}\approx 5.8$ and that $3-2\sqrt{2}\approx 0.2$. Now draw the sign chart:

Interval	Test Number	$k-5.8$	$k-0.2$	k^2-6k+1
$(-\infty, 3-2\sqrt{2})$	0	neg.	neg.	pos.
$(3-2\sqrt{2}, 3+2\sqrt{2})$	3	neg.	pos.	neg.
$(3+2\sqrt{2}, \infty)$	7	pos.	pos.	pos.

The expression is positive on the interval $(-\infty, 3-2\sqrt{2})\cup(3+2\sqrt{2}, \infty)$, and zero at $k=3-2\sqrt{2}$ and $k=3+2\sqrt{2}$. So the values of k are chosen from the interval $(-\infty, 3-2\sqrt{2}]\cup[3+2\sqrt{2}, \infty)$.

75. If $a=1$, we can find c and b:
$$c = r_1\bullet r_2 = \left(1-\sqrt{2}\right)\left(1+\sqrt{2}\right) = 1-2 = -1$$
$$b = -\left(r_1+r_2\right) = -\left(1-\sqrt{2}+1+\sqrt{2}\right) = -2$$
Thus the equation is $x^2-2x-1=0$.

77. If $a=1$, we can find c and b:
$$c = r_1\bullet r_2 = \left(3+\sqrt{6}\right)\left(3-\sqrt{6}\right) = 9-6 = 3$$
$$b = -\left(r_1+r_2\right) = -\left(3+\sqrt{6}+3-\sqrt{6}\right) = -6$$
Thus the equation is $x^2-6x+3=0$.

79. Let x and y represent the two numbers. So $x^3+y^3=2071$ while $x+y=19$. Solving for y yields $y=19-x$, now substitute:
$$x^3+y^3=2071$$
$$x^3+(19-x)^3=2071$$
$$x^3+6859-1083x+57x^2-x^3=2071$$
$$57x^2-1083x+4788=0$$
$$x^2-19x+84=0$$
$$(x-12)(x-7)=0$$
$$x=7,12$$
If $x=7$ then $y=19-x=12$, and if $x=12$ then $y=19-x=7$.
The two numbers are 7 and 12.

81. Let x and $11-x$ denote the units digit and tens digit, respectively. Then:
$$\text{original number}+27 = \text{new number}$$
$$10(11-x)+x+27 = 10x+(11-x)$$
$$110-10x+x+27 = 10x+11-x$$
$$-18x = -126$$
$$x = 7$$
$$11-x = 4$$
The original number is 47.

83.   First re-draw the figure:

Since $AE = FB$:
$$AE + EF + FB = 1$$
$$EF + 2AE = 1$$
$$AE = \frac{1 - EF}{2}$$

By the Pythagorean theorem and the fact that $AG = AE$ and $EF = GE$:
$$(AG)^2 + (AE)^2 = (GE)^2$$
$$(AE)^2 + (AE)^2 = (EF)^2$$
$$2(AE)^2 = (EF)^2$$
$$2\left(\frac{1 - EF}{2}\right)^2 = EF^2$$

Now, for convenience, let $EF = x$. Then:
$$2\left(\frac{1 - 2x + x^2}{4}\right) = x^2$$
$$1 - 2x + x^2 = 2x^2$$
$$0 = x^2 + 2x - 1$$

The quadratic formula then gives us:
$$x = \frac{-2 \pm \sqrt{4+4}}{2} = \frac{-2 \pm \sqrt{8}}{2} = \frac{-2 \pm 2\sqrt{2}}{2} = -1 \pm \sqrt{2}$$

Choose the positive root, since $x > 0$. Thus $x = EF = -1 + \sqrt{2}$ cm.

85.   Given that the length of a side is $\frac{x}{4}$, compute the area and the perimeter:
$$\text{area} = \left(\frac{x}{4}\right)^2 = \frac{x^2}{16}$$
$$\text{perimeter} = 4\left(\frac{x}{4}\right) = x$$

If the area is numerically greater than the perimeter, then:

$$\frac{x^2}{16} > x$$
$$x^2 > 16x$$
$$x^2 - 16x > 0$$
$$x(x - 16) > 0$$

Since x must be positive, this is satisfied only if $x - 16 > 0$, and hence $x > 16$. We conclude that the area will be numerically greater than the perimeter when $x > 16$ cm, which is the interval $(16 \text{ cm}, \infty)$.

87. Denoting the integers by $x - 1$, x and $x + 1$:

$$(x - 1)^2 + x^2 + (x + 1)^2 = 1454$$
$$x^2 - 2x + 1 + x^2 + x^2 + 2x + 1 = 1454$$
$$3x^2 = 1452$$
$$x^2 = 484$$
$$x = \sqrt{484} = 22$$

Thus $x - 1 = 21$ and $x + 1 = 23$, so the three integers are 21, 22, and 23.

89. Let Joan's speed for the first four miles be r mph. Then the time for those four miles is $\frac{4}{r}$. Since her rate for the next six miles is $r + \frac{1}{2}$, her time on that portion is $\frac{6}{r+1/2}$ or $\frac{12}{2r+1}$, after simplifying. Thus her time for the race is $\frac{4}{r} + \frac{12}{2r+1}$ hours. On the other hand, had she run the entire ten miles at the faster pace, her time would have been $\frac{10}{r+1/2}$ or $\frac{20}{2r+1}$, after simplifying. Since this time is $\frac{2}{60} = \frac{1}{30}$ of an hour faster than her actual time, we can write:

$$\frac{4}{r} + \frac{12}{2r+1} - \frac{1}{30} = \frac{20}{2r+1}$$
$$\frac{4}{r} - \frac{8}{2r+1} - \frac{1}{30} = 0$$
$$4(30)(2r + 1) - 8(30)r - r(2r + 1) = 0$$
$$120 - 2r^2 - r = 0$$
$$2r^2 + r - 120 = 0$$
$$(2r - 15)(r + 8) = 0$$

Choose the positive value here, since $r > 0$. Thus Joan's speed for the first four miles is 7.5 mph, and for the last six miles it is 8 mph. Her time for the entire race is $\left(\frac{4}{7.5} + \frac{6}{8}\right)$ hours. This simplifies to $\frac{77}{60}$ hours, or 1 hour and 17 minutes.

91. First draw a figure:

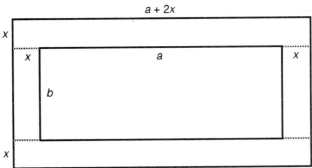

Let x denote the width of the path, as indicated in the figure. To compute the area of the path, divide it into four rectangular portions, as indicated by the dotted lines in the figure. Then the area of the path is:

$$2[x(a+2x)] + 2(bx) = 4x^2 + 2ax + 2bx$$

Since the area of the path and the flower garden are equal:

$$4x^2 + 2ax + 2bx = ab$$
$$4x^2 + 2(a+b)x - ab = 0$$
$$x = \frac{-2(a+b) \pm \sqrt{4(a+b)^2 + 16ab}}{8}$$

Choosing the positive root here (since $x > 0$):

$$x = \frac{-2(a+b) + 2\sqrt{(a+b)^2 + 4ab}}{8} = \frac{-(a+b) + \sqrt{a^2 + 6ab + b^2}}{4}$$

93. For each ball we will find at what time the height is 0. For the first ball we have $h = -16t^2 + 40t + 50$, so setting $h = 0$ yields $-16t^2 + 40t + 50 = 0$, or $8t^2 - 20t - 25 = 0$. Using the quadratic formula and choosing the positive root yields:

$$t = \frac{20 + \sqrt{400 + 32(25)}}{2(8)} = \frac{20 + 20\sqrt{3}}{16} = \frac{5 + 5\sqrt{3}}{4} \approx 3.4 \text{ sec}$$

For the second ball we have $h = -16t^2 + 5t + 100$, so setting $h = 0$ yields $-16t^2 + 5t + 100 = 0$, or $16t^2 - 5t - 100 = 0$. Using the quadratic formula and choosing the postive root yields:

$$t = \frac{5 + \sqrt{25 + 6400}}{2(16)} = \frac{5 + \sqrt{6425}}{32} = \frac{5 + 5\sqrt{257}}{32} \approx 2.7 \text{ sec}$$

Therefore the second ball (the one thrown from 100 ft) hits the ground first.

95. Re-draw the figure:

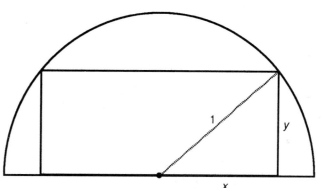

Let y be the height of the rectangle, so the area of the rectangle (using the Pythagorean theorem) is:

$$\text{area} = 2xy = 2x\sqrt{1-x^2}$$

If the area is 1 cm^2, then:

$$2x\sqrt{1-x^2} = 1$$
$$4x^2(1-x^2) = 1$$
$$4x^2 - 4x^4 - 1 = 0$$
$$4x^4 - 4x^2 + 1 = 0$$
$$\left(2x^2 - 1\right)^2 = 0$$
$$2x^2 - 1 = 0$$
$$x^2 = \tfrac{1}{2}$$
$$x = \sqrt{\tfrac{1}{2}} = \tfrac{\sqrt{2}}{2}$$

This value checks in the original equation.

97. Let x be the integer, so $x + 1$ is the next highest integer. Given some integer N, we must find where:

$$x^2 + x^3 = N(x+1)^2$$
$$x^2(1+x) = N(x+1)^2$$
$$\frac{x^2}{x+1} = N$$

But, by Exercise 85 of Section 1.6, $\dfrac{x^2}{x+1}$ cannot be a natural number (positive integer), thus there is no solution to the equation. This proves the desired result.

Chapter Three Test

1.   Multiplying by $6(1-x)(2-x)$:
$$\frac{1}{1-x} + \frac{4}{2-x} = \frac{11}{6}$$
$$6(2-x) + 4(6)(1-x) = 11(1-x)(2-x)$$
$$12 - 6x + 24 - 24x = 22 - 33x + 11x^2$$
$$-11x^2 + 3x + 14 = 0$$
$$11x^2 - 3x - 14 = 0$$
$$(11x - 14)(x+1) = 0$$
$$x = \tfrac{14}{11}, -1$$

2.   (a)   Solve by factoring:
$$x^2 + 4x = 5$$
$$x^2 + 4x - 5 = 0$$
$$(x+5)(x-1) = 0$$
$$x = -5, 1$$

(b)   Solve by writing the equation as $x^2 + 4x - 1 = 0$, then using the quadratic formula:
$$x = \frac{-4 \pm \sqrt{(4)^2 - 4(1)(-1)}}{2(1)} = \frac{-4 \pm \sqrt{20}}{2} = \frac{-4 \pm 2\sqrt{5}}{2} = -2 \pm \sqrt{5}$$

3.   Solve by factoring:
$$x^2\left(x^2 - 7\right) + 12 = 0$$
$$x^4 - 7x^2 + 12 = 0$$
$$\left(x^2 - 4\right)\left(x^2 - 3\right) = 0$$
So $x^2 = 4$ or $x^2 = 3$, and thus $x = \pm 2$ or $x = \pm\sqrt{3}$.

4.   Let $t = x^{2/3}$ and $t^2 = x^{4/3}$ so that the equation becomes $4t^2 - 13t + 9 = 0$. Factoring yields:
$$(4t - 9)(t - 1) = 0$$
$$t = \tfrac{9}{4}, 1$$
With $t = \tfrac{9}{4}$:
$$x^{2/3} = \tfrac{9}{4}$$
$$x^{1/3} = \pm\tfrac{3}{2}$$
$$x = \pm\tfrac{27}{8}$$
With $t = 1$ we have:
$$x^{2/3} = 1$$
$$x^{1/3} = \pm 1$$
$$x = \pm 1$$
So the solutions are ± 1 and $\pm\tfrac{27}{8}$.

5. Isolating $\sqrt{5-2x}$ and squaring:
$$\sqrt{5-2x} - \sqrt{2-x} - \sqrt{3-x} = 0$$
$$\sqrt{5-2x} = \sqrt{2-x} + \sqrt{3-x}$$
$$5 - 2x = 2 - x + 2\sqrt{2-x}\sqrt{3-x} + 3 - x$$
$$0 = 2\sqrt{2-x}\sqrt{3-x}$$

Thus $\sqrt{2-x} = 0$ and consequently $x = 2$ or $\sqrt{3-x} = 0$ and consequently $x = 3$. The value $x = 2$ checks in the original equation, but (considering only the real number system) the value $x = 3$ does not check.

6. Given $|3x - 1| = 2$, either $3x - 1 = 2$ or $3x - 1 = -2$. Therefore:

$$3x - 1 = 2 \qquad \text{or} \qquad 3x - 1 = -2$$
$$3x = 3 \qquad\qquad\qquad 3x = -1$$
$$x = 1 \qquad\qquad\qquad x = -\tfrac{1}{3}$$

The solutions are 1 and $-\tfrac{1}{3}$.

7. First divide by 2 to write the equation as $x^2 + 4x - \tfrac{9}{2} = 0$. Then the sum of the roots is given by $r_1 + r_2 = -b = -4$.

8. If $a = 1$, we can find c and b:
$$c = r_1 \cdot r_2 = \left(2 + 3\sqrt{7}\right)\left(2 - 3\sqrt{7}\right) = 4 - 63 = -59$$
$$b = -\left(r_1 + r_2\right) = -\left(2 + 3\sqrt{7} + 2 - 3\sqrt{7}\right) = -4$$

Thus the equation is $x^2 - 4x - 59 = 0$.

9. Using the quadratic formula:
$$x^2 = \frac{-(-3) \pm \sqrt{(-3)^2 - 4(1)(-1)}}{2(1)} = \frac{3 \pm \sqrt{9+4}}{2} = \frac{3 \pm \sqrt{13}}{2}$$

Since $\dfrac{3 - \sqrt{13}}{2} < 0$, we choose $x^2 = \dfrac{3 + \sqrt{13}}{2}$. Solving for x:

$$x = \pm\sqrt{\frac{3 + \sqrt{13}}{2}} \approx \pm 1.817$$

These values are consistent with the given graph.

10. Solve the inequality:
$$4(1 + x) - 3(2x - 1) \geq 1$$
$$4 + 4x - 6x + 3 \geq 1$$
$$-2x \geq -6$$
$$x \leq 3$$

The solution set is $(-\infty, 3]$.

11.   Solving the double inequality:

$$\frac{3}{5} < \frac{3 - 2x}{-4} < \frac{4}{5}$$
$$-12 > 15 - 10x > -16$$
$$-27 > -10x > -31$$
$$\frac{27}{10} < x < \frac{31}{10}$$

The solution set is $\left(\frac{27}{10}, \frac{31}{10}\right)$.

12.   Solving $|3x - 8| \leq 1$:

$$-1 \leq 3x - 8 \leq 1$$
$$7 \leq 3x \leq 9$$
$$\frac{7}{3} \leq x \leq 3$$

The solution set is $\left[\frac{7}{3}, 3\right]$.

13.   The key numbers are 4 and -8. Draw the sign chart:

Interval	Test Number	$(x+8)^3$	$(x-4)^2$	$(x-4)^2(x+8)^3$
$(-\infty, -8)$	-9	neg.	pos.	neg.
$(-8, 4)$	0	pos.	pos.	pos.
$(4, \infty)$	5	pos.	pos.	pos.

The product is positive on $(-8, 4) \cup (4, \infty)$ and zero at $x = -8$ and $x = 4$. So the inequality is satisfied on $[-8, 4] \cup [4, \infty)$, which combine as $[-8, \infty)$.

14.   Simplifying the inequality and factoring:

$$\frac{(x+1)(x+2) + x(x+2) + x(x+1)}{x(x+1)(x+2)} \geq 0$$
$$\frac{3x^2 + 6x + 2}{x(x+1)(x+2)} \geq 0$$

Three of the key numbers are 0, –1 and –2. The other two are found by using the quadratic formula to solve $3x^2 + 6x + 2 = 0$. The roots are found to be $\frac{-3\pm\sqrt{3}}{3}$. For purposes of picking appropriate test numbers, note that $\frac{-3+\sqrt{3}}{3} \approx -0.4$ and that $\frac{-3-\sqrt{3}}{3} \approx -1.6$. Draw the sign chart:

Interval	Test Number	x	$x+0.4$	$x+1$	$x+1.6$	$x+2$	$\frac{3x^2+6x+2}{x(x+1)(x+2)}$
$(-\infty,-2)$	-3	neg.	neg.	neg.	neg.	neg.	neg.
$(-2,-1.6)$	-1.8	neg.	neg.	neg.	neg.	pos.	pos.
$(-1.6,-1)$	-1.5	neg.	neg.	neg.	pos.	pos.	neg.
$(-1,-0.4)$	-0.5	neg.	neg.	pos.	pos.	pos.	pos.
$(-0.4,0)$	-0.1	neg.	pos.	pos.	pos.	pos.	neg.
$(0,\infty)$	2	pos.	pos.	pos.	pos.	pos.	pos.

So the solution set is $\left(-2,\frac{-3-\sqrt{3}}{3}\right] \cup \left(-1,\frac{-3+\sqrt{3}}{3}\right] \cup (0,\infty)$.

15. Let x and $280 - x$ denote the required number of pounds of Kona and Columbian beans, respectively. Organize the data in a chart:

type of coffee	pounds	cost / pound	total cost
Kona	x	$6.00	$6(x)$
Columbian	$280 - x$	$5.00	$5(280 - x)$
Blend	280	$5.60	$5.6(280)$

We therefore have the equation:
$$6x + 5(280 - x) = 5.6(280)$$
$$6x + 1400 - 5x = 1568$$
$$x = 168$$
$$280 - x = 112$$
Thus 168 pounds of Kona beans and 112 pounds of Columbian beans are required.

16. In the three hours from 6 a.m. to 9 a.m., the train from Los Angeles covers a distance of $(50)(3) = 150$ miles. Now let $t = 0$ correspond to 9 a.m. Then in t hours the train from Los Angeles has covered a total distance of $150 + 50t$, while the train from San Francisco has traveled $40t$ miles. When the trains pass one another the total distance must add up to 450 miles, so:
$$(150 + 50t) + 40t = 450$$
$$90t = 300$$
$$t = 3\tfrac{1}{3} \text{ hours}$$
So the trains meet 3 hours and 20 minutes after 9 A.M., which is 12:20 P.M.

17. The discriminant is $b^2 - 4ac = (3)^2 - 4(1)(k^2) = 9 - 4k^2$. For the equation to have real solutions, this discriminant must be non-negative:
$$9 - 4k^2 \geq 0$$
$$(3 + 2k)(3 - 2k) \geq 0$$
The key numbers are $-\frac{3}{2}$ and $\frac{3}{2}$. Draw the sign chart:

Interval	Test Number	$3 + 2k$	$3 - 2k$	$9 - 4k^2$
$\left(-\infty, -\frac{3}{2}\right)$	-2	neg.	pos.	neg.
$\left(-\frac{3}{2}, \frac{3}{2}\right)$	0	pos.	pos.	pos.
$\left(\frac{3}{2}, \infty\right)$	2	pos.	neg.	neg.

So the product is positive on $\left(-\frac{3}{2}, \frac{3}{2}\right)$, and zero at $k = -\frac{3}{2}$ and $k = \frac{3}{2}$. Thus the equation will have real solutions if k is in the interval $\left[-\frac{3}{2}, \frac{3}{2}\right]$.

18. (a) Since $h = 2r$, $r = \frac{h}{2}$. Substituting for r:
$$S = 2\pi\left(\tfrac{h}{2}\right)^2 + 2\pi\left(\tfrac{h}{2}\right) \cdot h = \frac{\pi h^2}{2} + \pi h^2 = \tfrac{3}{2}\pi h^2$$

(b) Since $h = 2r$, substituting for h:
$$S = 2\pi r^2 + 2\pi r \cdot (2r) = 2\pi r^2 + 4\pi r^2 = 6\pi r^2$$
Solving for r:
$$S = 6\pi r^2$$
$$r^2 = \frac{S}{2\pi}$$
$$r = \sqrt{\frac{S}{2\pi}}$$
We were able to choose the positive root here since $r \geq 0$.

19. Since (a, b) lies in the third quadrant, $a < 0$ and $b < 0$. Also, since (a, b) lies on the line $y = 2x + 1$, we have $b = 2a + 1$. Substituting into the distance formula:
$$\sqrt{a^2 + b^2} = \sqrt{65}$$
$$a^2 + b^2 = 65$$
$$a^2 + (2a + 1)^2 = 65$$
$$a^2 + 4a^2 + 4a + 1 = 65$$
$$5a^2 + 4a - 64 = 0$$
$$(5a - 16)(a + 4) = 0$$
$$a = -4, \tfrac{16}{5}$$
Since $a < 0$ (third quadrant), choose $a = -4$. Then $b = 2(-4) + 1 = -7$. The point $(a, b) = (-4, -7)$.

20. An extraneous solution of an equation is one that is a solution to a subsequent equation obtained by algebraic steps, but not a solution to the original equation. Extraneous solutions frequently arise in solving radical and fractional equations. Consider the solution of the first example, which is solved by squaring each side of the equation:

$$\sqrt{3x+7} = x-1$$
$$3x+7 = x^2 - 2x + 1$$
$$0 = x^2 - 5x - 6$$
$$0 = (x-6)(x+1)$$
$$x = -1, 6$$

Note that $x = 6$ is indeed a solution to the original equation. However, $x = -1$ results in the statement $\sqrt{4} = -2$, which is false. Thus $x = -1$ is an extraneous solution of the original equation, even though it is a solution to the subsequent equation $x^2 - 5x - 6 = 0$. Now consider the solution of the second example, which is solved by multiplying each side of the equation by $x^2 - 9 = (x+3)(x-3)$:

$$\frac{2x}{x^2-9} - \frac{1}{x-3} = \frac{2}{x+3}$$
$$2x - 1(x+3) = 2(x-3)$$
$$2x - x - 3 = 2x - 6$$
$$x - 3 = 2x - 6$$
$$3 = x$$

Note that $x = 3$ results in a 0 denominator of the original equation, thus $x = 3$ is an extraneous solution for that equation. Note that two ways extraneous solutions may arise are squaring both sides of an equation and multiplying both sides of an equation by a factor which may be 0 for certain values of x.

Chapter Four
Functions

4.1 The Definition of a Function

1. (a) The domain is all real numbers, or $(-\infty,\infty)$.

   (b) Find where the denominator is 0:
   $$-5x+1=0$$
   $$-5x=-1$$
   $$x=\tfrac{1}{5}$$
   The domain is all real numbers except $\tfrac{1}{5}$, or $\left(-\infty,\tfrac{1}{5}\right)\cup\left(\tfrac{1}{5},\infty\right)$.

   (c) Find where the radical is defined:
   $$-5x+1\geq 0$$
   $$-5x\geq -1$$
   $$x\leq \tfrac{1}{5}$$
   The domain is $x\leq\tfrac{1}{5}$, or $\left(-\infty,\tfrac{1}{5}\right]$.

   (d) The domain is all real numbers, or $(-\infty,\infty)$.

3.   (a)   The domain is all real numbers, or $(-\infty,\infty)$.

(b)   Find where the denominator is 0:
$$x^2 - 9 = 0$$
$$(x+3)(x-3) = 0$$
$$x = -3, 3$$
The domain is all real numbers except ± 3, or $(-\infty,-3)\cup(-3,3)\cup(3,\infty)$.

(c)   Find where the radical is defined:
$$x^2 - 9 \geq 0$$
$$(x+3)(x-3) \geq 0$$
The key numbers are -3 and 3. Construct the sign chart:

Interval	Test Number	$x+3$	$x-3$	$(x+3)(x-3)$
$(-\infty,-3)$	-4	neg.	neg.	pos.
$(-3,3)$	0	pos.	neg.	neg.
$(3,\infty)$	4	pos.	pos.	pos.

So $x^2 - 9 \geq 0$ on the set $(-\infty,-3]\cup[3,\infty)$. Thus the domain is $(-\infty,-3]\cup[3,\infty)$.

(d)   The domain is all real numbers, or $(-\infty,\infty)$.

5.   (a)   The domain is all real numbers, or $(-\infty,\infty)$.

(b)   Find where the denominator is 0:
$$t^2 - 8t + 15 = 0$$
$$(t-5)(t-3) = 0$$
$$t = 3, 5$$
The domain is all real numbers except 3 and 5, or $(-\infty,3)\cup(3,5)\cup(5,\infty)$.

(c)   Find where the radical is defined:
$$t^2 - 8t + 15 \geq 0$$
$$(t-5)(t-3) \geq 0$$
The key numbers are 3 and 5. Construct the sign chart:

Interval	Test Number	$t-3$	$t-5$	$t^2-8t+15$
$(-\infty,3)$	2	neg.	neg.	pos.
$(3,5)$	4	pos.	neg.	neg.
$(5,\infty)$	6	pos.	pos.	pos.

So $t^2 - 8t + 15 \geq 0$ on the set $(-\infty,3]\cup[5,\infty)$. Thus the domain is $(-\infty,3]\cup[5,\infty)$.

(d)   The domain is all real numbers, or $(-\infty,\infty)$.

7. (a) The denominator is 0 when $x = -3$, so the domain is all real numbers except -3, or $(-\infty, -3) \cup (-3, \infty)$.

(b) The radical is defined when $\dfrac{x-2}{2x+6} \geq 0$. The key numbers are -3 and 2. Construct the sign chart:

Interval	Test Number	$x - 2$	$2x + 6$	$\dfrac{x-2}{2x+6}$
$(-\infty, -3)$	-4	neg.	neg.	pos.
$(-3, 2)$	0	neg.	pos.	neg.
$(2, \infty)$	4	pos.	pos.	pos.

So $\dfrac{x-2}{2x+6} \geq 0$ on the set $(-\infty, -3) \cup [2, \infty)$. Notice that $x = -3$ results in a 0 denominator, so that endpoint is not included. Thus the domain is $(-\infty, -3) \cup [2, \infty)$.

(c) The denominator is 0 when $x = -3$, so the domain is all real numbers except -3, or $(-\infty, -3) \cup (-3, \infty)$.

9. (a) The domain is all real numbers, or $(-\infty, \infty)$.

(b) The domain is all real numbers, or $(-\infty, \infty)$.

(c) The radical is defined when $2(x-1)(x+2)(x-3) \geq 0$. The key numbers are -2, 1, and 3. Construct the sign chart:

Interval	Test Number	$x - 1$	$x + 2$	$x - 3$	$2(x-1)(x+2)(x-3)$
$(-\infty, -2)$	-3	neg.	neg.	neg.	neg.
$(-2, 1)$	0	neg.	pos.	neg.	pos.
$(1, 3)$	2	pos.	pos.	neg.	neg.
$(3, \infty)$	4	pos.	pos.	pos.	pos.

So $2(x-1)(x+2)(x-3) \geq 0$ on the set $[-2, 1] \cup [3, \infty)$. Thus the domain is $[-2, 1] \cup [3, \infty)$.

11. Find where the denominator is 0:
$$x^3 - 1 = 0$$
$$x^3 = 1$$
$$x = \sqrt[3]{1} = 1$$
The domain is all real numbers except 1, or $(-\infty, 1) \cup (1, \infty)$.

13. The domain is all real numbers, or $(-\infty, \infty)$. The range is all real numbers, or $(-\infty, \infty)$.

15. The domain is all real numbers, or $(-\infty,\infty)$. The range is all real numbers, or $(-\infty,\infty)$.

17. Find where the denominator is 0:
$$3x - 18 = 0$$
$$3x = 18$$
$$x = 6$$
The domain is all real numbers except 6, or $(-\infty,6)\cup(6,\infty)$.
To find the range, let $y = g(x)$ and solve for x:
$$y = \frac{4x - 20}{3x - 18}$$
$$3xy - 18y = 4x - 20$$
$$3xy - 4x = 18y - 20$$
$$x(3y - 4) = 18y - 20$$
$$x = \frac{18y - 20}{3y - 4}$$
Since the denominator is 0 when $y = \frac{4}{3}$, the range is all real numbers except $\frac{4}{3}$, or
$$\left(-\infty,\tfrac{4}{3}\right)\cup\left(\tfrac{4}{3},\infty\right).$$

19. (a)   The denominator is 0 when $x = 5$, so the domain is all real numbers except 5, or $(-\infty,5)\cup(5,\infty)$. To find the range, let $y = f(x)$ and solve for x:
$$y = \frac{x + 3}{x - 5}$$
$$xy - 5y = x + 3$$
$$xy - x = 5y + 3$$
$$x(y - 1) = 5y + 3$$
$$x = \frac{5y + 3}{y - 1}$$
Since the denominator is 0 when $y = 1$, the range is all real numbers except 1, or $(-\infty,1)\cup(1,\infty)$.

(b)   Find where the denominator is 0:
$$x^3 - 5 = 0$$
$$x^3 = 5$$
$$x = \sqrt[3]{5}$$

The domain is all real numbers except $\sqrt[3]{5}$, or $\left(-\infty, \sqrt[3]{5}\right) \cup \left(\sqrt[3]{5}, \infty\right)$. To find the range, let $y = F(x)$ and solve for x:

$$y = \frac{x^3 + 3}{x^3 - 5}$$
$$x^3 y - 5y = x^3 + 3$$
$$x^3 y - x^3 = 5y + 3$$
$$x^3 (y - 1) = 5y + 3$$
$$x^3 = \frac{5y + 3}{y - 1}$$
$$x = \sqrt[3]{\frac{5y + 3}{y - 1}}$$

Since the denominator is 0 when $y = 1$, the range is all real numbers except 1, or $(-\infty, 1) \cup (1, \infty)$.

21. The domain is all real numbers, or $(-\infty, \infty)$. Since $t^2 \geq 0$, $t^2 + 4 \geq 4$ and thus the range is $s \geq 4$, or $[4, \infty)$.

23. The denominator is 0 when $u = 0$, so the domain is all real numbers except 0, or $(-\infty, 0) \cup (0, \infty)$. To find the range, let $y = H(u)$ and solve for u:

$$y = \frac{au + b}{u}$$
$$yu = au + b$$
$$yu - au = b$$
$$u(y - a) = b$$
$$u = \frac{b}{y - a}$$

Since the denominator is 0 when $y = a$, the range is all real numbers except a, or $(-\infty, a) \cup (a, \infty)$.

25. Rules f, g, F and H are functions. Rule h is not a function since $h(x) = 1$ and $h(x) = 2$, which violates the definition of a function. Rule G is not a function, since $G(y)$ has not been assigned a value.

27. (a) The range of f is $\{1, 2, 3\}$, the range of g is $\{2, 3\}$, the range of F is $\{1\}$, and the range of H is $\{1, 2\}$.

    (b) The range of g is $\{i, j\}$, the range of F is $\{i, j\}$, and the range of G is $\{k\}$.

29. (a) $y = (x - 3)^2$
    (b) $y = x^2 - 3$
    (c) $y = (3x)^2$
    (d) $y = 3x^2$

31.   (a)   Compute $f(1)$:
$$f(1) = (1)^2 - 3(1) + 1 = 1 - 3 + 1 = -1$$

(b)   Compute $f(0)$:
$$f(0) = (0)^2 - 3(0) + 1 = 0 - 0 + 1 = 1$$

(c)   Compute $f(-1)$:
$$f(-1) = (-1)^2 - 3(-1) + 1 = 1 + 3 + 1 = 5$$

(d)   Compute $f\left(\frac{3}{2}\right)$:
$$f\left(\frac{3}{2}\right) = \left(\frac{3}{2}\right)^2 - 3\left(\frac{3}{2}\right) + 1 = \frac{9}{4} - \frac{9}{2} + 1 = -\frac{5}{4}$$

(e)   Compute $f(z)$:
$$f(z) = (z)^2 - 3(z) + 1 = z^2 - 3z + 1$$

(f)   Compute $f(x + 1)$:
$$\begin{aligned} f(x+1) &= (x+1)^2 - 3(x+1) + 1 \\ &= x^2 + 2x + 1 - 3x - 3 + 1 \\ &= x^2 - x - 1 \end{aligned}$$

(g)   Compute $f(a + 1)$:
$$\begin{aligned} f(a+1) &= (a+1)^2 - 3(a+1) + 1 \\ &= a^2 + 2a + 1 - 3a - 3 + 1 \\ &= a^2 - a - 1 \end{aligned}$$

(h)   Compute $f(-x)$:
$$f(-x) = (-x)^2 - 3(-x) + 1 = x^2 + 3x + 1$$

(i)   Using our result from (a):
$$|f(1)| = |-1| = 1$$

(j)   Compute $f(\sqrt{3})$:
$$f(\sqrt{3}) = \left(\sqrt{3}\right)^2 - 3\left(\sqrt{3}\right) + 1 = 3 - 3\sqrt{3} + 1 = 4 - 3\sqrt{3}$$

(k)   Compute $f\left(1 + \sqrt{2}\right)$:
$$f\left(1 + \sqrt{2}\right) = \left(1 + \sqrt{2}\right)^2 - 3\left(1 + \sqrt{2}\right) + 1 = 1 + 2\sqrt{2} + 2 - 3 - 3\sqrt{2} + 1 = 1 - \sqrt{2}$$

(l)   Compute $|1 - f(2)|$:
$$|1 - f(2)| = \left|1 - \left[(2)^2 - 3(2) + 1\right]\right| = \left|1 - [4 - 6 + 1]\right| = |1 - (-1)| = |1 + 1| = 2$$

33. (a) Compute $f(2x)$:
$$f(2x) = 3(2x)^2 = 3(4x^2) = 12x^2$$

(b) Compute $2f(x)$:
$$2f(x) = 2(3x^2) = 6x^2$$

(c) Compute $f(x^2)$:
$$f(x^2) = 3(x^2)^2 = 3x^4$$

(d) Compute $[f(x)]^2$:
$$[f(x)]^2 = (3x^2)^2 = 9x^4$$

(e) Compute $f\left(\dfrac{x}{2}\right)$:
$$f(\tfrac{x}{2}) = 3(\tfrac{x}{2})^2 = 3 \cdot \frac{x^2}{4} = \tfrac{3}{4}x^2$$

(f) Compute $\dfrac{f(x)}{2}$:
$$\frac{f(x)}{2} = \frac{3x^2}{2} = \tfrac{3}{2}x^2$$

35. (a) Compute $H(0)$:
$$H(0) = 1 - 2(0)^2 = 1 - 0 = 1$$

(b) Compute $H(2)$:
$$H(2) = 1 - 2(2)^2 = 1 - 2(4) = 1 - 8 = -7$$

(c) Compute $H(\sqrt{2})$:
$$H(\sqrt{2}) = 1 - 2(\sqrt{2})^2 = 1 - 2(2) = 1 - 4 = -3$$

(d) Compute $H(\tfrac{5}{6})$:
$$H(\tfrac{5}{6}) = 1 - 2(\tfrac{5}{6})^2 = 1 - 2(\tfrac{25}{36}) = 1 - \tfrac{25}{18} = -\tfrac{7}{18}$$

(e)   Compute $H(x+1)$:
$$H(x+1) = 1 - 2(x+1)^2$$
$$= 1 - 2\left(x^2 + 2x + 1\right)$$
$$= 1 - 2x^2 - 4x - 2$$
$$= -2x^2 - 4x - 1$$

(f)   Compute $H(x+h)$:
$$H(x+h) = 1 - 2(x+h)^2 = 1 - 2\left(x^2 + 2xh + h^2\right) = 1 - 2x^2 - 4xh - 2h^2$$

(g)   Using our result from part (f), we have:
$$H(x+h) - H(x) = \left(1 - 2x^2 - 4xh - 2h^2\right) - \left(1 - 2x^2\right)$$
$$= 1 - 2x^2 - 4xh - 2h^2 - 1 + 2x^2$$
$$= -4xh - 2h^2$$

(h)   Using our result from part (g), we have:
$$\frac{H(x+h) - H(x)}{h} = \frac{-4xh - 2h^2}{h} = \frac{h(-4x - 2h)}{h} = -4x - 2h$$

37.   (a)   For the domain, we must exclude those values which make $x - 2 = 0$, or $x = 2$. So the domain is all real numbers except 2, or $(-\infty, 2) \cup (2, \infty)$. For the range, solve for x:
$$y = \frac{2x - 1}{x - 2}$$
$$y(x - 2) = 2x - 1$$
$$yx - 2y = 2x - 1$$
$$yx - 2x = 2y - 1$$
$$x(y - 2) = 2y - 1$$
$$x = \frac{2y - 1}{y - 2}$$
Since the denominator cannot be zero, $y = 2$ is excluded. So the range is all real numbers except 2, or $(-\infty, 2) \cup (2, \infty)$.

(b)   Compute $R(0)$:
$$R(0) = \frac{2(0) - 1}{0 - 2} = \frac{-1}{-2} = \tfrac{1}{2}$$

(c)   Compute $R\left(\tfrac{1}{2}\right)$:
$$R\left(\tfrac{1}{2}\right) = \frac{2\left(\tfrac{1}{2}\right) - 1}{\tfrac{1}{2} - 2} = \frac{1 - 1}{-\tfrac{3}{2}} = 0$$

(d) Compute $R(-1)$:

$$R(-1) = \frac{2(-1)-1}{-1-2} = \frac{-2-1}{-3} = \frac{-3}{-3} = 1$$

(e) Compute $R(x^2)$:

$$R(x^2) = \frac{2(x^2)-1}{x^2-2} = \frac{2x^2-1}{x^2-2}$$

(f) Compute $R\left(\frac{1}{x}\right)$:

$$R\left(\frac{1}{x}\right) = \frac{2\left(\frac{1}{x}\right)-1}{\frac{1}{x}-2} = \frac{\frac{2}{x}-1}{\frac{1}{x}-2} = \frac{2-x}{1-2x}$$

(g) Compute $R(a)$:

$$R(a) = \frac{2(a)-1}{a-2} = \frac{2a-1}{a-2}$$

(h) Compute $R(x-1)$:

$$R(x-1) = \frac{2(x-1)-1}{(x-1)-2} = \frac{2x-2-1}{x-3} = \frac{2x-3}{x-3}$$

39. (a) Compute the indicated values:

$$d(1) = -16(1)^2 + 96(1) = -16 + 96 = 80$$

$$d\left(\tfrac{3}{2}\right) = -16\left(\tfrac{3}{2}\right)^2 + 96\left(\tfrac{3}{2}\right) = -16\left(\tfrac{9}{4}\right) + 144 = 108$$

$$d(2) = -16(2)^2 + 96(2) = -64 + 192 = 128$$

$$d(t_0) = -16t_0^2 + 96t_0$$

(b) Set $d(t) = 0$:

$$-16t^2 + 96t = 0$$
$$-16t(t-6) = 0$$
$$t = 0, 6$$

(c) Set $d(t) = 1$:

$$-16t^2 + 96t = 1$$
$$16t^2 - 96t + 1 = 0$$

Using the quadratic formula:

$$t = \frac{96 \pm \sqrt{(-96)^2 - 4(16)(1)}}{2(16)}$$

$$= \frac{96 \pm \sqrt{9216 - 64}}{32}$$

$$= \frac{96 \pm \sqrt{9152}}{32}$$

$$= \frac{96 \pm 8\sqrt{143}}{32}$$

$$= \frac{12 \pm \sqrt{143}}{4}$$

41. Compute the given values:

$$g(3) = |3 - 4| = |-1| = 1$$
$$g(x + 4) = |x + 4 - 4| = |x|$$

43. First compute:

$$f(x + h) = 8(x + h) - 3 = 8x + 8h - 3$$

Therefore:

$$\frac{f(x+h) - f(x)}{h} = \frac{(8x + 8h - 3) - (8x - 3)}{h} = \frac{8x + 8h - 3 - 8x + 3}{h} = \frac{8h}{h} = 8$$

45. (a) First compute:

$$f(x + h) = (x + h)^2 = x^2 + 2xh + h^2$$

Therefore:

$$\frac{f(x+h) - f(x)}{h} = \frac{\left(x^2 + 2xh + h^2\right) - \left(x^2\right)}{h} = \frac{2xh + h^2}{h} = \frac{h(2x + h)}{h} = 2x + h$$

(b) First compute:

$$f(x + h) = 2(x + h)^2 = 2x^2 + 4xh + 2h^2$$

Therefore:

$$\frac{f(x+h) - f(x)}{h} = \frac{\left(2x^2 + 4xh + 2h^2\right) - \left(2x^2\right)}{h}$$

$$= \frac{4xh + 2h^2}{h}$$

$$= \frac{h(4x + 2h)}{h}$$

$$= 4x + 2h$$

(c) First compute:

$$f(x+h) = 2(x+h)^3 = 2x^3 + 6x^2h + 6xh^2 + 2h^3$$

Therefore:

$$\frac{f(x+h) - f(x)}{h} = \frac{\left(2x^3 + 6x^2h + 6xh^2 + 2h^3\right) - \left(2x^3\right)}{h}$$

$$= \frac{6x^2h + 6xh^2 + 2h^3}{h}$$

$$= \frac{h\left(6x^2 + 6xh + 2h^2\right)}{h}$$

$$= 6x^2 + 6xh + 2h^2$$

47. First compute:

$$f(x+h) = 2(x+h)^2 - 3(x+h) + 1$$
$$= 2\left(x^2 + 2xh + h^2\right) - 3x - 3h + 1$$
$$= 2x^2 + 4xh + 2h^2 - 3x - 3h + 1$$

Therefore:

$$\frac{f(x+h) - f(x)}{h} = \frac{\left(2x^2 + 4xh + 2h^2 - 3x - 3h + 1\right) - \left(2x^2 - 3x + 1\right)}{h}$$

$$= \frac{2x^2 + 4xh + 2h^2 - 3x - 3h + 1 - 2x^2 + 3x - 1}{h}$$

$$= \frac{4xh + 2h^2 - 3h}{h}$$

$$= \frac{h(4x + 2h - 3)}{h}$$

$$= 4x + 2h - 3$$

49. Simplifying:

$$\frac{f(x+h) - f(x)}{h} = \frac{\frac{3}{x+h} - \frac{3}{x}}{h} \cdot \frac{x(x+h)}{x(x+h)}$$

$$= \frac{3x - 3(x+h)}{hx(x+h)}$$

$$= \frac{3x - 3x - 3h}{hx(x+h)}$$

$$= \frac{-3h}{hx(x+h)}$$

$$= \frac{-3}{x(x+h)}$$

51. Computing the difference quotient:

$$\frac{g(x) - g(a)}{x - a} = \frac{4x^2 - 4a^2}{x - a} = \frac{4(x+a)(x-a)}{x - a} = 4x + 4a$$

53. Computing the difference quotient:

$$\frac{g(x)-g(a)}{x-a} = \frac{(x^2-2x+4)-(a^2-2a+4)}{x-a}$$

$$= \frac{x^2-2x+4-a^2+2a-4}{x-a}$$

$$= \frac{x^2-a^2-2x+2a}{x-a}$$

$$= \frac{(x+a)(x-a)-2(x-a)}{x-a}$$

$$= \frac{(x-a)(x+a-2)}{x-a}$$

$$= x+a-2$$

55. Computing the difference quotient:

$$\frac{g(x)-g(a)}{x-a} = \frac{4x^3-4a^3}{x-a}$$

$$= \frac{4(x-a)(x^2+ax+a^2)}{x-a}$$

$$= 4(x^2+ax+a^2)$$

$$= 4x^2+4ax+4a^2$$

57. Computing the difference quotient:

$$\frac{g(x)-g(a)}{x-a} = \frac{\frac{5}{x^2}-\frac{5}{a^2}}{x-a}\cdot\frac{a^2x^2}{a^2x^2} = \frac{5a^2-5x^2}{a^2x^2(x-a)} = \frac{-5(x+a)(x-a)}{a^2x^2(x-a)} = \frac{-5(x+a)}{a^2x^2}$$

59. (a) Computing the difference quotient:

$$\frac{f(x)-f(a)}{x-a} = \frac{\frac{x}{x-1}-\frac{a}{a-1}}{x-a}\cdot\frac{(x-1)(a-1)}{(x-1)(a-1)}$$

$$= \frac{x(a-1)-a(x-1)}{(x-a)(x-1)(a-1)}$$

$$= \frac{ax-x-ax+a}{(x-a)(x-1)(a-1)}$$

$$= \frac{-(x-a)}{(x-a)(x-1)(a-1)}$$

$$= \frac{-1}{(x-1)(a-1)}$$

(b) Using our result from (a) with $a = 3$:

$$\frac{f(x)-f(3)}{x-3} = \frac{-1}{(x-1)(3-1)} = \frac{-1}{2(x-1)}$$

(c) Computing the difference quotient:

$$\frac{f(x+h)-f(x)}{h} = \frac{\frac{x+h}{x+h-1} - \frac{x}{x-1}}{h} \cdot \frac{(x-1)(x+h-1)}{(x-1)(x+h-1)}$$

$$= \frac{(x+h)(x-1) - x(x+h-1)}{h(x-1)(x+h-1)}$$

$$= \frac{x^2 + hx - x - h - x^2 - xh + x}{h(x-1)(x+h-1)}$$

$$= \frac{-h}{h(x-1)(x+h-1)}$$

$$= \frac{-1}{(x-1)(x+h-1)}$$

(d) Using our result from (c) with $x = 3$:
$$\frac{f(3+h)-f(3)}{h} = \frac{-1}{(3-1)(3+h-1)} = \frac{-1}{2(2+h)}$$

61. (a) Set $f(x_0) = g(x_0)$:
$$4x_0 - 3 = 8 - x_0$$
$$5x_0 = 11$$
$$x_0 = \tfrac{11}{5}$$

(b) Set $f(x_0) = g(x_0)$:
$$x_0^2 - 4 = 4 - x_0^2$$
$$2x_0^2 = 8$$
$$x_0^2 = 4$$
$$x_0 = \pm 2$$

(c) Set $f(x_0) = g(x_0)$:
$$x_0^2 = x_0^3$$
$$x_0^3 - x_0^2 = 0$$
$$x_0^2(x_0 - 1) = 0$$
$$x_0 = 0, 1$$

(d) Set $f(x_0) = g(x_0)$:
$$2x_0^2 - x_0 = 3$$
$$2x_0^2 - x_0 - 3 = 0$$
$$(2x_0 - 3)(x_0 + 1) = 0$$
$$x_0 = \tfrac{3}{2}, -1$$

63. (a) Compute $A(1)$ and $A(0)$:
$$A(1) = 1000\left(1 + \frac{0.12}{4}\right)^{4(1)} \approx \$1125.51$$
$$A(0) = 1000\left(1 + \frac{0.12}{4}\right)^{4(0)} = \$1000.00$$
So $A(1) - A(0) \approx 1125.51 - 1000 = \$125.51$.

(b) Compute $A(10)$ and $A(9)$:
$$A(10) = 1000\left(1 + \frac{0.12}{4}\right)^{4(10)} \approx \$3262.04$$
$$A(9) = 1000\left(1 + \frac{0.12}{4}\right)^{4(9)} \approx \$2898.28$$
So $A(10) - A(9) \approx 3262.04 - 2898.28 = \$363.76$.

65. (a) Complete the table:

n	2	3	4	5	6	7	8
$g(n)$	1.4142	1.4422	1.4142	1.3797	1.3480	1.3205	1.2968

(b) Computing $g(15) = 1.19786$ and $g(14) = 1.20744$, so 15 is the smallest natural number n such that $g(n) < 1.2$.

67. Let $a = 1$ and $b = 2$. Then:
$$f(a+b) = f(3) = 3^2 - 1 = 8$$
$$f(a) = f(1) = 1^2 - 1 = 0$$
$$f(b) = f(2) = 2^2 - 1 = 3$$
So $f(a+b) \neq f(a) + f(b)$.

69. Let $a = 2$. Then:
$$f\left(\tfrac{1}{a}\right) = f\left(\tfrac{1}{2}\right) = \left(\tfrac{1}{2}\right)^2 - 1 = \tfrac{1}{4} - 1 = -\tfrac{3}{4}$$
$$\frac{1}{f(a)} = \frac{1}{f(2)} = \frac{1}{2^2 - 1} = \frac{1}{3}$$
So $f\left(\tfrac{1}{a}\right) \neq \dfrac{1}{f(a)}$.

71. Simplifying the quantity:
$$f(f(n)) = f(1-n) = 1 - (1-n) = 1 - 1 + n = n$$

73. (a) Compute $f(a)$, $f(2a)$ and $f(3a)$:

$$f(a) = \frac{a-a}{a+a} = \frac{0}{2a} = 0$$

$$f(2a) = \frac{2a-a}{2a+a} = \frac{a}{3a} = \tfrac{1}{3}$$

$$f(3a) = \frac{3a-a}{3a+a} = \frac{2a}{4a} = \tfrac{1}{2}$$

Since $\tfrac{1}{2} \neq 0 + \tfrac{1}{3}$, $f(3a) \neq f(a) + f(2a)$.

(b) Compute $f(5a)$:

$$f(5a) = \frac{5a-a}{5a+a} = \frac{4a}{6a} = \tfrac{2}{3}$$

Since $f(2a) = \tfrac{1}{3}$, $f(5a) = 2f(2a)$.

75. Simplify $\phi(y^2)$ and $[\phi(y)]^2$:

$$\phi(y^2) = 2(y^2) - 3 = 2y^2 - 3$$

$$[\phi(y)]^2 = (2y-3)^2 = 4y^2 - 12y + 9$$

So $\phi(y^2) \neq [\phi(y)]^2$.

77. Simplify $f(ax + b) = 2(ax + b) + 3 = 2ax + 2b + 3$. Since $f(ax + b) = x$, we have $2ax + 2b + 3 = x$. Since a and b are constants, we must have:

$$2ax = x \qquad\qquad 2b + 3 = 0$$
$$2a = 1 \qquad\qquad 2b = -3$$
$$a = \tfrac{1}{2} \qquad\qquad b = -\tfrac{3}{2}$$

So $a = \tfrac{1}{2}$ and $b = -\tfrac{3}{2}$.

79. First compute:

$$f(x+y) = \frac{(x+y)-x}{(x+y)+y} = \frac{y}{x+2y} \text{ and } f(x-y) = \frac{(x-y)-x}{(x-y)+y} = \frac{-y}{x}$$

Therefore:

$$f(x+y) + f(x-y) = \frac{y}{x+2y} - \frac{y}{x}$$
$$= \frac{yx - y(x+2y)}{x(x+2y)}$$
$$= \frac{yx - yx - 2y^2}{x^2 + 2xy}$$
$$= \frac{-2y^2}{x^2 + 2xy}$$

81.   Compute $F\left(\frac{ax+b}{cx-a}\right)$:

$$F\left(\tfrac{ax+b}{cx-a}\right) = \frac{a\left(\tfrac{ax+b}{cx-a}\right)+b}{c\left(\tfrac{ax+b}{cx-a}\right)-a}$$

$$= \frac{a(ax+b)+b(cx-a)}{c(ax+b)-a(cx-a)}$$

$$= \frac{a^2x+ab+bcx-ab}{acx+bc-acx+a^2}$$

$$= \frac{a^2x+bcx}{a^2+bc}$$

$$= \frac{x(a^2+bc)}{a^2+bc}$$

$$= x$$

83.   Since $g(1)=(1)^2-3(1)k-4=1-3k-4=-3k-3$, if $g(1)=-2$:

$$-3k-3=-2$$
$$-3k=1$$
$$k=-\tfrac{1}{3}$$

85.   (a)   The value is $L(1)=0$, since $2^0=1$.

  (b)   The value is $L(2)=1$, since $2^1=2$.

  (c)   The value is $L(4)=2$, since $2^2=4$.

  (d)   The value is $L(64)=6$, since $2^6=64$.

  (e)   The value is $L\left(\tfrac{1}{2}\right)=-1$, since $2^{-1}=\tfrac{1}{2}$.

  (f)   The value is $L\left(\tfrac{1}{4}\right)=-2$, since $2^{-2}=\tfrac{1}{4}$.

  (g)   The value is $L\left(\tfrac{1}{64}\right)=-6$, since $2^{-6}=\tfrac{1}{64}$.

  (h)   The value is $L\left(\sqrt{2}\right)=\tfrac{1}{2}$, since $2^{1/2}=\sqrt{2}$.

87. Actually, we already know the answer to this question. Since $\dfrac{-b+\sqrt{b^2-4ac}}{2a}$ is one of

the roots to the quadratic equation $q(x)=0$, we know $q\left(\dfrac{-b+\sqrt{b^2-4ac}}{2a}\right)=0$. Let's

check our answer manually:

$$q\left(\frac{-b+\sqrt{b^2-4ac}}{2a}\right)=a\left(\frac{-b+\sqrt{b^2-4ac}}{2a}\right)^2+b\left(\frac{-b+\sqrt{b^2-4ac}}{2a}\right)+c$$

$$=a\left(\frac{b^2-2b\sqrt{b^2-4ac}+b^2-4ac}{4a^2}\right)+\left(\frac{-b^2+b\sqrt{b^2-4ac}}{2a}\right)+c$$

$$=\frac{2b^2-4ac-2b\sqrt{b^2-4ac}}{4a}+\frac{-b^2+b\sqrt{b^2-4ac}}{2a}+c$$

$$=\frac{b^2-2ac-b\sqrt{b^2-4ac}-b^2+b\sqrt{b^2-4ac}}{2a}+c$$

$$=\frac{-2ac}{2a}+c$$

$$=-c+c$$

$$=0$$

89. (a) Compute $f[f(x)]$:

$$f[f(x)]=f\left(\tfrac{3x-4}{x-3}\right)$$

$$=\frac{3\left(\frac{3x-4}{x-3}\right)-4}{\left(\frac{3x-4}{x-3}\right)-3}$$

$$=\frac{3(3x-4)-4(x-3)}{(3x-4)-3(x-3)}$$

$$=\frac{9x-12-4x+12}{3x-4-3x+9}$$

$$=\frac{5x}{5}$$

$$=x$$

(b) Since $f[f(x)]=x$, then $f\left[f\left(\tfrac{22}{7}\right)\right]=\tfrac{22}{7}$.

91. The problem here is the word "nearest." $G(4)=3$, but also $G(4)=5$, since both 3 and 5 are equally "near" 4. So G is not a function, since it assigns more than one value to $x=4$. To alter the definition of G, one could define G to assign the closest prime number greater than or equal to x (or, for that matter, greater than x). This would provide G with a way of "deciding" between 3 and 5, in the previous example.

93. Compute the values:
$$P(1) = (1)^2 - 1 + 17 = 17$$
$$P(2) = (2)^2 - 2 + 17 = 19$$
$$P(3) = (3)^2 - 3 + 17 = 23$$
$$P(4) = (4)^2 - 4 + 17 = 29$$
The first natural number x such that $P(x)$ is not prime is $x = 17$. Compute:
$$P(17) = (17)^2 - 17 + 17 = 17^2$$

95. Start by simplifying the left-hand side of the equation:
$$f\left(n^2 + k\right) = \left(n^2 + k\right)^2 - \left(n^2 + k\right) + 2$$
$$= n^4 + 2n^2k + k^2 - n^2 - k + 2$$
$$= n^4 + (2k - 1)n^2 + \left(k^2 - k + 2\right)$$
Now simplify:
$$f(n+1) = (n+1)^2 - (n+1) + 2 = n^2 + 2n + 1 - n - 1 + 2 = n^2 + n + 2$$
Therefore:
$$f(n) \bullet f(n+1) = \left(n^2 - n + 2\right)\left(n^2 + n + 2\right)$$
$$= n^4 + n^3 + 2n^2 - n^3 - n^2 - 2n + 2n^2 + 2n + 4$$
$$= n^4 + 3n^2 + 4$$
Thus the equation simplifies to:
$$n^4 + (2k - 1)n^2 + \left(k^2 - k + 2\right) = n^4 + 3n^2 + 4$$
Since this equation must hold for all values of n, we can equate coefficients of corresponding terms to result in the two equations $2k - 1 = 3$ and $k^2 - k + 2 = 4$.
Solving the first equation:
$$2k - 1 = 3$$
$$2k = 4$$
$$k = 2$$
Checking in the second equation:
$$k^2 - k + 2 = 2^2 - 2 + 2 = 4$$
Therefore $k = 2$ is the required value.

97. (a) Compute the function values:
$$S(0) = \frac{3^0 - 3^{-0}}{2} = \frac{1 - 1}{2} = \frac{0}{2} = 0$$
$$C(0) = \frac{3^0 + 3^{-0}}{2} = \frac{1 + 1}{2} = \frac{2}{2} = 1$$
$$S(1) = \frac{3^1 - 3^{-1}}{2} = \frac{3 - 1/3}{2} = \frac{8/3}{2} = \frac{4}{3}$$
$$C(1) = \frac{3^1 + 3^{-1}}{2} = \frac{3 + 1/3}{2} = \frac{10/3}{2} = \frac{5}{3}$$

(b) Working from the left-hand side:

$$[C(x)]^2 - [S(x)]^2 = \left(\frac{3^x + 3^{-x}}{2}\right)^2 - \left(\frac{3^x - 3^{-x}}{2}\right)^2$$

$$= \frac{3^{2x} + 2 \cdot 3^0 + 3^{-2x}}{4} - \frac{3^{2x} - 2 \cdot 3^0 + 3^{-2x}}{4}$$

$$= \frac{3^{2x} + 2 + 3^{-2x} - 3^{2x} + 2 - 3^{-2x}}{4}$$

$$= \frac{4}{4}$$

$$= 1$$

(c) Working from the left-hand sides:

$$S(-x) = \frac{3^{-x} - 3^{-(-x)}}{2} = \frac{3^{-x} - 3^x}{2} = -\frac{3^x - 3^{-x}}{2} = -S(x)$$

$$C(-x) = \frac{3^{-x} + 3^{-(-x)}}{2} = \frac{3^{-x} + 3^x}{2} = \frac{3^x + 3^{-x}}{2} = C(x)$$

(d) Working from the right-hand side:

$$S(x)C(y) + C(x)S(y)$$

$$= \left(\frac{3^x - 3^{-x}}{2}\right)\left(\frac{3^y + 3^{-y}}{2}\right) + \left(\frac{3^x + 3^{-x}}{2}\right)\left(\frac{3^y - 3^{-y}}{2}\right)$$

$$= \frac{3^x 3^y - 3^{-x} 3^y + 3^x 3^{-y} - 3^{-x} 3^{-y}}{4} + \frac{3^x 3^y + 3^{-x} 3^y - 3^x 3^{-y} - 3^{-x} 3^{-y}}{4}$$

$$= \frac{3^{x+y} - 3^{y-x} + 3^{x-y} - 3^{-(x+y)} + 3^{x+y} + 3^{y-x} - 3^{x-y} - 3^{-(x+y)}}{4}$$

$$= \frac{2[3^{x+y} - 3^{-(x+y)}]}{4}$$

$$= \frac{3^{x+y} - 3^{-(x+y)}}{2}$$

$$= S(x+y)$$

(e)   Working from the right-hand side:

$$C(x)C(y)+S(x)S(y)$$

$$=\left(\frac{3^x+3^{-x}}{2}\right)\left(\frac{3^y+3^{-y}}{2}\right)+\left(\frac{3^x-3^{-x}}{2}\right)\left(\frac{3^y-3^{-y}}{2}\right)$$

$$=\frac{3^x3^y+3^{-x}3^y+3^x3^{-y}+3^{-x}3^{-y}}{4}+\frac{3^x3^y-3^{-x}3^y-3^x3^{-y}+3^{-x}3^{-y}}{4}$$

$$=\frac{3^{x+y}+3^{y-x}+3^{x-y}+3^{-(x+y)}+3^{x+y}-3^{y-x}-3^{x-y}+3^{-(x+y)}}{4}$$

$$=\frac{2(3^{x+y}+3^{-(x+y)})}{4}$$

$$=\frac{3^{x+y}+3^{-(x+y)}}{2}$$

$$=C(x+y)$$

(f)   Working from the right-hand sides:

$$2S(x)C(x)=2\cdot\left(\frac{3^x-3^{-x}}{2}\right)\left(\frac{3^x+3^{-x}}{2}\right)$$

$$=\frac{3^x3^x-3^{-x}3^x+3^x3^{-x}-3^{-x}3^{-x}}{2}$$

$$=\frac{3^{2x}-3^{-2x}}{2}$$

$$=S(2x)$$

$$[C(x)]^2+[S(x)]^2$$

$$=\left(\frac{3^x+3^{-x}}{2}\right)^2+\left(\frac{3^x-3^{-x}}{2}\right)^2$$

$$=\frac{3^x3^x+3^x3^{-x}+3^x3^{-x}+3^{-x}3^{-x}}{4}+\frac{3^x3^x-3^x3^{-x}-3^{-x}3^x+3^{-x}3^{-x}}{4}$$

$$=\frac{3^{2x}+2+3^{-2x}+3^{2x}-2+3^{-2x}}{4}$$

$$=\frac{2(3^{2x}+3^{-2x})}{4}$$

$$=\frac{3^{2x}+3^{-2x}}{2}$$

$$=C(2x)$$

(g)   Working from the right-hand side:

$$3S(x)+4[S(x)]^3 = 3\left(\frac{3^x-3^{-x}}{2}\right)+4\left(\frac{3^x-3^{-x}}{2}\right)^3$$
$$= \tfrac{3}{2}(3^x-3^{-x})+\tfrac{4}{8}(3^x-3^{-x})(3^{2x}-2+3^{-2x})$$
$$= \tfrac{3}{2}(3^x-3^{-x})+\tfrac{1}{2}(3^{3x}-2\cdot 3^x+3^{-x}-3^x+2\cdot 3^{-x}-3^{-3x})$$
$$= \tfrac{3}{2}3^x-\tfrac{3}{2}3^{-x}+\tfrac{1}{2}3^{3x}-\tfrac{3}{2}3^x+\tfrac{3}{2}3^{-x}-\tfrac{1}{2}3^{-3x}$$
$$= \tfrac{1}{2}(3^{3x}-3^{-3x})$$
$$= S(3x)$$

4.2 The Graph of a Function

1.   Substituting $x=3$:
$$y=\sqrt{x}=\sqrt{3}\approx 1.732$$

3.   Substituting $x=\sqrt{5}$:
$$y=\frac{1}{x}=\frac{1}{\sqrt{5}}=\frac{\sqrt{5}}{5}\approx 0.447$$

5.   For point P, $x=4$ and thus $y=\sqrt[3]{4}$. For point Q, $y=\sqrt[3]{4}$ and thus, since Q lies on the line $y=x$, $x=\sqrt[3]{4}$. For point R, $x=\sqrt[3]{4}$ and thus $y=\sqrt[3]{\sqrt[3]{4}}=\sqrt[9]{4}$. Thus the coordinates of each point are given by $P\left(4,\sqrt[3]{4}\right)\approx P(4,1.587)$, $Q\left(\sqrt[3]{4},\sqrt[3]{4}\right)\approx Q(1.587,1.587)$, and $R\left(\sqrt[3]{4},\sqrt[9]{4}\right)\approx R(1.587,1.167)$.

7.   For point P, $x=\sqrt[3]{3}$ and thus:
$$y=\left(\sqrt[3]{3}\right)^3-3\sqrt[3]{3}=3-3\sqrt[3]{3}$$
For point Q, $y=3-3\sqrt[3]{3}$ and thus, since Q lies on the line $y=x$, $x=3-3\sqrt[3]{3}$.
For point R, $x=3-3\sqrt[3]{3}$ and thus:
$$y=\left(3-3\sqrt[3]{3}\right)^3-3\left(3-3\sqrt[3]{3}\right)$$
$$= 27-81\sqrt[3]{3}+81\sqrt[3]{9}-81-9+9\sqrt[3]{3}$$
$$= -63-72\sqrt[3]{3}+81\sqrt[3]{9}$$
Thus the coordinates of each point are given by $P\left(\sqrt[3]{3},3-3\sqrt[3]{3}\right)\approx P(1.442,-1.327)$, $Q\left(3-3\sqrt[3]{3},3-3\sqrt[3]{3}\right)\approx Q(-1.327,-1.327)$, and $R\left(3-3\sqrt[3]{3},-63-72\sqrt[3]{3}+81\sqrt[3]{9}\right)\approx R(-1.327,1.645)$.

9. The domain is $[-4, 2]$ and the range is $[-3, 3]$.

11. The domain is $[-4, -1) \cup (-1, 4]$ and the range is $[-2, 3)$.

13. The domain is $[-4, 4]$ and the range is $[-2, 2)$.

15. The domain is $[-4, 3]$ and the range is $\{2\}$.

17. (a) Since $(-5, 1)$ lies on the graph of F, $F(-5) = 1$.
   (b) Since $(2, -3)$ lies on the graph of F, $F(2) = -3$.
   (c) Since $F(1) \approx -2$, $F(1)$ is negative, not positive.
   (d) Since $(2, -3)$ lies on the graph of F, $F(x) = -3$ when $x = 2$.
   (e) Since $F(2) = -3$ and $F(-2) = -2$, $F(2) - F(-2) = -3 - (-2) = -1$.

19. (a) positive
   (b) $f(-2) = 4$; $f(1) = 1$; $f(2) = 2$; $f(3) = 0$
   (c) $f(2)$, since $f(2) > 0$ and $f(4) < 0$
   (d) $f(4) - f(1) = -2 - 1 = -3$
   (e) $\left| f(4) - f(1) \right| = \left| -3 \right| = 3$
   (f) domain $= [-2, 4]$; range $= [-2, 4]$

21. (a) $f(-2) = 0$ and $g(-2) = 1$, so $g(-2)$ is larger.

   (b) $f(0) - g(0) = 2 - (-3) = 2 + 3 = 5$

   (c) Compute the three values:
$$f(1) - g(1) = 1 - (-1) = 2$$
$$f(2) - g(2) = 1 - 0 = 1$$
$$f(3) - g(3) = 4 - 1 = 3$$
   So $f(2) - g(2)$ is the smallest.

   (d) Since $f(1) = 1$, we look for where $g(x) = 1$. This occurs at $x = -2$ or $x = 3$.

   (e) Since $(3, 4)$ is a point on the graph of f, 4 is in the range of f.

23. Complete the table:

Function	$\lvert x \rvert$	x^2	x^3
Domain	$(-\infty,\infty)$	$(-\infty,\infty)$	$(-\infty,\infty)$
Range	$[0,\infty)$	$[0,\infty)$	$(-\infty,\infty)$
Turning Point	$(0,0)$	$(0,0)$	none
Maximum Value	none	none	none
Minimum Value	0	0	none
Interval(s) where Increasing	$(0,\infty)$	$(0,\infty)$	$(-\infty,\infty)$
Intervals(s) where Decreasing	$(-\infty,0)$	$(-\infty,0)$	none

25. (a) The range is $[-1, 1]$.
   (b) The maximum value is 1 (occurring at $x = 1$).
   (c) The minimum value is -1 (occurring at $x = 3$).
   (d) The function is increasing on the intervals $[0, 1]$ and $[3, 4]$.
   (e) The function is decreasing on the interval $[1, 3]$.

27. (a) The range is $[-3, 0]$.
   (b) The maximum value is 0 (occurring at $x = 0$ and $x = 4$).
   (c) The minimum value is -3 (occurring at $x = 2$).
   (d) The function is increasing on the interval $[2, 4]$.
   (e) The function is decreasing on the interval $[0, 2]$.

29. (a) Since any vertical line drawn intersects the graph in at most one point, this is the graph of a function.

   (b) Since a vertical line can be drawn which intersects the graph in two points, this cannot be the graph of a function.

   (c) Since a vertical line can be drawn which intersects the graph in two points, this cannot be the graph of a function.

   (d) Since any vertical line drawn intersects the graph in only one point, this is the graph of a function.

31.   Graph the function:

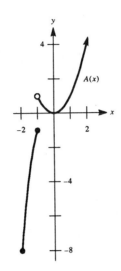

33.   Graph the function:

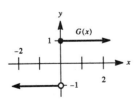

35.   (a)   Graph the function:

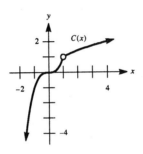

(b)   Graph the function:

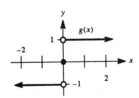

37.   Graph the function:

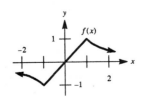

39. Graph the function:

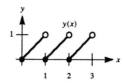

41. The domain is the set of all real numbers, or $(-\infty, \infty)$, and the range is the number 3, or $\{3\}$.

43. Since $f(3) = 9$ and $f(4) = 16$, find the slope:
$$m = \frac{f(4) - f(3)}{4 - 3} = \frac{16 - 9}{4 - 3} = \frac{7}{1} = 7$$
This represents the average rate of change of f on the interval $[3, 4]$.

45. Since $f(3) = 3^2 + 2(3) = 15$ and $f(5) = 5^2 + 2(5) = 35$, the average rate of change is given by:
$$\frac{f(5) - f(3)}{5 - 3} = \frac{35 - 15}{5 - 3} = \frac{20}{2} = 10$$

47. Since $g(-1) = 2(-1)^2 - 4(-1) = 6$ and $g(3) = 2(3)^2 - 4(3) = 6$, the average rate of change is given by:
$$\frac{g(3) - g(-1)}{3 - (-1)} = \frac{6 - 6}{3 + 1} = 0$$

49. Since $h(5) = 2(5) - 6 = 4$ and $h(12) = 2(12) - 6 = 18$, the average rate of change is given by:
$$\frac{h(12) - h(5)}{12 - 5} = \frac{18 - 4}{12 - 5} = \frac{14}{7} = 2$$

51. Since $f(a) = \frac{3}{a}$ and $f(b) = \frac{3}{b}$, the average rate of change is given by:
$$\frac{f(b) - f(a)}{b - a} = \frac{\frac{3}{b} - \frac{3}{a}}{b - a} \cdot \frac{ab}{ab} = \frac{3a - 3b}{ab(b - a)} = \frac{3(a - b)}{ab(b - a)} = -\frac{3}{ab}$$

53. Since $F(a) = -2a^3$ and $F(b) = -2b^3$, the average rate of change is given by:
$$\frac{F(b) - F(a)}{b - a} = \frac{-2b^3 + 2a^3}{b - a}$$
$$= \frac{2(a^3 - b^3)}{b - a}$$
$$= \frac{2(a - b)(a^2 + ab + b^2)}{b - a}$$
$$= -2(a^2 + ab + b^2)$$

55. (a) Since $G(0) = 22$ and $G(3) = 23$, the average rate of change is given by:
$$\frac{G(3) - G(0)}{3 - 0} = \frac{23 - 22}{3 - 0} = \frac{1}{3} \, °\text{C/min}$$

(b) Since $G(3) = 23$ and $G(6) = 27$, the average rate of change is given by:
$$\frac{G(6) - G(3)}{6 - 3} = \frac{27 - 23}{6 - 3} = \frac{4}{3} \, °\text{C/min}$$

(c) Since $G(6) = 27$ and $G(8) = 28$, the average rate of change is given by:
$$\frac{G(8) - G(6)}{8 - 6} = \frac{28 - 27}{8 - 6} = \frac{1}{2} \, °\text{C/min}$$

57. Since $f(1) = 1$ and $f(b) = \frac{1}{b}$, the average rate of change is given by:
$$\frac{f(b) - f(1)}{b - 1} = \frac{\frac{1}{b} - 1}{b - 1} \cdot \frac{b}{b} = \frac{1 - b}{b(b - 1)} = -\frac{1}{b}$$
Since this average rate of change is equal to $-\frac{1}{5}$:
$$-\frac{1}{b} = -\frac{1}{5}$$
$$b = 5$$

59. (a) To simplify calculations compute the average rate of change on the interval $[1, b]$:
$$\frac{s(b) - s(1)}{b - 1} = \frac{16b^2 - 16}{b - 1} = \frac{16(b + 1)(b - 1)}{b - 1} = 16(b + 1)$$
Now complete the table by evaluating this expression at $b = 1.1$, $b = 1.01$, $b = 1.001$, $b = 1.0001$, and $b = 1.00001$. The resulting values are displayed in the table:

Interval	[1,1.1]	[1,1.01]	[1,1.001]	[1,1.0001]	[1,1.00001]
$\Delta s / \Delta t$	33.6	32.16	32.016	32.0016	32.00016

(b) The average velocity $\Delta s / \Delta t$ seems to be approaching 32 ft/sec.

61. (a) Using the points $(0, 0)$ and $(1, 4)$:
$$m = \frac{4 - 0}{1 - 0} = \frac{4}{1} = 4 \, °\text{F/hour}$$

(b) Using the points $(1, 4)$ and $(2, 2)$:
$$m = \frac{2 - 4}{2 - 1} = \frac{-2}{1} = -2 \, °\text{F/hour}$$

(c) Using the points $(0, 0)$ and $(3, 0)$:
$$m = \frac{0 - 0}{3 - 0} = \frac{0}{3} = 0 \, °\text{F/hour}$$

(d) Using the points $(0, 0)$ and $(4, 4)$:
$$m = \frac{4-0}{4-0} = \frac{4}{4} = 1\,°F/hour$$

63. (a) Since the slope of the line segment in the 1983-1984 period is greater, the average rate of change of consumption is greater over that period.

(b) Computing the changes for each period:
   1983-1984:   change $= 304.1 - 298.4 = 6$
   1984-1985:   change $= 305.4 - 304.4 = 1$
   Note that the change in the 1983-1984 period is greater.

(c) Since the slope of the line segment in the 1989-1991 period is negative, the average rate of change of consumption is negative over that period.

(d) Computing the change for that period:
   1989-1991:   change $= 324.2 - 335.6 = -11.4$
   Note that the change in the 1989-1991 period is negative.

(e) Computing the average rate of change for each period:
   1982-1992:   slope $= \dfrac{330.6 - 294.4}{10} = 3.62$

   1991-1992:   slope $= \dfrac{330.6 - 324.2}{1} = 6.4$

   The average rate of change of consumption is greater over the 1991-1992 period.

65. (a) Simplifying the expression:
$$\frac{f(x+h) - f(x)}{h} = \frac{(x+h)^2 - x^2}{h} = \frac{x^2 + 2xh + h^2 - x^2}{h} = \frac{h(2x+h)}{h} = 2x + h$$

(b) Completing the tables:

h	2	1	0.1	0.01	0.001	0.0001
$\Delta f / \Delta x$ on the interval $[1, 1+h]$	4	3	2.1	2.01	2.001	2.0001

h	-2	-1	-0.1	-0.01	-0.001	-0.0001
$\Delta f / \Delta x$ on the interval $[1, 1+h]$	0	1	1.9	1.99	1.999	1.9999

(c) Both tables indicate $\Delta f / \Delta x$ is approaching 2 as h approaches 0.

(d) The slope of the line is 2. Using $(1, 1)$ in the point-slope formula:
$$y - 1 = 2(x - 1)$$
$$y - 1 = 2x - 2$$
$$y = 2x - 1$$

67. (a) Complete the table:

t	0	0.25	0.5	0.75	1	1.25	1.5	1.75	2	2.25	2.5	2.75	3
S(t)	0	0.25	0.5	0.75	1	1	1	1	1	1	1	1	1

Now graph the function $S(t)$ on the interval $0 \le t \le 3$:

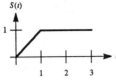

(b) Complete the table:

t	3	3.25	3.5	3.75	4	4.25	4.5	4.75	5
S(t)	1	0.75	0.5	0.25	0	−0.25	−0.5	−0.75	−1

Now graph the function $S(t)$ on the interval $3 \le t \le 5$:

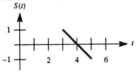

(c) Draw the graph:

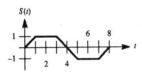

(d) Draw the graph:

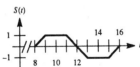

The graph is identical to that of part (c) except for the t-values. This is an example of a periodic function, that is, one which repeats its values over a specified period. In this example, the period is $t = 8$.

69. (a) Complete the table:

x	0	0.1	0.5	0.9	1.0
[x]	0	0	0	0	1

(b)   Graph $y = [x]$ on the interval $-2 \leq x < 3$:

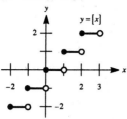

(c)   The range is all integers.

Graphing Utility Exercises for Section 4.2

1.   (a)   Graphing the function in the standard viewing rectangle:

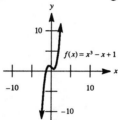

There appear to be two turning points.

(b)   Graphing using the new viewing rectangle:

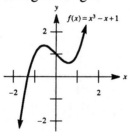

(c)   The x-coordinates of the turning points are approximately ± 0.568.

(d)   Our result agrees to one decimal place with $x = \pm 1/\sqrt{3} \approx \pm 0.577$.

(e) f is increasing on the intervals $\left(-\infty, -1/\sqrt{3}\right)$ and $\left(1/\sqrt{3}, \infty\right)$, while f is decreasing on the interval $\left(-1/\sqrt{3}, 1/\sqrt{3}\right)$.

3.   Graphing the function:

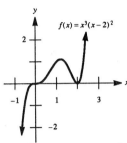

The x-coordinates of the turning points are approximately 1.2 and 2.0, which agree with the exact x-coordinates of 2 and 6/5. Note that f is increasing on the intervals $(-\infty, 6/5)$ and $(2, \infty)$, while f is decreasing on the interval $(6/5, 2)$.

5.   Graphing the function:

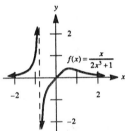

The x-coordinate of the turning point is approximately 0.630, which agrees with the exact x-coordinate of $\sqrt[3]{0.25}$. Note that f is increasing on the intervals $\left(-\infty, \sqrt[3]{-0.5}\right)$ and $\left(\sqrt[3]{-0.5}, \sqrt[3]{0.25}\right)$, while f is decreasing on the interval $\left(\sqrt[3]{0.25}, \infty\right)$.

7.   Graphing the function:

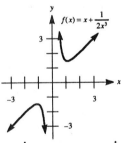

The x-coordinates of the turning points are approximately ± 1.107, which agree with the exact x-coordinates of $\pm\sqrt[4]{1.5}$. Note that y is increasing on the intervals $\left(-\infty, -\sqrt[4]{1.5}\right)$ and $\left(\sqrt[4]{1.5}, \infty\right)$, while y is decreasing on the intervals $\left(-\sqrt[4]{1.5}, 0\right)$ and $\left(0, \sqrt[4]{1.5}\right)$.

9. (a) The domain is all real numbers, or $(-\infty, \infty)$.

(b) Graphing the function:

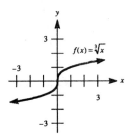

4.3 Techniques in Graphing

1. (a) C   (b) F   (c) I   (d) A   (e) J   (f) K
    (g) D   (h) B   (i) E   (j) H   (k) G

3. This graph will be that of $y = x^3$ translated down 3 units:

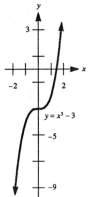

5. This graph will be that of $y = x^2$ translated to the left 4 units:

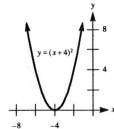

7.   This graph will be that of $y = x^2$ translated to the right 4 units:

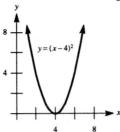

9.   This graph will be that of $y = x^2$ reflected across the x-axis:

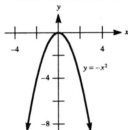

11.   This graph will be that of $y = x^2$ translated to the right 3 units, then reflected across the x-axis:

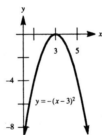

13.   This graph will be that of $y = \sqrt{x}$ translated to the right 3 units:

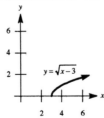

15.   This graph will be that of $y = \sqrt{x}$ translated to the left 1 unit, then reflected across the x-axis:

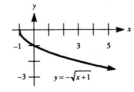

17. This graph will be that of $y = \dfrac{1}{x}$ translated to the left 2 units, then translated up 2 units:

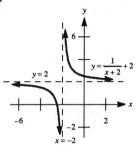

19. This graph will be that of $y = x^3$ translated to the right 2 units:

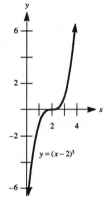

21. This graph will be that of $y = x^3$ reflected across the x-axis, then translated up 4 units:

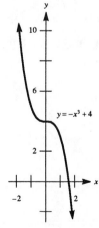

23. (a) This graph will be that of $y = |x|$ translated to the left 4 units:

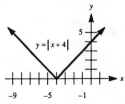

(b) This graph will be that of $y = |x|$ reflected across the y-axis, then translated to the right 4 units. Note that the reflection has no effect on the graph, since $y = |x|$ is symmetric about the y-axis:

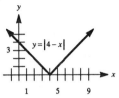

(c) This graph will be that of $y = |x|$ reflected across the y-axis, translated to the right 4 units, reflected across the x-axis, then translated up 1 unit:

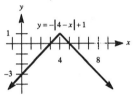

25. This is $f(x) = |x|$ translated to the right 5 units:

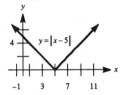

27. This is $f(x) = |x|$ reflected across the y-axis, then translated to the right 5 units. Note that the reflection has no effect on the graph, since $y = |x|$ is symmetric about the y-axis:

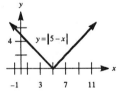

29. This is $f(x) = |x|$ translated to the right 5 units, reflected across the x-axis, then translated up 1 unit:

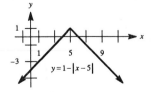

31. This is $F(x) = \dfrac{1}{x}$ translated to the left 3 units:

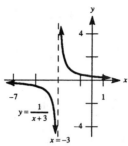

33. This is $F(x) = \dfrac{1}{x}$ translated to the left 3 units, then reflected across the x-axis:

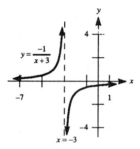

35. This is $g(x) = \sqrt{1 - x^2}$ translated to the right 2 units:

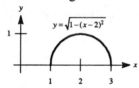

37. This is $g(x) = \sqrt{1 - x^2}$ translated to the right 2 units, reflected across the x-axis, then translated up 1 unit:

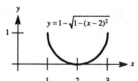

39. This is $g(x) = \sqrt{1 - x^2}$ reflected across the y-axis, then translated to the right 2 units:

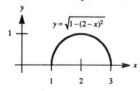

41.   (a)   This is $y = f(x)$ reflected across the x-axis:

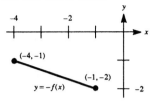

(b)   This is $y = f(x)$ reflected across the y-axis:

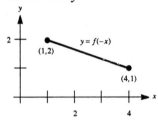

(c)   This is $y = f(x)$ reflected across the y-axis, then reflected across the x-axis:

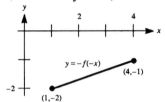

(d)   This is $y = f(x)$ displaced one unit to the right, then reflected across the x-axis:

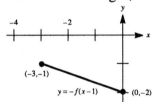

(e)   This is $y = f(x)$ displaced one unit to the left, reflected across the y-axis, then reflected across the x-axis:

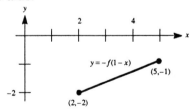

(f) This is $y = f(x)$ displaced one unit to the left, reflected across the y-axis, reflected across the x-axis, then displaced one unit up:

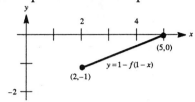

43. (a) This is $y = g(x)$ reflected across the y-axis:

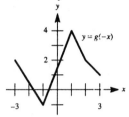

(b) This is $y = g(x)$ reflected across the x-axis:

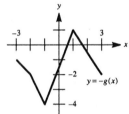

(c) This is $y = g(x)$ reflected across the y-axis, then reflected across the x-axis:

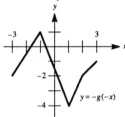

45. (a) Complete the table:

x	x^2	$x^2 - 1$	$x^2 + 1$
0	0	−1	1
±1	1	0	2
±2	4	3	5
±3	9	8	10

(b) Notice that the graph of $y = x^2 - 1$ is a vertical displacement down one unit (from $y = x^2$), while the graph of $y = x^2 + 1$ is a vertical displacement up one unit (from $y = x^2$):

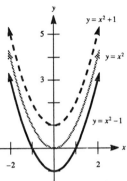

47. (a) Complete the table:

x	$\sqrt{x}$	$-\sqrt{x}$
0	0.0	0.0
1	1.0	−1.0
2	1.4	−1.4
3	1.7	−1.7
4	2.0	−2.0
5	2.2	−2.2

(b) Notice that the graph of $y = -\sqrt{x}$ is a reflection of $y = \sqrt{x}$ across the x-axis:

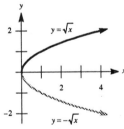

49. (a) Translating $y = \sqrt{x}$ two units to the right and one unit up:

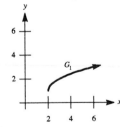

(b) Translating $y = \sqrt{x}$ one unit up and two units to the right:

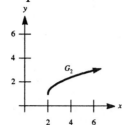

Notice that the graphs G_1 and G_2 are the same.

51. (a) Reflecting $y = \sqrt{x}$ in the x-axis, then in the y-axis:

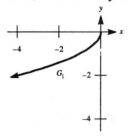

(b) Reflecting $y = \sqrt{x}$ in the y-axis, then in the x-axis:

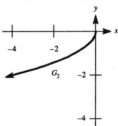

Notice that the graphs G_1 and G_2 are the same.

53. (a) Translating $y = \sqrt{x}$ two units to the right, then reflecting in the x-axis:

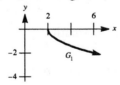

(b) Reflecting $y = \sqrt{x}$ in the x-axis, then translating two units to the right:

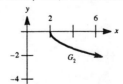

Notice that the graphs G_1 and G_2 are the same.

55. (a) Translating $y = \sqrt{x}$ one units up, then reflecting in the y-axis:

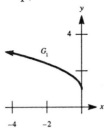

(b) Reflecting $y = \sqrt{x}$ in the y-axis, then translating one units up:

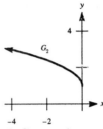

Notice that the graphs G_1 and G_2 are the same.

57. (a) Translating $y = \sqrt{x}$ two units up, then reflecting in the x-axis:

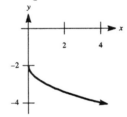

(b) Checking in the equation $y = -\sqrt{x} + 2$:
$$y = -\sqrt{1} + 2 = -1 + 2 = 1 \neq -3$$
So the point $(1, -3)$ does not satisfy the equation.

(c) The graph in part (a) is for the equation $y = -\sqrt{x} - 2$.

59. (a) Reflecting $y = \sqrt{x}$ in the y-axis, then translating two units to the left:

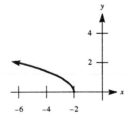

(b) Checking in the equation $y = \sqrt{-x+2}$:
$$y = \sqrt{-(-6)+2} = \sqrt{6+2} = \sqrt{8} \neq 2$$
So the point $(-6, 2)$ does not satisfy the equation.

(c) The graph in part (a) is for the equation $y = \sqrt{-x-2}$.

61. (a) $(-a, b+2)$, since $f(-(-a)) + 2 = f(a) + 2 = b + 2$
     (b) $(-a, -b+2)$, since $-f(-(-a)) + 2 = -f(a) + 2 = -b + 2$
     (c) $(a+3, -b)$, since $-f(a+3-3) = -f(a) = -b$
     (d) $(a-1, 1-b)$, since $1 - f(a-1+1) = 1 - f(a) = 1 - b$
     (e) $(-a+1, b)$, since $f(1-(-a+1)) = f(1+a-1) = f(a) = b$
     (f) $(-a+1, -b+1)$, since $-f(1-(-a+1)) + 1 = -f(a) + 1 = -b + 1$

63. Work from the right-hand side of the equality:
$$-(x-1)^2 + 3 = -(x^2 - 2x + 1) + 3 = -x^2 + 2x - 1 + 3 = -x^2 + 2x + 2$$
Since $h(x) = -(x-1)^2 + 3$, we graph $y = x^2$ displaced 1 unit to the right, reflected across the x-axis, then displaced up 3 units:

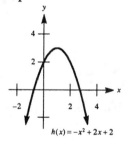

65. (a) Call $y = f(x)$. For an equation to be symmetric about the y-axis, then replacing x by $-x$ should not change the resulting equation: $y = f(-x) = f(x)$. So the resulting equation is $y = f(x)$, the original equation. Thus the graph of an even function is symmetric about the y-axis.

     (b) (i) Compute $f(-x)$:
$$f(-x) = (-x)^2 = x^2 = f(x)$$
          (ii) Compute $f(-x)$:
$$f(-x) = 2(-x)^4 - 6 = 2x^4 - 6 = f(x)$$
          (iii) Compute $f(-x)$:
$$f(-x) = 3(-x)^6 - \frac{4}{(-x)^2} + 1 = 3x^6 - \frac{4}{x^2} + 1 = f(x)$$

67. (a) Compute $f(-x)$:
$$f(-x) = \frac{1-(-x)^2}{2+(-x)^2} = \frac{1-x^2}{2+x^2} = f(x)$$
So $f(x)$ is an even function.

(b)   Compute $g(-x)$:

$$g(-x) = \frac{(-x)-(-x)^3}{2(-x)+(-x)^3} = \frac{-x+x^3}{-2x-x^3} = \frac{x-x^3}{2x+x^3} = g(x)$$

So $g(x)$ is an even function.

(c)   Compute $h(-x)$:

$$h(-x) = (-x)^2 + (-x) = x^2 - x$$

So $h(x)$ is neither even nor odd.

(d)   Compute $F(-x)$:

$$F(-x) = \left[(-x)^2 + (-x)\right]^2 = \left(x^2 - x\right)^2$$

So $F(x)$ is neither even nor odd.

(e)   Observe the graph (see below). Since this graph is symmetric to the origin (check some points if you don't see it), the function is odd.

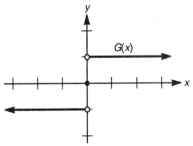

69.   If $1 < x_1 < x_2$, then $f(x_1) < f(x_2)$. Now let $g(x) = -f(-x)$, so $g(-x_2) = -f(x_2)$ and $g(-x_1) = -f(x_1)$. Now $-x_2 < -x_1 < -1$ and $-f(x_2) < -f(x_1)$, so $g(-x_2) < g(-x_1)$. So g is increasing on $(-\infty, -1)$. A similar argument shows that g is decreasing on $(-1, \infty)$.

Graphing Utility Exercises for Section 4.3

1.   (a)   Graphing the two functions:

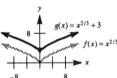

Notice that the graph of g is obtained by translating the graph of f up three units.

(b)   Graphing the two functions:

Notice that the graph of h is obtained by translating the graph of f down three units.

3. (a) Graphing the two functions:

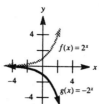

Notice that the graph of g is obtained by reflecting the graph of f in the x-axis.

(b) Graphing the two functions:

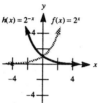

Notice that the graph of h is obtained by reflecting the graph of f in the y-axis.

5. (a) Draw the graph:

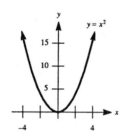

(b) This will be a translation of $y = x^2$ to the left 3 units:

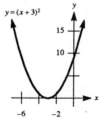

(c) This will be a translation of $y = x^2$ to the left 3 units, then a reflection across the x-axis:

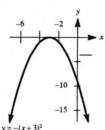

(d) This will be a translation of $y = x^2$ to the left 3 units, then a reflection across the y-axis:

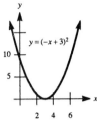

7.   (a) Draw the graph:

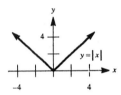

(b) This will be a translation of $y = |x|$ to the left 4 units:

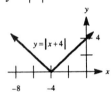

(c) This will be a translation of $y = |x|$ to the left 4 units, then a reflection across the y-axis:

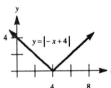

(d) This will be a translation of $y = |x|$ to the left 4 units, then a reflection across the y-axis, then a translation down 2 units:

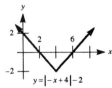

9.   (a)   Draw the graph:

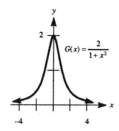

(b)   This will be a translation of $G(x)$ to the left 3 units. Compute $G(x + 3)$:

$$G(x+3) = \frac{2}{1+(x+3)^2}$$

Now draw the graph:

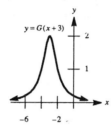

(c)   This will be a translation of $G(x)$ to the right 3 units. Compute $G(x - 3)$:

$$G(x-3) = \frac{2}{1+(x-3)^2}$$

Now draw the graph:

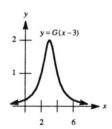

(d)   This will be a reflection of $G(x)$ across the x-axis, then a translation up 2 units. Compute $-G(x) + 2$:

$$-G(x)+2 = \frac{-2}{1+x^2} + 2$$

Now draw the graph:

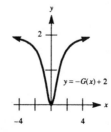

4.4 Methods of Combining Functions. Iteration

1.   (a)   Compute $(f + g)(x)$:
$$(f + g)(x) = f(x) + g(x) = (2x - 1) + (x^2 - 3x - 6) = x^2 - x - 7$$

(b)   Compute $(f - g)(x)$:
$$(f - g)(x) = f(x) - g(x) = (2x - 1) - (x^2 - 3x - 6) = -x^2 + 5x + 5$$

(c)   Using our answer from part (b), we have:
$$(f - g)(0) = -(0)^2 + 5(0) + 5 = 0 + 0 + 5 = 5$$

3.   (a)   Compute $(m - f)(x)$:
$$\begin{aligned}
(m - f)(x) &= m(x) - f(x) \\
&= (x^2 - 9) - (2x - 1) \\
&= x^2 - 9 - 2x + 1 \\
&= x^2 - 2x - 8
\end{aligned}$$

(b)   Compute $(f - m)(x)$:
$$\begin{aligned}
(f - m)(x) &= f(x) - m(x) \\
&= (2x - 1) - (x^2 - 9) \\
&= 2x - 1 - x^2 + 9 \\
&= -x^2 + 2x + 8
\end{aligned}$$

5.   (a)   Compute $(fk)(x)$:
$$(fk)(x) = f(x)k(x) = (2x - 1)(2) = 4x - 2$$

(b)   Compute $(kf)(x)$:
$$(kf)(x) = k(x)f(x) = 2(2x - 1) = 4x - 2$$

(c)   Using our results from parts (a) and (b):
$$(fk)(1) - (kf)(2) = [4(1) - 2] - [4(2) - 2] = 2 - 6 = -4$$

7.   (a)   Compute $(f/m)(x) - (m/f)(x)$:

$$\frac{f}{m}(x) - \frac{m}{f}(x) = \frac{f(x)}{m(x)} - \frac{m(x)}{f(x)}$$

$$= \frac{[f(x)]^2 - [m(x)]^2}{f(x)m(x)}$$

$$= \frac{(2x-1)^2 - (x^2-9)^2}{(x^2-9)(2x-1)}$$

$$= \frac{(4x^2 - 4x + 1) - (x^4 - 18x^2 + 81)}{2x^3 - x^2 - 18x + 9}$$

$$= \frac{-x^4 + 22x^2 - 4x - 80}{2x^3 - x^2 - 18x + 9}$$

   (b)   Using our result from part (a):

$$\frac{f}{m}(0) - \frac{m}{f}(0) = \frac{-0^4 + 22(0)^2 - 4(0) - 80}{2(0)^3 - 0^2 - 18(0) + 9} = -\frac{80}{9}$$

9.   (a)   Compute $[m \cdot (k - h)](x)$:

$$[m \cdot (k - h)](x) = m(x)[k(x) - h(x)]$$

$$= (x^2 - 9)(2 - x^3)$$

$$= 2x^2 - 18 - x^5 + 9x^3$$

$$= -x^5 + 9x^3 + 2x^2 - 18$$

   (b)   Compute $(mk)(x) - (mh)(x)$:

$$(mk)(x) - (mh)(x) = m(x)k(x) - m(x)h(x)$$

$$= (x^2 - 9)(2) - (x^2 - 9)(x^3)$$

$$= 2x^2 - 18 - x^5 + 9x^3$$

$$= -x^5 + 9x^3 + 2x^2 - 18$$

   (c)   Using our result from part (b):

$$(mk)(-1) - (mh)(-1) = -(-1)^5 + 9(-1)^3 + 2(-1)^2 - 18$$

$$= -(-1) + 9(-1) + 2(1) - 18$$

$$= 1 - 9 + 2 - 18$$

$$= -24$$

11.   (a)   Compute $(f \circ g)(x)$:

$$(f \circ g)(x) = f[g(x)] = f(-2x - 5) = 3(-2x - 5) + 1 = -6x - 15 + 1 = -6x - 14$$

   (b)   Using our result from part (a):

$$(f \circ g)(10) = -6(10) - 14 = -60 - 14 = -74$$

(c)   Compute $(g \circ f)(x)$:
$$(g \circ f)(x) = g[f(x)] = g(3x + 1) = -2(3x + 1) - 5 = -6x - 2 - 5 = -6x - 7$$

(d)   Using our result from part (a):
$$(g \circ f)(10) = -6(10) - 7 = -60 - 7 = -67$$

13.   (a)   Finding the indicated compositions:
$$(f \circ g)(x) = f[g(x)] = f(1 + x) = 1 - (1 + x) = 1 - 1 - x = -x$$
$$(f \circ g)(-2) = -(-2) = 2$$
$$(g \circ f)(x) = g[f(x)] = g(1 - x) = 1 + (1 - x) = 2 - x$$
$$(g \circ f)(-2) = 2 - (-2) = 2 + 2 = 4$$

(b)   Finding the indicated compositions:
$$(f \circ g)(x) = f[g(x)]$$
$$= f(2 - 3x)$$
$$= (2 - 3x)^2 - 3(2 - 3x) - 4$$
$$= 4 - 12x + 9x^2 - 6 + 9x - 4$$
$$= 9x^2 - 3x - 6$$
$$(f \circ g)(-2) = 9(-2)^2 - 3(-2) - 6 = 36 + 6 - 6 = 36$$
$$(g \circ f)(x) = g[f(x)]$$
$$= g(x^2 - 3x - 4)$$
$$= 2 - 3(x^2 - 3x - 4)$$
$$= 2 - 3x^2 + 9x + 12$$
$$= -3x^2 + 9x + 14$$
$$(g \circ f)(-2) = -2(-2)^2 + 9(-2) + 14 = -12 - 18 + 14 = -16$$

(c)   Finding the indicated compositions:
$$(f \circ g)(x) = f[g(x)] = f(1 - x^4) = \frac{1 - x^4}{3}$$
$$(f \circ g)(-2) = \frac{1 - (-2)^4}{3} = \frac{1 - 16}{3} = \frac{-15}{3} = -5$$
$$(g \circ f)(x) = g[f(x)] = g\left(\frac{x}{3}\right) = 1 - \left(\frac{x}{3}\right)^4 = 1 - \frac{x^4}{81}$$
$$(g \circ f)(-2) = 1 - \frac{(-2)^4}{81} = 1 - \frac{16}{81} = \frac{65}{81}$$

(d)   Finding the indicated compositions:
$$(f \circ g)(x) = f[g(x)] = f(x^2 + 1) = 2^{x^2 + 1}$$
$$(f \circ g)(-2) = 2^{(-2)^2 + 1} = 2^{4 + 1} = 2^5 = 32$$
$$(g \circ f)(x) = g[f(x)] = g(2^x) = (2^x)^2 + 1 = 2^{2x} + 1$$
$$(g \circ f)(-2) = 2^{2(-2)} + 1 = 2^{-4} + 1 = \frac{1}{16} + 1 = \frac{17}{16}$$

(e) Finding the indicated compositions:

$$(f \circ g)(x) = f[g(x)] = f(3x^5 - 4x^2) = 3x^5 - 4x^2$$
$$(f \circ g)(-2) = 3(-2)^5 - 4(-2)^2 = -96 - 16 = -112$$
$$(g \circ f)(x) = g[f(x)] = g(x) = 3x^5 - 4x^2$$
$$(g \circ f)(-2) = 3(-2)^5 - 4(-2)^2 = -96 - 16 = -112$$

(f) Finding the indicated compositions:

$$(f \circ g)(x) = f[g(x)] = f\left(\frac{x+4}{3}\right) = 3\left(\frac{x+4}{3}\right) - 4 = x + 4 - 4 = x$$
$$(f \circ g)(-2) = -2$$
$$(g \circ f)(x) = g[f(x)] = g(3x - 4) = \frac{3x - 4 + 4}{3} = \frac{3x}{3} = x$$
$$(g \circ f)(-2) = -2$$

15. (a) Compute $(F \circ G)(x)$:

$$(F \circ G)(x) = F[G(x)]$$
$$= F\left(\frac{x+1}{x-1}\right)$$
$$= \frac{3\left(\frac{x+1}{x-1}\right) - 4}{3\left(\frac{x+1}{x-1}\right) + 3}$$
$$= \frac{3(x+1) - 4(x-1)}{3(x+1) + 3(x-1)}$$
$$= \frac{3x + 3 - 4x + 4}{3x + 3 + 3x - 3}$$
$$= \frac{-x + 7}{6x}$$

(b) Using our result from part (a):

$$F[G(t)] = \frac{-t + 7}{6t}$$

(c) Using our result from part (a):

$$(F \circ G)(2) = F[G(2)] = \frac{-2 + 7}{6(2)} = \tfrac{5}{12}$$

(d) Compute $(G \circ F)(x)$:

$$\begin{aligned}
(G \circ F)(x) &= G[F(x)] \\
&= G\left(\frac{3x-4}{3x+3}\right) \\
&= \frac{\frac{3x-4}{3x+3}+1}{\frac{3x-4}{3x+3}-1} \\
&= \frac{(3x-4)+1(3x+3)}{(3x-4)-1(3x+3)} \\
&= \frac{3x-4+3x+3}{3x-4-3x-3} \\
&= \frac{6x-1}{-7} \\
&= \frac{1-6x}{7}
\end{aligned}$$

(e) Using our result from part (d):

$$G[F(y)] = \frac{1-6y}{7}$$

(f) Using our result from part (d):

$$(G \circ F)(2) = G[F(2)] = \frac{1-6(2)}{7} = -\frac{11}{7}$$

17. (a) Compute $M(7)$ and $M[M(7)]$:

$$M(7) = \frac{2(7)-1}{7-2} = \frac{14-1}{5} = \frac{13}{5}$$

$$M[M(7)] = M\left(\frac{13}{5}\right) = \frac{2\left(\frac{13}{5}\right)-1}{\frac{13}{5}-2} = \frac{\frac{26}{5}-1}{\frac{3}{5}} = \frac{\frac{21}{5}}{\frac{3}{5}} = 7$$

(b) Compute $(M \circ M)(x)$:

$$\begin{aligned}
(M \circ M)(x) &= M[M(x)] \\
&= M\left(\frac{2x-1}{x-2}\right) \\
&= \frac{2\left(\frac{2x-1}{x-2}\right)-1}{\left(\frac{2x-1}{x-2}\right)-2} \\
&= \frac{2(2x-1)-1(x-2)}{(2x-1)-2(x-2)} \\
&= \frac{4x-2-x+2}{2x-1-2x+4} \\
&= \frac{3x}{3} \\
&= x
\end{aligned}$$

(c) Using our result from part (b), we have $(M \circ M)(7) = M[M(7)] = 7$. This agrees with our answer from part (a).

19. (a) $f[g(3)] = f(0) = 1$
     (b) $g[f(3)] = g(4) = -3$
     (c) $f[h(3)] = f(2) = -1$
     (d) $(h \circ g)(2) = h[g(2)] = h(1) = 2$
     (e) $h\{f[g(3)]\} = h[f(0)] = h(1) = 2$
     (f) $(g \circ f \circ h \circ f)(2) = (g \circ f \circ h)(-1) = (g \circ f)(3) = g(4) = -3$

21. (a) Compute each composition:
$$(T \circ I)(x) = T[I(x)] = T(x) = 4x^3 - 3x^2 + 6x - 1$$
$$(I \circ T)(x) = I[T(x)] = I(4x^3 - 3x^2 + 6x - 1) = 4x^3 - 3x^2 + 6x - 1$$

     (b) Compute each composition:
$$(G \circ I)(x) = G[I(x)] = G(x) = ax^2 + bx + c$$
$$(I \circ G)(x) = I[G(x)] = I(ax^2 + bx + c) = ax^2 + bx + c$$

     (c) In general, given any function $f(x)$ and given $I(x) = x$, $(f \circ I)(x) = f(x)$ and $(I \circ f)(x) = f(x)$. The function $I(x) = x$ is called the identity function.

23. Compute the compositions:
$$(f \circ g)(0) = f[g(0)] = f(3) = 1$$
$$(f \circ g)(1) = f[g(1)] = f(2) = 3$$
$$(f \circ g)(2) = f[g(2)] = f(0) = 2$$
$$(f \circ g)(3) = f[g(3)] = f(4) = \text{undefined}$$
$$(f \circ g)(4) = f[g(4)] = f(-1) = 2$$
Thus we have the table:

x	0	1	2	3	4
$(f \circ g)(x)$	1	3	2	undef.	2

Compute the compositions:
$$(g \circ f)(-1) = g[f(-1)] = g(2) = 0$$
$$(g \circ f)(0) = g[f(0)] = g(2) = 0$$
$$(g \circ f)(1) = g[f(1)] = g(0) = 3$$
$$(g \circ f)(2) = g[f(2)] = g(3) = 4$$
$$(g \circ f)(3) = g[f(3)] = g(1) = 2$$
$$(g \circ f)(4) = g[f(4)] = \text{undefined}$$
Thus we have the table:

x	−1	0	1	2	3	4
$(g \circ f)(x)$	0	0	3	4	2	undef.

25.   (a)   Compute $(f \circ g)(x)$:
$$(f \circ g)(x) = f[g(x)] = f(3x - 4) = 2(3x - 4) + 1 = 6x - 8 + 1 = 6x - 7$$
Now draw the graph:

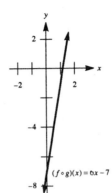

$(f \circ g)(x) = 6x - 7$

(b)   Compute $(g \circ f)(x)$:
$$(g \circ f)(x) = g[f(x)] = g(2x + 1) = 3(2x + 1) - 4 = 6x + 3 - 4 = 6x - 1$$
Now draw the graph:

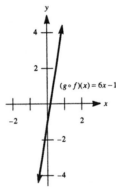

$(g \circ f)(x) = 6x - 1$

27.   (a)   The domain is $[0, \infty)$ and the range is $[-3, \infty)$:

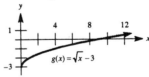

$g(x) = \sqrt{x} - 3$

(b)   The domain is $(-\infty, \infty)$ and the range is $(-\infty, \infty)$:

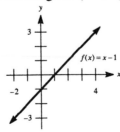

$f(x) = x - 1$

(c)   First compute $(f \circ g)(x)$:
$$(f \circ g)(x) = f[g(x)] = f(\sqrt{x} - 3) = (\sqrt{x} - 3) - 1 = \sqrt{x} - 4$$
The domain is $[0, \infty)$ and the range is $[-4, \infty)$:

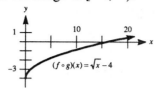

(d)   First compute $g[f(x)]$:
$$g[f(x)] = g(x - 1) = \sqrt{x - 1} - 3$$
The domain is $[1, \infty)$.

(e)   Draw the graph:

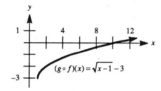

29.   Let $f(x) = x^4$ and $g(x) = 3x - 1$. Then $C(x) = (f \circ g)(x)$, since:
$$(f \circ g)(x) = f[g(x)] = f(3x - 1) = (3x - 1)^4$$

31.   (a)   Let $f(x) = \sqrt[3]{x}$ and $g(x) = 3x + 4$. Then $F(x) = (f \circ g)(x)$, since:
$$(f \circ g)(x) = f[g(x)] = f(3x + 4) = \sqrt[3]{3x + 4}$$

(b)   Let $f(x) = |x|$ and $g(x) = 2x - 3$. Then $G(x) = (f \circ g)(x)$, since:
$$(f \circ g)(x) = f[g(x)] = f(2x - 3) = |2x - 3|$$

(c)   Let $f(x) = x^5$ and $g(x) = ax + b$. Then $H(x) = (f \circ g)(x)$, since:
$$(f \circ g)(x) = f[g(x)] = f(ax + b) = (ax + b)^5$$

(d)   Let $f(x) = \frac{1}{x}$ and $g(x) = \sqrt{x}$. Then $T(x) = (f \circ g)(x)$, since:
$$(f \circ g)(x) = f[g(x)] = f(\sqrt{x}) = \frac{1}{\sqrt{x}}$$

33.   (a)   The composition is $f(x) = (b \circ c)(x)$, since:
$$(b \circ c)(x) = b[c(x)] = b(2x + 1) = \sqrt[3]{2x + 1}$$

(b)   The composition is $g(x) = (a \circ d)(x)$, since:
$$(a \circ d)(x) = a[d(x)] = a(x^2) = \frac{1}{x^2}$$

(c) The composition is $h(x) = (c \circ d)(x)$, since:
$$(c \circ d)(x) = c[d(x)] = c(x^2) = 2x^2 + 1$$

(d) The composition is $K(x) = (c \circ b)(x)$, since:
$$(c \circ b)(x) = c[b(x)] = c(\sqrt[3]{x}) = 2\sqrt[3]{x} + 1$$

(e) The composition is $l(x) = (c \circ a)(x)$, since:
$$(c \circ a)(x) = c[a(x)] = c\left(\tfrac{1}{x}\right) = 2\left(\tfrac{1}{x}\right) + 1 = \frac{2}{x} + 1$$

(f) The composition is $m(x) = (a \circ c)(x)$, since:
$$(a \circ c)(x) = a[c(x)] = a(2x + 1) = \frac{1}{2x + 1}$$

(g) The composition is $n(x) = (b \circ d)(x)$, since:
$$(b \circ d)(x) = b[d(x)] = b(x^2) = \sqrt[3]{x^2} = x^{2/3}$$
Note that we could also use $n(x) = (d \circ b)(x)$, since:
$$(d \circ b)(x) = d[b(x)] = d(\sqrt[3]{x}) = (\sqrt[3]{x})^2 = x^{2/3}$$

35. (a) The first six iterates for $x_0 = 1$ are:
$$x_1 = 2$$
$$x_2 = 4$$
$$x_3 = 8$$
$$x_4 = 16$$
$$x_5 = 32$$
$$x_6 = 64$$

(b) For $x_0 = 0$, all six iterates are 0.

(c) The first six iterates for $x_0 = -1$ are:
$$x_1 = -2$$
$$x_2 = -4$$
$$x_3 = -8$$
$$x_4 = -16$$
$$x_5 = -32$$
$$x_6 = -64$$

37. (a) The first six iterates for $x_0 = -2$ are:

$$x_1 = -3$$
$$x_2 = -5$$
$$x_3 = -9$$
$$x_4 = -17$$
$$x_5 = -33$$
$$x_6 = -65$$

(b) For $x_0 = -1$, all six iterates are -1.

(c) The first six iterates for $x_0 = 1$ are:

$$x_1 = 3$$
$$x_2 = 7$$
$$x_3 = 15$$
$$x_4 = 31$$
$$x_5 = 63$$
$$x_6 = 127$$

39. (a) The first six iterates for $x_0 = 0.9$ are:

$$x_1 = 0.81$$
$$x_2 \approx 0.656$$
$$x_3 \approx 0.430$$
$$x_4 \approx 0.185$$
$$x_5 \approx 0.034$$
$$x_6 \approx 0.001$$

(b) For $x_0 = 1$, all six iterates are 1.

(c) The first six iterates for $x_0 = 1.1$ are:

$$x_1 = 1.21$$
$$x_2 \approx 1.464$$
$$x_3 \approx 2.144$$
$$x_4 \approx 4.595$$
$$x_5 \approx 21.114$$
$$x_6 \approx 445.792$$

41. The first four iterates for $x_0 = 0.1$ are:

$$x_1 \approx 0.316$$
$$x_2 \approx 0.562$$
$$x_3 \approx 0.750$$
$$x_4 \approx 0.866$$

These values are consistent with the graph.

43. First compute $(C \circ f)(t)$:

$$(C \circ f)(t) = C[f(t)] = C\left(\frac{1}{t^2+1}\right) = 2\pi\left(\frac{1}{t^2+1}\right) = \frac{2\pi}{t^2+1}$$

When $t = 3$, we have:

$$(C \circ f)(3) = \frac{2\pi}{3^2+1} = \frac{2\pi}{10} = \frac{\pi}{5} \text{ ft}$$

45. (a) Compute $(C \circ f)(t)$:

$$(C \circ f)(t) = C[f(t)] = C(5t) = 100 + 90(5t) - (5t)^2 = 100 + 450t - 25t^2$$

(b) When $t = 3$ hr:

$$C[f(3)] = 100 + 450(3) - 25(3)^2 = \$1225$$

(c) When $t = 6$ hr:

$$C[f(6)] = 100 + 450(6) - 25(6)^2 = \$1900$$

No, the cost is not twice as much for 6 hours.

47. Call $y = f(x)$. Since $(g \circ f)(x) = g[f(x)] = g(y)$:

$$g(y) = x + 5$$
$$4y - 1 = x + 5$$
$$4y = x + 6$$
$$y = \frac{x+6}{4}$$

So $f(x) = \frac{x+6}{4}$.

49. Set $f[g(x)] = x$:

$$f(ax + b) = x$$
$$-2(ax + b) + 1 = x$$
$$-2ax - 2b + 1 = x$$

Since a and b are constants, we can equate components:

$$-2a = 1 \qquad \text{and} \qquad -2b + 1 = 0$$
$$a = -\tfrac{1}{2} \qquad\qquad\qquad -2b = -1$$
$$b = \tfrac{1}{2}$$

So $a = -\tfrac{1}{2}$ and $b = \tfrac{1}{2}$.

51. (a) Compute the difference quotient:

$$\frac{f[g(x)]-f[g(a)]}{g(x)-g(a)} = \frac{f(2x-1)-f(2a-1)}{(2x-1)-(2a-1)}$$

$$= \frac{(2x-1)^2-(2a-1)^2}{2x-1-2a+1}$$

$$= \frac{[(2x-1)+(2a-1)][(2x-1)-(2a-1)]}{2x-2a}$$

$$= \frac{(2x+2a-2)(2x-2a)}{2x-2a}$$

$$= 2x+2a-2$$

(b) Compute the difference quotient:

$$\frac{f[g(x)]-f[g(a)]}{x-a} = \frac{(2x+2a-2)(2x-2a)}{x-a} = \frac{4(x+a-1)(x-a)}{x-a} = 4x+4a-4$$

53. (a) Compute $(g \circ h \circ f)(x)$:

$$(g \circ h \circ f)(x) = g\{h[f(x)]\} = g[h(x^2)] = g\left(\frac{x^2}{2}\right) = \frac{x^2}{2}+1$$

(b) Compute $(h \circ f \circ g)(x)$:

$$(h \circ f \circ g)(x) = h\{f[g(x)]\} = h[f(x+1)] = h[(x+1)^2] = \frac{(x+1)^2}{2}$$

(c) Compute $(g \circ f \circ h)(x)$:

$$(g \circ f \circ h)(x) = g\{f[h(x)]\} = g\left[f\left(\tfrac{x}{2}\right)\right] = g\left[\left(\tfrac{x}{2}\right)^2\right] = g\left(\frac{x^2}{4}\right) = \frac{x^2}{4}+1$$

(d) Compute $(f \circ h \circ g)(x)$:

$$(f \circ h \circ g)(x) = f\{h[g(x)]\} = f[h(x+1)] = f\left(\frac{x+1}{2}\right) = \left(\frac{x+1}{2}\right)^2 = \frac{(x+1)^2}{4}$$

(e) Compute $(h \circ g \circ f)(x)$:

$$(h \circ g \circ f)(x) = h\{g[f(x)]\} = h[g(x^2)] = h(x^2+1) = \frac{x^2+1}{2}$$

55. (a) The composition is $p(x) = (g \circ f \circ h)(x)$, since:

$$(g \circ f \circ h)(x) = g\{f[h(x)]\} = g[f(3x)] = g[(3x)^2] = g(9x^2) = 1-9x^2$$

(b) The composition is $q(x) = (h \circ g \circ f)(x)$, since:

$$(h \circ g \circ f)(x) = h\{g[f(x)]\} = h[g(x^2)] = h(1-x^2) = 3(1-x^2) = 3-3x^2$$

(c)   The composition is $r(x) = (f \circ g \circ h)(x)$, since:
$$(f \circ g \circ h)(x) = f\{g[h(x)]\} = f[g(3x)] = f(1-3x) = (1-3x)^2 = 1 - 6x + 9x^2$$

(d)   The composition is $s(x) = (h \circ f \circ g)(x)$, since:
$$\begin{aligned}
(h \circ f \circ g)(x) &= h\{f[g(x)]\} \\
&= h[f(1-x)] \\
&= h\left[(1-x)^2\right] \\
&= h(1 - 2x + x^2) \\
&= 3(1 - 2x + x^2) \\
&= 3 - 6x + 3x^2
\end{aligned}$$

57.   (a)   Compute the function values:

$f(1) = 3(1) + 1 = 4$ $\qquad$ $f(2) = 2/2 = 1$

$f(3) = 3(3) + 1 = 10$ $\qquad$ $f(4) = 4/2 = 2$

$f(5) = 3(5) + 1 = 16$ $\qquad$ $f(6) = 6/2 = 3$

(b)   The first three iterates of $x_0 = 1$ are:

$f(1) = 3(1) + 1 = 4$

$f(4) = 4/2 = 2$

$f(2) = 2/2 = 1$

(c)   Compute the iterates of $x_0 = 3$:

$f(3) = 3(3) + 1 = 10$ $\qquad$ $f(10) = 10/2 = 5$

$f(5) = 3(5) + 1 = 16$ $\qquad$ $f(16) = 16/2 = 8$

$f(8) = 8/2 = 4$ $\qquad$ $f(4) = 4/2 = 2$

$f(2) = 2/2 = 1$

(d)   The iterates of $x_0 = 2$ are:

$f(2) = 2/2 = 1$

The iterates of $x_0 = 4$ are:

$f(4) = 4/2 = 2$ $\qquad$ $f(2) = 2/2 = 1$

The iterates of $x_0 = 5$ are:

$f(5) = 3(5) + 1 = 16$ $\qquad$ $f(16) = 16/2 = 8$

$f(8) = 8/2 = 4$ $\qquad$ $f(4) = 4/2 = 2$

$f(2) = 2/2 = 1$

The iterates of $x_0 = 6$ are:

$f(6) = 6/2 = 3$ $\qquad$ $f(3) = 3(3) + 1 = 10$

$f(10) = 10/2 = 5$ $\qquad$ $f(5) = 3(5) + 1 = 16$

$f(16) = 16/2 = 8$ $\qquad$ $f(8) = 8/2 = 4$

$f(4) = 4/2 = 2$ $\qquad$ $f(2) = 2/2 = 1$

The iterates of $x_0 = 7$ are:

$$f(7) = 3(7) + 1 = 22 \qquad f(22) = 22/2 = 11$$
$$f(11) = 3(11) + 1 = 34 \qquad f(34) = 34/2 = 17$$
$$f(17) = 3(17) + 1 = 52 \qquad f(52) = 52/2 = 26$$
$$f(26) = 26/2 = 13 \qquad f(13) = 3(13) + 1 = 40$$
$$f(40) = 40/2 = 20 \qquad f(20) = 20/2 = 10$$
$$f(10) = 10/2 = 5 \qquad f(5) = 3(5) + 1 = 16$$
$$f(16) = 16/2 = 8 \qquad f(8) = 8/2 = 4$$
$$f(4) = 4/2 = 2 \qquad f(2) = 2/2 = 1$$

The conjecture is valid for each of the given values of x_0.

59. First find $F[F(x)]$: (remember that $p + q = 1$)

$$F[F(x)] = F\left(p - \tfrac{1}{x+q}\right) = p - \frac{1}{\left(p - \tfrac{1}{x+q}\right) + q} = p - \frac{1}{\frac{x+q-1}{x+q}} = p - \frac{x+q}{x+q-1}$$

Therefore:

$$F\{F[F(x)]\} = F\left(p - \tfrac{x+q}{x+q-1}\right)$$

$$= p - \frac{1}{\left(p - \tfrac{x+q}{x+q-1}\right) + q}$$

$$= p - \frac{1}{1 - \tfrac{x+q}{x+q-1}}$$

$$= p - \frac{1}{\frac{x+q-1-x-q}{x+q-1}}$$

$$= p - \frac{x+q-1}{-1}$$

$$= p + x + q - 1$$

$$= 1 + x - 1$$

$$= x$$

Graphing Utility Exercises for Section 4.4

1. The completed table is:

x	0.1	0.2	0.3
$f(g(x))$	194.672	148.176	109.744
$g(f(x))$	4.992	4.936	4.784

3. The completed table is:

x	7	8	9
$f(g(x))$	1614	2218	2918
$g(f(x))$	−571	−747	−947

5. (a) The area function is given by:
$$A(r) = A(f(t)) = \pi(15 + t^{1.65})^2$$
Completing the table (rounding to the nearest whole number):

t	0	0.5	1	1.5	2	2.5	3	3.5	4	4.5	5
A	707	737	804	903	1034	1199	1402	1648	1940	2284	2685

(b) After one hour the area is approximately 800 m^2.

(c) Initially the area of the spill was 707 m^2. After approximately three hours this area has doubled.

(d) Computing the average rates of change:

0 to 2.5: $\dfrac{1199 - 707}{2.5} = 196.8 \text{ m}^2/\text{hr}$

2.5 to 5: $\dfrac{2685 - 1199}{2.5} = 594.4 \text{ m}^2/\text{hr}$

The area is increasing faster over the interval from $t = 2.5$ to $t = 5$.

7. (a) Computing the first ten iterates for $x_0 = 1$:
$$x_1 = 3$$
$$x_2 \approx 2.259259259$$
$$x_3 \approx 1.963308018$$
$$x_4 \approx 1.914212754$$
$$x_5 \approx 1.912932041$$
$$x_6 = \ldots = x_{10} \approx 1.912931183$$
Notice that the iterates converge to a number which is approximately 1.912931183.

(b) Note that $\sqrt[3]{7} \approx 1.912931183$. They are the same.

(c) The fifth iterate agrees with $\sqrt[3]{7}$ through the first three decimal places. The sixth iterate agrees with $\sqrt[3]{7}$ through the first eight decimal places.

4.5 Inverse Functions

1. (a) We must show that $f[g(x)] = x$ and $g[f(x)] = x$:
$$f[g(x)] = f\left(\tfrac{x}{3}\right) = 3\left(\tfrac{x}{3}\right) = x$$
$$g[f(x)] = g(3x) = \tfrac{3x}{3} = x$$
So $f(x)$ and $g(x)$ are inverse functions.

   (b) We must show that $f[g(x)] = x$ and $g[f(x)] = x$:
$$f[g(x)] = f\left(\tfrac{x+1}{4}\right) = 4\left(\tfrac{x+1}{4}\right) - 1 = x + 1 - 1 = x$$
$$g[f(x)] = g(4x - 1) = \tfrac{(4x-1)+1}{4} = \tfrac{4x}{4} = x$$
So $f(x)$ and $g(x)$ are inverse functions.

   (c) We must show that $g[h(x)] = x$ and $h[g(x)] = x$:
$$g[h(x)] = g\left(x^2\right) = \sqrt{x^2} = x, \quad \text{since } x \geq 0$$
$$h[g(x)] = h\left(\sqrt{x}\right) = \left(\sqrt{x}\right)^2 = x$$
So $g(x)$ and $h(x)$ are inverse functions.

3. Since $f(7) = 12$ and g is the inverse function for f, $g(12) = 7$.

5. (a) Since $f\left[f^{-1}(x)\right] = x$ for all x in the domain of f^{-1}, $f\left[f^{-1}(4)\right] = 4$ since the domain of f^{-1} is $(-\infty, \infty)$.

   (b) Since $f^{-1}[f(x)] = x$ as long as $f(x)$ is in the domain of f^{-1}, $f^{-1}[f(-1)] = -1$ since the domain of f^{-1} is $(-\infty, \infty)$.

   (c) Since $f\left[f^{-1}(x)\right] = x$ for all x in the domain of f^{-1}, $f\left[f^{-1}\left(\sqrt{2}\right)\right] = \sqrt{2}$ since the domain of f^{-1} is $(-\infty, \infty)$.

   (d) Since $f\left[f^{-1}(x)\right] = x$ for all x in the domain of f^{-1}, $f\left[f^{-1}(t+1)\right] = t + 1$ since the domain of f^{-1} is $(-\infty, \infty)$.

7. (a) Let $y = 2x + 1$. Switch the roles of x and y and solve the resulting equation for y:
$$x = 2y + 1$$
$$2y = x - 1$$
$$y = \frac{x-1}{2}$$
So the inverse is $f^{-1}(x) = \dfrac{x-1}{2}$.

(b) Calculate the required values:

$$f^{-1}(5) = \frac{5-1}{2} = \frac{4}{2} = 2$$

$$\frac{1}{f(5)} = \frac{1}{2(5)+1} = \frac{1}{11}$$

Note that the two answers are not the same. Remember that f^{-1} does not mean the reciprocal of f.

9. (a) Let $y = 3x - 1$. Switch the roles of x and y and solve the resulting equation for y:

$$x = 3y - 1$$
$$3y = x + 1$$
$$y = \frac{x+1}{3}$$

So the inverse is $f^{-1}(x) = \frac{x+1}{3}$.

(b) Verify that $f[f^{-1}(x)] = x$ and $f^{-1}[f(x)] = x$:

$$f[f^{-1}(x)] = f\left(\frac{x+1}{3}\right) = 3\left(\frac{x+1}{3}\right) - 1 = x + 1 - 1 = x$$

$$f^{-1}[f(x)] = f^{-1}(3x - 1) = \frac{(3x-1)+1}{3} = \frac{3x}{3} = x$$

(c) The graphs of each line are given below. Note the symmetry of the two lines about the line $y = x$:

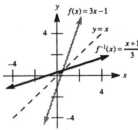

11. (a) Let $y = \sqrt{x-1}$. Switch the roles of x and y and solve the resulting equation for y:

$$x = \sqrt{y-1}$$
$$x^2 = y - 1$$
$$y = x^2 + 1$$

So the inverse is $f^{-1}(x) = x^2 + 1$ for $x \geq 0$.

(b) Verify that $f[f^{-1}(x)] = x$ and $f^{-1}[f(x)] = x$:

$$f[f^{-1}(x)] = f(x^2 + 1) = \sqrt{(x^2 + 1) - 1} = \sqrt{x^2} = x \quad (\text{since } x \geq 0)$$

$$f^{-1}[f(x)] = f^{-1}(\sqrt{x-1}) = (\sqrt{x-1})^2 + 1 = x - 1 + 1 = x$$

(c)   The graphs of each curve are given below. Note the symmetry of the curves about the line $y = x$:

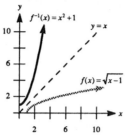

13.   (a)   Since the denominator is 0 when $x = 3$, the domain of f is the set of all real numbers except 3, or $(-\infty, 3) \cup (3, \infty)$. To find the range of f, first solve for x:

$$y = \frac{x+2}{x-3}$$
$$y(x-3) = x+2$$
$$yx - 3y = x+2$$
$$yx - x = 3y+2$$
$$x(y-1) = 3y+2$$
$$x = \frac{3y+2}{y-1}$$

Since the denominator is 0 when $y = 1$, the range of f is the set of all real numbers except 1, or $(-\infty, 1) \cup (1, \infty)$.

(b)   Let $y = \dfrac{x+2}{x-3}$. Switch the roles of x and y and solve the resulting equation for y:

$$x = \frac{y+2}{y-3}$$
$$x(y-3) = y+2$$
$$xy - 3x = y+2$$
$$xy - y = 3x+2$$
$$y(x-1) = 3x+2$$
$$y = \frac{3x+2}{x-1}$$

So the inverse is $f^{-1}(x) = \dfrac{3x+2}{x-1}$.

(c) Since the denominator is 0 when $x = 1$, the domain of f^{-1} is the set of all real numbers except 1, or $(-\infty, 1) \cup (1, \infty)$. To find the range of f^{-1}, first solve for x:

$$y = \frac{3x + 2}{x - 1}$$
$$y(x - 1) = 3x + 2$$
$$yx - y = 3x + 2$$
$$yx - 3x = y + 2$$
$$x(y - 3) = y + 2$$
$$x = \frac{y + 2}{y - 3}$$

Since the denominator is 0 when $y = 3$, the range of f^{-1} is the set of all real numbers except 3, or $(-\infty, 3) \cup (3, \infty)$. Observe that the domain of f is equal to the range of f^{-1}, and that the range of f is equal to the domain of f^{-1}.

15. Let $y = 2x^3 + 1$. Switch the roles of x and y and solve the resulting equation for y:

$$x = 2y^3 + 1$$
$$2y^3 = x - 1$$
$$y^3 = \frac{x-1}{2}$$
$$y = \sqrt[3]{\frac{x - 1}{2}}$$

So the inverse is $f^{-1}(x) = \sqrt[3]{\dfrac{x - 1}{2}}$.

17. Compute $f\left[f^{-1}(x)\right]$ and $f^{-1}\left[f(x)\right]$:

$$f\left[f^{-1}(x)\right] = f\left(\frac{x + 4}{3}\right) = 3\left(\frac{x + 4}{3}\right) - 4 = x + 4 - 4 = x$$

$$f^{-1}\left[f(x)\right] = f^{-1}(3x - 4) = \frac{(3x - 4) + 4}{3} = \frac{3x}{3} = x$$

This verfies that f and f^{-1} are inverse functions. Sketch the graph:

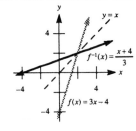

19. (a) Let $y = (x-3)^3 - 1$. Switch the roles of x and y and solve the resulting equation for y:

$$x = (y-3)^3 - 1$$
$$(y-3)^3 = x+1$$
$$y - 3 = \sqrt[3]{x+1}$$
$$y = \sqrt[3]{x+1} + 3$$

So the inverse is $f^{-1}(x) = \sqrt[3]{x+1} + 3$.

(b) Note that $f(x)$ and $f^{-1}(x)$ are symmetric about the line $y = x$:

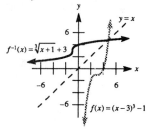

21. (a) The graph of $y = g^{-1}(x)$ is a reflection of $g(x)$ across the line $y = x$:

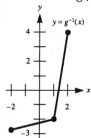

(b) The graph of $y = g^{-1}(x) - 1$ is a displacement of $g^{-1}(x)$ down 1 unit:

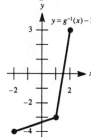

(c) The graph of $y = g^{-1}(x - 1)$ is a displacement of $g^{-1}(x)$ to the right 1 unit:

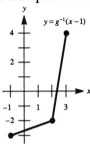

(d) The graph of $y = g^{-1}(-x)$ is a reflection of $g^{-1}(x)$ across the y-axis:

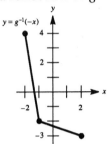

(e) The graph of $y = -g^{-1}(x)$ is a reflection of $g^{-1}(x)$ across the x-axis:

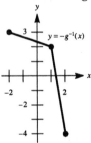

(f) The graph of $y = -g^{-1}(-x)$ is a reflection of $g^{-1}(x)$ across the y-axis and across the x-axis:

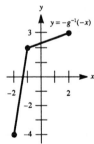

23. The graph of $y = x^2 + 1$ fails the horizontal line test, so it is not one-to-one:

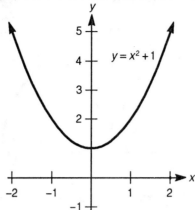

25. The graph of $y = \dfrac{1}{x}$ passes the horizontal line test, so it is one-to-one:

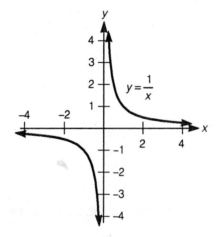

27. The graph of $y = x^3$ passes the horizontal line test, so it is one-to-one:

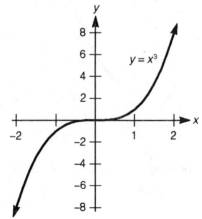

29. The graph of $y = \sqrt{1-x^2}$ fails the horizontal line test, so it is not one-to-one:

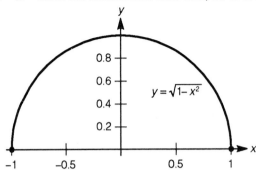

31. The graph of $g(x) = 5$ fails the horizontal line test (it <u>is</u> a horizontal line), so it is not one-to-one:

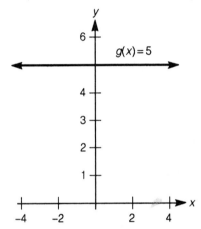

33. The graph of $f(x)$ passes the horizontal line test, so it is one-to-one:

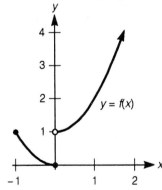

35. Assume $f(a) = f(b)$. To show that f is one-to-one, we must show that $a = b$:

$$f(a) = f(b)$$
$$4a - 9 = 4b - 9$$
$$4a = 4b$$
$$a = b$$

Thus f is one-to-one.

37. Assume $g(a) = g(b)$. To show that g is one-to-one, we must show that $a = b$:

$$g(a) = g(b)$$
$$\sqrt{2a+1} = \sqrt{2b+1}$$
$$2a+1 = 2b+1$$
$$2a = 2b$$
$$a = b$$

Thus g is one-to-one.

39. Assume $f(a) = f(b)$. To show that f is one-to-one, we must show that $a = b$:

$$f(a) = f(b)$$
$$\frac{a+4}{2a-7} = \frac{b+4}{2b-7}$$
$$(a+4)(2b-7) = (b+4)(2a-7)$$
$$2ab + 8b - 7a - 28 = 2ab + 8a - 7b - 28$$
$$8b - 7a = 8a - 7b$$
$$15b = 15a$$
$$b = a$$

Thus f is one-to-one.

41. To show that f is not one-to-one, we must demonstrate one case where $f(a) = f(b)$ but $a \neq b$. One approach here would be to factor:

$$f(x) = 2x^2 - 3x = x(2x-3)$$

Now let $a = 0$ and $b = \frac{3}{2}$. Then $f(a) = f(b) = 0$ while $a \neq b$.

43. Simplify $f[f(x)]$:

$$f[f(x)] = f\left(\tfrac{3x-2}{5x-3}\right)$$
$$= \frac{3\left(\frac{3x-2}{5x-3}\right) - 2}{5\left(\frac{3x-2}{5x-3}\right) - 3}$$
$$= \frac{3(3x-2) - 2(5x-3)}{5(3x-2) - 3(5x-3)}$$
$$= \frac{9x - 6 - 10x + 6}{15x - 10 - 15x + 9}$$
$$= \frac{-x}{-1}$$
$$= x$$

Thus $f^{-1}(x) = f(x)$.

45. (a)   Solve the equation:
$$7 + f^{-1}(x-1) = 9$$
$$f^{-1}(x-1) = 2$$
$$x - 1 = f(2)$$
$$x - 1 = 6$$
$$x = 7$$

(b)   Solve the equation:
$$4 + f(x+3) = -3$$
$$f(x+3) = -7$$
$$x + 3 = f^{-1}(-7)$$
$$x + 3 = 0$$
$$x = -3$$

47.   Solve the equation:
$$f^{-1}\left(\frac{t+1}{t-2}\right) = 12$$
$$\frac{t+1}{t-2} = f(12)$$
$$\frac{t+1}{t-2} = 13$$
$$t + 1 = 13t - 26$$
$$27 = 12t$$
$$t = \tfrac{27}{12} = \tfrac{9}{4}$$

49. (a)   Let $y = \sqrt{x}$. Switch the roles of x and y and solve the resulting equation for y:
$$x = \sqrt{y}$$
$$x^2 = y$$

So the inverse function is $f^{-1}(x) = x^2$. Since $x = \sqrt{y}$, $x \geq 0$. So the domain of $f^{-1}(x)$ is $[0, \infty)$.

(b) (i)   Since $2 = \sqrt{4}$, (4, 2) lies on the graph of f.

(ii)   Since (4, 2) lies on the graph of f, (2, 4) lies on the graph of f^{-1}.

(iii)   Since $\sqrt{5} = \sqrt{5}$, $(5, \sqrt{5})$ lies on the graph of f.

(iv)   Since $(5, \sqrt{5})$ lies on the graph of f, $(\sqrt{5}, 5)$ lies on the graph of f^{-1}.

(v)   Since $f(a) = f(a)$, $(a, f(a))$ lies on the graph of f.

(vi)   Since $(a, f(a))$ lies on the graph of f, $(f(a), a)$ lies on the graph of f^{-1}.

(vii)   Since the graph of f^{-1} is all points $(x, f^{-1}(x))$, $(b, f^{-1}(b))$ lies on the graph of f^{-1}.

(viii)   Since $b = f[f^{-1}(b)]$, $(f^{-1}(b), b)$ lies on the graph of f.

51. The coordinates of A are $(a, f(a))$. Since B lies on the line $y = x$, and its y-coordinate is $f(a)$, the coordinates of B are $(f(a), f(a))$. The y-coordinate of C is also $f(a)$, and so its x-coordinate is $f(f(a))$, since $f^{-1}(f(f(a))) = f(a)$, so the coordinates of C are $(f(f(a)), f(a))$. Since D is the reflection of C across $y = x$, the coordinates of D are $(f(a), f(f(a)))$.

53. (a) Computing the average rate of change:

$$\frac{f(9) - f(4)}{9 - 4} = \frac{\frac{3}{8} - 1}{5} = \frac{-\frac{5}{8}}{5} = -\frac{1}{8}$$

(b) To find $f^{-1}(x)$, call $y = f(x)$. Switch the roles of x and y and solve the resulting equation for y:

$$x = \frac{3}{y - 1}$$
$$xy - x = 3$$
$$xy = x + 3$$
$$y = \frac{x + 3}{x}$$

So $f^{-1}(x) = \dfrac{x + 3}{x}$. Now compute the average rate of change of $f^{-1}(x)$ on the interval $[f(4), f(9)] = \left[1, \frac{3}{8}\right]$:

$$\frac{f^{-1}\left(\frac{3}{8}\right) - f^{-1}(1)}{\frac{3}{8} - 1} = \frac{9 - 4}{-\frac{5}{8}} = \frac{5}{-\frac{5}{8}} = -8$$

Notice that our answer is the reciprocal of that found in part (a).

55. (a) Since the graph of $y = f(x) + 1$ is obtained by translating the graph of $y = f(x)$ up one unit, the point (a, b) is translated to the point $(a, b + 1)$. Thus E is the correct answer.

(b) Since the graph of $y = f(x + 1)$ is obtained by translating the graph of $y = f(x)$ to the left one unit, the point (a, b) is translated to the point $(a - 1, b)$. Thus C is the correct answer.

(c) Since the graph of $y = f(x - 1) + 1$ is obtained by translating the graph of $y = f(x)$ to the right one unit and up one unit, the point (a, b) is translated to the point $(a + 1, b + 1)$. Thus L is the correct answer.

(d) Since the graph of $y = f(-x)$ is obtained by reflecting the graph of $y = f(x)$ across the y-axis, the point (a, b) is reflected to the point $(-a, b)$. Thus A is the correct answer.

(e) Since the graph of $y = -f(x)$ is obtained by reflecting the graph of $y = f(x)$ across the x-axis, the point (a, b) is reflected to the point $(a, -b)$. Thus J is the correct answer.

(f)   Since the graph of $y = -f(-x)$ is obtained by reflecting the graph of $y = f(x)$ across the x-axis and across the y-axis, the point (a, b) is reflected to the point $(-a, -b)$. Thus G is the correct answer.

(g)   Since the graph of $y = f^{-1}(x)$ is obtained by reflecting the graph of $y = f(x)$ across the line $y = x$, the point (a, b) is reflected to the point (b, a). Thus B is the correct answer.

(h)   Since the graph of $y = f^{-1}(x) + 1$ is obtained by reflecting the graph of $y = f(x)$ across the line $y = x$, translating up one unit, then the point (a, b) is reflected to the point (b, a) then translated to the point $(b, a + 1)$. Thus M is the correct answer.

(i)   Since the graph of $y = f^{-1}(x - 1)$ is obtained by reflecting the graph of $y = f(x)$ across the line $y = x$, translating to the right one unit, the point (a, b) is reflected to the point (b, a) then translated to the point $(b + 1, a)$. Thus K is the correct answer.

(j)   Since the graph of $y = f^{-1}(-x) + 1$ is obtained by reflecting the graph of $y = f(x)$ across the line $y = x$, across the y-axis, then translating up one unit, the point (a, b) is reflected to the point (b, a), reflected to the point $(-b, a)$, then translated to the point $(-b, a + 1)$. Thus D is the correct answer.

(k)   Since the graph of $y = -f^{-1}(x)$ is obtained by reflecting the graph of $y = f(x)$ across the line $y = x$ then across the x-axis, the point (a, b) is reflected to the point (b, a) then reflected to the point $(b, -a)$. Thus I is the correct answer.

(l)   Since the graph of $y = -f^{-1}(-x) + 1$ is obtained by reflecting the graph of $y = f(x)$ across the line $y = x$, the x-axis, and the y-axis, then translating up one unit, the point (a, b) is reflected to the point (b, a), then reflected to the point $(-b, a)$ and then $(-b, -a)$, and finally translated to the point $(-b, -a + 1)$. Thus H is the correct answer.

(m)   Since the graph of $y = 1 - f^{-1}(x)$ is obtained by reflecting the graph of $y = f(x)$ across the line $y = x$ and across the x-axis, then translated up one unit, the point (a, b) is reflected to the point (b, a), then reflected to the point $(b, -a)$, then translated to the point $(b, -a + 1)$. Thus N is the correct answer.

(n)   Since the graph of $y = f(1 - x)$ is obtained by reflecting the graph of $y = f(x)$ across the y-axis, then translating to the right one unit, the point (a, b) is reflected to the point $(-a, b)$ then translated to the point $(-a + 1, b)$. Thus F is the correct answer.

Graphing Utility Exercises for Section 4.5

1. Draw the graphs, noting the symmetry about $y = x$:

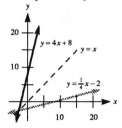

3. Draw the graphs, noting the symmetry about $y = x$:

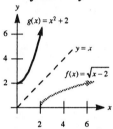

5. Draw the graphs, noting the symmetry about $y = x$:

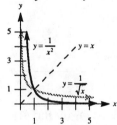

7. The function is not one-to-one. Draw the graph:

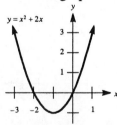

9. The function is not one-to-one. Draw the graph:

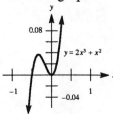

11. The function is one-to-one. Draw the graph:

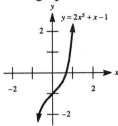

13. (a)   The function is one-to-one. Draw the graph:

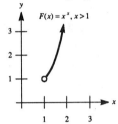

(b)   The function is not one-to-one. Draw the graph:

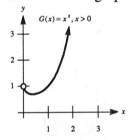

(c)   The function is not one-to-one. Draw the graph:

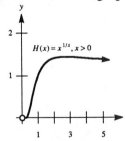

15. First find $F^{-1}(x)$. Let $y = F(x)$, then switch x and y and solve the resulting equation for y:

$$x = \sqrt{1 - y^2}$$
$$x^2 = 1 - y^2$$
$$y^2 = 1 - x^2$$
$$y = -\sqrt{1 - x^2}$$

Note that $y \leq 0$ since $x \leq 0$ in the original function $F(x)$. So $F^{-1}(x) = -\sqrt{1-x^2}$ with domain $0 \leq x \leq 1$. Draw the graph using a square viewing rectangle:

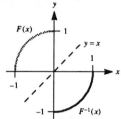

17. (a) Graph the function, noting that it passes the horizontal line test:

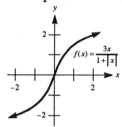

(b) If $x \geq 0$, then $y = \dfrac{3x}{1+x}$ since $|x| = x$. Switching the roles of x and y, and solving the resulting equation for y:

$$\frac{3y}{1+y} = x$$
$$3y = x + xy$$
$$y(3-x) = x$$
$$y = \frac{x}{3-x} \quad \text{(for } x < 3)$$

If $x < 0$, then $y = \dfrac{3x}{1-x}$ since $|x| = -x$. Switching the roles of x and y, and solving the resulting equation for y:

$$\frac{3y}{1-y} = x$$
$$3y = x - xy$$
$$y(3+x) = x$$
$$y = \frac{x}{3+x} \quad \text{(for } x > -3)$$

Thus the inverse function is:

$$f^{-1}(x) = \begin{cases} x/(3-x) & \text{if } 0 \leq x < 3 \\ x/(3+x) & \text{if } -3 < x < 0 \end{cases}$$

Now graph both functions, noting the symmetry about $y = x$:

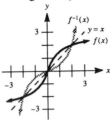

4.6 Variation

1. (a) $y = kx$

(b) $A = \dfrac{k}{B}$

3. (a) $x = kuv^2$
    (b) $z = kA^2B^3$

5. (a) $F = \dfrac{k}{r^2}$
    (b) $V^2 = k(U^2 + T^2)$

7. (a) Since A varies inversely as B, $A = \dfrac{k}{B}$. Find k by substituting $A = -1$ and $B = 2$:

$$-1 = \dfrac{k}{2}$$
$$k = -2$$

(b) $A = \dfrac{-2}{B}$

(c) When $B = \frac{5}{4}$, we have:
$$A = \dfrac{-2}{\frac{5}{4}} = -\dfrac{8}{5}$$

9. Since x varies jointly as y and z, $x = kyz$. Find k by substituting $x = 9$, $y = 2$, and $z = -3$:
$$9 = k(2)(-3)$$
$$9 = -6k$$
$$k = -\tfrac{3}{2}$$
So $x = -\tfrac{3}{2}yz$. Thus, when $y = 4$ and $z = 4$:
$$x = -\tfrac{3}{2}(4)(4) = -24$$

11. Since A varies jointly as B and C, $A = kBC$. Now substitute $3B$ and $2C$ in for B and C, respectively:

$$A = k(3B)(2C) = 6(kBC)$$

So A is six times its original value.

13. Since x varies jointly as B and C and inversely as $\sqrt{A}$, $x = \dfrac{kBC}{\sqrt{A}}$. Now substitute $4B$, $4C$ and $4A$ in for B, C and A, respectively:

$$x = \frac{k(4B)(4C)}{\sqrt{4A}} = \frac{16kBC}{2\sqrt{A}} = 8\left(\frac{kBC}{\sqrt{A}}\right)$$

So x is eight times its original value.

15. (a) Let S = surface area and r = radius. Then $S = kr^2$. When $r = 2$ and $S = 16\pi$:

$$16\pi = k(2)^2$$
$$16\pi = 4k$$
$$k = 4\pi$$

So $S = 4\pi r^2$.

   (b) When $r = \sqrt{3}$, we have:

$$S = 4\pi\left(\sqrt{3}\right)^2 = 12\pi \text{ cm}^2$$

17. Let m_1 and m_2 be the two masses, F = force, and d = distance. Then $F = \dfrac{km_1m_2}{d^2}$. Now substitute $3m_1$, $4m_2$ and $\frac{1}{2}d$ in for m_1, m_2 and d, respectively:

$$F = \frac{k(3m_1)(4m_2)}{\left(\frac{d}{2}\right)^2} = \frac{12m_1m_2}{\frac{d^2}{4}} = 48\left(\frac{km_1m_2}{d^2}\right)$$

So the force is 48 times its original value.

19. Let d = distance and t = time. Then $d = kt^2$. Find k by substituting $d = 490$ and $t = 10$:

$$490 = k(10)^2$$
$$490 = 100k$$
$$k = \frac{490}{100} = 4.9$$

So $d = 4.9t^2$. When $t = 5$, we have:

$$d = (4.9)(5)^2 = (4.9)(25) = 122.5 \text{ m}$$

21.   (a)   Let V = volume and P = pressure. Then $V = \dfrac{k}{P}$. Find k by substituting $V = 2$ and $P = 1.025$:

$$2 = \frac{k}{1.025}$$
$$k = 2.05$$

So $V = \dfrac{2.05}{P}$.

   (b)   When $P = 1$, we have $V = \frac{2.05}{1} = 2.05$ liters.

23.   (a)   Let E = kinetic energy, m = mass and r = radius. Then $E = \frac{km}{r}$. Now substitute $\frac{1}{2}r$ and $3m$ for r and m, respectively:

$$E = \frac{k(3m)}{\frac{1}{2}r} = \frac{6km}{r} = 6\left(\frac{km}{r}\right)$$

So the kinetic energy is six times its original value.

   (b)   Let V = velocity and r = radius. Then $V = \dfrac{k}{\sqrt{r}}$. Now substitute $\frac{1}{2}r$ for r:

$$V = \frac{k}{\sqrt{\frac{1}{2}r}} = \frac{k\sqrt{2}}{\sqrt{r}} = \sqrt{2}\left(\frac{k}{\sqrt{r}}\right)$$

So the velocity is $\sqrt{2}$ times its original value.

25.   Let W = weight and d = distance. Then $W = \dfrac{k}{d^2}$. Find k by substituting $W = 140$ and $d = 4000$:

$$140 = \frac{k}{(4000)^2}$$
$$k = 140(4000)^2 = 2.24 \times 10^9$$

When $d = 4500$, we have:

$$W = \frac{2.24 \times 10^9}{(4500)^2} = 110.6 \text{ lbs}$$

27.   (a)   Required is a graph which is linear with a negative slope. Such a graph is F.
   (b)   Required is a graph for $y = \frac{k}{x}$. Such a graph is C.
   (c)   Required is a graph which is a parabola pointed upward. Such a graph is D.
   (d)   Required is a graph which is linear with a positive slope. Such a graph is A.
   (e)   Required is a graph which is a parabola pointed downward. Such a graph is B.

29. Let P = period and L = length. Then $P = k\sqrt{L}$. Denote the new length and period by L' and P', respectively, where $P' = \frac{P}{2}$. Since $P' = k\sqrt{L'}$, we have:

$$\frac{P}{2} = k\sqrt{L'}$$

$$\frac{k\sqrt{L}}{2} = k\sqrt{L'}$$

$$\frac{k^2 L}{4} = k^2 L'$$

$$L' = \frac{L}{4}$$

So the new length L' should be $\frac{1}{4}$ of the original length. Thus it must be shortened by $\frac{3}{4}L$.

31. Let I = intensity and d = distance. Then $I = \dfrac{k}{d^2}$. When $d = 6$:

$$I = \frac{k}{6^2} = \frac{k}{36}$$

We are asked to find the value of d when $I = 2\left(\frac{k}{36}\right) = \frac{k}{18}$:

$$\frac{k}{18} = \frac{k}{d^2}$$
$$d^2 = 18$$
$$d = 3\sqrt{2} \approx 4.24$$

At a distance of 4.24 ft, the illumination will be twice as great.

33. The area of the triangle is:
$$A = \tfrac{1}{2}(\text{base})(\text{height}) = \tfrac{1}{2}(x)(y) = \tfrac{1}{2}(x)(mx) = \tfrac{1}{2}mx^2$$
Thus A varies jointly as m and x^2.

35. Let t = time and d = distance. Then $t^2 = kd^3$. First find k by using $d = 1.000000$ and $t = 365.2564$ (information for earth):
$$t^2 = kd^3$$
$$k = \frac{t^2}{d^3} = \frac{(365.2564)^2}{(1.000000)^3} = 133412.2377$$

So $t = \sqrt{kd^3} = \sqrt{133412.2377d^3}$. Now compute t for each planet:

Mercury: 87.9693 days
Venus: 224.7007 days
Earth: 365.2564 days
Mars: 686.9786 days
Jupiter: 4336.6159 days
Saturn: 10826.9994 days
Uranus: 30873.7244 days
Neptune: 60300.6863 days
Pluto: 91814.3739 days

37. (a) Let $x = k_1 z$ and $y = k_2 z$. Then:
$$x + y = k_1 z + k_2 z = (k_1 + k_2)z$$
Call $K = k_1 + k_2$. Then $x + y = Kz$. This shows that $x + y$ varies directly as z.

(b) Let $x = k_1 z$ and $y = k_2 z$. Then:
$$xy = (k_1 z)(k_2 z) = k_1 k_2 z^2$$
Call $K = k_1 k_2$. Then $xy = Kz^2$. So xy does <u>not</u> vary directly as z.

(c) Let $x = k_1 z$ and $y = k_2 z$. Then:
$$\sqrt{xy} = \sqrt{k_1 k_2 z^2} = \sqrt{k_1 k_2}\, z \text{ since } z > 0$$
Call $K = \sqrt{k_1 k_2}$. Then $\sqrt{xy} = Kz$. This shows that $\sqrt{xy}$ varies directly as z.

39. (a) Simplifying:
$$x + y = k(x - y)$$
$$x + y = kx - ky$$
$$y + ky = kx - x$$
$$y(1 + k) = x(k - 1)$$
$$y = \frac{k-1}{k+1} x$$
Setting $K = \frac{k-1}{k+1}$, we have $y = Kx$. Thus y varies directly as x.

(b) Simplifying:
$$x^2 + y^2 = x^2 + (Kx)^2 = x^2 + K^2 x^2 = (K^2 + 1)(x^2)$$

Since $y = Kx$, $x = \frac{y}{K}$, so:

$$x^2 + y^2 = (K^2 + 1)(x^2) = (K^2 + 1)\left(\frac{y}{k}\right)(x) = \frac{K^2 + 1}{K}(xy)$$

Setting $A = \frac{K^2 + 1}{K}$, we have $x^2 + y^2 = Axy$. Thus $x^2 + y^2$ varies jointly as x and y.

(c)   Simplifying:
$$x^3 + y^3 = (x + y)(x^2 - xy + y^2)$$
$$= (x + y)(x^2 + y^2 - xy)$$
$$= (x + y)(Axy - xy), \text{ from part (b)}$$
$$= (x + y)(xy)(A - 1)$$
$$= (A - 1)(x + y)(xy)$$

Let $B = A - 1$. Then $x^3 + y^3 = Bxy(x + y)$. So $x^3 + y^3$ varies jointly as x, y, and $x + y$.

41.   The area of the triangle is:
$$A = \tfrac{1}{2}(\text{base})(\text{height}) = \tfrac{1}{2}(x)(y) = \tfrac{1}{2}(x)\left(\frac{a^2}{x}\right) = \tfrac{1}{2}a^2$$

Thus A varies directly as a^2.

Chapter Four Review Exercises

1.   (a)   We must guarantee that the quantity within the radical is non-negative:
$$15 - 5x \geq 0$$
$$-5x \geq -15$$
$$x \leq 3$$
So the domain is $(-\infty, 3]$.

(b)   Solve $y = \dfrac{3 + x}{2x - 5}$ for x:
$$y(2x - 5) = 3 + x$$
$$2xy - 5y = 3 + x$$
$$2xy - x = 5y + 3$$
$$x(2y - 1) = 5y + 3$$
$$x = \frac{5y + 3}{2y - 1}$$
So the range is all real numbers except $\tfrac{1}{2}$, or $\left(-\infty, \tfrac{1}{2}\right) \cup \left(\tfrac{1}{2}, \infty\right)$.

3.   Yes. Let $f(x) = ax + b$ and $g(x) = cx + d$, then:
$$(f \circ g)(x) = f(cx + d) = a(cx + d) + b = acx + (ad + b)$$
So $(f \circ g)(x)$ is a linear function.

5.   (a)   Compute the difference quotient:
$$\frac{F(x) - F(a)}{x - a} = \frac{\frac{1}{x} - \frac{1}{a}}{x - a} \cdot \frac{ax}{ax} = \frac{a - x}{ax(x - a)} = -\frac{1}{ax}$$

(b) First compute:

$$g(x+h) = (x+h) - 2(x+h)^2 = x + h - 2x^2 - 4xh - 2h^2$$

So we have:

$$\frac{g(x+h) - g(x)}{h} = \frac{\left(x + h - 2x^2 - 4xh - 2h^2\right) - \left(x - 2x^2\right)}{h}$$

$$= \frac{x + h - 2x^2 - 4xh - 2h^2 - x + 2x^2}{h}$$

$$= \frac{h - 4xh - 2h^2}{h}$$

$$= 1 - 4x - 2h$$

7. (a) Switch the roles of x and y and solve the resulting equation for y:

$$x = \frac{1 - 5y}{3y}$$

$$3xy = 1 - 5y$$

$$3xy + 5y = 1$$

$$y(3x + 5) = 1$$

$$y = \frac{1}{3x + 5}$$

So $g^{-1}(x) = \dfrac{1}{3x + 5}$.

(b) Graph f^{-1}:

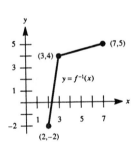

9. (a) Displace $y = |x|$ two units to the left and 3 units down. The x-intercepts are -5 and 1, and the y-intercept is -1:

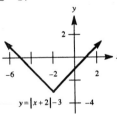

(b)   Displace $y = \frac{1}{x}$ two units to the left and 1 unit down. The x-intercept is -1 and the y-intercept is $-\frac{1}{2}$:

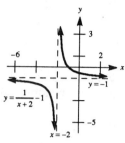

11.   (a)   Compute $f(-1)$:
$$f(-1) = 3(-1)^2 - 2(-1) = 3 + 2 = 5$$

(b)   Compute $f(1-\sqrt{2})$:
$$\begin{aligned} f(1-\sqrt{2}) &= 3(1-\sqrt{2})^2 - 2(1-\sqrt{2}) \\ &= 3(1 - 2\sqrt{2} + 2) - 2 + 2\sqrt{2} \\ &= 9 - 6\sqrt{2} - 2 + 2\sqrt{2} \\ &= 7 - 4\sqrt{2} \end{aligned}$$

13.   The slope is given by:
$$m = \frac{(5+h)^2 - 25}{5+h-5} = \frac{25 + 10h + h^2 - 25}{h} = \frac{10h + h^2}{h} = 10 + h$$
Using functional notation, we have $m(h) = 10 + h$.

15.   The graph of $y = f(-x)$ will result in a reflection across the y-axis:

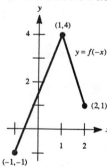

17.   (a)   The first three iterates are 2.5, 3.25, and 3.625.

(b)   The y-coordinate of point A is $f(1) = 2.5$. The x-coordinate of point B is 2.5, so the y-coordinate of point B is $f(2.5) = 3.25$. The x-coordinate of point C is 3.25, so the y-coordinate of point C is $f(3.25) = 3.625$.

(c) At the intersection point the y-coordinates must be equal, so:

$$x = \tfrac{1}{2}x + 2$$
$$2x = x + 4$$
$$x = 4$$

Since $y = x$, the y-coordinate is also 4.

(d) The fourth through tenth iterates are 3.813, 3.906, 3.953, 3.977, 3.988, 3.994, and 3.997. These iterates appear to be approaching 4.

19. (a) The first three iterates are approximately 0.4, 1.2, and 1.4.

(b) The calculated iterates are approximately 0.45, 1.20, and 1.48.

(c) Finding where $y = x$ intersects $y = \sqrt{x+1}$:

$$x = \sqrt{x+1}$$
$$x^2 = x + 1$$
$$x^2 - x - 1 = 0$$

Using the quadratic formula:

$$x = \frac{-(-1) \pm \sqrt{(-1)^2 - 4(1)(-1)}}{2(1)} = \frac{1 \pm \sqrt{1+4}}{2} = \frac{1 \pm \sqrt{5}}{2}$$

Since $x > 0$ (first quadrant), we choose $x = \dfrac{1 + \sqrt{5}}{2}$. Since $y = x$, $y = \dfrac{1 + \sqrt{5}}{2}$.

(d) Computing the first twelve iterates:

$$x_1 \approx 0.447$$
$$x_2 \approx 1.203$$
$$x_3 \approx 1.484$$
$$x_4 \approx 1.576$$
$$x_5 \approx 1.605$$
$$x_6 \approx 1.614$$
$$x_7 \approx 1.617$$
$$x_8 = \dots = x_{12} \approx 1.618$$

From part (c), we have $\dfrac{1 + \sqrt{5}}{2} \approx 1.618$, which confirms that the two values are equal.

21. To find the x-intercepts, let $y = 0$:

$$-(x-1)^2 + 2 = 0$$
$$-(x-1)^2 = -2$$
$$x - 1 = \pm\sqrt{2}$$
$$x = 1 \pm \sqrt{2}$$

To find the y-intercept, let $x = 0$:
$$y = -(0-1)^2 + 2 = -1 + 2 = 1$$

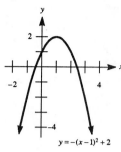

$$y = -(x-1)^2 + 2$$

23. To find the x-intercepts, let $y = 0$:
$$\frac{1}{x+1} = 0$$
$$1 = 0$$
Clearly this is impossible, so there are no x-intercepts. To find the y-intercept, let $x = 0$:
$$y = \frac{1}{0+1} = 1$$

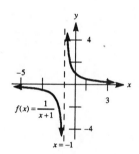

$$f(x) = \frac{1}{x+1}$$
$$x = -1$$

25. To find the x-intercept, let $y = 0$:
$$|x+3| = 0$$
$$x = -3$$
To find the y-intercept, let $x = 0$:
$$y = |0+3| = 3$$

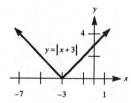

$$y = |x+3|$$

27. To find the x-intercepts, let $y = 0$:
$$\sqrt{1-x^2} = 0$$
$$1 - x^2 = 0$$
$$-x^2 = -1$$
$$x = \pm 1$$

To find the y-intercept, let $x = 0$:
$$y = \sqrt{1-0} = 1$$

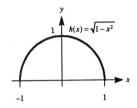

29.   To find the x-intercept, let $y = 0$:
$$1 - (x+1)^3 = 0$$
$$-(x+1)^3 = -1$$
$$(x+1)^3 = 1$$
$$x + 1 = 1$$
$$x = 0$$

To find the y-intercept, let $x = 0$:
$$y = 1 - (0+1)^3 = 1 - 1 = 0$$

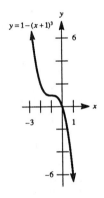

31.   The x-intercept is 0 and the y-intercept is 0:

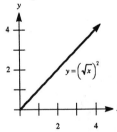

33. First compute $(f \circ g)(x)$:

$$(f \circ g)(x) = f(\sqrt{x-1}) = -(\sqrt{x-1})^2 = 1 - x \text{ for } x \geq 1$$

The x-intercept is 1 and there is no y-intercept:

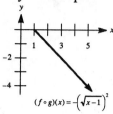

35. The x-intercepts are -1 and 0, and the y-intercept is 0:

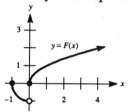

37. There are no x- or y-intercepts:

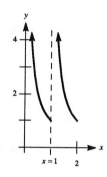

39. To find the inverse function, switch the roles of x and y and solve the resulting equation for y:

$$x = \frac{y+1}{2}$$
$$2x = y + 1$$
$$2x - 1 = y$$

So $f^{-1}(x) = 2x - 1$. To find the x-intercept, let $y = 0$:

$$2x - 1 = 0$$
$$2x = 1$$
$$x = \tfrac{1}{2}$$

To find the y-intercept, let $x = 0$:
$$y = 2(0) - 1 = -1$$

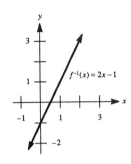

41. Compute $(f \circ f^{-1})(x) = x$ for $x \geq 0$. The x-intercept is 0 and the y-intercept is 0:

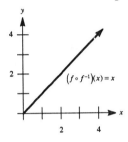

43. We must make sure that $x^2 - 9 \neq 0$. Find the points to exclude:
$$x^2 - 9 = 0$$
$$x^2 = 9$$
$$x = \pm 3$$
So the domain is all real numbers except 3 and –3, or $(-\infty, -3) \cup (-3, 3) \cup (3, \infty)$.

45. We must be sure that the quantity inside the radical is non-negative:
$$8 - 2x \geq 0$$
$$-2x \geq -8$$
$$x \leq 4$$
So the domain is $(-\infty, 4]$.

47. We must be sure that the quantity inside the radical is non-negative, so $|2 - 5x| \geq 0$.
But this is true for all real numbers, so the domain is all real numbers, or $(-\infty, \infty)$.

49. We must be sure that the quantity inside the radical is non-negative:

$$x^2 - 2x - 3 \geq 0$$
$$(x-3)(x+1) \geq 0$$

The key numbers are −1 and 3. Draw the sign chart:

Interval	Test Number	$x-3$	$x+1$	$(x-3)(x+1)$
$(-\infty, -1)$	−2	neg.	neg.	pos.
$(-1, 3)$	0	neg.	pos.	neg.
$(3, \infty)$	4	pos.	pos.	pos.

So the product is positive on the intervals $(-\infty, -1) \cup (3, \infty)$, and it is zero when $x = -1$ and $x = 3$. So the domain is $(-\infty, -1] \cup [3, \infty)$.

51. We must be sure that $x \neq 0$. So the domain is all real numbers except 0, or $(-\infty, 0) \cup (0, \infty)$.

53. Solve for x:

$$y = \frac{x+4}{3x-1}$$
$$y(3x-1) = x+4$$
$$3xy - y = x + 4$$
$$3xy - x = y + 4$$
$$x(3y-1) = y+4$$
$$x = \frac{y+4}{3y-1}$$

Now $3y - 1 = 0$ when $y = \frac{1}{3}$, so the range is all real numbers except $\frac{1}{3}$, or $\left(-\infty, \frac{1}{3}\right) \cup \left(\frac{1}{3}, \infty\right)$.

55. First compute $(f \circ g)(x)$:

$$(f \circ g)(x) = f(3x+4) = \frac{1}{3x+4}$$

Solve $y = \frac{1}{3x+4}$ for x:

$$y = \frac{1}{3x+4}$$
$$y(3x+4) = 1$$
$$3xy + 4y = 1$$
$$3xy = 1 - 4y$$
$$x = \frac{1-4y}{3y}$$

Now $3y = 0$ when $y = 0$, so the range is all real numbers except 0, or $(-\infty, 0) \cup (0, \infty)$.

57. Since the range of f^{-1} is the domain of f, exclude the values of x where $3x - 6 = 0$, or $x = 2$. So the range is all real numbers except 2, or $(-\infty, 2) \cup (2, \infty)$.

59. The composition is $a(x) = (f \circ g)(x)$, since:
$$(f \circ g)(x) = f[g(x)] = f(x - 1) = \frac{1}{x - 1}$$

61. The composition is $c(x) = (G \circ g)(x)$, since:
$$(G \circ g)(x) = G[g(x)] = G(x - 1) = \sqrt{x - 1}$$

63. The composition is $A(x) = (g \circ f \circ G)(x)$, since:
$$(g \circ f \circ G)(x) = (g \circ f)(\sqrt{x}) = g\left(\frac{1}{\sqrt{x}}\right) = \frac{1}{\sqrt{x}} - 1$$

65. The composition is $C(x) = (g \circ G \circ G)(x)$, since:
$$(g \circ G \circ G)(x) = (g \circ G)(\sqrt{x}) = g(\sqrt{\sqrt{x}}) = g(\sqrt[4]{x}) = \sqrt[4]{x} - 1$$

67. Using $f(x)$, compute $f(-3)$:
$$f(-3) = (-3)^2 - (-3) = 9 + 3 = 12$$

69. Using $F(x)$, compute $F\left(\frac{3}{4}\right)$:
$$F\left(\frac{3}{4}\right) = \frac{\frac{3}{4} - 3}{\frac{3}{4} + 4} = \frac{3 - 3(4)}{3 + 4(4)} = \frac{3 - 12}{3 + 16} = -\frac{9}{19}$$

71. Using $f(x)$, compute $f(-t)$:
$$f(-t) = (-t)^2 - (-t) = t^2 + t$$

73. Using $f(x)$, compute $f(x - 2)$:
$$f(x - 2) = (x - 2)^2 - (x - 2) = x^2 - 4x + 4 - x + 2 = x^2 - 5x + 6$$

75. Using $g(x)$, first compute $g(2)$ and $g(0)$:
$$g(2) = 1 - 2(2) = 1 - 4 = -3$$
$$g(0) = 1 - 2(0) = 1 - 0 = 1$$
So $g(2) - g(0) = -3 - 1 = -4$

77. Using $f(x)$, first compute $f(1)$ and $f(3)$:
$$f(1) = (1)^2 - 1 = 1 - 1 = 0$$
$$f(3) = (3)^2 - 3 = 9 - 3 = 6$$
So $|f(1) - f(3)| = |0 - 6| = |-6| = 6$.

79. Using $f(x)$, compute $f(x^2)$:
$$f(x^2) = (x^2)^2 - x^2 = x^4 - x^2$$

81. Compute $[f(x)][g(x)]$:
$$[f(x)][g(x)] = (x^2 - x)(1 - 2x) = x^2 - 2x^3 - x + 2x^2 = -2x^3 + 3x^2 - x$$

83. Compute $f[g(x)]$:
$$f[g(x)] = f(1 - 2x) = (1 - 2x)^2 - (1 - 2x) = 1 - 4x + 4x^2 - 1 + 2x = 4x^2 - 2x$$

85. Compute $(g \circ f)(x)$:
$$(g \circ f)(x) = g[f(x)] = g(x^2 - x) = 1 - 2(x^2 - x) = 2 - 2x^2 + 2x = -2x^2 + 2x + 1$$

87. Compute $(F \circ g)(x)$:
$$(F \circ g)(x) = F[g(x)] = F(1 - 2x) = \frac{(1 - 2x) - 3}{(1 - 2x) + 4} = \frac{-2x - 2}{-2x + 5} \text{ or } \frac{2x + 2}{2x - 5}$$

89. Find the difference quotient:
$$\frac{f(x + h) - f(x)}{h} = \frac{2xh + h^2 - h}{h} = \frac{h(2x + h - 1)}{h} = 2x + h - 1$$

91. Let $y = \dfrac{x - 3}{x + 4}$. To find $F^{-1}(x)$, switch the roles of x and y and solve the resulting equation for y:
$$x = \frac{y - 3}{y + 4}$$
$$x(y + 4) = y - 3$$
$$xy + 4x = y - 3$$
$$xy - y = -4x - 3$$
$$y(x - 1) = -4x - 3$$
$$y = \frac{-4x - 3}{x - 1}$$
So $F^{-1}(x) = \dfrac{-4x - 3}{x - 1} = \dfrac{4x + 3}{1 - x}$.

93. $F^{-1}[F(x)] = x$ for $x \neq -4$, by definition of $F^{-1}(x)$.

95. The value is x, since $(g \circ g^{-1})(x) = x$ for all x.

97. Since $g^{-1}(x) = \dfrac{1-x}{2}$, we have:

$$g^{-1}(-x) = \dfrac{1-(-x)}{2} = \dfrac{1+x}{2}$$

99. Since $\frac{22}{7}$ is in the domain of $F(x)$, $F^{-1}\left[F\left(\frac{22}{7}\right)\right] = \frac{22}{7}$.

101. Since $f(0) = -2$, it is negative.

103. Since the point $(-3, -1)$ lies on the graph, $f(-3) = -1$.

105. Since $f(0) = -2$ and $f(8) = -1$, $f(0) - f(8) = -2 - (-1) = -2 + 1 = -1$.

107. The coordinates of the turning points are $(0, -2)$ and $(5, 1)$.

109. $f(x)$ is decreasing on the intervals $[-6, 0]$ and $[5, 8]$.

111. Since $|x| \le 2$ corresponds to the interval $-2 \le x \le 2$, the largest value of $f(x)$ is 0, occurring at $x = 2$.

113. Since $f(x)$ is not a one-to-one function (it does not pass the horizontal line test) it does not possess an inverse function.

115. From the graph, note that $f(x) = g(x)$ when $x = 4$.

117. (a) From the graph, note that $f(x) = 0$ when $x = 10$.
   (b) From the graph, note that $g(x) = 0$ when $x = 0$.

119. (a) $(f + g)(8) = f(8) + g(8) = 1 + 4 = 5$
   (b) $(f - g)(8) = f(8) - g(8) = 1 - 4 = -3$
   (c) $fg(8) = f(8) \cdot g(8) = 1 \cdot 4 = 4$
   (d) $(f / g)(x) = \dfrac{f(8)}{g(8)} = \frac{1}{4}$

121. Compute each value:
$$(f \circ f)(10) = f[f(10)] = f(0) = 5$$
$$(g \circ g)(10) = g[g(10)] = g(3) = 1$$
So $(f \circ f)(10)$ is larger.

123. Note that $f(x) \ge 3$ for $0 \le x \le 4$, which is the interval $[0, 4]$.

125. The maximum point of $g(x)$ is $(6, 5)$, so the largest number in the range of g is 5.

127. Note that $g(x)$ is decreasing when $1 < x < 3$ and $6 < x < 10$, which are the intervals $(1,3) \cup (6,10)$.

129. Consider two cases:

If $4 < x < 5$, then $f(x) > f(5)$ and $x < 5$, thus $f(x) - f(5) > 0$ while $x - 5 < 0$.

Thus $\dfrac{f(x) - f(5)}{x - 5} < 0$.

If $5 < x < 7$, then $f(x) < f(5)$ and $x > 5$, thus $f(x) - f(5) < 0$ while $x - 5 > 0$.

Thus $\dfrac{f(x) - f(5)}{x - 5} < 0$.

Therefore, the quantity is negative for all x-values in the interval $(4, 7)$.

131. If A varies jointly as x and y^2, then $A = kxy^2$. Replace x by $3x$ and y by $2y$:

$$A = k(3x)(2y)^2 = k(3x)(4y^2) = 12(kxy^2)$$

So the value of A is 12 times its original value.

133. Let I = intensity and d = distance, so $I = \dfrac{k}{d^2}$. If $d = 10$ ft, then:

$$I = \frac{k}{10^2} = \frac{k}{100}$$

We are asked to find the value of d when $I = 5\left(\frac{k}{100}\right) = \frac{k}{20}$, so:

$$\frac{k}{20} = \frac{k}{d^2}$$

$$\frac{1}{20} = \frac{1}{d^2}$$

$$d^2 = 20$$

$$d = \sqrt{20} = 2\sqrt{5}$$

So the intensity will be five times as great when $d = 2\sqrt{5} = 4.47$ feet.

135. Let L = length and A = area, so $L = k\sqrt{A}$. Find k by substituting $A = 16\sqrt{3}$ and $L = 8$:

$$8 = k\sqrt{16\sqrt{3}}$$

$$8 = k \cdot 4\sqrt[4]{3}$$

$$k = \frac{8}{4\sqrt[4]{3}} = \frac{2}{\sqrt[4]{3}}$$

So $L = \frac{2}{\sqrt[4]{3}} \sqrt{A}$. When $A = 20$ we have:

$$L = \frac{2}{\sqrt[4]{3}} \sqrt{20} = \frac{4\sqrt{5}}{\sqrt[4]{3}} = \frac{4}{3}\sqrt{5}\,\sqrt[4]{27} \approx 6.80 \text{ cm}$$

Chapter Four Test

1.   We must be sure the quantity inside the radical is non-negative:
$$x^2 - 5x - 6 \geq 0$$
$$(x-6)(x+1) \geq 0$$
The key numbers are –1 and 6. Draw the sign chart:

Interval	Test Number	$x-6$	$x+1$	$(x-6)(x+1)$
$(-\infty, -1)$	–2	neg.	neg.	pos.
$(-1, 6)$	0	neg.	pos.	neg.
$(6, \infty)$	8	pos.	pos.	pos.

The product is positive on the intervals $(-\infty, -1) \cup (6, \infty)$ and 0 at $x = -1$ and $x = 6$. So the domain is $(-\infty, -1] \cup [6, \infty)$.

2.   Let $y = \dfrac{2x-8}{3x+5}$. Solve for x:
$$\frac{2x-8}{3x+5} = y$$
$$2x - 8 = 3yx + 5y$$
$$2x - 3yx = 5y + 8$$
$$x(2 - 3y) = 5y + 8$$
$$x = \frac{5y+8}{2-3y}$$

So $y \neq \frac{2}{3}$. The range is $\left(-\infty, \frac{2}{3}\right) \cup \left(\frac{2}{3}, \infty\right)$.

3.   (a)   Compute $(f-g)(x)$:
$$(f-g)(x) = f(x) - g(x) = (2x^2 - 3x) - (2 - x) = 2x^2 - 3x - 2 + x = 2x^2 - 2x - 2$$

   (b)   Compute $(f \circ g)(x)$:
$$
\begin{aligned}
(f \circ g)(x) &= f[g(x)] \\
&= f(2 - x) \\
&= 2(2-x)^2 - 3(2-x) \\
&= 8 - 8x + 2x^2 - 6 + 3x \\
&= 2x^2 - 5x + 2
\end{aligned}
$$

   (c)   Using our result from part (b):
$$f[g(-4)] = 2(-4)^2 - 5(-4) + 2 = 32 + 20 + 2 = 54$$

4.   Compute the difference quotient:
$$\frac{f(t) - f(a)}{t-a} = \frac{\frac{2}{t} - \frac{2}{a}}{t-a} \cdot \frac{at}{at} = \frac{2a - 2t}{at(t-a)} = \frac{-2(t-a)}{at(t-a)} = -\frac{2}{at}$$

5. First compute $g(x + h)$:

$$g(x + h) = 2(x + h)^2 - 5(x + h) = 2x^2 + 4xh + 2h^2 - 5x - 5h$$

Now compute the difference quotient:

$$\frac{g(x + h) - g(x)}{h} = \frac{2x^2 + 4xh + 2h^2 - 5x - 5h - 2x^2 + 5x}{h}$$

$$= \frac{4xh + 2h^2 - 5h}{h}$$

$$= 4x + 2h - 5$$

6. Let $y = \dfrac{-4x}{6x + 1}$. Switch the roles of x and y, then solve the resulting equation for y:

$$\frac{-4y}{6y + 1} = x$$

$$-4y = 6xy + x$$

$$-x = 6xy + 4y$$

$$-x = y(6x + 4)$$

$$\frac{-x}{6x + 4} = y$$

So $g^{-1}(x) = -\dfrac{x}{6x + 4}$.

7. Since $f^{-1}(x)$ will be the line segment joining the points $(1, -3)$ and $(6, 5)$, $y = -f^{-1}(x)$ will be the line segment joining the points $(1, 3)$ and $(6, -5)$. Compute the slope:

$$m = \frac{-5 - 3}{6 - 1} = -\frac{8}{5}$$

8. (a) The x-intercepts are 4 and 2, and the y-intercept is -2:

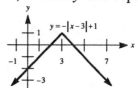

(b) The x-intercept is $-\frac{5}{2}$ and the y-intercept is $-\frac{5}{3}$:

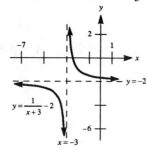

9.   (a)   The range of g is $[-3, 1]$.
   (b)   The turning point has coordinates $(-2, 1)$.
   (c)   The minimum value of g is -3, which occurs at $x = 2$.
   (d)   The maximum value of g is 1, occurring at $x = -2$.
   (e)   For $-2 < x < 2$, g is decreasing. So g is decreasing on $(-2, 2)$.
   (f)   On the interval $[-4, 2]$, we have:

$$\frac{\Delta g}{\Delta x} = \frac{-3-(-1)}{2-(-4)} = \frac{-2}{6} = -\frac{1}{3}$$

10.   (a)   Compute $f\left(-\frac{3}{2}\right)$:

$$f\left(-\tfrac{3}{2}\right) = \left(-\tfrac{3}{2}\right)^2 - 3\left(-\tfrac{3}{2}\right) - 1 = \tfrac{9}{4} + \tfrac{9}{2} - 1 = \tfrac{23}{4}$$

   (b)   Compute $f\left(\sqrt{3} - 2\right)$:

$$\begin{aligned} f\left(\sqrt{3} - 2\right) &= \left(\sqrt{3} - 2\right)^2 - 3\left(\sqrt{3} - 2\right) - 1 \\ &= 3 - 4\sqrt{3} + 4 - 3\sqrt{3} + 6 - 1 \\ &= 12 - 7\sqrt{3} \end{aligned}$$

11.   The domain is all real numbers except -1, or $(-\infty, -1) \cup (-1, \infty)$. Draw the graph:

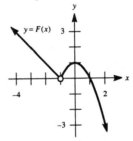

12.   The variation law is $z = \dfrac{kx^2}{\sqrt{t}}$. Given $x = 6, t = 9, z = 12$, we find k:

$$12 = \frac{k \cdot (6)^2}{\sqrt{9}}$$

$$12 = \frac{36k}{3}$$

$$1 = k$$

So the variation law is $z = \dfrac{x^2}{\sqrt{t}}$. Now substitute $x = 3$ and $t = 16$:

$$z = \frac{(3)^2}{\sqrt{16}} = \frac{9}{4}$$

13. (a) Assume $f(a) = f(b)$. We must show $a = b$:

$$f(a) = f(b)$$
$$3 - 2a = 3 - 2b$$
$$-2a = -2b$$
$$a = b$$

So f is a one-to-one function.

(b) We must give values for a and b so that $f(a) = f(b)$ but $a \neq b$. Let $a = 1$ and $b = -1$. Then $f(a) = f(b) = \sqrt{5}$ while $a \neq b$. So f is not a one-to-one function.

14. (a) The first six iterates are 0.4, 0.8, 0.4, 0.8, 0.4, and 0.8.

(b) The calculated six iterates are 0.4, 0.8, 0.4, 0.8, 0.4, and 0.8. These results are consistent with those found in part (a).

15. Since $y = f(-x)$ is a reflection of $y = f(x)$ across the y-axis, it will be the line segment joining the points $(-1, 3)$ and $(-5, -2)$:

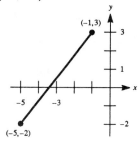

16. Solve the equation:

$$5 + f(4t - 3) = 2$$
$$f(4t - 3) = -3$$
$$4t - 3 = f^{-1}(-3)$$
$$4t - 3 = 1$$
$$4t = 4$$
$$t = 1$$

17. Since $f(1) = 1$ and $f(b) = \dfrac{1}{b}$:

$$\frac{\Delta f}{\Delta x} = \frac{f(b) - f(1)}{b - 1} = \frac{\frac{1}{b} - 1}{b - 1} \cdot \frac{b}{b} = \frac{1 - b}{b(b - 1)} = -\frac{1}{b}$$

Therefore:

$$-\frac{1}{b} = -\frac{1}{10}$$
$$b = 10$$

18.   (a)   Since $F(t_1) = 0.16t_1^2 - 1.6t_1 + 35$ and $F(5) = 0.16(5)^2 - 1.6(5) + 35 = 31$, the average value is given by:

$$\frac{\Delta F}{\Delta t} = \frac{F(5) - F(t_1)}{5 - t_1}$$

$$= \frac{31 - \left(0.16t_1^2 - 1.6t_1 + 35\right)}{5 - t_1}$$

$$= \frac{-0.16t_1^2 + 1.6t_1 - 4}{5 - t_1}$$

$$= \frac{0.16t_1^2 - 1.6t_1 + 4}{t_1 - 5}$$

$$= \frac{\left(t_1 - 5\right)\left(0.16t_1 - 0.8\right)}{t_1 - 5}$$

$$= 0.16t_1 - 0.8$$

(b)   Using the result from part (a) where $t_1 = 4$:

$$\frac{\Delta F}{\Delta t} = 0.16(4) - 0.8 = -0.16 \ ^\circ\text{C/hour}$$

Chapter Five
Polynomial and Rational Functions. Applications to Iteration and Optimization

5.1 Linear Functions

1. First find the slope between the points $(-1, 0)$ and $(5, 4)$:
$$m = \frac{4-0}{5-(-1)} = \tfrac{4}{6} = \tfrac{2}{3}$$
Now use the point $(-1, 0)$ in the point-slope formula:
$$y - 0 = \tfrac{2}{3}[x - (-1)]$$
$$y = \tfrac{2}{3}(x + 1)$$
$$y = \tfrac{2}{3}x + \tfrac{2}{3}$$
Using functional notation we have $f(x) = \tfrac{2}{3}x + \tfrac{2}{3}$.

3. First find the slope between the points $(0, 0)$ and $(1, \sqrt{2})$:
$$m = \frac{\sqrt{2} - 0}{1 - 0} = \sqrt{2}$$
Now, since $(0, 0)$ is the y-intercept, we use the slope-intercept formula to write $y = \sqrt{2}x$. Using functional notation we have $g(x) = \sqrt{2}x$.

5. Find the slope of $x - y = 1$:
$$-y = -x + 1$$
$$y = x - 1$$
So the parallel slope is 1. Use the point $\left(\frac{1}{2}, -3\right)$ in the point-slope formula:
$$y - (-3) = 1\left(x - \tfrac{1}{2}\right)$$
$$y + 3 = x - \tfrac{1}{2}$$
$$y = x - \tfrac{7}{2}$$
Using functional notation we have $f(x) = x - \frac{7}{2}$.

7. Set $x = 0$ to find the y-intercepts of the circle:
$$0 - 0 + y^2 - 3 = 0$$
$$y^2 = 3$$
$$y = \pm\sqrt{3}$$
So a horizontal line passing through $\left(0, \sqrt{3}\right)$ is $y = \sqrt{3}$. Using functional notation we have $f(x) = \sqrt{3}$.

9. Since the points $(-1, 2)$ and $(0, 4)$ lie on the graph of the inverse function, $(2, -1)$ and $(4, 0)$ must lie on the graph of the function. We find the slope:
$$m = \frac{0 - (-1)}{4 - 2} = \tfrac{1}{2}$$
Use the point $(4, 0)$ in the point-slope formula:
$$y - 0 = \tfrac{1}{2}(x - 4)$$
$$y = \tfrac{1}{2}x - 2$$
Using functional notation we have $f(x) = \frac{1}{2}x - 2$.

11. Compute $(f \circ g)(x)$:
$$(f \circ g)(x) = f[g(x)] = f(1 - 2x) = 3(1 - 2x) - 4 = 3 - 6x - 4 = -6x - 1$$
So $f \circ g$ is a linear function (it is in standard form).

13. Call $V(t)$ the value of the machine after t years. When $t = 0$ we have $V = 20{,}000$ and when $t = 8$ we have $V = 1{,}000$. Find the slope of the line between the points $(0, 20000)$ and $(8, 1000)$:
$$m = \frac{1000 - 20000}{8 - 0} = \frac{-19000}{8} = -2375$$
Since $(0, 20000)$ is the y-intercept, we use the slope-intercept formula to write $V = -2375t + 20000$. Using functional notation we have $V(t) = -2375t + 20000$.

15. (a) Call $V(t)$ the value of the machine after t years. Now $V = 60{,}000$ when $t = 0$ and $V = 0$ when $t = 5$. Find the slope of the line between $(0, 60000)$ and $(5, 0)$:
$$m = \frac{0 - 60000}{5 - 0} = \frac{-60000}{5} = -12000$$
Since $(0, 60000)$ is the y-intercept, we use the slope-intercept formula to write $V = -12000t + 60000$. Using functional notation we have $V(t) = -12000t + 60000$.

(b) The completed schedule is:

End of Year	Yearly Depreciation	Accumulated Depreciation	Value V
0	0	0	60,000
1	12,000	12,000	48,000
2	12,000	24,000	36,000
3	12,000	36,000	24,000
4	12,000	48,000	12,000
5	12,000	60,000	0

17. (a) Let $x = 10$, so $C(10) = 450 + 8x = 450 + 8(10) = \$530$.

(b) Let $x = 11$, so $C(11) = 450 + 8(11) = \$538$.

(c) There are two ways to find the marginal cost. One way is to recognize that the marginal cost will be the slope of the line, which is \$8/fan. Another way would be to use the definition of the marginal cost: it is cost of producing the <u>next</u> unit. Since $C(10)$ is the cost of producing 10 fans and $C(11)$ is the cost of producing 11 fans, the marginal cost would be $C(11) - C(10) = 538 - 530 = \$8$/fan. We get the same answer using either approach.

19. (a) First compute $C(n + 1)$:
$$C(n+1) = 400 + 50(n+1) = 400 + 50n + 50 = 450 + 50n$$
Therefore:
$$C(n+1) - C(n) = (450 + 50n) - (400 + 50n) = 450 + 50n - 400 - 50n = 50$$

(b) The marginal cost, which represents the cost to produce the next unit, is the slope of the line. So the marginal cost is \$50/player.

(c) The answers are the same. Notice that (a) directly computed marginal cost from its definition.

21. Since the velocity will be the slope of the distance line, compute the slope of each line.

(a) Let $(x_1, y_1) = (1, 4)$ and $(x_2, y_2) = (6, 8)$. Then:
$$m = \frac{8-4}{6-1} = \frac{4}{5}$$
So the velocity is $\frac{4}{5}$ ft/sec.

(b) Let $(x_1, y_1) = (2, 4)$ and $(x_2, y_2) = (5, 4)$. Then:
$$m = \frac{4-4}{5-2} = \frac{0}{3} = 0$$
So the velocity is 0 cm/sec.

(c) Let $(x_1, y_1) = (0, 0)$ and $(x_2, y_2) = (2, 16)$. Then:
$$m = \frac{16 - 0}{2 - 0} = \tfrac{16}{2} = 8$$
So the velocity is 8 mph.

23. (a) Since the velocity of A is 3 units/sec and the velocity of B is 20 units/sec, B is traveling faster.

(b) When $t = 0$, A is at $x = 100$ and B is at $x = -36$. So A is farther to the right.

(c) Set the two x-coordinates equal:
$$3t + 100 = 20t - 36$$
$$136 = 17t$$
$$8 = t$$
When $t = 8$ sec, A and B have the same x-coordinate.

25. (a) First find the slope:
$$m = \frac{23667764 - 19971069}{1980 - 1970} = \frac{3696695}{10} = 369669.5$$
Using the point (1970, 19971069) in the point-slope formula:
$$y - 19971069 = 369669.5(x - 1970)$$
$$y - 19971069 = 369669.5x - 728248915$$
$$y = 369669.5x - 708277846$$

(b) Substituting $x = 1990$:
$$y = 369669.5(1990) - 708277846 \approx 27364000$$
The population projection is 27,364,000.

(c) The linear function yields a projection that is too low.

(d) Calculating the percent error:
$$\text{percent error} = \frac{|29760021 - 27364000|}{29760021} \times 100 \approx 8.1\%$$

27. (a) First find the slope:
$$m = \frac{9746961 - 6791418}{1980 - 1970} = \frac{2955543}{10} = 295554.3$$
Using the point (1970, 6791418) in the point-slope formula:
$$y - 6791418 = 295554.3(x - 1970)$$
$$y - 6791418 = 295554.3x - 582241971$$
$$y = 295554.3x - 575450553$$

(b) Substituting $x = 1990$:
$$y = 295554.3(1990) - 575450553 \approx 12703000$$
The population projection is 12,703,000.

(c) The linear function yields a projection that is too low.

(d) Calculating the percent error:
$$\text{percent error} = \frac{|12937926 - 12703000|}{12937926} \times 100 \approx 1.8\%$$

29. (a) First find the slope:
$$m = \frac{205.50 - 67.76}{49 - 21} = \frac{137.74}{28} \approx 4.92$$
Using the point (21, 67.76) in the point-slope formula:
$$y - 67.76 = 4.92(x - 21)$$
$$y - 67.76 = 4.92x - 103.31$$
$$y = 4.92x - 35.55$$

(b) Substituting $x = 28$:
$$y = 4.92(28) - 35.55 = 102.21$$
The projection for average height is 102.21 cm.

(c) The estimate from part (b) is too high. Calculating the percent error:
$$\text{percent error} = \frac{|98.10 - 102.21|}{98.10} \times 100 \approx 4.2\%$$

(d) Substituting $x = 14$:
$$y = 4.92(14) - 35.55 = 33.33$$
The projection for average height is 33.33 cm. This estimate is too low. Calculating the percent error:
$$\text{percent error} = \frac{|36.36 - 33.33|}{36.36} \times 100 \approx 8.3\%$$

(e) Substituting $x = 84$:
$$y = 4.92(84) - 35.55 = 377.73$$
The projection for average height is 377.73 cm. This estimate is too high. Calculating the percent error:
$$\text{percent error} = \frac{|254.50 - 377.73|}{254.50} \times 100 \approx 48.4\%$$
Note the percent error is fairly large for this longer time period.

31. (a) Plotting the points:

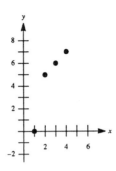

(b) The slope appears to be approximately 2.5 and the y-intercept is -2:

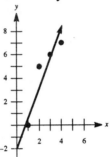

(c) Graphing the regression line $y = 2.4x - 2$:

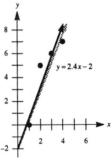

33. (a) Substituting $x = 2000$ into the regression line:
$$f(2000) = 37546.068(2000) - 71238863.429 \approx 3,853,000$$
Based on the projected year 2000 population, it is reasonable that the population of Los Angeles might reach 3.8 million.

(b) Let $y = f(x)$. Switching the roles of x and y and solving the resulting equation for y:
$$x = 37546.068y - 71238863.429$$
$$x + 71238863.429 = 37546.068y$$
$$y = \frac{x + 71238863.429}{37546.068}$$
So $f^{-1}(x) = \dfrac{x + 71238863.429}{37546.068}$.

(c) Substituting $x = 4{,}000{,}000$ into the inverse function:
$$f^{-1}(4000000) = \frac{4000000 + 71238863.429}{37546.068} \approx 2004$$
The population of Los Angeles might reach 4 million in the year 2004.

35. (a) Substitute $x = 1996$:
$$y = (0.3428)(1996) - 680.6114 = 3.6174$$
The value of the toy imports from China in 1996 will be approximately 3.6174 billion dollars.

(b) Substitute $x = 2000$:
$$y = (0.3428)(2000) - 680.6114 = 4.9886$$
The value of the toy imports will not exceed 5 billion dollars by the year 2000.

37. (a) Substituting $x = 1957$:
$$f(1957) = -0.370(1957) + 962.041 = 237.951$$
Our estimate is 238.0 seconds, or 3:58.0. This estimate is too high, but it is very close.

(b) Substituting $x = 1958$:
$$f(1958) = -0.370(1958) + 962.041 = 237.581$$
Our estimate is 237.6 seconds, or 3:57.6. This estimate is too high.

39. (a) Using the algebraic method:
$$f(x) = x$$
$$-\frac{3}{2}x + \frac{15}{2} = x$$
$$-3x + 15 = 2x$$
$$15 = 5x$$
$$3 = x$$
So 3 is a fixed point for the function.

(b) Graphing $f(x) = -\frac{3}{2}x + \frac{15}{2}$ and $y = x$:

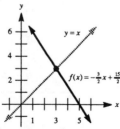

Note that $x = 3$ appears to be where the two curves intersect.

(c) The first five iterates of 3 are 3.

41. (a) Finding the fixed point of f:
$$f(x) = x$$
$$\tfrac{1}{2}x + \tfrac{3}{2} = x$$
$$x + 3 = 2x$$
$$3 = x$$
So 3 is a fixed point for the function.

   (b) The first five iterates of $x_0 = 1$ are 2, 2.5, 2.75, 2.875, and 2.9375.

   (c) Computing the first twelve iterates of $x_0 = 1$:
$$x_1 = 2$$
$$x_2 = 2.5$$
$$x_3 = 2.75$$
$$x_4 = 2.875$$
$$x_5 = 2.9375$$
$$x_6 \approx 2.9688$$
$$x_7 \approx 2.9844$$
$$x_8 \approx 2.9922$$
$$x_9 \approx 2.9961$$
$$x_{10} \approx 2.9980$$
$$x_{11} \approx 2.9990$$
$$x_{12} \approx 2.9995$$
Note that the first five iterates agree with those found in part (b).

   (d) The iterates are approaching the number 3, which is the fixed point of the function f.

43. (a) The average rate of change is given by:
$$\frac{f(b) - f(a)}{b - a} = \frac{(Ab + B) - (Aa + B)}{b - a} = \frac{Ab - Aa}{b - a} = \frac{A(b - a)}{b - a} = A$$

   (b) Since $(f \circ f)(x) = f(Ax + B) = A(Ax + B) + B = A^2 x + AB + B$, the average rate of change is given by:
$$\frac{(f \circ f)(b) - (f \circ f)(a)}{b - a} = \frac{\left(A^2 b + AB + B\right) - \left(A^2 a + AB + B\right)}{b - a}$$
$$= \frac{A^2 b - A^2 a}{b - a}$$
$$= \frac{A^2 (b - a)}{b - a}$$
$$= A^2$$

(c) Since $(g \circ f)(x) = g(Ax + B) = C(Ax + B) + D = ACx + BC + D$, the average rate of change is given by:

$$\frac{(g \circ f)(b) - (g \circ f)(a)}{b - a} = \frac{(ACb + BC + D) - (ACa + BC + D)}{b - a}$$
$$= \frac{ACb - ACa}{b - a}$$
$$= \frac{AC(b - a)}{b - a}$$
$$= AC$$

(d) Since $(f \circ g)(x) = f(Cx + D) = A(Cx + D) + B = ACx + AD + B$, the average rate of change is given by:

$$\frac{(f \circ g)(b) - (f \circ g)(a)}{b - a} = \frac{(ACb + AD + B) - (ACa + AD + B)}{b - a}$$
$$= \frac{ACb - ACa}{b - a}$$
$$= \frac{AC(b - a)}{b - a}$$
$$= AC$$

45. (a) Since $(f \circ g)(x) = ACx + AD + B$, we must find $(f \circ g)^{-1}$ by switching the roles of x and y, then solving for y:

$$ACy + AD + B = x$$
$$ACy = x - AD - B$$
$$y = \frac{x}{AC} - \frac{D}{C} - \frac{B}{AC}$$

So $(f \circ g)^{-1} = \frac{x}{AC} - \frac{D}{C} - \frac{B}{AC}$. The average rate of change is given by:

$$\frac{(f \circ g)^{-1}(b) - (f \circ g)^{-1}(a)}{b - a} = \frac{\left(\frac{b}{AC} - \frac{D}{C} - \frac{B}{AC}\right) - \left(\frac{a}{AC} - \frac{D}{C} - \frac{B}{AC}\right)}{b - a}$$
$$= \frac{\frac{b}{AC} - \frac{a}{AC}}{b - a}$$
$$= \frac{\frac{1}{AC}(b - a)}{b - a}$$
$$= \frac{1}{AC}$$

(b) Since $(g \circ f)(x) = ACx + BC + D$, we must find $(g \circ f)^{-1}$ by switching the roles of x and y, then solving for y:

$$ACy + BC + D = x$$
$$ACy = x - BC - D$$
$$y = \frac{x}{AC} - \frac{B}{A} - \frac{D}{AC}$$

So $(g \circ f)^{-1} = \dfrac{x}{AC} - \dfrac{B}{A} - \dfrac{D}{AC}$. The average rate of change is given by:

$$\frac{(g \circ f)^{-1}(b) - (g \circ f)^{-1}(a)}{b-a} = \frac{\left(\frac{b}{AC} - \frac{B}{A} - \frac{D}{AC}\right) - \left(\frac{a}{AC} - \frac{B}{A} - \frac{D}{AC}\right)}{b-a}$$

$$= \frac{\frac{b}{AC} - \frac{a}{AC}}{b-a}$$

$$= \frac{\frac{1}{AC}(b-a)}{b-a}$$

$$= \frac{1}{AC}$$

47. (a) Using $f(x) = mx$, we have:
$$f(a+b) = m(a+b) = ma + mb = f(a) + f(b)$$

(b) Using $f(x) = mx$, we have:
$$f(ax) = m(ax) = a(mx) = af(x)$$

49. First simplify $(f \circ f)(x)$:
$$(f \circ f)(x) = f(mx + b) = m(mx + b) + b = m^2x + (mb + b)$$
Since $(f \circ f)(x) = 9x + 4$, and since m and b are constants, we must have the two equations (equating x-coefficients and constants):
$$m^2 = 9 \quad \text{and} \quad mb + b = 4$$
Since $m > 0$, $m = 3$ and:
$$3b + b = 4$$
$$4b = 4$$
$$b = 1$$
So $f(x) = 3x + 1$.

51. (a) Compute the sums:
$$\sum x = 1 + 2 + 3 + 4 = 10$$
$$\sum y = 0 + 3 + 6 + 7 = 16$$

(b) Compute the sums:
$$\sum x^2 = 1 + 4 + 9 + 16 = 30$$
$$\sum xy = 0 + 6 + 18 + 28 = 52$$

(c)   Multiply the first equation by -3:
$$-12b - 30m = -48$$
$$10b + 30m = 52$$
Adding, we obtain:
$$-2b = 4$$
$$b = -2$$
Substituting into the first equation:
$$4b + 10m = 16$$
$$-8 + 10m = 16$$
$$10m = 24$$
$$m = 2.4$$
So the regression line is $y = 2.4x - 2$.

53.   First do the computations:
$$\sum x = 1 + 2 + 3 + 4 + 5 = 15$$
$$\sum y = 2 + 3 + 9 + 9 + 11 = 34$$
$$\sum x^2 = 1 + 4 + 9 + 16 + 25 = 55$$
$$\sum xy = 2 + 6 + 27 + 36 + 55 = 126$$
So the system of equations becomes:
$$5b + 15m = 34$$
$$15b + 55m = 126$$
We multiply the first equation by -3:
$$-15b - 45m = -102$$
$$15b + 55m = 126$$
Adding, we obtain:
$$10m = 24$$
$$m = 2.4$$
Substituting into the first equation:
$$5b + 36 = 34$$
$$5b = -2$$
$$b = -0.4$$
So the regression line is $y = 2.4x - 0.4$.

55.   First do the computations:
$$\sum x = 520 + 740 + 560 + 610 + 650 = 3080$$
$$\sum y = 81 + 98 + 83 + 88 + 95 = 445$$
$$\sum x^2 = 270400 + 547600 + 313600 + 372100 + 422500 = 1926200$$
$$\sum xy = 42120 + 72520 + 46480 + 53680 + 61750 = 276550$$
So the system of equations becomes:
$$5b + 3080m = 445$$
$$3080b + 1926200m = 276550$$
Divide the first equation by 5 and the second equation by 10:
$$b + 616m = 89$$
$$308b + 192620m = 27655$$

Multiply the first equation by -308:
$$-308b - 189728m = -27412$$
$$308b + 192620m = 27655$$
Adding, we obtain:
$$2892m = 243$$
$$m \approx 0.084$$
Substituting into the first equation:
$$5b + 258.8 = 445$$
$$5b = 186.2$$
$$b \approx 37.241$$
So the regression line is $y = 0.084x + 37.241$.

57. (a) Since f is a linear function, let $f(x) = ax + b$. Therefore:
$$f(f(x)) = a(ax + b) + b = a^2 x + (ab + b)$$
Since $f(f(x)) = 2x + 1$, we have $a^2 = 2$, thus $a = \pm\sqrt{2}$. Also $ab + b = 1$:
$$ab + b = 1$$
$$b(a + 1) = 1$$
$$b = \frac{1}{a + 1}$$
If $a = \sqrt{2}$, then $b = \dfrac{1}{\sqrt{2} + 1} = \sqrt{2} - 1$ and if $a = -\sqrt{2}$, then $b = \dfrac{1}{-\sqrt{2} + 1} = -\sqrt{2} - 1$.

The possible linear functions are $f(x) = \sqrt{2}\,x + \left(-1 + \sqrt{2}\right)$ and
$$f(x) = -\sqrt{2}\,x + \left(-1 - \sqrt{2}\right).$$

(b) Since f is a linear function, let $f(x) = ax + b$. From part (a) we have
$$f(f(x)) = a^2 x + (ab + b), \text{ so:}$$
$$f(f(f(x))) = f\left(a^2 x + (ab + b)\right)$$
$$= a^3 x + \left(a^2 b + ab\right) + b$$
$$= a^3 x + \left(a^2 b + ab + b\right)$$
Since $f(f(f(x))) = 2x + 1$, we have $a^3 = 2$, so $a = \sqrt[3]{2}$. Also $a^2 b + ab + b = 1$, so:
$$a^2 b + ab + b = 1$$
$$\sqrt[3]{4}\,b + \sqrt[3]{2}\,b + b = 1$$
$$b\left(\sqrt[3]{4} + \sqrt[3]{2} + 1\right) = 1$$
$$b = \frac{1}{1 + \sqrt[3]{2} + \sqrt[3]{4}} \cdot \frac{1 - \sqrt[3]{2}}{1 - \sqrt[3]{2}}$$
$$b = \frac{1 - \sqrt[3]{2}}{1 - 2} = -1 + \sqrt[3]{2}$$
The only possible linear function is $f(x) = \sqrt[3]{2}\,x + \left(-1 + \sqrt[3]{2}\right)$.

59. If $f^{-1}(x) = f(f(x))$, then $f(f(f(x))) = x$. From Exercise 57 (b), we know
$f(f(f(x))) = a^3 x + (a^2 b + ab + b)$. Thus $a^3 = 1$, so $a = 1$. Also $a^2 b + ab + b = 0$, so:

$$a^2 b + ab + b = 0$$
$$b + b + b = 0$$
$$3b = 0$$
$$b = 0$$

Thus $f(x) = x$ is the only linear function satisfying the conditions.

Graphing Utility Exercises for Section 5.1

1. (a) Graphing the function:

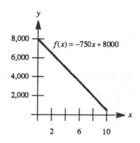

 (b) The y-coordinate is 4250, which matches the result from the text.

3. (a) Graphing the two functions:

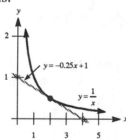

Note that the line does appear to be tangent to the curve at $\left(2, \frac{1}{2}\right)$.

 (b) Completing the tables:

x	1.9	1.99	1.999
$1/x$	0.5263	0.502513	0.5002501
$-0.25x + 1$	0.525	0.5025	0.50025

x	2.1	2.01	2.001
$1/x$	0.4762	0.497512	0.4997501
$-0.25x + 1$	0.475	0.4975	0.49975

5.   (a)   The regression line is $y = 1.0808x - 2080.1169$.

      (b)   Substituting $x = 1993$:
                    $y = 1.0808(1993) - 2080.1169 \approx 73.918$
            The estimate is 73.918 trillion cubic feet.

      (c)   The estimate is close, but it is too low.

      (d)   Dividing:
                    $$\frac{5016.2 \text{ tcf}}{74 \text{ tcf/year}} \approx 68 \text{ years}$$
            This would imply there are world reserves equivalent to 68 years of natural gas production.

5.2 Quadratic Functions

1.   The vertex is (–2, 0), the axis of symmetry is $x = -2$, the minimum value is 0, the x-intercept is –2, and the y-intercept is 4. Sketching the graph:

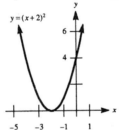

3.   The vertex is (–2, 0), the axis of symmetry is $x = -2$, the minimum value is 0, the x-intercept is –2, and the y-intercept is 8. Sketching the graph:

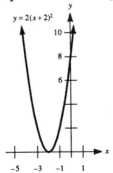

5. The vertex is (–2, 4), the axis of symmetry is $x = -2$, the maximum value is 4, the x-intercepts are $-2 \pm \sqrt{2}$, and the y-intercept is -4. Sketching the graph:

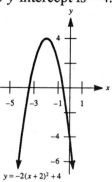

$y = -2(x+2)^2 + 4$

7. Completing the square:
$$f(x) = x^2 - 4x = \left(x^2 - 4x + 4\right) - 4 = (x-2)^2 - 4$$
The vertex is $(2, -4)$, the axis of symmetry is $x = 2$, the minimum value is -4, the x-intercepts are 0 and 4, and the y-intercept is 0. Sketching the graph:

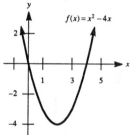

$f(x) = x^2 - 4x$

9. The vertex is (0, 1), the axis of symmetry is $x = 0$, the maximum value is 1, the x-intercepts are ± 1, and the y-intercept is 1. Sketching the graph:

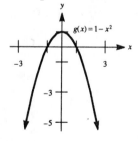

$g(x) = 1 - x^2$

11. Completing the square:
$$y = x^2 - 2x - 3 = \left(x^2 - 2x + 1\right) - 4 = (x-1)^2 - 4$$
The vertex is $(1, -4)$, the axis of symmetry is $x = 1$, the minimum value is -4, the x-intercepts are 3 and -1, and the y-intercept is -3. Sketching the graph:

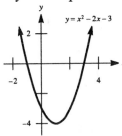

13. Completing the square:
$$y = -x^2 + 6x + 2 = -\left(x^2 - 6x\right) + 2 = -\left(x^2 - 6x + 9\right) + 11 = -(x-3)^2 + 11$$
The vertex is $(3, 11)$, the axis of symmetry is $x = 3$, the maximum value is 11, the x-intercepts are $3 \pm \sqrt{11}$, and the y-intercept is 2. Sketching the graph:

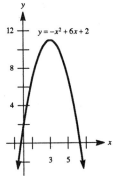

15. The vertex is $(0, 0)$, the axis of symmetry is $t = 0$, the minimum value is 0, the t-intercept is 0, and the s-intercept is 0. Sketching the graph:

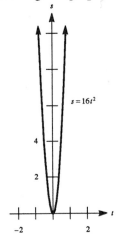

17. Completing the square:
$$s = -9t^2 + 3t + 2 = -9\left(t^2 - \tfrac{1}{3}t\right) + 2 = -9\left(t^2 - \tfrac{1}{3}t + \tfrac{1}{36}\right) + \tfrac{9}{4} = -9\left(t - \tfrac{1}{6}\right)^2 + \tfrac{9}{4}$$
The vertex is $\left(\tfrac{1}{6}, \tfrac{9}{4}\right)$, the axis of symmetry is $t = \tfrac{1}{6}$, the maximum value is $\tfrac{9}{4}$, the t-intercepts are $-\tfrac{1}{3}$ and $\tfrac{2}{3}$, and the s-intercept is 2. Sketching the graph:

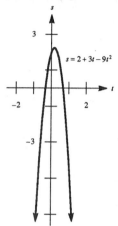

19. Completing the square:
$$y = 2x^2 - 4x + 11 = 2\left(x^2 - 2x\right) + 11 = 2\left(x^2 - 2x + 1\right) + 9 = 2(x - 1)^2 + 9$$
Since the vertex is $(1, 9)$ and the parabola will be pointed up, $x = 1$ will yield a minimum output value.

21. Completing the square:
$$g(x) = -6x^2 + 18x = -6\left(x^2 - 3x\right) = -6\left(x^2 - 3x + \tfrac{9}{4}\right) + \tfrac{27}{2} = -6\left(x - \tfrac{3}{2}\right)^2 + \tfrac{27}{2}$$
Since the vertex is $\left(\tfrac{3}{2}, \tfrac{27}{2}\right)$ and the parabola will be pointed down, $x = \tfrac{3}{2}$ will yield a maximum output value.

23. Since the vertex is $(0, -10)$ and the parabola will be pointed up, $x = 0$ will yield a minimum output value.

25. Completing the square:
$$y = x^2 - 8x + 3 = \left(x^2 - 8x + 16\right) - 13 = (x - 4)^2 - 13$$
Since the vertex is $(4, -13)$ and the parabola will be pointed up, the function has a minimum value of -13.

27. Completing the square:
$$y = -2x^2 - 3x + 2 = -2\left(x^2 + \tfrac{3}{2}x\right) + 2 = -2\left(x^2 + \tfrac{3}{2}x + \tfrac{9}{16}\right) + 2 + \tfrac{9}{8} = -2\left(x + \tfrac{3}{4}\right)^2 + \tfrac{25}{8}$$
Since the vertex is $\left(-\tfrac{3}{4}, \tfrac{25}{8}\right)$ and the parabola will be pointed down, the function has a maximum value of $\tfrac{25}{8}$.

29. We do not need to complete the square, since the vertex is (0, 1000) and the parabola will be pointed down. So the function has a maximum value of 1000.

31. Find the vertex of the parabola by completing the square:
$$y = x^2 - 6x + 13 = (x^2 - 6x + 9) + 4 = (x - 3)^2 + 4$$
So the vertex is (3, 4). We now use the distance formula with the points (0, 0) and (3, 4):
$$d = \sqrt{(3-0)^2 + (4-0)^2} = \sqrt{9+16} = \sqrt{25} = 5$$
So the vertex is 5 units from the origin.

33. Compute $(f \circ g)(x)$:
$$(f \circ g)(x) = 2(x^2 + 4x + 1) - 3 = 2x^2 + 8x + 2 - 3 = 2x^2 + 8x - 1$$
So $f \circ g$ is a quadratic function.

35. Compute $(g \circ h)(x)$:
$$\begin{aligned}(g \circ h)(x) &= (1 - 2x^2)^2 + 4(1 - 2x^2) + 1 \\ &= 1 - 4x^2 + 4x^4 + 4 - 8x^2 + 1 \\ &= 4x^4 - 12x^2 + 6\end{aligned}$$
So $g \circ h$ is neither linear nor quadratic.

37. Compute $(f \circ f)(x)$:
$$(f \circ f)(x) = 2(2x - 3) - 3 = 4x - 6 - 3 = 4x - 9$$
So $f \circ f$ is a linear function.

39. (a) First complete the square on $x^2 - 6x + 73$:
$$x^2 - 6x + 73 = (x^2 - 6x + 9) + 73 - 9 = (x - 3)^2 + 64$$
So $f(x) = \sqrt{(x-3)^2 + 64}$. This would achieve a minimum value at $(3, \sqrt{64}) = (3, 8)$.

    (b) Here $g(x) = \sqrt[3]{(x-3)^2 + 64}$, which would achieve a minimum value at
$$\left(3, \sqrt[3]{64}\right) = (3, 4).$$

    (c) Completing the square on $x^4 - 6x^2 + 73$:
$$x^4 - 6x^2 + 73 = (x^4 - 6x^2 + 9) + 73 - 9 = (x^2 - 3)^2 + 64$$
So $h(x) = (x^2 - 3)^2 + 64$, which would achieve a minimum value at $(\pm\sqrt{3}, 64)$.

41. (a) Completing the square on $-x^2 + 4x + 12$:
$$-x^2 + 4x + 12 = -(x^2 - 4x) + 12 = -(x^2 - 4x + 4) + 16 = -(x - 2)^2 + 16$$
So $f(x) = \sqrt{-(x-2)^2 + 16}$, which has a maximum value at $(2, \sqrt{16}) = (2, 4)$.

(b) Now $g(x) = \sqrt[3]{-(x-2)^2 + 16}$, which has a maximum value at $\left(2, \sqrt[3]{16}\right) = \left(2, 2\sqrt[3]{2}\right)$.

(c) Here $h(x) = -\left(x^2 - 2\right)^2 + 16$, which has a maximum value at $\left(\pm\sqrt{2}, 16\right)$.

43. The average rate of change is given by:

$$\frac{f(x_2) - f(x_1)}{x_2 - x_1} = \frac{\left(ax_2^2 + bx_2 + c\right) - \left(ax_1^2 + bx_1 + c\right)}{x_2 - x_1}$$

$$= \frac{ax_2^2 + bx_2 - ax_1^2 - bx_1}{x_2 - x_1}$$

$$= \frac{a\left(x_2^2 - x_1^2\right) + b\left(x_2 - x_1\right)}{x_2 - x_1}$$

$$= \frac{a\left(x_2 + x_1\right)\left(x_2 - x_1\right) + b\left(x_2 - x_1\right)}{x_2 - x_1}$$

$$= \frac{\left(x_2 - x_1\right)\left(ax_1 + ax_2 + b\right)}{x_2 - x_1}$$

$$= ax_1 + ax_2 + b$$

45. Completing the square:

$$y = ax^2 + bx + c$$

$$= a\left(x^2 + \frac{b}{a}x\right) + c$$

$$= a\left(x^2 + \frac{b}{a}x + \frac{b^2}{4a^2}\right) + c - \frac{ab^2}{4a^2}$$

$$= a\left(x + \frac{b}{2a}\right)^2 + \frac{4ac - b^2}{4a}$$

$$= a\left(x + \frac{b}{2a}\right)^2 - \frac{D}{4a}, \text{ where } D = b^2 - 4ac$$

So the vertex is $\left(-\frac{b}{2a}, -\frac{D}{4a}\right)$, where $D = b^2 - 4ac$.

47. (a) Compute $f\left(\dfrac{-b}{2a}+h\right)$:

$$f\left(\frac{-b}{2a}+h\right)=a\left(\frac{-b}{2a}+h\right)^{2}+b\left(\frac{-b}{2a}+h\right)+c$$

$$=a\left(\frac{b^{2}}{4a^{2}}-\frac{bh}{a}+h^{2}\right)-\frac{b^{2}}{2a}+bh+c$$

$$=\frac{b^{2}}{4a}-bh+ah^{2}-\frac{b^{2}}{2a}+bh+c$$

$$=ah^{2}-\frac{b^{2}}{4a}+c$$

$$=ah^{2}-\frac{b^{2}-4ac}{4a}$$

$$=ah^{2}-\frac{D}{4a},\text{ where }D=b^{2}-4ac$$

(b) Compute $f\left(\dfrac{-b}{2a}-h\right)$:

$$f\left(\frac{-b}{2a}-h\right)=a\left(\frac{-b}{2a}-h\right)^{2}+b\left(\frac{-b}{2a}-h\right)+c$$

$$=a\left(\frac{b^{2}}{4a^{2}}+\frac{bh}{a}+h^{2}\right)-\frac{b^{2}}{2a}-bh+c$$

$$=\frac{b^{2}}{4a}+bh+ah^{2}-\frac{b^{2}}{2a}-bh+c$$

$$=ah^{2}-\frac{b^{2}}{4a}+c$$

$$=ah^{2}-\frac{b^{2}-4ac}{4a}$$

$$=ah^{2}-\frac{D}{4a},\text{ where }D=b^{2}-4ac$$

49. (a) The y-coordinate of point A is $f(x_0)$, the first iterate of x_0. The x-coordinate of point B is $f(x_0)$, so the y-coordinate of point B is $f(f(x_0))$, the second iterate of x_0. The x-coordinate of point C is $f(f(x_0))$, so the y-coordinate of point C is $f(f(f(x_0)))$, the third iterate of x_0. The x-coordinate of point D is $f(f(f(x_0)))$, so the y-coordinate of point D is $f(f(f(f(x_0))))$, the fourth iterate of x_0.

(b) The first four iterates of $x_0=0.2$ are $-0.56,-0.2864,-0.518$, and -0.332.

51. Since the vertex is $(2, 2)$, its equation will be $y = A(x - 2)^2 + 2$. Since $(0, 0)$ must lie on this parabola, substitute the points to find A:

$$y = A(x - 2)^2 + 2$$
$$0 = A(0 - 2)^2 + 2$$
$$0 = 4A + 2$$
$$-2 = 4A$$
$$A = -\tfrac{1}{2}$$

So the parabola is $y = -\tfrac{1}{2}(x - 2)^2 + 2$.

53. Since the vertex is $(3, -1)$, its equation will be $y = A(x - 3)^2 - 1$. Since $(1, 0)$ must lie on this parabola, substitute the points to find A:

$$y = A(x - 3)^2 - 1$$
$$0 = A(1 - 3)^2 - 1$$
$$0 = 4A - 1$$
$$1 = 4A$$
$$A = \tfrac{1}{4}$$

So the parabola is $y = \tfrac{1}{4}(x - 3)^2 - 1$.

55. Find the vertex of the parabola by completing the square:

$$y = 2x^2 + 12x + 14 = 2(x^2 + 6x) + 14 = 2(x^2 + 6x + 9) - 4 = 2(x + 3)^2 - 4$$

Since the vertex is $(-3, -4)$, the circle would have an equation of $(x + 3)^2 + (y + 4)^2 = r^2$. Since $(0, 0)$ lies on the circle, we have:

$$(0 + 3)^2 + (0 + 4)^2 = r^2$$
$$r^2 = 9 + 16$$
$$r^2 = 25$$
$$r = 5, \text{ since } r > 0$$

So the circle is $(x + 3)^2 + (y + 4)^2 = 25$.

57. (a) First complete the square to find the vertex:

$$y = p(x^2 + x) + r = p\left(x^2 + x + \tfrac{1}{4}\right) + r - \tfrac{p}{4} = p\left(x + \tfrac{1}{2}\right)^2 + \tfrac{4r - p}{4}$$

If the vertex lies on the x-axis, then $\tfrac{4r - p}{4} = 0$, so $4r - p = 0$, thus $p = 4r$.

(b) If $p = 4r$, then $y = 4rx^2 + 4rx + r$. Completing the square:

$$y = 4r\left(x^2 + x\right) + r = 4r\left(x^2 + x + \tfrac{1}{4}\right) + r - r = 4r\left(x + \tfrac{1}{2}\right)^2$$

But then the vertex is $\left(-\tfrac{1}{2}, 0\right)$, and thus it lies on the x-axis.

59.   First simplify an expression for $f(f(x))$:

$$f(f(x)) = f\left(ax^2 + bx\right)$$
$$= a\left(ax^2 + bx\right)^2 + b\left(ax^2 + bx\right)$$
$$= \left(ax^2 + bx\right)\left[a\left(ax^2 + bx\right) + b\right]$$
$$= x(ax + b)\left(a^2x^2 + abx + b\right)$$

Since $f(f(x)) = 0$, we have $x = 0$ and $x = -\dfrac{b}{a}$, which are two distinct real solutions

since $b \neq 0$. Thus $a^2x^2 + abx + b = 0$ can have only one real root. Using the discriminant:

$$(ab)^2 - 4\left(a^2\right)(b) = 0$$
$$a^2b^2 - 4a^2b = 0$$
$$a^2b(b - 4) = 0$$

Since $a \neq 0$ and $b \neq 0$, we must have $b = 4$. Choosing $b = 4$ results in three real roots for the equation $f(f(x)) = 0$. Notice that the value of a doesn't matter as long as $a \neq 0$.

Graphing Utility Exercises for Section 5.2

1.   (a)   Graphing all four functions:

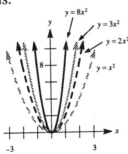

As the coefficient of x^2 increases from 1 to 8, the graph "narrows" and produces larger y-coordinates.

(b)   The graph should be "narrower" than the others:

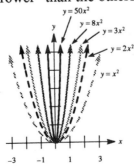

3.   (a)   Using the indicated settings, we graph both functions:

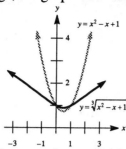

The graph indicates a minimum exists at $x = 0.5$.

(b)   When $x \approx 0.5$, the value of y is approximately 0.909. From Example 6, $\sqrt[3]{\frac{3}{4}} \approx 0.90856$, so our approximation is consistent with the text.

5.   (a)   The regression line is $y = 330.8t + 218.8$, where $t = 0$ corresponds to the year 1986.

(b)   Substituting $t = 4$ and $t = 7$ into the regression line:

$t = 4:$ $y = 330.8(4) + 218.8 = 1542$

$t = 7:$ $y = 330.8(7) + 218.8 = 2534.4$

The projections are 1542 thousand cases for 1990 and 2534 thousand cases for 1993. Both projections are too low.

(c)   The quadratic regression curve is $y = 62t^2 + 144.8t + 280.8$. Substituting $t = 4$ and $t = 7$ into the regression curve:

$t = 4:$ $y = 62(4)^2 + 144.8(4) + 280.8 = 1852$

$t = 7:$ $y = 62(7)^2 + 144.8(7) + 280.8 = 4332.4$

The projections are 1852 thousand cases for 1990 and 4332 thousand cases for 1993. The 1990 projection is closer to the actual value. Note that both projections are more accurate than those found in part (b), and that both projections are too low.

7.   (a)   The regression line is $y = 10.02x - 19598.75$, while the quadratic regression curve is $y = -2.54x^2 + 10114.14x - 10068145.15$.

(b)   Completing the table:

	1991	1992
Estimate for world energy production (in quadrillion Btu) using the linear model	351.07	361.09
Estimate for world energy production (in quadrillion Btu) using the quadratic model	341.85	339.17

Both projections are consistent with Figures A and B for 1991.

(c)   The quadratic model gives a closer estimate to 344.41 Btu for 1991.

(d)   The quadratic model gives a closer estimate to 346.33 Btu for 1992.

5.3 More On Iteration. Quadratics and Population Growth

1. Finding where $f(x) = x$:
$$-4x + 5 = x$$
$$5 = 5x$$
$$x = 1$$
The fixed point is $x = 1$.

3. Finding where $G(x) = x$:
$$\frac{1}{2} + x = x$$
$$\frac{1}{2} = 0$$
Since this last equation is false, there are no fixed points.

5. Finding where $h(x) = x$:
$$x^2 - 3x - 5 = x$$
$$x^2 - 4x - 5 = 0$$
$$(x - 5)(x + 1) = 0$$
$$x = -1, 5$$
The fixed points are $x = -1$ and $x = 5$.

7. Finding where $f(t) = t$:
$$t^2 - t + 1 = t$$
$$t^2 - 2t + 1 = 0$$
$$(t - 1)^2 = 0$$
$$t = 1$$
The fixed point is $t = 1$.

9. Finding where $k(t) = t$:
$$t^2 - 12 = t$$
$$t^2 - t - 12 = 0$$
$$(t - 4)(t + 3) = 0$$
$$t = -3, 4$$
The fixed points are $t = -3$ and $t = 4$.

11. Finding where $T(x) = x$:
$$1.8x(1 - x) = x$$
$$1.8x - 1.8x^2 = x$$
$$0.8x - 1.8x^2 = 0$$
$$0.2x(4 - 9x) = 0$$
$$x = 0, \tfrac{4}{9}$$
The fixed points are $x = 0$ and $x = \frac{4}{9}$.

13. Finding where $g(u) = u$:

$$2u^2 + 3u - 4 = u$$
$$2u^2 + 2u - 4 = 0$$
$$2(u^2 + u - 2) = 0$$
$$2(u + 2)(u - 1) = 0$$
$$u = -2, 1$$

The fixed points are $u = -2$ and $u = 1$.

15. Finding where $f(x) = x$:

$$7 + \sqrt{x - 1} = x$$
$$\sqrt{x - 1} = x - 7$$
$$x - 1 = x^2 - 14x + 49$$
$$0 = x^2 - 15x + 50$$
$$0 = (x - 5)(x - 10)$$
$$x = 5, 10$$

Note that $x = 5$ does not check in the original equation. The fixed point is $x = 10$.

17. Finding where $G(x) = x$:

$$\frac{x - 3}{x + 5} = x$$
$$x - 3 = x^2 + 5x$$
$$0 = x^2 + 4x + 3$$
$$0 = (x + 3)(x + 1)$$
$$x = -3, -1$$

The fixed points are $x = -3$ and $x = -1$.

19. (a) Finding where $g(x) = x$:

$$x^2 - 0.5 = x$$
$$2x^2 - 1 = 2x$$
$$2x^2 - 2x - 1 = 0$$

Using the quadratic formula:

$$x = \frac{-(-2) \pm \sqrt{(-2)^2 - 4(2)(-1)}}{2(2)} = \frac{2 \pm \sqrt{12}}{4} = \frac{2 \pm 2\sqrt{3}}{4} = \frac{1 \pm \sqrt{3}}{2}$$

Thus one of the fixed points is $x = \frac{1 - \sqrt{3}}{2} \approx -0.366$.

(b) Calculating the iterates of $x_0 = -1$:

$$x_1 = -0.49$$
$$x_2 = -0.2599$$
$$x_3 \approx -0.4325$$
$$x_4 \approx -0.3130$$

The fourth iterate is the first iterate to have the digit 3 in the first decimal place.

(c) Continuing the iterates:

$$x_5 \approx -0.4020$$
$$x_6 \approx -0.3384$$
$$x_7 \approx -0.3855$$
$$x_8 \approx -0.3514$$
$$x_9 \approx -0.3765$$
$$x_{10} \approx -0.3582$$
$$x_{11} \approx -0.3717$$
$$x_{12} \approx -0.3619$$

The twelfth iterate is the first iterate to have the digit 6 in the second decimal place.

21. (a) Completing the table:

	x_1	x_2	x_3	x_4	x_5	x_6	x_7	x_8
From graph	1.7	0.8	1.4	1.0	1.3	1.1	1.2	1.1
From calculator	1.72	0.796	1.443	0.990	1.307	1.085	1.240	1.132

(b) Finding where $f(x) = x$:

$$-0.7x + 2 = x$$
$$2 = 1.7x$$
$$x = \frac{2}{1.7} = \frac{20}{17}$$

The fixed point is $x = \frac{20}{17} \approx 1.176$.

(c) The eighth iterate has the same digit in the first decimal place.

23. (a) Completing the table:

	x_1	x_2	x_3	x_4	x_5	x_6	x_7	x_8	x_9
From graph	0.36	0.92	0.29	0.82	0.58	0.97	0.11	0.40	0.96
From calculator	0.36	0.922	0.289	0.822	0.585	0.971	0.113	0.402	0.962

(b) Evaluating the expressions:

$$x_1 = f\left(\frac{5+\sqrt{5}}{8}\right)$$

$$= 4\left(\frac{5+\sqrt{5}}{8}\right)\left(1-\frac{5+\sqrt{5}}{8}\right)$$

$$= \frac{5+\sqrt{5}}{2}\cdot\frac{3-\sqrt{5}}{8}$$

$$= \frac{15+3\sqrt{5}-5\sqrt{5}-5}{16}$$

$$= \frac{10-2\sqrt{5}}{16}$$

$$= \frac{5-\sqrt{5}}{8}$$

$$x_2 = f\left(\frac{5-\sqrt{5}}{8}\right)$$

$$= 4\left(\frac{5-\sqrt{5}}{8}\right)\left(1-\frac{5-\sqrt{5}}{8}\right)$$

$$= \frac{5-\sqrt{5}}{2}\cdot\frac{3+\sqrt{5}}{8}$$

$$= \frac{15-3\sqrt{5}+5\sqrt{5}-5}{16}$$

$$= \frac{10+2\sqrt{5}}{16}$$

$$= \frac{5+\sqrt{5}}{8}$$

Note that $x_2 = x_0$. Since $x_3 = f(x_2) = f(x_0) = x_1$ and $x_4 = f(x_3) = f(x_1) = x_2$, we have $x_3 = \dfrac{5-\sqrt{5}}{8}$ and $x_4 = \dfrac{5+\sqrt{5}}{8}$. The pattern is an alternating of iterate values.

(c) Calculating the first ten iterates of $x_0 = 0.905$:

$x_1 \approx 0.344$
$x_2 \approx 0.903$
$x_3 \approx 0.352$
$x_4 \approx 0.912$
$x_5 \approx 0.320$
$x_6 \approx 0.871$
$x_7 \approx 0.450$
$x_8 \approx 0.990$
$x_9 \approx 0.039$
$x_{10} \approx 0.151$

The behavior of the iterates is more like that in part (a).

25. (a) The graphical estimates and calculator estimates for $x_0 = 0.7$ are:

	x_1	x_2	x_3	x_4	x_5	x_6	x_7	x_8
From graph	0.6	0.8	0.4	0.8	0.4	0.8	0.4	0.8
From calculator	0.6	0.8	0.4	0.8	0.4	0.8	0.4	0.8

The pattern that emerges are alternating iterate values.

(b) The first eight iterates for $x_0 = 0.75$ are:

$x_1 = 0.5, x_2 = 1.0, x_3 = x_4 = ... = x_8 = 0.0$

The third iterate and all remaining iterates are 0.

(c) The first ten iterates for $x_0 = 2$ are:

$x_1 = -2, x_2 = -4, x_3 = -8, x_4 = -16, x_5 = -32$
$x_6 = -64, x_7 = -128, x_8 = -256, x_9 = -512, x_{10} = -1024$

The iterates are negative powers of 2.

27. (a) Since $x_{20} \approx 0.64594182$ and $f(x) = 2.9x(1-x)$, we can compute the required iterates:

$x_{21} \approx 0.6632$
$x_{22} \approx 0.6477$
$x_{23} \approx 0.6617$
$x_{24} \approx 0.6492$
$x_{25} \approx 0.6605$

(b) Multipliying each of these iterates by 500 results in the population of catfish in the pond:

n	20	21	22	23	24	25
Number of fish after n breeding seasons	323	332	324	331	325	330

Note that although the population is oscillating, the changes (sizes) of the oscillations are decreasing.

(c) Using $f(x) = 0.75x(1-x)$ and $x_0 = 0.1$, compute the iterates:

$x_1 = 0.0675$
$x_2 \approx 0.04721$
$x_3 \approx 0.03373$
$x_4 \approx 0.02445$
$x_5 \approx 0.01789$

The iterates appear to be approaching 0. Now calculate the fixed point of the function:

$$f(x) = x$$
$$0.75x(1-x) = x$$
$$0.75x - 0.75x^2 = x$$
$$-0.25x - 0.75x^2 = 0$$
$$-0.25x(1+3x) = 0$$
$$x = 0, -\tfrac{1}{3}$$

Since our iterates seem to be approaching 0, they are approaching a fixed point of the function. Since this fixed point is 0, eventually the population will decrease to 0.

29. (a) Using $f(x) = 3.1x(1-x)$ and $x_0 = 0.1$, complete the tables:

n	0	1	2	3	4	5
x_n	0.1	0.279	0.6236	0.7276	0.6143	0.7345
Number of fish after n breeding seasons	50	140	312	364	307	367

n	6	7	8	9	10
x_n	0.6046	0.7411	0.5948	0.7471	0.5857
Number of fish after n breeding seasons	302	371	297	374	293

Using $x_{20} \approx 0.56140323$, complete the table:

n	21	22	23	24	25	26
x_n	0.7633	0.5601	0.7638	0.5592	0.7641	0.5587
Number of fish after n breeding seasons	382	280	382	280	382	279

(b) Let $f(x) = x$:
$$3.1x(1-x) = x$$
$$3.1x - 3.1x^2 = x$$
$$2.1x - 3.1x^2 = 0$$
$$0.1x(21 - 31x) = 0$$
$$x = 0, \tfrac{21}{31}$$

The nonzero fixed point is $x = \tfrac{21}{31} \approx 0.6774$. The iterates are not approaching this fixed point.

(c) Evaluating these expressions when $k = 3.1$:
$$a = \frac{1 + 3.1 + \sqrt{(3.1-3)(3.1+1)}}{2(3.1)} = \frac{4.1 + \sqrt{(0.1)(4.1)}}{6.2} \approx 0.7646$$
$$b = \frac{1 + 3.1 - \sqrt{(3.1-3)(3.1+1)}}{2(3.1)} = \frac{4.1 - \sqrt{(0.1)(4.1)}}{6.2} \approx 0.5580$$

These iterates are consistent with the table from part (a).

(d) These correspond to the populations of 382 and 279 fish.

31. (a) The first six iterates of $x_0 = c$ are:
$$x_1 = g(c) = d$$
$$x_2 = g(x_1) = g(d) = c$$
$$x_3 = g(x_2) = g(c) = d$$
$$x_4 = g(x_3) = g(d) = c$$
$$x_5 = g(x_4) = g(c) = d$$
$$x_6 = g(x_5) = g(d) = c$$

Similarly, the first six iterates of $x_0 = d$ are:
$$x_1 = g(d) = c$$
$$x_2 = g(x_1) = g(c) = d$$
$$x_3 = g(x_2) = g(d) = c$$
$$x_4 = g(x_3) = g(c) = d$$
$$x_5 = g(x_4) = g(d) = c$$
$$x_6 = g(x_5) = g(c) = d$$

(b) The first six iterates of 0.4 are 0.8, 0.4, 0.8, 0.4, 0.8, and 0.4.

(c) The work in part (b) shows that {0.4, 0.8} is a 2-cycle for the function T.

(d)   Computing the first two iterates:
$$x_1 = T(0.4) = 1 - |0.8 - 1| = 1 - 0.2 = 0.8$$
$$x_2 = T(0.8) = 1 - |1.6 - 1| = 1 - 0.6 = 0.4$$
These answers are consistent with the work in part (b).

33.   (a)   Subtracting the two equations:
$$4a(1-a) - 4b(1-b) = b - a$$
$$4a - 4a^2 - 4b + 4b^2 = b - a$$
$$4b^2 - 4a^2 + 4a - 4b = b - a$$
$$4(b+a)(b-a) - 4(b-a) = b - a$$
$$4(b-a)(b+a-1) = b - a$$

(b)   Since $a \neq b$, we can divide by $b - a$ to obtain:
$$4(b+a-1) = 1$$
$$b + a - 1 = \tfrac{1}{4}$$
$$b + a = \tfrac{5}{4}$$
$$b = \tfrac{5}{4} - a$$

(c)   Substituting $b = \tfrac{5}{4} - a$:
$$4a(1-a) = \tfrac{5}{4} - a$$
$$16a - 16a^2 = 5 - 4a$$
$$0 = 16a^2 - 20a + 5$$

(d)   Using the quadratic formula:
$$a = \frac{-(-20) \pm \sqrt{(-20)^2 - 4(16)(5)}}{2(16)} = \frac{20 \pm \sqrt{80}}{32} = \frac{20 \pm 4\sqrt{5}}{32} = \frac{5 \pm \sqrt{5}}{8}$$

(e)   Choosing the positive root, we have $a = \tfrac{1}{8}\left(5 + \sqrt{5}\right)$. Substituting to find b:
$$b = \tfrac{5}{4} - \tfrac{1}{8}\left(5 + \sqrt{5}\right) = \tfrac{5}{4} - \tfrac{5}{8} - \tfrac{1}{8}\sqrt{5} = \tfrac{5}{8} - \tfrac{1}{8}\sqrt{5} = \tfrac{1}{8}\left(5 - \sqrt{5}\right)$$
Now checking these values:
$$f\left[\tfrac{1}{8}\left(5+\sqrt{5}\right)\right] = 4 \cdot \tfrac{1}{8}\left(5+\sqrt{5}\right)\left[1 - \tfrac{1}{8}\left(5+\sqrt{5}\right)\right]$$
$$= \tfrac{1}{2}\left(5+\sqrt{5}\right)\left(1 - \tfrac{5}{8} - \tfrac{1}{8}\sqrt{5}\right)$$
$$= \tfrac{1}{2}\left(5+\sqrt{5}\right)\left(\tfrac{3}{8} - \tfrac{1}{8}\sqrt{5}\right)$$
$$= \tfrac{1}{16}\left(5+\sqrt{5}\right)\left(3 - \sqrt{5}\right)$$
$$= \tfrac{1}{16}\left(10 - 2\sqrt{5}\right)$$
$$= \tfrac{1}{8}\left(5 - \sqrt{5}\right)$$

$$f\left[\tfrac{1}{8}\left(5-\sqrt{5}\right)\right]=4\cdot\tfrac{1}{8}\left(5-\sqrt{5}\right)\left[1-\tfrac{1}{8}\left(5-\sqrt{5}\right)\right]$$

$$=\tfrac{1}{2}\left(5-\sqrt{5}\right)\left(1-\tfrac{5}{8}+\tfrac{1}{8}\sqrt{5}\right)$$

$$=\tfrac{1}{2}\left(5-\sqrt{5}\right)\left(\tfrac{3}{8}+\tfrac{1}{8}\sqrt{5}\right)$$

$$=\tfrac{1}{16}\left(5-\sqrt{5}\right)\left(3+\sqrt{5}\right)$$

$$=\tfrac{1}{16}\left(10+2\sqrt{5}\right)$$

$$=\tfrac{1}{8}\left(5+\sqrt{5}\right)$$

Therefore $f(a)=b$ and $f(b)=a$.

35. (a) Substituting $k=3.5$ yields:

$$a=\frac{1+3.5+\sqrt{(3.5+1)(3.5-3)}}{2(3.5)}=\frac{4.5+\sqrt{(4.5)(0.5)}}{7}=\frac{\tfrac{9}{2}+\sqrt{\tfrac{9}{4}}}{7}=\frac{\tfrac{9}{2}+\tfrac{3}{2}}{7}=\frac{6}{7}$$

$$b=\frac{\tfrac{9}{2}-\tfrac{3}{2}}{7}=\frac{3}{7}$$

(b) The coordinates are $P\left(\tfrac{3}{7},\tfrac{3}{7}\right)$, $Q\left(\tfrac{3}{7},\tfrac{6}{7}\right)$, $R\left(\tfrac{6}{7},\tfrac{6}{7}\right)$, and $S\left(\tfrac{6}{7},\tfrac{3}{7}\right)$.

37. (a) Let $f(a)=b$ and $f(b)=a$. Therefore:

$$a^2-3=b$$
$$b^2-3=a$$

Subtracting yields:

$$a^2-b^2=b-a$$
$$(a+b)(a-b)=-(a-b)$$

Since $a\neq b$, dividing by $a-b$ yields:

$$a+b=-1$$
$$b=-1-a$$

Substituting:

$$a^2-3=b$$
$$a^2-3=-1-a$$
$$a^2+a-2=0$$
$$(a+2)(a-1)=0$$
$$a=-2,1$$
$$b=1,-2$$

Choosing the larger value for a, the values are $a=1$ and $b=-2$.

(b) Let $f(a) = b$ and $f(b) = a$. Therefore:

$$a^2 - c = b$$
$$b^2 - c = a$$

Subtracting yields:

$$a^2 - b^2 = b - a$$
$$(a+b)(a-b) = -(a-b)$$

Since $a \neq b$, dividing by $a - b$ yields:

$$a + b = -1$$
$$b = -1 - a$$

Substituting:

$$a^2 - c = b$$
$$a^2 - c = -1 - a$$
$$a^2 + a + (1-c) = 0$$

Using the quadratic formula:

$$a = \frac{-1 \pm \sqrt{(1)^2 - 4(1)(1-c)}}{2(1)} = \frac{-1 \pm \sqrt{1 - 4 + 4c}}{2} = \frac{-1 \pm \sqrt{4c - 3}}{2}$$

Choosing the positive root, we have $a = \dfrac{-1 + \sqrt{4c-3}}{2}$. Substituting to find b:

$$b = -\frac{-1 + \sqrt{4c-3}}{2} - 1 = \frac{1 - \sqrt{4c-3}}{2} - \frac{2}{2} = \frac{-1 - \sqrt{4c-3}}{2}$$

To check, let $c = 3$. We have:

$$a = \frac{-1 + \sqrt{12-3}}{2} = \frac{-1 + 3}{2} = 1$$
$$b = \frac{-1 - \sqrt{12-3}}{2} = \frac{-1 - 3}{2} = -2$$

These values check our result from part (a).

Graphing Utility Exercises for Section 5.3

1. (a) There is one fixed point for the function:

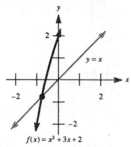

$f(x) = x^3 + 3x + 2$

(b) The fixed point is approximately -0.771.

3. (a) There are three fixed points for the function:

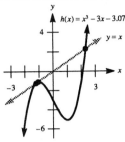

(b) The fixed points are approximately -1.206, -1.103, and 2.309.

5. (a) There are two fixed points for the function:

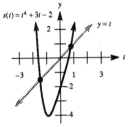

(b) The fixed points are approximately -1.495 and 0.798.

7. (a) There are two fixed points for the function:

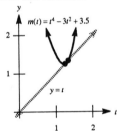

(b) The fixed points are approximately 1.256 and 1.344.

9. (a) There are four fixed points for the function:

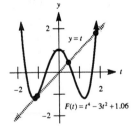

(b) The fixed points are approximately -1.177, -1.083, 0.463, and 1.797.

11. (a) There are two fixed points for the function:

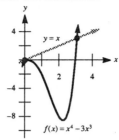

(b) The fixed points are 0 (exact) and 3.104 (approximate).

13. (a) Graphing the function and line:

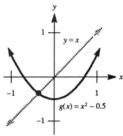

(b) Evaluating, we have $\dfrac{1 - \sqrt{3}}{2} \approx -0.36602540$.

(c) (i) The twelfth iterate is –0.36186, which agrees with the fixed point to two decimal places. Yes, the next iterate (–0.36906) also agrees with the fixed point to two decimal places.

(ii) The 27th iterate is –0.36606, which agrees with the fixed point to four decimal places. Yes, the next iterate (–0.36600) also agrees with the fixed point to four decimal places.

5.4 Applied Functions: Setting Up Equations

1. (a) Call P the perimeter and A the area. We are asked to come up with a formula for A in terms of x. Since $P = 16$ and $P = 2x + 2l$, we have:
$$2x + 2l = 16$$
$$2l = 16 - 2x$$
$$l = 8 - x$$
Now $A = xl = x(8 - x) = 8x - x^2$. Using functional notation we have $A(x) = 8x - x^2$.

(b) Since $A = 85$ and $A = xl$, we have:

$$xl = 85$$

$$l = \frac{85}{x}$$

Now $P = 2x + 2l = 2x + 2\left(\dfrac{85}{x}\right) = 2x + \dfrac{170}{x}$. Using functional notation we have

$$P(x) = 2x + \frac{170}{x}.$$

3.   (a) First draw the figure:

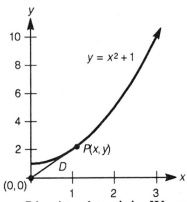

Call D the distance from $P(x,y)$ to the origin. We are asked to come up with a formula for D in terms of x. By the distance formula:

$$D = \sqrt{(x-0)^2 + (y-0)^2} = \sqrt{x^2 + y^2}$$

Since $P(x, y)$ lies on the curve, $y = x^2 + 1$. Substituting this for y in our equation for D:

$$D = \sqrt{x^2 + y^2} = \sqrt{x^2 + \left(x^2 + 1\right)^2} = \sqrt{x^2 + x^4 + 2x^2 + 1} = \sqrt{x^4 + 3x^2 + 1}$$

Using functional notation we have $D(x) = \sqrt{x^4 + 3x^2 + 1}$.

(b) Let m denote the slope of the line segment from the origin to $P(x, y)$. Then:

$$m = \frac{y - 0}{x - 0} = \frac{y}{x}$$

Substituting $y = x^2 + 1$ in for y in this equation:

$$m = \frac{y}{x} = \frac{x^2 + 1}{x}$$

Using functional notation we have $m(x) = \dfrac{x^2 + 1}{x}$.

5. (a) First draw the figure:

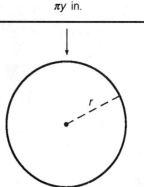

πy in.

Call A the area of the circle and r its radius. We are asked to come up with a formula for A in terms of y. Since πy is the circumference C of the circle, and $C = 2\pi r$, we have:

$$2\pi r = \pi y$$
$$r = \frac{y}{2}$$

Now $A = \pi r^2 = \pi \left(\frac{y}{2}\right)^2 = \frac{\pi y^2}{4}$. Using functional notation we have $A(y) = \frac{\pi y^2}{4}$.

(b) Draw the figure:

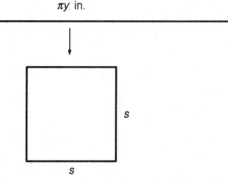

πy in.

s

s

Let A denote the area of the square, and s the length of the side. Since the perimeter P is πy and $P = 4s$, we have:

$$4s = \pi y$$
$$s = \frac{\pi y}{4}$$

Now $A = s^2 = \left(\frac{\pi y}{4}\right)^2 = \frac{\pi^2 y^2}{16}$. Using functional notation we have $A(y) = \frac{\pi^2 y^2}{16}$.

7. (a) Let the two numbers be x and $16 - x$. Then the product P would be:
$$P = x(16 - x) = 16x - x^2$$

Using functional notation we have $P(x) = 16x - x^2$.

(b) Since the two numbers are x and $16 - x$, the sum of squares S would be:
$$S = (x)^2 + (16 - x)^2 = x^2 + 256 - 32x + x^2 = 2x^2 - 32x + 256$$
Using functional notation we have $S(x) = 2x^2 - 32x + 256$.

(c) There are two ways to set this up. Since the two numbers are x and $16 - x$, the difference of the cubes D could be:
$$D = x^3 - (16 - x)^3 \quad \text{or} \quad D = (16 - x)^3 - x^3$$
Using functional notation we have:
$$D(x) = x^3 - (16 - x)^3 \quad \text{or} \quad D(x) = (16 - x)^3 - x^3$$

(d) Let A denote the average of the two numbers. Since the two numbers are x and $16 - x$, we have:
$$A = \frac{x + 16 - x}{2} = \frac{16}{2} = 8$$
So $A(x) = 8$. Notice that the average does not depend on what the two numbers are!

9.   Let R be the revenue, x be the number of units sold, and p be the demand (price). Then:
$$R = xp = x\left(-\tfrac{1}{4}x + 8\right) = -\tfrac{1}{4}x^2 + 8x$$
Using functional notation we have $R(x) = -\tfrac{1}{4}x^2 + 8x$.

11. (a) Complete the table:

x	1	2	3	4	5	6	7
$P(x)$	17.88	19.49	20.83	21.86	22.49	22.58	21.75

(b) The largest value for $P(x)$ is 22.58, corresponding to $x = 6$.

(c) Compute $P(4\sqrt{2})$:
$$\begin{aligned}
P(4\sqrt{2}) &= 2(4\sqrt{2}) + 2\sqrt{64 - (4\sqrt{2})^2} \\
&= 8\sqrt{2} + 2\sqrt{64 - 32} \\
&= 8\sqrt{2} + 2\sqrt{32} \\
&= 8\sqrt{2} + 8\sqrt{2} \\
&= 16\sqrt{2} \\
&\approx 22.63
\end{aligned}$$
This is indeed larger than any of our table values.

13. (a) First draw the figure:

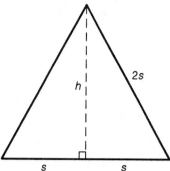

Let h denote the height and $2s$ denote the sides. Note that the height (called the altitude) bisects the base into the lengths of s and s. We are asked to find h in terms of s, so we use the Pythagorean theorem on the right triangle:

$$h^2 + s^2 = (2s)^2$$
$$h^2 + s^2 = 4s^2$$
$$h^2 = 3s^2$$
$$h = \sqrt{3s^2} = \sqrt{3}s$$

Using functional notation we have $h(s) = \sqrt{3}s$.

(b) Let A denote the area of the triangle. Then:

$$A = \tfrac{1}{2}(\text{base})(\text{height}) = \tfrac{1}{2}(2s)(\sqrt{3}s) = \sqrt{3}s^2$$

Using functional notation we have $A(s) = \sqrt{3}s^2$.

(c) If each side is 8 cm, then:

$$2s = 8$$
$$s = 4$$

Using the function from (a), we have:

$$h(4) = \sqrt{3} \cdot 4 = 4\sqrt{3} \text{ cm}$$

(d) If each side is 5 in., then:

$$2s = 5$$
$$s = \tfrac{5}{2}$$

Using the function from (b), we have:

$$A\left(\tfrac{5}{2}\right) = \sqrt{3}\left(\tfrac{5}{2}\right)^2 = \frac{25\sqrt{3}}{4} \text{ in.}^2$$

15. Let h be the height, r be the radius, and V be the volume. We know that:

$$V = \pi r^2 h$$

We are also given that $h = 2r$, so we substitute into the formula for V:

$$V = \pi r^2 (2r) = 2\pi r^3$$

Using functional notation we have $V(r) = 2\pi r^3$.

17. (a) Let h be the height, r be the radius and V be the volume. We know that $V = 12\pi$ and $V = \pi r^2 h$, so:

$$\pi r^2 h = 12\pi$$

$$h = \frac{12\pi}{\pi r^2} = \frac{12}{r^2}$$

Using functional notation we have $h(r) = \dfrac{12}{r^2}$.

(b) Let S be the total surface area. Then:

$$S = 2\pi r^2 + 2\pi rh = 2\pi r^2 + 2\pi r\left(\frac{12}{r^2}\right) = 2\pi r^2 + \frac{24\pi}{r}$$

Using functional notation we have $S(r) = 2\pi r^2 + \dfrac{24\pi}{r}$.

19. Solve $S = 4\pi r^2$ for r:

$$4\pi r^2 = S$$

$$r^2 = \frac{S}{4\pi}$$

$$r = \sqrt{\frac{S}{4\pi}}$$

Now since $V = \frac{4}{3}\pi r^3$, we have:

$$V = \frac{4}{3}\pi\left(\sqrt{\frac{S}{4\pi}}\right)^3 = \frac{4\pi S\sqrt{S}}{3(4\pi)\sqrt{4\pi}} = \frac{S\sqrt{S}}{3\sqrt{4\pi}} = \frac{S\sqrt{S\pi}}{6\pi}$$

Using functional notation we have $V(S) = \dfrac{S\sqrt{S\pi}}{6\pi}$.

21. Draw a figure:

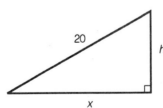

Let A be the area of the triangle and let x and h be its two legs. By the Pythagorean theorem:

$$x^2 + h^2 = 20^2$$

$$h^2 = 400 - x^2$$

$$h = \sqrt{400 - x^2}$$

Therefore:

$$A = \tfrac{1}{2}(\text{base})(\text{height}) = \tfrac{1}{2}(x)(h) = \tfrac{1}{2}x\sqrt{400 - x^2}$$

Using functional notation we have $A(x) = \tfrac{1}{2}x\sqrt{400 - x^2}$.

23. Draw the figure:

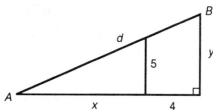

Using similar triangles, we have:

$$\frac{x}{5} = \frac{x+4}{y}$$
$$xy = 5(x+4)$$
$$y = \frac{5(x+4)}{x}$$

Therefore:

$$d^2 = (x+4)^2 + y^2$$

$$= (x+4)^2 + \left(\frac{5(x+4)}{x}\right)^2$$

$$= (x+4)^2 + \frac{25(x+4)^2}{x^2}$$

$$= \frac{x^2(x+4)^2 + 25(x+4)^2}{x^2}$$

$$= \frac{(x+4)^2(x^2+25)}{x^2}$$

So $d = \frac{(x+4)\sqrt{x^2+25}}{x}$. Using functional notation we have $d(x) = \frac{(x+4)\sqrt{x^2+25}}{x}$.

25. Substitute into $A(x) = 50x - x^2$:

x	5	10	20	24	24.8	24.9	25	25.1	25.2	45
$A(x)$	225	400	600	624	624.96	624.99	625	624.99	624.96	225

So $x = 25$ yields the largest area. Since $L = 50 - x = 50 - 25 = 25$, $L = 25$ is the corresponding value.

27. (a) Substitute into $A(x) = 8x - \frac{1}{2}x^3$:

Table 1:

x	1	2	3	4
A	7.5	12	10.5	0

Here $x = 2$ yields the largest area.

Table 2:

x	1.75	2.00	2.25	2.50	2.75
A	11.3203	12.0000	12.3047	12.1875	11.6016

Here $x = 2.25$ yields the largest area.

Table 3:

x	2.15	2.20	2.25	2.30	2.35
A	12.2308	12.2760	12.3047	12.3165	12.3111

Here $x = 2.30$ yields the largest area.

(b) Since $x = \frac{4\sqrt{3}}{3} = 2.309$ to four significant places, $x = 2.30$ is the closest x-value. This yields an area of 12.3168. Notice that $x = 2.30$ agrees with this to five significant digits.

29. For the square, the side is $\frac{x}{4}$. For the rectangle which has a perimeter of $3 - x$, we have:

$$2w + 2l = 3 - x$$
$$2 \cdot \tfrac{1}{2}l + 2l = 3 - x$$
$$3l = 3 - x$$
$$l = 1 - \tfrac{1}{3}x$$
$$w = \tfrac{1}{2} - \tfrac{1}{6}x$$

So the combined area of the square and rectangle is:

$$A = \left(\tfrac{x}{4}\right)^2 + \left(1 - \tfrac{1}{3}x\right)\left(\tfrac{1}{2} - \tfrac{1}{6}x\right) = \tfrac{1}{16}x^2 + \tfrac{1}{2} - \tfrac{1}{3}x + \tfrac{1}{18}x^2 = \tfrac{17}{144}x^2 - \tfrac{1}{3}x + \tfrac{1}{2}$$

Using functional notation we have $A(x) = \tfrac{17}{144}x^2 - \tfrac{1}{3}x + \tfrac{1}{2}$.

31. (a) Since $V = \tfrac{1}{3}\pi r^2 h$ and $h = \sqrt{3}r$, we have:

$$V = \tfrac{1}{3}\pi r^2 \left(\sqrt{3}r\right) = \tfrac{\sqrt{3}}{3}\pi r^3$$

Using functional notation we have $V(r) = \tfrac{\sqrt{3}}{3}\pi r^3$.

(b) Since $S = \pi r\sqrt{r^2 + h^2}$ and $h = \sqrt{3}r$, we have:

$$S = \pi r\sqrt{r^2 + h^2} = \pi r\sqrt{r^2 + \left(\sqrt{3}r\right)^2} = \pi r\sqrt{r^2 + 3r^2} = \pi r\sqrt{4r^2} = \pi r(2r) = 2\pi r^2$$

Using functional notation we have $S(r) = 2\pi r^2$.

33.   (a)   Since $V = \frac{1}{3}\pi r^2 h$ and $S = \pi r\sqrt{r^2 + h^2}$, and $V = S$, we have:

$$\frac{1}{3}\pi r^2 h = \pi r\sqrt{r^2 + h^2}$$
$$rh = 3\sqrt{r^2 + h^2}$$
$$r^2 h^2 = 9\left(r^2 + h^2\right)$$
$$r^2 h^2 = 9r^2 + 9h^2$$
$$r^2 h^2 - 9r^2 = 9h^2$$
$$r^2\left(h^2 - 9\right) = 9h^2$$
$$r^2 = \frac{9h^2}{h^2 - 9}$$
$$r = \sqrt{\frac{9h^2}{h^2 - 9}} = \frac{3h}{\sqrt{h^2 - 9}}$$

Using functional notation we have $r(h) = \dfrac{3h}{\sqrt{h^2 - 9}}$.

(b)   After squaring in (a), we had:

$$r^2 h^2 = 9r^2 + 9h^2$$
$$r^2 h^2 - 9h^2 = 9r^2$$
$$h^2\left(r^2 - 9\right) = 9r^2$$
$$h^2 = \frac{9r^2}{r^2 - 9}$$

Taking roots:

$$h = \sqrt{\frac{9r^2}{r^2 - 9}} = \frac{3r}{\sqrt{r^2 - 9}}$$

Using functional notation we have $h(r) = \dfrac{3r}{\sqrt{r^2 - 9}}$.

35.   Let x be the length of wire used for the circle, so $14 - x$ is the length of wire used for the square. For the circle, we have $2\pi r = x$, so $r = \frac{x}{2\pi}$ and thus the area is given by:

$$\pi r^2 = \pi\left(\frac{x}{2\pi}\right)^2 = \frac{x^2}{4\pi}$$

For the square, we have $4s = 14 - x$, so $s = \frac{14-x}{4}$ and thus the area is given by:

$$\left(\frac{14 - x}{4}\right)^2 = \frac{(14 - x)^2}{16}$$

So the total combined area is:

$$A = \frac{x^2}{4\pi} + \frac{(14 - x)^2}{16} = \frac{4x^2 + \pi(14 - x)^2}{16\pi}$$

Using functional notation we have $A(x) = \dfrac{4x^2 + \pi(14 - x)^2}{16\pi}$.

37. The perimeter of each semi-circle is $\frac{1}{2}(2\pi r) = \pi r$, so the total perimeter P is given by:

$P = \pi r + \pi r + l + l = 2\pi r + 2l$, where l is the length of the rectangle

Since $P = \frac{1}{4}$, we have:

$$2\pi r + 2l = \frac{1}{4}$$
$$2l = \frac{1}{4} - 2\pi r$$
$$2l = \frac{1 - 8\pi r}{4}$$
$$l = \frac{1 - 8\pi r}{8}$$

Now find the area A. The area of each semicircle is $\frac{1}{2}\pi r^2$, and the area of the rectangle is length $\bullet$ width, where $w = 2r$, so:

$$A = \frac{1}{2}\pi r^2 + \frac{1}{2}\pi r^2 + lw$$
$$= \pi r^2 + \left(\frac{1 - 8\pi r}{8}\right)(2r)$$
$$= \pi r^2 + \frac{r - 8\pi r^2}{4}$$
$$= \frac{4\pi r^2 + r - 8\pi r^2}{4}$$
$$= \frac{r - 4\pi r^2}{4}$$

Using functional notation we have $A(r) = \dfrac{r(1 - 4\pi r)}{4}$.

39. Draw the figure:

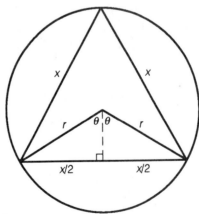

Using geometry, we see that:
$$2\theta = \frac{1}{3}(360°)$$
$$2\theta = 120°$$
$$\theta = 60°$$

Since $\theta = 60°$, we have:

$$\frac{x}{2} = \frac{\sqrt{3}}{2}r$$

$$x = \sqrt{3}r$$

$$r = \frac{x}{\sqrt{3}}$$

So the area of the circle A is:

$$A = \pi r^2 = \pi\left(\frac{x}{\sqrt{3}}\right)^2 = \frac{\pi x^2}{3}$$

Using functional notation we have $A(x) = \frac{\pi}{3}x^2$.

41. (a)   Since $V = l \cdot w \cdot h$ and $l = 8 - 2w$, $w = 6 - 2x$, $h = x$, we have:

$$V = (8 - 2x)(6 - 2x)(x) = \left(48 - 28x + 4x^2\right)(x) = 4x^3 - 28x^2 + 48x$$

Using functional notation we have $V(x) = 4x^3 - 28x^2 + 48x$.

(b)   Completing the table:

x (in.)	0	0.5	1.0	1.5	2.0	2.5	3.0
volume (in.^3)	0	17.5	24	22.5	16	7.5	0

(c)   The value $x = 1.0$ appears to yield the largest volume.

(d)   Completing the table:

x (in.)	0.8	0.9	1.0	1.1	1.2	1.3	1.4
volume (in.^3)	22.5	23.4	24	24.2	24.2	23.9	23.3

(e)   The value $x = 1.1$ appears to yield the largest volume.

43. (a)   The area of the window would be:

$$A = \tfrac{1}{2}\left(\pi r^2\right) + lw$$

It remains to find l and w in terms of r. We see that $w = 2r$, and the perimeter $P = 32$, so:

$$P = \tfrac{1}{2}(2\pi r) + 2l + w$$

Therefore:

$$\tfrac{1}{2}(2\pi r) + 2l + w = 32$$

$$\pi r + 2l + 2r = 32$$

$$2l = 32 - \pi r - 2r$$

$$l = \frac{32 - \pi r - 2r}{2}$$

Now find the area:
$$A = \tfrac{1}{2}\left(\pi r^2\right) + lw$$
$$= \tfrac{1}{2}\left(\pi r^2\right) + \left(\frac{32 - \pi r - 2r}{2}\right)(2r)$$
$$= \frac{\pi r^2}{2} + 32r - \pi r^2 - 2r^2$$
$$= 32r - 2r^2 - \frac{\pi r^2}{2}$$

Using functional notation we write $A(r) = 32r - 2r^2 - \dfrac{\pi r^2}{2}$.

(b)   We have $A(r) = -\left(\frac{4+\pi}{2}\right)r^2 + 32r$, which will open downward. Since $A(0) = 0$, it does pass through the origin. Complete the square:
$$A(r) = -\left(\frac{4+\pi}{2}\right)\left(r^2 - \frac{64}{4+\pi}r\right)$$
$$= -\left(\frac{4+\pi}{2}\right)\left[r^2 - \frac{64}{4+\pi}r + \left(\frac{32}{4+\pi}\right)^2\right] + \left(\frac{4+\pi}{2}\right)\left(\frac{32}{4+\pi}\right)^2$$
$$= -\left(\frac{4+\pi}{2}\right)\left(r - \frac{32}{4+\pi}\right)^2 + \frac{512}{4+\pi}$$

So the vertex is $\left(\frac{32}{4+\pi}, \frac{512}{4+\pi}\right)$.

45.   (a)   Use the Pythagorean theorem:
$$3^2 + y^2 = z^2$$
Taking roots:
$$z = \sqrt{y^2 + 9}$$

Now $s = \dfrac{y}{z} = \dfrac{y}{\sqrt{y^2+9}}$, so $s\sqrt{y^2+9} = y$. Squaring yields:
$$s^2\left(y^2 + 9\right) = y^2$$
$$s^2 y^2 + 9s^2 = y^2$$
$$y^2 - s^2 y^2 = 9s^2$$
$$y^2\left(1 - s^2\right) = 9s^2$$
$$y^2 = \frac{9s^2}{1 - s^2}$$
$$y = \frac{3s}{\sqrt{1 - s^2}}$$

Using functional notation we have $y(s) = \dfrac{3s}{\sqrt{1 - s^2}}$.

(b) From part (a) we had $s(y) = \dfrac{y}{\sqrt{y^2 + 9}}$.

(c) Since $s = \frac{y}{z}$, then $z = \frac{y}{s}$. Using our result from (a), we have:

$$z = \frac{y}{s} = \frac{\frac{3s}{\sqrt{1-s^2}}}{s} = \frac{3}{\sqrt{1-s^2}}$$

Using functional notation we have $z(s) = \dfrac{3}{\sqrt{1-s^2}}$.

(d) Using our answer from (c), we have:

$$z = \frac{3}{\sqrt{1-s^2}}$$
$$z\sqrt{1-s^2} = 3$$

Squaring each side, we obtain:

$$z^2(1-s^2) = 9$$
$$z^2 - z^2 s^2 = 9$$
$$-z^2 s^2 = 9 - z^2$$
$$s^2 = \frac{z^2 - 9}{z^2}$$
$$s = \frac{\sqrt{z^2 - 9}}{z}$$

Using functional notation we have $s(z) = \dfrac{\sqrt{z^2 - 9}}{z}$.

47. (a) Using the points $(0, -1)$ and (a, a^2), we have:

$$m(a) = \frac{a^2 - (-1)}{a - 0} = \frac{a^2 + 1}{a}$$

(b) If x_0 is the x-intercept, then the area of the triangle A is:

$$A = \tfrac{1}{2}(\text{base})(\text{height}) = \tfrac{1}{2}(a - x_0)(a^2)$$

To find the x-intercept, we must find the equation of the line. Use $m = \dfrac{a^2 + 1}{a}$ (from (a) above) and $(0, -1)$ in the slope-intercept formula to obtain:

$$y = \frac{a^2 + 1}{a}x - 1$$

We find x_0 by letting $y = 0$:

$$0 = \frac{a^2 + 1}{a} x_0 - 1$$

$$\frac{a^2 + 1}{a} x_0 = 1$$

$$x_0 = \frac{a}{a^2 + 1}$$

Therefore:

$$A = \tfrac{1}{2}\left(a - x_0\right)a^2$$

$$= \tfrac{1}{2}\left(a - \frac{a}{a^2 + 1}\right)a^2$$

$$= \frac{a^2}{2}\left[\frac{a\left(a^2 + 1\right) - a}{a^2 + 1}\right]$$

$$= \frac{a^2\left(a^3 + a - a\right)}{2\left(a^2 + 1\right)}$$

$$= \frac{a^2\left(a^3\right)}{2\left(a^2 + 1\right)}$$

$$= \frac{a^5}{2\left(a^2 + 1\right)}$$

49. Re-draw the figure (differently):

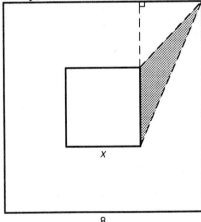

Extend the triangle to form a right triangle as pictured. Now find the areas of the large and small right triangles:

$$A_{\text{large}} = \tfrac{1}{2}(\text{base})(\text{height})$$

$$= \tfrac{1}{2}\left(x + \frac{8-x}{2}\right)\left(\frac{8-x}{2}\right)$$

$$= \tfrac{1}{2}\left(\frac{2x+8-x}{2}\right)\left(\frac{8-x}{2}\right)$$

$$= \frac{(x+8)(8-x)}{8}$$

$$= \frac{64-x^2}{8}$$

$$A_{\text{small}} = \tfrac{1}{2}(\text{base})(\text{height}) = \tfrac{1}{2}\left(\frac{8-x}{2}\right)\left(\frac{8-x}{2}\right) = \frac{64-16x+x^2}{8}$$

Therefore:

$$A = A_{\text{large}} - A_{\text{small}}$$

$$= \frac{64-x^2}{8} - \frac{64-16x+x^2}{8}$$

$$= \frac{64-x^2-64+16x-x^2}{8}$$

$$= \frac{16x-2x^2}{8}$$

$$= \frac{8x-x^2}{4}$$

Using functional notation we write $A(x) = \dfrac{8x-x^2}{4}$.

Note: A much easier approach is to realize that the altitude of the triangle need not lie on the triangle. That is:

$$\text{Area} = \tfrac{1}{2}(\text{base})(\text{altitude}) = \tfrac{1}{2}(x)\left(\frac{8-x}{2}\right) = \frac{8x-x^2}{4}$$

Both approaches are correct.

51. Draw the diagram:

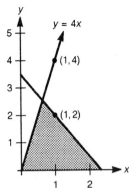

We are asked to find the area A of the shaded triangle. Since the line has slope m and passes through $(1, 2)$, by the point-slope formula we have:

$$y - 2 = m(x - 1)$$
$$y - 2 = mx - m$$
$$y = mx + (2 - m)$$

Its x-intercept is where $y = 0$:

$$0 = mx + (2 - m)$$
$$mx = m - 2$$
$$x = \frac{m - 2}{m}$$

This is the base of a triangle. To find its height, we must find the value of y where this line and $y = 4x$ intersect. Set the two y-values equal:

$$mx + 2 - m = 4x$$
$$mx - 4x = m - 2$$
$$x(m - 4) = m - 2$$
$$x = \frac{m - 2}{m - 4}$$

Since this point lies on $y = 4x$, its y-coordinate is:

$$y = 4x = 4\left(\frac{m - 2}{m - 4}\right)$$

Finally, find the area:

$$A = \tfrac{1}{2}(\text{base})(\text{height})$$
$$= \tfrac{1}{2}\left(\frac{m - 2}{m}\right)(4)\left(\frac{m - 2}{m - 4}\right)$$
$$= 2\frac{(m - 2)^2}{m(m - 4)}$$
$$= \frac{2\left(m^2 - 4m + 4\right)}{m^2 - 4m}$$
$$= \frac{2m^2 - 8m + 8}{m^2 - 4m}$$

Using functional notation we have $A(m) = \dfrac{2m^2 - 8m + 8}{m^2 - 4m}$.

53. Draw a figure:

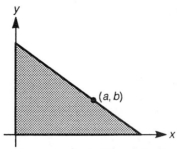

The line has the equation $y - b = m(x - a)$. Since the base and height are the x- and y-intercepts, respectively, find each intercept:

      base: $y = 0$
$$-b = m(x - a)$$
$$-b = mx - ma$$
$$mx = ma - b$$
$$x = \frac{ma - b}{m}$$

    height: $x = 0$
$$y - b = m(-a)$$
$$y = b - ma$$

So the area of the triangle is $A = \frac{1}{2}\left(\dfrac{ma - b}{m}\right)(b - ma) = \dfrac{(ma - b)^2}{-2m}$.

Graphing Utility Exercises for Section 5.4

1. Graphing the function:

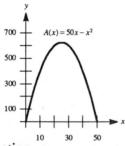

(a) Yes, this is a quadratic function.
(b) There is one turning point.
(c) Yes, there is a maximum value of 625, which occurs at $x = 25$.
(d) No, there is no minimum value for the function.

3.   Graphing the function:

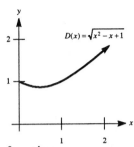

$$D(x) = \sqrt{x^2 - x + 1}$$

(a) No, this is not a quadratic function.
(b) There is one turning point.
(c) No, there is no maximum value for the function.
(d) Yes, there is a minimum value of $\sqrt{0.75} \approx 0.87$, which occurs at $x = 1/2$.

5.   Graphing the function:

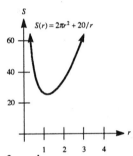

$$S(r) = 2\pi r^2 + 20/r$$

(a) No, this is not a quadratic function.
(b) There is one turning point.
(c) No, there is no maximum value for the function.
(d) Yes, there is a minimum value of approximately 25.69, which occurs at approximately $x \approx 1.17$.

7.   (a) Graphing the function:

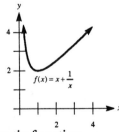

$$f(x) = x + \frac{1}{x}$$

(a) No, this is not a quadratic function.
(b) There is one turning point.
(c) No, there is no maximum value for the function.
(d) Yes, there is a minimum value of 2, which occurs at $x = 1$.

(b) Graphing the function:

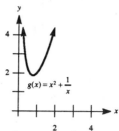

$g(x) = x^2 + \dfrac{1}{x}$

(a) No, this is not a quadratic function.
(b) There is one turning point.
(c) No, there is no maximum value for the function.
(d) Yes, there is a minimum value of approximately 1.890, which occurs at $x \approx 0.794$.

9. (a) Let x and y represent the two numbers. Then $x + y = \sqrt{11}$, so $y = \sqrt{11} - x$. The product is given by:
$$P = xy = x\left(\sqrt{11} - x\right) = -x^2 + \sqrt{11}\, x$$
The product is given by $P(x) = -x^2 + \sqrt{11}\, x$.

(b) Graphing the function:

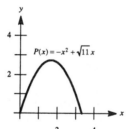

$P(x) = -x^2 + \sqrt{11}\, x$

The product has a maximum value of 2.75.

5.5 Maximum and Minimum Problems

1. Call the two numbers x and y. Then $x + y = 5$, so $y = 5 - x$. So the product can be written as:
$$P = xy = x(5 - x) = 5x - x^2$$
Completing the square:
$$P = -\left(x^2 - 5x\right) = -\left(x^2 - 5x + \tfrac{25}{4}\right) + \tfrac{25}{4} = -\left(x - \tfrac{5}{2}\right)^2 + \tfrac{25}{4}$$
Since this is a parabola opening downward, it will have a maximum value of $\tfrac{25}{4}$.

3. Call the two numbers x and y. Then $y - x = 1$, so $y = x + 1$. The sum of their squares can be written as:
$$S = x^2 + y^2 = x^2 + (x+1)^2 = x^2 + x^2 + 2x + 1 = 2x^2 + 2x + 1$$
Completing the square:
$$S = 2(x^2 + x) + 1 = 2(x^2 + x + \tfrac{1}{4}) + 1 - \tfrac{1}{2} = 2(x + \tfrac{1}{2})^2 + \tfrac{1}{2}$$
Since this is a parabola opening upward, it will have a minimum value of $\tfrac{1}{2}$.

5. Let w and l be the width and length, respectively. Since $P = 2w + 2l$, we have:
$$2w + 2l = 25$$
$$2l = 25 - 2w$$
$$l = \frac{25 - 2w}{2}$$
So the area is given by:
$$A = wl = w\left(\frac{25 - 2w}{2}\right) = \tfrac{1}{2}(-2w^2 + 25)$$
Completing the square:
$$A = -\left(w^2 - \tfrac{25}{2}w\right) = -\left(w^2 - \tfrac{25}{2}w + \tfrac{625}{16}\right) = -\left(w - \tfrac{25}{4}\right)^2 + \tfrac{625}{16}$$
This is a parabola opening downward, so it will achieve a maximum value when $w = \tfrac{25}{4}$.
Substitute to find l:
$$l = \frac{25 - 2\left(\tfrac{25}{4}\right)}{2} = \frac{25 - \tfrac{25}{2}}{2} = \tfrac{25}{4}$$
So the largest such rectangle is a square of dimensions $\tfrac{25}{4}$ m by $\tfrac{25}{4}$ m.

7. Let x and y be the lengths of the two shorter sides, so $x + y = 100$, and $y = 100 - x$. Then the area is given by:
$$A = \tfrac{1}{2}xy = \tfrac{1}{2}x(100 - x) = \tfrac{1}{2}(-x^2 + 100x)$$
Completing the square:
$$A = -\tfrac{1}{2}(x^2 - 100x) = -\tfrac{1}{2}(x^2 - 100x + 2500) + 1250 = -\tfrac{1}{2}(x - 50)^2 + 1250$$
This is a parabola opening downward, so it will achieve a maximum value of 1250 in.^{2}.

9. Let x and y be the two numbers, so $x + y = 6$ and thus $y = 6 - x$.

(a) Simplifying:
$$T = x^2 + y^2 = x^2 + (6 - x)^2 = x^2 + 36 - 12x + x^2 = 2x^2 - 12x + 36$$
Completing the square:
$$T = 2(x^2 - 6x) + 36 = 2(x^2 - 6x + 9) + 36 - 18 = 2(x - 3)^2 + 18$$
This is a parabola opening upward, so it will have a minimum value of 18.

(b) Simplifying:
$$S = x + y^2 = x + (6-x)^2 = x + 36 - 12x + x^2 = x^2 - 11x + 36$$
Completing the square:
$$S = \left(x^2 - 11x\right) + 36 = \left(x^2 - 11x + \tfrac{121}{4}\right) + 36 - \tfrac{121}{4} = \left(x - \tfrac{11}{2}\right)^2 + \tfrac{23}{4}$$
This is a parabola opening upward, so it will have a minimum value of $\tfrac{23}{4}$.

(c) Simplifying:
$$U = x + 2y^2 = x + 2(6-x)^2 = x + 72 - 24x + 2x^2 = 2x^2 - 23x + 72$$
Completing the square:
$$U = 2\left(x^2 - \tfrac{23}{2}x\right) + 72 = 2\left(x^2 - \tfrac{23}{2}x + \tfrac{529}{16}\right) + 72 - \tfrac{529}{8} = 2\left(x - \tfrac{23}{4}\right)^2 + \tfrac{47}{8}$$
This is a parabola opening upward, so it will have a minimum value of $\tfrac{47}{8}$.

(d) Substituting:
$$V = x + (2y)^2$$
$$= x + 4y^2$$
$$= x + 4(6-x)^2$$
$$= x + 144 - 48x + 4x^2$$
$$= 4x^2 - 47x + 144$$
Completing the square:
$$V = 4\left(x^2 - \tfrac{47}{4}x\right) + 144 = 4\left(x^2 - \tfrac{47}{4}x + \tfrac{2209}{64}\right) + 144 - \tfrac{2209}{16} = 4\left(x - \tfrac{47}{8}\right)^2 + \tfrac{95}{16}$$
This is a parabola opening upward, so it will have a minimum value of $\tfrac{95}{16}$.

11. (a) Compute the heights:
$$h(1) = -16(1)^2 + 32(1) = -16 + 32 = 16 \text{ ft}$$
$$h\left(\tfrac{3}{2}\right) = -16\left(\tfrac{3}{2}\right)^2 + 32\left(\tfrac{3}{2}\right) = -16\left(\tfrac{9}{4}\right) + 48 = -36 + 48 = 12 \text{ ft}$$

(b) Completing the square:
$$h = -16t^2 + 32t = -16\left(t^2 - 2t\right) = -16\left(t^2 - 2t + 1\right) + 16 = -16(t-1)^2 + 16$$
This is a parabola opening downward, so it will have a maximum height of 16 ft, attained after 1 second.

(c) Set $h = 7$ and solve for t:
$$7 = -16t^2 + 32t$$
$$16t^2 - 32t + 7 = 0$$
$$(4t - 7)(4t - 1) = 0$$
$$t = \tfrac{7}{4}, \tfrac{1}{4}$$
So $h = 7$ when $t = \tfrac{7}{4}$ sec or $t = \tfrac{1}{4}$ sec.

13. Every point on the given curve has coordinates of the form $(x, \sqrt{x-2}+1)$, and using the distance formula gives:
$$d = \sqrt{(4-x)^2 + (1-\sqrt{x-2}-1)^2} = \sqrt{16-8x+x^2+x-2} = \sqrt{x^2-7x+14}$$
Completing the square:
$$(x^2-7x)+14 = \left(x^2-7x+\tfrac{49}{4}\right)+14-\tfrac{49}{4} = \left(x-\tfrac{7}{2}\right)^2+\tfrac{7}{4}$$
This is a parabola opening upward which will achieve a minimum value of $\sqrt{\tfrac{7}{4}} = \tfrac{\sqrt{7}}{2}$ at $x = \tfrac{7}{2}$. Then:
$$y = \sqrt{\tfrac{7}{2}-2}+1 = \sqrt{\tfrac{3}{2}}+1 = \tfrac{2+\sqrt{6}}{2}$$
So the point is $\left(\tfrac{7}{2}, \tfrac{2+\sqrt{6}}{2}\right)$ and the distance is $\tfrac{\sqrt{7}}{2}$.

15. (a) We must find the value of x such that $x - x^2$ is as large as possible.
Call $f(x) = -x^2 + x$. Completing the square:
$$f(x) = -\left(x^2-x\right) = -\left(x^2-x+\tfrac{1}{4}\right)+\tfrac{1}{4} = -\left(x-\tfrac{1}{2}\right)^2+\tfrac{1}{4}$$
This is a parabola opening downward, so it will achieve a maximum value when $x = \tfrac{1}{2}$. So the number is $\tfrac{1}{2}$.

(b) We must find the value of x such that $x - 2x^2$ is as large as possible.
Call $f(x) = -2x^2 + x$. Completing the square:
$$f(x) = -2\left(x^2-\tfrac{1}{2}x\right) = -2\left(x^2-\tfrac{1}{2}x+\tfrac{1}{16}\right)+\tfrac{1}{8} = -2\left(x-\tfrac{1}{4}\right)^2+\tfrac{1}{8}$$
This is a parabola opening downward, so it will achieve a maximum value when $x = \tfrac{1}{4}$. So the number is $\tfrac{1}{4}$.

17. If we choose x for the depth of the pasture, then $500 - 2x$ is the length parallel to the river. The area of the pasture will then be given by:
$$A = x(500-2x) = -2x^2 + 500x$$
Completing the square:
$$A = -2\left(x^2-250x\right) = -2\left(x^2-250x+125^2\right)+2(125)^2 = -2(x-125)^2+31250$$
This is a parabola opening downward, so it will achieve a maximum value at $x = 125$. Then the length is $500 - 2(125) = 500 - 250 = 250$. So the dimensions are 125 ft by 250 ft.

19. Simplifying:
$$\begin{aligned}
R - C &= \left(0.4x^2+10x+5\right)-\left(0.5x^2+2x+101\right) \\
&= 0.4x^2+10x+5-0.5x^2-2x-101 \\
&= -0.1x^2+8x-96
\end{aligned}$$

Completing the square:
$$R - C = -0.1(x^2 - 80x) - 96$$
$$= -0.1(x^2 - 80x + 1600) - 96 + 160$$
$$= -0.1(x - 40)^2 + 64$$

This is a parabola opening downward, so it will achieve a maximum value when $x = 40$.

21. Recall that revenue R is $x \cdot p$. So:
$$R = x\left(-\tfrac{1}{4}x + 30\right) = -\tfrac{1}{4}x^2 + 30x$$

Completing the square:
$$R = -\tfrac{1}{4}x^2 + 30x$$
$$= -\tfrac{1}{4}\left(x^2 - 120x\right)$$
$$= -\tfrac{1}{4}\left(x^2 - 120x + 3600\right) + 900$$
$$= -\tfrac{1}{4}(x - 60)^2 + 900$$

This is a parabola opening downward, so it will achieve a maximum value at $x = 60$. The maximum revenue is \$900. The corresponding unit price p is:
$$p = -\tfrac{1}{4}(60) + 30 = -15 + 30 = \$15$$

23. (a) To use max/min methods, we need to substitute in the quantity $x^2 + y^2$ and write it strictly in terms of x or y. So take $2x + 3y = 6$ and solve for y:
$$3y = 6 - 2x$$
$$y = \frac{6 - 2x}{3}$$

Then substitute, and the quantity $x^2 + y^2$ becomes:
$$Q = x^2 + \left(\frac{6 - 2x}{3}\right)^2 = x^2 + \frac{36 - 24x + 4x^2}{9} = \tfrac{13}{9}x^2 - \tfrac{8}{3}x + 4$$

Completing the square:
$$Q = \tfrac{13}{9}\left(x^2 - \tfrac{24}{13}x\right) + 4 = \tfrac{13}{9}\left(x^2 - \tfrac{24}{13}x + \tfrac{144}{169}\right) + 4 - \tfrac{13}{9}\left(\tfrac{144}{169}\right) = \tfrac{13}{9}\left(x - \tfrac{12}{13}\right)^2 + \tfrac{36}{13}$$

This is a parabola opening up, so it will achieve a minimum value of $\tfrac{36}{13}$.

(b) The equation of a circle with its center at the origin is $x^2 + y^2 = r^2$ where r is the radius. The line $2x + 3y = 6$ will intersect the circle in two points wherever r is sufficiently large. As we reduce r, we gradually reach a position where the circle and line are tangent and this is the minimum value of r or $\sqrt{x^2 + y^2}$. In this case, it is $\sqrt{\tfrac{36}{13}} = \tfrac{6\sqrt{13}}{13}$. This is the square root of the answer from (a).

25. **(a)** Substitute $y = 15 - x$:
$$Q = x^2 + y^2 = x^2 + (15 - x)^2 = x^2 + 225 - 30x + x^2 = 2x^2 - 30x + 225$$
Completing the square:
$$\begin{aligned}Q &= 2x^2 - 30x + 225 \\ &= 2(x^2 - 15x) + 225 \\ &= 2\left(x^2 - 15x + \tfrac{225}{4}\right) + 225 - \tfrac{225}{2} \\ &= 2\left(x - \tfrac{15}{2}\right)^2 + \tfrac{225}{2}\end{aligned}$$
This is a parabola opening upward, so it will achieve a minimum value of $\tfrac{225}{2}$.

(b) Substitute $y = C - x$:
$$Q = x^2 + y^2 = x^2 + (C - x)^2 = x^2 + C^2 - 2Cx + x^2 = 2x^2 - 2Cx + C^2$$
Completing the square:
$$\begin{aligned}Q &= 2x^2 - 2Cx + C^2 \\ &= 2(x^2 - Cx) + C^2 \\ &= 2\left(x^2 - Cx + \tfrac{C^2}{4}\right) + C^2 - \tfrac{C^2}{2} \\ &= 2\left(x - \tfrac{C}{2}\right)^2 + \tfrac{C^2}{2}\end{aligned}$$
This is a parabola opening upward, so it will achieve a minimum value of $\tfrac{C^2}{2}$.
When $C = 15$, the result from (a) is verified.

27. Let the other two sides of each of the four triangles be t and $1 - t$, respectively. Then the area of the square will be a minimum when the area of these triangles is a maximum. Now write an expression for the total area of the four triangles:
$$A = 4\left(\tfrac{1}{2}\right)(t)(1 - t) = 2t - 2t^2 = -2t^2 + 2t$$
Completing the square:
$$A = -2t^2 + 2t = -2(t^2 - t) = -2\left(t^2 - t + \tfrac{1}{4}\right) + \tfrac{1}{2} = -2\left(t - \tfrac{1}{2}\right)^2 + \tfrac{1}{2}$$
This is a parabola opening downward, so it will achieve a maximum area of $\tfrac{1}{2}$ when $t = \tfrac{1}{2}$. Since the large square has area $= 1$, the smaller square has area $1 - A = 1 - \tfrac{1}{2} = \tfrac{1}{2}$. Using the Pythagorean theorem to find x:
$$x^2 = t^2 + (1 - t)^2 = \tfrac{1}{4} + \tfrac{1}{4} = \tfrac{1}{2}$$
So $x = \sqrt{\tfrac{1}{2}} = \tfrac{\sqrt{2}}{2}$.

29. Draw the figure:

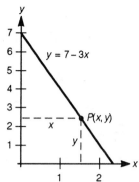

We have $A = xy = x(7 - 3x) = -3x^2 + 7x$. Completing the square:

$$A = -3\left(x^2 - \tfrac{7}{3}x\right) = -3\left(x^2 - \tfrac{7}{3}x + \tfrac{49}{36}\right) + \tfrac{49}{12} = -3\left(x - \tfrac{7}{6}\right)^2 + \tfrac{49}{12}$$

Since this parabola opens downward, the largest possible area is $\tfrac{49}{12}$.

31. Draw the figure:

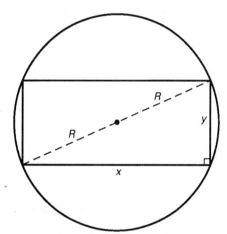

Using the Pythagorean theorem, we obtain $y = \sqrt{4R^2 - x^2}$. The area A of the rectangle is:

$$A = xy = x\sqrt{4R^2 - x^2}$$

Therefore:

$$A^2 = x^2\left(4R^2 - x^2\right) = 4R^2x^2 - x^4$$

In order to maximize the expression $4R^2x^2 - x^4$, first let $t = x^2$, so that the expression becomes $4R^2t - t^2$. Completing the square:

$$A^2 = -t^2 + 4R^2t$$
$$= -\left(t^2 - 4R^2t\right)$$
$$= -\left(t^2 - 4R^2t + 4R^4\right) + 4R^4$$
$$= -\left(t - 2R^2\right)^2 + 4R^4$$

This parabola opens downward, so the maximum value of A^2 is $4R^4$. Thus the maximum area is $\sqrt{4R^4} = 2R^2$.

33. Let x = east-west dimension, and y = north-south dimension. So the cost is given by
$C = 12(2x) + 8(2y) = 24x + 16y$. Since this cost is \$4800, we have $24x + 16y = 4800$, so:
$$y = \frac{4800 - 24x}{16} = \frac{600 - 3x}{2}$$
Now the area is $A = xy = x\left(\dfrac{600 - 3x}{2}\right) = -\frac{3}{2}x^2 + 300x$. So $A(x) = -\frac{3}{2}x^2 + 300x$.

This will be a parabola opening downward, so it will have a maximum value.
Completing the square:
$$A(x) = -\tfrac{3}{2}x^2 + 300x$$
$$= -\tfrac{3}{2}\left(x^2 - 200x\right)$$
$$= -\tfrac{3}{2}\left(x^2 - 200x + 100^2\right) + 15000$$
$$= -\tfrac{3}{2}(x - 100)^2 + 15000$$
So $x = 100$ will maximize area, which is 15,000 yd^2. We find y:
$$y = \frac{600 - 3(100)}{2} = \frac{600 - 300}{2} = 150 \text{ yd}$$
So the dimensions are 100 yd by 150 yd.

35. The given function can be rewritten:
$$y = \left(a_1 + a_2\right)x^2 - 2\left(a_1 x_1 + a_2 x_2\right)x + \left(a_1 x_1^2 + a_2 x_2^2\right)$$
Completing the square:
$$y = \left(a_1 + a_2\right)\left[x^2 - \frac{2\left(a_1 x_1 + a_2 x_2\right)}{a_1 + a_2}\right] + \left(a_1 x_1^2 + a_2 x_2^2\right)$$
$$= \left(a_1 + a_2\right)\left(x - \frac{a_1 x_1 + a_2 x_2}{a_1 + a_2}\right)^2 + \left(a_1 x_1^2 + a_2 x_2^2\right) - \frac{\left(a_1 x_1 + a_2 x_2\right)^2}{a_1 + a_2}$$
Since a_1 and a_2 are both positive, $a_1 + a_2 > 0$ and thus this parabola opens upward.
So the minimum must occur where $x = \frac{a_1 x_1 + a_2 x_2}{a_1 + a_2}$.

37. (a) We have $\frac{\Delta p}{\Delta x} = \frac{10}{-5} = -2$. Also $p = 200$ when $x = 150$. Using the point-slope formula
with the point $(150, 200)$:
$$p - 200 = -2(x - 150)$$
$$p - 200 = -2x + 300$$
$$p = -2x + 500$$
Using functional notation we have $p(x) = -2x + 500$.

   (b) Since $R = xp$, we have $R = x(-2x + 500) = -2x^2 + 500x$. Completing the square:
$$R = -2x^2 + 500x$$
$$= -2\left(x^2 - 250x\right)$$
$$= -2\left(x^2 - 250x + 15625\right) + 31250$$
$$= -2(x - 125)^2 + 31250$$

Using functional notation we have $R(x) = -2(x-125)^2 + 31250$. Since this parabola opens downward, we have a maximum revenue of \$31,250 when $x = 125$. We find $p = -2(125) + 500 = \$250$.

39. Let $x = t^2$, so $f(x) = x - x^2 = -x^2 + x$. Completing the square:
$$f(x) = -\left(x^2 - x\right) = -\left(x^2 - x + \tfrac{1}{4}\right) + \tfrac{1}{4} = -\left(x - \tfrac{1}{2}\right)^2 + \tfrac{1}{4}$$
So $f(x)$ has a maximum value when $x = \tfrac{1}{2}$. Then $t^2 = \tfrac{1}{2}$, so $t = \pm\tfrac{\sqrt{2}}{2}$.

41. Let $x = t^2$. Then $y = -t^4 + 6t^2 - 6 = -x^2 + 6x - 6$. Completing the square:
$$y = -x^2 + 6x - 6 = -\left(x^2 - 6x\right) - 6 = -\left(x^2 - 6x + 9\right) - 6 + 9 = -(x-3)^2 + 3$$
So $x = 3$ will yield the largest output. Since $x = t^2$, we have:
$$t^2 = 3$$
$$t = \pm\sqrt{3}$$
So $t = \sqrt{3}$ or $t = -\sqrt{3}$ will yield the largest output.

43. (a) Since $x^2 - x + 1 = \left(x - \tfrac{1}{2}\right)^2 + \tfrac{3}{4}$, the quantity must be positive. We know that
$\left(x - \tfrac{1}{2}\right)^2 \geq 0$, thus $\left(x - \tfrac{1}{2}\right)^2 + \tfrac{3}{4} \geq \tfrac{3}{4} > 0$.

(b) Computing the discriminant:
$$D = (-1)^2 - 4(1)(1) = 1 - 4 = -3$$
Since $D < 0$, there are no x-intercepts (because no real solutions exist). Since the graph is a parabola opening up, and there are no x-intercepts, the y-values must be positive for all values of x.

45. (a) For the circle, we have $2\pi r = x$, so $r = \tfrac{x}{2\pi}$, and thus the area is:
$$A = \pi r^2 = \pi\left(\frac{x}{2\pi}\right)^2 = \frac{x^2}{4\pi}$$
For the square, we have $4s = 16 - x$, so $s = \tfrac{16-x}{4}$, and thus the area is:
$$A = s^2 = \left(\frac{16-x}{4}\right)^2 = \frac{256 - 32x + x^2}{16} = 16 - 2x + \tfrac{1}{16}x^2$$
So the total combined area is given by:
$$A(x) = \frac{x^2}{4\pi} + 16 - 2x + \tfrac{1}{16}x^2 = \tfrac{4+\pi}{16\pi}x^2 - 2x + 16$$

(b) This is a parabola opening upward, so it will have a minimum value at:
$$x = \frac{2}{2 \cdot \frac{4+\pi}{16\pi}} = \frac{16\pi}{4 + \pi}$$

(c) This ratio is given by:

$$\frac{\frac{16\pi}{4+\pi}}{16-\frac{16\pi}{4+\pi}}\cdot\frac{4+\pi}{4+\pi}=\frac{16\pi}{64+16\pi-16\pi}=\frac{16\pi}{64}=\frac{\pi}{4}$$

47. (a) For the circle, we have $2\pi r = x$, so $r = \frac{x}{2\pi}$, and thus the area is:

$$A = \pi r^2 = \pi\left(\frac{x}{2\pi}\right)^2 = \frac{x^2}{4\pi}$$

For the square, we have $4s = L - x$, so $s = \frac{L-x}{4}$, and thus the area is:

$$A = \left(\frac{L-x}{4}\right)^2 = \frac{L^2 - 2Lx + x^2}{16} = \frac{1}{16}x^2 - \frac{L}{8}x + \frac{L^2}{16}$$

So the total combined area is given by:

$$A(x) = \left(\frac{1}{4\pi} + \frac{1}{16}\right)x^2 - \frac{L}{8}x + \frac{L^2}{16} = \frac{4+\pi}{16\pi}x^2 - \frac{L}{8}x + \frac{L^2}{16}$$

(b) This will have a minimum value when $x = -\frac{b}{2a}$, so:

$$x = \frac{\frac{L}{8}}{\frac{4+\pi}{8\pi}}\cdot\frac{8\pi}{8\pi} = \frac{\pi L}{4+\pi}$$

(c) This ratio is:

$$\frac{\frac{\pi L}{4+\pi}}{L - \frac{\pi L}{4+\pi}}\cdot\frac{4+\pi}{4+\pi}=\frac{\pi L}{4L+\pi L-\pi L}=\frac{\pi L}{4L}=\frac{\pi}{4}$$

It is interesting to note that this ratio does not depend on L, the length of the wire.

49. Re-draw the figure and label additional sides:

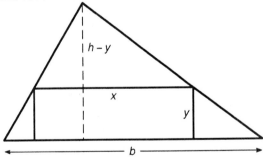

By similar triangles:

$$\frac{h-y}{h} = \frac{x}{b}$$
$$bh - by = xh$$
$$by = bh - xh$$
$$y = h - \frac{h}{b}x$$

So the area of the rectangle is given by:

$$A = xy = x\left(h - \frac{h}{b}x\right) = -\frac{h}{b}x^2 + hx$$

This is a parabola opening downward, so it will have a maximum value when:

$$x = \frac{-h}{2\left(-\frac{h}{b}\right)} = \frac{b}{2}$$

So the maximum area is given by:

$$A\left(\frac{b}{2}\right) = -\frac{h}{b} \cdot \frac{b^2}{4} + h \cdot \frac{b}{2} = -\frac{hb}{4} + \frac{hb}{2} = \frac{hb}{4}$$

Since the triangle has an area of $\frac{hb}{2}$, the desired ratio is:

$$\frac{\frac{hb}{2}}{\frac{hb}{4}} = 2$$

51. Call h the height of the triangle (see figure), and call y the indicated value:

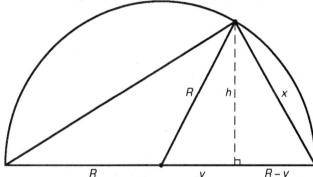

By the Pythagorean theorem, we have the following relationships:

$$y^2 + h^2 = R^2 \qquad \text{and} \qquad (R-y)^2 + h^2 = x^2$$
$$y^2 = R^2 - h^2$$
$$y = \sqrt{R^2 - h^2}$$

Substituting, we have:

$$\left(R - \sqrt{R^2 - h^2}\right)^2 + h^2 = x^2$$
$$R^2 - 2R\sqrt{R^2 - h^2} + R^2 - h^2 + h^2 = x^2$$
$$2R^2 - 2R\sqrt{R^2 - h^2} = x^2$$

Solve for h:

$$-2R\sqrt{R^2 - h^2} = x^2 - 2R^2$$

$$\sqrt{R^2 - h^2} = R - \frac{x^2}{2R}$$

$$R^2 - h^2 = R^2 - x^2 + \frac{x^4}{4R^2}$$

$$-h^2 = \frac{x^4}{4R^2} - x^2$$

$$h^2 = -\frac{x^4}{4R^2} + x^2$$

Now use the hint. Since the area of the triangle is given by $A = \frac{1}{2}(2R)(h) = Rh$, we maximize:

$$A^2 = R^2 h^2 = R^2\left(\frac{-x^4}{4R^2} + x^2\right) = -\frac{1}{4}x^4 + R^2 x^2$$

This will have a maximum value when:

$$x^2 = \frac{-b}{2a} = \frac{-R^2}{-\frac{1}{2}} = 2R^2$$

Therefore we have:

$$A^2 = -\frac{1}{4}\left(2R^2\right)^2 + R^2\left(2R^2\right) = -R^4 + 2R^4 = R^4$$

So the maximum value is $A^2 = R^4$, or $A = R^2$. Thus the shaded area is:

$$\frac{1}{2}\left(\pi R^2\right) - R^2 = \frac{\pi R^2}{2} - R^2 = R^2\left(\frac{\pi - 2}{2}\right)$$

53. We wish to minimize the sum $x + \frac{1}{x}$. Completing the square:

$$x + \frac{1}{x} = \left(\sqrt{x} - \sqrt{\frac{1}{x}}\right)^2 + 2$$

The smallest possible value of 2 occurs if:

$$\sqrt{x} = \sqrt{\frac{1}{x}}$$

$$x = \frac{1}{x}$$

$$x^2 = 1$$

$$x = 1$$

If both numbers are 1, the minimum sum of 2 occurs.

55. Using the hint, we have:

$$G(x) = \frac{(a+x)(b+x)}{x}$$

$$= \frac{ab + (a+b)x + x^2}{x}$$

$$= \frac{ab}{x} + (a+b) + x$$

$$= (a+b) + \left(\frac{ab}{x} + x\right)$$

We can rewrite $G(x)$ as:

$$G(x) = \left(\sqrt{x} - \sqrt{\frac{ab}{x}}\right)^2 + \left(a + b - 2\sqrt{ab}\right) = \left(\sqrt{x} - \sqrt{\frac{ab}{x}}\right)^2 + \left(\sqrt{a} - \sqrt{b}\right)^2$$

Since $G(x)$ is the sum of two squares, we have $G(x) \geq 0$. Thus the minimum value occurs when:

$$\sqrt{x} - \sqrt{\frac{ab}{x}} = 0$$

$$\sqrt{x} = \frac{\sqrt{ab}}{\sqrt{x}}$$

$$x = \sqrt{ab}$$

So the minimum value is:

$$G\left(\sqrt{ab}\right) = (a+b) + \left(\frac{ab}{\sqrt{ab}} + \sqrt{ab}\right) = a + b + 2\sqrt{ab} = \left(\sqrt{a} + \sqrt{b}\right)^2$$

This proves the desired result.

57. (a) The largest value of $f(x)$ is approximately 1.143.

(b) Since $f(x) = \dfrac{1}{2x^2 - x + 1}$, its maximum value will be the same as the minimum value of $2x^2 - x + 1$. This occurs when:

$$x = \frac{-b}{2a} = \frac{1}{4}$$

Thus the maximum value of $f(x)$ is:

$$f\left(\tfrac{1}{4}\right) = \frac{1}{2\left(\tfrac{1}{4}\right)^2 - \tfrac{1}{4} + 1} = \frac{1}{\tfrac{1}{8} - \tfrac{1}{4} + 1} = \frac{1}{\tfrac{7}{8}} = \tfrac{8}{7}$$

So $f(x) \leq \tfrac{8}{7}$ for every value of x.

Graphing Utility Exercises for Section 5.5

1. (a) Call x and y the two sides, so $xy = 3200$ and thus $y = \dfrac{3200}{x}$. Then the cost is given by:

$$C = 1x + 2y = x + 2\left(\frac{3200}{x}\right) = x + \frac{6400}{x}$$

(b) Graphing the function:

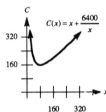

The minimum value is 160 when $x = 80$ and $y = 40$. The dimensions are 80 ft ($1 per foot fencing) and 40 ft ($2 per foot fencing).

3. (a) Let x and y represent the dimensions of the printed matter, so $x + 2$ and $y + 2$ represent the dimensions of the page. Since $xy = 30$, we have $y = \dfrac{30}{x}$. The area of the page is:

$$A = (x+2)(y+2) = (x+2)\left(\frac{30}{x} + 2\right) = 30 + \frac{60}{x} + 2x + 4 = 34 + 2x + \frac{60}{x}$$

(b) Graphing the function:

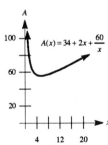

The minimum value is approximately 55.91 when $x \approx 5.48$ and $y \approx 5.48$, thus $x + 2 \approx 7.48$ and $y + 2 \approx 7.48$. The dimensions of the page are both approximately 7.48 in.

5. (a) Let $\left(x, x^2\right)$ represent a point on the parabola. The distance from this point to the point $(3, 0)$ is given by the distance formula:

$$d = \sqrt{(x-3)^2 + \left(x^2 - 0\right)^2} = \sqrt{x^2 - 6x + 9 + x^4} = \sqrt{x^4 + x^2 - 6x + 9}$$

(b) Graphing the function:

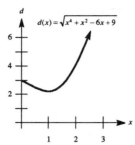

The minimum value is $\sqrt{5} \approx 2.236$ when $x = 1$. The closest point on the parabola is $(1, 1)$.

7.   (a) Since the perimeter of the rectangle is 36, we have $2r + 2h = 36$, thus $r + h = 18$ and $h = 18 - r$. The volume of the cylinder is given by:
$$V = \pi r^2 h = \pi r^2 (18 - r) = 18 \pi r^2 - \pi r^3$$

(b) Graphing the function:

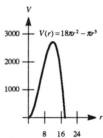

The maximum value is approximately 2714 when $r = 12$ and $h = 6$. The maximum volume is:
$$V = \pi (12)^2 (6) = 864 \pi \approx 2714$$

9.   (a) First draw the figure:

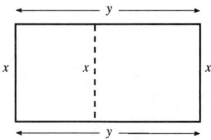

Since the area is 5000 ft^2, we have $xy = 5000$, so $y = \dfrac{5000}{x}$. Thus the cost is given by:

$$
\begin{aligned}
C &= 150(2x + 2y) + 100(x) \\
&= 300x + 300y + 100x \\
&= 400x + 300\left(\frac{5000}{x}\right) \\
&= 400x + \frac{1500000}{x}
\end{aligned}
$$

(b) Graphing the function:

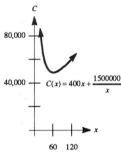

$C(x) = 400x + \dfrac{1500000}{x}$

The minimum value is approximately 48,990 when $x \approx 61.2$ and $y \approx 81.7$. The minimum cost occurs when the dimensions are 61.2 ft by 81.7 ft.

11. (a) Let x represent the width of the window (which is the diameter of the circle) and y represent the rectangular height. Since the perimeter is 30, we have:

$$
\begin{aligned}
x + 2y + \tfrac{1}{2}(2)(\pi)\left(\frac{x}{2}\right) &= 30 \\
x + 2y + \tfrac{\pi}{2}x &= 30 \\
\left(1 + \tfrac{\pi}{2}\right)x + 2y &= 30 \\
2y &= 30 - \left(1 + \tfrac{\pi}{2}\right)x \\
y &= 15 - \tfrac{1}{2}\left(1 + \tfrac{\pi}{2}\right)x
\end{aligned}
$$

The greatest possible amount of light admitted occurs when the area of the window is maximized. The area is given by:

$$
\begin{aligned}
A &= xy + \tfrac{1}{2}(\pi)\left(\frac{x}{2}\right)^2 \\
&= x\left[15 - \tfrac{1}{2}\left(1 + \tfrac{\pi}{2}\right)x\right] + \tfrac{\pi}{2}\left(\frac{x^2}{4}\right) \\
&= 15x - \tfrac{1}{2}x^2 - \tfrac{\pi}{4}x^2 + \tfrac{\pi}{8}x^2 \\
&= 15x - \left(\tfrac{1}{2} + \tfrac{\pi}{4} - \tfrac{\pi}{8}\right)x^2 \\
&= 15x - \left(\tfrac{1}{2} + \tfrac{\pi}{8}\right)x^2
\end{aligned}
$$

(b) Graphing the function:

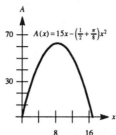

The maximum value is approximately 63.01 when $x \approx 8.40$ and $y \approx 4.20$. The maximum area occurs with approximate dimensions of 8.40 ft by 4.20 ft.

(c) Completing the square:

$$A(x) = -\left(\tfrac{1}{2} + \tfrac{\pi}{8}\right)x^2 + 15x$$
$$= -\left(\frac{4+\pi}{8}\right)\left[x^2 - \frac{120}{4+\pi}x\right]$$
$$= -\left(\frac{4+\pi}{8}\right)\left[x^2 - \frac{120}{4+\pi}x + \left(\frac{60}{4+\pi}\right)^2\right] + \frac{450}{4+\pi}$$
$$= -\left(\frac{4+\pi}{8}\right)\left(x - \frac{60}{4+\pi}\right)^2 + \frac{450}{4+\pi}$$

This will have a maximum value of $\dfrac{450}{4+\pi} \approx 63.01$ when $x = \dfrac{60}{4+\pi} \approx 8.40$, which agrees with our results from part (b).

5.6 Polynomial Functions

1. This graph has 4 turning points, but a polynomial function of degree 3 can have at most 2 turning points.

3. As $|x|$ gets very large, our function should be similar to $f(x) = a_3 x^3$. But $f(x)$ does not have a parabolic shape like the given graph.

5. As $|x|$ gets very large with x negative, then the graph should resemble $2x^5$. But the y-values of $2x^5$ are always negative when x is negative, contrary to the given graph.

7. This graph has a corner, which cannot occur in the graph of a polynomial function.

9.   This is $y = x^2$ translated 2 units to the right and 1 unit up. There are no x-intercepts and the y-intercept is 5. Sketching the graph:

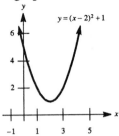

11.   This is $y = x^4$ reflected across the x-axis and translated 1 unit to the right. The x-intercept is 1 and the y-intercept is –1. Sketching the graph:

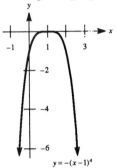

13.   This is $y = x^3$ translated 4 units to the right and 2 units down. The x-intercept is $4 + \sqrt[3]{2}$ and the y-intercept is – 66. Sketching the graph:

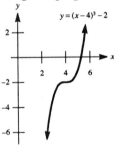

15.   This is $y = 2x^4$ translated 5 units to the right and reflected across the x-axis. The x-intercept is –5 and the y-intercept is –1250. Sketching the graph:

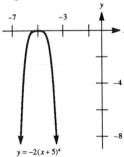

17. This is $y = \frac{1}{2}x^5$ translated 1 unit to the left. The x-intercept is -1 and the y-intercept is $\frac{1}{2}$. Sketching the graph:

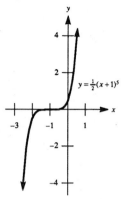

19. This is $y = x^3$ translated 1 unit to the right and 1 unit down, then reflected across the x-axis. The x- and y-intercepts are both 0. Sketching the graph:

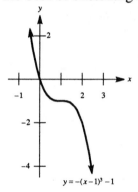

21. (a) The x-intercepts are 2, 1, and -1, and the y-intercept is 2. The signs of y are given by:

Interval	$y = (x-2)(x-1)(x+1)$
$(-\infty, -1)$	neg.
$(-1, 1)$	pos.
$(1, 2)$	neg.
$(2, \infty)$	pos.

Sketching the excluded regions:

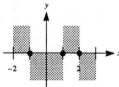

(b)   Now graph the function:

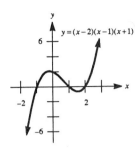

23. (a)   The x-intercepts are 0, 2, and 1, and the y-intercept is 0. The signs of y are given by:

Interval	$y = 2x(x-2)(x-1)$
$(-\infty,0)$	neg.
$(0,1)$	pos.
$(1,2)$	neg.
$(2,\infty)$	pos.

Sketching the excluded regions:

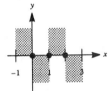

(b)   Now graph the function:

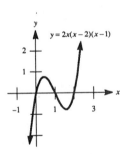

25. (a)   First factor to find the x-intercepts:
$$y = x\left(x^2 - 4x - 5\right) = x(x-5)(x+1)$$
The x-intercepts are 0, 5, and –1, and the y-intercept is 0. The signs of y are given by:

Interval	$y = x(x-5)(x+1)$
$(-\infty,-1)$	neg.
$(-1,0)$	pos.
$(0,5)$	neg.
$(5,\infty)$	pos.

Sketching the excluded regions:

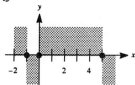

(b) Now graph the function:

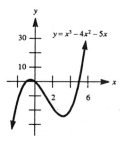

27. (a) First factor (by grouping) to find the *x*-intercepts:

$$y = x^2(x+3) - 4(x+3) = (x+3)(x^2-4) = (x+3)(x+2)(x-2)$$

The *x*-intercepts are –3, –2, and 2, and the *y*-intercept is –12. The signs of *y* are given by:

Interval	$y = (x+3)(x+2)(x-2)$
$(-\infty, -3)$	neg.
$(-3, -2)$	pos.
$(-2, 2)$	neg.
$(2, \infty)$	pos.

Sketching the excluded regions:

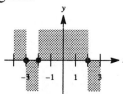

(b) Now graph the function:

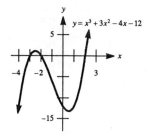

29. (a)   The x-intercepts are 0 and –2, and the y-intercept is 0. The signs of y are given by:

Interval	$y = x^3(x+2)$
$(-\infty, -2)$	pos.
$(-2, 0)$	neg.
$(0, \infty)$	pos.

Sketching the excluded regions:

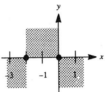

(b)   Near $x = 0$, we have the approximation $y \approx x^3(0 + 2) = 2x^3$. So in the immediate vicinity of $x = 0$, the graph of y resembles $y = 2x^3$:

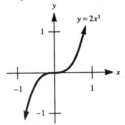

(c)   Now graph the function:

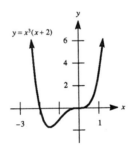

31. (a)   The x-intercepts are 1 and 4, and the y-intercept is 128. The signs of y are given by:

Interval	$y = 2(x-1)(x-4)^3$
$(-\infty, 1)$	pos.
$(1, 4)$	neg.
$(4, \infty)$	pos.

Sketching the excluded regions:

(b) Near $x = 4$, we have the approximation $y \approx 2(4-1)(x-4)^3 = 6(x-4)^3$. So in the immediate vicinity of $x = 4$, the graph of y resembles $y = 6(x-4)^3$:

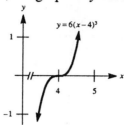

(c) Now graph the function:

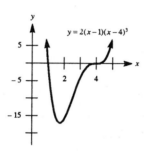

33. (a) The x-intercepts are -1, 1, and 3, and the y-intercept is 3. The signs of y are given by:

Interval	$y = (x+1)^2(x-1)(x-3)$
$(-\infty, -1)$	pos.
$(-1, 1)$	pos.
$(1, 3)$	neg.
$(3, \infty)$	pos.

Sketching the excluded regions:

(b) Near $x = -1$, we have the approximation $y \approx (x+1)^2(-2)(-4) = 8(x+1)^2$. So in the immediate vicinity of $x = -1$, the graph of y resembles $y = 8(x+1)^2$:

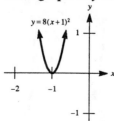

(c) Now graph the function:

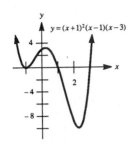

35. (a) The x-intercepts are 0, 4, and –2, and the y-intercept is 0. The signs of y are given by:

Interval	$y = -x^3(x-4)(x+2)$
$(-\infty, -2)$	pos.
$(-2, 0)$	neg.
$(0, 4)$	pos.
$(4, \infty)$	neg.

Sketching the excluded regions:

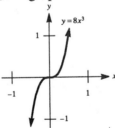

(b) Near $x = 0$, we have the approximation $y \approx -x^3(0-4)(0+2) = 8x^3$. So in the immediate vicinity of $x = 0$, the graph resembles $y = 8x^3$:

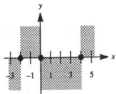

(c) Now graph the function:

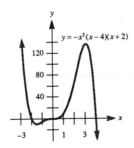

37. (a) The x-intercepts are 0, 2, and –2, and the y-intercept is 0. The signs of y are given by:

Interval	$y = -4x(x-2)^2(x+2)^3$
$(-\infty, -2)$	neg.
$(-2, 0)$	pos.
$(0, 2)$	neg.
$(2, \infty)$	neg.

Sketching the excluded regions:

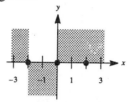

(b) Near $x = 2$, we have the approximation $y \approx -4(2)(x-2)^2(2+2)^3 = -512(x-2)^2$. So in the immediate vicinity of $x = 2$, the graph resembles $y = -512(x-2)^2$:

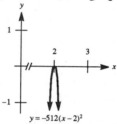

Near $x = -2$, we have the approximation $y \approx -4(-2)(-2-2)^2(x+2)^3 = 128(x+2)^3$. So in the immediate vicinity of $x = -2$, the graph resembles $y = 128(x+2)^3$:

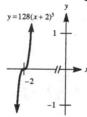

(c) Now graph the function:

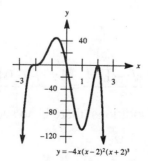

39. Finding the x-intercepts:
$$x^3 - 3x^2 - 5x = 0$$
$$x\left(x^2 - 3x - 5\right) = 0$$
So $x = 0$ is one x-intercept. Find the other two using the quadratic formula:
$$x = \frac{-(-3) \pm \sqrt{(-3)^2 - 4(1)(-5)}}{2(1)} = \frac{3 \pm \sqrt{29}}{2} \approx -1.193, 4.193$$
Note that the x-intercepts are consistent with the given graph.

41. Finding the x-intercepts:
$$x^3 + 6x^2 - 3x - 18 = 0$$
$$x^2(x + 6) - 3(x + 6) = 0$$
$$(x + 6)\left(x^2 - 3\right) = 0$$
$$x = -6, \pm\sqrt{3} \approx \pm 1.732$$
Note that the x-intercepts are consistent with the given graph.

43. Finding the x-intercepts:
$$x^7 + 8x^4 + 16x = 0$$
$$x\left(x^6 + 8x^3 + 16\right) = 0$$
$$x\left(x^3 + 4\right)^2 = 0$$
$$x = 0, -\sqrt[3]{4} \approx -1.587$$
Note that the x-intercepts are consistent with the given graph.

45. From left to right, they are $f(x) = x$, $g(x) = x^2$, $h(x) = x^3$, $F(x) = x^4$, $G(x) = x^5$, and $H(x) = x^6$.

47. Find where $0 \le H(x) < 0.1$:
$$0 \le x^6 < 0.1$$
$$0 \le x < 0.68$$
When x lies in the interval $[0, 0.68)$, then $H(x)$ will lie in the interval $[0, 0.1)$.

49. Find where:
$$g(t) - F(t) = 0.26$$
$$t^2 - t^4 = 0.26$$
$$t^4 - t^2 + 0.26 = 0$$
Using the quadratic formula:
$$t^2 = \frac{1 \pm \sqrt{1 - 4(0.26)}}{2} = \frac{1 \pm \sqrt{1 - 1.04}}{2} = \frac{1 \pm \sqrt{-0.04}}{2}$$
Since this equation has no real solutions, there is no such value of t.

51. Set the two y-coordinates equal:
$$x = \tfrac{1}{100} x^2$$
$$0 = \tfrac{1}{100} x^2 - x$$
$$0 = \tfrac{1}{100}\left(x^2 - 100x\right)$$
$$0 = \tfrac{1}{100} x(x - 100)$$
The graphs intersect at the origin but also at the point (100, 100).

53. (a) First factor to obtain:
$$D(x) = x^2\left(1 - x^2\right) = x^2(1 + x)(1 - x)$$
So the x-intercepts are 0, –1, and 1. Sketching the graph:

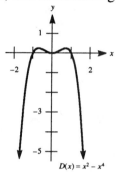

$$D(x) = x^2 - x^4$$

(b) Complete the square:
$$D(x) = x^2 - x^4 = -\left(x^4 - x^2\right) = -\left(x^4 - x^2 + \tfrac{1}{4}\right) + \tfrac{1}{4} = -\left(x^2 - \tfrac{1}{2}\right)^2 + \tfrac{1}{4}$$
The turning points are at $\left(\pm\tfrac{1}{\sqrt{2}},\tfrac{1}{4}\right) = \left(\pm\tfrac{\sqrt{2}}{2},\tfrac{1}{4}\right)$, which yield maximum values.
Note that the graph also has a turning point at $(0, 0)$, which is a minimum value.

(c) Graph the two functions:

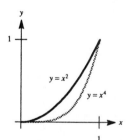

$$y = x^2$$
$$y = x^4$$

From part (b), we know that the maximum vertical distance between the two curves is $\tfrac{1}{4}$.

55. (a) Drawing a diagonal between two corners where the cylinder touches the circle yields the right triangle:

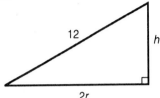

Using the Pythagorean theorem:
$$(2r)^2 + h^2 = 12^2$$
$$4r^2 + h^2 = 144$$
$$h^2 = 144 - 4r^2$$
$$h = \sqrt{144 - 4r^2} = 2\sqrt{36 - r^2}$$

(b) Compute the volume:
$$V = \pi r^2 h = \pi r^2 \left(2\sqrt{36 - r^2}\right) = 2\pi r^2 \sqrt{36 - r^2}$$

(c) We must have $36 - r^2 > 0$, so $0 < r < 6$. So the domain is $(0, 6)$.

(d) Complete the table:

r	$f(r)$
0.0	0
0.5	88
1.0	1382
1.5	6745
2.0	20213
2.5	45878
3.0	86339
3.5	140700
4.0	202129
4.5	254971
5.0	271414
5.5	207720
6.0	0

Graph $f(r) = 4\pi^2 r^4 (36 - r^2)$:

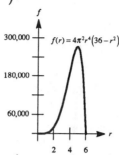

We can approximate the maximum value to be 273000. Thus the maximum possible volume is approximately $\sqrt{273000} \approx 522$ cm^3.

Graphing Utility Exercises for Section 5.6

1. (a) Graphing the curve using the standard viewing rectangle:

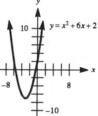

Now graphing the curve using the suggested settings:

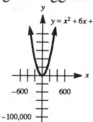

(b) Graphing the curve using the standard viewing rectangle:

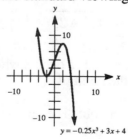

Changing the viewing rectangle and adding the curve $y = -0.25x^3$, note that the curves appear identical:

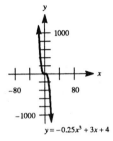

$$y = -0.25x^3 + 3x + 4$$

(c)   Graphing the curve using the suggested settings:

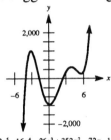

$$y = 2x^5 - 16x^4 - 26x^3 + 352x^2 - 72x - 1140$$

Adding the curve $y = 2x^5$ and using the suggested settings:

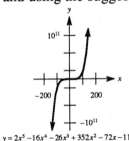

$$y = 2x^5 - 16x^4 - 26x^3 + 352x^2 - 72x - 1140$$

3.   Let $P(x, x^3)$ represent a point on the curve. The distance from this point to the point $(4, 5)$ is given by:

$$D(x) = \sqrt{(x-4)^2 + (x^3 - 5)^2}$$
$$= \sqrt{x^2 - 8x + 16 + x^6 - 10x^3 + 25}$$
$$= \sqrt{x^6 - 10x^3 + x^2 - 8x + 41}$$

The minimum point on the resulting curve occurs when $x \approx 1.738$, and $y \approx 5.250$. The point $(1.738, 5.250)$ is closest to the point $(4, 5)$.

5.   Let x be the length of the squares cut out. The box will have dimensions of $9 - 2x$ by $12 - 2x$ by x, so the volume is given by:

$$V(x) = x(9 - 2x)(12 - 2x) = x(108 - 42x + 4x^2) = 4x^3 - 42x^2 + 108x$$

The maximum point on the resulting curve occurs when $x \approx 1.697$, with a maximum possible volume of 81.872 cubic inches.

5.7 Rational Functions

1. For the domain, we must exclude those values of x where:
$$4x - 12 = 0$$
$$4x = 12$$
$$x = 3$$
So the domain is all real numbers except 3, or $(-\infty, 3) \cup (3, \infty)$.
For the x-intercepts, we must find where:
$$3x + 15 = 0$$
$$3x = -15$$
$$x = -5$$
So the x-intercept is -5. For the y-intercept, we let $x = 0$ to obtain:
$$y = \frac{0 + 15}{0 - 12} = -\frac{5}{4}$$
So the y-intercept is $-\frac{5}{4}$.

3. For the domain, we must exclude those values of x where:
$$x^2 - x - 6 = 0$$
$$(x - 3)(x + 2) = 0$$
$$x = 3, -2$$
So the domain is all real numbers except 3 and -2, or $(-\infty, -2) \cup (-2, 3) \cup (3, \infty)$.
For the x-intercepts, we must find where:
$$x^2 - 8x - 9 = 0$$
$$(x - 9)(x + 1) = 0$$
$$x = 9, -1$$
So the x-intercepts are 9 and -1. For the y-intercept, we let $x = 0$ to obtain:
$$y = \frac{-9}{-6} = \frac{3}{2}$$
So the y-intercept is $\frac{3}{2}$.

5. For the domain, we must exclude those values of x where:
$$x^6 = 0$$
$$x = 0$$
So the domain is all real numbers except 0, or $(-\infty, 0) \cup (0, \infty)$.
For the x-intercepts, we must find where:
$$\left(x^2 - 4\right)\left(x^3 - 1\right) = 0$$
$$x^2 - 4 = 0 \qquad \text{or} \qquad x^3 - 1 = 0$$
$$x^2 = 4 \qquad\qquad\qquad x^3 = 1$$
$$x = \pm 2 \qquad\qquad\qquad x = 1$$
So the x-intercepts are -2, 2, and 1. For the y-intercept, we let $x = 0$. But $x = 0$ is not in the domain of the function, so there are no y-intercepts.

7.   There is no x-intercept, the y-intercept is $\frac{1}{4}$, the horizontal asymptote is $y = 0$, and the
      vertical asymptote is $x = -4$. Sketching the graph:

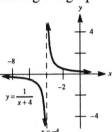

9.   There is no x-intercept, the y-intercept is $\frac{3}{2}$, the horizontal asymptote is $y = 0$, and the
      vertical asymptote is $x = -2$. Sketching the graph:

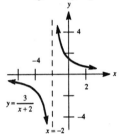

11.   There is no x-intercept, the y-intercept is $\frac{2}{3}$, the horizontal asymptote is $y = 0$, and the
      vertical asymptote is $x = 3$. Sketching the graph:

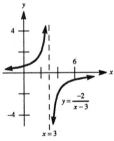

13.   Using long division, we have:
      $$\frac{x-3}{x-1} = 1 - \frac{2}{x-1}$$
      The x-intercept is 3, the y-intercept is 3, the horizontal asymptote is $y = 1$, and the
      vertical asymptote is $x = 1$. Sketching the graph:

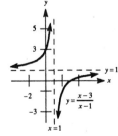

15. Using long division, we have:
$$\frac{4x-2}{2x+1} = 2 - \frac{4}{2x+1}$$
The x-intercept is $\frac{1}{2}$, the y-intercept is -2, the horizontal asymptote is $y = 2$, and the vertical asymptote is $x = -\frac{1}{2}$. Sketching the graph:

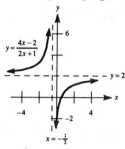

17. There is no x-intercept, the y-intercept is $\frac{1}{4}$, the horizontal asymptote is $y = 0$, and the vertical asymptote is $x = 2$. Sketching the graph:

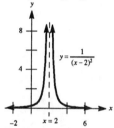

19. There is no x-intercept, the y-intercept is 3, the horizontal asymptote is $y = 0$, and the vertical asymptote is $x = -1$. Sketching the graph:

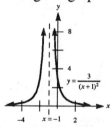

21. There is no x-intercept, the y-intercept is $\frac{1}{8}$, the horizontal asymptote is $y = 0$, and the vertical asymptote is $x = -2$. Sketching the graph:

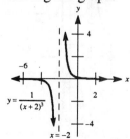

23. There is no x-intercept, the y-intercept is $-\frac{4}{125}$, the horizontal asymptote is $y = 0$, and the vertical asymptote is $x = -5$. Sketching the graph:

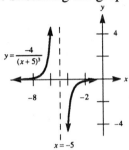

25. The x- and y-intercepts are 0, the horizontal asymptote is $y = 0$, and the vertical asymptotes are $x = -2$ and $x = 2$. Sketching the graph:

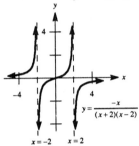

27. (a) The x-intercept is 0, the y-intercept is 0, the horizontal asymptote is $y = 0$, and the vertical asymptotes are $x = 1$ and $x = -3$. Sketching the graph:

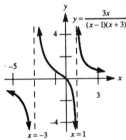

   (b) The x-intercept is 0, the y-intercept is 0, the horizontal asymptote is $y = 3$, and the vertical asymptotes are $x = 1$ and $x = -3$. Sketching the graph:

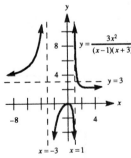

29. (a) The *x*-intercepts are 2 and 4, there is no *y*-intercept, the horizontal asymptote is *y* = 1, and the vertical asymptotes are *x* = 0 and *x* = 1. Sketching the graph:

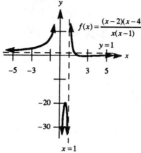

An expanded view of the graph near *x* = 3 illustrates the behavior of the curve for *x* ≥ 2:

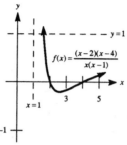

(b) The *x*-intercepts are 2 and 4, there is no *y*-intercept, the horizontal asymptote is *y* = 1, and the vertical asymptotes are *x* = 0 and *x* = 3. Sketching the graph:

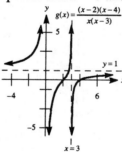

31. Finding where the curve crosses the horizontal asymptote:

$$\frac{(x-4)(x+2)}{(x-1)(x-3)} = 1$$
$$(x-4)(x+2) = (x-1)(x-3)$$
$$x^2 - 2x - 8 = x^2 - 4x + 3$$
$$-2x - 8 = -4x + 3$$
$$2x = 11$$
$$x = \tfrac{11}{2}$$

The x-intercepts are -2 and 4, the y-intercept is $-\frac{8}{3}$, the horizontal asymptote is $y = 1$, and the vertical asymptotes are $x = 1$ and $x = 3$. Sketching the graph:

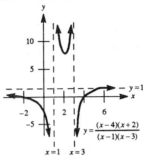

33.   Finding where the curve crosses the horizontal asymptote:

$$\frac{(x+1)^2}{(x-1)(x-3)} = 1$$
$$(x+1)^2 = (x-1)(x-3)$$
$$x^2 + 2x + 1 = x^2 - 4x + 3$$
$$2x + 1 = -4x + 3$$
$$6x = 2$$
$$x = \frac{1}{3}$$

The x-intercept is -1, the y-intercept is $\frac{1}{3}$, the horizontal asymptote is $y = 1$, and the vertical asymptotes are $x = 1$ and $x = 3$. Sketching the graph:

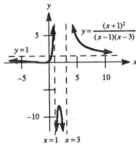

35.   (a)   For x near -2, we have:
$$y \approx \frac{(-5)(x+2)}{(-1)(-4)} = -\frac{5}{4}(x+2)$$

(b)   For x near -1, we have:
$$y \approx \frac{(-4)(1)}{(x+1)(-3)} = \frac{\frac{4}{3}}{x+1}$$

(c)   For x near 2, we have:
$$y \approx \frac{(-1)(4)}{(3)(x-2)} = \frac{-\frac{4}{3}}{x-2}$$

37. **(a)** For $x \neq -3$, we have:
$$y = \frac{x^2 - 9}{x+3} = \frac{(x+3)(x-3)}{x+3} = x - 3$$
So this is the graph of $y = x - 3$, without the point at $(-3, -6)$:

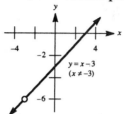

(b) For $x \neq 3$, we have:
$$y = \frac{x^2 - 5x + 6}{x^2 - 2x - 3} = \frac{(x-2)(x-3)}{(x+1)(x-3)} = \frac{x-2}{x+1}$$
So this is the graph of $y = \frac{x-2}{x+1}$, without the point at $\left(3, \frac{1}{4}\right)$:

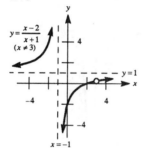

(c) For $x \neq 1, 2, 3$, we have:
$$y = \frac{(x-1)(x-2)(x-3)}{(x-1)(x-2)(x-3)(x-4)} = \frac{1}{x-4}$$
So this is the graph of $y = \frac{1}{x-4}$, without the points at $\left(1, -\frac{1}{3}\right), \left(2, -\frac{1}{2}\right), (3, -1)$:

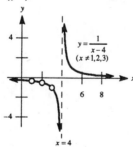

39. The horizontal asymptote is $y = 0$ and the vertical asymptote is $x = 3$. We find the value of k where $k = \frac{x}{(x-3)^2}$:
$$k(x-3)^2 = x$$
$$kx^2 - 6kx + 9k = x$$
$$kx^2 - (6k+1)x + 9k = 0$$

Since this equation must have only one solution, we set the discriminant equal to zero:

$$[-(6k+1)]^2 - 4k(9k) = 0$$
$$(6k+1)^2 - 36k^2 = 0$$
$$(6k+1+6k)(6k+1-6k) = 0$$
$$(12k+1)(1) = 0$$
$$k = -\tfrac{1}{12}$$

Thus $y = -\tfrac{1}{12}$. Find x:

$$-\tfrac{1}{12} = \frac{x}{(x-3)^2}$$
$$-(x-3)^2 = 12x$$
$$(x-3)^2 = -12x$$
$$x^2 - 6x + 9 = -12x$$
$$x^2 + 6x + 9 = 0$$
$$(x+3)^2 = 0$$
$$x = -3$$

So the low point is $\left(-3, -\tfrac{1}{12}\right)$. Sketching the graph:

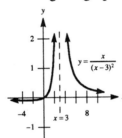

41.   (a)   Use long division:

$$
\begin{array}{r}
x+4 \\
x-3 \enclose{longdiv}{x^2+\ x-6} \\
\underline{x^2-3x} \\
4x-6 \\
\underline{4x-12} \\
6
\end{array}
$$

So $\dfrac{x^2+x-6}{x-3} = (x+4) + \dfrac{6}{x-3}$.

(b) Complete the tables:

x	$x+4$	$\dfrac{x^2+x-6}{x-3}$
10	14	14.8571
100	104	104.0619
1000	1004	1004.0060

x	$x+4$	$\dfrac{x^2+x-6}{x-3}$
−10	−6	−6.4615
−100	−96	−96.0583
−1000	−996	−996.0600

(c) The vertical asymptote is $x=3$, the x-intercepts are −3 and 2, and the y-intercept is 2.

(d) Sketching the graph:

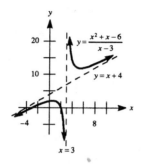

(e) Find where:

$$\frac{x^2+x-6}{x-3}=k$$
$$x^2+x-6=kx-3k$$
$$x^2+(1-k)x+(3k-6)=0$$

Setting the discriminant equal to 0, we have:

$$(1-k)^2-4(1)(3k-6)=0$$
$$1-2k+k^2-12k+24=0$$
$$k^2-14k+25=0$$
$$(k-7)^2=-25+49$$
$$(k-7)^2=24$$
$$k-7=\pm\sqrt{24}$$
$$k=7\pm2\sqrt{6}$$

So either $y=7+2\sqrt{6}$ or $y=7-2\sqrt{6}$. For each of these values, we find x:

$$\frac{x^2+x-6}{x-3}=7+2\sqrt{6}$$
$$x^2+x-6=(7+2\sqrt{6})x-21-6\sqrt{6}$$
$$x^2+(-6-2\sqrt{6})x+(15+6\sqrt{6})=0$$
$$\left(x-(3+\sqrt{6})\right)^2=0$$
$$x=3+\sqrt{6}$$

So one point is $\left(3+\sqrt{6}, 7+2\sqrt{6}\right)$. For $y = 7 - 2\sqrt{6}$, we find x:

$$\frac{x^2 + x - 6}{x - 3} = 7 - 2\sqrt{6}$$

$$x^2 + x - 6 = \left(7 - 2\sqrt{6}\right)x - 21 + 6\sqrt{6}$$

$$x^2 + \left(-6 + 2\sqrt{6}\right)x + \left(15 - 6\sqrt{6}\right) = 0$$

$$\left(x - \left(3 - \sqrt{6}\right)\right)^2 = 0$$

$$x = 3 - \sqrt{6}$$

So the other point is $\left(3 - \sqrt{6}, 7 - 2\sqrt{6}\right)$.

43. Note that:

$$\frac{-x^2 + 1}{x} = -x + \frac{1}{x}$$

Thus $y = -x$ is a slant asymptote. Sketching the graph:

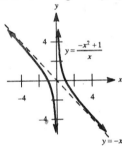

Graphing Utility Exercises for Section 5.7

1. The equation of the line can be found using the point-slope formula:

$$y - 1 = m(x - 2)$$
$$y = mx - 2m + 1$$

We now find the x- and y-intercepts. If $x = 0$, then $y = -2m + 1$, and if $y = 0$, we have:

$$0 = mx - 2m + 1$$

$$2m - 1 = mx$$

$$\frac{2m - 1}{m} = x$$

So the area of the triangle is given by:

$$A(m) = \tfrac{1}{2} \cdot (-2m + 1)\left(\frac{2m - 1}{m}\right) = \frac{-4m^2 + 4m - 1}{2m}$$

The graph of $A(m)$ appears as:

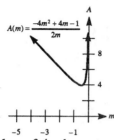

This function has a minimum value of 4 when $m \approx -0.5$.

3.   Let the coordinates of P be $P(x, x^2)$. We find the slope between P and $(0, -1)$:

$$m(x) = \frac{x^2 - (-1)}{x - 0} = \frac{x^2 + 1}{x}$$

The graph of $m(x)$ appears as:

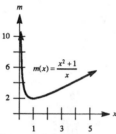

This function has a minimum value of 2 when $x \approx 1$. Thus the point P is $P(1, 1)$. We draw the parabola and the line through P and $(0, -1)$:

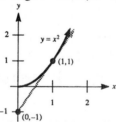

The line appears to be tangent to the parabola at $(1, 1)$.

5.   **(a)**   Since the area is 6 m^2, $xy = 6$, so $y = \frac{6}{x}$ where x is the width and y is the length. Thus the perimeter is given by:

$$P(x) = 2x + 2\left(\frac{6}{x}\right) = 2x + \frac{12}{x} = \frac{2x^2 + 12}{x}$$

(b)   The graph of $P(x)$ appears as:

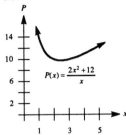

This function has a minimum value of 9.80 when $x \approx 2.45$. Substituting into $y = \frac{6}{x}$, we obtain $y \approx 2.45$. So the length and width are approximately 2.45 m.

7.   Since $a + b = 1$, $b = 1 - a$. Substituting into the quantity:

$$Q = \left(a + a^{-1}\right)^2 + \left(b + b^{-1}\right)^2$$

$$= a^2 + 2 + \frac{1}{a^2} + b^2 + 2 + \frac{1}{b^2}$$

$$= a^2 + \frac{1}{a^2} + 4 + (1-a)^2 + \frac{1}{(1-a)^2}$$

Graphing the function:

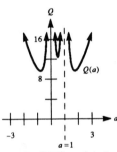

The minimum value is approximately 9.73, which occurs when $a \approx 1.75$ and $a \approx -0.75$.

9.   (a)   Graphing the function:

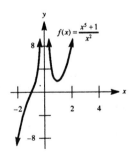

(b) Adding the curve $y = x^3$ to the graph:

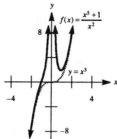

Using the viewing rectangle $-4 \le x \le 4$ and $-20 \le y \le 20$:

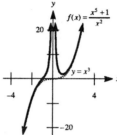

Using the viewing rectangle $-10 \le x \le 10$ and $-100 \le y \le 100$:

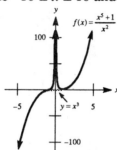

Note that as $|x|$ increases, the curve $f(x)$ approaches the curve $y = x^3$.

(c) Completing the tables:

x	5	10	50	100	500
d	0.04	0.01	0.0004	0.0001	0.000004

x	-5	-10	-50	-100	-500
d	0.04	0.01	0.0004	0.0001	0.000004

Note that the values of d decrease as $|x|$ increases.

(d) Working from the left-hand side:

$$\frac{x^5 + 1}{x^2} = \frac{x^5}{x^2} + \frac{1}{x^2} = x^3 + \frac{1}{x^2}$$

As $|x|$ increases, the quantity $\dfrac{1}{x^2}$ approaches 0, and thus $f(x) \approx x^3$ when $|x|$ gets very large.

Chapter Five Review Exercises

1. Find the slope between the points $(1, -2)$ and $(-2, -11)$:
$$m = \frac{-11 - (-2)}{-2 - 1} = \frac{-9}{-3} = 3$$
Now use the point-slope formula:
$$y - (-2) = 3(x - 1)$$
$$y + 2 = 3x - 3$$
$$y = 3x - 5$$
So $G(x) = 3x - 5$, and thus $G(0) = -5$.

3. First compute the revenue:
$$R(x) = xp = x\left(-\tfrac{1}{8}x + 100\right) = -\tfrac{1}{8}x^2 + 100x$$
Now complete the square:
$$R(x) = -\tfrac{1}{8}\left(x^2 - 800x\right) = -\tfrac{1}{8}\left(x^2 - 800x + 160000\right) + 20000 = -\tfrac{1}{8}(x - 400)^2 + 20000$$
So the maximum possible revenue is \$20,000.

5. Graph the function:

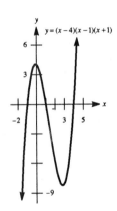

7. We have the points $(0, 1000)$ and $(5, 100)$. Find the slope:
$$m = \frac{100 - 1000}{5 - 0} = \frac{-900}{5} = -180$$
So $V(t) = -180t + 1000$.

9. The x-intercept is $-\tfrac{5}{3}$, the y-intercept is $\tfrac{5}{2}$, the horizontal asymptote is $y = 3$, and the vertical asymptote is $x = -2$. Graph the function:

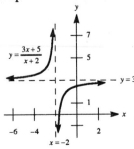

11. Graph the function:

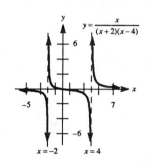

13. Draw the figure:

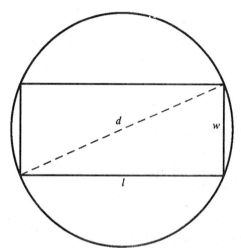

We can find r, since the circumference of the circle is 12 cm:

$$2\pi r = 12, \text{ so } r = \frac{6}{\pi} \text{ and thus } d = \frac{12}{\pi}$$

Using the Pythagorean theorem:

$$w^2 + l^2 = \left(\frac{12}{\pi}\right)^2$$

$$w^2 + l^2 = \frac{144}{\pi^2}$$

$$l^2 = \frac{144}{\pi^2} - w^2$$

$$l = \sqrt{\frac{144 - \pi^2 w^2}{\pi^2}} = \frac{\sqrt{144 - \pi^2 w^2}}{\pi}$$

Since the perimeter is given by $P = 2w + 2l$, we have:

$$P(w) = 2w + \frac{2\sqrt{144 - \pi^2 w^2}}{\pi}$$

15. Find the slope between the points $(3, 5)$ and $(-2, 0)$:
$$m = \frac{0-5}{-2-3} = \frac{-5}{-5} = 1$$
Use the point $(3, 5)$ in the point-slope formula:
$$y - 5 = 1(x - 3)$$
$$y - 5 = x - 3$$
$$y = x + 2$$
Using functional notation we have $f(x) = x + 2$.

17. Find the slope of the line $3x - 8y = 16$:
$$-8y = -3x + 16$$
$$y = \tfrac{3}{8}x - 2$$
Use $m = \tfrac{3}{8}$ and the point $(4, -1)$ in the point-slope formula:
$$y - (-1) = \tfrac{3}{8}(x - 4)$$
$$y + 1 = \tfrac{3}{8}x - \tfrac{3}{2}$$
$$y = \tfrac{3}{8}x - \tfrac{5}{2}$$
Using functional notation we have $f(x) = \tfrac{3}{8}x - \tfrac{5}{2}$.

19. If the graph of the inverse function passes through $(2, 1)$, then $(1, 2)$ must lie on the graph of the function. Find the slope between the points $(1, 2)$ and $(-3, 5)$:
$$m = \frac{5-2}{-3-1} = \frac{3}{-4} = -\tfrac{3}{4}$$
Use the point $(1, 2)$ in the point-slope formula:
$$y - 2 = -\tfrac{3}{4}(x - 1)$$
$$y - 2 = -\tfrac{3}{4}x + \tfrac{3}{4}$$
$$y = -\tfrac{3}{4}x + \tfrac{11}{4}$$
Using functional notation we have $f(x) = -\tfrac{3}{4}x + \tfrac{11}{4}$.

21. Completing the square:
$$y = x^2 + 2x - 3 = \left(x^2 + 2x + 1\right) - 4 = (x+1)^2 - 4$$
The vertex is $(-1, -4)$, the x-intercepts are 1 and -3, and the y-intercept is -3. Graphing the function:

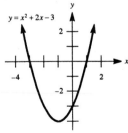

23. Completing the square:
$$y = -x^2 + 2\sqrt{3}x + 3 = -\left(x^2 - 2\sqrt{3}x\right) + 3 = -\left(x^2 - 2\sqrt{3}x + 3\right) + 6 = -\left(x - \sqrt{3}\right)^2 + 6$$
The vertex is $\left(\sqrt{3}, 6\right)$, the x-intercepts are $\sqrt{3} + \sqrt{6}$ and $\sqrt{3} - \sqrt{6}$, and the y-intercept is 3. Graphing the function:

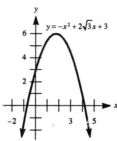

25. Completing the square:
$$y = -3x^2 + 12x = -3\left(x^2 - 4x\right) = -3\left(x^2 - 4x + 4\right) + 12 = -3(x - 2)^2 + 12$$
The vertex is (2, 12), the x-intercepts are 0 and 4, and the y-intercept is 0. Drawing the graph:

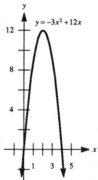

27. Find the two vertices:

$$
\begin{aligned}
y &= x^2 - 4x + 6 \\
y - 6 &= x^2 - 4x \\
y - 6 + 4 &= x^2 - 4x + 4 \\
y - 2 &= (x - 2)^2 \\
\text{vertex:} &\ (2, 2)
\end{aligned}
\qquad
\begin{aligned}
y &= -x^2 - 4x - 5 \\
y + 5 &= -\left(x^2 + 4x\right) \\
y + 5 - 4 &= -\left(x^2 + 4x + 4\right) \\
y + 1 &= -(x + 2)^2 \\
\text{vertex:} &\ (-2, -1)
\end{aligned}
$$

Now find the distance between (2, 2) and (–2, –1):
$$d = \sqrt{(-2 - 2)^2 + (-1 - 2)^2} = \sqrt{16 + 9} = \sqrt{25} = 5$$

29. Call x and y the two numbers, so $x + y = \sqrt{3}$, and thus $y = \sqrt{3} - x$. Find their product:
$$P = xy = x\left(\sqrt{3} - x\right) = -x^2 + \sqrt{3}x$$
Now complete the square:
$$P = -\left(x^2 - \sqrt{3}x\right) = -\left(x^2 - \sqrt{3}x + \tfrac{3}{4}\right) + \tfrac{3}{4} = -\left(x - \tfrac{\sqrt{3}}{2}\right)^2 + \tfrac{3}{4}$$
The maximum product is $\tfrac{3}{4}$.

31. (a) Completing the square:

$$h = -16t^2 + v_0 t$$

$$= -16\left(t^2 - \frac{v_0}{16}t\right)$$

$$= -16\left(t^2 - \frac{v_0}{16}t + \frac{v_0^2}{1024}\right) + \frac{v_0^2}{64}$$

$$= -16\left(t - \frac{v_0}{32}\right)^2 + \frac{v_0^2}{64}$$

So the maximum height of $\frac{v_0^2}{64}$ ft. is obtained when $t = \frac{v_0}{32}$ sec.

(b) The object will strike the ground when $h(t) = 0$:

$$-16t^2 + v_0 t = 0$$

$$t\left(-16t + v_0\right) = 0$$

$$t = 0 \text{ or } t = \frac{v_0}{16}$$

So the object will strike the ground when $t = \frac{v_0}{16}$ sec.

33. (a) Find the distance between $(0, 2)$ and (x, x^2):

$$d = \sqrt{(x-0)^2 + (x^2-2)^2} = \sqrt{x^2 + x^4 - 4x^2 + 4} = \sqrt{x^4 - 3x^2 + 4}$$

(b) We want to minimize d. This will occur at the same x-coordinate as the minimum of d^2. Completing the square:

$$d^2 = x^4 - 3x^2 + 4 = \left(x^4 - 3x^2 + \tfrac{9}{4}\right) + 4 - \tfrac{9}{4} = \left(x^2 - \tfrac{3}{2}\right)^2 + \tfrac{7}{4}$$

So the minimum occurs where $x^2 = \frac{3}{2}$, so $x = \pm\sqrt{\frac{3}{2}} = \frac{\pm\sqrt{6}}{2}$. Since the point is in the second quadrant, we know $x < 0$, and thus the point on the parabola is $\left(\frac{-\sqrt{6}}{2}, \frac{3}{2}\right)$.

35. Find the distance between $(2, 0)$ and $\left(x, \frac{4}{3}x + b\right)$:

$$d = \sqrt{(x-2)^2 + \left(\tfrac{4}{3}x + b\right)^2}$$

$$= \sqrt{x^2 - 4x + 4 + \tfrac{16}{9}x^2 + \tfrac{8b}{3}x + b^2}$$

$$= \sqrt{\tfrac{25}{9}x^2 + \left(\tfrac{8b}{3} - 4\right)x + \left(4 + b^2\right)}$$

Completing the square on d^2:

$$d^2 = \frac{25}{9}\left[x^2 + \frac{9}{25}\left(\frac{8b}{3} - 4\right)x\right] + \left(4 + b^2\right)$$

$$= \frac{25}{9}\left[x^2 + \frac{12}{25}(2b - 3)x\right] + \left(4 + b^2\right)$$

$$= \frac{25}{9}\left[x^2 + \frac{12}{25}(2b - 3)x + \frac{36}{625}(2b - 3)^2\right] + \left(4 + b^2\right) - \frac{4}{25}(2b - 3)^2$$

$$= \frac{25}{9}\left[x + \frac{6}{25}(2b - 3)\right]^2 + \frac{25\left(4 + b^2\right) - 4\left(4b^2 - 12b + 9\right)}{25}$$

$$= \frac{25}{9}\left[x + \frac{6}{25}(2b - 3)\right]^2 + \frac{(3b + 8)^2}{25}$$

Since the minimum distance is 5, $5 = \left|\frac{3b+8}{5}\right|$. Now solve for b:

$$|3b + 8| = 25$$

$3b + 8 = 25$	or	$3b + 8 = -25$
$3b = 17$		$3b = -33$
$b = \frac{17}{3}$		$b = -11$

37. Since $x + y = \sqrt{2}$, $y = \sqrt{2} - x$. Therefore:

$$s = x^2 + y^2 = x^2 + \left(\sqrt{2} - x\right)^2 = x^2 + 2 - 2\sqrt{2}x + x^2 = 2x^2 - 2\sqrt{2}x + 2$$

Completing the square:

$$s = 2\left(x^2 - \sqrt{2}x\right) + 2 = 2\left(x^2 - \sqrt{2}x + \frac{1}{2}\right) + 2 - 1 = 2\left(x - \frac{\sqrt{2}}{2}\right)^2 + 1$$

So the minimum value of s is 1.

39. Let x and h be the two legs. We have:

$$x^2 + h^2 = 15^2$$

$$h^2 = 225 - x^2$$

$$h = \sqrt{225 - x^2}$$

So $A = \frac{1}{2}(\text{base})(\text{height}) = \frac{1}{2}x\sqrt{225 - x^2}$. We find $A^2 = \frac{1}{4}x^2\left(225 - x^2\right) = -\frac{1}{4}x^4 + \frac{225}{4}x^2$.

Now complete the square on A^2:

$$A^2 = -\frac{1}{4}\left(x^4 - 225x^2\right) = -\frac{1}{4}\left[x^4 - 225x^2 + \left(\frac{225}{2}\right)^2\right] + \left(\frac{225}{4}\right)^2 = -\frac{1}{4}\left(x^2 - \frac{225}{2}\right)^2 + \left(\frac{225}{4}\right)^2$$

So the maximum of $A^2 = \left(\frac{225}{4}\right)^2$, thus $A = \frac{225}{4}$ cm^2.

41. Factor $f(x)$ as:

$$f(x) = x^2 - \left(a^2 + 2a\right)x + 2a^3 = \left(x - a^2\right)(x - 2a)$$

So the x-intercepts are a^2 and $2a$. Since $2a > a^2$ when $0 < a < 2$, the distance between $(a^2, 0)$ and $(2a, 0)$ will be $D = 2a - a^2 = -a^2 + 2a$. Now complete the square:

$$D = -\left(a^2 - 2a\right) = -\left(a^2 - 2a + 1\right) + 1 = -(a - 1)^2 + 1$$

So D is maximum when $a = 1$.

43.   Compute R:
$$R = xp = x\left(160 - \tfrac{1}{5}x\right) = -\tfrac{1}{5}x^2 + 160x$$
Completing the square:
$$R = -\tfrac{1}{5}\left(x^2 - 800x\right) = -\tfrac{1}{5}\left(x^2 - 800x + 160000\right) + 32000 = -\tfrac{1}{5}(x - 400)^2 + 32000$$
So $x = 400$ units will maximize the revenue. Then $p = 160 - \tfrac{1}{5}(400) = \$80$.

45.   (a)   The first five iterates are $3, 4\tfrac{1}{4}, 5\tfrac{1}{2}, 6\tfrac{1}{4},$ and $6\tfrac{3}{4}$. Using a calculator, the iterates for

   $x_0 = 2$ are:
$$x_1 = 3$$
$$x_2 \approx 4.243$$
$$x_3 \approx 5.402$$
$$x_4 \approx 6.294$$
$$x_5 \approx 6.903$$

   (b)   Finding the fixed point:
$$f(x) = x$$
$$3\sqrt{x-1} = x$$
$$9(x-1) = x^2$$
$$9x - 9 = x^2$$
$$0 = x^2 - 9x + 9$$
   Using the quadratic formula:
$$x = \frac{-(-9) \pm \sqrt{(-9)^2 - 4(1)(9)}}{2(1)} = \frac{9 \pm \sqrt{81 - 36}}{2} = \frac{9 \pm 3\sqrt{5}}{2} \approx 1.146, 7.854$$

   Choosing the larger value, the fixed point is $\dfrac{9 + 3\sqrt{5}}{2} \approx 7.854$.

47.   Since the x-coordinate of point A is a, the coordinates of P are $(a, f(a))$. Similarly, the
coordinates of point R are $(b, f(b))$. But, since $PQRS$ is a square with diagonal $y = x$,
the points P and R are reflections of one another about the line $y = x$. So their
coordinates are just interchanges of each other, thus $a = f(b)$ and $b = f(a)$.

49. The *x*-intercepts are – 4 and 2 and the *y*-intercept is – 8. Drawing the graph:

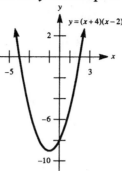

51. The *x*-intercept is –5 and the *y*-intercept is –125. Drawing the graph:

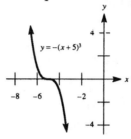

53. The *x*-intercepts are –1 and 0, and the *y*-intercept is 0. Drawing the graph:

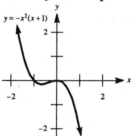

55. The *x*-intercepts are 0, 2, and –2, and the *y*-intercept is 0. Drawing the graph:

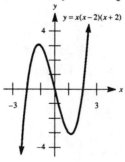

57.   The x-intercept is $-\frac{1}{3}$, there is no y-intercept, the horizontal asymptote is $y = 3$, and the vertical asymptote is $x = 0$. Drawing the graph:

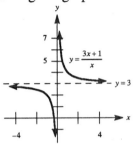

59.   There is no x-intercept, the y-intercept is -1, the horizontal asymptote is $y = 0$, and the vertical asymptote is $x = 1$. Drawing the graph:

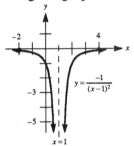

61.   The x-intercept is 2, the y-intercept is $\frac{2}{3}$, the horizontal asymptote is $y = 1$, and the vertical asymptote is $x = 3$. Drawing the graph:

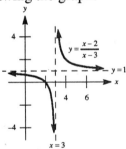

63.   The x-intercept is 1, the y-intercept is $\frac{1}{4}$, the horizontal asymptote is $y = 1$, and the vertical asymptote is $x = 2$. Drawing the graph:

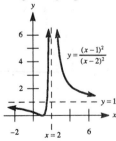

65. (a) If $b = 1$, then $f(x) = x^2 + 2x + 1 = (x + 1)^2$, so the vertex is $(-1, 0)$. This point is 1 unit from the origin.

(b) If $b = 2$, then $f(x) = x^2 + 4x + 1$. Completing the square:
$$f(x) = (x^2 + 4x + 4) + 1 - 4 = (x + 2)^2 - 3$$
So the vertex is $(-2, -3)$. By the distance formula:
$$d = \sqrt{(-2 - 0)^2 + (-3 - 0)^2} = \sqrt{4 + 9} = \sqrt{13}$$

(c) Completing the square:
$$f(x) = x^2 + 2bx + 1 = (x^2 + 2bx + b^2) + 1 - b^2 = (x + b)^2 + 1 - b^2$$
So the vertex is $(-b, 1 - b^2)$. By the distance formula:
$$d = \sqrt{(-b)^2 + (1 - b^2)^2} = \sqrt{b^2 + 1 - 2b^2 + b^4} = \sqrt{b^4 - b^2 + 1}$$
Now complete the square on d^2:
$$d^2 = b^4 - b^2 + 1 = \left(b^4 - b^2 + \tfrac{1}{4}\right) + 1 - \tfrac{1}{4} = \left(b^2 - \tfrac{1}{2}\right)^2 + \tfrac{3}{4}$$
So d^2 (and thus d) is minimized when $b^2 = \tfrac{1}{2}$, thus $b = \pm\tfrac{1}{\sqrt{2}} = \tfrac{\pm\sqrt{2}}{2}$.

67. We need to find the vertex of the parabola. Completing the square:
$$y = x^2 - 2x + k$$
$$y - k = x^2 - 2x$$
$$y - k + 1 = x^2 - 2x + 1$$
$$y = (x - 1)^2 + (k - 1)$$
Since the vertex is $(1, k - 1)$ and the parabola is opening upward, $k - 1 = 5$, so $k = 6$.

69. Solve for x:
$$y = \frac{(x - 1)(x - 3)}{x - 4}$$
$$y(x - 4) = (x - 1)(x - 3)$$
$$yx - 4y = x^2 - 4x + 3$$
$$0 = x^2 - (4 + y)x + (4y + 3)$$
Using the quadratic formula:
$$x = \frac{4 + y \pm \sqrt{(4 + y)^2 - 4(4y + 3)}}{2}$$
$$= \frac{4 + y \pm \sqrt{16 + 8y + y^2 - 16y - 12}}{2}$$
$$= \frac{4 + y \pm \sqrt{y^2 - 8y + 4}}{2}$$
So we must make sure that $y^2 - 8y + 4 \geq 0$.
Find the key numbers by using the quadratic formula:
$$y = \frac{8 \pm \sqrt{64 - 16}}{2} = \frac{8 \pm 4\sqrt{3}}{2} = 4 \pm 2\sqrt{3}$$

From a sign chart, we see that the range is $y \le 4 - 2\sqrt{3}$ or $y \ge 4 + 2\sqrt{3}$. We write this as $(-\infty, 4 - 2\sqrt{3}] \cup [4 + 2\sqrt{3}, \infty)$.

71. First draw a figure:

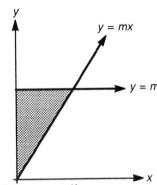

Find the intersection point of these two lines:

$$mx = m$$
$$x = 1$$
$$y = m$$

So the point is $(1, m)$. Since these are the base and height, respectively, of the triangle, we have:

$$A = \tfrac{1}{2}(1)(m) = \frac{m}{2}$$

Using functional notation we have $A(m) = \dfrac{m}{2}$.

73. Re-draw the figure and label essential parts:

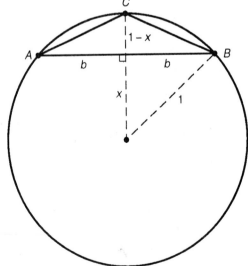

By the labeled parts of the figure, we have:

$$x^2 + b^2 = 1^2, \text{ so } b = \sqrt{1 - x^2}$$

Thus the base of the triangle is $2b = 2\sqrt{1-x^2}$ and the height is $1-x$, thus the area is given by:

$$A = \tfrac{1}{2} \cdot 2\sqrt{1-x^2} \cdot (1-x) = (1-x)\sqrt{1-x^2}$$

Using functional notation we have $A(x) = (1-x)\sqrt{1-x^2}$.

75. (a) Assume $f(x_0+x) = f(x_0-x)$. Then:

$$a(x_0+x)^2 + b(x_0+x) + c = a(x_0-x)^2 + b(x_0-x) + c$$
$$ax_0^2 + 2ax_0x + ax^2 + bx_0 + bx = ax_0^2 - 2ax_0x + ax^2 + bx_0 - bx$$
$$4ax_0x + 2bx = 0$$
$$2ax_0x + bx = 0$$
$$x(2ax_0 + b) = 0$$
$$x = 0 \quad \text{or} \quad 2ax_0 + b = 0$$
$$x_0 = -\frac{b}{2a}$$

We can discard $x = 0$ since the relationship must hold for all x, thus $x_0 = -\dfrac{b}{2a}$.

(b) Since $x_0 = -\dfrac{b}{2a}$ is the x-coordinate of the vertex, the identity $f(x_0+x) = f(x_0-x)$ shows that the graph of f is symmetric about the vertical line $x = -\dfrac{b}{2a}$, which passes through the vertex.

Chapter Five Test

1. First find the slope between $(-2, -4)$ and $(5, 1)$:
$$m = \frac{-4-1}{-2-5} = \frac{-5}{-7} = \tfrac{5}{7}$$
Using the point-slope formula, we have:
$$y - 1 = \tfrac{5}{7}(x-5)$$
$$y - 1 = \tfrac{5}{7}x - \tfrac{25}{7}$$
$$y = \tfrac{5}{7}x - \tfrac{18}{7}$$
So $L(x) = \tfrac{5}{7}x - \tfrac{18}{7}$, and thus $L(0) = -\tfrac{18}{7}$.

2. (a) Completing the square:
$$F(x) = -2x^2 + 4x = -2(x^2 - 2x) = -2(x^2 - 2x + 1) + 2 = -2(x-1)^2 + 2$$
So the graph of $F(x)$ is a parabola pointing downward with vertex $(1, 2)$, and thus it is increasing on the interval $(-\infty, 1)$. The maximum value of $F(x)$ is 2.

(b) Since $9t^4 \geq 0$ and $6t^2 \geq 0$, $G(t)$ will achieve a minimum value of 2 when $t = 0$.

3. (a) The x-intercepts are 3 and -4, and the y-intercept is -48. The signs of $f(x)$ are given by:

Interval	$f(x) = (x-3)(x+4)^2$
$(-\infty, -4)$	neg.
$(-4, 3)$	neg.
$(3, \infty)$	pos.

Sketch the excluded regions for $f(x)$:

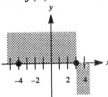

(b) For $x \approx -4$, we have $f(x) \approx (-4-3)(x+4)^2 = -7(x+4)^2$. So $f(x)$ is approximately $-7(x+4)^2$ when x is close to -4. Sketch $y = -7(x+4)^2$:

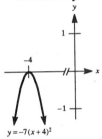

$$y = -7(x+4)^2$$

(c) Graph the function:

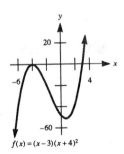

$$f(x) = (x-3)(x+4)^2$$

4. (a) Sketch the graph:

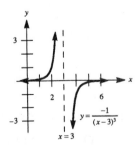

$$y = \frac{-1}{(x-3)^3}$$

$$x = 3$$

(b)   Sketch the graph:

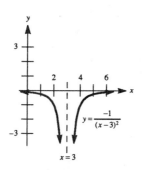

5.   First complete the square:

$$y = -x^2 + 7x + 6 = -\left(x^2 - 7x\right) + 6 = -\left(x^2 - 7x + \tfrac{49}{4}\right) + \tfrac{73}{4} = -\left(x - \tfrac{7}{2}\right)^2 + \tfrac{73}{4}$$

The turning point (vertex) is $\left(\tfrac{7}{2}, \tfrac{73}{4}\right)$, the x-intercepts are $\dfrac{7 \pm \sqrt{73}}{2}$, the y-intercept is 6, and the axis of symmetry is $x = \tfrac{7}{2}$. Draw the graph:

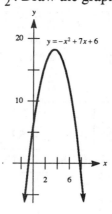

6.   Set up the revenue function:

$$R = xp = x\left(-\tfrac{1}{6}x + 80\right) = -\tfrac{1}{6}x^2 + 80x$$

Now complete the square:

$$R = -\tfrac{1}{6}\left(x^2 - 480x\right) = -\tfrac{1}{6}\left(x^2 - 480x + 57600\right) + 9600 = -\tfrac{1}{6}\left(x - 240\right)^2 + 9600$$

The maximum possible revenue is $9600, which occurs when $x = 240$.
Thus $p = -\tfrac{1}{6}(240) + 80 = \$40 \,/\, \text{unit}$.

7.   Find the slope between the points (0, 14000) and (10, 750):

$$m = \frac{750 - 14000}{10 - 0} = \frac{-13250}{10} = -1325$$

So the value of the machine is $V(t) = -1325t + 14000$.

8.   (a) Finding the fixed points:

$$f(x) = x$$
$$2x(1-x) = x$$
$$2x - 2x^2 = x$$
$$x - 2x^2 = 0$$
$$x(1-2x) = 0$$
$$x = 0, \tfrac{1}{2}$$

     (b) Finding the fixed points:

$$g(x) = x$$
$$\frac{4x-1}{3x+6} = x$$
$$4x - 1 = 3x^2 + 6x$$
$$0 = 3x^2 + 2x + 1$$

Using the quadratic formula:

$$x = \frac{-2 \pm \sqrt{(2)^2 - 4(3)(1)}}{2(3)} = \frac{-2 \pm \sqrt{4-12}}{6} = \frac{-2 \pm \sqrt{-8}}{6}$$

Since there are no real solutions, there are no fixed points for the function.

9.   (a) Completing the table:

	x_1	x_2	x_3	x_4	x_5	x_6
From graph	0.56	0.28	0.52	0.33	0.49	0.36
From calculator	0.56	0.286	0.518	0.332	0.490	0.360

     (b) Finding the fixed points:

$$f(x) = x$$
$$0.6 - x^2 = x$$
$$0 = x^2 + x - 0.6$$
$$0 = 10x^2 + 10x - 6$$
$$0 = 5x^2 + 5x - 3$$

Using the quadratic formula:

$$x = \frac{-5 \pm \sqrt{(5)^2 - 4(5)(-3)}}{2(5)} = \frac{-5 \pm \sqrt{25+60}}{10} = \frac{-5 \pm \sqrt{85}}{10}$$

The fixed points are $\dfrac{-5 - \sqrt{85}}{10} \approx -1.4220$ and $\dfrac{-5 + \sqrt{85}}{10} \approx 0.4220$. The iterates from part (a) are approaching 0.4220.

(c) Computing the first six iterates of $x_0 = 1$:

$$x_1 = -0.4$$
$$x_2 = 0.44$$
$$x_3 = 0.4064$$
$$x_4 \approx 0.4348$$
$$x_5 \approx 0.4109$$
$$x_6 \approx 0.4311$$

Yes, these iterates are approaching the fixed point 0.4220.

(d) Computing the first six iterates of $x_0 = 2$:

$$x_1 = -3.4$$
$$x_2 = -10.96$$
$$x_3 \approx -119.52$$
$$x_4 \approx -14285$$
$$x_5 \approx -204055878$$
$$x_6 \approx -4.16 \times 10^{16}$$

The iterates are not approaching either of the fixed points determined in part (b).

10. Since $y = -\frac{1}{2}(3 - x)^3$ will be a reflection of $y = -\frac{1}{2}(x + 3)^3$ across the y-axis, we have the graph:

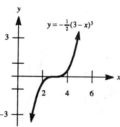

The x-intercept is 3 and the y-intercept is $-\frac{27}{2}$.

11. The x-intercept is $\frac{3}{2}$, the y-intercept is -3, the vertical asymptote is $x = -1$, and the horizontal asymptote is $y = 2$. Draw the graph:

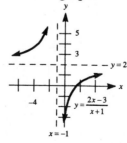

12. (a) Using the distance formula, we have:
$$L(x) = \sqrt{(x+1)^2 + (y-3)^2}$$
$$= \sqrt{(x+1)^2 + (3x-1-3)^2}$$
$$= \sqrt{x^2 + 2x + 1 + 9x^2 - 24x + 16}$$
$$= \sqrt{10x^2 - 22x + 17}$$

(b) Complete the square inside the radical:
$$10x^2 - 22x + 17 = 10(x^2 - 2.2x) + 17$$
$$= 10(x^2 - 2.2x + 1.21) + 4.9$$
$$= 10(x - 1.1)^2 + 4.9$$

So the length will be a minimum when L^2 is a minimum, which occurs when $x = 1.1$.

13. (a) The vertical asymptotes are $x = 3$ and $x = -3$, and the horizontal asymptote is $y = 1$.

(b) Near $x = 0, f(x)$ will look like:
$$y = \frac{x(0-2)}{0-9} = \tfrac{2}{9}x$$
Near $x = 2, f(x)$ will look like:
$$y = \frac{2(x-2)}{4-9} = -\tfrac{2}{5}(x-2)$$
Sketch the graph:

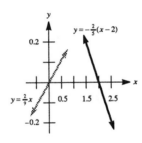

(c) Near $x = 3, f(x)$ will look like:
$$y = \frac{3(3-2)}{(3+3)(x-3)} = \frac{1}{2(x-3)}$$
Near $x = -3, f(x)$ will look like:
$$y = \frac{-3(-3-2)}{(x+3)(-3-3)} = -\frac{5}{2(x+3)}$$

Sketch the graph:

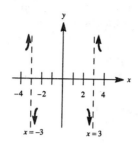

(d) Graph the function:

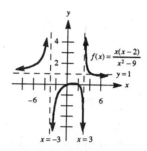

14. Let w and l represent the width and length, respectively. Draw the figure:

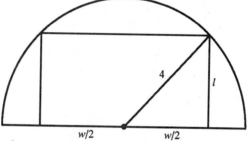

From the Pythagorean theorem, we have:

$$\left(\frac{w}{2}\right)^2 + l^2 = 4^2$$

$$\frac{w^2}{4} + l^2 = 16$$

$$l^2 = \frac{64 - w^2}{4}$$

$$l = \frac{\sqrt{64 - w^2}}{2}$$

So the area of the rectangle is given by:

$$A(w) = w \cdot l = w \cdot \frac{\sqrt{64 - w^2}}{2} = \tfrac{1}{2} w \sqrt{64 - w^2}$$

15.   (a)   As $|x|$ increases in size for x positive, $-x^3$ increases in the negative direction, which the pictured function does not.

      (b)   This graph has four turning points, and a polynomial function with highest degree term $-x^3$ can have at most two turning points.

16.   (a)   Sketching the points and the regression line:

      (b)   When $x = 1991$, we have $y = 32.6$. The estimate is \$32.6 billion. Note that the estimate is too high, compared to the actual value of \$25.5 billion.

Chapter Six
Exponential and Logarithmic Functions

6.1 Exponential Functions

1. (a) Since $2^{10} \approx 10^3$, $2^{30} = \left(2^{10}\right)^3 \approx \left(10^3\right)^3 = 10^9$.

(b) Since $2^{10} \approx 10^3$, $2^{50} = \left(2^{10}\right)^5 \approx \left(10^3\right)^5 = 10^{15}$.

3. Using the property $\left(b^m\right)^n = b^{mn}$:

$$\left(5^{\sqrt{3}}\right)^{\sqrt{3}} = 5^{\sqrt{3}\cdot\sqrt{3}} = 5^3 = 125$$

5. Using the property $b^n b^m = b^{n+m}$:

$$\left(4^{1+\sqrt{2}}\right)\left(4^{1-\sqrt{2}}\right) = 4^{1+\sqrt{2}+1-\sqrt{2}} = 4^2 = 16$$

7. Using the property $\dfrac{b^m}{b^n} = b^{m-n}$:

$$\frac{2^{4+\pi}}{2^{1+\pi}} = 2^{4+\pi-1-\pi} = 2^3 = 8$$

9.   Using the property $\left(b^m\right)^n = b^{mn}$:

$$\left(\sqrt{5}^{\sqrt{2}}\right)^2 = \sqrt{5}^{2\sqrt{2}} = \left(5^{1/2}\right)^{2\sqrt{2}} = 5^{\sqrt{2}}$$

11.   (a) Solve for x:
$$3^x = 27$$
$$3^x = 3^3$$
$$x = 3$$

(b) Solve for t:
$$9^t = 27$$
$$\left(3^2\right)^t = 3^3$$
$$3^{2t} = 3^3$$
$$2t = 3$$
$$t = \tfrac{3}{2}$$

(c) Solve for y:
$$3^{1-2y} = \sqrt{3}$$
$$3^{1-2y} = 3^{1/2}$$
$$1 - 2y = \tfrac{1}{2}$$
$$-2y = -\tfrac{1}{2}$$
$$y = \tfrac{1}{4}$$

(d) Solve for z:
$$3^z = 9\sqrt{3}$$
$$3^z = 3^2 \cdot 3^{1/2}$$
$$3^z = 3^{5/2}$$
$$z = \tfrac{5}{2}$$

13.   The domain is all real numbers, or $(-\infty, \infty)$.

15.   Since $2^{x-1} \neq 0$ (even if $x = 1$), the domain is all real numbers, or $(-\infty, \infty)$.

17.   Graph $y = 2^x$ and $y = 2^{-x}$:

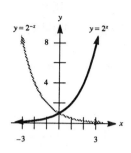

19.   Graph $y = 3^x$ and $y = -3^x$ (note that $y = -3^x$ is a reflection of $y = 3^x$ about the x-axis):

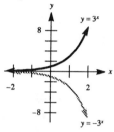

21.   Graph $y = 2^x$ and $y = 3^x$:

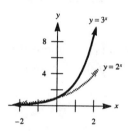

23.   Graph $y = \left(\frac{1}{2}\right)^x = 2^{-x}$ and $y = \left(\frac{1}{3}\right)^x = 3^{-x}$:

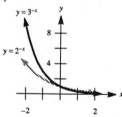

25.   The domain is $(-\infty, \infty)$, the range is $(-\infty, 1)$, the x- and y-intercepts are 0, and the asymptote is $y = 1$. Graphing the function:

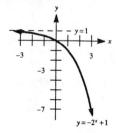

27. The domain is $(-\infty, \infty)$, the range is $(1, \infty)$, there is no x-intercept, the y-intercept is 2, and the asymptote is $y = 1$. Graphing the function:

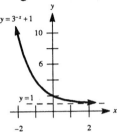

29. The domain is $(-\infty, \infty)$, the range is $(0, \infty)$, there is no x-intercept, the y-intercept is $\frac{1}{2}$, and the asymptote is $y = 0$. Graphing the function:

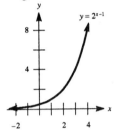

31. The domain is $(-\infty, \infty)$, the range is $(1, \infty)$, there is no x-intercept, the y-intercept is 4, and the asymptote is $y = 1$. Graphing the function:

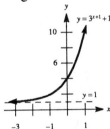

33. Solve for x:
$$3x(10^x) + 10^x = 0$$
$$10^x(3x + 1) = 0$$
$$x = -\tfrac{1}{3} \ (\text{since } 10^x \neq 0)$$

35. Solve for x:
$$3(3^x) - 5x(3^x) + 2x^2(3^x) = 0$$
$$3^x(3 - 5x + 2x^2) = 0$$
$$3^x(3 - 2x)(1 - x) = 0$$
$$x = \tfrac{3}{2}, 1 \ (\text{since } 3^x \neq 0)$$

37. (a) Graphing the two functions:

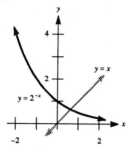

(b) The fixed point lies between 0 and 1.

(c) Solving the quadratic equation:
$$\tfrac{1}{8}x^2 - \tfrac{5}{8}x + 1 = x$$
$$x^2 - 5x + 8 = 8x$$
$$x^2 - 13x + 8 = 0$$
Using the quadratic formula:
$$x = \frac{-(-13) \pm \sqrt{(-13)^2 - 4(1)(8)}}{2(1)} = \frac{13 \pm \sqrt{169 - 32}}{2} = \frac{13 \pm \sqrt{137}}{2}$$
The two roots are $\dfrac{13 - \sqrt{137}}{2} \approx 0.648$ and $\dfrac{13 + \sqrt{137}}{2} \approx 12.352$. Based on our

answer from part (b), the root is $\dfrac{13 - \sqrt{137}}{2} \approx 0.648$.

(d) The approximation from part (c) agrees with this to two decimal places.

39. Simplify the difference quotient:
$$\frac{f(x+h) - f(x)}{h} = \frac{2^{x+h} - 2^x}{h} = \frac{2^x 2^h - 2^x}{h} = \frac{2^x(2^h - 1)}{h} = 2^x\left(\frac{2^h - 1}{h}\right)$$

41. (a) We know the graph of $y = 2^x$ and the graph of g, the inverse of f, should contain these points with the x and y coordinates interchanged. So $g(x)$ will be the reflection of $f(x)$ across the line $y = x$:

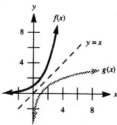

(b) The domain is $(0, \infty)$, the range is $(-\infty, \infty)$, the x-intercept is 1, there is no y-intercept, and the asymptote is $x = 0$.

43. (a) From the graph, we have $2^{1/2} \approx 1.4$.
  (b) Using a calculator, we have $\sqrt{2} \approx 1.41$.

45. (a) From the graph, we have $2^{3/5} = 2^{0.6} \approx 1.5$.
  (b) Using a calculator, we have $2^{3/5} \approx 1.52$.

47. (a) From the graph, we have $\sqrt{3} = 3^{1/2} \approx 1.7$.
  (b) Using a calculator, we have $\sqrt{3} \approx 1.73$.

49. (a) From the graph, we have $5^{3/10} = 5^{0.3} \approx 1.6$.
  (b) Using a calculator, we have $5^{3/10} \approx 1.62$.

51. From the graph, we have $x \approx 0.3$.

53. From the graph, we have $x \approx 0.7$.

55. (a) Since $10^0 = 1$, the entry in the table corresponding to $x = 1$ is 0. Since $10^1 = 10$, the entry corresponding to $x = 10$ is 1.

  (b) Since $10^{0.3} \approx 2$, we have $\left(10^{0.3}\right)^2 \approx 2^2$. That is, $10^{0.6} \approx 4$. Therefore the entry in the table corresponding to $x = 4$ is 0.6. Similarly, by cubing both sides of the approximation $10^{0.3} \approx 2$, we obtain $10^{0.9} \approx 8$. Thus the entry in the table corresponding to $x = 8$ is 0.9.

  (c) Using the hint that is given, we have $5 \approx \dfrac{10}{10^{0.3}} = 10^{0.7}$. Thus, the entry in the table corresponding to $x = 5$ is 0.7.

  (d) We have $7^2 \approx (5)(10) \approx \left(10^{0.7}\right)\left(10^1\right) = 10^{1.7}$. Therefore, $\left(7^2\right)^{1/2} \approx \left(10^{1.7}\right)^{1/2}$, or $7 \approx 10^{0.85}$. Thus the power to which 10 must be raised to yield 7 is approximately 0.85.

  (e) We have $3^4 \approx (8)(10) \approx \left(10^{0.9}\right)\left(10^1\right) = 10^{1.9}$. Therefore $\left(3^4\right)^{1/4} \approx \left(10^{1.9}\right)^{1/4} \approx 10^{0.475}$. Thus the power to which 10 must be raised to yield 3 is approximately 0.48.

  (f) We have $6 = (2)(3) \approx \left(10^{0.3}\right)\left(10^{0.48}\right) = 10^{0.78}$. Thus the power to which 10 must be raised to yield 6 is approximately 0.78. We also have:
  $$9 = (3)(3) \approx \left(10^{0.475}\right)\left(10^{0.475}\right) = 10^{0.95}$$
  Thus the power to which 10 must be raised to yield 9 is approximately 0.95.

Complete the table:

x	$\log_{10} x$
1	0.00
2	0.30
3	0.48
4	0.60
5	0.70
6	0.78
7	0.85
8	0.90
9	0.95
10	1.00

Graphing Utility Exercises for Section 6.1

1. Graphing the function and the line:

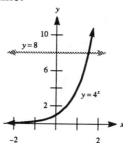

The x-coordinate of the intersection point is $x = 1.5$, which matches the value given in the text.

3. Graphing the equation:

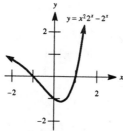

The x-intercepts are $x = -1$ and $x = 1$. These are the solutions to the equation $x^2 2^x - 2^x = 0$ in Example 6.

5. (a) There is no x-intercept, the y-intercept is $-\frac{1}{9}$, and the asymptote is $y = 0$.

    (b) Graphing the function:

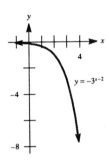

7. (a) The x-intercept is -1, the y-intercept is -3, and the asymptote is $y = -4$.

    (b) Graphing the function:

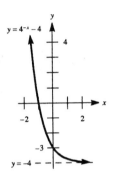

9. (a) There is no x-intercept, the y-intercept is $\frac{1}{10}$, and the asymptote is $y = 0$.

    (b) Graphing the function:

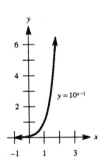

11. (a) Drawing the graph:

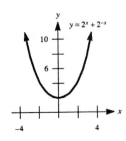

(b)   Adding $y = x^2$, draw the graph:

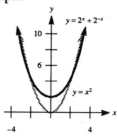

(c)   Sketching the graph of both functions:

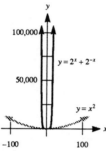

13.   (a)   Graphing the two functions:

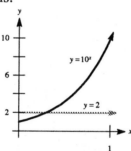

(b)   The point $(0.3, 2)$ is the approximate point of intersection, and thus $10^{0.3} \approx 2$.

6.2   The Exponential Function $y = e^x$

1.   The domain is $(-\infty, \infty)$, the range is $(0, \infty)$, there is no x-intercept, the y-intercept is 1, and the asymptote is $y = 0$. Drawing the graph:

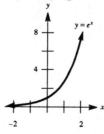

3.   The domain is $(-\infty, \infty)$, the range is $(-\infty, 0)$, there is no x-intercept, the y-intercept is -1, and the asymptote is $y = 0$. Drawing the graph:

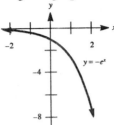

5.   The domain is $(-\infty, \infty)$, the range is $(1, \infty)$, there is no x-intercept, the y-intercept is 2, and the asymptote is $y = 1$. Drawing the graph:

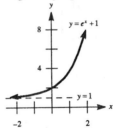

7.   The domain is $(-\infty, \infty)$, the range is $(1, \infty)$, there is no x-intercept, the y-intercept is $e + 1$, and the asymptote is $y = 1$. Drawing the graph:

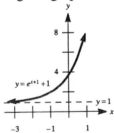

9.   The domain is $(-\infty, \infty)$, the range is $(-\infty, 0)$, there is no x-intercept, the y-intercept is $-\frac{1}{e^2}$, and the asymptote is $y = 0$. Drawing the graph:

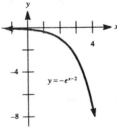

11. The domain is $(-\infty, \infty)$, the range is $(-\infty, e)$, the x-intercept is 1, the y-intercept is $e - 1$, and the asymptote is $y = e$. Drawing the graph:

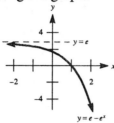

13. Graphing the three functions:

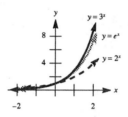

15. False (since $e \approx 2.7$)

17. False (since $\sqrt{2.7} > 1$)

19. True

21. False (since $2.7^{-1} > 0$)

23. From the graph, we have $e^{0.1} \approx 1.1$. Using a calculator, we have $e^{0.1} \approx 1.105$.

25. From the graph, we have $e^{-0.3} \approx 0.75$. Using a calculator, we have $e^{-0.3} \approx 0.741$.

27. From the graph, we have $e^{-1} \approx 0.35$. Using a calculator, we have $e^{-1} \approx 0.368$.

29. From the graph, we have $\dfrac{1}{\sqrt{e}} = e^{-0.5} \approx 0.6$. Using a calculator, we have

$$\frac{1}{\sqrt{e}} = e^{-0.5} \approx 0.607.$$

31. (a) When $y = 1.5$, the value of x is $x \approx 0.4$.
    (b) The value is $\ln 1.5 \approx 0.405$, which is consistent with the answer from part (a).

33. (a) When $y = 1.8$, the value of x is $x \approx 0.6$.
    (b) The value is $\ln 1.8 \approx 0.588$, which is consistent with the answer from part (a).

35. (a) On the interval [2, 3]:

$$\frac{\Delta f}{\Delta x} = \frac{f(3) - f(2)}{3 - 2} = \frac{3^4 - 2^4}{3 - 2} = \frac{81 - 16}{1} = 65$$

$$\frac{\Delta g}{\Delta x} = \frac{g(3) - g(2)}{3 - 2} = \frac{2^3 - 2^2}{3 - 2} = \frac{8 - 4}{1} = 4$$

$$\frac{\Delta h}{\Delta x} = \frac{h(3) - h(2)}{3 - 2} = \frac{e^3 - e^2}{1} \approx 12.70$$

Note that $\Delta f / \Delta x$ is 5 times as large as $\Delta h / \Delta x$.

(b) On the interval [8, 9]:

$$\frac{\Delta f}{\Delta x} = \frac{f(9) - f(8)}{9 - 8} = \frac{9^4 - 8^4}{9 - 8} = \frac{6561 - 4096}{1} = 2465$$

$$\frac{\Delta g}{\Delta x} = \frac{g(9) - g(8)}{9 - 8} = \frac{2^9 - 2^8}{9 - 8} = \frac{512 - 256}{1} = 256$$

$$\frac{\Delta h}{\Delta x} = \frac{h(9) - h(8)}{9 - 8} = \frac{e^9 - e^8}{1} \approx 5122.13$$

Note that $\Delta h / \Delta x$ is more than twice as large as $\Delta f / \Delta x$.

(c) On the interval [14, 15]:

$$\frac{\Delta f}{\Delta x} = \frac{f(15) - f(14)}{15 - 14} = \frac{15^4 - 14^4}{15 - 14} = \frac{50625 - 38416}{1} = 12,209$$

$$\frac{\Delta g}{\Delta x} = \frac{g(15) - g(14)}{15 - 14} = \frac{2^{15} - 2^{14}}{15 - 14} = \frac{32768 - 16384}{1} = 16,384$$

$$\frac{\Delta h}{\Delta x} = \frac{h(15) - h(14)}{15 - 14} = \frac{e^{15} - e^{14}}{1} \approx 2,066,413$$

Note that $\Delta h / \Delta x$ is 126 times as large as $\Delta g / \Delta x$.

37. (a) Based on the graph, $e^\pi \approx 23.4$ while $\pi^e \approx 22.2$, so e^π is slightly larger.
(b) Since $e^\pi \approx 23.14$ while $\pi^e \approx 22.46$, e^π is larger.

39. (a) Completing the table:

	x_1	x_2	x_3	x_4	x_5
From graph	0.37	0.70	0.50	0.60	0.55
From calculator	0.3679	0.6922	0.5005	0.6062	0.5454

(b) Computing additional iterates:

$x_6 \approx 0.5796$

$x_7 \approx 0.5601$

$x_8 \approx 0.5711$

$x_9 \approx 0.5649$

The seventh iterate is the first iterate to have 56 in the first two decimal places. The next iterate (the eighth iterate) does not begin with 56 in the first two decimal places, however the iterate after that (the ninth iterate) does have a 56 in the first two decimal places.

(c) No. Every input appears to produce iterates which converge to the fixed point from part (b).

41. (a) The domain is all real numbers, or $(-\infty, \infty)$.

(b) Evaluate when $x = 0$:

$$C(0) = \frac{e^0 + e^0}{2} = \frac{1+1}{2} = 1$$

(c) Compute $C(-x)$:

$$C(-x) = \frac{e^{-x} + e^x}{2} = C(x)$$

The graph must be symmetric about the y-axis.

(d) Drawing the graph:

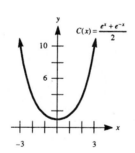

43. (a) Complete the table:

x	1000	10^4	10^5	10^6	10^7
$\left(1+\frac{2}{x}\right)^x$	7.374312	7.387579	7.388908	7.389041	7.389055

(b) The value of e^2 is approximately 7.389056. Note that the values in the table are approaching e^2.

45. (a) Draw the graphs, noting that $L(x)$ will be the reflection of $f(x)$ across $y = x$:

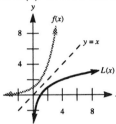

   (b) The domain is $(0, \infty)$, the range is $(-\infty, \infty)$, the x-intercept is 1, and the asymptote is $x = 0$.

   (c) (i) This will be a reflection of $L(x)$ across the x-axis. The x-intercept is 1 and the asymptote is $x = 0$:

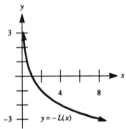

       (ii) This will be a reflection of $L(x)$ across the y-axis. The x-intercept is -1 and the asymptote is $x = 0$:

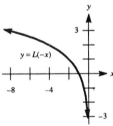

       (iii) This will be a displacement of $L(x)$ to the right one unit. The x-intercept is 2 and the asymptote is $x = 1$:

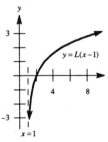

47. The function possesses properties A, D, E, and G.

49. The function possesses properties B, D, E, and G.

51.   The function possesses properties A, D, F, G, and H.

53.   The function possesses properties A, D, F, G, and H.

55.   (a)   First compute $[C(x)]^2$ and $[S(x)]^2$:

$$[C(x)]^2 = \left(\frac{e^x + e^{-x}}{2}\right)^2 = \frac{e^{2x} + 2 + e^{-2x}}{4}$$

$$[S(x)]^2 = \left(\frac{e^x - e^{-x}}{2}\right)^2 = \frac{e^{2x} - 2 + e^{-2x}}{4}$$

Therefore:

$$[C(x)]^2 - [S(x)]^2 = \frac{e^{2x} + 2 + e^{-2x} - e^{2x} + 2 - e^{-2x}}{4} = \frac{4}{4} = 1$$

(b)   Working from the left-hand side, we have:

$$S(-x) = \frac{e^{-x} - e^{-(-x)}}{2} = \frac{e^{-x} - e^x}{2} = -\frac{e^x - e^{-x}}{2} = -S(x)$$

(c)   Working from the left-hand side, we have:

$$C(-x) = \frac{e^{-x} + e^{-(-x)}}{2} = \frac{e^{-x} + e^x}{2} = \frac{e^x + e^{-x}}{2} = C(x)$$

(d)   Working from the right-hand side, we have:

$$S(x)C(y) + C(x)S(y)$$

$$= \left(\frac{e^x - e^{-x}}{2}\right)\left(\frac{e^y + e^{-y}}{2}\right) + \left(\frac{e^x + e^{-x}}{2}\right)\left(\frac{e^y - e^{-y}}{2}\right)$$

$$= \frac{e^{x+y} - e^{-x+y} + e^{x-y} - e^{-x-y} + e^{x+y} + e^{-x+y} - e^{x-y} - e^{-x-y}}{4}$$

$$= \frac{2e^{x+y} - 2e^{-x-y}}{4}$$

$$= \frac{e^{x+y} - e^{-(x+y)}}{2}$$

$$= S(x + y)$$

(e) Working from the right-hand side, we have:

$C(x)C(y) + S(x)S(y)$

$$= \left(\frac{e^x + e^{-x}}{2}\right)\left(\frac{e^y + e^{-y}}{2}\right) + \left(\frac{e^x - e^{-x}}{2}\right)\left(\frac{e^y - e^{-y}}{2}\right)$$

$$= \frac{e^{x+y} + e^{-x+y} + e^{x-y} + e^{-x-y} + e^{x+y} - e^{-x+y} - e^{x-y} + e^{-x-y}}{4}$$

$$= \frac{2e^{x+y} + 2e^{-x-y}}{4}$$

$$= \frac{e^{x+y} + e^{-(x+y)}}{2}$$

$$= C(x + y)$$

(f) Working from the right-hand side, we have:

$$2S(x)C(x) = 2\left(\frac{e^x - e^{-x}}{2}\right)\left(\frac{e^x + e^{-x}}{2}\right) = \frac{e^{2x} - e^{-2x}}{2} = S(2x)$$

(g) Working from the right-hand side, we have:

$$[C(x)]^2 + [S(x)]^2 = \left(\frac{e^x + e^{-x}}{2}\right)^2 + \left(\frac{e^x - e^{-x}}{2}\right)^2$$

$$= \frac{e^{2x} + 2 + e^{-2x} + e^{2x} - 2 + e^{-2x}}{4}$$

$$= \frac{2e^{2x} + 2e^{-2x}}{4}$$

$$= \frac{e^{2x} + e^{-2x}}{2}$$

$$= C(2x)$$

Graphing Utility Exercises for Section 6.2

1. (a) Sketch the graphs using the standard viewing rectangle:

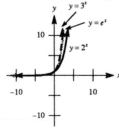

(b) Sketch the graphs using the indicated settings:

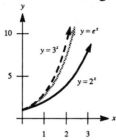

Note that $2^x < e^x < 3^x$ when $x > 0$.

(c) Sketch the graphs using the indicated settings:

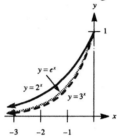

Note that $3^x < e^x < 2^x$ when $x < 0$.

3. The graph is a reflection of $y = e^x$ across the y-axis:

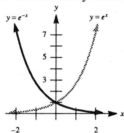

5. The graph is a reflection of $y = e^x$ across the x- and y-axes.

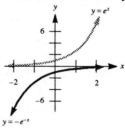

7.   (a)   The graph is a translation of $y = e^x$ to the right 1 unit:

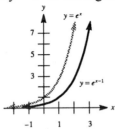

   (b)   The graph is a translation of $y = e^x$ to the right 1 unit, then a reflection across the y-axis:

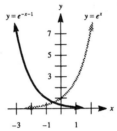

9.   (a)   Graphing the functions:

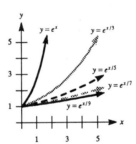

The y-intercept for each graph is 1. As k increases, the graph of $y = e^{x/k}$ flattens.

   (b)   It should be much flatter than the others. Graphing the functions:

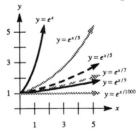

(c)   Graphing the function:

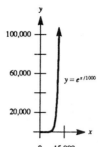

Note that the graph takes on the more familiar exponential shape.

11.   (a)   Graphing the line and the curve:

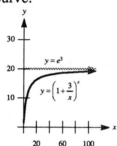

Note that the line appears to be an asymptote for the curve.

(b)   Using the suggested range settings:

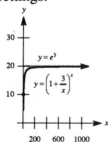

6.3 Logarithmic Functions

1.   (a)   No, because a horizontal line can intersect the parabola twice.
     (b)   Yes, because any horizontal line intersects the line only once.
     (c)   Yes, because any horizontal line intersects the curve only once.

3. (a) Let $y = \dfrac{2x-1}{3x+4}$. Switch the roles of x and y and solve the resulting equation for y:

$$x = \frac{2y-1}{3y+4}$$
$$x(3y+4) = 2y-1$$
$$3xy+4x = 2y-1$$
$$3xy-2y = -1-4x$$
$$y(3x-2) = -1-4x$$
$$y = \frac{-1-4x}{3x-2} = \frac{4x+1}{2-3x}$$

So $f^{-1}(x) = \dfrac{4x+1}{2-3x}$.

(b) Compute $\dfrac{1}{f(x)}$:
$$\frac{1}{f(x)} = \frac{3x+4}{2x-1}$$

(c) Compute $f^{-1}(0)$:
$$f^{-1}(0) = \frac{4(0)+1}{2-3(0)} = \tfrac{1}{2}$$

(d) Compute $\dfrac{1}{f(0)}$:
$$\frac{1}{f(0)} = \frac{3(0)+4}{2(0)-1} = \frac{4}{-1} = -4$$

5. Since $y = f^{-1}(x)$ would join the points $(-2, 3)$ and $(5, -1)$, $y = f^{-1}(x-1)$ will join the points $(-1, 3)$ and $(6, -1)$.

7. (a) $\log_3 9 = 2$
(b) $\log_{10} 1000 = 3$
(c) $\log_7 343 = 3$
(d) $\log_2 \sqrt{2} = 1/2$

9. (a) $2^5 = 32$
(b) $10^0 = 1$
(c) $e^{1/2} = \sqrt{e}$
(d) $3^{-4} = 1/81$
(e) $t^v = u$

11. Since $\log_5 30$ represents the power to which 5 must be raised to get 30, it is clearly greater than 2, since $5^2 = 25$. But $\log_8 60$ is less than 2, since $8^2 = 64$. Hence $\log_5 30$ is larger.

13. (a) Since $\log_9 27$ is the power to which 9 must be raised to get 27, we can see it is between 1 and 2, since $9^1 = 9$ while $9^2 = 81$. To find it let $\log_9 27 = n$ then $9^n = 27$ in exponential form and $3^{2n} = 3^3$. So $2n = 3$ and $n = \frac{3}{2}$.

   (b) If $\log_4 \frac{1}{32} = n$ then $4^n = \frac{1}{32}$, thus $2^{2n} = 2^{-5}$. So $2n = -5$, and $n = -\frac{5}{2}$. So $\log_4 \frac{1}{32} = -\frac{5}{2}$.

   (c) Follow the same steps. If $\log_5 5\sqrt{5} = n$, then:
$$5^n = 5\sqrt{5}$$
$$5^n = 5^{3/2}$$
$$n = \frac{3}{2}$$

15. (a) Writing $\log_4 x = -2$ in expontential form, we have $x = 4^{-2} = \frac{1}{16}$.

   (b) Writing $\ln x = -2$ in exponential form, we have $x = e^{-2} \approx 0.14$.

17. (a) We must have $5x > 0$, so $x > 0$. So the domain is $(0, \infty)$.

   (b) We must have $3 - 4x > 0$, so $3 > 4x$ and $x < \frac{3}{4}$. So the domain is $\left(-\infty, \frac{3}{4}\right)$.

   (c) We must have $x^2 > 0$, so $x \neq 0$. So the domain is all real numbers except 0, or $(-\infty, 0) \cup (0, \infty)$.

   (d) We must have $x > 0$. So the domain is $(0, \infty)$.

   (e) Solve the inequality:
$$x^2 - 25 > 0$$
$$(x + 5)(x - 5) > 0$$
   The key numbers are −5 and 5. Construct the sign chart:

Inteval	Test Number	$x + 5$	$x - 5$	$(x + 5)(x - 5)$
$(-\infty, -5)$	−6	neg.	neg.	pos.
$(-5, 5)$	0	pos.	neg.	neg.
$(5, \infty)$	6	pos.	pos.	pos.

   So the product is positive on the intervals $(-\infty, -5) \cup (5, \infty)$, which is the domain.

Note: This inequality can also be solved without using a sign chart:

$$x^2 - 25 > 0$$
$$x^2 > 25$$
$$|x| > 5$$
$$x > 5 \text{ or } x < -5$$

19. The coordinates for each point are:

    A: $(0, 1)$
    B: $(1, 0)$
    C: $(4, \log_2 4) = (4, 2)$
    D: $(2, 4)$

21. (a) The domain is $(0, \infty)$, the range is $(-\infty, \infty)$, the *x*-intercept is 1, there is no *y*-intercept, and the asymptote is $x = 0$:

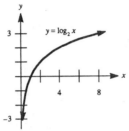

(b) The domain is $(0, \infty)$, the range is $(-\infty, \infty)$, the *x*-intercept is 1, there is no *y*-intercept, and the asymptote is $x = 0$:

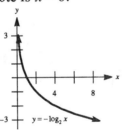

(c) The domain is $(-\infty, 0)$, the range is $(-\infty, \infty)$, the *x*-intercept is -1, there is no *y*-intercept, and the asymptote is $x = 0$:

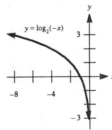

(d) The domain is $(-\infty, 0)$, the range is $(-\infty, \infty)$, the x-intercept is -1, there is no y-intercept, and the asymptote is $x = 0$:

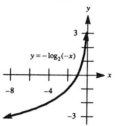

23. Translate $y = \log_3 x$ two units to the right, reflect across the x-axis, and translate 1 unit up. The domain is $(2, \infty)$, the range is $(-\infty, \infty)$, the x-intercept is 5, there is no y-intercept, and the asymptote is $x = 2$:

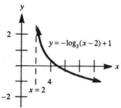

25. Translate $y = \ln x$ to the left e units. The domain is $(-e, \infty)$, the range is $(-\infty, \infty)$, the x-intercept is $-e + 1$, the y-intercept is 1, and the asymptote is $x = -e$:

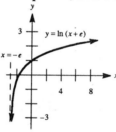

27. (a) Writing $x = \ln e^4$ in exponential form:
$$e^x = e^4$$
$$x = 4$$

(b) Writing $x = \ln \frac{1}{e}$ in exponential form:
$$e^x = \frac{1}{e}$$
$$e^x = e^{-1}$$
$$x = -1$$

(c)   Writing $x = \ln\sqrt{e}$ in exponential form:
$$e^x = \sqrt{e}$$
$$e^x = e^{1/2}$$
$$x = \tfrac{1}{2}$$

29.   Solve for x:
$$10^x = 25$$
$$x = \log_{10} 25 \approx 1.40$$

31.   Solve for x:
$$10^{(x^2)} = 40$$
$$x^2 = \log_{10} 40$$
$$x = \pm\sqrt{\log_{10} 40} = \pm\sqrt{1 + \log_{10} 4} \approx \pm 1.27$$

33.   Solve for t:
$$e^{2t+3} = 10$$
$$2t + 3 = \ln 10$$
$$2t = -3 + \ln 10$$
$$t = \frac{-3 + \ln 10}{2} \approx -0.35$$

35.   Solve for t:
$$e^{1-4t} = 12.405$$
$$1 - 4t = \ln 12.405$$
$$-4t = -1 + \ln 12.405$$
$$t = \frac{1 - \ln 12.405}{4} \approx -0.38$$

37.   (a)   No, the value is not exact, since $-\log(0.4) \approx 0.39794$.

(b)   Completing the table:

	x_1	x_2	x_3	x_4	x_5	x_6
From graph	0.22	0.65	0.20	0.75	0.15	0.85
From calculator	0.2218	0.6539	0.1845	0.7341	0.1342	0.8721

(c)   Computing the iterates:
$$x_7 \approx 0.0594$$
$$x_8 \approx 1.2260$$
$$x_9 \approx -0.0885$$
No, these iterates do not appear to be approaching the fixed point of the function.

(d)   The error occurs because $\log(-0.0885)$ does not exist. Recall that the domain of $\log x$ is $x > 0$.

39. Let $x = \ln 5$, so $e^x = 5$. We find the point $(x, 5)$ on the graph, and estimate $x \approx 1.61$. Now:

$$\% \text{ error} = \frac{|1.61 - \ln 5|}{\ln 5} \cdot 100 \approx 0.035\% \text{ error}$$

41. (a) The coordinates of A are $(1.6, \ln(1.6)) = (1.6, 0.470)$.
    (b) The coordinates of B are $(0.470, 0.470)$.
    (c) The coordinates of C are $(0.470, \ln(0.470)) = (0.470, -0.755)$.
    (d) The coordinates of D are $(-0.755, -0.755)$.
    (e) The coordinates of E are $\left(-0.755, \ln|-0.755|\right) = (-0.755, -0.281)$.

43. Since $\sqrt{5000} \approx 70$ while $\ln 5000 \approx 9$, A is the curve $g(x) = \ln x$ and B is the curve $f(x) = \sqrt{x}$.

45. Since $\sqrt[10]{7 \times 10^{15}} \approx 38$ while $\ln\left(7 \times 10^{15}\right) \approx 36$, A is the curve $g(x) = \ln x$ and B is the curve $f(x) = \sqrt[10]{x}$.

47. Since $\log_3 3 = 1$ while $\log_2 3 > 1$, the upper curve must be $y = \log_2 x$ and the lower curve must be $y = \log_3 x$.

49. Let $y = e^{x+1}$. We switch the roles of x and y and solve the resulting equation for y:

$$x = e^{y+1}$$
$$\ln x = \ln\left(e^{y+1}\right)$$
$$\ln x = y + 1$$
$$y = -1 + \ln x$$
$$f^{-1}(x) = -1 + \ln x$$

The x-intercept is e and the asymptote is $x = 0$:

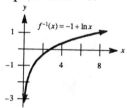

51. Graphing the region:

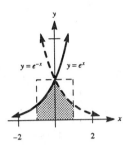

The area of the shaded region is less than that of the rectangle (defined on the graph), which has an area of two square units.

53. Solve for x:

$$\left(e^x\right)^2 - 5e^x - 6 = 0$$
$$\left(e^x - 6\right)\left(e^x + 1\right) = 0$$
$$e^x = 6, -1$$

Since $e^x > 0$, $e^x \neq -1$. Then $e^x = 6$, so $x = \ln 6$.

55. Using the approximation $2^{10} \approx 10^3$, then writing $\log_2 x = 100$ in exponential form:

$$x = 2^{100} = \left(2^{10}\right)^{10} \approx \left(10^3\right)^{10} = 10^{30}$$

57. (a) Compute $pH = -\log_{10}\left(3 \times 10^{-4}\right) \approx 3.5$. This would be an acid.

   (b) Compute $pH = -\log_{10}(1) = 0$. This would be an acid.

59. (a) We have $P(10) = 4$, since 2, 3, 5, 7 do not exceed 10.
   We have $P(18) = 7$, since 2, 3, 5, 7, 11, 13, 17 do not exceed 18.
   We have $P(19) = 8$, since 2, 3, 5, 7, 11, 13, 17, 19 do not exceed 19.

   (b) Complete the table:

x	$P(x)$	$\dfrac{x}{\ln x}$	$\dfrac{P(x)}{x / \ln x}$
10^2	25	22	1.151
10^4	1229	1086	1.132
10^6	78498	72382	1.084
10^8	5761455	5428681	1.061
10^9	50847534	48254942	1.054
10^{10}	455052512	434294482	1.048

(c) Complete the table:

x	$P(x)$	$\dfrac{x}{\ln x - 1.08366}$	$\dfrac{P(x)}{x\,/\,(\ln x - 1.08366)}$
10^2	25	28	0.8804
10^4	1229	1231	0.9988
10^6	78498	78543	0.9994
10^8	5761455	5768004	0.9989
10^9	50847534	50917519	0.9986
10^{10}	455052512	455743004	0.9985

61. The function possesses properties A, D, E, and H.

63. The function possesses properties B, D, E, and H.

65. The function possesses properties A, D, E, and H.

67. The function possesses properties A, D, E, and H.

69. Taking logarithms (base e), we must find a natural number n such that:
$$\ln\left(e^n\right) < \ln\left(10^{12}\right) < \ln\left(e^{n+1}\right)$$
$$n < 12\ln 10 < n+1$$
$$n < 27.63 < n+1$$
So $n = 27$.

71. The easiest approach to use is to convert the equation to exponential form:
$$\ln y = x, \text{ so } y = e^x$$
Thus the graph is identical to $y = e^x$:

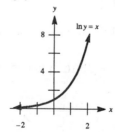

Graphing Utility Exercises for Section 6.3

1. (a) Since $y = e^x$ and $y = \ln x$ are inverse functions, we would expect their graphs to be symmetric about the line $y = x$.

    (b) Graphing the curves using the suggested settings:

    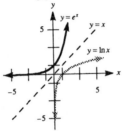

3. (a) For $F(x) = e^x$, the domain is $(-\infty, \infty)$:

    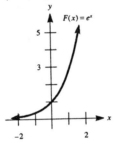

    (b) For $G(x) = \ln x$, the domain is $(0, \infty)$:

    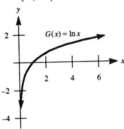

    (c) For $(G \circ F)(x) = \ln e^x = x$, the domain is $(-\infty, \infty)$:

    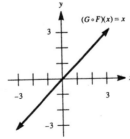

(d)　For $(F \circ G)(x) = e^{\ln x} = x$ for $x > 0$, the domain is $(0, \infty)$:

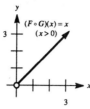

(e)　For $(G \circ H)(x) = \ln(x - 2)$, the domain is $(2, \infty)$:

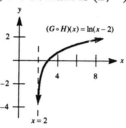

(f)　For $(H \circ G)(x) = \ln x - 2$, the domain is $(0, \infty)$:

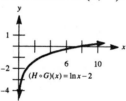

(g)　For $(G \circ K)(x) = \ln x^2$, the domain is $(-\infty, 0) \cup (0, \infty)$:

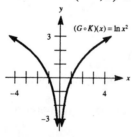

(h)　For $(K \circ G)(x) = (\ln x)^2$, the domain is $(0, \infty)$:

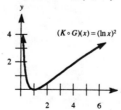

5.   (a)   Graphing the function using the suggested settings:

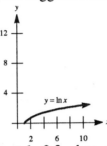

The *y*-coordinate is approximately 2.3 when *x* = 10.

(b)   Graphing the two curves using the suggested settings:

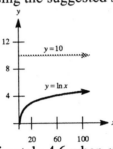

The *y*-coordinate is approximately 4.6 when *x* = 100.

(c)   Graphing the two curves using the suggested settings:

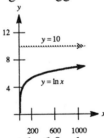

The *y*-coordinate is approximately 6.9 when *x* = 1000.

(d)   Graphing the two curves using the suggested settings:

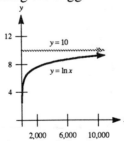

The *y*-coordinate is approximately 9.2 when *x* = 10,000.

(e) Graphing the two curves using the suggested settings:

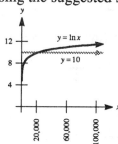

The y-coordinate is approximately 11.5 when $x = 100{,}000$. It appears $\ln x = 10$ when $x \approx 22{,}000$. To find the exact value, we write $\ln x = 10$ in exponential form to obtain $x = e^{10} \approx 22026$.

7. (a) Completing the table:

Planet	x Average Distance from the Sun (millions of miles)	y Average Orbital Velocity (miles/sec)	$\ln x$	$\ln y$
Mercury	35.98	29.75	3.583	3.393
Venus	67.08	21.76	4.206	3.080
Earth	92.96	18.51	4.532	2.918
Mars	141.64	14.99	4.953	2.707

(b) The values are $A = -0.500$ and $B = 5.184$, so $\ln y = -0.500 \ln x + 5.184$.

(c) Solving for y:
$$\ln y = -0.500 \ln x + 5.184$$
$$y = e^{-0.500 \ln x + 5.184}$$
$$y = e^{-\frac{1}{2} \ln x + 5.184}$$
$$y = e^{\ln x^{-1/2}} \cdot e^{5.184}$$
$$y = x^{-1/2} \cdot e^{5.184}$$
$$y = \frac{e^{5.184}}{\sqrt{x}}$$

(d) Completing the table:

Planet	x Average Distance from the Sun (millions of miles)	y Average Orbital Velocity (miles/sec) from (2)	y Average Orbital Velocity (miles/sec) from observation
Jupiter	483.63	8.11	8.12
Saturn	888.22	5.99	5.99
Uranus	1786.55	4.22	4.23
Neptune	2799.06	3.37	3.38
Pluto	3700.75	2.93	2.95

Note that the equation fits the observed data very accurately.

6.4 Properties of Logarithms

1. Using properties of logarithms:
$$\log_{10} 70 - \log_{10} 7 = \log_{10} \tfrac{70}{7} = \log_{10} 10 = 1$$

3. Using properties of logarithms:
$$\log_7 \sqrt{7} = \log_7\left(7^{1/2}\right) = \tfrac{1}{2}$$

5. Using properties of logarithms:
$$\log_3 108 + \log_3 \tfrac{3}{4} = \log_3\left(\tfrac{108 \cdot 3}{4}\right) = \log_3 81 = \log_3 3^4 = 4$$

7. Using properties of logarithms:
$$-\tfrac{1}{2} + \ln \sqrt{e} = -\tfrac{1}{2} + \ln e^{1/2} = -\tfrac{1}{2} + \tfrac{1}{2} = 0$$

9. Using properties of logarithms:
$$2^{\log_2 5} - 3\log_5 \sqrt[3]{5} = 5 - 3\log_5 5^{1/3} = 5 - 3\left(\tfrac{1}{3}\right) = 5 - 1 = 4$$

11. Using properties of logarithms:
$$\log_{10} 30 + \log_{10} 2 = \log_{10}(30 \cdot 2) = \log_{10} 60$$

13. Using properties of logarithms:
$$\log_5 6 + \log_5 \tfrac{1}{3} + \log_5 10 = \log_5\left(6 \cdot \tfrac{1}{3} \cdot 10\right) = \log_5 20$$

15. (a) Using properties of logarithms:
$$\ln 3 - 2\ln 4 + \ln 32 = \ln 3 - \ln\left(4^2\right) + \ln 32 = \ln\left(\frac{3 \cdot 32}{4^2}\right) = \ln 6$$

(b) Using properties of logarithms:

$$\ln 3 - 2(\ln 4 + \ln 32) = \ln 3 - 2\big[\ln(4 \cdot 32)\big]$$
$$= \ln 3 - 2\ln 128$$
$$= \ln 3 - \ln\big(128^2\big)$$
$$= \ln\left(\frac{3}{128^2}\right)$$
$$= \ln\left(\tfrac{3}{16384}\right)$$

17. Using properties of logarithms:

$$\log_b 4 + 3\Big[\log_b(1+x) - \tfrac{1}{2}\log_b(1-x)\Big] = \log_b 4 + 3\Big(\log_b(1+x) - \log_b\sqrt{1-x}\Big)$$
$$= \log_b 4 + \log_b(1+x)^3 - \log_b(1-x)^{3/2}$$
$$= \log_b\left[\frac{4(1+x)^3}{(1-x)^{3/2}}\right]$$

19. Using properties of logarithms:

$$4\log_{10} 3 - 6\log_{10}\big(x^2+1\big) + \tfrac{1}{2}\Big[\log_{10}(x+1) - 2\log_{10} 3\Big]$$
$$= \log_{10} 3^4 - \log_{10}\big(x^2+1\big)^6 + \tfrac{1}{2}\Big[\log_{10}(x+1) - \log_{10} 3^2\Big]$$
$$= \log_{10} 81 - \log_{10}\big(x^2+1\big)^6 + \tfrac{1}{2}\log_{10}\left(\frac{x+1}{9}\right)$$
$$= \log_{10} 81 - \log_{10}\big(x^2+1\big)^6 + \log_{10}\left(\frac{x+1}{9}\right)^{1/2}$$
$$= \log_{10}\left[\frac{81\frac{\sqrt{x+1}}{3}}{\big(x^2+1\big)^6}\right]$$
$$= \log_{10}\left[\frac{27\sqrt{x+1}}{\big(x^2+1\big)^6}\right]$$

21. (a) Using properties of logarithms:

$$\log_{10}\left(\frac{x^2}{1+x^2}\right) = \log_{10} x^2 - \log_{10}\big(1+x^2\big) = 2\log_{10} x - \log_{10}\big(1+x^2\big)$$

(b) Using properties of logarithms:

$$\ln\left(\frac{x^2}{\sqrt{1+x^2}}\right) = \ln x^2 - \ln\sqrt{1+x^2} = 2\ln x - \tfrac{1}{2}\ln\big(1+x^2\big)$$

23. (a) Using properties of logarithms:
$$\log_{10}\sqrt{9-x^2} = \tfrac{1}{2}\log_{10}(9-x^2)$$
$$= \tfrac{1}{2}\log_{10}[(3+x)(3-x)]$$
$$= \tfrac{1}{2}\log_{10}(3+x)+\tfrac{1}{2}\log_{10}(3-x)$$

(b) Using properties of logarithms:
$$\ln\left[\frac{\sqrt{4-x^2}}{(x-1)(x+1)^{3/2}}\right] = \tfrac{1}{2}\ln(4-x^2)-\ln(x-1)-\ln(x+1)^{3/2}$$
$$= \tfrac{1}{2}\ln[(2+x)(2-x)]-\ln(x-1)-\tfrac{3}{2}\ln(x+1)$$
$$= \tfrac{1}{2}\ln(2+x)+\tfrac{1}{2}\ln(2-x)-\ln(x-1)-\tfrac{3}{2}\ln(x+1)$$

25. (a) Using properties of logarithms:
$$\log_b\sqrt{\frac{x}{b}} = \tfrac{1}{2}\log_b\frac{x}{b} = \tfrac{1}{2}\log_b x-\tfrac{1}{2}\log_b b = \tfrac{1}{2}\log_b x-\tfrac{1}{2}$$

(b) Using properties of logarithms:
$$2\ln\sqrt{(1+x^2)(1+x^4)(1+x^6)} = \ln\left[(1+x^2)(1+x^4)(1+x^6)\right]$$
$$= \ln(1+x^2)+\ln(1+x^4)+\ln(1+x^6)$$

27. (a) Using properties of logarithms:
$$\log_b 6 = \log_b(2\cdot 3) = \log_b 2+\log_b 3 = A+B$$

(b) Using properties of logarithms:
$$\log_b\tfrac{1}{6} = \log_b(6^{-1}) = -\log_b 6 = -(A+B) = -A-B$$

(c) Using properties of logarithms:
$$\log_b 27 = \log_b(3^3) = 3\log_b 3 = 3B$$

(d) Using properties of logarithms:
$$\log_b\tfrac{1}{27} = \log_b(27^{-1}) = -\log_b 27 = -3B$$

29. (a) Using properties of logarithms:
$$\log_b\tfrac{5}{3} = \log_b 5-\log_b 3 = C-B$$

(b) Using properties of logarithms:
$$\log_b 0.6 = \log_b\tfrac{3}{5} = \log_b 3-\log_b 5 = B-C$$

(c) Using properties of logarithms:
$$\log_b\tfrac{5}{9} = \log_b 5-\log_b(3^2) = \log_b 5-2\log_b 3 = C-2B$$

(d) Using properties of logarithms:

$$\log_b \tfrac{5}{16} = \log_b 5 - \log_b\left(2^4\right) = \log_b 5 - 4\log_b 2 = C - 4A$$

31. (a) Using the change of base formula:

$$\log_3 b = \frac{\log_b b}{\log_b 3} = \frac{1}{B}$$

(b) Using the change of base formula:

$$\log_3 (10b) = \frac{\log_b(10b)}{\log_b 3} = \frac{\log_b(2 \cdot 5b)}{\log_b 3} = \frac{\log_b 2 + \log_b 5 + \log_b b}{\log_b 3} = \frac{A + C + 1}{B}$$

33. (a) Using the change of base formula:

$$\log_{3b} 2 = \frac{\log_b 2}{\log_b(3b)} = \frac{\log_b 2}{\log_b 3 + \log_b b} = \frac{A}{B+1}$$

(b) Using the change of base formula:

$$\log_{3b} 15 = \frac{\log_b 15}{\log_b(3b)} = \frac{\log_b 3 + \log_b 5}{\log_b 3 + \log_b b} = \frac{B+C}{B+1}$$

35. (a) Using the change of base formula:

$$\left(\log_b 5\right)\left(\log_5 b\right) = \log_b 5 \cdot \frac{\log_b b}{\log_b 5} = \log_b b = 1$$

(b) Using the change of base formula:

$$\left(\log_b 6\right)\left(\log_6 b\right) = \log_b 6 \cdot \frac{\log_b b}{\log_b 6} = \log_b b = 1$$

37. (a) Using properties of logarithms:

$$\log_{10}\left(AB^2C^3\right) = \log_{10} A + \log_{10} B^2 + \log_{10} C^3$$
$$= \log_{10} A + 2\log_{10} B + 3\log_{10} C$$
$$= a + 2b + 3c$$

(b) Using properties of logarithms:

$$\log_{10} 10\sqrt{A} = \log_{10} 10 + \log_{10} \sqrt{A} = 1 + \tfrac{1}{2}\log_{10} A = 1 + \tfrac{1}{2}a$$

(c) Using properties of logarithms:

$$\log_{10} \sqrt{10ABC} = \tfrac{1}{2}\log_{10}(10ABC)$$
$$= \tfrac{1}{2}\log_{10} 10 + \tfrac{1}{2}\log_{10} A + \tfrac{1}{2}\log_{10} B + \tfrac{1}{2}\log_{10} C$$
$$= \tfrac{1}{2}(1) + \tfrac{1}{2}a + \tfrac{1}{2}b + \tfrac{1}{2}c$$
$$= \tfrac{1}{2}(1 + a + b + c)$$

(d)   Using properties of logarithms:
$$\log_{10}\left(\tfrac{10A}{\sqrt{BC}}\right) = \log_{10}(10A) - \tfrac{1}{2}\log_{10}(BC)$$
$$= \log_{10}10 + \log_{10}A - \tfrac{1}{2}\log_{10}B - \tfrac{1}{2}\log_{10}C$$
$$= 1 + a - \tfrac{1}{2}b - \tfrac{1}{2}c$$

39.   (a)   Using properties of logarithms:
$$\ln(ex) = \ln e + \ln x = 1 + t$$

(b)   Using properties of logarithms:
$$\ln(xy) - \ln\left(x^2\right) = \ln\frac{xy}{x^2} = \ln\frac{y}{x} = \ln y - \ln x = u - t$$

(c)   Using properties of logarithms:
$$\ln\sqrt{xy} + \ln\frac{x}{e} = \tfrac{1}{2}(\ln x + \ln y) + \ln x - \ln e = \tfrac{1}{2}(t+u) + t - 1 = \tfrac{3}{2}t + \tfrac{1}{2}u - 1$$

(d)   Using properties of logarithms:
$$\ln\left(e^2x\sqrt{y}\right) = \ln e^2 + \ln x + \ln\sqrt{y} = 2\ln e + \ln x + \tfrac{1}{2}\ln y = 2 + t + \tfrac{1}{2}u$$

41.   (a)   To find the x-intercept, set $y = 0$:
$$0 = 2^x - 5$$
$$2^x = 5$$
$$\ln 2^x = \ln 5$$
$$x\ln 2 = \ln 5$$
$$x = \frac{\ln 5}{\ln 2} \approx 2.32$$
Now sketch the graph:

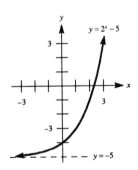

(b) To find the x-intercept, set $y = 0$:

$$0 = 2^{x/2} - 5$$
$$2^{x/2} = 5$$
$$\ln 2^{x/2} = \ln 5$$
$$\frac{x}{2} \ln 2 = \ln 5$$
$$x = \frac{2 \ln 5}{\ln 2} \approx 4.64$$

Now sketch the graph:

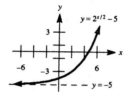

43. Taking the natural log of each side:

$$\ln 5 = \ln 2 + \ln e^{2x-1}$$
$$\ln 5 = \ln 2 + (2x - 1)$$
$$\ln 5 = \ln 2 + 2x - 1$$
$$2x = \ln 5 - \ln 2 + 1$$
$$x = \frac{\ln 5 - \ln 2 + 1}{2}$$

45. Taking the natural logarithm of each side:

$$2^x = 13$$
$$\ln 2^x = \ln 13$$
$$x \ln 2 = \ln 13$$
$$x = \frac{\ln 13}{\ln 2}$$

47. Taking the natural logarithm of each side:

$$10^x = e$$
$$\ln 10^x = \ln e$$
$$x \ln 10 = 1$$
$$x = \frac{1}{\ln 10}$$

49. Taking the natural logarithm of each side:

$$3^{x^2-1} = 12$$
$$\log 3^{x^2-1} = \log 12$$
$$(x^2 - 1)\log 3 = \log 12$$
$$x^2 - 1 = \frac{\log 12}{\log 3}$$
$$x^2 = 1 + \frac{\log 12}{\log 3}$$
$$x = \pm\sqrt{1 + \frac{\log 12}{\log 3}}$$
$$x \approx \pm 1.806$$

51. Using the change of base formula:

$$\log_2 5 = \frac{\log_{10} 5}{\log_{10} 2}$$

53. Using the change of base formula:

$$\ln 3 = \log_e 3 = \frac{\log_{10} 3}{\log_{10} e}$$

55. Using the change of base formula:

$$\log_b 2 = \frac{\log_{10} 2}{\log_{10} b}$$

57. Using the change of base formula:

$$\log_{10} 6 = \frac{\ln 6}{\ln 10}$$

59. Using the change of base formula:

$$\log_{10} e = \frac{\ln e}{\ln 10} = \frac{1}{\ln 10}$$

61. Using the change of base formula:

$$\log_{10}(\log_{10} x) = \log_{10}\left(\frac{\ln x}{\ln 10}\right) = \frac{\ln\left(\frac{\ln x}{\ln 10}\right)}{\ln 10} = \frac{\ln(\ln x) - \ln(\ln 10)}{\ln 10}$$

63. (a) true
    (b) true
    (c) true
    (d) false, since $\ln x^3 = 3\ln x$, not $\ln 3x$
    (e) true
    (f) false, since $\ln 2x^3 = \ln 2 + 3\ln x \neq 3\ln 2x = \ln(2x)^3$

(g) true

(h) false, since $\log_5 24$ is between 1 and 2, not 5^1 and 5^2

(i) true

(j) false, since $\log_5 24$ is close to 2 ($2 = \log_5 25$)

(k) false, since the domain is $(0, \infty)$

(l) true

(m) true

65. (a) We have:
$$\log_{10} \pi^7 \approx 3.48005$$
$$7\log_{10} \pi \approx 7(0.49715) = 3.48005$$
So $\log_{10} \pi^7 = 7\log_{10} \pi$.

(b) We have:
$$\log_b P^n = \log_{10} \pi^7 \approx 3.48005$$
$$\left(\log_b P\right)^n = \left(\log_{10} \pi\right)^7 \approx 0.00751$$
So $\log_b \pi^7 \neq \left(\log_b \pi\right)^7$.

(c) We have:
$$b^{\log_b P} = 10^{\log_{10} 1776} = 10^{3.24944} = 1776$$
So $10^{\log_{10} 1776} = 1776$.

(d) We have:
$$\ln 2 + \ln 3 + \ln 4 \approx 0.69315 + 1.09861 + 1.38629 \approx 3.17805$$
$$\ln 24 \approx 3.17805$$
So $\ln 2 + \ln 3 + \ln 4 = \ln 24$.

(e) We have:
$$\log_{10} A + \log_{10} B + \log_{10} C = \log_{10} 11 + \log_{10} 12 + \log_{10} 13$$
$$\approx 1.04139 + 1.07918 + 1.11394$$
$$= 3.23451$$
$$\log_{10}(ABC) = \log_{10}(11 \cdot 12 \cdot 13) = \log_{10} 1716 \approx 3.23451$$
So $\log_{10} A + \log_{10} B + \log_{10} C = \log_{10}(ABC)$.

(f) Compute $f[g(2345.6)]$:
$$f[g(2345.6)] = f(\ln 2345.6) \approx f(7.76030) \approx e^{7.76030} \approx 2345.6$$

(g) Compute $g[f(0.123456)]$:
$$g[f(0.123456)] = g\left(10^{0.123456}\right) = \log_{10}(1.32879) = 0.123456$$

67. Complete the table:

x	0.1	0.05	0.005	0.0005
$\ln(1+x)$	0.095310	0.048790	0.004987	0.000499

Note that the values of $\ln(1 + x)$ are close to x, when x is close to 0.

69. Since $(-2, 324)$ and $\left(\frac{1}{2}, \frac{4}{3}\right)$ lie on the curve $y = ae^{bx}$:

$$324 = ae^{-2b} \quad \text{and} \quad \tfrac{4}{3} = ae^{\frac{1}{2}b}$$

Dividing:

$$\frac{324}{\frac{4}{3}} = \frac{ae^{-2b}}{ae^{\frac{1}{2}b}}$$

$$243 = e^{-2.5b}$$

$$\ln 243 = -2.5b$$

$$b = -\frac{\ln 243}{2.5}$$

Substituting into $324 = ae^{-2b}$:

$$324 = ae^{(2\ln 243)/2.5}$$

$$324 = ae^{0.8\ln 243}$$

$$324 = a\left(e^{\ln 243}\right)^{0.8}$$

$$324 = a \cdot 243^{0.8}$$

$$a = \frac{324}{243^{0.8}} = \frac{324}{81} = 4$$

71. (a) Solve the equation:

$$2^x - 3 = 0$$
$$2^x = 3$$
$$\log_{10} 2^x = \log_{10} 3$$
$$x\log_{10} 2 = \log_{10} 3$$
$$x = \frac{\log_{10} 3}{\log_{10} 2}$$

(b) We have $\dfrac{\log_{10} 3}{\log_{10} 2} \approx 1.585$ and $\dfrac{\ln 3}{\ln 2} \approx 1.585$.

(c) Using the change of base formula:

$$\frac{\log_{10} 3}{\log_{10} 2} = \frac{\frac{\ln 3}{\ln 10}}{\frac{\ln 2}{\ln 10}} = \frac{\ln 3}{\ln 2}$$

73. Working from the left-hand side:

$$\log_b\left(\frac{\sqrt{3}+\sqrt{2}}{\sqrt{3}-\sqrt{2}}\right) = \log_b\left(\frac{\sqrt{3}+\sqrt{2}}{\sqrt{3}-\sqrt{2}} \cdot \frac{\sqrt{3}+\sqrt{2}}{\sqrt{3}+\sqrt{2}}\right)$$

$$= \log_b\left[\frac{\left(\sqrt{3}+\sqrt{2}\right)^2}{3-2}\right]$$

$$= \log_b\left(\sqrt{3}+\sqrt{2}\right)^2$$

$$= 2\log_b\left(\sqrt{3}+\sqrt{2}\right)$$

75. Using properties of logarithms:

$$b^{3\log_b x} = b^{\log_b x^3} = x^3$$

77. Using the change of base formula:

$$\log_b a = \frac{\log_a a}{\log_a b} = \frac{1}{\log_a b}$$

79. (a) Using the hint:

$$\frac{1}{\log_2 \pi} + \frac{1}{\log_5 \pi} = \frac{1}{\frac{\log_\pi \pi}{\log_\pi 2}} + \frac{1}{\frac{\log_\pi \pi}{\log_\pi 5}} = \log_\pi 2 + \log_\pi 5 = \log_\pi 10$$

Since $\pi^2 < 10$, $\log_\pi 10 > \log_\pi \pi^2 = 2$, which completes the proof.

(b) Since $\log_\pi 2 > \log_\pi 1 = 0$, $\log_\pi 2$ is positive. Using the inequality $a + b \geq 2\sqrt{ab}$ with $a = \log_\pi 2$ and $b = \frac{1}{\log_\pi 2}$:

$$\log_\pi 2 + \frac{1}{\log_\pi 2} \geq 2\sqrt{1} = 2$$

The equality occurs only if $a = b$, which occurs when:

$$\log_\pi 2 = \frac{1}{\log_\pi 2}$$

$$\left(\log_\pi 2\right)^2 = 1, \text{ which is impossible}$$

Thus $\log_\pi 2 + \frac{1}{\log_\pi 2} > 2$.

(c)   The two quantities are:
$$\frac{1}{\log_2 \pi} + \frac{1}{\log_5 \pi} = \log_\pi 10 = \frac{\ln 10}{\ln \pi} \approx 2.0115$$

$$\log_\pi 2 + \frac{1}{\log_\pi 2} = \frac{\ln 2}{\ln \pi} + \frac{\ln \pi}{\ln 2} \approx 2.2570$$

Thus $\log_\pi 2 + \dfrac{1}{\log_\pi 2}$ is the larger quantity.

81.   Using the change of base formula:
$$\log_{ab} x = \frac{\log_a x}{\log_a ab} = \frac{\log_a x}{\log_a a + \log_a b} = \frac{\log_a x}{1 + \log_a b}$$

Therefore:
$$\frac{\log_a x}{\log_{ab} x} = \frac{\log_a x}{\frac{\log_a x}{1 + \log_a b}} = 1 + \log_a b$$

83.   Work from the right-hand side:
$$\tfrac{1}{2}(\log a + \log b) = \tfrac{1}{2}\log(ab) = \log \sqrt{ab}$$
Proving the desired equality is equivalent to proving:
$$\tfrac{1}{3}(a + b) = \sqrt{ab}, \text{ since log is a one-to-one function.}$$
If a and b are both positive, we can square each side:
$$\tfrac{1}{9}\left(a^2 + 2ab + b^2\right) = ab$$
Now work with the left-hand side:
$$\tfrac{1}{9}\left(a^2 + 2ab + b^2\right) = \tfrac{1}{9}(7ab + 2ab) \text{ by our assumption } a^2 + b^2 = 7ab$$
$$= \tfrac{1}{9}(9ab)$$
$$= ab$$

85.   (a)   Using properties of logarithms:
$$f(2) + f(3) + f(4) = \ln\left(1 - \tfrac{1}{4}\right) + \ln\left(1 - \tfrac{1}{9}\right) + \ln\left(1 - \tfrac{1}{16}\right)$$
$$= \ln\tfrac{3}{4} + \ln\tfrac{8}{9} + \ln\tfrac{15}{16}$$
$$= \ln\left(\tfrac{3}{4} \cdot \tfrac{8}{9} \cdot \tfrac{15}{16}\right)$$
$$= \ln\tfrac{5}{8}$$

(b)   Both sides evaluate to approximately –0.470.

6.5 Equations and Inequalities with Logs and Exponents

1. Taking natural logarithms:
$$5^x = 3^{2x-1}$$
$$\ln 5^x = \ln 3^{2x-1}$$
$$x \ln 5 = (2x-1)\ln 3$$
$$x \ln 5 = 2x \ln 3 - \ln 3$$
$$x(\ln 5 - 2\ln 3) = -\ln 3$$
$$x = \frac{-\ln 3}{\ln 5 - 2\ln 3}$$
$$x = \frac{\ln 3}{2\ln 3 - \ln 5} \approx 1.869$$

3. Converting to exponential form:
$$\ln(\ln x) = 1.5$$
$$\ln x = e^{1.5}$$
$$x = e^{e^{1.5}} \approx 88.384$$

5. Converting to exponential form:
$$\log_{10}\left(x^2 + 36\right) = 2$$
$$x^2 + 36 = 10^2$$
$$x^2 + 36 = 100$$
$$x^2 = 64$$
$$x = \pm 8$$

7. Converting to exponential form:
$$\log_{10}\left(2x^2 - 3x\right) = 2$$
$$2x^2 - 3x = 10^2$$
$$2x^2 - 3x = 100$$
$$2x^2 - 3x - 100 = 0$$
Using the quadratic formula:
$$x = \frac{-(-3) \pm \sqrt{(-3)^2 - 4(2)(-100)}}{2(2)} = \frac{3 \pm \sqrt{9 + 800}}{4} = \frac{3 \pm \sqrt{809}}{4}$$
The solutions are $x = \dfrac{3 - \sqrt{809}}{4} \approx -6.361$ or $x = \dfrac{3 + \sqrt{809}}{4} \approx 7.861$.

9. Let $u = 10^x$. Then the equation becomes:
$$\left(10^x\right)^2 + 3\left(10^x\right) - 10 = 0$$
$$u^2 + 3u - 10 = 0$$
$$(u+5)(u-2) = 0$$
$$u = -5, 2$$

So either $10^x = -5$, which is impossible, or $10^x = 2$, so $x = \log 2 \approx 0.301$.

11. (a) Since $\ln\left(x^3\right) = 3\ln x$, the equation is true for all $x > 0$.

(b) Let $u = \ln x$. Then the equation becomes:
$$(\ln x)^3 = 3\ln x$$
$$u^3 = 3u$$
$$u^3 - 3u = 0$$
$$u\left(u^2 - 3\right) = 0$$
$$u = 0, -\sqrt{3}, \sqrt{3}$$

So either $\ln x = 0$, so $x = 1$, $\ln x = -\sqrt{3}$, so $x = e^{-\sqrt{3}} \approx 0.177$, or $\ln x = \sqrt{3}$, so $x = e^{\sqrt{3}} \approx 5.652$.

13. (a) Since $\log_3 6x = \log_3 6 + \log_3 x$, the equation is true for all $x > 0$.

(b) Using properties of logarithms:
$$\log_3 6x = 6\log_3 x$$
$$\log_3 6 + \log_3 x = 6\log_3 x$$
$$\log_3 6 = 5\log_3 x$$
$$\tfrac{1}{5}\log_3 6 = \log_3 x$$
$$\log_3 6^{1/5} = \log_3 x$$
$$x = 6^{1/5} \approx 1.431$$

15. Since $7^{\log_7 2x} = 2x$, the equation is true for all $x > 0$.

17. Converting to exponential form:
$$\log_2\left(\log_3 x\right) = -1$$
$$\log_3 x = 2^{-1}$$
$$\log_3 x = \tfrac{1}{2}$$
$$x = 3^{1/2}$$
$$x = \sqrt{3} \approx 1.732$$

19. Solving the equation:

$$\ln 4 - \ln x = \frac{\ln 4}{\ln x}$$

$$(\ln 4)(\ln x) - (\ln x)^2 = \ln 4$$

Let $u = \ln x$. Then the equation becomes:

$$(\ln 4)u - u^2 = \ln 4$$

$$0 = u^2 - (\ln 4)u + \ln 4$$

Using the quadratic formula:

$$u = \frac{\ln 4 \pm \sqrt{(\ln 4)^2 - 4\ln 4}}{2} = \frac{\ln 4 \pm \sqrt{(\ln 4)(\ln 4 - 4)}}{2}$$

Since $\ln 4 > 0$ and $\ln 4 - 4 < 0$, the quantity inside the radical is negative and thus the equation has no real solutions.

21. Using properties of logarithms:

$$\ln\left(3x^2\right) = 2\ln(3x)$$

$$\ln\left(3x^2\right) = \ln(3x)^2$$

$$3x^2 = (3x)^2$$

$$3x^2 = 9x^2$$

$$0 = 6x^2$$

$$x = 0$$

Since $\ln 0$ is undefined, the equation has no real solutions.

23. Converting to exponential form:

$$\log_{16} \frac{x+3}{x-1} = \tfrac{1}{2}$$

$$\frac{x+3}{x-1} = 16^{1/2}$$

$$\frac{x+3}{x-1} = 4$$

$$x + 3 = 4x - 4$$

$$7 = 3x$$

$$x = \tfrac{7}{3}$$

25. (a) Let $u = e^x$. Then the equation becomes:

$$\left(e^x\right)^2 + 2e^x + 1 = 0$$

$$u^2 + 2u + 1 = 0$$

$$(u+1)^2 = 0$$

$$u = -1$$

So $e^x = -1$, which is impossible. The equation has no real solutions.

(b) Let $u = e^x$. Then the equation becomes:

$$\left(e^x\right)^2 - 2e^x + 1 = 0$$
$$u^2 - 2u + 1 = 0$$
$$(u-1)^2 = 0$$
$$u = 1$$

So $e^x = 1$, thus $x = 0$.

(c) Let $u = e^x$. Then the equation becomes:

$$\left(e^x\right)^2 - 2e^x - 3 = 0$$
$$u^2 - 2u - 3 = 0$$
$$(u-3)(u+1) = 0$$
$$u = -1, 3$$

So either $e^x = -1$, which is impossible, or $e^x = 3$, so $x = \ln 3 \approx 1.099$.

(d) Let $u = e^x$. Then the equation becomes:

$$\left(e^x\right)^2 - 2e^x - 4 = 0$$
$$u^2 - 2u - 4 = 0$$

Using the quadratic formula:

$$u = \frac{-(-2) \pm \sqrt{(-2)^2 - 4(1)(-4)}}{2(1)} = \frac{2 \pm \sqrt{4+16}}{2} = \frac{2 \pm 2\sqrt{5}}{2} = 1 \pm \sqrt{5}$$

So either $e^x = 1 - \sqrt{5}$, which is impossible (since $1 - \sqrt{5} < 0$), or $e^x = 1 + \sqrt{5}$, so $x = \ln\left(1 + \sqrt{5}\right) \approx 1.174$.

27. Using the hint:

$$e^x - e^{-x} = 1$$
$$e^{2x} - 1 = e^x$$
$$e^{2x} - e^x - 1 = 0$$

Let $u = e^x$. Then the equation becomes $u^2 - u - 1 = 0$. Using the quadratic formula:

$$u = \frac{-(-1) \pm \sqrt{(-1)^2 - 4(1)(-1)}}{2(1)} = \frac{1 \pm \sqrt{1+4}}{2} = \frac{1 \pm \sqrt{5}}{2}$$

So either $e^x = \dfrac{1-\sqrt{5}}{2}$, which is impossible, or $e^x = \dfrac{1+\sqrt{5}}{2}$, so $x = \ln\left(\dfrac{1+\sqrt{5}}{2}\right) \approx 0.481$.

29. Taking natural logarithms:

$$2^{5x} = 3^x \cdot 5^{x+3}$$

$$\ln\left(2^{5x}\right) = \ln\left(3^x \cdot 5^{x+3}\right)$$

$$(5x)\ln 2 = \ln\left(3^x\right) + \ln\left(5^{x+3}\right)$$

$$(5\ln 2)x = (\ln 3)x + (x+3)\ln 5$$

$$(5\ln 2)x = (\ln 3)x + (\ln 5)x + 3\ln 5$$

$$(5\ln 2 - \ln 3 - \ln 5)x = 3\ln 5$$

$$x = \frac{3\ln 5}{5\ln 2 - \ln 3 - \ln 5} \approx 6.372$$

31. Using properties of logarithms:

$$\log_6 x + \log_6 (x+1) = 1$$

$$\log_6 \left(x(x+1)\right) = 1$$

$$x(x+1) = 6$$

$$x^2 + x = 6$$

$$x^2 + x - 6 = 0$$

$$(x+3)(x-2) = 0$$

$$x = -3, 2$$

But $x = -3$ is an extraneous root $\left(\log_6(-3) \text{ is undefined}\right)$, so $x = 2$ is the only solution.

33. Solving for x:

$$\log_9 (x+1) = \tfrac{1}{2} + \log_9 x$$

$$\log_9 (x+1) - \log_9 x = \tfrac{1}{2}$$

$$\log_9 \left(\frac{x+1}{x}\right) = \tfrac{1}{2}$$

$$9^{1/2} = \frac{x+1}{x}$$

$$3 = \frac{x+1}{x}$$

$$3x = x+1$$

$$2x = 1$$

$$x = \tfrac{1}{2}$$

35. Solving for x:

$$\log_{10}(2x+4) + \log_{10}(x-2) = 1$$

$$\log_{10}\left[(2x+4)(x-2)\right] = 1$$

$$(2x+4)(x-2) = 10^1$$

$$2x^2 - 8 = 10$$

$$2x^2 = 18$$

$$x^2 = 9$$

$$x = \pm 3$$

Since $\log_{10}(-2)$ is undefined, $x = -3$ is an extraneous root. So the solution is $x = 3$.

37.   Solving for x:

$$\log_{10}(x+3) - \log_{10}(x-2) = 2$$

$$\log_{10}\left(\frac{x+3}{x-2}\right) = 2$$

$$10^2 = \frac{x+3}{x-2}$$

$$100(x-2) = x+3$$

$$100x - 200 = x+3$$

$$99x = 203$$

$$x = \frac{203}{99}$$

39.   Solving for x:

$$\log_b(x+1) = 2\log_b(x-1)$$

$$\log_b(x+1) = \log_b(x-1)^2$$

$$x+1 = (x-1)^2$$

$$x+1 = x^2 - 2x + 1$$

$$x^2 - 3x = 0$$

$$x(x-3) = 0$$

$$x = 0, 3$$

Since $\log_b(-1)$ is undefined, $x = 0$ is an extraneous root. So the solution is $x = 3$.

41.   Solving for x:

$$\log_{10}(x-6) + \log_{10}(x+3) = 1$$

$$\log_{10}[(x-6)(x+3] = 1$$

$$10^1 = (x-6)(x+3)$$

$$x^2 - 3x - 18 = 10$$

$$x^2 - 3x - 28 = 0$$

$$(x-7)(x+4) = 0$$

$$x = 7, -4$$

Since $\log_{10}(-10)$ is undefined, $x = -4$ is an extraneous root. So the solution is $x = 7$.

43. (a) Solving for x:

$$\log_{10} x - y = \log_{10}(3x - 1)$$
$$\log_{10} x - \log_{10}(3x - 1) = y$$
$$\log_{10}\left(\frac{x}{3x - 1}\right) = y$$
$$10^y = \frac{x}{3x - 1}$$
$$10^y(3x - 1) = x$$
$$3(10^y)x - 10^y = x$$
$$3(10^y)x - x = 10^y$$
$$x[3(10^y) - 1] = 10^y$$
$$x = \frac{10^y}{3(10^y) - 1}$$

(b) Solving for x:

$$\log_{10}(x - y) = \log_{10}(3x - 1)$$
$$x - y = 3x - 1$$
$$-2x = y - 1$$
$$x = \frac{y - 1}{-2} \text{ or } \frac{1 - y}{2}$$

45. Solving the inequality:

$$3(2 - 0.6^x) \le 1$$
$$2 - 0.6^x \le \tfrac{1}{3}$$
$$-0.6^x \le -\tfrac{5}{3}$$
$$0.6^x \ge \tfrac{5}{3}$$
$$\left(\tfrac{3}{5}\right)^x \ge \tfrac{5}{3}$$
$$\left(\tfrac{5}{3}\right)^{-x} \ge \tfrac{5}{3}$$
$$-x \ge 1$$
$$x \le -1$$

47. Solving the inequality:
$$4\left(10 - e^x\right) \le -3$$
$$10 - e^x \le -\tfrac{3}{4}$$
$$-e^x \le -\tfrac{43}{4}$$
$$e^x \ge \tfrac{43}{4}$$
$$x \ge \ln\tfrac{43}{4}$$
$$x \ge 2.375$$

49. Solving the inequality:
$$\ln(2 - 5x) > 2$$
$$2 - 5x > e^2$$
$$-5x > e^2 - 2$$
$$5x < 2 - e^2$$
$$x < \frac{2 - e^2}{5}$$
$$x < -1.078$$

51. Solving the inequality:
$$e^{2+x} \ge 100$$
$$2 + x \ge \ln 100$$
$$x \ge \ln 100 - 2$$
$$x \ge 2.605$$

53. Since $2^x > 0$ for all real numbers x, the solution set is all real numbers.

55. Solving the inequality:
$$\log_2 \frac{2x-1}{x-2} < 0$$
$$0 < \frac{2x-1}{x-2} < 1$$

Now $\dfrac{2x-1}{x-2} > 0$ for $x < \tfrac{1}{2}$ or $x > 2$ (using a sign chart). Solving the second inequality:
$$\frac{2x-1}{x-2} < 1$$
$$\frac{2x-1}{x-2} - \frac{x-2}{x-2} < 0$$
$$\frac{2x-1-x+2}{x-2} < 0$$
$$\frac{x+1}{x-2} < 0$$

This inequality is satisfied (using a sign chart) when $-1 < x < 2$. Combining answers, the solution set is $-1 < x < \tfrac{1}{2}$.

57. Solving the inequality:

$$e^{x^2-4x} \geq e^5$$
$$x^2 - 4x \geq 5$$
$$x^2 - 4x - 5 \geq 0$$
$$(x-5)(x+1) \geq 0$$

This inequality is satisfied (using a sign chart) when $x \leq -1$ or $x \geq 5$.

59. Solving the inequality:

$$e^{1/x-1} > 1$$
$$\frac{1}{x} - 1 > 0$$
$$\frac{1}{x} > 1$$
$$0 < x < 1$$

61. (a) We must have both $x > 0$ and $x - 4 > 0$, which occurs when $x > 4$. The domain is $(4, \infty)$.

(b) Solving the inequality:

$$\ln x + \ln(x-4) \leq \ln 21$$
$$\ln\left(x^2 - 4x\right) \leq \ln 21$$
$$x^2 - 4x \leq 21$$
$$x^2 - 4x - 21 \leq 0$$
$$(x-7)(x+3) \leq 0$$

The inequality is satisfied (using a sign chart) when $-3 \leq x \leq 7$. Combined with the domain from part (a), the solution set is $4 < x \leq 7$.

63. First note that the domain requires $x > 0$, $x + 1 > 0$, and $2x + 6 > 0$, which occurs when $x > 0$. Solving the inequality:

$$\log_2 x + \log_2 (x+1) - \log_2 (2x+6) < 0$$
$$\log_2 \frac{x(x+1)}{2x+6} < 0$$
$$\frac{x(x+1)}{2x+6} < 1$$
$$\frac{x^2 + x}{2x+6} - \frac{2x+6}{2x+6} < 0$$
$$\frac{x^2 - x - 6}{2x+6} < 0$$
$$\frac{(x-3)(x+2)}{2x+6} < 0$$

This inequality is satisfied (using a sign chart) when $x < -3$ or $-2 < x < 3$. Combined with the domain $x > 0$, the solution set is $0 < x < 3$.

65. (a) The domain of $y = \log_2(2x - 1)$ is $x > \frac{1}{2}$, while the domain of $y = \log_4 x$ is $x > 0$.

(b) Solving the inequality:
$$\log_2(2x - 1) > \log_4 x$$
$$4^{\log_2(2x-1)} > 4^{\log_4 x}$$
$$2^{2\log_2(2x-1)} > x$$
$$2^{\log_2(2x-1)^2} > x$$
$$(2x - 1)^2 > x$$
$$4x^2 - 4x + 1 > x$$
$$4x^2 - 5x + 1 > 0$$
$$(4x - 1)(x - 1) > 0$$

This inequality is satisfied (using a sign chart) when $x < \frac{1}{4}$ or $x > 1$. Since the combined domain from part (a) is $x > \frac{1}{2}$, the solution set is $x > 1$.

67. Let $u = \ln x$. Then the equation becomes:
$$3(\ln x)^2 - \ln(x^2) - 8 = 0$$
$$3(\ln x)^2 - 2\ln x - 8 = 0$$
$$3u^2 - 2u - 8 = 0$$
$$(3u + 4)(u - 2) = 0$$
$$u = -\tfrac{4}{3}, 2$$
Either $\ln x = -\frac{4}{3}$, so $x = e^{-4/3} \approx 0.264$, or $\ln x = 2$, so $x = e^2 \approx 7.389$.

69. Changing all logarithms to natural logarithms:
$$\log_6 x = \frac{1}{\frac{1}{\log_2 x} + \frac{1}{\log_3 x}}$$
$$\frac{\ln x}{\ln 6} = \frac{1}{\frac{\ln 2}{\ln x} + \frac{\ln 3}{\ln x}}$$
$$\frac{\ln x}{\ln 6} = \frac{\ln x}{\ln 2 + \ln 3}$$
$$\ln 6 = \ln 2 + \ln 3$$
Since this last equation is true for all real x, the solution set is $x > 0$ (note that the domain of $\ln x$ requires that $x > 0$).

71. Solving for x:

$$\alpha \ln x + \ln \beta = 0$$
$$\alpha \ln x = -\ln \beta$$
$$\ln x = -\frac{\ln \beta}{\alpha}$$
$$x = e^{-(\ln \beta)/\alpha}$$
$$x = e^{\ln \beta^{-1/\alpha}}$$
$$x = \beta^{-1/\alpha}$$

73. Solving for x:

$$y = Ae^{kx}$$
$$\frac{y}{A} = e^{kx}$$
$$\ln \frac{y}{A} = kx$$
$$x = \frac{1}{k} \ln \frac{y}{A}$$

75. Solving for x:

$$y = \frac{a}{1 + be^{-kx}}$$
$$y + bye^{-kx} = a$$
$$bye^{-kx} = a - y$$
$$e^{-kx} = \frac{a - y}{by}$$
$$-kx = \ln\left(\frac{a - y}{by}\right)$$
$$x = -\frac{1}{k} \ln\left(\frac{a - y}{by}\right)$$

77. (a) Starting with $x = \dfrac{\ln 3}{\ln 4 - 2\ln 3}$, we have:

$$x = \frac{\ln 3}{\ln 4 - 2\ln 3}$$
$$x(\ln 4 - 2\ln 3) = \ln 3$$
$$x\ln 4 - 2x\ln 3 = \ln 3$$
$$x\ln 4 = (2\ln 3)x + \ln 3$$
$$x\ln 4 = (2x + 1)\ln 3$$
$$\ln 4^x = \ln 3^{2x+1}$$
$$4^x = 3^{2x+1}$$

(b) The solution is approximately -1.355, which checks in the original equation.

79. (a) (a) Starting with $x = \pm\sqrt{1002}$, we have:
$$\log_{10}\left(x^2 - 2\right) = \log_{10}(1002 - 2) = \log_{10} 1000 = 3$$

(b) The solutions are approximately ± 31.654, which check in the original equation.

(b) (a) Starting with $x = 1 + \sqrt{1001}$, we have:
$$\log_{10}\left(x^2 - 2x\right) = \log_{10}\left[\left(1 + \sqrt{1001}\right)^2 - 2\left(1 + \sqrt{1001}\right)\right]$$
$$= \log_{10}\left(1 + 2\sqrt{1001} + 1001 - 2 - 2\sqrt{1001}\right)$$
$$= \log_{10} 1000$$
$$= 3$$
Starting with $x = 1 - \sqrt{1001}$, we have:
$$\log_{10}\left(x^2 - 2x\right) = \log_{10}\left[\left(1 - \sqrt{1001}\right)^2 - 2\left(1 - \sqrt{1001}\right)\right]$$
$$= \log_{10}\left(1 - 2\sqrt{1001} + 1001 - 2 + 2\sqrt{1001}\right)$$
$$= \log_{10} 1000$$
$$= 3$$

(b) The solutions are approximately -30.639 and 32.639, which check in the original equation.

81. Since $e > 0$, $2 - e < 2$, thus $\dfrac{2-e}{3} < \dfrac{2}{3}$.

83. Since $\log_\pi\left[\log_4\left(x^2 - 5\right)\right] < 0$, we must have:
$$0 < \log_4\left(x^2 - 5\right) < 1$$
$$4^0 < x^2 - 5 < 4^1$$
$$1 < x^2 - 5 < 4$$
$$6 < x^2 < 9$$
This inequality is satisfied on the set $\left(-3, -\sqrt{6}\right) \cup \left(\sqrt{6}, 3\right)$. Note that the original domain requires that $x^2 > 5$, which is satisfied by our solution set.

85. Taking natural logarithms of each side:

$$x^{(x^x)} = \left(x^x\right)^x$$

$$\ln x^{(x^x)} = \ln\left(x^x\right)^x$$

$$x^x \ln x = x \ln\left(x^x\right)$$

$$x^x \ln x = x^2 \ln x$$

$$(\ln x)\left(x^x - x^2\right) = 0$$

So either $\ln x = 0$, thus $x = 1$, or $x^x - x^2 = 0$:

$$x^x - x^2 = 0$$

$$x^x = x^2$$

$$\ln x^x = \ln x^2$$

$$x \ln x = 2 \ln x$$

$$(\ln x)(x - 2) = 0$$

$$x = 1, 2$$

The solutions are $x = 1$ or $x = 2$.

87. Solving for x:

$$\left(a^4 - 2a^2b^2 + b^4\right)^{x-1} = (a-b)^{2x}(a+b)^{-2}$$

$$\left(a^2 - b^2\right)^{2x-2} = \frac{(a-b)^{2x}}{(a+b)^2}$$

$$(a+b)^{2x-2}(a-b)^{2x-2} = \frac{(a-b)^{2x}}{(a+b)^2}$$

$$(a+b)^{2x}(a-b)^{-2} = 1$$

$$(a+b)^{2x} = (a-b)^2$$

$$2x \ln(a+b) = 2 \ln(a-b)$$

$$x = \frac{\ln(a-b)}{\ln(a+b)}$$

89. Solving for x:

$$6^x = \tfrac{10}{3} - 6^{-x}$$

$$6^x + 6^{-x} = \tfrac{10}{3}$$

$$6^{2x} + 1 = \tfrac{10}{3} 6^x$$

$$3 \cdot 6^{2x} - 10 \cdot 6^x + 3 = 0$$

$$\left(3 \cdot 6^x - 1\right)\left(6^x - 3\right) = 0$$

Setting each factor equal to 0, we have:

$$3 \cdot 6^x - 1 = 0 \qquad\qquad\qquad 6^x - 3 = 0$$
$$6^x = \tfrac{1}{3} \qquad\qquad\qquad\qquad 6^x = 3$$
$$x \log_{10} 6 = -\log_{10} 3 \qquad\qquad x \log_{10} 6 = \log_{10} 3$$
$$x = -\frac{\log_{10} 3}{\log_{10} 6} \qquad\qquad\qquad x = \frac{\log_{10} 3}{\log_{10} 6}$$

Now, since $\log_{10} 6 = \log_{10}(2 \cdot 3) = \log_{10} 2 + \log_{10} 3$, and since $\log_{10} 2 = a$ and $\log_{10} 3 = b$, we have:

$$x = \frac{-b}{a+b} \ \text{ or } \ x = \frac{b}{a+b}$$

So $x = \dfrac{\pm b}{a+b}$.

Graphing Utility Exercises for Section 6.5

1. (a) Graphing the two functions:

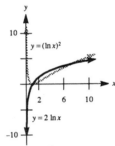

Note that the two graphs appear to intersect at $x = 1$.

   (b) The root is approximately $x \approx 7.39$, to two decimal places.

   (c) Solving the equation:
$$(\ln x)^2 = 2 \ln x$$
$$(\ln x)^2 - 2 \ln x = 0$$
$$(\ln x)(\ln x - 2) = 0$$

Either $\ln x = 0$, so $x = 1$, or $\ln x = 2$, so $x = e^2 \approx 7.39$. These answers check our results from part (a).

3.   (a)   Graphing the two functions:

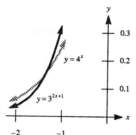

The x-coordinate of the intersection point is approximately $x \approx -1.355$.

(b)   Using a calculator, $x = \dfrac{\ln 3}{\ln 4 - 2\ln 3} \approx -1.355$, which checks our estimation from part (a).

(c)   Graphing the function:

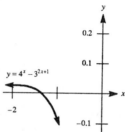

The x-intercept is approximately $x \approx -1.355$, which is consistent with our answer from part (a).

5.   (a)   Graphing the function and the line:

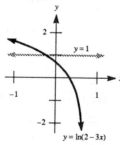

(b)   The x-coordinate of the intersection point is the left endpoint of the solution set from Example 8(a) in the text.

(c)   The x-coordinate of the intersection point is $x \approx -0.24$.

(d)   Using a calculator, $x = \dfrac{2-e}{3} \approx -0.24$, which checks our solution from part (c).

6.6 Compound Interest

1.   For annual compounding of money use the formula $A = P(1 + r)^t$. Now find A when $P = \$800$, $r = 0.06$ and $t = 4$ yrs:
$$A = 800(1 + 0.06)^4 = 800(1.06)^4 \approx \$1009.98$$

3.   For annual compounding of money use the formula $A = P(1 + r)^t$. Now find r when $A = \$6000$, $P = \$4000$ and $t = 5$:
$$6000 = 4000(1 + r)^5$$
$$1.5 = (1 + r)^5$$
$$(1.5)^{1/5} = 1 + r$$
$$r = (1.5)^{1/5} - 1 \approx 0.0845$$
So the interest rate is 8.45%.

5.   In the first bank, we have $P = \$500$, $r = 0.05$, and $t = 4$:
$$A = 500(1 + 0.05)^4 = 500(1.05)^4 \approx 607.75$$
Now deposit $P = \$607.75$, $r = 0.06$, and $t = 4$:
$$A = 607.75(1 + 0.06)^4 = 607.75(1.06)^4 \approx 767.27$$
So the new balance will be $767.27.

7.   (a) Use $A = P(1 + r)^t$ with $P = 1000$, $r = 0.07$, and $t = 20$:
$$A = 1000(1 + 0.07)^{20} = 1000(1.07)^{20} \approx 3869.68$$
The new balance is $3869.68.

(b) Use $A = P\left(1 + \frac{r}{N}\right)^t$ with $P = 1000$, $r = 0.07$, $N = 4$, and $t = 20$:
$$A = 1000\left(1 + \frac{0.07}{4}\right)^{4(20)} = 1000(1.0175)^{80} \approx 4006.39$$
The new balance is $4006.39.

9.   For compounding quarterly use the formula:
$$A = P\left(1 + \frac{r}{N}\right)^{Nt}$$
So here $P = \$100$, $r = 6\%$ and we have $N = 4$ compoundings per year. We find the value of t for which $A \geq \$120$:
$$120 \leq 100\left(1 + \frac{0.06}{4}\right)^{4t}$$
$$1.2 \leq (1.015)^{4t}$$
$$\ln 1.2 \leq 4t \ln 1.015$$
$$4t \geq \frac{\ln 1.2}{\ln 1.015}$$
$$t \geq \frac{\ln 1.2}{4 \ln 1.015} \approx 3.06$$
This is slightly over 3 years, and so 13 quarters will be required.

11. Use $A = P\left(1 + \frac{r}{N}\right)^{Nt}$ where $r = 0.055$, $N = 2$, $A = 6000$, and $t = 10$:

$$6000 = P\left(1 + \frac{0.55}{2}\right)^{2(10)}$$
$$6000 = P(1.0275)^{20}$$
$$P = \frac{6000}{(1.0275)^{20}} \approx 3487.50$$

You must deposit a principal of \$3487.50.

13. Use $A = Pe^{rt}$ where $A = 5000$, $t = 10$ and $r = 0.065$:

$$5000 = Pe^{(0.065)(10)}$$
$$5000 = Pe^{0.65}$$
$$P = \frac{5000}{e^{0.65}} \approx 2610.23$$

A principal of \$2610.23 will grow to \$5000 in 10 yrs.

15. Since the effective rate is $r = 0.06$, we have:

$$A = P(1 + 0.06)^1 = 1.06P$$

The nominal rate r would yield a balance of:

$$A = Pe^{r(1)} = Pe^r$$

Setting these equal:

$$Pe^r = P(1.06)$$
$$e^r = 1.06$$
$$r = \ln 1.06 \approx 0.0583$$

So the nominal rate is 5.83%.

17. Let's take the 6% investment first:

$$A = 10000(1 + 0.06)^5 = 10000(1.3382) \approx \$13382.26$$

The second choice will be:

$$A = 10000e^{0.05(5)} = 10000e^{0.25} \approx \$12840.25, \text{ considerably less.}$$

19. (a) We have $T_2 \approx \frac{0.7}{r} = \frac{0.7}{0.05} = 14$ yrs.

(b) We have $T_2 = \frac{\ln 2}{r} = \frac{\ln 2}{0.05} = 13.86$ yrs.

(c) Here $d_1 = 13.86$, $d_2 = 14$, so $d = |13.86 - 14| = 0.14$.

This represents $\frac{0.14}{13.86}(100) \approx 1.01\%$ of the actual doubling time.

21. Compute:

$$A = 1000e^{(0.08)(300)} = 1000e^{24} = 1000(2.65 \times 10^{10}) = \$2.65 \times 10^{13}$$

That's \$26.5 trillion, a nice inheritance.

23. (a) We have $T_2 \approx \frac{0.7}{0.05} = 14$ yrs.

(b) Sketch the graph:

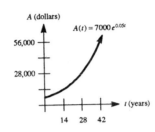

25. (a) For account #1, we have $P = 2000$, $r = 0.04$, and $N = 4$, so:
$$A_1(t) = 2000\left(1 + \frac{0.04}{4}\right)^{4t} = 2000(1.01)^{4t}$$
For account #2, we have $P = 2000$ and $r = 0.04$, so:
$$A_2(t) = 2000e^{0.04t}$$
Now complete the table:

t (years)	1	2	3	4	10
A (account #1)	2081	2166	2254	2345	2978
A (account #2)	2082	2167	2255	2347	2984

(b) For account #1, we find when $A_1(t) = 4000$:
$$4000 = 2000(1.01)^{4t}$$
$$2 = (1.01)^{4t}$$
$$\ln 2 = 4t \ln 1.01$$
$$t = \frac{\ln 2}{4 \ln 1.01} \approx 17.4 \text{ years}$$
For account #2, we find when $A_2(t) = 4000$:
$$4000 = 2000e^{0.04t}$$
$$2 = e^{0.04t}$$
$$\ln 2 = 0.04t$$
$$t = \frac{\ln 2}{0.04} \approx 17.3 \text{ years}$$
The doubling time for account #1 is 17.4 years while that of account #2 is 17.3 years.

(c) Using the table values:
$$\frac{\Delta A}{\Delta t} = \frac{A_1(3) - A_1(2)}{3 - 2} = \frac{2254 - 2166}{3 - 2} = \$88 \text{ per year}$$

(d) Using the table values:
$$\frac{\Delta A}{\Delta t} = \frac{A_2(3) - A_2(2)}{3 - 2} = \frac{2255 - 2167}{3 - 2} = \$88 \text{ per year}$$

27. (a) The quadratic model appears to fit the four data points best.

(b) Completing the table:

	1991	1992
Producer Price Index (logarithmic model)	116.5	117.9
Producer Price Index (quadratic model)	116.2	116.7
Producer Price Index (power model)	117.0	118.7
Producer Price Index (actual)	116.5	117.2

(c) The logarithmic model provides the best estimate for the actual 1991 producer price index.

(d) The quadratic model provides the best estimate for the actual 1992 producer price index.

(e) Using $t = 25$:

        logarithmic: 127.0
        quadratic: 107.1
        power: 129.8

The highest projection is provided by the power model, while the lowest projection is provided by the quadratic model.

(f) The quadratic model is probably the least likely to be useful in making long-range predictions about the producer price index. Since the parabola will eventually turn downward, and it is unlikely the index will decrease substantially, it is probably only useful in making short-range projections.

6.7 Exponential Growth and Decay

1. (a) Since 2000 bacteria are present initially, we have $N_0 = 2000$. Now $N = 3800$ when $t = 2$, so:

$$3800 = 2000e^{k \cdot 2}$$
$$1.9 = e^{2k}$$
$$\ln 1.9 = 2k$$
$$k = \frac{\ln 1.9}{2} \approx 0.3209$$

(b) Using the formula $N(t) = 2000e^{0.3209t}$, we find $N(5)$:

$$N(5) = 2000e^{(0.3209)(5)} \approx 9951 \text{ bacteria}$$

(c) Find when $N = 10000$:

$$10000 = 2000e^{0.3209t}$$
$$5 = e^{0.3209t}$$
$$\ln 5 = 0.3209t$$
$$t = \frac{\ln 5}{0.3209} \approx 5.0 \text{ hours}$$

3. Since $N(0) = 2000$, we have $N_0 = 2000$. Using $N(t) = 2000e^{kt}$, we find when $N(3) = 3400$:

$$3400 = 2000e^{k \cdot 3}$$
$$1.7 = e^{3k}$$
$$\ln 1.7 = 3k$$
$$k = \frac{\ln 1.7}{3} \approx 0.1769$$

5. The growth problems are solved by assuming an exponential growth rate:
$$N = N_0 e^{kt}$$
First we find the percent distribution by dividing the total population into each of the two regions:
$$\frac{1.169}{5.702} \approx 20.5\% \text{ for more developed regions}$$
Applying the growth formula to the total world population, we have:
$$N = N_0 e^{kt} = 5.702e^{0.015(5)} = 5.702e^{0.075} \approx 6.146 \text{ billion}$$
The other calculations are similar to these two. The completed table is:

Region	1995 Population (billions)	Percent of Population in 1995	Relative Growth Rate (percent per year)	Year 2000 Population (billions)	Percent of World Population in 2000
World	5.702	100	1.5	6.146	100
More dev.	1.169	20.5	0.2	1.181	19.22
Less dev.	4.533	79.5	1.9	4.985	81.11

Note that some discrepency in the last two columns is due to round-off error.

7. (a) Completing the table:

Country	1995 Population (millions)	Relative Growth Rate (percent per year)	Year 2000 Population (millions)	Percent Increase in Population
Iraq	20.6	3.7	24.8	20.4
United Kingdom	58.6	0.2	59.2	1.0

(b) Solving the equation:

$$20.6e^{0.037t} = 58.6e^{0.002t}$$

$$\frac{e^{0.037t}}{e^{0.002t}} = \frac{58.6}{20.6}$$

$$e^{0.035t} \approx 2.84$$

$$0.035t \approx \ln 2.84$$

$$t \approx \frac{\ln 2.84}{0.035} \approx 30$$

Iraq's population will be equal to the United Kingdom's in approximately 30 years, which is the year 2025.

9.   (a) For Cyprus, we use $N_0 = 0.7$ million, $k = 0.011$, and $t = 8$:

$$N(8) = 0.7e^{(0.011)(8)} \approx 0.8 \text{ million}$$

For Gaza, we use $N_0 = 0.7$ million, $k = 0.046$, and $t = 8$:

$$N(8) = 0.7e^{(0.046)(8)} \approx 1.0 \text{ million}$$

(b) For Cyprus, we find when $N = 1$ million:

$$1 = 0.7e^{0.011t}$$

$$1.43 = e^{0.011t}$$

$$\ln 1.43 = 0.011t$$

$$t = \frac{\ln 1.43}{0.011} \approx 33 \text{ years}$$

Cyprus will reach 1 million people in 33 years, which is the year 2025. At that time, the population of Gaza will be:

$$N(33) = 0.7e^{(0.046)(33)} \approx 3.2 \text{ million}$$

11.   (a) Use $N_0 = 1.3$, $t = 16$, and $N = 2$:

$$2 = 1.3e^{k \cdot 16}$$

$$1.54 \approx e^{16k \cdot}$$

$$\ln 1.54 \approx 16k$$

$$k \approx \frac{\ln 1.54}{16} \approx 0.0269$$

(b) Use $N_0 = 1.3$, $t = 106$, and $N = 2$:

$$2 = 1.3e^{k \cdot 106}$$

$$1.54 \approx e^{106k}$$

$$\ln 1.54 \approx 106k$$

$$k \approx \frac{\ln 1.54}{106} \approx 0.0041$$

13. (a) Use $N = N_0 e^{kt}$:
$$62947714 = 23191876 e^{k(50)}$$
$$2.7142 = e^{50k}$$
$$\ln 2.7142 = 50k$$
$$k = \frac{\ln 2.7142}{50} \approx 0.0200$$

(b) In 1950, we have:
$$N = (62947714)e^{(0.02)(50)} \approx 170853155$$
Your answer may differ slightly due to round-off error.

(c) The actual growth over that period was slower than that predicted from (b) with a constant growth rate.

15. (a) Use $N = N_0 e^{kt}$:
$$13479142 = 12588066 e^{k(10)}$$
$$1.0708 = e^{10k}$$
$$\ln 1.0708 = 10k$$
$$k = \frac{\ln 1.0708}{10} \approx 0.00684$$
The growth rate is 0.684%.

(b) Use $N = N_0 e^{kt}$:
$$14830192 = 13479142 e^{k(10)}$$
$$1.1002 = e^{10k}$$
$$\ln 1.1002 = 10k$$
$$k = \frac{\ln 1.1002}{10} \approx 0.00955$$
The growth rate is 0.955%.

(c) Use $N = N_0 e^{kt}$, $N_0 = 14830192$, $k = 0.00955$, and $t = 40$:
$$N = 14830192 e^{0.00955(40)} \approx 21731258$$

(d) Our prediction is higher than the actual population.

17. (a) Complete the table:

Region	1990 Population (millions)	Growth Rate (%)	2025 Population
North America	275.2	0.7	351.6
Soviet Union	291.3	0.7	372.2
Europe	499.5	0.2	535.7
Nigeria	113.3	3.1	335.3

(b) It will be 335.3 mil – 113.3 mil = 222.0 million.

(c) The net increases are given by:

North America:	351.6 mil – 275.2 mil = 76.4 mil
Soviet Union:	372.2 mil – 291.3 mil = 80.9 mil
Europe:	535.7 mil – 499.5 mil = 36.2 mil
combined:	193.5 million

(d) For Nigeria, our results support this projection.

19. (a) Complete the table:

t(seconds)	0	550	1100	1650	2200
N(grams)	8	4	2	1	0.5

(b) Complete the table:

t(years)	0	4.9×10^9	9.8×10^9	14.7×10^9	19.6×10^9
N(grams)	10	5	2.5	1.25	0.625

21. Given the decay law $N = N_0 e^{kt}$, and $N = \frac{1}{2}N_0$ when $t = 8$, we have:

$$\frac{1}{2}N_0 = N_0 e^{8k}$$

$$\frac{1}{2} = e^{8k}$$

$$\ln\frac{1}{2} = 8k$$

$$k = \frac{\ln\frac{1}{2}}{8} \approx -0.0866$$

So when $t = 7$ and $N_0 = 1$, we have:

$$N(7) = 1e^{(-0.0866)(7)} \approx 0.55 \text{ g}$$

23. (a) Using $N = \frac{1}{2}N_0$ when $t = 14.9$ hours:

$$\frac{1}{2}N_0 = N_0 e^{14.9k}$$

$$\frac{1}{2} = e^{14.9k}$$

$$\ln\frac{1}{2} = 14.9k$$

$$k = \frac{\ln\frac{1}{2}}{14.9} \approx -0.0465$$

Now using $t = 48$ hours and $N_0 = 40$ g:

$$N(48) = 40e^{(-0.0465)(48)} \approx 4.29 \text{ g}$$

Approximately 4.29 g of the sample will remain after 48 hours.

(b) Find when $N(t) = 1$:
$$1 = 40e^{-0.0465t}$$
$$\frac{1}{40} = e^{-0.0465t}$$
$$\ln\frac{1}{40} = -0.0465t$$
$$t = \frac{\ln\frac{1}{40}}{-0.0465} \approx 79 \text{ hours}$$
The isotope will decay to 1 gram in approximately 79 hours.

25. (a) Drawing the graph:

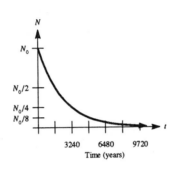

(b) Drawing the graph:

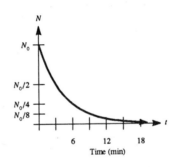

27. (a) First find the decay constant k:
$$\tfrac{1}{2}N_0 = N_0e^{13k}$$
$$\tfrac{1}{2} = e^{13k}$$
$$\ln\tfrac{1}{2} = 13k$$
$$k = \frac{\ln\frac{1}{2}}{13} \approx -0.0533$$
Now using $N_0 = 2$ g and $t = 5$ years:
$$N(5) = 2e^{(-0.0533)(5)} \approx 1.53 \text{ grams}$$

(b) Find when $N(t) = 0.2$ g:
$$0.2 = 2e^{-0.0533t}$$
$$0.1 = e^{-0.0533t}$$
$$\ln 0.1 = -0.0533t$$
$$t = \frac{\ln 0.1}{-0.0533} \approx 43 \text{ years}$$

29. (a) The decay law is $N = N_0 e^{kt}$, where $k = \frac{\ln 0.5}{7340}$. We want to determine t when $\frac{N_0}{1000} = N_0 e^{kt}$. After dividing by N_0, and then taking the logarithm of both sides, we obtain $\ln 0.001 = kt$. Thus $t = \frac{\ln 0.001}{k}$, where $k = \frac{\ln 0.5}{7340}$. A calculator then shows that it is approximately 73,000 yrs.

(b) The answer 73,000 yrs is approximately 10 half-lives.

(c) After one half-life, we have $N = \frac{N_0}{2}$. After two half-lives, we have $N = \frac{N_0}{2^2}$. After three half-lives, we have $N = \frac{N_0}{2^3}$. Continuing the pattern, we see that after 10 half-lives, $N = \frac{N_0}{2^{10}} \approx \frac{N_0}{1000}$. This agrees with the result in part (b).

31. (a) We have $k = \frac{\ln 0.5}{28} \approx -0.0248$.

(b) Solve for t:
$$\frac{N_0}{1000} = N_0 e^{kt}$$
$$0.001 = e^{-0.0248t}$$
$$\ln 0.001 = -0.0248t$$
$$t = \frac{\ln 0.001}{-0.0248} \approx 279 \text{ years}$$

(c) Since $2^{10} \approx 1000$, after 10 half-lives, it should be reduced to $\frac{N_0}{1000}$. Since each half-life is 28 years, this is approximately 280 years. Note that this is close to our answer from (b).

33. (a) Let T represent the half-life. Since $T = \frac{\ln \frac{1}{2}}{k} = \frac{-\ln 2}{k}$, $k = \frac{-\ln 2}{T}$. Therefore:
$$\mu = \frac{-1}{k} = \frac{-1}{\frac{-\ln 2}{T}} = \frac{T}{\ln 2}$$

(b) The mean life is $\mu = \frac{10}{\ln 2} \approx 14.4 \text{ seconds}$.

(c)   After 3 mean lives we have $3(14.4) = 43.2$ seconds passed. Since $k = \dfrac{-\ln 2}{10}$,

we have:

$$N(43.2) = 1e^{\left(\frac{-\ln 2}{10}\right)(43.2)} \approx 0.05$$

Approximately 0.05 g of the sample remains.

35.   First compute the decay constant:

$$\tfrac{1}{2}N_0 = N_0 e^{\left(7 \times 10^8\right)k}$$

$$\tfrac{1}{2} = e^{\left(7 \times 10^8\right)k}$$

$$\ln \tfrac{1}{2} = \left(7 \times 10^8\right)k$$

$$k = \frac{\ln \tfrac{1}{2}}{7 \times 10^8} \approx -9.90 \times 10^{-10}$$

Now using the decay law:

$$N = 0.028 e^{\left(-9.76 \times 10^{-10}\right)(1000)} \approx 0.02799997 \text{ ounces}$$

37.   (a)   We use $N = N_0 e^{kt}$ with $N = 10 \times 10^9$, $N_0 = 3.6 \times 10^9$, and $k = 0.02$:

$$10 \times 10^9 = \left(3.6 \times 10^9\right)e^{0.02t}$$

$$\tfrac{10}{3.6} = e^{0.02t}$$

$$\ln \tfrac{10}{3.6} = 0.02t$$

$$t = \frac{\ln \tfrac{10}{3.6}}{0.02} \approx 51 \text{ years}$$

So 51 years after 1969, or 2020, the carrying capacity will be reached.

(b)   We use $N = N_0 e^{kt}$ with $N = 10 \times 10^9$, $N_0 = 5.7 \times 10^9$, and $k = 0.015$:

$$10 \times 10^9 = \left(5.7 \times 10^9\right)e^{0.015t}$$

$$\tfrac{10}{5.7} = e^{0.015t}$$

$$\ln \tfrac{10}{5.7} = 0.015t$$

$$t = \frac{\ln \tfrac{10}{5.7}}{0.015} \approx 37 \text{ years}$$

So 37 years after 1995, or 2032, the carrying capacity will be reached. Using more recent data, the carrying capacity will be reached 12 years later.

39.   (a)   Substituting $t = 7$:

$$y = (2224.955)e^{(0.168)(7)} \approx 7212$$

The estimate is $7212 million, which is lower than the actual figure of $8590 million.

(b) Substituting $t = 15$:
$$y = (2224.955)e^{(0.168)(15)} \approx 27,653$$
The estimate is $27.653 billion.

(c) Since $(1, 2633)$ lies on the curve, we have:
$$2633 = ae^{b \cdot 1}$$
Since $(4, 4414)$ lies on the curve, we have:
$$4414 = ae^{b \cdot 4}$$
Now dividing:
$$\frac{4414}{2633} = \frac{ae^{4b}}{ae^b}$$
$$\frac{4414}{2633} = e^{3b}$$
$$3b = \ln \frac{4414}{2633}$$
$$b = \tfrac{1}{3} \ln \frac{4414}{2633} \approx 0.1722$$
Substituting to find a:
$$2633 = ae^{0.1722}$$
$$a = \frac{2633}{e^{0.1722}} \approx 2216.445$$

The exponential curve is $y = (2216.445)e^{0.1722t}$.

(d) Substituting $t = 7$:
$$y = (2216.445)e^{(0.1722)(7)} \approx 7399$$
This equation comes closer to predicting the actual 1992 value.

41. (a) First multiply both sides by $\dfrac{k}{A_0}$ to get:
$$\frac{Ak}{A_0} = e^{kT} - 1$$
$$e^{kT} = \frac{Ak}{A_0} + 1$$
Now taking the natural logarithm of both sides we have:
$$kT = \ln\left(\frac{Ak}{A_0} + 1\right)$$
$$T = \frac{\ln\left(\dfrac{Ak}{A_0} + 1\right)}{k}$$

(b) We have:

$$T = \frac{\ln\left[\frac{(124)(0.016)}{3.090} + 1\right]}{0.016} \approx 31 \text{ yrs}$$

(c) Computing the life expectancy:

$$\frac{136.7 \text{ billion metric tons}}{3.15 \text{ billion metric tons/year}} \approx 43 \text{ years}$$

43. (a) Computing the life expectancy:

$$T = \frac{\ln\left[\frac{(321000)(0.022)}{10773.2} + 1\right]}{0.022} \approx 23 \text{ yrs}$$

The corresponding depletion date is 2013.

(b) Doubling the world reserves and halving the growth rate:

$$T = \frac{\ln\left[\frac{(642000)(0.011)}{10773.2} + 1\right]}{0.011} \approx 46 \text{ yrs}$$

The corresponding depletion date is 2036.

(c) Computing the life expectancy:

$$\frac{321000 \text{ thousand metric tons}}{10773.2 \text{ thousand metric tons/year}} \approx 30 \text{ years}$$

The corresponding depletion date is 2020.

45. (a) Computing the life expectancy:

$$T = \frac{\ln\left[\frac{(70000)(0.004)}{5544.5} + 1\right]}{0.004} \approx 12 \text{ yrs}$$

The corresponding depletion date is 2002.

(b) Doubling the world reserves and halving the growth rate:

$$T = \frac{\ln\left[\frac{(140000)(0.002)}{5544.5} + 1\right]}{0.002} \approx 25 \text{ yrs}$$

The corresponding depletion date is 2015.

(c) Computing the life expectancy:

$$\frac{70,000 \text{ thousand metric tons}}{5,544.5 \text{ thousand metric tons/year}} \approx 13 \text{ years}$$

The corresponding depletion date is 2003.

47. **(a)** **(i)** Using $N = 600$, $N_0 = 346.83$, and $k = 0.01$:

$$600 = 346.83e^{0.01t}$$

$$\frac{600}{346.83} = e^{0.01t}$$

$$0.01t = \ln\frac{600}{346.83}$$

$$t = 100\ln\frac{600}{346.83} \approx 55 \text{ years}$$

(ii) Using $N = 600$, $N_0 = 346.83$, and $k = 0.02$:

$$600 = 346.83e^{0.02t}$$

$$\frac{600}{346.83} = e^{0.02t}$$

$$0.02t = \ln\frac{600}{346.83}$$

$$t = 50\ln\frac{600}{346.83} \approx 27 \text{ years}$$

(b) **(i)** Computing the life expectancy:

$$T = \frac{\ln\left[\frac{(225,000)(0.01)}{346.83} + 1\right]}{0.01} \approx 201 \text{ yrs}$$

(ii) Computing the life expectancy:

$$T = \frac{\ln\left[\frac{(225,000)(0.02)}{346.83} + 1\right]}{0.02} \approx 132 \text{ yrs}$$

(c) Computing the life expectancy:

$$\frac{155,000 \text{ quadrillion Btu}}{346.83 \text{ quadrillion Btu/year}} \approx 447 \text{ years}$$

49. **(a)** Solving for N_0:

$$N_r = N_0e^{kT}$$

$$\frac{N_r}{e^{kT}} = N_0$$

$$N_0 = N_re^{-kT}$$

So $N_s = N_0 - N_r = N_re^{-kT} - N_r$.

(b)   Solving for T:
$$N_s + N_r = N_r e^{-kT}$$
$$\frac{N_s}{N_r} + 1 = e^{-kT}$$
$$\ln\left(\frac{N_s}{N_r} + 1\right) = -kT$$
$$T = \frac{\ln\left(\frac{N_s}{N_r} + 1\right)}{-k}$$

51.   Compute T:
$$T = \frac{\ln\left(\frac{N_s}{N_r} + 1\right)}{-k} = \frac{\ln(0.0636 + 1)}{-\left(-1.4748 \times 10^{-11}\right)} \approx 4.181 \times 10^9 \text{ yrs}$$
The rock is approximately 4.181 billion years old.

53.   Compute T:
$$T = \frac{5730 \ln\frac{N}{920}}{\ln\frac{1}{2}} = \frac{5730 \ln\frac{141}{920}}{\ln\frac{1}{2}} \approx 15505 \text{ yrs}$$
The two paintings are 15,505 years old.

55.   Compute T:
$$T = \frac{5730 \ln\frac{N}{920}}{\ln\frac{1}{2}} = \frac{5730 \ln\frac{348}{920}}{\ln\frac{1}{2}} \approx 8000 \text{ yrs}$$
The site is older than which was originally estimated.

57.   Compute T:
$$T = \frac{5730 \ln\left(\frac{N}{920}\right)}{\ln\left(\frac{1}{2}\right)} = \frac{5730 \ln\left(\frac{723}{920}\right)}{\ln\left(\frac{1}{2}\right)} \approx 1992$$
This would correspond to the years 10-5 B.C., which fit in the historical range.

59.   (a)   Completing the table:

	$N(-1)$	$N(0)$	$N(1)$	$N(4)$	$N(5)$
From graph	0.25	0.5	1.0	3.5	3.75
From calculator	0.176	0.444	1.014	3.489	3.795

(b) Computing the values, we have:

$$N(10) = \frac{4}{1+8e^{-10}} \approx 3.99854773$$

$$N(15) = \frac{4}{1+8e^{-15}} \approx 3.99999021$$

$$N(20) = \frac{4}{1+8e^{-20}} \approx 3.99999993$$

Note that the values are approaching $N = 4$.

(c) The value is approximately $t \approx 3$.

(d) Let $N = 3$ and solve for t:

$$3 = \frac{4}{1+8e^{-t}}$$

$$1+8e^{-t} = \tfrac{4}{3}$$

$$8e^{-t} = \tfrac{1}{3}$$

$$e^{-t} = \tfrac{1}{24}$$

$$e^{t} = 24$$

$$t = \ln 24 \approx 3.178$$

The answer is consistent with the answer from part (c).

(e) Switch the roles of N and t, and solve the resulting equation for N:

$$t = \frac{4}{1+8e^{-N}}$$

$$1+8e^{-N} = \frac{4}{t}$$

$$8e^{-N} = \frac{4}{t} - 1$$

$$e^{-N} = \frac{4-t}{8t}$$

$$e^{N} = \frac{8t}{4-t}$$

$$N = \ln\left(\frac{8t}{4-t}\right)$$

Thus $N^{-1}(t) = \ln\left(\dfrac{8t}{4-t}\right)$.

(f) Computing $N^{-1}(3) = \ln\left(\dfrac{8(3)}{4-3}\right) = \ln 24 \approx 3.178$. The answer is the same as in part (d).

61. (a) Since $N = 0.24$ when $t = 0$, we have:

$$0.24 = \frac{50}{1 + ae^0}$$

$$0.24 = \frac{50}{1 + a}$$

$$1 + a = \frac{50}{0.24}$$

$$1 + a = \frac{625}{3}$$

$$a = \frac{622}{3} \approx 207.33$$

So the logistic function is $N(t) = \dfrac{50}{1 + 207.33e^{-bt}}$. Since $N = 13.53$ when $t = 2$, we have:

$$13.53 = \frac{50}{1 + 207.33e^{-b(2)}}$$

$$1 + 207.33e^{-2b} = \frac{50}{13.53}$$

$$207.33e^{-2b} \approx 2.695$$

$$e^{-2b} \approx 0.013$$

$$-2b \approx \ln 0.013$$

$$b \approx \frac{\ln 0.013}{-2} \approx 2.17$$

So the logistic function is $N(t) = \dfrac{50}{1 + 207.33e^{-2.17t}}$.

(b) Computing the function values:

$$N(1) = \frac{50}{1 + 207.33e^{-2.17(1)}} \approx 2.0$$

$$N(3) = \frac{50}{1 + 207.33e^{-2.17(3)}} \approx 38.2$$

$$N(4) = \frac{50}{1 + 207.33e^{-2.17(4)}} \approx 48.3$$

$$N(5) = \frac{50}{1 + 207.33e^{-2.17(5)}} \approx 49.8$$

Note that these values agree fairly closely with Thornton's data.

(c)   Finding when $N = 10$:

$$10 = \frac{50}{1 + 207.33e^{-2.17t}}$$

$$1 + 207.33e^{-2.17t} = 5$$

$$207.33e^{-2.17t} = 4$$

$$e^{-2.17t} = \frac{4}{207.33}$$

$$-2.17t = \ln\frac{4}{207.33}$$

$$t = \frac{\ln\frac{4}{207.33}}{-2.17} \approx 1.82$$

Thus the area is 10 cm^2 when $t \approx 1.82$ days, which is 1 day 19.5 hours.

63.   (a)   Compute $\dfrac{\Delta N}{\Delta t}$ over the interval $[t, t + \Delta t]$:

$$\frac{\Delta N}{\Delta t} = \frac{N(t + \Delta t) - N(t)}{t + \Delta t - t}$$

$$= \frac{N_0 e^{k(t+\Delta t)} - N_0 e^{kt}}{\Delta t}$$

$$= \frac{N_0 e^{kt} e^{k\Delta t} - N_0 e^{kt}}{\Delta t}$$

$$= \frac{N_0 e^{kt}\left(e^{k\Delta t} - 1\right)}{\Delta t}$$

$$= \frac{N\left(e^{k\Delta t} - 1\right)}{\Delta t}$$

(b)   Substituting:

$$\frac{\Delta N}{\Delta t} = \frac{N\left(e^{k\Delta t} - 1\right)}{\Delta t} \approx \frac{N(1 + k\Delta t - 1)}{\Delta t} = \frac{kN\Delta t}{\Delta t} = kN$$

Thus $\dfrac{\Delta N}{\Delta t} \approx kN$ when Δt is close to 0.

Chapter Six Review Exercises

1.   If $x = \log_5 126$, then $5^x = 126$. Since $5^3 = 125$, $x > 3$.
If $y = \log_{10} 999$, then $10^y = 999$. Since $10^3 = 1000$, $y < 3$. So $\log_5 126$ is larger.

3. Since $N(t) = N_0 e^{kt}$ and $N_0 = 8000$, $N(t) = 8000 e^{kt}$. Since $N(4) = 10000$, we can find k:
$$N(4) = 10000$$
$$8000 e^{4k} = 10000$$
$$e^{4k} = 1.25$$
$$k = \frac{\ln 1.25}{4}$$

Now find t such that $N(t) = 12000$:
$$N(t) = 12000$$
$$8000 e^{kt} = 12000$$
$$e^{kt} = 1.5$$
$$kt = \ln 1.5$$
$$t = \frac{\ln 1.5}{k} = \frac{\ln 1.5}{\frac{\ln 1.25}{4}} = \frac{4 \ln 1.5}{\ln 1.25} \text{ hours}$$

5. Graphing $f(x)$:

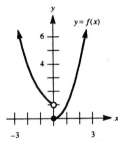

f is not one-to-one since it does not pass the horizontal line test.

7. Solve for x:
$$\ln(x+1) - 1 = \ln(x-1)$$
$$\ln(x+1) - \ln(x-1) = 1$$
$$\ln \frac{x+1}{x-1} = 1$$
$$\frac{x+1}{x-1} = e$$
$$x + 1 = ex - e$$
$$e + 1 = ex - x$$
$$e + 1 = x(e-1)$$
$$x = \frac{e+1}{e-1}$$

9. For $y = e^x$, the domain is $(-\infty, \infty)$ and the range is $(0, \infty)$. For $y = \ln x$, the domain is $(0, \infty)$ and the range is $(-\infty, \infty)$. Graphing both functions:

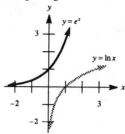

11. Simplify the logarithm:
$$\log_9 \tfrac{1}{27} = \log_9\left(3^{-3}\right) = \log_9\left(9^{1/2}\right)^{-3} = \log_9 9^{-3/2} = -\tfrac{3}{2}$$

13. Using the decay equation $N = N_0 e^{kt}$, find k when $t = 13$ and $N = \tfrac{1}{2}N_0$:
$$\tfrac{1}{2}N_0 = N_0 e^{13k}$$
$$\tfrac{1}{2} = e^{13k}$$
$$\ln\tfrac{1}{2} = 13k$$
$$k = \frac{\ln\tfrac{1}{2}}{13} \approx -0.05$$

15. Solve for x:
$$5e^{2-x} = 12$$
$$e^{2-x} = \tfrac{12}{5}$$
$$2 - x = \ln\tfrac{12}{5}$$
$$-x = -2 + \ln\tfrac{12}{5}$$
$$x = 2 - \ln\tfrac{12}{5}$$

17. Use $N = N_0 e^{kt}$ where $N_0 = 2$ and $k = 0.02$, so $N(t) = 2e^{0.02t}$. Find t when $N = 3$:
$$3 = 2e^{0.02t}$$
$$1.5 = e^{0.02t}$$
$$\ln 1.5 = 0.02t$$
$$t = \frac{\ln 1.5}{0.02} \approx 20 \text{ yrs}$$
We would expect the population to reach 3 million in the year 2015.

19. Use $A = Pe^{rt}$ where $P = \$1000$ and $r = 0.10$, so $A(t) = 1000e^{0.1t}$. The doubling time is approximately $\frac{70}{10} = 7$ years. Sketching the graph:

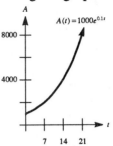

21. The horizontal asymptote is $y = 0$, there is no vertical asymptote, there is no x-intercept, and the y-intercept is 1. Drawing the graph:

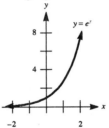

23. There is no horizontal asymptote, the vertical asymptote is $x = 0$, the x-intercept is 1, and there is no y-intercept. Drawing the graph:

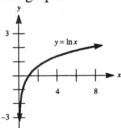

25. The horizontal asymptote is $y = 1$, there is no vertical asymptote, there is no x-intercept, and the y-intercept is 3. Drawing the graph:

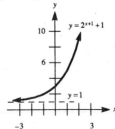

27. The horizontal asymptote is $y = 0$, there is no vertical asymptote, there is no x-intercept, and the y-intercept is 1. Drawing the graph:

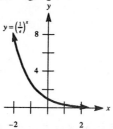

29. The horizontal asymptote is $y = 1$, there is no vertical asymptote, there is no x-intercept, and the y-intercept is $e + 1$. Drawing the graph:

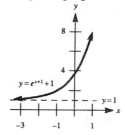

31. There are no asymptotes, and the x- and y-intercepts are both 0. Drawing the graph:

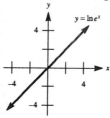

33. Using properties of logarithms:
$$\log_4 x + \log_4(x - 3) = 1$$
$$\log_4[x(x - 3)] = 1$$
$$x(x - 3) = 4^1$$
$$x^2 - 3x = 4$$
$$x^2 - 3x - 4 = 0$$
$$(x - 4)(x + 1) = 0$$
$$x = 4, -1$$
Since $\log_4(-1)$ is undefined, $x = -1$ is an extraneous root. So the solution is $x = 4$.

35. Using properties of logarithms:
$$\ln x + \ln(x+2) = \ln 15$$
$$\ln[x(x+2)] = \ln 15$$
$$x(x+2) = 15$$
$$x^2 + 2x = 15$$
$$x^2 + 2x - 15 = 0$$
$$(x+5)(x-3) = 0$$
$$x = -5, 3$$
Since $\ln(-5)$ is undefined, $x = -5$ is an extraneous root. So the solution is $x = 3$.

37. Using properties of logarithms:
$$\log_2 x + \log_2(3x+10) - 3 = 0$$
$$\log_2[x(3x+10)] = 3$$
$$x(3x+10) = 2^3$$
$$3x^2 + 10x = 8$$
$$3x^2 + 10x - 8 = 0$$
$$(3x-2)(x+4) = 0$$
$$x = \tfrac{2}{3}, -4$$
Since $\log_2(-4)$ is undefined, $x = -4$ is an extraneous root. So the solution is $x = \tfrac{2}{3}$.

39. Solve for x:
$$3\log_9 x = \tfrac{1}{2}$$
$$\log_9 x = \tfrac{1}{6}$$
$$x = 9^{1/6} = \sqrt[6]{9} = \sqrt[3]{3}$$

41. Solve for x:
$$e^{1-5x} = 3\sqrt{e}$$
$$1 - 5x = \ln 3\sqrt{e}$$
$$1 - 5x = \ln 3 + \tfrac{1}{2}\ln e$$
$$1 - 5x = \ln 3 + \tfrac{1}{2}$$
$$-5x = \ln 3 - \tfrac{1}{2}$$
$$x = \frac{1 - 2\ln 3}{10}$$

43. Using properties of logarithms:
$$\log_{10} x - 2 = \log_{10}(x - 2)$$
$$\log_{10} x - \log_{10}(x - 2) = 2$$
$$\log_{10} \frac{x}{x - 2} = 2$$
$$\frac{x}{x - 2} = 100$$
$$x = 100(x - 2)$$
$$x = 100x - 200$$
$$200 = 99x$$
$$x = \tfrac{200}{99}$$

45. Using properties of logarithms:
$$\ln(x + 2) = \ln x + \ln 2$$
$$\ln(x + 2) = \ln 2x$$
$$x + 2 = 2x$$
$$x = 2$$

47. Since $\ln(x^4) = 4 \ln x$ is an identity, the solution is all real numbers $x > 0$, or $(0, \infty)$.

49. Using properties of logarithms:
$$\log_{10} x = \ln x$$
$$\frac{\ln x}{\ln 10} = \ln x$$
$$\ln x = (\ln 10)(\ln x)$$
$$\ln x - (\ln 10)(\ln x) = 0$$
$$\ln x(1 - \ln 10) = 0$$
$$\ln x = 0, \text{ so } x = 1$$

51. Using properties of logarithms:
$$\log_{10} \sqrt{10} = \log_{10}\left(10^{1/2}\right) = \tfrac{1}{2}$$

53. Using properties of logarithms:
$$\ln\left(\sqrt[5]{e}\right) = \ln\left(e^{1/5}\right) = \tfrac{1}{5}$$

55. Using properties of logarithms:
$$\log_{10} \pi - \log_{10} 10\pi = \log_{10} \pi - \left(\log_{10} 10 + \log_{10} \pi\right)$$
$$= \log_{10} \pi - \log_{10} 10 - \log_{10} \pi$$
$$= -\log_{10} 10$$
$$= -1$$

57. Using properties of logarithms:
$$10^{\log_{10} 16} = 16$$

59. Evaluate $\ln\left(e^4\right) = 4$.

61. Using properties of logarithms:
$$\log_{12} 2 + \log_{12} 18 + \log_{12} 4 = \log_{12}(2 \cdot 18 \cdot 4) = \log_{12} 144 = \log_{12}\left(12^2\right) = 2$$

63. Using properties of logarithms:
$$\frac{\ln 100}{\ln 10} = \frac{\ln\left(10^2\right)}{\ln 10} = \frac{2 \ln 10}{\ln 10} = 2$$

65. Using properties of logarithms:
$$\log_2 \sqrt[7]{16\sqrt[3]{2\sqrt{2}}} = \tfrac{1}{7}\log_2 16\sqrt[3]{2\sqrt{2}}$$
$$= \tfrac{1}{7}\left(\log_2 16 + \log_2 \sqrt[3]{2\sqrt{2}}\right)$$
$$= \tfrac{1}{7}\left[\log_2\left(2^4\right) + \tfrac{1}{3}\log_2 2\sqrt{2}\right]$$
$$= \tfrac{1}{7}\left[4 + \tfrac{1}{3}\left(\log_2 2 + \log_2 \sqrt{2}\right)\right]$$
$$= \tfrac{1}{7}\left[4 + \tfrac{1}{3}\left(1 + \log_2\left(2^{1/2}\right)\right)\right]$$
$$= \tfrac{1}{7}\left(4 + \tfrac{1}{3} \cdot \tfrac{3}{2}\right)$$
$$= \tfrac{1}{7}\left(4 + \tfrac{1}{2}\right)$$
$$= \tfrac{1}{7} \cdot \tfrac{9}{2}$$
$$= \tfrac{9}{14}$$

67. Using properties of logarithms:
$$\log_{10}\left(A^2 B^3 \sqrt{C}\right) = \log_{10} A^2 + \log_{10} B^3 + \log_{10} \sqrt{C}$$
$$= 2\log_{10} A + 3\log_{10} B + \tfrac{1}{2}\log_{10} C$$
$$= 2a + 3b + \tfrac{c}{2}$$

69. Using properties of logarithms:
$$16\log_{10} \sqrt{A}\ \sqrt[4]{B} = 16\left(\log_{10} \sqrt{A} + \log_{10} \sqrt[4]{B}\right)$$
$$= 16\left(\tfrac{1}{2}\log_{10} A + \tfrac{1}{4}\log_{10} B\right)$$
$$= 8\log_{10} A + 4\log_{10} B$$
$$= 8a + 4b$$

71. Since $\log_{10} 100 = 2$ and $\log_{10} 1000 = 3$, $\log_{10} 209$ lies between 2 and 3.

73. Since $\log_6 36 = 2$ and $\log_6 216 = 3$, $\log_6 100$ lies between 2 and 3.

75. Since $\log_{10} 0.010 = -2$ and $\log_{10} 0.001 = -3$, $\log_{10} 0.003$ lies between -2 and -3.

77. (a) Graph $y = \ln (x + 2)$ and $y = \ln (-x) - 1$:

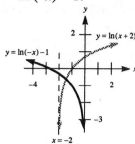

The graph shows these two curves intersect in the third quadrant.

(b) Setting the y-coordinates equal:
$$\ln(x + 2) = \ln(-x) - 1$$
$$\ln(x + 2) - \ln(-x) = -1$$
$$\ln\frac{x + 2}{-x} = -1$$
$$\frac{x + 2}{-x} = e^{-1}$$
$$e(x + 2) = -x$$
$$ex + 2e = -x$$
$$ex + x = -2e$$
$$x(e + 1) = -2e$$
$$x = \frac{-2e}{e + 1} \approx -1.46$$

79. We know $N = N_0 e^{kt}$, and $N = \tfrac{1}{2}N_0$ when $t = T$:
$$\tfrac{1}{2}N_0 = N_0 e^{kT}$$
$$\tfrac{1}{2} = e^{kT}$$
$$\ln\tfrac{1}{2} = kT$$
$$k = \frac{\ln\tfrac{1}{2}}{T}$$

81. Since 4 half-lives have passed, we will have $\left(\tfrac{1}{2}\right)^4 N_0 = \tfrac{1}{16}N_0$ of the substance left, or 6.25% remaining.

83. We have $k = \dfrac{\ln\frac{1}{2}}{d}$ and $N_0 = b$. We want to find the value of t when $N = c$:

$$N = N_0 e^{kt}$$

$$c = be^{kt}$$

$$\frac{c}{b} = e^{kt}$$

$$\ln\frac{c}{b} = kt$$

$$\ln\frac{c}{b} = \frac{\ln\frac{1}{2}}{d} \bullet t$$

$$t = \frac{d\ln\frac{c}{b}}{\ln\frac{1}{2}} \text{ days}$$

85. Using properties of logarithms:

$$\log_{10} 8 + \log_{10} 3 - \log_{10} 12 = \log_{10}\left(\frac{8 \bullet 3}{12}\right) = \log_{10} 2$$

87. Using properties of logarithms:

$$\ln 5 - 3\ln 2 + \ln 16 = \ln 5 - \ln 2^3 + \ln 16 = \ln 5 - \ln 8 + \ln 16 = \ln\left(\frac{5 \bullet 16}{8}\right) = \ln 10$$

89. Using properties of logarithms:

$$a\ln x + b\ln y = \ln x^a + \ln y^b = \ln x^a y^b$$

91. Using properties of logarithms:

$$\ln\sqrt{(x-3)(x+4)} = \tfrac{1}{2}\ln[(x-3)(x+4)] = \tfrac{1}{2}\ln(x-3) + \tfrac{1}{2}\ln(x+4)$$

93. Using properties of logarithms:

$$\log_{10}\frac{x^3}{\sqrt{1+x}} = \log_{10} x^3 - \log_{10}\sqrt{1+x}$$

$$= \log_{10} x^3 - \log_{10}(1+x)^{1/2}$$

$$= 3\log_{10} x - \tfrac{1}{2}\log_{10}(1+x)$$

95. Using properties of logarithms:

$$\log_{10}\sqrt[3]{\frac{x}{100}} = \log_{10}\left(\frac{x}{100}\right)^{1/3}$$

$$= \tfrac{1}{3}\log_{10}\frac{x}{100}$$

$$= \tfrac{1}{3}\log_{10} x - \tfrac{1}{3}\log_{10} 100$$

$$= \tfrac{1}{3}\log_{10} x - \tfrac{2}{3}$$

97. Using properties of logarithms:

$$\ln\left(\frac{1+2e}{1-2e}\right)^3 = 3\ln\frac{1+2e}{1-2e} = 3\ln(1+2e) - 3\ln(1-2e)$$

99. Use $B = P(1+r)^t$, where $P = A$, $B = 2A$, and $r = \frac{R}{100}$:

$$2A = A\left(1+\frac{R}{100}\right)^t$$
$$2 = \left(1+\frac{R}{100}\right)^t$$
$$\ln 2 = t\ln\left(1+\frac{R}{100}\right)$$
$$t = \frac{\ln 2}{\ln\left(1+\frac{R}{100}\right)} \text{ years}$$

101. The balance after 1 year will be $A = P\left(1+\frac{0.095}{12}\right)^{12}$. We must find the effective interest rate r where $A = P(1 + r)^1$. Set these equal:

$$P(1+r) = P\left(1+\frac{0.095}{12}\right)^{12}$$
$$1+r = \left(1+\frac{0.095}{12}\right)^{12}$$
$$r = \left(1+\frac{0.095}{12}\right)^{12} - 1 \approx 0.0992$$

So the effective interest rate is 9.92%.

103. Use $A = Pe^{rt}$ where $P = D$, $A = 2D$, and $r = \frac{R}{100}$:

$$2D = De^{(R/100)t}$$
$$2 = e^{(R/100)t}$$
$$\ln 2 = \frac{R}{100}(t)$$
$$t = \frac{100\ln 2}{R} \text{ years}$$

105. (a) Use $A = P\left(1+\frac{r}{N}\right)^{Nt}$ where $P = \$660$, $r = 0.055$, $N = 4$, and $A = \$1000$:

$$1000 = 660\left(1+\frac{0.055}{4}\right)^{4t}$$
$$\frac{50}{33} = (1.01375)^{4t}$$
$$\ln\frac{50}{33} = 4t\ln 1.01375$$
$$t = \frac{\ln\frac{50}{33}}{4\ln 1.01375} \approx 7.61 \text{ yrs}$$

So the balance will reach $1000 after $7\frac{3}{4}$ years.

(b) Use $A = P\left(1+\frac{r}{N}\right)^{Nt}$ where $P = D$, $r = \frac{R}{100}$, $N = 4$, and $A = nD$:

$$nD = D\left(1+\frac{R}{400}\right)^{t}$$

$$n = \left(1+\frac{R}{400}\right)^{t}$$

$$\ln n = t\ln\left(1+\frac{R}{400}\right)$$

$$t = \frac{\ln n}{\ln\left(1+\frac{R}{400}\right)} \text{ years}$$

107. (a) The domain is $x > 0$, or $(0, \infty)$.

(b) We must make sure $\log_{10} x \geq 0$, so $x \geq 1$. So the domain is $x \geq 1$, or $[1, \infty)$.

109. (a) We must make sure that the expression in the log is non-zero:

$$x^2 - 2x - 15 \neq 0$$
$$(x-5)(x+3) \neq 0$$
$$x \neq 5, -3$$

So the domain is all real numbers except 5 and –3, or $(-\infty, -3) \cup (-3, 5) \cup (5, \infty)$.

(b) We must make sure that the expression in the log is positive:

$$x^2 - 2x - 15 > 0$$
$$(x-5)(x+3) > 0$$

The key numbers are –3 and 5. Draw a sign chart:

Interval	Test Number	$x-5$	$x+3$	$(x-5)(x+3)$
$(-\infty, -3)$	–5	neg.	neg.	pos.
$(-3, 5)$	0	neg.	pos.	neg.
$(5, \infty)$	6	pos.	pos.	pos.

So the product is positive on the intervals $(-\infty, -3) \cup (5, \infty)$. Thus the domain is $(-\infty, -3) \cup (5, \infty)$.

111. Solve for x:

$$y = \frac{e^x + 1}{e^x - 1}$$
$$y(e^x - 1) = e^x + 1$$
$$ye^x - y = e^x + 1$$
$$ye^x - e^x = y + 1$$
$$e^x(y - 1) = y + 1$$
$$e^x = \frac{y+1}{y-1}$$
$$x = \ln\frac{y+1}{y-1}$$

We must solve the inequality $\dfrac{y+1}{y-1} > 0$. The key numbers are -1 and 1. Draw a sign chart:

Interval	Test Number	$y+1$	$y-1$	$\frac{y+1}{y-1}$
$(-\infty, -1)$	-4	neg.	neg.	pos.
$(-1, 1)$	0	pos.	neg.	neg.
$(1, \infty)$	4	pos.	pos.	pos.

The quotient is positive on the intervals $(-\infty, -1) \cup (1, \infty)$. So the range is $(-\infty, -1) \cup (1, \infty)$.

113. Using properties of logarithms:
$$\ln 0.5 = \ln\left(2^{-1}\right) = -\ln 2 \approx -0.7$$

115. Using properties of logarithms:
$$\ln \tfrac{1}{9} = \ln\left(3^{-2}\right) = -2\ln 3 \approx -2(1.1) = -2.2$$

117. Using properties of logarithms:
$$\begin{aligned}
\ln 72 &= \ln\left(3^2 \cdot 2^3\right) \\
&= \ln\left(3^2\right) + \ln\left(2^3\right) \\
&= 2\ln 3 + 3\ln 2 \\
&\approx 2(1.1) + 3(0.7) \\
&= 2.2 + 2.1 \\
&= 4.3
\end{aligned}$$

119. Let $x = e$, then $\ln x = 1$. From the graph, note that $x \approx 2.7$.

121. Using properties of logarithms:
$$\log_2 3 = \frac{\ln 3}{\ln 2} \approx \frac{1.1}{0.7} \approx 1.6$$

123. (a) The corresponding x-coordinate when $y = 3.000$ is $x \approx 1.0986$.

(b) The percent error is given by:
$$\% \text{ error} = \frac{|1.0986 - \ln 3|}{\ln 3} \cdot 100 = 0.00112\%$$

(c) Estimate the quantities:
$$\begin{aligned}
\ln\sqrt{3} &= \tfrac{1}{2}\ln 3 \approx \tfrac{1}{2}(1.0986) \approx 0.5493 \\
\ln 9 &= \ln\left(3^2\right) = 2\ln 3 \approx 2(1.0986) \approx 2.1972 \\
\ln\tfrac{1}{3} &= \ln\left(3^{-1}\right) = -\ln 3 \approx -1.0986
\end{aligned}$$

Chapter Six Test

1. The domain is $(-\infty, \infty)$, the range is $(-3, \infty)$, the x-intercept is $-\frac{\ln 3}{\ln 2}$, the y-intercept is -2, and the asymptote is $y = -3$. Drawing the graph:

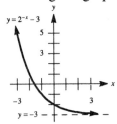

2. Using $N_0 = 6000$, $N = 6200$, and $t = 1$, find the growth constant k:
    $$6200 = 6000e^{k \cdot 1}$$
    $$1.033 = e^k$$
    $$k = \ln 1.033 \approx 0.03279$$
    Now find the value of t when $N = 10000$:
    $$10000 = 6000e^{kt}$$
    $$\tfrac{5}{3} = e^{kt}$$
    $$kt = \ln \tfrac{5}{3}$$
    $$t = \frac{\ln \tfrac{5}{3}}{\ln 1.033} \approx 16 \text{ hrs}$$

3. Since $2^4 = 16$, $\log_2 17 > 4$. Since $3^4 = 81$, $\log_3 80 < 4$. So $\log_2 17$ is larger.

4. Using the change-of-base formula:
    $$\log_2 15 = \frac{\ln 15}{\ln 2}$$

5. Since $2^{10} \approx 10^3$:
    $$2^{40} = \left(2^{10}\right)^4 \approx \left(10^3\right)^4 = 10^{12}$$

6. Using $A = P(1 + r)^t$ for interest compounded annually, with $P = \$9500$, $A = \$12,000$ and $r = 0.06$:
    $$12000 = 9500(1.06)^t$$
    $$1.2632 = (1.06)^t$$
    $$\ln 1.2632 = t \ln 1.06$$
    $$t = \frac{\ln 1.2632}{\ln 1.06} \approx 4 \text{ yrs}$$

7. (a) The identity is valid only if $\ln x$ is defined, which is the interval $(0, \infty)$.

   (b) Graphing the two functions:

   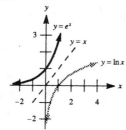

8. Using properties of logarithms:

$$\log_{10}\frac{A^3}{\sqrt{B}} = \log_{10} A^3 - \log_{10}\sqrt{B} = 3\log_{10} A - \tfrac{1}{2}\log_{10} B = 3a - \tfrac{1}{2}b$$

9. (a) Using properties of logarithms:
$$\log_5\tfrac{1}{\sqrt{5}} = \log_5 5^{-1/2} = -\tfrac{1}{2}$$

   (b) Using properties of logarithms:
$$\ln e^2 + \ln 1 - e^{\ln 3} = 2 + 0 - 3 = -1$$

10. Factoring:
$$e^x\left(x^3 - 4x\right) = 0$$
$$xe^x\left(x^2 - 4\right) = 0$$
$$xe^x(x+2)(x-2) = 0$$
$$x = 0, -2, 2 \quad (\text{since } e^x \neq 0)$$

11. (a) Using the formula $T = \dfrac{\ln\frac{1}{2}}{k}$:

$$4 = \frac{\ln\frac{1}{2}}{k}$$

$$k = \frac{\ln\frac{1}{2}}{4} \approx -0.1733$$

   (b) Using $N = N_0 e^{kt}$ with $N_0 = 2$, $k = -0.1733$, and $t = 10$:
$$N = 2e^{-0.1733(10)} \approx 0.35 \text{ grams}$$

12. Using properties of logarithms:

$$2\ln x - \ln\sqrt[3]{x^2+1} = \ln x^2 - \ln\left(x^2+1\right)^{1/3} = \ln\frac{x^2}{\left(x^2+1\right)^{1/3}}$$

13. Solve for x:

$$-2e^{3x-1} = 9$$
$$e^{3x-1} = -4.5$$

But this is impossible $\left(e^{3x-1} > 0 \text{ for all } x\right)$, so there is no solution.

14. Let $y = \log_{10}(x - 1)$. Switching the roles of x and y and solving the resulting equation for y yields:

$$x = \log_{10}(y - 1)$$
$$10^x = y - 1$$
$$y = 10^x + 1$$

So $g^{-1}(x) = 10^x + 1$. The range of $g^{-1}(x)$ is $(1, \infty)$.

15. (a) Using the formula $T = \frac{\ln 2}{r}$:

$$T = \frac{\ln 2}{0.06} \approx 12 \text{ years}$$

(b) Graphing $A = 12000e^{0.06t}$:

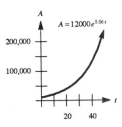

16. (a) Since $\ln x$ must be defined, this is true on the interval $(0, \infty)$. Note that $\ln(x^2) = 2\ln x$ is an identity on the interval.

(b) Solve for x:

$$(\ln x)^2 = 2\ln x$$
$$(\ln x)^2 - 2\ln x = 0$$
$$(\ln x)(\ln x - 2) = 0$$
$$\ln x = 0 \quad \text{or} \quad \ln x = 2$$
$$x = 1 \qquad\qquad x = e^2$$

17. (a) Graphing $f(x)$:

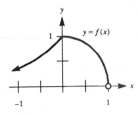

(b) By the horizontal line test, $f(x)$ is not one-to-one.

18. Using properties of logarithms, solve for x:
$$\tfrac{1}{2} - \log_{16}(x-3) = \log_{16} x$$
$$\tfrac{1}{2} = \log_{16} x + \log_{16}(x-3)$$
$$\tfrac{1}{2} = \log_{16}[x(x-3)]$$
$$16^{1/2} = x(x-3)$$
$$4 = x^2 - 3x$$
$$0 = x^2 - 3x - 4$$
$$0 = (x-4)(x+1)$$
$$x = -1, 4$$

Since $\log_{10}(-1)$ is undefined, $x = -1$ is an extraneous root. So the solution is $x = 4$.

19. Let $u = e^x$. Then the equation becomes:
$$6(e^x)^2 - 5e^x = 6$$
$$6u^2 - 5u - 6 = 0$$
$$(3u+2)(2u-3) = 0$$
$$u = -\tfrac{2}{3}, \tfrac{3}{2}$$

So either $e^x = -\tfrac{2}{3}$, which is impossible, or $e^x = \tfrac{3}{2}$, so $x = \ln\tfrac{3}{2} \approx 0.405$.

20. (a) Solving the inequality:
$$5(4 - 0.3^x) > 12$$
$$4 - 0.3^x > 2.4$$
$$-0.3^x > -1.6$$
$$0.3^x < 1.6$$
$$x \ln 0.3 < \ln 1.6$$
$$x > \frac{\ln 1.6}{\ln 0.3} \approx -0.39 \qquad \text{(since } \ln 0.3 < 0\text{)}$$

(b) Solving the inequality:
$$\ln x + \ln(x-3) \le \ln 4$$
$$\ln(x^2 - 3x) \le \ln 4$$
$$x^2 - 3x \le 4$$
$$x^2 - 3x - 4 \le 0$$
$$(x-4)(x+1) \le 0$$

Using a sign chart, the solution is $-1 \le x \le 4$. For the original domains, we must have $x > 0$ and $x > 3$, so the combined solution is $3 < x \le 4$, or $(3,4]$.

Chapter Seven
Trigonometric Functions of Angles

7.1 Trigonometric Functions of Acute Angles

1.   (a)   Use the definitions:

$$\sin\theta = \frac{\text{opposite}}{\text{hypotenuse}} = \frac{15}{17} \qquad\qquad \cos\theta = \frac{\text{adjacent}}{\text{hypotenuse}} = \frac{8}{17}$$

$$\tan\theta = \frac{\text{opposite}}{\text{adjacent}} = \frac{15}{8} \qquad\qquad \cot\theta = \frac{\text{adjacent}}{\text{opposite}} = \frac{8}{15}$$

$$\sec\theta = \frac{\text{hypotenuse}}{\text{adjacent}} = \frac{17}{8} \qquad\qquad \csc\theta = \frac{\text{hypotenuse}}{\text{opposite}} = \frac{17}{15}$$

    (b)   Use the definitions:

$$\sin\beta = \frac{\text{opposite}}{\text{hypotenuse}} = \frac{8}{17} \qquad\qquad \cos\beta = \frac{\text{adjacent}}{\text{hypotenuse}} = \frac{15}{17}$$

$$\tan\beta = \frac{\text{opposite}}{\text{adjacent}} = \frac{8}{15} \qquad\qquad \cot\beta = \frac{\text{adjacent}}{\text{opposite}} = \frac{15}{8}$$

$$\sec\beta = \frac{\text{hypotenuse}}{\text{adjacent}} = \frac{17}{15} \qquad\qquad \csc\beta = \frac{\text{hypotenuse}}{\text{opposite}} = \frac{17}{8}$$

3.   (a) Use the definitions:

$$\sin\theta = \frac{\text{opposite}}{\text{hypotenuse}} = \frac{3}{3\sqrt{5}} = \frac{1}{\sqrt{5}} = \frac{\sqrt{5}}{5}$$

$$\cos\theta = \frac{\text{adjacent}}{\text{hypotenuse}} = \frac{6}{3\sqrt{5}} = \frac{2}{\sqrt{5}} = \frac{2\sqrt{5}}{5}$$

$$\tan\theta = \frac{\text{opposite}}{\text{adjacent}} = \frac{3}{6} = \frac{1}{2}$$

$$\cot\theta = \frac{\text{adjacent}}{\text{opposite}} = \frac{6}{3} = 2$$

$$\sec\theta = \frac{\text{hypotenuse}}{\text{adjacent}} = \frac{3\sqrt{5}}{6} = \frac{\sqrt{5}}{2}$$

$$\csc\theta = \frac{\text{hypotenuse}}{\text{opposite}} = \frac{3\sqrt{5}}{3} = \sqrt{5}$$

(b) Use the definitions:

$$\sin\beta = \frac{\text{opposite}}{\text{hypotenuse}} = \frac{6}{3\sqrt{5}} = \frac{2}{\sqrt{5}} = \frac{2\sqrt{5}}{5}$$

$$\cos\beta = \frac{\text{adjacent}}{\text{hypotenuse}} = \frac{3}{3\sqrt{5}} = \frac{1}{\sqrt{5}} = \frac{\sqrt{5}}{5}$$

$$\tan\beta = \frac{\text{opposite}}{\text{adjacent}} = \frac{6}{3} = 2$$

$$\cot\beta = \frac{\text{adjacent}}{\text{opposite}} = \frac{3}{6} = \frac{1}{2}$$

$$\sec\beta = \frac{\text{hypotenuse}}{\text{adjacent}} = \frac{3\sqrt{5}}{3} = \sqrt{5}$$

$$\csc\beta = \frac{\text{hypotenuse}}{\text{opposite}} = \frac{3\sqrt{5}}{6} = \frac{\sqrt{5}}{2}$$

5.   First draw $\triangle ABC$ and label $AC = 3$ and $BC = 2$:

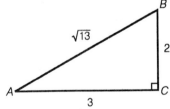

Now find AB by using the Pythagorean theorem:

$$(AC)^2 + (BC)^2 = (AB)^2$$

$$(3)^2 + (2)^2 = (AB)^2$$

$$9 + 4 = (AB)^2$$

$$\sqrt{13} = AB$$

(a) Find $\cos A$, $\sin A$ and $\tan A$:

$$\cos A = \frac{\text{adjacent}}{\text{hypotenuse}} = \frac{3}{\sqrt{13}} = \frac{3\sqrt{13}}{13}$$

$$\sin A = \frac{\text{opposite}}{\text{hypotenuse}} = \frac{2}{\sqrt{13}} = \frac{2\sqrt{13}}{13}$$

$$\tan A = \frac{\text{opposite}}{\text{adjacent}} = \frac{2}{3}$$

(b) Find $\sec B$, $\csc B$ and $\cot B$:

$$\sec B = \frac{\text{hypotenuse}}{\text{adjacent}} = \frac{\sqrt{13}}{2}$$

$$\csc B = \frac{\text{hypotenuse}}{\text{opposite}} = \frac{\sqrt{13}}{3}$$

$$\cot B = \frac{\text{adjacent}}{\text{opposite}} = \frac{2}{3}$$

7.   Draw a sketch with $AB = 13$ and $BC = 5$:

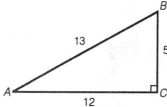

Now find AC by using the Pythagorean theorem:

$$(AC)^2 + (BC)^2 = (AB)^2$$
$$(AC)^2 + (5)^2 = (13)^2$$
$$(AC)^2 + 25 = 169$$
$$(AC)^2 = 144$$
$$AC = 12$$

Now find the six trigonometric functions of angle B:

$$\sin B = \frac{\text{opposite}}{\text{hypotenuse}} = \frac{12}{13} \qquad \cos B = \frac{\text{adjacent}}{\text{hypotenuse}} = \frac{5}{13}$$

$$\tan B = \frac{\text{opposite}}{\text{adjacent}} = \frac{12}{5} \qquad \cot B = \frac{\text{adjacent}}{\text{opposite}} = \frac{5}{12}$$

$$\sec B = \frac{\text{hypotenuse}}{\text{adjacent}} = \frac{13}{5} \qquad \csc B = \frac{\text{hypotenuse}}{\text{opposite}} = \frac{13}{12}$$

9.   Draw a sketch where $AC = 1$ and $BC = \frac{3}{4}$:

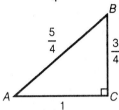

Now find AB by using the Pythagorean theorem:
$$(AC)^2 + (BC)^2 = (AB)^2$$
$$(1)^2 + \left(\tfrac{3}{4}\right)^2 = (AB)^2$$
$$1 + \tfrac{9}{16} = (AB)^2$$
$$\tfrac{25}{16} = (AB)^2$$
$$\tfrac{5}{4} = AB$$

(a) Find $\sin B$ and $\cos A$:
$$\sin B = \frac{\text{opposite}}{\text{hypotenuse}} = \frac{1}{\frac{5}{4}} = \tfrac{4}{5}$$
$$\cos A = \frac{\text{adjacent}}{\text{hypotenuse}} = \frac{1}{\frac{5}{4}} = \tfrac{4}{5}$$

(b) Find $\sin A$ and $\cos B$:
$$\sin A = \frac{\text{opposite}}{\text{hypotenuse}} = \frac{\frac{3}{4}}{\frac{5}{4}} = \tfrac{3}{5}$$
$$\cos B = \frac{\text{adjacent}}{\text{hypotenuse}} = \frac{\frac{3}{4}}{\frac{5}{4}} = \tfrac{3}{5}$$

(c) First find $\tan A$ and $\tan B$:
$$\tan A = \frac{\text{opposite}}{\text{adjacent}} = \frac{\frac{3}{4}}{1} = \tfrac{3}{4}$$
$$\tan B = \frac{\text{opposite}}{\text{adjacent}} = \frac{1}{\frac{3}{4}} = \tfrac{4}{3}$$

Now find $(\tan A)(\tan B)$:
$$(\tan A)(\tan B) = \tfrac{3}{4} \cdot \tfrac{4}{3} = 1$$

11. Draw a sketch where $AB = 25$ and $AC = 24$:

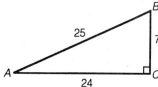

Now find BC by using the Pythagorean theorem:

$$(AC)^2 + (BC)^2 = (AB)^2$$
$$(24)^2 + (BC)^2 = (25)^2$$
$$576 + (BC)^2 = 625$$
$$(BC)^2 = 49$$
$$BC = 7$$

(a) Find $\cos A$, $\sin A$ and $\tan A$:

$$\cos A = \frac{\text{adjacent}}{\text{hypotenuse}} = \frac{24}{25}$$

$$\sin A = \frac{\text{opposite}}{\text{hypotenuse}} = \frac{7}{25}$$

$$\tan A = \frac{\text{opposite}}{\text{adjacent}} = \frac{7}{24}$$

(b) Find $\cos B$, $\sin B$ and $\tan B$:

$$\cos B = \frac{\text{adjacent}}{\text{hypotenuse}} = \frac{7}{25}$$

$$\sin B = \frac{\text{opposite}}{\text{hypotenuse}} = \frac{24}{25}$$

$$\tan B = \frac{\text{opposite}}{\text{adjacent}} = \frac{24}{7}$$

(c) Using the values obtained in parts (a) and (b):

$$(\tan A)(\tan B) = \frac{7}{24} \cdot \frac{24}{7} = 1$$

13. The calculated values for $\theta = 65°$ are approximately:

$$\sin \theta \approx 0.906 \qquad \cos \theta \approx 0.423 \qquad \tan \theta \approx 2.145$$

15. The calculated values for $\theta = 38.5°$ are approximately:

$$\sin \theta \approx 0.623 \qquad \cos \theta \approx 0.783 \qquad \tan \theta \approx 0.795$$

17. The calculated values for $\theta = 80.06°$ are approximately:

$$\sin \theta \approx 0.985 \qquad \cos \theta \approx 0.173 \qquad \tan \theta \approx 5.706$$

19. The calculated values for $\theta = 20°$ are approximately:
$$\sec\theta \approx 1.064 \qquad \csc\theta \approx 2.924 \qquad \cot\theta \approx 2.747$$

21. The calculated values for $\theta = 17.5°$ are approximately:
$$\sec\theta \approx 1.049 \qquad \csc\theta \approx 3.326 \qquad \cot\theta \approx 3.172$$

23. The calculated values for $\theta = 1°$ are approximately:
$$\sec\theta \approx 1.000 \qquad \csc\theta \approx 57.299 \qquad \cot\theta \approx 57.290$$

25. Compute each side of the equation:
$$\cos 60° = \tfrac{1}{2}$$
$$\cos^2 30° - \sin^2 30° = \left(\tfrac{\sqrt{3}}{2}\right)^2 - \left(\tfrac{1}{2}\right)^2 = \tfrac{3}{4} - \tfrac{1}{4} = \tfrac{2}{4} = \tfrac{1}{2}$$
So $\cos 60° = \cos^2 30° - \sin^2 30°$.

27. Compute the left-hand side of the equation:
$$\sin^2 30° + \sin^2 45° + \sin^2 60° = \left(\tfrac{1}{2}\right)^2 + \left(\tfrac{\sqrt{2}}{2}\right)^2 + \left(\tfrac{\sqrt{3}}{2}\right)^2 = \tfrac{1}{4} + \tfrac{2}{4} + \tfrac{3}{4} = \tfrac{6}{4} = \tfrac{3}{2}$$
So $\sin^2 30° + \sin^2 45° + \sin^2 60° = \tfrac{3}{2}$.

29. Compute each side of the equation:
$$2\sin 30°\cos 30° = 2\left(\tfrac{1}{2}\right)\left(\tfrac{\sqrt{3}}{2}\right) = \tfrac{2\sqrt{3}}{4} = \tfrac{\sqrt{3}}{2}$$
$$\sin 60° = \tfrac{\sqrt{3}}{2}$$
So $2\sin 30°\cos 30° = \sin 60°$.

31. Compute each side of the equation:
$$\sin 30° = \tfrac{1}{2}$$
$$\sqrt{\tfrac{1}{2}(1 - \cos 60°)} = \sqrt{\tfrac{1}{2}\left(1 - \tfrac{1}{2}\right)} = \sqrt{\tfrac{1}{2}\left(\tfrac{1}{2}\right)} = \sqrt{\tfrac{1}{4}} = \tfrac{1}{2}$$
So $\sin 30° = \sqrt{\tfrac{1}{2}(1 - \cos 60°)}$.

33. Compute each side of the equation:
$$\tan 30° = \frac{1}{\sqrt{3}} = \frac{\sqrt{3}}{3}$$
$$\frac{\sin 60°}{1 + \cos 60°} = \frac{\tfrac{\sqrt{3}}{2}}{1 + \tfrac{1}{2}} = \frac{\tfrac{\sqrt{3}}{2}}{\tfrac{3}{2}} = \frac{\sqrt{3}}{3}$$
So $\tan 30° = \dfrac{\sin 60°}{1 + \cos 60°}$.

35. Compute each side of the equation:
$$1+\tan^2 45° = 1+(1)^2 = 1+1 = 2$$
$$\sec^2 45° = \left(\tfrac{\sqrt{2}}{1}\right)^2 = 2$$
So $1+\tan^2 45° = \sec^2 45°$.

37. (a) Using a calculator, the approximate values are:
$$\cos 30° \approx 0.8660254038$$
$$\cos 45° \approx 0.7071067812$$

(b) Using the values given in Table 1, the approximate values are:
$$\cos 30° = \tfrac{\sqrt{3}}{2} \approx 0.8660254038$$
$$\cos 45° = \tfrac{\sqrt{2}}{2} \approx 0.7071067812$$
Note that these results agree with those in part (a).

39. Using the values in Table 1 to evaluate each side of the statement:
$$2[S(30°)] = 2\sin 30° = 2\left(\tfrac{1}{2}\right) = 1$$
$$S(60°) = \sin 60° = \frac{\sqrt{3}}{2}$$
The statement is false.

41. Using the values in Table 1 to evaluate the left-hand side of the statement:
$$(T \circ D)(30°) = \tan[2 \cdot 30°] = \tan 60° = \sqrt{3} > 1$$
The statement is true.

43. Using the values in Table 1 to evaluate the left-hand side of the statement:
$$S(45°) - C(45°) = \sin 45° - \cos 45° = \frac{\sqrt{2}}{2} - \frac{\sqrt{2}}{2} = 0$$
The statement is true.

45. Using the values in Table 1 to evaluate each side of the statement:
$$(C \circ D)(30°) = \cos[2 \cdot 30°] = \cos 60° = \tfrac{1}{2}$$
$$S(30°) = \sin 30° = \tfrac{1}{2}$$
The statement is true.

47. (a) The expressions can be written as:
$$\sin 20° = \frac{RC}{r}$$
$$\sin 40° = \frac{QB}{r}$$
$$\sin 60° = \frac{PA}{r}$$
But $RC < QB < PA$, so $\sin 20° < \sin 40° < \sin 60°$.

(b) Using a calculator:
$$\sin 20° \approx 0.3420$$
$$\sin 40° \approx 0.6428$$
$$\sin 60° \approx 0.8660$$
Thus $\sin 20° < \sin 40° < \sin 60°$.

49. (a) $\cos \theta$ is larger

(b) $\sec \beta$ is larger

51. Extending AB to D an equal distance to AB guarantees $\triangle DBC$ congruent to $\triangle ABC$. Now $\triangle ADC$ is equilateral as each angle is 60°. By construction $AD = 2AB$ and since the triangle is equilateral, $AC = AD$, hence $AC = 2AB$.

53. The values are equal. Computed to ten decimal places, the values are:

$\sin 3° \approx 0.0523359562$ $\sin 6° \approx 0.1045284633$

$\sin 9° \approx 0.1564344650$ $\sin 12° \approx 0.2079116908$

$\sin 15° \approx 0.2588190451$ $\sin 18° \approx 0.3090169944$

7.2 Algebra and the Trigonometric Functions

1. (a) Combining like terms:
$$-SC + 12SC = 11SC$$

(b) Combining as in (a):
$$-\sin \theta \cos \theta + 12 \sin \theta \cos \theta = 11 \sin \theta \cos \theta$$

3. (a) Combining like terms:
$$4C^3 S - 12C^3 S = -8C^3 S$$

(b) Combining as in (a):
$$4\cos^3 \theta \sin \theta - 12 \cos^3 \theta \sin \theta = -8 \cos^3 \theta \sin \theta$$

5. (a) Squaring:
$$(1+T)^2 = 1 + 2T + T^2$$

(b) Squaring as in (a):
$$(1 + \tan \theta)^2 = 1 + 2 \tan \theta + \tan^2 \theta$$

7. (a) Using the distributive property:
$$(T+3)(T-2) = T^2 + 3T - 2T - 6 = T^2 + T - 6$$

(b) Using the distributive property as in (a):
$$(\tan \theta + 3)(\tan \theta - 2) = \tan^2 \theta + 3 \tan \theta - 2 \tan \theta - 6 = \tan^2 \theta + \tan \theta - 6$$

9. (a) Factoring then simplifying:

$$\frac{S-C}{C-S} = \frac{-1(C-S)}{C-S} = -1$$

(b) Factoring then simplifying as in (a):

$$\frac{\sin\theta - \cos\theta}{\cos\theta - \sin\theta} = \frac{-1(\cos\theta - \sin\theta)}{\cos\theta - \sin\theta} = -1$$

11. (a) Obtaining common denominators then adding:

$$C + \frac{2}{S} = \frac{CS}{S} + \frac{2}{S} = \frac{CS+2}{S}$$

(b) Obtaining common denominators then adding as in (a):

$$\cos A + \frac{2}{\sin A} = \frac{\cos A \sin A}{\sin A} + \frac{2}{\sin A} = \frac{\cos A \sin A + 2}{\sin A}$$

13. (a) The expression factors as:

$$T^2 + 8T - 9 = (T-1)(T+9)$$

(b) Factoring as in (a):

$$\tan^2 \beta + 8\tan\beta - 9 = (\tan\beta - 1)(\tan\beta + 9)$$

15. (a) Factoring as a difference of squares:

$$4C^2 - 1 = (2C+1)(2C-1)$$

(b) Factoring as in (a):

$$4\cos^2 B - 1 = (2\cos B + 1)(2\cos B - 1)$$

17. (a) Factoring the greatest common factor:

$$9S^2T^3 + 6ST^2 = 3ST^2(3ST+2)$$

(b) Factoring as in (a):

$$9\sec^2 B \tan^3 B + 6\sec B \tan^2 B = 3\sec B \tan^2 B(3\sec B \tan B + 2)$$

19. Since $\sin\theta = \dfrac{\text{opposite}}{\text{hypotenuse}} = \frac{3}{4}$, construct the triangle:

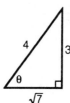

Using the Pythagorean theorem:

$$a^2 + 3^2 = 4^2$$
$$a^2 + 9 = 16$$
$$a^2 = 7$$
$$a = \sqrt{7}$$

The six trigonometric functions are:

$$\sin\theta = \frac{\text{opposite}}{\text{hypotenuse}} = \frac{3}{4} \qquad\qquad \cos\theta = \frac{\text{adjacent}}{\text{hypotenuse}} = \frac{\sqrt{7}}{4}$$

$$\tan\theta = \frac{\text{opposite}}{\text{adjacent}} = \frac{3}{\sqrt{7}} = \frac{3\sqrt{7}}{7} \qquad\qquad \cot\theta = \frac{\text{adjacent}}{\text{opposite}} = \frac{\sqrt{7}}{3}$$

$$\sec\theta = \frac{\text{hypotenuse}}{\text{adjacent}} = \frac{4}{\sqrt{7}} = \frac{4\sqrt{7}}{7} \qquad\qquad \csc\theta = \frac{\text{hypotenuse}}{\text{opposite}} = \frac{4}{3}$$

21. Since $\cos\beta = \dfrac{\text{adjacent}}{\text{hypotenuse}} = \dfrac{\sqrt{3}}{5}$, construct the triangle:

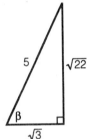

Using the Pythagorean theorem:

$$\left(\sqrt{3}\right)^2 + b^2 = 5^2$$
$$3 + b^2 = 25$$
$$b^2 = 22$$
$$b = \sqrt{22}$$

The six trigonometric functions are:

$$\sin\beta = \frac{\text{opposite}}{\text{hypotenuse}} = \frac{\sqrt{22}}{5} \qquad\qquad \cos\beta = \frac{\text{adjacent}}{\text{hypotenuse}} = \frac{\sqrt{3}}{5}$$

$$\tan\beta = \frac{\text{opposite}}{\text{adjacent}} = \frac{\sqrt{22}}{\sqrt{3}} = \frac{\sqrt{66}}{3} \qquad\qquad \cot\beta = \frac{\text{adjacent}}{\text{opposite}} = \frac{\sqrt{3}}{\sqrt{22}} = \frac{\sqrt{66}}{22}$$

$$\sec\beta = \frac{\text{hypotenuse}}{\text{adjacent}} = \frac{5}{\sqrt{3}} = \frac{5\sqrt{3}}{3} \qquad\qquad \csc\beta = \frac{\text{hypotenuse}}{\text{opposite}} = \frac{5}{\sqrt{22}} = \frac{5\sqrt{22}}{22}$$

23.   Since $\sin A = \dfrac{\text{opposite}}{\text{hypotenuse}} = \frac{5}{13}$, construct the triangle:

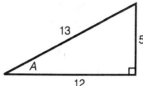

Using the Pythagorean theorem:
$$a^2 + 5^2 = 13^2$$
$$a^2 + 25 = 169$$
$$a^2 = 144$$
$$a = 12$$
The six trigonometric functions are:

$$\sin A = \frac{\text{opposite}}{\text{hypotenuse}} = \frac{5}{13} \qquad\qquad \cos A = \frac{\text{adjacent}}{\text{hypotenuse}} = \frac{12}{13}$$

$$\tan A = \frac{\text{opposite}}{\text{adjacent}} = \frac{5}{12} \qquad\qquad \cot A = \frac{\text{adjacent}}{\text{opposite}} = \frac{12}{5}$$

$$\sec A = \frac{\text{hypotenuse}}{\text{adjacent}} = \frac{13}{12} \qquad\qquad \csc A = \frac{\text{hypotenuse}}{\text{opposite}} = \frac{13}{5}$$

25.   Since $\tan B = \dfrac{\text{opposite}}{\text{adjacent}} = \frac{4}{3}$, construct the triangle:

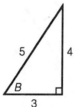

Using the Pythagorean theorem:
$$3^2 + 4^2 = c^2$$
$$9 + 16 = c^2$$
$$25 = c^2$$
$$5 = c$$
The six trigonometric functions are:

$$\sin B = \frac{\text{opposite}}{\text{hypotenuse}} = \frac{4}{5} \qquad\qquad \cos B = \frac{\text{adjacent}}{\text{hypotenuse}} = \frac{3}{5}$$

$$\tan B = \frac{\text{opposite}}{\text{adjacent}} = \frac{4}{3} \qquad\qquad \cot B = \frac{\text{adjacent}}{\text{opposite}} = \frac{3}{4}$$

$$\sec B = \frac{\text{hypotenuse}}{\text{adjacent}} = \frac{5}{3} \qquad\qquad \csc B = \frac{\text{hypotenuse}}{\text{opposite}} = \frac{5}{4}$$

27. Since $\sec C = \dfrac{\text{hypotenuse}}{\text{adjacent}} = \dfrac{3}{2}$, construct the triangle:

Using the Pythagorean theorem:
$$2^2 + b^2 = 3^2$$
$$4 + b^2 = 9$$
$$b^2 = 5$$
$$b = \sqrt{5}$$

The six trigonometric functions are:

$$\sin C = \frac{\text{opposite}}{\text{hypotenuse}} = \frac{\sqrt{5}}{3}$$ $$\cos C = \frac{\text{adjacent}}{\text{hypotenuse}} = \frac{2}{3}$$

$$\tan C = \frac{\text{opposite}}{\text{adjacent}} = \frac{\sqrt{5}}{2}$$ $$\cot C = \frac{\text{adjacent}}{\text{opposite}} = \frac{2}{\sqrt{5}} = \frac{2\sqrt{5}}{5}$$

$$\sec C = \frac{\text{hypotenuse}}{\text{adjacent}} = \frac{3}{2}$$ $$\csc C = \frac{\text{hypotenuse}}{\text{opposite}} = \frac{3}{\sqrt{5}} = \frac{3\sqrt{5}}{5}$$

29. Since $\cot \alpha = \dfrac{\text{adjacent}}{\text{opposite}} = \dfrac{\sqrt{3}}{3}$, construct the triangle:

Using the Pythagorean theorem:
$$\left(\sqrt{3}\right)^2 + (3)^2 = c^2$$
$$3 + 9 = c^2$$
$$12 = c^2$$
$$c = \sqrt{12} = 2\sqrt{3}$$

The six trigonometric functions are:

$$\sin \alpha = \frac{\text{opposite}}{\text{hypotenuse}} = \frac{3}{2\sqrt{3}} = \frac{3\sqrt{3}}{6} = \frac{\sqrt{3}}{2}$$ $$\cos \alpha = \frac{\text{adjacent}}{\text{hypotenuse}} = \frac{\sqrt{3}}{2\sqrt{3}} = \frac{1}{2}$$

$$\tan \alpha = \frac{\text{opposite}}{\text{adjacent}} = \frac{3}{\sqrt{3}} = \frac{3\sqrt{3}}{3} = \sqrt{3}$$ $$\cot \alpha = \frac{\text{adjacent}}{\text{opposite}} = \frac{\sqrt{3}}{3}$$

$$\sec \alpha = \frac{\text{hypotenuse}}{\text{adjacent}} = \frac{2\sqrt{3}}{\sqrt{3}} = 2$$ $$\csc \alpha = \frac{\text{hypotenuse}}{\text{opposite}} = \frac{2\sqrt{3}}{3}$$

31. From the identity $\sin^2\theta + \cos^2\theta = 1$, find $\sin\theta$:

$$\sin^2\theta + (0.4626)^2 = 1$$
$$\sin^2\theta + 0.21399876 = 1$$
$$\sin^2\theta = 0.78600124$$
$$\sin\theta \approx 0.887$$

Using trigonometric identities:

$$\tan\theta = \frac{\sin\theta}{\cos\theta} \approx \frac{0.887}{0.4626} \approx 1.916$$

$$\cot\theta = \frac{\cos\theta}{\sin\theta} \approx \frac{0.4626}{0.887} \approx 0.522$$

$$\sec\theta = \frac{1}{\cos\theta} = \frac{1}{0.4626} \approx 2.162$$

$$\csc\theta = \frac{1}{\sin\theta} \approx \frac{1}{0.887} \approx 1.128$$

33. Since $\tan\theta = \dfrac{\text{opposite}}{\text{adjacent}} = \dfrac{1.1998}{1}$, construct the triangle:

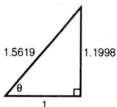

Using the Pythagorean theorem:

$$1^2 + (1.1998)^2 = c^2$$
$$1 + 1.43952004 = c^2$$
$$2.43952004 = c^2$$
$$c \approx 1.5619$$

The remaining trigonometric functions are:

$$\sin\theta = \frac{\text{opposite}}{\text{hypotenuse}} \approx \frac{1.1998}{1.5619} \approx 0.768$$

$$\cos\theta = \frac{\text{adjacent}}{\text{hypotenuse}} \approx \frac{1}{1.5619} \approx 0.640$$

$$\cot\theta = \frac{1}{\tan\theta} = \frac{1}{1.1998} \approx 0.833$$

$$\sec\theta = \frac{1}{\cos\theta} \approx \frac{1}{0.640} \approx 1.562$$

$$\csc\theta = \frac{1}{\sin\theta} \approx \frac{1}{0.768} \approx 1.302$$

35.   By factoring the numerator:

$$\frac{\sin^2 A - \cos^2 A}{\sin A - \cos A} = \frac{(\sin A - \cos A)(\sin A + \cos A)}{\sin A - \cos A} = \sin A + \cos A$$

37.   By using the identities for csc θ and sec θ:

$$\sin^2 \theta \cos \theta \csc^3 \theta \sec \theta = \sin^2 \theta \cdot \cos \theta \cdot \frac{1}{\sin^3 \theta} \cdot \frac{1}{\cos \theta} = \frac{1}{\sin \theta} = \csc \theta$$

39.   By using the identity for cot B:

$$\cot B \sin^2 B \cot B = \frac{\cos B}{\sin B} \cdot \sin^2 B \cdot \frac{\cos B}{\sin B} = \cos^2 B$$

41.   By factoring the numerator:

$$\frac{\cos^2 A + \cos A - 12}{\cos A - 3} = \frac{(\cos A - 3)(\cos A + 4)}{\cos A - 3} = \cos A + 4$$

43.   By using the identities for tan θ and sec θ:

$$\frac{\tan \theta}{\sec \theta - 1} + \frac{\tan \theta}{\sec \theta + 1} = \frac{\frac{\sin \theta}{\cos \theta}}{\frac{1}{\cos \theta} - 1} + \frac{\frac{\sin \theta}{\cos \theta}}{\frac{1}{\cos \theta} + 1}$$

$$= \frac{\sin \theta}{1 - \cos \theta} + \frac{\sin \theta}{1 + \cos \theta}$$

$$= \frac{\sin \theta (1 + \cos \theta) + \sin \theta (1 - \cos \theta)}{1 - \cos^2 \theta}$$

$$= \frac{2 \sin \theta}{\sin^2 \theta}$$

$$= \frac{2}{\sin \theta}$$

$$= 2 \csc \theta$$

45.   By using the identities for sec A, csc A, tan A and cot A:

$$\sec A \csc A - \tan A - \cot A = \frac{1}{\cos A} \cdot \frac{1}{\sin A} - \frac{\sin A}{\cos A} - \frac{\cos A}{\sin A}$$

$$= \frac{1 - \sin^2 A - \cos^2 A}{\sin A \cos A}$$

$$= \frac{1 - \left(\sin^2 A + \cos^2 A \right)}{\sin A \cos A}$$

$$= \frac{1 - 1}{\sin A \cos A}$$

$$= 0$$

47. By using the identities for $\cot\theta$, $\tan\theta$, $\csc\theta$ and $\sec\theta$:

$$\frac{\cot^2\theta}{\csc^2\theta}+\frac{\tan^2\theta}{\sec^2\theta}=\frac{\frac{\cos^2\theta}{\sin^2\theta}}{\frac{1}{\sin^2\theta}}+\frac{\frac{\sin^2\theta}{\cos^2\theta}}{\frac{1}{\cos^2\theta}}=\cos^2\theta+\sin^2\theta=1$$

49. By using the identity $\cos(90°-\theta)=\sin\theta$:

$$\frac{\cos(90°-\theta)}{\cos\theta}=\frac{\sin\theta}{\cos\theta}=\tan\theta$$

51. By using the identities $\cos(90°-A)=\sin A$ and $\sin(90°-A)=\cos A$:

$$\frac{\cos^2(90°-A)}{\sin^2(90°-A)}-\frac{1}{\cos^2 A}=\frac{\sin^2 A}{\cos^2 A}-\frac{1}{\cos^2 A}$$
$$=\frac{\sin^2 A-1}{\cos^2 A}$$
$$=-\frac{1-\sin^2 A}{\cos^2 A}$$
$$=-\frac{\cos^2 A}{\cos^2 A}$$
$$=-1$$

53. By using the identity $\sin(90°-\theta)=\cos\theta$:

$$1-\sin(90°-\theta)\cos\theta=1-\cos\theta\cdot\cos\theta=1-\cos^2\theta=\sin^2\theta$$

55. By using only algebraic simplification:

$$\frac{\frac{\cos\theta+1}{\cos\theta}+1}{\frac{\cos\theta-1}{\cos\theta}-1}=\frac{\cos\theta+1+\cos\theta}{\cos\theta-1-\cos\theta}=\frac{2\cos\theta+1}{-1}=-2\cos\theta-1$$

57. Solve the first equation for A:
$$A\sin\theta+\cos\theta=1$$
$$A\sin\theta=1-\cos\theta$$
$$A=\frac{1-\cos\theta}{\sin\theta}$$
Now solve the second equation for B:
$$B\sin\theta-\cos\theta=1$$
$$B\sin\theta=1+\cos\theta$$
$$B=\frac{1+\cos\theta}{\sin\theta}$$
Now compute the product AB:
$$AB=\frac{1-\cos\theta}{\sin\theta}\cdot\frac{1+\cos\theta}{\sin\theta}=\frac{1-\cos^2\theta}{\sin^2\theta}=\frac{\sin^2\theta}{\sin^2\theta}=1$$

59. Recall that $\sin^2\theta + \cos^2\theta = 1$, so we have the system of equations:
$$a\sin^2\theta + b\cos^2\theta = 1$$
$$\sin^2\theta + \cos^2\theta = 1$$
Multiply the second equation by $-b$:
$$a\sin^2\theta + b\cos^2\theta = 1$$
$$-b\sin^2\theta - b\cos^2\theta = -b$$
Adding yields:
$$(a-b)\sin^2\theta = 1-b$$
$$\sin^2\theta = \frac{1-b}{a-b}$$
Therefore:
$$\cos^2\theta = 1 - \sin^2\theta = 1 - \frac{1-b}{a-b} = \frac{(a-b)-(1-b)}{a-b} = \frac{a-1}{a-b}$$
Using the identity for $\tan\theta$:
$$\tan^2\theta = \frac{\sin^2\theta}{\cos^2\theta} = \frac{\frac{1-b}{a-b}}{\frac{a-1}{a-b}} = \frac{1-b}{a-1} = \frac{b-1}{1-a}$$
This proves the desired results.

61. Solving the identity $\sin^2\beta + \cos^2\beta = 1$ for $\cos\beta$ yields $\cos\beta = \sqrt{1-\sin^2\beta}$, so:
$$\cos\beta = \sqrt{1-\sin^2\beta}$$
$$= \sqrt{1 - \frac{m^4 - 2m^2n^2 + n^4}{\left(m^2+n^2\right)^2}}$$
$$= \frac{\sqrt{m^4 + 2m^2n^2 + n^4 - m^4 + 2m^2n^2 - n^4}}{m^2+n^2}$$
$$= \frac{\sqrt{4m^2n^2}}{m^2+n^2}$$
$$= \frac{2mn}{m^2+n^2} \quad \text{(since } m > 0, n > 0\text{)}$$
Now using the identity for $\tan\beta$:
$$\tan\beta = \frac{\sin\beta}{\cos\beta} = \frac{\frac{m^2-n^2}{m^2+n^2}}{\frac{2mn}{m^2+n^2}} = \frac{m^2-n^2}{2mn}$$

7.3 Right-Triangle Applications

1. Draw the figure:

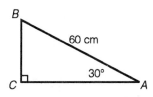

Since $\sin 30° = \dfrac{BC}{60}$:

$BC = 60 \sin 30° = 60 \cdot \frac{1}{2} = 30$ cm

Since $\cos 30° = \dfrac{AC}{60}$:

$AC = 60 \cos 30° = 60 \cdot \frac{\sqrt{3}}{2} = 30\sqrt{3}$ cm

3. Draw the figure:

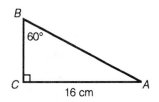

Since $\sin 60° = \dfrac{16}{AB}$:

$AB = \dfrac{16}{\sin 60°} = \dfrac{16}{\frac{\sqrt{3}}{2}} = \dfrac{32}{\sqrt{3}} = \dfrac{32\sqrt{3}}{3}$ cm

Since $\tan 60° = \dfrac{16}{BC}$:

$BC = \dfrac{16}{\tan 60°} = \dfrac{16}{\sqrt{3}} = \dfrac{16\sqrt{3}}{3}$ cm

5. Draw the figure:

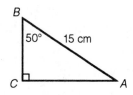

Since $\sin 50° = \dfrac{AC}{15}$:

$AC = 15 \sin 50° \approx 11.5$ cm

Since $\cos 50° = \dfrac{BC}{15}$:

$BC = 15 \cos 50° \approx 9.6$ cm

7.   Draw a figure:

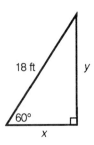

(a)   Now find y:

$\sin 60° = \frac{y}{18}$

$y = 18\sin 60° = 18\left(\frac{\sqrt{3}}{2}\right) = 9\sqrt{3}$ ft

Using a calculator, this is approximately 15.59 ft.

(b)   Now find x:

$\cos 60° = \frac{x}{18}$

$x = 18\cos 60° = 18\left(\frac{1}{2}\right) = 9$ ft

9.   Using the sine function, $\sin(\angle SEM) = \dfrac{MS}{SE}$, so:

$MS = SE\sin(\angle SEM) = 93\sin 21.16° \approx 34$

Thus the distance MS is approximately 34 million miles.

11.   Draw a figure:

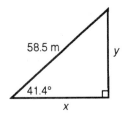

First find x and y:

$\cos 41.4° = \dfrac{x}{58.5}$, so $x = 58.5\cos 41.4° \approx 43.9$ m

$\sin 41.4° = \dfrac{y}{58.5}$, so $y = 58.5\sin 41.4° \approx 38.7$ m

So the total length of fencing required is:

43.9 m $+ 38.7$ m $+ 58.5$ m ≈ 141.1 m

13.   (a)   First draw the triangle:

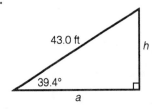

Now find h:

$$\sin 39.4° = \frac{h}{43.0}, \text{ so } h = 43.0 \sin 39.4° \approx 27.3 \text{ ft}$$

(b) First find a:

$$\cos 39.4° = \frac{a}{43.0}, \text{ so } a = 43.0 \cos 39.4° \approx 33.2 \text{ ft}$$

Now the gable has a base of 2(33.2) = 66.4 ft and a height of 27.3 ft, so its area is given by:

$$\tfrac{1}{2}(\text{base})(\text{height}) \approx \tfrac{1}{2}(66.4)(27.3) \approx 906.9 \text{ ft}^2$$

Note: This calculation was done with the full calculator approximation, not the values rounded to one decimal place.

15. Using the formula $A = \tfrac{1}{2}ab\sin\theta$:

$$A = \tfrac{1}{2}(2)(3)\sin 30° = 1.5 \text{ in.}^2$$

17. Start by finding the measure of a central angle of a triangle drawn from the center out to two adjacent vertices:

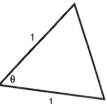

Since the polygon is 7-sided, $7\theta = 360°$, so $\theta = \frac{360°}{7}$. The area of this triangle is thus:

$$\tfrac{1}{2}ab\sin\theta = \tfrac{1}{2}(1)(1)\sin\tfrac{360°}{7} = \tfrac{1}{2}\sin\tfrac{360°}{7}$$

The area of the septagon (7-sided figure) is therefore:

$$7\left(\tfrac{1}{2}\sin\tfrac{360°}{7}\right) = \tfrac{7}{2}\sin\tfrac{360°}{7} \approx 2.736 \text{ square units}$$

19. Start by finding the measure of a central angle of a triangle drawn from the center out to two adjacent vertices:

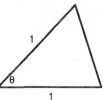

Since the polygon is 8-sided, $8\theta = 360°$, so $\theta = 45°$. The area of this triangle is thus:

$$\tfrac{1}{2}ab\sin\theta = \tfrac{1}{2}(1)(1)\sin 45° = \tfrac{\sqrt{2}}{4}$$

The area of the octagon (8-sided figure) is therefore:

$$8\left(\tfrac{\sqrt{2}}{4}\right) = 2\sqrt{2}$$

The shaded area is obtained by subtracting this area from the circular area, therefore:

$$\text{Area} = \pi(1)^2 - 2\sqrt{2} = \pi - 2\sqrt{2} \approx 0.313 \text{ square units}$$

21. Start by finding the measure of a central angle of a triangle drawn from the center out to two adjacent vertices:

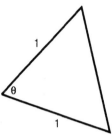

Since the polygon is 6-sided, $6\theta = 360°$, so $\theta = 60°$. The area of this triangle is thus:

$\frac{1}{2}ab\sin\theta = \frac{1}{2}(1)(1)\sin 60° = \frac{\sqrt{3}}{4}$

Since the shaded area is comprised of four such triangles, the shaded area is:

$4\left(\frac{\sqrt{3}}{4}\right) = \sqrt{3} \approx 1.732$ square units

23. Using the hint, draw the figure:

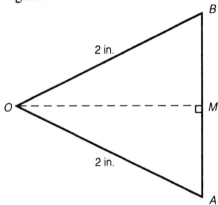

Now $\angle AOB = \frac{360°}{5} = 72°$, so $\angle AOM = 36°$. Now find AM:

$\sin 36° = \dfrac{AM}{2}$

$AM = 2\sin 36°$

Thus $AB = 4\sin 36°$, and since there are five sides, the perimeter is:

$5(AB) = 5(4\sin 36°) = 20\sin 36°$ inches

25. Draw the figure:

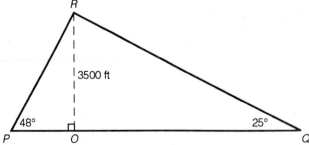

Now compute:

$$\tan 48° = \frac{3500}{PO} \text{ and } \tan 25° = \frac{3500}{OQ}$$

So $PO = \dfrac{3500}{\tan 48°}$ and $OQ = \dfrac{3500}{\tan 25°}$. Thus $PQ = \dfrac{3500}{\tan 48°} + \dfrac{3500}{\tan 25°} \approx 10,660$ ft.

27. Draw a figure:

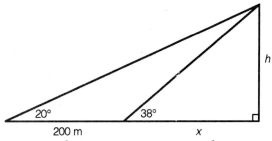

Now $\tan 38° = \dfrac{h}{x}$, so $x = \dfrac{h}{\tan 38°}$. Also $\tan 20° = \dfrac{h}{200 + x}$, so $h = (200 + x)\tan 20°$.

Substituting:

$$h = \left(200 + \frac{h}{\tan 38°}\right)\tan 20°$$
$$h\tan 38° = 200\tan 38°\tan 20° + h\tan 20°$$
$$h(\tan 38° - \tan 20°) = 200\tan 38°\tan 20°$$
$$h = \frac{200\tan 38°\tan 20°}{\tan 38° - \tan 20°} \approx 136 \text{ m}$$

29. Draw the figure:

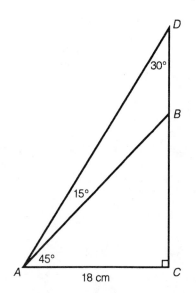

Now compute:

$$\tan 45° = \frac{BC}{18}, \text{ so } BC = 18\tan 45°$$

$$\tan 60° = \frac{CD}{18}, \text{ so } CD = 18\tan 60°$$

Therefore:

$$BD = CD - BC = 18(\tan 60° - \tan 45°) = 18(\sqrt{3} - 1) \text{ cm}$$

31. (a) First, note that $\angle BOA = 90° - \theta$ since it forms a right angle with θ. Also, $\angle OAB = \theta$ since both angles are complemetary to the same angle ($\angle AOB$). Now, $\angle BAP = 90° - \theta$ since it forms a right angle with $\angle OAB$. Finally, $\angle BPA = \theta$ since $\angle BPA$ and $\angle OAB$ are both complementary to the same angle ($\angle BAP$).

(b) Using $\triangle AOP$:

$$\sin \theta = \frac{AO}{OP} = \frac{AO}{1}, \text{ so } AO = \sin \theta$$

$$\cos \theta = \frac{AP}{OP} = \frac{AP}{1}, \text{ so } AP = \cos \theta$$

Using $\triangle AOB$:

$$\sin \theta = \frac{OB}{OA} = \frac{OB}{\sin \theta}, \text{ so } OB = \sin^2 \theta$$

Using $\triangle ABP$:

$$\cos \theta = \frac{BP}{AP} = \frac{BP}{\cos \theta}, \text{ so } BP = \cos^2 \theta$$

33. (a) Examine the similar triangles having $\overline{AB}$ and $\overline{BC}$ as hypotenuses, and notice that θ is also the angle at B in the smaller triangle. Thus:

$$\sin \theta = \frac{5}{BC}, \text{ so } BC = \frac{5}{\sin \theta}$$

(b) Using the smaller triangle:

$$\cos \theta = \frac{4}{AB}, \text{ so } AB = \frac{4}{\cos \theta}$$

(c) Since $AC = AB + BC$:

$$AC = \frac{4}{\cos \theta} + \frac{5}{\sin \theta} = 4\sec \theta + 5\csc \theta$$

35. First observe that the figure $x^2 + y^2 = 1$ is a circle with a radius of one. Since OA, OD and OF all represent the radius, they are each equal to 1. In each case, look for a trigonometric relationship involving the required segment:

(a) $\sin\theta = \dfrac{DE}{OD}$, so $DE = \sin\theta$

(b) $\cos\theta = \dfrac{OE}{OD}$, so $OE = \cos\theta$

(c) $\tan\theta = \dfrac{CF}{OF}$, so $CF = \tan\theta$

(d) $\sec\theta = \dfrac{OC}{OF}$, so $OC = \sec\theta$

Going to $\triangle OAB$, $\angle ABO = \theta$, thus:

(e) $\cot\theta = \dfrac{AB}{OA}$, so $AB = \cot\theta$

(f) $\csc\theta = \dfrac{OB}{OA}$, so $OB = \csc\theta$

37. (a) Use $\sin\theta$ to set up a trigonometric relationship:

$$\sin\theta = \frac{r}{PS + r}$$
$$PS\sin\theta + r\sin\theta = r$$
$$PS\sin\theta = r - r\sin\theta$$
$$PS\sin\theta = r(1 - \sin\theta)$$
$$r = \left(\frac{\sin\theta}{1 - \sin\theta}\right)PS$$

(b) Using a calculator:

$$r = \left(\frac{\sin 0.257°}{1 - \sin 0.257°}\right)(238{,}857) \approx 1080 \text{ miles}$$

39. (a) Draw the figure:

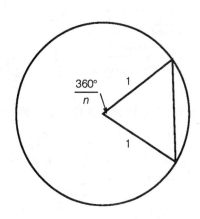

The area of the triangle is given by:
$$\tfrac{1}{2}ab\sin\theta = \tfrac{1}{2}(1)(1)\sin\tfrac{360°}{n} = \tfrac{1}{2}\sin\tfrac{360°}{n}$$
Since there are n congruent triangles in the n-gon, its area is $\tfrac{n}{2}\sin\tfrac{360°}{n}$.

(b) Complete the table:

n	5	10	50	100	1,000	5,000	10,000
A_n	2.38	2.94	3.1333	3.1395	3.141572	3.1415918	3.1415924

(c) As n gets larger, A_n becomes closer to the area of the circle, which is π.

41. Draw the figure:

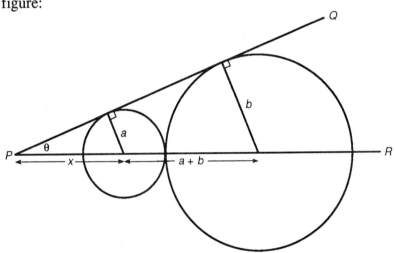

Using the smaller triangle:
$$\sin\theta = \frac{a}{x}, \text{ so } x = \frac{a}{\sin\theta}$$
Using the larger triangle and substituting for x:
$$\sin\theta = \frac{b}{x+a+b}$$
$$\sin\theta = \frac{b}{\frac{a}{\sin\theta}+a+b}$$
$$\sin\theta = \frac{b\sin\theta}{a+(a+b)\sin\theta}$$
$$a\sin\theta + (a+b)\sin^2\theta = b\sin\theta$$
$$a + (a+b)\sin\theta = b \quad (\text{since } \theta \neq 0 \text{ and } \theta \neq 180°, \text{ then } \sin\theta \neq 0)$$
$$(a+b)\sin\theta = b - a$$
$$\sin\theta = \frac{b-a}{a+b}$$

Since $\cos^2\theta + \sin^2\theta = 1$:

$$\cos\theta = \sqrt{1 - \sin^2\theta}$$

$$= \sqrt{1 - \frac{(b-a)^2}{(a+b)^2}}$$

$$= \sqrt{\frac{a^2 + 2ab + b^2 - b^2 + 2ab - a^2}{(a+b)^2}}$$

$$= \sqrt{\frac{4ab}{(a+b)^2}}$$

$$= \frac{2\sqrt{ab}}{a+b}$$

43. Draw the figure:

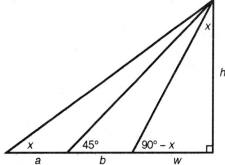

Use the following relationships from the three triangles:

small: $\tan x = \dfrac{w}{h}$

middle: $\tan 45° = \dfrac{h}{b+w}$, so $h = b + w$ and thus $w = h - b$

large: $\tan x = \dfrac{h}{a+b+w}$

Substituting:

$$\frac{w}{h} = \frac{h}{a+b+w}$$

Since $w = h - b$:

$$\frac{h-b}{h} = \frac{h}{a+b+h-b}$$

$$\frac{h-b}{h} = \frac{h}{h+a}$$

$$(h-b)(h+a) = h^2$$

$$h^2 - bh + ah - ab = h^2$$

$$ah - bh = ab$$

$$h(a-b) = ab$$

$$h = \frac{ab}{a-b}$$

The tower is $\dfrac{ab}{a-b}$ feet high.

7.4 Trigonometric Functions of Angles

1.   (a)   The reference angle for 110° is 70°:

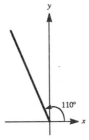

   (b)   The reference angle for –110° is 70°:

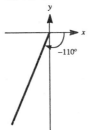

3.   (a)   The reference angle for 200° is 20°:

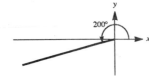

(b)   The reference angle for –200° is 20°:

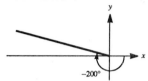

5.   (a)   The reference angle for 300° is 60°:

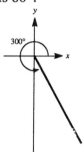

(b)   The reference angle for –300° is 60°:

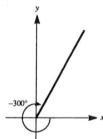

7.   (a)   The reference angle for 60° is 60°:

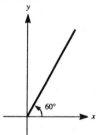

(b)   The reference angle for – 60° is 60°:

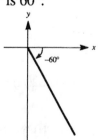

9. Since $270°$ corresponds to the point $(0, -1)$ on the unit circle, using $x = 0$ and $y = -1$ in the definitions yields:

$$\cos 270° = x = 0 \qquad\qquad \sec 270° = \frac{1}{x} = \frac{1}{0}, \text{ which is undefined}$$

$$\sin 270° = y = -1 \qquad\qquad \csc 270° = \frac{1}{y} = \frac{1}{-1} = -1$$

$$\tan 270° = \frac{y}{x} = \frac{-1}{0}, \text{ which is undefined} \qquad \cot 270° = \frac{x}{y} = \frac{0}{-1} = 0$$

11. Since $-270°$ corresponds to the point $(0, 1)$ on the unit circle, using $x = 0$ and $y = 1$ in the definitions yields:

$$\cos(-270°) = x = 0 \qquad\qquad \sec(-270°) = \frac{1}{x} = \frac{1}{0}, \text{ which is undefined}$$

$$\sin(-270°) = y = 1 \qquad\qquad \csc(-270°) = \frac{1}{y} = \frac{1}{1} = 1$$

$$\tan(-270°) = \frac{y}{x} = \frac{1}{0}, \text{ which is undefined} \qquad \cot(-270°) = \frac{x}{y} = \frac{0}{1} = 0$$

13. Since $810°$ results in the same point as $90°$ on the unit circle, which corresponds to the point $(0, 1)$, using $x = 0$ and $y = 1$ in the definitions yields:

$$\cos 810° = x = 0 \qquad\qquad \sec 810° = \frac{1}{x} = \frac{1}{0}, \text{ which is undefined}$$

$$\sin 810° = y = 1 \qquad\qquad \csc 810° = \frac{1}{y} = \frac{1}{1} = 1$$

$$\tan 810° = \frac{y}{x} = \frac{1}{0}, \text{ which is undefined} \qquad \cot 810° = \frac{x}{y} = \frac{0}{1} = 0$$

15. Using Figure 8, $\sin 10° \approx 0.2$ and $\sin(-10°) \approx -0.2$. Using a calculator, $\sin 10° \approx 0.17$ and $\sin(-10°) \approx -0.17$.

17. Using Figure 8, $\cos 80° \approx 0.2$ and $\cos(-80°) \approx 0.2$. Using a calculator, $\cos 80° \approx 0.17$ and $\cos(-80°) \approx 0.17$.

19. Using Figure 8, $\sin 120° \approx 0.9$ and $\sin(-120°) \approx -0.9$. Using a calculator, $\sin 120° \approx 0.87$ and $\sin(-120°) \approx -0.87$.

21. Using Figure 8, $\sin 150° = 0.5$ and $\sin(-150°) = -0.5$. Using a calculator, $\sin 150° = 0.5$ and $\sin(-150°) = -0.5$.

23. Using Figure 8, $\cos 220° \approx -0.8$ and $\cos(-220°) \approx -0.8$. Using a calculator, $\cos 220° \approx -0.77$ and $\cos(-220°) \approx -0.77$.

25. Using Figure 8, $\cos 310° \approx 0.6$ and $\cos(-310°) \approx 0.6$. Using a calculator, $\cos 310° \approx 0.64$ and $\cos(-310°) \approx 0.64$.

27. Since $40° + 360°$ is one full revolution on the unit circle past $40°$, $\sin(40° + 360°) = \sin 40°$. Using Figure 8, $\sin(40° + 360°) \approx 0.6$. Using a calculator, $\sin(40° + 360°) \approx 0.64$.

29. Using Figure 8, $\sin 70° \approx 0.9$ and $\cos 70° \approx 0.3$, so $\sin 70°$ is larger. Using a calculator, $\sin 70° \approx 0.94$ and $\cos 70° \approx 0.34$, so $\sin 70°$ is larger.

31. Using Figure 8, $\cos 170° \approx -1.0$ and $\cos 160° \approx -0.9$, so $\cos 160°$ is larger. Using a calculator, $\cos 170° \approx -0.98$ and $\cos 160° \approx -0.94$, so $\cos 160°$ is larger.

33. Using Figure 8, $\cos 280° \approx 0.2$ and $\cos 290° \approx 0.3$, so $\cos 290°$ is larger. Using a calculator, $\cos 280° \approx 0.17$ and $\cos 290° \approx 0.34$, so $\cos 290°$ is larger.

35. Using Figure 8, $\sin 10° \approx 0.2$ and $\sin(-10°) \approx -0.2$, so $\sin 10°$ is larger. Using a calculator, $\sin 10° \approx 0.17$ and $\sin(-10°) \approx -0.17$, so $\sin 10°$ is larger.

37. Using Figure 8, $\sin 80° \approx 1.0$ and $\sin 110° \approx 0.9$, so $\sin 80°$ is larger. Using a calculator, $\sin 80° \approx 0.98$ and $\sin 110° \approx 0.94$, so $\sin 80°$ is larger.

39. Using Figure 8, $\sin(-80°) \approx -1.0$ and $\cos(-110°) \approx -0.3$, so $\cos(-110°)$ is larger. Using a calculator, $\sin(-80°) \approx -0.98$ and $\cos(-110°) \approx -0.34$, so $\cos(-110°)$ is larger.

41. (a) The reference angle for $315°$ is $45°$, and $\cos 45° = \frac{\sqrt{2}}{2}$. Since $\cos \theta$ is the x-coordinate, and $\theta = 315°$ lies in the fourth quadrant where the x-coordinates are positive, $\cos 315° = \frac{\sqrt{2}}{2}$.

(b) The reference angle for $-315°$ is $45°$, and $\cos 45° = \frac{\sqrt{2}}{2}$. Since $\cos \theta$ is the x-coordinate, and $\theta = -315°$ lies in the first quadrant where the x-coordinates are positive, $\cos(-315°) = \frac{\sqrt{2}}{2}$.

(c) The reference angle for $315°$ is $45°$, and $\sin 45° = \frac{\sqrt{2}}{2}$. Since $\sin \theta$ is the y-coordinate, and $\theta = 315°$ lies in the fourth quadrant where the y-coordinates are negative, $\sin 315° = -\frac{\sqrt{2}}{2}$.

(d) The reference angle for $-315°$ is $45°$, and $\sin 45° = \frac{\sqrt{2}}{2}$. Since $\sin \theta$ is the y-coordinate, and $\theta = -315°$ lies in the first quadrant where the y-coordinates are positive, $\sin(-315°) = \frac{\sqrt{2}}{2}$.

43. (a) The reference angle for $300°$ is $60°$, and $\cos 60° = \frac{1}{2}$. Since $\cos \theta$ is the x-coordinate, and $\theta = 300°$ lies in the fourth quadrant where the x-coordinates are positive, $\cos 300° = \frac{1}{2}$.

   (b) The reference angle for $-300°$ is $60°$, and $\cos 60° = \frac{1}{2}$. Since $\cos \theta$ is the x-coordinate, and $\theta = -300°$ lies in the first quadrant where the x-coordinates are positive, $\cos(-300°) = \frac{1}{2}$.

   (c) The reference angle for $300°$ is $60°$, and $\sin 60° = \frac{\sqrt{3}}{2}$. Since $\sin \theta$ is the y-coordinate, and $\theta = 300°$ lies in the fourth quadrant where the y-coordinates are negative, $\sin 300° = -\frac{\sqrt{3}}{2}$.

   (d) The reference angle for $-300°$ is $60°$, and $\sin 60° = \frac{\sqrt{3}}{2}$. Since $\sin \theta$ is the y-coordinate, and $\theta = -300°$ lies in the first quadrant where the y-coordinates are positive, $\sin(-300°) = \frac{\sqrt{3}}{2}$.

45. (a) The reference angle for $210°$ is $30°$, and $\cos 30° = \frac{\sqrt{3}}{2}$. Since $\cos \theta$ is the x-coordinate, and $\theta = 210°$ lies in the third quadrant where the x-coordinates are negative, $\cos 210° = -\frac{\sqrt{3}}{2}$.

   (b) The reference angle for $-210°$ is $30°$, and $\cos 30° = \frac{\sqrt{3}}{2}$. Since $\cos \theta$ is the x-coordinate, and $\theta = -210°$ lies in the second quadrant where the x-coordinates are negative, $\cos(-210°) = -\frac{\sqrt{3}}{2}$.

   (c) The reference angle for $210°$ is $30°$, and $\sin 30° = \frac{1}{2}$. Since $\sin \theta$ is the y-coordinate, and $\theta = 210°$ lies in the third quadrant where the y-coordinates are negative, $\sin 210° = -\frac{1}{2}$.

   (d) The reference angle for $-210°$ is $30°$, and $\sin 30° = \frac{1}{2}$. Since $\sin \theta$ is the y-coordinate, and $\theta = -210°$ lies in the second quadrant where the y-coordinates are positive, $\sin(-210°) = \frac{1}{2}$.

47. (a) The reference angle for $390°$ is $30°$ ($390° = 30° + 360°$), and $\cos 30° = \frac{\sqrt{3}}{2}$. Since $\cos \theta$ is the x-coordinate, and $\theta = 390°$ lies in the first quadrant where the x-coordinates are positive, $\cos 390° = \frac{\sqrt{3}}{2}$.

(b) The reference angle for $-390°$ is $30°$ ($-390° = -30° - 360°$), and $\cos 30° = \frac{\sqrt{3}}{2}$. Since $\cos \theta$ is the x-coordinate, and $\theta = -390°$ lies in the fourth quadrant where the x-coordinates are positive, $\cos(-390°) = \frac{\sqrt{3}}{2}$.

(c) The reference angle for $390°$ is $30°$, and $\sin 30° = \frac{1}{2}$. Since $\sin \theta$ is the y-coordinate, and $\theta = 390°$ lies in the first quadrant where the y-coordinates are positive, $\sin 390° = \frac{1}{2}$.

(d) The reference angle for $-390°$ is $30°$, and $\sin 30° = \frac{1}{2}$. Since $\sin \theta$ is the y-coordinate, and $\theta = -390°$ lies in the fourth quadrant where the y-coordinates are negative, $\sin(-390°) = -\frac{1}{2}$.

49. (a) The reference angle for $600°$ is $60°$ ($600° = 240° + 360°$), and $\sec 60° = 2$. Since $\sec \theta$ is the reciprocal of the x-coordinate, and $\theta = 600°$ lies in the third quadrant where the x-coordinates are negative, $\sec 600° = -2$.

(b) The reference angle for $-600°$ is $60°$ ($-600° = -240° - 360°$), and $\csc 60° = \frac{2\sqrt{3}}{3}$. Since $\csc \theta$ is the reciprocal of the y-coordinate, and $\theta = -600°$ lies in the second quadrant where the y-coordinates are positive, $\csc(-600°) = \frac{2\sqrt{3}}{3}$.

(c) The reference angle for $600°$ is $60°$, and $\tan 60° = \sqrt{3}$. Since $\tan \theta = \frac{y}{x}$, and $\theta = 600°$ lies in the third quadrant where both the x- and y-coordinates are negative, $\tan 600° = \sqrt{3}$.

(d) The reference angle for $-600°$ is $60°$, and $\cot 60° = \frac{\sqrt{3}}{3}$. Since $\cot \theta = \frac{x}{y}$ and $\theta = -600°$ lies in the second quadrant where the y-coordinates are positive and the x-coordinates are negative, $\cot(-600°) = -\frac{\sqrt{3}}{3}$.

51. Complete the table:

θ	$\sin\theta$	$\cos\theta$	$\tan\theta$
0°	0	1	0
30°	$\frac{1}{2}$	$\frac{\sqrt{3}}{2}$	$\frac{\sqrt{3}}{3}$
45°	$\frac{\sqrt{2}}{2}$	$\frac{\sqrt{2}}{2}$	1
60°	$\frac{\sqrt{3}}{2}$	$\frac{1}{2}$	$\sqrt{3}$
90°	1	0	undefined
120°	$\frac{\sqrt{3}}{2}$	$-\frac{1}{2}$	$-\sqrt{3}$
135°	$\frac{\sqrt{2}}{2}$	$-\frac{\sqrt{2}}{2}$	-1
150°	$\frac{1}{2}$	$-\frac{\sqrt{3}}{2}$	$-\frac{\sqrt{3}}{3}$
180°	0	-1	0

53. Draw the figure:

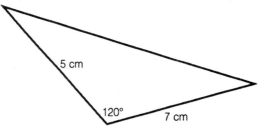

Now using the area formula:

$$\text{Area} = \tfrac{1}{2}ab\sin\theta = \tfrac{1}{2}(7)(5)\sin 120° = \left(\tfrac{35}{2}\right)\left(\tfrac{\sqrt{3}}{2}\right) = \tfrac{35\sqrt{3}}{4}\text{ cm}^2$$

55. Draw the figure:

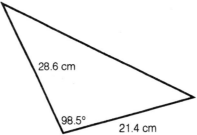

Now using the area formula:

$$\text{Area} = \tfrac{1}{2}ab\sin\theta = \tfrac{1}{2}(21.4)(28.6)\sin 98.5° \approx 302.7\text{ cm}^2$$

57. Start by finding the measure of a central angle of a triangle drawn from the center out to two adjacent vertices:

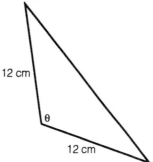

12 cm

12 cm

θ

Since a triangle is 3-sided, $3\theta = 360°$, so $\theta = 120°$. The area of this triangle is thus:
$$\tfrac{1}{2}ab\sin\theta = \tfrac{1}{2}(12)(12)\sin 120° = 72 \cdot \tfrac{\sqrt{3}}{2} = 36\sqrt{3}$$
The area of the shaded figure (3 triangles) is therefore:
$$3(36\sqrt{3}) = 108\sqrt{3} \approx 187.06 \text{ cm}^2$$

59. (a) Complete the table:

Terminal side of angle θ lies in

	Quadrant I	Quadrant II	Quadrant III	Quadrant IV
$\sin\theta$	positive	positive	negative	negative
$\cos\theta$	positive	negative	negative	positive
$\tan\theta$	positive	negative	positive	negative

(b) It works!

61. (a) Sketch the figure showing the obtuse angle $180° - \theta$ in standard position:

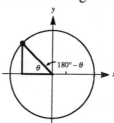

y

$180° - \theta$

θ

x

Note that the terminal side of the angle $180° - \theta$ lies in the second quadrant.

(b) The reference angle for $180° - \theta$ is θ.

(c) Since $180° - \theta$ lies in the second quadrant where all y-coordinates are positive, $\sin(180° - \theta) = \sin\theta$.

63. Since the height of the triangle is the y-coordinate of point P, the height is $\sin\theta$. Thus the area is:

$$\text{area} = \tfrac{1}{2}(\text{base})(\text{height}) = \tfrac{1}{2}(2)(\sin\theta) = \sin\theta$$

65. (a) Using $\theta = 20°$, calculate:

$$\log_{10}\left(\sin^2 20°\right) \approx -0.9319$$
$$2\log_{10}\left(\sin 20°\right) \approx -0.9319$$

The two expressions are equal.

(b) They are equal as long as $\sin\theta > 0$, thus ensuring $\log_{10}\left(\sin\theta\right)$ is defined. Thus they are equal on the interval $0° < \theta < 180°$.

67. Since $(L \circ S)(\theta) = \ln(\sin\theta)$, we must have $\sin\theta > 0$. This occurs on the interval $0° < \theta < 180°$, which (in interval notation) is $(0°, 180°)$.

69. (a) Using $\theta = 20°$, calculate:

$$\ln\sqrt{1 - \cos 20°} + \ln\sqrt{1 + \cos 20°} \approx -1.0729$$
$$\ln\left(\sin 20°\right) \approx -1.0729$$

The two expressions are equal.

(b) Working from the left-hand side:

$$\ln\sqrt{1 - \cos\theta} + \ln\sqrt{1 + \cos\theta} = \ln\sqrt{1 - \cos^2\theta} = \ln\sqrt{\sin^2\theta} = \ln(\sin\theta)$$

(c) For the equation to be valid, we must ensure $\sin\theta > 0$. This occurs on the interval $0° < \theta < 180°$, which (in interval notation) is $(0°, 180°)$.

71. (a) In $\triangle ADC$, $\sec\alpha = \dfrac{AC}{AD} = \dfrac{AC}{1} = AC$.

Similarly, in $\triangle ADB$ $\sec\beta = \dfrac{AB}{AD} = \dfrac{AB}{1} = AB$.

(b) Compute the areas:

Area $\triangle ADC = \tfrac{1}{2}(AC)(AD)\sin\alpha = \tfrac{1}{2}\sec\alpha\sin\alpha$

Area $\triangle ADB = \tfrac{1}{2}(AD)(AB)\sin\beta = \tfrac{1}{2}\sec\beta\sin\beta$

Area $\triangle ABC = \tfrac{1}{2}(AC)(AB)\sin(\alpha + \beta) = \tfrac{1}{2}\sec\alpha\sec\beta\sin(\alpha + \beta)$

(c) Since the sum of the areas of the two smaller triangles equals the area of $\triangle ABC$:

$$\tfrac{1}{2}\sec\alpha\sec\beta\sin(\alpha + \beta) = \tfrac{1}{2}\sec\alpha\sin\alpha + \tfrac{1}{2}\sec\beta\sin\beta$$

Now multiply both sides of this equation by the quantity $2\cos\alpha\cos\beta$. This yields:

$$\sin(\alpha + \beta) = \sin\alpha\cos\beta + \cos\alpha\sin\beta$$

(d)   Using the hint:

$$\sin 75° = \sin(30°+45°)$$
$$= \sin 30° \cos 45° + \cos 30° \sin 45°$$
$$= \frac{1}{2} \cdot \frac{\sqrt{2}}{2} + \frac{\sqrt{3}}{2} \cdot \frac{\sqrt{2}}{2}$$
$$= \frac{\sqrt{2}+\sqrt{6}}{4}$$

(e)   Compute $\sin 75° = \dfrac{\sqrt{2}+\sqrt{6}}{4} \approx \dfrac{1.4+2.4}{4} = \dfrac{3.8}{4} < 1.$
But $\sin 30° + \sin 45° \approx 0.5 + 0.7 = 1.2 > 1$. Since $\sin 75°$ is less than 1, whereas $\sin 30° + \sin 45°$ is greater than 1, the quantities cannot be equal. We can also argue with exact values (not approximations):

$$\sin 30° + \sin 45° = \frac{1}{2} + \frac{\sqrt{2}}{2} = \frac{2+2\sqrt{2}}{4}$$
$$\sin 75° = \frac{\sqrt{2}+\sqrt{6}}{4}$$

So, if these were to be equal then $2 + 2\sqrt{2} = \sqrt{2} + \sqrt{6}$, thus $2 + \sqrt{2} = \sqrt{6}$. But $2 + \sqrt{2} > 3$ while $\sqrt{6} < 3$, thus they cannot be equal.

(f)   Compute $\sin 105° \approx 0.9659$, $\sin 45° \approx 0.7071$, and $\sin 60° \approx 0.8660$. So $\sin 45° + \sin 60° \approx 1.5731 \neq \sin 105°$.
Note: An argument similar to (e) using exact values can be made here.

7.5 Trigonometric Identities

1.   (a)   Substituting $\sin \theta = \frac{1}{5}$ into $\cos^2 \theta = 1 - \sin^2 \theta$ yields:

$$\cos^2 \theta = 1 - \left(\tfrac{1}{5}\right)^2 = 1 - \tfrac{1}{25} = \tfrac{24}{25}$$

Since the terminal side of θ lies in quadrant 2, $\cos \theta < 0$ and thus:

$$\cos \theta = -\sqrt{\tfrac{24}{25}} = -\frac{2\sqrt{6}}{5}$$

Now compute the remaining four trigonometric values:

$$\tan \theta = \frac{\sin \theta}{\cos \theta} = \frac{\frac{1}{5}}{-\frac{2\sqrt{6}}{5}} = -\frac{1}{2\sqrt{6}} = -\frac{\sqrt{6}}{12}$$

$$\cot \theta = \frac{\cos \theta}{\sin \theta} = \frac{-\frac{2\sqrt{6}}{5}}{\frac{1}{5}} = -2\sqrt{6}$$

$$\sec \theta = \frac{1}{\cos \theta} = \frac{1}{-\frac{2\sqrt{6}}{5}} = -\frac{5}{2\sqrt{6}} = -\frac{5\sqrt{6}}{12}$$

$$\csc \theta = \frac{1}{\sin \theta} = \frac{1}{\frac{1}{5}} = 5$$

(b) Substituting $\sin\theta = -\frac{1}{5}$ into $\cos^2\theta = 1 - \sin^2\theta$ yields:

$$\cos^2\theta = 1 - \left(\tfrac{1}{5}\right)^2 = 1 - \tfrac{1}{25} = \tfrac{24}{25}$$

Since the terminal side of θ lies in quadrant 3, $\cos\theta < 0$ and thus:

$$\cos\theta = -\sqrt{\tfrac{24}{25}} = -\tfrac{2\sqrt{6}}{5}$$

Now compute the remaining four trigonometric values:

$$\tan\theta = \frac{\sin\theta}{\cos\theta} = \frac{-\frac{1}{5}}{-\frac{2\sqrt{6}}{5}} = \frac{1}{2\sqrt{6}} = \frac{\sqrt{6}}{12}$$

$$\cot\theta = \frac{\cos\theta}{\sin\theta} = \frac{-\frac{2\sqrt{6}}{5}}{-\frac{1}{5}} = 2\sqrt{6}$$

$$\sec\theta = \frac{1}{\cos\theta} = \frac{1}{-\frac{2\sqrt{6}}{5}} = -\frac{5}{2\sqrt{6}} = -\frac{5\sqrt{6}}{12}$$

$$\csc\theta = \frac{1}{\sin\theta} = \frac{1}{-\frac{1}{5}} = -5$$

3. (a) Substituting $\cos\theta = \frac{5}{13}$ into $\sin^2\theta = 1 - \cos^2\theta$ yields:

$$\sin^2\theta = 1 - \left(\tfrac{5}{13}\right)^2 = 1 - \tfrac{25}{169} = \tfrac{144}{169}$$

Since the terminal side of θ lies in quadrant 1, $\sin\theta > 0$ and thus:

$$\sin\theta = \sqrt{\tfrac{144}{169}} = \tfrac{12}{13}$$

Now compute the remaining four trigonometric values:

$$\tan\theta = \frac{\sin\theta}{\cos\theta} = \frac{\frac{12}{13}}{\frac{5}{13}} = \frac{12}{5} \qquad \cot\theta = \frac{\cos\theta}{\sin\theta} = \frac{\frac{5}{13}}{\frac{12}{13}} = \frac{5}{12}$$

$$\sec\theta = \frac{1}{\cos\theta} = \frac{1}{\frac{5}{13}} = \frac{13}{5} \qquad \csc\theta = \frac{1}{\sin\theta} = \frac{1}{\frac{12}{13}} = \frac{13}{12}$$

(b) Substituting $\cos\theta = -\frac{5}{13}$ into $\sin^2\theta = 1 - \cos^2\theta$ yields:

$$\sin^2\theta = 1 - \left(-\tfrac{5}{13}\right)^2 = 1 - \tfrac{25}{169} = \tfrac{144}{169}$$

Since the terminal side of θ lies in quadrant 3, $\sin\theta < 0$ and thus:

$$\sin\theta = -\sqrt{\tfrac{144}{169}} = -\tfrac{12}{13}$$

Now compute the remaining four trigonometric values:

$$\tan\theta = \frac{\sin\theta}{\cos\theta} = \frac{-\frac{12}{13}}{-\frac{5}{13}} = \frac{12}{5} \qquad \cot\theta = \frac{\cos\theta}{\sin\theta} = \frac{-\frac{5}{13}}{-\frac{12}{13}} = \frac{5}{12}$$

$$\sec\theta = \frac{1}{\cos\theta} = \frac{1}{-\frac{5}{13}} = -\frac{13}{5} \qquad \csc\theta = \frac{1}{\sin\theta} = \frac{1}{-\frac{12}{13}} = -\frac{13}{12}$$

5. Since $\csc A = -3$, $\sin A = -\frac{1}{3}$. Substituting into $\cos^2 A = 1 - \sin^2 A$ yields:

$$\cos^2 A = 1 - \left(-\tfrac{1}{3}\right)^2 = 1 - \tfrac{1}{9} = \tfrac{8}{9}$$

Since the terminal side of A lies in quadrant 4, $\cos A > 0$ and thus:

$$\cos A = \sqrt{\tfrac{8}{9}} = \tfrac{2\sqrt{2}}{3}$$

Now compute the remaining three trigonometric values:

$$\tan A = \frac{\sin A}{\cos A} = \frac{-\frac{1}{3}}{\frac{2\sqrt{2}}{3}} = -\frac{1}{2\sqrt{2}} = -\frac{\sqrt{2}}{4}$$

$$\cot A = \frac{\cos A}{\sin A} = \frac{\frac{2\sqrt{2}}{3}}{-\frac{1}{3}} = -2\sqrt{2}$$

$$\sec A = \frac{1}{\cos A} = \frac{1}{\frac{2\sqrt{2}}{3}} = \frac{3}{2\sqrt{2}} = \frac{3\sqrt{2}}{4}$$

7. Since $\sec B = -\frac{3}{2}$, then $\cos B = -\frac{2}{3}$. Substituting into $\sin^2 B = 1 - \cos^2 B$ yields:

$$\sin^2 B = 1 - \left(-\tfrac{2}{3}\right)^2 = 1 - \tfrac{4}{9} = \tfrac{5}{9}$$

Since the terminal side of B lies in quadrant 3, $\sin B < 0$ and thus:

$$\sin B = -\sqrt{\tfrac{5}{9}} = -\tfrac{\sqrt{5}}{3}$$

Now compute the remaining three trigonometric values:

$$\tan B = \frac{\sin B}{\cos B} = \frac{-\frac{\sqrt{5}}{3}}{-\frac{2}{3}} = \frac{\sqrt{5}}{2}$$

$$\cot B = \frac{\cos B}{\sin B} = \frac{-\frac{2}{3}}{-\frac{\sqrt{5}}{3}} = \frac{2}{\sqrt{5}} = \frac{2\sqrt{5}}{5}$$

$$\csc B = \frac{1}{\sin B} = \frac{1}{-\frac{\sqrt{5}}{3}} = -\frac{3}{\sqrt{5}} = -\frac{3\sqrt{5}}{5}$$

9. Substituting $\cos\theta = \dfrac{t}{3}$ into $\sin^2\theta = 1 - \cos^2\theta$ yields:

$$\sin^2\theta = 1 - \left(\frac{t}{3}\right)^2 = 1 - \frac{t^2}{9} = \frac{9 - t^2}{9}$$

Since the terminal side of θ lies in quadrant 4, $\sin\theta < 0$ and thus:

$$\sin\theta = -\sqrt{\frac{9 - t^2}{9}} = -\frac{\sqrt{9 - t^2}}{3}$$

Now compute the remaining four trigonometric values:

$$\tan \theta = \frac{\sin \theta}{\cos \theta} = \frac{-\frac{\sqrt{9-t^2}}{3}}{\frac{t}{3}} = -\frac{\sqrt{9-t^2}}{t}$$

$$\cot \theta = \frac{\cos \theta}{\sin \theta} = \frac{\frac{t}{3}}{-\frac{\sqrt{9-t^2}}{3}} = -\frac{t}{\sqrt{9-t^2}} = -\frac{t\sqrt{9-t^2}}{9-t^2}$$

$$\sec \theta = \frac{1}{\cos \theta} = \frac{1}{\frac{t}{3}} = \frac{3}{t}$$

$$\csc \theta = \frac{1}{\sin \theta} = \frac{1}{-\frac{\sqrt{9-t^2}}{3}} = -\frac{3}{\sqrt{9-t^2}} = -\frac{3\sqrt{9-t^2}}{9-t^2}$$

11.   Substituting $\sin \theta = -3u$ into $\cos^2 \theta = 1 - \sin^2 \theta$ yields:

$$\cos^2 \theta = 1 - (-3u)^2 = 1 - 9u^2$$

Since the terminal side of θ lies in quadrant 3, $\cos \theta < 0$ and thus:

$$\cos \theta = -\sqrt{1 - 9u^2}$$

Now compute the remaining four trigonometric values:

$$\tan \theta = \frac{\sin \theta}{\cos \theta} = \frac{-3u}{-\sqrt{1-9u^2}} = \frac{3u\sqrt{1-9u^2}}{1-9u^2}$$

$$\cot \theta = \frac{\cos \theta}{\sin \theta} = \frac{-\sqrt{1-9u^2}}{-3u} = \frac{\sqrt{1-9u^2}}{3u}$$

$$\sec \theta = \frac{1}{\cos \theta} = \frac{1}{-\sqrt{1-9u^2}} = -\frac{\sqrt{1-9u^2}}{1-9u^2}$$

$$\csc \theta = \frac{1}{\sin \theta} = \frac{1}{-3u} = -\frac{1}{3u}$$

13.   Substituting $\cos \theta = \frac{u}{\sqrt{3}}$ into $\sin^2 \theta = 1 - \cos^2 \theta$ yields:

$$\sin^2 \theta = 1 - \left(\frac{u}{\sqrt{3}}\right)^2 = 1 - \frac{u^2}{3} = \frac{3-u^2}{3}$$

Since the terminal side of θ lies in quadrant 1, $\sin \theta > 0$ and thus:

$$\sin \theta = \sqrt{\frac{3-u^2}{3}} = \frac{\sqrt{3-u^2}}{\sqrt{3}} = \frac{\sqrt{9-3u^2}}{3}$$

Now compute the remaining four trigonometric values:

$$\tan \theta = \frac{\sin \theta}{\cos \theta} = \frac{\frac{\sqrt{9-3u^2}}{3}}{\frac{u}{\sqrt{3}}} = \frac{\sqrt{9-3u^2}}{u\sqrt{3}} = \frac{\sqrt{3-u^2}}{u}$$

$$\cot \theta = \frac{\cos \theta}{\sin \theta} = \frac{\frac{u}{\sqrt{3}}}{\frac{\sqrt{9-3u^2}}{3}} = \frac{u\sqrt{3}}{\sqrt{9-3u^2}} = \frac{u}{\sqrt{3-u^2}} = \frac{u\sqrt{3-u^2}}{3-u^2}$$

$$\sec \theta = \frac{1}{\cos \theta} = \frac{1}{\frac{u}{\sqrt{3}}} = \frac{\sqrt{3}}{u}$$

$$\csc \theta = \frac{1}{\sin \theta} = \frac{1}{\frac{\sqrt{9-3u^2}}{3}} = \frac{3}{\sqrt{9-3u^2}} = \frac{3\sqrt{9-3u^2}}{9-3u^2} = \frac{\sqrt{9-3u^2}}{3-u^2}$$

15. Using the identities for $\sec \theta$ and $\csc \theta$:

$$\sin \theta \cos \theta \sec \theta \csc \theta = \frac{\sin \theta \cos \theta}{\sin \theta \cos \theta} = 1$$

17. Using the identities for $\sec \theta$ and $\tan \theta$:

$$\frac{\sin \theta \sec \theta}{\tan \theta} = \frac{\sin \theta \cdot \frac{1}{\cos \theta}}{\frac{\sin \theta}{\cos \theta}} = \frac{\frac{\sin \theta}{\cos \theta}}{\frac{\sin \theta}{\cos \theta}} = 1$$

19. Working from the right-hand side and using the identities for $\sec x$ and $\tan x$:

$$\sec x - 5 \tan x = \frac{1}{\cos x} - \frac{5 \sin x}{\cos x} = \frac{1 - 5 \sin x}{\cos x}$$

21. Using the identity for $\sec A$:

$$\cos A(\sec A - \cos A) = \cos A\left(\frac{1}{\cos A} - \cos A\right) = 1 - \cos^2 A = \sin^2 A$$

23. Multiplying out parentheses and using the identities for $\sec \theta$ and $\tan \theta$:

$$(1 - \sin \theta)(\sec \theta + \tan \theta) = \sec \theta + \tan \theta - \sin \theta \sec \theta - \sin \theta \tan \theta$$

$$= \frac{1}{\cos \theta} + \frac{\sin \theta}{\cos \theta} - \sin \theta \cdot \frac{1}{\cos \theta} - \sin \theta \cdot \frac{\sin \theta}{\cos \theta}$$

$$= \frac{1 + \sin \theta - \sin \theta - \sin^2 \theta}{\cos \theta}$$

$$= \frac{1 - \sin^2 \theta}{\cos \theta}$$

$$= \frac{\cos^2 \theta}{\cos \theta}$$

$$= \cos \theta$$

25. Multiplying out parentheses and using the identities for $\sec \alpha$ and $\tan \alpha$:

$$
\begin{aligned}
(\sec \alpha - \tan \alpha)^2 &= \left(\frac{1}{\cos \alpha} - \frac{\sin \alpha}{\cos \alpha} \right)^2 \\
&= \frac{(1 - \sin \alpha)^2}{\cos^2 \alpha} \\
&= \frac{(1 - \sin \alpha)^2}{1 - \sin^2 \alpha} \\
&= \frac{(1 - \sin \alpha)^2}{(1 + \sin \alpha)(1 - \sin \alpha)} \\
&= \frac{1 - \sin \alpha}{1 + \sin \alpha}
\end{aligned}
$$

27. Working from the right-hand side and using identities for $\cot A$ and $\tan A$:

$$
\begin{aligned}
\frac{\sin A}{1 - \cot A} - \frac{\cos A}{\tan A - 1} &= \frac{\sin A}{1 - \frac{\cos A}{\sin A}} - \frac{\cos A}{\frac{\sin A}{\cos A} - 1} \\
&= \frac{\sin^2 A}{\sin A - \cos A} - \frac{\cos^2 A}{\sin A - \cos A} \\
&= \frac{(\sin A - \cos A)(\sin A + \cos A)}{\sin A - \cos A} \\
&= \sin A + \cos A
\end{aligned}
$$

29. Working from the left-hand side and using identities for $\csc \theta$ and $\sec \theta$:

$$
\begin{aligned}
\csc^2 \theta + \sec^2 \theta &= \frac{1}{\sin^2 \theta} + \frac{1}{\cos^2 \theta} \\
&= \frac{\cos^2 \theta + \sin^2 \theta}{\sin^2 \theta \cos^2 \theta} \\
&= \frac{1}{\sin^2 \theta \cos^2 \theta} \\
&= \frac{1}{\sin^2 \theta} \cdot \frac{1}{\cos^2 \theta} \\
&= \csc^2 \theta \sec^2 \theta
\end{aligned}
$$

31. Using the identity for $\tan A$ and $\sin^2 A = 1 - \cos^2 A$:

$$
\sin A \tan A = \sin A \cdot \frac{\sin A}{\cos A} = \frac{\sin^2 A}{\cos A} = \frac{1 - \cos^2 A}{\cos A}
$$

33. Working from the right-hand side:

$$-\cot^4 A + \csc^4 A = \frac{-\cos^4 A}{\sin^4 A} + \frac{1}{\sin^4 A}$$

$$= \frac{\left(1 - \cos^2 A\right)\left(1 + \cos^2 A\right)}{\sin^4 A}$$

$$= \frac{\sin^2 A\left(1 + \cos^2 A\right)}{\sin^4 A}$$

$$= \frac{1 + \cos^2 A}{\sin^2 A}$$

$$= \frac{1}{\sin^2 A} + \frac{\cos^2 A}{\sin^2 A}$$

$$= \csc^2 A + \cot^2 A$$

35. Working from the left-hand side:

$$\frac{\sin A - \cos A}{\sin A} + \frac{\cos A - \sin A}{\cos A}$$

$$= \frac{\cos A(\sin A - \cos A) + \sin A(\cos A - \sin A)}{\sin A \cos A}$$

$$= \frac{\cos A \sin A - \cos^2 A + \sin A \cos A - \sin^2 A}{\sin A \cos A}$$

$$= \frac{2\cos A \sin A - \left(\cos^2 A + \sin^2 A\right)}{\sin A \cos A}$$

$$= 2 - \frac{1}{\sin A \cos A}$$

$$= 2 - \sec A \csc A$$

37. (a) Start by testing the value $\alpha = 30°$. Since $\cos 30° = \frac{\sqrt{3}}{2}$ and $\csc 30° = 2$, the equation states that:

$$\frac{2^2 - 1}{2^2} = \frac{\sqrt{3}}{2}$$

$$\frac{3}{4} = \frac{\sqrt{3}}{2}$$

Since the left-hand side does not equal the right-hand side, this is not an identity.

(b)   Proceeding as in (a), test the value $\alpha = 30°$. Since $\sec 30° = \frac{2}{\sqrt{3}}$ and $\csc 30° = 2$, the equation states that:

$$\left[\left(\tfrac{2}{\sqrt{3}}\right)^2 - 1\right]\left[(2)^2 - 1\right] = 1$$

$$\left[\tfrac{4}{3} - 1\right]\left[4 - 1\right] = 1$$

$$\tfrac{1}{3} \cdot 3 = 1$$

Since this is true, proceed by proving the identity:

$$(\sec^2 \alpha - 1)(\csc^2 \alpha - 1) = \left(\frac{1}{\cos^2 \alpha} - 1\right)\left(\frac{1}{\sin^2 \alpha} - 1\right)$$

$$= \frac{1 - \cos^2 \alpha}{\cos^2 \alpha} \cdot \frac{1 - \sin^2 \alpha}{\sin^2 \alpha}$$

$$= \frac{\sin^2 \alpha}{\cos^2 \alpha} \cdot \frac{\cos^2 \alpha}{\sin^2 \alpha}$$

$$= 1$$

39.   (a)   Starting with the left-hand side, multiply the numerator and denominator by $1 + \cos \theta$ to obtain:

$$\frac{\sin \theta}{1 - \cos \theta} \cdot \frac{1 + \cos \theta}{1 + \cos \theta} = \frac{\sin \theta(1 + \cos \theta)}{1 - \cos^2 \theta} = \frac{\sin \theta(1 + \cos \theta)}{\sin^2 \theta} = \frac{1 + \cos \theta}{\sin \theta}$$

(b)   Starting with the left-hand side, multiply the numerator and denominator by $\sin \theta$ to obtain:

$$\frac{\sin \theta}{1 - \cos \theta} \cdot \frac{\sin \theta}{\sin \theta} = \frac{\sin^2 \theta}{\sin \theta(1 - \cos \theta)}$$

$$= \frac{1 - \cos^2 \theta}{\sin \theta(1 - \cos \theta)}$$

$$= \frac{(1 + \cos \theta)(1 - \cos \theta)}{\sin \theta(1 - \cos \theta)}$$

$$= \frac{1 + \cos \theta}{\sin \theta}$$

41.   Starting with the left-hand side, use the identities for $\sec \theta$ and $\csc \theta$, then multiply the resulting fraction by $\sin \theta$ to obtain:

$$\frac{\sec \theta - \csc \theta}{\sec \theta + \csc \theta} = \frac{\frac{1}{\cos \theta} - \frac{1}{\sin \theta}}{\frac{1}{\cos \theta} + \frac{1}{\sin \theta}} = \frac{\frac{\sin \theta}{\cos \theta} - \frac{\sin \theta}{\sin \theta}}{\frac{\sin \theta}{\cos \theta} + \frac{\sin \theta}{\sin \theta}} = \frac{\tan \theta - 1}{\tan \theta + 1}$$

43. Starting with the left-hand side, use the identity for $\cot\theta$ to obtain:

$$\sin^2\theta\left(1+n\cot^2\theta\right)=\sin^2\theta\left(1+n\cdot\frac{\cos^2\theta}{\sin^2\theta}\right)$$
$$=\sin^2\theta+n\cos^2\theta$$
$$=\cos^2\theta\left(\frac{\sin^2\theta}{\cos^2\theta}+n\right)$$
$$=\cos^2\theta\left(n+\tan^2\theta\right)$$

45. (a) Using the difference of cubes formula $A^3-B^3=(A-B)\left(A^2+AB+B^2\right)$:

$$\cos^3\theta-\sin^3\theta=(\cos\theta-\sin\theta)\left(\cos^2\theta+\cos\theta\sin\theta+\sin^2\theta\right)$$
$$=(\cos\theta-\sin\theta)(1+\cos\theta\sin\theta)$$

(b) Starting with the left-hand side, use the identities for $\cot\phi$, $\tan\phi$, $\sec\phi$ and $\csc\phi$ to obtain:

$$\frac{\cos\phi\cot\phi-\sin\phi\tan\phi}{\csc\phi-\sec\phi}=\frac{\cos\phi\cdot\frac{\cos\phi}{\sin\phi}-\sin\phi\cdot\frac{\sin\phi}{\cos\phi}}{\frac{1}{\sin\phi}-\frac{1}{\cos\phi}}$$
$$=\frac{\frac{\cos^2\phi}{\sin\phi}-\frac{\sin^2\phi}{\cos\phi}}{\frac{\cos\phi-\sin\phi}{\sin\phi\cos\phi}}$$
$$=\frac{\cos^3\phi-\sin^3\phi}{\cos\phi-\sin\phi}$$
$$=\frac{(\cos\phi-\sin\phi)(1+\cos\phi\sin\phi)}{\cos\phi-\sin\phi}$$
$$=1+\sin\phi\cos\phi$$

47. Since $\tan\alpha\tan\beta=1$, $\tan\alpha=\dfrac{1}{\tan\beta}$. Using the identities from Exercise 46:

$$\sec^2\alpha=1+\tan^2\alpha=1+\frac{1}{\tan^2\beta}=1+\cot^2\beta=\csc^2\beta$$

Since α and β are acute angles, $\sec\alpha>0$ and $\csc\beta>0$, so $\sec^2\alpha=\csc^2\beta$ implies $\sec\alpha=\csc\beta$.

49. Using the identity $\cos^2 \theta = 1 - \sin^2 \theta$:

$$\cos^2 \theta = 1 - \frac{(p-q)^2}{(p+q)^2}$$

$$= 1 - \frac{p^2 - 2pq + q^2}{p^2 + 2pq + q^2}$$

$$= \frac{p^2 + 2pq + q^2 - p^2 + 2pq - q^2}{p^2 + 2pq + q^2}$$

$$= \frac{4pq}{(p+q)^2}$$

Since the terminal side of θ lies in quadrant 2, $\cos \theta < 0$ and thus:

$$\cos \theta = -\sqrt{\frac{4pq}{(p+q)^2}} = -\frac{2\sqrt{pq}}{p+q} \quad \text{(since } p > 0, q > 0\text{)}$$

Now find $\tan \theta$:

$$\tan \theta = \frac{\sin \theta}{\cos \theta} = \frac{\dfrac{p-q}{p+q}}{-\dfrac{2\sqrt{pq}}{p+q}} = \frac{p-q}{-2\sqrt{pq}} = \frac{q-p}{2\sqrt{qp}}$$

Chapter Seven Review Exercises

1. Since the terminal side of $135°$ lies in quadrant 2, $\sin 135° > 0$. Since the reference angle for $135°$ is $45°$:

$$\sin 135° = \sin 45° = \tfrac{\sqrt{2}}{2}$$

3. Since the terminal side of $-240°$ lies in quadrant 2, $\tan(-240°) < 0$. Since the reference angle for $-240°$ is $60°$:

$$\tan(-240°) = -\tan 60° = -\sqrt{3}$$

5. Since the terminal side of $210°$ lies in quadrant 3, $\csc(210°) < 0$. Since the reference angle for $210°$ is $30°$:

$$\csc 210° = -\csc 30° = -2$$

7. Since $270°$ corresponds to the point $(0, -1)$ on the unit circle, $\sin 270° = -1$, which is the y-coordinate.

9. Since the terminal side of $-315°$ lies in quadrant 1, $\cos(-315°) > 0$. Since the reference angle for $-315°$ is $45°$:

$$\cos(-315°) = \cos 45° = \tfrac{\sqrt{2}}{2}$$

11. Since $1800°$ corresponds to the point $(1, 0)$ on the unit circle, $\cos 1800° = 1$, which is the x-coordinate.

13. Since the terminal side of $240°$ lies in quadrant 3, $\csc 240° < 0$. Since the reference angle for $240°$ is $60°$:

$$\csc 240° = -\csc 60° = -\frac{2\sqrt{3}}{3}$$

15. Since the terminal side of $780°$ lies in quadrant 1, $\sec 780° > 0$. Since the reference angle for $780°$ is $60°$:

$$\sec 780° = \sec 60° = 2$$

17. Since $\sin^2 17° + \cos^2 17° = 1$ while $\ln e^\pi = \pi$:

$$\ln\left(\sin^2 17° + \cos^2 17°\right) - \cos\left(\ln e^\pi\right) = \ln(1) - \cos(\pi) = 0 - (-1) = 1$$

19. Draw the triangle:

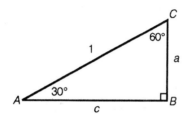

Now compute a and c:

$$\sin 30° = \frac{a}{1}, \text{ so } a = \sin 30° = \tfrac{1}{2}$$
$$\cos 30° = \frac{c}{1}, \text{ so } c = \cos 30° = \frac{\sqrt{3}}{2}$$

21. Draw a triangle:

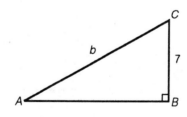

Now compute b:

$$\sin A = \frac{7}{b}, \text{ so } b = \frac{7}{\sin A} = \frac{7}{\tfrac{2}{5}} = \frac{35}{2}$$

23. Draw a triangle:

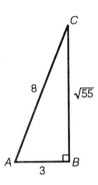

Using the Pythagorean theorem:
$$3^2 + b^2 = 8^2$$
$$9 + b^2 = 64$$
$$b^2 = 55$$
$$b = \sqrt{55}$$
From the triangle:
$$\sin A = \frac{\sqrt{55}}{8} \text{ and } \cot A = \frac{3}{\sqrt{55}} = \frac{3\sqrt{55}}{55}$$

25. Draw the triangle:

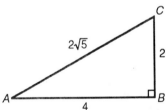

Now compute the area:
$$\tfrac{1}{2}ab\sin\theta = \tfrac{1}{2}(5)(4)\sin 150° = 10 \cdot \tfrac{1}{2} = 5 \text{ square units}$$

27. Draw a triangle:

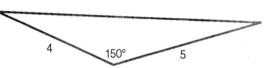

Using the Pythagorean theorem:
$$4^2 + 2^2 = c^2$$
$$16 + 4 = c^2$$
$$c^2 = 20$$
$$c = 2\sqrt{5}$$
From the triangle:
$$\sin A = \frac{2}{2\sqrt{5}} = \frac{1}{\sqrt{5}} = \frac{\sqrt{5}}{5} \text{ and } \cos B = 0$$
Therefore:
$$\sin^2 A + \cos^2 B = \left(\frac{\sqrt{5}}{5}\right)^2 + (0)^2 = \frac{5}{25} + 0 = \frac{1}{5}$$

29. Draw a triangle:

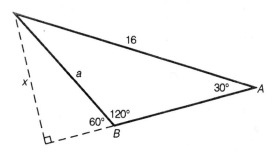

First find x:

$$\sin 30° = \frac{x}{16}, \text{ so } x = 16\sin 30° = \tfrac{1}{2}(16) = 8$$

Now, using the smaller triangle:

$$\sin 60° = \frac{8}{a}, \text{ so } a = \frac{8}{\sin 60°} = \frac{8}{\frac{\sqrt{3}}{2}} = \frac{16}{\sqrt{3}} = \frac{16\sqrt{3}}{3}$$

31. Draw a triangle:

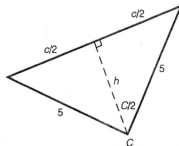

Notice that the bisector drawn for C bisects the opposite side (of length c) into two pieces of length $\frac{c}{2}$. This is true since $a = b = 5$. Using one of the smaller triangles:

$$\sin\left(\tfrac{1}{2}C\right) = \frac{\frac{c}{2}}{5}$$

$$\frac{9}{10} = \frac{c}{10}$$

$$9 = c$$

To find the height h of the triangle, use the Pythagorean theorem:

$$\left(\tfrac{c}{2}\right)^2 + h^2 = 5^2$$

$$\left(\tfrac{9}{2}\right)^2 + h^2 = 25$$

$$\tfrac{81}{4} + h^2 = 25$$

$$h^2 = \tfrac{19}{4}$$

$$h = \tfrac{\sqrt{19}}{2}$$

So the area of $\triangle ABC$ is given by:

$$\tfrac{1}{2}(\text{base})(\text{height}) = \tfrac{1}{2}(9)\left(\tfrac{\sqrt{19}}{2}\right) = \tfrac{9\sqrt{19}}{4}$$

33. Letting z be the side opposite the $90° - \beta$ angle (adjacent to β):

$$\cot \alpha = \frac{x+z}{y} \text{ and } \cot \beta = \frac{z}{y}$$

Thus:

$$\cot \alpha - \cot \beta = \frac{x+z}{y} - \frac{z}{y}$$

$$\cot \alpha - \cot \beta = \frac{x}{y}$$

$$y(\cot \alpha - \cot \beta) = x$$

$$y = \frac{x}{\cot \alpha - \cot \beta}$$

35. Since $\sin^2 \theta + \cos^2 \theta = 1$:

$$\left(\frac{2p^2q^2}{p^4 + q^4} \right)^2 + \cos^2 \theta = 1$$

$$\frac{4p^4q^4}{\left(p^4 + q^4\right)^2} + \cos^2 \theta = 1$$

$$\cos^2 \theta = \frac{\left(p^4 + q^4\right)^2 - 4p^4q^4}{\left(p^4 + q^4\right)^2}$$

$$\cos^2 \theta = \frac{p^8 - 2p^4q^4 + q^8}{\left(p^4 + q^4\right)^2}$$

$$\cos^2 \theta = \frac{\left(p^4 - q^4\right)^2}{\left(p^4 + q^4\right)^2}$$

$$\cos \theta = \frac{p^4 - q^4}{p^4 + q^4}$$

Now find $\tan \theta$:

$$\tan \theta = \frac{\sin \theta}{\cos \theta} = \frac{2p^2q^2}{p^4 - q^4}$$

37. Using the identities for $\sec A$ and $\csc A$:

$$\frac{\sin A + \cos A}{\sec A + \csc A} = \frac{\sin A + \cos A}{\frac{1}{\cos A} + \frac{1}{\sin A}} = \frac{\sin A \cos A(\sin A + \cos A)}{\sin A + \cos A} = \sin A \cos A$$

39. Using the identities for $\sec A$, $\tan A$ and $\cot A$:

$$\frac{\sin A \sec A}{\tan A + \cot A} = \frac{\frac{\sin A}{\cos A}}{\frac{\sin A}{\cos A} + \frac{\cos A}{\sin A}} = \frac{\sin^2 A}{\sin^2 A + \cos^2 A} = \sin^2 A$$

41. Using the identities for $\tan A$ and $\cot A$:

$$\frac{\cos A}{1 - \tan A} + \frac{\sin A}{1 - \cot A} = \frac{\cos A}{1 - \frac{\sin A}{\cos A}} + \frac{\sin A}{1 - \frac{\cos A}{\sin A}}$$

$$= \frac{\cos^2 A}{\cos A - \sin A} + \frac{\sin^2 A}{\sin A - \cos A}$$

$$= \frac{\cos^2 A}{\cos A - \sin A} - \frac{\sin^2 A}{\cos A - \sin A}$$

$$= \frac{\cos^2 A - \sin^2 A}{\cos A - \sin A}$$

$$= \frac{(\cos A + \sin A)(\cos A - \sin A)}{\cos A - \sin A}$$

$$= \cos A + \sin A$$

43. Using the identities for $\sec A$ and $\csc A$:

$$(\sec A + \csc A)^{-1}\left[(\sec A)^{-1} + (\csc A)^{-1}\right] = \frac{1}{\sec A + \csc A} \cdot \left[\frac{1}{\sec A} + \frac{1}{\csc A}\right]$$

$$= \frac{1}{\sec A + \csc A} \cdot \frac{\csc A + \sec A}{\sec A \csc A}$$

$$= \frac{1}{\sec A \csc A}$$

$$= \frac{1}{\frac{1}{\cos A} \cdot \frac{1}{\sin A}}$$

$$= \sin A \cos A$$

45. Simplifying the complex fraction:

$$\frac{\frac{\sin A + \cos A}{\sin A - \cos A} - \frac{\sin A - \cos A}{\sin A + \cos A}}{\frac{\sin A + \cos A}{\sin A - \cos A} + \frac{\sin A - \cos A}{\sin A + \cos A}} = \frac{(\sin A + \cos A)^2 - (\sin A - \cos A)^2}{(\sin A + \cos A)^2 + (\sin A - \cos A)^2}$$

$$= \frac{1 + 2\sin A \cos A - 1 + 2\sin A \cos A}{1 + 2\sin A \cos A + 1 - 2\sin A \cos A}$$

$$= \frac{4\sin A \cos A}{2}$$

$$= 2\sin A \cos A$$

47. The x-coordinate of P is approximately 0.9848, so $\cos 10° \approx 0.9848$.

49. The y-coordinate of P is approximately 0.1736, so $\sin 10° \approx 0.1736$.

51. Draw a triangle:

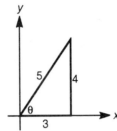

Using the Pythagorean theorem:
$$3^2 + b^2 = 5^2$$
$$9 + b^2 = 25$$
$$b^2 = 16$$
$$b = 4$$

Since $0° < \theta < 90°$, both $\sin \theta$ and $\tan \theta$ are positive, so:
$$\sin \theta = \tfrac{4}{5} \text{ and } \tan \theta = \tfrac{4}{3}$$

53. Draw a triangle:

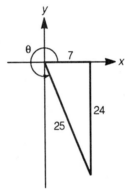

Using the Pythagorean theorem:
$$7^2 + b^2 = 25^2$$
$$49 + b^2 = 625$$
$$b^2 = 576$$
$$b = 24$$

Since $270° < \theta < 360°$, $\tan \theta$ is negative, so:
$$\tan \theta = -\tfrac{24}{7}$$

55.   Draw a triangle:

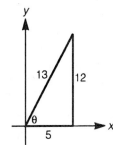

Using the Pythagorean theorem:
$$a^2 + 12^2 = 13^2$$
$$a^2 + 144 = 169$$
$$a^2 = 25$$
$$a = 5$$

Since $0° < \theta < 90°$, $\cot \theta$ is positive, so:
$$\cot \theta = \frac{5}{12}$$

57.   Draw a triangle:

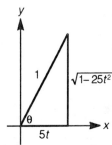

Using the Pythagorean theorem:
$$(5t)^2 + b^2 = 1^2$$
$$25t^2 + b^2 = 1$$
$$b^2 = 1 - 25t^2$$
$$b = \sqrt{1 - 25t^2}$$

Since $0° < \theta < 90°$, $\tan(90° - \theta)$ is positive, so:
$$\tan(90° - \theta) = \cot \theta = \frac{5t}{\sqrt{1 - 25t^2}} = \frac{5t\sqrt{1 - 25t^2}}{1 - 25t^2}$$

59. Draw a triangle:

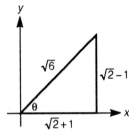

Using the Pythagorean theorem:
$$\left(\sqrt{2}+1\right)^2 + \left(\sqrt{2}-1\right)^2 = c^2$$
$$2 + 2\sqrt{2} + 1 + 2 - 2\sqrt{2} + 1 = c^2$$
$$6 = c^2$$
$$\sqrt{6} = c$$

Since $0° < \theta < 90°$, $\sin\theta$ is positive, so:
$$\sin\theta = \frac{\sqrt{2}-1}{\sqrt{6}} = \frac{2\sqrt{3}-\sqrt{6}}{6}$$

61. Using the identities for $\tan\theta$ and $\cot\theta$:
$$\tan\theta + \cot\theta = 2$$
$$\frac{\sin\theta}{\cos\theta} + \frac{\cos\theta}{\sin\theta} = 2$$
$$\frac{\sin^2\theta + \cos^2\theta}{\sin\theta\cos\theta} = 2$$
$$1 = 2\sin\theta\cos\theta$$

Now $(\sin\theta + \cos\theta)^2 = \sin^2\theta + 2\sin\theta\cos\theta + \cos^2\theta = 1 + 1 = 2$, so $\sin\theta + \cos\theta = \sqrt{2}$ since $0° < \theta < 90°$.

63. Draw the figure:

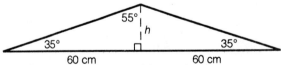

Note that the altitude of the triangle must bisect the upper vertex angle, which is $110°$ $(180° - 35° - 35°)$. Now find the height h:
$$\tan 35° = \frac{h}{60}, \text{ so } h = 60\tan 35°$$

Thus the area is given by:
$$\tfrac{1}{2}(\text{base})(\text{height}) = \tfrac{1}{2}(120)(60\tan 35°) = 3600\tan 35° \approx 2521 \text{ cm}^2$$

65. Draw the figure:

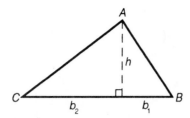

Now:

$$\cot B = \frac{b_1}{h}, \text{ so } b_1 = h\cot B$$

$$\cot C = \frac{b_2}{h}, \text{ so } b_2 = h\cot C$$

Therefore:

$$b_1 + b_2 = a$$
$$h\cot B + h\cot C = a$$
$$h(\cot B + \cot C) = a$$

$$h = \frac{a}{\cot B + \cot C}$$

67. Draw the figure:

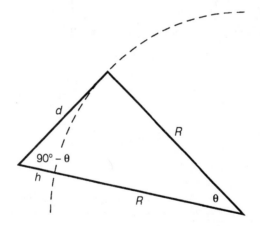

Thus:

$$d^2 + R^2 = (R+h)^2$$
$$d^2 = h^2 + 2Rh$$
$$d = \sqrt{2Rh + h^2}$$

Therefore:

$$\cot\theta = \frac{R}{d} = \frac{R}{\sqrt{2Rh + h^2}}$$

69. Using the identities $\sin(90° - \theta) = \cos\theta$ and $\tan(90° - \theta) = \cot\theta$:

$$\sin^2(90° - \theta)\csc\theta - \tan^2(90° - \theta)\sin\theta = \cos^2\theta \cdot \frac{1}{\sin\theta} - \cot^2\theta\sin\theta$$

$$= \frac{\cos^2\theta}{\sin\theta} - \frac{\cos^2\theta}{\sin^2\theta} \cdot \sin\theta$$

$$= \frac{\cos^2\theta}{\sin\theta} - \frac{\cos^2\theta}{\sin\theta}$$

$$= 0$$

71. Working from the right-hand side and using the identity for csc A:

$$\csc A - \sin A = \frac{1}{\sin A} - \sin A = \frac{1 - \sin^2 A}{\sin A} = \frac{\cos^2 A}{\sin A} = \frac{\cos A}{\sin A} \cdot \cos A = \cos A \cot A$$

73. Using the identity for cot A:

$$\frac{\cot A - 1}{\cot A + 1} = \frac{\frac{\cos A}{\sin A} - 1}{\frac{\cos A}{\sin A} + 1} = \frac{\cos A - \sin A}{\cos A + \sin A}$$

75. Simplifying the left-hand side:

$$\frac{1}{1 - \cos A} + \frac{1}{1 + \cos A} = \frac{1 + \cos A + 1 - \cos A}{(1 - \cos A)(1 + \cos A)} = \frac{2}{1 - \cos^2 A} = \frac{2}{\sin^2 A}$$

Simplifying the right-hand side using the identity for cot A:

$$2 + 2\cot^2 A = 2 + 2 \cdot \frac{\cos^2 A}{\sin^2 A} = \frac{2\sin^2 A + 2\cos^2 A}{\sin^2 A} = \frac{2(\sin^2 A + \cos^2 A)}{\sin^2 A} = \frac{2}{\sin^2 A}$$

Since both sides of the equality simplify to the same quantity, the original equality is an identity.

77. First simplifying two of the fractions using the identities for sec A and csc A:

$$\frac{1}{1 + \sec^2 A} = \frac{1}{1 + \frac{1}{\cos^2 A}} = \frac{\cos^2 A}{\cos^2 A + 1}$$

$$\frac{1}{1 + \csc^2 A} = \frac{1}{1 + \frac{1}{\sin^2 A}} = \frac{\sin^2 A}{\sin^2 A + 1}$$

Replacing the original fractions with these simplified fractions, the left-hand side becomes:

$$\frac{1}{1 + \sin^2 A} + \frac{1}{1 + \cos^2 A} + \frac{\cos^2 A}{1 + \cos^2 A} + \frac{\sin^2 A}{1 + \sin^2 A} = \frac{1 + \sin^2 A}{1 + \sin^2 A} + \frac{1 + \cos^2 A}{1 + \cos^2 A} = 2$$

79. (a) Since $OP = 1$, $PN = \sin\theta$.

(b) Since $OP = 1$, $ON = \cos\theta$.

(c) Since $\cos\theta = \dfrac{PN}{PT}$ (note $\theta = \angle TPN$), $\cos\theta = \dfrac{\sin\theta}{PT}$. Hence $PT = \tan\theta$.

(d)   Here $\cos\theta = \dfrac{1}{OT}$ (using $\triangle OPT$), so $OT = \dfrac{1}{\cos\theta} = \sec\theta$.

(e)   Simplify $NA = 1 - ON = 1 - \cos\theta$.

(f)   Since $NT = OT - ON$:

$$NT = \sec\theta - \cos\theta$$
$$= \frac{1}{\cos\theta} - \cos\theta$$
$$= \frac{1 - \cos^2\theta}{\cos\theta}$$
$$= \frac{\sin^2\theta}{\cos\theta}$$
$$= \sin\theta \cdot \frac{\sin\theta}{\cos\theta}$$
$$= \sin\theta\tan\theta$$

81.   Re-draw the figure (note the labels):

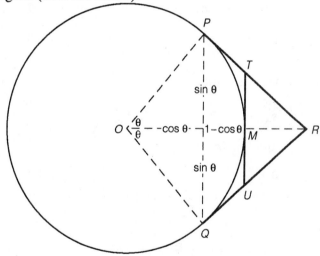

Using $\triangle QOR$:

$$\cos\theta = \frac{1}{1 + MR}$$
$$1 + MR = \frac{1}{\cos\theta}$$
$$MR = \frac{1}{\cos\theta} - 1$$

Since $\triangle RTU$ and $\triangle RPQ$ are similar, the corresponding sides are proportional:

$$\frac{TU}{PQ} = \frac{MR}{1 - \cos\theta + MR}$$

But $PQ = 2\sin\theta$ and $MR = \dfrac{1}{\cos\theta} - 1$, so:

$$\frac{TU}{2\sin\theta} = \frac{\frac{1}{\cos\theta} - 1}{1 - \cos\theta + \frac{1}{\cos\theta} - 1} = \frac{1 - \cos\theta}{1 - \cos^2\theta} = \frac{1 - \cos\theta}{\sin^2\theta}$$

$$TU = \frac{2(1 - \cos\theta)}{\sin\theta}$$

Thus the area is given by:

$$A = \frac{1}{2}(TU)(MR)$$

$$= \frac{1}{2}\cdot\frac{2(1 - \cos\theta)}{\sin\theta}\cdot\left(\frac{1}{\cos\theta} - 1\right)$$

$$= \frac{1 - \cos\theta}{\sin\theta}\cdot\frac{1 - \cos\theta}{\cos\theta}$$

$$= \frac{(1 - \cos\theta)^2}{\sin\theta\cos\theta}$$

83. (a) Using ΔODC, $OD = \cos\theta$ and $DC = \sin\theta$ (recall that $OC = 1$).

Thus, using ΔADC, $\tan\dfrac{\theta}{2} = \dfrac{CD}{AD} = \dfrac{\sin\theta}{1 + \cos\theta}$, since $AO = 1$.

(b) Compute $\tan 15°$ using the formula:

$$\tan 15° = \frac{\sin 30°}{1 + \cos 30°} = \frac{\frac{1}{2}}{1 + \frac{\sqrt{3}}{2}} = \frac{1}{2 + \sqrt{3}}\cdot\frac{2 - \sqrt{3}}{2 - \sqrt{3}} = \frac{2 - \sqrt{3}}{4 - 3} = 2 - \sqrt{3}$$

Compute $\tan 22.5°$ using the formula:

$$\tan 22.5° = \frac{\sin 45°}{1 + \cos 45°} = \frac{\frac{\sqrt{2}}{2}}{1 + \frac{\sqrt{2}}{2}} = \frac{\sqrt{2}}{2 + \sqrt{2}}\cdot\frac{2 - \sqrt{2}}{2 - \sqrt{2}} = \frac{2\sqrt{2} - 2}{4 - 2} = \sqrt{2} - 1$$

Chapter Seven Test

1. (a) Since the terminal side of $30°$ lies in the first quadrant, $\tan 30° > 0$, so $\tan 30° = \frac{\sqrt{3}}{3}$.

(b) Since the terminal side of $45°$ lies in the first quadrant, $\sec 45° > 0$, so $\sec 45° = \sqrt{2}$.

(c) Since $\sin^2\theta + \cos^2\theta = 1$ for all angles θ, $\sin^2 25° + \cos^2 25° = 1$.

(d) Since $\sin 53° = \cos(90° - 53°) = \cos 37°$:

$$\sin 53° - \cos 37° = \cos 37° - \cos 37° = 0$$

2. (a) Since $\cos\theta$ is negative, $\sec\theta$ is negative.

(b) Since $\sin\theta$ is negative, $\csc\theta$ is negative.

(c) Since both $\sin\theta$ and $\cos\theta$ are negative, $\cot\theta$ is positive.

3. (a) Since –270° corresponds to the point (0, 1) on the unit circle, sin (–270°) = 1.
    (b) Since 180° corresponds to the point (–1, 0) on the unit circle, cos 180° = –1.
    (c) Since 720° corresponds to the point (1, 0) on the unit circle:
    $$\tan 720° = \frac{y}{x} = \frac{0}{1} = 0$$

4. Factoring as a trinomial:
    $$2\cot^2\theta + 11\cot\theta + 12 = (2\cot\theta + 3)(\cot\theta + 4)$$

5. Using the formula for area:
    $$\tfrac{1}{2}ab\sin\theta = \tfrac{1}{2}(8)(9)\sin 150° = 36 \cdot \tfrac{1}{2} = 18 \text{ cm}^2$$

6. (a) Since the terminal side of –225° lies in quadrant 2, sin (–225°) > 0. Using a reference angle of 45°:
    $$\sin(-225°) = \sin 45° = \tfrac{\sqrt{2}}{2}$$

    (b) Since the terminal side of 330° lies in quadrant 4, tan 330° < 0. Using a reference angle of 30°:
    $$\tan 330° = -\tan 30° = -\tfrac{\sqrt{3}}{3}$$

    (c) Since the terminal side of 120° lies in quadrant 2, sec 120° < 0. Using a reference angle of 60°:
    $$\sec 120° = -\sec 60° = -2$$

7. Draw a triangle:

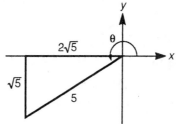

Using the Pythagorean theorem:
$$a^2 + \left(\sqrt{5}\right)^2 = 5^2$$
$$a^2 + 5 = 25$$
$$a^2 = 20$$
$$a = 2\sqrt{5}$$

Since 180° < θ < 270°, cos θ < 0 and tan θ > 0, so:
$$\cos\theta = -\tfrac{2\sqrt{5}}{5} \quad \text{and} \quad \tan\theta = \tfrac{\sqrt{5}}{2\sqrt{5}} = \tfrac{1}{2}$$

8.   Draw a triangle:

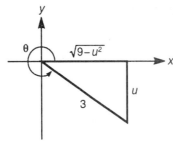

Using the Pythagorean theorem:
$$a^2 + u^2 = 3^2$$
$$a^2 + u^2 = 9$$
$$a^2 = 9 - u^2$$
$$a = \sqrt{9 - u^2}$$

Since $270° < \theta < 360°$, $\cos\theta > 0$ and $\cot\theta < 0$, so:
$$\cos\theta = \frac{\sqrt{9-u^2}}{3} \quad \text{and} \quad \cot\theta = -\frac{\sqrt{9-u^2}}{u}$$

9.   Multiplying the numerator and denominator by $\cos\theta$:
$$\frac{\frac{\cos\theta+1}{\cos\theta}+1}{\frac{\cos\theta-1}{\cos\theta}-1} = \frac{\cos\theta+1+\cos\theta}{\cos\theta-1-\cos\theta} = \frac{2\cos\theta+1}{-1} = -2\cos\theta-1$$

10.   The central angle of a triangle drawn from the center to two adjacent vertices is given by $9\theta = 360°$, so $\theta = 40°$. The area of this triangle is therefore:
$$\tfrac{1}{2}ab\sin\theta = \tfrac{1}{2}(3)(3)\sin 40° = \tfrac{9}{2}\sin 40°$$
Since there are nine such triangles which form the polygon, its area is:
$$9\left(\tfrac{9}{2}\sin 40°\right) = \tfrac{81}{2}\sin 40° \approx 26.033 \text{ m}^2$$

11.   Using the identities for $\cot\theta$, $\tan\theta$, $\csc\theta$ and $\sec\theta$:
$$\frac{\cot^2\theta}{\csc^2\theta} + \frac{\tan^2\theta}{\sec^2\theta} = \frac{\frac{\cos^2\theta}{\sin^2\theta}}{\frac{1}{\sin^2\theta}} + \frac{\frac{\sin^2\theta}{\cos^2\theta}}{\frac{1}{\cos^2\theta}} = \cos^2\theta + \sin^2\theta = 1$$

12.   First find BD and BC. Using the smaller right triangle:
$$\tan 25° = \frac{BD}{50}, \text{ so } BD = 50\tan 25°$$
Using the larger right triangle:
$$\tan 55° = \frac{BC}{50}, \text{ so } BC = 50\tan 55°$$
Since $CD = BC - BD$:
$$CD = 50\tan 55° - 50\tan 25° = 50(\tan 55° - \tan 25°)$$

13. (a) Since 85° has a much larger y-coordinate than 5°, sin 85° is larger.

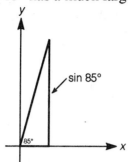

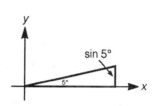

(b) Since the x-coordinate is much larger at 5° than the y-coordinate, cos 5° is larger.

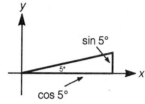

(c) Since the terminal side for 175° lies in quadrant 2, tan 175° < 0. Since the terminal side for 185° lies in quadrant 3, tan 185° > 0. Thus tan 185° is larger. Note that a sketch isn't necessary to make this comparison.

14. (a) To find AD, use $\triangle ADC$:
$$\cos 20° = \frac{AD}{2.75}, \text{ so } AD = 2.75\cos 20°$$
To find CD, use $\triangle ADC$:
$$\sin 20° = \frac{CD}{2.75}, \text{ so } CD = 2.75\sin 20°$$
To find DB, note that $DB = AB - AD$, so:
$$DB = 3.25 - 2.75\cos 20°$$

(b) Using the Pythagorean theorem in $\triangle CDB$:
$$(CB)^2 = (DB)^2 + (CD)^2$$
$$(CB)^2 = (3.25 - 2.75\cos 20°)^2 + (2.75\sin 20°)^2$$
$$(CB)^2 = 10.5625 - 17.875\cos 20° + 7.5625\cos^2 20° + 7.5625\sin^2 20°$$
$$(CB)^2 = 10.5625 - 17.875\cos 20° + 7.5625\left(\cos^2 20° + \sin^2 20°\right)$$
$$(CB)^2 = 10.5625 - 17.875\cos 20° + 7.5625$$
$$(CB)^2 = 18.125 - 17.875\cos 20°$$
$$CB = \sqrt{18.125 - 17.875\cos 20°} \approx 1.15 \text{ cm}$$

Chapter Eight
Trigonometric Functions of Real Numbers

8.1 Radian Measure

1. Using $\theta = \dfrac{s}{r}$, the radian measure is:

    $$\theta = \frac{5 \text{ cm}}{2 \text{ cm}} = 2.5 \text{ radians}$$

3. Using $\theta = \dfrac{s}{r}$, the radian measure is:

    $$\theta = \frac{200 \text{ cm}}{1 \text{ m}} = \frac{200 \text{ cm}}{100 \text{ cm}} = 2 \text{ radians}$$

5. (a) Multiplying by the conversion factor $\frac{\pi}{180°}$:
    $$45° \cdot \frac{\pi}{180°} = \frac{\pi}{4} \approx 0.79 \text{ radians}$$

    (b) Multiplying by the conversion factor $\frac{\pi}{180°}$:
    $$90° \cdot \frac{\pi}{180°} = \frac{\pi}{2} \approx 1.57 \text{ radians}$$

    (c) Multiplying by the conversion factor $\frac{\pi}{180°}$:
    $$135° \cdot \frac{\pi}{180°} = \frac{3\pi}{4} \approx 2.36 \text{ radians}$$

7.   (a)   Multiplying by the conversion factor $\frac{\pi}{180°}$:

$$0° \cdot \frac{\pi}{180°} = 0 \text{ radians}$$

(b)   Multiplying by the conversion factor $\frac{\pi}{180°}$:

$$360° \cdot \frac{\pi}{180°} = 2\pi \approx 6.28 \text{ radians}$$

(c)   Multiplying by the conversion factor $\frac{\pi}{180°}$:

$$450° \cdot \frac{\pi}{180°} = \frac{5\pi}{2} \approx 7.85 \text{ radians}$$

9.   (a)   Multiplying by the conversion factor $\frac{180°}{\pi}$:

$$\frac{\pi}{12} \cdot \frac{180°}{\pi} = 15°$$

(b)   Multiplying by the conversion factor $\frac{180°}{\pi}$:

$$\frac{\pi}{6} \cdot \frac{180°}{\pi} = 30°$$

(c)   Multiplying by the conversion factor $\frac{180°}{\pi}$:

$$\frac{\pi}{4} \cdot \frac{180°}{\pi} = 45°$$

11.   (a)   Multiplying by the conversion factor $\frac{180°}{\pi}$:

$$\frac{\pi}{3} \cdot \frac{180°}{\pi} = 60°$$

(b)   Multiplying by the conversion factor $\frac{180°}{\pi}$:

$$\frac{5\pi}{3} \cdot \frac{180°}{\pi} = 300°$$

(c)   Multiplying by the conversion factor $\frac{180°}{\pi}$:

$$4\pi \cdot \frac{180°}{\pi} = 720°$$

13.   (a)   Multiplying by the conversion factor $\frac{180°}{\pi}$:

$$2 \cdot \frac{180°}{\pi} = \frac{360°}{\pi} \approx 114.59°$$

(b)   Multiplying by the conversion factor $\frac{180°}{\pi}$:

$$3 \cdot \frac{180°}{\pi} = \frac{540°}{\pi} \approx 171.89°$$

(c)   Multiplying by the conversion factor $\frac{180°}{\pi}$:

$$\pi^2 \cdot \frac{180°}{\pi} = 180\pi° \approx 565.49°$$

15. Since a right angle has radian measure of $\frac{\pi}{2}$, which is larger than $\frac{3}{2}$ (since π is larger than 3), this angle is smaller than a right angle.

17. Multiplying by the conversion factor $\frac{\pi}{180°}$:

$30° = 30° \cdot \frac{\pi}{180°} = \frac{\pi}{6}$ radians

$45° = 45° \cdot \frac{\pi}{180°} = \frac{\pi}{4}$ radians

$60° = 60° \cdot \frac{\pi}{180°} = \frac{\pi}{3}$ radians

$120° = 120° \cdot \frac{\pi}{180°} = \frac{2\pi}{3}$ radians

$135° = 135° \cdot \frac{\pi}{180°} = \frac{3\pi}{4}$ radians

$150° = 150° \cdot \frac{\pi}{180°} = \frac{5\pi}{6}$ radians

19. Since the angles correspond to 0°, 90°, 180°, 270°, and 360°, the completed table is:

θ	$\sin\theta$	$\cos\theta$	$\tan\theta$
0	0	1	0
$\frac{\pi}{2}$	1	0	undefined
π	0	−1	0
$\frac{3\pi}{2}$	−1	0	undefined
2π	0	1	0

21. From the figure, $0.5 < \cos 1 < 0.6$ and $0.8 < \sin 1 < 0.9$. Using a calculator, $\cos 1 \approx 0.54$ and $\sin 1 \approx 0.84$.

23. From the figure, $0.5 < \cos(-1) < 0.6$ and $-0.9 < \sin(-1) < -0.8$. Using a calculator, $\cos(-1) \approx 0.54$ and $\sin(-1) \approx -0.84$.

25. From the figure, $-0.7 < \cos 4 < -0.6$ and $-0.8 < \sin 4 < -0.7$. Using a calculator, $\cos 4 \approx -0.65$ and $\sin 4 \approx -0.76$.

27. From the figure, $-0.7 < \cos(-4) < -0.6$ and $0.7 < \sin(-4) < 0.8$. Using a calculator, $\cos(-4) \approx -0.65$ and $\sin(-4) \approx 0.76$.

29. Since $\sin(1 + 2\pi) = \sin 1$, use the figure to obtain $0.8 < \sin(1 + 2\pi) < 0.9$. Using a calculator, $\sin(1 + 2\pi) \approx 0.84$.

31. (a) The reference angle for $\frac{5\pi}{6}$ is $\frac{\pi}{6}$, and $\cos\frac{\pi}{6} = \frac{\sqrt{3}}{2}$. The terminal side of $\frac{5\pi}{6}$ lies in the second quadrant where the x-coordinates are negative, so:

$$\cos\frac{5\pi}{6} = -\cos\frac{\pi}{6} = -\frac{\sqrt{3}}{2}$$

(b) The reference angle for $-\frac{5\pi}{6}$ is $\frac{\pi}{6}$, and $\cos\frac{\pi}{6} = \frac{\sqrt{3}}{2}$. The terminal side of $-\frac{5\pi}{6}$ lies in the third quadrant where the x-coordinates are negative, so:

$$\cos\left(-\frac{5\pi}{6}\right) = -\cos\frac{\pi}{6} = -\frac{\sqrt{3}}{2}$$

(c) The reference angle for $\frac{5\pi}{6}$ is $\frac{\pi}{6}$, and $\sin\frac{\pi}{6} = \frac{1}{2}$. The terminal side of $\frac{5\pi}{6}$ lies in the second quadrant where the y-coordinates are positive, so:

$$\sin\frac{5\pi}{6} = \sin\frac{\pi}{6} = \frac{1}{2}$$

(d) The reference angle for $-\frac{5\pi}{6}$ is $\frac{\pi}{6}$, and $\sin\frac{\pi}{6} = \frac{1}{2}$. The terminal side of $-\frac{5\pi}{6}$ lies in the third quadrant where the y-coordinates are negative, so:

$$\sin\left(-\frac{5\pi}{6}\right) = -\sin\frac{\pi}{6} = -\frac{1}{2}$$

33. (a) The reference angle for $\frac{4\pi}{3}$ is $\frac{\pi}{3}$, and $\cos\frac{\pi}{3} = \frac{1}{2}$. The terminal side of $\frac{4\pi}{3}$ lies in the third quadrant where the x-coordinates are negative, so:

$$\cos\frac{4\pi}{3} = -\cos\frac{\pi}{3} = -\frac{1}{2}$$

(b) The reference angle for $-\frac{4\pi}{3}$ is $\frac{\pi}{3}$, and $\cos\frac{\pi}{3} = \frac{1}{2}$. The terminal side of $-\frac{4\pi}{3}$ lies in the second quadrant where the x-coordinates are negative, so:

$$\cos\left(-\frac{4\pi}{3}\right) = -\cos\frac{\pi}{3} = -\frac{1}{2}$$

(c) The reference angle for $\frac{4\pi}{3}$ is $\frac{\pi}{3}$, and $\sin\frac{\pi}{3} = \frac{\sqrt{3}}{2}$. The terminal side of $\frac{4\pi}{3}$ lies in the third quadrant where the y-coordinates are negative, so:

$$\sin\frac{4\pi}{3} = -\sin\frac{\pi}{3} = -\frac{\sqrt{3}}{2}$$

(d) The reference angle for $-\frac{4\pi}{3}$ is $\frac{\pi}{3}$, and $\sin\frac{\pi}{3} = \frac{\sqrt{3}}{2}$. The terminal side of $-\frac{4\pi}{3}$ lies in the second quadrant where the y-coordinates are positive, so:

$$\sin\left(-\frac{4\pi}{3}\right) = \sin\frac{\pi}{3} = \frac{\sqrt{3}}{2}$$

35. (a) The reference angle for $\frac{5\pi}{4}$ is $\frac{\pi}{4}$, and $\cos\frac{\pi}{4} = \frac{\sqrt{2}}{2}$. The terminal side of $\frac{5\pi}{4}$ lies in the third quadrant where the x-coordinates are negative, so:

$$\cos\frac{5\pi}{4} = -\cos\frac{\pi}{4} = -\frac{\sqrt{2}}{2}$$

(b) The reference angle for $-\frac{5\pi}{4}$ is $\frac{\pi}{4}$, and $\cos\frac{\pi}{4} = \frac{\sqrt{2}}{2}$. The terminal side of $-\frac{5\pi}{4}$ lies in the second quadrant where the x-coordinates are negative, so:

$$\cos\left(-\frac{5\pi}{4}\right) = -\cos\frac{\pi}{4} = -\frac{\sqrt{2}}{2}$$

(c) The reference angle for $\frac{5\pi}{4}$ is $\frac{\pi}{4}$, and $\sin\frac{\pi}{4} = \frac{\sqrt{2}}{2}$. The terminal side of $\frac{5\pi}{4}$ lies in the third quadrant where the y-coordinates are negative, so:

$$\sin\frac{5\pi}{4} = -\sin\frac{\pi}{4} = -\frac{\sqrt{2}}{2}$$

(d) The reference angle for $-\frac{5\pi}{4}$ is $\frac{\pi}{4}$, and $\sin\frac{\pi}{4} = \frac{\sqrt{2}}{2}$. The terminal side of $-\frac{5\pi}{4}$ lies in the second quadrant where the y-coordinates are positive, so:

$$\sin\left(-\frac{5\pi}{4}\right) = \sin\frac{\pi}{4} = \frac{\sqrt{2}}{2}$$

37. (a) The reference angle for $\frac{4\pi}{3}$ is $\frac{\pi}{3}$, and $\sec\frac{\pi}{3} = 2$. The terminal side of $\frac{4\pi}{3}$ lies in the third quadrant where the x-coordinates are negative, so:
$$\sec\frac{4\pi}{3} = -\sec\frac{\pi}{3} = -2$$

(b) The reference angle for $-\frac{4\pi}{3}$ is $\frac{\pi}{3}$, and $\csc\frac{\pi}{3} = \frac{2\sqrt{3}}{3}$. The terminal side of $-\frac{4\pi}{3}$ lies in the second quadrant where the y-coordinates are positive, so:

$$\csc\left(-\frac{4\pi}{3}\right) = \csc\frac{\pi}{3} = \frac{2\sqrt{3}}{3}$$

(c) The reference angle for $\frac{4\pi}{3}$ is $\frac{\pi}{3}$, and $\tan\frac{\pi}{3} = \sqrt{3}$. The terminal side of $\frac{4\pi}{3}$ lies in the third quadrant where both the x- and y-coordinates are negative, so:
$$\tan\frac{4\pi}{3} = \tan\frac{\pi}{3} = \sqrt{3}$$

(d) The reference angle for $-\frac{4\pi}{3}$ is $\frac{\pi}{3}$, and $\cot\frac{\pi}{3} = \frac{\sqrt{3}}{3}$. The terminal side of $-\frac{4\pi}{3}$ lies in the second quadrant where the x-coordinates are negative while the y-coordinates are positive, so:

$$\cot\left(-\frac{4\pi}{3}\right) = -\cot\frac{\pi}{3} = -\frac{\sqrt{3}}{3}$$

39. (a) The reference angle for $\frac{17\pi}{6}$ is $\frac{\pi}{6}$, and $\sec\frac{\pi}{6} = \frac{2\sqrt{3}}{3}$. The terminal side of $\frac{17\pi}{6}$ lies in the second quadrant where the x-coordinates are negative, so:

$$\sec\frac{17\pi}{6} = -\sec\frac{\pi}{6} = -\frac{2\sqrt{3}}{3}$$

(b) The reference angle for $-\frac{17\pi}{6}$ is $\frac{\pi}{6}$, and $\csc\frac{\pi}{6} = 2$. The terminal side of $-\frac{17\pi}{6}$ lies in the third quadrant where the y-coordinates are negative, so:

$$\csc\left(-\frac{17\pi}{6}\right) = -\csc\frac{\pi}{6} = -2$$

(c) The reference angle for $\frac{17\pi}{6}$ is $\frac{\pi}{6}$, and $\tan\frac{\pi}{6} = \frac{\sqrt{3}}{3}$. The terminal side of $\frac{17\pi}{6}$ lies in the second quadrant where the x-coordinates are negative while the y-coordinates are positive, so:

$$\tan\frac{17\pi}{6} = -\tan\frac{\pi}{6} = -\frac{\sqrt{3}}{3}$$

(d) The reference angle for $-\frac{17\pi}{6}$ is $\frac{\pi}{6}$, and $\cot\frac{\pi}{6} = \sqrt{3}$. The terminal side of $-\frac{17\pi}{6}$ lies in the third quadrant where both the x- and y-coordinates are negative, so:

$$\cot\left(-\frac{17\pi}{6}\right) = \cot\frac{\pi}{6} = \sqrt{3}$$

41. Since the triangle is equilateral with sides 1 ft, the circle formed will have a radius of 1 ft and arc length of 1 ft, so:

$$\theta = \frac{s}{r} = \frac{1}{1} = 1 \text{ radian}$$

Converting to degrees, the angle at A will have a measure of:

$$1 \cdot \frac{180°}{\pi} = \frac{180°}{\pi} \approx 57.30°$$

43. Since $x° = \frac{\pi x}{180}$ radians, the following equation must hold:

$$x = \frac{\pi x}{180}$$

$$180x = \pi x$$

The only solution to this equation is $x = 0$. The real number $x = 0$ satisfies the property that x degrees equals x radians.

8.2 Radian Measure and Geometry

1. Using the formula $s = r\theta$ where $r = 3$ ft and $\theta = \frac{4\pi}{3}$:

$$s = r\theta = 3 \cdot \frac{4\pi}{3} = 4\pi \text{ ft}$$

3.   First convert $45°$ to radian measure:

$$45° = 45° \cdot \tfrac{\pi}{180°} = \tfrac{\pi}{4} \text{ radians}$$

Now using the formula $s = r\theta$ where $r = 2$ cm and $\theta = \tfrac{\pi}{4}$:

$$s = r\theta = 2 \cdot \tfrac{\pi}{4} = \tfrac{\pi}{2} \text{ cm}$$

5.   (a)   The area is given by:

$$A = \tfrac{1}{2}r^2\theta = \tfrac{1}{2}(6)^2\left(\tfrac{2\pi}{3}\right) = 12\pi \text{ cm}^2 \approx 37.70 \text{ cm}^2$$

(b)   First convert θ to radian measure:

$$\theta = 80° \cdot \tfrac{\pi}{180°} = \tfrac{4\pi}{9} \text{ radians}$$

The area is given by:

$$A = \tfrac{1}{2}r^2\theta = \tfrac{1}{2}(5)^2\left(\tfrac{4\pi}{9}\right) = \tfrac{50\pi}{9} \text{ m}^2 \approx 17.45 \text{ m}^2$$

(c)   The area is given by:

$$A = \tfrac{1}{2}r^2\theta = \tfrac{1}{2}(24)^2\left(\tfrac{\pi}{20}\right) = \tfrac{72\pi}{5} \text{ m}^2 \approx 45.24 \text{ m}^2$$

(d)   First convert θ to radian measure:

$$\theta = 144° \cdot \tfrac{\pi}{180°} = \tfrac{4\pi}{5} \text{ radians}$$

The area is given by:

$$A = \tfrac{1}{2}r^2\theta = \tfrac{1}{2}(1.8)^2\left(\tfrac{4\pi}{5}\right) = \tfrac{12.96\pi}{10} = 1.296\pi \text{ cm}^2 \approx 4.07 \text{ cm}^2$$

7.   We have $r = 1$ cm and $A = \tfrac{\pi}{5}$ cm^2. Substituting into $A = \tfrac{1}{2}r^2\theta$:

$$\tfrac{\pi}{5} = \tfrac{1}{2}(1)^2\theta$$
$$\tfrac{\pi}{5} = \tfrac{\theta}{2}$$
$$\theta = \tfrac{2\pi}{5} \text{ radians}$$

9.   (a)   First convert $30° = \tfrac{\pi}{6}$ radians. The arc length is given by:

$$s = r\theta = (5 \text{ in.})\left(\tfrac{\pi}{6}\right) = \tfrac{5\pi}{6} \text{ in.}$$

Therefore the perimeter is:

$$5 \text{ in.} + 5 \text{ in.} + \tfrac{5\pi}{6} \text{ in.} = \left(10 + \tfrac{5\pi}{6}\right) \text{ in.} \approx 12.62 \text{ in.}$$

(b)   The area is given by:

$$A = \tfrac{1}{2}r^2\theta = \tfrac{1}{2}(5 \text{ in.})^2\left(\tfrac{\pi}{6}\right) = \tfrac{25\pi}{12} \text{ in.}^2 \approx 6.54 \text{ in.}^2$$

11. The area of the sector is given by:
$$A_s = \tfrac{1}{2}(6 \text{ cm})^2 (1.4) = 25.2 \text{ cm}^2$$
The area of the triangle is given by:
$$A_t = \tfrac{1}{2}(6 \text{ cm})(6 \text{ cm})(\sin 1.4) = 18\sin 1.4 \text{ cm}^2$$
Thus the shaded area is:
$$A_s - A_t = 25.2 \text{ cm}^2 - 18\sin 1.4 \text{ cm}^2 \approx 7.46 \text{ cm}^2$$

13. (a) Each revolution of the wheel is 2π radians, so in 6 revolutions there are
$\theta = 6(2\pi) = 12\pi$ radians. Consequently:
$$\omega = \tfrac{\theta}{t} = \tfrac{12\pi \text{ radians}}{1 \text{ sec}} = 12\pi \tfrac{\text{radians}}{\text{sec}}$$

(b) Using the formula $s = r\theta$, where $r = 12$ cm and $\theta = 12\pi$ radians,
$s = (12 \text{ cm})(12\pi \text{ radians}) = 144\pi$ cm. The linear speed, therefore, is:
$$v = \tfrac{d}{t} = \tfrac{144\pi \text{ cm}}{1\text{sec}} = 144\pi \tfrac{\text{cm}}{\text{sec}}$$

(c) Using the formula $s = r\theta$, where $r = 6$ cm and $\theta = 12\pi$ radians,
$s = (6 \text{ cm})(12\pi \text{ radians}) = 72\pi$ cm. Thus:
$$v = \tfrac{d}{t} = \tfrac{72\pi \text{ cm}}{1\text{sec}} = 72\pi \tfrac{\text{cm}}{\text{sec}}$$

15. (a) In 1 second, the wheel has rotated $1080° \bullet \tfrac{\pi \text{ radians}}{180°} = 6\pi$ radians. Consequently:
$$\omega = \tfrac{\theta}{t} = \tfrac{6\pi \text{ radians}}{1\text{sec}} = 6\pi \tfrac{\text{radians}}{\text{sec}}$$

(b) Using $\theta = 6\pi$ radians and $r = 25$ cm, then:
$$s = r\theta = (25 \text{ cm})(6\pi \text{ radians}) = 150\pi \text{ cm}$$
Thus:
$$v = \tfrac{150\pi \text{ cm}}{1\text{sec}} = 150\pi \tfrac{\text{cm}}{\text{sec}}$$

(c) Using $\theta = 6\pi$ radians and $r = \tfrac{25}{2}$ cm, then:
$$s = r\theta = \left(\tfrac{25}{2} \text{ cm}\right)(6\pi \text{ radians}) = 75\pi \text{ cm}$$
Thus:
$$v = \tfrac{75\pi \text{ cm}}{1\text{sec}} = 75\pi \tfrac{\text{cm}}{\text{sec}}$$

17. **(a)** In 1 minute, the wheel has rotated 500 rev. Since each revolution is equal to 2π radians:

$$\theta = (500)(2\pi) = 1000\pi \text{ radians}$$

Consequently:

$$\omega = \frac{\theta}{t} = \frac{1000\pi \text{ radians}}{60 \text{ sec}} = \frac{50\pi}{3} \frac{\text{radians}}{\text{sec}}$$

(b) Using $\theta = 1000\pi$ radians and $r = 45$ cm, then:

$$s = r\theta = (45 \text{ cm})(1000\pi \text{ radians}) = 45000\pi \text{ cm}$$

Thus:

$$v = \frac{45000\pi \text{ cm}}{60 \text{ sec}} = 750\pi \frac{\text{cm}}{\text{sec}}$$

(c) Using $\theta = 1000\pi$ radians and $r = \frac{45}{2}$ cm, then:

$$s = r\theta = \left(\frac{45}{2} \text{ cm}\right)(1000\pi \text{ radians}) = 22500\pi \text{ cm}$$

Thus:

$$v = \frac{22500\pi \text{ cm}}{60 \text{ sec}} = 375\pi \frac{\text{cm}}{\text{sec}}$$

19. **(a)** Each revolution is 2π radians, and 24 hr = 86400 sec, so:

$$\omega = \frac{\theta}{t} = \frac{2\pi \text{ radians}}{86400 \text{ sec}} \approx 0.000073 \frac{\text{radians}}{\text{sec}}$$

(b) Using $\theta = 2\pi$ radians and $r = 3960$ mi, then:

$$s = r\theta = (3960 \text{ mi})(2\pi \text{ radians}) = 7920\pi \text{ mi}$$

Thus:

$$v = \frac{s}{t} = \frac{7920\pi \text{ mi}}{24 \text{ hr}} = 330\pi \frac{\text{mi}}{\text{hr}} \approx 1040 \text{ mph}$$

21. **(a)** The angular speed of the larger wheel is given by:

$$\omega = 100 \frac{\text{rev}}{\text{min}} \cdot 2\pi \frac{\text{rad}}{\text{rev}} = 200\pi \frac{\text{rad}}{\text{min}}$$

(b) In 1 minute, a point on the larger wheel has traveled:

$$s = R\theta = (10 \text{ cm})(200\pi \text{ rad}) = 2000\pi \text{ cm}$$

So the linear speed is given by:

$$v = \frac{2000\pi \text{ cm}}{1 \text{ min}} = 2000\pi \frac{\text{cm}}{\text{min}}$$

(c) Since the linear speed on the smaller wheel must also be $2000\pi \frac{\text{cm}}{\text{min}}$, and the radius of the smaller wheel is $r = 6$ cm, its angular speed is:

$$\omega = \frac{v}{r} = \frac{2000\pi \text{ cm/min}}{6 \text{ cm}} = \frac{1000}{3}\pi \frac{\text{radians}}{\text{min}}$$

(d) In rpm, the angular speed of the smaller wheel is:

$$\frac{1 \text{ rev}}{2\pi \text{ rad}} \cdot \frac{1000}{3}\pi \frac{\text{rad}}{\text{min}} = \frac{500}{3} \text{ rpm}$$

23. First convert θ to radian measure:

$$\theta = 71°23' = 71\tfrac{23°}{60} \cdot \tfrac{\pi}{180°} = \tfrac{4283\pi}{10800}$$

Now compute the arc length:

$$s = r\theta = 3960 \text{ mi} \cdot \tfrac{4283\pi}{10800} = \tfrac{47113\pi}{30} \text{ mi} \approx 4930 \text{ mi}$$

25. First convert θ to radian measure:

$$\theta = 21°19' = 21\tfrac{19°}{60} \cdot \tfrac{\pi}{180°} = \tfrac{1279\pi}{10800}$$

Now compute the arc length:

$$s = r\theta = 3960 \text{ mi} \cdot \tfrac{1279\pi}{10800} = \tfrac{14069\pi}{30} \text{ mi} \approx 1470 \text{ mi}$$

27. First convert θ to radian measure:

$$\theta = 38°54' = 38\tfrac{54°}{60} \cdot \tfrac{\pi}{180°} = \tfrac{2334\pi}{10800}$$

Now compute the arc length:

$$s = r\theta = 3960 \text{ mi} \cdot \tfrac{2334\pi}{10800} = \tfrac{8558\pi}{10} \text{ mi} \approx 2690 \text{ mi}$$

29. (a) Since each angle of the equilateral triangle is $60° = \tfrac{\pi}{3}$ radians, each arc length is given by $r\theta = \tfrac{\pi}{3}s$. Therefore the perimeter is given by:

$$P = \tfrac{\pi}{3}s + \tfrac{\pi}{3}s + s = \left(\tfrac{2\pi}{3} + 1\right)s$$

(b) The area of sectors ABC and CAB are both given by:

$$A_s = \tfrac{1}{2}r^2\theta = \tfrac{1}{2}(s)^2 \cdot \tfrac{\pi}{3} = \tfrac{\pi}{6}s^2$$

The area of the triangle is given by:

$$A_t = \tfrac{1}{2}(s)(s)\sin\tfrac{\pi}{3} = \left(\tfrac{1}{2}s^2\right)\left(\tfrac{\sqrt{3}}{2}\right) = \tfrac{\sqrt{3}}{4}s^2$$

Therefore the area of the equilateral arch is given by:

$$A = A_s + A_s - A_t = \tfrac{\pi}{6}s^2 + \tfrac{\pi}{6}s^2 - \tfrac{\sqrt{3}}{4}s^2 = \left(\tfrac{\pi}{3} - \tfrac{\sqrt{3}}{4}\right)s^2$$

31. (a) From Exercise 29(b), the area of equilateral arch ABC is given by:

$$A_{ABC} = \left(\tfrac{\pi}{3} - \tfrac{\sqrt{3}}{4}\right)s^2$$

For arch ADF, use $\dfrac{s}{2}$ as the radius of the circle. The area of sectors DAF and AFD are each given by:

$$A_s = \tfrac{1}{2}r^2\theta = \tfrac{1}{2}\left(\tfrac{s}{2}\right)^2 \cdot \tfrac{\pi}{3} = \tfrac{\pi}{24}s^2$$

The area of the triangle is given by:

$$A_t = \frac{1}{2}\left(\frac{s}{2}\right)\left(\frac{s}{2}\right)\sin\frac{\pi}{3} = \left(\frac{1}{8}s^2\right)\left(\frac{\sqrt{3}}{2}\right) = \frac{\sqrt{3}}{16}s^2$$

Therefore the area of arch ADF is given by:

$$A_{ADF} = A_s + A_s - A_t = \frac{\pi}{24}s^2 + \frac{\pi}{24}s^2 - \frac{\sqrt{3}}{16}s^2 = \left(\frac{\pi}{12} - \frac{\sqrt{3}}{16}\right)s^2$$

Finally, the required ratio is given by:

$$\frac{A_{ADF}}{A_{ABC}} = \frac{\left(\frac{\pi}{12} - \frac{\sqrt{3}}{16}\right)s^2}{\left(\frac{\pi}{3} - \frac{\sqrt{3}}{4}\right)s^2} = \frac{\frac{\pi}{12} - \frac{\sqrt{3}}{16}}{\frac{\pi}{3} - \frac{\sqrt{3}}{4}} \cdot \frac{48}{48} = \frac{4\pi - 3\sqrt{3}}{16\pi - 12\sqrt{3}} = \frac{1}{4}$$

(b) From Exercise 30(a), use $\frac{s}{2}$ as the radius of the circle. The area of each sector is given by:

$$A_s = \frac{1}{2}r^2\theta = \frac{1}{2}\left(\frac{s}{2}\right)^2\left(\frac{\pi}{3}\right) = \frac{\pi}{24}s^2$$

The area of the triangle is given by:

$$A_t = \frac{1}{2}\left(\frac{s}{2}\right)\left(\frac{s}{2}\right)\sin\frac{\pi}{3} = \left(\frac{1}{8}s^2\right)\left(\frac{\sqrt{3}}{2}\right) = \frac{\sqrt{3}}{16}s^2$$

Therefore the area of the equilateral curved triangle DBE is given by:

$$A_{DBE} = 3A_s - 2A_t = 3\left(\frac{\pi}{24}s^2\right) - 2\left(\frac{\sqrt{3}}{16}s^2\right) = \left(\frac{\pi}{8} - \frac{\sqrt{3}}{8}\right)s^2$$

(c) The area of triangle DEF is given by:

$$A_t = \frac{1}{2}\left(\frac{s}{2}\right)\left(\frac{s}{2}\right)\sin\frac{\pi}{3} = \left(\frac{1}{8}s^2\right)\left(\frac{\sqrt{3}}{2}\right) = \frac{\sqrt{3}}{16}s^2$$

The area of each sector to be removed is given by:

$$A_s = \frac{1}{2}\left(\frac{s}{2}\right)^2\left(\frac{\pi}{3}\right) - \frac{\sqrt{3}}{16}s^2 = \frac{\pi}{24}s^2 - \frac{\sqrt{3}}{16}s^2 = \left(\frac{\pi}{24} - \frac{\sqrt{3}}{16}\right)s^2$$

Therefore the area of the curved figure DEF is given by:

$$A_t - 3A_s = \frac{\sqrt{3}}{16}s^2 - 3\left(\frac{\pi}{24} - \frac{\sqrt{3}}{16}\right)s^2 = \left(\frac{\sqrt{3}}{16} - \frac{\pi}{8} + \frac{3\sqrt{3}}{16}\right)s^2 = \left(\frac{\sqrt{3}}{4} - \frac{\pi}{8}\right)s^2$$

(d) From part (c), the sectors BE and EC each have areas of $A_s = \left(\dfrac{\pi}{24} - \dfrac{\sqrt{3}}{16}\right)s^2$. The large sector BAC has an area of:

$$A_{BAC} = \tfrac{1}{2}(s)^2\left(\tfrac{\pi}{3}\right) - \tfrac{1}{2}(s)^2 \sin\tfrac{\pi}{3} = \tfrac{\pi}{6}s^2 - \tfrac{\sqrt{3}}{4}s^2 = \left(\tfrac{\pi}{6} - \tfrac{\sqrt{3}}{4}\right)s^2$$

Therefore the area of the curved figure BEC is given by:

$$A_{BAC} - 2A_s = \left(\frac{\pi}{6} - \frac{\sqrt{3}}{4}\right)s^2 - 2\left(\frac{\pi}{24} - \frac{\sqrt{3}}{16}\right)s^2$$

$$= \left(\frac{\pi}{6} - \frac{\sqrt{3}}{4} - \frac{\pi}{12} + \frac{\sqrt{3}}{8}\right)s^2$$

$$= \left(\frac{\pi}{12} - \frac{\sqrt{3}}{8}\right)s^2$$

33. (a) The area of the sector is given by:
$$A_s = \tfrac{1}{2}r^2\alpha$$
The area of the triangle is given by:
$$A_t = \tfrac{1}{2}r^2 \sin\alpha$$
Thus the area of the segment is given by:
$$A = A_s - A_t = \tfrac{1}{2}r^2\alpha - \tfrac{1}{2}r^2 \sin\alpha = \tfrac{1}{2}r^2(\alpha - \sin\alpha)$$

(b) The two central angles used are $45° = \tfrac{\pi}{4}$ radians and $135° = \tfrac{3\pi}{4}$ radians. Adding the two areas together:
$$A = \tfrac{1}{2}(1)^2\left(\tfrac{\pi}{4} - \sin\tfrac{\pi}{4}\right) + \tfrac{1}{2}(1)^2\left(\tfrac{3\pi}{4} - \sin\tfrac{3\pi}{4}\right)$$
$$= \tfrac{1}{2}\left(\tfrac{\pi}{4} - \tfrac{\sqrt{2}}{2}\right) + \tfrac{1}{2}\left(\tfrac{3\pi}{4} - \tfrac{\sqrt{2}}{2}\right)$$
$$= \tfrac{1}{2}\left(\pi - \sqrt{2}\right)$$
$$\approx 0.86$$

35. (a) The coordinates of point A are $A(\cos\theta, \sin\theta)$.

(b) Using the distance formula:
$$d = \sqrt{(\cos\theta - 1)^2 + (\sin\theta - 0)^2}$$
$$= \sqrt{\cos^2\theta - 2\cos\theta + 1 + \sin^2\theta}$$
$$= \sqrt{\cos^2\theta + \sin^2\theta + 1 - 2\cos\theta}$$
$$= \sqrt{1 + 1 - 2\cos\theta}$$
$$= \sqrt{2 - 2\cos\theta}$$

37. (a) The arc length is $s = r\theta$. Since the perimeter is 12 cm:

$$r + r + r\theta = 12$$
$$2r + r\theta = 12$$
$$r(2 + \theta) = 12$$
$$r = \frac{12}{2 + \theta}$$

Using functional notation, the function is $r(\theta) = \dfrac{12}{2 + \theta}$.

(b) The area is given by:

$$A = \tfrac{1}{2}r^2(\theta) = \tfrac{1}{2}\left(\frac{12}{2 + \theta}\right)^2 (\theta) = \frac{72\theta}{(2 + \theta)^2}$$

Using functional notation, the function is $A(\theta) = \dfrac{72\theta}{(2 + \theta)^2}$. This is not a quadratic function.

(c) Solving the expression in part (a) for θ:

$$r = \frac{12}{2 + \theta}$$
$$2 + \theta = \frac{12}{r}$$
$$\theta = \frac{12}{r} - 2$$

Using functional notation, the function is $\theta(r) = \dfrac{12}{r} - 2$.

(d) The area is given by:

$$A = \tfrac{1}{2}r^2(\theta) = \tfrac{1}{2}r^2\left(\frac{12}{r} - 2\right) = 6r - r^2$$

Using functional notation, the function is $A(r) = 6r - r^2$. Yes, this is a quadratic function.

(e) Completing the square on A:

$$A = 6r - r^2$$
$$= -\left(r^2 - 6r\right)$$
$$= -\left(r^2 - 6r + 9\right) + 9$$
$$= -(r - 3)^2 + 9$$

The area will be a maximum of 9 cm^2 when $r = 3$ cm. The corresponding value of θ is $\theta = 2$ radians.

8.3 Trigonometric Functions of Real Numbers

1.   (a)   The reference angle for $\frac{11\pi}{6}$ is $\frac{\pi}{6}$, and $\cos\frac{\pi}{6} = \frac{\sqrt{3}}{2}$. The terminal side of $\frac{11\pi}{6}$ lies in the fourth quadrant where the x-coordinate is positive, so:

$$\cos\frac{11\pi}{6} = \cos\frac{\pi}{6} = \frac{\sqrt{3}}{2}$$

     (b)   The reference angle for $-\frac{11\pi}{6}$ is $\frac{\pi}{6}$, and $\cos\frac{\pi}{6} = \frac{\sqrt{3}}{2}$. The terminal side of $-\frac{11\pi}{6}$ lies in the first quadrant where the x-coordinate is positive, so:

$$\cos\left(-\frac{11\pi}{6}\right) = \cos\frac{\pi}{6} = \frac{\sqrt{3}}{2}$$

     (c)   The reference angle for $\frac{11\pi}{6}$ is $\frac{\pi}{6}$, and $\sin\frac{\pi}{6} = \frac{1}{2}$. The terminal side of $\frac{11\pi}{6}$ lies in the fourth quadrant where the y-coordinate is negative, so:

$$\sin\frac{11\pi}{6} = -\sin\frac{\pi}{6} = -\frac{1}{2}$$

     (d)   The reference angle for $-\frac{11\pi}{6}$ is $\frac{\pi}{6}$, and $\sin\frac{\pi}{6} = \frac{1}{2}$. The terminal side of $-\frac{11\pi}{6}$ lies in the first quadrant where the y-coordinate is positive, so:

$$\sin\left(-\frac{11\pi}{6}\right) = \sin\frac{\pi}{6} = \frac{1}{2}$$

3.   (a)   The reference angle for $\frac{\pi}{6}$ is $\frac{\pi}{6}$, and $\cos\frac{\pi}{6} = \frac{\sqrt{3}}{2}$. The terminal side of $\frac{\pi}{6}$ lies in the first quadrant where the x-coordinate is positive, so:

$$\cos\frac{\pi}{6} = \frac{\sqrt{3}}{2}$$

     (b)   The reference angle for $-\frac{\pi}{6}$ is $\frac{\pi}{6}$, and $\cos\frac{\pi}{6} = \frac{\sqrt{3}}{2}$. The terminal side of $-\frac{\pi}{6}$ lies in the fourth quadrant where the x-coordinate is positive, so:

$$\cos\left(-\frac{\pi}{6}\right) = \cos\frac{\pi}{6} = \frac{\sqrt{3}}{2}$$

     (c)   The reference angle for $\frac{\pi}{6}$ is $\frac{\pi}{6}$, and $\sin\frac{\pi}{6} = \frac{1}{2}$. The terminal side of $\frac{\pi}{6}$ lies in the first quadrant where the y-coordinate is positive, so:

$$\sin\frac{\pi}{6} = \frac{1}{2}$$

(d) The reference angle for $-\frac{\pi}{6}$ is $\frac{\pi}{6}$, and $\sin\frac{\pi}{6} = \frac{1}{2}$. The terminal side of $-\frac{\pi}{6}$ lies in the fourth quadrant where the y-coordinate is negative, so:

$$\sin\left(-\frac{\pi}{6}\right) = -\sin\frac{\pi}{6} = -\frac{1}{2}$$

5. (a) The reference angle for $\frac{5\pi}{4}$ is $\frac{\pi}{4}$, and $\cos\frac{\pi}{4} = \frac{\sqrt{2}}{2}$. The terminal side of $\frac{5\pi}{4}$ lies in the third quadrant where the x-coordinate is negative, so:

$$\cos\frac{5\pi}{4} = -\cos\frac{\pi}{4} = -\frac{\sqrt{2}}{2}$$

(b) The reference angle for $-\frac{5\pi}{4}$ is $\frac{\pi}{4}$, and $\cos\frac{\pi}{4} = \frac{\sqrt{2}}{2}$. The terminal side of $-\frac{5\pi}{4}$ lies in the second quadrant where the x-coordinate is negative, so:

$$\cos\left(-\frac{5\pi}{4}\right) = -\cos\frac{\pi}{4} = -\frac{\sqrt{2}}{2}$$

(c) The reference angle for $\frac{5\pi}{4}$ is $\frac{\pi}{4}$, and $\sin\frac{\pi}{4} = \frac{\sqrt{2}}{2}$. The terminal side of $\frac{5\pi}{4}$ lies in the third quadrant where the y-coordinate is negative, so:

$$\sin\frac{5\pi}{4} = -\sin\frac{\pi}{4} = -\frac{\sqrt{2}}{2}$$

(d) The reference angle for $-\frac{5\pi}{4}$ is $\frac{\pi}{4}$, and $\sin\frac{\pi}{4} = \frac{\sqrt{2}}{2}$. The terminal side of $-\frac{5\pi}{4}$ lies in the second quadrant where the y-coordinate is positive, so:

$$\sin\left(-\frac{5\pi}{4}\right) = \sin\frac{\pi}{4} = \frac{\sqrt{2}}{2}$$

7. (a) The reference angle for $\frac{5\pi}{3}$ is $\frac{\pi}{3}$, and $\sec\frac{\pi}{3} = 2$. The terminal side of $\frac{5\pi}{3}$ lies in the fourth quadrant where the x-coordinate is positive, so:

$$\sec\frac{5\pi}{3} = \sec\frac{\pi}{3} = 2$$

(b) The reference angle for $-\frac{5\pi}{3}$ is $\frac{\pi}{3}$, and $\csc\frac{\pi}{3} = \frac{2\sqrt{3}}{3}$. The terminal side of $-\frac{5\pi}{3}$ lies in the first quadrant where the y-coordinate is positive, so:

$$\csc\left(-\frac{5\pi}{3}\right) = \csc\frac{\pi}{3} = \frac{2\sqrt{3}}{3}$$

(c) The reference angle for $\frac{5\pi}{3}$ is $\frac{\pi}{3}$, and $\tan\frac{\pi}{3}=\sqrt{3}$. The terminal side of $\frac{5\pi}{3}$ lies in the fourth quadrant where the x-coordinate is positive while the y-coordinate is negative, so:

$$\tan\tfrac{5\pi}{3}=-\tan\tfrac{\pi}{3}=-\sqrt{3}$$

(d) The reference angle for $-\frac{5\pi}{3}$ is $\frac{\pi}{3}$, and $\cot\frac{\pi}{3}=\dfrac{\sqrt{3}}{3}$. The terminal side of $-\frac{5\pi}{3}$ lies in the first quadrant where both the x- and y-coordinates are positive, so:

$$\cot\left(-\tfrac{5\pi}{3}\right)=\cot\tfrac{\pi}{3}=\dfrac{\sqrt{3}}{3}$$

9. (a) Choose positive radian values such that the x-coordinate of the point on the unit circle is 0. Four such values are $t=\frac{\pi}{2},\frac{3\pi}{2},\frac{5\pi}{2}$, and $\frac{7\pi}{2}$ (other answers are possible).

(b) Choose negative radian values such that the x-coordinate of the point on the unit circle is 0. Four such values are $t=-\frac{\pi}{2},-\frac{3\pi}{2},-\frac{5\pi}{2}$, and $-\frac{7\pi}{2}$ (other answers are possible).

11. (a) Using a calculator to two decimal place accuracy:

$\sin 2.06\approx 0.88$	$\cos 2.06\approx -0.47$	$\tan 2.06\approx -1.88$
$\sec 2.06\approx -2.13$	$\csc 2.06\approx 1.13$	$\cot 2.06\approx -0.53$

(b) Using a calculator to two decimal place accuracy:

$\sin(-2.06)\approx -0.88$	$\cos(-2.06)\approx -0.47$	$\tan(-2.06)\approx 1.88$
$\sec(-2.06)\approx -2.13$	$\csc(-2.06)\approx -1.13$	$\cot(-2.06)\approx 0.53$

13. (a) Using a calculator to two decimal place accuracy:

$\sin\frac{\pi}{6}\approx 0.50$	$\cos\frac{\pi}{6}\approx 0.87$	$\tan\frac{\pi}{6}\approx 0.58$
$\sec\frac{\pi}{6}\approx 1.15$	$\csc\frac{\pi}{6}\approx 2.00$	$\cot\frac{\pi}{6}\approx 1.73$

(b) Since $\frac{\pi}{6}+2\pi$ will intersect the unit circle at the same location as in (a), all six trigonometric functions will have the same values as in (a):

$\sin\left(\frac{\pi}{6}+2\pi\right)\approx 0.50$	$\cos\left(\frac{\pi}{6}+2\pi\right)\approx 0.87$	$\tan\left(\frac{\pi}{6}+2\pi\right)\approx 0.58$
$\sec\left(\frac{\pi}{6}+2\pi\right)\approx 1.15$	$\csc\left(\frac{\pi}{6}+2\pi\right)\approx 2.00$	$\cot\left(\frac{\pi}{6}+2\pi\right)\approx 1.73$

15. (a) Checking the identity $\sin^2 t+\cos^2 t=1$:

$$\sin^2\tfrac{\pi}{3}+\cos^2\tfrac{\pi}{3}=\left(\tfrac{\sqrt{3}}{2}\right)^2+\left(\tfrac{1}{2}\right)^2=\tfrac{3}{4}+\tfrac{1}{4}=1$$

(b) Checking the identity $\sin^2 t + \cos^2 t = 1$:

$$\sin^2 \tfrac{5\pi}{4} + \cos^2 \tfrac{5\pi}{4} = \left(-\frac{\sqrt{2}}{2}\right)^2 + \left(-\frac{\sqrt{2}}{2}\right)^2 = \tfrac{2}{4} + \tfrac{2}{4} = 1$$

(c) Checking the identity $\sin^2 t + \cos^2 t = 1$ and using a calculator:
$$\sin^2(-53) + \cos^2(-53) \approx (-0.3959)^2 + (-0.9183)^2 \approx 0.1568 + 0.8432 = 1$$

17. (a) Checking the identity $\cot^2 t + 1 = \csc^2 t$:
$$\cot^2\left(-\tfrac{\pi}{6}\right) + 1 = \left(-\sqrt{3}\right)^2 + 1 = 3 + 1 = 4$$
$$\csc^2\left(-\tfrac{\pi}{6}\right) = (-2)^2 = 4$$

(b) Checking the identity $\cot^2 t + 1 = \csc^2 t$:
$$\cot^2\left(\tfrac{7\pi}{4}\right) + 1 = (-1)^2 + 1 = 1 + 1 = 2$$
$$\csc^2\left(\tfrac{7\pi}{4}\right) = \left(-\sqrt{2}\right)^2 = 2$$

(c) Checking the identity $\cot^2 t + 1 = \csc^2 t$ and using a calculator:
$$\cot^2(0.12) + 1 \approx 68.7787 + 1 = 69.7787$$
$$\csc^2(0.12) \approx 69.7787$$

19. (a) Checking the identity $\sin(-t) = -\sin t$:
$$\sin\left(-\tfrac{3\pi}{2}\right) = 1$$
$$-\sin\tfrac{3\pi}{2} = -(-1) = 1$$

(b) Checking the identity $\sin(-t) = -\sin t$:
$$\sin\tfrac{5\pi}{6} = \tfrac{1}{2}$$
$$-\sin\left(-\tfrac{5\pi}{6}\right) = -\left(-\tfrac{1}{2}\right) = \tfrac{1}{2}$$

(c) Checking the identity $\sin(-t) = -\sin t$, and using a calculator:
$$\sin(-13.24) \approx -0.6238$$
$$-\sin 13.24 \approx -(0.6238) = -0.6238$$

21. (a) Checking the identity $\sin(t + 2\pi) = \sin t$:

$$\sin\left(\tfrac{5\pi}{3} + 2\pi\right) = \sin\left(\tfrac{11\pi}{3}\right) = -\frac{\sqrt{3}}{2}$$

$$\sin\tfrac{5\pi}{3} = -\frac{\sqrt{3}}{2}$$

(b) Checking the identity $\sin(t + 2\pi) = \sin t$:

$$\sin\left(-\tfrac{3\pi}{2} + 2\pi\right) = \sin\tfrac{\pi}{2} = 1$$
$$\sin\left(-\tfrac{3\pi}{2}\right) = 1$$

(c) Checking the identity $\sin(t + 2\pi) = \sin t$ and using a calculator:

$$\sin(\sqrt{19} + 2\pi) \approx \sin(10.6421) \approx -0.9382$$
$$\sin(\sqrt{19}) \approx \sin(4.3589) \approx -0.9382$$

23. Evaluating each side of the "equality" when $t = \tfrac{\pi}{6}$:

$$\cos\left(2 \cdot \tfrac{\pi}{6}\right) = \cos\tfrac{\pi}{3} = \tfrac{1}{2}$$
$$2\cos\tfrac{\pi}{6} = 2 \cdot \tfrac{\sqrt{3}}{2} = \sqrt{3}$$

Since these results are unequal, $\cos 2t = 2\cos t$ is not an identity.

25. Using $\sin^2 t + \cos^2 t = 1$:

$$\cos t = \pm\sqrt{1 - \sin^2 t} = \pm\sqrt{1 - \tfrac{9}{25}} = \pm\tfrac{4}{5}$$

Since $\pi < t < \tfrac{3\pi}{2}$, $\cos t = -\tfrac{4}{5}$ and thus:

$$\tan t = \frac{\sin t}{\cos t} = \frac{-\tfrac{3}{5}}{-\tfrac{4}{5}} = \tfrac{3}{4}$$

27. Since $\tfrac{\pi}{2} < t < \pi$, $\cos t < 0$ and thus:

$$\cos t = -\sqrt{1 - \sin^2 t} = -\sqrt{1 - \tfrac{3}{16}} = -\sqrt{\tfrac{13}{16}} = -\frac{\sqrt{13}}{4}$$

Thus:

$$\tan t = \frac{\sin t}{\cos t} = \frac{\tfrac{\sqrt{3}}{4}}{-\tfrac{\sqrt{13}}{4}} = -\frac{\sqrt{3}}{\sqrt{13}} = -\frac{\sqrt{39}}{13}$$

29. Using the identity $1 + \tan^2\alpha = \sec^2\alpha$:

$$1 + \left(\tfrac{12}{5}\right)^2 = \sec^2\alpha$$
$$1 + \tfrac{144}{25} = \sec^2\alpha$$
$$\tfrac{169}{25} = \sec^2\alpha$$
$$\pm\tfrac{13}{5} = \sec\alpha$$

Since $\cos\alpha > 0$, $\sec\alpha > 0$, pick $\sec\alpha = \tfrac{13}{5}$. Thus $\cos\alpha = \tfrac{5}{13}$, and use $\sin^2\alpha + \cos^2\alpha = 1$ to get:

$$\sin^2\alpha = 1 - \left(\tfrac{5}{13}\right)^2 = \tfrac{144}{169}, \text{ thus } \sin\alpha = \pm\tfrac{12}{13}$$

Pick the positive value since, if both the tangent and cosine are positive, so is the sine. Thus $\sin\alpha = \tfrac{12}{13}$.

31. For $0 < \theta < \frac{\pi}{2}$:
$$\sqrt{9 - x^2} = \sqrt{9 - (3\sin\theta)^2} = \sqrt{9(1 - \sin^2\theta)} = \sqrt{9\cos^2\theta} = 3\cos\theta$$
Choose the positive root since $\cos\theta > 0$ for $0 < \theta < \frac{\pi}{2}$.

33. Since $0 < \theta < \frac{\pi}{2}$, $\tan\theta > 0$. Thus:
$$\frac{1}{\left(u^2 - 25\right)^{3/2}} = \frac{1}{\left(25\sec^2\theta - 25\right)^{3/2}} = \frac{1}{125\tan^3\theta} = \frac{\cot^3\theta}{125}$$

35. Since $\sec\theta > 0$:
$$\frac{1}{\sqrt{u^2 + 7}} = \frac{1}{\sqrt{7\tan^2\theta + 7}} = \frac{1}{\sqrt{7}\sec\theta} = \frac{\cos\theta}{\sqrt{7}} = \frac{\sqrt{7}\cos\theta}{7}$$

37. (a) Since $\sin(-t) = -\sin t$:
$$\sin(-t) = -\sin t = -\tfrac{2}{3}$$

(b) Since $\sin(-\phi) = -\sin\phi$:
$$\sin(-\phi) = -\sin\phi = -\left(-\tfrac{1}{4}\right) = \tfrac{1}{4}$$

(c) Since $\cos(-\alpha) = \cos\alpha$:
$$\cos(-\alpha) = \cos\alpha = \tfrac{1}{5}$$

(d) Since $\cos(-s) = \cos s$:
$$\cos(-s) = \cos s = -\tfrac{1}{5}$$

39. (a) Since $\cos t = -\tfrac{1}{3}$ and $\frac{\pi}{2} < t < \pi$, $\sin t > 0$ and thus:
$$\sin t = \sqrt{1 - \tfrac{1}{9}} = \sqrt{\tfrac{8}{9}} = \frac{2\sqrt{2}}{3}$$
Therefore the values are:
$$\sin(-t) = -\sin t = -\frac{2\sqrt{2}}{3}$$
$$\cos(-t) = \cos t = -\tfrac{1}{3}$$
Thus:
$$\sin(-t) + \cos(-t) = -\frac{2\sqrt{2}}{3} - \frac{1}{3} = -\frac{1 + 2\sqrt{2}}{3}$$

(b) Note that $\sin^2(-t) + \cos^2(-t) = 1$, regardless of the value of t.

41. (a) Using the identity $\cos(t + 2\pi k) = \cos t$:

$$\cos\left(\tfrac{\pi}{4} + 2\pi\right) = \cos\tfrac{\pi}{4} = \frac{\sqrt{2}}{2}$$

(b) Using the identity $\sin(t + 2\pi k) = \sin t$:

$$\sin\left(\tfrac{\pi}{3} + 2\pi\right) = \sin\tfrac{\pi}{3} = \frac{\sqrt{3}}{2}$$

(c) Using the identity $\sin(t + 2\pi k) = \sin t$:

$$\sin\left(\tfrac{\pi}{2} - 6\pi\right) = \sin\tfrac{\pi}{2} = 1$$

43. Using the Pythagorean identities:

$$\frac{\sin^2 t + \cos^2 t}{\tan^2 t + 1} = \frac{1}{\sec^2 t} = \cos^2 t$$

45. Using the Pythagorean identities:

$$\frac{\sec^2 \theta - \tan^2 \theta}{1 + \cot^2 \theta} = \frac{\tan^2 \theta + 1 - \tan^2 \theta}{\csc^2 \theta} = \frac{1}{\csc^2 \theta} = \sin^2 \theta$$

47. Working from the right-hand side and using the identity for $\cot t$:

$$\sin t + \cot t \cos t = \sin t + \frac{\cos t}{\sin t}(\cos t) = \frac{\sin^2 t}{\sin t} + \frac{\cos^2 t}{\sin t} = \frac{\sin^2 t + \cos^2 t}{\sin t} = \frac{1}{\sin t} = \csc t$$

49. Combining fractions on the left-hand side:

$$\frac{1}{1 + \sec s} + \frac{1}{1 - \sec s} = \frac{(1 - \sec s) + (1 + \sec s)}{(1 - \sec s)(1 + \sec s)} = \frac{2}{1 - \sec^2 s} = \frac{-2}{\tan^2 s} = -2\cot^2 s$$

51. Denoting $\sin\theta$ by S and $\cos\theta$ by C:

$$\frac{\cot\theta}{1-\tan\theta}+\frac{\tan\theta}{1-\cot\theta}=\frac{\frac{C}{S}}{1-\frac{S}{C}}+\frac{\frac{S}{C}}{1-\frac{C}{S}}$$

$$=\frac{C^2}{S(C-S)}-\frac{S^2}{C(C-S)}$$

$$=\frac{C^3-S^3}{SC(C-S)}$$

$$=\frac{(C-S)(C^2+SC+S^2)}{SC(C-S)}$$

$$=\frac{1+SC}{SC}$$

$$=\frac{1}{SC}+1$$

Now replace C with $\cos\theta$ and S with $\sin\theta$:

$$\frac{1}{SC}+1=\frac{\cos^2\theta+\sin^2\theta}{\sin\theta\cos\theta}+1=\frac{\cos\theta}{\sin\theta}+\frac{\sin\theta}{\cos\theta}+1=\cot\theta+\tan\theta+1$$

53. The expression on the left-hand side becomes:

$$\tan\theta-(\tan\theta\cot\theta)(\cot\theta)+\cot\theta-(\cot\theta\tan\theta)(\tan\theta)$$

This multiplies out to become:

$$\tan\theta-\cot\theta+\cot\theta-\tan\theta=0$$

55. If $\sec t=\frac{13}{5}$, then $\cos t=\frac{5}{13}$. Given $\frac{3\pi}{2}<t<2\pi$, then $\sin t<0$ (fourth quadrant), thus:

$$\sin t=-\sqrt{1-\cos^2 t}=-\sqrt{1-\frac{25}{169}}=-\sqrt{\frac{144}{169}}=-\frac{12}{13}$$

Therefore:

$$\frac{2\sin t-3\cos t}{4\sin t-9\cos t}=\frac{2\left(-\frac{12}{13}\right)-3\left(\frac{5}{13}\right)}{4\left(-\frac{12}{13}\right)-9\left(\frac{5}{13}\right)}=\frac{-24-15}{-48-45}=\frac{-39}{-93}=\frac{13}{31}$$

57. Assuming that the radius of the circle is 1, the coordinates of the point labeled (x, y) are $(\cos t, \sin t)$, and the coordinates of the point labeled $(-x, -y)$ are $(\cos(t+\pi), \sin(t+\pi))$. So $y=\sin t$ and $-y=\sin(t+\pi)$, from which it follows that $\sin(t+\pi)=-\sin t$. Similarly, $-x=\cos(t+\pi)$ and $x=\cos t$, from which it follows that $\cos(t+\pi)=-\cos t$. Since $t-\pi$ results in the same intersection point with the unit circle as $t+\pi$, identities (ii) and (iv) follow in a similar manner.

59. The substitutions are $x = \dfrac{\sqrt{2}}{2}(X - Y)$ and $y = \dfrac{\sqrt{2}}{2}(X + Y)$. Thus:

$$x^2 = \tfrac{1}{2}\left(X^2 - 2XY + Y^2\right) \text{ and } y^2 = \tfrac{1}{2}\left(X^2 + 2XY + Y^2\right)$$

Squaring again, we obtain:

$$x^4 = \tfrac{1}{4}\left(X^4 - 4X^3Y + 6X^2Y^2 - 4XY^3 + Y^4\right)$$
$$y^4 = \tfrac{1}{4}\left(X^4 + 4X^3Y + 6X^2Y^2 + 4XY^3 + Y^4\right)$$

Also find $x^2y^2 = \tfrac{1}{4}\left(X^2 - Y^2\right)^2 = \tfrac{1}{4}\left(X^4 - 2X^2Y^2 + Y^4\right)$. Now use the expressions that have been found for x^2, y^2, x^4, and y^4 to substitute in the expression $x^4 + 6x^2y^2 + y^4$ to obtain:

$$\tfrac{X^4}{4} - X^3Y + \tfrac{3}{2}X^2Y^2 - XY^3 + \tfrac{Y^4}{4} + \tfrac{3}{2}X^4 - 3X^2Y^2 + \tfrac{3}{2}Y^4 + \tfrac{X^4}{4} + X^3Y + \tfrac{3}{2}X^2Y^2 + XY^3 + \tfrac{Y^4}{4}$$

which is $2X^4 + 2Y^4$. In light of this result, the equation $x^4 + 6x^2y^2 + y^4 = 32$ is equivalent to $2X^4 + 2Y^4 = 32$, or $X^4 + Y^4 = 16$, as required.

61. (a) Complete the table:

t	0.2	0.4	0.6	0.8	1.0	1.2	1.4
$f(t)$	219.07	50.53	19.70	9.55	6.14	7.98	33.88

(b) The smallest output is 6.14, which occurs at $t = 1.0$.

(c) Work from the right-hand side:

$$(\tan t - 3\cot t)^2 + 6 = \tan^2 t - 6\tan t \cot t + 9\cot^2 t + 6$$
$$= \tan^2 t - 6 + 9\cot^2 t + 6$$
$$= \tan^2 t + 9\cot^2 t$$

(d) Since $(\tan t - 3\cot t)^2 \geq 0$:

$$\tan^2 t + 9\cot^2 t = (\tan t - 3\cot t)^2 + 6 \geq 0 + 6 \geq 6$$

(e) The minimum will occur when:

$$\tan t - 3\cot t = 0$$
$$\tan t = 3\cot t$$
$$\tan t = \frac{3}{\tan t}$$
$$\tan^2 t = 3$$
$$\tan t = \sqrt{3} \text{ (since } 0 < t < \pi/2, \text{ then } \tan t > 0)$$
$$t = \tfrac{\pi}{3} \approx 1.05$$

The answer from part (b) is consistent with this result.

63. (a) When $t = \frac{\pi}{6}$:

$$2\sin^2\frac{\pi}{6} - \sin\frac{\pi}{6} = 2\left(\frac{1}{2}\right)^2 - \frac{1}{2} = \frac{1}{2} - \frac{1}{2} = 0$$

$$2\sin\frac{\pi}{6}\cos\frac{\pi}{6} - \cos\frac{\pi}{6} = 2\cdot\frac{1}{2}\cdot\frac{\sqrt{3}}{2} - \frac{\sqrt{3}}{2} = \frac{\sqrt{3}}{2} - \frac{\sqrt{3}}{2} = 0$$

(b) When $t = \frac{\pi}{4}$:

$$2\sin^2\frac{\pi}{4} - \sin\frac{\pi}{4} = 2\left(\frac{\sqrt{2}}{2}\right)^2 - \frac{\sqrt{2}}{2} = 1 - \frac{\sqrt{2}}{2} = \frac{2-\sqrt{2}}{2}$$

$$2\sin\frac{\pi}{4}\cos\frac{\pi}{4} - \cos\frac{\pi}{4} = 2\cdot\frac{\sqrt{2}}{2}\cdot\frac{\sqrt{2}}{2} - \frac{\sqrt{2}}{2} = 1 - \frac{\sqrt{2}}{2} = \frac{2-\sqrt{2}}{2}$$

(c) No. Using the value $t = 0$:

$$2\sin^2 0 - \sin 0 = 2(0)^2 - 0 = 0 - 0 = 0$$

$$2\sin 0\cos 0 - \cos 0 = 2(0)(1) - 1 = 0 - 1 = -1$$

Since the two sides of the equation are not equal, the given equation is not an identity.

65. Complete the table:

t	$1 - \frac{1}{2}t^2$	$\cos\theta$
0.02	0.9998	0.999800
0.05	0.99875	0.998750
0.1	0.995	0.995004
0.2	0.980	0.980067
0.3	0.955	0.955336

67. Complete the table:

x	$\frac{1}{3}x^3 + x$	$\frac{2}{15}x^5 + \frac{1}{3}x^3 + x$	$\tan x$
0.1	0.100333	0.100335	0.100335
0.2	0.202667	0.202709	0.202710
0.3	0.309	0.309324	0.309336
0.4	0.421333	0.422699	0.422793
0.5	0.541667	0.545833	0.546302

69. The y-coordinates of points $P, Q, R, S,$ and T are just the sines of the radian measures of the corresponding arcs. Thus the y-coordinates are:

$$P : \sin\tfrac{\pi}{12} \approx 0.259$$

$$Q : \sin\tfrac{\pi}{6} = \tfrac{1}{2} = 0.5$$

$$R : \sin\tfrac{\pi}{4} = \frac{\sqrt{2}}{2} \approx 0.707$$

$$S : \sin\tfrac{\pi}{3} = \frac{\sqrt{3}}{2} \approx 0.866$$

$$T : \sin\tfrac{5\pi}{12} \approx 0.966$$

8.4 Graphs of the Sine and Cosine Functions

1. A cycle is completed every 2 units, so the period is 2. The curve has high and low points of 1 and –1, respectively, so the amplitude is 1.

3. A cycle is completed every 4 units, so the period is 4. The curve has high and low points of 6 and – 6, respectively, so the amplitude is 6.

5. A cycle is completed every 4 units, so the period is 4. The curve has high and low points of 6 and 2, respectively, so the amplitude is $\frac{6-2}{2} = 2$.

7. A cycle is completed every 6 units, so the period is 6. The curve has high and low points of –3 and – 6, respectively, so the amplitude is $\frac{-3-(-6)}{2} = \frac{3}{2}$.

9. The coordinates of point C are $\left(-\tfrac{7\pi}{2}, 1\right) \approx (-10.996, 1)$.

11. The coordinates of point G are $\left(\tfrac{5\pi}{2}, 1\right) \approx (7.854, 1)$.

13. The coordinates of point B are $(-4\pi, 0) \approx (-12.566, 0)$.

15. The coordinates of point D are $(-3\pi, 0) \approx (-9.425, 0)$.

17. The coordinates of point E are $(-\pi, 0) \approx (-3.142, 0)$.

19. Referring to the graph of $y = \sin x$, note that $y = \sin x$ is increasing on the interval $\tfrac{3\pi}{2} < x < 2\pi$.

21. Referring to the graph of $y = \sin x$, note that $y = \sin x$ is decreasing on the interval $\tfrac{5\pi}{2} < x < \tfrac{7\pi}{2}$.

23. The coordinates of point J are $\left(\frac{9\pi}{2},0\right) \approx (14.137,0)$.

25. The coordinates of point A are $(-4\pi,1) \approx (-12.566,1)$.

27. The coordinates of point E are $\left(\frac{\pi}{2},0\right) \approx (1.571,0)$.

29. The coordinates of point I are $(4\pi,1) \approx (12.566,1)$.

31. The coordinates of point B are $\left(-\frac{5\pi}{2},0\right) \approx (-7.854,0)$.

33. Referring to the graph of $y = \cos x$, note that $y = \cos x$ is decreasing on the interval $0 < x < \pi$.

35. Referring to the graph of $y = \cos x$, note that $y = \cos x$ is increasing on the interval $-\frac{\pi}{2} < x < 0$.

37. (a) From the figure, the root is $x \approx 0.8$.
  (b) Using a calculator, the root is $x \approx 0.7954$.
  (c) Another root would be $2\pi - x \approx 5.4878$.
  (d) These roots would be $\pi - x \approx 2.3462$ and $\pi + x \approx 3.9370$.

39. (a) From the figure, the root is $x \approx 1.15$.
  (b) Using a calculator, the root is $x \approx 1.1593$.
  (c) Another root would be $2\pi - x \approx 5.1239$.
  (d) These roots would be $\pi - x \approx 1.9823$ and $\pi + x \approx 4.3009$.

41. (a) From the figure, the root is $x \approx 0.9$.
  (b) Using a calculator, the root is $x \approx 0.9273$.
  (c) Another root would be $2\pi - x \approx 5.3559$.
  (d) These roots would be $\pi - x \approx 2.2143$ and $\pi + x \approx 4.0689$.

43. (a) From the figure, the root is $x \approx 0.1$.
  (b) Using a calculator, the root is $x \approx 0.1002$.
  (c) Another root would be $\pi - x \approx 3.0414$.
  (d) These roots would be $\pi + x \approx 3.2418$ and $2\pi - x \approx 6.1830$.

45. (a) From the figure, the root is $x \approx 0.4$.
  (b) Using a calculator, the root is $x \approx 0.4115$.
  (c) Another root would be $\pi - x \approx 2.7301$.
  (d) These roots would be $\pi + x \approx 3.5531$ and $2\pi - x \approx 5.8717$.

47. (a) From the figure, the root is $x \approx 1.0$.
    (b) Using a calculator, the root is $x \approx 1.0160$.
    (c) Another root would be $\pi - x \approx 2.1256$.
    (d) These roots would be $\pi + x \approx 4.1576$ and $2\pi - x \approx 5.2672$.

49. (a) Since C lies on the unit circle, its coordinates are $C(\cos\theta, \sin\theta)$.
    (b) Since $\angle AOB = \theta = \angle COD$ and $AO = CO$, the two triangles are congruent.
    (c) Since $AB = CD$ and $OB = OD$, the coordinates of A are $A(-\sin\theta, \cos\theta)$.
    (d) Matching up x- and y-coordinates:
    $$\cos\left(\theta + \tfrac{\pi}{2}\right) = -\sin\theta \text{ and } \sin\left(\theta + \tfrac{\pi}{2}\right) = \cos\theta$$

51. (a) The corresponding value of $\cos x$ is 0.
    (b) The corresponding value of $\sin x$ is 0.

53. Both functions are decreasing on the open interval $\left(\tfrac{\pi}{2}, \pi\right)$.

55. (a) The completed table values using $x_0 = \tfrac{50}{500} = 0.1$ are:

	x_1	x_2	x_3	x_4
From graph	1.0	0.55	0.85	0.65
From calculator	0.99500	0.54450	0.85539	0.65593

	x_5	x_6	x_7
From graph	0.80	0.70	0.75
From calculator	0.79248	0.70208	0.76350

(b) Multiplying the results by 500, the completed table is:

n	0	1	2	3	4	5	6	7
Number of fish after n breeding seasons	50	498	272	428	328	396	351	382

(c) The corresponding equilibrium population is $(0.7391)(500) \approx 370$ fish.

Graphing Utility Exercises for Section 8.4

1. (a) The x-intercept is π.

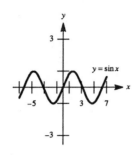

    (b) There are four turning points. The turning point between 1 and 2 has an x-coordinate of $\frac{\pi}{2}$.

    (c) There are five turning points for $y = \cos x$. The turning point between 6 and 7 has an x-coordinate of 2π.

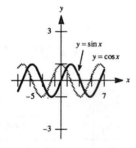

    (d) If $y = \sin x$ is shifted $\frac{\pi}{2}$ units to the left, its graph will coincide with $y = \cos x$.

    (e) The results from parts (a) through (c) are confirmed.

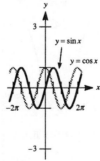

3.   (a)   Between $-\frac{\pi}{2}$ and $\frac{\pi}{2}$, the curve does appear similar to that of a parabola:

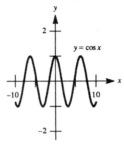

(b)   Graphing the two curves:

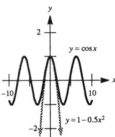

In the vicinity of $x = 0$, the graphs of $y = \cos x$ and $y = 1 - 0.5x^2$ are very similar.

(c)   Completing the table:

x	1	0.5	0.1	0.01	0.001
$\cos x$	0.54	0.8776	0.9950	0.99995	0.9999995
$1 - 0.5x^2$	0.50	0.8750	0.9950	0.99995	0.9999995

5.   Graphing the two curves:

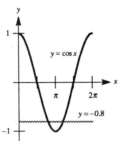

The x-coordinates of the intersection points are $x \approx 2.498$ and $x \approx 3.785$, matching the results in the text.

7. **(a)** The roots to $\sin x = 0.687$ are approximately $x \approx 0.7574$ and $x \approx 2.3842$.

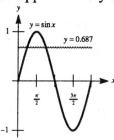

(b) The roots to $\sin x = -0.687$ are approximately $x \approx 3.8989$ and $x \approx 5.5258$.

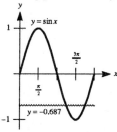

8.5 Graphs of $y = A \sin (Bx - C)$ and $y = A \cos (Bx - C)$

1. **(a)** The amplitude is 2, the period is 2π, and the x-intercepts are 0, π, 2π. The function is increasing on the intervals $\left(0, \frac{\pi}{2}\right)$ and $\left(\frac{3\pi}{2}, 2\pi\right)$.

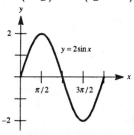

(b) The amplitude is 1 and the period is $\frac{2\pi}{2} = \pi$. The x-intercepts are 0, $\frac{\pi}{2}$, π, and the function is increasing on the interval $\left(\frac{\pi}{4}, \frac{3\pi}{4}\right)$. Notice that the graph is a reflection of $y = \sin 2x$ across the x-axis.

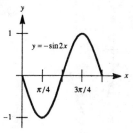

3. (a) The amplitude is 1 and the period is $\frac{2\pi}{2} = \pi$. The x-intercepts are $\frac{\pi}{4}, \frac{3\pi}{4}$, and the function is increasing on the interval $\left(\frac{\pi}{2}, \pi\right)$.

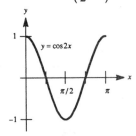

(b) The amplitude is 2 and the period is $\frac{2\pi}{2} = \pi$. The x-intercepts are $\frac{\pi}{4}, \frac{3\pi}{4}$, and the function is increasing on the interval $\left(\frac{\pi}{2}, \pi\right)$.

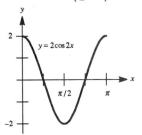

5. (a) The amplitude is 3 and the period is $\frac{2\pi}{\pi/2} = 4$. The x-intercepts are 0, 2, 4, and the function is increasing on the intervals (0, 1) and (3, 4).

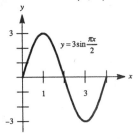

(b) The amplitude is 3 and the period is $\frac{2\pi}{\pi/2} = 4$. The x-intercepts are 0, 2, 4, and the function is increasing on the interval (1, 3). Notice that the graph is a reflection of $y = 3\sin\frac{\pi x}{2}$ across the x-axis.

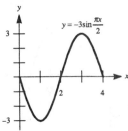

7. (a) The amplitude is 1 and the period is $\frac{2\pi}{2\pi} = 1$. The x-intercepts are $\frac{1}{4}, \frac{3}{4}$, and the function is increasing on the interval $\left(\frac{1}{2}, 1\right)$.

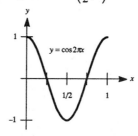

(b) The amplitude is 4 and the period is $\frac{2\pi}{2\pi} = 1$. The x-intercepts are $\frac{1}{4}, \frac{3}{4}$, and the function is increasing on the interval $\left(0, \frac{1}{2}\right)$. Notice that the graph is a reflection of $y = 4 \cos 2\pi x$ across the x-axis.

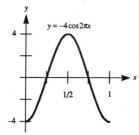

9. The amplitude is 1 and the period is $\frac{2\pi}{2} = \pi$. The x-intercept is $\frac{3\pi}{4}$, and the function is increasing on the intervals $\left(0, \frac{\pi}{4}\right)$ and $\left(\frac{3\pi}{4}, \pi\right)$. Notice that the graph is a displacement of $y = \sin 2x$ up 1 unit.

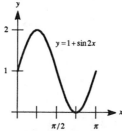

11. The amplitude is 1 and the period is $\frac{2\pi}{\pi/3} = 6$. The x-intercepts are 0, 6, and the function is increasing on the interval (0, 3). Notice that the graph is a reflection of $y = \cos\frac{\pi x}{3}$ across the x-axis, then a displacement up 1 unit.

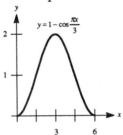

13. The amplitude is 1, the period is 2π, and the phase shift is $\frac{\pi}{6}$. The x-intercepts are $\frac{\pi}{6}, \frac{7\pi}{6}, \frac{13\pi}{6}$, the high point is $\left(\frac{2\pi}{3}, 1\right)$ and the low point is $\left(\frac{5\pi}{3}, -1\right)$.

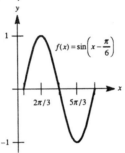

15. The amplitude is 1, the period is 2π, and the phase shift is $-\frac{\pi}{4}$. The x-intercepts are $\frac{\pi}{4}, \frac{5\pi}{4}$, the high point is $\left(\frac{3\pi}{4}, 1\right)$, and the low points are $\left(-\frac{\pi}{4}, -1\right)$ and $\left(\frac{7\pi}{4}, -1\right)$. Notice that the graph is a reflection of $y = \cos\left(x + \frac{\pi}{4}\right)$ across the x-axis.

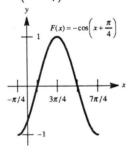

17. The amplitude is 1, the period is $\frac{2\pi}{2} = \pi$, and the phase shift is $\frac{\pi/2}{2} = \frac{\pi}{4}$. The x-intercepts are $\frac{\pi}{4}, \frac{3\pi}{4}, \frac{5\pi}{4}$, the high point is $\left(\frac{\pi}{2},1\right)$, and the low point is $(\pi, -1)$.

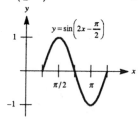

19. The amplitude is 1, the period is $\frac{2\pi}{2} = \pi$, and the phase shift is $\frac{\pi}{2}$. The x-intercepts are $\frac{3\pi}{4}, \frac{5\pi}{4}$, the high points are $\left(\frac{\pi}{2},1\right)$ and $\left(\frac{3\pi}{2},1\right)$, and the low point is $(\pi, -1)$.

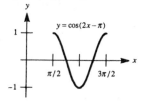

21. The amplitude is 3, the period is $\frac{2\pi}{1/2} = 4\pi$, and the phase shift is $\frac{-\pi/6}{1/2} = -\frac{\pi}{3}$. The x-intercepts are $-\frac{\pi}{3}, \frac{5\pi}{3}, \frac{11\pi}{3}$, the high point is $\left(\frac{2\pi}{3},3\right)$, and the low point is $\left(\frac{8\pi}{3},-3\right)$.

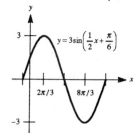

23. The amplitude is 4, the period is $\frac{2\pi}{3}$, and the phase shift is $\frac{\pi/4}{3} = \frac{\pi}{12}$. The x-intercepts are $\frac{\pi}{4}, \frac{7\pi}{12}$, the high points are $\left(\frac{\pi}{12},4\right)$ and $\left(\frac{3\pi}{4},4\right)$, and the low point is $\left(\frac{5\pi}{12},-4\right)$.

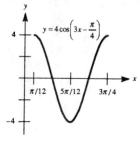

25. The amplitude is $\frac{1}{2}$, the period is $\frac{2\pi}{\pi/2} = 4$, and the phase shift is $\frac{\pi^2}{\pi/2} = 2\pi$. The x-intercepts are 2π, $2\pi + 2$, $2\pi + 4$, the high point is $\left(2\pi + 1, \frac{1}{2}\right)$, and the low point is $\left(2\pi + 3, -\frac{1}{2}\right)$.

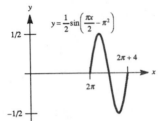

27. The amplitude is 1, the period is $\frac{2\pi}{2} = \pi$, and the phase shift is $\frac{\pi/3}{2} = \frac{\pi}{6}$. The x-intercepts are $\frac{\pi}{6}$ and $\frac{7\pi}{6}$, the high point is $\left(\frac{2\pi}{3}, 2\right)$, and the low points are $\left(\frac{\pi}{6}, 0\right)$ and $\left(\frac{7\pi}{6}, 0\right)$. Notice that the graph is a reflection of $y = \cos\left(2x - \frac{\pi}{3}\right)$ across the x-axis, then a displacement up 1 unit.

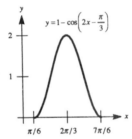

29. This is a sine function where the amplitude is 2, so $A = 2$. Since the period is 4π:

$$\frac{2\pi}{B} = 4\pi$$
$$2\pi = 4\pi B$$
$$\frac{1}{2} = B$$

The equation is $y = 2\sin\frac{1}{2}x$.

31. This is a sine function (reflected across the x-axis) where the amplitude is 3, so $A = -3$. Since the period is 2:

$$\frac{2\pi}{B} = 2$$
$$2\pi = 2B$$
$$\pi = B$$

The equation is $y = -3\sin\pi x$.

33. This is a cosine function (reflected across the x-axis) where the amplitude is 4, so $A = -4$. Since the period is 10π:

$$\frac{2\pi}{B} = 10\pi$$
$$2\pi = 10\pi B$$
$$\frac{1}{5} = B$$

The equation is $y = -4\cos\frac{1}{5}x$.

35. Graph $y = \frac{1}{2} - \frac{1}{2}\cos 2x$, which has an amplitude of $\frac{1}{2}$ and a period of $\frac{2\pi}{2} = \pi$. The graph will be a reflection of $y = \frac{1}{2}\cos 2x$ across the x-axis, then a displacement up $\frac{1}{2}$ unit.

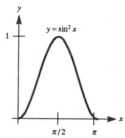

37. Graph $y = \frac{1}{2}\sin 2x$, which has an amplitude of $\frac{1}{2}$ and a period of $\frac{2\pi}{2} = \pi$.

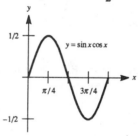

39. The curve $y = \cos x$ begins a period when $x = 0$ and completes that period when $x = 2\pi$. Thus $y = A\cos Bx$ will begin a period when $Bx = 0$ and complete that period when $Bx = 2\pi$, which yields $x = 0$ and $x = \frac{2\pi}{B}$. Thus the period of $y = A\cos Bx$ is $\frac{2\pi}{B}$.

Graphing Utility Exercises for Section 8.5

1. The amplitudes and periods are:

$y = \sin x:$ amplitude $= 1$, period $= 2\pi$
$y = 2\sin x:$ amplitude $= 2$, period $= 2\pi$
$y = 3\sin x:$ amplitude $= 3$, period $= 2\pi$
$y = 4\sin x:$ amplitude $= 4$, period $= 2\pi$

The graphs of the four functions are:

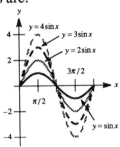

3.   (a)   For $y = 2\sin \pi x$, the amplitude is 2 and the period is $\dfrac{2\pi}{\pi} = 2$. For $y = \sin 2\pi x$, the

amplitude is 1 and the period is $\dfrac{2\pi}{2\pi} = 1$.

(b)   The amplitudes and periods agree with the graphs:

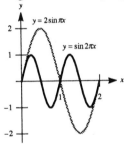

5.   (a)   The amplitude is 2.5, the period is $\frac{2\pi}{3\pi} = \frac{2}{3}$, and the phase shift is $-\frac{4}{3\pi}$.

(b)   Graphing the curve:

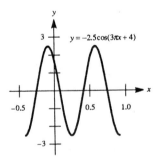

(c)   The high points are approximately $(-0.09, 2.5)$ and $(0.58, 2.5)$, while the low points
are approximately $(-0.42, -2.5)$, $(0.24, -2.5)$, and $(0.91, -2.5)$.

(d)   The high points are $\left(-\frac{4}{3\pi} + \frac{1}{3}, 2.5\right)$ and $\left(-\frac{4}{3\pi} + 1, 2.5\right)$, while the low points are
$\left(-\frac{4}{3\pi}, -2.5\right)$, $\left(-\frac{4}{3\pi} + \frac{2}{3}, -2.5\right)$, and $\left(-\frac{4}{3\pi} + \frac{4}{3}, -2.5\right)$.

7.   (a)   The amplitude is 2.5, the period is 6, and the phase shift is $-\frac{12}{\pi}$.

(b)   Graphing the curve:

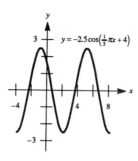

(c)   The high points are approximately (–0.82, 2.5) and (5.18, 2.5), while the low points are approximately (–3.82, –2.5), (2.18, –2.5), and (8.18, –2.5).

(d)   The high points are $\left(-\frac{12}{\pi}+3, 2.5\right)$ and $\left(-\frac{12}{\pi}+9, 2.5\right)$, while the low points are $\left(-\frac{12}{\pi}, -2.5\right)$, $\left(-\frac{12}{\pi}+6, -2.5\right)$, and $\left(-\frac{12}{\pi}+12, -2.5\right)$.

9.   (a)   The amplitude is 1, the period is $\frac{2\pi}{0.5} = 4\pi$, and the phase shift is $\frac{-0.75}{0.5} = -1.5$.

(b)   Graphing the curve:

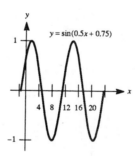

(c)   The high points are approximately (1.64, 1) and (14.21, 1), while the low points are approximately (7.92, –1) and (20.49, –1).

(d)   The high points are $(-1.5 + \pi, 1)$ and $(-1.5 + 5\pi, 1)$, while the low points are $(-1.5 + 3\pi, -1)$ and $(-1.5 + 7\pi, -1)$.

11. (a) The amplitude is 0.02, the period is $\frac{2\pi}{0.01\pi} = 200$, and the phase shift is $\frac{4\pi}{0.01\pi} = 400$.

(b) Graphing the curve:

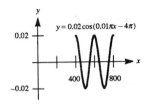

(c) The high points are (400, 0.02), (600, 0.02), and (800, 0.02), while the low points are (500, –0.02) and (700, –0.02).

(d) The high points are (400, 0.02), (600, 0.02), and (800, 0.02), while the low points are (500, –0.02) and (700, –0.02).

13. (a) Graphing the two curves in the standard viewing rectangle:

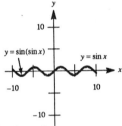

Graphing using the suggested settings:

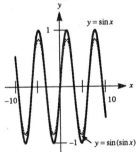

Although the two graphs have the same period, the second graph appears to have a smaller amplitude.

(b) The highest point on $y = \sin(\sin x)$ is approximately (1.58, 0.84), so the amplitude is approximately 0.84.

(c) Since $\sin x$ is largest when $x = \frac{\pi}{2}$ (where $\sin x = 1$), and $\sin x$ is increasing for $0 < x < \frac{\pi}{2}$, $\sin(\sin x)$ should have its largest value when $x = \frac{\pi}{2}$. Its value then is $\sin\left(\sin \frac{\pi}{2}\right) = \sin 1$. Since $\sin 1 \approx 0.84$, the result from part (b) is confirmed.

15. (a) Graphing the curve:

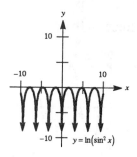

$y = \ln(\sin^2 x)$

(b) Graphing the curve:

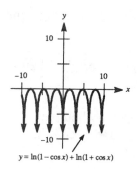

$y = \ln(1 - \cos x) + \ln(1 + \cos x)$

(c) Using properties of logarithms:

$$\ln(1 - \cos x) + \ln(1 + \cos x) = \ln(1 - \cos^2 x) = \ln(\sin^2 x)$$

Since $\ln(1 - \cos x) + \ln(1 + \cos x) = \ln(\sin^2 x)$ using properties of logarithms, the two graphs are identical.

8.6 Simple Harmonic Motion

1. (a) Evaluating $s = 4\cos\dfrac{\pi t}{2}$ at the given values of t:

$t = 0$: $s = 4\cos 0 = 4$ cm

$t = 0.5$: $s = 4\cos\frac{\pi}{4} = 2\sqrt{2}$ cm ≈ 2.83 cm

$t = 1$: $s = 4\cos\frac{\pi}{2} = 0$ cm

$t = 2$: $s = 4\cos\pi = -4$ cm

(b)   The amplitude is 4 cm, the period is $\frac{2\pi}{\pi/2} = 4$ sec, and the frequency is $\frac{1}{4}$ cycles/sec. Sketch the graph over the interval $0 \le t \le 8$:

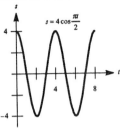

(c)   The mass is farthest from the origin at high and low points, which occur at $t = 0$, $t = 2$, $t = 4$, $t = 6$ and $t = 8$ sec.

(d)   The mass is passing through the origin when $s = 0$, which occurs at $t = 1$, $t = 3$, $t = 5$ and $t = 7$ sec.

(e)   The mass is moving to the right when the s-coordinate is increasing, which occurs during the intervals $2 < t < 4$ and $6 < t < 8$.

3.   (a)   The amplitude is 3 feet, the period is $\frac{2\pi}{\pi/3} = 6$ sec, and the frequency is $\frac{1}{6}$ cycles/sec. Sketch the graph over the interval $0 \le t \le 12$:

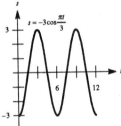

(b)   The mass is moving upward when the s-coordinate is increasing, which occurs during the intervals $0 < t < 3$ and $6 < t < 9$.

(c)   The mass is moving downward when the s-coordinate is decreasing, which occurs during the intervals $3 < t < 6$ and $9 < t < 12$.

(d)   Graph the velocity function for $0 \le t \le 12$, noting its period is $\frac{2\pi}{\pi/3} = 6$ sec and its amplitude is π feet/sec.

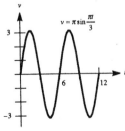

(e) Note that $v = 0$ when $t = 0$, $t = 3$, $t = 6$, $t = 9$, and $t = 12$ sec. At these times the mass (s-coordinate) is at -3, 3, -3, 3 and -3 feet, respectively.

(f) The velocity is a maximum at the high points of this graph, which occur at $t = 1.5$ and $t = 7.5$ sec. At these times the mass is at 0 feet.

(g) The velocity is a minimum at the low points of this graph, which occur at $t = 4.5$ and $t = 10.5$ sec. At these times the mass is at 0 feet.

(h) Graph the velocity function and position function on the same set of axes for $0 \le t \le 12$:

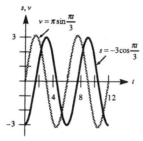

5. (a) The amplitude is 170 volts and the period is $\frac{2\pi}{120\pi} = \frac{1}{60}$ sec, so the frequency is 60 cycles/sec.

(b) Graph the voltage for $0 \le t \le \frac{1}{30}$ sec:

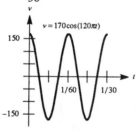

(c) The voltage is a maximum at the high points of this graph, which occur at $t = 0$, $t = \frac{1}{60}$ and $t = \frac{1}{30}$ sec.

7. (a) Complete the table:

t(sec)	0	1	2	3	4	5	6	7
θ(radians)	0	$\frac{\pi}{3}$	$\frac{2\pi}{3}$	π	$\frac{4\pi}{3}$	$\frac{5\pi}{3}$	2π	$\frac{7\pi}{3}$

(b) Since the x-coordinate of the point Q is $\cos \theta$, the corresponding x-coordinates are $\cos 0 = 1$, $\cos\frac{\pi}{3} = \frac{1}{2}$, $\cos\frac{2\pi}{3} = -\frac{1}{2}$, $\cos \pi = -1$, $\cos\frac{4\pi}{3} = -\frac{1}{2}$, $\cos\frac{5\pi}{3} = \frac{1}{2}$, $\cos 2\pi = 1$, and $\cos\frac{7\pi}{3} = \frac{1}{2}$.

(c)   Sketch P and Q when $t = 1$ second, thus $\theta = \frac{\pi}{3}$ radians:

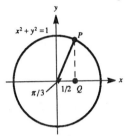

(d)   Sketch P and Q when $t = 2$ seconds, thus $\theta = \frac{2\pi}{3}$ radians:

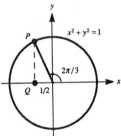

Sketch P and Q when $t = 3$ seconds, thus $\theta = \pi$ radians. Note that P and Q coincide at the same point, since P lies on the x-axis:

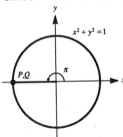

Sketch P and Q when $t = 4$ seconds, thus $\theta = \frac{4\pi}{3}$ radians:

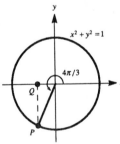

(e)   The x-coordinate of Q is the same as the x-coordinate of P. But P is a point on the unit circle and θ is the radian measure to that point, thus the x-coordinate is $\cos\theta$.

(f) The period of this function is $\frac{2\pi}{\pi/3} = 6$, the amplitude is 1, and the frequency is $\frac{1}{6}$. Graph the function for two complete cycles, or $0 \le t \le 12$:

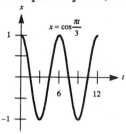

(g) The period is $\frac{2\pi}{\pi/3} = 6$ and the amplitude is $\frac{\pi}{3}$. Graph the velocity for two complete cycles, or $0 \le t \le 12$:

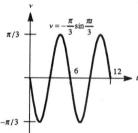

(h) The velocity is 0 when $t = 0, 3, 6, 9$ and 12 sec. The corresponding x-coordinates are 1, –1, 1, –1 and 1, respectively.

(i) The velocity is a maximum at the high points of the curve, which occur at $t = 4.5$ and $t = 10.5$ sec. The x-coordinate of Q is 0 at each of these times, so Q is located at the origin.

(j) The velocity is a minimum at the low points of the curve, which occur at $t = 1.5$ and $t = 7.5$ sec. The x-coordinate of Q is 0 at each of these times, so Q is located at the origin.

8.7 Graphs of the Tangent and the Reciprocal Functions

1. (a) The x-intercept is $-\frac{\pi}{4}$, the y-intercept is 1, and the asymptotes are $x = -\frac{3\pi}{4}$ and $x = \frac{\pi}{4}$. Notice that this graph is $y = \tan x$ displaced $\frac{\pi}{4}$ units to the left.

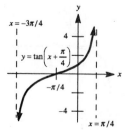

(b) The x-intercept is $-\frac{\pi}{4}$, the y-intercept is -1, and the asymptotes are $x = -\frac{3\pi}{4}$ and $x = \frac{\pi}{4}$. Notice that this graph is $y = \tan\left(x + \frac{\pi}{4}\right)$ reflected across the x-axis.

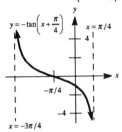

3. (a) The x- and y-intercepts are both 0, and the asymptotes are $x = -\frac{3\pi}{2}$ and $x = \frac{3\pi}{2}$. Notice that the period is $\frac{\pi}{1/3} = 3\pi$.

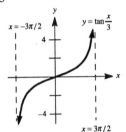

(b) The x- and y-intercepts are both 0, and the asymptotes are $x = -\frac{3\pi}{2}$ and $x = \frac{3\pi}{2}$. Notice that this graph is $y = \tan\frac{x}{3}$ reflected across the x-axis.

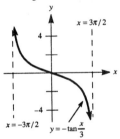

5. The x- and y-intercepts are both 0, and the asymptotes are $x = -1$ and $x = 1$. Notice that the period is $\frac{\pi}{\pi/2} = 2$.

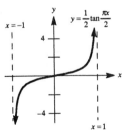

7.   The x-intercept is 1, there is no y-intercept, and the asymptotes are $x = 0$ and $x = 2$. Notice that the period is $\frac{\pi}{\pi/2} = 2$.

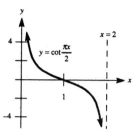

9.   The x-intercept is $\frac{3\pi}{4}$, the y-intercept is 1, and the asymptotes are $x = \frac{\pi}{4}$ and $x = \frac{5\pi}{4}$. Notice that this graph is $y = \cot x$ displaced $\frac{\pi}{4}$ units to the right, then reflected across the x-axis.

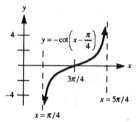

11.   The x-intercept is $\frac{\pi}{4}$, there is no y-intercept, and the asymptotes are $x = 0$ and $x = \frac{\pi}{2}$. Notice that the period is $\frac{\pi}{2}$.

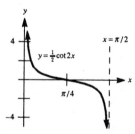

13.   There is no x-intercept, the y-intercept is $-\sqrt{2}$, and the asymptotes are $x = -\frac{3\pi}{4}$, $x = \frac{\pi}{4}$, and $x = \frac{5\pi}{4}$. Notice that this graph is $y = \csc x$ displaced $\frac{\pi}{4}$ units to the right.

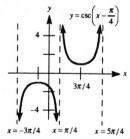

15. There are no x- or y-intercepts, and the asymptotes are $x = -2\pi$, $x = 0$ and $x = 2\pi$. Notice that the period is $\frac{2\pi}{1/2} = 4\pi$, and that this graph is $y = \csc\frac{x}{2}$ reflected across the x-axis.

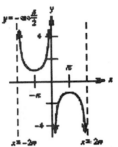

17. There are no x- or y-intercepts, and the asymptotes are $x = -1$, $x = 0$, and $x = 1$. Notice that the period is $\frac{2\pi}{\pi} = 2$.

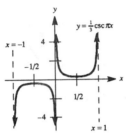

19. There is no x-intercept, the y-intercept is -1, and the asymptotes are $x = -\frac{\pi}{2}$, $x = \frac{\pi}{2}$, and $x = \frac{3\pi}{2}$. Notice that this graph is $y = \sec x$ reflected across the x-axis.

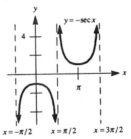

21. There is no x-intercept, the y-intercept is -1, and the asymptotes are $x = \frac{\pi}{2}$, $x = \frac{3\pi}{2}$, and $x = \frac{5\pi}{2}$. Notice that this graph is $y = \sec x$ displaced π units to the right.

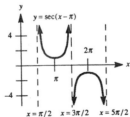

23. There is no x-intercept, the y-intercept is 3, and the asymptotes are $x = -1$, $x = 1$, and $x = 3$. Notice that the period is $\frac{2\pi}{\pi/2} = 4$.

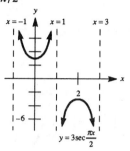

25. **(a)** The x-intercepts are $-\frac{11}{18}$, $-\frac{5}{18}$, $\frac{1}{18}$, $\frac{7}{18}$ and $\frac{13}{18}$, and the y-intercept is -1, and there are no asymptotes. Notice that the period is $\frac{2\pi}{3\pi} = \frac{2}{3}$ and the phase shift is $\frac{\pi/6}{3\pi} = \frac{1}{18}$.

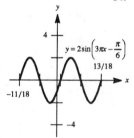

(b) There are no x-intercepts, the y-intercept is -4, and the asymptotes are $x = -\frac{11}{18}$, $x = -\frac{5}{18}$, $x = \frac{1}{18}$, $x = \frac{7}{18}$ and $x = \frac{13}{18}$. Notice that the period is $\frac{2\pi}{3\pi} = \frac{2}{3}$ and the phase shift is $\frac{\pi/6}{3\pi} = \frac{1}{18}$.

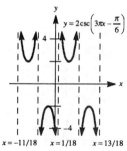

27.   (a)   The x-intercepts are $-\frac{5}{8}, -\frac{1}{8}, \frac{3}{8}$, and $\frac{7}{8}$, and the y-intercept is $-\frac{3\sqrt{2}}{2} \approx -2.12$, and there are no asymptotes. Notice that the period is $\frac{2\pi}{2\pi} = 1$, the phase shift is $\frac{\pi/4}{2\pi} = \frac{1}{8}$, and that this graph is $y = 3\cos\left(2\pi x - \frac{\pi}{4}\right)$ reflected across the x-axis.

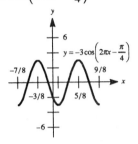

(b)   There are no x-intercepts, the y-intercept is $-3\sqrt{2}$, and the asymptotes are $x = -\frac{5}{8}$, $x = -\frac{1}{8}$, $x = \frac{3}{8}$ and $x = \frac{7}{8}$. Notice that the period is $\frac{2\pi}{2\pi} = 1$, the phase shift is $\frac{\pi/4}{2\pi} = \frac{1}{8}$, and that this graph is $y = 3\sec\left(2\pi x - \frac{\pi}{4}\right)$ reflected across the x-axis.

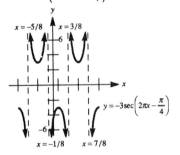

29.   (a)   First find the composition:
$$(f \circ h)(x) = f\left(\pi x - \frac{\pi}{6}\right) = \sin\left(\pi x - \frac{\pi}{6}\right)$$
The period is $\frac{2\pi}{\pi} = 2$, the amplitude is 1, and the phase shift is $\frac{\pi/6}{\pi} = \frac{1}{6}$.

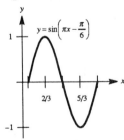

(b) First find the composition:

$$(g \circ h)(x) = g\left(\pi x - \tfrac{\pi}{6}\right) = \csc\left(\pi x - \tfrac{\pi}{6}\right)$$

The period is $\frac{2\pi}{\pi} = 2$, the phase shift is $\frac{\pi/6}{\pi} = \tfrac{1}{6}$, and the asymptotes are $x = \tfrac{1}{6}$, $x = \tfrac{7}{6}$ and $x = \tfrac{13}{6}$.

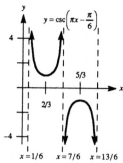

31. (a) First find the composition:

$$(f \circ H)(x) = f\left(\pi x + \tfrac{\pi}{4}\right) = \sin\left(\pi x + \tfrac{\pi}{4}\right)$$

The period is $\frac{2\pi}{\pi} = 2$, the amplitude is 1, and the phase shift is $\frac{-\pi/4}{\pi} = -\tfrac{1}{4}$.

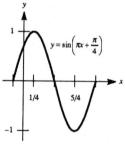

(b) First find the composition:

$$(g \circ H)(x) = g\left(\pi x + \tfrac{\pi}{4}\right) = \csc\left(\pi x + \tfrac{\pi}{4}\right)$$

The period is $\frac{2\pi}{\pi} = 2$, the phase shift is $\frac{-\pi/4}{\pi} = -\tfrac{1}{4}$, and the asymptotes are $x = -\tfrac{1}{4}$, $x = \tfrac{3}{4}$ and $x = \tfrac{7}{4}$.

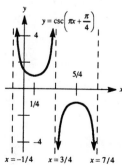

33.   First find the composition:
$$(A \circ T)(x) = A(\tan x) = |\tan x|$$
Now draw the graph, noting that values of x where $\tan x < 0$ will be reflected across the x-axis.

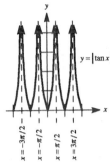

35.   First find the composition:
$$(A \circ f)(x) = A(\csc x) = |\csc x|$$
Now draw the graph, noting that values of x where $\csc x < 0$ will be reflected across the x-axis.

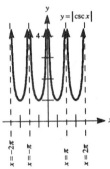

37.   (a)   Since P and Q are both points on the unit circle, the coordinates are $P(\cos s, \sin s)$ and $Q\left(\cos\left(s - \frac{\pi}{2}\right), \sin\left(s - \frac{\pi}{2}\right)\right)$.

   (b)   Since $\triangle OAP$ is congruent to $\triangle OBQ$ (labeling the third vertex B), $OA = OB$ and $AP = BQ$. Because the y-coordinate at Q is negative, we have concluded what was required.

   (c)   Restating part (b), we have shown that:
$$\cos\left(s - \tfrac{\pi}{2}\right) = \sin s \quad \text{and} \quad \sin\left(s - \tfrac{\pi}{2}\right) = -\cos s$$

   (d)   Compute $\cot s$:
$$\cot s = \frac{\cos s}{\sin s} = \frac{-\sin\left(s - \frac{\pi}{2}\right)}{\cos\left(s - \frac{\pi}{2}\right)} = -\tan\left(s - \tfrac{\pi}{2}\right)$$

Graphing Utility Exercises for Section 8.7

1. Graphing the function:

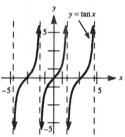

The x-intercepts are $-\pi$, 0, and π.

3. The two graphs are identical, verifying the identity $\cot x = -\tan\left(x - \frac{\pi}{2}\right)$.

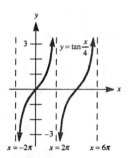

5. (a) Graphing the function:

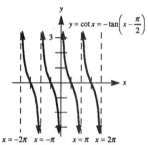

(b) Graphing the function:

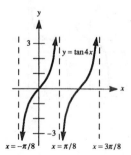

7.   (a)   Graphing the function:

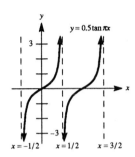

(b)   Graphing the function:

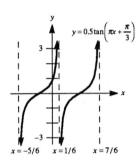

(c)   Graphing the function:

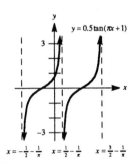

9.   (a)   Graphing the function:

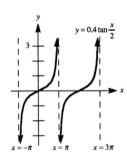

(b)   Graphing the function:

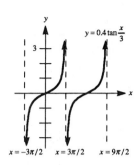

(c)   Graphing the function:

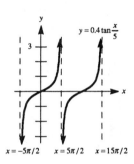

11.   It appears that $\sin x = \csc x$ at odd multiples of $\frac{\pi}{2}$, such as $-\frac{3\pi}{2}$, $-\frac{\pi}{2}$, $\frac{\pi}{2}$, and $\frac{3\pi}{2}$. Since $\sin x$ and $\csc x$ have the same sign, there are no points in which $\sin x = -\csc x$. Using the standard viewing rectangle, the graphs appear as:

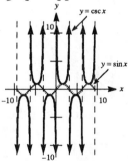

13.   Graphing the two functions:

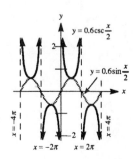

15.   Graphing the two functions:

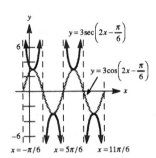

17.   The two graphs are identical. This demonstrates that $\tan^2 x = \sec^2 x - 1$ is a trigonometric identity. Graphing the two functions:

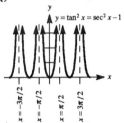

19.   (a)   Using the standard viewing rectangle, the graphs appear as:

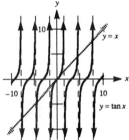

Note that the two curves are close to each other when $x \approx 0$.

(b)   Graphing the two functions using the indicated settings:

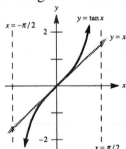

Again note that $\tan x \approx x$ when $x \approx 0$. Completing the table:

x	0.000123	0.01	0.05	0.1
$\tan x$	0.000123	0.010000	0.050042	0.100335
x	0.2	0.3	0.4	0.5
$\tan x$	0.202710	0.309336	0.422793	0.546302

(c) Graphing the three functions, note that $y = x + \frac{1}{3}x^3$ is an even better approximation to $y = \tan x$:

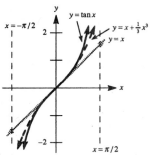

(d) Complete the table:

x	0.000123	0.01	0.05	0.1
$\tan x$	0.000123	0.010000	0.050042	0.100335
$x + \frac{x^3}{3}$	0.000123	0.010000	0.050042	0.100333
x	0.2	0.3	0.4	0.5
$\tan x$	0.202710	0.309336	0.422793	0.546302
$x + \frac{x^3}{3}$	0.202667	0.309	0.421333	0.541667

Note that the values of $x + \frac{x^3}{3}$ are much closer to $\tan x$ than are those for x.

21. (a) Graphing the two functions:

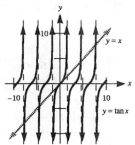

The two graphs appear to intersect at approximately $x \approx 4.5$.

(b) More accurately, the root is $x \approx 4.4934$.

(c) No, since $\tan(r + \pi) = \tan r = r$, the value does not equal $r + \pi$.

Chapter Eight Review Exercises

1. (a) Since the terminal side of $\frac{5\pi}{3}$ lies in the fourth quadrant, $\sin\frac{5\pi}{3} < 0$. Using a reference angle of $\frac{\pi}{3}$:

$$\sin\frac{5\pi}{3} = -\sin\frac{\pi}{3} = -\frac{\sqrt{3}}{2}$$

   (b) Since the terminal side of $\frac{11\pi}{6}$ lies in the fourth quadrant, $\cot\frac{11\pi}{6} < 0$. Using a reference angle of $\frac{\pi}{6}$:

$$\cot\frac{11\pi}{6} = -\cot\frac{\pi}{6} = -\frac{\cos\frac{\pi}{6}}{\sin\frac{\pi}{6}} = -\frac{\frac{\sqrt{3}}{2}}{\frac{1}{2}} = -\sqrt{3}$$

3. Using the identities $\cos(-t) = \cos t$ and $\sin(-t) = -\sin t$:
$$\cos t - \cos(-t) + \sin t - \sin(-t) = \cos t - \cos t + \sin t - (-\sin t) = 2\sin t$$

5. The period is $\frac{2\pi}{2\pi} = 1$ and the phase shift is $\frac{3}{2\pi}$. The asymptotes will occur at $x = -\frac{1}{4} + \frac{3}{2\pi}$, $x = \frac{1}{4} + \frac{3}{2\pi}$ and $x = \frac{3}{4} + \frac{3}{2\pi}$.

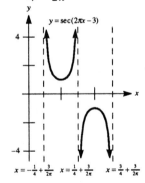

7. The amplitude is 1, the period is $\frac{2\pi}{2} = \pi$, and the phase shift is $\frac{\pi}{2}$. Notice that this graph is $y = \sin(2x - \pi)$ reflected across the x-axis.

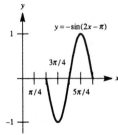

9. Since the terminal side of π radians intersects the unit circle at the point $(-1, 0)$, $\cos \pi = -1$.

11. Since the terminal side of $\frac{2\pi}{3}$ radians lies in the second quadrant, $\csc \frac{2\pi}{3} > 0$. Using a reference angle of $\frac{\pi}{3}$ radians:

$$\csc \frac{2\pi}{3} = \csc \frac{\pi}{3} = \frac{1}{\sin \frac{\pi}{3}} = \frac{2}{\sqrt{3}} = \frac{2\sqrt{3}}{3}$$

13. Since the terminal side of $\frac{11\pi}{6}$ radians lies in the fourth quadrant, $\tan \frac{11\pi}{6} < 0$. Using a reference angle of $\frac{\pi}{6}$ radians:

$$\tan \frac{11\pi}{6} = -\tan \frac{\pi}{6} = -\frac{\sin \frac{\pi}{6}}{\cos \frac{\pi}{6}} = -\frac{\frac{1}{2}}{\frac{\sqrt{3}}{2}} = -\frac{1}{\sqrt{3}} = -\frac{\sqrt{3}}{3}$$

15. Since the terminal side of $\frac{\pi}{6}$ radians lies in the first quadrant, $\sin \frac{\pi}{6} > 0$. Thus $\sin \frac{\pi}{6} = \frac{1}{2}$.

17. Since the terminal side of $\frac{5\pi}{4}$ radians lies in the third quadrant, $\cot \frac{5\pi}{4} > 0$. Using a reference angle of $\frac{\pi}{4}$ radians:

$$\cot \frac{5\pi}{4} = \cot \frac{\pi}{4} = \frac{\cos \frac{\pi}{4}}{\sin \frac{\pi}{4}} = \frac{\frac{\sqrt{2}}{2}}{\frac{\sqrt{2}}{2}} = 1$$

19. Since the terminal side of $-\frac{5\pi}{6}$ radians lies in the third quadrant, $\csc\left(-\frac{5\pi}{6}\right) < 0$. Using a reference angle of $\frac{\pi}{6}$ radians:

$$\csc\left(-\frac{5\pi}{6}\right) = -\csc \frac{\pi}{6} = -\frac{1}{\sin \frac{\pi}{6}} = -\frac{1}{\frac{1}{2}} = -2$$

21. Using a calculator, we find $\sin 1 \approx 0.841$.

23. Since $\frac{3\pi}{2}$ radians intersects the unit circle at the point $(0, -1)$, $\sin \frac{3\pi}{2} = -1$. A calculator also verifies this value.

25. Using a calculator, we find $\sin (\sin 0.0123) \approx 0.0123$.

27. Since $\sin 1776 \approx -0.842$ and $\cos 1776 \approx -0.540$:
$$\sin^2 1776 + \cos^2 1776 \approx (-0.842)^2 + (-0.540)^2 = 1$$

29. Since $\cos (0.25) \approx 0.969$ and $\sin (0.25) \approx 0.247$:
$$\cos^2 (0.25) - \sin^2 (0.25) \approx (0.969)^2 - (0.247)^2 \approx 0.878$$
Since $\cos (0.5) \approx 0.878$, the equality is verified.

31. Since $0 < \theta < \frac{\pi}{2}$, $\cos \theta > 0$ and thus:
$$\sqrt{25 - x^2} = \sqrt{25 - 25\sin^2 \theta} = 5\sqrt{1 - \sin^2 \theta} = 5\sqrt{\cos^2 \theta} = 5\cos \theta$$

33. Since $0 < \theta < \frac{\pi}{2}$, $\tan \theta > 0$ and thus:
$$\left(x^2 - 100\right)^{1/2} = \left(100\sec^2 \theta - 100\right)^{1/2} = 10\left(\sec^2 \theta - 1\right)^{1/2} = 10\left(\tan^2 \theta\right)^{1/2} = 10\tan \theta$$

35. Since $0 < \theta < \frac{\pi}{2}$, $\sec \theta > 0$ and thus:
$$\left(x^2 + 5\right)^{-1/2} = \left(5\tan^2 \theta + 5\right)^{-1/2} = \frac{\sqrt{5}}{5}\left(\sec^2 \theta\right)^{-1/2} = \frac{\sqrt{5}}{5}\cos \theta$$

37. Since $\sin \theta$ is negative:
$$\sin \theta = -\sqrt{1 - \cos^2 \theta} = -\sqrt{1 - \left(\tfrac{8}{17}\right)^2} = -\sqrt{1 - \tfrac{64}{289}} = -\sqrt{\tfrac{225}{289}} = -\tfrac{15}{17}$$
Thus:
$$\tan \theta = \frac{\sin \theta}{\cos \theta} = \frac{-\frac{15}{17}}{\frac{8}{17}} = -\frac{15}{8}$$

39. Since $\angle BPA = \theta$, $\angle APC = \pi - \theta$, and the area of the sector formed by $\angle APC$ is:
$$\tfrac{1}{2}r^2\theta = \tfrac{1}{2}\left(\sqrt{2}\right)^2(\pi - \theta) = \pi - \theta$$
Now using the area formula for $\triangle APC$:
$$\text{Area}_{\triangle APC} = \tfrac{1}{2}ab\sin(\pi - \theta) = \tfrac{1}{2}\left(\sqrt{2}\right)\left(\sqrt{2}\right)\sin \theta = \sin \theta$$
Thus, the area of the shaded region is given by:
$$A(\theta) = \pi - \theta - \sin \theta$$

41. Using the hint, we have two congruent shaded regions each with $r = 1$ cm and $\theta = \frac{\pi}{2}$, and thus the total area is:
$$2 \cdot \tfrac{1}{2}\left(\tfrac{\pi}{2} - 1\right) \text{cm}^2 = \frac{\pi - 2}{2} \text{ cm}^2$$

43. This is a sine function where the amplitude is 4, so $A = 4$. Since the period is 2π.
$$\frac{2\pi}{B} = 2\pi$$
$$2\pi = 2\pi B$$
$$1 = B$$
So the equation is $y = 4\sin x$.

45. This is a cosine function (reflected across the x-axis) where the amplitude is 2, so $A = -2$. Since the period is $\frac{\pi}{2}$:

$$\frac{2\pi}{B} = \frac{\pi}{2}$$
$$4\pi = \pi B$$
$$4 = B$$

So the equation is $y = -2\cos 4x$.

47. The x-intercepts are $\frac{\pi}{8}$ and $\frac{3\pi}{8}$, the high point is $\left(\frac{\pi}{4}, 3\right)$, and the low points are $(0, -3)$ and $\left(\frac{\pi}{2}, -3\right)$. Notice that the period is $\frac{2\pi}{4} = \frac{\pi}{2}$, and that this graph is $y = 3\cos 4x$ reflected across the x-axis.

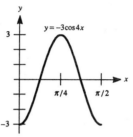

49. The x-intercepts are $\frac{1}{2}$, $\frac{5}{2}$ and $\frac{9}{2}$, the high point is $\left(\frac{3}{2}, 2\right)$, and the low point is $\left(\frac{7}{2}, -2\right)$. Notice that the period is $\frac{2\pi}{\pi/2} = 4$, and the phase shift is $\frac{\pi/4}{\pi/2} = \frac{1}{2}$.

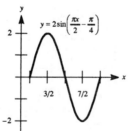

51. The x-intercepts are $\frac{5}{2}$ and $\frac{11}{2}$, the high points are $(1, 3)$ and $(7, 3)$, and the low point is $(4, -3)$. Notice that the period is $\frac{2\pi}{\pi/6} = 6$, and the phase shift is $\frac{\pi/3}{\pi/3} = 1$.

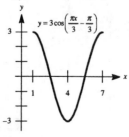

53. (a) Notice that the period is $\frac{\pi}{\pi/4} = 4$, and the asymptotes occur at $x = 2$ and $x = -2$.

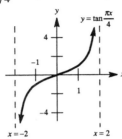

(b) Notice that the period is $\frac{\pi}{\pi/4} = 4$, and the asymptotes occur at $x = 0$ and $x = 4$.

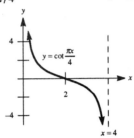

55. (a) Notice that the period is $\frac{2\pi}{1/4} = 8\pi$, and the asymptotes occur at $x = -2\pi$, $x = 2\pi$ and $x = 6\pi$.

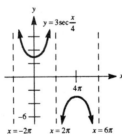

(b) Notice that the period is $\frac{2\pi}{1/4} = 8\pi$, and the asymptotes occur at $x = -4\pi$, $x = 0$ and $x = 4\pi$.

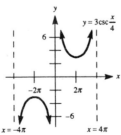

57. (a) The amplitude is 2.5 cm, the period is $\frac{2\pi}{\pi/8} = 16$ sec, and the frequency is $\frac{1}{16}$ cycles/sec. Sketch the graph over the interval $0 \le t \le 32$:

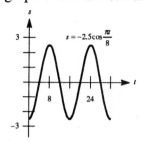

(b) The mass is farthest from the origin at high and low points, which occur at $t = 0$, $t = 8$, $t = 16$, $t = 24$ and $t = 32$ sec.

(c) The mass is passing through the equilibrium position when $s = 0$, which occurs at $t = 4$, $t = 12$, $t = 20$ and $t = 28$ sec.

59. First find the area of each region, noting that $AD = \tan\alpha$ and $AE = \tan\beta$:

$$A_{ABDA} = \tfrac{1}{2}(1)(\tan\alpha) - \tfrac{1}{2}(1)^2\alpha = \tfrac{1}{2}(\tan\alpha - \alpha)$$

$$A_{ABCEDA} = \tfrac{1}{2}(1)(\tan\beta) - \tfrac{1}{2}(1)^2\beta = \tfrac{1}{2}(\tan\beta - \beta)$$

Since $A_{ABDA} < A_{ABCEDA}$, the resulting inequality is:

$$\tfrac{1}{2}(\tan\alpha - \alpha) < \tfrac{1}{2}(\tan\beta - \beta)$$

$$\tan\alpha - \alpha < \tan\beta - \beta$$

$$\tan\alpha - \tan\beta < \alpha - \beta$$

$$\tan\beta - \tan\alpha > \beta - \alpha$$

This proves the desired inequality.

Chapter Eight Test

1. (a) Since the terminal side of $\frac{4\pi}{3}$ radians lies in the third quadrant, $\cos\frac{4\pi}{3} < 0$. Using a reference angle of $\frac{\pi}{3}$ radians:

$$\cos\frac{4\pi}{3} = -\cos\frac{\pi}{3} = -\tfrac{1}{2}$$

(b) Since the terminal side of $-\frac{5\pi}{6}$ radians lies in the third quadrant, $\csc\left(-\frac{5\pi}{6}\right) < 0$. Using a reference angle of $\frac{\pi}{6}$ radians:

$$\csc\left(-\frac{5\pi}{6}\right) = -\csc\frac{\pi}{6} = -\frac{1}{\sin\frac{\pi}{6}} = -\frac{1}{\frac{1}{2}} = -2$$

(c)   Since $\sin^2 t + \cos^2 t = 1$ for all values of t:
$$\sin^2 \tfrac{3\pi}{4} + \cos^2 \tfrac{3\pi}{4} = 1$$

2.   Substituting $t = 4\sin u$ and noting that $\sin u > 0$ when $0 < u < \tfrac{\pi}{2}$:
$$\frac{1}{\sqrt{16 - t^2}} = \frac{1}{\sqrt{16 - 16\sin^2 u}} = \frac{1}{\sqrt{16\left(1 - \sin^2 u\right)}} = \frac{1}{\sqrt{16\cos^2 u}} = \frac{1}{4\cos u} = \tfrac{1}{4}\sec u$$

3.   Noting that the period is $\tfrac{2\pi}{4\pi} = \tfrac{1}{2}$, and the phase shift is $\tfrac{1}{4\pi}$, then the asymptotes are
$x = -\tfrac{1}{8} + \tfrac{1}{4\pi}$, $x = \tfrac{1}{8} + \tfrac{1}{4\pi}$, and $x = \tfrac{3}{8} + \tfrac{1}{4\pi}$.

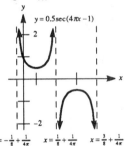

4.   The amplitude is 1, the period is $\tfrac{2\pi}{3}$, and the phase shift is $\tfrac{\pi/4}{3} = \tfrac{\pi}{12}$. Notice that this graph is $y = \sin\left(3x - \tfrac{\pi}{4}\right)$ reflected across the x-axis.

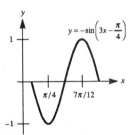

5.   Notice that the period is $\tfrac{\pi}{\pi/4} = 4$, and that the asymptote occurs at $x = 2$.

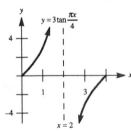

6.   (a)   Multiplying by the conversion factor $\frac{\pi}{180°}$:
$$175° \cdot \frac{\pi}{180°} = \frac{35}{36}\pi \text{ radians}$$

(b)   Multiplying by the conversion factor $\frac{180°}{\pi}$:
$$5 \cdot \frac{180°}{\pi} = \frac{900°}{\pi}$$

7.   (a)   First converting $\theta = \angle BAC$ to radians:
$$\theta = 75° \cdot \frac{\pi}{180°} = \frac{5\pi}{12} \text{ radians}$$
Thus the arc length is given by:
$$s = r\theta = 5\,\text{cm} \cdot \frac{5\pi}{12} = \frac{25\pi}{12}\,\text{cm}$$

(b)   The area of the sector is given by:
$$A = \tfrac{1}{2}r^2\theta = \tfrac{1}{2}(5)^2 \cdot \frac{5\pi}{12} = \frac{125\pi}{24}\,\text{cm}^2$$

8.   (a)   Converting the angular speed to radians:
$$\omega = 25\,\frac{\text{rev}}{\text{sec}} \cdot 2\pi\,\frac{\text{rad}}{\text{rev}} = 50\pi\,\frac{\text{rad}}{\text{sec}}$$

(b)   In 1 second, the wheel has rotated 50π radians, so the total distance traveled by a point 5 cm from the center is:
$$d = (5\text{ cm})(50\pi\text{ rad}) = 250\pi\text{ cm}$$
Thus the linear velocity is given by:
$$v = \frac{d}{t} = \frac{250\pi\,\text{cm}}{1\text{ sec}} = 250\pi\,\frac{\text{cm}}{\text{sec}}$$

9. Simplify each side of the equation, using the identities $\sin(-\theta) = -\sin\theta$ and $\cos(-\theta) = \cos\theta$:

$$\frac{\cot\theta}{1+\tan(-\theta)} + \frac{\tan\theta}{1+\cot(-\theta)} = \frac{\frac{\cos\theta}{\sin\theta}}{1+\frac{\sin(-\theta)}{\cos(-\theta)}} + \frac{\frac{\sin\theta}{\cos\theta}}{1+\frac{\cos(-\theta)}{\sin(-\theta)}}$$

$$= \frac{\frac{\cos\theta}{\sin\theta}}{1-\frac{\sin\theta}{\cos\theta}} + \frac{\frac{\sin\theta}{\cos\theta}}{1-\frac{\cos\theta}{\sin\theta}}$$

$$= \frac{\cos^2\theta}{\sin\theta\cos\theta - \sin^2\theta} + \frac{\sin^2\theta}{\sin\theta\cos\theta - \cos^2\theta}$$

$$= \frac{\cos^2\theta}{\sin\theta(\cos\theta - \sin\theta)} + \frac{\sin^2\theta}{\cos\theta(\sin\theta - \cos\theta)}$$

$$= \frac{-\cos^3\theta}{\sin\theta\cos\theta(\sin\theta - \cos\theta)} + \frac{\sin^3\theta}{\sin\theta\cos\theta(\sin\theta - \cos\theta)}$$

$$= \frac{\sin^3\theta - \cos^3\theta}{\sin\theta\cos\theta(\sin\theta - \cos\theta)}$$

$$= \frac{(\sin\theta - \cos\theta)(\sin^2\theta + \sin\theta\cos\theta + \cos^2\theta)}{\sin\theta\cos\theta(\sin\theta - \cos\theta)}$$

$$= \frac{1+\sin\theta\cos\theta}{\sin\theta\cos\theta}$$

$$\cot\theta + \tan\theta + 1 = \frac{\cos\theta}{\sin\theta} + \frac{\sin\theta}{\cos\theta} + 1 = \frac{\cos^2\theta + \sin^2\theta + \sin\theta\cos\theta}{\sin\theta\cos\theta} = \frac{1+\sin\theta\cos\theta}{\sin\theta\cos\theta}$$

Since both sides of the equation simplify to the same quantity, the equation is an identity.

10. (a) The period is $\frac{2\pi}{\pi/3} = 6$ sec, and the amplitude is 10 cm. Graph the function over the interval $0 \le t \le 12$:

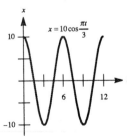

(b) The point is passing through the origin when $x = 0$, which occurs when $t = 1.5$, $t = 4.5$, $t = 7.5$ and $t = 10.5$ sec. The point is farthest from the origin at the high and low points of this graph, which occur when $t = 0$, $t = 3$, $t = 6$, $t = 9$ and $t = 12$ sec.

11. (a) $\sin^2 13 + \cos^2 13 = 1$, since $\sin^2 x + \cos^2 x = 1$ regardless of the value of x.

(b) $\sin 5 + \sin (-5) = \sin 5 - \sin 5 = 0$

(c) $\tan 1 + \tan (-1 - 2\pi) = \tan 1 + \tan (-1) = \tan 1 - \tan 1 = 0$

12. (a) Using the graph, a root is $x \approx 1.1$.

(b) Using a calculator, a root is $x \approx 1.1198$.

(c) Another root would be $\pi - x \approx 2.0218$.

Chapter Nine
Analytical Trigonometry

9.1 The Addition Formulas

1. Using the identity $\sin(s+t) = \sin s \cos t + \cos s \sin t$ where $s = \theta$ and $t = 2\theta$:
$$\sin\theta\cos 2\theta + \cos\theta\sin 2\theta = \sin(\theta + 2\theta) = \sin 3\theta$$

3. Using the identity $\sin(s-t) = \sin s \cos t - \cos s \sin t$ where $s = 3\theta$ and $t = \theta$:
$$\sin 3\theta\cos\theta - \cos 3\theta\sin\theta = \sin(3\theta - \theta) = \sin 2\theta$$

5. Using the identity $\cos(s+t) = \cos s \cos t - \sin s \sin t$ where $s = 2u$ and $t = 3u$:
$$\cos 2u\cos 3u - \sin 2u\sin 3u = \cos(2u + 3u) = \cos 5u$$

7. Using the identity $\cos(s-t) = \cos s \cos t + \sin s \sin t$ where $s = \frac{2\pi}{9}$ and $t = \frac{\pi}{18}$:
$$\cos\frac{2\pi}{9}\cos\frac{\pi}{18} + \sin\frac{2\pi}{9}\sin\frac{\pi}{18} = \cos\left(\frac{2\pi}{9} - \frac{\pi}{18}\right) = \cos\frac{3\pi}{18} = \cos\frac{\pi}{6} = \frac{\sqrt{3}}{2}$$

9. Using the identity $\sin(s-t) = \sin s \cos t - \cos s \sin t$ where $s = A + B$ and $t = A$:
$$\sin(A+B)\cos A - \cos(A+B)\sin A = \sin(A + B - A) = \sin B$$

11. Using the identity $\sin(s-t) = \sin s \cos t - \cos s \sin t$ with $s = \theta$ and $t = \frac{3\pi}{2}$:
$$\sin\left(\theta - \frac{3\pi}{2}\right) = \sin\theta\cos\frac{3\pi}{2} - \cos\theta\sin\frac{3\pi}{2} = \sin\theta \cdot 0 - \cos\theta \cdot (-1) = \cos\theta$$

13. Using the identity $\cos (s + t) = \cos s \cos t - \sin s \sin t$ with $s = \theta$ and $t = \pi$:

$$\cos(\theta + \pi) = \cos \theta \cos \pi - \sin \theta \sin \pi = \cos \theta \cdot (-1) - \sin \theta \cdot 0 = -\cos \theta$$

15. Using the identity $\sin (s + t) = \sin s \cos t + \cos s \sin t$ with $s = t$ and $t = 2\pi$:

$$\sin(t + 2\pi) = \sin t \cos 2\pi + \cos t \sin 2\pi = \sin t \cdot 1 + \cos t \cdot 0 = \sin t$$

Notice that this verifies the formula $\sin (t + 2\pi) = \sin t$.

17. Using the identity $\cos (s + t) = \cos s \cos t - \sin s \sin t$:

$$\begin{aligned}
\cos 75° &= \cos(45° + 30°) \\
&= \cos 45° \cos 30° - \sin 45° \sin 30° \\
&= \frac{\sqrt{2}}{2} \cdot \frac{\sqrt{3}}{2} - \frac{\sqrt{2}}{2} \cdot \frac{1}{2} \\
&= \frac{\sqrt{6} - \sqrt{2}}{4}
\end{aligned}$$

19. Using the identity $\sin (s + t) = \sin s \cos t + \cos s \sin t$:

$$\sin \tfrac{7\pi}{12} = \sin\left(\tfrac{\pi}{3} + \tfrac{\pi}{4}\right) = \sin \tfrac{\pi}{3} \cos \tfrac{\pi}{4} + \sin \tfrac{\pi}{4} \cos \tfrac{\pi}{3} = \frac{\sqrt{3}}{2} \cdot \frac{\sqrt{2}}{2} + \frac{\sqrt{2}}{2} \cdot \frac{1}{2} = \frac{\sqrt{6} + \sqrt{2}}{4}$$

21. Using the identities $\sin (s + t) = \sin s \cos t + \cos s \sin t$ and
$\sin (s - t) = \sin s \cos t - \cos s \sin t$:

$$\begin{aligned}
\sin\left(\tfrac{\pi}{4} + s\right) - \sin\left(\tfrac{\pi}{4} - s\right) &= \left(\sin \tfrac{\pi}{4} \cos s + \cos \tfrac{\pi}{4} \sin s\right) - \left(\sin \tfrac{\pi}{4} \cos s - \cos \tfrac{\pi}{4} \sin s\right) \\
&= \frac{\sqrt{2}}{2} \cos s + \frac{\sqrt{2}}{2} \sin s - \frac{\sqrt{2}}{2} \cos s + \frac{\sqrt{2}}{2} \sin s \\
&= \sqrt{2} \sin s
\end{aligned}$$

23. Using the identities $\cos (s + t) = \cos s \cos t - \sin s \sin t$ and
$\cos (s - t) = \cos s \cos t + \sin s \sin t$:

$$\begin{aligned}
\cos\left(\tfrac{\pi}{3} - \theta\right) - \cos\left(\tfrac{\pi}{3} + \theta\right) &= \left(\cos \tfrac{\pi}{3} \cos \theta + \sin \tfrac{\pi}{3} \sin \theta\right) - \left(\cos \tfrac{\pi}{3} \cos \theta - \sin \tfrac{\pi}{3} \sin \theta\right) \\
&= \tfrac{1}{2} \cos \theta + \frac{\sqrt{3}}{2} \sin \theta - \tfrac{1}{2} \cos \theta + \frac{\sqrt{3}}{2} \sin \theta \\
&= \sqrt{3} \sin \theta
\end{aligned}$$

25. First compute $\cos \alpha$ and $\sin \beta$. Since $\frac{\pi}{2} < \alpha < \pi$, $\cos \alpha < 0$ and thus:

$$\cos \alpha = -\sqrt{1 - \sin^2 \alpha} = -\sqrt{1 - \left(\tfrac{12}{13}\right)^2} = -\sqrt{1 - \tfrac{144}{169}} = -\sqrt{\tfrac{25}{169}} = -\tfrac{5}{13}$$

Since $\pi < \beta < \frac{3\pi}{2}$, $\sin \beta < 0$ and thus:

$$\sin \beta = -\sqrt{1 - \cos^2 \beta} = -\sqrt{1 - \left(-\tfrac{3}{5}\right)^2} = -\sqrt{1 - \tfrac{9}{25}} = -\sqrt{\tfrac{16}{25}} = -\tfrac{4}{5}$$

(a) Using the identity for sin $(\alpha + \beta)$:

$$\sin(\alpha + \beta) = \sin\alpha\cos\beta + \cos\alpha\sin\beta$$
$$= \left(\tfrac{12}{13}\right)\cdot\left(-\tfrac{3}{5}\right) + \left(-\tfrac{5}{13}\right)\cdot\left(-\tfrac{4}{5}\right)$$
$$= -\tfrac{36}{65} + \tfrac{20}{65}$$
$$= -\tfrac{16}{65}$$

(b) Using the identity for cos $(\alpha + \beta)$:

$$\cos(\alpha + \beta) = \cos\alpha\cos\beta - \sin\alpha\sin\beta$$
$$= \left(-\tfrac{5}{13}\right)\cdot\left(-\tfrac{3}{5}\right) - \left(\tfrac{12}{13}\right)\cdot\left(-\tfrac{4}{5}\right)$$
$$= \tfrac{15}{65} + \tfrac{48}{65}$$
$$= \tfrac{63}{65}$$

27. Use the value for cos α computed in Exercise 26. Since $-2\pi < \theta < -\tfrac{3\pi}{2}$, $\sin\theta > 0$ and thus:

$$\sin\theta = \sqrt{1 - \cos^2\theta} = \sqrt{1 - \left(\tfrac{7}{25}\right)^2} = \sqrt{1 - \tfrac{49}{625}} = \sqrt{\tfrac{576}{625}} = \tfrac{24}{25}$$

(a) Using the identity for sin $(\theta - \beta)$:

$$\sin(\theta - \beta) = \sin\theta\cos\beta - \cos\theta\sin\beta$$
$$= \left(\tfrac{24}{25}\right)\cdot\left(-\tfrac{3}{5}\right) - \left(\tfrac{7}{25}\right)\cdot\left(-\tfrac{4}{5}\right)$$
$$= -\tfrac{72}{125} + \tfrac{28}{125}$$
$$= -\tfrac{44}{125}$$

(b) Using the identity for sin $(\theta + \beta)$:

$$\sin(\theta + \beta) = \sin\theta\cos\beta + \cos\theta\sin\beta$$
$$= \left(\tfrac{24}{25}\right)\cdot\left(-\tfrac{3}{5}\right) + \left(\tfrac{7}{25}\right)\cdot\left(-\tfrac{4}{5}\right)$$
$$= -\tfrac{72}{125} - \tfrac{28}{125}$$
$$= -\tfrac{100}{125}$$
$$= -\tfrac{4}{5}$$

29. (a) Since $0 < \theta < \tfrac{\pi}{2}$, $\cos\theta > 0$ and thus:

$$\cos\theta = \sqrt{1 - \sin^2\theta} = \sqrt{1 - \left(\tfrac{1}{5}\right)^2} = \sqrt{1 - \tfrac{1}{25}} = \sqrt{\tfrac{24}{25}} = \frac{2\sqrt{6}}{5}$$

(b) Since $\sin 2\theta = \sin(\theta + \theta)$, use the addition formula for sin $(s + t)$ to obtain:

$$\sin 2\theta = \sin\theta\cos\theta + \sin\theta\cos\theta = 2\sin\theta\cos\theta = 2\cdot\frac{1}{5}\cdot\frac{2\sqrt{6}}{5} = \frac{4\sqrt{6}}{25}$$

31. First find $\sin\theta$, $\cos\theta$, $\sin\beta$ and $\cos\beta$. Since $\frac{\pi}{2} < \theta < \pi$, $\sec\theta < 0$, so using the identity $\sec^2\theta = 1 + \tan^2\theta$:

$$\sec\theta = -\sqrt{1 + \tan^2\theta} = -\sqrt{1 + \left(-\tfrac{2}{3}\right)^2} = -\sqrt{1 + \tfrac{4}{9}} = -\sqrt{\tfrac{13}{3}} = -\frac{\sqrt{13}}{3}$$

Thus $\cos\theta = -\frac{3}{\sqrt{13}} = -\frac{3\sqrt{13}}{13}$. Since $\frac{\pi}{2} < \theta < \pi$, $\sin\theta > 0$ and thus:

$$\sin\theta = \sqrt{1 - \cos^2\theta} = \sqrt{1 - \left(-\frac{3\sqrt{13}}{13}\right)^2} = \sqrt{1 - \tfrac{9}{13}} = \sqrt{\tfrac{4}{13}} = \frac{2}{\sqrt{13}} = \frac{2\sqrt{13}}{13}$$

Since $\csc\beta = 2$, $\sin\beta = \frac{1}{2}$. Since $0 < \beta < \frac{\pi}{2}$, $\cos\beta > 0$ and thus:

$$\cos\beta = \sqrt{1 - \sin^2\beta} = \sqrt{1 - \left(\tfrac{1}{2}\right)^2} = \sqrt{1 - \tfrac{1}{4}} = \sqrt{\tfrac{3}{4}} = \frac{\sqrt{3}}{2}$$

Now using the addition formula for $\sin(s + t)$:

$$\sin(\theta + \beta) = \sin\theta\cos\beta + \cos\theta\sin\beta$$
$$= \frac{2\sqrt{13}}{13} \cdot \frac{\sqrt{3}}{2} - \frac{3\sqrt{13}}{13} \cdot \frac{1}{2}$$
$$= \frac{2\sqrt{39}}{26} - \frac{3\sqrt{13}}{26}$$
$$= \frac{2\sqrt{39} - 3\sqrt{13}}{26}$$

Finally using the addition formula for $\cos(s - t)$:

$$\cos(\beta - \theta) = \cos\beta\cos\theta + \sin\beta\sin\theta$$
$$= \frac{\sqrt{3}}{2} \cdot \left(-\frac{3\sqrt{13}}{13}\right) + \frac{1}{2} \cdot \frac{2\sqrt{13}}{13}$$
$$= -\frac{3\sqrt{39}}{26} + \frac{2\sqrt{13}}{26}$$
$$= \frac{2\sqrt{13} - 3\sqrt{39}}{26}$$

33. Using the addition formula for $\sin(s + t)$:

$$\sin\left(t + \tfrac{\pi}{4}\right) = \sin t\cos\tfrac{\pi}{4} + \cos t\sin\tfrac{\pi}{4} = \tfrac{1}{\sqrt{2}}\sin t + \tfrac{1}{\sqrt{2}}\cos t = \frac{\sin t + \cos t}{\sqrt{2}}$$

35. Using the results from Exercises 33 and 34:

$$\sin\left(t + \tfrac{\pi}{4}\right) + \cos\left(t + \tfrac{\pi}{4}\right) = \frac{\sin t + \cos t}{\sqrt{2}} + \frac{\cos t - \sin t}{\sqrt{2}} = \frac{2\cos t}{\sqrt{2}} = \sqrt{2}\cos t$$

37. Using the addition formulas for tan $(s + t)$ and tan $(s - t)$:

$$\tan(s+t) = \frac{\tan s + \tan t}{1 - \tan s \tan t} = \frac{2+3}{1-2\cdot 3} = \frac{5}{-5} = -1$$

$$\tan(s-t) = \frac{\tan s - \tan t}{1 + \tan s \tan t} = \frac{2-3}{1+2\cdot 3} = \frac{-1}{7} = -\frac{1}{7}$$

39. Using the addition formulas for tan $(s + t)$ and tan $(s - t)$:

$$\tan(s+t) = \frac{\tan s + \tan t}{1 - \tan s \tan t} = \frac{\tan \frac{3\pi}{4} + (-4)}{1 - \tan \frac{3\pi}{4} \cdot (-4)} = \frac{-1-4}{1-(-1)(-4)} = \frac{-5}{-3} = \frac{5}{3}$$

$$\tan(s-t) = \frac{\tan s - \tan t}{1 + \tan s \tan t} = \frac{\tan \frac{3\pi}{4} - (-4)}{1 + \tan \frac{3\pi}{4} \cdot (-4)} = \frac{-1+4}{1+(-1)(-4)} = \frac{3}{5}$$

41. Since $\tan(s+t) = \dfrac{\tan s + \tan t}{1 - \tan s \tan t}$, applying this formula with $s = t$ and $t = 2t$:

$$\frac{\tan t + \tan 2t}{1 - \tan t \tan 2t} = \tan(t + 2t) = \tan 3t$$

43. Since $\tan(s-t) = \dfrac{\tan s - \tan t}{1 + \tan s \tan t}$, applying this formula with $s = 70°$ and $t = 10°$:

$$\frac{\tan 70° - \tan 10°}{1 + \tan 70° \tan 10°} = \tan(70° - 10°) = \tan 60° = \sqrt{3}$$

45. Since $\tan(s+t) = \dfrac{\tan s + \tan t}{1 - \tan s \tan t}$, applying this formula with $s = x - y$ and $t = y$:

$$\frac{\tan(x-y) + \tan y}{1 - \tan(x-y)\tan y} = \tan(x - y + y) = \tan x$$

47. Using the addition formula for tan $(s + t)$ with $s = \frac{\pi}{3}$ and $t = \frac{\pi}{4}$:

$$\tan \frac{7\pi}{12} = \frac{\tan \frac{\pi}{3} + \tan \frac{\pi}{4}}{1 - \tan \frac{\pi}{3} \cdot \tan \frac{\pi}{4}} = \frac{\sqrt{3}+1}{1-\sqrt{3}\cdot 1} \cdot \frac{1+\sqrt{3}}{1+\sqrt{3}} = \frac{1+2\sqrt{3}+3}{1-3} = \frac{4+2\sqrt{3}}{-2} = -2 - \sqrt{3}$$

49. Using the addition formula for sin $(s + t)$:

$$\frac{\sin(s+t)}{\cos s \cos t} = \frac{\sin s \cos t + \cos s \sin t}{\cos s \cos t} = \frac{\sin s}{\cos s} + \frac{\sin t}{\cos t} = \tan s + \tan t$$

51. Using the addition formulas for cos $(s - t)$ and cos $(s + t)$:

$$\cos(A-B) - \cos(A+B) = (\cos A \cos B + \sin A \sin B) - (\cos A \cos B - \sin A \sin B)$$
$$= \cos A \cos B + \sin A \sin B - \cos A \cos B + \sin A \sin B$$
$$= 2 \sin A \sin B$$

53. Using the addition formulas for $\cos (s + t)$ and $\cos (s - t)$, as well as the identities $\sin^2 A = 1 - \cos^2 A$ and $\sin^2 B = 1 - \cos^2 B$:

$$\cos(A + B)\cos(A - B) = (\cos A \cos B - \sin A \sin B)(\cos A \cos B + \sin A \sin B)$$
$$= (\cos A \cos B)^2 - (\sin A \sin B)^2$$
$$= \cos^2 A \cos^2 B - \sin^2 A \sin^2 B$$
$$= \cos^2 A(1 - \sin^2 B) - (1 - \cos^2 A)\sin^2 B$$
$$= \cos^2 A - \cos^2 A \sin^2 B - \sin^2 B + \cos^2 A \sin^2 B$$
$$= \cos^2 A - \sin^2 B$$

55. Using the addition formulas for $\cos (s + t)$ and $\sin (s + t)$, as well as the identity $\cos^2 \beta + \sin^2 \beta = 1$:

$$\cos(\alpha + \beta)\cos\beta + \sin(\alpha + \beta)\sin\beta$$
$$= (\cos\alpha\cos\beta - \sin\alpha\sin\beta)\cos\beta + (\sin\alpha\cos\beta + \cos\alpha\sin\beta)\sin\beta$$
$$= \cos\alpha\cos^2\beta - \sin\alpha\sin\beta\cos\beta + \sin\alpha\sin\beta\cos\beta + \cos\alpha\sin^2\beta$$
$$= \cos\alpha\cos^2\beta + \cos\alpha\sin^2\beta$$
$$= \cos\alpha(\cos^2\beta + \sin^2\beta)$$
$$= \cos\alpha$$

57. Writing $2t = t + t$:

$$\tan 2t = \frac{\tan t + \tan t}{1 - \tan t \cdot \tan t} = \frac{2\tan t}{1 - \tan^2 t}$$

59. The average rate of change is given by:

$$\frac{\Delta f}{\Delta x} = \frac{f(x + h) - f(x)}{h}$$
$$= \frac{\sin(x + h) - \sin x}{h}$$
$$= \frac{\sin x \cos h + \cos x \sin h - \sin x}{h}$$
$$= \frac{(\sin x)(\cos h - 1) + (\cos x)(\sin h)}{h}$$
$$= (\sin x)\left(\frac{\cos h - 1}{h}\right) + (\cos x)\left(\frac{\sin h}{h}\right)$$

61. The average rate of change is given by:

$$\frac{\Delta T}{\Delta x} = \frac{T(x+h) - T(x)}{h}$$

$$= \frac{\tan(x+h) - \tan x}{h}$$

$$= \frac{\dfrac{\tan x + \tan h}{1 - \tan x \tan h} - \tan x}{h}$$

$$= \frac{\tan x + \tan h - \tan x + \tan^2 x \tan h}{h(1 - \tan x \tan h)}$$

$$= \frac{(\tan h)(1 + \tan^2 x)}{h(1 - \tan x \tan h)}$$

$$= \frac{\tan h}{h} \cdot \frac{\sec^2 h}{1 - \tan x \tan h}$$

63. (a) Since θ is as pictured, $\sin\theta = \dfrac{b}{\sqrt{a^2 + b^2}}$ and $\cos\theta = \dfrac{a}{\sqrt{a^2 + b^2}}$. Using an addition formula:

$$\sqrt{a^2 + b^2}\,\sin(x + \theta) = \sqrt{a^2 + b^2}\,(\sin x \cos\theta + \cos x \sin\theta)$$

$$= \sqrt{a^2 + b^2}\left(\frac{a}{\sqrt{a^2 + b^2}}\sin x + \frac{b}{\sqrt{a^2 + b^2}}\cos x\right)$$

$$= a\sin x + b\cos x$$

(b) Since $\sin(x + \theta) \leq 1$, a maximum value of $f(x) = a\sin x + b\cos x$ is $\sqrt{a^2 + b^2}$.

65. (a) Using an addition formula:

$$\sqrt{2}\cos\left(x - \tfrac{\pi}{4}\right) = \sqrt{2}\left(\cos x \cos\tfrac{\pi}{4} + \sin x \sin\tfrac{\pi}{4}\right)$$

$$= \sqrt{2}\left(\frac{1}{\sqrt{2}}\cos x + \frac{1}{\sqrt{2}}\sin x\right)$$

$$= \cos x + \sin x$$

(b) Graphing $f(x) = \sqrt{2}\cos\left(x - \tfrac{\pi}{4}\right) = \cos x + \sin x$:

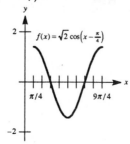

67. To prove this identity, instead show an alternate identity:

$$1 - \cos^2 C - \cos^2 A - \cos^2 B - 2\cos A \cos B \cos C$$

$$= \sin^2 C - \cos^2 A - \cos^2 B - 2\cos A \cos B \cos[\pi - (A+B)]$$

$$= \sin^2[\pi - (A+B)] - \cos^2 A - \cos^2 B + 2\cos A \cos B \cos(A+B)$$

$$= \sin^2(A+B) - \cos^2 A - \cos^2 B + 2\cos A \cos B(\cos A \cos B - \sin A \sin B)$$

$$= (\sin A \cos B + \sin B \cos A)^2 - \cos^2 A - \cos^2 B + 2\cos^2 A \cos^2 B$$

$$\qquad -2\cos A \cos B \sin A \sin B$$

$$= \sin^2 A \cos^2 B + \sin^2 B \cos^2 A - \cos^2 A - \cos^2 B + 2\cos^2 A \cos^2 B$$

$$= \cos^2 B(\sin^2 A - 1) + \cos^2 A(\sin^2 B - 1) + 2\cos^2 A \cos^2 B$$

$$= -\cos^2 B \cos^2 A - \cos^2 A \cos^2 B + 2\cos^2 A \cos^2 B$$

$$= 0$$

Hence $\cos^2 A + \cos^2 B + \cos^2 C + 2\cos A \cos B \cos C = 1$.

69. Since $a^2 + b^2 = 1$ and $c^2 + d^2 = 1$, there is some angle θ for which $a = \cos\theta$ and $b = \sin\theta$ and some angle φ for which $c = \cos\varphi$ and $d = \sin\varphi$. Thus:

$$|ac + bd| = |\cos\theta\cos\varphi + \sin\theta\sin\varphi| = |\cos(\theta - \varphi)| \le 1$$

71. Let $A = \frac{\pi}{3} - t$ and $B = \frac{\pi}{3} + t$, then:

$$\sin A \cos B + \cos A \sin B = \sin(A+B)$$

But $A + B = \frac{2\pi}{3}$, so $\sin(A+B) = \sin\frac{2\pi}{3} = \dfrac{\sqrt{3}}{2}$.

73. If $\alpha + \beta = \frac{\pi}{4}$, then $\tan(\alpha + \beta) = \tan\frac{\pi}{4} = 1$, so:

$$1 = \tan(\alpha + \beta)$$

$$1 = \frac{\tan\alpha + \tan\beta}{1 - \tan\alpha\tan\beta}$$

$$1 - \tan\alpha\tan\beta = \tan\alpha + \tan\beta$$

$$1 = \tan\alpha + \tan\beta + \tan\alpha\tan\beta$$

Working from the left-hand side:

$$(1 + \tan\alpha)(1 + \tan\beta) = 1 + \tan\alpha + \tan\beta + \tan\alpha\tan\beta = 1 + 1 = 2$$

75. (a) Using $\triangle ABH$, $\cos(\alpha + \beta) = \dfrac{AB}{1}$, so $\cos(\alpha + \beta) = AB$.

(b) Using $\triangle ACF$, $\cos\alpha = \dfrac{AC}{AF} = \dfrac{AC}{\cos\beta}$ from Exercise 74(e), so $AC = \cos\alpha\cos\beta$.

(c) Using $\triangle EFH$, $\sin(\angle EHF) = \dfrac{EF}{HF}$. But $\angle EHF = \alpha$ from Exercise 74(c), and

$HF = \sin\beta$ from Exercise 74(b), so $\sin\alpha = \dfrac{EF}{\sin\beta}$, and thus $EF = \sin\alpha\sin\beta$.

(d) From part (a), $\cos(\alpha+\beta) = AB = AC - BC$. But $AC = \cos\alpha\cos\beta$ from part (b), and $BC = EF = \sin\alpha\sin\beta$ from part (c), so $\cos(\alpha+\beta) = \cos\alpha\cos\beta - \sin\alpha\sin\beta$.

77. Working from the left-hand side:

$$\frac{\sin(A+B)}{\sin(A-B)} = \frac{\sin A\cos B + \cos A\sin B}{\sin A\cos B - \cos A\sin B} \div \frac{\cos A\cos B}{\cos A\cos B}$$

$$= \frac{\dfrac{\sin A}{\cos A} + \dfrac{\sin B}{\cos B}}{\dfrac{\sin A}{\cos A} - \dfrac{\sin B}{\cos B}}$$

$$= \frac{\tan A + \tan B}{\tan A - \tan B}$$

79. Working from the left-hand side:

$$\cot(A+B) = \frac{\cos(A+B)}{\sin(A+B)}$$

$$= \frac{\cos A\cos B - \sin A\sin B}{\sin A\cos B + \cos A\sin B} \div \frac{\sin A\sin B}{\sin A\sin B}$$

$$= \frac{\dfrac{\cos A}{\sin A}\cdot\dfrac{\cos B}{\sin B} - 1}{\dfrac{\cos B}{\sin B} + \dfrac{\cos A}{\sin A}}$$

$$= \frac{\cot A\cot B - 1}{\cot B + \cot A}$$

81. (a) Complete the table:

t	1	2	3	4
$f(t)$	1.5	1.5	1.5	1.5

(b) Conjecture: $f(t) = 1.5$
To prove this, first simplify the expressions:

$$\cos\left(t + \tfrac{2\pi}{3}\right) = \cos t\cos\tfrac{2\pi}{3} - \sin t\sin\tfrac{2\pi}{3} = -\tfrac{1}{2}\cos t - \tfrac{\sqrt{3}}{2}\sin t$$

$$\cos\left(t - \tfrac{2\pi}{3}\right) = \cos t\cos\tfrac{2\pi}{3} + \sin t\sin\tfrac{2\pi}{3} = -\tfrac{1}{2}\cos t + \tfrac{\sqrt{3}}{2}\sin t$$

Therefore:

$$\cos^2\left(t + \tfrac{2\pi}{3}\right) = \left(-\tfrac{1}{2}\cos t - \tfrac{\sqrt{3}}{2}\sin t\right)^2 = \tfrac{1}{4}\cos^2 t + \tfrac{\sqrt{3}}{2}\sin t\cos t + \tfrac{3}{4}\sin^2 t$$

$$\cos^2\left(t - \tfrac{2\pi}{3}\right) = \left(-\tfrac{1}{2}\cos t + \tfrac{\sqrt{3}}{2}\sin t\right)^2 = \tfrac{1}{4}\cos^2 t - \tfrac{\sqrt{3}}{2}\sin t\cos t + \tfrac{3}{4}\sin^2 t$$

Thus:

$$f(t) = \cos^2 t + \left(\tfrac{1}{4}\cos^2 t + \tfrac{\sqrt{3}}{2}\sin t \cos t + \tfrac{3}{4}\sin^2 t\right)$$
$$+ \left(\tfrac{1}{4}\cos^2 t - \tfrac{\sqrt{3}}{2}\sin t \cos t + \tfrac{3}{4}\sin^2 t\right)$$
$$= \tfrac{3}{2}\cos^2 t + \tfrac{3}{2}\sin^2 t$$
$$= \tfrac{3}{2}\left(\cos^2 t + \sin^2 t\right)$$
$$= \tfrac{3}{2}$$

83. Working from the left-hand side:

$$\tan(A - B) = \frac{\tan A - \dfrac{n\sin A \cos A}{1 - n\sin^2 A}}{1 + \tan A \cdot \dfrac{n\sin A \cos A}{1 - n\sin^2 A}}$$

$$= \frac{\tan A\left(1 - n\sin^2 A\right) - n\sin A \cos A}{1 - n\sin^2 A + n\tan A \sin A \cos A}$$

$$= \frac{\tan A - n\tan A \sin^2 A - n\sin A \cos A}{1 - n\sin^2 A + n\tan A \sin A \cos A}$$

$$= \frac{\tan A - n\tan A\left(\sin^2 A + \cos^2 A\right)}{1 - n\sin^2 A + n\sin^2 A}$$

$$= \frac{\tan A - n\tan A}{1}$$

$$= (1 - n)\tan A$$

85. **(a)** Using a calculator:
$$\tan A + \tan B + \tan C = \tan 20^\circ + \tan 50^\circ + \tan 110^\circ \approx -1.1918$$
$$\tan A \tan B \tan C = \tan 20^\circ \tan 50^\circ \tan 110^\circ \approx -1.1918$$
It appears that the two values are equal.

(b) Using a calculator:
$$\tan\alpha + \tan\beta + \tan\gamma = \tan\tfrac{\pi}{10} + \tan\tfrac{3\pi}{10} + \tan\tfrac{3\pi}{5} \approx -1.3764$$
$$\tan\alpha \tan\beta \tan\gamma = \tan\tfrac{\pi}{10} \cdot \tan\tfrac{3\pi}{10} \cdot \tan\tfrac{3\pi}{5} \approx -1.3764$$
It appears that the two values are equal.

(c) Since $A + B + C = \pi$, $\tan(A + B) = \tan(\pi - C) = -\tan C$. Since
$$\tan(A + B) = \frac{\tan A + \tan B}{1 - \tan A \tan B}:$$
$$\frac{\tan A + \tan B}{1 - \tan A \tan B} = -\tan C$$
$$\tan A + \tan B = -\tan C + \tan A \tan B \tan C$$
$$\tan A + \tan B + \tan C = \tan A \tan B \tan C$$

9.2 The Double-Angle Formulas

1. Since $0° < \phi < 90°$, $\sin \phi > 0$ and thus:
$$\sin \phi = \sqrt{1 - \cos^2 \phi} = \sqrt{1 - \left(\frac{7}{25}\right)^2} = \sqrt{1 - \frac{49}{625}} = \sqrt{\frac{576}{625}} = \frac{24}{25}$$

(a) Using the double-angle identity for $\sin 2\phi$:
$$\sin 2\phi = 2 \sin \phi \cos \phi = 2 \cdot \frac{24}{25} \cdot \frac{7}{25} = \frac{336}{625}$$

(b) Using the double-angle identity for $\cos 2\phi$:
$$\cos 2\phi = \cos^2 \phi - \sin^2 \phi = \left(\frac{7}{25}\right)^2 - \left(\frac{24}{25}\right)^2 = \frac{49}{625} - \frac{576}{625} = -\frac{527}{625}$$

(c) Using parts (a) and (b):
$$\tan 2\phi = \frac{\sin 2\phi}{\cos 2\phi} = \frac{\frac{336}{625}}{-\frac{527}{625}} = -\frac{336}{527}$$

3. Since $\frac{3\pi}{2} < u < 2\pi$, $\sec u > 0$ and thus:
$$\sec u = \sqrt{1 + \tan^2 u} = \sqrt{1 + (-4)^2} = \sqrt{1 + 16} = \sqrt{17}$$
Thus $\cos u = \frac{1}{\sqrt{17}}$. Since $\frac{3\pi}{2} < u < 2\pi$, $\sin u < 0$ and thus:
$$\sin u = -\sqrt{1 - \cos^2 u} = -\sqrt{1 - \frac{1}{17}} = -\sqrt{\frac{16}{17}} = -\frac{4}{\sqrt{17}}$$

(a) Using the double-angle identity for $\sin 2u$:
$$\sin 2u = 2 \sin u \cos u = 2 \cdot \left(-\frac{4}{\sqrt{17}}\right) \cdot \left(\frac{1}{\sqrt{17}}\right) = -\frac{8}{17}$$

(b) Using the double-angle identity for $\cos 2u$:
$$\cos 2u = \cos^2 u - \sin^2 u = \left(\frac{1}{\sqrt{17}}\right)^2 - \left(-\frac{4}{\sqrt{17}}\right)^2 = \frac{1}{17} - \frac{16}{17} = -\frac{15}{17}$$

(c) Using parts (a) and (b):
$$\tan 2u = \frac{\sin 2u}{\cos 2u} = \frac{-\frac{8}{17}}{-\frac{15}{17}} = \frac{8}{15}$$

5. Since $0° < \alpha < 90°$, $\cos \alpha > 0$ and thus:
$$\cos \alpha = \sqrt{1 - \sin^2 \alpha} = \sqrt{1 - \left(\frac{\sqrt{3}}{2}\right)^2} = \sqrt{1 - \frac{3}{4}} = \sqrt{\frac{1}{4}} = \frac{1}{2}$$

(a) Since $0° < \alpha < 90°$, $0° < \frac{\alpha}{2} < 45°$ and thus $\sin \frac{\alpha}{2} > 0$. Therefore:
$$\sin \frac{\alpha}{2} = \sqrt{\frac{1 - \cos \alpha}{2}} = \sqrt{\frac{1 - \frac{1}{2}}{2}} = \sqrt{\frac{1}{4}} = \frac{1}{2}$$

(b)   Since $0° < \alpha < 90°$, $0° < \frac{\alpha}{2} < 45°$ and thus $\cos\frac{\alpha}{2} > 0$. Therefore:

$$\cos\frac{\alpha}{2} = \sqrt{\frac{1+\cos\alpha}{2}} = \sqrt{\frac{1+\frac{1}{2}}{2}} = \sqrt{\frac{3}{4}} = \frac{\sqrt{3}}{2}$$

(c)   Using parts (a) and (b):

$$\tan\frac{\alpha}{2} = \frac{\sin\frac{\alpha}{2}}{\cos\frac{\alpha}{2}} = \frac{\frac{1}{2}}{\frac{\sqrt{3}}{2}} = \frac{1}{\sqrt{3}} = \frac{\sqrt{3}}{3}$$

Notice that an alternate solution is to spot that since $\sin\alpha = \frac{\sqrt{3}}{2}$ and $0° < \alpha < 90°$, $\alpha = 60°$ and thus $\frac{\alpha}{2} = 30°$. Thus $\sin\frac{\alpha}{2} = \frac{1}{2}$, $\cos\frac{\alpha}{2} = \frac{\sqrt{3}}{2}$ and $\tan\frac{\alpha}{2} = \frac{\sqrt{3}}{3}$.

7.   (a)   Since $\frac{\pi}{2} < \theta < \pi$, $\frac{\pi}{4} < \frac{\theta}{2} < \frac{\pi}{2}$ and thus $\sin\frac{\theta}{2} > 0$. Therefore:

$$\sin\frac{\theta}{2} = \sqrt{\frac{1-\cos\theta}{2}} = \sqrt{\frac{1+\frac{7}{9}}{2}} = \sqrt{\frac{8}{9}} = \frac{2\sqrt{2}}{3}$$

(b)   Since $\frac{\pi}{2} < \theta < \pi$, $\frac{\pi}{4} < \frac{\theta}{2} < \frac{\pi}{2}$ and thus $\cos\frac{\theta}{2} > 0$. Therefore:

$$\cos\frac{\theta}{2} = \sqrt{\frac{1+\cos\theta}{2}} = \sqrt{\frac{1-\frac{7}{9}}{2}} = \sqrt{\frac{1}{9}} = \frac{1}{3}$$

(c)   Using parts (a) and (b):

$$\tan\frac{\theta}{2} = \frac{\sin\frac{\theta}{2}}{\cos\frac{\theta}{2}} = \frac{\frac{2\sqrt{2}}{3}}{\frac{1}{3}} = 2\sqrt{2}$$

9.   Since $\frac{\pi}{2} < \theta < \pi$, $\cos\theta < 0$ and thus:

$$\cos\theta = -\sqrt{1-\sin^2\theta} = -\sqrt{1-\left(\frac{3}{4}\right)^2} = -\sqrt{1-\frac{9}{16}} = -\sqrt{\frac{7}{16}} = -\frac{\sqrt{7}}{4}$$

(a)   Using the double-angle identity for $\sin 2\theta$:

$$\sin 2\theta = 2\sin\theta\cos\theta = 2\cdot\left(\frac{3}{4}\right)\cdot\left(-\frac{\sqrt{7}}{4}\right) = -\frac{3\sqrt{7}}{8}$$

(b)   Using the double-angle identity for $\cos 2\theta$:

$$\cos 2\theta = \cos^2\theta - \sin^2\theta = \left(-\frac{\sqrt{7}}{4}\right)^2 - \left(\frac{3}{4}\right)^2 = \frac{7}{16} - \frac{9}{16} = -\frac{1}{8}$$

(c)   Since $\frac{\pi}{4} < \frac{\theta}{2} < \frac{\pi}{2}$, $\sin\frac{\theta}{2} > 0$ and thus:

$$\sin\frac{\theta}{2} = \sqrt{\frac{1-\cos\theta}{2}} = \sqrt{\frac{1-\left(-\frac{\sqrt{7}}{4}\right)}{2}} = \sqrt{\frac{4+\sqrt{7}}{8}} = \frac{\sqrt{8+2\sqrt{7}}}{4}$$

(d)   Since $\frac{\pi}{4} < \frac{\theta}{2} < \frac{\pi}{2}$, $\cos\frac{\theta}{2} > 0$ and thus:

$$\cos\frac{\theta}{2} = \sqrt{\frac{1+\cos\theta}{2}} = \sqrt{\frac{1-\frac{\sqrt{7}}{4}}{2}} = \sqrt{\frac{4-\sqrt{7}}{8}} = \frac{\sqrt{8-2\sqrt{7}}}{4}$$

11.   Since $180° < \theta < 270°$, $\sin\theta < 0$ and thus:

$$\sin\theta = -\sqrt{1-\cos^2\theta} = -\sqrt{1-\left(-\tfrac{1}{3}\right)^2} = -\sqrt{1-\tfrac{1}{9}} = -\sqrt{\tfrac{8}{9}} = -\frac{2\sqrt{2}}{3}$$

(a)   Using the double-angle identity for $\sin 2\theta$:

$$\sin 2\theta = 2\sin\theta\cos\theta = 2\cdot\left(-\frac{2\sqrt{2}}{3}\right)\cdot\left(-\tfrac{1}{3}\right) = \frac{4\sqrt{2}}{9}$$

(b)   Using the double-angle identity for $\cos 2\theta$:

$$\cos 2\theta = \cos^2\theta - \sin^2\theta = \left(-\tfrac{1}{3}\right)^2 - \left(-\tfrac{2\sqrt{2}}{3}\right)^2 = \tfrac{1}{9} - \tfrac{8}{9} = -\tfrac{7}{9}$$

(c)   Since $90° < \frac{\theta}{2} < 135°$, $\sin\frac{\theta}{2} > 0$ and thus:

$$\sin\frac{\theta}{2} = \sqrt{\frac{1-\cos\theta}{2}} = \sqrt{\frac{1+\frac{1}{3}}{2}} = \sqrt{\tfrac{2}{3}} = \frac{\sqrt{6}}{3}$$

(d)   Since $90° < \frac{\theta}{2} < 135°$, $\cos\frac{\theta}{2} < 0$ and thus:

$$\cos\frac{\theta}{2} = -\sqrt{\frac{1+\cos\theta}{2}} = -\sqrt{\frac{1-\frac{1}{3}}{2}} = -\sqrt{\tfrac{1}{3}} = -\frac{\sqrt{3}}{3}$$

13.   (a)   Since $0 < \frac{\pi}{12} < \frac{\pi}{2}$, $\sin\frac{\pi}{12} > 0$. Using the half-angle formula for $\sin\frac{1}{2}x$:

$$\sin\frac{\pi}{12} = \sqrt{\frac{1-\cos\frac{\pi}{6}}{2}} = \sqrt{\frac{1-\frac{\sqrt{3}}{2}}{2}} = \sqrt{\frac{2-\sqrt{3}}{4}} = \frac{\sqrt{2-\sqrt{3}}}{2}$$

(b)   Since $0 < \frac{\pi}{12} < \frac{\pi}{2}$, $\cos\frac{\pi}{12} > 0$. Using the half-angle formula for $\cos\frac{1}{2}x$:

$$\cos\frac{\pi}{12} = \sqrt{\frac{1+\cos\frac{\pi}{6}}{2}} = \sqrt{\frac{1+\frac{\sqrt{3}}{2}}{2}} = \frac{\sqrt{2+\sqrt{3}}}{2}$$

(c)   Using the half-angle formula for $\tan\frac{1}{2}x$:

$$\tan\frac{\pi}{12} = \frac{\sin\frac{\pi}{6}}{1+\cos\frac{\pi}{6}} = \frac{\frac{1}{2}}{1+\frac{\sqrt{3}}{2}} = \frac{1}{2+\sqrt{3}}\cdot\frac{2-\sqrt{3}}{2-\sqrt{3}} = \frac{2-\sqrt{3}}{4-3} = 2-\sqrt{3}$$

Note: We could also use parts (a) and (b), as follows:

$$\tan\frac{\pi}{12} = \frac{\sin\frac{\pi}{12}}{\cos\frac{\pi}{12}}$$

$$= \frac{\frac{\sqrt{2-\sqrt{3}}}{2}}{\frac{\sqrt{2+\sqrt{3}}}{2}}$$

$$= \frac{\sqrt{2-\sqrt{3}}}{\sqrt{2+\sqrt{3}}} \cdot \frac{\sqrt{2+\sqrt{3}}}{\sqrt{2+\sqrt{3}}}$$

$$= \frac{\sqrt{4-3}}{2+\sqrt{3}} \cdot \frac{2-\sqrt{3}}{2-\sqrt{3}}$$

$$= \frac{2-\sqrt{3}}{4-3}$$

$$= 2-\sqrt{3}$$

15. (a) Since $90° < 105° < 180°$, $\sin 105° > 0$. Using the half-angle formula for $\sin\frac{1}{2}x$:

$$\sin 105° = \sqrt{\frac{1-\cos 210°}{2}} = \sqrt{\frac{1+\frac{\sqrt{3}}{2}}{2}} = \sqrt{\frac{2+\sqrt{3}}{4}} = \frac{\sqrt{2+\sqrt{3}}}{2}$$

(b) Since $90° < 105° < 180°$, $\cos 105° < 0$. Using the half-angle formula for $\cos\frac{1}{2}x$:

$$\cos 105° = -\sqrt{\frac{1+\cos 210°}{2}} = -\sqrt{\frac{1-\frac{\sqrt{3}}{2}}{2}} = -\sqrt{\frac{2-\sqrt{3}}{4}} = -\frac{\sqrt{2-\sqrt{3}}}{2}$$

(c) Using the half-angle formula for $\tan\frac{1}{2}x$:

$$\tan 105° = \frac{\sin 210°}{1+\cos 210°} = \frac{-\frac{1}{2}}{1-\frac{\sqrt{3}}{2}} = -\frac{1}{2-\sqrt{3}} \cdot \frac{2+\sqrt{3}}{2+\sqrt{3}} = -\frac{2+\sqrt{3}}{4-3} = -2-\sqrt{3}$$

17. From the first triangle we have $\sin\theta = \frac{3}{5}$, $\cos\theta = \frac{4}{5}$ and $\tan\theta = \frac{3}{4}$.

(a) By the double-angle identity for $\sin 2\theta$:

$$\sin 2\theta = 2\sin\theta\cos\theta = 2\cdot\frac{3}{5}\cdot\frac{4}{5} = \frac{24}{25}$$

(b) By the double-angle identity for $\cos 2\theta$:

$$\cos 2\theta = \cos^2\theta - \sin^2\theta = \left(\frac{4}{5}\right)^2 - \left(\frac{3}{5}\right)^2 = \frac{16}{25} - \frac{9}{25} = \frac{7}{25}$$

(c) By the double-angle identity for $\tan 2\theta$:

$$\tan 2\theta = \frac{2\tan\theta}{1-\tan^2\theta} = \frac{2\cdot\frac{3}{4}}{1-\left(\frac{3}{4}\right)^2} = \frac{\frac{3}{2}}{\frac{7}{16}} = \frac{24}{7}$$

Note: An easier approach, after doing (a) and (b), would be to say:

$$\tan 2\theta = \frac{\sin 2\theta}{\cos 2\theta} = \frac{\frac{24}{25}}{\frac{7}{25}} = \frac{24}{7}$$

19. From the first triangle we have $\sin\beta = \frac{4}{5}$ and $\cos\beta = \frac{3}{5}$.

(a) By the double-angle identity for $\sin 2\beta$:
$$\sin 2\beta = 2\sin\beta\cos\beta = 2 \cdot \tfrac{4}{5} \cdot \tfrac{3}{5} = \tfrac{24}{25}$$

(b) By the double-angle identity for $\cos 2\beta$:
$$\cos 2\beta = \cos^2\beta - \sin^2\beta = \left(\tfrac{3}{5}\right)^2 - \left(\tfrac{4}{5}\right)^2 = \tfrac{9}{25} - \tfrac{16}{25} = -\tfrac{7}{25}$$

(c) Using the identity for $\tan x$:
$$\tan 2\beta = \frac{\sin 2\beta}{\cos 2\beta} = \frac{\frac{24}{25}}{-\frac{7}{25}} = -\frac{24}{7}$$

Note: We could also have used the double-angle formula for $\tan 2\beta$.

21. From the first triangle we have $\sin\theta = \frac{3}{5}$ and $\cos\theta = \frac{4}{5}$.

(a) Since $\sin\frac{\theta}{2} > 0$, use the half-angle formula for $\sin\frac{\theta}{2}$ to obtain:
$$\sin\frac{\theta}{2} = \sqrt{\frac{1-\cos\theta}{2}} = \sqrt{\frac{1-\frac{4}{5}}{2}} = \sqrt{\tfrac{1}{10}} = \frac{\sqrt{10}}{10}$$

(b) Since $\cos\frac{\theta}{2} > 0$, use the half-angle formula for $\cos\frac{\theta}{2}$ to obtain:
$$\cos\frac{\theta}{2} = \sqrt{\frac{1+\cos\theta}{2}} = \sqrt{\frac{1+\frac{4}{5}}{2}} = \sqrt{\tfrac{9}{10}} = \frac{3\sqrt{10}}{10}$$

(c) Using the half-angle formula for $\tan\frac{\theta}{2}$ to obtain:
$$\tan\frac{\theta}{2} = \frac{\sin\theta}{1+\cos\theta} = \frac{\frac{3}{5}}{1+\frac{4}{5}} = \frac{3}{9} = \frac{1}{3}$$

Note: We could also have computed this directly after parts (a) and (b) as:
$$\tan\frac{\theta}{2} = \frac{\sin\frac{\theta}{2}}{\cos\frac{\theta}{2}} = \frac{\frac{\sqrt{10}}{10}}{\frac{3\sqrt{10}}{10}} = \frac{1}{3}$$

23.   From the first triangle we have $\cos \beta = \frac{3}{5}$.

(a)   Since $\sin \frac{\beta}{2} > 0$, use the half-angle formula for $\sin \frac{\beta}{2}$ to obtain:

$$\sin \frac{\beta}{2} = \sqrt{\frac{1 - \cos \beta}{2}} = \sqrt{\frac{1 - \frac{3}{5}}{2}} = \sqrt{\frac{1}{5}} = \frac{\sqrt{5}}{5}$$

(b)   Since $\cos \frac{\beta}{2} > 0$, use the half-angle formula for $\cos \frac{\beta}{2}$ to obtain:

$$\cos \frac{\beta}{2} = \sqrt{\frac{1 + \cos \beta}{2}} = \sqrt{\frac{1 + \frac{3}{5}}{2}} = \sqrt{\frac{4}{5}} = \frac{2\sqrt{5}}{5}$$

(c)   Using the identity for $\tan x$ and parts (a) and (b):

$$\tan \frac{\beta}{2} = \frac{\sin \frac{\beta}{2}}{\cos \frac{\beta}{2}} = \frac{\frac{\sqrt{5}}{5}}{\frac{2\sqrt{5}}{5}} = \frac{1}{2}$$

25.   Since $0 < \theta < \frac{\pi}{2}$, $\cos \theta > 0$. Since $\sin \theta = \frac{x}{5}$:

$$\cos \theta = \sqrt{1 - \sin^2 \theta} = \sqrt{1 - \left(\frac{x}{5}\right)^2} = \sqrt{1 - \frac{x^2}{25}} = \frac{\sqrt{25 - x^2}}{5}$$

Now apply the double angle formulas:

$$\sin 2\theta = 2 \sin \theta \cos \theta = 2 \cdot \frac{x}{5} \cdot \frac{\sqrt{25 - x^2}}{5} = \frac{2x\sqrt{25 - x^2}}{25}$$

$$\cos 2\theta = \cos^2 \theta - \sin^2 \theta = \left(\frac{\sqrt{25 - x^2}}{5}\right)^2 - \left(\frac{x}{5}\right)^2 = \frac{25 - x^2}{25} - \frac{x^2}{25} = \frac{25 - 2x^2}{25}$$

27.   Since $0 < \theta < \frac{\pi}{2}$, $\cos \theta > 0$. Since $\sin \theta = \frac{x - 1}{2}$:

$$\cos \theta = \sqrt{1 - \sin^2 \theta} = \sqrt{1 - \left(\frac{x - 1}{2}\right)^2} = \sqrt{1 - \frac{x^2 - 2x + 1}{4}} = \frac{\sqrt{3 + 2x - x^2}}{2}$$

Now apply the double-angle formulas:

$$\sin 2\theta = 2 \sin \theta \cos \theta = 2 \cdot \frac{x - 1}{2} \cdot \frac{\sqrt{3 + 2x - x^2}}{2} = \frac{(x - 1)\sqrt{3 + 2x - x^2}}{2}$$

$$\cos 2\theta = \cos^2 \theta - \sin^2 \theta$$

$$= \left(\frac{\sqrt{3 + 2x - x^2}}{2}\right)^2 - \left(\frac{x - 1}{2}\right)^2$$

$$= \frac{3 + 2x - x^2}{4} - \frac{x^2 - 2x + 1}{4}$$

$$= \frac{2 + 4x - 2x^2}{4}$$

$$= \frac{1 + 2x - x^2}{2}$$

29. Using the identity $\sin^2\theta = \dfrac{1-\cos 2\theta}{2}$:

$$\sin^4\theta = \left(\sin^2\theta\right)^2$$
$$= \left(\frac{1-\cos 2\theta}{2}\right)^2$$
$$= \frac{1-2\cos 2\theta + (\cos 2\theta)^2}{4}$$
$$= \frac{1-2\cos 2\theta + \frac{1+\cos 4\theta}{2}}{4}$$
$$= \frac{2-4\cos 2\theta + 1 + \cos 4\theta}{8}$$
$$= \frac{3-4\cos 2\theta + \cos 4\theta}{8}$$

31. Using the identities $\sin^2\dfrac{\theta}{2} = \dfrac{1-\cos\theta}{2}$ and $\cos^2\theta = \dfrac{1+\cos 2\theta}{2}$:

$$\sin^4\frac{\theta}{2} = \left(\sin^2\frac{\theta}{2}\right)^2$$
$$= \left(\frac{1-\cos\theta}{2}\right)^2$$
$$= \frac{1-2\cos\theta + \cos^2\theta}{4}$$
$$= \frac{1-2\cos\theta + \frac{1+\cos 2\theta}{2}}{4}$$
$$= \frac{2-4\cos\theta + 1 + \cos 2\theta}{8}$$
$$= \frac{3-4\cos\theta + \cos 2\theta}{8}$$

33. (a) Replacing 2θ with $\theta + \theta$ and using the addition formula for $\cos(s+t)$:
$$\cos 2\theta = \cos(\theta+\theta) = \cos\theta\cos\theta - \sin\theta\sin\theta = \cos^2\theta - \sin^2\theta$$

(b) Replacing 2θ with $\theta + \theta$ and using the addition formula for $\tan(s+t)$:
$$\tan 2\theta = \tan(\theta+\theta) = \frac{\tan\theta + \tan\theta}{1-\tan\theta\tan\theta} = \frac{2\tan\theta}{1-\tan^2\theta}$$

35. Working from the right-hand side:
$$\frac{1-\tan^2 s}{1+\tan^2 s} = \frac{1-\frac{\sin^2 s}{\cos^2 s}}{1+\frac{\sin^2 s}{\cos^2 s}} = \frac{\cos^2 s - \sin^2 s}{\cos^2 s + \sin^2 s} = \frac{\cos 2s}{1} = \cos 2s$$

37. Writing $\theta = 2 \cdot \frac{\theta}{2}$, apply the double-angle formula:

$$\cos\theta = \cos\left(2 \cdot \frac{\theta}{2}\right) = \cos^2\frac{\theta}{2} - \sin^2\frac{\theta}{2} = \cos^2\frac{\theta}{2} - \left(1 - \cos^2\frac{\theta}{2}\right) = 2\cos^2\frac{\theta}{2} - 1$$

39. Using the identities $\sin^2\frac{\theta}{2} = \dfrac{1-\cos\theta}{2}$ and $\cos^2\theta = \dfrac{1+\cos 2\theta}{2}$:

$$\sin^4\theta = \left(\sin^2\theta\right)^2$$
$$= \left(\frac{1-\cos 2\theta}{2}\right)^2$$
$$= \frac{1 - 2\cos 2\theta + \cos^2 2\theta}{4}$$
$$= \frac{1 - 2\cos 2\theta + \frac{1+\cos 4\theta}{2}}{4}$$
$$= \frac{2 - 4\cos 2\theta + 1 + \cos 4\theta}{8}$$
$$= \frac{3 - 4\cos 2\theta + \cos 4\theta}{8}$$

41. Working from the right-hand side:

$$\frac{2\tan\theta}{1+\tan^2\theta} = \frac{2 \cdot \frac{\sin\theta}{\cos\theta}}{1 + \frac{\sin^2\theta}{\cos^2\theta}} = \frac{2\sin\theta\cos\theta}{\cos^2\theta + \sin^2\theta} = \frac{\sin 2\theta}{1} = \sin 2\theta$$

43. Working from the right-hand side:

$$2\sin^3\theta\cos\theta + 2\sin\theta\cos^3\theta = (2\sin\theta\cos\theta)(\sin^2\theta + \cos^2\theta) = \sin 2\theta \cdot 1 = \sin 2\theta$$

45. Working from the left-hand side:

$$\frac{1+\tan\frac{\theta}{2}}{1-\tan\frac{\theta}{2}} = \frac{1 + \frac{\sin\frac{\theta}{2}}{\cos\frac{\theta}{2}}}{1 - \frac{\sin\frac{\theta}{2}}{\cos\frac{\theta}{2}}}$$
$$= \frac{\cos\frac{\theta}{2} + \sin\frac{\theta}{2}}{\cos\frac{\theta}{2} - \sin\frac{\theta}{2}}$$
$$= \frac{\left(\cos\frac{\theta}{2} + \sin\frac{\theta}{2}\right)^2}{\cos^2\frac{\theta}{2} - \sin^2\frac{\theta}{2}}$$
$$= \frac{1 + 2\cos\frac{\theta}{2}\sin\frac{\theta}{2}}{\cos\theta}$$
$$= \frac{1 + \sin\theta}{\cos\theta}$$
$$= \tan\theta + \sec\theta$$

47. Working from the left-hand side and using the addition formula for $\sin(s-t)$:

$$2\sin^2(45°-\theta) = 2(\sin 45°\cos\theta - \cos 45°\sin\theta)^2$$

$$= 2\left(\frac{\cos\theta - \sin\theta}{\sqrt{2}}\right)^2$$

$$= 2 \cdot \frac{\cos^2\theta - 2\sin\theta\cos\theta + \sin^2\theta}{2}$$

$$= 1 - 2\cos\theta\sin\theta$$

$$= 1 - \sin 2\theta$$

49. Simplifying the left-hand side:

$$1 + \tan\theta\tan 2\theta = 1 + \tan\theta \cdot \frac{2\tan\theta}{1-\tan^2\theta}$$

$$= 1 + \frac{2\tan^2\theta}{1-\tan^2\theta}$$

$$= \frac{1 - \tan^2\theta + 2\tan^2\theta}{1-\tan^2\theta}$$

$$= \frac{1 + \tan^2\theta}{1-\tan^2\theta}$$

Simplifying the right-hand side:

$$\tan 2\theta\cot\theta - 1 = \frac{2\tan\theta}{1-\tan^2\theta} \cdot \frac{1}{\tan\theta} - 1 = \frac{2}{1-\tan^2\theta} - 1 = \frac{2-1+\tan^2\theta}{1-\tan^2\theta} = \frac{1+\tan^2\theta}{1-\tan^2\theta}$$

Since both sides simplify to the same quantity, the original equation is an identity.

51. Using the addition formula for $\tan(\alpha+\beta)$:

$$\tan(\alpha+\beta) = \frac{\tan\alpha + \tan\beta}{1-\tan\alpha\tan\beta} = \frac{\frac{1}{11} + \frac{5}{6}}{1 - \frac{1}{11}\cdot\frac{5}{6}} = \frac{\frac{61}{66}}{\frac{61}{66}} = 1$$

But if $\tan(\alpha+\beta) = 1$ and $0 < \alpha+\beta < \pi$, then $\alpha+\beta = \frac{\pi}{4}$.

53. (a) Substituting $x = \cos 20° \approx 0.9397$ verifies that it is a root of the equation.

   (b) Substituting $\theta = 20°$ in the given identity:

$$\cos 60° = 4\cos^3 20° - 3\cos 20°$$

$$\tfrac{1}{2} = 4\cos^3 20° - 3\cos 20°$$

$$1 = 8\cos^3 20° - 6\cos 20°$$

$$0 = 8\cos^3 20° - 6\cos 20° - 1$$

Thus $\cos 20°$ is a root of the cubic equation $8x^3 - 6x - 1 = 0$.

55. (a) Using $\triangle ODC$ and the fact that $OC = 1$, $OD = \cos\theta$ and $DC = \sin\theta$. Now using $\triangle ADC$ and the fact that $AO = 1$:

$$\tan\frac{\theta}{2} = \frac{CD}{AO + OD} = \frac{\sin\theta}{1 + \cos\theta}$$

(b) Using the formula from (a) and rationalizing denominators:

$$\tan 15° = \frac{\sin 30°}{1 + \cos 30°} = \frac{\frac{1}{2}}{1 + \frac{\sqrt{3}}{2}} = \frac{1}{2 + \sqrt{3}} \cdot \frac{2 - \sqrt{3}}{2 - \sqrt{3}} = \frac{2 - \sqrt{3}}{4 - 3} = 2 - \sqrt{3}$$

Using the formula from (a) and rationalizing denominators:

$$\tan\frac{\pi}{8} = \frac{\sin\frac{\pi}{4}}{1 + \cos\frac{\pi}{4}} = \frac{\frac{\sqrt{2}}{2}}{1 + \frac{\sqrt{2}}{2}} = \frac{\sqrt{2}}{2 + \sqrt{2}} \cdot \frac{2 - \sqrt{2}}{2 - \sqrt{2}} = \frac{2\sqrt{2} - 2}{4 - 2} = \sqrt{2} - 1$$

57. (a) Following the suggestion and applying the double-angle formulas for $\sin 4\theta$ and $\sin 2\theta$:

$$\frac{\sin 4\theta}{4\sin\theta} = \frac{2\sin 2\theta\cos 2\theta}{4\sin\theta} = \frac{4\sin\theta\cos\theta\cos 2\theta}{4\sin\theta} = \cos\theta\cos 2\theta$$

(b) Following the suggestion and applying the double-angle formulas for $\sin 8\theta$, $\sin 4\theta$ and $\sin 2\theta$:

$$\frac{\sin 8\theta}{8\sin\theta} = \frac{2\sin 4\theta\cos 4\theta}{8\sin\theta}$$
$$= \frac{4\sin 2\theta\cos 2\theta\cos 4\theta}{8\sin\theta}$$
$$= \frac{8\sin\theta\cos\theta\cos 2\theta\cos 4\theta}{8\sin\theta}$$
$$= \cos\theta\cos 2\theta\cos 4\theta$$

(c) Following the suggestion and applying the double-angle formulas for $\sin 16\theta$, $\sin 8\theta$, $\sin 4\theta$ and $\sin 2\theta$:

$$\frac{\sin 16\theta}{16\sin\theta} = \frac{2\sin 8\theta\cos 8\theta}{16\sin\theta}$$
$$= \frac{4\sin 4\theta\cos 4\theta\cos 8\theta}{16\sin\theta}$$
$$= \frac{8\sin 2\theta\cos 2\theta\cos 4\theta\cos 8\theta}{16\sin\theta}$$
$$= \frac{16\sin\theta\cos\theta\cos 2\theta\cos 4\theta\cos 8\theta}{16\sin\theta}$$
$$= \cos\theta\cos 2\theta\cos 4\theta\cos 8\theta$$

59. (a) Using a calculator, $\cos 72° + \cos 144° = -0.5$.

(b) Using the observation and the addition formulas for cosine:

$$\cos 72° + \cos 144°$$
$$= \cos(108° - 36°) + \cos(108° + 36°)$$
$$= (\cos 108° \cos 36° + \sin 108° \sin 36°) + (\cos 108° \cos 36° - \sin 108° \sin 36°)$$
$$= 2\cos 108° \cos 36°$$

(c) Since $108° = 180° - 72°$, $\cos 108° = -\cos 72°$. Similarly, since $36° = 180° - 144°$, $\cos 36° = -\cos 144°$. Therefore:

$$\cos 108° \cos 36° = (-\cos 72°)(-\cos 144°) = \cos 72° \cos 144°$$

(d) Using the above relationships:

$$\cos 72° + \cos 144° = 2\cos 108° \cos 36° = 2\cos 72° \cos 144° = 2\left(-\tfrac{1}{4}\right) = -\tfrac{1}{2}$$

61. The coordinates of points A_1, A_2, and A_3 are:

$$A_1(1,0)$$
$$A_2(\cos 120°, \sin 120°) = A_2\left(-\tfrac{1}{2}, \tfrac{\sqrt{3}}{2}\right)$$
$$A_3(\cos 240°, \sin 240°) = A_3\left(-\tfrac{1}{2}, -\tfrac{\sqrt{3}}{2}\right)$$

Using the distance formula from the point $P(x, 0)$:

$$PA_1 = \sqrt{(x-1)^2 + (0-0)^2} = \sqrt{(x-1)^2} = |x-1| = 1-x, \text{ since } x < 1$$
$$PA_2 = \sqrt{\left(x+\tfrac{1}{2}\right)^2 + \left(0 - \tfrac{\sqrt{3}}{2}\right)^2} = \sqrt{x^2 + x + \tfrac{1}{4} + \tfrac{3}{4}} = \sqrt{x^2 + x + 1}$$
$$PA_3 = \sqrt{\left(x+\tfrac{1}{2}\right)^2 + \left(0 + \tfrac{\sqrt{3}}{2}\right)^2} = \sqrt{x^2 + x + \tfrac{1}{4} + \tfrac{3}{4}} = \sqrt{x^2 + x + 1}$$

Now compute the product:

$$\left(PA_1\right)\left(PA_2\right)\left(PA_3\right) = (1-x)\sqrt{x^2 + x + 1}\sqrt{x^2 + x + 1} = (1-x)\left(1 + x + x^2\right) = 1 - x^3$$

63. (a) The result is verified, since $\sin 18° \sin 54° = 0.25$.

(b) (i) This is true because of the double-angle formula for sine, and the fact that $\cos 36° = \sin 54°$.

(ii) This is true because of the double-angle formula for sine, and the fact that $\cos 18° = \sin 72°$.

(iii) Dividing each side by $4 \sin 72°$ produces this result.

65. (a) Using two addition formulas:
$$\cos(60° - \theta)\cos(60° + \theta)$$
$$= (\cos 60° \cos\theta + \sin 60° \sin\theta)(\cos 60° \cos\theta - \sin 60° \sin\theta)$$
$$= \left(\tfrac{1}{2}\cos\theta + \tfrac{\sqrt{3}}{2}\sin\theta\right)\left(\tfrac{1}{2}\cos\theta - \tfrac{\sqrt{3}}{2}\sin\theta\right)$$
$$= \tfrac{1}{4}\cos^2\theta - \tfrac{3}{4}\sin^2\theta$$
$$= \tfrac{1}{4}\cos^2\theta - \tfrac{3}{4}\left(1 - \cos^2\theta\right)$$
$$= \tfrac{1}{4}\cos^2\theta - \tfrac{3}{4} + \tfrac{3}{4}\cos^2\theta$$
$$= \cos^2\theta - \tfrac{3}{4}$$
$$= \frac{4\cos^2\theta - 3}{4}$$

(b) From part (a):
$$\cos\theta\cos(60° - \theta)\cos(60° + \theta) = \cos\theta \cdot \frac{4\cos^2\theta - 3}{4} = \frac{4\cos^3\theta - 3\cos\theta}{4}$$

From Example 3:
$$\frac{\cos 3\theta}{4} = \frac{4\cos^3\theta - 3\cos\theta}{4}$$
The two expressions are equal.

(c) Using $\theta = 20°$ in the identity:
$$\cos 20°\cos(60° - 20°)\cos(60° + 20°) = \frac{\cos 60°}{4}$$
$$\cos 20°\cos 40°\cos 80° = \frac{1/2}{4} = \tfrac{1}{8}$$

(d) A calculator verifies the result.

67. (a) Using the addition formulas and double-angle formulas:
$$\cos 3\theta = \cos\theta\cos 2\theta - \sin\theta\sin 2\theta$$
$$= \cos\theta\left(\cos^2\theta - \sin^2\theta\right) - \sin\theta(2\sin\theta\cos\theta)$$
$$= \cos^3\theta - \cos\theta\left(1 - \cos^2\theta\right) - 2\cos\theta\left(1 - \cos^2\theta\right)$$
$$= \cos^3\theta - \cos\theta + \cos^3\theta - 2\cos\theta + 2\cos^3\theta$$
$$= 4\cos^3\theta - 3\cos\theta$$
$$\cos 4\theta = \cos^2 2\theta - \sin^2 2\theta$$
$$= \left(\cos^2\theta - \sin^2\theta\right)^2 - 4\sin^2\theta\cos^2\theta$$
$$= \cos^4\theta - 2\cos^2\theta\sin^2\theta + \sin^4\theta - 4\cos^2\theta\sin^2\theta$$
$$= \cos^4\theta - 2\cos^2\theta\left(1 - \cos^2\theta\right) + \left(1 - \cos^2\theta\right)^2 - 4\cos^2\theta\left(1 - \cos^2\theta\right)$$
$$= \cos^4\theta - 2\cos^2\theta + 2\cos^4\theta + 1 - 2\cos^2\theta + \cos^4\theta - 4\cos^2\theta + 4\cos^4\theta$$
$$= 8\cos^4\theta - 8\cos^2\theta + 1$$

(b) Since $\cos 3\theta = \cos\frac{6\pi}{7}$ and $\cos 4\theta = \cos\frac{8\pi}{7}$, both angles have a negative cosine with a reference angle of $\frac{\pi}{7}$. So $\cos\frac{6\pi}{7} = \cos\frac{8\pi}{7}$.

(c) Since $\cos 4\theta = \cos 3\theta$:
$$8\cos^4\theta - 8\cos^2\theta + 1 = 4\cos^3\theta - 3\cos\theta$$
$$8\cos^4\theta - 4\cos^3\theta - 8\cos^2\theta + 3\cos\theta + 1 = 0$$

(d) The factoring $(\cos\theta - 1)(8\cos^3\theta + 4\cos^2\theta - 4\cos\theta - 1) = 0$ checks, and since $\theta = \frac{2\pi}{7}$ is not a root to $\cos\theta - 1 = 0$, it must be a root to $8\cos^3\theta + 4\cos^2\theta - 4\cos\theta - 1 = 0$. Thus $\cos\frac{2\pi}{7}$ is a solution to the equation $8x^3 + 4x^2 - 4x - 1 = 0$.

(e) The value $x = \cos\frac{2\pi}{7}$ checks in the equation.

9.3 The Product-to-Sum and Sum-to-Product Formulas

1. Using the product-to-sum formula for $\cos A \cos B$:
$$\cos 70°\cos 20° = \tfrac{1}{2}\left[\cos(70°-20°) + \cos(70°+20°)\right]$$
$$= \tfrac{1}{2}\left[\cos 50° + \cos 90°\right]$$
$$= \tfrac{1}{2}\left[\cos 50° + 0\right]$$
$$= \tfrac{1}{2}\cos 50°$$

3. Using the product-to-sum formula for $\sin A \sin B$:
$$\sin 5°\sin 85° = \tfrac{1}{2}\left[\cos(5°-85°) - \cos(5°+85°)\right]$$
$$= \tfrac{1}{2}\left[\cos(-80°) - \cos 90°\right]$$
$$= \tfrac{1}{2}\left[\cos 80° - 0\right]$$
$$= \tfrac{1}{2}\cos 80°$$

5. Using the product-to-sum formula for $\sin A \cos B$:
$$\sin 20°\cos 10° = \tfrac{1}{2}\left[\sin(20°-10°) + \sin(20°+10°)\right]$$
$$= \tfrac{1}{2}\left[\sin 10° + \sin 30°\right]$$
$$= \tfrac{1}{2}\left[\sin 10° + \tfrac{1}{2}\right]$$
$$= \tfrac{1}{2}\sin 10° + \tfrac{1}{4}$$

7.   Using the product-to-sum formula for $\cos A \cos B$:

$$\cos\tfrac{\pi}{5}\cos\tfrac{4\pi}{5} = \tfrac{1}{2}\left[\cos\left(\tfrac{\pi}{5}-\tfrac{4\pi}{5}\right)+\cos\left(\tfrac{\pi}{5}+\tfrac{4\pi}{5}\right)\right]$$
$$= \tfrac{1}{2}\left[\cos\left(-\tfrac{3\pi}{5}\right)+\cos\pi\right]$$
$$= \tfrac{1}{2}\left[\cos\tfrac{3\pi}{5}-1\right]$$
$$= \tfrac{1}{2}\cos\tfrac{3\pi}{5}-\tfrac{1}{2}$$

9.   Using the product-to-sum formula for $\sin A \sin B$:

$$\sin\tfrac{2\pi}{7}\sin\tfrac{5\pi}{7} = \tfrac{1}{2}\left[\cos\left(\tfrac{2\pi}{7}-\tfrac{5\pi}{7}\right)-\cos\left(\tfrac{2\pi}{7}+\tfrac{5\pi}{7}\right)\right]$$
$$= \tfrac{1}{2}\left[\cos\left(-\tfrac{3\pi}{7}\right)-\cos\pi\right]$$
$$= \tfrac{1}{2}\left[\cos\tfrac{3\pi}{7}-(-1)\right]$$
$$= \tfrac{1}{2}\cos\tfrac{3\pi}{7}+\tfrac{1}{2}$$

11.   Using the product-to-sum formula for $\sin A \cos B$:

$$\sin\tfrac{7\pi}{12}\cos\tfrac{\pi}{12} = \tfrac{1}{2}\left[\sin\left(\tfrac{7\pi}{12}-\tfrac{\pi}{12}\right)+\sin\left(\tfrac{7\pi}{12}+\tfrac{\pi}{12}\right)\right]$$
$$= \tfrac{1}{2}\left[\sin\tfrac{\pi}{2}+\sin\tfrac{2\pi}{3}\right]$$
$$= \tfrac{1}{2}\left[1+\tfrac{\sqrt{3}}{2}\right]$$
$$= \tfrac{1}{2}+\tfrac{\sqrt{3}}{4}$$

13.   Using the product-to-sum formula for $\sin A \sin B$:

$$\sin 3x\sin 4x = \tfrac{1}{2}\left[\cos(3x-4x)-\cos(3x+4x)\right]$$
$$= \tfrac{1}{2}\left[\cos(-x)-\cos(7x)\right]$$
$$= \tfrac{1}{2}\left[\cos x-\cos 7x\right]$$
$$= \tfrac{1}{2}\cos x-\tfrac{1}{2}\cos 7x$$

15.   Using the product-to-sum formula for $\sin A \cos B$:

$$\sin 6\theta\cos 5\theta = \tfrac{1}{2}\left[\sin(6\theta-5\theta)+\sin(6\theta+5\theta)\right]$$
$$= \tfrac{1}{2}\left[\sin\theta+\sin 11\theta\right]$$
$$= \tfrac{1}{2}\sin\theta+\tfrac{1}{2}\sin 11\theta$$

17.   Using the product-to-sum formula for $\sin A \cos B$:

$$\sin\tfrac{3\theta}{2}\cos\tfrac{\theta}{2} = \tfrac{1}{2}\left[\sin\left(\tfrac{3\theta}{2}-\tfrac{\theta}{2}\right)+\sin\left(\tfrac{3\theta}{2}+\tfrac{\theta}{2}\right)\right] = \tfrac{1}{2}\left[\sin\theta+\sin 2\theta\right] = \tfrac{1}{2}\sin\theta+\tfrac{1}{2}\sin 2\theta$$

19. Using the product-to-sum formula for $\sin A \sin B$:
$$\sin(2x+y)\sin(2x-y) = \tfrac{1}{2}\left[\cos(2x+y-2x+y)-\cos(2x+y+2x-y)\right]$$
$$= \tfrac{1}{2}\left[\cos 2y - \cos 4x\right]$$
$$= \tfrac{1}{2}\cos 2y - \tfrac{1}{2}\cos 4x$$

21. Using the product-to-sum formula for $\sin A \cos B$:
$$\sin 2t \cos(s-t) = \tfrac{1}{2}\left[\sin(2t-s+t)+\sin(2t+s-t)\right]$$
$$= \tfrac{1}{2}\left[\sin(3t-s)+\sin(t+s)\right]$$
$$= \tfrac{1}{2}\sin(3t-s)+\tfrac{1}{2}\sin(t+s)$$

23. Using the sum-to-product formula for $\cos\alpha + \cos\beta$:
$$\cos 35° + \cos 55° = 2\cos\tfrac{35°+55°}{2}\cos\tfrac{35°-55°}{2}$$
$$= 2\cos 45°\cos(-10°)$$
$$= 2\cdot\tfrac{\sqrt{2}}{2}\cdot\cos 10°$$
$$= \sqrt{2}\cos 10°$$

25. Using the sum-to-product formula for $\sin\alpha - \sin\beta$:
$$\sin\tfrac{\pi}{5} - \sin\tfrac{3\pi}{10} = 2\cos\tfrac{\frac{\pi}{5}+\frac{3\pi}{10}}{2}\sin\tfrac{\frac{\pi}{5}-\frac{3\pi}{10}}{2}$$
$$= 2\cos\tfrac{\pi}{4}\sin\left(-\tfrac{\pi}{20}\right)$$
$$= -2\cdot\tfrac{\sqrt{2}}{2}\sin\tfrac{\pi}{20}$$
$$= -\sqrt{2}\sin\tfrac{\pi}{20}$$

27. Using the sum-to-product formula for $\cos\alpha - \cos\beta$:
$$\cos 5\theta - \cos 3\theta = -2\sin\tfrac{5\theta+3\theta}{2}\sin\tfrac{5\theta-3\theta}{2} = -2\sin 4\theta\sin\theta$$

29. Using the hint, $\cos 65° = \sin(90°-65°) = \sin 25°$. Now using the sum-to-product formula for $\sin\alpha + \sin\beta$:
$$\sin 35° + \sin 25° = 2\sin\tfrac{35°+25°}{2}\cos\tfrac{35°-25°}{2} = 2\sin 30°\cos 5° = 2\cdot\tfrac{1}{2}\cos 5° = \cos 5°$$

31. Using the sum-to-product formula for $\sin\alpha - \sin\beta$:
$$\sin\left(\tfrac{\pi}{3}+2\theta\right) - \sin\left(\tfrac{\pi}{3}-2\theta\right) = 2\cos\tfrac{2\pi/3}{2}\sin\tfrac{4\theta}{2} = 2\cos\tfrac{\pi}{3}\sin 2\theta = 2\cdot\tfrac{1}{2}\sin 2\theta = \sin 2\theta$$

33. Simplify the numerator and denominator separately. For the numerator, first use the identity $\cos\theta = \sin\left(\frac{\pi}{2} - \theta\right)$ so that $\cos\frac{5\pi}{12} = \sin\frac{\pi}{12}$. Now using the sum-to-product formula for $\sin\alpha + \sin\beta$:

$$\sin\frac{\pi}{12} + \sin\frac{5\pi}{12} = 2\sin\frac{\frac{\pi}{12}+\frac{5\pi}{12}}{2}\cos\frac{\frac{\pi}{12}-\frac{5\pi}{12}}{2} = 2\sin\frac{\pi}{4}\cos\left(-\frac{\pi}{6}\right) = 2\cdot\frac{\sqrt{2}}{2}\cdot\frac{\sqrt{3}}{2} = \frac{\sqrt{6}}{2}$$

For the denominator, first use the identity $\cos\theta = \sin\left(\frac{\pi}{2} - \theta\right)$ so that $\cos\frac{\pi}{12} = \sin\frac{5\pi}{12}$. Now using the sum-to-product formula for $\sin\alpha - \sin\beta$:

$$\sin\frac{5\pi}{12} - \sin\frac{\pi}{12} = 2\cos\frac{\frac{5\pi}{12}+\frac{\pi}{12}}{2}\sin\frac{\frac{5\pi}{12}-\frac{\pi}{12}}{2} = 2\cos\frac{\pi}{4}\sin\frac{\pi}{6} = 2\cdot\frac{\sqrt{2}}{2}\cdot\frac{1}{2} = \frac{\sqrt{2}}{2}$$

Thus, the original problem becomes:

$$\frac{\cos\frac{5\pi}{12} + \sin\frac{5\pi}{12}}{\cos\frac{\pi}{12} - \sin\frac{\pi}{12}} = \frac{\frac{\sqrt{6}}{2}}{\frac{\sqrt{2}}{2}} = \frac{\sqrt{6}}{\sqrt{2}} = \sqrt{3}$$

35. Using the sum-to-product formulas for $\sin s + \sin t$ and $\cos s + \cos t$:

$$\frac{\sin s + \sin t}{\cos s + \cos t} = \frac{2\sin\frac{s+t}{2}\cos\frac{s-t}{2}}{2\cos\frac{s+t}{2}\cos\frac{s-t}{2}} = \frac{\sin\frac{s+t}{2}}{\cos\frac{s+t}{2}} = \tan\frac{s+t}{2}$$

37. Using the sum-to-product formulas for $\sin 2x + \sin 2y$ and $\cos 2x + \cos 2y$:

$$\frac{\sin 2x + \sin 2y}{\cos 2x + \cos 2y} = \frac{2\sin\frac{2x+2y}{2}\cos\frac{2x-2y}{2}}{2\cos\frac{2x+2y}{2}\cos\frac{2x-2y}{2}} = \frac{\sin(x+y)}{\cos(x+y)} = \tan(x+y)$$

Note that we can also use the result from Exercise 35 where $s = 2x$ and $t = 2y$.

39. Using the hint, use the sum-to-product formulas to obtain:

$$\cos 7\theta + \cos 5\theta = 2\cos\frac{7\theta+5\theta}{2}\cos\frac{7\theta-5\theta}{2} = 2\cos 6\theta\cos\theta$$

$$\cos 3\theta + \cos\theta = 2\cos\frac{3\theta+\theta}{2}\cos\frac{3\theta-\theta}{2} = 2\cos 2\theta\cos\theta$$

Thus:

$$\cos 7\theta + \cos 5\theta + \cos 3\theta + \cos\theta = 2\cos 6\theta\cos\theta + 2\cos 2\theta\cos\theta$$
$$= (2\cos\theta)(\cos 6\theta + \cos 2\theta)$$
$$= (2\cos\theta)\left(2\cos\frac{6\theta+2\theta}{2}\cos\frac{6\theta-2\theta}{2}\right)$$
$$= 4\cos\theta\cos 4\theta\cos 2\theta$$

41. **(a)** Since $\cos\frac{x}{2} = \sin\left(\frac{\pi}{2} - \frac{x}{2}\right)$, use the sum-to-product identity for $\sin\alpha + \sin\beta$ to obtain:

$$\sqrt{2}\left[\sin\frac{x}{2} + \cos\frac{x}{2}\right] = \sqrt{2}\left[\sin\frac{x}{2} + \sin\left(\frac{\pi}{2} - \frac{x}{2}\right)\right]$$

$$= \sqrt{2}\left[2\sin\frac{\frac{\pi}{2}}{2}\cos\frac{x - \frac{\pi}{2}}{2}\right]$$

$$= \sqrt{2}\left[2\sin\frac{\pi}{4}\cos\left(\frac{x}{2} - \frac{\pi}{4}\right)\right]$$

$$= \sqrt{2}\cdot 2\cdot\frac{\sqrt{2}}{2}\cos\left(\frac{x}{2} - \frac{\pi}{4}\right)$$

$$= 2\cos\left(\frac{x}{2} - \frac{\pi}{4}\right)$$

(b) Since $f(x) = 2\cos\left(\frac{x}{2} - \frac{\pi}{4}\right)$, the amplitude is 2, the period is $\frac{2\pi}{1/2} = 4\pi$, and the phase shift is $\frac{\pi/4}{1/2} = \frac{\pi}{2}$. Sketch the graph:

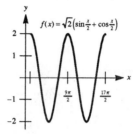

43. Begin with the product-to-sum formula:

$$\cos A\cos B = \tfrac{1}{2}\left[\cos(A - B) + \cos(A + B)\right]$$

If we let $A + B = \alpha$ and $A - B = \beta$, then:

$$A = \frac{\alpha + \beta}{2} \text{ and } B = \frac{\alpha - \beta}{2}$$

Substituting:

$$\cos\frac{\alpha + \beta}{2}\cos\frac{\alpha - \beta}{2} = \tfrac{1}{2}\left[\cos\beta + \cos\alpha\right]$$

Multiplying by 2, we have the desired identity:

$$\cos\alpha + \cos\beta = 2\cos\frac{\alpha + \beta}{2}\cos\frac{\alpha - \beta}{2}$$

45. Using the identity for $\sin\alpha + \sin\beta$ (derived in Exercise 44), we replace β with $-\beta$ and note that $\sin(-\beta) = -\sin\beta$:

$$\sin\alpha + \sin(-\beta) = 2\sin\frac{\alpha - \beta}{2}\cos\frac{\alpha + \beta}{2}$$

$$\sin\alpha - \sin\beta = 2\cos\frac{\alpha + \beta}{2}\sin\frac{\alpha - \beta}{2}$$

47. Obtaining a common denominator in the expression, and noting that $\sin 10° = \cos(90° - 10°) = \cos 80°$:

$$\frac{1}{2\sin 10°} - 2\sin 70° = \frac{1 - 4\sin 10°\sin 70°}{2\sin 10°} = \frac{1 - 4\sin 10°\sin 70°}{2\cos 80°}$$

Now using the product-to-sum formula for $\sin A \sin B$:

$$\sin 10° \sin 70° = \tfrac{1}{2}\left[\cos(10°-70°)-\cos(10°+70°)\right]$$
$$= \tfrac{1}{2}\left[\cos(-60°)-\cos 80°\right]$$
$$= \tfrac{1}{2}\left[\tfrac{1}{2}-\cos 80°\right]$$
$$= \tfrac{1}{4}-\tfrac{1}{2}\cos 80°$$

Thus:

$$\frac{1}{2\sin 10°}-2\sin 70° = \frac{1-4\sin 10° \sin 70°}{2\cos 80°}$$
$$= \frac{1-4\left(\tfrac{1}{4}-\tfrac{1}{2}\cos 80°\right)}{2\cos 80°}$$
$$= \frac{1-1+2\cos 80°}{2\cos 80°}$$
$$= \frac{2\cos 80°}{2\cos 80°}$$
$$= 1$$

49. First, note that $A + B + C = 180°$ implies that $A + B = 180° - C, B + C = 180° - A$ and $A + C = 180° - B$. Also, note the identities $\sin(180° - \alpha) = \sin \alpha$ and $\cos(180° - \alpha) = -\cos \alpha$. Working from the right-hand side, note that:

$$\sin \tfrac{A}{2}\sin \tfrac{B}{2} = \tfrac{1}{2}\cos\left(\tfrac{A}{2}-\tfrac{B}{2}\right)-\tfrac{1}{2}\cos\left(\tfrac{A}{2}+\tfrac{B}{2}\right)$$

Thus:

$$\sin \tfrac{A}{2}\sin \tfrac{B}{2}\sin \tfrac{C}{2}$$
$$= \tfrac{1}{2}\cos\left(\tfrac{A}{2}-\tfrac{B}{2}\right)\sin \tfrac{C}{2} - \tfrac{1}{2}\cos\left(\tfrac{A}{2}+\tfrac{B}{2}\right)\sin \tfrac{C}{2}$$
$$= \tfrac{1}{2}\sin\left[90°-\tfrac{A}{2}+\tfrac{B}{2}\right]\sin \tfrac{C}{2} - \tfrac{1}{2}\sin\left[90°-\tfrac{A}{2}-\tfrac{B}{2}\right]\sin \tfrac{C}{2}$$
$$= \tfrac{1}{4}\left[\cos\left(90°-\tfrac{A}{2}+\tfrac{B}{2}-\tfrac{C}{2}\right)-\cos\left(90°-\tfrac{A}{2}+\tfrac{B}{2}+\tfrac{C}{2}\right)\right]$$
$$\quad -\tfrac{1}{4}\left[\cos\left(90°-\tfrac{A}{2}-\tfrac{B}{2}-\tfrac{C}{2}\right)-\cos\left(90°-\tfrac{A}{2}-\tfrac{B}{2}+\tfrac{C}{2}\right)\right]$$
$$= \tfrac{1}{4}\left[\cos\frac{180°+B-(A+C)}{2}-\cos\frac{180°-A+(B+C)}{2}-\cos\frac{180°-(A+B+C)}{2}+\cos\frac{180°+C-(A+B)}{2}\right]$$
$$= \tfrac{1}{4}\left[\cos\frac{180°+B-(180°-B)}{2}-\cos\frac{180°-A+(180°-A)}{2}-\cos\frac{180°-180°}{2}+\cos\frac{180°+C-(180°-C)}{2}\right]$$
$$= \tfrac{1}{4}\left[\cos B - \cos(180°-A)-\cos 0 + \cos C\right]$$
$$= \tfrac{1}{4}\left[\cos B + \cos A - 1 + \cos C\right]$$

Now prove the identity:

$$1+4\sin \tfrac{A}{2}\sin \tfrac{B}{2}\sin \tfrac{C}{2} = 1 + \cos B + \cos A - 1 + \cos C = \cos A + \cos B + \cos C$$

51. (a) Calculating the sum of the cosines for each triangle:

$$\cos 30° + \cos 70° + \cos 80° \approx 1.38$$
$$\cos 40° + \cos 25° + \cos 115° \approx 1.25$$
$$\cos 55° + \cos 55° + \cos 70° \approx 1.49$$

In each case the sum of the cosines is less than $\frac{3}{2}$.

   (b) Since each angle of an equilateral triangle is 60°, the sum is:

$$\cos 60° + \cos 60° + \cos 60° = \tfrac{1}{2} + \tfrac{1}{2} + \tfrac{1}{2} = \tfrac{3}{2}$$

   (c) (i) This is the sum-to-product formula for $\cos A + \cos B$.

   (ii) This is true since $\cos \dfrac{A-B}{2} \leq 1$.

   (iii) This is true since $A + B = 180° - C$.

   (iv) This is just division by 2.

   (v) The identities used are $\cos(90° - \theta) = \sin \theta$ and $\cos \theta = 1 - 2\sin^2 \frac{\theta}{2}$.

   (vi) Multiplying out this expression shows they are equal.

   (vii) Since $2\left(\sin \frac{C}{2} - \frac{1}{2}\right)^2 \geq 0$, the expression is at most $\frac{3}{2}$.

9.4 Trigonometric Equations

1. For $\theta = \frac{\pi}{2}$, $2\cos^2 \theta - 3\cos \theta = 2(0)^2 - 3(0) = 0$, so $\theta = \frac{\pi}{2}$ is a solution.

3. For $x = \frac{3\pi}{4}$, $\tan^2 x - 3\tan x + 2 = (-1)^2 + 3 + 2 = 6$, so $x = \frac{3\pi}{4}$ is not a solution.

5. Since $\sin \theta = \frac{\sqrt{3}}{2}$, $\theta = \frac{\pi}{3}$ and $\theta = \frac{2\pi}{3}$ are the primary solutions. All solutions will be of the form $\theta = \frac{\pi}{3} + 2\pi k$ or $\theta = \frac{2\pi}{3} + 2\pi k$, where k is any integer.

7. Since $\sin \theta = -\frac{1}{2}$, $\theta = \frac{7\pi}{6}$ and $\theta = \frac{11\pi}{6}$ are the primary solutions. All solutions will be of the form $\theta = \frac{7\pi}{6} + 2\pi k$ or $\theta = \frac{11\pi}{6} + 2\pi k$, where k is any integer.

9. Since $\cos \theta = -1$, $\theta = \pi$ is the primary solution. All solutions will be of the form $\theta = \pi + 2\pi k$, where k is any integer.

11. Since $\tan \theta = \sqrt{3}$, $\theta = \frac{\pi}{3}$ is the primary solution. All solutions will be of the form $\theta = \frac{\pi}{3} + \pi k$, where k is any integer.

13. Since $\tan x = 0$, $x = 0$ is the primary solution. All solutions will be of the form $x = 0 + \pi k = \pi k$, where k is any integer.

15. Since $2\cos^2\theta + \cos\theta = \cos\theta(2\cos\theta + 1) = 0$, the primary solutions are the solutions of $\cos\theta = 0$ or $\cos\theta = -\frac{1}{2}$, which are $\theta = \frac{\pi}{2}$, $\theta = \frac{3\pi}{2}$, $\theta = \frac{2\pi}{3}$ or $\theta = \frac{4\pi}{3}$. All solutions will be of the form $\theta = \frac{\pi}{2} + \pi k$, $\theta = \frac{2\pi}{3} + 2\pi k$ or $\theta = \frac{4\pi}{3} + 2\pi k$, where k is any integer.

17. Since $\cos^2 t\sin t - \sin t = \sin t(\cos^2 t - 1) = 0$, $\sin t = 0$ or $\cos t = \pm 1$. Therefore primary solutions are $t = 0$ or $t = \pi$. All solutions are of the form $t = \pi k$, where k is any integer.

19. Using the identity $\cos^2 x = 1 - \sin^2 x$:
$$2\cos^2 x - \sin x - 1 = 2(1 - \sin^2 x) - \sin x - 1$$
$$= -2\sin^2 x - \sin x + 1$$
$$= (-2\sin x + 1)(\sin x + 1)$$
So $\sin x = \frac{1}{2}$ or $\sin x = -1$. Thus the primary solutions are $x = \frac{\pi}{6}$, $x = \frac{5\pi}{6}$ or $x = \frac{3\pi}{2}$. All solutions are of the form $x = \frac{\pi}{6} + 2\pi k$, $x = \frac{5\pi}{6} + 2\pi k$ or $x = \frac{3\pi}{2} + 2\pi k$, where k is any integer.

21. Since $\sqrt{3}\sin t - \sqrt{1 + \sin^2 t} = 0$ is equivalent to $3\sin^2 t = 1 + \sin^2 t$ by squaring each side, $2\sin^2 t = 1$ and thus $\sin^2 t = \frac{1}{2}$ and $\sin t = \pm\frac{\sqrt{2}}{2}$. This would have primary solutions of $\frac{\pi}{4}$, $\frac{3\pi}{4}$, $\frac{5\pi}{4}$ or $\frac{7\pi}{4}$, but $\frac{5\pi}{4}$ and $\frac{7\pi}{4}$ do not work in the original equation. So the primary solutions are $t = \frac{\pi}{4}$ or $t = \frac{3\pi}{4}$. All solutions are of the form $t = \frac{\pi}{4} + 2\pi k$ or $t = \frac{3\pi}{4} + 2\pi k$, where k is any integer.

23. Since $\cos x = 0.184$, $x \approx 1.39$ or $x \approx 4.90$.

25. Since $\sin x = \frac{1}{\sqrt{5}}$, $x \approx 0.46$ or $x \approx 2.68$.

27. Since $\tan x = 6$, $x \approx 1.41$ or $x \approx 4.55$.

29. Dividing through by $\cos t$ results in $\tan t = 5$, thus $t \approx 1.37$ or $t \approx 4.51$.

31. Since $\sec t = 2.24$, $\cos t = \dfrac{1}{2.24} \approx 0.45$. Thus $t \approx 1.11$ or $t \approx 5.18$.

33. Factoring the equation:
$$\tan^2 x + \tan x - 12 = 0$$
$$(\tan x + 4)(\tan x - 3) = 0$$
$$\tan x = -4 \text{ or } 3$$
If $\tan x = -4$, then $x \approx 1.82$ or $x \approx 4.96$. If $\tan x = 3$, then $x \approx 1.25$ or $x \approx 4.39$. The solutions are approximately $x \approx 1.25, 1.82, 4.39, 4.96$.

35.   Factoring the equation by grouping:
$$16\sin^3 x - 12\sin^2 x + 36\sin x - 27 = 0$$
$$4\sin^2 x(4\sin x - 3) + 9(4\sin x - 3) = 0$$
$$(4\sin x - 3)\left(4\sin^2 x + 9\right) = 0$$
$$\sin x = \tfrac{3}{4}$$
Note that $4\sin^2 x + 9 \neq 0$. If $\sin x = \tfrac{3}{4}$, then $x \approx 0.85$ or $x \approx 2.29$. The solutions are approximately $x \approx 0.85, 2.29$.

37.   If $\sin \theta = \tfrac{1}{4}$, then $\theta \approx 14.5°, 165.5°$.

39.   Factoring the equation:
$$9\tan^2 \theta - 16 = 0$$
$$(3\tan \theta + 4)(3\tan \theta - 4) = 0$$
$$\tan \theta = -\tfrac{4}{3}, \tfrac{4}{3}$$
If $\tan \theta = -\tfrac{4}{3}$, then $\theta \approx 126.9°$ or $\theta \approx 306.9°$. If $\tan \theta = \tfrac{4}{3}$, then $\theta \approx 53.1°$ or $\theta \approx 233.1°$. The solutions are approximately $\theta \approx 53.1°, 126.9°, 233.1°, 306.9°$.

41.   Using the quadratic formula:
$$\cos \theta = \frac{-(-1) \pm \sqrt{(-1)^2 - 4(1)(-1)}}{2(1)} = \frac{1 \pm \sqrt{1 + 4}}{2} = \frac{1 \pm \sqrt{5}}{2} \approx -0.6180, 1.6180$$
Note that $\cos \theta = 1.6180$ is impossible. If $\cos \theta = -0.6180$, then $\theta \approx 128.2°$ or $\theta \approx 231.8°$. The solutions are approximately $\theta \approx 128.2°, 231.8°$.

43.   Since $\cos 3\theta = 1$, $3\theta = 2\pi k$ for any integer k. So $\theta = \frac{2\pi k}{3}$. Thus the values of θ in the interval $[0°, 360°)$ are $0°$, $120°$ and $240°$.

45.   Since $\sin 3\theta = -\frac{\sqrt{2}}{2}$, $3\theta = \frac{5\pi}{4} + 2\pi k$ or $\frac{7\pi}{4} + 2\pi k$. Thus $\theta = \frac{5\pi}{12} + \frac{2\pi k}{3}$ or $\frac{7\pi}{12} + \frac{2\pi k}{3}$. So the primary solutions are $75°$, $105°$, $195°$, $225°$, $315°$ or $345°$.

47.   Using the hint:
$$\sin \theta = \cos \tfrac{\theta}{2}$$
$$2\sin \tfrac{\theta}{2} \cos \tfrac{\theta}{2} = \cos \tfrac{\theta}{2}$$
$$\left(\cos \tfrac{\theta}{2}\right)\left(2\sin \tfrac{\theta}{2} - 1\right) = 0$$
Thus $\cos \tfrac{\theta}{2} = 0$ or $\sin \tfrac{\theta}{2} = \tfrac{1}{2}$. Now $\sin \tfrac{\theta}{2} = \tfrac{1}{2}$ when $\tfrac{\theta}{2} = 30°$ or $\tfrac{\theta}{2} = 150°$, and therefore when $\theta = 60°$ or $300°$. When $\cos \tfrac{\theta}{2} = 0$, $\tfrac{\theta}{2} = 90° + 360°k$ or $270° + 360°k$, so $\theta = 180°$. Combining yields $\theta = 60°$, $180°$ or $300°$.

49. Dividing each side by $\cos 2\theta$ results in $\tan 2\theta = \sqrt{3}$, thus $2\theta = 60° + 180°k$ and so $\theta = 30° + 90°k$. So the solutions in the interval $[0°, 360°)$ are $\theta = 30°, 120°, 210°$ or $300°$.

51. Following the hint, use the addition formula and double-angle formula for tangent to obtain:

$$\tan 3x = \tan(2x + x) = \frac{\tan 2x + \tan x}{1 - \tan x \tan 2x} = \frac{\frac{2\tan x}{1-\tan^2 x} + \tan x}{1 - \tan x \frac{2\tan x}{1-\tan^2 x}} = \frac{3\tan x - \tan^3 x}{1 - 3\tan^2 x}$$

So the left-hand side of the equation becomes:

$$\tan 3x - \tan x = \frac{3\tan x - \tan^3 x}{1 - 3\tan^2 x} - \tan x$$

$$= \frac{3\tan x - \tan^3 x - \tan x + 3\tan^3 x}{1 - 3\tan^2 x}$$

$$= \frac{2\tan x + 2\tan^3 x}{1 - 3\tan^2 x}$$

$$= \frac{2\tan x\left(1 + \tan^2 x\right)}{1 - 3\tan^2 x}$$

Since $1 + \tan^2 x \neq 0$, it must be that $\tan x = 0$ and thus $x = 0$ or $x = \pi$.

53. By the half-angle formula for cosine:

$$\cos\tfrac{x}{2} = \pm\sqrt{\frac{1 + \cos x}{2}}$$

Writing the original equation, then squaring:

$$\pm\sqrt{\frac{1 + \cos x}{2}} = 1 + \cos x$$

$$\frac{1 + \cos x}{2} = 1 + 2\cos x + \cos^2 x$$

$$1 + \cos x = 2 + 4\cos x + 2\cos^2 x$$

$$0 = 2\cos^2 x + 3\cos x + 1$$

$$0 = (2\cos x + 1)(\cos x + 1)$$

$$\cos x = -\tfrac{1}{2} \text{ or } \cos x = -1$$

These equations have solutions of $x = \frac{2\pi}{3}, \frac{4\pi}{3}$ or π. Upon checking $x = \frac{4\pi}{3}$ is not a solution, and thus the solutions are $x = \frac{2\pi}{3}$ or $x = \pi$.

55. Using the identity for $\sec 4\theta$ and the double-angle identity for sine:

$$\sec 4\theta + 2\sin 4\theta = 0$$

$$\frac{1}{\cos 4\theta} + 2\sin 4\theta = 0$$

$$1 + 2\sin 4\theta \cos 4\theta = 0$$

$$\sin 8\theta = -1$$

Thus $8\theta = \frac{3\pi}{2} + 2\pi k$, and so $\theta = \frac{3\pi}{16} + \frac{k\pi}{4}$, where k is any integer.

57. Following the hint:

$$4\sin\theta - 3\cos\theta = 2$$

$$4\sin\theta = 3\cos\theta + 2$$

$$16\sin^2\theta = 9\cos^2\theta + 12\cos\theta + 4$$

$$16(1 - \cos^2\theta) = 9\cos^2\theta + 12\cos\theta + 4$$

$$16 - 16\cos^2\theta = 9\cos^2\theta + 12\cos\theta + 4$$

$$25\cos^2\theta + 12\cos\theta - 12 = 0$$

This will not factor, so use the quadratic formula:

$$\cos\theta = \frac{-12 \pm \sqrt{(12)^2 - 4(25)(-12)}}{2(25)} = \frac{-12 \pm \sqrt{1344}}{50} = \frac{-12 \pm 8\sqrt{21}}{50} = \frac{-6 \pm 4\sqrt{21}}{25}$$

So $\cos\theta = 0.4932$ or $\cos\theta = -0.9732$, and thus $\theta = 60.45°$ (the other solution is not in the required interval).

59. (a) Squaring each side:

$$\sin^2 x \cos^2 x = 1$$

$$\sin^2 x(1 - \sin^2 x) = 1$$

$$\sin^2 x - \sin^4 x = 1$$

$$\sin^4 x - \sin^2 x + 1 = 0$$

(b) Using the quadratic formula:

$$\sin^2 x = \frac{1 \pm \sqrt{(-1)^2 - 4(1)(1)}}{2(1)} = \frac{1 \pm \sqrt{1-4}}{2} = \frac{1 \pm \sqrt{-3}}{2}$$

Thus the original equation has no real-number solutions.

61. (a) Since $\cos x = 0.412$, $x \approx 1.146$ or $x \approx -1.146$. Now add multiples of 2π on to these values until we reach an x-value greater than 1000. Note that $\frac{1000}{2\pi} \approx 159$ so check $x = 1.146 + 159(2\pi)$ as a starting point. The first such value is at $x \approx 1000.173$.

(b) Since $\cos x = -0.412$, $x \approx 1.995$ or $x \approx -1.995$. Again, add multiples of 2π on to these values until we reach an x-value greater than 1000. See the note from part (a). The first such value is at $x \approx 1001.022$.

63. (a) Solving the equation $f(x) = x$:
$$x + 0.4\sin(2\pi x) = x$$
$$0.4\sin(2\pi x) = 0$$
$$\sin(2\pi x) = 0$$
$$2\pi x = 0, \pi, 2\pi$$
$$x = 0, 0.5, 1$$
Note that $x = 0.5$ is a fixed point for the function.

(b) Computing the first ten iterates of $x_0 = 0.25$:

x	x_1	x_2	x_3	x_4	x_5
$f(x)$	0.65	0.3264	0.6812	0.3180	0.6820

x	x_6	x_7	x_8	x_9	x_{10}
$f(x)$	0.3180	0.6820	0.3180	0.6820	0.3180

The iterates are alternating between 0.3180 and 0.6820. They are not approaching the fixed point $x = 0.5$.

(c) Computing the first ten iterates of $x_0 = 0.45$:

x	x_1	x_2	x_3	x_4	x_5
$f(x)$	0.5736	0.3951	0.6400	0.3318	0.6801

x	x_6	x_7	x_8	x_9	x_{10}
$f(x)$	0.3181	0.6820	0.3180	0.6820	0.3180

The iterates are alternating between 0.3180 and 0.6820. They are not approaching the fixed point $x = 0.5$.

Graphing Utility Exercises for Section 9.4

1. (a) Graphing the equation:

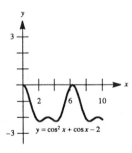

$$y = \cos^2 x + \cos x - 2$$

(b) The x-intercept is approximately $x \approx 6.28$.

(c) The exact value is $x = 2\pi$.

3.   (a)   Graphing the function:

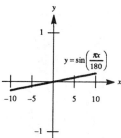

The graph is a sine function with amplitude 1 and period 360°.

(b)   Changing the viewing rectangle:

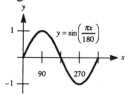

(c)   Graphing the two equations:

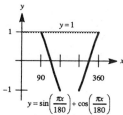

Note that there are two intersection points.

(d) $x = 90°$ is an intersection point of the two curves.

5.   Graphing the two equations:

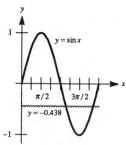

The x-coordinates of the intersection points are $x \approx 3.595, 5.830$.

7.   Graphing the two equations:

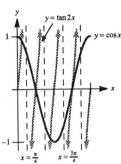

The x-coordinates of the intersection points are $x \approx 0.375, 1.571, 2.767, 4.712$.

9.   Graphing the two equations:

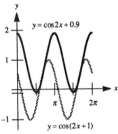

The x-coordinates of the intersection points are $x \approx 0.401, 3.542$.

11.   Graphing the two equations:

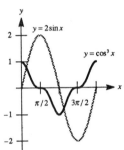

The x-coordinates of the intersection points are $x \approx 2.282, 5.424$.

13.   Graphing the two equations:

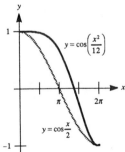

The x-coordinates of the intersection points are $x = 0, 6$, $x \approx 6.187$.

15. Graphing the two equations:

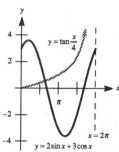

The x-coordinate of the intersection point is $x \approx 2.006$.

17. Graphing the two equations:

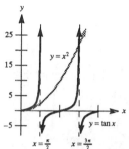

The x-coordinates of the intersection points are $x = 0$ and $x \approx 4.667$.

19. Graphing the two equations:

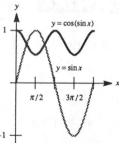

The x-coordinates of the intersection points are $x \approx 0.832, 2.310$.

21. Graphing the two equations:

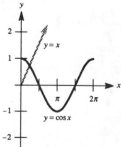

The x-coordinate of the intersection point is $x \approx 0.739$.

23. Graphing the two equations:

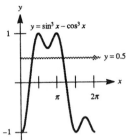

The x-coordinates of the intersection points are $x \approx 1.034, 3.679$.

25. Graphing the two equations:

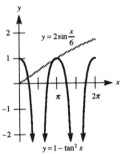

The x-coordinates of the intersection points are $x \approx 0.717, 2.864, 3.142$.

9.5 The Inverse Trigonometric Functions

1. We are asked to find the number x in the interval $\left[-\frac{\pi}{2}, \frac{\pi}{2}\right]$ such that $\sin x = \frac{\sqrt{3}}{2}$. Since $x = \frac{\pi}{3}$ is that number, $\sin^{-1}\left(\frac{\sqrt{3}}{2}\right) = \frac{\pi}{3}$.

3. We are asked to find the number x in the interval $\left(-\frac{\pi}{2}, \frac{\pi}{2}\right)$ such that $\tan x = \sqrt{3}$. Since $x = \frac{\pi}{3}$ is that number, $\tan^{-1}\sqrt{3} = \frac{\pi}{3}$.

5. We are asked to find the number x in the interval $\left(-\frac{\pi}{2}, \frac{\pi}{2}\right)$ such that $\tan x = -\frac{1}{\sqrt{3}}$. Since $x = -\frac{\pi}{6}$ is that number, $\arctan\left(-\frac{1}{\sqrt{3}}\right) = -\frac{\pi}{6}$.

7. We are asked to find the number x in the interval $\left(-\frac{\pi}{2}, \frac{\pi}{2}\right)$ such that $\tan x = 1$. Since $x = \frac{\pi}{4}$ is that number, $\tan^{-1} 1 = \frac{\pi}{4}$.

9. We are asked to find the number x in the interval $[0, \pi]$ such that $\cos x = 2\pi$. Since $\cos x \leq 1$ for all x, this value for x does not exist, thus $\cos^{-1} 2\pi$ is undefined.

11. If $x = \sin^{-1}\left(\frac{1}{4}\right)$, then $\sin x = \frac{1}{4}$. Thus $\sin\left[\sin^{-1}\frac{1}{4}\right] = \frac{1}{4}$.

13. If $x = \cos^{-1}\left(\frac{3}{4}\right)$, then $\cos x = \frac{3}{4}$. Thus $\cos\left[\cos^{-1}\frac{3}{4}\right] = \frac{3}{4}$.

15. We are asked to find the number x in the interval $\left(-\frac{\pi}{2}, \frac{\pi}{2}\right)$ such that $\tan x = \tan\left(-\frac{\pi}{7}\right)$. Since $x = -\frac{\pi}{7}$ is that number, $\arctan\left[\tan\left(-\frac{\pi}{7}\right)\right] = -\frac{\pi}{7}$.

17. Since $\sin\frac{\pi}{2} = 1$, we must find $\arcsin 1$. We are asked to find the number x in the interval $\left[-\frac{\pi}{2}, \frac{\pi}{2}\right]$ such that $\sin x = 1$. Since $x = \frac{\pi}{2}$ is that number, $\arcsin\left[\sin\frac{\pi}{2}\right] = \frac{\pi}{2}$.

19. Since $\cos 2\pi = 1$, we must find $\arccos 1$. We are asked to find the number x in the interval $[0, \pi]$ such that $\cos x = 1$. Since $x = 0$ is that number, $\arccos(\cos 2\pi) = 0$.

21. If $x = \sin^{-1}\left(\frac{4}{5}\right)$, then $\sin x = \frac{4}{5}$ and thus $\cos x = \frac{3}{5}$. Therefore:
$$\tan\left[\sin^{-1}\left(\frac{4}{5}\right)\right] = \tan x = \frac{\sin x}{\cos x} = \frac{\frac{4}{5}}{\frac{3}{5}} = \frac{4}{3}$$

23. Let $x = \tan^{-1} 1$, so $\tan x = 1$ and thus $x = \frac{\pi}{4}$. Therefore:
$$\sin(\tan^{-1} 1) = \sin\frac{\pi}{4} = \frac{\sqrt{2}}{2}$$

25. If $x = \arccos\frac{5}{13}$, then $\cos x = \frac{5}{13}$ and thus $\sin x = \frac{12}{13}$. Therefore:
$$\tan\left(\arccos\frac{5}{13}\right) = \tan x = \frac{\sin x}{\cos x} = \frac{\frac{12}{13}}{\frac{5}{13}} = \frac{12}{5}$$

27. Since $\arctan\sqrt{3} = \frac{\pi}{3}$:
$$\cos(\arctan\sqrt{3}) = \cos\left(\frac{\pi}{3}\right) = \frac{1}{2}$$

29. If $x = \arccos\left(-\frac{1}{3}\right)$, then $\cos x = -\frac{1}{3}$ and since x is in the interval $[0, \pi]$, x must lie in the second quadrant, so:
$$\sin x = \sqrt{1 - \cos^2 x} = \sqrt{1 - \left(-\frac{1}{3}\right)^2} = \sqrt{1 - \frac{1}{9}} = \sqrt{\frac{8}{9}} = \frac{2\sqrt{2}}{3}$$
Therefore:
$$\sin\left[\arccos\left(-\frac{1}{3}\right)\right] = \sin x = \frac{2\sqrt{2}}{3}$$

31. (a) $\sin^{-1}\left(\frac{3}{4}\right) \approx 0.84$ radians or $48.59°$

(b) $\cos^{-1}\left(\frac{2}{3}\right) \approx 0.84$ radians or $48.19°$

(c) $\tan^{-1}\pi \approx 1.26$ radians or $72.34°$

(d) $\tan^{-1}\left(\tan^{-1}\pi\right) \approx \tan^{-1}1.26 \approx 0.90$ radians or $51.57°$

33. First compute $\cos^{-1}\left(\frac{\sqrt{2}}{2}\right)$ and $\sin^{-1}(-1)$:

$$\cos^{-1}\left(\frac{\sqrt{2}}{2}\right) = \frac{\pi}{4} \text{ since } \cos\frac{\pi}{4} = \frac{\sqrt{2}}{2} \text{ and } \frac{\pi}{4} \text{ is in the interval } [0,\pi]$$

$$\sin^{-1}(-1) = -\frac{\pi}{2} \text{ since } \sin\left(-\frac{\pi}{2}\right) = -1 \text{ and } -\frac{\pi}{2} \text{ is in the interval } \left[-\frac{\pi}{2},\frac{\pi}{2}\right]$$

Therefore:

$$\sec\left[\cos^{-1}\left(\frac{\sqrt{2}}{2}\right) + \sin^{-1}(-1)\right] = \sec\left(\frac{\pi}{4} - \frac{\pi}{2}\right) = \sec\left(-\frac{\pi}{4}\right) = \frac{1}{\cos\left(-\frac{\pi}{4}\right)} = \frac{1}{\frac{\sqrt{2}}{2}} = \sqrt{2}$$

35. Let $u = \sin^{-1}x$, so $\sin u = x$ and u is in the interval $-\frac{\pi}{2} \le u \le \frac{\pi}{2}$. Since $\cos u \ge 0$:

$$\cos u = \sqrt{1-\sin^2 u} = \sqrt{1-x^2}$$

Therefore:

$$\cos\left(\sin^{-1}x\right) = \cos u = \sqrt{1-x^2}$$

37. Since $\sin 2\theta = 2\sin\theta\cos\theta$ by the double-angle identity for sine, we must find $\cos\theta$. Since $0 < \theta < \frac{\pi}{2}$, $\cos\theta > 0$ and thus:

$$\cos\theta = \sqrt{1-\sin^2\theta} = \sqrt{1-\left(\frac{3x}{2}\right)^2} = \sqrt{1-\frac{9x^2}{4}} = \frac{\sqrt{4-9x^2}}{2}$$

Since $\sin\theta = \frac{3x}{2}$, $\theta = \sin^{-1}\left(\frac{3x}{2}\right)$, and therefore:

$$\frac{\theta}{4} - \sin 2\theta = \frac{1}{4}\theta - 2\sin\theta\cos\theta$$

$$= \frac{1}{4}\sin^{-1}\left(\frac{3x}{2}\right) - 2\cdot\frac{3x}{2}\cdot\frac{\sqrt{4-9x^2}}{2}$$

$$= \frac{1}{4}\sin^{-1}\left(\frac{3x}{2}\right) - \frac{3x\sqrt{4-9x^2}}{2}$$

39. Given $\tan\theta = \frac{x-1}{2}$, construct the triangle:

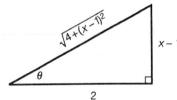

Therefore:

$$\theta - \cos\theta = \tan^{-1}\left(\tfrac{x-1}{2}\right) - \frac{2}{\sqrt{4+(x-1)^2}} = \tan^{-1}\left(\tfrac{x-1}{2}\right) - \frac{2}{\sqrt{5-2x+x^2}}$$

41. (a) Reflect $y = \sin^{-1} x$ across the x-axis:

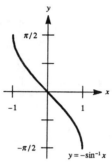

(b) Reflect $y = \sin^{-1} x$ across the y-axis:

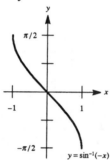

(c) Reflect $y = \sin^{-1} x$ across both the x- and y-axes:

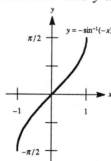

43.   (a)   Displace $y = \arccos x$ to the left 1 unit:

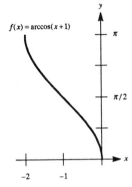

(b)   Displace $y = \arccos x$ up $\frac{\pi}{2}$ units:

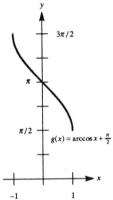

45.   (a)   Displace $y = \arcsin x$ to the left 2 units, then reflect across the y-axis, then displace up $\frac{\pi}{2}$ units:

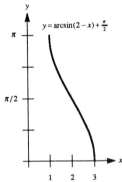

(b) Displace $y = \arcsin x$ to the left 2 units, then reflect across both the x- and y-axes, then displace up $\frac{\pi}{2}$ units:

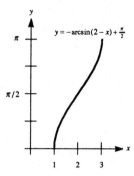

47. (a) Reflect $y = \tan^{-1} x$ across the x-axis:

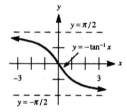

(b) Reflect $y = \tan^{-1} x$ across the y-axis:

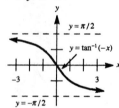

(c) Reflect $y = \tan^{-1} x$ across both the x- and y-axes:

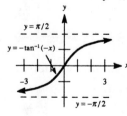

49. Displace $y = \arctan x$ to the left 1 unit, then reflect across both the x- and y-axes, then displace down $\frac{\pi}{2}$ units:

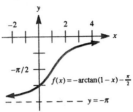

51. Let $\theta = \tan^{-1} 4$, so $\tan \theta = 4$. Draw the triangle:

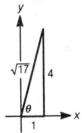

Using the double-angle formula for $\sin 2\theta$:

$$\sin\left(2\tan^{-1} 4\right) = \sin 2\theta = 2\sin\theta\cos\theta = 2 \cdot \frac{4}{\sqrt{17}} \cdot \frac{1}{\sqrt{17}} = \tfrac{8}{17}$$

53. Let $s = \arccos\frac{3}{5}$, so $\cos s = \frac{3}{5}$. Drawing a triangle:

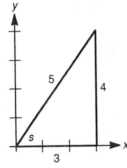

Let $t = \arctan\frac{7}{13}$, so $\tan t = \frac{7}{13}$. Drawing a triangle:

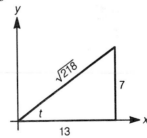

Using the addition formula for $\sin(s-t)$:

$$\sin\left(\arccos\tfrac{3}{5} - \arctan\tfrac{7}{13}\right) = \sin(s-t)$$

$$= \sin s\cos t - \cos s\sin t$$

$$= \tfrac{4}{5}\cdot\frac{13}{\sqrt{218}} - \tfrac{3}{5}\cdot\frac{7}{\sqrt{218}}$$

$$= \frac{31}{5\sqrt{218}}$$

$$= \frac{31\sqrt{218}}{1090}$$

55. (a) Since $\alpha = \sin^{-1}x$, $-\tfrac{\pi}{2} \le \alpha \le \tfrac{\pi}{2}$ and since $\beta = \cos^{-1}x$, $0 \le \beta \le \pi$. Thus:

$$-\tfrac{\pi}{2} + 0 \le \alpha + \beta \le \tfrac{\pi}{2} + \pi$$

$$-\tfrac{\pi}{2} \le \alpha + \beta \le \tfrac{3\pi}{2}$$

(b) Since $\alpha = \sin^{-1}x$, $\sin\alpha = x$ and thus:

$$\cos\alpha = \sqrt{1 - \sin^2\alpha} = \sqrt{1 - x^2}$$

Similarly, since $\beta = \cos^{-1}x$, $\cos\beta = x$ and thus:

$$\sin\beta = \sqrt{1 - \cos^2\beta} = \sqrt{1 - x^2}$$

Using the addition formula for $\sin(\alpha + \beta)$:

$$\sin(\alpha + \beta) = \sin\alpha\cos\beta + \cos\alpha\sin\beta$$

$$= x\cdot x + \sqrt{1-x^2}\cdot\sqrt{1-x^2}$$

$$= x^2 + 1 - x^2$$

$$= 1$$

But if $-\tfrac{\pi}{2} \le \alpha + \beta \le \tfrac{3\pi}{2}$ (from part (a)) and $\sin(\alpha+\beta) = 1$, then $\alpha + \beta = \tfrac{\pi}{2}$.

57. Take the sine of each side to get:

$$\sin\left(\sin^{-1}(3t-2)\right) = \sin\left(\sin^{-1}t - \cos^{-1}t\right)$$

$$3t - 2 = \sin\left(\sin^{-1}t - \cos^{-1}t\right)$$

Let $u = \sin^{-1}t$, so $\sin u = t$ and $\cos u = \sqrt{1-t^2}$. Also, let $v = \cos^{-1}t$, so $\cos v = t$ and $\sin v = \sqrt{1-t^2}$. Thus:

$$\sin\left(\sin^{-1}t - \cos^{-1}t\right) = \sin(u-v)$$

$$= \sin u\cos v - \cos u\sin v$$

$$= t\cdot t - \sqrt{1-t^2}\,\sqrt{1-t^2}$$

$$= t^2 - 1 + t^2$$

$$= 2t^2 - 1$$

Thus, solve the equation:

$$3t - 2 = 2t^2 - 1$$
$$0 = 2t^2 - 3t + 1$$
$$0 = (2t - 1)(t - 1)$$
$$t = \tfrac{1}{2}, 1$$

59.   Let $s = \arctan x$ and $t = \arctan y$, so $\tan s = x$ and $\tan t = y$. Using the addition formula for $\tan (s + t)$:

$$\tan(s + t) = \frac{\tan s + \tan t}{1 - \tan s \tan t} = \frac{x + y}{1 - xy}$$

Then:

$$\arctan(\tan(s + t)) = \arctan\left(\frac{x+y}{1-xy}\right)$$

So $s + t = \arctan x + \arctan y = \arctan\left(\frac{x+y}{1-xy}\right)$.

61.   Following the hint, take the tangent of each side to obtain:

$$\tan\left(2 \tan^{-1} x\right) = \tan\left[\tan^{-1}\left(\tfrac{1}{4x}\right)\right]$$
$$\tan\left(2 \tan^{-1} x\right) = \tfrac{1}{4x}$$

Let $\alpha = \tan^{-1} x$, so using the double-angle formula for $\tan 2\alpha$:

$$\frac{2 \tan \alpha}{1 - \tan^2 \alpha} = \frac{1}{4x}$$
$$\frac{2x}{1 - x^2} = \frac{1}{4x}$$
$$8x^2 = 1 - x^2$$
$$9x^2 = 1$$
$$x^2 = \tfrac{1}{9}$$
$$x = \pm\tfrac{1}{3}$$

63.   Take the tangent of each side of the equation to obtain:

$$\tan\left(2 \tan^{-1} \sqrt{t - t^2}\right) = \tan\left(\tan^{-1} t + \tan^{-1}(1 - t)\right)$$

Let $\alpha = \tan^{-1} \sqrt{t - t^2}$, so $\tan \alpha = \sqrt{t - t^2}$. Let $\beta = \tan^{-1} t$, so $\tan \beta = t$. Let $\gamma = \tan^{-1}(1 - t)$, so $\tan \gamma = 1 - t$. Now simplify each side of the equation, using the double-angle and sum-angle identities for tangent:

$$\tan\left(2 \tan^{-1} \sqrt{t - t^2}\right) = \tan 2\alpha = \frac{2 \tan \alpha}{1 - \tan^2 \alpha} = \frac{2\sqrt{t - t^2}}{1 - \left(t - t^2\right)} = \frac{2\sqrt{t - t^2}}{1 - t + t^2}$$

$$\tan\left(\tan^{-1} + \tan^{-1}(1 - t)\right) = \tan(\beta + \gamma) = \frac{\tan \beta + \tan \gamma}{1 - \tan \beta \tan \gamma} = \frac{t + 1 - t}{1 - t(1 - t)} = \frac{1}{1 - t + t^2}$$

So the original equation becomes:

$$\frac{2\sqrt{t-t^2}}{1-t+t^2} = \frac{1}{1-t+t^2}$$
$$2\sqrt{t-t^2} = 1$$

Squaring each side:

$$4\left(t-t^2\right) = 1$$
$$4t - 4t^2 = 1$$
$$0 = 4t^2 - 4t + 1$$
$$0 = (2t-1)^2$$
$$t = \tfrac{1}{2}$$

65. (a) Graph $y = \sec x$ on the domain $\left[0, \frac{\pi}{2}\right) \cup \left[\pi, \frac{3\pi}{2}\right)$, noting that it is one-to-one:

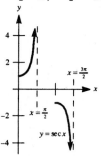

(b) Graph $y = \sec^{-1} x$:

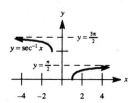

(c) Let $x = \sec^{-1}\left(\frac{2}{\sqrt{3}}\right)$, so $\sec x = \frac{2}{\sqrt{3}}$. Then $\cos x = \frac{\sqrt{3}}{2}$, so $x = \frac{\pi}{6}$. Let $y = \sec^{-1}\left(-\frac{2}{\sqrt{3}}\right)$, so $\sec y = -\frac{2}{\sqrt{3}}$. Then $\cos y = -\frac{\sqrt{3}}{2}$, so $y = \frac{7\pi}{6}$. Thus $\sec^{-1}\left(\frac{2}{\sqrt{3}}\right) = \frac{\pi}{6}$ and $\sec^{-1}\left(-\frac{2}{\sqrt{3}}\right) = \frac{7\pi}{6}$.

(d) Let $x = \sec^{-1}\left(\sqrt{2}\right)$, so $\sec x = \sqrt{2}$. Then $\cos x = \frac{1}{\sqrt{2}}$, so $x = \frac{\pi}{4}$. Let $y = \sec^{-1}\left(-\sqrt{2}\right)$, so $\sec y = -\sqrt{2}$. Then $\cos y = -\frac{1}{\sqrt{2}}$, so $y = \frac{5\pi}{4}$. Thus $\sec^{-1}\left(\sqrt{2}\right) = \frac{\pi}{4}$ and $\sec^{-1}\left(-\sqrt{2}\right) = \frac{5\pi}{4}$.

(e) Since 2 is in the domain of $y = \sec^{-1} x$, $\sec\left(\sec^{-1} 2\right) = 2$. Since 0 is in the domain of the restricted secant function, $\sec^{-1}(\sec 0) = 0$.

67. (a)   If A corresponds to $A\,(0, 0)$, then the coordinates of B and C are $B\,(3, -2)$ and $C\,(5, 1)$. Following the hint, compute the slopes:

$$m_{\overline{AB}} = \frac{-2-0}{3-0} = -\frac{2}{3}$$

$$m_{\overline{BC}} = \frac{1-(-2)}{5-3} = \frac{3}{2}$$

Since the product of these slopes is $\left(-\frac{2}{3}\right)\left(\frac{3}{2}\right) = -1$, $\overline{AB}$ is perpendicular to $\overline{BC}$, and thus $\angle ABC = \frac{\pi}{2}$.

(b)   Compute each distance using the coordinates specified in part (a) and the distance formula:

$$AB = \sqrt{(3-0)^2 + (-2-0)^2} = \sqrt{9+4} = \sqrt{13}$$

$$BC = \sqrt{(5-3)^2 + (1+2)^2} = \sqrt{4+9} = \sqrt{13}$$

Thus $AB = BC$. Since $AB = BC$ and $\angle ABC = \frac{\pi}{2}$, $\triangle ABC$ is an isosceles triangle and thus $\angle BAC = \angle ACB = \frac{\pi}{4}$.

(c)   The relationships $\alpha = \tan^{-1}\left(\frac{2}{3}\right)$ and $\beta = \tan^{-1}\left(\frac{1}{5}\right)$ follow directly from the right triangles that contain α and β as acute angles. Since $\alpha + \beta = \frac{\pi}{4}$,

$$\tan^{-1}\left(\frac{2}{3}\right) + \tan^{-1}\left(\frac{1}{5}\right) = \frac{\pi}{4}.$$

Graphing Utility Exercises for Section 9.5

1. (a)   Graphing the function:

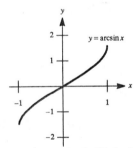

The maximum value is approximately 1.57 (when $x = 1$), and the minimum value is approximately -1.57 (when $x = -1$).

(b)   The maximum value is $\frac{\pi}{2}$ (when $x = 1$), and the minimum value is $-\frac{\pi}{2}$ (when $x = -1$).

3.   Graphing the function:

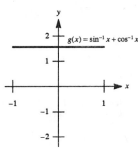

The graph is a horizontal line with a y-intercept of $\frac{\pi}{2}$. This demonstrates the identity $\sin^{-1} x + \cos^{-1} x = \frac{\pi}{2}$, for $-1 \leq x \leq 1$.

5.   Note that the graphs appear to be identical:

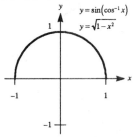

7.   Note that the graphs appear to be identical:

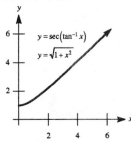

9.   The solution is $x \approx 0.74$.

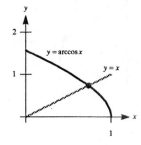

11.   (a)   The solution is $x \approx 0.96$.

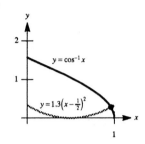

(b)   The solution is $x \approx 0.96$.

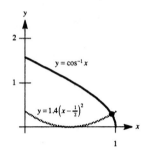

13.   (a)   The solution is $x \approx 0.24$.

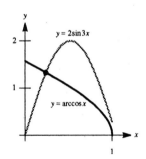

(b)   The solutions are $x \approx 0.19$ and $x \approx 0.68$.

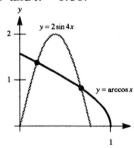

15. (a) The solution is $x \approx 0.56$.

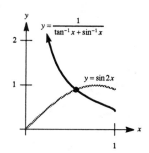

(b) The solutions are $x \approx 0.51$ and $x \approx 0.84$.

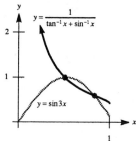

17. The solution is $x \approx 0.71$.

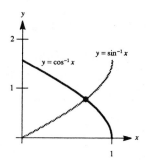

19. The solution is $x \approx 0.74$.

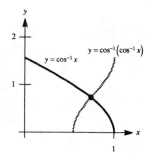

21.   The solution is $x \approx 0.94$.

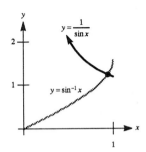

23.   The solution is $x \approx 0.93$.

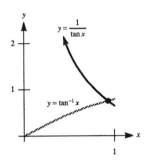

Chapter Nine Review Exercises

1.   Using the identity for $\cot x$ and the addition formula for $\tan (s + t)$:

$$\cot(x+y) = \frac{1}{\tan(x+y)} = \frac{1-\tan x \tan y}{\tan x + \tan y} \cdot \frac{\cot x \cot y}{\cot x \cot y} = \frac{\cot x \cot y - 1}{\cot y + \cot x}$$

3.   Working from the right-hand side:

$$\frac{2\tan x}{1+\tan^2 x} = \frac{\frac{2\sin x}{\cos x}}{1+\frac{\sin^2 x}{\cos^2 x}} = \frac{2\sin x \cos x}{\cos^2 x + \sin^2 x} = \frac{\sin 2x}{1} = \sin 2x$$

5.   Using the addition formulas for $\sin (x + y)$ and $\sin (x - y)$:

$$\frac{\sin(x+y)\sin(x-y)}{\cos^2 x \cos^2 y} = \frac{(\sin x \cos y + \cos x \sin y)(\sin x \cos y - \cos x \sin y)}{\cos^2 x \cos^2 y}$$

$$= \frac{\sin^2 x \cos^2 y - \cos^2 x \sin^2 y}{\cos^2 x \cos^2 y}$$

$$= \frac{\sin^2 x}{\cos^2 x} - \frac{\sin^2 y}{\cos^2 y}$$

$$= \tan^2 x - \tan^2 y$$

7. Using the half-angle formula for $\tan \frac{x}{2}$:

$$\sin x\left(\tan \tfrac{x}{2} + \cot \tfrac{x}{2}\right) = \sin x\left[\frac{\sin x}{1+\cos x} + \frac{1+\cos x}{\sin x}\right]$$

$$= \sin x\left[\frac{\sin^2 x + 1 + 2\cos x + \cos^2 x}{(\sin x)(1+\cos x)}\right]$$

$$= \sin x\left[\frac{2+2\cos x}{\sin x(1+\cos x)}\right]$$

$$= \frac{2(1+\cos x)}{1+\cos x}$$

$$= 2$$

9. Using the addition formula for $\tan(s-t)$:

$$\tan\left(\tfrac{\pi}{4} - x\right) = \frac{\tan \frac{\pi}{4} - \tan x}{1 + \tan x \tan \frac{\pi}{4}} = \frac{1 - \tan x}{1 + \tan x}$$

Now using the result of Exercise 8:

$$\tan\left(\tfrac{\pi}{4} + x\right) - \tan\left(\tfrac{\pi}{4} - x\right) = \frac{1 + \tan x}{1 - \tan x} - \frac{1 - \tan x}{1 + \tan x} = \frac{4\tan x}{1 - \tan^2 x} = 2\tan 2x$$

11. Using the double-angle formula for $\sin 2\theta$:

$$2\sin\left(\tfrac{\pi}{4} - \tfrac{x}{2}\right)\cos\left(\tfrac{\pi}{4} - \tfrac{x}{2}\right) = \sin\left[2 \cdot \left(\tfrac{\pi}{4} - \tfrac{x}{2}\right)\right] = \sin\left(\tfrac{\pi}{2} - x\right) = \cos x$$

13. First simplify $\tan\left(\tfrac{\pi}{4} - t\right)$ using the addition formula for $\tan(s-t)$:

$$\tan\left(\tfrac{\pi}{4} - t\right) = \frac{\tan \frac{\pi}{4} - \tan t}{1 + \tan \frac{\pi}{4} \tan t} = \frac{1 - \tan t}{1 + \tan t}$$

Therefore:

$$\frac{1 - \tan\left(\tfrac{\pi}{4} - t\right)}{1 + \tan\left(\tfrac{\pi}{4} - t\right)} = \frac{1 - \frac{1-\tan t}{1+\tan t}}{1 + \frac{1-\tan t}{1+\tan t}} = \frac{1 + \tan t - 1 + \tan t}{1 + \tan t + 1 - \tan t} = \frac{2\tan t}{2} = \tan t$$

15. Using the addition formula for $\tan(s+t)$:

$$\frac{\tan(\alpha - \beta) + \tan \beta}{1 - \tan(\alpha - \beta)\tan \beta} = \tan\left[(\alpha - \beta) + \beta\right] = \tan \alpha$$

17. Using the addition formula for tan $(s+t)$ and the double-angle formula for tan 2θ:

$$\tan 3\theta = \tan(\theta + 2\theta)$$
$$= \frac{\tan\theta + \tan 2\theta}{1 - \tan\theta\tan 2\theta}$$
$$= \frac{\tan\theta + \frac{2\tan\theta}{1-\tan^2\theta}}{1 - \tan\theta \cdot \frac{2\tan\theta}{1-\tan^2\theta}}$$
$$= \frac{\tan\theta\left(1 - \tan^2\theta\right) + 2\tan\theta}{1 - \tan^2\theta - 2\tan^2\theta}$$
$$= \frac{3\tan\theta - \tan^3\theta}{1 - 3\tan^2\theta}$$
$$= \frac{3t - t^3}{1 - 3t^2}, \text{ where } t = \tan\theta$$

19. Working from the right-hand side:

$$\frac{\cos x + \sin x}{\cos x - \sin x} = \frac{\cos x + \sin x}{\cos x - \sin x} \cdot \frac{\cos x + \sin x}{\cos x + \sin x}$$
$$= \frac{\cos^2 x + 2\sin x\cos x + \sin^2 x}{\cos^2 x - \sin^2 x}$$
$$= \frac{1 + \sin 2x}{\cos 2x}$$
$$= \frac{1}{\cos 2x} + \frac{\sin 2x}{\cos 2x}$$
$$= \tan 2x + \sec 2x$$

21. Using the double-angle formula for sin $2x$:

$$2\sin x + \sin 2x = 2\sin x + 2\sin x\cos x$$
$$= 2\sin x(1 + \cos x)$$
$$= 2\sin x(1 + \cos x) \cdot \frac{1 - \cos x}{1 - \cos x}$$
$$= \frac{2\sin x\left(1 - \cos^2 x\right)}{1 - \cos x}$$
$$= \frac{2\sin^3 x}{1 - \cos x}$$

23. Working from the right-hand side:

$$\frac{1-\cos x + \sin x}{1+\cos x + \sin x} = \frac{1+\sin x - \cos x}{1+\sin x + \cos x} \cdot \frac{1+\sin x + \cos x}{1+\sin x + \cos x}$$

$$= \frac{(1+\sin x)^2 - \cos^2 x}{\left[(1+\sin x) + \cos x\right]^2}$$

$$= \frac{1 + 2\sin x + \sin^2 x - \cos^2 x}{\left(1 + 2\sin x + \sin^2 x\right) + 2\cos x(1+\sin x) + \cos^2 x}$$

$$= \frac{2\sin x + 2\sin^2 x}{2 + 2\sin x + 2\cos x(1+\sin x)}$$

$$= \frac{2\sin x(1+\sin x)}{2(1+\sin x) + 2\cos x(1+\sin x)}$$

$$= \frac{2\sin x(1+\sin x)}{2(1+\sin x)(1+\cos x)}$$

$$= \frac{\sin x}{1+\cos x}$$

$$= \tan \tfrac{x}{2}$$

25. Using the addition formula for $\sin(s-t)$:

$$\sin(x+y)\cos y - \cos(x+y)\sin y = \sin\left[(x+y) - y\right] = \sin x$$

27. Working from the left-hand side:

$$\frac{1-\tan^2 \frac{x}{2}}{1+\tan^2 \frac{x}{2}} = \frac{1 - \dfrac{\sin^2 \frac{x}{2}}{\cos^2 \frac{x}{2}}}{1 + \dfrac{\sin^2 \frac{x}{2}}{\cos^2 \frac{x}{2}}} = \frac{\cos^2 \frac{x}{2} - \sin^2 \frac{x}{2}}{\cos^2 \frac{x}{2} + \sin^2 \frac{x}{2}} = \cos^2 \tfrac{x}{2} - \sin^2 \tfrac{x}{2} = \cos\left(2 \cdot \tfrac{x}{2}\right) = \cos x$$

29. Using the double-angle formulas for $\sin 2\theta$ and $\cos 2\theta$:

$$\sin 4x = 2\sin 2x \cos 2x$$

$$= 2(2\sin x \cos x)\left(\cos^2 x - \sin^2 x\right)$$

$$= 4\sin x \cos x\left(1 - 2\sin^2 x\right)$$

$$= 4\sin x \cos x - 8\sin^3 x \cos x$$

31. Using the addition formula for $\sin(s + t)$:
$$\sin 5x = \sin(4x + x) = \sin 4x \cos x + \cos 4x \sin x$$

Simplifying each of these products using the double-angle formulas for $\sin 2\theta$ and $\cos 2\theta$, we have:
$$\sin 4x \cos x = 2\sin 2x \cos 2x \cos x$$
$$= 2(2\sin x \cos x)(\cos^2 x - \sin^2 x)\cos x$$
$$= 4\sin x \cos^2 x(1 - 2\sin^2 x)$$
$$= 4\sin x(1 - \sin^2 x)(1 - 2\sin^2 x)$$
$$= 4\sin x - 12\sin^3 x + 8\sin^5 x$$
$$\cos 4x \sin x = (\cos^2 2x - \sin^2 2x)(\sin x)$$
$$= \left[(\cos^2 x - \sin^2 x)^2 - (2\sin x \cos x)^2\right]\sin x$$
$$= \left[(1 - 2\sin^2 x)^2 - 4\sin^2 x \cos^2 x\right]\sin x$$
$$= \left[1 - 4\sin^2 x + 4\sin^4 x - 4\sin^2 x(1 - \sin^2 x)\right]\sin x$$
$$= \left[1 - 8\sin^2 x + 8\sin^4 x\right]\sin x$$
$$= \sin x - 8\sin^3 x + 8\sin^5 x$$

Therefore:
$$\sin 5x = \sin 4x \cos x + \cos 4x \sin x$$
$$= 4\sin x - 12\sin^3 x + 8\sin^5 x + \sin x - 8\sin^3 x + 8\sin^5 x$$
$$= 16\sin^5 x - 20\sin^3 x + 5\sin x$$

33. Using the sum-to-product formula for $\sin\alpha - \sin\beta$:
$$\sin 80° - \sin 20° = 2\cos\tfrac{80°+20°}{2}\sin\tfrac{80°-20°}{2} = 2\cos 50°\sin 30° = 2\cos 50°\cdot\tfrac{1}{2} = \cos 50°$$

35. Using the sum-to-product formulas for $\cos\alpha - \cos\beta$ and $\sin\alpha + \sin\beta$:
$$\cos x - \cos 3x = -2\sin\tfrac{x+3x}{2}\sin\tfrac{x-3x}{2} = -2\sin 2x\sin(-x) = 2\sin 2x\sin x$$
$$\sin x + \sin 3x = 2\sin\tfrac{x+3x}{2}\cos\tfrac{x-3x}{2} = 2\sin 2x\cos(-x) = 2\sin 2x\cos x$$

Therefore:
$$\frac{\cos x - \cos 3x}{\sin x + \sin 3x} = \frac{2\sin 2x\sin x}{2\sin 2x\cos x} = \frac{\sin x}{\cos x} = \tan x$$

37. Using the sum-to-product formula for $\sin\alpha + \sin\beta$:
$$\sin\tfrac{5\pi}{12} + \sin\tfrac{\pi}{12} = 2\sin\tfrac{\frac{5\pi}{12}+\frac{\pi}{12}}{2}\cos\tfrac{\frac{5\pi}{12}-\frac{\pi}{12}}{2} = 2\sin\tfrac{\pi}{4}\cos\tfrac{\pi}{6} = 2\cdot\tfrac{\sqrt{2}}{2}\cdot\tfrac{\sqrt{3}}{2} = \tfrac{\sqrt{6}}{2}$$

39. Using the sum-to-product formulas for $\cos \alpha + \cos \beta$ and $\sin \alpha + \sin \beta$:

$$\cos 3y + \cos(2x - 3y) = 2\cos\tfrac{2x}{2}\cos\tfrac{6y-2x}{2} = 2\cos x\cos(3y - x)$$

$$\sin 3y + \sin(2x - 3y) = 2\sin\tfrac{2x}{2}\cos\tfrac{6y-2x}{2} = 2\sin x\cos(3y - x)$$

Therefore:

$$\frac{\cos 3y + \cos(2x - 3y)}{\sin 3y + \sin(2x - 3y)} = \frac{2\cos x\cos(3y - x)}{2\sin x\cos(3y - x)} = \frac{\cos x}{\sin x} = \cot x$$

41. Using the sum-to-product formulas for $\sin \alpha - \sin \beta$ and $\cos \alpha - \cos \beta$:

$$\sin 40° - \sin 20° = 2\cos\tfrac{40°+20°}{2}\sin\tfrac{40°-20°}{2}$$

$$= 2\cos 30°\sin 10°$$

$$= 2 \cdot \tfrac{\sqrt{3}}{2}\sin 10°$$

$$= \sqrt{3}\sin 10°$$

$$\cos 20° - \cos 40° = -2\sin\tfrac{40°+20°}{2}\sin\tfrac{40°-20°}{2}$$

$$= -2\sin 30°\sin(-10°)$$

$$= 2 \cdot \tfrac{1}{2}\sin 10°$$

$$= \sin 10°$$

Therefore:

$$\frac{\sin 40° - \sin 20°}{\cos 20° - \cos 40°} = \frac{\sqrt{3}\sin 10°}{\sin 10°} = \sqrt{3}$$

Using the result of Exercise 40, we have shown the required identity.

43. (a) Since $a = 1$, the area is $\tan^{-1}1 = \tfrac{\pi}{4}$.

(b) (i)   Since $\tan^{-1} a = 1.5$, $a = \tan 1.5 \approx 14$.
(ii)   Since $\tan^{-1} a = 1.56$, $a = \tan 1.56 \approx 93$.
(iii)   Since $\tan^{-1} a = 1.57$, $a = \tan 1.57 \approx 1256$.

45. The principal solution is $x = \tan^{-1} 4.26 \approx 1.34$. Since $\tan x$ is also positive in the third quadrant, the other solution in the interval $[0, 2\pi]$ is $1.34 + \pi \approx 4.48$.

47. Since $\csc x = 2.24$, $\sin x = \tfrac{1}{2.24} \approx 0.45$. The principal solution is $x = \sin^{-1} 0.45 \approx 0.46$. Since $\sin x$ is also positive in the second quadrant, the other solution in the interval $[0, 2\pi)$ is $\pi - 0.46 \approx 2.68$.

49. Given $\tan^2 x - 3 = 0$, so $\tan^2 x = 3$ and $\tan x = \pm\sqrt{3}$. If $\tan x = \sqrt{3}$ then $x = \tfrac{\pi}{3}$ or $\tfrac{4\pi}{3}$ while if $\tan x = -\sqrt{3}$ then $x = \tfrac{2\pi}{3}$ or $\tfrac{5\pi}{3}$. So the solutions in the interval $[0, 2\pi)$ are $x = \tfrac{\pi}{3}, \tfrac{2\pi}{3}, \tfrac{4\pi}{3}$ or $\tfrac{5\pi}{3}$.

51. Squaring each side of the equation:
$$(1+\sin x)^2 = \cos^2 x$$
$$1+2\sin x+\sin^2 x = 1-\sin^2 x$$
$$2\sin^2 x+2\sin x = 0$$
$$2\sin x(\sin x+1) = 0$$

So $\sin x = 0$ or $\sin x = -1$, thus $x = 0$, π or $\frac{3\pi}{2}$. Upon checking we find that $x = \pi$ is not a solution (recall that squaring an equation can produce extraneous roots), so the solutions are $x = 0$ or $\frac{3\pi}{2}$.

53. Using the double-angle formula for $\cos 2x$:
$$\sin x - \left(\cos^2 x - \sin^2 x\right)+1 = 0$$
$$\sin x - \left(1-2\sin^2 x\right)+1 = 0$$
$$2\sin^2 x+\sin x = 0$$
$$\sin x(2\sin x+1) = 0$$

So $\sin x = 0$ or $\sin x = -\frac{1}{2}$. Thus the solutions are $x = 0$, π, $\frac{7\pi}{6}$ or $\frac{11\pi}{6}$.

55. Solving the equation:
$$3\csc x - 4\sin x = 0$$
$$\frac{3}{\sin x} = 4\sin x$$
$$\sin^2 x = \frac{3}{4}$$
$$\sin x = \pm\frac{\sqrt{3}}{2}$$

So the solutions are $x = \frac{\pi}{3}, \frac{2\pi}{3}, \frac{4\pi}{3}$ or $\frac{5\pi}{3}$.

57. Factoring:
$$2\sin^4 x - 3\sin^2 x+1 = 0$$
$$\left(2\sin^2 x-1\right)\left(\sin^2 x-1\right) = 0$$
$$\sin^2 x = \tfrac{1}{2} \qquad \text{or} \qquad \sin^2 x = 1$$
$$\sin x = \pm\tfrac{\sqrt{2}}{2} \qquad \text{or} \qquad \sin x = \pm 1$$

So the solutions are $x = \frac{\pi}{4}, \frac{\pi}{2}, \frac{3\pi}{4}, \frac{5\pi}{4}, \frac{3\pi}{2}$ or $\frac{7\pi}{4}$.

59. Using the identity $\sin^2 x = 1 - \cos^2 x$:
$$\left(1-\cos^2 x\right)^2 + \cos^4 x = \tfrac{5}{8}$$
$$1-2\cos^2 x+2\cos^4 x = \tfrac{5}{8}$$
$$16\cos^4 x - 16\cos^2 x+3 = 0$$
$$\left(4\cos^2 x-3\right)\left(4\cos^2 x-1\right) = 0$$
$$\cos^2 x = \tfrac{3}{4} \qquad \text{or} \qquad \cos^2 x = \tfrac{1}{4}$$
$$\cos x = \pm\tfrac{\sqrt{3}}{2} \qquad \text{or} \qquad \cos x = \pm\tfrac{1}{2}$$

So the solutions are $x = \frac{\pi}{6}, \frac{\pi}{3}, \frac{2\pi}{3}, \frac{5\pi}{6}, \frac{7\pi}{6}, \frac{4\pi}{3}, \frac{5\pi}{3}$ or $\frac{11\pi}{6}$.

61. Using the suggestion, re-write the equation in terms of sines and cosines:

$$\cot x + \csc x + \sec x = \tan x$$

$$\frac{\cos x}{\sin x} + \frac{1}{\sin x} + \frac{1}{\cos x} = \frac{\sin x}{\cos x}$$

Multiplying each side of the equation by $\sin x \cos x$ yields:

$$\cos^2 x + \cos x + \sin x = \sin^2 x$$

$$\cos^2 x - \sin^2 x + \cos x + \sin x = 0$$

$$(\cos x + \sin x)(\cos x - \sin x + 1) = 0$$

$$\cos x + \sin x = 0 \quad \text{or} \quad \cos x - \sin x + 1 = 0$$

From the first equation $\sin x = -\cos x$, so $\tan x = -1$ and thus $x = \frac{3\pi}{4}$ or $\frac{7\pi}{4}$.

From the second equation, isolate $\cos x$ and square each side:

$$\cos x = \sin x - 1$$

$$\cos^2 x = \sin^2 x - 2\sin x + 1$$

$$1 - \sin^2 x = \sin^2 x - 2\sin x + 1$$

$$0 = 2\sin^2 x - 2\sin x$$

$$0 = 2\sin x(\sin x - 1)$$

Now $\sin x = 0$ when $x = 0$ or π, but then $\csc x$ is undefined. Also $\sin x = 1$ when $x = \frac{\pi}{2}$, but then $\tan x$ is undefined. So the only solutions are $x = \frac{3\pi}{4}$ or $\frac{7\pi}{4}$.

63. Let $\theta = \tan^{-1}\left(\frac{\sqrt{2}}{2}\right)$, so $\tan\theta = \frac{\sqrt{2}}{2}$. Draw the triangle:

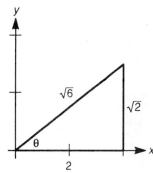

Thus $\sin\theta = \frac{\sqrt{2}}{\sqrt{6}} = \frac{\sqrt{3}}{3}$, so:

$$\cos\left\{\tan^{-1}\left[\sin\left(\tan^{-1}\left(\frac{\sqrt{2}}{2}\right)\right)\right]\right\} = \cos\left\{\tan^{-1}[\sin\theta]\right\} = \cos\left\{\tan^{-1}\left(\frac{\sqrt{3}}{3}\right)\right\} = \cos\frac{\pi}{6} = \frac{\sqrt{3}}{2}$$

65. We are asked to find the number x in $\left(-\frac{\pi}{2}, \frac{\pi}{2}\right)$ such that $\tan x = \frac{\sqrt{3}}{3}$. Since $x = \frac{\pi}{6}$ is that number, $\arctan\frac{\sqrt{3}}{3} = \frac{\pi}{6}$.

67. We are asked to find the number x in $\left[-\frac{\pi}{2}, \frac{\pi}{2}\right]$ such that $\sin x = \frac{1}{2}$. Since $x = \frac{\pi}{6}$ is that number, $\arcsin\frac{1}{2} = \frac{\pi}{6}$.

69. We are asked to find the number x in $[0, \pi]$ such that $\cos x = \frac{1}{2}$. Since $x = \frac{\pi}{3}$ is that number, $\cos^{-1}\left(\frac{1}{2}\right) = \frac{\pi}{3}$.

71. We are asked to find the number x in $[0, \pi]$ such that $\cos x = -\frac{1}{2}$. Since $x = \frac{2\pi}{3}$ is that number, $\cos^{-1}\left(-\frac{1}{2}\right) = \frac{2\pi}{3}$.

73. Since $\cos\left(\cos^{-1} x\right) = x$ for every x in the interval $[-1, 1]$, $\cos\left[\cos^{-1}\left(\frac{2}{7}\right)\right] = \frac{2}{7}$.

75. Let $\theta = \tan^{-1}(-1)$, so $\tan\theta = -1$ and θ is in the interval $\left(-\frac{\pi}{2}, \frac{\pi}{2}\right)$, thus $\theta = -\frac{\pi}{4}$. Therefore:

$$\sin\left[\tan^{-1}(-1)\right] = \sin\left(-\frac{\pi}{4}\right) = -\frac{\sqrt{2}}{2}$$

77. Let $\theta = \cos^{-1}\left(\frac{\sqrt{2}}{3}\right)$, so $\cos\theta = \frac{\sqrt{2}}{3}$ and θ is in the interval $[0, \pi]$. Therefore:

$$\sec\left[\cos^{-1}\left(\frac{\sqrt{2}}{3}\right)\right] = \sec\theta = \frac{1}{\cos\theta} = \frac{1}{\frac{\sqrt{2}}{3}} = \frac{3}{\sqrt{2}} = \frac{3\sqrt{2}}{2}$$

79. Let $\theta = \sin^{-1}\left(\frac{5}{13}\right)$, so $\sin\theta = \frac{5}{13}$ and θ is in the interval $\left[-\frac{\pi}{2}, \frac{\pi}{2}\right]$. Draw the triangle:

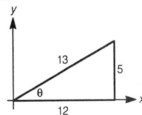

Using the addition formula for $\tan(s + t)$:

$$\tan\left(\frac{\pi}{4} + \theta\right) = \frac{\tan\frac{\pi}{4} + \tan\theta}{1 - \tan\frac{\pi}{4}\tan\theta} = \frac{1 + \frac{5}{12}}{1 - 1 \cdot \frac{5}{12}} = \frac{\frac{17}{12}}{\frac{7}{12}} = \frac{17}{7}$$

81. Let $\theta = \tan^{-1} 2$, so $\tan\theta = 2$ and θ is in the interval $\left(-\frac{\pi}{2}, \frac{\pi}{2}\right)$. Using the double-angle formula for $\tan 2\theta$:

$$\tan(2\theta) = \frac{2\tan\theta}{1 - \tan^2\theta} = \frac{2 \cdot 2}{1 - (2)^2} = -\frac{4}{3}$$

83. Let $\theta = \cos^{-1}\left(\frac{4}{5}\right)$, so $\cos\theta = \frac{4}{5}$ and θ lies in the first quadrant. Thus $\frac{\theta}{2}$ lies in the first quadrant, so using the half-angle formula for $\cos\frac{\theta}{2}$:

$$\cos\frac{\theta}{2} = \sqrt{\frac{1+\cos\theta}{2}} = \sqrt{\frac{1+\frac{4}{5}}{2}} = \sqrt{\frac{9}{10}} = \frac{3}{\sqrt{10}} = \frac{3\sqrt{10}}{10}$$

85. Let $\theta = \sec^{-1}\left(-\sqrt{2}\right)$, so $\sec\theta = -\sqrt{2}$ and θ is in the interval $\left[0,\frac{\pi}{2}\right)\cup\left(\frac{\pi}{2},\pi\right]$. Since $\cos\theta = -\frac{\sqrt{2}}{2}$, $\theta = \frac{3\pi}{4}$. So $\sec^{-1}\left(-\sqrt{2}\right) = \frac{3\pi}{4}$.

87. Since $\sec\left(\sec^{-1}x\right) = x$ for every x in $(-\infty,-1]\cup[1,\infty)$, $\sec\left(\sec^{-1}\sqrt{6}\right) = \sqrt{6}$.

89. Let $\alpha = \tan^{-1}x$ and $\beta = \tan^{-1}y$, so $\tan\alpha = x$ and $\tan\beta = y$. Using the addition formula for $\tan(s+t)$:

$$\tan\left(\tan^{-1}x + \tan^{-1}y\right) = \tan(\alpha+\beta) = \frac{\tan\alpha + \tan\beta}{1 - \tan\alpha\tan\beta} = \frac{x+y}{1-xy}$$

91. Let $\theta = \arctan x$, so $\tan\theta = x$. Draw the triangle:

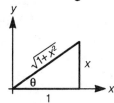

Using the double-angle formula for $\sin 2\theta$:

$$\sin(2\arctan x) = \sin 2\theta = 2\sin\theta\cos\theta = 2\cdot\frac{x}{\sqrt{x^2+1}}\cdot\frac{1}{\sqrt{x^2+1}} = \frac{2x}{x^2+1}$$

93. Let $\theta = \sin^{-1}\left(x^2\right)$, so $\sin\theta = x^2$. Since $\sin\theta \geq 0$, θ must lie in the first quadrant. Draw the triangle:

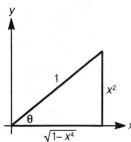

Noting that $\frac{\theta}{2}$ must also lie in the first quadrant, use the half-angle formula for $\sin\frac{\theta}{2}$ to obtain:

$$\sin\left[\frac{1}{2}\sin^{-1}\left(x^2\right)\right] = \sin\frac{\theta}{2} = \sqrt{\frac{1-\cos\theta}{2}} = \sqrt{\frac{1-\sqrt{1-x^4}}{2}} = \sqrt{\frac{1}{2} - \frac{1}{2}\sqrt{1-x^4}}$$

95. Let $\alpha = \arcsin\frac{4\sqrt{41}}{41}$ and $\beta = \arcsin\frac{\sqrt{82}}{82}$, so $\sin\alpha = \frac{4\sqrt{41}}{41} = \frac{4}{\sqrt{41}}$ and $\sin\beta = \frac{\sqrt{82}}{82} = \frac{1}{\sqrt{82}}$.
Draw the triangles:

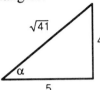 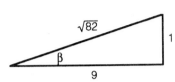

Now find $\tan(\alpha + \beta)$ by the addition formula for $\tan(s+t)$:

$$\tan(\alpha + \beta) = \frac{\tan\alpha + \tan\beta}{1 - \tan\alpha\tan\beta} = \frac{\frac{4}{5} + \frac{1}{9}}{1 - \frac{4}{5}\cdot\frac{1}{9}} = \frac{\frac{41}{45}}{\frac{41}{45}} = 1$$

Since $\tan(\alpha + \beta) = 1$ and $0 < \alpha + \beta < \pi$, $\alpha + \beta = \frac{\pi}{4}$. Therefore:

$$\arcsin\frac{4\sqrt{41}}{41} + \arcsin\frac{\sqrt{82}}{82} = \frac{\pi}{4}$$

97. (a) Using a calculator, $\cos 20° \cos 40° \cos 60° \cos 80° = 0.0625$.

(b) Since $\cos 60° = \frac{1}{2}$:

$$\cos 20°\cos 40°\cos 60°\cos 80° = \tfrac{1}{2}\cos 20°\cos 40°\cos 80°$$

Using the product-to-sum formula for $\cos A \cos B$:

$$\cos 20°\cos 40° = \tfrac{1}{2}\left[\cos 60° + \cos(-20°)\right] = \tfrac{1}{2}\left[\tfrac{1}{2} + \cos 20°\right] = \tfrac{1}{4} + \tfrac{1}{2}\cos 20°$$

Therefore:

$$\cos 20°\cos 40°\cos 60°\cos 80° = \tfrac{1}{2}\cos 20°\cos 40°\cos 80°$$
$$= \tfrac{1}{2}\left(\tfrac{1}{4} + \tfrac{1}{2}\cos 20°\right)\cos 80°$$
$$= \tfrac{1}{8}\cos 80° + \tfrac{1}{4}\cos 20°\cos 80°$$

Using the product-to-sum formula for $\cos A \cos B$, and the identity
$\cos\theta = -\cos(180° - \theta)$:

$$\cos 20°\cos 80° = \tfrac{1}{2}\left[\cos 100° + \cos(-60°)\right] = \tfrac{1}{2}\left[-\cos 80° + \tfrac{1}{2}\right] = -\tfrac{1}{2}\cos 80° + \tfrac{1}{4}$$

Therefore:

$$\cos 20°\cos 40°\cos 60°\cos 80° = \tfrac{1}{8}\cos 80° + \tfrac{1}{4}\cos 20°\cos 80°$$
$$= \tfrac{1}{8}\cos 80° + \tfrac{1}{4}\left(-\tfrac{1}{2}\cos 80° + \tfrac{1}{4}\right)$$
$$= \tfrac{1}{8}\cos 80° - \tfrac{1}{8}\cos 80° + \tfrac{1}{16}$$
$$= \tfrac{1}{16}$$
$$= 0.0625$$

99. Working from the right-hand side:

$$\frac{2}{\cot\theta + \tan\theta} = \frac{2}{\frac{\cos\theta}{\sin\theta} + \frac{\sin\theta}{\cos\theta}} = \frac{2}{\frac{\cos^2\theta + \sin^2\theta}{\sin\theta\cos\theta}} = \frac{2}{\frac{1}{\sin\theta\cos\theta}} = 2\sin\theta\cos\theta = \sin 2\theta$$

101. Using the double-angle formula for $\tan 2\theta$:

$$\frac{1}{1+\tan 2\theta \tan \theta} = \frac{1}{1+\frac{2\tan \theta}{1-\tan^2 \theta}\cdot \tan \theta}$$

$$= \frac{1}{1+\frac{2\tan^2 \theta}{1-\tan^2 \theta}}$$

$$= \frac{1-\tan^2 \theta}{\left(1-\tan^2 \theta\right)+2\tan^2 \theta}$$

$$= \frac{1-\tan^2 \theta}{1+\tan^2 \theta}$$

$$= \frac{1-\frac{\sin^2 \theta}{\cos^2 \theta}}{1+\frac{\sin^2 \theta}{\cos^2 \theta}}$$

$$= \frac{\cos^2 \theta - \sin^2 \theta}{\cos^2 \theta + \sin^2 \theta}$$

$$= \frac{\cos 2\theta}{1}$$

$$= \cos 2\theta$$

103. (a) Working from the right-hand side:

$$\frac{2\tan \theta}{1+\tan^2 \theta} = \frac{2\cdot \frac{\sin \theta}{\cos \theta}}{1+\frac{\sin^2 \theta}{\cos^2 \theta}} = \frac{2\sin \theta \cos \theta}{\cos^2 \theta + \sin^2 \theta} = \frac{\sin 2\theta}{1} = \sin 2\theta$$

(b) Working from the right-hand side:

$$\frac{1-\tan^2 \theta}{1+\tan^2 \theta} = \frac{1-\frac{\sin^2 \theta}{\cos^2 \theta}}{1+\frac{\sin^2 \theta}{\cos^2 \theta}} = \frac{\cos^2 \theta - \sin^2 \theta}{\cos^2 \theta + \sin^2 \theta} = \frac{\cos 2\theta}{1} = \cos 2\theta$$

105. Using the product-to-sum formula for $\sin A \sin B$:

$$\sin \theta \sin[(n-1)\theta] = \tfrac{1}{2}\left[\cos(2\theta - n\theta) - \cos(n\theta)\right]$$

$$= \tfrac{1}{2}\cos[(2-n)\theta] - \tfrac{1}{2}\cos n\theta$$

$$= \tfrac{1}{2}\cos[(n-2)\theta] - \tfrac{1}{2}\cos n\theta$$

Therefore, the right-hand side of the original identity becomes:

$$\cos[(n-2)\theta] - 2\sin \theta \sin[(n-1)\theta] = \cos[(n-2)\theta] - \cos[(n-2)\theta] + \cos n\theta$$

$$= \cos n\theta$$

Chapter Nine Test

1. Using the addition formula for $\sin(s + t)$:
$$\sin\left(\theta + \tfrac{3\pi}{2}\right) = \sin\theta\cos\tfrac{3\pi}{2} + \cos\theta\sin\tfrac{3\pi}{2} = \sin\theta \cdot 0 + \cos\theta \cdot (-1) = -\cos\theta$$

2. Since $\tfrac{3\pi}{2} < t < 2\pi$, draw the triangle:

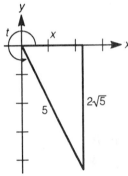

By the Pythagorean theorem:
$$x^2 + \left(2\sqrt{5}\right)^2 = (5)^2$$
$$x^2 + 20 = 25$$
$$x^2 = 5$$
$$x = \sqrt{5}$$

So $\cos t = \tfrac{\sqrt{5}}{5}$, now use the double-angle formula for $\cos 2t$:
$$\cos 2t = \cos^2 t - \sin^2 t = \left(\tfrac{\sqrt{5}}{5}\right)^2 - \left(-\tfrac{2\sqrt{5}}{5}\right)^2 = \tfrac{1}{5} - \tfrac{4}{5} = -\tfrac{3}{5}$$

3. Since $\pi < \theta < \tfrac{3\pi}{2}$, draw the triangle:

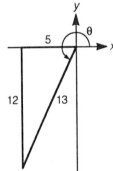

So $\sin\theta = -\tfrac{12}{13}$, now use the half-angle formula for $\tan\tfrac{\theta}{2}$:
$$\tan\tfrac{\theta}{2} = \frac{\sin\theta}{1 + \cos\theta} = \frac{-\tfrac{12}{13}}{1 - \tfrac{5}{13}} = \frac{-\tfrac{12}{13}}{\tfrac{8}{13}} = -\tfrac{3}{2}$$

4.   Dividing each side of the equation by $\cos x$ yields $\tan x = 3$, which has a principal solution of $x = \tan^{-1} 3 \approx 1.25$. Since $\tan x$ is also positive in the third quadrant, the other solution in the interval $(0, 2\pi)$ is $1.25 + \pi \approx 4.39$.

5.   Factoring:
$$2\sin^2 x + 7\sin x + 3 = 0$$
$$(2\sin x + 1)(\sin x + 3) = 0$$
$$\sin x = -\tfrac{1}{2} \text{ or } \sin x = -3 \text{ (impossible)}$$
So $x = \frac{7\pi}{6}, \frac{11\pi}{6}$.

6.   Since $\cos\alpha = \frac{2}{\sqrt{5}}$ and $\frac{3\pi}{2} < \alpha < 2\pi$ (fourth quadrant):
$$\sin\alpha = -\sqrt{1 - \cos^2\alpha} = -\sqrt{1 - \tfrac{4}{5}} = -\tfrac{1}{\sqrt{5}}$$
Similarly, since $\sin\beta = \frac{4}{5}$ and $\frac{\pi}{2} < \beta < \pi$ (second quadrant):
$$\cos\beta = -\sqrt{1 - \sin^2\beta} = -\sqrt{1 - \tfrac{16}{25}} = -\sqrt{\tfrac{9}{25}} = -\tfrac{3}{5}$$
Using the addition formula for $\sin(\beta - \alpha)$:
$$\sin(\beta - \alpha) = \sin\beta\cos\alpha - \cos\beta\sin\alpha = \tfrac{4}{5}\cdot\tfrac{2}{\sqrt{5}} - \left(-\tfrac{3}{5}\right)\left(-\tfrac{1}{\sqrt{5}}\right) = \tfrac{8}{5\sqrt{5}} - \tfrac{3}{5\sqrt{5}} = \tfrac{5}{5\sqrt{5}} = \tfrac{\sqrt{5}}{5}$$

7.   First, simplify the left-hand side using the addition formula for $\sin(\alpha + \beta)$:
$$\sin(x + 30°) = \sin x\cos 30° + \cos x\sin 30° = \tfrac{\sqrt{3}}{2}\sin x + \tfrac{1}{2}\cos x$$
Thus the equation is:
$$\tfrac{\sqrt{3}}{2}\sin x + \tfrac{1}{2}\cos x = \sqrt{3}\sin x$$
$$\sqrt{3}\sin x + \cos x = 2\sqrt{3}\sin x$$
$$\cos x = \sqrt{3}\sin x$$
$$\tan x = \tfrac{1}{\sqrt{3}}$$
$$x = 30°$$

8.   If $\csc\theta = -3$, then $\sin\theta = -\frac{1}{3}$. Since $\pi < \theta < \frac{3\pi}{2}$ (third quadrant):
$$\cos\theta = -\sqrt{1 - \sin^2\theta} = -\sqrt{1 - \tfrac{1}{9}} = -\sqrt{\tfrac{8}{9}} = -\tfrac{2\sqrt{2}}{3}$$
Now $\frac{\pi}{2} < \frac{\theta}{2} < \frac{3\pi}{4}$ (second quadrant), so using the half-angle formula for $\sin\frac{\theta}{2}$:
$$\sin\tfrac{\theta}{2} = \sqrt{\frac{1 - \cos\theta}{2}} = \sqrt{\frac{1 + \frac{2\sqrt{2}}{3}}{2}} = \sqrt{\frac{3 + 2\sqrt{2}}{6}} = \frac{\sqrt{18 + 12\sqrt{2}}}{6}$$

9. For the restricted sine function, the domain is $\left[-\frac{\pi}{2},\frac{\pi}{2}\right]$ and the range is $[-1,1]$. For the function $y = \sin^{-1} x$, the domain is $[-1,1]$ and the range is $\left[-\frac{\pi}{2},\frac{\pi}{2}\right]$.

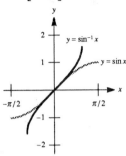

10. (a) Since $\sin^{-1}(\sin x) = x$ for every x in the interval $\left[-\frac{\pi}{2},\frac{\pi}{2}\right]$, $\sin^{-1}\left(\sin\frac{\pi}{10}\right) = \frac{\pi}{10}$.

(b) Since $\sin 2\pi = 0$, we are asked to find $x = \sin^{-1} 0$. Then $\sin x = 0$ and x is in the interval $\left[-\frac{\pi}{2},\frac{\pi}{2}\right]$, thus $x = 0$. So $\sin^{-1}(\sin 2\pi) = 0$. Notice that we cannot use the identity $\sin^{-1}(\sin x) = x$, since 2π is not in the interval $\left[-\frac{\pi}{2},\frac{\pi}{2}\right]$.

11. Let $\theta = \arcsin\frac{3}{4}$, so $\sin\theta = \frac{3}{4}$ and θ is in the interval $\left[-\frac{\pi}{2},\frac{\pi}{2}\right]$. Draw the triangle:

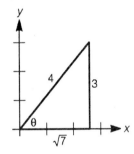

Therefore:
$$\cos\left(\arcsin\tfrac{3}{4}\right) = \cos\theta = \frac{\sqrt{7}}{4}$$

12. Using the addition formula for $\tan(s+t)$:
$$\tan\left(\tfrac{\pi}{4}+\tfrac{\theta}{2}\right) = \frac{\tan\frac{\pi}{4}+\tan\frac{\theta}{2}}{1-\tan\frac{\pi}{4}\tan\frac{\theta}{2}} = \frac{1+\tan\frac{\theta}{2}}{1-\tan\frac{\theta}{2}}$$

Now using the half-angle formula for $\tan\frac{\theta}{2}$:
$$\tan\left(\tfrac{\pi}{4}+\tfrac{\theta}{2}\right) = \frac{1+\tan\frac{\theta}{2}}{1-\tan\frac{\theta}{2}} = \frac{1+\frac{\sin\theta}{1+\cos\theta}}{1-\frac{\sin\theta}{1+\cos\theta}} = \frac{1+\cos\theta+\sin\theta}{1+\cos\theta-\sin\theta}$$

13. Using the product-to-sum formula for $\sin A \cos B$:

$$\sin\tfrac{7\pi}{24}\cos\tfrac{\pi}{24} = \tfrac{1}{2}\left[\sin\left(\tfrac{7\pi}{24}+\tfrac{\pi}{24}\right)+\sin\left(\tfrac{7\pi}{24}-\tfrac{\pi}{24}\right)\right]$$

$$= \tfrac{1}{2}\left[\sin\tfrac{\pi}{3}+\sin\tfrac{\pi}{4}\right]$$

$$= \frac{1}{2}\left(\frac{\sqrt{3}}{2}+\frac{\sqrt{2}}{2}\right)$$

$$= \frac{\sqrt{3}+\sqrt{2}}{4}$$

14. Using the sum-to-product formulas for $\sin\alpha+\sin\beta$ and $\cos\alpha+\cos\beta$:

$$\sin 3\theta + \sin 5\theta = 2\sin\tfrac{3\theta+5\theta}{2}\cos\tfrac{3\theta-5\theta}{2} = 2\sin 4\theta\cos(-\theta) = 2\sin 4\theta\cos\theta$$

$$\cos 3\theta + \cos 5\theta = 2\cos\tfrac{3\theta+5\theta}{2}\cos\tfrac{3\theta-5\theta}{2} = 2\cos 4\theta\cos(-\theta) = 2\cos 4\theta\cos\theta$$

Therefore:

$$\frac{\sin 3\theta + \sin 5\theta}{\cos 3\theta + \cos 5\theta} = \frac{2\sin 4\theta\cos\theta}{2\cos 4\theta\cos\theta} = \frac{\sin 4\theta}{\cos 4\theta} = \tan 4\theta$$

15. (a) Let $\theta = \arctan\sqrt{x^2-1}$, so $\tan\theta = \sqrt{x^2-1}$. Draw the triangle:

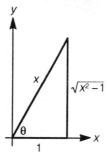

Therefore:

$$\sec\left[\arctan\sqrt{x^2-1}\right] = \sec\theta = \tfrac{x}{1} = x$$

(b) Let $\alpha = \sec^{-1}\left(\tfrac{5}{3}\right)$ and $\beta = \tan^{-1}\left(\tfrac{3}{4}\right)$, so $\sec\alpha = \tfrac{5}{3}$ and $\tan\beta = \tfrac{3}{4}$. Draw the triangles:

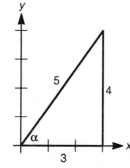
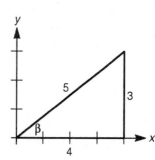

Since α and β are both in the first quadrant, $\sin\alpha > 0$, $\sin\beta > 0$ and $\cos\beta > 0$.

Now use the addition formula for $\sin(s+t)$:

$$\sin\left[\sec^{-1}\left(\tfrac{5}{3}\right)+\tan^{-1}\left(\tfrac{3}{4}\right)\right]=\sin(\alpha+\beta)$$
$$=\sin\alpha\cos\beta+\cos\alpha\sin\beta$$
$$=\tfrac{4}{5}\cdot\tfrac{4}{5}+\tfrac{3}{5}\cdot\tfrac{3}{5}$$
$$=\tfrac{16}{25}+\tfrac{9}{25}$$
$$=\tfrac{25}{25}$$
$$=1$$

Notice that since $\sin(\alpha+\beta)=1$, $\alpha+\beta=\tfrac{\pi}{2}$ and thus α and β are complementary angles.

16. The domain of $y=\tan^{-1}x$ is $(-\infty,\infty)$, and the range is $\left(-\tfrac{\pi}{2},\tfrac{\pi}{2}\right)$.

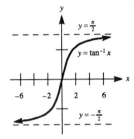

Chapter Ten
Additional Topics in Trigonometry

10.1 The Law of Sines and the Law of Cosines

1. Draw the triangle:

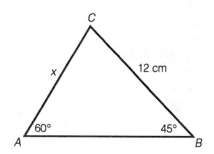

Using the law of sines:
$$\frac{\sin 45°}{x} = \frac{\sin 60°}{12}$$

Thus:

$$x = 12 \cdot \frac{\sin 45°}{\sin 60°} = 12 \cdot \frac{\sqrt{2}}{2} \cdot \frac{2}{\sqrt{3}} = \frac{12\sqrt{2}}{\sqrt{3}} = 4\sqrt{6} \text{ cm}$$

3.   Draw the triangle:

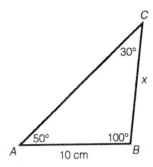

Using the law of sines:
$$\frac{\sin 50°}{x} = \frac{\sin 30°}{10}$$
Thus:
$$x = \frac{10\sin 50°}{\sin 30°} = \frac{10\sin 50°}{\frac{1}{2}} = 20\sin 50° \text{ cm}$$

5.   Draw a triangle:

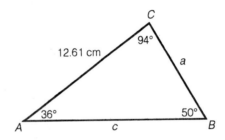

Using the law of sines:
$$\frac{\sin 36°}{a} = \frac{\sin 50°}{12.61}$$
Thus:
$$a = \frac{12.61\sin 36°}{\sin 50°} \approx 9.7 \text{ cm}$$
Using the law of sines:
$$\frac{\sin 94°}{c} = \frac{\sin 50°}{12.61}$$
Thus:
$$c = \frac{12.61\sin 94°}{\sin 50°} \approx 16.4 \text{ cm}$$

7. Draw a triangle:

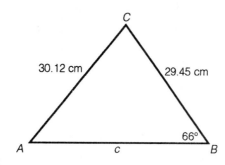

Using the law of sines:
$$\frac{\sin A}{29.45 \text{ cm}} = \frac{\sin 66°}{30.12 \text{ cm}}$$
Thus:
$$\sin A = \frac{29.45 \sin 66°}{30.12} \approx 0.8932, \text{ so } A \approx 63.3°$$
Then $C = 180° - 66° - 63.3° \approx 50.7°$. Using the law of sines:
$$\frac{\sin 50.7°}{c} = \frac{\sin 66°}{30.12}$$
Thus:
$$c = \frac{30.12 \sin 50.7°}{\sin 66°} \approx 25.5 \text{ cm}$$

9. (a) Since $\sin B = \frac{\sqrt{2}}{2}$, $\angle B = 45°$ or $\angle B = 135°$.

(b) Since $\cos E = \frac{\sqrt{2}}{2}$, $\angle E = 45°$. Note that $\angle E \neq 135°$.

(c) Since $\sin H = \frac{1}{4}$, $\angle H \approx 14.5°$ or $\angle H \approx 165.5°$.

(d) Since $\cos K = -\frac{2}{3}$, $\angle K \approx 131.8°$.

11. (a) Using the law of sines:
$$\frac{\sin 23.1°}{2.0} = \frac{\sin B}{6.0}$$
Thus:
$$\sin B = \frac{6.0 \sin 23.1°}{2.0} \approx 1.18$$
But $\sin B \leq 1$, so no such triangle exists.

(b) First find $\angle B$:
$$\frac{\sin 23.1°}{2.0} = \frac{\sin B}{3.0}$$
Thus:
$$\sin B = \frac{3.0 \sin 23.1°}{2.0} \approx 0.5885, \text{ so } B \approx 36.05° \text{ or } B \approx 143.95°$$

Since $\angle B$ is obtuse, $\angle B \approx 143.95°$. Therefore
$\angle C = 180° - 23.1° - 143.95° \approx 12.95°$, so by the law of sines:
$$\frac{\sin 12.95°}{c} = \frac{\sin 23.1°}{2.0}$$
Thus:
$$c = \frac{2.0 \sin 12.95°}{\sin 23.1°} \approx 1.1 \text{ feet}$$

13. (a) Using the law of sines:
$$\frac{\sin A}{\sqrt{2}} = \frac{\sin 30°}{1}$$
Thus:
$$\sin A = \sqrt{2} \sin 30° = \sqrt{2} \cdot \tfrac{1}{2} = \tfrac{\sqrt{2}}{2}$$
Therefore $\angle A = 45°$ or $\angle A = 135°$.

(b) If $\angle A = 45°$, then $\angle C = 180° - 45° - 30° = 105°$. Using the law of sines:
$$\frac{\sin 105°}{c} = \frac{\sin 30°}{1}$$
Thus:
$$c = \frac{\sin 105°}{\sin 30°} \approx 1.93$$

(c) If $\angle A = 135°$, then $\angle C = 180° - 135° - 30° = 15°$. Using the law of sines:
$$\frac{\sin 15°}{c} = \frac{\sin 30°}{1}$$
$$c = \frac{\sin 15°}{\sin 30°} \approx 0.52$$

(d) Find the area of each triangle:
$$A = \tfrac{1}{2} ab \sin C = \tfrac{1}{2} \cdot \sqrt{2} \cdot 1 \sin 105° \approx 0.68$$
$$A = \tfrac{1}{2} ab \sin C = \tfrac{1}{2} \cdot \sqrt{2} \sin 15° \approx 0.18$$

15. Using the law of sines:
$$\frac{\sin 20°}{2} = \frac{\sin 100°}{a} = \frac{\sin 50°}{b}$$
$$\frac{\sin 70°}{c} = \frac{\sin 95°}{b} = \frac{\sin 15°}{d}$$

Hence:

$$a = \frac{2\sin 110°}{\sin 20°} = \frac{2\sin 70°}{\sin 20°}\,\text{cm}$$

$$b = \frac{2\sin 50°}{\sin 20°}\,\text{cm}$$

$$c = \frac{b\sin 70°}{\sin 95°} = \frac{2\sin 50°\sin 70°}{\sin 20°\sin 85°}\,\text{cm}$$

$$d = \frac{b\sin 15°}{\sin 95°} = \frac{2\sin 50°\sin 15°}{\sin 20°\sin 85°}\,\text{cm}$$

17. Sketch the triangle:

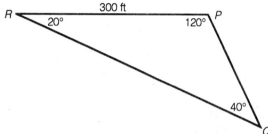

Applying the law of sines:

$$\frac{\sin 40°}{300} = \frac{\sin 20°}{PQ}$$

So $PQ = \dfrac{300\sin 20°}{\sin 40°} \approx 160$ ft.

19. (a) Using the law of cosines:

$$x^2 = 5^2 + 8^2 - 2(5)(8)\cos 60° = 25 + 64 - 80 \cdot \tfrac{1}{2} = 49$$

So $x = \sqrt{49} = 7$ cm.

(b) Using the law of cosines:

$$x^2 = 5^2 + 8^2 - 2(5)(8)\cos 120° = 25 + 64 - 80 \cdot \left(-\tfrac{1}{2}\right) = 129$$

So $x = \sqrt{129}$ cm.

21. (a) Using the law of cosines:

$$x^2 = (7.3)^2 + (11.5)^2 - 2(7.3)(11.5)\cos 40°$$
$$= 53.29 + 132.25 - 167.9\cos 40°$$
$$= 185.54 - 167.9\cos 40°$$
$$\approx 56.92$$

So $x \approx \sqrt{56.92} \approx 7.5$ cm.

(b)   Using the law of cosines:
$$x^2 = (7.3)^2 + (11.5)^2 - 2(7.3)(11.5)\cos 140°$$
$$= 53.29 + 132.25 - 167.9\cos 140°$$
$$= 185.54 - 167.9\cos 140°$$
$$\approx 314.16$$
So $x \approx \sqrt{314.16} \approx 17.7$ cm.

23.   This is incorrect because x is not the side opposite the $130°$ angle. The correct equation is:
$$6^2 = x^2 + 3^2 - 2(x)(3)\cos 130°$$

25.   Using $a = 6$, $b = 7$, $c = 10$ and the law of cosines:
$$6^2 = 7^2 + 10^2 - 2(7)(10)\cos A, \text{ so } \cos A = \tfrac{113}{140}$$
$$7^2 = 6^2 + 10^2 - 2(6)(10)\cos B, \text{ so } \cos B = \tfrac{87}{120} = \tfrac{29}{40}$$
$$10^2 = 6^2 + 7^2 - 2(6)(7)\cos C, \text{ so } \cos C = -\tfrac{15}{84} = -\tfrac{5}{28}$$

27.   Using the law of cosines to find angle A:
$$7^2 = 8^2 + 13^2 - 2(8)(13)\cos A$$
$$49 = 64 + 169 - 208\cos A$$
$$-184 = -208\cos A$$
$$\cos A = \tfrac{184}{208}$$
$$A \approx 27.8°$$
Now use the law of cosines to find angle B:
$$8^2 = 7^2 + 13^2 - 2(7)(13)\cos B$$
$$64 = 49 + 169 - 182\cos B$$
$$-154 = -182\cos B$$
$$\cos B = \tfrac{154}{182}$$
$$B \approx 32.2°$$
Since $A + B + C = 180°$:
$$C = 180° - A - B \approx 180° - 27.8° - 32.2° \approx 120°$$

29.   Using the law of cosines to find angle A:
$$\left(\tfrac{2}{\sqrt{3}}\right)^2 = \left(\tfrac{2}{\sqrt{3}}\right)^2 + 2^2 - 2 \cdot \tfrac{2}{\sqrt{3}} \cdot 2\cos A$$
$$\tfrac{4}{3} = \tfrac{4}{3} + 4 - \tfrac{8}{\sqrt{3}}\cos A$$
$$-4 = -\tfrac{8}{\sqrt{3}}\cos A$$
$$\cos A = \tfrac{\sqrt{3}}{2}$$
$$A = 30°$$
Since $a = b$, $B = 30°$. Since $A + B + C = 180°$:
$$C = 180° - A - B = 180° - 30° - 30° = 120°$$

31. First draw a figure, noting that the central angle is $\frac{360°}{5} = 72°$:

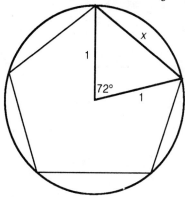

Apply the law of cosines to find x:

$$x^2 = 1^2 + 1^2 - 2(1)(1)\cos 72° = 2 - 2\cos 72° \approx 1.382$$

So $x \approx \sqrt{1.382} \approx 1.18$, and thus the perimeter is $5(1.18) \approx 5.9$ units.

33. (a) Using the law of cosines:

$$\begin{aligned}
a^2 &= (6.1)^2 + (3.2)^2 - 2(6.1)(3.2)\cos 40° \\
&= 37.21 + 10.24 - 39.04\cos 40° \\
&= 47.45 - 39.04\cos 40° \\
&\approx 17.54
\end{aligned}$$

So $a \approx \sqrt{17.54} \approx 4.2$ cm.

(b) Using the law of sines:

$$\frac{\sin C}{3.2} = \frac{\sin 40°}{4.2}$$

$$\sin C = \frac{3.2\sin 40°}{4.2}$$

$$\sin C \approx 0.49$$

$$C \approx 29.3°$$

(c) Since $A + B + C = 180°$:

$$B = 180° - A - C \approx 180° - 40° - 29.3° \approx 110.7°$$

35. First draw a figure indicating the relationship between Town A, Town B and Town C, where Town A is centered at the origin:

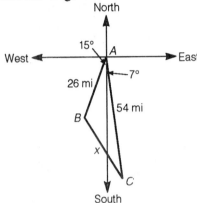

Using $\triangle ABC$ and the law of cosines:

$$x^2 = 26^2 + 54^2 - 2(26)(54)\cos 22°$$
$$= 676 + 2916 - 2808\cos 22°$$
$$= 3592 - 2808\cos 22°$$
$$\approx 988.5$$

So $x \approx \sqrt{988.5} \approx 31$. The distance between Towns B and C is approximately 31 miles.

37. Using P to denote the plane, note the following figure:

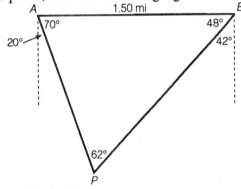

Using the law of sines to find AP:

$$\frac{AP}{\sin 48°} = \frac{1.5}{\sin 62°}$$
$$AP = \frac{1.5\sin 48°}{\sin 62°} \approx 1.26 \text{ mi}$$

Using the law of sines to find BP:

$$\frac{BP}{\sin 70°} = \frac{1.5}{\sin 62°}$$
$$BP = \frac{1.5\sin 70°}{\sin 62°} \approx 1.60 \text{ mi}$$

So the distance from the plane to lighthouse A is 1.26 miles and to lighthouse B is 1.60 miles.

39. Using the law of cosines:
$$D^2 = d^2 + d^2 - 2(d)(d)\cos\tfrac{32°}{60}$$
$$= 2d^2 - 2d^2\cos\tfrac{32°}{60}$$
$$= 2d^2\left(1 - \cos\tfrac{32°}{60}\right)$$
$$= 2(92{,}690{,}000)^2\left(1 - \cos\tfrac{32°}{60}\right)$$
$$\approx 744{,}414{,}483{,}000$$
So $D \approx \sqrt{744{,}414{,}483{,}000} \approx 860{,}000$ miles.
Note: That is about 100 times the diameter of Earth!

41. (a) Using the law of cosines:
$$\left(m^2 + n^2 + mn\right)^2 = \left(2mn + n^2\right)^2 + \left(m^2 - n^2\right)^2 - 2\left(2mn + n^2\right)\left(m^2 - n^2\right)\cos C$$
After carrying out the indicated squaring operations, and then combining like terms, the equation becomes:
$$2m^3 n - 2mn^3 + m^2 n^2 - n^4 = -2\left(2mn + n^2\right)\left(m^2 - n^2\right)\cos C$$
$$2mn\left(m^2 - n^2\right) + n^2\left(m^2 - n^2\right) = -2\left(2mn + n^2\right)\left(m^2 - n^2\right)\cos C$$
$$\left(m^2 - n^2\right)\left(2mn + n^2\right) = -2\left(2mn + n^2\right)\left(m^2 - n^2\right)\cos C$$
Therefore $\cos C = -\tfrac{1}{2}$, and consequently $C = 120°$.

(b) Let $m = 2$ and $n = 1$. Then by means of the expressions in part (a), we obtain $a = 5$, $b = 3$, $c = 7$.

43. Draw the figure:

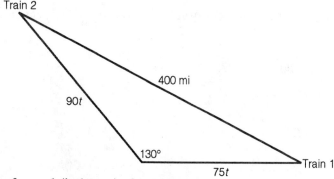

If t is the time of travel (in hours), then the trains have traveled $75t$ mi and $90t$ mi, as indicated in the figure. Use the law of cosines:
$$(400)^2 = (90t)^2 + (75t)^2 - 2(90t)(75t)\cos 130°$$
$$160000 = 8100t^2 + 5625t^2 - 13500\cos 130° t^2$$
$$t^2 = \frac{160000}{13725 - 13500\cos 130°} \approx 7.142$$
$$t \approx 2.672 \text{ hr} \approx 160 \text{ minutes}$$
The trains are 400 mi apart after 160 minutes, which occurs at 2:40 P.M.

45.   Since $\angle AEB = 180° - 15° - 15° = 150°$, using the law of sines:

$$\frac{\sin \angle AEB}{AB} = \frac{\sin 15°}{AE}$$

$$\frac{\sin 150°}{1} = \frac{\sin 15°}{AE}$$

$$\tfrac{1}{2} AE = \sin 15°$$

$$AE = 2\sin 15°$$

Now $\angle DAE = 90° - 15° = 75°$, so using the law of cosines:

$$(DE)^2 = (AD)^2 + (AE)^2 - 2(AD)(AE)\cos 75°$$

$$= 1 + (2\sin 15°)^2 - 2(1)(2\sin 15°)\cos 75°$$

$$= 1 + 4\sin^2 15° - 4\sin 15° \cos 75°$$

$$= 1 + 4\sin^2 15° - 4\sin 15° \sin 15°$$

$$= 1$$

So $DE = 1$, and since $CE = DE$ (congruent triangles), $CE = 1$. Because $DC = 1$, $\triangle CDE$ is equilateral.

47.   (a)   The measure of the angle drawn from two vertices of a triangle to the center of the circumscribed circle is twice the measure of the remaining angle of the triangle at the third vertex.

(b)   Since $\overline{OT}$ is the perpendicular bisector of $\overline{AC}$, $AT = TC = \dfrac{b}{2}$.

(c)   By the side-angle-side postulate in geometry, the two triangles are congruent.

(d)   Since the two triangles are congruent, $\angle AOT = \angle COT$. Since $\angle AOC = 2\angle B$, then $\angle AOT = \angle COT = \angle B$.

(e)   Since $OC = R$ and $TC = b/2$:

$$\sin \angle TOC = \frac{TC}{R}$$

$$\sin B = \frac{b/2}{R}$$

$$R = \frac{b}{2\sin B}$$

From the law of sines:

$$R = \frac{a}{2\sin A} = \frac{b}{2\sin B} = \frac{c}{2\sin C}$$

(f) Consider the figure:

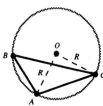

All of the above steps are valid if O lies outside of the triangle, so the proof is still valid.

49. The hint that is given results in the two equations:
$$\lambda^2 = a^2 + d^2 - 2ad\cos(180° - \theta) \qquad (1)$$
$$\lambda^2 = b^2 + c^2 - 2bc\cos\theta \qquad (2)$$
Equation (1) can be rewritten:
$$\lambda^2 = a^2 + d^2 + 2ad\cos\theta$$

Solving this last equation for $\cos\theta$:
$$\cos\theta = \frac{\lambda^2 - a^2 - d^2}{2ad}$$

Now use this expression for $\cos\theta$ in equation (2) to obtain:
$$\lambda^2 = b^2 + c^2 - 2bc\left(\frac{\lambda^2 - a^2 - d^2}{2ad}\right)$$
$$\lambda^2 ad = b^2 ad + c^2 ad - bc\lambda^2 + a^2 bc + bcd^2$$
$$\lambda^2(ad + bc) = b^2 ad + c^2 ad + a^2 bc + d^2 bc$$
$$\lambda^2(ad + bc) = \left(c^2 ad + a^2 bc\right) + \left(b^2 ad + d^2 bc\right)$$
$$\lambda^2(ad + bc) = ac(cd + ab) + bd(ab + cd)$$
$$\lambda^2(ad + bc) = (ab + cd)(ac + bd)$$
$$\lambda^2 = \frac{(ab + cd)(ac + bd)}{ad + bc}$$

51. Using the hint:
$$a^4 + b^4 + c^4 = 2\left(a^2 + b^2\right)c^2$$
$$c^4 - 2\left(a^2 + b^2\right)c^2 + \left(a^4 + b^4\right) = 0$$
Using the quadratic formula:
$$c^2 = \frac{2\left(a^2 + b^2\right) \pm \sqrt{4\left(a^2 + b^2\right)^2 - 4\left(a^4 + b^4\right)}}{2}$$
$$= a^2 + b^2 \pm \sqrt{a^4 + 2a^2 b^2 + b^4 - a^4 - b^4}$$
$$= a^2 + b^2 \pm \sqrt{2a^2 b^2}$$
$$= a^2 + b^2 \pm ab\sqrt{2}$$
By the law of cosines:
$$c^2 = a^2 + b^2 - 2ab\cos C$$

Setting these expressions equal:
$$a^2 + b^2 - 2ab\cos C = a^2 + b^2 \pm ab\sqrt{2}$$
$$-2ab\cos C = \pm ab\sqrt{2}$$
$$\cos C = \pm\frac{\sqrt{2}}{2}$$
$$C = 45° \text{ or } 135°$$

53. (a) Simplifying each of the fractions:
$$\frac{\sin A}{a} = \frac{\frac{\sqrt{T}}{2bc}}{a} = \frac{\sqrt{T}}{2abc}$$
$$\frac{\sin B}{b} = \frac{\frac{\sqrt{T}}{2ac}}{b} = \frac{\sqrt{T}}{2abc}$$
$$\frac{\sin C}{c} = \frac{\frac{\sqrt{T}}{2ab}}{c} = \frac{\sqrt{T}}{2abc}$$

(b) Since each of the fractions is equal to $\dfrac{\sqrt{T}}{2abc}$:
$$\frac{\sin A}{a} = \frac{\sin B}{b} = \frac{\sin C}{c}$$

55. (a) Following the hint, drop a perpendicular from O to $\overline{AD}$, and call the intersection point P. Using $\triangle APO$:
$$\cos\alpha = \frac{AP}{AO} = \frac{AP}{1} = AP$$
Since $AD = 2 \cdot AP$, $AD = 2\cos\alpha$.

(b) Following steps similar to part (a), notice that $\triangle AOE$ is an isosceles triangle. Drop a perpendicular from O to $\overline{AE}$, and call the intersection point Q. Using $\triangle AQO$:
$$\cos\beta = \frac{AQ}{AO} = \frac{AQ}{1} = AQ$$
Since $AE = 2 \cdot AQ$, $AE = 2\cos\beta$.

(c) Since $\angle AFC = 60°$, $\angle AFB = 180° - 60° = 120°$. Since the angles of $\triangle AFB$ must sum to $180°$:
$$\alpha + 120° + \angle B = 180°$$
$$\angle B = 60° - \alpha$$
Similarly, since the angles of $\triangle AFC$ must sum to $180°$:
$$\beta + 60° + \angle C = 180°$$
$$\angle C = 120° - \beta$$

(d) Using the law of sines on $\triangle ABC$:

$$\frac{\sin(\angle A)}{BC} = \frac{\sin(\angle B)}{AC} = \frac{\sin(\angle C)}{AB}$$

Since $\angle A = \alpha + \beta$, $\angle B = 60° - \alpha$ and $\angle C = 120° - \beta$:

$$\frac{\sin(\alpha+\beta)}{BC} = \frac{\sin(60°-\alpha)}{AC}, \text{ so } AC = \frac{BC \cdot \sin(60°-\alpha)}{\sin(\alpha+\beta)}$$

$$\frac{\sin(\alpha+\beta)}{BC} = \frac{\sin(120°-\beta)}{AB}, \text{ so } AB = \frac{BC \cdot \sin(120°-\beta)}{\sin(\alpha+\beta)}$$

(e) Using the results in parts (a), (b) and (d):

$$AD \cdot AB - AE \cdot AC$$

$$= (2\cos\alpha) \cdot \frac{BC\sin(120°-\beta)}{\sin(\alpha+\beta)} - (2\cos\beta) \cdot \frac{BC\sin(60°-\alpha)}{\sin(\alpha+\beta)}$$

$$= BC \cdot \left[\frac{2\cos\alpha\sin(120°-\beta) - 2\cos\beta\sin(60°-\alpha)}{\sin(\alpha+\beta)} \right]$$

(f) Using the addition formulas for sine:

$$\sin(120°-\beta) = \sin 120° \cos\beta - \cos 120° \sin\beta = \tfrac{\sqrt{3}}{2}\cos\beta + \tfrac{1}{2}\sin\beta$$

$$\sin(60°-\alpha) = \sin 60° \cos\alpha - \cos 60° \sin\alpha = \tfrac{\sqrt{3}}{2}\cos\alpha - \tfrac{1}{2}\sin\alpha$$

Therefore:

$$2\cos\alpha\sin(120°-\beta) - 2\cos\beta\sin(60°-\alpha)$$

$$= 2\cos\alpha\left(\tfrac{\sqrt{3}}{2}\cos\beta + \tfrac{1}{2}\sin\beta\right) - 2\cos\beta\left(\tfrac{\sqrt{3}}{2}\cos\alpha - \tfrac{1}{2}\sin\alpha\right)$$

$$= \sqrt{3}\cos\alpha\cos\beta + \cos\alpha\sin\beta - \sqrt{3}\cos\alpha\cos\beta + \sin\alpha\cos\beta$$

$$= \sin\alpha\cos\beta + \cos\alpha\sin\beta$$

$$= \sin(\alpha+\beta)$$

This completes the proof of the result.

57. (a) Computing areas of each triangle:

$$\text{Area}_{\text{left}} = \tfrac{1}{2}(\text{base})(\text{height}) = \tfrac{1}{2}af\sin\tfrac{C}{2}$$

$$\text{Area}_{\text{right}} = \tfrac{1}{2}(\text{base})(\text{height}) = \tfrac{1}{2}bf\sin\tfrac{C}{2}$$

$$\text{Area}_{\text{entire}} = \tfrac{1}{2}(\text{base})(\text{height}) = \tfrac{1}{2}ab\sin C$$

Adding the left and right triangles:

$$\tfrac{1}{2}af\sin\tfrac{C}{2} + \tfrac{1}{2}bf\sin\tfrac{C}{2} = \tfrac{1}{2}ab\sin C$$

(b) Since $\sin C = 2\sin\frac{C}{2}\cos\frac{C}{2}$:

$$\tfrac{1}{2}af\sin\frac{C}{2} + \tfrac{1}{2}bf\sin\frac{C}{2} = ab\sin\frac{C}{2}\cos\frac{C}{2}$$

Multiplying by $\frac{2}{\sin(C/2)}$:

$$af + bf = 2ab\cos\frac{C}{2}$$
$$f(a+b) = 2ab\cos\frac{C}{2}$$
$$f = \frac{2ab\cos\frac{C}{2}}{a+b}$$

(c) Using the half-angle formula:

$$\cos\frac{C}{2} = \sqrt{\frac{1+\cos C}{2}}$$
$$= \sqrt{\frac{1+\frac{a^2+b^2-c^2}{2ab}}{2}}$$
$$= \sqrt{\frac{a^2+2ab+b^2-c^2}{4ab}}$$
$$= \frac{1}{2}\sqrt{\frac{(a+b)^2-c^2}{ab}}$$
$$= \frac{1}{2}\sqrt{\frac{(a+b-c)(a+b+c)}{ab}}$$

(d) Combining our results from (b) and (c):

$$f = \frac{2ab}{a+b}\cos\frac{C}{2} = \frac{ab}{a+b}\sqrt{\frac{(a+b-c)(a+b+c)}{ab}} = \frac{\sqrt{ab}}{a+b}\sqrt{(a+b-c)(a+b+c)}$$

59. From the law of sines, note that $a\sin B = b\sin A$. Now, using the double-angle identity for $\sin 2B$ and $\sin 2A$:

$$\frac{a^2\sin 2B + b^2\sin 2A}{4} = \frac{2a^2\sin B\cos B + 2b^2\sin A\cos A}{4}$$
$$= \frac{(a\sin B)(a\cos B) + (b\sin A)(b\cos A)}{2}$$
$$= \frac{(b\sin A)(a\cos B) + (a\sin B)(b\cos A)}{2}$$
$$= \frac{ab(\sin A\cos B + \cos A\sin B)}{2}$$
$$= \frac{ab\sin(A+B)}{2}$$
$$= \frac{ab\sin C}{2}$$
$$= \tfrac{1}{2}ab\sin C$$

Note that the fact that $\sin(A+B) = \sin(180° - (A+B)) = \sin C$ was used. Since this last expression is the area of $\triangle ABC$, the proof is complete.

10.2 Vectors in the Plane, A Geometric Approach

1. Graph the vector:

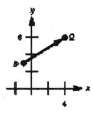

The magnitude is given by:

$$|\overrightarrow{PQ}| = \sqrt{(4-(-1))^2 + (6-3)^2} = \sqrt{25+9} = \sqrt{34}$$

3. Graph the vector:

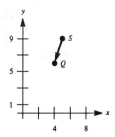

The magnitude is given by:

$$|\overrightarrow{SQ}| = \sqrt{(4-5)^2 + (6-9)^2} = \sqrt{1+9} = \sqrt{10}$$

5. Graph the vector:

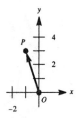

The magnitude is given by:

$$|\overrightarrow{OP}| = \sqrt{(-1-0)^2 + (3-0)^2} = \sqrt{1+9} = \sqrt{10}$$

7. Graph the vector sum:

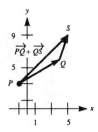

The magnitude is given by:

$$|\overrightarrow{PQ}+\overrightarrow{QS}| = |\overrightarrow{PS}| = \sqrt{(5-(-1))^2 + (9-3)^2} = \sqrt{36+36} = 6\sqrt{2}$$

9.   Graph the vector sum:

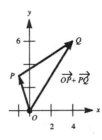

The magnitude is given by:

$$|\overrightarrow{OP} + \overrightarrow{PQ}| = |\overrightarrow{OQ}| = \sqrt{(4-0)^2 + (6-0)^2} = \sqrt{16+36} = 2\sqrt{13}$$

11.   Graph the vector sum:

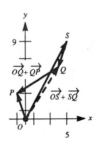

The magnitude is given by:

$$|\overrightarrow{OS} + \overrightarrow{SQ} + \overrightarrow{QP}| = |\overrightarrow{OQ} + \overrightarrow{QP}| = |\overrightarrow{OP}| = \sqrt{(-1-0)^2 + (3-0)^2} = \sqrt{1+9} = \sqrt{10}$$

13.   Graph the vector sum:

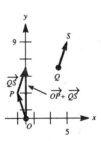

The magnitude is given by:

$$|\overrightarrow{OP} + \overrightarrow{QS}| = \sqrt{(0-0)^2 + (6-0)^2} = \sqrt{0+36} = 6$$

15.   Graph the vector sum:

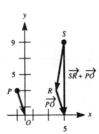

The magnitude is given by:

$$|\overrightarrow{SR} + \overrightarrow{PO}| = \sqrt{(5-5)^2 + (9-0)^2} = \sqrt{0+81} = 9$$

17. Graph the vector sum:

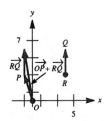

The magnitude is given by:

$$|\overrightarrow{OP} + \overrightarrow{RQ}| = \sqrt{(-1-0)^2 + (6-0)^2} = \sqrt{1+36} = \sqrt{37}$$

19. Graph the vector sum:

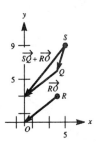

The magnitude is given by:

$$|\overrightarrow{SQ} + \overrightarrow{RO}| = \sqrt{(5-0)^2 + (9-3)^2} = \sqrt{25+36} = \sqrt{61}$$

21. Graph the vector sum:

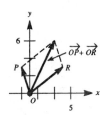

The magnitude is given by:

$$|\overrightarrow{OP} + \overrightarrow{OR}| = \sqrt{(3-0)^2 + (6-0)^2} = \sqrt{9+36} = 3\sqrt{5}$$

23. Graph the vector sum:

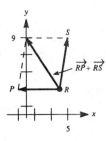

The magnitude is given by:

$$|\overrightarrow{RP} + \overrightarrow{RS}| = \sqrt{(0-4)^2 + (9-3)^2} = \sqrt{16+36} = 2\sqrt{13}$$

25.   Graph the vector sum:

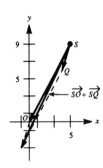

The magnitude is given by:

$$\left|\vec{SO} + \vec{SQ}\right| = \sqrt{(-1-5)^2 + (-3-9)^2} = \sqrt{36 + 144} = 6\sqrt{5}$$

27.   Draw the figure:

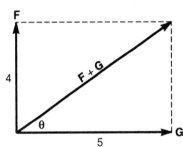

Now compute the magnitude and direction of **F** + **G**:

$$\left|\mathbf{F} + \mathbf{G}\right| = \sqrt{4^2 + 5^2} = \sqrt{16 + 25} = \sqrt{41} \text{ N}$$

$$\theta = \tan^{-1}\left(\tfrac{4}{5}\right) \approx 38.7°$$

29.   Draw the figure:

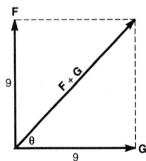

Now compute the magnitude and direction of **F** + **G**:

$$\left|\mathbf{F} + \mathbf{G}\right| = \sqrt{9^2 + 9^2} = 9\sqrt{2} \text{ N}$$

$$\theta = \tan^{-1}\left(\tfrac{9}{9}\right) = \tan^{-1}1 = 45°$$

31. Draw the figure:

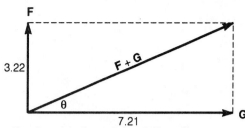

Now compute the magnitude and direction of $\mathbf{F} + \mathbf{G}$:

$$|\mathbf{F} + \mathbf{G}| = \sqrt{3.22^2 + 7.21^2} = \sqrt{62.3525} \approx 7.90 \text{ N}$$

$$\theta = \tan^{-1}\left(\tfrac{3.22}{7.21}\right) \approx 24.1°$$

33. Draw the parallelogram:

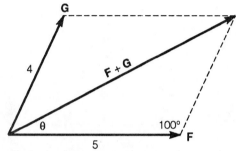

Let $d = |\mathbf{F} + \mathbf{G}|$. Then using the law of cosines:

$$d^2 = 5^2 + 4^2 - 2(5)(4)\cos 100°$$

$$d^2 = 41 - 40\cos 100°$$

$$d = \sqrt{41 - 40\cos 100°} \approx 6.92 \text{ N}$$

Find θ by the law of sines:

$$\frac{\sin \theta}{4} = \frac{\sin 100°}{d}$$

$$\sin \theta = \frac{4\sin 100°}{\sqrt{41 - 40\cos 100°}} \approx 0.5689$$

$$\theta \approx 34.67°$$

35. Draw the parallelogram:

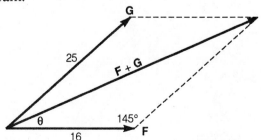

Let $d = |\mathbf{F} + \mathbf{G}|$. Then using the law of cosines:

$$d^2 = 16^2 + 25^2 - 2(16)(25)\cos 145°$$
$$d^2 = 881 - 800\cos 145°$$
$$d = \sqrt{881 - 800\cos 145°} \approx 39.20 \text{ N}$$

Find θ by the law of sines:

$$\frac{\sin \theta}{25} = \frac{\sin 145°}{d}$$
$$\sin \theta = \frac{25\sin 145°}{\sqrt{881 - 800\cos 145°}} \approx 0.3658$$
$$\theta \approx 21.46°$$

37. Draw the parallelogram:

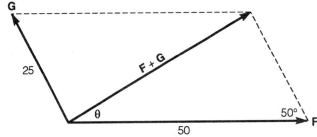

Let $d = |\mathbf{F} + \mathbf{G}|$. Then using the law of cosines:

$$d^2 = 50^2 + 25^2 - 2(50)(25)\cos 50°$$
$$d^2 = 3125 - 2500\cos 50°$$
$$d = \sqrt{3125 - 2500\cos 50°} \approx 38.96 \text{ N}$$

Find θ by the law of sines:

$$\frac{\sin \theta}{25} = \frac{\sin 50°}{d}$$
$$\sin \theta = \frac{25\sin 50°}{\sqrt{3125 - 2500\cos 50°}} \approx 0.4915$$
$$\theta \approx 29.44°$$

39. Compute the horizontal and vertical components:

$$V_x = 16\cos 30° \approx 13.86 \text{ cm/sec}$$
$$V_y = 16\sin 30° = 8 \text{ cm/sec}$$

41. Compute the horizontal and vertical components:

$$F_x = 14\cos 75° \approx 3.62 \text{ N}$$
$$F_y = 14\sin 75° \approx 13.52 \text{ N}$$

43. Compute the horizontal and vertical components:

$$V_x = 1\cos 135° \approx -0.71 \text{ cm/sec}$$
$$V_y = 1\sin 135° \approx 0.71 \text{ cm/sec}$$

45. Compute the horizontal and vertical components:
$$F_x = 1.25\cos 145° \approx -1.02 \text{ N}$$
$$F_y = 1.25\sin 145° \approx 0.72 \text{ N}$$

47. Draw the vectors:

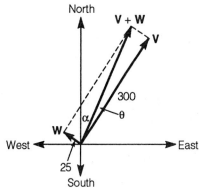

Let θ be the drift angle. Then:
$$\tan\theta = \tfrac{25}{300} = \tfrac{1}{12}$$
$$\theta = \tan^{-1}\left(\tfrac{1}{12}\right) \approx 4.76°$$
The ground speed is given by:
$$|V + W| = \sqrt{25^2 + 300^2} = \sqrt{90625} \approx 301.04 \text{ mph}$$
Let α be the bearing. Then:
$$\alpha = 30° - \theta \approx 30° - 4.76° \approx 25.24°$$

49. Draw the vectors:

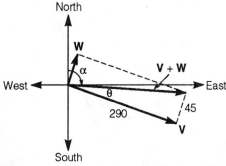

Let θ be the drift angle. Then:
$$\tan\theta = \tfrac{45}{290} = \tfrac{9}{58}$$
$$\theta = \tan^{-1}\left(\tfrac{9}{58}\right) \approx 8.82°$$
The ground speed is given by:
$$|V + W| = \sqrt{290^2 + 45^2} = \sqrt{86125} \approx 293.47 \text{ mph}$$
Let α be the bearing. Then:
$$\alpha = 100° - \theta \approx 100° - 8.82° \approx 91.18°$$

51.   Draw a figure:

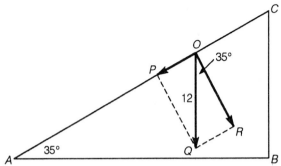

Notice that $\angle QOR = 35°$ since it is complementary to $\angle POQ$, but $\angle POQ = \angle ACB$. So the desired components are given by:

$$|\overrightarrow{OR}| = 12\cos 35° \approx 9.83\,\text{lb}$$

$$|\overrightarrow{OP}| = 12\sin 35° \approx 6.88\,\text{lb}$$

53.   Using the same approach as in Exercise 51:
perpendicular: $12\cos 10° \approx 11.82\,\text{lb}$
parallel: $12\sin 10° \approx 2.08\,\text{lb}$

55.   (a)   Draw the vector sum $(\mathbf{A} + \mathbf{B}) + \mathbf{C}$:

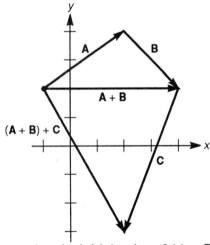

From the diagram, note that the initial point of $(\mathbf{A} + \mathbf{B}) + \mathbf{C}$ is $(-1, 2)$ and the terminal point is $(2, -3)$.

(b)   Draw the vector sum **A** + (**B** + **C**):

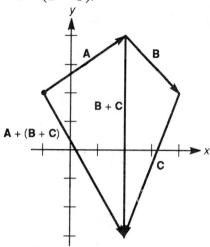

From the diagram, note that the initial point of **A** + (**B** + **C**) is (–1, 2) and the terminal point is (2, –3).

10.3 Vectors in the Plane, An Algebraic Approach

1.   Compute the length of the vector:

$$\left|\langle 4,3\rangle\right| = \sqrt{4^2 + 3^2} = \sqrt{25} = 5$$

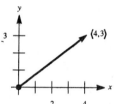

3.   Compute the length of the vector:

$$\left|\langle -4,2\rangle\right| = \sqrt{(-4)^2 + 2^2} = \sqrt{20} = 2\sqrt{5}$$

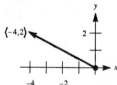

5.   Compute the length of the vector:

$$\left|\left\langle \tfrac{3}{4}, -\tfrac{1}{2}\right\rangle\right| = \sqrt{\left(\tfrac{3}{4}\right)^2 + \left(-\tfrac{1}{2}\right)^2} = \sqrt{\tfrac{9}{16} + \tfrac{1}{4}} = \tfrac{\sqrt{13}}{4}$$

7.   Subtracting components:

$$\overrightarrow{PQ} = \langle 3 - 2, 7 - 3\rangle = \langle 1, 4\rangle$$

9.   Subtracting components:

$$\overrightarrow{PQ} = \langle -3 - (-2), -2 - (-3)\rangle = \langle -3 + 2, -2 + 3\rangle = \langle -1, 1\rangle$$

11.   Subtracting components:

$$\overrightarrow{PQ} = \langle 3 - (-5), -4 - 1\rangle = \langle 3 + 5, -5\rangle = \langle 8, -5\rangle$$

13. $\mathbf{a} + \mathbf{b} = \langle 2 + 5, 3 + 4\rangle = \langle 7, 7\rangle$

15. $2\mathbf{a} + 4\mathbf{b} = \langle 4, 6\rangle + \langle 20, 16\rangle = \langle 24, 22\rangle$

17.   Since $\mathbf{b} + \mathbf{c} = \langle 5 + 6, 4 - 1\rangle = \langle 11, 3\rangle$:

$$|\mathbf{b} + \mathbf{c}| = \sqrt{11^2 + 3^2} = \sqrt{130}$$

19.   Since $\mathbf{a} + \mathbf{c} = \langle 2 + 6, 3 - 1\rangle = \langle 8, 2\rangle$:

$$|\mathbf{a} + \mathbf{c}| = \sqrt{8^2 + 2^2} = \sqrt{68} = 2\sqrt{17}$$
$$|\mathbf{a}| = \sqrt{2^2 + 3^2} = \sqrt{13}$$
$$|\mathbf{c}| = \sqrt{6^2 + (-1)^2} = \sqrt{37}$$

So $|\mathbf{a} + \mathbf{c}| - |\mathbf{a}| - |\mathbf{c}| = 2\sqrt{17} - \sqrt{13} - \sqrt{37}$.

21.   Since $\mathbf{b} + \mathbf{c} = \langle 5 + 6, 4 - 1\rangle = \langle 11, 3\rangle$:

$$\mathbf{a} + (\mathbf{b} + \mathbf{c}) = \langle 2, 3\rangle + \langle 11, 3\rangle = \langle 13, 6\rangle$$

23. $3\mathbf{a} + 4\mathbf{a} = \langle 6, 9\rangle + \langle 8, 12\rangle = \langle 14, 21\rangle$

25. $\mathbf{a} - \mathbf{b} = \langle 2, 3\rangle - \langle 5, 4\rangle = \langle -3, -1\rangle$

27. $3\mathbf{b} - 4\mathbf{d} = \langle 15, 12\rangle - \langle -8, 0\rangle = \langle 15 + 8, 12 - 0\rangle = \langle 23, 12\rangle$

29. Since $\mathbf{b} + \mathbf{c} = \langle 11, 3 \rangle$:
$$\mathbf{a} - (\mathbf{b} + \mathbf{c}) = \langle 2, 3 \rangle - \langle 11, 3 \rangle = \langle -9, 0 \rangle$$

31. Since $\mathbf{c} + \mathbf{d} = \langle 4, -1 \rangle$ and $\mathbf{c} - \mathbf{d} = \langle 8, -1 \rangle$:
$$|\mathbf{c} + \mathbf{d}| = \sqrt{16 + 1} = \sqrt{17} \text{ and } |\mathbf{c} - \mathbf{d}| = \sqrt{64 + 1} = \sqrt{65}$$
Therefore:
$$|\mathbf{c} + \mathbf{d}|^2 - |\mathbf{c} - \mathbf{d}|^2 = 17 - 65 = -48$$

33. Separating individual components:
$$\langle 3, 8 \rangle = \langle 3, 0 \rangle + \langle 0, 8 \rangle = 3\mathbf{i} + 8\mathbf{j}$$

35. Separating individual components:
$$\langle -8, -6 \rangle = \langle -8, 0 \rangle + \langle 0, -6 \rangle = -8\mathbf{i} - 6\mathbf{j}$$

37. Separating individual components:
$$3\langle 5, 3 \rangle + 2\langle 2, 7 \rangle = 3(5\mathbf{i} + 3\mathbf{j}) + 2(2\mathbf{i} + 7\mathbf{j}) = 15\mathbf{i} + 9\mathbf{j} + 4\mathbf{i} + 14\mathbf{j} = 19\mathbf{i} + 23\mathbf{j}$$

39. $\mathbf{i} + \mathbf{j} = \langle 1, 1 \rangle$

41. $5\mathbf{i} - 4\mathbf{j} = \langle 5, -4 \rangle$

43. First compute the length of the vector:
$$|\langle 4, 8 \rangle| = \sqrt{4^2 + 8^2} = \sqrt{80} = 4\sqrt{5}$$
So a unit vector would be given by:
$$\frac{1}{4\sqrt{5}}\langle 4, 8 \rangle = \left\langle \frac{1}{\sqrt{5}}, \frac{2}{\sqrt{5}} \right\rangle = \left\langle \frac{\sqrt{5}}{5}, \frac{2\sqrt{5}}{5} \right\rangle$$

45. First compute the length of the vector:
$$|\langle 6, -3 \rangle| = \sqrt{6^2 + (-3)^2} = \sqrt{45} = 3\sqrt{5}$$
So a unit vector would be given by:
$$\frac{1}{3\sqrt{5}}\langle 6, -3 \rangle = \left\langle \frac{2}{\sqrt{5}}, -\frac{1}{\sqrt{5}} \right\rangle = \left\langle \frac{2\sqrt{5}}{5}, -\frac{\sqrt{5}}{5} \right\rangle$$

47. First compute the length of the vector:
$$|8\mathbf{i} - 9\mathbf{j}| = \sqrt{8^2 + (-9)^2} = \sqrt{145}$$
So a unit vector would be given by:
$$\frac{1}{\sqrt{145}}(8\mathbf{i} - 9\mathbf{j}) = \frac{8}{\sqrt{145}}\mathbf{i} - \frac{9}{\sqrt{145}}\mathbf{j} = \frac{8\sqrt{145}}{145}\mathbf{i} - \frac{9\sqrt{145}}{145}\mathbf{j}$$

49. Compute the components u_1 and u_2:
$$u_1 = \cos\frac{\pi}{6} = \frac{\sqrt{3}}{2}$$
$$u_2 = \sin\frac{\pi}{6} = \frac{1}{2}$$

51.   Compute the components u_1 and u_2:
$$u_1 = \cos\frac{2\pi}{3} = -\frac{1}{2}$$
$$u_2 = \sin\frac{2\pi}{3} = \frac{\sqrt{3}}{2}$$

53.   Compute the components u_1 and u_2:
$$u_1 = \cos\frac{5\pi}{6} = -\frac{\sqrt{3}}{2}$$
$$u_2 = \sin\frac{5\pi}{6} = \frac{1}{2}$$

55.   Verify property 1:
$$\begin{aligned}
\mathbf{u} + (\mathbf{v}+\mathbf{w}) &= \langle u_1, u_2 \rangle + \left(\langle v_1, v_2 \rangle + \langle w_1, w_2 \rangle\right) \\
&= \langle u_1, u_2 \rangle + \langle v_1 + w_1, v_2 + w_2 \rangle \\
&= \langle u_1 + v_1 + w_1, u_2 + v_2 + w_2 \rangle \\
&= \langle u_1 + v_1, u_2 + v_2 \rangle + \langle w_1, w_2 \rangle \\
&= \left(\langle u_1, u_2 \rangle + \langle v_1, v_2 \rangle\right) + \langle w_1, w_2 \rangle \\
&= (\mathbf{u}+\mathbf{v}) + \mathbf{w}
\end{aligned}$$

Verify property 2:
$$\mathbf{0} + \mathbf{v} = \langle 0,0 \rangle + \langle v_1, v_2 \rangle = \langle 0+v_1, 0+v_2 \rangle = \langle v_1+0, v_2+0 \rangle = \langle v_1, v_2 \rangle + \langle 0,0 \rangle = \mathbf{v} + \mathbf{0}$$
$$\mathbf{v} + \mathbf{0} = \langle v_1, v_2 \rangle + \langle 0,0 \rangle = \langle v_1+0, v_2+0 \rangle = \langle v_1, v_2 \rangle = \mathbf{v}$$

57.   Verify property 5:
$$\begin{aligned}
a(\mathbf{u}+\mathbf{v}) &= a\left(\langle u_1, u_2 \rangle + \langle v_1, v_2 \rangle\right) \\
&= a\langle u_1 + v_1, u_2 + v_2 \rangle \\
&= \left\langle a(u_1+v_1), a(u_2+v_2) \right\rangle \\
&= \langle au_1 + av_1, au_2 + av_2 \rangle \\
&= \langle au_1, au_2 \rangle + \langle av_1, av_2 \rangle \\
&= a\langle u_1, u_2 \rangle + a\langle v_1, v_2 \rangle \\
&= a\mathbf{u} + a\mathbf{v}
\end{aligned}$$

Verify property 6:
$$\begin{aligned}
(a+b)\mathbf{v} &= (a+b)\langle v_1, v_2 \rangle \\
&= \left\langle (a+b)v_1, (a+b)v_2 \right\rangle \\
&= \langle av_1 + bv_1, av_2 + bv_2 \rangle \\
&= \langle av_1, av_2 \rangle + \langle bv_1, bv_2 \rangle \\
&= a\langle v_1, v_2 \rangle + b\langle v_1, v_2 \rangle \\
&= a\mathbf{v} + b\mathbf{v}
\end{aligned}$$

59.   (a)   Compute the dot products:
$$\mathbf{u} \cdot \mathbf{v} = \langle -4, 5 \rangle \cdot \langle 3, 4 \rangle = (-4)(3) + (5)(4) = -12 + 20 = 8$$
$$\mathbf{v} \cdot \mathbf{u} = \langle 3, 4 \rangle \cdot \langle -4, 5 \rangle = (3)(-4) + (4)(5) = -12 + 20 = 8$$

(b) Compute the dot products:
$$\mathbf{v} \cdot \mathbf{w} = \langle 3,4 \rangle \cdot \langle 2,-5 \rangle = (3)(2) + (4)(-5) = 6 - 20 = -14$$
$$\mathbf{w} \cdot \mathbf{v} = \langle 2,-5 \rangle \cdot \langle 3,4 \rangle = (2)(3) + (-5)(4) = 6 - 20 = -14$$

(c) Let $\mathbf{A} = \langle x_1, y_1 \rangle$ and $\mathbf{B} = \langle x_2, y_2 \rangle$. Compute each dot product:
$$\mathbf{A} \cdot \mathbf{B} = \langle x_1, y_1 \rangle \cdot \langle x_2, y_2 \rangle = x_1 x_2 + y_1 y_2$$
$$\mathbf{B} \cdot \mathbf{A} = \langle x_2, y_2 \rangle \cdot \langle x_1, y_1 \rangle = x_2 x_1 + y_2 y_1 = x_1 x_2 + y_1 y_2$$
Thus $\mathbf{A} \cdot \mathbf{B} = \mathbf{B} \cdot \mathbf{A}$.

61. (a) Compute each quantity:
$$\mathbf{v} \cdot \mathbf{v} = \langle 3,4 \rangle \cdot \langle 3,4 \rangle = (3)(3) + (4)(4) = 9 + 16 = 25$$
$$|\mathbf{v}| = \sqrt{3^2 + 4^2} = \sqrt{9 + 16} = \sqrt{25} = 5, \text{ so } |\mathbf{v}|^2 = 25$$

(b) Compute each quantity:
$$\mathbf{w} \cdot \mathbf{w} = \langle 2,-5 \rangle \cdot \langle 2,-5 \rangle = (2)(2) + (-5)(-5) = 4 + 25 = 29$$
$$|\mathbf{w}| = \sqrt{2^2 + (-5)^2} = \sqrt{4 + 25} = \sqrt{29}, \text{ so } |\mathbf{w}|^2 = 29$$

63. First do the computations:
$$|\mathbf{A}| = \sqrt{16 + 1} = \sqrt{17}$$
$$|\mathbf{B}| = \sqrt{4 + 36} = \sqrt{40} = 2\sqrt{10}$$
$$\mathbf{A} \cdot \mathbf{B} = 8 + 6 = 14$$
Thus:
$$\cos\theta = \frac{14}{\sqrt{17} \cdot 2\sqrt{10}} = \frac{7}{\sqrt{170}}$$
So $\theta \approx 57.53°$ or $\theta \approx 1.00$ radian.

65. First do the computations:
$$|\mathbf{A}| = \sqrt{25 + 36} = \sqrt{61}$$
$$|\mathbf{B}| = \sqrt{9 + 49} = \sqrt{58}$$
$$\mathbf{A} \cdot \mathbf{B} = -15 - 42 = -57$$
Thus:
$$\cos\theta = \frac{-57}{\sqrt{61} \cdot \sqrt{58}} = \frac{-57}{\sqrt{3538}}$$
So $\theta \approx 163.39°$ or $\theta \approx 2.85$ radians.

67. (a) First do the computations:
$$|\mathbf{A}| = \sqrt{64 + 4} = \sqrt{68} = 2\sqrt{17}$$
$$|\mathbf{B}| = \sqrt{1 + 9} = \sqrt{10}$$
$$\mathbf{A} \cdot \mathbf{B} = -8 - 6 = -14$$
Thus:
$$\cos\theta = \frac{-14}{2\sqrt{17} \cdot \sqrt{10}} = \frac{-7}{\sqrt{170}}$$
So $\theta \approx 122.47°$ or $\theta \approx 2.14$ radians.

(b) Again, do the computations:
$$|\mathbf{A}| = \sqrt{64+4} = \sqrt{68} = 2\sqrt{17}$$
$$|\mathbf{B}| = \sqrt{1+9} = \sqrt{10}$$
$$\mathbf{A} \cdot \mathbf{B} = 8+6 = 14$$
Thus:
$$\cos\theta = \frac{14}{2\sqrt{17} \cdot \sqrt{10}} = \frac{7}{\sqrt{170}}$$
So $\theta \approx 57.53°$ or $\theta \approx 1.00$ radian.

69. (a) First compute:
$$|\langle 2,5 \rangle| = \sqrt{4+25} = \sqrt{29}$$
$$|\langle -5,2 \rangle| = \sqrt{25+4} = \sqrt{29}$$
$$\langle 2,5 \rangle \cdot \langle -5,2 \rangle = -10+10 = 0$$
So $\cos\theta = \frac{0}{29} = 0$.

(b) Since the angle between the vectors is 90°, the vectors must be perpendicular.

(c) Draw the sketch:

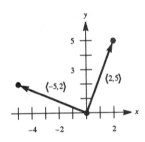

71. Since $\cos\theta = \frac{\mathbf{A} \cdot \mathbf{B}}{|\mathbf{A}||\mathbf{B}|}$, $\mathbf{A} \cdot \mathbf{B} = 0$ implies $\cos\theta = 0$, and thus $\theta = 90°$. So the vectors are perpendicular.

73. Call such a vector $\langle x,y \rangle$, so $x^2 + y^2 = 1$. Now $\langle x,y \rangle \cdot \langle -12,5 \rangle = 0$, so:
$$-12x + 5y = 0$$
$$5y = 12x$$
$$y = \tfrac{12}{5}x$$
Substituting into $x^2 + y^2 = 1$:
$$x^2 + \tfrac{144}{25}x^2 = 1$$
$$\tfrac{169}{25}x^2 = 1$$
$$x^2 = \tfrac{25}{169}$$
$$x = \pm\tfrac{5}{13}$$
$$y = \pm\tfrac{12}{13}$$
So the two unit vectors are $\left\langle \tfrac{5}{13}, \tfrac{12}{13} \right\rangle$ and $\left\langle -\tfrac{5}{13}, -\tfrac{12}{13} \right\rangle$.

75. **(a)** First find:
$$\mathbf{C} = \mathbf{B} - \mathbf{A} = \langle x_2, y_2 \rangle - \langle x_1, y_1 \rangle = \langle x_2 - x_1, y_2 - y_1 \rangle$$
Therefore:
$$|\mathbf{C}| = \sqrt{(x_2 - x_1)^2 + (y_2 - y_1)^2} = \sqrt{x_1^2 + y_1^2 + x_2^2 + y_2^2 - 2x_1 x_2 - 2y_1 y_2}$$

(b) Working from the left-hand side:
$$|\mathbf{C}|^2 = x_1^2 + y_1^2 + x_2^2 + y_2^2 - 2x_1 x_2 - 2y_1 y_2 = |\mathbf{A}|^2 + |\mathbf{B}|^2 - 2(\mathbf{A} \cdot \mathbf{B})$$

(c) Using the suggestion given:
$$|\mathbf{A}|^2 + |\mathbf{B}|^2 - 2|\mathbf{A}||\mathbf{B}|\cos\theta = |\mathbf{A}|^2 + |\mathbf{B}|^2 - 2(\mathbf{A} \cdot \mathbf{B})$$
$$-2|\mathbf{A}||\mathbf{B}|\cos\theta = -2(\mathbf{A} \cdot \mathbf{B})$$
$$|\mathbf{A}||\mathbf{B}|\cos\theta = \mathbf{A} \cdot \mathbf{B}$$
$$\cos\theta = \frac{\mathbf{A} \cdot \mathbf{B}}{|\mathbf{A}||\mathbf{B}|}$$

10.4 Parametric Equations

1. Find the x- and y-coordinates corresponding to $t = 0$:
$$x = 2 - 4(0) = 2$$
$$y = 3 - 5(0) = 3$$
The point corresponding to $t = 0$ is (2, 3).

3. Find the x- and y-coordinates corresponding to $t = \frac{\pi}{6}$:
$$x = 5\cos\frac{\pi}{6} = 5 \cdot \frac{\sqrt{3}}{2} = \frac{5\sqrt{3}}{2}$$
$$y = 2\sin\frac{\pi}{6} = 2 \cdot \frac{1}{2} = 1$$
The point corresponding to $t = \frac{\pi}{6}$ is $\left(\frac{5\sqrt{3}}{2}, 1\right)$.

5. Find the x- and y-coordinates corresponding to $t = \frac{\pi}{4}$:
$$x = 3\sin^3\frac{\pi}{4} = 3\left(\frac{\sqrt{2}}{2}\right)^3 = 3 \cdot \frac{\sqrt{2}}{4} = \frac{3\sqrt{2}}{4}$$
$$y = 3\cos^3\frac{\pi}{4} = 3\left(\frac{\sqrt{2}}{2}\right)^3 = 3 \cdot \frac{\sqrt{2}}{4} = \frac{3\sqrt{2}}{4}$$
The point corresponding to $t = \frac{\pi}{4}$ is $\left(\frac{3\sqrt{2}}{4}, \frac{3\sqrt{2}}{4}\right)$.

7.   Solving $x = t + 1$ for t yields $t = x - 1$, now substituting:
$$y = t^2 = (x - 1)^2$$
Graph the parabola which has a vertex at $(1, 0)$:

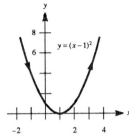

9.   Solving $y = t + 1$ for t yields $t = y - 1$, now substituting:
$$x = (y - 1)^2 - 1$$
$$x + 1 = (y - 1)^2$$
Graph the parabola which has a vertex at $(-1, 1)$:

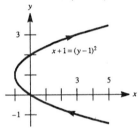

11.   Multiplying the first equation by 2 and the second equation by 5 yields $2x = 10 \cos t$ and $5y = 10 \sin t$, so:
$$(2x)^2 + (5y)^2 = 100 \cos^2 t + 100 \sin^2 t$$
$$4x^2 + 25y^2 = 100$$
$$\frac{x^2}{25} + \frac{y^2}{4} = 1$$
Graph the ellipse which is centered at the origin:

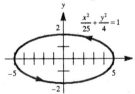

13.   Multiplying the first equation by 3 and the second equation by 2 yields $3x = 12 \cos 2t$ and $2y = 12 \sin 2t$, so:
$$(3x)^2 + (2y)^2 = 144 \cos^2 2t + 144 \sin^2 2t$$
$$9x^2 + 4y^2 = 144$$
$$\frac{x^2}{16} + \frac{y^2}{36} = 1$$

Graph the ellipse which is centered at the origin:

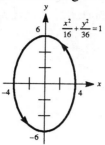

15. **(a)** Compute:
$$x^2 + y^2 = 4\cos^2 t + 4\sin^2 t$$
$$x^2 + y^2 = 4$$
Graph the circle which is centered at the origin:

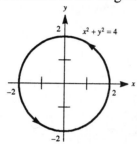

(b) Multiplying the second equation by 2 yields $x = 4\cos t$ and $2y = 4\sin t$, so:
$$x^2 + (2y)^2 = 16\cos^2 t + 16\sin^2 t$$
$$x^2 + 4y^2 = 16$$
$$\frac{x^2}{16} + \frac{y^2}{4} = 1$$
Graph the ellipse which is centered at the origin:

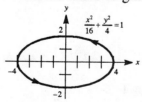

17. **(a)** When $t = 1$, compute the x- and y-coordinates:
$$x = (100\cos 70°) \cdot 1 = 100\cos 70° \approx 34.2$$
$$y = 5 + (100\sin 70°) \cdot 1 - 16(1)^2 = 100\sin 70° - 11 \approx 83.0$$
When $t = 2$, compute the x- and y-coordinates:
$$x = (100\cos 70°) \cdot 2 = 200\cos 70° \approx 68.4$$
$$y = 5 + (100\sin 70°) \cdot 2 - 16(2)^2 = 200\sin 70° - 59 \approx 128.9$$

When $t = 3$, compute the x- and y-coordinates:
$$x = (100\cos 70°) \cdot 3 = 300\cos 70° \approx 102.6$$
$$y = 5 + (100\sin 70°) \cdot 3 - 16(3)^2 = 300\sin 70° - 139 \approx 142.9$$

(b)   Find the value of t when $y = 0$:
$$5 + (100\sin 70°)t - 16t^2 = 0$$
$$16t^2 - (100\sin 70°)t - 5 = 0$$
Using the quadratic formula:
$$t = \frac{100\sin 70° \pm \sqrt{10000\sin^2 70° + 320}}{32} \approx 5.93, -0.05$$
Discard the negative value and conclude that the ball is in flight for approximately 5.93 seconds. When $t \approx 5.93$:
$$x \approx (100\cos 70°)(5.93) = 593\cos 70° \approx 203$$
The total horizontal distance traveled is approximately 203 feet. Note that this is consistent with the figure.

19.   Solve for t:
$$6 + (88\sin 35°)t - 16t^2 = 0$$
$$16t^2 - (88\sin 35°) - 6 = 0$$
Using the quadratic formula:
$$t = \frac{88\sin 35° \pm \sqrt{7744\sin^2 35° + 384}}{32} \approx 3.27, -0.11$$
The solutions are verified.

21.   Using the hint, raise each side to the 2/3 power to obtain $x^{2/3} = \cos^2 t$ and $y^{2/3} = \sin^2 t$, so:
$$x^{2/3} + y^{2/3} = \cos^2 t + \sin^2 t = 1$$
The x-y equation for the curve is $x^{2/3} + y^{2/3} = 1$.

Graphing Utility Exercises for Section 10.4

1.   (a)   Graphing the equations for $0 \le t \le 1$:

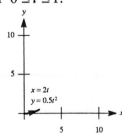

Graphing the equations for $0 \leq t \leq 3$:

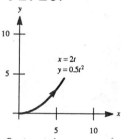

Graphing the equations for $0 \leq t \leq 4$:

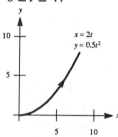

As the interval for t gets larger, the curve resembles a parabola.

(b) Graphing the equations for $-5 \leq t \leq 5$:

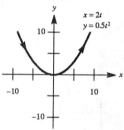

The restrictions on t in Figure 1(b) are $0 \leq t \leq 5$.

3. (a) Graphing the equations for $0 \leq t \leq 2\pi$:

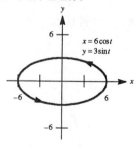

(b) Graphing the equations for $0 \le t \le 2\pi$:

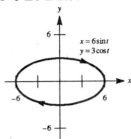

$$x = 6\sin t$$
$$y = 3\cos t$$

(c) Graphing the equations for $0 \le t \le \pi$:

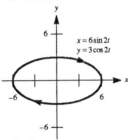

$$x = 6\sin 2t$$
$$y = 3\cos 2t$$

Note that the graphs appear identical.

5.   (a) Graphing the equations for $\frac{2}{3} < t \le 1$:

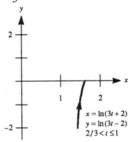

$$x = \ln(3t + 2)$$
$$y = \ln(3t - 2)$$
$$2/3 < t \le 1$$

(b) Graphing the equations for $\frac{2}{3} < t \le 10$:

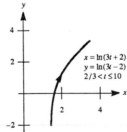

$$x = \ln(3t + 2)$$
$$y = \ln(3t - 2)$$
$$2/3 < t \le 10$$

(c) Graphing the equations for $\frac{2}{3} < t \leq 100$:

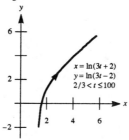

$x = \ln(3t + 2)$
$y = \ln(3t - 2)$
$2/3 < t \leq 100$

(d) Graphing the equations for $\frac{2}{3} < t \leq 1000$:

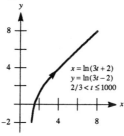

$x = \ln(3t + 2)$
$y = \ln(3t - 2)$
$2/3 < t \leq 1000$

7.   Graphing the equations for $0 \leq t \leq 2\pi$:

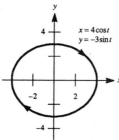

$x = 4\cos t$
$y = -3\sin t$

9.   Graphing the equations for $0 \leq t \leq 2\pi$:

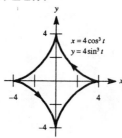

$x = 4\cos^3 t$
$y = 4\sin^3 t$

11.   Graphing the equations for $0 \le t \le 2\pi$:

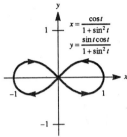

$$x = \frac{\cos t}{1 + \sin^2 t}$$

$$y = \frac{\sin t \cos t}{1 + \sin^2 t}$$

13.   Graphing the equations for $-10 \le t \le 10$:

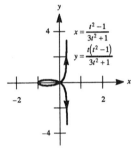

$$x = \frac{t^2 - 1}{3t^2 + 1}$$

$$y = \frac{t(t^2 - 1)}{3t^2 + 1}$$

15.   Graphing the equations for $0 \le t \le 2\pi$:

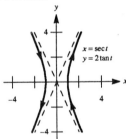

$$x = \sec t$$
$$y = 2 \tan t$$

17.   Graphing the equations for $0 \le t \le 2\pi$:

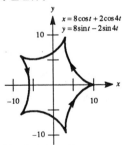

$$x = 8\cos t + 2\cos 4t$$
$$y = 8\sin t - 2\sin 4t$$

19. (a) Graphing the equations for $-\pi \le t \le \pi$:

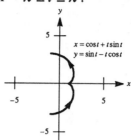

(b) Graphing the equations for $-2\pi \le t \le 2\pi$:

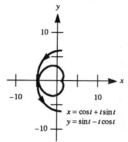

(c) Graphing the equations for $-4\pi \le t \le 4\pi$:

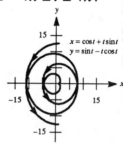

10.5 Introduction to Polar Coordinates

1. (a) Using $x = r\cos\theta$ and $y = r\sin\theta$:

$$x = 3\cos\frac{2\pi}{3} = 3\left(-\frac{1}{2}\right) = -\frac{3}{2}$$

$$y = 3\sin\frac{2\pi}{3} = 3\left(\frac{\sqrt{3}}{2}\right) = \frac{3\sqrt{3}}{2}$$

So the rectangular coordinates are $\left(-\frac{3}{2}, \frac{3\sqrt{3}}{2}\right)$.

(b) Using $x = r\cos\theta$ and $y = r\sin\theta$:

$$x = 4\cos\frac{11\pi}{6} = 4\left(\frac{\sqrt{3}}{2}\right) = 2\sqrt{3}$$

$$y = 4\sin\frac{11\pi}{6} = 4\left(-\frac{1}{2}\right) = -2$$

So the rectangular coordinates are $\left(2\sqrt{3}, -2\right)$.

(c) Using $x = r\cos\theta$ and $y = r\sin\theta$.
$$x = 4\cos\left(-\tfrac{\pi}{6}\right) = 4\left(\tfrac{\sqrt{3}}{2}\right) = 2\sqrt{3}$$
$$y = 4\sin\left(-\tfrac{\pi}{6}\right) = 4\left(-\tfrac{1}{2}\right) = -2$$
So the rectangular coordinates are $\left(2\sqrt{3}, -2\right)$.

3.   (a) Using $x = r\cos\theta$ and $y = r\sin\theta$.
$$x = 1\cos\tfrac{\pi}{2} = 1(0) = 0$$
$$y = 1\sin\tfrac{\pi}{2} = 1(1) = 1$$
So the rectangular coordinates are $(0, 1)$.

(b) Using $x = r\cos\theta$ and $y = r\sin\theta$.
$$x = 1\cos\tfrac{5\pi}{2} = 1(0) = 0$$
$$y = 1\sin\tfrac{5\pi}{2} = 1(1) = 1$$
So the rectangular coordinates are $(0, 1)$.

(c) Using $x = r\cos\theta$ and $y = r\sin\theta$, and using the half-angle formulas for $\sin\theta$ and $\cos\theta$.
$$x = -1\cos\tfrac{\pi}{8} = -1\left(\sqrt{\frac{1+\frac{\sqrt{2}}{2}}{2}}\right) = -\sqrt{\frac{2+\sqrt{2}}{4}} = -\frac{\sqrt{2+\sqrt{2}}}{2}$$
$$y = -1\sin\tfrac{\pi}{8} = -1\left(\sqrt{\frac{1-\frac{\sqrt{2}}{2}}{2}}\right) = -\sqrt{\frac{2-\sqrt{2}}{4}} = -\frac{\sqrt{2-\sqrt{2}}}{2}$$
So the rectangular coordinates are $\left(-\frac{\sqrt{2+\sqrt{2}}}{2}, -\frac{\sqrt{2-\sqrt{2}}}{2}\right)$.

5.   Computing:
$$r^2 = 1 + 1 = 2, \text{ so } r = \sqrt{2}$$
$$\theta = \tan^{-1}\left(\tfrac{-1}{-1}\right) + \pi = \tfrac{\pi}{4} + \pi = \tfrac{5\pi}{4}$$
So the polar form is $\left(\sqrt{2}, \tfrac{5\pi}{4}\right)$.

7.   If we first multiply by r:
$$r^2 = 2r\cos\theta$$
$$x^2 + y^2 = 2x$$
$$x^2 - 2x + y^2 = 0$$
$$x^2 - 2x + 1 + y^2 = 1$$
$$(x-1)^2 + y^2 = 1$$

9. Substituting for r and $\tan \theta$:

$$\sqrt{x^2 + y^2} = \frac{y}{x}$$

$$x^2 + y^2 = \frac{y^2}{x^2}$$

$$x^4 + x^2 y^2 = y^2$$

$$x^4 + x^2 y^2 - y^2 = 0$$

11. Using the double-angle formula for $\cos 2\theta$:

$$r = 3\left(\cos^2 \theta - \sin^2 \theta\right)$$

Multiplying by r^2:

$$r^3 = 3\left(r^2 \cos^2 \theta - r^2 \sin^2 \theta\right)$$

$$\left(x^2 + y^2\right)^{3/2} = 3\left(x^2 - y^2\right)$$

$$\left(x^2 + y^2\right)^3 = 9\left(x^2 - y^2\right)^2$$

13. Multiplying each side by $2 - \sin^2 \theta$ yields:

$$2r^2 - r^2 \sin^2 \theta = 8$$

$$2\left(x^2 + y^2\right) - y^2 = 8$$

$$2x^2 + 2y^2 - y^2 = 8$$

$$2x^2 + y^2 = 8$$

$$\frac{x^2}{4} + \frac{y^2}{8} = 1$$

15. Using the addition formula for $\cos(s - t)$:

$$r\cos\left(\theta - \tfrac{\pi}{6}\right) = 2$$

$$r\left(\cos\theta\cos\tfrac{\pi}{6} + \sin\theta\sin\tfrac{\pi}{6}\right) = 2$$

$$\tfrac{\sqrt{3}}{2} r\cos\theta + \tfrac{1}{2} r\sin\theta = 2$$

$$\tfrac{\sqrt{3}}{2} x + \tfrac{1}{2} y = 2$$

$$\sqrt{3}x + y = 4$$

$$y = -\sqrt{3}x + 4$$

17. Substituting $x = r\cos\theta$ and $y = r\sin\theta$:

$$3r\cos\theta - 4r\sin\theta = 2$$

$$r(3\cos\theta - 4\sin\theta) = 2$$

$$r = \frac{2}{3\cos\theta - 4\sin\theta}$$

19.   Substituting $x = r\cos\theta$ and $y = r\sin\theta$:

$$r^2\sin^2\theta = r^3\cos^3\theta$$
$$\sin^2\theta = r\cos^3\theta$$
$$r = \frac{\sin^2\theta}{\cos^3\theta}$$
$$r = \tan^2\theta\sec\theta$$

21.   Substituting $x = r\cos\theta$ and $y = r\sin\theta$:

$$2(r\cos\theta)(r\sin\theta) = 1$$
$$r^2(2\sin\theta\cos\theta) = 1$$
$$r^2\sin 2\theta = 1$$
$$r^2 = \frac{1}{\sin 2\theta}$$
$$r^2 = \csc 2\theta$$

23.   Substituting $x = r\cos\theta$ and $y = r\sin\theta$:

$$9r^2\cos^2\theta + r^2\sin^2\theta = 9$$
$$r^2(9\cos^2\theta + \sin^2\theta) = 9$$
$$r^2 = \frac{9}{9\cos^2\theta + \sin^2\theta}$$

25. A: Since $\theta = \frac{\pi}{6}$:

$$r = \frac{4}{1 + \sin\frac{\pi}{6}} = \frac{4}{1 + \frac{1}{2}} = \frac{4}{\frac{3}{2}} = \frac{8}{3}$$

B: Since $\theta = \frac{5\pi}{6}$:

$$r = \frac{4}{1 + \sin\frac{5\pi}{6}} = \frac{4}{1 + \frac{1}{2}} = \frac{4}{\frac{3}{2}} = \frac{8}{3}$$

C: Since $\theta = \pi$:

$$r = \frac{4}{1 + \sin\pi} = \frac{4}{1 + 0} = 4$$

D: Since $\theta = \frac{7\pi}{6}$:

$$r = \frac{4}{1 + \sin\frac{7\pi}{6}} = \frac{4}{1 - \frac{1}{2}} = \frac{4}{\frac{1}{2}} = 8$$

The coordinates of the points are $A\left(\frac{8}{3}, \frac{\pi}{6}\right)$, $B\left(\frac{8}{3}, \frac{5\pi}{6}\right)$, $C(4, \pi)$, and $D\left(8, \frac{7\pi}{6}\right)$.

27. *A:*   Since $\theta = \frac{\pi}{6}$:
$$r = 2\cos\frac{\pi}{3} = 2 \cdot \frac{1}{2} = 1$$
   B: Since $\theta = \frac{5\pi}{6}$:
$$r = 2\cos\frac{5\pi}{3} = 2 \cdot \frac{1}{2} = 1$$
   C: Since $\theta = \frac{7\pi}{6}$:
$$r = 2\cos\frac{7\pi}{3} = 2 \cdot \frac{1}{2} = 1$$
   D: Since $\theta = \frac{\pi}{2}$:
$$r = 2\cos\pi = 2 \cdot (-1) = -2$$

The coordinates of the points are $A\left(1, \frac{\pi}{6}\right)$, $B\left(1, \frac{5\pi}{6}\right)$, $C\left(1, \frac{7\pi}{6}\right)$, and $D\left(-2, \frac{\pi}{2}\right)$.

29. *A:*   Since $\theta = 0$:
$$r = e^{0/6} = 1$$
   B: Since $\theta = \frac{\pi}{4}$:
$$r = e^{\pi/24} \approx 1.14$$
   C: Since $\theta = \frac{3\pi}{4}$:
$$r = e^{\pi/8} \approx 1.48$$
   D: Since $\theta = \pi$:
$$r = e^{\pi/6} \approx 1.69$$
   E: Since $\theta = \frac{5\pi}{4}$:
$$r = e^{5\pi/24} \approx 1.92$$
   F: Since $\theta = \frac{3\pi}{2}$:
$$r = e^{\pi/4} \approx 2.19$$
   G: Since $\theta = \frac{7\pi}{4}$:
$$r = e^{7\pi/24} \approx 2.50$$
   H: Since $\theta = 2\pi$:
$$r = e^{\pi/3} \approx 2.85$$
   I: Since $\theta = \frac{9\pi}{4}$:
$$r = e^{3\pi/8} \approx 3.25$$
   J: Since $\theta = \frac{11\pi}{4}$:
$$r = e^{11\pi/24} \approx 4.22$$
   K: Since $\theta = \frac{13\pi}{4}$:
$$r = e^{13\pi/24} \approx 5.48$$

The coordinates of the points are $A(1,0)$, $B\left(1.14,\frac{\pi}{4}\right)$, $C\left(1.48,\frac{3\pi}{4}\right)$, $D(1.69,\pi)$, $E\left(1.92,\frac{5\pi}{4}\right)$, $F\left(2.19,\frac{3\pi}{2}\right)$, $G\left(2.50,\frac{7\pi}{4}\right)$, $H(2.85,2\pi)$, $I\left(3.25,\frac{9\pi}{4}\right)$, $J\left(4.22,\frac{11\pi}{4}\right)$, and $K\left(5.48,\frac{13\pi}{4}\right)$.

31. Use the distance formula, taking the points $\left(r_1,\theta_1\right)$ and $\left(r_2,\theta_2\right)$ to be $\left(2,\frac{2\pi}{3}\right)$ and $\left(4,\frac{\pi}{6}\right)$:

$$
\begin{aligned}
d^2 &= r_1^2 + r_2^2 - 2r_1r_2\cos\left(\theta_2 - \theta_1\right) \\
&= 2^2 + 4^2 - 2(2)(4)\cos\left(\tfrac{2\pi}{3} - \tfrac{\pi}{6}\right) \\
&= 20 - 16\cos\tfrac{\pi}{2} \\
&= 20 - 16\cdot 0 \\
&= 20
\end{aligned}
$$

So $d = \sqrt{20} = 2\sqrt{5}$.

33. Use the distance formula, taking the points $\left(r_1,\theta_1\right)$ and $\left(r_2,\theta_2\right)$ to be $\left(4,\frac{4\pi}{3}\right)$ and $(1,0)$:

$$
\begin{aligned}
d^2 &= r_1^2 + r_2^2 - 2r_1r_2\cos\left(\theta_2 - \theta_1\right) \\
&= 4^2 + 1^2 - 2(4)(1)\cos\left(\tfrac{4\pi}{3} - 0\right) \\
&= 17 - 8\cos\tfrac{4\pi}{3} \\
&= 17 - 8\cdot\left(-\tfrac{1}{2}\right) \\
&= 21
\end{aligned}
$$

So $d = \sqrt{21}$.

35. (a) Using the equation $r^2 + r_0^2 - 2rr_0\cos\left(\theta - \theta_0\right) = a^2$:

$$
\begin{aligned}
r^2 + 4^2 - 2(r)(4)\cos(\theta - 0) &= 2^2 \\
r^2 + 16 - 8r\cos\theta &= 4 \\
r^2 - 8r\cos\theta &= -12
\end{aligned}
$$

(b) Using the equation $r^2 + r_0^2 - 2rr_0\cos\left(\theta - \theta_0\right) = a^2$:

$$
\begin{aligned}
r^2 + 4^2 - 2(r)(4)\cos\left(\theta - \tfrac{2\pi}{3}\right) &= 2^2 \\
r^2 + 16 - 8r\cos\left(\theta - \tfrac{2\pi}{3}\right) &= 4 \\
r^2 - 8r\cos\left(\theta - \tfrac{2\pi}{3}\right) &= -12
\end{aligned}
$$

(c) Using the equation $r^2 + r_0^2 - 2rr_0\cos\left(\theta - \theta_0\right) = a^2$:

$$
\begin{aligned}
r^2 + 0^2 - 2(r)(0)\cos(\theta - 0) &= 2^2 \\
r^2 &= 4 \\
r &= 2
\end{aligned}
$$

37. (a) Using the equation $r^2 + r_0^2 - 2rr_0 \cos(\theta - \theta_0) = a^2$:

$$r^2 + 1^2 - 2(r)(1)\cos\left(\theta - \tfrac{3\pi}{2}\right) = 1^2$$
$$r^2 + 1 - 2r\cos\left(\theta - \tfrac{3\pi}{2}\right) = 1$$
$$r^2 = 2r\cos\left(\theta - \tfrac{3\pi}{2}\right)$$
$$r = 2\cos\left(\theta - \tfrac{3\pi}{2}\right)$$

(b) Using the equation $r^2 + r_0^2 - 2rr_0 \cos(\theta - \theta_0) = a^2$:

$$r^2 + 1^2 - 2(r)(1)\cos\left(\theta - \tfrac{\pi}{4}\right) = 1^2$$
$$r^2 + 1 - 2r\cos\left(\theta - \tfrac{\pi}{4}\right) = 1$$
$$r^2 = 2r\cos\left(\theta - \tfrac{\pi}{4}\right)$$
$$r = 2\cos\left(\theta - \tfrac{\pi}{4}\right)$$

39. (a) Since the line is of the form $r\cos(\theta - \alpha) = d$, $\alpha = \tfrac{\pi}{6}$ and $d = 2$. So the desired perpendicular distance is 2.

(b) When $\theta = 0$:

$$r\cos\left(0 - \tfrac{\pi}{6}\right) = 2$$
$$r\left(\tfrac{\sqrt{3}}{2}\right) = 2$$
$$r = \frac{4}{\sqrt{3}} = \frac{4\sqrt{3}}{3}$$

When $\theta = \tfrac{\pi}{2}$:

$$r\cos\left(\tfrac{\pi}{2} - \tfrac{\pi}{6}\right) = 2$$
$$r \cdot \tfrac{1}{2} = 2$$
$$r = 4$$

The required polar points are $\left(\dfrac{4\sqrt{3}}{3}, 0\right)$ and $\left(4, \tfrac{\pi}{2}\right)$.

(c) These coordinates are (d, α), which is the polar point $\left(2, \tfrac{\pi}{6}\right)$.

(d) Sketch the line:

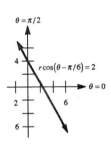

41. (a) Since the line is of the form $r\cos(\theta - \alpha) = d$, $\alpha = -\frac{2\pi}{3}$ and $d = 4$. So the desired perpendicualr distance is 4.

(b) When $\theta = 0$:
$$r\cos\left(0 + \frac{2\pi}{3}\right) = 4$$
$$r\left(-\frac{1}{2}\right) = 4$$
$$r = -8$$
When $\theta = \frac{\pi}{2}$:
$$r\cos\left(\frac{\pi}{2} + \frac{2\pi}{3}\right) = 4$$
$$r\cos\frac{7\pi}{6} = 4$$
$$r \cdot \left(-\frac{\sqrt{3}}{2}\right) = 4$$
$$r = -\frac{8}{\sqrt{3}} = -\frac{8\sqrt{3}}{3}$$

The required polar points are $(-8, 0)$ and $\left(-\frac{8}{3}\sqrt{3}, \frac{\pi}{2}\right)$.

(c) These coordinates are (d, α), which is the polar point $\left(4, -\frac{2\pi}{3}\right)$.

(d) Sketch the line:

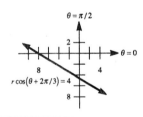

43. (a) Since $x^2 + y^2 = r^2$:

$$r = \sqrt{x^2 + y^2} = \sqrt{\left(-\sqrt{3}\right)^2 + (1)^2} = \sqrt{4} = 2$$

Now using $x = r\cos\theta$ and $y = r\sin\theta$:

$2\cos\theta = -\sqrt{3}$, so $\cos\theta = -\frac{\sqrt{3}}{2}$

$2\sin\theta = 1$, so $\sin\theta = \frac{1}{2}$

Therefore a value of θ is $\frac{5\pi}{6}$. The polar coordinates of P are $\left(2, \frac{5\pi}{6}\right)$.

(b) In the general equation $r\cos(\theta - \alpha) = d$, use the values $\alpha = \frac{5\pi}{6}$ and $d = 2$ to obtain:

$$r\cos\left(\theta - \frac{5\pi}{6}\right) = 2$$

This is the polar equation for the tangent line.

(c) The x-axis corresponds to $\theta = 0$, therefore:

$$r\cos\left(0 - \frac{5\pi}{6}\right) = 2$$

$$r\left(-\frac{\sqrt{3}}{2}\right) = 2$$

$$r = -\frac{4}{\sqrt{3}} = -\frac{4\sqrt{3}}{3}$$

So the x-intercept of the line is $-\dfrac{4\sqrt{3}}{3}$. The y-axis corresponds to $\theta = \frac{\pi}{2}$, so:

$$r\cos\left(\frac{\pi}{2} - \frac{5\pi}{6}\right) = 2$$

$$r\cos\left(-\frac{\pi}{3}\right) = 2$$

$$r \cdot \frac{1}{2} = 2$$

$$r = 4$$

So the y-intercept of the line is 4.

45. (a) Since $r\cos\theta = x$, this is the graph of $x = 3$:

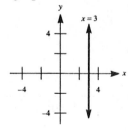

(b)   Since $r \sin \theta = y$, this is the graph of $y = 3$:

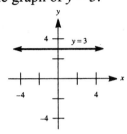

47.   (a)   Since d is the perpendicular distance from the origin to L, and (d, α) is the foot of the perpendicular on L, the polar equation for L must be $r \cos(\theta - \alpha) = d$.

(b)   Using the formula $\cos(s - t) = \cos s \cos t + \sin s \sin t$:

$$r \cos(\theta - \alpha) = d$$
$$r(\cos\theta \cos\alpha + \sin\theta \sin\alpha) = d$$
$$(r\cos\theta)\cos\alpha + (r\sin\theta)\sin\alpha = d$$
$$x \cos\alpha + y \sin\alpha = d$$

49.   First establish an identity for $\sin 3\theta$:

$$\sin 3\theta = \sin(2\theta + \theta)$$
$$= \sin 2\theta \cos\theta + \cos 2\theta \sin\theta$$
$$= (2\sin\theta \cos\theta)\cos\theta + (\cos^2\theta - \sin^2\theta)\sin\theta$$
$$= 3\sin\theta \cos^2\theta - \sin^3\theta$$

Now multiply the equation $r = a \sin 3\theta$ by r^3 and substitute:

$$r^4 = a\left((3r\sin\theta)(r^2\cos^2\theta) - r^3\sin^3\theta\right)$$
$$\left(x^2 + y^2\right)^2 = a\left(3yx^2 - y^3\right)$$
$$\left(x^2 + y^2\right)^2 = ay\left(3x^2 - y^2\right)$$

51.   Substituting $x = r\cos\theta$ and $y = r\sin\theta$ results in the equation:

$$\frac{r^2\cos^2\theta}{a^2} - \frac{r^2\sin^2\theta}{b^2} = 1$$
$$\frac{r^2\left(b^2\cos^2\theta - a^2\sin^2\theta\right)}{a^2b^2} = 1$$
$$r^2 = \frac{a^2b^2}{b^2\cos^2\theta - a^2\sin^2\theta}$$

53. (a) *A* is the point $(1, 0)$, *B* is the point $\left(1, \frac{2\pi}{3}\right)$, and *C* is the point $\left(1, \frac{4\pi}{3}\right)$.

(b) Use the distance formula $d^2 = r_1^2 + r_2^2 - 2r_1 r_2 \cos(\theta_2 - \theta_1)$ to find each distance:
$$(PA)^2 = r^2 + 1^2 - 2(r)(1)\cos(0-0) = r^2 + 1 - 2r$$
$$(PB)^2 = r^2 + 1^2 - 2(r)(1)\cos\left(0 - \frac{2\pi}{3}\right) = r^2 + 1 + r$$
$$(PC)^2 = r^2 + 1^2 - 2(r)(1)\cos\left(0 - \frac{4\pi}{3}\right) = r^2 + 1 + r$$

Therefore:
$$(PA)^2 (PB)^2 (PC)^2 = \left(r^2 - 2r + 1\right)\left(r^2 + r + 1\right)^2$$
$$= (r-1)^2\left(r^2 + r + 1\right)^2$$
$$= \left(r^3 - 1\right)^2$$

So:
$$(PA)(PB)(PC) = \sqrt{\left(r^3 - 1\right)^2} = 1 - r^3, \text{ since } r < 1$$

10.6 Curves in Polar Coordinates

1. Sketch the polar curve:

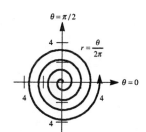

3. Sketch the polar curve:

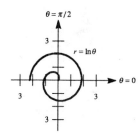

5.   Sketch the polar curve:

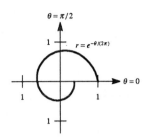

7.   Sketch the polar curve:

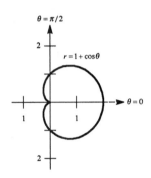

9.   Sketch the polar curve:

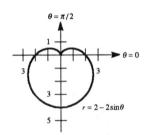

11.   Sketch the polar curve:

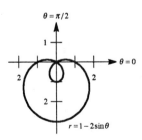

13.   Sketch the polar curve:

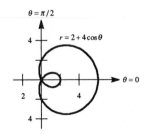

15.   Sketch the polar curve:

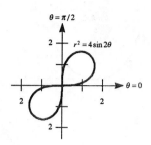

17.   Sketch the polar curve:

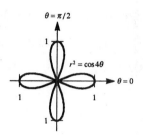

19.   Sketch the polar curve:

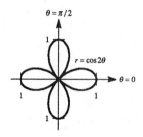

21.   Sketch the polar curve:

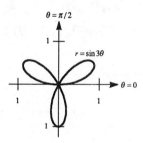

23.   Sketch the polar curve:

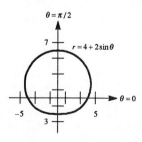

25.   Sketch the polar curve:

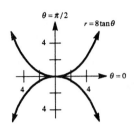

27.   (a)   The graph is C, since $\theta = \frac{\pi}{2}$ corresponds to $r = 6$.

   (b)   The graph is B, since $\theta = \frac{\pi}{2}$ corresponds to $r = 0$.

   (c)   The graph is D, since $\theta = \frac{\pi}{2}$ corresponds to $r = 3$.

   (d)   The graph is A, since $\theta = 0$ corresponds to $r = 6$.

29.   (a)   For A, let $\theta = 0$:

   $\ln r = 0$, so $r = 1$

   For B, let $\theta = \frac{\pi}{2}$:

$$\ln r = \frac{a\pi}{2}, \text{ so } r = e^{a\pi/2}$$

   For C, let $\theta = \pi$:

   $\ln r = a\pi$, so $r = e^{a\pi}$

   For D, let $\theta = \frac{3\pi}{2}$:

$$\ln r = \frac{3a\pi}{2}, \text{ so } r = e^{3a\pi/2}$$

   So the polar coordinates are $A(1, 0)$, $B\left(e^{a\pi/2}, \frac{\pi}{2}\right)$, $C\left(e^{a\pi}, \pi\right)$, and $D\left(e^{3a\pi/2}, \frac{3\pi}{2}\right)$.

   Since $OA = 1$, $OB = e^{a\pi/2}$, $OC = e^{a\pi}$, and $OD = e^{3a\pi/2}$, the required ratios are therefore:

$$\frac{OD}{OC} = \frac{e^{3a\pi/2}}{e^{a\pi}} = e^{3a\pi/2 - a\pi} = e^{a\pi/2}$$

$$\frac{OC}{OB} = \frac{e^{a\pi}}{e^{a\pi/2}} = e^{a\pi - a\pi/2} = e^{a\pi/2}$$

$$\frac{OB}{OA} = \frac{e^{a\pi/2}}{1} = e^{a\pi/2}$$

   Thus:

$$\frac{OD}{OC} = \frac{OC}{OB} = \frac{OB}{OA} = e^{a\pi/2}$$

(b) Using the hint, the rectangular coordinates are $A(1,0)$, $B\left(0,e^{a\pi/2}\right)$, $C\left(-e^{a\pi},0\right)$, and $D\left(0,-e^{3a\pi/2}\right)$. Now compute the slope of line segments $\overline{AB}$, $\overline{BC}$, and $\overline{CD}$:

$$m_{\overline{AB}} = \frac{e^{a\pi/2}-0}{0-1} = -e^{a\pi/2}$$

$$m_{\overline{BC}} = \frac{0-e^{a\pi/2}}{-e^{a\pi}-0} = \frac{1}{e^{a\pi/2}}$$

$$m_{\overline{CD}} = \frac{-e^{3a\pi/2}-0}{0+e^{a\pi}} = -e^{a\pi/2}$$

Since $m_{\overline{AB}} \cdot m_{\overline{BC}} = -1$ and $m_{\overline{BC}} \cdot m_{\overline{CD}} = -1$, $\angle ABC$ and $\angle BCD$ are right angles.

31. (a) Graph the curves. Note that the curves are identical.

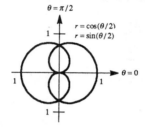

(b) Square each side of the polar curve $r = \cos\dfrac{\theta}{2}$ and apply the half-angle identity:

$$r^2 = \cos^2\frac{\theta}{2}$$

$$r^2 = \frac{1+\cos\theta}{2}$$

$$2r^2 = 1+\cos\theta$$

$$2\left(x^2+y^2\right) = 1+\frac{x}{\sqrt{x^2+y^2}}$$

$$2\left(x^2+y^2\right)^{3/2} = \left(x^2+y^2\right)^{1/2}+x$$

$$\left(x^2+y^2\right)^{1/2}\left(2x^2+2y^2-1\right) = x$$

$$\left(x^2+y^2\right)\left(2x^2+2y^2-1\right)^2 = x^2$$

$$\left(x^2+y^2\right)\left(2x^2+2y^2-1\right)^2 - x^2 = 0$$

Now square each side of the polar curve $r = \sin\dfrac{\theta}{2}$ and apply the half-angle identity:

$$r^2 = \sin^2\frac{\theta}{2}$$

$$r^2 = \frac{1-\cos\theta}{2}$$

$$2r^2 = 1-\cos\theta$$

$$2\left(x^2+y^2\right) = 1 - \frac{x}{\sqrt{x^2+y^2}}$$

$$2\left(x^2+y^2\right)^{3/2} = \left(x^2+y^2\right)^{1/2} - x$$

$$\left(x^2+y^2\right)^{1/2}\left(2x^2+2y^2-1\right) = -x$$

$$\left(x^2+y^2\right)\left(2x^2+2y^2-1\right)^2 = x^2$$

$$\left(x^2+y^2\right)\left(2x^2+2y^2-1\right)^2 - x^2 = 0$$

So both curves convert to the same rectangular equation.

Graphing Utility Exercises for Section 10.6

1. Graphing the equations:

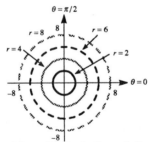

Notice the graphs are circles with radii 2, 4, 6, and 8.

3. (a) Replacing θ with $-\theta$:
    $$r = \cos^2(-\theta) - 2\cos(-\theta) = \cos^2\theta - 2\cos\theta$$
    Since the equation is unchanged, the graph is symmetric about the x-axis.

    (b) Graphing the curve, note the symmetry about the x-axis:

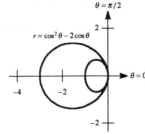

5. (a) Graph the polar curve:

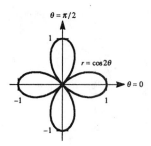

(b) Graph the polar curve:

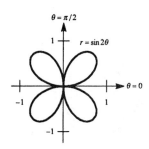

7. (a) Graph the polar curve:

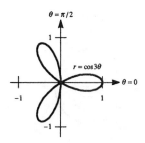

(b) Graph the polar curve:

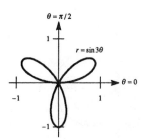

9. Graph the polar curve:

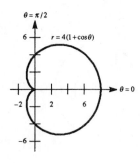

11. Graphing the polar curve:

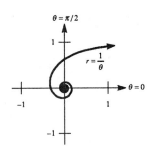

13. Graphing the polar curve:

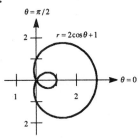

15. (a) Graphing the polar curve:

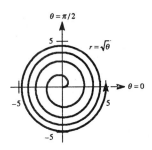

(b) Graphing the polar curve:

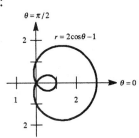

(c) Graphing the polar curve:

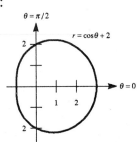

(d) Graphing the polar curve:

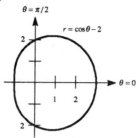

(e) Graphing the polar curve:

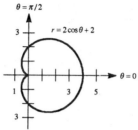

(f) Graphing the polar curve:

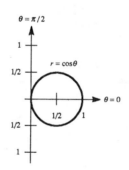

17. Graphing the polar curve:

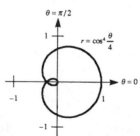

Zooming in several times, note that the inner loop near the origin is not simple, but rather a cardioid type shape which passes through both the first and fourth quadrants.

19. Graphing the polar curve:

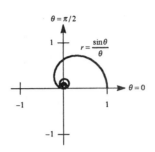

Chapter Ten Review Exercises

1. Draw the triangle:

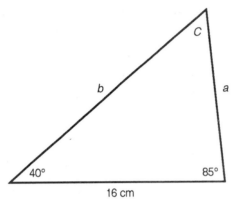

First note that $\angle C = 180° - 40° - 85° = 55°$. Using the law of sines:

$$\frac{\sin 40°}{a} = \frac{\sin 55°}{16}, \text{ so } a = \frac{16 \sin 40°}{\sin 55°} \approx 12.6 \text{ cm}$$

Using the law of sines:

$$\frac{\sin 85°}{b} = \frac{\sin 55°}{16}, \text{ so } b = \frac{16 \sin 85°}{\sin 55°} \approx 19.5 \text{ cm}$$

3. (a) Draw the triangle:

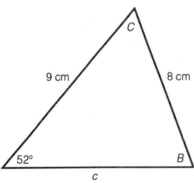

Using the law of sines:

$$\frac{\sin B}{9} = \frac{\sin 52°}{8}, \text{ so } \sin B = \frac{9 \sin 52°}{8} \approx 0.8865$$

Thus $\angle B \approx 62.4°$. Now note that $\angle C \approx 180° - 52° - 62.4° \approx 65.6°$. Using the law of sines:

$$\frac{\sin 65.6°}{c} = \frac{\sin 52°}{8}, \text{ so } c = \frac{8 \sin 65.6°}{\sin 52°} \approx 9.2 \text{ cm}$$

(b) Note that $\sin B \approx 0.8865$, thus $\angle B \approx 117.6°$. Now note that $\angle C = 180° - 52° - 117.6° \approx 10.4°$. Using the law of sines:

$$\frac{\sin 10.4°}{c} = \frac{\sin 52°}{8}, \text{ so } c = \frac{8 \sin 10.4°}{\sin 52°} \approx 1.8 \text{ cm}$$

5.   Draw the triangle:

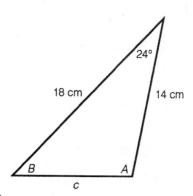

Using the law of cosines:

$$c^2 = 18^2 + 14^2 - 2(18)(14) \cos 24°, \text{ so } c \approx 7.7 \text{ cm}$$

Using the law of sines:

$$\frac{\sin A}{18} = \frac{\sin 24°}{7.7}, \text{ so } \sin A = \frac{18 \sin 24°}{7.7} \approx 0.9486$$

Then either $\angle A \approx 71.5°$ or $\angle A \approx 108.5°$. If $\angle A \approx 71.5°$, then $\angle B \approx 180° - 24° - 71.5° \approx 84.5°$. But this is impossible since $\angle B < \angle A$. If $\angle A \approx 108.5°$, then $\angle B \approx 180° - 24° - 108.5° \approx 47.5°$.

7.   Draw the triangle:

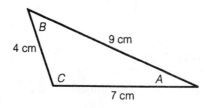

Using the law of cosines:

$$9^2 = 4^2 + 7^2 - 2(4)(7) \cos C, \text{ so } \cos C \approx -0.2857, \text{ thus } C \approx 106.6°$$

Using the law of sines:

$$\frac{\sin 106.6°}{9} = \frac{\sin B}{7}, \text{ so } \sin B = \frac{7 \sin 106.6°}{9} \approx 0.7454$$

Thus $B \approx 48.2°$. Therefore $\angle A \approx 180° - 106.6° - 48.2° \approx 25.2°$.

9.   Note that $\angle AEB = 86°$. Using the law of sines:
$$\frac{\sin 50°}{BE} = \frac{\sin 86°}{12}, \text{ so } BE = \frac{12\sin 50°}{\sin 86°} \approx 9.21 \text{ cm}$$

11.   Using the area formula:
$$A = \tfrac{1}{2}(BC)(BE)\sin 36° = \tfrac{1}{2}(12 \text{ cm})(9.21 \text{ cm})(\sin 36°) \approx 32.48 \text{ cm}^2$$

13.   Using the area formula:
$$A = \tfrac{1}{2}(12 \text{ cm})(BD\sin 44°) = \tfrac{1}{2}(12 \text{ cm})(13.25 \text{ cm})(\sin 44°) \approx 55.23 \text{ cm}^2$$

15.   Using the law of cosines:
$$(CD)^2 = (BC)^2 + (BD)^2 - 2(BC)(BD)\cos 36°$$
$$\approx (12)^2 + (13.25)^2 - 2(12)(13.25)\cos 36°$$
$$\approx 319.56 - 318\cos 36°$$
$$\approx 62.29$$
So $CD \approx \sqrt{62.29} \approx 7.89$ cm.

17.   Using the law of sines:
$$\frac{\sin 80°}{AC} = \frac{\sin 50°}{12}, \text{ so } AC = \frac{12\sin 80°}{\sin 50°} \approx 15.43 \text{ cm}$$

19.   Re-draw the figure:

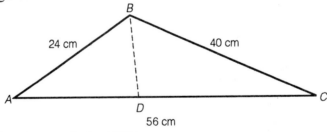

Use the law of cosines to find $\angle ABC$:
$$56^2 = 40^2 + 24^2 - 2(40)(24)\cos(\angle ABC)$$
$$-0.5 = \cos(\angle ABC)$$
$$\angle ABC = 120°$$
Now use the law of sines to find $\angle A$:
$$\frac{\sin A}{40} = \frac{\sin 120°}{56}, \text{ so } \sin A \approx 0.6186, \text{ thus } A \approx 38.21°$$
Then $\angle ADB \approx 180° - 60° - 38.21° \approx 81.79°$. Find BD by the law of sines:
$$\frac{\sin 81.79°}{24} = \frac{\sin 38.21°}{BD}, \text{ so } BD = \frac{24\sin 38.21°}{\sin 81.79°} \approx 15 \text{ cm}$$

21. Re-draw the figure:

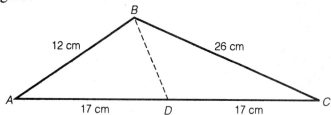

Using the law of cosines, find $\angle A$:

$$26^2 = 12^2 + 34^2 - 2(12)(34)\cos A$$
$$\cos A \approx 0.7647$$
$$A \approx 40.12°$$

Now find BD by using the law of cosines (on $\triangle ADB$):

$$(BD)^2 = 12^2 + 17^2 - 2(12)(17)\cos 40.12°$$
$$(BD)^2 = 121$$
$$BD = 11 \text{ cm}$$

23. (a) Using the law of cosines:

$$a^2 = b^2 + c^2 - 2bc\cos A$$
$$4^2 = 5^2 + 6^2 - 2(5)(6)\cos A$$
$$16 = 25 + 36 - 60\cos A$$
$$-45 = -60\cos A$$
$$\tfrac{3}{4} = \cos A$$

Also using the law of cosines:

$$c^2 = a^2 + b^2 - 2ab\cos C$$
$$6^2 = 4^2 + 5^2 - 2(4)(5)\cos C$$
$$36 = 16 + 25 - 40\cos C$$
$$-5 = -40\cos C$$
$$\tfrac{1}{8} = \cos C$$

(b) Since $\cos A = \tfrac{3}{4}$:

$$\sin A = \sqrt{1 - \cos^2 A} = \sqrt{1 - \tfrac{9}{16}} = \sqrt{\tfrac{7}{16}} = \tfrac{\sqrt{7}}{4}$$

Therefore:

$$\cos 2A = \cos^2 A - \sin^2 A = \left(\tfrac{3}{4}\right)^2 - \left(\tfrac{\sqrt{7}}{4}\right)^2 = \tfrac{9}{16} - \tfrac{7}{16} = \tfrac{1}{8}$$

But since $\cos C = \tfrac{1}{8}$, $C = 2A$.

25. Since $BE = BD$, $\triangle BDE$ is isosceles and thus its base angles are congruent. Therefore, calling $\angle BED = \theta$:

$$\theta + \theta + 108° = 180°$$
$$2\theta = 72°$$
$$\theta = 36°$$

So $\angle BED = 36°$.

27. Note that from Exercise 26 $\angle ABE = 36°$. Using the law of sines on $\triangle ABE$:

$$\frac{\sin \angle AEB}{16} = \frac{\sin 36°}{11}$$
$$\sin \angle AEB = \frac{16 \sin 36°}{11} \approx 0.8550$$
$$\angle AEB \approx 58.76° \text{ or } 121.24°$$

Since the given figure indicates that $\angle AEB$ is an obtuse angle, $\angle AEB \approx 121.24°$.

29. Note that from Exercise 28 $\angle BAE \approx 22.76°$. Using the law of sines on $\triangle ABE$:

$$\frac{\sin 22.76°}{BE} = \frac{\sin 36°}{11}$$
$$BE = \frac{11 \sin 22.76°}{\sin 36°} \approx 7.24 \text{ m}$$

31. (a) Let O be the center of the circle, so that $OA = OB = 10$ cm. Since $\angle AOB = \frac{360°}{5} = 72°$, now find AB using the law of cosines on $\triangle AOB$:

$$(AB)^2 = (OA)^2 + (OB)^2 - 2(OA)(OB)\cos 72°$$
$$= (10)^2 + (10)^2 - 2(10)(10)\cos 72°$$
$$= 200 - 200\cos 72°$$

So $AB = \sqrt{200 - 200\cos 72°} \approx 11.76$ cm.

(b) The interior angles of the pentagon are $\frac{180°(5-2)}{5} = 108°$, so $\angle ABC = 108°$. Using the law of cosines on $\triangle ABC$ (using the exact value from (a)):

$$(AC)^2 = (AB)^2 + (BC)^2 - 2(AB)(BC)\cos 108°$$
$$\approx (11.76)^2 + (11.76)^2 - 2(11.76)(11.76)\cos 108°$$
$$\approx 361.80$$

So $AC \approx \sqrt{361.80} \approx 19.02$ cm.

33. For the triangle, note the following figure:

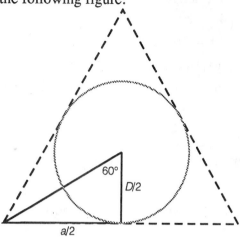

So $\tan 60° = \frac{a/2}{D/2}$, or $\sqrt{3} = \frac{a}{D}$, thus $a = D\sqrt{3}$. For the hexagon, note the following figure:

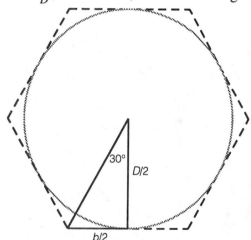

So $\tan 30° = \frac{b/2}{D/2}$, or $\frac{1}{\sqrt{3}} = \frac{b}{D}$, thus $b = \frac{D}{\sqrt{3}}$. Thus:

$$ab = \left(D\sqrt{3}\right)\left(\frac{D}{\sqrt{3}}\right) = D^2$$

35. The area of the triangle is given by:

$$A = \tfrac{1}{2}(\text{base})(\text{height}) = \tfrac{1}{2}(a)(b\sin 60°) = \tfrac{\sqrt{3}}{4}ab$$

Since the area is $10\sqrt{3}$ cm^2:

$$\tfrac{\sqrt{3}}{4}ab = 10\sqrt{3}$$
$$ab = 40$$

Find the third side d in terms of a and b using the law of cosines:

$$d^2 = a^2 + b^2 - 2ab\cos 60°$$
$$d^2 = a^2 + b^2 - 2(40)\left(\tfrac{1}{2}\right)$$
$$d = \sqrt{a^2 + b^2 - 40}$$

Since the perimeter of the triangle is 20 cm:
$$a + b + \sqrt{a^2 + b^2 - 40} = 20$$
$$\sqrt{a^2 + b^2 - 40} = 20 - (a + b)$$
$$a^2 + b^2 - 40 = 400 - 40(a + b) + (a + b)^2$$
$$a^2 + b^2 - 40 = 400 - 40a - 40b + a^2 + 2ab + b^2$$
$$-440 = -40a - 40b + 2(40)$$
$$-520 = -40a - 40b$$
$$13 = a + b$$

So, we have the system of equations:
$$a + b = 13$$
$$ab = 40$$

Solving the first equation for b yields $b = 13 - a$, now substitute:
$$a(13 - a) = 40$$
$$13a - a^2 = 40$$
$$a^2 - 13a + 40 = 0$$
$$(a - 8)(a - 5) = 0$$
$$a = 8, 5$$

Since a is the smaller of the two numbers, $a = 5$ and $b = 8$.

37. Draw a figure:

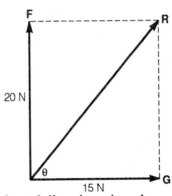

The resultant has a magnitude and direction given by:
$$|\mathbf{R}| = \sqrt{15^2 + 20^2} = \sqrt{625} = 25 \text{ N}$$
$$\theta = \tan^{-1}\left(\tfrac{20}{15}\right) \approx 53.1°$$

39. The components are given by:
$$v_x = 50\cos 35° \approx 41.0 \text{ cm/sec}$$
$$v_y = 50\sin 35° \approx 28.7 \text{ cm/sec}$$

41. Draw the figure:

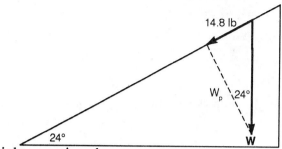

So the desired weights are given by:

$$\tan 24° = \frac{14.8}{\left|\mathbf{W}_p\right|}, \text{ so } \left|\mathbf{W}_p\right| = \frac{14.8}{\tan 24°} \approx 33.2 \text{ lb}$$

$$\sin 24° = \frac{14.8}{\left|\mathbf{W}\right|}, \text{ so } \left|\mathbf{W}\right| = \frac{14.8}{\sin 24°} \approx 36.4 \text{ lb}$$

43. First find the lengths of $\langle 2,6 \rangle$ and $\langle -5,b \rangle$:

$$\left|\langle 2,6 \rangle\right| = \sqrt{2^2 + 6^2} = \sqrt{40}$$

$$\left|\langle -5,b \rangle\right| = \sqrt{(-5)^2 + b^2} = \sqrt{b^2 + 25}$$

Find b by solving the equation:

$$\sqrt{b^2 + 25} = \sqrt{40}$$
$$b^2 + 25 = 40$$
$$b^2 = 15$$
$$b = \pm\sqrt{15}$$

45. $\mathbf{a} + \mathbf{b} = \langle 3,5 \rangle + \langle 7,4 \rangle = \langle 10,9 \rangle$

47. Compute the quantity:

$$3\mathbf{c} + 2\mathbf{a} = 3\langle 2,-1 \rangle + 2\langle 3,5 \rangle = \langle 6,-3 \rangle + \langle 6,10 \rangle = \langle 12,7 \rangle$$

49. Since $\mathbf{b} + \mathbf{d} = \langle 7,7 \rangle$ and $\mathbf{b} - \mathbf{d} = \langle 7,1 \rangle$:

$$\left|\mathbf{b}+\mathbf{d}\right|^2 - \left|\mathbf{b}-\mathbf{d}\right|^2 = \left(7^2 + 7^2\right) - \left(7^2 + 1^2\right) = 98 - 50 = 48$$

51. $(\mathbf{a} + \mathbf{b}) + \mathbf{c} = \langle 10,9 \rangle + \langle 2,-1 \rangle = \langle 12,8 \rangle$

53. $(\mathbf{a} - \mathbf{b}) - \mathbf{c} = \langle -4,1 \rangle - \langle 2,-1 \rangle = \langle -6,2 \rangle$

55. $4\mathbf{c} + 2\mathbf{a} - 3\mathbf{b} = \langle 8,-4 \rangle + \langle 6,10 \rangle - \langle 21,12 \rangle = \langle -7,-6 \rangle$

57. $\langle 7,-6 \rangle = 7\mathbf{i} - 6\mathbf{j}$

59. First compute the length of $\langle 6, 4 \rangle$:
$$|\langle 6, 4 \rangle| = \sqrt{36 + 16} = \sqrt{52} = 2\sqrt{13}$$
Therefore a unit vector in the same direction as $\langle 6, 4 \rangle$ would be:
$$\frac{1}{2\sqrt{13}}\langle 6, 4 \rangle = \left\langle \frac{3}{\sqrt{13}}, \frac{2}{\sqrt{13}} \right\rangle = \left\langle \frac{3\sqrt{13}}{13}, \frac{2\sqrt{13}}{13} \right\rangle$$

61. Use the distance formula, taking the points (r_1, θ_1) and (r_2, θ_2) to be $\left(3, \frac{\pi}{12}\right)$ and $\left(2, \frac{17\pi}{18}\right)$:
$$\begin{aligned} d^2 &= r_1^2 + r_2^2 - 2r_1 r_2 \cos\left(\theta_2 - \theta_1\right) \\ &= 3^2 + 2^2 - 2(3)(2)\cos\left(\frac{\pi}{12} - \frac{17\pi}{18}\right) \\ &= 13 - 12\cos\left(-\frac{31\pi}{36}\right) \\ &\approx 23.8757 \end{aligned}$$
So $d \approx \sqrt{23.8757} \approx 4.89$.

63. Using the equation $r^2 + r_0^2 - 2rr_0 \cos\left(\theta - \theta_0\right) = a^2$:
$$\begin{aligned} r^2 + (5)^2 - 2(r)(5)\cos\left(\theta - \frac{\pi}{6}\right) &= 3^2 \\ r^2 + 25 - 10r\cos\left(\theta - \frac{\pi}{6}\right) &= 9 \\ r^2 - 10r\cos\left(\theta - \frac{\pi}{6}\right) &= -16 \end{aligned}$$

65. (a) Since the line is of the form $r\cos(\theta - \alpha) = d$, $\alpha = \frac{\pi}{3}$ and $d = 3$. So the desired perpendicular distance is 3.

(b) When $\theta = 0$:
$$\begin{aligned} r\cos\left(0 - \frac{\pi}{3}\right) &= 3 \\ r\left(\frac{1}{2}\right) &= 3 \\ r &= 6 \end{aligned}$$
When $\theta = \frac{\pi}{2}$:
$$\begin{aligned} r\cos\left(\frac{\pi}{2} - \frac{\pi}{3}\right) &= 3 \\ r\cos\frac{\pi}{6} &= 3 \\ r \cdot \frac{\sqrt{3}}{2} &= 3 \\ r &= \frac{6}{\sqrt{3}} = 2\sqrt{3} \end{aligned}$$
The required polar points are $(6, 0)$ and $\left(2\sqrt{3}, \frac{\pi}{2}\right)$.

(c) These coordinates are (d, α), which is the polar point $\left(3, \frac{\pi}{3}\right)$.

(d) Sketch the line:

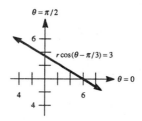

67. (a) Plot points in polar form:

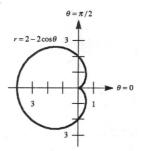

(b) Plot points in polar form:

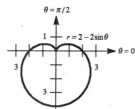

69. (a) Plot points in polar form:

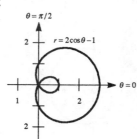

(b) Plot points in polar form:

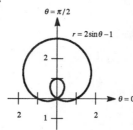

71.   (a)   Plot points in polar form:

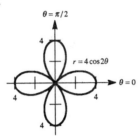

(b)   Plot points in polar form:

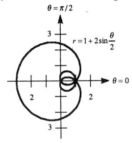

73.   (a)   Plot points in polar form (note that we must plot $0 \le \theta \le 4\pi$):

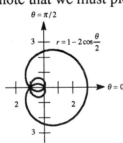

(b)   Plot points in polar form (note that we must plot $0 \le \theta \le 4\pi$):

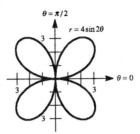

75. Solving $y = 1 + t$ for t yields $t = y - 1$, now substituting:
$$x = 3 - 5(y - 1)$$
$$x = 3 - 5y + 5$$
$$x + 5y = 8$$
So the given parametric equations determine a line.

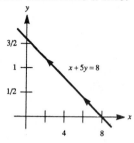

77. Multiplying the first equation by 2 yields $2x = 6 \sin t$ and $y = 6 \cos t$, so:
$$(2x)^2 + y^2 = 36 \sin^2 t + 36 \cos^2 t$$
$$4x^2 + y^2 = 36$$
$$\frac{x^2}{9} + \frac{y^2}{36} = 1$$
So the given parametric equations determine an ellipse.

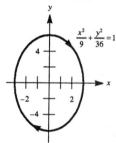

79. Multiplying the first equation by 3 and the second equation by 4 yields $3x = 12 \sec t$ and $4y = 12 \tan t$, so:
$$(3x)^2 - (4y)^2 = 144 \sec^2 t - 144 \tan^2 t$$
$$9x^2 - 16y^2 = 144$$
$$\frac{x^2}{16} - \frac{y^2}{9} = 1$$
So the given parametric equations determine a hyperbola.

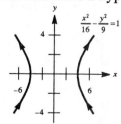

81.   (a) P can be written as $\left(\dfrac{\sin\alpha}{\alpha}, \alpha \right)$, while Q can be written as $\left(\dfrac{\sin 2\alpha}{2\alpha}, 2\alpha \right)$.

(b)   For point P:

$$x = r\cos\theta = \frac{\sin\alpha}{\alpha} \cdot \cos\alpha = \frac{\sin\alpha\cos\alpha}{\alpha}$$

$$y = r\sin\theta = \frac{\sin\alpha}{\alpha} \cdot \sin\alpha = \frac{\sin^2\alpha}{\alpha}$$

So the rectangular coordinates are $P\left(\dfrac{\sin\alpha\cos\alpha}{\alpha}, \dfrac{\sin^2\alpha}{\alpha} \right)$. For point Q:

$$x = r\cos\theta = \frac{\sin 2\alpha}{2\alpha} \cdot \cos 2\alpha = \frac{\sin 2\alpha\cos 2\alpha}{2\alpha}$$

$$y = r\sin\theta = \frac{\sin 2\alpha}{2\alpha} \cdot \sin 2\alpha = \frac{\sin^2 2\alpha}{2\alpha}$$

So the rectangular coordinates are $Q\left(\dfrac{\sin 2\alpha\cos 2\alpha}{2\alpha}, \dfrac{\sin^2 2\alpha}{2\alpha} \right)$.

(c)   The slope of $\overline{OQ}$ is given by:

$$\frac{\dfrac{\sin^2 2\alpha}{2\alpha} - 0}{\dfrac{\sin 2\alpha\cos 2\alpha}{2\alpha} - 0} = \frac{\sin^2 2\alpha}{\sin 2\alpha\cos 2\alpha} = \frac{\sin 2\alpha}{\cos 2\alpha} = \tan 2\alpha$$

(d)   The slope of $\overline{PQ}$ is given by:

$$\frac{\dfrac{\sin^2 2\alpha}{2\alpha} - \dfrac{\sin^2\alpha}{\alpha}}{\dfrac{\sin 2\alpha\cos 2\alpha}{2\alpha} - \dfrac{\sin\alpha\cos\alpha}{\alpha}} = \frac{\sin^2 2\alpha - 2\sin^2\alpha}{\sin 2\alpha\cos 2\alpha - 2\sin\alpha\cos\alpha}$$

(e) Making the substitution $\sin^2 2\alpha = 1 - \cos^2 2\alpha$ and using the double-angle identities for sine and cosine:

$$\frac{\sin^2 2\alpha - 2\sin^2 \alpha}{\sin 2\alpha \cos 2\alpha - 2\sin \alpha \cos \alpha} = \frac{1 - \cos^2 2\alpha - 2\sin^2 \alpha}{\sin 2\alpha \cos 2\alpha - \sin 2\alpha}$$

$$= \frac{\left(1 - 2\sin^2 \alpha\right) - \cos^2 2\alpha}{\sin 2\alpha(\cos 2\alpha - 1)}$$

$$= \frac{\cos 2\alpha - \cos^2 2\alpha}{\sin 2\alpha(\cos 2\alpha - 1)}$$

$$= \frac{\cos 2\alpha(1 - \cos 2\alpha)}{\sin 2\alpha(\cos 2\alpha - 1)}$$

$$= -\frac{\cos 2\alpha}{\sin 2\alpha}$$

$$= -\cot 2\alpha$$

(f) Since the slope of $\overline{OQ}$ is $\tan 2\alpha$ and the slope of $\overline{PQ}$ is $-\cot 2\alpha$:

$$m_{\overline{OQ}} \cdot m_{\overline{PQ}} = (\tan 2\alpha)(-\cot 2\alpha) = \tan 2\alpha \cdot \frac{-1}{\tan 2\alpha} = -1$$

Since the product of these slopes is -1, $\overline{OQ} \perp \overline{PQ}$.

83. (a) If $\dfrac{a}{b} = \dfrac{x}{y}$, then:

$$\frac{a-b}{a+b} \div \frac{b}{b} = \frac{\frac{a}{b} - 1}{\frac{a}{b} + 1} = \frac{\frac{x}{y} - 1}{\frac{x}{y} + 1} \cdot \frac{y}{y} = \frac{x-y}{x+y}$$

(b) Since $\dfrac{a}{b} = \dfrac{\sin A}{\sin B}$, using the result in part (a) with $x = \sin A$ and $y = \sin B$:

$$\frac{a-b}{a+b} = \frac{x-y}{x+y} = \frac{\sin A - \sin B}{\sin A + \sin B}$$

(c) Using the sum-to-product formulas:

$$\sin A - \sin B = 2\cos \tfrac{A+B}{2} \sin \tfrac{A-B}{2}$$

$$\sin A + \sin B = 2\sin \tfrac{A+B}{2} \cos \tfrac{A-B}{2}$$

Therefore our result from part (b) becomes:

$$\frac{a-b}{a+b} = \frac{2\cos \frac{A+B}{2}\sin \frac{A-B}{2}}{2\sin \frac{A+B}{2}\cos \frac{A-B}{2}} = \frac{\sin \frac{A-B}{2}}{\cos \frac{A-B}{2}} \div \frac{\sin \frac{A+B}{2}}{\cos \frac{A+B}{2}} = \frac{\tan \frac{1}{2}(A-B)}{\tan \frac{1}{2}(A+B)}$$

<u>Chapter Ten Test</u>

1. Draw the triangle:

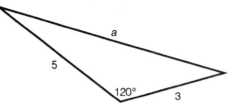

Using the law of cosines:

$$a^2 = 3^2 + 5^2 - 2(3)(5)\cos 120° = 9 + 25 - 30\left(-\tfrac{1}{2}\right) = 49$$

So $a = 7$ cm.

2. Draw the triangle:

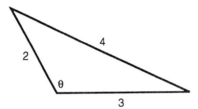

Using the law of cosines:

$$4^2 = 2^2 + 3^2 - 2(2)(3)\cos\theta$$
$$16 = 4 + 9 - 12\cos\theta$$
$$3 = -12\cos\theta$$
$$\cos\theta = -\tfrac{1}{4}$$

The angle opposite the 4 cm side must be obtuse (not acute), since its cosine is negative.

3. Using the law of sines:

$$\frac{\sin 45°}{20\sqrt{2}} = \frac{\sin 30°}{x}$$

$$x = \frac{20\sqrt{2}\sin 30°}{\sin 45°} = \frac{20\sqrt{2}\cdot\tfrac{1}{2}}{\tfrac{\sqrt{2}}{2}} = 20 \text{ cm}$$

4.   Draw the figure:

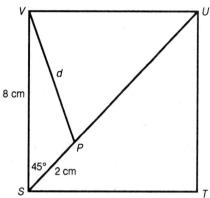

Note that the diagonals of a square bisect the vertex angles. Use the law of cosines with $\triangle SPV$:

$$d^2 = 8^2 + 2^2 - 2(8)(2)\cos 45° = 64 + 4 - 32\left(\tfrac{\sqrt{2}}{2}\right) = 68 - 16\sqrt{2}$$

$$d = \sqrt{68 - 16\sqrt{2}} = 2\sqrt{17 - 4\sqrt{2}}$$

So $PV = 2\sqrt{17 - 4\sqrt{2}}$ cm.

5.   Using the law of cosines to find a:

$$a^2 = (5.8)^2 + (3.2)^2 - 2(5.8)(3.2)\cos 27° = 43.88 - 37.12\cos 27° \approx 10.81$$

So $a \approx \sqrt{10.81} \approx 3.3$ cm. Now use the law of sines to find $\angle C$:

$$\frac{\sin C}{3.2} = \frac{\sin 27°}{3.3}$$

$$\sin C = \frac{3.2\sin 27°}{3.3} \approx 0.442$$

$$C \approx 26.2°$$

Finally, find $B \approx 180° - 27° - 26.2° \approx 126.8°$.

6.   Draw the figure:

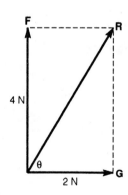

(a)   Find the magnitude of the resultant:

$$|\mathbf{R}| = \sqrt{2^2 + 4^2} = \sqrt{20} = 2\sqrt{5}\ \text{N}$$

(b) $\tan \theta = \frac{4}{2} = 2$

7.   Draw the figure:

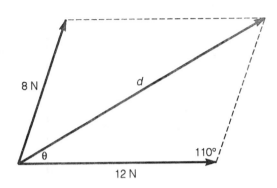

(a)   Using the law of cosines:
$$d^2 = 12^2 + 8^2 - 2(12)(8)\cos 110° = 208 - 192\cos 110°$$
So $d = \sqrt{208 - 192\cos 110°} = 4\sqrt{13 - 12\cos 110°}$ N.

(b)   Using the law of sines:
$$\frac{\sin\theta}{8} = \frac{\sin 110°}{4\sqrt{13 - 12\cos 110°}}$$
$$\sin\theta = \frac{2\sin 110°}{\sqrt{13 - 12\cos 110°}}$$

8.   Draw the figure:

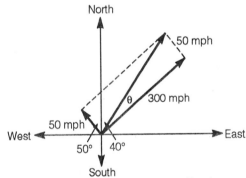

The heading vector and the wind vector are perpendicular to each other, so:
ground speed $= \sqrt{300^2 + 50^2} = 50\sqrt{37}$ mph
$\tan\theta = \frac{50}{300} = \frac{1}{6}$

9.   (a) $2\mathbf{A} + 3\mathbf{B} = 2\langle 2,4\rangle + 3\langle 3,-1\rangle = \langle 13,5\rangle$

(b)   Compute the length:
$$|2\mathbf{A} + 3\mathbf{B}| = |\langle 13,5\rangle| = \sqrt{194}$$

(c) $\mathbf{C} - \mathbf{B} = \langle 4,-4\rangle - \langle 3,-1\rangle = \langle 1,-3\rangle = \mathbf{i} - 3\mathbf{j}$

10. Subtracting components:

$$\vec{PQ} = \langle -7,2 \rangle - \langle 4,5 \rangle = \langle -11,-3 \rangle$$

The required unit vector is:

$$\frac{\vec{PQ}}{|\vec{PQ}|} = \frac{1}{\sqrt{130}} \langle -11,-3 \rangle = \left\langle \frac{-11\sqrt{130}}{130}, \frac{-3\sqrt{130}}{130} \right\rangle$$

11. Using the double-angle formula for $\cos 2\theta$ and multiplying by r^2 yields:

$$r^2 = \cos^2 \theta - \sin^2 \theta$$
$$r^4 = r^2 \cos^2 \theta - r^2 \sin^2 \theta$$
$$\left(x^2 + y^2\right)^2 = x^2 - y^2$$

12. Plotting points in polar form:

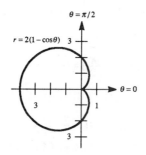

13. Multiplying the second equation by 2 yields $x = 4 \sin t$ and $2y = 4 \cos t$, so:

$$x^2 + (2y)^2 = 16 \sin^2 t + 16 \cos^2 t$$
$$x^2 + 4y^2 = 16$$
$$\frac{x^2}{16} + \frac{y^2}{4} = 1$$

So the given parametric equations determine an ellipse.

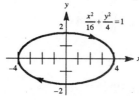

14.   Use the distance formula, taking the points (r_1, θ_1) and (r_2, θ_2) to be $\left(4, \frac{10\pi}{21}\right)$ and $\left(1, \frac{\pi}{7}\right)$:

$$\begin{aligned}
d^2 &= r_1^2 + r_2^2 - 2r_1r_2 \cos(\theta_2 - \theta_1) \\
&= 4^2 + 1^2 - 2(4)(1)\cos\left(\frac{\pi}{7} - \frac{10\pi}{21}\right) \\
&= 17 - 8\cos\left(-\frac{\pi}{3}\right) \\
&= 17 - 8 \cdot \frac{1}{2} \\
&= 13
\end{aligned}$$

So $d = \sqrt{13}$.

15.   Using the equation $r^2 + r_0^2 - 2rr_0 \cos(\theta - \theta_0) = a^2$:

$$\begin{aligned}
r^2 + (5)^2 - 2(r)(5)\cos\left(\theta - \frac{\pi}{2}\right) &= 2^2 \\
r^2 + 25 - 10r\cos\left(\theta - \frac{\pi}{2}\right) &= 4 \\
r^2 - 10r\cos\left(\theta - \frac{\pi}{2}\right) &= -21
\end{aligned}$$

Now substitute the polar point $\left(2, \frac{\pi}{6}\right)$ in the equation:

$$2^2 - 10(2)\cos\left(\frac{\pi}{6} - \frac{\pi}{2}\right) = 4 - 20\cos\left(-\frac{\pi}{3}\right) = 4 - 20 \cdot \frac{1}{2} = -6$$

Since we did not obtain -21, the point does not lie on the circle.

16.   (a)   The line will be of the form $r\cos(\theta - \alpha) = d$. Since $\alpha = \frac{5\pi}{6}$ and $d = 4$, the equation of the line is $r\cos\left(\theta - \frac{5\pi}{6}\right) = 4$.

      (b)   Using the addition formula for cosine:

$$\begin{aligned}
r\cos\left(\theta - \frac{5\pi}{6}\right) &= 4 \\
r\left(\cos\theta\cos\frac{5\pi}{6} + \sin\theta\sin\frac{5\pi}{6}\right) &= 4 \\
r\left[\cos\theta \cdot \left(-\frac{\sqrt{3}}{2}\right) + \sin\theta \cdot \frac{1}{2}\right] &= 4 \\
-r\sqrt{3}\cos\theta + r\sin\theta &= 8 \\
-r\sqrt{3}\cos\theta + r\sin\theta - 8 &= 0
\end{aligned}$$

Chapter Eleven
Systems of Equations

11.1 Systems of Two Linear Equations in Two Unknowns

1. (a) yes
    (b) no--the xy term makes it non-linear
    (c) yes
    (d) yes

3. Test $(5, 1)$ in the two equations and see if it produces true statements:
    $$2(5) - 8(1) = 2$$
    $$10 - 8 = 2 \qquad \text{true}$$
    $$3(5) + 7(1) = 22$$
    $$15 + 7 = 22 \qquad \text{true}$$
    So $(5, 1)$ is a solution to the given system.

5. Test $(0, -4)$ in the two equations and see if it produces true statements:
    $$\tfrac{1}{6}(0) + \tfrac{1}{2}(-4) = -2$$
    $$0 - 2 = -2 \qquad \text{true}$$
    $$\tfrac{2}{3}(0) + \tfrac{3}{4}(-4) = 2$$
    $$0 - 3 = 2 \qquad \text{false}$$
    So $(0, -4)$ is not a solution to the given system.

7. Test $(3, -2)$ in the two equations and see if it produces true statements:

$$\tfrac{2}{7}(3) - \tfrac{1}{5}(-2) = \tfrac{44}{35}$$
$$\tfrac{6}{7} + \tfrac{2}{5} = \tfrac{44}{35}$$
$$\tfrac{30}{35} + \tfrac{14}{35} = \tfrac{44}{35} \qquad \text{true}$$
$$\tfrac{1}{3}(3) - \tfrac{5}{4}(-2) = \tfrac{7}{2}$$
$$1 + \tfrac{5}{2} = \tfrac{7}{2}$$
$$\tfrac{2}{2} + \tfrac{5}{2} = \tfrac{7}{2} \qquad \text{true}$$

So $(3, -2)$ is a solution to the given system.

9. Solve the second equation for x:

$$x + 4y = -4$$
$$x = -4 - 4y$$

Now substitute $x = -4 - 4y$ for x in the first equation, and solve for y:

$$3x - 2y = -19$$
$$3(-4 - 4y) - 2y = -19$$
$$-12 - 12y - 2y = -19$$
$$-14y - 12 = -19$$
$$-14y = -7$$
$$y = \tfrac{1}{2}$$

Substitute this back into the second equation:

$$x = -4 - 4y = -4 - 4 \cdot \tfrac{1}{2} = -4 - 2 = -6$$

So the solution is $\left(-6, \tfrac{1}{2}\right)$.

11. Solve the first equation for y:

$$4x + 2y = 3$$
$$2y = 3 - 4x$$
$$y = \frac{3 - 4x}{2}$$

Now substitute $y = \frac{3-4x}{2}$ for y in the second equation, and solve for x:

$$10x + 4y = 1$$
$$10x + 4\left(\frac{3 - 4x}{2}\right) = 1$$
$$10x + 2(3 - 4x) = 1$$
$$10x + 6 - 8x = 1$$
$$2x + 6 = 1$$
$$2x = -5, \text{ so } x = -\tfrac{5}{2}$$

Substitute this back into the first equation:

$$y = \frac{3 - 4x}{2} = \frac{3 - 4\left(-\tfrac{5}{2}\right)}{2} = \frac{3 + 10}{2} = \tfrac{13}{2}$$

The solution is $\left(-\tfrac{5}{2}, \tfrac{13}{2}\right)$.

13. Solve the second equation for y:

$$-7x + 2y = 0$$
$$2y = 7x$$
$$y = \frac{7x}{2}$$

Now substitute $y = \frac{7x}{2}$ for y in the first equation, and solve for x:

$$13x - 8y = -3$$
$$13x - 8 \cdot \frac{7x}{2} = -3$$
$$13x - 28x = -3$$
$$-15x = -3$$
$$x = \tfrac{1}{5}$$

Substitute this back into the second equation:

$$y = \frac{7x}{2} = \frac{7 \cdot \frac{1}{5}}{2} = \tfrac{7}{10}$$

The solution is $\left(\tfrac{1}{5}, \tfrac{7}{10} \right)$.

15. First multiply each equation by 20 to clear fractions:

$$20\left(-\tfrac{2}{5}x + \tfrac{1}{4}y\right) = 20(3)$$
$$20\left(\tfrac{1}{4}x - \tfrac{2}{5}y\right) = 20(-3)$$

Therefore we have the system:

$$-8x + 5y = 60$$
$$5x - 8y = -60$$

Solve the first equation for y:

$$-8x + 5y = 60$$
$$5y = 8x + 60$$
$$y = \frac{8x + 60}{5}$$

Now substitute $y = \frac{8x+60}{5}$ for y in the second equation and solve for x:

$$5x - 8y = -60$$
$$5x - 8\left(\frac{8x + 60}{5}\right) = -60$$

Multiply by 5 to clear fractions:

$$25x - 8(8x + 60) = -300$$
$$25x - 64x - 480 = -300$$
$$-39x = 180$$
$$x = -\frac{180}{39} = -\frac{60}{13}$$

Substitute this back into the first equation:

$$y = \frac{8x + 60}{5} = \frac{8\left(-\frac{60}{13}\right) + 60}{5} = \frac{-480 + 780}{65} = \frac{300}{65} = \frac{60}{13}$$

The solution is $\left(-\tfrac{60}{13}, \tfrac{60}{13}\right)$.

17. Solve the first equation for x:
$$\sqrt{2}x - \sqrt{3}y = \sqrt{3}$$
$$\sqrt{2}x = \sqrt{3}y + \sqrt{3}$$
$$x = \frac{\sqrt{3}y + \sqrt{3}}{\sqrt{2}}$$

Now substitute $x = \frac{\sqrt{3}y+\sqrt{3}}{\sqrt{2}}$ for x in the second equation and solve for y:
$$\sqrt{3}x - \sqrt{8}y = \sqrt{2}$$
$$\sqrt{3}\left(\frac{\sqrt{3}y + \sqrt{3}}{\sqrt{2}}\right) - \sqrt{8}y = \sqrt{2}$$
$$\frac{3y + 3}{\sqrt{2}} - \sqrt{8}y = \sqrt{2}$$

Multiply by $\sqrt{2}$ to clear the fractions:
$$3y + 3 - 4y = 2$$
$$3 - y = 2$$
$$-y = -1$$
$$y = 1$$

Substitute this back into the first equation:
$$x = \frac{\sqrt{3}y + \sqrt{3}}{\sqrt{2}} = \frac{\sqrt{3} + \sqrt{3}}{\sqrt{2}} = \frac{2\sqrt{3}}{\sqrt{2}} = \frac{2\sqrt{6}}{2} = \sqrt{6}$$

The solution is $\left(\sqrt{6}, 1\right)$.

19. Multiply the second equation by 2:
$$5x + 6y = 4$$
$$4x - 6y = -6$$
Adding:
$$9x = -2$$
$$x = -\tfrac{2}{9}$$

Substitute $x = -\tfrac{2}{9}$ into the first equation:
$$5x + 6y = 4$$
$$5\left(-\tfrac{2}{9}\right) + 6y = 4$$
$$-\tfrac{10}{9} + 6y = 4$$
$$6y = \tfrac{46}{9}$$
$$y = \tfrac{46}{54} = \tfrac{23}{27}$$

The solution is $\left(-\tfrac{2}{9}, \tfrac{23}{27}\right)$.

21. Multiply the second equation by -2:
$$4x + 13y = -5$$
$$-4x + 108y = 2$$
Adding:
$$121y = -3$$
$$y = -\tfrac{3}{121}$$

Substitute $y = -\frac{3}{121}$ into the first equation:

$$4x + 13y = -5$$
$$4x + 13\left(-\frac{3}{121}\right) = -5$$
$$4x - \frac{39}{121} = -5$$
$$4x = -\frac{566}{121}$$
$$x = -\frac{283}{242}$$

The solution is $\left(-\frac{283}{242}, -\frac{3}{121}\right)$.

23. Multiply the first equation by 12 and the second equation by 70 to clear fractions:

$$12\left(\tfrac{1}{4}x - \tfrac{1}{3}y\right) = 12(4)$$
$$70\left(\tfrac{2}{7}x - \tfrac{1}{7}y\right) = 70\left(\tfrac{1}{10}\right)$$

Therefore we have the system:

$$3x - 4y = 48$$
$$20x - 10y = 7$$

Multiply the first equation by -5 and the second equation by 2:

$$-15x + 20y = -240$$
$$40x - 20y = 14$$

Adding:

$$25x = -226$$
$$x = -\frac{226}{25}$$

Substitute $x = -\frac{226}{25}$ into the second equation:

$$20x - 10y = 7$$
$$20\left(-\frac{226}{25}\right) - 10y = 7$$
$$-\frac{904}{5} - 10y = 7$$
$$-10y = \frac{939}{5}$$
$$y = -\frac{939}{50}$$

The solution is $\left(-\frac{226}{25}, -\frac{939}{50}\right)$.

25. Multiply the second equation by -4:

$$8x + 16y = 5$$
$$-8x - 20y = -5$$

Adding:

$$-4y = 0, \text{ so } y = 0$$

Substitute $y = 0$ into the first equation:

$$8x + 16y = 5$$
$$8x = 5$$
$$x = \tfrac{5}{8}$$

The solution is $\left(\tfrac{5}{8}, 0\right)$.

27. Divide the first equation by 5 and multiply the second equation by 10 to get:

$$25x - 8y = 9$$
$$x + y = 3$$

Multiply the second equation by 8:

$$25x - 8y = 9$$
$$8x + 8y = 24$$

Adding:

$$33x = 33$$
$$x = 1$$

Now substitute $x = 1$ into the second equation:

$$x + y = 3$$
$$1 + y = 3$$
$$y = 2$$

The solution is $(1, 2)$.

29. The given points must satisfy the equation, so:

$$4 = 0^2 = b \cdot 0 + c$$
$$4 = c$$

For the other point:

$$14 = 2^2 + b \cdot 2 + c$$
$$10 = 2b + c$$

This system, $c = 4$ and $2b + c = 10$, is easily solved by substitution:

$$2b + c = 10$$
$$2b + 4 = 10$$
$$2b = 6$$
$$b = 3$$
$$y = x^2 + 3x + 4$$

So the equation of the parabola is $y = x^2 + 3x + 4$.

31. Again, the points satisfy the equation:

$$Ax + By = 2$$
$$A(-4) + B(5) = 2$$
$$A(7) + B(-9) = 2$$

So we have the system:

$$-4A + 5B = 2$$
$$7A - 9B = 2$$

Multiply the first equation by 7 and the second equation by 4:

$$-28A + 35B = 14$$
$$28A - 36B = 8$$
$$-B = 22$$
$$B = -22$$

Substituting for B:

$$7A - 9(-22) = 2$$
$$7A + 198 = 2$$
$$7A = -196$$
$$A = -28$$

33. First graph the region:

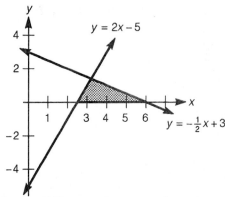

The base of the triangle is the difference between the x-intercepts, so:

$$\text{base} = 6 - \tfrac{5}{2} = \tfrac{7}{2}$$

The height is the y-coordinate of the intersection of the two lines $y = -\tfrac{1}{2}x + 3$ and $y = 2x - 5$, so:

$$-\tfrac{1}{2}x + 3 = 2x - 5$$
$$8 = \tfrac{5}{2}x$$
$$\tfrac{16}{5} = x$$

So $y = 2\left(\tfrac{16}{5}\right) - 5 = \tfrac{32}{5} - 5 = \tfrac{7}{5}$. So the area is:

$$\tfrac{1}{2}(\text{base})(\text{height}) = \tfrac{1}{2}\left(\tfrac{7}{2}\right)\left(\tfrac{7}{5}\right) = \tfrac{49}{20} \text{ square units}$$

35. Let $x =$ the amount of the 10% solution and $y =$ the amount of the 35% solution. Then $x + y = 200$ cc. Also we know that the amount of acid in each separate solution, 10% of x and 35% of y, must add to equal the acid in the mixture, 25% of $(x + y)$. This second equation is usually in need of simplifying:

$$0.10x + 0.35y = 0.25(x + y)$$
$$10x + 35y = 25(x + y)$$
$$2x + 7y = 5x + 5y$$
$$-3x + 2y = 0$$

This system is now solved by either method:

$$x + \ y = 200$$
$$-3x + 2y = 0$$

Multiply the first equation by 3:

$$3x + 3y = 600$$
$$-3x + 2y = 0$$

Adding:

$$5y = 600$$
$$y = 120$$

So $x = 80$, and we need 80 cc of the 10% solution and 120 cc of the 35% solution.

37. Let x be the amount of \$5.20 coffee and y be the amount of \$5.80 coffee. Then $x + y = 16$ pounds. The total value of each bean is $5.20x$ and $5.80y$, respectively, and that of the mixture is $5.50(x + y)$, so we have:
$$5.20x + 5.80y = 5.50(x + y)$$
$$52x + 58y = 55(x + y)$$
$$52x + 58y = 55x + 55y$$
$$-3x + 3y = 0$$
$$-x + y = 0$$
Solve the system:
$$x + y = 16$$
$$-x + y = 0$$
Adding:
$$2y = 16$$
$$y = 8$$
Since $x = 16 - y$, $x = 16 - 8 = 8$. So we mix 8 pounds of \$5.20 coffee and 8 pounds of \$5.80 coffee.

39. Eliminating fractions:
$$bx + ay = ab$$
$$ax + by = ab$$
Therefore:
$$abx + a^2y = a^2b$$
$$-abx - b^2y = -ab^2$$
Adding:
$$(a^2 - b^2)y = a^2b - ab^2$$
$$y = \frac{ab(a - b)}{a^2 - b^2}$$
If we factor and reduce, we have $y = \dfrac{ab}{a + b}$. Now substitute to solve for x:
$$\frac{x}{a} + \frac{\frac{ab}{a+b}}{b} = 1$$
$$\frac{x}{a} + \frac{a}{a + b} = 1$$
$$(a + b)x + a^2 = a(a + b)$$
$$(a + b)x = a^2 + ab - a^2$$
$$x = \frac{ab}{a + b}$$
The solution is $\left(\dfrac{ab}{a + b}, \dfrac{ab}{a + b} \right)$. Note that we cannot have $a = \pm b$ for this solution.

41. Multiply the first equation by b and the second equation by $-a$:

$$abx + a^2by = b$$
$$-abx - ab^2y = -a$$

Adding:

$$(a^2b - ab^2)y = b - a$$

$$y = \frac{b-a}{a^2b - ab^2} = \frac{-1}{ab}$$

Substitute $y = \frac{-1}{ab}$ into the first equation:

$$ax + a^2y = 1$$

$$ax - \frac{a}{b} = 1$$

$$ax = \frac{a+b}{b}$$

$$x = \frac{a+b}{ab}$$

So the solution is $\left(\dfrac{a+b}{ab}, \dfrac{-1}{ab}\right)$. Note that we cannot have $ab = 0$, so $a \neq 0$ and $b \neq 0$.

43. Let $x = \frac{1}{s}$ and $y = \frac{1}{t}$. Then we have the system:

$$\tfrac{1}{2}x - \tfrac{1}{2}y = -10$$
$$2x + 3y = 5$$

Multiply the first equation by -4:

$$-2x + 2y = 40$$
$$2x + 3y = 5$$

Adding:

$$5y = 45$$
$$y = 9$$

Substitute $y = 9$ into the second equation, and solve for x:

$$2x + 3y = 5$$
$$2x + 3(9) = 5$$
$$2x + 27 = 5$$
$$2x = -22$$
$$x = -11$$

Since $s = \frac{1}{x}$ and $t = \frac{1}{y}$, then $s = -\frac{1}{11}$ and $t = \frac{1}{9}$. So the solution is $\left(-\frac{1}{11}, \frac{1}{9}\right)$.

45. First clear fractions and put into a standard form:

$$\frac{2w-1}{3} + \frac{z+2}{4} = 4$$

Multiplying by 12 we have the equation:

$$4(2w-1) + 3(z+2) = 12 \cdot 4$$
$$8w - 4 + 3z + 6 = 48$$
$$8w + 3z = 46$$

The second equation will be $w + 2z = 9$. So we have the system:
$$8w + 3z = 46$$
$$w + 2z = 9$$
Multiply the second equation by -8:
$$8w + 3z = 46$$
$$-8w - 16z = -72$$
Adding:
$$-13z = -26$$
$$z = 2$$
$$w = 9 - 2(2) = 5$$
So the solution is $(w, z) = (5, 2)$.

47.   Using substitution, solve the first equation for x:
$$1.03x - 2.54y = 5.47$$
$$1.03x = 2.54y + 5.47$$
$$x = 2.4660y + 5.3107$$
Now substitute into the second equation:
$$3.85x + 4.29y = -1.84$$
$$3.85(2.466y + 5.3107) + 4.29y = -1.84$$
$$13.7841y = -22.2862$$
$$y \approx -1.62$$
Substituting back for x yields:
$$x = 2.4660(-1.62) + 5.3107 \approx 1.32$$
So the solution is $(1.32, -1.62)$.

49.   Letting $u = \ln x$ and $v = \ln y$, we have the system:
$$2u - 5v = 11$$
$$u + v = -5$$
Using substitution, solve the second equation for v to obtain $v = -5 - u$.
Now substitute into the first equation:
$$2u - 5(-5 - u) = 11$$
$$7u + 25 = 11$$
$$7u = -14$$
$$u = -2$$
$$v = -5 + 2 = -3$$
So $\ln x = -2$ thus $x = e^{-2}$, and $\ln y = -3$ thus $y = e^{-3}$. The solution is (e^{-2}, e^{-3}), which is approximately $(0.14, 0.05)$.

51. Letting $u = e^x$ and $v = e^y$, we have the system:
$$u - 3v = 2$$
$$3u + v = 16$$
Using substitution, solve the first equation for u to obtain $u = 2 + 3v$.
Now substitute into the second equation:
$$3(2 + 3v) + v = 16$$
$$10v + 6 = 16$$
$$10v = 10$$
$$v = 1$$
$$u = 5$$
So $e^x = 5$ thus $x = \ln 5$, and $e^y = 1$ thus $y = 0$. The solution is $(\ln 5, 0)$, which is approximately $(1.61, 0)$.

53. Letting $u = \sqrt{x^2 - 3x}$ and $v = \sqrt{y^2 + 6y}$, we have the system:
$$4u - 3v = -4$$
$$\tfrac{1}{2}u + \tfrac{1}{2}v = 3$$
Multiplying the second equation by 6, we have the system:
$$4u - 3v = -4$$
$$3u + 3v = 18$$
Adding we obtain $7u = 14$, so $u = 2$. Substitute into the first equation:
$$4(2) - 3v = -4$$
$$8 - 3v = -4$$
$$-3v = -12$$
$$v = 4$$
So $u = 2$ and $v = 4$. Since $u = \sqrt{x^2 - 3x}$ and $v = \sqrt{y^2 + 6y}$, we have the two equations:

$$\sqrt{x^2 - 3x} = 2 \qquad\qquad \sqrt{y^2 + 6y} = 4$$
$$x^2 - 3x = 4 \qquad\qquad y^2 + 6y = 16$$
$$x^2 - 3x - 4 = 0 \qquad\qquad y^2 + 6y - 16 = 0$$
$$(x - 4)(x + 1) = 0 \qquad\qquad (y + 8)(y - 2) = 0$$
$$x = 4, -1 \qquad\qquad y = -8, 2$$
So the solutions are $(4, -8)$, $(4, 2)$, $(-1, -8)$, and $(-1, 2)$.

55. If we take tu as our number, then it is important to distinguish between its value, $10t + u$, and the sum of its digits $t + u$. Here we are told that $t + u = 14$ and that $2t = u + 1$. We solve this system:
$$t + u = 14$$
$$2t - u = 1$$
Adding:
$$3t = 15, \text{ so } t = 5$$
Thus $u = 9$ and our original two-digit number was 59.

57. Let l and w be the length and width, respectively, of the rectangle. Since the perimeter is 34 in., $2l + 2w = 34$, or $l + w = 17$. We are also given that $l = 2 + 2w$, so we substitute:

$$l + w = 17$$
$$(2 + 2w) + w = 17$$
$$2 + 3w = 17$$
$$3w = 15, \text{ so } w = 5$$

Thus $l = 17 - w = 17 - 5 = 12$. Thus the length is 12 in. and the width is 5 in.

59. (a) Solve the first equation for y:

$$x + y = 100$$
$$y = 100 - x$$

Now substituting into the second equation:

$$-2x + 3y = 0$$
$$-2x + 3(100 - x) = 0$$
$$-2x + 300 - 3x = 0$$
$$300 - 5x = 0$$
$$300 = 5x, \text{ so } x = 60$$

Then $y = 100 - 60 = 40$. Thus the solution is $(60, 40)$.

(b) Multiply the first equation by 2:

$$2x + 2y = 200$$
$$-2x + 3y = 0$$

Adding:

$$5y = 200, \text{ so } y = 40$$

Substituting back into the first equation:

$$x + y = 100$$
$$x + 40 = 100$$
$$x = 60$$

Again, the solution is $(60, 40)$.

61. Solve each equation for x:

$$x = by - ab$$
$$x = cy - ac$$

Setting these equal:

$$by - ab = cy - ac$$
$$by - cy = ab - ac$$
$$y(b - c) = a(b - c)$$
$$y = a, \text{ as long as } b \neq c$$

Substituting into the second equation:

$$x = cy - ac = ac - ac = 0$$

So the solution is $(0, a)$.

63. Let $u = \frac{1}{x}$ and $v = \frac{1}{y}$. Then:

$$\frac{a}{b}u + \frac{b}{a}v = a + b$$
$$bu + av = a^2 + b^2$$

Multiply the first equation by ab to clear the fractions:

$$a^2u + b^2v = a^2b + ab^2$$
$$bu + av = a^2 + b^2$$

Multiply the first equation by $-b$ and the second equation by a^2:

$$-a^2bu - b^3v = -a^2b^2 - ab^3$$
$$a^2bu + a^3v = a^4 + a^2b^2$$

Adding:

$$\left(a^3 - b^3\right)v = a^4 - ab^3$$
$$\left(a^3 - b^3\right)v = a\left(a^3 - b^3\right)$$
$$v = a$$

Substituting $v = a$ into the second equation:

$$bu + av = a^2 + b^2$$
$$bu + a^2 = a^2 + b^2$$
$$bu = b^2$$
$$u = b$$

Since $x = \frac{1}{u}$ and $y = \frac{1}{v}$, then $x = \frac{1}{b}$ and $y = \frac{1}{a}$. So the solution is $\left(\dfrac{1}{b}, \dfrac{1}{a}\right)$.

65. Since the lines are concurrent, the following two systems must possess the same solution:

$$\begin{array}{ll}
I. \quad 7x + 5y = 4 & \qquad II. \quad x + ky = 3 \\
 x + ky = 3 & \qquad 5x + y = -k
\end{array}$$

Using either method of this section, we find that the solution of system I is $\left(\frac{4k-15}{7k-5}, \frac{17}{7k-5}\right)$.

Also the solution of system II is found to be $\left(\frac{-3-k^2}{5k-1}, \frac{k+15}{5k-1}\right)$. Since the two solutions are the same, the corresponding x- and y-coordinates must be equal. Equating the y-coordinates gives us:

$$\frac{17}{7k-5} = \frac{k+15}{5k-1}$$

After clearing fractions and simplifying, this equation becomes $7k^2 + 15k - 58 = 0$. We factor to get $(7k + 29)(k - 2) = 0$ and therefore $k = -\frac{29}{7}$ or $k = 2$. These are the required values of k. If we equate the x-coordinates rather than the y-coordinates we obtain the equation $7k^3 + 15k^2 - 58k = 0$. We factor to get $k(7k + 29)(k - 2) = 0$, which has solutions $k = 0, -\frac{29}{7}$, and 2. The root $k = 0$ is extraneous, for in that case the two y-coordinates are not equal.

67.   (a)   First draw the triangle:

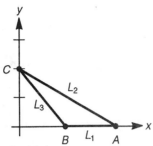

Find the equations for each side:

L_1: $y = 0$

L_2: $m = \dfrac{0-2}{2a-0} = -\dfrac{1}{a}$

Therefore:

$$y = -\frac{1}{a}x + 2$$
$$ay = -x + 2a$$
$$x + ay = 2a$$

L_3: $m = \dfrac{0-2}{2b-0} = -\dfrac{1}{b}$

Therefore:

$$y = -\frac{1}{b}x + 2$$
$$by = -x + 2b$$
$$x + by = 2b$$

(b)   First find the midpoints of each side:

L_1: $\text{mid} = \left(\dfrac{2b+2a}{2}, \dfrac{0}{2}\right) = (a+b, 0)$

L_2: $\text{mid} = \left(\dfrac{0+2a}{2}, \dfrac{2+0}{2}\right) = (a, 1)$

L_3: $\text{mid} = \left(\dfrac{0+2b}{2}, \dfrac{2+0}{2}\right) = (b, 1)$

Now find the slopes from each vertex to the adjacent midpoint:

CL_1: $m = \dfrac{0-2}{a+b-0} = \dfrac{-2}{a+b}$

BL_2: $m = \dfrac{1-0}{a-2b} = \dfrac{1}{a-2b}$

AL_3: $m = \dfrac{1-0}{b-2a} = \dfrac{1}{b-2a}$

Finally, use these slopes and vertices in the point-slope formula:

For CL_1:

$$y - 2 = \frac{-2}{a+b}(x - 0)$$

$$y = \frac{-2}{a+b}x + 2$$

$$(a+b)y = -2x + 2(a+b)$$

$$2x + (a+b)y = 2(a+b)$$

For BL_2:

$$y - 0 = \frac{1}{a-2b}(x - 2b)$$

$$y = \frac{1}{a-2b}x - \frac{2b}{a-2b}$$

$$(a-2b)y = x - 2b$$

$$-x + (a-2b)y = -2b$$

$$x + (2b-a)y = 2b$$

For AL_3:

$$y - 0 = \frac{1}{b-2a}(x - 2a)$$

$$y = \frac{1}{b-2a}x - \frac{2a}{b-2a}$$

$$(b-2a)y = x - 2a$$

$$-x + (b-2a)y = -2a$$

$$x + (2a-b)y = 2a$$

(c) Find the intersection of the three lines:

$$2x + (a+b)y = 2(a+b)$$
$$x + (2b-a)y = 2b$$
$$x + (2a-b)y = 2a$$

Solve the second and third equations for x:

$$x = 2b - (2b-a)y$$
$$x = 2a - (2a-b)y$$

Setting these equal:

$$2b - (2b-a)y = 2a - (2a-b)y$$
$$(2a-b)y - (2b-a)y = 2a - 2b$$
$$(2a - b - 2b + a)y = 2a - 2b$$
$$(3a - 3b)y = 2a - 2b$$
$$y = \frac{2a - 2b}{3a - 3b} = \frac{2(a-b)}{3(a-b)} = \frac{2}{3}$$

Substitute $y = \frac{2}{3}$ into the second equation:

$$x = 2b - \frac{2}{3}(2b - a) = \frac{6b - 4b + 2a}{3} = \frac{2b + 2a}{3} = \frac{2(a+b)}{3}$$

We must still show that this point $\left(\frac{2(a+b)}{3}, \frac{2}{3}\right)$ lies on the first line:

$$2x + (a+b)y = 2(a+b)$$

$$4\left(\frac{a+b}{3}\right) + (a+b)\frac{2}{3} = 2(a+b)$$

$$6\frac{(a+b)}{3} = 2(a+b) \quad \text{true}$$

Thus the point $\left(\frac{2(a+b)}{3}, \frac{2}{3}\right)$ is the intersection of all three medians.

(d)   The altitudes will be perpendicular to the slopes of each side, which we found in (a). So:

For BL_2:
$$m = a$$
$$y - 0 = a(x - 2b)$$
$$y = ax - 2ab$$
$$-ax + y = -2ab$$
$$ax - y = 2ab$$

For CL_2:
$$m = \text{undefined}$$
$$x = 0$$

For AL_3:
$$m = b$$
$$y - 0 = b(x - 2a)$$
$$y = bx - 2ab$$
$$-bx + y = -2ab$$
$$bx - y = 2ab$$

(e)   Find the intersection of the three lines:
$$ax - y = 2ab$$
$$bx - y = 2ab$$
$$x = 0$$

Since $x = 0$, substituting into the first equation yields:
$$-y = 2ab$$
$$y = -2ab$$

We must still show that this point $(0, -2ab)$ lies on the second line:
$$bx - y = 2ab$$
$$b(0) - (-2ab) = 2ab$$
$$2ab = 2ab \quad \text{true}$$

Thus, the point $(0, -2ab)$ is the intersection of all three altitudes.

(f) Recall that the perpendicular bisectors must be perpendicular and pass through the midpoints (which we found in (b)):

For L_1:
$$\text{mid} = (a+b, 0)$$
$$m = \text{undefined}$$
$$x = a+b$$

For L_2:
$$\text{mid} = (a, 1)$$
$$m = a$$
$$y - 1 = a(x - a)$$
$$y - 1 = ax - a^2$$
$$-ax + y = 1 - a^2$$
$$ax - y = a^2 - 1$$

For L_3:
$$\text{mid} = (b, 1)$$
$$m = b$$
$$y - 1 = b(x - b)$$
$$y - 1 = bx - b^2$$
$$-bx + y = 1 - b^2$$
$$bx - y = b^2 - 1$$

(g) Find the intersection of the three lines:
$$ax - y = a^2 - 1$$
$$bx - y = b^2 - 1$$
$$x = a+b$$

Substituting into the first equation yields:
$$a(a+b) - y = a^2 - 1$$
$$a^2 + ab - y = a^2 - 1$$
$$-y = -ab - 1$$
$$y = ab + 1$$

We must still show that the point $(a + b, ab + 1)$ lies on the second line:
$$bx - y = b^2 - 1$$
$$b(a+b) - (ab+1) = b^2 - 1$$
$$ab + b^2 - ab - 1 = b^2 - 1$$
$$b^2 - 1 = b^2 - 1 \quad \text{true}$$

Thus the point $(a + b, ab + 1)$ is the intersection of all three perpendicular bisectors.

(h) Find the line that passes through $\left(\frac{2(a+b)}{3}, \frac{2}{3}\right)$, $(0, -2ab)$, and $(a + b, ab + 1)$. We work with the last two points:
$$m = \frac{ab + 1 - (-2ab)}{a + b - 0} = \frac{3ab + 1}{a + b}$$

Since $(0, -2ab)$ is the y-intercept, the line should be:

$$y = \frac{3ab + 1}{a + b}x - 2ab$$

We must still show that the point $\left(\frac{2(a+b)}{3}, \frac{2}{3}\right)$ lies on this line:

$$\frac{2}{3} = \left(\frac{3ab + 1}{a + b}\right)\left(\frac{2(a + b)}{3}\right) - 2ab$$

$$\frac{2}{3} = \frac{2(3ab + 1)}{3} - 2ab$$

$$\frac{2}{3} = \frac{6ab + 2 - 6ab}{3}$$

$$\frac{2}{3} = \frac{2}{3} \quad \text{true}$$

Thus $y = \dfrac{3ab + 1}{a + b}x - 2ab$ is the Euler line.

Graphing Utility Exercises for Section 11.1

1. Graphing the two equations:

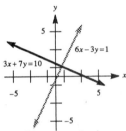

The system is consistent with exactly one solution.

3. Graphing the two equations:

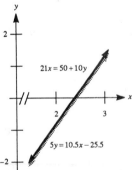

The system is inconsistent with no solution.

5. (a) Graphing the system:

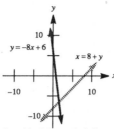

The solution is approximately $(1.56, -6.44)$.

(b) Substituting $x = 8 + y$:
$$y = -8(8 + y) + 6$$
$$y = -64 - 8y + 6$$
$$9y = -58$$
$$y = -\frac{58}{9}$$
$$x = 8 - \frac{58}{9} = \frac{14}{9}$$

The solution is $\left(\frac{14}{9}, -\frac{58}{9}\right)$, which is consistent with the estimate from part (a).

7. (a) Graphing the system:

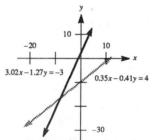

The solution is approximately $(-7.95, -16.54)$.

(b) Multiply the first equation by -3.02 and the second equation by 0.35:
$$-1.057x + 1.2382y = -12.08$$
$$1.057x - 0.4445y = -1.05$$
Adding yields:
$$0.7937y = -13.13$$
$$y = -\frac{131300}{7937}$$

Substituting into the first equation:

$$0.35x - 0.41\left(-\frac{131300}{7937}\right) = 4$$

$$\frac{7}{20}x - \frac{41}{100}\left(-\frac{131300}{7937}\right) = 4$$

$$\frac{7}{20}x + \frac{53833}{7937} = 4$$

$$\frac{7}{20}x = -\frac{22085}{7937}$$

$$x = -\frac{63100}{7937}$$

The solution is $\left(-\frac{63100}{7937}, -\frac{131300}{7937}\right)$, which is consistent with the estimate from part (a).

9. (a) Graphing the two lines:

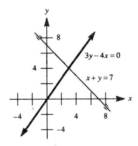

(b) The intersection point is (3, 4). Solving for y yields $y = 7 - x$, now substitute:

$$3(7 - x) - 4x = 0$$

$$21 - 3x - 4x = 0$$

$$21 = 7x$$

$$x = 3$$

$$y = 7 - 3 = 4$$

The solution is (3, 4), which checks our graphical method.

11.2 Gaussian Elimination

1. We are given the system:

$$2x + y + z = -9$$
$$3y - 2z = -4$$
$$8z = -8$$

Solve for z in the third equation:

$$8z = -8$$
$$z = -1$$

Substitute into the second equation:

$$3y - 2z = -4$$
$$3y - 2(-1) = -4$$
$$3y + 2 = -4$$
$$3y = -6$$
$$y = -2$$

Substitute into the first equation:
$$2x + y + z = -9$$
$$2x + (-2) + (-1) = -9$$
$$2x - 3 = 9$$
$$2x = -6$$
$$x = -3$$
So the solution is $(-3, -2, -1)$.

3. We are given the system:
$$8x + 5y + 3z = 1$$
$$3y + 4z = 2$$
$$5z = 3$$
Solve for z in the third equation:
$$5z = 3$$
$$z = \tfrac{3}{5}$$
Substitute into the second equation:
$$3y + 4z = 2$$
$$3y + (4)\tfrac{3}{5} = 2$$
$$3y + \tfrac{12}{5} = 2$$
$$3y = -\tfrac{2}{5}$$
$$y = -\tfrac{2}{15}$$
Substitute into the first equation:
$$8x + 5y + 3z = 1$$
$$8x + 5\left(-\tfrac{2}{15}\right) + (3)\tfrac{3}{5} = 1$$
$$8x - \tfrac{2}{3} + \tfrac{9}{5} = 1$$
$$8x - \tfrac{10}{15} + \tfrac{27}{15} = 1$$
$$8x = -\tfrac{2}{15}$$
$$x = -\tfrac{1}{60}$$
So the solution is $\left(-\tfrac{1}{60}, -\tfrac{2}{15}, \tfrac{3}{5}\right)$.

5. We are given the system:
$$-4x + 5y = 0$$
$$3y + 2z = 1$$
$$3z = -1$$
Solve for z in the third equation:
$$3z = -1$$
$$z = -\tfrac{1}{3}$$

Substitute into the second equation:
$$3y + 2z = 1$$
$$3y + 2\left(-\tfrac{1}{3}\right) = 1$$
$$3y - \tfrac{2}{3} = 1$$
$$3y = \tfrac{5}{3}$$
$$y = \tfrac{5}{9}$$

Substitute into the first equation:
$$-4x + 5y = 0$$
$$-4x + (5)\tfrac{5}{9} = 0$$
$$-4x + \tfrac{25}{9} = 0$$
$$-4x = -\tfrac{25}{9}$$
$$x = \tfrac{25}{36}$$

So the solution is $\left(\tfrac{25}{36}, \tfrac{5}{9}, -\tfrac{1}{3}\right)$.

7.   We are given the system:
$$-x + 8y + 3z = 0$$
$$2z = 0$$

Solve for z in the second equation:
$$2z = 0$$
$$z = 0$$

Substitute into the first equation:
$$-x + 8y + 3z = 0$$
$$-x + 8y + 3(0) = 0$$
$$-x + 8y = 0$$
$$8y = x$$
$$y = \frac{x}{8}$$

So the solution is $\left(x, \tfrac{x}{8}, 0\right)$, where x is any real number.

9.   We are given the system:
$$2x + 3y + z + w = -6$$
$$y + 3z - 4w = 23$$
$$6z - 5w = 31$$
$$-2w = 10$$

Solve for w in the fourth equation:
$$-2w = 10$$
$$w = -5$$

Substitute into the third equation:
$$6z - 5w = 31$$
$$6z - 5(-5) = 31$$
$$6z + 25 = 31$$
$$6z = 6$$
$$z = 1$$

Substitute into the second equation:
$$y + 3z - 4w = 23$$
$$y + 3(1) - 4(-5) = 23$$
$$y + 23 = 23$$
$$y = 0$$
Substitute into the first equation:
$$2x + 3y + z + w = -6$$
$$2x + 3(0) + 1 + (-5) = -6$$
$$2x - 4 = -6$$
$$2x = -2$$
$$x = -1$$
So the solution is $(-1, 0, 1, -5)$.

11. We must arrange this according to echelon form. First add -2 times the first equation to the second one, and -3 times it to the third:
$$x + y + z = 12$$
$$-3y - 3z = -25$$
$$-y - 2z = -14$$
We want to eliminate the y term from the third equation so interchange the second and third equations:
$$x + y + z = 12$$
$$-y - 2z = -14$$
$$-3y - 3z = -25$$
Add -3 times the second equation to the third:
$$x + y + z = 12$$
$$-y - 2z = -14$$
$$3z = 17$$
Solve for z in the third equation:
$$3z = 17$$
$$z = \tfrac{17}{3}$$
Substitute into the second equation:
$$-y - 2z = -14$$
$$-y - (2)\tfrac{17}{3} = -14$$
$$-y - \tfrac{34}{3} = -14$$
$$-y = -\tfrac{8}{3}$$
$$y = \tfrac{8}{3}$$
Substitute into the first equation:
$$x + y + z = 12$$
$$x + \tfrac{8}{3} + \tfrac{17}{3} = 12$$
$$x + \tfrac{25}{3} = 12$$
$$x = \tfrac{11}{3}$$
So the solution is $\left(\tfrac{11}{3}, \tfrac{8}{3}, \tfrac{17}{3}\right)$.

13. We must arrange the system in echelon form. Multiply the first equation by -2 and add it to the second equation:
$$-4x + 6y - 4z = -8$$
$$4x + 2y + 3z = 7$$
Adding:
$$8y - z = -1$$
Multiply the first equation by $-\frac{5}{2}$ and add it to the third equation:
$$-5x + \frac{15}{2}y - 5z = -10$$
$$5x + 4y + 2z = 7$$
Adding:
$$\frac{23}{2}y - 3z = -3$$
$$23y - 6z = -6$$
So we have the system:
$$2x - 3y + 2z = 4$$
$$8y - z = -1$$
$$23y - 6z = -6$$
Multiply the second equation by $-\frac{23}{8}$ and add it to the third equation:
$$-23y + \frac{23}{8}z = \frac{23}{8}$$
$$23y - 6z = -6$$
Adding:
$$-\frac{25}{8}z = -\frac{25}{8}$$
So we have the system in echelon form:
$$2x - 3y + 2z = 4$$
$$8y - z = -1$$
$$-\frac{25}{8}z = -\frac{25}{8}$$
Solve for z in the third equation:
$$-\frac{25}{8}z = -\frac{25}{8}$$
$$z = 1$$
Substitute into the second equation:
$$8y - z = -1$$
$$8y - 1 = -1$$
$$8y = 0$$
$$y = 0$$
Substitute into the first equation:
$$2x - 3y + 2z = 4$$
$$2x - 3(0) + 2(1) = 4$$
$$2x + 2 = 4$$
$$2x = 2$$
$$x = 1$$
So the solution is $(1, 0, 1)$.

15. We must arrange the system in echelon form. Multiply the first equation by -2 and add it to the second equation:

$$-6x - 6y + 4z = -26$$
$$6x + 2y - 5z = 13$$

Adding:

$$-4y - z = -13$$

Multiply the first equation by $-\frac{7}{3}$ and add it to the third equation:

$$-7x - 7y + \tfrac{14}{3}z = -\tfrac{91}{3}$$
$$7x + 5y - 3z = 26$$

Adding:

$$-2y + \tfrac{5}{3}z = -\tfrac{13}{3}$$
$$-6y + 5z = -13$$

So we have the system:

$$3x + 3y - 2z = 13$$
$$-4y - z = -13$$
$$-6y + 5z = -13$$

Multiply the second equation by $-\frac{3}{2}$ and add it to the third equation:

$$6y + \tfrac{3}{2}z = \tfrac{39}{2}$$
$$-6y + 5z = -13$$

Adding:

$$\tfrac{13}{2}z = \tfrac{13}{2}$$

So we have the system in echelon form:

$$3x + 3y - 2z = 13$$
$$-4y - z = -13$$
$$\tfrac{13}{2}z = \tfrac{13}{2}$$

Solve for z in the third equation:

$$\tfrac{13}{2}z = \tfrac{13}{2}$$
$$z = 1$$

Substitute into the second equation:

$$-4y - z = -13$$
$$-4y - 1 = -13$$
$$-4y = -12$$
$$y = 3$$

Substitute into the first equation:

$$3x + 3y - 2z = 13$$
$$3x + 3(3) - 2(1) = 13$$
$$3x + 7 = 13$$
$$3x = 6$$
$$x = 2$$

So the solution is $(2, 3, 1)$.

17.   We must arrange the system in echelon form. Multiply the first equation by 2 and add it to the second equation:

$$2x + 2y + 2z = 2$$
$$-2x + y + z = -2$$

Adding:

$$3y + 3z = 0$$
$$y + z = 0$$

Multiply the first equation by -3 and add it to the third equation:

$$-3x - 3y - 3z = -3$$
$$3x + 6y + 6z = 5$$

Adding:

$$3y + 3z = 2$$

So we have the system:

$$x + y + z = 1$$
$$y + z = 0$$
$$3y + 3z = 2$$

Multiply the second equation by -3 and add it to the third equation:

$$-3y - 3z = 0$$
$$3y + 3z = 2$$

Adding:

$$0 = 2$$

Since this equation is false, there is no solution. This system is inconsistent.

19.   First re-arrange the system as:

$$x + 3y - 2z = 2$$
$$2x - y + z = -1$$
$$-5x + 6y - 5z = 5$$

Multiply the first equation by -2 and add it to the second equation:

$$-2x - 6y + 4z = -4$$
$$2x - y + z = -1$$

Adding:

$$-7y + 5 = -5$$

Multiply the first equation by 5 and add it to the third equation:

$$5x + 15y - 10z = 10$$
$$-5x + 6y - 5z = 5$$

Adding:

$$21y - 15z = 15$$
$$7y - 5z = 5$$

So we have the system:

$$x + 3y - 2z = 2$$
$$-7y + 5z = -5$$
$$7y - 5z = 5$$

Adding the second and third equations yields $0 = 0$, so the system is dependent. Solving for x and y in terms of z yields the solution $\left(\frac{z+1}{-7}, \frac{5(z+1)}{7}, z \right)$, where z is any real number.

21. First re-arrange the system as:
$$x + 3y + 2z = -1$$
$$2x - y + z = 4$$
$$7x + 5z = 11$$

Multiply the first equation by –2 and add it to the second equation:
$$-2x - 6y - 4z = 2$$
$$2x - y + z = 4$$

Adding:
$$-7y - 3z = 6$$

Multiply the first equation by –7 and add it to the third equation:
$$-7x - 21y - 14z = 7$$
$$7x + 5z = 11$$

Adding:
$$-21y - 9z = 18$$
$$-7y - 3z = 6$$

So we have the system:
$$x + 3y + 2z = -1$$
$$-7y - 3z = 6$$
$$-7y - 3z = 6$$

Multiply the second equation by –1 and add it to the third equation:
$$7y + 3z = -6$$
$$-7y - 3z = 6$$

Adding:
$$0 = 0$$

So we have the system:
$$x + 3y + 2z = -1$$
$$-7y - 3z = 6$$

Solve the second equation for y:
$$-7y - 3z = 6$$
$$-7y = 3z + 6$$
$$y = \frac{-3z - 6}{7}$$

Substitute into the first equation:
$$x + 3y + 2z = -1$$
$$x + 3\left(\frac{-3z - 6}{7}\right) + 2z = -1$$
$$x + \frac{-9z - 18 + 14z}{7} = -1$$
$$x + \frac{5z - 18}{7} = -1$$
$$x = \frac{-7 - 5z + 18}{7} = \frac{-5z + 11}{7}$$

So the solution is $\left(\frac{11 - 5z}{7}, \frac{-3z - 6}{7}, z\right)$, where z is any real number.

23.   Multiply the first equation by -1 and add it to the second equation:
$$-x - y - z - w = -4$$
$$x - 2y - z - w = 3$$
Adding:
$$-3y - 2z - 2w = -1$$
$$3y + 2z + 2w = 1$$
Multiply the first equation by -2 and add it to the third equation:
$$-2x - 2y - 2z - 2w = -8$$
$$2x - y + z - w = 2$$
Adding:
$$-3y - z - 3w = -6$$
$$3y + z + 3w = 6$$
Multiply the first equation by -1 and add it to the fourth equation:
$$-x - y - z - w = -4$$
$$x - y + 2z - 2w = -7$$
Adding:
$$-2y + z - 3w = -11$$
$$2y - z + 3w = 11$$
So we have the system:
$$x + y + z + w = 4$$
$$3y + 2z + 2w = 1$$
$$3y + z + 3w = 6$$
$$2y - z + 3w = 11$$
Multiply the second equation by -1 and add it to the third equation:
$$-3y - 2z - 2w = -1$$
$$3y + z + 3w = 6$$
Adding:
$$-z + w = 5$$
$$z - w = -5$$
Multiply the second equation by $-\frac{2}{3}$ and add it to the fourth equation:
$$-2y - \frac{4}{3}z - \frac{4}{3}w = -\frac{2}{3}$$
$$2y - z + 3w = 11$$
Adding:
$$-\frac{7}{3}z + \frac{5}{3}w = \frac{31}{3}$$
$$-7z + 5w = 31$$
So we have the system:
$$x + y + z + w = 4$$
$$3y + 2z + 2w = 1$$
$$z - w = -5$$
$$-7z + 5w = 31$$
Multiply the third equation by 7 and add it to the fourth equation:
$$7z - 7w = -35$$
$$-7z + 5w = 31$$
Adding:
$$-2w = -4$$

So we have the system:
$$x + y + z + w = 4$$
$$3y + 2z + 2w = 1$$
$$z - w = -5$$
$$-2w = -4$$
Solve the fourth equation for w:
$$-2w = -4$$
$$w = 2$$
Substitute into the third equation:
$$z - w = -5$$
$$z - 2 = -5$$
$$z = -3$$
Substitute into the second equation:
$$3y + 2z + 2w = 1$$
$$3y + 2(-3) + 2(2) = 1$$
$$3y - 2 = 1$$
$$3y = 3$$
$$y = 1$$
Substitute into the first equation:
$$x + y + z + w = 4$$
$$x + 1 - 3 + 2 = 4$$
$$x = 4$$
So the solution is $(4, 1, -3, 2)$.

25.   First re-arrange the system as:
$$x + 4y - 3z = 1$$
$$2x + 3y + 2z = 5$$
Multiply the first equation by -2 and add it to the second equation:
$$-2x - 8y + 6z = -2$$
$$2x + 3y + 2z = 5$$
Adding:
$$-5y + 8z = 3$$
So we have the system:
$$x + 4y - 3z = 1$$
$$-5y + 8z = 3$$
Solve the second equation for y:
$$-5y + 8z = 3$$
$$-5y = 3 - 8z$$
$$y = \frac{8z - 3}{5}$$

Substitute into the first equation:

$$x + 4y - 3z = 1$$

$$x + 4\left(\frac{8z-3}{5}\right) - 3z = 1$$

$$x + \frac{17z - 12}{5} = 1$$

$$x = \frac{17 - 17z}{5}$$

So the solution is $\left(\frac{17-17z}{5}, \frac{8z-3}{5}, z\right)$, where z is any real number.

27. Multiply the first equation by -3 and add it to the second equation:

$$-3x + 6y + 6z - 6w = 30$$
$$3x + 4y - z - 3w = 11$$

Adding:

$$10y + 5z - 9w = 41$$

Multiply the first equation by 4 and add it to the third equation:

$$4x - 8y - 8z + 8w = -40$$
$$-4x - 3y - 3z + 8w = -21$$

Adding:

$$-11y - 11z + 16w = -61$$

So we have the system:

$$x - 2y - 2z + 2w = -10$$
$$10y + 5z - 9w = 41$$
$$-11y - 11z + 16w = -61$$

Add the second and third equations:

$$10y + 5z - 9w = 41$$
$$-11y - 11z + 16w = -61$$

Adding:

$$-y - 6z + 7w = -20$$
$$y + 6z - 7w = 20$$

So we have the system:

$$x - 2y - 2z + 2w = -10$$
$$y + 6z - 7w = 20$$
$$-11y - 11z + 16w = -61$$

Multiply the second equation by 11 and add it to the third equation:

$$11y + 66z - 77w = 220$$
$$-11y - 11z + 16w = -61$$

Adding:

$$55z - 61w = 159$$

So we have the system:

$$x - 2y - 2z + 2w = -10$$
$$y + 6z - 7w = 20$$
$$55z - 61w = 159$$

Solve the third equation for z:
$$55z - 61w = 159$$
$$55z = 61w + 159$$
$$z = \frac{61w + 159}{55}$$

Substitute into the second equation:
$$y + 6z - 7w = 20$$
$$y + 6\left(\frac{61w + 159}{55}\right) - 7w = 20$$
$$y + \frac{954 - 19w}{55} = 20$$
$$y = \frac{146 + 19w}{55}$$

Substitute into the first equation:
$$x - 2y - 2z + 2w = -10$$
$$x - 2\left(\frac{146 + 19w}{55}\right) - 2\left(\frac{61w + 159}{55}\right) + 2w = -10$$
$$x + \frac{-122 - 10w}{11} = -10$$
$$x = \frac{12 + 10w}{11}$$

So the solution is $\left(\frac{12+10w}{11}, \frac{146+19w}{55}, \frac{61w+159}{55}, w\right)$, where w is any real number.

29. Solving the second equation for y yields $y = \frac{2z+3}{3}$. We substitute this into the first equation to get $x = -\frac{5z}{12}$. So the solution is $\left(-\frac{5z}{12}, \frac{2z+3}{3}, z\right)$, where z is any real number.

31. (a) Substituting the coordinates of each point, we have the system of equations:
$$A + B + C = -2$$
$$A - B + C = 0$$
$$4A + 2B + C = 3$$
Adding -1 times the first equation to the second equation, and -4 times the first equation to the third equation:
$$A + B + C = -2$$
$$-2B = 2$$
$$-2B - 3C = 11$$
Dividing the second equation by -2, and adding 2 times the (new) second equation to the third equation:
$$A + B + C = -2$$
$$B = -1$$
$$-3C = 9$$
So $C = -3$ and $B = -1$. Substituting into the first equation yields $A - 4 = -2$, so $A = 2$. The constants are $A = 2$, $B = -1$, and $C = -3$.

(b)   Substituting the coordinates of each point, we have the system of equations:
$$1 + a + b + c = -2$$
$$-1 + a - b + c = 0$$
$$8 + 4a + 2b + c = 3$$
Rewrite the system as:
$$a + b + c = -3$$
$$a - b + c = 1$$
$$4a + 2b + c = -5$$
Adding -1 times the first equation to the second equation, and -4 times the first equation to the third equation:
$$a + b + c = -3$$
$$-2b = 4$$
$$-2b - 3c = 7$$
Dividing the second equation by -2, and adding 2 times the (new) second equation to the third equation:
$$a + b + c = -3$$
$$b = -2$$
$$-3c = 3$$
So $c = -1$ and $b = -2$. Substituting into the first equation yields $a - 3 = -3$, so $a = 0$.
The constants are $a = 0$, $b = -2$, and $c = -1$.

33.   Represent the equation of the parabola as $y = ax^2 + bx + c$. Substituting the three points results in the system:
$$a - b + c = 1$$
$$a + b + c = -2$$
$$16a + 4b + c = 1$$
Adding -1 times the first equation to the second equation, and -16 times the first equation to the third equation:
$$a - b + c = 1$$
$$2b = -3$$
$$20b - 15c = -15$$
Dividing the second equation by 2, and adding -20 times the (new) second equation to the third equation:
$$a - b + c = 1$$
$$b = -\tfrac{3}{2}$$
$$-15c = 15$$
So $c = -1$ and $b = -\tfrac{3}{2}$. Substituting into the first equation:
$$a + \tfrac{3}{2} - 1 = 1$$
$$a + \tfrac{1}{2} = 1$$
$$a = \tfrac{1}{2}$$

So the equation of the parabola is $y = \frac{1}{2}x^2 - \frac{3}{2}x - 1$. Set $y = 0$ to find the x-intercepts:

$$\frac{1}{2}x^2 - \frac{3}{2}x - 1 = 0$$
$$x^2 - 3x - 2 = 0$$

Using the quadratic formula:

$$x = \frac{-(-3) \pm \sqrt{(-3)^2 - 4(1)(-2)}}{2(1)} = \frac{3 \pm \sqrt{9+8}}{2} = \frac{3 \pm \sqrt{17}}{2}$$

The x-intercepts are $\dfrac{3 \pm \sqrt{17}}{2}$. The x-coordinate of the vertex is given by:

$$x = -\frac{b}{2a} = \frac{-(-3/2)}{2(1/2)} = \frac{3}{2}$$

Substitute to find the y-coordinate:

$$y = \frac{1}{2}\left(\frac{3}{2}\right)^2 - \frac{3}{2}\left(\frac{3}{2}\right) - 1 = \frac{9}{8} - \frac{9}{4} - 1 = -\frac{17}{8}$$

The vertex is $\left(\frac{3}{2}, -\frac{17}{8}\right)$.

35. We know its form is $(x-h)^2 + (y-k)^2 = r^2$, so:

$$(-1-h)^2 + (-5-k)^2 = r^2$$
$$(1-h)^2 + (2-k)^2 = r^2$$
$$(-2-h)^2 + k^2 = r^2$$

Therefore:

$$1 + 2h + h^2 + 25 + 10k + k^2 = r^2$$
$$1 - 2h + h^2 + 4 - 4k + k^2 = r^2$$
$$4 + 4h + h^2 + k^2 = r^2$$

Setting the first two equations equal yields $4h + 14k = -21$ and setting the last two equations equal yields $6h + 4k = 1$. So we solve the system:

$$4h + 14k = -21$$
$$6h + 4k = 1$$

Multiplying the first equation by -3 and the second equation by 2, then adding, yields $k = -\frac{65}{34}$, and substituting into the second equation yields $h = \frac{49}{34}$. Find r by the original third equation:

$$r^2 = 4 + 4h + h^2 + k^2 = 4 + \frac{98}{17} + \left(\frac{49}{34}\right)^2 + \left(-\frac{65}{34}\right)^2 = \frac{17914}{34^2} = \frac{8957}{578}$$

So we have $\left(x - \frac{49}{34}\right)^2 + \left(y + \frac{65}{34}\right)^2 = \frac{8957}{578}$, which simplifies to:

$$17x^2 + 17y^2 - 49x + 65y - 166 = 0$$

37. Using the hint, we let $A = e^x$, $B = e^y$, and $C = e^z$. Then we have the system:

$$A + B - 2C = 2a$$
$$A + 2B - 4C = 3a$$
$$\tfrac{1}{2}A - 3B + C = -5a$$

Multiply the first equation by -1 and add it to the second equation:

$$-A - B + 2C = -2a$$
$$A + 2B - 4C = 3a$$

Adding:

$$B - 2C = a$$

Multiply the first equation by -1 and the third equation by 2, then add the resulting equations:

$$-A - B + 2C = -2a$$
$$A - 6B + 2C = -10a$$

Adding:

$$-7B + 4C = -12a$$

So we have the system:

$$A + B - 2C = 2a$$
$$B - 2C = a$$
$$-7B + 4C = -12a$$

Multiply the second equation by 7 and add it to the third equation:

$$7B - 14C = 7a$$
$$-7B + 4C = -12a$$

Adding:

$$-10C = -5a$$
$$C = \frac{a}{2}$$

Substituting into the second equation:

$$B - 2C = a$$
$$B - 2\left(\frac{a}{2}\right) = a$$
$$B - a = a$$
$$B = 2a$$

Substituting into the first equation:

$$A + B - 32C = 2a$$
$$A + 2a - a = 2a$$
$$A + a = 2a$$
$$A = a$$

So $A = a$, $B = 2a$, and $C = \frac{a}{2}$. Since $A = e^x, B = e^y, C = e^z, x = \ln a, y = \ln 2a$ and $z = \ln \frac{a}{2}$.

39.   From the figure in the text, we obtain the system of three equations:

$$r_1 + r_2 = a$$
$$r_2 + r_3 = b$$
$$r_1 + r_3 = c$$

This system can be solved by repeated substitution, or by using the following technique. Add the three equations and then divide by 2 to obtain $r_1 + r_2 + r_3 = \dfrac{a+b+c}{2}$. Now replace $r_2 + r_3$ in this last equation by b. The result is $r_1 = \dfrac{a+b+c}{2} - b$. Thus $r_1 = \dfrac{a-b+c}{2}$. Similarly, we find that $r_2 = \dfrac{a+b-c}{2}$ and $r_3 = \dfrac{b+c-a}{2}$.

41. Let α be the distance from A to B, β be the distance from B to C, and γ be the distance from C to A. The walking, riding and driving rates are, respectively, $\frac{1}{a}$ mi / min, $\frac{1}{b}$ mi / min and $\frac{1}{c}$ mi / min. From the statement of the problem we can obtain the following system of three equations:

$$\alpha a + \beta b + \gamma c = 60(a + c - b)$$
$$\beta a + \gamma b + \alpha c = 60(b + a - c)$$
$$\gamma a + \alpha b + \beta c = 60(c + b - a)$$

By adding all three equations:

$$\alpha(a + b + c) + \beta(a + b + c) + \gamma(a + b + c) = 60(a + b + c)$$

Now divide through by the quantity $a + b + c$. This yields $\alpha + \beta + \gamma = 60$ miles.

11.3 Matrices

1. (a) two by three (2×3)
(b) three by two (3×2)

3. five by four (5×4)

5. The coefficient matrix is $\begin{pmatrix} 2 & 3 & 4 \\ 5 & 6 & 7 \\ 8 & 9 & 10 \end{pmatrix}$ and the augmented matrix is $\begin{pmatrix} 2 & 3 & 4 & 10 \\ 5 & 6 & 7 & 9 \\ 8 & 9 & 10 & 8 \end{pmatrix}$.

7. The coefficient matrix is $\begin{pmatrix} 1 & 0 & 1 & 1 \\ 1 & 1 & 0 & 2 \\ 0 & 1 & 1 & 1 \\ 2 & -1 & -1 & 0 \end{pmatrix}$ and the augmented matrix is

$\begin{pmatrix} 1 & 0 & 1 & 1 & -1 \\ 1 & 1 & 0 & 2 & 0 \\ 0 & 1 & 1 & 1 & 1 \\ 2 & -1 & -1 & 0 & 2 \end{pmatrix}$.

9. Form the augmented matrix:

$\begin{pmatrix} 1 & -1 & 2 & 7 \\ 3 & 2 & -1 & -10 \\ -1 & 3 & 1 & -2 \end{pmatrix}$

Adding –3 times row 1 to row 2 and adding row 1 to row 3 yields:

$$\begin{pmatrix} 1 & -1 & 2 & 7 \\ 0 & 5 & -7 & -31 \\ 0 & 2 & 3 & 5 \end{pmatrix}$$

Multiplying row 3 by –3 and adding to row 2 yields:

$$\begin{pmatrix} 1 & -1 & 2 & 7 \\ 0 & -1 & -16 & -46 \\ 0 & 2 & 3 & 5 \end{pmatrix}$$

Multiplying row 2 by 2 and adding to row 3 yields:

$$\begin{pmatrix} 1 & -1 & 2 & 7 \\ 0 & -1 & -16 & -46 \\ 0 & 0 & -29 & -87 \end{pmatrix}$$

So we have the system of equations:

$$x - y + 2z = 7$$
$$-y - 16z = -46$$
$$-29z = -87$$

Solve equation three for z:

$$-29z = -87$$
$$z = 3$$

Substitute into equation two:

$$-y - 48 = -46$$
$$-y = 2$$
$$y = -2$$

Substitute into equation one:

$$x + 2 + 6 = 7$$
$$x = -1$$

So the solution is $(-1, -2, 3)$.

11. Form the augmented matrix:

$$\begin{pmatrix} 1 & 0 & 1 & -2 \\ -3 & 2 & 0 & 17 \\ 1 & -1 & -1 & -9 \end{pmatrix}$$

Adding 3 times row 1 to row 2, and –1 times row 1 to row 3 yields:

$$\begin{pmatrix} 1 & 0 & 1 & -2 \\ 0 & 2 & 3 & 11 \\ 0 & -1 & -2 & -7 \end{pmatrix}$$

Switching row 2 and row 3:

$$\begin{pmatrix} 1 & 0 & 1 & -2 \\ 0 & -1 & -2 & -7 \\ 0 & 2 & 3 & 11 \end{pmatrix}$$

Multiplying row 2 by 2 and adding it to row 3 yields:

$$\begin{pmatrix} 1 & 0 & 1 & -2 \\ 0 & -1 & -2 & -7 \\ 0 & 0 & -1 & -3 \end{pmatrix}$$

So we have the system:

$$\begin{aligned} x \quad + \quad z &= -2 \\ -y - 2z &= -7 \\ -z &= -3 \end{aligned}$$

Solve equation three for z:

$$-z = -3$$
$$z = 3$$

Substitute into equation two:

$$-y - 6 = -7$$
$$-y = -1$$
$$y = 1$$

Substitute into equation one:

$$x + 3 = -2$$
$$x = -5$$

So the solution is $(-5, 1, 3)$.

13. Form the augmented matrix:

$$\begin{pmatrix} 1 & 1 & 1 & -4 \\ 2 & -3 & 1 & -1 \\ 4 & 2 & -3 & 33 \end{pmatrix}$$

Adding -2 times row 1 to row 2, and -4 times row 1 to row 3 yields:

$$\begin{pmatrix} 1 & 1 & 1 & -4 \\ 0 & -5 & -1 & 7 \\ 0 & -2 & -7 & 49 \end{pmatrix}$$

Adding -3 times row 3 to row 2 (to get a 1 entry) yields:

$$\begin{pmatrix} 1 & 1 & 1 & -4 \\ 0 & 1 & 20 & -140 \\ 0 & -2 & -7 & 49 \end{pmatrix}$$

Adding 2 times row 2 to row 3 yields:

$$\begin{pmatrix} 1 & 1 & 1 & -4 \\ 0 & 1 & 20 & -140 \\ 0 & 0 & 33 & -231 \end{pmatrix}$$

So we have the system of equations:

$$\begin{aligned} x + y + \quad z &= -4 \\ y + 20z &= -140 \\ 33z &= -231 \end{aligned}$$

Solving the last equation for z:
$$33z = -231$$
$$z = -7$$
Substituting into the second equation:
$$y + 20z = -140$$
$$y - 140 = -140$$
$$y = 0$$
Substituting into the first equation:
$$x + y + z = -4$$
$$x + 0 - 7 = -4$$
$$x = 3$$
So the solution is $(3, 0, -7)$.

15.   Form the augmented matrix:
$$\begin{pmatrix} 3 & -2 & 6 & 0 \\ 1 & 3 & 20 & 15 \\ 10 & -11 & -10 & -9 \end{pmatrix}$$
Switching row 1 and row 2 yields:
$$\begin{pmatrix} 1 & 3 & 20 & 15 \\ 3 & -2 & 6 & 0 \\ 10 & -11 & -10 & -9 \end{pmatrix}$$
Adding -3 times row 1 to row 2, and -10 times row 1 to row 3 yields:
$$\begin{pmatrix} 1 & 3 & 20 & 15 \\ 0 & -11 & -54 & -45 \\ 0 & -41 & -210 & -159 \end{pmatrix}$$
Multiplying row 2 by -4 and adding it to row 3 yields:
$$\begin{pmatrix} 1 & 3 & 20 & 15 \\ 0 & -11 & -54 & -45 \\ 0 & 3 & 6 & 21 \end{pmatrix}$$
Dividing row 3 by 3 and then switching it with row 2 yields:
$$\begin{pmatrix} 1 & 3 & 20 & 15 \\ 0 & 1 & 2 & 7 \\ 0 & -11 & -54 & -45 \end{pmatrix}$$
Multiplying row 2 by 11 and adding it to row 3 yields:
$$\begin{pmatrix} 1 & 3 & 20 & 15 \\ 0 & 1 & 2 & 7 \\ 0 & 0 & -32 & 32 \end{pmatrix}$$
So we have the system of equations:
$$x + 3y + 20z = 15$$
$$y + 2z = 7$$
$$-32z = 32$$

Solve equation three for z:
$$-32z = 32$$
$$z = -1$$
Substitute into equation two:
$$y - 2 = 7$$
$$y = 9$$
Substitute into equation one:
$$x + 27 - 20 = 15$$
$$x + 7 = 15$$
$$x = 8$$
So the solution is $(8, 9, -1)$.

17. Form the augmented matrix:
$$\begin{pmatrix} 4 & -3 & 3 & 2 \\ 5 & 1 & -4 & 1 \\ 9 & -2 & -1 & 3 \end{pmatrix}$$
Subtracting row 1 from row 2 yields:
$$\begin{pmatrix} 4 & -3 & 3 & 2 \\ 1 & 4 & -7 & -1 \\ 9 & -2 & -1 & 3 \end{pmatrix}$$
Switching row 1 and row 2 yields:
$$\begin{pmatrix} 1 & 4 & -7 & -1 \\ 4 & -3 & 3 & 2 \\ 9 & -2 & -1 & 3 \end{pmatrix}$$
Adding -4 times row 1 to row 2, and -9 times row 1 to row 3 yields:
$$\begin{pmatrix} 1 & 4 & -7 & -1 \\ 0 & -19 & 31 & 6 \\ 0 & -38 & 62 & 12 \end{pmatrix}$$
Multiplying row 2 by -2 and adding it to row 3 yields:
$$\begin{pmatrix} 1 & 4 & -7 & -1 \\ 0 & -19 & 31 & 6 \\ 0 & 0 & 0 & 0 \end{pmatrix}$$
So we have the system of equations:
$$x + 4y - 7z = -1$$
$$-19y + 31z = 6$$
Solve equation two for y:
$$-19y = 6 - 31z$$
$$y = \frac{31z - 6}{19}$$

Substitute into equation one:

$$x + 4\left(\frac{31z - 6}{19}\right) - 7z = -1$$

$$x + \frac{-9z - 24}{19} = -1$$

$$x = \frac{9z + 5}{19}$$

So the solution is $\left(\frac{9z+5}{19}, \frac{31z-6}{19}, z\right)$, for any real number z.

19. Form the augmented matrix:

$$\begin{pmatrix} 1 & -1 & 1 & 1 & 6 \\ 1 & 1 & -1 & 1 & 4 \\ 1 & 1 & 1 & -1 & -2 \\ -1 & 1 & 1 & 1 & 0 \end{pmatrix}$$

Adding -1 times row 1 to both row 2 and row 3, and adding row 1 to row 4 yields:

$$\begin{pmatrix} 1 & -1 & 1 & 1 & 6 \\ 0 & 2 & -2 & 0 & -2 \\ 0 & 2 & 0 & -2 & -8 \\ 0 & 0 & 2 & 2 & 6 \end{pmatrix}$$

Dividing rows 2, 3, and 4 by 2 and subtracting row 2 from row 3 yields:

$$\begin{pmatrix} 1 & -1 & 1 & 1 & 6 \\ 0 & 1 & -1 & 0 & -1 \\ 0 & 0 & 1 & -1 & -3 \\ 0 & 0 & 1 & 1 & 3 \end{pmatrix}$$

Multiplying row 3 by -1 and adding it to row 4 yields:

$$\begin{pmatrix} 1 & -1 & 1 & 1 & 6 \\ 0 & 1 & -1 & 0 & -1 \\ 0 & 0 & 1 & -1 & -3 \\ 0 & 0 & 0 & 2 & 6 \end{pmatrix}$$

So we have the system of equations:

$$x - y + z + w = 6$$
$$y - z = -1$$
$$z - w = -3$$
$$2w = 6$$

Solve equation four for w:

$$2w = 6$$
$$w = 3$$

Substitute into equation three:

$$z - 3 = -3$$
$$z = 0$$

Substitute into equation two:
$$y - 0 = -1$$
$$y = -1$$
Substitute into equation one:
$$x + 1 + 0 + 3 = 6$$
$$x + 4 = 6$$
$$x = 2$$
So the solution is $(2, -1, 0, 3)$.

21. Form the augmented matrix:
$$\begin{pmatrix} 15 & 14 & 26 & 1 \\ 18 & 17 & 32 & -1 \\ 21 & 20 & 38 & 0 \end{pmatrix}$$
Reduce coefficients by subtracting row 2 from row 3, and row 1 from row 2:
$$\begin{pmatrix} 15 & 14 & 26 & 1 \\ 3 & 3 & 6 & -2 \\ 3 & 3 & 6 & 1 \end{pmatrix}$$
Subtracting row 2 from row 3 yields:
$$\begin{pmatrix} 15 & 14 & 26 & 1 \\ 3 & 3 & 6 & -2 \\ 0 & 0 & 0 & 3 \end{pmatrix}$$
But this last row corresponds to the equation $0 = 3$, which is false. So the system has no solution.

23. Adding the corresponding terms, $A + B$ will be:
$$\begin{pmatrix} 2 & 3 \\ -1 & 4 \end{pmatrix} + \begin{pmatrix} 1 & -1 \\ 3 & 0 \end{pmatrix} = \begin{pmatrix} 3 & 2 \\ 2 & 4 \end{pmatrix}$$

25. Multiplying:
$$2A = \begin{pmatrix} 4 & 6 \\ -2 & 8 \end{pmatrix} \text{ and } 2B = \begin{pmatrix} 2 & -2 \\ 6 & 0 \end{pmatrix}$$
Therefore:
$$2A + 2B = \begin{pmatrix} 6 & 4 \\ 4 & 8 \end{pmatrix}$$

27. The multiplication is defined, since # cols in A = # rows in B:
row 1, col 1: $2(1) + 3(3) = 2 + 9 = 11$
row 1, col 2: $2(-1) + 3(0) = -2 + 0 = -2$
row 2, col 1: $-1(1) + 4(3) = -1 + 12 = 11$
row 2, col 2: $-1(-1) + 4(0) = 1 + 0 = 1$
So $AB = \begin{pmatrix} 11 & -2 \\ 11 & 1 \end{pmatrix}$.

29. The multiplication is defined, since # cols in A = # rows in C:
row 1, col 1: $2(1) + 3(0) = 2 + 0 = 2$
row 1, col 2: $2(0) + 3(1) = 0 + 3 = 3$
row 2, col 1: $-1(1) + 4(0) = -1 + 0 = -1$
row 2, col 2: $-1(0) + 4(1) = 0 + 4 = 4$

So $AC = \begin{pmatrix} 2 & 3 \\ -1 & 4 \end{pmatrix}$.

31. This operation is not defined, since D and E do not have the same size.

33. Computing:
$$2F - 3G = \begin{pmatrix} 10 & -2 \\ -8 & 0 \\ 4 & 6 \end{pmatrix} - \begin{pmatrix} 0 & 0 \\ 0 & 0 \\ 0 & 0 \end{pmatrix} = \begin{pmatrix} 10 & -2 \\ -8 & 0 \\ 4 & 6 \end{pmatrix}$$

35. The multiplication is defined, since # cols of E = # rows of D.
row 1, col 1: $2(-1) + 1(4) = -2 + 4 = 2$
row 1, col 2: $2(2) + 1(0) = 4 + 0 = 4$
row 1, col 3: $2(3) + 1(5) = 6 + 5 = 11$
row 2, col 1: $8(-1) - 1(4) = -8 - 4 = -12$
row 2, col 2: $8(2) - 1(0) = 16 - 0 = 16$
row 2, col 3: $8(3) - 1(5) = 24 - 5 = 19$
row 3, col 1: $6(-1) + 5(4) = -6 + 20 = 14$
row 3, col 2: $6(2) + 5(0) = 12 + 0 = 12$
row 3, col 3: $6(3) + 5(5) = 18 + 25 = 43$

So $ED = \begin{pmatrix} 2 & 4 & 11 \\ -12 & 16 & 19 \\ 14 & 12 & 43 \end{pmatrix}$.

37. The multiplication is defined, since # cols of F = # rows of D:
row 1, col 1: $5(-1) - 1(4) = -5 - 4 = -9$
row 1, col 2: $5(2) - 1(0) = 10 - 0 = 10$
row 1, col 3: $5(3) - 1(5) = 15 - 5 = 10$
row 2, col 1: $-4(-1) + 0(4) = 4 + 0 = 4$
row 2, col 2: $-4(2) + 0(0) = -8 + 0 = -8$
row 2, col 3: $-4(3) + 0(5) = -12 + 0 = -12$
row 3, col 1: $2(-1) + 3(4) = -2 + 12 = 10$
row 3, col 2: $2(2) + 3(0) = 4 + 0 = 4$
row 3, col 3: $2(3) + 3(5) = 6 + 15 = 21$

So $FD = \begin{pmatrix} -9 & 10 & 10 \\ 4 & -8 & -12 \\ 10 & 4 & 21 \end{pmatrix}$.

39. The operation is not defined, since G and A are not the same size.

41. The multiplication is defined, since # cols of G = # rows of D:

$$GD = \begin{pmatrix} 0 & 0 & 0 \\ 0 & 0 & 0 \\ 0 & 0 & 0 \end{pmatrix}$$

43. Adding:

$$A + (B + C) = \begin{pmatrix} 2 & 3 \\ -1 & 4 \end{pmatrix} + \begin{pmatrix} 2 & -1 \\ 3 & 1 \end{pmatrix} = \begin{pmatrix} 4 & 2 \\ 2 & 5 \end{pmatrix}$$

45. The multiplication is not defined, since # cols of $D \neq$ # rows of C.

47. The multiplication is defined, since # cols of A = # rows of A.
row 1, col 1: $2(2) + 3(-1) = 4 - 3 = 1$
row 1, col 2: $2(3) + 3(4) = 6 + 12 = 18$
row 2, col 1: $-1(2) + 4(-1) = -2 - 4 = -6$
row 2, col 2: $-1(3) + 4(4) = -3 + 16 = 13$

So $A^2 = \begin{pmatrix} 1 & 18 \\ -6 & 13 \end{pmatrix}$.

49. Multiplying:

$$AA^2 = \begin{pmatrix} 2 & 3 \\ -1 & 4 \end{pmatrix}\begin{pmatrix} 1 & 18 \\ -6 & 13 \end{pmatrix} = \begin{pmatrix} -16 & 75 \\ -25 & 34 \end{pmatrix}$$

51. (a) Computing:

$$A(B+C) = \begin{pmatrix} -1 & 3 & 4 \\ 3 & 2 & -3 \\ 9 & 1 & 6 \end{pmatrix}\begin{pmatrix} 11 & 6 & 2 \\ 2 & 1 & 6 \\ -2 & 1 & 6 \end{pmatrix} = \begin{pmatrix} -13 & 1 & 40 \\ 43 & 17 & 0 \\ 89 & 61 & 60 \end{pmatrix}$$

(b) Computing:

$$AB + AC = \begin{pmatrix} -1 & 3 & 4 \\ 3 & 2 & -3 \\ 9 & 1 & 6 \end{pmatrix}\begin{pmatrix} 7 & 0 & 1 \\ 0 & 0 & 3 \\ -1 & 2 & 4 \end{pmatrix} + \begin{pmatrix} -1 & 3 & 4 \\ 3 & 2 & -3 \\ 9 & 1 & 6 \end{pmatrix}\begin{pmatrix} 4 & 6 & 1 \\ 2 & 1 & 3 \\ -1 & -1 & 2 \end{pmatrix}$$

$$= \begin{pmatrix} -11 & 8 & 24 \\ 24 & -6 & -3 \\ 57 & 12 & 36 \end{pmatrix} + \begin{pmatrix} -2 & -7 & 16 \\ 19 & 23 & 3 \\ 32 & 49 & 24 \end{pmatrix}$$

$$= \begin{pmatrix} -13 & 1 & 40 \\ 43 & 17 & 0 \\ 89 & 61 & 60 \end{pmatrix}$$

(c) Computing:

$$(AB)C = \left[\begin{pmatrix} -1 & 3 & 4 \\ 3 & 2 & -3 \\ 9 & 1 & 6 \end{pmatrix}\begin{pmatrix} 7 & 0 & 1 \\ 0 & 0 & 3 \\ -1 & 2 & 4 \end{pmatrix}\right]\begin{pmatrix} 4 & 6 & 1 \\ 2 & 1 & 3 \\ -1 & -1 & 2 \end{pmatrix}$$

$$= \begin{pmatrix} -11 & 8 & 24 \\ 24 & -6 & -3 \\ 57 & 12 & 36 \end{pmatrix}\begin{pmatrix} 4 & 6 & 1 \\ 2 & 1 & 3 \\ -1 & -1 & 2 \end{pmatrix}$$

$$= \begin{pmatrix} -52 & -82 & 61 \\ 87 & 141 & 0 \\ 216 & 318 & 165 \end{pmatrix}$$

(d) Computing:

$$A(BC) = \begin{pmatrix} -1 & 3 & 4 \\ 3 & 2 & -3 \\ 9 & 1 & 6 \end{pmatrix}\begin{pmatrix} 27 & 41 & 9 \\ -3 & -3 & 6 \\ -4 & -8 & 13 \end{pmatrix} = \begin{pmatrix} -52 & -82 & 61 \\ 87 & 141 & 0 \\ 216 & 318 & 165 \end{pmatrix}$$

53. (a) Compute A^2 and B^2:

$$A^2 = \begin{pmatrix} 3 & 5 \\ 7 & 9 \end{pmatrix}\begin{pmatrix} 3 & 5 \\ 7 & 9 \end{pmatrix} = \begin{pmatrix} 44 & 60 \\ 84 & 116 \end{pmatrix}$$

$$B^2 = \begin{pmatrix} 2 & 4 \\ 6 & 8 \end{pmatrix}\begin{pmatrix} 2 & 4 \\ 6 & 8 \end{pmatrix} = \begin{pmatrix} 28 & 40 \\ 60 & 88 \end{pmatrix}$$

So $A^2 - B^2 = \begin{pmatrix} 16 & 20 \\ 24 & 28 \end{pmatrix}$.

(b) We have $A - B = \begin{pmatrix} 1 & 1 \\ 1 & 1 \end{pmatrix}$ and $A + B = \begin{pmatrix} 5 & 9 \\ 13 & 17 \end{pmatrix}$, so:

$$(A - B)(A + B) = \begin{pmatrix} 1 & 1 \\ 1 & 1 \end{pmatrix}\begin{pmatrix} 5 & 9 \\ 13 & 17 \end{pmatrix} = \begin{pmatrix} 18 & 26 \\ 18 & 26 \end{pmatrix}$$

(c) Compute the product:

$$(A + B)(A - B) = \begin{pmatrix} 5 & 9 \\ 13 & 17 \end{pmatrix}\begin{pmatrix} 1 & 1 \\ 1 & 1 \end{pmatrix} = \begin{pmatrix} 14 & 14 \\ 30 & 30 \end{pmatrix}$$

(d) Compute AB and BA:

$$AB = \begin{pmatrix} 3 & 5 \\ 7 & 9 \end{pmatrix}\begin{pmatrix} 2 & 4 \\ 6 & 8 \end{pmatrix} = \begin{pmatrix} 36 & 52 \\ 68 & 100 \end{pmatrix}$$

$$BA = \begin{pmatrix} 2 & 4 \\ 6 & 8 \end{pmatrix}\begin{pmatrix} 3 & 5 \\ 7 & 9 \end{pmatrix} = \begin{pmatrix} 34 & 46 \\ 74 & 102 \end{pmatrix}$$

Therefore:

$A^2 + AB - BA - B^2 =$

$$\begin{pmatrix} 44 & 60 \\ 84 & 116 \end{pmatrix} + \begin{pmatrix} 36 & 52 \\ 68 & 100 \end{pmatrix} - \begin{pmatrix} 34 & 46 \\ 74 & 102 \end{pmatrix} - \begin{pmatrix} 28 & 40 \\ 60 & 88 \end{pmatrix} = \begin{pmatrix} 18 & 26 \\ 18 & 26 \end{pmatrix}$$

55. (a) The product AZ is given by:

$$AZ = \begin{pmatrix} 1 & 0 \\ 0 & -1 \end{pmatrix}\begin{pmatrix} x \\ y \end{pmatrix} = \begin{pmatrix} x \\ -y \end{pmatrix}$$

(b) The product BZ is given by:

$$BZ = \begin{pmatrix} -1 & 0 \\ 0 & 1 \end{pmatrix}\begin{pmatrix} x \\ y \end{pmatrix} = \begin{pmatrix} -x \\ y \end{pmatrix}$$

(c) The product AB is given by:

$$AB = \begin{pmatrix} 1 & 0 \\ 0 & -1 \end{pmatrix}\begin{pmatrix} -1 & 0 \\ 0 & 1 \end{pmatrix} = \begin{pmatrix} -1 & 0 \\ 0 & -1 \end{pmatrix}$$

So $(AB)Z = \begin{pmatrix} -1 & 0 \\ 0 & -1 \end{pmatrix}\begin{pmatrix} x \\ y \end{pmatrix} = \begin{pmatrix} -x \\ -y \end{pmatrix}$. This would represent a reflection about the origin.

57. (a) First compute the product:

$$AB = \begin{pmatrix} 1 & 2 \\ 3 & 4 \end{pmatrix}\begin{pmatrix} 3 & -1 \\ 5 & 8 \end{pmatrix} = \begin{pmatrix} 13 & 15 \\ 29 & 29 \end{pmatrix}$$

Now compute the function values:

$f(A) = 1(4) - 2(3) = 4 - 6 = -2$
$f(B) = 3(8) - (-1)(5) = 24 + 5 = 29$
$f(AB) = 13(29) - 15(29) = -2(29) = -58$

So $f(A) \cdot f(B) = f(AB)$.

(b) First compute the product:

$$AB = \begin{pmatrix} a & b \\ c & d \end{pmatrix}\begin{pmatrix} e & f \\ g & h \end{pmatrix} = \begin{pmatrix} ae + bg & af + bh \\ ce + dg & cf + dh \end{pmatrix}$$

Now compute the function values:

$f(A) = ad - bc$
$f(B) = eh - fg$
$f(AB) = (ae + bg)(cf + dh) - (af + bh)(ce + dg)$
$\qquad = (acef + bcfg + adeh + bdgh) - (acef + bceh + adfg + bdgh)$
$\qquad = bcfg + adeh - bceh - adfg$

So $f(A) \cdot f(B) = f(AB)$.

59. (a) Let $A = \begin{pmatrix} a & b \\ c & d \end{pmatrix}$ and $B = \begin{pmatrix} e & f \\ g & h \end{pmatrix}$. Then $A + B = \begin{pmatrix} a+e & b+f \\ c+g & d+h \end{pmatrix}$, so

$(A+B)^T = \begin{pmatrix} a+e & c+g \\ b+f & d+h \end{pmatrix}$. Now $A^T = \begin{pmatrix} a & c \\ b & d \end{pmatrix}$ and $B^T = \begin{pmatrix} e & g \\ f & h \end{pmatrix}$, so we have

$A^T + B^T = \begin{pmatrix} a+e & c+g \\ b+f & d+h \end{pmatrix}$. Thus $(A+B)^T = A^T + B^T$.

(b) Since $A^T = \begin{pmatrix} a & c \\ b & d \end{pmatrix}$, $\left(A^T\right)^T = \begin{pmatrix} a & b \\ c & d \end{pmatrix}$. Thus $\left(A^T\right)^T = A$.

(c) Since $AB = \begin{pmatrix} ae+bg & af+bh \\ ce+dg & cf+dh \end{pmatrix}$, $(AB)^T = \begin{pmatrix} ae+bg & ce+dg \\ af+bh & cf+dh \end{pmatrix}$. Also

$B^T A^T = \begin{pmatrix} e & g \\ f & h \end{pmatrix}\begin{pmatrix} a & c \\ b & d \end{pmatrix} = \begin{pmatrix} ae+bg & ce+dg \\ af+bh & cf+dh \end{pmatrix}$. So we see that $(AB)^T = B^T A^T$.

11.4 The Inverse of a Square Matrix

1. Compute AI_2 and $I_2 A$:
$$AI_2 = \begin{pmatrix} 4 & -1 \\ -5 & 2 \end{pmatrix}\begin{pmatrix} 1 & 0 \\ 0 & 1 \end{pmatrix} = \begin{pmatrix} 4 & -1 \\ -5 & 2 \end{pmatrix} = A$$
$$I_2 A = \begin{pmatrix} 1 & 0 \\ 0 & 1 \end{pmatrix}\begin{pmatrix} 4 & -1 \\ -5 & 2 \end{pmatrix} = \begin{pmatrix} 4 & -1 \\ -5 & 2 \end{pmatrix} = A$$

3. Compute CI_3 and $I_3 C$:
$$CI_3 = \begin{pmatrix} 3 & 0 & -2 \\ 0 & 5 & 6 \\ 1 & 4 & -7 \end{pmatrix}\begin{pmatrix} 1 & 0 & 0 \\ 0 & 1 & 0 \\ 0 & 0 & 1 \end{pmatrix} = \begin{pmatrix} 3 & 0 & -2 \\ 0 & 5 & 6 \\ 1 & 4 & -7 \end{pmatrix} = C$$
$$I_3 C = \begin{pmatrix} 1 & 0 & 0 \\ 0 & 1 & 0 \\ 0 & 0 & 1 \end{pmatrix}\begin{pmatrix} 3 & 0 & -2 \\ 0 & 5 & 6 \\ 1 & 4 & -7 \end{pmatrix} = \begin{pmatrix} 3 & 0 & -2 \\ 0 & 5 & 6 \\ 1 & 4 & -7 \end{pmatrix} = C$$

5. We need to find numbers a, b, c and d such that:
$$\begin{pmatrix} 7 & 9 \\ 4 & 5 \end{pmatrix}\begin{pmatrix} a & b \\ c & d \end{pmatrix} = \begin{pmatrix} 1 & 0 \\ 0 & 1 \end{pmatrix} \text{ and } \begin{pmatrix} a & b \\ c & d \end{pmatrix}\begin{pmatrix} 7 & 9 \\ 4 & 5 \end{pmatrix} = \begin{pmatrix} 1 & 0 \\ 0 & 1 \end{pmatrix}$$
Compute the left-hand product:
$$\begin{pmatrix} 7 & 9 \\ 4 & 5 \end{pmatrix}\begin{pmatrix} a & b \\ c & d \end{pmatrix} = \begin{pmatrix} 7a+9c & 7b+9d \\ 4a+5c & 4b+5d \end{pmatrix} = \begin{pmatrix} 1 & 0 \\ 0 & 1 \end{pmatrix}$$

So we have the following systems of equations:

$$7a + 9c = 1 \qquad \text{and} \qquad 7b + 9d = 0$$
$$4a + 5c = 0 \qquad\qquad\qquad 4b + 5d = 1$$

Multiplying the first equation by 5 and the second equation by -9 yields:

$$35a + 45c = 5 \qquad \text{and} \qquad 35b + 45d = 0$$
$$-36a - 45c = 0 \qquad\qquad\qquad -36b - 45d = -9$$

Adding:

$$-a = 5 \qquad\qquad\qquad -b = -9$$
$$a = -5 \qquad\qquad\qquad b = 9$$

Substituting for c and d, respectively:

$$4a + 5c = 0 \qquad\qquad\qquad 4b + 5d = 1$$
$$-20 + 5c = 0 \qquad\qquad\qquad 36 + 5d = 1$$
$$5c = 20 \qquad\qquad\qquad 5d = -35$$
$$c = 4 \qquad\qquad\qquad d = -7$$

So the inverse matrix is $A^{-1} = \begin{pmatrix} -5 & 9 \\ 4 & -7 \end{pmatrix}$.

7.   We need to find numbers a, b, c, and d such that:

$$\begin{pmatrix} -3 & 1 \\ 5 & 6 \end{pmatrix} \begin{pmatrix} a & b \\ c & d \end{pmatrix} = \begin{pmatrix} 1 & 0 \\ 0 & 1 \end{pmatrix}$$

Compute the left-hand product:

$$\begin{pmatrix} -3a + c & -3b + d \\ 5a + 6c & 5b + 6d \end{pmatrix} = \begin{pmatrix} 1 & 0 \\ 0 & 1 \end{pmatrix}$$

So we have the following systems of equations:

$$-3a + c = 1 \qquad \text{and} \qquad -3b + d = 0$$
$$5a + 6c = 0 \qquad\qquad\qquad 5b + 6d = 1$$

Solving for c and d respectively, then substituting:

$$-3a + c = 1 \qquad\qquad\qquad -3b + d = 0$$
$$c = 3a + 1 \qquad\qquad\qquad d = 3b$$

Substituting:

$$5a + 6(3a + 1) = 0 \qquad\qquad\qquad 5b + 6(3b) = 1$$
$$23a + 6 = 0 \qquad\qquad\qquad 23b = 1$$
$$23a = -6 \qquad\qquad\qquad b = \tfrac{1}{23}$$
$$a = -\tfrac{6}{23}$$

Now substituting to find c and d:

$$c = 3a + 1 \qquad\qquad\qquad d = 3b$$
$$c = -\tfrac{18}{23} + 1 \qquad\qquad\qquad d = \tfrac{3}{23}$$
$$c = \tfrac{5}{23}$$

So the inverse matrix is $A^{-1} = \begin{pmatrix} -\tfrac{6}{23} & \tfrac{1}{23} \\ \tfrac{5}{23} & \tfrac{3}{23} \end{pmatrix}$.

9. We need to find numbers a, b, c, and d such that:
$$\begin{pmatrix} -2 & 3 \\ -4 & 6 \end{pmatrix}\begin{pmatrix} a & b \\ c & d \end{pmatrix} = \begin{pmatrix} 1 & 0 \\ 0 & 1 \end{pmatrix}$$
Compute the left-hand product:
$$\begin{pmatrix} -2a+3c & -2b+3d \\ -4a+6c & -4b+6d \end{pmatrix} = \begin{pmatrix} 1 & 0 \\ 0 & 1 \end{pmatrix}$$
So we have the following systems of equations:

$$\begin{array}{lcl} -2a+3c = 1 & \text{and} & -2b+3d = 0 \\ -4a+6c = 0 & & -4b+6d = 1 \end{array}$$

Multiplying the top equation by -2 yields:

$$\begin{array}{lcl} 4a-6c = -2 & & 4b-6d = 0 \\ -4a+6c = 0 & & -4b+6d = 1 \end{array}$$

Adding:

$$\begin{array}{lcl} 0 = -2 & & 0 = 1 \end{array}$$

Since both of these equations are false, no A^{-1} exists.

11. We need to find numbers a, b, c, and d such that:
$$\begin{pmatrix} \frac{1}{3} & \frac{1}{3} \\ -\frac{1}{9} & \frac{2}{9} \end{pmatrix}\begin{pmatrix} a & b \\ c & d \end{pmatrix} = \begin{pmatrix} 1 & 0 \\ 0 & 1 \end{pmatrix}$$
Compute the left-hand product:
$$\begin{pmatrix} \frac{1}{3}a+\frac{1}{3}c & \frac{1}{3}b+\frac{1}{3}d \\ -\frac{1}{9}a+\frac{2}{9}c & -\frac{1}{9}b+\frac{2}{9}d \end{pmatrix} = \begin{pmatrix} 1 & 0 \\ 0 & 1 \end{pmatrix}$$
So we have the following systems of equations:

$$\begin{array}{lcl} \frac{1}{3}a+\frac{1}{3}c = 1 & \text{and} & \frac{1}{3}b+\frac{1}{3}d = 0 \\ -\frac{1}{9}a+\frac{2}{9}c = 0 & & -\frac{1}{9}b+\frac{2}{9}d = 1 \end{array}$$

Multiplying the first equation by 3 and the second by 9 yields:

$$\begin{array}{lcl} a+\ \ c = 3 & & b+\ \ d = 0 \\ -a+2c = 0 & & -b+2d = 9 \end{array}$$

Adding:

$$\begin{array}{lcl} 3c = 3 & & 3d = 9 \\ c = 1 & & d = 3 \end{array}$$

Now substituting to find a and b:

$$\begin{array}{lcl} a+c = 3 & & b+d = 0 \\ a+1 = 3 & & b+3 = 0 \\ a = 2 & & b = -3 \end{array}$$

So the inverse is $A^{-1} = \begin{pmatrix} 2 & -3 \\ 1 & 3 \end{pmatrix}$.

13. Form the matrix:
$$\begin{pmatrix} 2 & 1 & 1 & 0 \\ 3 & 2 & 0 & 1 \end{pmatrix}$$
Multiply row 1 by -1 and add to row 2:
$$\begin{pmatrix} 2 & 1 & 1 & 0 \\ 1 & 1 & -1 & 1 \end{pmatrix}$$
Switch rows 1 and 2:
$$\begin{pmatrix} 1 & 1 & -1 & 1 \\ 2 & 1 & 1 & 0 \end{pmatrix}$$
Multiply row 1 by -2 and add to row 2:
$$\begin{pmatrix} 1 & 1 & -1 & 1 \\ 0 & -1 & 3 & -2 \end{pmatrix}$$
Add row 2 to row 1:
$$\begin{pmatrix} 1 & 0 & 2 & -1 \\ 0 & -1 & 3 & -2 \end{pmatrix}$$
Multiply row 2 by -1:
$$\begin{pmatrix} 1 & 0 & 2 & -1 \\ 0 & 1 & -3 & 2 \end{pmatrix}$$
So the inverse is $\begin{pmatrix} 2 & -1 \\ -3 & 2 \end{pmatrix}$.

15. Form the matrix:
$$\begin{pmatrix} 0 & -11 & 1 & 0 \\ 1 & 6 & 0 & 1 \end{pmatrix}$$
Switch rows 1 and 2:
$$\begin{pmatrix} 1 & 6 & 0 & 1 \\ 0 & -11 & 1 & 0 \end{pmatrix}$$
Multiply row 2 by $-\frac{1}{11}$:
$$\begin{pmatrix} 1 & 6 & 0 & 1 \\ 0 & 1 & -\frac{1}{11} & 0 \end{pmatrix}$$
Multiply row 2 by -6 and add to row 1:
$$\begin{pmatrix} 1 & 0 & \frac{6}{11} & 1 \\ 0 & 1 & -\frac{1}{11} & 0 \end{pmatrix}$$
So the inverse is $\begin{pmatrix} \frac{6}{11} & 1 \\ -\frac{1}{11} & 0 \end{pmatrix}$.

17. Form the matrix:

$$\begin{pmatrix} \frac{2}{3} & -\frac{1}{4} & 1 & 0 \\ -8 & 3 & 0 & 1 \end{pmatrix}$$

Multiply row 1 by 12:

$$\begin{pmatrix} 8 & -3 & 12 & 0 \\ -8 & 3 & 0 & 1 \end{pmatrix}$$

Add row 1 to row 2:

$$\begin{pmatrix} 8 & -3 & 12 & 0 \\ 0 & 0 & 12 & 1 \end{pmatrix}$$

So the inverse does not exist.

19. Form the matrix:

$$\begin{pmatrix} -5 & 4 & -3 & 1 & 0 & 0 \\ 10 & -7 & 6 & 0 & 1 & 0 \\ 8 & -6 & 5 & 0 & 0 & 1 \end{pmatrix}$$

Multiply row 1 by 2 and add to row 2:

$$\begin{pmatrix} -5 & 4 & -3 & 1 & 0 & 0 \\ 0 & 1 & 0 & 2 & 1 & 0 \\ 8 & -6 & 5 & 0 & 0 & 1 \end{pmatrix}$$

Add row 3 to row 1 (to reduce the numbers):

$$\begin{pmatrix} 3 & -2 & 2 & 1 & 0 & 1 \\ 0 & 1 & 0 & 2 & 1 & 0 \\ 8 & -6 & 5 & 0 & 0 & 1 \end{pmatrix}$$

Multiply row 2 by 2 and add to row 1, and also multiply row 2 by 6 and add to row 3:

$$\begin{pmatrix} 3 & 0 & 2 & 5 & 2 & 1 \\ 0 & 1 & 0 & 2 & 1 & 0 \\ 8 & 0 & 5 & 12 & 6 & 1 \end{pmatrix}$$

Multiply row 1 by $-\frac{8}{3}$ and add to row 3:

$$\begin{pmatrix} 3 & 0 & 2 & 5 & 2 & 1 \\ 0 & 1 & 0 & 2 & 1 & 0 \\ 0 & 0 & -\frac{1}{3} & -\frac{4}{3} & \frac{2}{3} & -\frac{5}{3} \end{pmatrix}$$

Multiply row 3 by -3:

$$\begin{pmatrix} 3 & 0 & 2 & 5 & 2 & 1 \\ 0 & 1 & 0 & 2 & 1 & 0 \\ 0 & 0 & 1 & 4 & -2 & 5 \end{pmatrix}$$

Multiply row 3 by –2 and add to row 1:

$$\begin{pmatrix} 3 & 0 & 0 & -3 & 6 & -9 \\ 0 & 1 & 0 & 2 & 1 & 0 \\ 0 & 0 & 1 & 4 & -2 & 5 \end{pmatrix}$$

Multiply row 1 by $\frac{1}{3}$:

$$\begin{pmatrix} 1 & 0 & 0 & -1 & 2 & -3 \\ 0 & 1 & 0 & 2 & 1 & 0 \\ 0 & 0 & 1 & 4 & -2 & 5 \end{pmatrix}$$

So the inverse is $\begin{pmatrix} -1 & 2 & -3 \\ 2 & 1 & 0 \\ 4 & -2 & 5 \end{pmatrix}$.

21. Form the matrix:

$$\begin{pmatrix} 1 & 2 & -1 & 1 & 0 & 0 \\ 0 & 3 & 0 & 0 & 1 & 0 \\ -4 & 0 & 5 & 0 & 0 & 1 \end{pmatrix}$$

Multiply row 1 by 4 and add to row 3:

$$\begin{pmatrix} 1 & 2 & -1 & 1 & 0 & 0 \\ 0 & 3 & 0 & 0 & 1 & 0 \\ 0 & 8 & 1 & 4 & 0 & 1 \end{pmatrix}$$

Multiply row 2 by $\frac{1}{3}$:

$$\begin{pmatrix} 1 & 2 & -1 & 1 & 0 & 0 \\ 0 & 1 & 0 & 0 & \frac{1}{3} & 0 \\ 0 & 8 & 1 & 4 & 0 & 1 \end{pmatrix}$$

Multiply row 2 by –2 and add to row 1, and multiply row 2 by – 8 and add to row 3:

$$\begin{pmatrix} 1 & 0 & -1 & 1 & -\frac{2}{3} & 0 \\ 0 & 1 & 0 & 0 & \frac{1}{3} & 0 \\ 0 & 0 & 1 & 4 & -\frac{8}{3} & 1 \end{pmatrix}$$

Add row 3 to row 1:

$$\begin{pmatrix} 1 & 0 & 0 & 5 & -\frac{10}{3} & 1 \\ 0 & 1 & 0 & 0 & \frac{1}{3} & 0 \\ 0 & 0 & 1 & 4 & -\frac{8}{3} & 1 \end{pmatrix}$$

So the inverse is $\begin{pmatrix} 5 & -\frac{10}{3} & 1 \\ 0 & \frac{1}{3} & 0 \\ 4 & -\frac{8}{3} & 1 \end{pmatrix}$.

23. Form the matrix:

$$\begin{pmatrix} -7 & 5 & 3 & 1 & 0 & 0 \\ 3 & -2 & -2 & 0 & 1 & 0 \\ 3 & -2 & -1 & 0 & 0 & 1 \end{pmatrix}$$

Multiply row 2 by −1 and add to row 3:

$$\begin{pmatrix} -7 & 5 & 3 & 1 & 0 & 0 \\ 3 & -2 & -2 & 0 & 1 & 0 \\ 0 & 0 & 1 & 0 & -1 & 1 \end{pmatrix}$$

Multiply row 2 by 2 and add to row 1:

$$\begin{pmatrix} -1 & 1 & -1 & 1 & 2 & 0 \\ 3 & -2 & -2 & 0 & 1 & 0 \\ 0 & 0 & 1 & 0 & -1 & 1 \end{pmatrix}$$

Multiply row 1 by 3 and add to row 2:

$$\begin{pmatrix} -1 & 1 & -1 & 1 & 2 & 0 \\ 0 & 1 & -5 & 3 & 7 & 0 \\ 0 & 0 & 1 & 0 & -1 & 1 \end{pmatrix}$$

Multiply row 3 by 5 and add to row 2, and add row 3 to row 1:

$$\begin{pmatrix} -1 & 1 & 0 & 1 & 1 & 1 \\ 0 & 1 & 0 & 3 & 2 & 5 \\ 0 & 0 & 1 & 0 & -1 & 1 \end{pmatrix}$$

Multiply row 1 by −1:

$$\begin{pmatrix} 1 & -1 & 0 & -1 & -1 & -1 \\ 0 & 1 & 0 & 3 & 2 & 5 \\ 0 & 0 & 1 & 0 & -1 & 1 \end{pmatrix}$$

Add row 2 to row 1:

$$\begin{pmatrix} 1 & 0 & 0 & 2 & 1 & 4 \\ 0 & 1 & 0 & 3 & 2 & 5 \\ 0 & 0 & 1 & 0 & -1 & 1 \end{pmatrix}$$

So the inverse is $\begin{pmatrix} 2 & 1 & 4 \\ 3 & 2 & 5 \\ 0 & -1 & 1 \end{pmatrix}$.

25. Form the matrix:

$$\begin{pmatrix} 1 & 2 & 3 & 1 & 0 & 0 \\ 4 & 5 & 6 & 0 & 1 & 0 \\ 7 & 8 & 9 & 0 & 0 & 1 \end{pmatrix}$$

Multiply row 1 by -4 and add to row 2, and multiply row 1 by -7 and add to row 3:

$$\begin{pmatrix} 1 & 2 & 3 & 1 & 0 & 0 \\ 0 & -3 & -6 & -4 & 1 & 0 \\ 0 & -6 & -12 & -7 & 0 & 1 \end{pmatrix}$$

Multiply row 2 by -2 and add to row 3:

$$\begin{pmatrix} 1 & 2 & 3 & 1 & 0 & 0 \\ 0 & -3 & -6 & -4 & 1 & 0 \\ 0 & 0 & 0 & 1 & -2 & 1 \end{pmatrix}$$

So the inverse does not exist.

27. (a) Since the system can be written as $A \cdot X = B$, where $X = \begin{pmatrix} x \\ y \end{pmatrix}$ and $B = \begin{pmatrix} 5 \\ 7 \end{pmatrix}$,

$X = A^{-1} \cdot B$. So $X = \begin{pmatrix} 11 & -8 \\ -4 & 3 \end{pmatrix}\begin{pmatrix} 5 \\ 7 \end{pmatrix} = \begin{pmatrix} -1 \\ 1 \end{pmatrix}$, thus $x = -1$ and $y = 1$.

(b) Again this system can be written as $A \cdot X = B$, where $X = \begin{pmatrix} x \\ y \end{pmatrix}$ and $B = \begin{pmatrix} -12 \\ 0 \end{pmatrix}$, then

$X = A^{-1} \cdot B$. So $X = \begin{pmatrix} 11 & -8 \\ -4 & 3 \end{pmatrix}\begin{pmatrix} -12 \\ 0 \end{pmatrix} = \begin{pmatrix} -132 \\ 48 \end{pmatrix}$, thus $x = -132$ and $y = 48$.

29. (a) Since the system can be written as $A \cdot X = B$, where $X = \begin{pmatrix} x \\ y \\ z \end{pmatrix}$ and $B = \begin{pmatrix} 28 \\ 9 \\ 22 \end{pmatrix}$,

$X = A^{-1} \cdot B$. So $X = \begin{pmatrix} 1 & 2 & -2 \\ -1 & 3 & 0 \\ 0 & -2 & 1 \end{pmatrix}\begin{pmatrix} 28 \\ 9 \\ 22 \end{pmatrix} = \begin{pmatrix} 2 \\ -1 \\ 4 \end{pmatrix}$, thus $x = 2$, $y = -1$, $z = 4$.

(b) Again this system can be written as $A \cdot X = B$, where $X = \begin{pmatrix} x \\ y \\ z \end{pmatrix}$ and $B = \begin{pmatrix} -7 \\ -2 \\ -6 \end{pmatrix}$, then

$X = A^{-1} \cdot B$. So $X = \begin{pmatrix} 1 & 2 & -2 \\ -1 & 3 & 0 \\ 0 & -2 & 1 \end{pmatrix}\begin{pmatrix} -7 \\ -2 \\ -6 \end{pmatrix} = \begin{pmatrix} 1 \\ 1 \\ -2 \end{pmatrix}$, thus $x = 1$, $y = 1$, $z = -2$.

31. (a) Compute AA:

$$AA = \begin{pmatrix} 1 & -6 & 3 \\ 2 & -7 & 3 \\ 4 & -12 & 5 \end{pmatrix}\begin{pmatrix} 1 & -6 & 3 \\ 2 & -7 & 3 \\ 4 & -12 & 5 \end{pmatrix} = \begin{pmatrix} 1 & 0 & 0 \\ 0 & 1 & 0 \\ 0 & 0 & 1 \end{pmatrix}$$

The product is the identity matrix I_3, and thus $A^{-1} = A$.

(b) Since this can be written as $A \cdot X = B$, where $X = \begin{pmatrix} x \\ y \\ z \end{pmatrix}$ and $B = \begin{pmatrix} \frac{19}{2} \\ 11 \\ 19 \end{pmatrix}$,

$X = A^{-1} \cdot B$. Recall from part (a) that $A^{-1} = A$, so $X = \begin{pmatrix} 1 & -6 & 3 \\ 2 & -7 & 3 \\ 4 & -12 & 5 \end{pmatrix}\begin{pmatrix} \frac{19}{2} \\ 11 \\ 19 \end{pmatrix} = \begin{pmatrix} \frac{1}{2} \\ -1 \\ 1 \end{pmatrix}$,

thus $x = \frac{1}{2}$, $y = -1$, and $z = 1$.

33. We find $\begin{pmatrix} a & b \\ c & d \end{pmatrix}$ where $\begin{pmatrix} 2 & 5 \\ 6 & 15 \end{pmatrix}\begin{pmatrix} a & b \\ c & d \end{pmatrix} = \begin{pmatrix} 1 & 0 \\ 0 & 1 \end{pmatrix}$. Carrying out the multiplication:

$$\begin{pmatrix} 2a + 5c & 2b + 5d \\ 6a + 15c & 6b + 15d \end{pmatrix} = \begin{pmatrix} 1 & 0 \\ 0 & 1 \end{pmatrix}$$

So we have the following systems of equations:

$$\begin{array}{ll} 2a + 5c = 1 \quad \text{and} & 2b + 5d = 0 \\ 6a + 15c = 0 & 6b + 15d = 1 \end{array}$$

Multiplying the first equation by -3 yields:

$$\begin{array}{ll} -6a - 15c = -3 & -6b - 15d = 0 \\ 6a + 15c = 0 & 6b + 15d = 1 \end{array}$$

Adding these equations:

$$\begin{array}{ll} 0 = -3 & 0 = 1 \end{array}$$

Neither of these systems has a solution, thus the matrix has no inverse.

35. (a) Form the matrix:

$$\begin{pmatrix} 2 & 3 & 1 & 0 \\ 4 & 5 & 0 & 1 \end{pmatrix}$$

Multiply row 1 by -2 and add to row 2:

$$\begin{pmatrix} 2 & 3 & 1 & 0 \\ 0 & -1 & -2 & 1 \end{pmatrix}$$

Multiply row 2 by 3 and add to row 1:

$$\begin{pmatrix} 2 & 0 & -5 & 3 \\ 0 & -1 & -2 & 1 \end{pmatrix}$$

Multiply row 1 by $\frac{1}{2}$ and row 2 by -1:

$$\begin{pmatrix} 1 & 0 & -\frac{5}{2} & \frac{3}{2} \\ 0 & 1 & 2 & -1 \end{pmatrix}$$

So $A^{-1} = \begin{pmatrix} -\frac{5}{2} & \frac{3}{2} \\ 2 & -1 \end{pmatrix}$. Now form the matrix:

$$\begin{pmatrix} 7 & 8 & 1 & 0 \\ 6 & 7 & 0 & 1 \end{pmatrix}$$

Subtract row 2 from row 1:

$$\begin{pmatrix} 1 & 1 & 1 & -1 \\ 6 & 7 & 0 & 1 \end{pmatrix}$$

Multiply row 1 by -6 and add to row 2:

$$\begin{pmatrix} 1 & 1 & 1 & -1 \\ 0 & 1 & -6 & 7 \end{pmatrix}$$

Subtract row 2 from row 1:

$$\begin{pmatrix} 1 & 0 & 7 & -8 \\ 0 & 1 & -6 & 7 \end{pmatrix}$$

So $B^{-1} = \begin{pmatrix} 7 & -8 \\ -6 & 7 \end{pmatrix}$.

Then $B^{-1}A^{-1} = \begin{pmatrix} 7 & -8 \\ -6 & 7 \end{pmatrix}\begin{pmatrix} -\frac{5}{2} & \frac{3}{2} \\ 2 & -1 \end{pmatrix} = \begin{pmatrix} -\frac{67}{2} & \frac{37}{2} \\ 29 & -16 \end{pmatrix}$.

(b) First find AB:

$$AB = \begin{pmatrix} 2 & 3 \\ 4 & 5 \end{pmatrix}\begin{pmatrix} 7 & 8 \\ 6 & 7 \end{pmatrix} = \begin{pmatrix} 32 & 37 \\ 58 & 67 \end{pmatrix}$$

Now form the matrix:

$$\begin{pmatrix} 32 & 37 & 1 & 0 \\ 58 & 67 & 0 & 1 \end{pmatrix}$$

Subtract row 1 from row 2 (to reduce numbers):

$$\begin{pmatrix} 32 & 37 & 1 & 0 \\ 26 & 30 & -1 & 1 \end{pmatrix}$$

Subtract row 2 from row 1 (to reduce numbers):

$$\begin{pmatrix} 6 & 7 & 2 & -1 \\ 26 & 30 & -1 & 1 \end{pmatrix}$$

Multiply row 1 by -4 and add to row 2:

$$\begin{pmatrix} 6 & 7 & 2 & -1 \\ 2 & 2 & -9 & 5 \end{pmatrix}$$

Multiply row 2 by -3 and add to row 1:

$$\begin{pmatrix} 0 & 1 & 29 & -16 \\ 2 & 2 & -9 & 5 \end{pmatrix}$$

Multiply row 1 by -2 and add to row 2:

$$\begin{pmatrix} 0 & 1 & 29 & -16 \\ 2 & 0 & -67 & 37 \end{pmatrix}$$

Multiply row 2 by $\frac{1}{2}$, then switch rows 1 and 2:

$$\begin{pmatrix} 1 & 0 & -\frac{67}{2} & \frac{37}{2} \\ 0 & 1 & 29 & -16 \end{pmatrix}$$

So $(AB)^{-1} = \begin{pmatrix} -\frac{67}{2} & \frac{37}{2} \\ 29 & -16 \end{pmatrix}$, which is the same as $B^{-1}A^{-1}$ from part (a).

Graphing Utility Exercises for Sections 11.3 and 11.4

For Exercises 1-33, first enter the matrices and store them under the variable names A through G, respectively.

1. The product is $AB = \begin{pmatrix} 23 & -101 \\ 18 & -79 \end{pmatrix}$.

3. The product AE is undefined.

5. The product is $EC = \begin{pmatrix} 21 & 16 & 43 \\ 17 & 18 & 43 \\ -4 & -4 & -14 \end{pmatrix}$.

7. The product is $A(A+B) = \begin{pmatrix} 139 & -36 \\ 109 & -28 \end{pmatrix}$.

9. The product is $AA + AB = \begin{pmatrix} 139 & -36 \\ 109 & -28 \end{pmatrix}$.

11. The product EA is undefined.

13. The product is $FA = \begin{pmatrix} -18 & -10 \\ 34 & 19 \\ 59 & 33 \end{pmatrix}$.

15. The product is $DC = \begin{pmatrix} 39 & 30 & 82 \\ 10 & 8 & 21 \\ -10 & -7 & -21 \end{pmatrix}$.

17. The product is $CD - CE = \begin{pmatrix} 0 & -42 & 23 \\ 1 & -14 & 5 \\ -1 & -35 & 15 \end{pmatrix}$.

19. The product CG is undefined.

21. The product GA is undefined.

23. The inverse is $B^{-1} = \begin{pmatrix} -4 & 9 \\ -1 & 2 \end{pmatrix}$.

25. The inverse is $(BA)^{-1} = \begin{pmatrix} -11 & 26 \\ 19 & -45 \end{pmatrix}$.

27. The inverse is $A^{-1}B^{-1} = \begin{pmatrix} -11 & 26 \\ 19 & -45 \end{pmatrix}$.

29. The inverse G^{-1} does not exist.

31. The inverse is $(DC)^{-1} = \begin{pmatrix} -21 & 56 & -26 \\ 0 & 1 & 1 \\ 10 & -27 & 12 \end{pmatrix}$.

33. The inverse $(FG)^{-1}$ does not exist.

35. Enter $A = \begin{pmatrix} 8 & -5 \\ 3 & 4 \end{pmatrix}$, so $A^{-1} = \begin{pmatrix} 0.085 & 0.106 \\ -0.064 & 0.170 \end{pmatrix}$.

   (a) Let $b = \begin{pmatrix} -13 \\ 48 \end{pmatrix}$, so the solution is $A^{-1}b = \begin{pmatrix} 4 \\ 9 \end{pmatrix}$. The solution to the system is $(4, 9)$.

   (b) Let $b = \begin{pmatrix} 5 \\ -2 \end{pmatrix}$, so the solution is $A^{-1}b = \begin{pmatrix} 0.213 \\ -0.660 \end{pmatrix}$. The solution to the system is $(0.213, -0.660)$.

37.   Enter $A = \begin{pmatrix} 5 & -2 & -2 \\ 3 & 1 & 0 \\ 1 & 1 & 1 \end{pmatrix}$, so $A^{-1} = \begin{pmatrix} 0.143 & 0 & 0.286 \\ -0.429 & 1 & -0.857 \\ 0.286 & -1 & 1.571 \end{pmatrix}$.

(a)   Let $b = \begin{pmatrix} 15 \\ 4 \\ -4 \end{pmatrix}$, so the solution is $A^{-1}b = \begin{pmatrix} 1 \\ 1 \\ -6 \end{pmatrix}$. The solution to the system is

$(1, 1, -6)$.

(b)   Let $b = \begin{pmatrix} 0 \\ 6 \\ 1 \end{pmatrix}$, so the solution is $A^{-1}b = \begin{pmatrix} 0.286 \\ 5.143 \\ -4.429 \end{pmatrix}$. The solution to the system is

$(0.286, 5.143, -4.429)$.

11.5 Determinants and Cramer's Rule

1.   (a)   Evaluate the determinant:
$$\begin{vmatrix} 2 & -17 \\ 1 & 6 \end{vmatrix} = 2(6) - (-17)(1) = 12 + 17 = 29$$

(b)   Evaluate the determinant:
$$\begin{vmatrix} 1 & 6 \\ 2 & -17 \end{vmatrix} = 1(-17) - (6)(2) = -17 - 12 = -29$$

3.   (a)   Evaluate the determinant:
$$\begin{vmatrix} 7 & 7 \\ 500 & 700 \end{vmatrix} = 100 \begin{vmatrix} 5 & 7 \\ 5 & 7 \end{vmatrix} = 100[5(7) - 7(5)] = 100(35 - 35) = 0$$

(b)   Evaluate the determinant:
$$\begin{vmatrix} 5 & 500 \\ 7 & 700 \end{vmatrix} = 100 \begin{vmatrix} 5 & 5 \\ 7 & 7 \end{vmatrix} = 100[5(7) - 5(7)] = 100(35 - 35) = 0$$

5.   Evaluate the determinant:
$$\begin{vmatrix} \sqrt{2} - 1 & \sqrt{2} \\ \sqrt{2} & \sqrt{2} + 1 \end{vmatrix} = (\sqrt{2} - 1)(\sqrt{2} + 1) - \sqrt{2}(\sqrt{2}) = (2 - 1) - 2 = -1$$

7.   Evaluate the minor of 3:
$$\begin{vmatrix} 5 & 1 \\ 10 & -10 \end{vmatrix} = 10 \begin{vmatrix} 5 & 1 \\ 1 & -1 \end{vmatrix} = 10[5(-1) - 1(1)] = 10(-6) = -60$$

9. Evaluate the minor of -10:

$$\begin{vmatrix} -6 & 3 \\ 5 & -4 \end{vmatrix} = 3\begin{vmatrix} -2 & 1 \\ 5 & -4 \end{vmatrix} = 3[-2(-4) - 1(5)] = 3(3) = 9$$

11. (a) Compute:

$$-6\begin{vmatrix} -4 & 1 \\ 9 & -10 \end{vmatrix} + 3\begin{vmatrix} 5 & 1 \\ 10 & -10 \end{vmatrix} + 8\begin{vmatrix} 5 & -4 \\ 10 & 9 \end{vmatrix} = -6(31) + 3(-60) + 8(85) = 314$$

(b) Compute:

$$-6\begin{vmatrix} -4 & 1 \\ 9 & -10 \end{vmatrix} - 3\begin{vmatrix} 5 & 1 \\ 10 & -10 \end{vmatrix} + 8\begin{vmatrix} 5 & -4 \\ 10 & 9 \end{vmatrix} = -6(31) - 3(-60) + 8(85) = 674$$

(c) The answer in part (b) would be the determinant.

13. (a) Expand along the second row:

$$-4\begin{vmatrix} 2 & 3 \\ 8 & 9 \end{vmatrix} + 5\begin{vmatrix} 1 & 3 \\ 7 & 9 \end{vmatrix} - 6\begin{vmatrix} 1 & 2 \\ 7 & 8 \end{vmatrix} = -4(-6) + 5(-12) - 6(-6) = 0$$

(b) Expand along the third row:

$$7\begin{vmatrix} 2 & 3 \\ 5 & 6 \end{vmatrix} - 8\begin{vmatrix} 1 & 3 \\ 4 & 6 \end{vmatrix} + 9\begin{vmatrix} 1 & 2 \\ 4 & 5 \end{vmatrix} = 7(-3) - 8(-6) + 9(-3) = 0$$

(c) Expand along the first column:

$$1\begin{vmatrix} 5 & 6 \\ 8 & 9 \end{vmatrix} - 4\begin{vmatrix} 2 & 3 \\ 8 & 9 \end{vmatrix} + 7\begin{vmatrix} 2 & 3 \\ 5 & 6 \end{vmatrix} = 1(-3) - 4(-6) + 7(-3) = 0$$

(d) Expand along the third column:

$$3\begin{vmatrix} 4 & 5 \\ 7 & 8 \end{vmatrix} - 6\begin{vmatrix} 1 & 2 \\ 7 & 8 \end{vmatrix} + 9\begin{vmatrix} 1 & 2 \\ 4 & 5 \end{vmatrix} = 3(-3) - 6(-6) + 9(-3) = 0$$

15. Factoring 5 out of the first row:

$$\begin{vmatrix} 5 & 10 & 15 \\ 1 & 2 & 3 \\ -9 & 11 & 7 \end{vmatrix} = 5\begin{vmatrix} 1 & 2 & 3 \\ 1 & 2 & 3 \\ -9 & 11 & 7 \end{vmatrix} = 5\begin{vmatrix} 0 & 0 & 0 \\ 1 & 2 & 3 \\ -9 & 11 & 7 \end{vmatrix} = 0$$

17. Expanding along the third row:

$$\begin{vmatrix} 1 & 2 & -3 \\ 4 & 5 & -9 \\ 0 & 0 & 1 \end{vmatrix} = 0\begin{vmatrix} 2 & -3 \\ 5 & -9 \end{vmatrix} - 0\begin{vmatrix} 1 & -3 \\ 4 & -9 \end{vmatrix} + 1\begin{vmatrix} 1 & 2 \\ 4 & 5 \end{vmatrix} = 1(-3) = -3$$

19. Factoring 2 out of the first row and 4 out of the second column:

$$\begin{vmatrix} -6 & -8 & 18 \\ 25 & 12 & 15 \\ -9 & 4 & 13 \end{vmatrix} = 2 \begin{vmatrix} -3 & -4 & 9 \\ 25 & 12 & 15 \\ -9 & 4 & 13 \end{vmatrix} = 8 \begin{vmatrix} -3 & -1 & 9 \\ 25 & 3 & 15 \\ -9 & 1 & 13 \end{vmatrix}$$

Now expanding along the first row:

$$\begin{vmatrix} -3 & -1 & 9 \\ 25 & 3 & 15 \\ -9 & 1 & 13 \end{vmatrix} = -3 \begin{vmatrix} 3 & 15 \\ 1 & 13 \end{vmatrix} + 1 \begin{vmatrix} 25 & 15 \\ -9 & 13 \end{vmatrix} + 9 \begin{vmatrix} 25 & 3 \\ -9 & 1 \end{vmatrix} = -3(24) + 1(460) + 9(52) = 856$$

Therefore:

$$8 \begin{vmatrix} -3 & -1 & 9 \\ 25 & 3 & 15 \\ -9 & 1 & 13 \end{vmatrix} = 8(856) = 6848$$

21. Factoring 16 out of the first row and 10 out of the third row:

$$\begin{vmatrix} 16 & 0 & -64 \\ -8 & 15 & -12 \\ 30 & -20 & 10 \end{vmatrix} = 16 \begin{vmatrix} 1 & 0 & -4 \\ -8 & 15 & -12 \\ 30 & -20 & 10 \end{vmatrix} = 160 \begin{vmatrix} 1 & 0 & -4 \\ -8 & 15 & -12 \\ 3 & -2 & 1 \end{vmatrix}$$

Now expanding along the first row:

$$\begin{vmatrix} 1 & 0 & -4 \\ -8 & 15 & -12 \\ 3 & -2 & 1 \end{vmatrix} = 1 \begin{vmatrix} 15 & -12 \\ -2 & 1 \end{vmatrix} - 0 \begin{vmatrix} -8 & -12 \\ 3 & 1 \end{vmatrix} - 4 \begin{vmatrix} -8 & 15 \\ 3 & 2 \end{vmatrix} = 1(-9) - 4(-29) = 107$$

Therefore:

$$160 \begin{vmatrix} 1 & 0 & -4 \\ -8 & 15 & -12 \\ 3 & -2 & 1 \end{vmatrix} = 160(107) = 17120$$

23. Evaluate the determinant:

$$\begin{vmatrix} 1 & x & x^2 \\ 1 & y & y^2 \\ 1 & z & z^2 \end{vmatrix} = \begin{vmatrix} 1 & x & x^2 \\ 0 & y-x & y^2-x^2 \\ 0 & z-x & z^2-x^2 \end{vmatrix}$$

$$= \begin{vmatrix} y-x & y^2-x^2 \\ z-x & z^2-x^2 \end{vmatrix}$$

$$= \begin{vmatrix} y-x & (y+x)(y-x) \\ z-x & (z+x)(z-x) \end{vmatrix}$$

$$= (y-x)(z-x) \begin{vmatrix} 1 & y+x \\ 1 & z+x \end{vmatrix}$$

$$= (y-x)(z-x)[z+x-y-x]$$

$$= (y-x)(z-x)(z-y)$$

25. Substracting row 1 from row 2 and row 1 from row 3:

$$\begin{vmatrix} 1 & 1 & 1 \\ 1 & 1+x & 1 \\ 1 & 1 & 1+y \end{vmatrix} = \begin{vmatrix} 1 & 1 & 1 \\ 0 & x & 0 \\ 0 & 0 & y \end{vmatrix} = 1\begin{vmatrix} x & 0 \\ 0 & y \end{vmatrix} = xy$$

27. Adding column 2 to column 3:

$$\begin{vmatrix} 1 & -1 & -1 & 2 \\ 0 & 1 & 0 & 0 \\ 2 & 1 & i & -1 \\ -2 & 2 & 3 & 1 \end{vmatrix} = 1\begin{vmatrix} 1 & -1 & 2 \\ 2 & 1 & -1 \\ -2 & 3 & 1 \end{vmatrix}$$

Adding twice column 3 to column 1:

$$\begin{vmatrix} 5 & -1 & 2 \\ 0 & 1 & -1 \\ 0 & 3 & 1 \end{vmatrix} = 5\begin{vmatrix} 1 & -1 \\ 3 & 1 \end{vmatrix} = 5(4) = 20$$

29. Evaluate the determinant:

$$\begin{vmatrix} 2 & 0 & 0 & 0 \\ 0 & 3 & 0 & 0 \\ 0 & 0 & 4 & 0 \\ 0 & 0 & 0 & 5 \end{vmatrix} = 2\begin{vmatrix} 3 & 0 & 0 \\ 0 & 4 & 0 \\ 0 & 0 & 5 \end{vmatrix} = 2(3)\begin{vmatrix} 4 & 0 \\ 0 & 5 \end{vmatrix} = 6(20) = 120$$

31. (a) Subtract the first column from the second and add twice the first column to the third. The result is:

$$\begin{vmatrix} -20 & 22 & -43 \\ 6 & -10 & 13 \\ -1 & 0 & 0 \end{vmatrix} = -1[22(13) - (-43)(-10)] = 144$$

(b) Multiply the second row by 2 and subtract it from the first row, and multiply the second row by 4 and subtract it from the third row. This yields:

$$\begin{vmatrix} 0 & -32 & -5 \\ 1 & 6 & 1 \\ 0 & -25 & -2 \end{vmatrix} = -1[(-32)(-2) - (-5)(-25)] = 61$$

(c) Subtract the first column from the second and add 10 times the first column to the third. This yields:

$$\begin{vmatrix} 2 & 0 & 0 \\ 1 & -5 & 16 \\ 4 & -5 & 39 \end{vmatrix} = 2[(-5)(39) - 16(-5)] = -230$$

33. Compute D:

$$D = \begin{vmatrix} 3 & 4 & -1 \\ 1 & -3 & 2 \\ 5 & 0 & -6 \end{vmatrix} = 5 \begin{vmatrix} 4 & -1 \\ -3 & 2 \end{vmatrix} - 6 \begin{vmatrix} 3 & 4 \\ 1 & -3 \end{vmatrix} = 5(5) - 6(-13) = 103$$

Now compute D_x:

$$D_x = \begin{vmatrix} 5 & 4 & -1 \\ 2 & -3 & 2 \\ -7 & 0 & -6 \end{vmatrix} = -7 \begin{vmatrix} 4 & -1 \\ -3 & 2 \end{vmatrix} - 6 \begin{vmatrix} 5 & 4 \\ 2 & -3 \end{vmatrix} = -7(5) - 6(-23) = 103$$

For D_y, adding -2 times column 1 to both column 2 and column 3 yields:

$$D_y = \begin{vmatrix} 3 & 5 & -1 \\ 1 & 2 & 2 \\ 5 & -7 & -6 \end{vmatrix} = \begin{vmatrix} 3 & -1 & -7 \\ 1 & 0 & 0 \\ 5 & -17 & -16 \end{vmatrix} = -1 \begin{vmatrix} -1 & -7 \\ -17 & -16 \end{vmatrix} = -(-103) = 103$$

For D_z, adding 3 times column 1 to column 2, and -2 times column 1 to column 3 yields:

$$D_z = \begin{vmatrix} 3 & 4 & 5 \\ 1 & -3 & 2 \\ 5 & 0 & -7 \end{vmatrix} = \begin{vmatrix} 3 & 13 & -1 \\ 1 & 0 & 0 \\ 5 & 15 & -17 \end{vmatrix} = -1 \begin{vmatrix} 13 & 1 \\ 15 & -17 \end{vmatrix} = -(-206) = 206$$

So $x = \frac{D_x}{D} = 1$, $y = \frac{D_y}{D} = 1$, and $z = \frac{D_z}{D} = 2$. So the solution is $(1, 1, 2)$.

35. Subtracting column 1 from both column 2 and column 3 yields:

$$D = \begin{vmatrix} 3 & 2 & -1 \\ 2 & -3 & -4 \\ 1 & 1 & 1 \end{vmatrix} = \begin{vmatrix} 3 & -1 & -4 \\ 2 & -5 & -6 \\ 1 & 0 & 0 \end{vmatrix} = 1 \begin{vmatrix} -1 & -4 \\ -5 & -6 \end{vmatrix} = -14$$

Adding -5 times column 3 to column 1, and subtracting column 3 from column 2 yields:

$$D_x = \begin{vmatrix} -6 & 2 & -1 \\ -11 & -3 & -4 \\ 5 & 1 & 1 \end{vmatrix} = \begin{vmatrix} -1 & 3 & -1 \\ 9 & 1 & -4 \\ 0 & 0 & 1 \end{vmatrix} = 1 \begin{vmatrix} -1 & 3 \\ 9 & 1 \end{vmatrix} = -28$$

Adding row 3 to row 1, and 4 times row 3 to row 2 yields:

$$D_y = \begin{vmatrix} 3 & -6 & -1 \\ 2 & -11 & -4 \\ 1 & 5 & 1 \end{vmatrix} = \begin{vmatrix} 4 & -1 & 0 \\ 6 & 9 & 0 \\ 1 & 5 & 1 \end{vmatrix} = 1 \begin{vmatrix} 4 & -1 \\ 6 & 9 \end{vmatrix} = 42$$

Adding -2 times row 3 to row 1, and 3 times row 3 to row 2 yields:

$$D_z = \begin{vmatrix} 3 & 2 & -6 \\ 2 & 3 & -11 \\ 1 & 1 & 5 \end{vmatrix} = \begin{vmatrix} 1 & 0 & -16 \\ 5 & 0 & 4 \\ 1 & 1 & 5 \end{vmatrix} = -1 \begin{vmatrix} 1 & -16 \\ 5 & 4 \end{vmatrix} = -(84) = -84$$

So $x = \frac{D_x}{D} = 2$, $y = \frac{D_y}{D} = -3$, and $z = \frac{D_z}{D} = 6$. So the solution is $(2, -3, 6)$.

37. Adding −2 times row 3 to row 1, and −3 times row 3 to row 2 yields:

$$D = \begin{vmatrix} 2 & 5 & 2 \\ 3 & -1 & -4 \\ 1 & 2 & -3 \end{vmatrix} = \begin{vmatrix} 0 & 1 & 8 \\ 0 & -7 & 5 \\ 1 & 2 & -3 \end{vmatrix} = 1\begin{vmatrix} 1 & 8 \\ -7 & 5 \end{vmatrix} = 61$$

Now compute:

$$D_x = \begin{vmatrix} 0 & 5 & 2 \\ 0 & -1 & -4 \\ 0 & 2 & 3 \end{vmatrix} = 0$$

$$D_y = \begin{vmatrix} 2 & 0 & 2 \\ 3 & 0 & -4 \\ 1 & 0 & -3 \end{vmatrix} = 0$$

$$D_z = \begin{vmatrix} 2 & 5 & 0 \\ 3 & -1 & 0 \\ 1 & 2 & 0 \end{vmatrix} = 0$$

So $x = \frac{D_x}{D} = 0$, $y = \frac{D_y}{D} = 0$, and $z = \frac{D_z}{D} = 0$. So the solution is $(0, 0, 0)$.

39. Factoring 6 out of column 1 and 2 out of column 2:

$$D = \begin{vmatrix} 12 & 0 & -11 \\ 6 & 6 & -4 \\ 6 & 2 & -5 \end{vmatrix} = 6\begin{vmatrix} 2 & 0 & -11 \\ 1 & 6 & -4 \\ 1 & 2 & -5 \end{vmatrix} = 12\begin{vmatrix} 2 & 0 & -11 \\ 1 & 3 & -4 \\ 1 & 1 & -5 \end{vmatrix}$$

Adding −3 times row 3 to row 2 yields:

$$D = 12\begin{vmatrix} 2 & 0 & -11 \\ -2 & 0 & 11 \\ 1 & 1 & -5 \end{vmatrix} = 12(-1)\begin{vmatrix} 2 & -11 \\ -2 & 11 \end{vmatrix} = -12(0) = 0$$

So Cramer's Rule will not work. We form the augmented matrix (using equation two as row 1):

$$\begin{pmatrix} 6 & 6 & -4 & 26 \\ 12 & 0 & -11 & 13 \\ 6 & 2 & -5 & 13 \end{pmatrix}$$

Adding −2 times row 1 to row 2 and −1 times row 1 to row 3 yields:

$$\begin{pmatrix} 6 & 6 & -4 & 26 \\ 0 & -12 & -3 & -39 \\ 0 & -4 & -1 & -13 \end{pmatrix}$$

Multiply row 1 by $\frac{1}{2}$ and row 2 by $\frac{1}{3}$:

$$\begin{pmatrix} 3 & 3 & -2 & 13 \\ 0 & 4 & 1 & 13 \\ 0 & -4 & -1 & -13 \end{pmatrix}$$

Adding row 2 to row 3 yields:

$$\begin{pmatrix} 3 & 3 & -2 & 13 \\ 0 & 4 & 1 & 13 \\ 0 & 0 & 0 & 0 \end{pmatrix}$$

So we have the system:

$$3x + 3y - 2z = 13$$
$$4y + z = 13$$

Solve equation 2 for z:

$$4y + z = 13$$
$$z = 13 - 4y$$

Substitute into equation 1:

$$3x + 3y - 26 + 8y = 13$$
$$3x = 39 - 11y$$
$$x = 13 - \tfrac{11}{3}y$$

So the solution is $\left(13 - \tfrac{11}{3}y, y, 13 - 4y\right)$, for any real number y.

41. Adding row 1 to row 2, 3 times row 1 to row 3, and row 1 to row 4 yields:

$$D = \begin{vmatrix} 1 & 1 & 1 & 1 \\ 1 & -1 & 1 & -1 \\ 2 & -2 & -3 & -3 \\ 3 & 2 & 1 & -1 \end{vmatrix} = \begin{vmatrix} 1 & 1 & 1 & 1 \\ 2 & 0 & 2 & 0 \\ 5 & 1 & 0 & 0 \\ 4 & 3 & 2 & 0 \end{vmatrix} = -1 \begin{vmatrix} 2 & 0 & 2 \\ 5 & 1 & 0 \\ 4 & 3 & 2 \end{vmatrix}$$

Subtracting column 3 from column 1 yields:

$$D = -1 \begin{vmatrix} 0 & 0 & 2 \\ 5 & 1 & 0 \\ 2 & 3 & 2 \end{vmatrix} = -2 \begin{vmatrix} 5 & 1 \\ 2 & 3 \end{vmatrix} = -2(13) = -26$$

Adding row 1 to row 2, 3 times row 1 to row 3, and row 1 to row 4 yields:

$$D_x = \begin{vmatrix} -7 & 1 & 1 & 1 \\ -11 & -1 & 1 & -1 \\ 26 & -2 & -3 & -3 \\ -9 & 2 & 1 & -1 \end{vmatrix} = \begin{vmatrix} -7 & 1 & 1 & 1 \\ -18 & 0 & 2 & 0 \\ 5 & 1 & 0 & 0 \\ -16 & 3 & 2 & 0 \end{vmatrix} = -1 \begin{vmatrix} -18 & 0 & 2 \\ 5 & 1 & 0 \\ -16 & 3 & 2 \end{vmatrix}$$

Subtracting row 3 from row 1 yields:

$$D_x = -1 \begin{vmatrix} -2 & -3 & 0 \\ 5 & 1 & 0 \\ -16 & 3 & 2 \end{vmatrix} = -2 \begin{vmatrix} -2 & -3 \\ 5 & 1 \end{vmatrix} = -2(13) = -26$$

Adding row 1 to row 2, 3 times row 1 to row 3, and row 1 to row 4 yields:

$$D_y = \begin{vmatrix} 1 & -7 & 1 & 1 \\ 1 & -11 & 1 & -1 \\ 2 & 26 & -3 & -3 \\ 3 & -9 & 1 & -1 \end{vmatrix} = \begin{vmatrix} 1 & -7 & 1 & 1 \\ 2 & -18 & 2 & 0 \\ 5 & 5 & 0 & 0 \\ 4 & -16 & 2 & 0 \end{vmatrix} = -1 \begin{vmatrix} 2 & -18 & 2 \\ 5 & 5 & 0 \\ 4 & -16 & 2 \end{vmatrix}$$

Subtracting row 3 from row 1 yields:

$$D_y = -1 \begin{vmatrix} -2 & -2 & 0 \\ 5 & 5 & 0 \\ 4 & -16 & 2 \end{vmatrix} = -2 \begin{vmatrix} -2 & -2 \\ 5 & 5 \end{vmatrix} = -2(0) = 0$$

Adding row 1 to row 2, 3 times row 1 to row 3, and row 1 to row 4 yields:

$$D_z = \begin{vmatrix} 1 & 1 & -7 & 1 \\ 1 & -1 & -11 & -1 \\ 2 & -2 & 26 & -3 \\ 3 & 2 & -9 & -1 \end{vmatrix} = \begin{vmatrix} 1 & 1 & -7 & 1 \\ 2 & 0 & -18 & 0 \\ 5 & 1 & 5 & 0 \\ 4 & 3 & -16 & 0 \end{vmatrix} = -1 \begin{vmatrix} 2 & 0 & -18 \\ 5 & 1 & 5 \\ 4 & 3 & -16 \end{vmatrix}$$

Adding –3 times row 2 to row 3 yields:

$$D_z = -1 \begin{vmatrix} 2 & 0 & -18 \\ 5 & 1 & 5 \\ 11 & 0 & -31 \end{vmatrix} = -1 \begin{vmatrix} 2 & -18 \\ -11 & -31 \end{vmatrix} = -1(-260) = 260$$

Adding row 1 to row 2, 2 times row 1 to row 3, and –2 times row 1 to row 4 yields:

$$D_w = \begin{vmatrix} 1 & 1 & 1 & -7 \\ 1 & -1 & 1 & -11 \\ 2 & -2 & -3 & 26 \\ 3 & 2 & 1 & -9 \end{vmatrix} = \begin{vmatrix} 1 & 1 & 1 & -7 \\ 2 & 0 & 2 & -18 \\ 4 & 0 & -1 & 12 \\ 1 & 0 & -1 & 5 \end{vmatrix} = -1 \begin{vmatrix} 2 & 2 & -18 \\ 4 & -1 & 12 \\ 1 & -1 & 5 \end{vmatrix} = -2 \begin{vmatrix} 1 & 1 & -9 \\ 4 & -1 & 12 \\ 1 & -1 & 5 \end{vmatrix}$$

Adding row 1 to both row 2 and row 3 yields:

$$D_w = -2 \begin{vmatrix} 1 & 1 & -9 \\ 5 & 0 & 3 \\ 2 & 0 & -4 \end{vmatrix} = 2 \begin{vmatrix} 5 & 3 \\ 2 & -4 \end{vmatrix} = 2(-26) = -52$$

So $x = \frac{D_x}{D} = 1$, $y = \frac{D_y}{D} = 0$, $z = \frac{D_z}{D} = -10$, and $w = \frac{D_w}{D} = 2$. So the solution is $(1, 0, -10, 2)$.

43. Expanding the determinant:

$$\begin{vmatrix} x-4 & 0 & 0 \\ 0 & x+4 & 0 \\ 0 & 0 & x+1 \end{vmatrix} = (x-4) \begin{vmatrix} x+4 & 0 \\ 0 & x+1 \end{vmatrix} = (x-4)(x+4)(x+1)$$

This will equal 0 when $x = 4$, $x = -4$, or $x = -1$.

45. Expanding along the first column:

$$\begin{vmatrix} a & b & c \\ a & b & c \\ d & e & f \end{vmatrix} = a \begin{vmatrix} b & c \\ e & f \end{vmatrix} - a \begin{vmatrix} b & c \\ e & f \end{vmatrix} + d \begin{vmatrix} b & c \\ b & c \end{vmatrix}$$

$$= a(bf - ec) - a(bf - ec) + d(bc - bc)$$
$$= abf - aec - abf + aec + 0$$
$$= 0$$

47. Expanding the determinant:

$$\begin{vmatrix} a_1 + A_1 & b_1 & c_1 \\ a_2 + A_2 & b_2 & c_2 \\ a_3 + A_3 & b_3 & c_3 \end{vmatrix} = (a_1 + A_1)\begin{vmatrix} b_2 & c_2 \\ b_3 & c_3 \end{vmatrix} - (a_2 + A_2)\begin{vmatrix} b_1 & c_1 \\ b_3 & c_3 \end{vmatrix} + (a_3 + A_3)\begin{vmatrix} b_1 & c_1 \\ b_2 & c_2 \end{vmatrix}$$

$$= \left\{ a_1\begin{vmatrix} b_2 & c_2 \\ b_3 & c_3 \end{vmatrix} - a_2\begin{vmatrix} b_1 & c_1 \\ b_3 & c_3 \end{vmatrix} + a_3\begin{vmatrix} b_1 & c_1 \\ b_2 & c_2 \end{vmatrix} \right\}$$

$$+ \left\{ A_1\begin{vmatrix} b_2 & c_2 \\ b_3 & c_3 \end{vmatrix} - A_2\begin{vmatrix} b_1 & c_1 \\ b_3 & c_3 \end{vmatrix} + A_3\begin{vmatrix} b_1 & c_1 \\ b_2 & c_2 \end{vmatrix} \right\}$$

Now observe that the expression in the first set of braces is:

$$\begin{vmatrix} a_1 & b_1 & c_1 \\ a_2 & b_2 & c_2 \\ a_3 & b_3 & c_3 \end{vmatrix}$$

The expression in the second set of braces is:

$$\begin{vmatrix} A_1 & b_1 & c_1 \\ A_2 & b_2 & c_2 \\ A_3 & b_3 & c_3 \end{vmatrix}$$

49. Expanding the determinant on the left-hand side of the given equation along its first row:

$$\begin{vmatrix} a_1 & b_1 & c_1 \\ a_2 & b_2 & c_2 \\ a_3 & b_3 & c_3 \end{vmatrix} = a_1\begin{vmatrix} b_2 & c_2 \\ b_3 & c_3 \end{vmatrix} - b_1\begin{vmatrix} a_2 & c_2 \\ a_3 & c_3 \end{vmatrix} + c_1\begin{vmatrix} a_2 & b_2 \\ a_3 & b_3 \end{vmatrix}$$

Next, expanding the determinant on the right-hand side of the given equation along its second row:

$$\begin{vmatrix} a_2 & b_2 & c_2 \\ a_1 & b_1 & c_1 \\ a_3 & b_3 & c_3 \end{vmatrix} = -a_1\begin{vmatrix} b_2 & c_2 \\ b_3 & c_3 \end{vmatrix} + b_1\begin{vmatrix} a_2 & c_2 \\ a_3 & c_3 \end{vmatrix} - c_1\begin{vmatrix} a_2 & b_2 \\ a_3 & b_3 \end{vmatrix}$$

By inspection now, we observe that the two expressions for the determinants are negatives of one another.

51. Subtract the fourth row from each of the other three rows. Therefore:

$$\begin{vmatrix} a & 0 & 0 & -d \\ 0 & b & 0 & -d \\ 0 & 0 & c & -d \\ 1 & 1 & 1 & 1+d \end{vmatrix} = abcd \begin{vmatrix} 1 & 0 & 0 & -1 \\ 0 & 1 & 0 & -1 \\ 0 & 0 & 1 & -1 \\ \frac{1}{a} & \frac{1}{b} & \frac{1}{c} & 1+\frac{1}{d} \end{vmatrix}$$

$$= abcd \begin{vmatrix} 1 & 0 & 0 & 0 \\ 0 & 1 & 0 & -1 \\ 0 & 0 & 1 & -1 \\ \frac{1}{a} & \frac{1}{b} & \frac{1}{c} & 1+\frac{1}{a}+\frac{1}{d} \end{vmatrix}$$

$$= abcd \begin{vmatrix} 1 & 0 & -1 \\ 0 & 1 & -1 \\ \frac{1}{b} & \frac{1}{c} & 1+\frac{1}{a}+\frac{1}{d} \end{vmatrix}$$

$$= abcd \begin{vmatrix} 1 & 0 & 0 \\ 0 & 1 & -1 \\ \frac{1}{b} & \frac{1}{c} & 1+\frac{1}{a}+\frac{1}{b}+\frac{1}{d} \end{vmatrix}$$

$$= abcd \begin{vmatrix} 1 & -1 \\ \frac{1}{c} & 1+\frac{1}{a}+\frac{1}{b}+\frac{1}{d} \end{vmatrix}$$

$$= abcd \left(1+\frac{1}{a}+\frac{1}{b}+\frac{1}{c}+\frac{1}{d}\right)$$

53. By expanding D along its first column, we obtain the equation:

$$a_1 D = a_1 \left[a_1(b_2 c_3 - b_3 c_2) - a_2(b_1 c_3 - b_3 c_1) + a_3(b_1 c_2 - b_2 c_1) \right] \quad **$$

On the other hand $\begin{vmatrix} B_2 & C_2 \\ B_3 & C_3 \end{vmatrix}$ is equal to:

$$B_2 C_3 - B_3 C_2 = (a_1 c_3 - a_3 c_1)(a_1 b_2 - a_2 b_1) - (a_2 c_1 - a_1 c_2)(a_3 b_1 - a_1 b_3)$$

$$= a_1^2 b_2 c_3 - a_1 a_3 b_2 c_1 - a_1 a_2 b_1 c_3 + a_2 a_3 b_1 c_1 - a_1^2 b_3 c_2$$
$$+ a_1 a_3 b_1 c_2 + a_1 a_2 b_3 c_1 - a_2 a_3 b_1 c_1$$

$$= a_1 \left[a_1 b_2 c_3 - a_3 b_2 c_1 - a_2 b_1 c_3 - a_1 b_3 c_2 + a_3 b_1 c_2 + a_2 b_3 c_1 \right]$$

$$= a_1 \left[a_1(b_2 c_3 - b_3 c_2) - a_2(b_1 c_3 - b_3 c_1) + a_3(b_1 c_2 - b_2 c_1) \right]$$

By inspection now, we see that this last expression agrees with the right-hand side of equation **. This proves that:

$$\begin{vmatrix} B_2 & C_2 \\ B_3 & C_3 \end{vmatrix} = a_1 D$$

55. (a) Expanding along the first row:

$$\begin{vmatrix} 1 & 0 & 0 \\ x & 1 & 0 \\ x & y & 1 \end{vmatrix} = 1 \begin{vmatrix} 1 & 0 \\ y & 1 \end{vmatrix} = 1(1) = 1$$

(b) Expanding along the first row:

$$\begin{vmatrix} 1 & 0 & 0 & 0 \\ x & 1 & 0 & 0 \\ x & y & 1 & 0 \\ x & y & z & 1 \end{vmatrix} = 1 \begin{vmatrix} 1 & 0 & 0 \\ y & 1 & 0 \\ y & z & 1 \end{vmatrix} = 1 \begin{vmatrix} 1 & 0 \\ z & 1 \end{vmatrix} = 1(1) = 1$$

57. Subtracting row 1 from each of row 2, row 3, and row 4 yields:

$$\begin{vmatrix} 1 & a & a & a \\ 1 & b & a & a \\ 1 & a & b & a \\ 1 & a & a & b \end{vmatrix} = \begin{vmatrix} 1 & a & a & a \\ 0 & b-a & 0 & 0 \\ 0 & 0 & b-a & 0 \\ 0 & 0 & 0 & b-a \end{vmatrix}$$

$$= 1 \begin{vmatrix} b-a & 0 & 0 \\ 0 & b-a & 0 \\ 0 & 0 & b-a \end{vmatrix}$$

$$= (b-a) \begin{vmatrix} b-a & 0 \\ 0 & b-a \end{vmatrix}$$

$$= (b-a)(b-a)^2$$

$$= (b-a)^3$$

59. Form the augmented matrix:

$$\begin{pmatrix} a & b & c & k \\ a^2 & b^2 & c^2 & k^2 \\ a^3 & b^3 & c^3 & k^3 \end{pmatrix}$$

Adding $-a$ times row 1 to row 2 and $-a^2$ times row 1 to row 3:

$$\begin{pmatrix} a & b & c & k \\ 0 & b^2-ab & c^2-ac & k^2-ak \\ 0 & b^3-a^2b & c^3-a^2c & k^3-a^2k \end{pmatrix}$$

Factoring:

$$\begin{pmatrix} a & b & c & k \\ 0 & b(b-a) & c(c-a) & k(k-a) \\ 0 & b(b+a)(b-a) & c(c+a)(c-a) & k(k+a)(k-a) \end{pmatrix}$$

Adding $-(b+a)$ times row 2 to row 3:

$$\begin{pmatrix} a & b & c & k \\ 0 & b(b-a) & c(c-a) & k(k-a) \\ 0 & 0 & c(c-a)(c-b) & k(k-a)(k-b) \end{pmatrix}$$

So we have the system:

$$ax + by + cz = k$$
$$b(b-a)y + c(c-a)z = k(k-a)$$
$$c(c-a)(c-b)z = k(k-a)(k-b)$$

Solving the third equation for z:

$$c(c-a)(c-b)z = k(k-a)(k-b)$$
$$z = \frac{k(k-a)(k-b)}{c(c-a)(c-b)}$$

Substitute into the second equation:

$$b(b-a)y + \frac{k(k-a)(k-b)}{c-b} = k(k-a)$$
$$b(b-a)(c-b)y = k(k-a)(c-b-k+b)$$
$$y = \frac{k(k-a)(k-c)}{b(b-a)(b-c)}$$

Substitute into the first equation:

$$ax + \frac{k(k-a)(k-c)}{(b-a)(b-c)} + \frac{k(k-a)(k-b)}{(c-a)(c-b)} = k$$
$$ax = \frac{k(k-b)(k-c)}{(b-a)(c-a)}$$
$$x = \frac{k(k-b)(k-c)}{a(a-b)(a-c)}$$

So the solution is $\left(\dfrac{k(k-b)(k-c)}{a(a-b)(a-c)}, \dfrac{k(k-a)(k-c)}{b(b-a)(b-c)}, \dfrac{k(k-a)(k-b)}{c(c-a)(c-b)} \right)$.

61. Evaluate the determinant by expanding along the first row:

$$\begin{vmatrix} x & y & 1 \\ -3 & -1 & 1 \\ 2 & 9 & 1 \end{vmatrix} = x(-1-9) - y(-3-2) + 1(-27+2) = -10x + 5y - 25$$

Setting this equal to 0:

$$-10x + 5y - 25 = 0$$
$$5y = 10x + 25$$
$$y = 2x + 5$$

63.   Re-draw the figure:

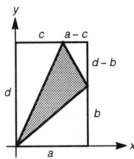

The area of the rectangle is ad, and the three triangles have areas of $\frac{1}{2}(ab)$, $\frac{1}{2}(a-c)(d-b)$, and $\frac{1}{2}(cd)$, so the area of the shaded triangle is:

$$ad - \left[\tfrac{1}{2}ab + \tfrac{1}{2}ad - \tfrac{1}{2}cd - \tfrac{1}{2}ab + \tfrac{1}{2}bc + \tfrac{1}{2}cd\right] = \tfrac{1}{2}ad - \tfrac{1}{2}bc = \tfrac{1}{2}(ad - bc) = \tfrac{1}{2}\begin{vmatrix} a & b \\ c & d \end{vmatrix}$$

65.   **(a)**   From Exercise 36 of the previous section, we found $A^{-1} = \begin{pmatrix} \frac{d}{ad-bc} & \frac{-b}{ad-bc} \\ \frac{-c}{ad-bc} & \frac{a}{ad-bc} \end{pmatrix}$. To

check this, note that A times this matrix equals I_2. Since $D = ad - bc$,

$$A^{-1} = \tfrac{1}{D}\begin{pmatrix} d & -b \\ -c & a \end{pmatrix}.$$

(b)   We have $D = -6(9) - 7(1) = -54 - 7 = -61$, so the inverse is:

$$-\tfrac{1}{61}\begin{pmatrix} 9 & -7 \\ -1 & -6 \end{pmatrix} = \begin{pmatrix} -\frac{9}{61} & \frac{7}{61} \\ \frac{1}{61} & \frac{6}{61} \end{pmatrix}$$

Graphing Utility Exercises for Section 11.5

1.   Entering $A = \begin{pmatrix} 1 & 2 & 3 \\ 4 & 5 & 6 \\ 7 & 8 & 9 \end{pmatrix}$, we find det $A = 0$. This checks the value in the text.

3.   Entering $A = \begin{pmatrix} 25 & 40 & 5 & 10 \\ 9 & 0 & 3 & 6 \\ -2 & 3 & 11 & -17 \\ -3 & 4 & 7 & 2 \end{pmatrix}$, we find det $A = -65520$. This checks the value in the text.

5. (a) The right-hand determinant should be 10 times the left-hand determinant.

(b) Entering $A = \begin{pmatrix} 1 & 2 & 3 \\ -7 & -4 & 5 \\ 9 & 2 & 6 \end{pmatrix}$ and $B = \begin{pmatrix} 10 & 20 & 30 \\ -7 & -4 & 5 \\ 9 & 2 & 6 \end{pmatrix}$, we find det $A = 206$ and

det $B = 2060$. Thus det $B = 10 \times$ det A.

7. (a) Entering each of the matrices and computing determinants, we have $D = -219$, $D_x = -287$, $D_y = 124$, and $D_z = -294$.

(b) Using Cramer's rule:

$$x = \frac{D_x}{D} = \frac{-287}{-219} \approx 1.31$$

$$y = \frac{D_y}{D} = \frac{124}{-219} \approx -0.57$$

$$z = \frac{D_z}{D} = \frac{-294}{-219} \approx 1.34$$

The solution is $(1.31, -0.57, 1.34)$.

11.6 Nonlinear Systems of Equations

1. Substituting:
$$x^2 = 3x$$
$$x^2 - 3x = 0$$
$$x(x - 3) = 0$$
From which we get $x = 0$ or $x = 3$. If $x = 0$, $y = 3 \cdot 0 = 0$ and if $x = 3$, $y = 9$. Our solutions are $(0, 0)$ and $(3, 9)$. Note that we have found the points of intersection of a line and a parabola.

3. Since $x^2 = 24y$, $x^2 + y^2 = 25$ becomes:
$$24y + y^2 = 25$$
$$y^2 + 24y - 25 = 0$$
$$(y + 25)(y - 1) = 0$$
$$y = -25 \text{ and } y = 1$$
When $y = 1$ we get $x^2 = 24$ or $x = \pm 2\sqrt{6}$, but $y = -25$ means $x^2 = -600$ which is impossible. Our two solutions are $\left(2\sqrt{6}, 1\right)$ and $\left(-2\sqrt{6}, 1\right)$.

5. Substitute into the first equation:
$$x\left(-x^2\right) = 1$$
$$-x^3 = 1$$
$$x = -1$$
Since $y = \frac{1}{x}$, $y = -1$. So the only solution is $(-1, -1)$.

7. Multiply the first equation by –2:
$$-4x^2 - 2y^2 = -34$$
$$x^2 + 2y^2 = 22$$
Adding:
$$-3x^2 = -12$$
$$x^2 = 4$$
$$x = \pm 2$$
When $x = 2$:
$$2(4) + y^2 = 17$$
$$y^2 = 9$$
$$y = \pm 3$$
When $x = -2$:
$$2(4) + y^2 = 17$$
$$y^2 = 9$$
$$y = \pm 3$$
So the solutions are $(2, 3)$, $(2, -3)$, $(-2, 3)$, and $(-2, -3)$.

9. Substitute into the first equation:
$$x^2 - 1 = 1 - x^2$$
$$2x^2 = 2$$
$$x^2 = 1$$
$$x = \pm 1$$
So $y = 1 - 1 = 0$ for each value of x. So the solutions are $(1, 0)$ and $(-1, 0)$.

11. Substitute into the first equation:
$$x(4x + 1) = 4$$
$$4x^2 + x - 4 = 0$$
$$x = \frac{-1 \pm \sqrt{1 + 64}}{8} = \frac{-1 \pm \sqrt{65}}{8}$$
When $x = \frac{-1+\sqrt{65}}{8}$:
$$y = \frac{-1 + \sqrt{65}}{2} + 1 = \frac{1 + \sqrt{65}}{2}$$
When $x = \frac{-1-\sqrt{65}}{8}$:
$$y = \frac{-1 - \sqrt{65}}{2} + 1 = \frac{1 - \sqrt{65}}{2}$$
So the solutions are $\left(\frac{-1+\sqrt{65}}{8}, \frac{1+\sqrt{65}}{2}\right)$ and $\left(\frac{-1-\sqrt{65}}{8}, \frac{1-\sqrt{65}}{2}\right)$.

13. Let $a = \frac{1}{x^2}$ and $b = \frac{1}{y^2}$, so:

$$a - 3b = 14$$
$$2a + b = 35$$

Multiply the first equation by -2:

$$-2a + 6b = -28$$
$$2a + b = 35$$

Adding:

$$7b = 7$$
$$b = 1$$

So $a - 3 = 14$, and $a = 17$. Since $a = \frac{1}{x^2}$ and $b = \frac{1}{y^2}$, we have $x^2 = \frac{1}{17}$ and $y^2 = 1$.

So $x = \frac{\pm\sqrt{17}}{17}$ and $y = \pm 1$. So the solutions are $\left(\frac{\sqrt{17}}{17}, 1\right), \left(\frac{\sqrt{17}}{17}, -1\right), \left(\frac{-\sqrt{17}}{17}, 1\right)$, and $\left(\frac{-\sqrt{17}}{17}, -1\right)$.

15. Substitute into the second equation:

$$(x-3)^2 + \left(-\sqrt{x-1}\right)^2 = 4$$
$$x^2 - 6x + 9 + x - 1 = 4$$
$$x^2 - 5x + 4 = 0$$
$$(x-1)(x-4) = 0$$
$$x = 1 \text{ or } x = 4$$

When $x = 1$, $y = -\sqrt{1-1} = 0$ and when $x = 4$, $y = -\sqrt{4-1} = -\sqrt{3}$. So the solutions are $(1, 0)$ and $\left(4, -\sqrt{3}\right)$.

17. Since $y = 2^{2x} - 12 = \left(2^x\right)^2 - 12$, substitute into the second equation:

$$y = y^2 - 12$$
$$0 = y^2 - y - 12$$
$$0 = (y-4)(y+3)$$
$$y = 4 \text{ or } y = -3$$

When $y = 4$, we have $2^x = 4$, or $x = 2$. But $y = -3$ will not yield a solution. So the only solution is $(2, 4)$.

19. Let $u = \log_{10} x$ and $v = \log_{10} y$, so:

$$2u^2 - v^2 = -1$$
$$4u^2 - 3v^2 = -11$$

Multiply the first equation by -2:

$$-4u^2 + 2v^2 = 2$$
$$4u^2 - 3v^2 = -11$$

Adding:

$$-v^2 = -9$$
$$v^2 = 9$$
$$v = \pm 3$$

Substitute into the first equation:
$$2u^2 - 9 = -1$$
$$2u^2 = 8$$
$$u^2 = 4$$
$$u = \pm 2$$

Since $u = \log_{10} x$, $x = 10^{\pm 2}$. Similarly, $y = 10^{\pm 3}$. So the solutions are $(100, 1000)$, $\left(100, \frac{1}{1000}\right)$, $\left(\frac{1}{100}, 1000\right)$, and $\left(\frac{1}{100}, \frac{1}{1000}\right)$.

21. First take the logarithm of each side of the first equation:
$$\ln(2^x 3^y) = \ln 4$$
$$\ln(2^x) + \ln(3^y) = \ln 2^2$$
$$(\ln 2)x + (\ln 3)y = 2\ln 2$$

Multiply the second equation by $-\ln 2$:
$$(\ln 2)x + (\ln 3)y = 2\ln 2$$
$$(-\ln 2)x - (\ln 2)y = -5\ln 2$$

Adding:
$$(\ln 3 - \ln 2)y = -3\ln 2$$
$$y = \frac{3\ln 2}{\ln 2 - \ln 3}$$

Substitute into the second equation:
$$x = 5 - \frac{3\ln 2}{\ln 2 - \ln 3} = \frac{2\ln 2 - 5\ln 3}{\ln 2 - \ln 3}$$

The solution is $\left(\dfrac{2\ln 2 - 5\ln 3}{\ln 2 - \ln 3}, \dfrac{3\ln 2}{\ln 2 - \ln 3}\right)$.

23. Substituting x yields:
$$y = 3x + 1 = \frac{-3 + 3\sqrt{13}}{6} + 1 = \frac{3 + 3\sqrt{13}}{6} = \frac{1 + \sqrt{13}}{2}$$
$$y = \frac{1}{x} = \frac{6}{-1 + \sqrt{13}} \cdot \frac{-1 - \sqrt{13}}{-1 - \sqrt{13}} = \frac{-6(1 + \sqrt{13})}{1 - 13} = \frac{-6(1 + \sqrt{13})}{-12} = \frac{1 + \sqrt{13}}{2}$$

They both yield the same y-value.

25. Since $ax + by = 2$, $by = 2 - ax$. Substitute into the second equation:
$$ax(by) = 1$$
$$ax(2 - ax) = 1$$
$$2ax - a^2 x^2 = 1$$
$$a^2 x^2 - 2ax + 1 = 0$$
$$(ax - 1)^2 = 0$$
$$ax = 1$$
$$x = \frac{1}{a}$$

When $ax = 1$, $by = 2 - 1 = 1$, so $y = \frac{1}{b}$. So the solution is $\left(\frac{1}{a}, \frac{1}{b}\right)$.

27. Solve the second equation for y to get $y = 23 - x$. Substitute into the first equation:
$$x^3 + (23 - x)^3 = 3473$$
$$x^3 + 12167 - 1587x + 69x^2 - x^3 = 3473$$
$$69x^2 - 1587x + 8694 = 0$$
$$x^2 - 23x + 126 = 0$$
$$(x - 9)(x - 14) = 0$$
$$x = 9 \text{ or } x = 14$$
When $x = 9$, $y = 14$ and when $x = 14$, $y = 9$. So the solutions are $(9, 14)$ and $(14, 9)$.

29. First draw the rectangle:

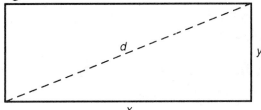

Now $x^2 + y^2 = d^2$ and $2x + 2y = 2p$. Solve the second equation for y:
$$2x + 2y = 2p$$
$$2y = 2p - 2x$$
$$y = p - x$$
Substitute into the first equation:
$$x^2 + (p - x)^2 = d^2$$
$$x^2 + p^2 - 2px + x^2 = d^2$$
$$2x^2 - 2px + p^2 - d^2 = 0$$
Using the quadratic formula:
$$x = \frac{2p \pm \sqrt{4p^2 - 8(p^2 - d^2)}}{4} = \frac{2p \pm 2\sqrt{2d^2 - p^2}}{4} = \frac{p \pm \sqrt{2d^2 - p^2}}{2}$$
When $x = \dfrac{p + \sqrt{2d^2 - p^2}}{2}$:
$$y = p - \frac{p + \sqrt{2d^2 - p^2}}{2} = \frac{p - \sqrt{2d^2 - p^2}}{2}$$
When $x = \dfrac{p - \sqrt{2d^2 - p^2}}{2}$:
$$y = p - \frac{p - \sqrt{2d^2 - p^2}}{2} = \frac{p + \sqrt{2d^2 - p^2}}{2}$$
So the rectangle has dimensions $\dfrac{p - \sqrt{2d^2 - p^2}}{2}$ by $\dfrac{p - \sqrt{2d^2 - p^2}}{2}$.

31. If we follow the hint:
$$\left(\sqrt{u+v} + \sqrt{u-v}\right)^2 = 4^2$$
$$u + v + 2\sqrt{u^2 - v^2} + u - v = 16$$
But since $u^2 - v^2 = 9$ from the first equation:
$$2u + 2\sqrt{9} = 16$$
$$2u + 6 = 16$$
$$2u = 10, \text{ so } u = 5$$
When $u = 5$:
$$5^2 - v^2 = 9$$
$$v^2 = 16$$
$$v = \pm 4$$
So the solutions are $(5, 4)$ and $(5, -4)$.

33. (a) Using the point $(2, 3)$, we have $3 = N_0 e^{2k}$. Using the point $(8, 24)$, we have $24 = N_0 e^{8k}$. Dividing these two yields:
$$\frac{24}{3} = \frac{N_0 e^{8k}}{N_0 e^{2k}}$$
$$8 = e^{6k}$$
$$\ln 8 = 6k$$
$$k = \frac{\ln 8}{6}$$
Substituting into the first equation:
$$3 = N_0 e^{\frac{1}{3}\ln 8}$$
$$3 = N_0 e^{\ln(8^{1/3})}$$
$$3 = N_0 \cdot 2$$
$$N_0 = \frac{3}{2}$$

(b) Using the point $\left(\frac{1}{2}, 1\right)$, we have $1 = N_0 e^{\frac{1}{2}k}$. Using the point $(4, 10)$, we have $10 = N_0 e^{4k}$. Dividing these two yields:
$$\frac{10}{1} = \frac{N_0 e^{4k}}{N_0 e^{\frac{1}{2}k}}$$
$$10 = e^{\frac{7}{2}k}$$
$$\ln 10 = \frac{7}{2}k$$
$$k = \frac{2}{7}\ln 10$$

Substituting into the first equation:

$$1 = N_0 e^{\frac{1}{7}\ln 10}$$
$$1 = N_0 e^{\ln(10^{1/7})}$$
$$1 = N_0 \cdot 10^{1/7}$$
$$N_0 = 10^{-1/7}$$

35. Let $w = x + y + z$, so:

$$xw = p^2$$
$$yw = q^2$$
$$zw = r^2$$

Adding the three equations:

$$(x + y + z)w = p^2 + q^2 + r^2$$
$$w^2 = p^2 + q^2 + r^2$$
$$w = \pm\sqrt{p^2 + q^2 + r^2}$$

Substitute into the first equation:

$$x = \frac{p^2}{\pm\sqrt{p^2 + q^2 + r^2}}$$

Similarly $y = \frac{q^2}{\pm A}$ and $z = \frac{r^2}{\pm A}$ where $A = \sqrt{p^2 + q^2 + r^2}$. So the solutions are $\left(\frac{p^2}{A}, \frac{q^2}{A}, \frac{r^2}{A}\right)$ and $\left(\frac{-p^2}{A}, \frac{-q^2}{A}, \frac{-r^2}{A}\right)$, where $A = \sqrt{p^2 + q^2 + r^2}$.

37. The area is given by $A = \frac{1}{2}bh$, where b and h are the missing legs. Therefore:

$$180 = \frac{1}{2}bh, \text{ so } bh = 360 \text{ thus } b = \frac{360}{h}$$

Also from the Pythagorean theorem $b^2 + h^2 = 41^2$, so:

$$\left(\frac{360}{h}\right)^2 + h^2 = 41^2$$
$$\frac{129600}{h^2} + h^2 = 1681$$
$$129600 + h^4 = 1681h^2$$
$$h^4 - 1681h^2 + 129600 = 0$$
$$\left(h^2 - 1600\right)\left(h^2 - 81\right) = 0$$
$$h^2 = 1600 \text{ or } h^2 = 81$$
$$h = \pm 40 \text{ or } h = \pm 9$$

Since these must be the length of the sides, we can neglect the negative values. So $h = 40$ and $h = 9$ are solutions. Since $b = \frac{360}{h}$ we obtain $b = 9$ when $h = 40$, and $b = 40$ when $h = 9$. So the legs are 9 cm and 40 cm.

39.   We have $lw = 60$ and $2l + 2w = 46$, so $l + w = 23$ and $l = 23 - w$. Substitute this into the first equation:

$$(23 - w)(w) = 60$$
$$23w - w^2 = 60$$
$$w^2 - 23w + 60 = 0$$
$$(w - 20)(w - 3) = 0$$
$$w = 20 \ \text{ or } \ w = 3$$

When $w = 20$ we have $l = 3$, and when $w = 3$ we have $l = 20$. So the rectangle must be 3 cm by 20 cm.

41.   Solve $xy = 2$ to get $y = \frac{2}{x}$. Now substitute into the first equation:

$$x^2 + \frac{4}{x^2} = 5$$
$$x^4 + 4 = 5x^2$$
$$x^4 - 5x^2 + 4 = 0$$
$$\left(x^2 - 1\right)\left(x^2 - 4\right) = 0$$
$$x^2 = 1 \ \text{ or } \ x^2 = 4$$
$$x = \pm 1 \qquad x = \pm 2$$

When $x = 1$ we have $y = 2$, when $x = -1$ we have $y = -2$, when $x = 2$ we have $y = 1$, and when $x = -2$ we have $y = -1$. So the solutions are $(1, 2)$, $(-1, -2)$, $(2, 1)$, and $(-2, -1)$.

43.   Multiply the second equation by 2:

$$2xy = 6$$

Adding to the first equation:

$$x^2 + 2xy + y^2 = 13$$
$$(x + y)^2 = 13$$
$$x + y = \pm\sqrt{13}$$

Subtracting from the first equation:

$$x^2 - 2xy + y^2 = 1$$
$$(x - y)^2 = 1$$
$$x - y = \pm 1$$

Solve the four systems of equations:

$$
\begin{array}{llll}
x + y = \sqrt{13} & x + y = \sqrt{13} & x + y = -\sqrt{13} & x + y = -\sqrt{13} \\
x - y = 1 & x - y = -1 & x - y = 1 & x - y = -1 \\
\hline
2x = 1 + \sqrt{13} & 2x = -1 + \sqrt{13} & 2x = 1 - \sqrt{13} & 2x = -1 - \sqrt{13} \\
x = \dfrac{1 + \sqrt{13}}{2} & x = \dfrac{-1 + \sqrt{13}}{2} & x = \dfrac{1 - \sqrt{13}}{2} & x = \dfrac{-1 - \sqrt{13}}{2} \\
y = \dfrac{-1 + \sqrt{13}}{2} & y = \dfrac{1 + \sqrt{13}}{2} & y = \dfrac{-1 - \sqrt{13}}{2} & y = \dfrac{1 - \sqrt{13}}{2}
\end{array}
$$

So the solutions are $\left(\frac{1+\sqrt{13}}{2}, \frac{-1+\sqrt{13}}{2}\right)$, $\left(\frac{-1+\sqrt{13}}{2}, \frac{1+\sqrt{13}}{2}\right)$, $\left(\frac{1-\sqrt{13}}{2}, \frac{-1-\sqrt{13}}{2}\right)$, and $\left(\frac{-1-\sqrt{13}}{2}, \frac{1-\sqrt{13}}{2}\right)$.

45. Solving by factoring:
$$2m^2 - 7m + 6 = 0$$
$$(2m - 3)(m - 2) = 0$$
$$m = \tfrac{3}{2} \text{ or } m = 2$$

So we have the equations:

$$x^2\left(\tfrac{9}{2} - 4\right) = 2 \qquad \text{or} \qquad x^2(6 - 4) = 2$$
$$\begin{array}{ll} x^2 = 4 & \qquad x^2 = 1 \\ x = \pm 2 & \qquad x = \pm 1 \\ y = \pm 3 & \qquad y = \pm 2 \end{array}$$

So the solutions are (2, 3), (–2, –3), (1, 2), and (–1, -2).

47. Using the hint, square the first equation:
$$x^2 y^2 + 2pqxy + p^2 q^2 = 4p^2 x^2$$
$$x^2 y^2 \qquad\quad + p^2 q^2 = 2q^2 y^2$$

Subtracting the second equation from the first:
$$2pqxy = 4p^2 x^2 - 2q^2 y^2$$
$$pqxy = 2p^2 x^2 - q^2 y^2$$
$$0 = 2p^2 x^2 - pqxy - q^2 y^2$$
$$0 = (2px + qy)(px - qy)$$

So we have the equations:
$$2px + qy = 0, \text{ so } y = -\tfrac{2p}{q}x$$
$$px - qy = 0, \text{ so } y = \tfrac{p}{q}x$$

Substituting into the first equation:
$$x\left(-\tfrac{2p}{q}x\right) + pq = 2px$$
$$x\left(\tfrac{p}{q}x\right) + pq = 2px$$

Multiplying each side by $\tfrac{q}{p}$:
$$-2x^2 - 2qx + q^2 = 0$$
$$x^2 - 2qx + q^2 = 0$$

The second equation factors to $(x - q)^2 = 0$, so $x = q$ and thus $y = \tfrac{p}{q} \cdot q = p$. Thus one solution is (q, p). This solution checks in the original (non-squared) first equation. For the first equation, use the quadratic formula:
$$x = \frac{2q \pm \sqrt{4q^2 + 8q^2}}{-4} - \frac{2q \pm 2q\sqrt{3}}{-4} = \frac{-1 \pm \sqrt{3}}{2}q$$
$$y = -\frac{2p}{q}\left(\frac{-1 \pm \sqrt{3}}{2}\right)q = (1 \pm \sqrt{3})p$$

This results in two other solutions, $\left(\tfrac{-1+\sqrt{3}}{2}q, (1 - \sqrt{3})p\right)$ and $\left(\tfrac{-1-\sqrt{3}}{2}q, (1 + \sqrt{3})p\right)$. Both of these solutions check in the original (non-squared) first equation.

49.   Taking logs in the first equation:

$$\ln(x^4) = \ln(y^6)$$
$$4\ln x = 6\ln y$$
$$2\ln x = 3\ln y$$

The second equation is:

$$\ln x - \ln y = \frac{\ln x}{\ln y}$$

$$\ln x \ln y - (\ln y)^2 = \ln x$$

Let $u = \ln x$ and $v = \ln y$, so we have the equations $2u = 3v$ and $uv - v^2 = u$. Solving the first equation for u yields $u = \frac{3v}{2}$, and substituting into the second equation yields:

$$\left(\frac{3v}{2}\right)v - v^2 = \frac{3v}{2}$$
$$3v^2 - 2v^2 = 3v$$
$$v^2 - 3v = 0$$
$$v(v-3) = 0$$
$$v = 0 \text{ or } v = 3$$

When $v = 0$, $u = 0$ and when $v = 3$, $u = \frac{9}{2}$. Since $v = \ln y$, $v = 0$ cannot be a solution to the original second equation ($\ln y$ is the denominator). Thus $u = \ln x$ and $v = \ln y$ yields:

$$\ln x = \frac{9}{2} \text{ so } x = e^{9/2}$$
$$\ln y = 3 \text{ so } y = e^3$$

So the only solution is $\left(e^{9/2}, e^3\right)$.

Graphing Utility Exercises for Section 11.6

1.   **(a)** Graphing the two curves:

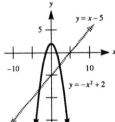

The intersection points are (2.193, –2.807) and (–3.193, –8.193).

(b) Substituting $y = x - 5$:

$$x - 5 = -x^2 + 2$$
$$x^2 + x - 7 = 0$$

Using the quadratic formula:

$$x = \frac{-1 \pm \sqrt{(1)^2 - 4(1)(-7)}}{2(1)} = \frac{-1 \pm \sqrt{1+28}}{2} = \frac{-1 \pm \sqrt{29}}{2}$$

If $x = \dfrac{-1+\sqrt{29}}{2}$, $y = \dfrac{-1+\sqrt{29}}{2} - 5 = \dfrac{-11+\sqrt{29}}{2}$. If $x = \dfrac{-1-\sqrt{29}}{2}$,

$y = \dfrac{-1-\sqrt{29}}{2} - 5 = \dfrac{-11-\sqrt{29}}{2}$. The intersection points are

$\left(\dfrac{-1+\sqrt{29}}{2}, \dfrac{-11+\sqrt{29}}{2} \right) \approx (2.1926, -2.8074)$ and

$\left(\dfrac{-1-\sqrt{29}}{2}, \dfrac{-11-\sqrt{29}}{2} \right) \approx (-3.1926, -8.1926)$.

3.   (a)   Graphing the two curves:

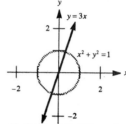

The intersection points are $(0.316, 0.949)$ and $(-0.316, -0.949)$.

(b)   Substituting $y = 3x$:

$$x^2 + (3x)^2 = 1$$
$$x^2 + 9x^2 = 1$$
$$10x^2 = 1$$
$$x^2 = \tfrac{1}{10}$$
$$x = \pm\sqrt{\tfrac{1}{10}} = \pm\dfrac{\sqrt{10}}{10}$$

If $x = \dfrac{\sqrt{10}}{10}$, $y = \dfrac{3\sqrt{10}}{10}$. If $x = -\dfrac{\sqrt{10}}{10}$, $y = -\dfrac{3\sqrt{10}}{10}$. The intersection points are

$\left(\dfrac{\sqrt{10}}{10}, \dfrac{3\sqrt{10}}{10} \right) \approx (0.3162, 0.9487)$ and $\left(-\dfrac{\sqrt{10}}{10}, -\dfrac{3\sqrt{10}}{10} \right) \approx (-0.3162, -0.9487)$.

5.   (a)   Graphing the two curves:

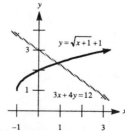

The intersection point is $(0.852, 2.361)$.

(b)   Substituting $y = \sqrt{x+1} + 1$:

$$3x + 4\left(\sqrt{x+1} + 1\right) = 12$$
$$3x + 4\sqrt{x+1} + 4 = 12$$
$$4\sqrt{x+1} = 8 - 3x$$
$$16(x+1) = 64 - 48x + 9x^2$$
$$16x + 16 = 64 - 48x + 9x^2$$
$$0 = 9x^2 - 64x + 48$$

Using the quadratic formula:

$$x = \frac{-(-64) \pm \sqrt{(-64)^2 - 4(9)(48)}}{2(9)}$$
$$= \frac{64 \pm \sqrt{4096 - 1728}}{18}$$
$$= \frac{64 \pm 8\sqrt{37}}{18}$$
$$= \frac{32 \pm 4\sqrt{37}}{9}$$

Since $3x + 4y = 12$, $y = \dfrac{12 - 3x}{4}$. Substituting:

$$x = \frac{32 - 4\sqrt{37}}{9}: \quad y = \tfrac{1}{4}\left(12 - \frac{96 - 12\sqrt{37}}{9}\right) = \frac{12 + 12\sqrt{37}}{36} = \frac{1 + \sqrt{37}}{3}$$

$$x = \frac{32 + 4\sqrt{37}}{9}: \quad y = \tfrac{1}{4}\left(12 - \frac{96 + 12\sqrt{37}}{9}\right) = \frac{12 - 12\sqrt{37}}{36} = \frac{1 - \sqrt{37}}{3}$$

Since $y \geq 0$, this second point is discarded. The intersection point is

$$\left(\frac{32 - 4\sqrt{37}}{9}, \frac{1 + \sqrt{37}}{3}\right) \approx (0.8521, 2.3609).$$

7.   (a)   Graphing the two curves:

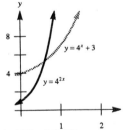

The intersection point is $(0.602, 5.303)$.

(b)   Setting the two equations equal:

$$4^{2x} = 4^x + 3$$

Let $u = 4^x$. Then the equation becomes:

$$u^2 = u + 3$$

$$u^2 - u - 3 = 0$$

Using the quadratic formula:

$$u = \frac{-(-1) \pm \sqrt{(-1)^2 - 4(1)(-3)}}{2(1)} = \frac{1 \pm \sqrt{1+12}}{2} = \frac{1 \pm \sqrt{13}}{2}$$

Since $u > 0$, choose $u = \dfrac{1 + \sqrt{13}}{2}$, thus:

$$4^x = \frac{1 + \sqrt{13}}{2}$$

$$x \ln 4 = \ln\left(\frac{1 + \sqrt{13}}{2}\right)$$

$$x = \frac{\ln\left(\dfrac{1 + \sqrt{13}}{2}\right)}{\ln 4}$$

Then $y = 4^x + 3 = \dfrac{1 + \sqrt{13}}{2} + 3 = \dfrac{7 + \sqrt{13}}{2}$. The solution is

$$\left(\frac{\ln\left(\dfrac{1 + \sqrt{13}}{2}\right)}{\ln 4}, \frac{7 + \sqrt{13}}{2}\right) \approx (0.6017, 5.3028).$$

9.   Sketch the graphs:

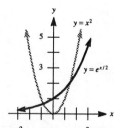

The intersection points are approximately (–0.816, 0.665) and (1.430, 2.044).

11.   Sketch the graphs:

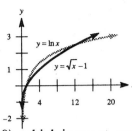

The intersection points are $(1, 0)$, which is exact, and approximately $(12.340, 2.513)$.

13.   Sketch the graphs:

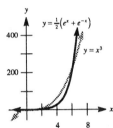

The intersection points are approximately $(1.229, 1.855)$ and $(6.135, 230.949)$.

11.7 Systems of Inequalities. Linear Programming

1.   (a)   Substitute the pair $(1, 2)$:
$$4(1) - 6(2) + 3 \geq 0$$
$$-5 \geq 0$$
Then $(1, 2)$ is not a solution.

   (b)   Substitute the pair $\left(0, \tfrac{1}{2}\right)$:
$$4(0) - 6\left(\tfrac{1}{2}\right) + 3 \geq 0$$
$$0 \geq 0$$
Then $\left(0, \tfrac{1}{2}\right)$ is a solution.

3.   Graph the region:

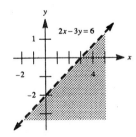

5. Graph the region:

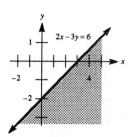

7. Graph the region:

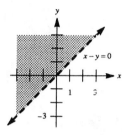

9. Graph the region:

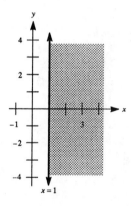

11. Graph the region:

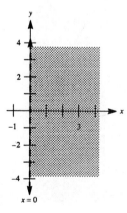

13.   Graph the region:

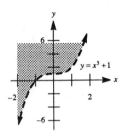

15.   Graph the region:

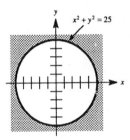

17.   Graph the system of inequalities:

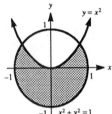

19.   Graph the system of inequalities:

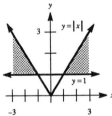

21.   Graph the system of inequalities:

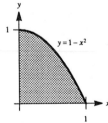

23. The region is convex and bounded. The vertices are $(0, 0)$, $(7, 0)$, $(3, 8)$, and $(0, 5)$. Graph the system of inequalities:

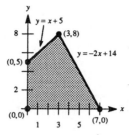

25. The region is convex and bounded. The vertices are $(0, 0)$, $(0, 4)$, $(3, 5)$, and $(8, 0)$. Graph the system of inequalities:

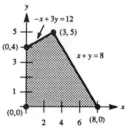

27. The region is convex but not bounded. The vertices are $(2, 7)$ and $(8, 5)$. Graph the system of inequalities:

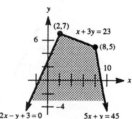

29. The region is convex but not bounded. The only vertex is $(6, 0)$. Graph the system of inequalities:

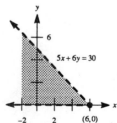

31. The region is convex and bounded. The vertices are $(0, 0)$, $(0, 5)$, and $(6, 0)$. Graph the system of inequalities:

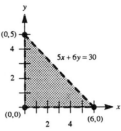

33. The region is convex and bounded. The vertices are $(5, 30)$, $(10, 30)$, $(20, 15)$, and $(20, 20)$. Graph the system of inequalities:

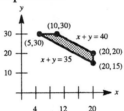

35. First graph the conditions:

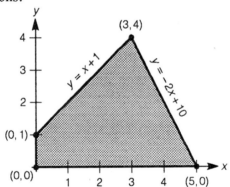

Now set up the table:

Vertices	$C = 3y - x$
$(0,0)$	0
$(5,0)$	-5
$(3,4)$	9
$(0,1)$	3

Thus we see that C has a maximum of 9 at $(3, 4)$ and a minimum of -5 at $(5, 0)$.

37. First graph the conditions:

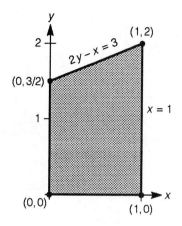

Now set up the table:

Vertices	$C = 2x + y$
$(0,0)$	0
$(0,\frac{3}{2})$	$\frac{3}{2}$
$(1,0)$	2
$(1,2)$	4

The maximum value is 4 at (1, 2) and the minimum value is 0 at (0, 0).

39. First graph the conditions:

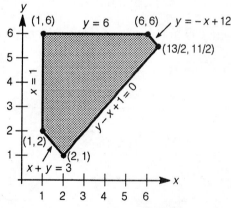

Now set up the table:

Vertices	$C = 10y + 9x - 1$
$(1,2)$	28
$(2,1)$	27
$(\frac{13}{2},\frac{11}{2})$	112.5
$(6,6)$	113
$(1,6)$	68

The maximum value is 113 at (6, 6) and the minimum value is 27 at (2, 1).

41.   First graph the conditions:

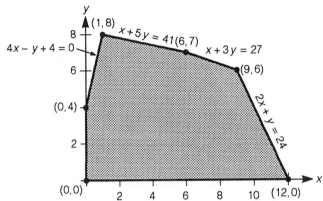

Now set up the table:

Vertices	$C = 10y + 3x + 100$
(0,0)	100
(0,4)	140
(1,8)	183
(6,7)	188
(9,6)	187
(12,0)	136

The maximum value is 188 at (6, 7), and the minimum value is 100 at (0, 0).

43.   Using the same constraints as in Exercise 41, set up a table containing both objective functions:

Vertices	(a) $C = 19x + 100y$	(b) $C = 21x + 100y$
(0,0)	0	0
(0,4)	400	400
(1,8)	819	821
(6,7)	814	826
(9,6)	771	789
(12,0)	228	252

(a)   The maximum value is 819 at (1, 8), and the minimum value is 0 at (0, 0).
(b)   The maximum value is 826 at (6, 7), and the minimum value is 0 at (0, 0).

45. First organize the data in tables:

Shipping	Cost
W_1 to D_1	180
W_1 to D_2	150
W_2 to D_1	160
W_2 to D_2	170

Supply	Demand
W_1 has 30	D_1 wants 40
W_2 has 50	D_2 wants 25

Let $x = W_1$ to D_1, so $40 - x = W_2$ to D_1. Let $y = W_1$ to D_2, so $25 - y = W_2$ to D_2. The cost function is:

$$C = 180(x) + 150(y) + 160(40 - x) + 175(25 - y)$$
$$= 180x + 150y + 6400 - 160x + 4250 - 170y$$
$$= 10650 + 20x - 20y$$

The constraints are:

(1) $x \geq 0$
(2) $y \geq 0$
(3) $x \leq 40$
(4) $y \leq 25$
(5) $x + y \leq 30$
(6) $(40 - x) + (25 - y) \leq 50$, or $x + y \geq 15$

Graph these constraints:

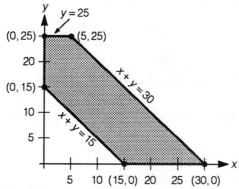

Set up a table of values:

Vertices	$C = 10650 + 20x - 20y$
(0,15)	10350
(0,25)	10150
(5,25)	10250
(30,0)	11250

So the minimum cost is $10,150 when $x = 0$ and $y = 25$.
Shipping instructions:
W_1 to D_1: 0 cars
W_2 to D_1: 40 cars
W_1 to D_2: 25 cars
W_2 to D_2: 0 cars

47. Let x = number of A units produced and y = number of B units produced. We have
$P = 0.60x + 0.80 \, y$.
The constraints are:
   (1) $x \geq 0$
   (2) $y \geq 0$
   (3) $4x + 5y \leq 5000$
   (4) $x + 2y \leq 1500$
Graph these constraints:

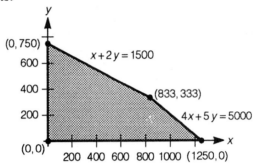

Now set up a table:

Vertices	$P = 0.60x + 0.80y$
(0,0)	0
(0,750)	600
(1250,0)	750
(833,333)	766.20

So the maximum profit is \$766.20 when 833 units of A and 333 units of B are produced.

49. Let x = acres of cherry tomatoes and y = acres of regular tomatoes. Then $P = 50x + 36y$.
The constraints are:
   (1) $x \geq 0$
   (2) $y \geq 0$
   (3) $x + y \leq 600$
   (4) $3x + 2y \leq 1350$
Graph the constraints:

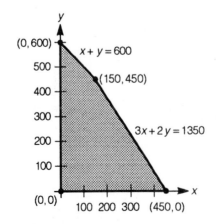

Set up a table:

Vertices	$P = 50x + 36y$
$(0,0)$	0
$(0,600)$	21600
$(450,0)$	22500
$(150,450)$	23700

So the maximum profit is $23,700 when 150 acres of cherry tomatoes and 450 acres of regular tomatoes are planted.

51. Let x = amount to invest in A and y = amount to invest in B. Then $R = 0.06x + 0.08y$. The constraints are:

(1) $x \geq 0$
(2) $y \geq 2000$
(3) $x + y \leq 12000$
(4) $x \leq 6000$
(5) $y \leq 2x$

Graph the constraints:

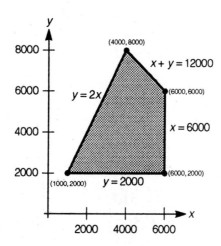

Set up a table:

Vertices	$P = 0.06x + 0.08y$
$(1000,2000)$	220
$(4000,8000)$	880
$(6000,6000)$	840
$(6000,2000)$	520

Your maximum return will be $880 when you invest $4000 in stock A and $8000 in stock B.

53. Using the same constraints as from Exercise 52:

(1) $x \geq 1000$

(2) $y \geq 1000$

(3) $10000 - x - y \geq 1000,\ x + y \leq 9000$

(4) $x + y \geq 5000$

(5) $y \leq 5x$

The profit function is:

$P = 0.05x + 0.04y + 0.06(10000 - x - y) = 600 - 0.01x - 0.02y$

Graph the constraints:

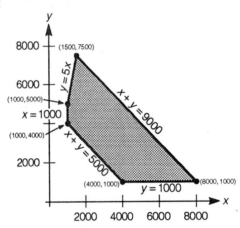

Set up a table:

Vertices	$P = 600 - 0.01x - 0.02y$
$(4000, 1000)$	540
$(8000, 1000)$	500
$(1500, 7500)$	435
$(1000, 5000)$	490
$(1000, 4000)$	510

Invest $4000 in A, $1000 in B and $5000 in C for a maximum return of $540.

55. Let x = pounds of A and y = pounds of B. Then $C = 0.44x + 0.80y$. The constraints are:

(1) $x \geq 0$

(2) $y \geq 0$

(3) $3x + 5y \geq 50$

(4) $2x + 4y \geq 36$

(5) $2x + 8y \geq 40$

Graph the constraints:

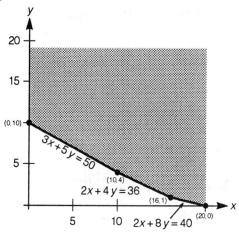

Set up a table:

Vertices	$C = 0.44x + 0.08y$
$(0,10)$	8
$(10,4)$	7.6
$(16,1)$	7.84
$(20,0)$	8.8

So 10 pounds of A and 4 pounds of B should be mixed to produce a mixture with minimum cost of $7.60.

57. The domain results from the inequality $x + y + 2 \geq 0$, or $x + y \geq -2$. Sketch the graph:

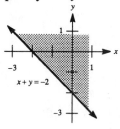

59. The domain results from the inequality $25 - x^2 - y^2 \geq 0$, or $x^2 + y^2 \leq 25$. Sketch the graph:

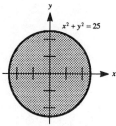

61.   The domain results from the inequality $xy > 0$. This consists of two quadrants: $x > 0$ and $y > 0$ (first quadrant) or $x < 0$ and $y < 0$ (third quadrant). Sketch the graph:

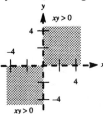

Chapter Eleven Review Exercises

1.   Adding the two equations yields:
$$2x = 6$$
$$x = 3$$
Substitute into equation 1:
$$3xy = -2$$
$$y = -5$$
So the solution is $(3, -5)$.

3.   Multiply the first equation by -2:
$$-4x - 2y = -4$$
$$x + 2y = 7$$
Adding:
$$-3x = 3$$
$$x = -1$$
Substitute into equation 1:
$$-2 + y = 2$$
$$y = 4$$
So the solution is $(-1, 4)$.

5.   Multiply equation 1 by -5 and equation 2 by 2:
$$-35x - 10y = -45$$
$$8x + 10y = 126$$
Adding:
$$-27x = 81$$
$$x = -3$$
Substitute into equation 1:
$$-21 + 2y = 9$$
$$2y = 30$$
$$y = 15$$
So the solution is $(-3, 15)$.

7. Multiply the first equation by 2 and the second equation by 24 to clear fractions:
$$4x - y = -16$$
$$8x + 3y = -24$$
Multiply equation 1 by –2:
$$-8x + 2y = 32$$
$$8x + 3y = -24$$
Adding:
$$5y = 8$$
$$y = \tfrac{8}{5}$$
Substitute into equation 1:
$$4x - \tfrac{8}{5} = -16$$
$$4x = -\tfrac{72}{5}$$
$$x = -\tfrac{18}{5}$$
So the solution is $\left(-\tfrac{18}{5}, \tfrac{8}{5}\right)$.

9. Multiply the first equation by 2:
$$6x + 10y = 2$$
$$9x - 10y = 8$$
Adding:
$$15x = 10$$
$$x = \tfrac{2}{3}$$
Substitute into equation 1:
$$2 + 5y = 1$$
$$5y = -1$$
$$y = -\tfrac{1}{5}$$
So the solution is $\left(\tfrac{2}{3}, -\tfrac{1}{5}\right)$.

11. Multiply the first equation by 6 and the second equation by 2 to clear fractions:
$$4x + 3y = -72$$
$$x - 2y = 4$$
Multiply the second equation by – 4:
$$4x + 3y = -72$$
$$-4x + 8y = -16$$
Adding:
$$11y = -88$$
$$y = -8$$
Substitute into equation 2:
$$x + 16 = 4$$
$$x = -12$$
So the solution is $(-12, -8)$.

13. Let $a = \frac{1}{x}$ and $b = \frac{1}{y}$. So:
$$a + b = -1$$
$$2a + 5b = -14$$
Multiply the first equation by -2:
$$-2a - 2b = 2$$
$$2a + 5b = -14$$
Adding:
$$3b = -12$$
$$b = -4$$
Substitute into equation 1:
$$a - 4 = -1$$
$$a = 3$$
Since $x = \frac{1}{a}$ and $y = \frac{1}{b}$, $x = \frac{1}{3}$ and $y = -\frac{1}{4}$. So the solution is $\left(\frac{1}{3}, -\frac{1}{4}\right)$.

15. Multiply the second equation by $a - 1$:
$$ax + (1-a)y = 1$$
$$\left(-a^2 + 2a - 1\right)x + (a-1)y = 0$$
Adding:
$$\left(-a^2 + 3a - 1\right)x = 1$$
$$x = \frac{-1}{a^2 - 3a + 1}$$
Substitute into equation 2:
$$\frac{a-1}{a^2 - 3a + 1} + y = 0$$
$$y = \frac{1-a}{a^2 - 3a + 1}$$
So the solution is $\left(\frac{-1}{a^2-3a+1}, \frac{1-a}{a^2-3a+1}\right)$. We must assume that $a^2 - 3a + 1 \neq 0$, or $a \neq \frac{3 \pm \sqrt{5}}{2}$.

17. Multiply the first equation by 2:
$$4x - 2y = 6a^2 - 2$$
$$x + 2y = -a^2 + 2$$
Adding:
$$5x = 5a^2$$
$$x = a^2$$
Substitute into equation 2:
$$2y + a^2 = 2 - a^2$$
$$2y = 2 - 2a^2$$
$$y = 1 - a^2$$
So the solution is $\left(a^2, 1 - a^2\right)$.

19. Multiply the first equation by 3:

$$15x - 3y = 12a^2 - 18b^2$$
$$2x + 3y = 5a^2 + b^2$$

Adding:

$$17x = 17a^2 - 17b^2$$
$$x = a^2 - b^2$$

Substitute into equation 1:

$$5a^2 - 5b^2 - y = 4a^2 - 6b^2$$
$$-y = -a^2 - b^2$$
$$y = a^2 + b^2$$

So the solution is $\left(a^2 - b^2, a^2 + b^2\right)$.

21. Multiply the first equation by p and the second equation by q:

$$p^2x - pqy = pq^2$$
$$q^2x + pqy = p^2q$$

Adding:

$$\left(p^2 + q^2\right)x = pq(p+q)$$
$$x = \frac{pq(p+q)}{p^2 + q^2}$$

Re-solve the system to find y. Multiply the first equation by $-q$ and the second equation by p:

$$-pqx + q^2y = -q^3$$
$$pqx + p^2y = p^3$$

Adding:

$$\left(p^2 + q^2\right)y = p^3 - q^3$$
$$y = \frac{p^3 - q^3}{p^2 + q^2}$$

So the solution is $\left(\dfrac{pq(p+q)}{p^2 + q^2}, \dfrac{p^3 - q^3}{p^2 + q^2}\right)$. We must assume that p and q are not both 0.

23. Let $u = \frac{1}{x}$ and $v = \frac{1}{y}$, so:

$$4au - 3bv = a - 7b$$
$$2a^2u - 2b^2v = 3a^2 - 5ab - 2b^2$$

Multiply the first equation by $-2b$ and the second equation by 3:

$$-8abu + 6b^2v = 14b^2 - 2ab$$
$$9a^2u - 6b^2v = 9a^2 - 15ab - 6b^2$$

Adding:

$$\left(9a^2 - 8ab\right)u = 9a^2 - 17ab + 8b^2$$
$$u = \frac{9a^2 - 17ab + 8b^2}{9a^2 - 8ab} = \frac{(9a - 8b)(a - b)}{a(9a - 8b)} = \frac{a - b}{a}$$

Substitute into equation 1:
$$4a\left(\frac{a-b}{a}\right) - 3bv = a - 7b$$
$$4a - 4b - 3bv = a - 7b$$
$$-3bv = -3a - 3b$$
$$v = \frac{a+b}{b}$$

Since $x = \frac{1}{u}$ and $y = \frac{1}{v}$, we have $x = \frac{a}{a-b}$ and $y = \frac{b}{a+b}$. So the solution is $\left(\dfrac{a}{a-b}, \dfrac{b}{a+b}\right)$.
We must assume that $ab \neq 0$ and $a \neq \pm b$.

25. Form the augmented matrix:
$$\begin{pmatrix} 1 & 1 & 1 & 9 \\ 1 & -1 & -1 & -5 \\ 2 & 1 & -2 & -1 \end{pmatrix}$$
Add -1 times row 1 to row 2 and -2 times row 1 to row 3:
$$\begin{pmatrix} 1 & 1 & 1 & 9 \\ 0 & -2 & -2 & -14 \\ 0 & -1 & -4 & -19 \end{pmatrix}$$
Switch row 2 and row 3, multiply each by -1:
$$\begin{pmatrix} 1 & 1 & 1 & 9 \\ 0 & 1 & 4 & 19 \\ 0 & 2 & 2 & 14 \end{pmatrix}$$
Add -2 times row 2 to row 3:
$$\begin{pmatrix} 1 & 1 & 1 & 9 \\ 0 & 1 & 4 & 19 \\ 0 & 0 & -6 & -24 \end{pmatrix}$$
So we have the system:
$$x + y + z = 9$$
$$y + 4z = 19$$
$$-6z = -24$$
Solve equation 3 for z:
$$-6z = -24$$
$$z = 4$$
Substitute into equation 2:
$$y + 16 = 19$$
$$y = 3$$
Substitute into equation 1:
$$x + 3 + 4 = 9$$
$$x = 2$$
So the solution is $(2, 3, 4)$.

27. Switching equations 1 and 3, form the augmented matrix:

$$\begin{pmatrix} 1 & 1 & 1 & -3 \\ 2 & 3 & 3 & -8 \\ 4 & -4 & 1 & 4 \end{pmatrix}$$

Add -2 times row 1 to row 2 and -4 times row 1 to row 3:

$$\begin{pmatrix} 1 & 1 & 1 & -3 \\ 0 & 1 & 1 & -2 \\ 0 & -8 & -3 & 16 \end{pmatrix}$$

Add 8 times row 2 to row 3:

$$\begin{pmatrix} 1 & 1 & 1 & -3 \\ 0 & 1 & 1 & -2 \\ 0 & 0 & 5 & 0 \end{pmatrix}$$

So we have the system:

$$x + y + z = -3$$
$$y + z = -2$$
$$5z = 0$$

Solve equation 3 for z:

$$5z = 0$$
$$z = 0$$

Substitute into equation 2:

$$y + 0 = -2$$
$$y = -2$$

Substitute into equation 1:

$$x - 2 + 0 = -3$$
$$x = -1$$

So the solution is $(-1, -2, 0)$.

29. Using equation 3 as row 1, form the augmented matrix:

$$\begin{pmatrix} 1 & 1 & -2 & 4 \\ 1 & -2 & 1 & -2 \\ -2 & 1 & 1 & 1 \end{pmatrix}$$

Add -1 times row 1 to row 2 and 2 times row 1 to row 3:

$$\begin{pmatrix} 0 & 1 & -2 & 4 \\ 0 & -3 & 3 & -6 \\ 0 & 3 & -3 & 9 \end{pmatrix}$$

Multiply row 2 by $\frac{1}{3}$ and row 3 by $\frac{1}{3}$:

$$\begin{pmatrix} 1 & 1 & -2 & 4 \\ 0 & -1 & 1 & -2 \\ 0 & 1 & -1 & 3 \end{pmatrix}$$

Add row 2 to row 3:

$$\begin{pmatrix} 1 & 1 & -2 & 4 \\ 0 & -1 & 1 & -2 \\ 0 & 0 & 0 & 1 \end{pmatrix}$$

Since $0 = 1$ is false, there is no solution to the system.

31. Multiply the second equation by -2:

$$4x + 2y - 3z = 15$$
$$-4x - 2y - 6z = -6$$

Adding:

$$-9z = 9$$
$$z = -1$$

Substitute into the original equation:

$$2x + y - 3 = 3$$
$$2x + y = 6$$
$$y = 6 - 2x$$

So the solution is $(x, 6 - 2x, -1)$, for any real number x.

33. Form the augmented matrix:

$$\begin{pmatrix} 1 & 2 & -3 & -2 \\ 2 & -1 & 1 & 1 \\ 3 & -4 & 5 & 1 \end{pmatrix}$$

Add -2 times row 1 to row 2 and -3 times row 1 to row 3:

$$\begin{pmatrix} 1 & 2 & -3 & -2 \\ 0 & -5 & 7 & 5 \\ 0 & -10 & 14 & 7 \end{pmatrix}$$

Add -2 times row 2 to row 3:

$$\begin{pmatrix} 1 & 2 & -3 & -2 \\ 0 & -5 & 7 & 5 \\ 0 & 0 & 0 & -3 \end{pmatrix}$$

Since $0 = -3$ is false, there is no solution to the system.

35. Form the augmented matrix:

$$\begin{pmatrix} 1 & 1 & 1 & a+b \\ 2 & -1 & 2 & -a+5b \\ 1 & -2 & 1 & -2a+4b \end{pmatrix}$$

Add -2 times row 1 to row 2 and -1 times row 1 to row 3:

$$\begin{pmatrix} 1 & 1 & 1 & a+b \\ 0 & -3 & 0 & -3a+3b \\ 0 & -3 & 0 & -3a+3b \end{pmatrix}$$

Add -1 times row 2 to row 3:

$$\begin{pmatrix} 1 & 1 & 1 & a+b \\ 0 & -3 & 0 & -3a+3b \\ 0 & 0 & 0 & 0 \end{pmatrix}$$

So we have the system:

$$x + y + z = a+b$$
$$-3y \phantom{{}+z} = -3a+3b$$

Solve equation 2 for y:

$$-3y = -3a+3b$$
$$y = a-b$$

Substitute into equation 1:

$$x + a - b + z = a + b$$
$$x = 2b - z$$

So the solution is $(2b - z, a - b, z)$, for any real number z.

37. Form the augmented matrix:

$$\begin{pmatrix} 1 & 1 & 1 & 1 & 8 \\ 3 & 3 & -1 & -1 & 20 \\ 4 & -1 & -1 & 2 & 18 \\ 2 & 5 & 5 & -5 & 8 \end{pmatrix}$$

Add -3 times row 1 to row 2, -4 times row 1 to row 3, and -2 times row 1 to row 4:

$$\begin{pmatrix} 1 & 1 & 1 & 1 & 8 \\ 0 & 0 & -4 & -4 & -4 \\ 0 & -5 & -5 & -2 & -14 \\ 0 & 3 & 3 & -7 & -8 \end{pmatrix}$$

Switch row 2 and row 3, and multiply row 3 by $-\frac{1}{4}$:

$$\begin{pmatrix} 1 & 1 & 1 & 1 & 8 \\ 0 & -5 & -5 & -2 & -14 \\ 0 & 0 & 1 & 1 & 1 \\ 0 & 3 & 3 & -7 & -8 \end{pmatrix}$$

Add 2 times row 4 to row 2:

$$\begin{pmatrix} 1 & 1 & 1 & 1 & 8 \\ 0 & 1 & 1 & -16 & -30 \\ 0 & 0 & 1 & 1 & 1 \\ 0 & 3 & 3 & -7 & -8 \end{pmatrix}$$

Add -3 times row 2 to row 4:
$$\begin{pmatrix} 1 & 1 & 1 & 1 & 8 \\ 0 & 1 & 1 & -16 & -30 \\ 0 & 0 & 1 & 1 & 1 \\ 0 & 0 & 0 & 41 & 82 \end{pmatrix}$$

So we have the system:
$$\begin{aligned} x + y + z + w &= 8 \\ y + z - 16w &= -30 \\ z + w &= 1 \\ 41w &= 82 \end{aligned}$$

Solve equation 4 for w:
$$41w = 82$$
$$w = 2$$

Substitute into equation 3:
$$z + 2 = 1$$
$$z = -1$$

Substitute into equation 2:
$$y - 1 - 32 = -30$$
$$y = 3$$

Substitute into equation 1:
$$x + 3 - 1 + 2 = 8$$
$$x = 4$$

So the solution is $(4, 3, -1, 2)$.

39. Evaluate the determinant:
$$\begin{vmatrix} 1 & 5 \\ -6 & 4 \end{vmatrix} = 1(4) - 5(-6) = 4 + 30 = 34$$

41. Adding -2 times row 2 to row 3 yields:
$$\begin{vmatrix} 4 & 0 & 3 \\ -2 & 1 & 5 \\ 0 & 2 & -1 \end{vmatrix} = \begin{vmatrix} 4 & 0 & 3 \\ -2 & 1 & 5 \\ 4 & 0 & -11 \end{vmatrix} = 1 \cdot \begin{vmatrix} 4 & 3 \\ 4 & -11 \end{vmatrix} = -56$$

43. Subtracting row 2 from row 1 yields:
$$\begin{vmatrix} 1 & 5 & 7 \\ 1 & 5 & 7 \\ 17 & 19 & 21 \end{vmatrix} = \begin{vmatrix} 0 & 0 & 0 \\ 1 & 5 & 7 \\ 17 & 19 & 21 \end{vmatrix} = 0$$

45. Evaluate the determinant:

$$\begin{vmatrix} 1 & 0 & 0 & 0 \\ 0 & 2 & 0 & 0 \\ 0 & 0 & 3 & 0 \\ 0 & 0 & 0 & 4 \end{vmatrix} = 1\begin{vmatrix} 2 & 0 & 0 \\ 0 & 3 & 0 \\ 0 & 0 & 4 \end{vmatrix} = 2\begin{vmatrix} 3 & 0 \\ 0 & 4 \end{vmatrix} = 2(12) = 24$$

47. By expanding along column 1:

$$\begin{vmatrix} a & b & c \\ b & c & a \\ c & a & b \end{vmatrix} = a\begin{vmatrix} c & a \\ a & b \end{vmatrix} - b\begin{vmatrix} b & c \\ a & b \end{vmatrix} + c\begin{vmatrix} b & c \\ c & a \end{vmatrix}$$

$$= a\left(bc - a^2\right) - b\left(b^2 - ac\right) + c\left(ab - c^2\right)$$
$$= abc - a^3 - b^3 + abc + abc - c^3$$
$$= 3abc - a^3 - b^3 - c^3$$

49. Adding b times column 2 to column 1 yields:

$$\begin{vmatrix} a^2 + x & b & c & d \\ -b & 1 & 0 & 0 \\ -c & 0 & 1 & 0 \\ -d & 0 & 0 & 1 \end{vmatrix} = \begin{vmatrix} a^2 + b^2 + x & b & c & d \\ 0 & 1 & 0 & 0 \\ -c & 0 & 1 & 0 \\ -d & 0 & 0 & 1 \end{vmatrix} = 1\begin{vmatrix} a^2 + b^2 + x & c & d \\ -c & 1 & 0 \\ -d & 0 & 1 \end{vmatrix}$$

Adding c times column 2 to column 1 yields:

$$\begin{vmatrix} a^2 + b^2 + c^2 + x & c & d \\ 0 & 1 & 0 \\ -d & 0 & 1 \end{vmatrix} = 1\begin{vmatrix} a^2 + b^2 + c^2 + x & d \\ -d & 1 \end{vmatrix}$$

$$= a^2 + b^2 + c^2 + x - \left(-d^2\right)$$
$$= a^2 + b^2 + c^2 + d^2 + x$$

51. Substituting $(x, y) = (-2, 5)$ and $(x, y) = (2, 9)$:

$$5 = 4a - 2b - 1$$
$$9 = 4a + 2b - 1$$

Adding:

$$14 = 8a - 2$$
$$16 = 8a$$
$$a = 2$$

Substitute into the first equation:

$$5 = 8 - 2b - 1$$
$$-2 = -2b$$
$$b = 1$$

So $a = 2$ and $b = 1$.

53. (a) First graph the triangle:

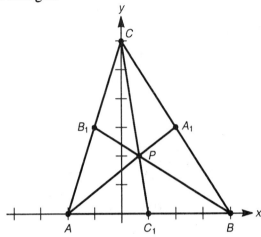

By the midpoint formula:

$$A_1 = \left(\frac{4+0}{2}, \frac{0+6}{2}\right) = (2,3)$$

$$B_1 = \left(\frac{-2+0}{2}, \frac{0+6}{2}\right) = (-1,3)$$

$$C_1 = \left(\frac{-2+4}{2}, \frac{0+0}{2}\right) = (1,0)$$

For AA_1:

$$\text{slope} = \frac{3-0}{2-(-2)} = \tfrac{3}{4}$$
$$\text{point} = (-2,0)$$
$$y - 0 = \tfrac{3}{4}(x+2)$$
$$y = \tfrac{3}{4}x + \tfrac{3}{2}$$

For BB_1:

$$\text{slope} = \frac{3-0}{-1-4} = -\tfrac{3}{5}$$
$$\text{point} = (4,0)$$
$$y - 0 = -\tfrac{3}{5}(x-4)$$
$$y = -\tfrac{3}{5}x + \tfrac{12}{5}$$

Solve $\tfrac{3}{4}x + \tfrac{3}{2} = -\tfrac{3}{5}x + \tfrac{12}{5}$. Multiply by 20 to clear fractions:

$$15x + 30 = -12x + 48$$
$$27x = 18$$
$$x = \tfrac{2}{3}$$

So $y = \tfrac{3}{4} \cdot \tfrac{2}{3} + \tfrac{3}{2} = 2$. Thus the point of intersection is $\left(\tfrac{2}{3}, 2\right)$.

(b) For CC_1:
$$\text{slope} = \frac{6-0}{0-1} = -6$$
$$y = -6x + 6$$
Solve $-6x + 6 = -\frac{3}{5}x + \frac{12}{5}$.
Multiply by 5 to clear fractions:
$$-30x + 30 = -3x + 12$$
$$18 = 27x$$
$$\tfrac{2}{3} = x$$
So $y = -6\left(\frac{2}{3}\right) + 6 = 2$. Thus the point of intersection is $\left(\frac{2}{3}, 2\right)$.

(c) Solve $-6x + 6 = \frac{3}{4}x + \frac{3}{2}$. Multiply by 4 to clear fractions:
$$-24x + 24 = 3x + 6$$
$$18 = 27x$$
$$\tfrac{2}{3} = x$$
So $y = -6\left(\frac{2}{3}\right) + 6 = 2$. Thus the point of intersection is $\left(\frac{2}{3}, 2\right)$.

(d) Use the distance formula:
$$AP = \sqrt{\left(-2 - \tfrac{2}{3}\right)^2 + (0-2)^2} = \sqrt{\tfrac{64}{9} + \tfrac{36}{9}} = \tfrac{10}{3}$$
$$PA_1 = \sqrt{\left(\tfrac{2}{3} - 2\right)^2 + (2-3)^2} = \sqrt{\tfrac{16}{9} + \tfrac{9}{9}} = \tfrac{5}{3}$$
So $\dfrac{AP}{PA_1} = \dfrac{\tfrac{10}{3}}{\tfrac{5}{3}} = 2$. Again using the distance formula:
$$BP = \sqrt{\left(4 - \tfrac{2}{3}\right)^2 + (0-2)^2} = \sqrt{\tfrac{100}{9} + \tfrac{36}{9}} = \tfrac{2\sqrt{34}}{3}$$
$$PB_1 = \sqrt{\left(\tfrac{2}{3} + 1\right)^2 + (2-3)^2} = \sqrt{\tfrac{25}{9} + \tfrac{9}{9}} = \tfrac{\sqrt{34}}{3}$$
So $\dfrac{BP}{PB_1} = \dfrac{\tfrac{2\sqrt{34}}{3}}{\tfrac{\sqrt{34}}{3}} = 2$. Again using the distance formula:
$$CP = \sqrt{\left(0 - \tfrac{2}{3}\right)^2 + (6-2)^2} = \sqrt{\tfrac{4}{9} + \tfrac{144}{9}} = \tfrac{2\sqrt{37}}{3}$$
$$PC_1 = \sqrt{\left(\tfrac{2}{3} - 1\right)^2 + (2-0)^2} = \sqrt{\tfrac{1}{9} + \tfrac{36}{9}} = \tfrac{\sqrt{37}}{3}$$
So $\dfrac{CP}{PC_1} = \dfrac{\tfrac{2\sqrt{37}}{3}}{\tfrac{\sqrt{37}}{3}} = 2$. These ratios are all equal.

55. See the figure:

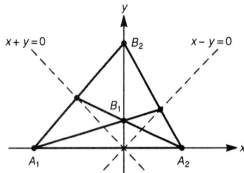

Start by finding the equation for $A_2 B_2$:
$$m = -\frac{b_2}{a_2}$$
point: (a, a)

By the point-slope formula:
$$a_2 y + b_2 x = a a_2 + a b_2$$

Since $(a_2, 0)$ lies on this curve, we have $a_2 b_2 = a a_2 + a b_2$. (*)

Now find the equation for $A_1 B_1$:
$$m = -\frac{b_1}{a_1}$$
point: (a, a)

By the point-slope formula:
$$a_1 y + b_1 x = a a_1 + a b_1$$

Since $(a_1, 0)$ lies on this curve, we have $a_1 b_1 = a a_1 + a b_1$. (**)

Now look at $A_1 B_2$:
$$m = -\frac{b_2}{a_1}$$
$$b_2 x + a_1 y = a_1 b_2$$

Also look at $A_2 B_1$:
$$m = -\frac{b_1}{a_2}, \text{ so:}$$
$$b_1 x + a_2 y = a_2 b_1$$

Solve the system:
$$b_2 x + a_1 y = a_1 b_2$$
$$b_1 x + a_2 y = a_2 b_1$$

Multiply equation 1 by $-b_1$ and equation 2 by b_2:
$$-b_1 b_2 x - a_1 b_1 y = -a_1 b_1 b_2$$
$$b_1 b_2 x + a_2 b_2 y = a_2 b_1 b_2$$

Adding:

$$\left(a_2 b_2 - a_1 b_1\right) y = a_2 b_1 b_2 - a_1 b_1 b_2$$

$$y = \frac{b_1 b_2\left(a_2 - a_1\right)}{a_2 b_2 - a_1 b_1}$$

Substituting into equation 1:

$$b_2 x + \frac{a_1 b_1 b_2\left(a_2 - a_1\right)}{a_2 b_2 - a_1 b_1} = a_1 b_2$$

$$x = \frac{a_1 a_2\left(b_2 - b_1\right)}{a_2 b_2 - a_1 b_1}$$

Now, we are asked to show that $x + y = 0$, so:

$$x + y = \frac{a_1 a_2\left(b_2 - b_1\right) + b_1 b_2\left(a_2 - a_1\right)}{a_2 b_2 - a_1 b_1} = \frac{a_1 a_2 b_2 - a_1 a_2 b_1 + a_2 b_1 b_2 - a_1 b_1 b_2}{a_2 b_2 - a_1 b_1}$$

Replacing $a_1 b_1$ and $a_2 b_2$ by equations (*) and (**):

$$x + y = \frac{a_1\left(aa_2 + ab_2\right) - a_2\left(aa_1 + ab_1\right) + b_1\left(aa_2 + ab_2\right) - b_2\left(aa_1 + ab_1\right)}{a_2 b_2 - a_1 b_1}$$

$$= \frac{a\left(a_1 a_2 + a_1 b_2 - a_1 a_2 - a_2 b_1 + a_2 b_1 + b_1 b_2 - a_1 b_2 - b_1 b_2\right)}{a_2 b_2 - a_1 b_1}$$

$$= 0$$

57. Find each intersection point:

$y = x - 1$	$y = x - 1$	$y = -x - 2$
$y = -x - 2$	$y = 2x + 3$	$y = 2x + 3$
$2y = -3$	$x - 1 = 2x + 3$	$-x - 2 = 2x + 3$
$y = -\frac{3}{2}$	$-4 = x$	$-5 = 3x$
		$x = -\frac{5}{3}$
$x = -\frac{1}{2}$	$y = -5$	$y = -\frac{1}{3}$
$\left(-\frac{1}{2}, -\frac{3}{2}\right)$	$\left(-4, -5\right)$	$\left(-\frac{5}{3}, -\frac{1}{3}\right)$

Now substitute into the equation $(x - h)^2 + (y - k)^2 = r^2$:

$$\left(-\tfrac{1}{2} - h\right)^2 + \left(-\tfrac{3}{2} - k\right)^2 = r^2$$

$$\left(-4 - h\right)^2 + \left(-5 - k\right)^2 = r^2$$

$$\left(-\tfrac{5}{3} - h\right)^2 + \left(-\tfrac{1}{3} - k\right)^2 = r^2$$

These equations, when multiplied out, become:

$$\tfrac{5}{2} + h + h^2 + 3k + k^2 = r^2$$

$$41 + 8h + h^2 + 10k + k^2 = r^2$$

$$\tfrac{26}{9} + \tfrac{10}{3} h + h^2 + \tfrac{2}{3} k + k^2 = r^2$$

Subtracting the first equation from the other two yields:

$$\tfrac{77}{2} + 7h + 7k = 0$$

$$\tfrac{7}{18} + \tfrac{7}{3} h - \tfrac{7}{3} k = 0$$

These simplify to:

$$6h + 6k = -33$$

$$6h - 6k = -1$$

Adding:

$$12h = -34$$

$$h = -\frac{17}{6}$$

To find k, we subtract the two equations:

$$12k = -32$$

$$k = -\frac{8}{3}$$

Finally, find r:

$$r^2 = (-4 - h)^2 + (-5 - k)^2 = \left(-\frac{7}{6}\right)^2 + \left(-\frac{7}{3}\right)^2 = \frac{245}{36}$$

The equation of the circle is:

$$\left(x + \frac{17}{6}\right)^2 + \left(y + \frac{8}{3}\right)^2 = \frac{245}{36}$$

59. $2A + 2B = \begin{pmatrix} 6 & -4 \\ 2 & 10 \end{pmatrix} + \begin{pmatrix} 4 & 2 \\ 2 & 16 \end{pmatrix} = \begin{pmatrix} 10 & -2 \\ 4 & 26 \end{pmatrix}$

61. $4B = \begin{pmatrix} 8 & 4 \\ 4 & 32 \end{pmatrix}$

63. $AB = \begin{pmatrix} 4 & -13 \\ 7 & 41 \end{pmatrix}$

65. $AB - BA = \begin{pmatrix} 7 & -13 \\ 7 & 41 \end{pmatrix} - \begin{pmatrix} 7 & 1 \\ 11 & 38 \end{pmatrix} = \begin{pmatrix} -3 & -14 \\ -4 & 3 \end{pmatrix}$

67. $C + B = \begin{pmatrix} 1 & 1 \\ 1 & 7 \end{pmatrix}$

69. $AB + AC = \begin{pmatrix} 4 & -13 \\ 7 & 41 \end{pmatrix} + \begin{pmatrix} -3 & 2 \\ -1 & -5 \end{pmatrix} = \begin{pmatrix} 1 & -11 \\ 6 & 36 \end{pmatrix}$

71. $BA + CA = \begin{pmatrix} 7 & 1 \\ 11 & 38 \end{pmatrix} + \begin{pmatrix} -3 & 2 \\ -1 & -5 \end{pmatrix} = \begin{pmatrix} 4 & 3 \\ 10 & 33 \end{pmatrix}$

73. $DE = \begin{pmatrix} -42 & 58 \\ 5 & 20 \end{pmatrix}$

75. The multiplication is undefined.

77. The multiplication is undefined.

79. $(A+B)+C = \begin{pmatrix} 5 & -1 \\ 2 & 13 \end{pmatrix} + \begin{pmatrix} -1 & 0 \\ 0 & -1 \end{pmatrix} = \begin{pmatrix} 4 & -1 \\ 2 & 12 \end{pmatrix}$

81. $(AB)C = \begin{pmatrix} 4 & -13 \\ 7 & 41 \end{pmatrix} \begin{pmatrix} -1 & 0 \\ 0 & -1 \end{pmatrix} = \begin{pmatrix} -4 & 13 \\ -7 & -41 \end{pmatrix}$

83. Compute A^2 and then A^3:

$$A^2 = \begin{pmatrix} 1 & 1 \\ 0 & 1 \end{pmatrix}\begin{pmatrix} 1 & 1 \\ 0 & 1 \end{pmatrix} = \begin{pmatrix} 1 & 2 \\ 0 & 1 \end{pmatrix}$$

$$A^3 = A \cdot A^2 = \begin{pmatrix} 1 & 1 \\ 0 & 1 \end{pmatrix}\begin{pmatrix} 1 & 2 \\ 0 & 1 \end{pmatrix} = \begin{pmatrix} 1 & 3 \\ 0 & 1 \end{pmatrix}$$

85. (a) Form the matrix:

$$\begin{pmatrix} 1 & 5 & 1 & 0 \\ 2 & 9 & 0 & 1 \end{pmatrix}$$

Multiply row 1 by –2 and add to row 2:

$$\begin{pmatrix} 1 & 5 & 1 & 0 \\ 0 & -1 & -2 & 1 \end{pmatrix}$$

Multiply row 2 by 5 and add to row 1:

$$\begin{pmatrix} 1 & 0 & -9 & 5 \\ 0 & -1 & -2 & 1 \end{pmatrix}$$

Multiply row 2 by –1:

$$\begin{pmatrix} 1 & 0 & -9 & 5 \\ 0 & 1 & 2 & -1 \end{pmatrix}$$

So the inverse is $\begin{pmatrix} -9 & 5 \\ 2 & -1 \end{pmatrix}$.

(b) Since the system is equivalent to $A \cdot X = B$, where $A = \begin{pmatrix} 1 & 5 \\ 2 & 9 \end{pmatrix}$,

$X = \begin{pmatrix} x \\ y \end{pmatrix}$, and $B = \begin{pmatrix} 3 \\ -4 \end{pmatrix}$:

$$X = A^{-1} \cdot B = \begin{pmatrix} -9 & 5 \\ 2 & -1 \end{pmatrix}\begin{pmatrix} 3 \\ -4 \end{pmatrix} = \begin{pmatrix} -47 \\ 10 \end{pmatrix}$$

So the solution is $(-47, 10)$.

87. (a) Form the matrix:

$$\begin{pmatrix} 1 & -2 & 3 & 1 & 0 & 0 \\ 2 & -5 & 10 & 0 & 1 & 0 \\ -1 & 2 & -2 & 0 & 0 & 1 \end{pmatrix}$$

Multiply row 1 by -2 and add to row 2, then add row 1 to row 3:

$$\begin{pmatrix} 1 & -2 & 3 & 1 & 0 & 0 \\ 0 & -1 & 4 & -2 & 1 & 0 \\ 0 & 0 & 1 & 1 & 0 & 1 \end{pmatrix}$$

Multiply row 2 by -1:

$$\begin{pmatrix} 1 & -2 & 3 & 1 & 0 & 0 \\ 0 & 1 & -4 & 2 & -1 & 0 \\ 0 & 0 & 1 & 1 & 0 & 1 \end{pmatrix}$$

Multiply row 2 by 2 and add to row 1:

$$\begin{pmatrix} 1 & 0 & -5 & 5 & -2 & 0 \\ 0 & 1 & -4 & 2 & -1 & 0 \\ 0 & 0 & 1 & 1 & 0 & 1 \end{pmatrix}$$

Multiply row 3 by 4 and add to row 2, then multiply row 3 by 5 and add to row 1:

$$\begin{pmatrix} 1 & 0 & 0 & 10 & -2 & 5 \\ 0 & 1 & 0 & 6 & -1 & 4 \\ 0 & 0 & 1 & 1 & 0 & 1 \end{pmatrix}$$

So the inverse is $\begin{pmatrix} 10 & -2 & 5 \\ 6 & -1 & 4 \\ 1 & 0 & 1 \end{pmatrix}$.

(b) Since the system is equivalent to $A \cdot X = B$, where $A = \begin{pmatrix} 1 & -2 & 3 \\ 2 & -5 & 10 \\ -1 & 2 & -2 \end{pmatrix}$,

$X = \begin{pmatrix} x \\ y \\ z \end{pmatrix}$, and $B = \begin{pmatrix} -2 \\ -3 \\ 6 \end{pmatrix}$:

$$X = A^{-1} \cdot B = \begin{pmatrix} 10 & -2 & 5 \\ 6 & -1 & 4 \\ 1 & 0 & 1 \end{pmatrix}\begin{pmatrix} -2 \\ -3 \\ 6 \end{pmatrix} = \begin{pmatrix} 16 \\ 15 \\ 4 \end{pmatrix}$$

So the solution is $(16, 15, 4)$.

89. Form the matrix:

$$\begin{pmatrix} 5 & 3 & 6 & -7 & 1 & 0 & 0 & 0 \\ 3 & -4 & 0 & -9 & 0 & 1 & 0 & 0 \\ 0 & 1 & -1 & -1 & 0 & 0 & 1 & 0 \\ 2 & 2 & 3 & -2 & 0 & 0 & 0 & 1 \end{pmatrix}$$

Multiply row 4 by –2 and add to row 1 (to get a 1):

$$\begin{pmatrix} 1 & -1 & 0 & -3 & 1 & 0 & 0 & -2 \\ 3 & -4 & 0 & -9 & 0 & 1 & 0 & 0 \\ 0 & 1 & -1 & -1 & 0 & 0 & 1 & 0 \\ 2 & 2 & 3 & -2 & 0 & 0 & 0 & 1 \end{pmatrix}$$

Multiply row 1 by –3 and add to row 2, and multiply row 1 by –2 and add to row 4:

$$\begin{pmatrix} 1 & -1 & 0 & -3 & 1 & 0 & 0 & -2 \\ 0 & -1 & 0 & 0 & -3 & 1 & 0 & 6 \\ 0 & 1 & -1 & -1 & 0 & 0 & 1 & 0 \\ 0 & 4 & 3 & 4 & -2 & 0 & 0 & 5 \end{pmatrix}$$

Add row 2 to row 3, then mulitply row 2 by 4 and add to row 4:

$$\begin{pmatrix} 1 & -1 & 0 & -3 & 1 & 0 & 0 & -2 \\ 0 & -1 & 0 & 0 & -3 & 1 & 0 & 6 \\ 0 & 0 & -1 & -1 & -3 & 1 & 1 & 6 \\ 0 & 0 & 3 & 4 & -14 & 4 & 0 & 29 \end{pmatrix}$$

Multiply row 3 by 3 and add to row 4:

$$\begin{pmatrix} 1 & -1 & 0 & -3 & 1 & 0 & 0 & -2 \\ 0 & -1 & 0 & 0 & -3 & 1 & 0 & 6 \\ 0 & 0 & -1 & -1 & -3 & 1 & 1 & 6 \\ 0 & 0 & 0 & 1 & -23 & 7 & 3 & 47 \end{pmatrix}$$

Add row 4 to row 3, then multiply row 4 by 3 and add to row 1:

$$\begin{pmatrix} 1 & -1 & 0 & 0 & -68 & 21 & 9 & 139 \\ 0 & -1 & 0 & 0 & -3 & 1 & 0 & 6 \\ 0 & 0 & -1 & 0 & -26 & 8 & 4 & 53 \\ 0 & 0 & 0 & 1 & -23 & 7 & 3 & 47 \end{pmatrix}$$

Multiply row 2 by –1 and row 3 by –1:

$$\begin{pmatrix} 1 & -1 & 0 & 0 & -68 & 21 & 9 & 139 \\ 0 & 1 & 0 & 0 & 3 & -1 & 0 & -6 \\ 0 & 0 & 1 & 0 & 26 & -8 & -4 & -53 \\ 0 & 0 & 0 & 1 & -23 & 7 & 3 & 47 \end{pmatrix}$$

Add row 2 to row 1:

$$\begin{pmatrix} 1 & 0 & 0 & 0 & -65 & 20 & 9 & 133 \\ 0 & 1 & 0 & 0 & 3 & -1 & 0 & -6 \\ 0 & 0 & 1 & 0 & 26 & -8 & -4 & -53 \\ 0 & 0 & 0 & 1 & -23 & 7 & 3 & 47 \end{pmatrix}$$

So the inverse is $\begin{pmatrix} -65 & 20 & 9 & 133 \\ 3 & -1 & 0 & -6 \\ 26 & -8 & -4 & -53 \\ -23 & 7 & 3 & 47 \end{pmatrix}$.

91. Adding 2 times column 2 to column 1 and column 2 to column 3 yields:

$$D = \begin{vmatrix} 2 & -1 & 1 \\ 3 & 2 & 2 \\ 1 & -5 & -3 \end{vmatrix} = \begin{vmatrix} 0 & -1 & 0 \\ 7 & 2 & 4 \\ -9 & -5 & -8 \end{vmatrix} = -(-1)\begin{vmatrix} 7 & 4 \\ -9 & -8 \end{vmatrix} = 1(-20) = -20$$

Adding 2 times row 1 to row 3 yields:

$$D_x = \begin{vmatrix} 1 & -1 & 1 \\ 0 & 2 & 2 \\ -2 & -5 & -3 \end{vmatrix} = \begin{vmatrix} 1 & -1 & 1 \\ 0 & 2 & 2 \\ 0 & -7 & -1 \end{vmatrix} = 1\begin{vmatrix} 2 & 2 \\ -7 & -1 \end{vmatrix} = 12$$

Adding 2 times row 1 to row 3 yields:

$$D_y = \begin{vmatrix} 2 & 1 & 1 \\ 3 & 0 & 2 \\ 1 & -2 & -3 \end{vmatrix} = \begin{vmatrix} 2 & 1 & 1 \\ 3 & 0 & 2 \\ 5 & 0 & -1 \end{vmatrix} = -1\begin{vmatrix} 3 & 2 \\ 5 & -1 \end{vmatrix} = -(-13) = 13$$

Adding 2 times row 1 to row 3 yields:

$$D_z = \begin{vmatrix} 2 & -1 & 1 \\ 3 & 2 & 0 \\ 1 & -5 & -2 \end{vmatrix} = \begin{vmatrix} 2 & -1 & 1 \\ 3 & 2 & 0 \\ 5 & -7 & 0 \end{vmatrix} = 1\begin{vmatrix} 3 & 2 \\ 5 & -7 \end{vmatrix} = -31$$

So $x = \frac{D_x}{D} = -\frac{12}{20} = -\frac{3}{5}$, $y = \frac{D_y}{D} = -\frac{13}{20}$, and $z = \frac{D_z}{D} = \frac{31}{20}$. So the solution is $\left(-\frac{3}{5}, -\frac{13}{20}, \frac{31}{20}\right)$.

93. Subtracting column 2 from column 3 and column 1 from column 2 yields:

$$D = \begin{vmatrix} 1 & 2 & 3 \\ 4 & 5 & 6 \\ 7 & 8 & 9 \end{vmatrix} = \begin{vmatrix} 1 & 1 & 1 \\ 4 & 1 & 1 \\ 7 & 1 & 1 \end{vmatrix}$$

Subtracting column 3 from column 2 yields:

$$D = \begin{vmatrix} 1 & 0 & 1 \\ 4 & 0 & 1 \\ 7 & 0 & 1 \end{vmatrix} = 0$$

So Cramer's Rule will not work. Form the augmented matrix:

$$\begin{pmatrix} 1 & 2 & 3 & -1 \\ 4 & 5 & 6 & 2 \\ 7 & 8 & 9 & -3 \end{pmatrix}$$

Add -4 times row 1 to row 2 and -7 times row 1 to row 3:

$$\begin{pmatrix} 1 & 2 & 3 & -1 \\ 0 & -3 & -6 & 6 \\ 0 & -6 & -12 & 4 \end{pmatrix}$$

Add -2 times row 2 to row 3:

$$\begin{pmatrix} 1 & 2 & 3 & -1 \\ 0 & -3 & -6 & 6 \\ 0 & 0 & 0 & -8 \end{pmatrix}$$

So $0 = -8$, which is false. The system has no solution.

95. Since $D = 0$ (See Exercise 94), form the augmented matrix:

$$\begin{pmatrix} 3 & 2 & -2 & 1 \\ 2 & 3 & -1 & -2 \\ 8 & 7 & -5 & 0 \end{pmatrix}$$

Subtract row 2 from row 1:

$$\begin{pmatrix} 1 & -1 & -1 & 3 \\ 2 & 3 & -1 & -2 \\ 8 & 7 & -5 & 0 \end{pmatrix}$$

Add -2 times row 1 to row 2 and -8 times row 1 to row 3:

$$\begin{pmatrix} 1 & -1 & -1 & 3 \\ 0 & 5 & 1 & -8 \\ 0 & 15 & 3 & -24 \end{pmatrix}$$

Add -3 times row 2 to row 3:

$$\begin{pmatrix} 1 & -1 & -1 & 3 \\ 0 & 5 & 1 & -8 \\ 0 & 0 & 0 & 0 \end{pmatrix}$$

So we have the system:
$$x - y - z = 3$$
$$5y + z = -8$$

Solve equation 2 for z:
$$z = -8 - 5y$$

Substitute into equation 1:
$$x - y + 8 + 5y = 3$$
$$x = -5 - 4y$$

So the solution is $(-5 - 4y, y, -8 - 5y)$, for any real number y.

97. Adding 4 times column 2 to both column 1 and column 3 yields:

$$D = \begin{vmatrix} 2 & -1 & 1 & 3 \\ 1 & 2 & 0 & 2 \\ 0 & 3 & 3 & 4 \\ -4 & 1 & -4 & 0 \end{vmatrix} = \begin{vmatrix} -2 & -1 & -3 & 3 \\ 9 & 2 & 8 & 2 \\ 12 & 3 & 15 & 4 \\ 0 & 1 & 0 & 0 \end{vmatrix} = 1 \begin{vmatrix} -2 & -3 & 3 \\ 9 & 8 & 2 \\ 12 & 15 & 4 \end{vmatrix}$$

Subtracting row 2 from row 3 yields:

$$D = \begin{vmatrix} -2 & -3 & 3 \\ 9 & 8 & 2 \\ 3 & 7 & 2 \end{vmatrix}$$

Adding $\frac{2}{3}$ times column 3 to column 1, and column 3 to column 2 yields:

$$D = \begin{vmatrix} 0 & 0 & 3 \\ \frac{31}{3} & 10 & 2 \\ \frac{13}{3} & 9 & 2 \end{vmatrix} = 3 \begin{vmatrix} \frac{31}{3} & 10 \\ \frac{13}{3} & 9 \end{vmatrix} = \begin{vmatrix} 31 & 10 \\ 13 & 9 \end{vmatrix} = 149$$

Adding 11 times column 2 to column 1 and 4 times column 2 to column 3 yields:

$$D_x = \begin{vmatrix} 15 & -1 & 1 & 3 \\ 12 & 2 & 0 & 2 \\ 12 & 3 & 3 & 4 \\ -11 & 1 & -4 & 0 \end{vmatrix} = \begin{vmatrix} 4 & -1 & -3 & 3 \\ 34 & 2 & 8 & 2 \\ 45 & 3 & 15 & 4 \\ 0 & 1 & 0 & 0 \end{vmatrix} = 1 \begin{vmatrix} 4 & -3 & 3 \\ 34 & 8 & 2 \\ 45 & 15 & 4 \end{vmatrix}$$

Subtracting row 2 from row 3 yields:

$$D_x = \begin{vmatrix} 4 & -3 & 3 \\ 34 & 8 & 2 \\ 11 & 7 & 2 \end{vmatrix}$$

Subtracting row 3 from row 2 yields:

$$D_x = \begin{vmatrix} 4 & -3 & 3 \\ 23 & 1 & 0 \\ 11 & 7 & 2 \end{vmatrix}$$

Adding –23 times column 2 to column 1:

$$D_x = \begin{vmatrix} 73 & -3 & 3 \\ 0 & 1 & 0 \\ -150 & 7 & 2 \end{vmatrix} = \begin{vmatrix} 73 & 3 \\ -150 & 2 \end{vmatrix} = 596$$

Adding –2 times row 2 to row 1 and 4 times row 2 to row 4 yields:

$$D_y = \begin{vmatrix} 2 & 15 & 1 & 3 \\ 1 & 12 & 0 & 2 \\ 0 & 12 & 3 & 4 \\ -4 & -11 & -4 & 0 \end{vmatrix} = \begin{vmatrix} 0 & -9 & 1 & -1 \\ 1 & 12 & 0 & 2 \\ 0 & 12 & 3 & 4 \\ 0 & 37 & -4 & 8 \end{vmatrix} = -1 \begin{vmatrix} -9 & 1 & -1 \\ 12 & 3 & 4 \\ 37 & -4 & 8 \end{vmatrix}$$

Adding 9 times column 2 to column 1 and column 2 to column 3 yields:

$$D_y = -1 \begin{vmatrix} 0 & 1 & 0 \\ 39 & 3 & 7 \\ 1 & -4 & 4 \end{vmatrix} = 1 \begin{vmatrix} 39 & 7 \\ 1 & 4 \end{vmatrix} = 149$$

Adding -2 times row 2 and row 1 to 4 times row 2 to row 4 yields:

$$D_z = \begin{vmatrix} 2 & -1 & 15 & 3 \\ 1 & 2 & 12 & 2 \\ 0 & 3 & 12 & 4 \\ -4 & 1 & -11 & 0 \end{vmatrix} = \begin{vmatrix} 0 & -5 & -9 & -1 \\ 1 & 2 & 12 & 2 \\ 0 & 3 & 12 & 4 \\ 0 & 9 & 37 & 8 \end{vmatrix} = -1 \begin{vmatrix} -5 & -9 & -1 \\ 3 & 12 & 4 \\ 9 & 37 & 8 \end{vmatrix}$$

Adding 4 times row 1 to row 2 and 8 times row 1 to row 3:

$$D_z = -1 \begin{vmatrix} -5 & -9 & -1 \\ -17 & -24 & 0 \\ -31 & -25 & 0 \end{vmatrix} = 1 \begin{vmatrix} -17 & -24 \\ -31 & -35 \end{vmatrix} = -149$$

Adding -2 times row 2 to row 1 and 4 times row 2 to row 4:

$$D_w = \begin{vmatrix} 2 & -1 & 1 & 15 \\ 1 & 2 & 0 & 12 \\ 0 & 3 & 3 & 12 \\ -4 & 1 & -4 & -11 \end{vmatrix} = \begin{vmatrix} 0 & -5 & 1 & -9 \\ 1 & 2 & 0 & 12 \\ 0 & 3 & 3 & 12 \\ 0 & 9 & -4 & 37 \end{vmatrix} = -1 \begin{vmatrix} -5 & 1 & -9 \\ 3 & 3 & 12 \\ 9 & -4 & 37 \end{vmatrix}$$

Factoring 3 out of row 2 yields:

$$D_w = -3 \begin{vmatrix} -5 & 1 & -9 \\ 1 & 1 & 4 \\ 9 & -4 & 37 \end{vmatrix}$$

Adding -1 times row 2 to row 1 and 4 times row 2 to row 3:

$$D_w = -3 \begin{vmatrix} -6 & 0 & -13 \\ 1 & 1 & 4 \\ 13 & 0 & 53 \end{vmatrix} = -3 \begin{vmatrix} -6 & -13 \\ 13 & 53 \end{vmatrix} = -3(-149) = 447$$

So $x = \frac{D_x}{D} = \frac{596}{149} = 4$, $y = \frac{D_y}{D} = \frac{149}{149} = 1$, $z = \frac{D_z}{D} = -\frac{149}{149} = -1$, and $w = \frac{D_w}{D} = \frac{447}{149} = 3$.
So the solution is $(4, 1, -1, 3)$.

99. Substitute to obtain:
$$x^2 = 6x$$
$$x^2 - 6x = 0$$
$$x(x - 6) = 0$$
$$x = 0 \text{ or } x = 6$$
When $x = 0$, $y = 0$ and when $x = 6$, $y = 36$. So the solutions are $(0,0)$ and $(6, 36)$.

101. Substitute to obtain:
$$x^2 - 9 = 9 - x^2$$
$$2x^2 = 18$$
$$x^2 = 9$$
$$x = \pm 3$$
When $x = \pm 3$, $y = 0$. So the solutions are $(3, 0)$ and $(-3, 0)$.

103. Adding the two equations yields:
$$2x^2 = 25$$
$$x^2 = \frac{25}{2}$$
$$x = \frac{\pm 5\sqrt{2}}{2}$$
Substituting for x yields:
$$\tfrac{25}{2} + y^2 = 16$$
$$y^2 = \tfrac{7}{2}$$
$$y = \frac{\pm\sqrt{14}}{2}$$
So the solutions are $\left(\frac{5\sqrt{2}}{2}, \frac{\sqrt{14}}{2}\right)$, $\left(-\frac{5\sqrt{2}}{2}, \frac{\sqrt{14}}{2}\right)$, $\left(\frac{5\sqrt{2}}{2}, -\frac{\sqrt{14}}{2}\right)$, and $\left(-\frac{5\sqrt{2}}{2}, -\frac{\sqrt{14}}{2}\right)$.

105. Substitute to obtain:
$$x^2 + x = 1$$
$$x^2 + x - 1 = 0$$
$$x = \frac{-1 \pm \sqrt{1+4}}{2} = \frac{-1 \pm \sqrt{5}}{2}$$
Now $x = \frac{-1-\sqrt{5}}{2}$ is impossible, since $x \geq 0$. So the only solution is
$\left(\frac{-1+\sqrt{5}}{2}, \sqrt{\frac{-1+\sqrt{5}}{2}}\right)$ or $\left(\frac{-1+\sqrt{5}}{2}, \frac{\sqrt{-2+2\sqrt{5}}}{2}\right)$.

107. Substitute to obtain:
$$x^2 + \left(2x^2\right)^2 = 1$$
$$x^2 + 4x^4 = 1$$
$$4x^4 + x^2 - 1 = 0$$
$$x^2 = \frac{-1 \pm \sqrt{1+16}}{8} = \frac{-1 \pm \sqrt{17}}{8}$$
Since $x^2 \neq \frac{-1-\sqrt{17}}{8}$, we have $x^2 = \frac{-1+\sqrt{17}}{8}$, so $x = \frac{\pm\sqrt{-2+2\sqrt{17}}}{4}$. Since $y = 2x^2$, we have
$y = \frac{-1+\sqrt{17}}{4}$. So the solutions are $\left(\frac{\sqrt{-2+2\sqrt{17}}}{4}, \frac{-1+\sqrt{17}}{4}\right)$ and $\left(\frac{-\sqrt{-2+2\sqrt{17}}}{4}, \frac{-1+\sqrt{17}}{4}\right)$.

109. Multiply the second equation by -3:
$$-9x^2 + 3xy - 3y^2 = -54$$
$$x^2 + 2xy + 3y^2 = 68$$
Adding:
$$-8x^2 + 5xy = 14$$
$$-5xy = 8x^2 + 14$$
$$y = \frac{8x^2 + 14}{5x}$$
Now substitute into the second equation:
$$3x^2 - \frac{8x^2 + 14}{5} + \frac{\left(8x^2 + 14\right)^2}{25x^2} = 18$$
Multiply by $25x^2$:
$$75x^4 - 5x^2\left(8x^2 + 14\right) + \left(8x^2 + 14\right)^2 = 450x^2$$
$$75x^4 - 40x^4 - 70x^2 + 64x^4 + 224x^2 + 196 = 450x^2$$
$$99x^4 - 296x^2 + 196 = 0$$
$$\left(99x^2 - 98\right)\left(x^2 - 2\right) = 0$$
Therefore we have the equations:
$$x^2 = \frac{98}{99} \qquad \text{or} \qquad x^2 = 2$$
$$x = \frac{\pm 7\sqrt{22}}{33} \qquad \text{or} \qquad x = \pm\sqrt{2}$$
Now $y = \dfrac{8x^2 + 14}{5x}$, so:

When $x = \sqrt{2}$, $y = \dfrac{16 + 14}{5\sqrt{2}} = 3\sqrt{2}$

When $x = -\sqrt{2}$, $y = \dfrac{16 + 14}{-5\sqrt{2}} = -3\sqrt{2}$

When $x = \dfrac{7\sqrt{22}}{33}$, $y = \dfrac{31\sqrt{22}}{33}$

When $x = \dfrac{-7\sqrt{22}}{33}$, $y = \dfrac{-31\sqrt{22}}{33}$

So the solutions are $\left(\sqrt{2}, 3\sqrt{2}\right), \left(-\sqrt{2}, -3\sqrt{2}\right), \left(\frac{7\sqrt{22}}{33}, \frac{31\sqrt{22}}{33}\right)$, and $\left(\frac{-7\sqrt{22}}{33}, \frac{-31\sqrt{22}}{33}\right)$.

111. Let $u = x - 3$ and $v = y + 1$, so:
$$2u^2 - v^2 = -1$$
$$-3u^2 + 2v^2 = 6$$
Multiply the first equation by 2 and add:
$$u^2 = 4$$
$$u = \pm 2$$

Substituting:
$$8 - v^2 = -1$$
$$-v^2 = -9$$
$$v = \pm 3$$
So the solutions (u, v) are $(2, 3)$, $(2, -3)$, $(-2, 3)$, and $(-2, -3)$. Since $u = x - 3$ and $v = y + 1$, $x = u + 3$ and $y = v - 1$. So the solutions are $(5, 2)$, $(5, -4)$, $(1, 2)$, and $(1, -4)$.

113. Call the numbers x and y. So $x + y = s$ and $\frac{x}{y} = \frac{a}{b}$. So $y = s - x$, and substituting:

$$\frac{x}{s - x} = \frac{a}{b}$$
$$bx = as - ax$$
$$(a + b)x = as$$
$$x = \frac{as}{a + b}$$

Thus $y = s - \frac{as}{a+b} = \frac{bs}{a+b}$. So the two numbers are $\frac{as}{a+b}$ and $\frac{bs}{a+b}$.

115. Given the equations:
$$\tfrac{1}{2}x + \tfrac{1}{3}y + \tfrac{1}{4}z = 62$$
$$\tfrac{1}{3}x + \tfrac{1}{4}y + \tfrac{1}{5}z = 47$$
$$\tfrac{1}{4}x + \tfrac{1}{5}y + \tfrac{1}{6}z = 38$$
Multiply the first equation by 12, the second equation by 60, and the third equation by 60 to clear the fractions:
$$6x + 4y + 3z = 744$$
$$20x + 15y + 12z = 2820$$
$$15x + 12y + 10z = 2280$$
Use Cramer's Rule. Subtracting row 3 from row 2 yields:
$$D = \begin{vmatrix} 6 & 4 & 3 \\ 20 & 15 & 12 \\ 15 & 12 & 10 \end{vmatrix} = \begin{vmatrix} 6 & 4 & 3 \\ 5 & 3 & 2 \\ 15 & 12 & 10 \end{vmatrix}$$
Subtracting row 2 from row 1 and adding -3 times row 2 to row 3 yields:
$$D = \begin{vmatrix} 1 & 1 & 1 \\ 5 & 3 & 2 \\ 0 & 3 & 4 \end{vmatrix}$$
Subtracting column 1 from column 2 and column 3 yields:
$$D = \begin{vmatrix} 1 & 0 & 0 \\ 5 & -2 & -3 \\ 0 & 3 & 4 \end{vmatrix} = 1 \begin{vmatrix} -2 & -3 \\ 3 & 4 \end{vmatrix} = 1(1) = 1$$

Subtracting row 3 from row 2 yields:

$$D_x = \begin{vmatrix} 744 & 4 & 3 \\ 2820 & 15 & 12 \\ 2280 & 12 & 10 \end{vmatrix} = \begin{vmatrix} 744 & 4 & 3 \\ 540 & 3 & 2 \\ 2280 & 12 & 10 \end{vmatrix}$$

Subtracting row 2 from row 1 and adding -4 times row 2 to row 3 yields:

$$D_x = \begin{vmatrix} 204 & 1 & 1 \\ 540 & 3 & 2 \\ 120 & 0 & 2 \end{vmatrix}$$

Adding -3 times row 1 to row 2 yields:

$$D_x = \begin{vmatrix} 204 & 1 & 1 \\ -72 & 0 & -1 \\ 120 & 0 & 2 \end{vmatrix} = -1 \begin{vmatrix} -72 & -1 \\ 120 & 2 \end{vmatrix} = -(-24) = 24$$

Subtracting row 3 from row 2 yields:

$$D_y = \begin{vmatrix} 6 & 744 & 3 \\ 20 & 2820 & 12 \\ 15 & 2280 & 10 \end{vmatrix} = \begin{vmatrix} 6 & 744 & 3 \\ 5 & 540 & 2 \\ 15 & 2280 & 10 \end{vmatrix}$$

Subtracting row 2 from row 1 and adding -3 times row 2 to row 3 yields:

$$D_y = \begin{vmatrix} 1 & 204 & 1 \\ 5 & 540 & 2 \\ 0 & 660 & 4 \end{vmatrix}$$

Adding -5 times row 1 to row 2 yields:

$$D_y = \begin{vmatrix} 1 & 204 & 1 \\ 0 & -480 & -3 \\ 0 & 660 & 4 \end{vmatrix} = 1 \begin{vmatrix} -480 & -3 \\ 660 & 4 \end{vmatrix} = 1(60) = 60$$

Subtracting row 3 from row 2 yields:

$$D_z = \begin{vmatrix} 6 & 4 & 744 \\ 20 & 15 & 2820 \\ 15 & 12 & 2280 \end{vmatrix} = \begin{vmatrix} 6 & 4 & 744 \\ 5 & 3 & 540 \\ 15 & 12 & 2280 \end{vmatrix}$$

Subtracting row 2 from row 1 and adding -3 times row 2 to row 3 yields:

$$D_z = \begin{vmatrix} 1 & 1 & 204 \\ 5 & 3 & 540 \\ 0 & 3 & 660 \end{vmatrix}$$

Adding -5 times row 1 to row 2 yields:

$$D_z = \begin{vmatrix} 1 & 1 & 204 \\ 0 & -2 & -480 \\ 0 & 3 & 660 \end{vmatrix} = 1 \begin{vmatrix} -2 & -480 \\ 3 & 660 \end{vmatrix} = 1(120) = 120$$

So $x = \dfrac{D_x}{D} = 24$, $y = \dfrac{D_y}{D} = 60$, and $z = \dfrac{D_z}{D} = 120$. So the numbers are 24, 60, and 120.

117. Given the equations:
$$xy = m$$
$$x^2 + y^2 = n$$
Add twice the first equation to the second to obtain:
$$x^2 + 2xy + y^2 = n + 2m$$
$$(x + y)^2 = n + 2m$$
$$x + y = \pm\sqrt{n + 2m}$$
Subtract twice the first equation from the second to obtain:
$$x^2 - 2xy + y^2 = n - 2m$$
$$(x - y)^2 = n - 2m$$
$$x - y = \pm\sqrt{n - 2m}$$
Solve the systems:

$$x + y = \sqrt{n + 2m}$$
$$x - y = \sqrt{n - 2m}$$
$$2x = \sqrt{n + 2m} + \sqrt{n - 2m}$$
$$x = \frac{\sqrt{n + 2m} + \sqrt{n - 2m}}{2}$$
$$y = \frac{\sqrt{n + 2m} - \sqrt{n - 2m}}{2}$$

$$x + y = \sqrt{n + 2m}$$
$$x - y = -\sqrt{n - 2m}$$
$$2x = \sqrt{n + 2m} - \sqrt{n - 2m}$$
$$x = \frac{\sqrt{n + 2m} - \sqrt{n - 2m}}{2}$$
$$y = \frac{\sqrt{n + 2m} + \sqrt{n - 2m}}{2}$$

And also the systems:

$$x + y = -\sqrt{n + 2m}$$
$$x - y = \sqrt{n - 2m}$$
$$2x = \sqrt{n - 2m} - \sqrt{n + 2m}$$
$$x = \frac{\sqrt{n - 2m} - \sqrt{n + 2m}}{2}$$
$$y = \frac{-\sqrt{n - 2m} - \sqrt{n + 2m}}{2}$$

$$x + y = -\sqrt{n + 2m}$$
$$x - y = -\sqrt{n - 2m}$$
$$2x = -\sqrt{n + 2m} - \sqrt{n - 2m}$$
$$x = \frac{-\sqrt{n + 2m} - \sqrt{n - 2m}}{2}$$
$$y = \frac{\sqrt{n - 2m} - \sqrt{n + 2m}}{2}$$

So the possible pairs of numbers are:
$$\frac{\sqrt{n + 2m} + \sqrt{n - 2m}}{2} \text{ and } \frac{\sqrt{n + 2m} - \sqrt{n - 2m}}{2},$$
$$\frac{\sqrt{n - 2m} - \sqrt{n + 2m}}{2} \text{ and } \frac{-\sqrt{n - 2m} - \sqrt{n + 2m}}{2}$$

119. The region is neither convex nor bounded. Graph the system of inequalities:

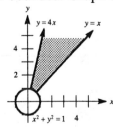

121. The region is convex and bounded. Graph the system of inequalities:

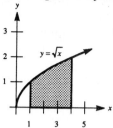

123. The region is neither convex nor bounded. Graph the system of inequalities:

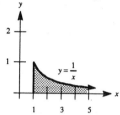

125. Graph the constraints:

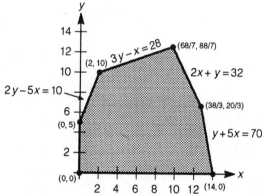

By solving systems of equations, we find the vertices to be $(2, 10)$, $\left(\frac{68}{7}, \frac{88}{7}\right)$, $\left(\frac{38}{3}, \frac{20}{3}\right)$, $(14, 0)$, $(0, 0)$, and $(0, 5)$. Complete the table:

Vertices	$C = 7x + 6y + 6$
$(2,10)$	80
$\left(\frac{68}{7}, \frac{88}{7}\right)$	$\frac{1046}{7} \approx 149.4$
$\left(\frac{38}{3}, \frac{20}{3}\right)$	$\frac{404}{3} \approx 134.7$
$(14,0)$	104
$(0,0)$	6
$(0,5)$	36

The maximum value is $\frac{1046}{7}$ at $\left(\frac{68}{7}, \frac{88}{7}\right)$, and the minimum value is 6 at $(0, 0)$.

Chapter Eleven Test

1.   Substitute $y = x^2 + 2x + 3$ into the first equation:

$$3x + 4y = 12$$
$$3x + 4(x^2 + 2x + 3) = 12$$
$$3x + 4x^2 + 8x + 12 = 12$$
$$4x^2 + 11x = 0$$
$$x(4x + 11) = 0$$
$$x = 0, -\frac{11}{4}$$

When $x = 0$, $y = 3$ and when $x = -\frac{11}{4}$, $y = \frac{81}{16}$. So the solutions are $(0, 3)$ and $\left(-\frac{11}{4}, \frac{81}{16}\right)$.

2.   Multiply the first equation by -3:

$$-3x + 6y = -39$$
$$3x + 5y = -16$$

Adding:

$$11y = -55$$
$$y = -5$$

Substitute into the first equation:

$$x - 2y = 13$$
$$x - 2(-5) = 13$$
$$x + 10 = 13$$
$$x = 3$$

So the solution is $(3, -5)$.

3.   (a)   Adding -3 times the first equation to the second equation, and also adding -4 times the first equation to the third equation results in the system:

$$x + 4y - z = 0$$
$$-11y + 4z = -1$$
$$-20y + 9z = -7$$

To reduce coefficients, multiply the second equation by -2 and add the third equation to the second equation:

$$x + 4y - z = 0$$
$$2y + z = -5$$
$$-20y + 9z = -7$$

Multiply the second equation by 10 and add it to the third equation:

$$x + 4y - z = 0$$
$$2y + z = -5$$
$$19z = -57$$

Solving the third equation for z yields $z = -3$. Substitute into the second equation:

$$2y - 3 = -5$$
$$2y = -2$$
$$y = -1$$

Substitute into the first equation:
$$x + 4(-1) - (-3) = 0$$
$$x - 4 + 3 = 0$$
$$x = 1$$
So the solution is $(1, -1, -3)$.

(b)   Adding row 1 to row 2, and 5 times row 1 to row 3:
$$D = \begin{vmatrix} 1 & 4 & -1 \\ 3 & 1 & 1 \\ 4 & -4 & 5 \end{vmatrix} = \begin{vmatrix} 1 & 4 & -1 \\ 4 & 5 & 0 \\ 9 & 16 & 0 \end{vmatrix}$$

Now expand along column 3:
$$D = -1 \begin{vmatrix} 4 & 5 \\ 9 & 16 \end{vmatrix} = -(64 - 45) = -19$$

Adding column 1 to columns 2 and 3:
$$D_x = \begin{vmatrix} 0 & 4 & -1 \\ -1 & 1 & 1 \\ -7 & -4 & 5 \end{vmatrix} = \begin{vmatrix} 0 & 4 & -1 \\ -1 & 0 & 0 \\ -7 & -11 & -2 \end{vmatrix}$$

Now expand along row 2:
$$D_x = -(-1) \begin{vmatrix} 4 & -1 \\ -11 & -2 \end{vmatrix} = 1(-8 - 11) = -19$$

Adding column 1 to column 3:
$$D_y = \begin{vmatrix} 1 & 0 & -1 \\ 3 & -1 & 1 \\ 4 & -7 & 5 \end{vmatrix} = \begin{vmatrix} 1 & 0 & 0 \\ 3 & -1 & 4 \\ 4 & -7 & 9 \end{vmatrix}$$

Now expand along row 1:
$$D_y = 1 \begin{vmatrix} -1 & 4 \\ -7 & 9 \end{vmatrix} = -9 + 28 = 19$$

Adding -4 times column 1 to column 2:
$$D_z = \begin{vmatrix} 1 & 4 & 0 \\ 3 & 1 & -1 \\ 4 & -4 & -7 \end{vmatrix} = \begin{vmatrix} 1 & 0 & 0 \\ 3 & -11 & -1 \\ 4 & -20 & -7 \end{vmatrix}$$

Now expand along row 1:
$$D_z = 1 \begin{vmatrix} -11 & -1 \\ -20 & -7 \end{vmatrix} = 77 - 20 = 57$$

So $x = \frac{D_x}{D} = \frac{-19}{-19} = 1$, $y = \frac{D_y}{D} = \frac{19}{-19} = -1$, and $z = \frac{D_z}{D} = \frac{57}{-19} = -3$. So the solution is $(1, -1, -3)$, which verifies our answer from (a).

4.   (a) $2A - B = \begin{pmatrix} 2 & -6 \\ 4 & -2 \end{pmatrix} - \begin{pmatrix} 0 & 4 \\ 1 & 3 \end{pmatrix} = \begin{pmatrix} 2 & -10 \\ 3 & -5 \end{pmatrix}$

(b) $BA = \begin{pmatrix} 0 & 4 \\ 1 & 3 \end{pmatrix}\begin{pmatrix} 1 & -3 \\ 2 & -1 \end{pmatrix} = \begin{pmatrix} 8 & -4 \\ 7 & -6 \end{pmatrix}$

5.   First graph the region:

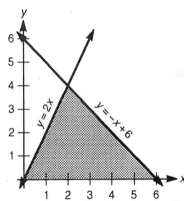

The base of the triangle is the x-intercept of $y = -x + 6$, which is 6. To find the height, find the intersection points of the lines $y = 2x$ and $y = -x + 6$:

$$2x = -x + 6$$
$$3x = 6$$
$$x = 2$$
$$y = 4$$

So the height is 4. Find the area:

$$\text{Area} = \tfrac{1}{2}(\text{base})(\text{height}) = \tfrac{1}{2}(6)(4) = 12 \text{ sq. units}$$

6.   Let $u = \tfrac{1}{x}$ and $v = \tfrac{1}{y}$, so we have the system:

$$\tfrac{1}{2}u + \tfrac{1}{3}v = 10$$
$$-5u - 4v = -4$$

Multiply the first equation by 10:

$$5u + \tfrac{10}{3}v = 100$$
$$-5u - \ 4v = -4$$

Adding:

$$-\tfrac{2}{3}v = 96$$
$$v = -144$$

Substitute into the first equation to find u:

$$\tfrac{1}{2}u + \tfrac{1}{3}(-144) = 10$$
$$\tfrac{1}{2}u - 48 = 10$$
$$\tfrac{1}{2}u = 58$$
$$u = 116$$

So $\tfrac{1}{x} = 116$ thus $x = \tfrac{1}{116}$, and $\tfrac{1}{y} = -144$ thus $y = -\tfrac{1}{144}$. So the solution is $\left(\tfrac{1}{116}, -\tfrac{1}{144}\right)$.

7.   The coefficient matrix is $\begin{pmatrix} 1 & 1 & -1 \\ 2 & -1 & 2 \\ 1 & -2 & 1 \end{pmatrix}$, and the augmented matrix is

$$\begin{pmatrix} 1 & 1 & -1 & -1 \\ 2 & -1 & 2 & 11 \\ 1 & -2 & 1 & 10 \end{pmatrix}.$$

8.   Using the augmented matrix formed in the preceeding exercise, add −2 times row 1 to row 2 and −1 times row 1 to row 3:

$$\begin{pmatrix} 1 & 1 & -1 & -1 \\ 0 & -3 & 4 & 13 \\ 0 & -3 & 2 & 11 \end{pmatrix}$$

Multiply row 2 by −1 and add to row 3:

$$\begin{pmatrix} 1 & 1 & -1 & -1 \\ 0 & -3 & 4 & 13 \\ 0 & 0 & -2 & -2 \end{pmatrix}$$

So we have the system of equations:
$$x + y - z = -1$$
$$-3y + 4z = 13$$
$$-2z = -2$$

Solving the third equation for z yields $z = 1$. Substitute into the second equation:
$$-3y + 4(1) = 13$$
$$-3y = 9$$
$$y = -3$$

Substitute into the first equation:
$$x - 3 - 1 = -1$$
$$x - 4 = -1$$
$$x = 3$$

So the solution is $(3, -3, 1)$.

9.   First find the point of intersection:
$$x + y = 11$$
$$3x + 2y = 7$$

Multiply the first equation by −2:
$$-2x - 2y = -22$$
$$3x + 2y = 7$$

Adding:
$$x = -15$$

Substitute into the first equation:
$$-15 + y = 11$$
$$y = 26$$

So the point of intersection is (–15, 26). Now find the slope of the line $2x - 4y = 7$:

$$-4y = -2x + 7$$
$$y = \tfrac{1}{2}x - \tfrac{7}{4}$$

Since this slope is $\tfrac{1}{2}$, the perpendicular slope must be –2. Use this slope and the point (–15, 26) in the point-slope formula:

$$y - 26 = -2(x + 15)$$
$$y - 26 = -2x - 30$$
$$y = -2x - 4$$

10. (a) The minor is $\begin{vmatrix} 2 & -1 \\ 0 & 4 \end{vmatrix} = 8.$

   (b) The cofactor is $-1\begin{vmatrix} 2 & -1 \\ 0 & 4 \end{vmatrix} = -1(8) = -8.$

11. Adding 2 times row 1 to row 2, and – 4 times row 1 to row 3:

$$\begin{vmatrix} 4 & -5 & 0 \\ 0 & 0 & 7 \\ 0 & 40 & 14 \end{vmatrix}$$

Expanding along column 1:

$$4\begin{vmatrix} 0 & 7 \\ 49 & 14 \end{vmatrix} = 4(0 - 280) = -1120$$

12. If we multiply the second equation by 2 we have $2xy = 10$. Adding the two equations, we obtain:

$$x^2 + 2xy + y^2 = 25$$
$$(x + y)^2 = 25$$
$$x + y = \pm 5$$

Subtracting the two equations:

$$x^2 - 2xy + y^2 = 5$$
$$(x - y)^2 = 5$$
$$x - y = \pm\sqrt{5}$$

So we have the following four systems of equations:

$x + y = 5$	$x + y = 5$	$x + y = -5$	$x + y = -5$
$x - y = \sqrt{5}$	$x - y = -\sqrt{5}$	$x - y = \sqrt{5}$	$x - y = -\sqrt{5}$

Adding results in the four equations:

$2x = 5 + \sqrt{5}$	$2x = 5 - \sqrt{5}$	$2x = -5 + \sqrt{5}$	$2x = -5 - \sqrt{5}$
$x = \dfrac{5 + \sqrt{5}}{2}$	$x = \dfrac{5 - \sqrt{5}}{2}$	$x = \dfrac{-5 + \sqrt{5}}{2}$	$x = \dfrac{-5 - \sqrt{5}}{2}$
$y = \dfrac{5 - \sqrt{5}}{2}$	$y = \dfrac{5 + \sqrt{5}}{2}$	$y = \dfrac{-5 - \sqrt{5}}{2}$	$y = \dfrac{-5 + \sqrt{5}}{2}$

So the solutions are $\left(\tfrac{5+\sqrt{5}}{2}, \tfrac{5-\sqrt{5}}{2}\right)$, $\left(\tfrac{5-\sqrt{5}}{2}, \tfrac{5+\sqrt{5}}{2}\right)$, $\left(\tfrac{-5+\sqrt{5}}{2}, \tfrac{-5-\sqrt{5}}{2}\right)$, and $\left(\tfrac{-5-\sqrt{5}}{2}, \tfrac{-5+\sqrt{5}}{2}\right)$.

13. Multiply the first equation by –2:
$$-2A - 4B - 6C = -2$$
$$2A - B - C = 2$$
Adding:
$$-5B - 7C = 0$$
$$-5B = 7C$$
$$B = -\tfrac{7}{5}C$$
Substitute into the (original) first equation:
$$A + 2B + 3C = 1$$
$$A - \tfrac{14}{5}C + 3C = 1$$
$$A + \tfrac{1}{5}C = 1$$
$$A = 1 - \tfrac{1}{5}C$$
So the solution is $\left(1 - \tfrac{1}{5}C, -\tfrac{7}{5}C, C\right)$, where $C =$ any real number.

14. (a) Form the matrix:
$$\begin{pmatrix} 10 & -2 & 5 & 1 & 0 & 0 \\ 6 & -1 & 4 & 0 & 1 & 0 \\ 1 & 0 & 1 & 0 & 0 & 1 \end{pmatrix}$$
Switching rows:
$$\begin{pmatrix} 1 & 0 & 1 & 0 & 0 & 1 \\ 6 & -1 & 4 & 0 & 1 & 0 \\ 10 & -2 & 5 & 1 & 0 & 0 \end{pmatrix}$$
Adding – 6 times row 1 to row 2, and –10 times row 1 to row 3:
$$\begin{pmatrix} 1 & 0 & 1 & 0 & 0 & 1 \\ 0 & -1 & -2 & 0 & 1 & -6 \\ 0 & -2 & -5 & 1 & 0 & -10 \end{pmatrix}$$
Multiply row 2 by –1, and adding 2 times this new row 2 to row 3:
$$\begin{pmatrix} 1 & 0 & 1 & 0 & 0 & 1 \\ 0 & 1 & 2 & 0 & -1 & 6 \\ 0 & 0 & -1 & 1 & -2 & 2 \end{pmatrix}$$
Adding 2 times row 3 to row 2, row 3 to row 1, then multiplying row 3 by –1:
$$\begin{pmatrix} 1 & 0 & 0 & 1 & -2 & 3 \\ 0 & 1 & 0 & 2 & -5 & 10 \\ 0 & 0 & 1 & -1 & 2 & -2 \end{pmatrix}$$
So the inverse is $\begin{pmatrix} 1 & -2 & 3 \\ 2 & -5 & 10 \\ -1 & 2 & -2 \end{pmatrix}$.

(b) Calling A the coefficient matrix and $B = \begin{pmatrix} -1 \\ -2 \\ 3 \end{pmatrix}$, we know the solution will be given

by $A^{-1}B$:

$$\begin{pmatrix} u \\ v \\ w \end{pmatrix} = \begin{pmatrix} 1 & -2 & 3 \\ 2 & -5 & 10 \\ -1 & 2 & -2 \end{pmatrix} \begin{pmatrix} -1 \\ -2 \\ 3 \end{pmatrix} = \begin{pmatrix} 12 \\ 38 \\ -9 \end{pmatrix}$$

So the solution is $(12, 38, -9)$.

15. Graph $5x - 6y \geq 30$:

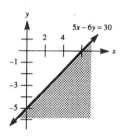

16. Substituting the two points into $y = Px^2 + Qx - 5$:
$$-1 = 4P - 2Q - 5$$
$$-2 = P - Q - 5$$
So we have the system:
$$4P - 2Q = 4$$
$$P - Q = 3$$
Divide the first equation by 2, and multiply the second equation by -1:
$$2P - Q = 2$$
$$-P + Q = -3$$
Adding, we obtain $P = -1$. Substitute into the second equation:
$$-1 - Q = 3$$
$$-Q = 4$$
$$Q = -4$$
So $P = -1$ and $Q = -4$.

17. Graph the inequality:

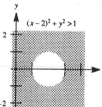

The solution set is neither bounded nor convex.

18. The vertices are $(0, 0)$, $(0, 7)$, $(6, 10)$, $\left(\frac{261}{26}, \frac{225}{26}\right)$, and $(11, 0)$. Sketch the graph:

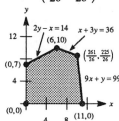

19. Let x = number of small motors produced and y = number of large motors produced. Therefore:

Revenue $= 40x + 104y$
Cost $= 15x + 30y + 8(2x + 6y) = 31x + 78y$
Profit $= (40x + 104y) - (31x + 78y) = 9x + 26y$

The constraints are given by:
   (1) $x \geq 0$
   (2) $y \geq 0$
   (3) $15x + 30y \leq 1500$, or $x + 2y \leq 100$
   (4) $16x + 48y \leq 2000$, or $x + 3y \leq 125$

Graph the constraints:

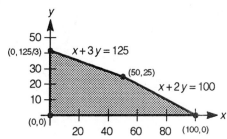

Set up the table:

Vertices	$P = 9x + 26y$
$(0,0)$	0
$(100,0)$	900
$(50,25)$	1100
$(0, \frac{125}{3})$	1083.33

The maximum profit is $1100 with 50 small motors and 25 large motors being produced.

20. Since the solution is $A^{-1}b$ for $(\ln x, \ln y, \ln z)$:

$$A^{-1}b = \begin{pmatrix} -3 & 2 & -4 \\ -1 & 1 & -1 \\ 8 & -5 & 10 \end{pmatrix} \begin{pmatrix} 3 \\ -1 \\ 2 \end{pmatrix} = \begin{pmatrix} -19 \\ -6 \\ 49 \end{pmatrix}$$

So $(\ln x, \ln y, \ln z) = (-19, -6, 49)$, and thus $(x, y, z) = \left(e^{-19}, e^{-6}, e^{49}\right)$.

Chapter Twelve
Roots of Polynomial Equations

12.1 The Complex Number System

1. Complete the table:

i^2	i^3	i^4	i^5	i^6	i^7	i^8
-1	$-i$	1	i	-1	$-i$	1

3. (a) The real part is 4 and the imaginary part is 5.
   (b) The real part is 4 and the imaginary part is -5.
   (c) The real part is $\frac{1}{2}$ and the imaginary part is -1.
   (d) The real part is 0 and the imaginary part is 16.

5. Equating the real parts gives $2c = 8$, and therefore $c = 4$. Similarly, equating the imaginary parts yields $d = -3$.

7. (a) $(5 - 6i) + (9+2i) = (5 + 9) + (-6 + 2)i = 14 - 4i$
   (b) $(5 - 6i) - (9+2i) = (5 - 9) + (-6 - 2)i = -4 - 8i$

9. (a) $(3 - 4i)(5 + i) = 15 - 17i - 4i^2 = 19 - 17i$

   (b) $(5 + i)(3 - 4i) = 19 - 17i$, from part (a)

(c) Compute the quotient:

$$\frac{3-4i}{5+i} \cdot \frac{5-i}{5-i} = \frac{15-23i+4i^2}{25-i^2} = \frac{11-23i}{26} = \tfrac{11}{26} - \tfrac{23}{26}i$$

(d) Compute the quotient:

$$\frac{5+i}{3-4i} \cdot \frac{3+4i}{3+4i} = \frac{15+23i+4i^2}{9-16i^2} = \frac{11+23i}{25} = \tfrac{11}{25} + \tfrac{23}{25}i$$

11. (a) $z+w = (2+3i)+(9-4i) = 11-i$

(b) $\bar{z}+w = (2-3i)+(9-4i) = 11-7i$

(c) $z+\bar{z} = (2+3i)+(2-3i) = 4$

13. Evaluate the expression:

$$(z+w)+w_1 = \left[(2+3i)+(9-4i)\right]+(-7-i) = (11-i)+(-7-i) = 4-2i$$

15. Compute the product:

$$zw = (2+3i)(9-4i) = 18+19i-12i^2 = 30+19i$$

17. $z\bar{z} = (2+3i)(2-3i) = 4-9i^2 = 13$

19. First compute the product:

$$ww_1 = (9-4i)(-7-i) = -63+19i+4i^2 = -67+19i$$

Therefore:

$$z(ww_1) = (2+3i)(-67+19i) = -134-163i+57i^2 = -191-163i$$

21. First compute the sum:

$$w+w_1 = (9-4i)+(-7-i) = 2-5i$$

Therefore:

$$z(w+w_1) = (2+3i)(2-5i) = 4-4i-15i^2 = 19-4i$$

23. First compute the powers:

$$z^2 = (2+3i)(2+3i) = 4+12i+9i^2 = -5+12i$$
$$w^2 = (9-4i)(9-4i) = 81-72i+16i^2 = 65-72i$$

Therefore:

$$z^2-w^2 = (-5+12i)-(65-72i) = -70+84i$$

25. Since $zw = 30+19i$ (from Exercise 15), we have:

$$(zw)^2 = (30+19i)(30+19i) = 900+1140i+361i^2 = 539+1140i$$

27. Since $z^2 = -5+12i$ (from Exercise 23), we have:

$$z^3 = z \cdot z^2 = (2+3i)(-5+12i) = -10+9i+36i^2 = -46+9i$$

29. Compute the quotient:
$$\frac{z}{w} = \frac{2+3i}{9-4i} \cdot \frac{9+4i}{9+4i} = \frac{18+35i+12i^2}{81-16i^2} = \frac{6+35i}{97} = \frac{6}{97} + \frac{35}{97}i$$

31. Using the values of $\bar{z}$ and $\bar{w}$, we have:
$$\frac{\bar{z}}{\bar{w}} = \frac{2-3i}{9+4i} \cdot \frac{9-4i}{9-4i} = \frac{18-35i+12i^2}{81-16i^2} = \frac{6-35i}{97} = \frac{6}{97} - \frac{35}{97}i$$

33. Using the value of $\bar{z}$, we have:
$$\frac{z}{\bar{z}} = \frac{2+3i}{2-3i} \cdot \frac{2+3i}{2+3i} = \frac{4+12i+9i^2}{4-9i^2} = \frac{-5+12i}{13} = -\frac{5}{13} + \frac{12}{13}i$$

35. Since $w - \bar{w} = (9-4i)-(9+4i) = -8i$, we have:
$$\frac{w-\bar{w}}{2i} = \frac{-8i}{2i} = -4$$

37. Compute the quotient:
$$\frac{i}{5+i} \cdot \frac{5-i}{5-i} = \frac{5i-i^2}{25-i^2} = \frac{1+5i}{26} = \frac{1}{26} + \frac{5}{26}i$$

39. Compute the quotient:
$$\frac{1}{i} \cdot \frac{i}{i} = \frac{i}{i^2} = \frac{i}{-1} = -i$$

41. By writing the numbers in complex form, we have:
$$\sqrt{-49} + \sqrt{-9} + \sqrt{-4} = 7i + 3i + 2i = 12i$$

43. By writing the numbers in complex form, we have:
$$\sqrt{-20} - 3\sqrt{-45} + \sqrt{-80} = \sqrt{4}\sqrt{-5} - 3\sqrt{9}\sqrt{-5} + \sqrt{16}\sqrt{-5}$$
$$= 2\sqrt{5}i - 9\sqrt{5}i + 4\sqrt{5}i$$
$$= -3\sqrt{5}i$$

45. By writing the numbers in complex form, we have:
$$1 + \sqrt{-36}\sqrt{-36} = 1 + (6i)(6i) = 1 + 36i^2 = -35$$

47. By writing the numbers in complex form, we have:
$$3\sqrt{-128} - 4\sqrt{-18} = 3\sqrt{-64}\sqrt{2} - 4\sqrt{-9}\sqrt{2}$$
$$= 3(8i)\sqrt{2} - 4(3i)\sqrt{2}$$
$$= 24\sqrt{2}i - 12\sqrt{2}i$$
$$= 12\sqrt{2}i$$

49. (a) Computing the discriminant:
$$b^2 - 4ac = (-1)^2 - 4(1)(1) = 1 - 4 = -3$$
Since the discriminant is negative, the quadratic will have complex roots.

(b) Using the quadratic formula:
$$x = \frac{-(-1) \pm \sqrt{-3}}{2(1)} = \frac{1 \pm i\sqrt{3}}{2} = \frac{1}{2} \pm \frac{\sqrt{3}}{2}i$$

51. (a) Computing the discriminant:
$$b^2 - 4ac = (2)^2 - 4(5)(2) = 4 - 40 = -36$$
Since the discriminant is negative, the quadratic will have complex roots.

(b) Using the quadratic formula:
$$z = \frac{-2 \pm \sqrt{-36}}{2(5)} = \frac{-2 \pm 6i}{10} = -\frac{1}{5} \pm \frac{3}{5}i$$

53. (a) Computing the discriminant:
$$b^2 - 4ac = (3)^2 - 4(2)(4) = 9 - 32 = -23$$
Since the discriminant is negative, the quadratic will have complex roots.

(b) Using the quadratic formula:
$$z = \frac{-3 \pm \sqrt{-23}}{2(2)} = \frac{-3 \pm i\sqrt{23}}{4} = -\frac{3}{4} \pm \frac{\sqrt{23}}{4}i$$

55. (a) Computing the discriminant:
$$b^2 - 4ac = \left(-\frac{1}{4}\right)^2 - 4\left(\frac{1}{6}\right)(1) = \frac{1}{16} - \frac{2}{3} = -\frac{29}{48}$$
Since the discriminant is negative, the quadratic will have complex roots.

(b) Using the quadratic formula:
$$z = \frac{-(-3) \pm \sqrt{-87}}{2(2)} = \frac{3 \pm i\sqrt{87}}{4} = \frac{3}{4} \pm \frac{\sqrt{87}}{4}i$$

57. (a) Substituting $x = 2 - i\sqrt{2}$:
$$\begin{aligned}
x^2 - 4x + 6 &= \left(2 - i\sqrt{2}\right)^2 - 4\left(2 - i\sqrt{2}\right) + 6 \\
&= 4 - 4i\sqrt{2} - 2 - 8 + 4i\sqrt{2} + 6 \\
&= 0
\end{aligned}$$

(b) This verifies the solution found in Example 6.

59. (a) Notice that the result does agree with the definition:
$$z + w = (a + bi) + (c + di) = (a + c) + (b + d)i$$

(b) Notice that the result does agree with the definition:
$$z - w = (a + bi) - (c + di) = (a - c) + (b - d)i$$

(c) Notice that the result does agree with the definition:
$$zw = (a + bi)(c + di) = ac + bci + adi + bdi^2 = (ac - bd) + (bc + ad)i$$

(d) Notice that the result does agree with the definition:
$$\frac{z}{w} = \frac{a + bi}{c + di} \cdot \frac{c - di}{c - di} = \frac{ac + bci - adi - bdi^2}{c^2 - d^2 i^2} = \frac{ac + bd}{c^2 + d^2} + \frac{bc - ad}{c^2 + d^2} i$$

61. (a) Compute each power:

$$z^3 = \left(\frac{-1 + \sqrt{3}i}{2}\right)^3$$

$$= \frac{\left(-1 + \sqrt{3}i\right)^3}{8}$$

$$= \frac{\left(-1 + \sqrt{3}i\right)\left(-1 + \sqrt{3}i\right)^2}{8}$$

$$= \frac{\left(-1 + \sqrt{3}i\right)\left(1 - 2\sqrt{3}i + 3i^2\right)}{8}$$

$$= \frac{\left(-1 + \sqrt{3}i\right)\left(-2 - 2\sqrt{3}i\right)}{8}$$

$$= \frac{2 - 2\sqrt{3}i + 2\sqrt{3}i - 6i^2}{8}$$

$$= \frac{8}{8}$$

$$= 1$$

$$w^3 = \left(\frac{-1 - \sqrt{3}i}{2}\right)^3$$

$$= \frac{\left(-1 - \sqrt{3}i\right)^3}{8}$$

$$= \frac{\left(-1 - \sqrt{3}i\right)\left(-1 - \sqrt{3}i\right)^2}{8}$$

$$= \frac{\left(-1 - \sqrt{3}i\right)\left(1 + 2\sqrt{3}i + 3i^2\right)}{8}$$

$$= \frac{\left(-1 - \sqrt{3}i\right)\left(-2 + 2\sqrt{3}i\right)}{8}$$

$$= \frac{2 + 2\sqrt{3}i - 2\sqrt{3}i - 6i^2}{8}$$

$$= \frac{8}{8}$$

$$= 1$$

(b) Compute the product:
$$zw = \left(\frac{-1 + \sqrt{3}i}{2}\right)\left(\frac{-1 - \sqrt{3}i}{2}\right) = \frac{1 - \sqrt{3}i + \sqrt{3}i - 3i^2}{4} = \frac{4}{4} = 1$$

(c) Compute each power:

$$w^2 = \left(\frac{-1 - \sqrt{3}i}{2}\right)^2 = \frac{1 + 2\sqrt{3}i + 3i^2}{4} = \frac{-2 + 2\sqrt{3}i}{4} = \frac{-1 + \sqrt{3}i}{2} = z$$

$$z^2 = \left(\frac{-1 + \sqrt{3}i}{2}\right)^2 = \frac{1 - 2\sqrt{3}i + 3i^2}{4} = \frac{-2 - 2\sqrt{3}i}{4} = \frac{-1 - \sqrt{3}i}{2} = w$$

Note: Another approach (using parts (a) and (b)) is to recognize:

$$z^3 = zw, \text{ so } z^2 = \frac{z^3}{z} = \frac{zw}{z} = w$$

$$w^3 = zw, \text{ so } w^2 = \frac{w^3}{w} = \frac{zw}{w} = z$$

(d) Compute the product:

$$\left(1 - z + z^2\right)\left(1 + z - z^2\right) = (1 - z + w)(1 + z - w)$$

$$= \left(\frac{2}{2} - \frac{-1 + \sqrt{3}i}{2} + \frac{-1 - \sqrt{3}i}{2}\right)\left(\frac{2}{2} + \frac{-1 + \sqrt{3}i}{2} - \frac{-1 - \sqrt{3}i}{2}\right)$$

$$= \left(\frac{2 - 2\sqrt{3}i}{2}\right)\left(\frac{2 + 2\sqrt{3}i}{2}\right)$$

$$= \left(1 - \sqrt{3}i\right)\left(1 + \sqrt{3}i\right)$$

$$= 1 - 3i^2$$

$$= 4$$

63. (a) Let $z = a + bi$, then:

$$0 + z = (0 + 0i) + (a + bi) = a + bi = z$$

$$z + 0 = (a + bi) + (0 + 0i) = a + bi = z$$

(b) Let $z = a + bi$, then:

$$0 \bullet z = (0 + 0i)(a + bi) = 0 + 0i = 0$$

$$z \bullet 0 = (a + bi)(0 + 0i) = 0$$

65. (a) Compute each sum:

$$z + w = (a + bi) + (c + di) = (a + c) + (b + d)i$$

$$w + z = (c + di) + (a + bi) = (c + a) + (d + b)i = z + w$$

(b) Compute each product:

$$zw = (a + bi)(c + di) = ac + bci + adi + bdi^2 = (ac - bd) + (bc + ad)i$$

$$wz = (c + di)(a + bi) = ac + adi + bci + bdi^2 = (ac - bd) + (bc + ad)i = zw$$

67. Factoring by grouping:

$$x^3 - 3x^2 + 4x - 12 = 0$$

$$x^2(x - 3) + 4(x - 3) = 0$$

$$(x - 3)\left(x^2 + 4\right) = 0$$

So $x = 3$ or $x^2 + 4 = 0$, thus $x^2 = -4$ and $x = \pm 2i$. The roots are $3, \pm 2i$.

69. Factoring by grouping:
$$x^6 - 9x^4 + 16x^2 - 144 = 0$$
$$x^4(x^2 - 9) + 16(x^2 - 9) = 0$$
$$(x^2 - 9)(x^4 + 16) = 0$$

So $x^2 = 9$, or $x = \pm 3$, or $x^4 = -16$, so $x^2 = \pm 4i$, so $x = \pm\sqrt{4i} = \pm 2\sqrt{i}$ or $x = \pm\sqrt{-4i} = \pm 2i\sqrt{i}$. The roots are ± 3, $\pm 2\sqrt{i}$, and $\pm 2i\sqrt{i}$.

71. Compute each quotient then add the complex numbers:
$$\frac{a+bi}{a-bi} + \frac{a-bi}{a+bi} = \frac{a+bi}{a-bi}\cdot\frac{a+bi}{a+bi} + \frac{a-bi}{a+bi}\cdot\frac{a-bi}{a-bi}$$
$$= \frac{(a+bi)^2}{a^2+b^2} + \frac{(a-bi)^2}{a^2+b^2}$$
$$= \frac{a^2 + 2abi - b^2 + a^2 - 2abi - b^2}{a^2+b^2}$$
$$= \frac{2a^2 - 2b^2}{a^2+b^2}$$

Thus the real part is $\dfrac{2a^2 - 2b^2}{a^2+b^2}$, and the imaginary part is 0.

73. Compute each quotient then add the complex numbers:
$$\frac{(a+bi)^2}{a-bi} - \frac{(a-bi)^2}{a+bi} = \frac{(a+bi)^3 - (a-bi)^3}{(a-bi)(a+bi)}$$
$$= \frac{\left(a^3 + 3a^2bi + 3ab^2i^2 + b^3i^3\right) - \left(a^3 - 3a^2bi + 3ab^2i^2 - b^3i^3\right)}{a^2 - b^2i^2}$$
$$= \frac{\left(a^3 - 3ab^2\right) + \left(3a^2b - b^3\right)i - \left(a^3 - 3ab^2\right) + \left(3a^2b - b^3\right)}{a^2+b^2}$$
$$= \frac{6a^2b - 2b^3}{a^2+b^2}\,i$$

Thus the real part is 0.

75. First compute zw:
$$zw = (a+bi)(c+di) = ac + bci + adi + bdi^2 = (ac - bd) + (bc + ad)i$$
Since $zw = 0$, $ac - bd = 0$ and $bc + ad = 0$. Now assuming $a \neq 0$, we can solve the first equation for c to obtain $c = \frac{bd}{a}$. Substituting into the second equation yields:
$$b\left(\tfrac{bd}{a}\right) + ad = 0$$
$$b^2d + a^2d = 0$$
$$(b^2 + a^2)d = 0$$
But if $a \neq 0$, then $b^2 + a^2 \neq 0$ and thus $d = 0$. Since $c = \frac{bd}{a}$, $c = 0$. Thus $w = 0$.

12.2 More on Division of Polynomials

1. Using synthetic division:

    $$\begin{array}{r|rrr} 5 & 1 & -6 & -2 \\ & & 5 & -5 \\ \hline & 1 & -1 & -7 \end{array}$$

    The quotient is $x - 1$ and the remainder is -7. So $x^2 - 6x - 2 = (x - 5)(x - 1) - 7$.

3. Using synthetic division:

    $$\begin{array}{r|rrr} -1 & 4 & -1 & -5 \\ & & -4 & 5 \\ \hline & 4 & -5 & 0 \end{array}$$

    The quotient is $4x - 5$ and the remainder is 0. So $4x^2 - x - 5 = (x + 1)(4x - 5) + 0$.

5. Using synthetic division:

    $$\begin{array}{r|rrrr} 4 & 6 & -5 & 2 & 1 \\ & & 24 & 76 & 312 \\ \hline & 6 & 19 & 78 & 313 \end{array}$$

    The quotient is $6x^2 + 19x + 78$ and the remainder is 313.
    So $6x^3 - 5x^2 + 2x + 1 = (x - 4)(6x^2 + 19x + 78) + 313$.

7. Using synthetic division:

    $$\begin{array}{r|rrrr} 2 & 1 & 0 & 0 & -1 \\ & & 2 & 4 & 8 \\ \hline & 1 & 2 & 4 & 7 \end{array}$$

    The quotient is $x^2 + 2x + 4$ and the remainder is 7. So $x^3 - 1 = (x - 2)(x^2 + 2x + 4) + 7$.

9. Using synthetic division:

    $$\begin{array}{r|rrrrrr} -2 & 1 & 0 & 0 & 0 & 0 & -1 \\ & & -2 & 4 & -8 & 16 & -32 \\ \hline & 1 & -2 & 4 & -8 & 16 & -33 \end{array}$$

    The quotient is $x^4 - 2x^3 + 4x^2 - 8x + 16$ and the remainder is -33.
    So $x^5 - 1 = (x + 2)(x^4 - 2x^3 + 4x^2 - 8x + 16) - 33$.

11. Using synthetic division:

    $$\begin{array}{r|rrrrr} -4 & 1 & -6 & 0 & 0 & 2 \\ & & -4 & 40 & -160 & 640 \\ \hline & 1 & -10 & 40 & -160 & 642 \end{array}$$

    The quotient is $x^3 - 10x^2 + 40x - 160$ and the remainder is 642.
    So $x^4 - 6x^3 + 2 = (x + 4)(x^3 - 10x^2 + 40x - 160) + 642$.

13. Using synthetic division:

$$\begin{array}{r|rrrr} 10 & 1 & -4 & -3 & 6 \\ & & 10 & 60 & 570 \\ \hline & 1 & 6 & 57 & 576 \end{array}$$

The quotient is $x^2 + 6x + 57$ and the remainder is 576.
So $x^3 - 4x^2 - 3x + 6 = (x - 10)(x^2 + 6x + 57) + 576$.

15. Using synthetic division:

$$\begin{array}{r|rrrr} -5 & 1 & -1 & 0 & 0 \\ & & -5 & 30 & -150 \\ \hline & 1 & -6 & 30 & -150 \end{array}$$

The quotient is $x^2 - 6x + 30$ and the remainder is -150.
So $x^3 - x^2 = (x + 5)(x^2 - 6x + 30) - 150$.

17. Using synthetic division:

$$\begin{array}{r|rrrr} 2/3 & 54 & -27 & -27 & 14 \\ & & 36 & 6 & -14 \\ \hline & 54 & 9 & -21 & 0 \end{array}$$

The quotient is $54x^2 + 9x - 21$ and the remainder is 0.
So $54x^3 - 27x^2 - 27x + 14 = \left(x - \frac{2}{3}\right)\left(54x^2 + 9x - 21\right) + 0$.

19. Using synthetic division:

$$\begin{array}{r|rrrrr} 3 & 1 & 0 & 3 & 0 & 12 \\ & & 3 & 9 & 36 & 108 \\ \hline & 1 & 3 & 12 & 36 & 120 \end{array}$$

The quotient is $x^3 + 3x^2 + 12x + 36$ and the remainder is 120.
So $x^4 + 3x^2 + 12 = (x - 3)(x^3 + 3x^2 + 12x + 36) + 120$.

21. Since $\sqrt[5]{32} = 2$ is a root, we use synthetic division:

$$\begin{array}{r|rrrrrr} 2 & 1 & 0 & 0 & 0 & 0 & -32 \\ & & 2 & 4 & 8 & 16 & 32 \\ \hline & 1 & 2 & 4 & 8 & 16 & 0 \end{array}$$

So $x^5 - 32 = (x - 2)\left(x^4 + 2x^3 + 4x^2 + 8x + 16\right)$.

23. Since $\sqrt[4]{81} = 3$ is a root, we use synthetic division:

$$\begin{array}{r|rrrrr} 3 & 1 & 0 & 0 & 0 & -81 \\ & & 3 & 9 & 27 & 81 \\ \hline & 1 & 3 & 9 & 27 & 0 \end{array}$$

So $z^4 - 81 = (z - 3)\left(z^3 + 3z^2 + 9z + 27\right)$.

25.   Since the two quotients are the same, we can now perform synthetic division:

$$\begin{array}{r|rrr} 4/3 & 2 & -8/3 & 1/3 \\ & & 8/3 & 0 \\ \hline & 2 & 0 & 1/3 \end{array}$$

The quotient is $2x$ and the remainder is $\frac{1}{3}$. So the original problem has a quotient of $2x$ and a remainder of $3\left(\frac{1}{3}\right) = 1$.

27.   Adapting the hint from Exercise 25, we first divide the numerator and denominator by 2 to form the quotient:

$$\frac{3x^3 + \frac{1}{2}}{x + \frac{1}{2}}$$

Using synthetic division:

$$\begin{array}{r|rrrr} -1/2 & 3 & 0 & 0 & 1/2 \\ & & -3/2 & 3/4 & -3/8 \\ \hline & 3 & -3/2 & 3/4 & 1/8 \end{array}$$

The quotient is $3x^2 - \frac{3}{2}x + \frac{3}{4}$ and the remainder is $\frac{1}{8}$. So the original problem has a quotient of $3x^2 - \frac{3}{2}x + \frac{3}{4}$ and a remainder of $2\left(\frac{1}{8}\right) = \frac{1}{4}$.

29.   Using synthetic division:

$$\begin{array}{r|rrrr} -1 & 1 & 0 & k & 1 \\ & & -1 & 1 & -k-1 \\ \hline & 1 & -1 & k+1 & -k \end{array}$$

So $-k = -4$, and thus $k = 4$.

31.   Using synthetic division:

$$\begin{array}{r|rrr} p & 1 & 2p & -3q^2 \\ & & p & 3p^2 \\ \hline & 1 & 3p & 3p^2 - 3q^2 \end{array}$$

Since the remainder is 0, we have:

$$3p^2 - 3q^2 = 0$$
$$3p^2 = 3q^2$$
$$p^2 = q^2$$

33.   By the division algorithm, we have:

$$f(x) = (x - a)(x - b)q(x) + (Ax + B)$$

Therefore:

$$f(a) = Aa + B$$
$$f(b) = Ab + B$$

We now use these values to simplify the expressions:

$$\frac{f(a) - f(b)}{a - b} = \frac{(Aa + B) - (Ab + B)}{a - b} = \frac{Aa - Ab}{a - b} = \frac{A(a - b)}{a - b} = A$$

$$\frac{bf(a) - af(b)}{b - a} = \frac{b(Aa + B) - a(Ab + B)}{b - a} = \frac{abA + bB - abA - aB}{b - a} = \frac{B(b - a)}{b - a} = B$$

This proves the desired results.

12.3 Roots of Polynomial Equations: The Remainder Theorem and the Factor Theorem

1. Substitute $x = 10$:
$$12(10) - 8 = 112$$
$$120 - 8 = 112$$
Yes, it is a root.

3. Substitute $x = 1 - \sqrt{5}$:
$$\left(1 - \sqrt{5}\right)^2 - 2\left(1 - \sqrt{5}\right) - 4 = 0$$
$$1 - 2\sqrt{5} + 5 - 2 + 2\sqrt{5} - 4 = 0$$
$$4 - 4 = 0$$
Yes, it is a root.

5. Substitute $x = \frac{1}{2}$:
$$2\left(\tfrac{1}{2}\right)^2 - 3\left(\tfrac{1}{2}\right) + 1 = 0$$
$$\tfrac{1}{2} - \tfrac{3}{2} + 1 = 0$$
$$-1 + 1 = 0$$
Yes, it is a root.

7. Compute $f\left(\frac{2}{3}\right)$:
$$f\left(\tfrac{2}{3}\right) = 3\left(\tfrac{2}{3}\right) - 2 = 2 - 2 = 0$$
So $x = \frac{2}{3}$ is a zero of $f(x)$.

9. Compute $h(-1)$:
$$h(-1) = 5(-1)^3 - (-1)^2 + 2(-1) + 8 = -5 - 1 - 2 + 8 = 0$$
So $x = -1$ is a zero of $h(x)$.

11. Compute $f(2)$:
$$f(2) = 1 + 2(2) + (2)^3 - (2)^5 = 1 + 4 + 8 - 32 = -19$$
So $t = 2$ is not a zero of $f(t)$.

13. (a) Compute $f\left(\frac{\sqrt{3}-1}{2}\right)$:
$$f\left(\frac{\sqrt{3}-1}{2}\right) = 2\left(\frac{\sqrt{3}-1}{2}\right)^3 - 3\left(\frac{\sqrt{3}-1}{2}\right) + 1 = \frac{3\sqrt{3}-5}{2} - \frac{3\sqrt{3}-3}{2} + 1 = -1 + 1 = 0$$
So $x = \frac{\sqrt{3}-1}{2}$ is a zero of $f(x)$.

(b) Compute $f\left(\frac{\sqrt{3}+1}{2}\right)$:

$$f\left(\frac{\sqrt{3}+1}{2}\right)=2\left(\frac{\sqrt{3}+1}{2}\right)^{3}-3\left(\frac{\sqrt{3}+1}{2}\right)+1=\frac{5+3\sqrt{3}}{2}-\frac{3\sqrt{3}-3}{2}+1=4+1=5$$

So $x=\frac{\sqrt{3}+1}{2}$ is not a zero of $f(x)$.

15. (a) 1, 2 (multiplicity 3), 3
    (b) 1 (multiplicity 3)
    (c) 5 (multiplicity 6), −1 (multiplicity 4)
    (d) 0 (multiplicity 5), 1

17. Using synthetic division:

$$
\begin{array}{r|rrrr}
-3 & 4 & -6 & 1 & -5 \\
   &   & -12 & 54 & -165 \\
\hline
   & 4 & -18 & 55 & -170 \\
\end{array}
$$

So $f(-3)=-170$.

19. Using synthetic division:

$$
\begin{array}{r|rrrrr}
1/2 & 6 & 5 & -8 & -10 & -3 \\
    &   & 3 & 4 & -2 & -6 \\
\hline
    & 6 & 8 & -4 & -12 & -9 \\
\end{array}
$$

So $f\left(\frac{1}{2}\right)=-9$.

21. Using synthetic division:

$$
\begin{array}{r|rrr}
-\sqrt{2} & 1 & 3 & -4 \\
          &   & -\sqrt{2} & -3\sqrt{2}+2 \\
\hline
          & 1 & 3-\sqrt{2} & -3\sqrt{2}-2 \\
\end{array}
$$

So $f\left(-\sqrt{2}\right)=-3\sqrt{2}-2$.

23. Using synthetic division:

$$
\begin{array}{r|rrrr}
12 & 1/2 & -5 & -13 & -10 \\
   &   & 6 & 12 & -12 \\
\hline
   & 1/2 & 1 & -1 & -22 \\
\end{array}
$$

So $f(12)=-22$.

25. Using synthetic division:

$$
\begin{array}{r|rrrr}
-3 & 1 & -4 & -9 & 36 \\
   &   & -3 & 21 & -36 \\
\hline
   & 1 & -7 & 12 & 0 \\
\end{array}
$$

So $x^{3}-4x^{2}-9x+36=(x+3)\left(x^{2}-7x+12\right)=(x+3)(x-4)(x-3)$.
So the roots are ±3, 4.

27. Using synthetic division:

$$\begin{array}{r|rrrr} 1 & 1 & 1 & -7 & 5 \\ & & 1 & 2 & -5 \\ \hline & 1 & 2 & -5 & 0 \end{array}$$

So $x^3 + x^2 - 7x + 5 = (x-1)(x^2 + 2x - 5)$. So $x = 1$ is a root. We use the quadratic formula:

$$x = \frac{-2 \pm \sqrt{4 + 20}}{2} = \frac{-2 \pm 2\sqrt{6}}{2} = -1 \pm \sqrt{6}$$

So the roots are $1, -1 \pm \sqrt{6}$.

29. Using synthetic division:

$$\begin{array}{r|rrrr} -2 & 3 & -5 & -16 & 12 \\ & & -6 & 22 & -12 \\ \hline & 3 & -11 & 6 & 0 \end{array}$$

So $3x^3 - 5x^2 - 16x + 12 = (x + 2)(3x^2 - 11x + 6) = (x + 2)(3x - 2)(x - 3)$.

So the roots are $-2, \frac{2}{3}$, and 3.

31. Using synthetic division:

$$\begin{array}{r|rrrr} -3/2 & 2 & 1 & -5 & -3 \\ & & -3 & 3 & 3 \\ \hline & 2 & -2 & -2 & 0 \end{array}$$

So $2x^3 + x^2 - 5x - 3 = \left(x + \frac{3}{2}\right)(2x^2 - 2x - 2) = 2\left(x + \frac{3}{2}\right)(x^2 - x - 1)$.

So $x = -\frac{3}{2}$ is a root. Use the quadratic formula:

$$x = \frac{1 \pm \sqrt{1 + 4}}{2} = \frac{1 \pm \sqrt{5}}{2}$$

So the roots are $-\frac{3}{2}, \frac{1 \pm \sqrt{5}}{2}$.

33. Using synthetic division:

$$\begin{array}{r|rrrrr} 5 & 1 & -15 & 75 & -125 & 0 \\ & & 5 & -50 & 125 & 0 \\ \hline & 1 & -10 & 25 & 0 & 0 \end{array}$$

Therefore:

$$\begin{aligned} x^4 - 15x^3 + 75x^2 - 125x &= (x - 5)(x^3 - 10x^2 + 25x) \\ &= x(x - 5)(x^2 - 10x + 25) \\ &= x(x - 5)^3 \end{aligned}$$

So the roots are 0 and 5.

35. Using synthetic division:

$$\begin{array}{r|rrrrr} -4 & 1 & 2 & -23 & -24 & 144 \\ & & -4 & 8 & 60 & -144 \\ \hline & 1 & -2 & -15 & 36 & 0 \end{array}$$

Use synthetic division again:

$$
\begin{array}{r|rrrr}
3 & 1 & -2 & -15 & 36 \\
  &   & 3 & 3   & -36 \\
\hline
  & 1 & 1 & -12 & 0
\end{array}
$$

So $x^4 + 2x^3 - 23x^2 - 24x + 144 = (x+4)(x-3)(x^2 + x - 12) = (x+4)^2(x-3)^2$.
So the roots are -4 and 3.

37. Using synthetic division:

$$
\begin{array}{r|rrrr}
-9 & 1 & 7 & -19 & -9 \\
   &   & -9 & 18 & 9 \\
\hline
   & 1 & -2 & -1 & 0
\end{array}
$$

So $x^3 + 7x^2 - 19x - 9 = (x+9)(x^2 - 2x - 1)$. So $x = -9$ is a root. Use the quadratic formula:

$$
x = \frac{2 \pm \sqrt{4+4}}{2} = \frac{2 \pm 2\sqrt{2}}{2} = 1 \pm \sqrt{2}
$$

So the roots are -9 and $1 \pm \sqrt{2}$.

39. Using long division:

$$
\begin{array}{r}
2x^2 - 3x - 1 \\
2x^2 - 3x - 1 \enclose{longdiv}{4x^4 - 12x^3 + 5x^2 + 6x + 1} \\
\underline{4x^4 - 6x^3 - 2x^2} \\
-6x^3 + 7x^2 + 6x \\
\underline{-6x^3 + 9x^2 + 3x} \\
-2x^2 + 3x + 1 \\
\underline{-2x^2 + 3x + 1} \\
0
\end{array}
$$

Use the quadratic formula to solve $2x^2 - 3x - 1 = 0$:

$$
x = \frac{3 \pm \sqrt{9+8}}{4} = \frac{3 \pm \sqrt{17}}{4}
$$

The solutions are $\dfrac{3 \pm \sqrt{17}}{4}$ (each of multiplicity 2).

41. (a) $R\left(\tfrac{1}{2}\right) = f\left(\tfrac{1}{2}\right) = 1.125$

    (b) $R(1.25) = f(1.25) = -0.046875$

    (c) Since $f(1) = 0$, $t - 1$ is a linear factor of $f(t)$.

    (d) Since 1 is a solution, use synthetic division:

$$
\begin{array}{r|rrrr}
1 & 1 & 0 & -4 & 3 \\
  &   & 1 & 1 & -3 \\
\hline
  & 1 & 1 & -3 & 0
\end{array}
$$

Using the quadratic formula to solve $t^2 + t - 3 = 0$:

$$t = \frac{-1 \pm \sqrt{1+12}}{2} = \frac{-1 \pm \sqrt{13}}{2}$$

So the solutions are 1, $\frac{-1 \pm \sqrt{13}}{2}$.

43. We must have $f(x) = (x-3)(x-5)(x+4) = 0$. Multiplying out, we have $x^3 - 4x^2 - 17x + 60 = 0$.

45. We must have $f(x) = (x+1)^2(x+6) = 0$. Multiplying out, we have $x^3 + 8x^2 + 13x + 6 = 0$.

47. Since $\frac{1}{2}$ is not a root of $x^2 - 3x - 4$, we know that such an equation must have

$f(x) = \left(x^2 - 3x - 4\right)\left(x - \frac{1}{2}\right)^3$, which has degree 5. Thus no such polynomial of degree 4 exists.

49. We must have $x^2 - 3x + 1$ and $x + 6$ as factors, and thus we can write $f(x) = (x^2 - 3x + 1)(x + 6)(ax + b) = 0$, for some real numbers a and b, where $a \neq 0$. Multiplying out, we have $ax^4 + (3a + b)x^3 + (-17a + 3b)x^2 + (6a - 17b)x + 6b = 0$.

51. Using synthetic division:

1.16	1	−3	12	9
		1.16	−2.1344	11.44
	1	−1.84	9.8656	20.44

So $f(1.16)$ is approximately equal to 20.44.

53. (a) Using synthetic division:

2.41	1	0	−5	−2
		2.41	5.8081	1.9475
	1	2.41	0.8081	−0.05

So $f(2.41)$ is approximately equal to -0.05.

(b) Using synthetic division:

2.42	1	0	−5	−2
		2.42	5.8564	2.0725
	1	2.42	0.8564	0.07

So $f(2.42)$ is approximately equal to 0.07.

55. (a) Since −5 is a zero, use synthetic division:

−5	1	1	−18	10
		−5	20	−10
	1	−4	2	0

So the remaining factor is $x^2 - 4x + 2 = 0$. Using the quadratic formula:

$$x = \frac{4 \pm \sqrt{16 - 4(2)}}{2(1)} = \frac{4 \pm \sqrt{8}}{2} = \frac{4 \pm 2\sqrt{2}}{2} = 2 \pm \sqrt{2}$$

So $x_2 = 2 - \sqrt{2}$ and $x_3 = 2 + \sqrt{2}$.

(b) The values are the same.

57. Evaluate the function:

$$
\begin{aligned}
f\left(-b - \sqrt{b^2 - 2c}\right) &= \tfrac{1}{2}\left(-b - \sqrt{b^2 - 2c}\right)^2 + b\left(-b - \sqrt{b^2 - 2c}\right) + c \\
&= \tfrac{1}{2}\left(b^2 + 2b\sqrt{b^2 - 2c} + b^2 - 2c\right) - b^2 - b\sqrt{b^2 - 2c} + c \\
&= b^2 + b\sqrt{b^2 - 2c} - c - b^2 - b\sqrt{b^2 - 2c} + c \\
&= 0
\end{aligned}
$$

So $x = -b - \sqrt{b^2 - 2c}$ is a zero of $f(x)$.

59. Evaluate the function:

$$
\begin{aligned}
F\left(\frac{-\sqrt{2} + \sqrt{2\sqrt{2} - 2}}{2}\right) &= 2\left(\frac{-\sqrt{2} + \sqrt{2\sqrt{2} - 2}}{2}\right)^4 + 4\left(\frac{-\sqrt{2} + \sqrt{2\sqrt{2} - 2}}{2}\right) + 1 \\
&= -1 + 2\sqrt{2} - 2\sqrt{2\sqrt{2} - 2} - 2\sqrt{2} + \sqrt{2\sqrt{2} - 2} + 1 \\
&= 0
\end{aligned}
$$

Yes, it is a zero.

61. Using synthetic division:

$$
\begin{array}{r|cccc}
1 & 1 & 1 & a & b \\
  &   & 1 & 2 & a+2 \\
\hline
  & 1 & 2 & a+2 & a+b+2
\end{array}
$$

Using synthetic division again:

$$
\begin{array}{r|cccc}
1 & 1 & -1 & -a & b \\
  &   & 1 & 0 & -a \\
\hline
  & 1 & 0 & -a & -a+b
\end{array}
$$

If 1 is a root, then:

$a + b + 2 = 0$

$-a + b = 0$

Adding these, we get $2b + 2 = 0$, so $b = -1$. Substituting we get $a = -1$.
So $a = -1$ and $b = -1$.

63. Let r_1 and r_2 be the roots, so $r_2 = 2r_1$. Then:

$$x^2 + bx + 1 = \left(x - r_1\right)\left(x - 2r_1\right) = x^2 - 3r_1 x + 2r_1^2$$

Since r_1 is a constant, $-3r_1 = b$ and $2r_1^2 = 1$, so $r_1^2 = \tfrac{1}{2}$ and thus $r_1 = \frac{\pm\sqrt{2}}{2}$. So $b = \frac{\pm 3\sqrt{2}}{2}$.

65. Let r denote the root with multiplicity 2. Using synthetic division:

$$
\begin{array}{r|cccc}
r & 1 & 0 & -12 & 16 \\
  &   & r & r^2 & r^3 - 12r \\
\hline
  & 1 & r & r^2 - 12 & r^3 - 12r + 16
\end{array}
$$

Since r is a double root, the equation $x^2 + rx + (r^2 - 12) = 0$ must also have r as a root. Substituting $x = r$, we have:

$$r^2 + r^2 + r^2 - 12 = 0$$
$$3r^2 = 12$$
$$r^2 = 4$$
$$r = \pm 2$$

But $r = -2$ does not check in the original equation, thus $r = 2$. Thus the quadratic equation becomes $x^2 + 2x - 8 = 0$, which factors to $(x + 4)(x - 2) = 0$, and thus $x = -4$ is also a solution. So the solutions are 2 (multiplicity 2) and -4.

Graphing Utility Exercises for Section 12.3

1. Sketching the graph:

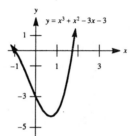

Note that $b = 2$ is not an x-intercept for the graph.

3. Sketching the graph:

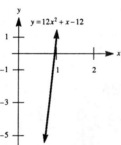

Note that $b = 0.9$ is not an x-intercept for the graph.

5. (a) Sketching the graph:

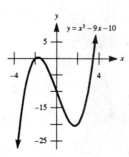

   (b) The x-intercept is $x \approx 3.449$.

(c)   The value is consistent with the x-intercept.

(d)   Computing $f(3.449) \approx 0$.

(e)   Substituting $r = 1 + \sqrt{6}$:
$$f\left(1+\sqrt{6}\right) = \left(1+\sqrt{6}\right)^3 - 9\left(1+\sqrt{6}\right) - 10$$
$$= 1 + 3\sqrt{6} + 18 + 6\sqrt{6} - 9 - 9\sqrt{6} - 10$$
$$= 0$$

7.   (a)   Sketching the graph:

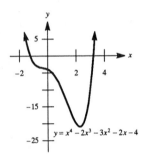

(b)   The x-intercept is $x \approx -1.236$.

(c)   The value is consistent with the x-intercept.

(d)   Computing $f(-1.236) \approx 0$.

(e)   Substituting $r = 1 - \sqrt{5}$:
$$f\left(1-\sqrt{5}\right) = \left(1-\sqrt{5}\right)^4 - 2\left(1-\sqrt{5}\right)^3 - 3\left(1-\sqrt{5}\right)^2 - 2\left(1-\sqrt{5}\right) - 4$$
$$= \left(56 - 24\sqrt{5}\right) - 2\left(16 - 8\sqrt{5}\right) - 3\left(6 - 2\sqrt{5}\right) - 2\left(1 - \sqrt{5}\right) - 4$$
$$= 56 - 24\sqrt{5} - 32 + 16\sqrt{5} - 18 + 6\sqrt{5} - 2 + 2\sqrt{5} - 4$$
$$= 0$$

9.   (a)   There appear to be three x-intercepts for the graph:

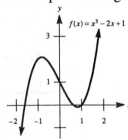

(b)   The roots are approximately -1.618, 0.618, and 1 (exact).

(c)   These values are consistent with those obtained in Example 5.

11. **(a)** The graph has only two x-intercepts:

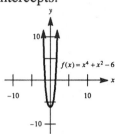

(b) The x-intercept between 1 and 2 is approximately 1.414.

(c) Since $\sqrt{2} \approx 1.414$, our work verifies that $\sqrt{2}$ is a root of the equation $x^4 + x^2 - 6 = 0$.

12.4 The Fundamental Theorem of Algebra

1. **(a)** yes
  (b) yes
  (c) yes
  (d) no--not a polynomial equation

3. $x^2 - 2x - 3 = (x + 1)(x - 3) = [x - (-1)](x - 3)$

5. $4x^2 + 23x - 6 = (4x - 1)(x + 6) = 4\left(x - \frac{1}{4}\right)[x - (-6)]$

7. $x^2 - 5 = (x + \sqrt{5})(x - \sqrt{5}) = [x - (-\sqrt{5})](x - \sqrt{5})$

9. Using the quadratic formula for $x^2 - 10x + 26 = 0$, we have:
$$x = \frac{10 \pm \sqrt{100 - 104}}{2} = \frac{10 \pm 2i}{2} = 5 \pm i$$
So $x^2 - 10x + 26 = [x - (5 + i)][x - (5 - i)]$

11. $f(x) = (x - 1)^2(x + 3) = x^3 + x^2 - 5x + 3$

13. $f(x) = (x - 2)(x + 2)(x - 2i)(x + 2i) = x^4 - 16$

15. $f(x) = (x - \sqrt{3})^2[x - (-\sqrt{3})]^2(x - 4i)[x - (-4i)] = x^6 + 10x^4 - 87x^2 + 144$

17. We can write $f(x)$ as $f(x) = a(x + 4)(x - 9)$, for some constant a. Since $f(3) = 5$, we have:
$$5 = a(7)(-6)$$
$$a = -\tfrac{5}{42}$$
So $f(x) = -\tfrac{5}{42}(x^2 - 5x - 36) = -\tfrac{5}{42}x^2 + \tfrac{25}{42}x + \tfrac{30}{7}$.

19. We can write $f(x)$ as $f(x) = a(x+5)(x-2)(x-3)$, for some constant a. Since $f(0) = 1$:

$$1 = a(5)(-2)(-3)$$
$$a = \tfrac{1}{30}$$

So $f(x) = \tfrac{1}{30}\left(x^3 - 19x + 30\right) = \tfrac{1}{30}x^3 - \tfrac{19}{30}x + 1$.

21. We have $-b = -i - \sqrt{3}$, so $b = i + \sqrt{3}$. Also $c = -i\left(-\sqrt{3}\right) = i\sqrt{3}$.
So $x^2 + \left(i + \sqrt{3}\right)x + i\sqrt{3} = 0$.

23. We have $-b = 3$, so $b = -3$. Also $c = (9)(-6) = -54$. So $x^2 - 3x - 54 = 0$.

25. We have $-b = 2$, so $b = -2$. Also $c = \left(1 + \sqrt{5}\right)\left(1 - \sqrt{5}\right) = -4$. So $x^2 - 2x - 4 = 0$.

27. We have $-B = 2a$, so $B = -2a$. Also $C = \left(a + \sqrt{b}\right)\left(a - \sqrt{b}\right) = a^2 - b$.
So $x^2 - 2ax + a^2 - b = 0$.

29. Using the hint, we have:

$$x^4 + 64 = x^4 + 16x^2 + 64 - 16x^2$$
$$= \left(x^2 + 8\right)^2 - (4x)^2$$
$$= \left(x^2 + 8 + 4x\right)\left(x^2 + 8 - 4x\right)$$
$$= \left(x^2 + 4x + 8\right)\left(x^2 - 4x + 8\right)$$

Now $x^2 + 4x + 8 = 0$ when $x = \dfrac{-4 \pm \sqrt{16 - 32}}{2} = \dfrac{-4 \pm 4i}{2} = -2 \pm 2i$.

Also $x^2 - 4x + 8 = 0$ when $x = \dfrac{4 \pm \sqrt{16 - 32}}{2} = \dfrac{4 \pm 4i}{2} = 2 \pm 2i$.

Therefore:

$$x^4 + 64 = \left(x^2 + 4x + 8\right)\left(x^2 - 4x + 8\right)$$
$$= \left[x - (-2 + 2i)\right]\left[x - (-2 - 2i)\right]\left[x - (2 + 2i)\right]\left[x - (2 - 2i)\right]$$
$$= (x + 2 - 2i)(x + 2 + 2i)(x - 2 - 2i)(x - 2 + 2i)$$

31. We know:

$$x^3 + bx^2 + cx + d = \left(x - r_1\right)\left(x - r_2\right)\left(x - r_3\right)$$
$$= x^3 - \left(r_1 + r_2 + r_3\right)x^2 + \left(r_1 r_2 + r_1 r_3 + r_2 r_3\right)x - r_1 r_2 r_3$$

Therefore:

$$r_1 + r_2 + r_3 = -b$$
$$r_1 r_2 + r_1 r_3 + r_2 r_3 = c$$
$$r_1 r_2 r_3 = -d$$

33. Let r_1, r_2 and r_3 be the roots and assume $r_2 = -r_1$ (their sum is therefore 0). The identities from Exercise 31 give us:

$$r_1 + r_2 + r_3 = 4 \qquad\qquad r_1 r_2 r_3 = -36$$
$$r_1 - r_1 + r_3 = 4 \qquad\qquad -r_1^2 \cdot 4 = -36$$
$$r_3 = 4 \qquad\qquad r_1^2 = 9$$
$$r_1 = \pm 3$$

So $r_1 = 3$ and $r_2 = -3$ (or switch them, it doesn't matter). So the roots are 4, 3, and –3.

35. (a) Substituting $x = \tan 15°$ into the equation:

$$\tan^3 15° - 3\tan^2 15° - 3\tan 15° + 1$$
$$= \left(\tan^3 15° - 3\tan 15°\right) - \left(3\tan^2 15° - 1\right)$$
$$= (\tan 45°)\left(3\tan^2 15° - 1\right) - \left(3\tan^2 15° - 1\right)$$
$$= \left(3\tan^2 15° - 1\right)(\tan 45° - 1)$$
$$= \left(3\tan^2 15° - 1\right)(1 - 1)$$
$$= 0$$

Thus $x = \tan 15°$ is a root of the cubic equation.

(b) Substituting $\tan 15° \approx 0.2679$ verifies it is a root of the equation.

(c) Factoring by grouping:

$$x^3 - 3x^2 - 3x + 1 = \left(x^3 + 1\right) - 3x^2 - 3x$$
$$= (x+1)\left(x^2 - x + 1\right) - 3x(x+1)$$
$$= (x+1)\left(x^2 - 4x + 1\right)$$

Since $\tan 15° + 1 \neq 0$, it must be the case that $\tan 15°$ is a root of $x^2 - 4x + 1 = 0$.

(d) Substituting $x = \tan 75°$ into the equation:

$$\tan^3 75° - 3\tan^2 75° - 3\tan 75° + 1$$
$$= \left(\tan^3 75° - 3\tan 75°\right) - \left(3\tan^2 75° - 1\right)$$
$$= (\tan 225°)\left(3\tan^2 75° - 1\right) - \left(3\tan^2 75° - 1\right)$$
$$= \left(3\tan^2 75° - 1\right)(\tan 225° - 1)$$
$$= \left(3\tan^2 75° - 1\right)(1 - 1)$$
$$= 0$$

Thus $x = \tan 75°$ is a root of $x^3 - 3x^2 - 3x + 1 = 0$. Since $\tan 75° + 1 \neq 0$, it must also be a root of $x^2 - 4x + 1 = 0$.

(e) Using Table 2:

$$\tan 15° + \tan 75° = -(-4) = 4$$
$$\tan 15° \tan 75° = 1$$

(f)   Using the quadratic formula:

$$x = \frac{-(-4) \pm \sqrt{(-4)^2 - 4(1)(1)}}{2(1)} = \frac{4 \pm \sqrt{16-4}}{2} = \frac{4 \pm 2\sqrt{3}}{2} = 2 \pm \sqrt{3}$$

Since $\tan 75° > \tan 15°$, it must be the case that $\tan 15° = 2 - \sqrt{3}$ and $\tan 75° = 2 + \sqrt{3}$.

37. (a) Substituting $\theta = 20°$ yields the identity $\cos 60° = 4\cos^3 20° - 3\cos 20°$. Since $\cos 60° = \frac{1}{2}$, we have $4\cos^3 20° - 3\cos 20° = \frac{1}{2}$. Letting $x = \cos 20°$ and multiplying the equation by 2 yields $8x^3 - 6x = 1$, or $8x^3 - 6x - 1 = 0$.

(b) Substituting $\theta = 100°$ results in $\cos 300° = 4\cos^3 100° - 3\cos 100°$. Since $\cos 300° = \frac{1}{2}$, the equation $8x^3 - 6x - 1 = 0$ again results. Substituting $\theta = 140°$ results in $\cos 420° = \cos 60° = \frac{1}{2}$, so by the above argument again $8x^3 - 6x - 1 = 0$ results.

(c) Since $r_1 = \cos 20°$, $r_2 = \cos 100°$, and $r_3 = \cos 140°$ are three roots to the equation $x^3 - \frac{3}{4}x - \frac{1}{8} = 0$ (note that we divided the equation in part (b) by 8), by Table 2 $r_1 r_2 r_3 = -d$, so $(\cos 20°)(\cos 100°)(\cos 140°) = \frac{1}{8}$.

(d) Since $\cos 140° = -\cos 40°$ and $\cos 100° = -\cos 80°$, we have:

$$(\cos 20°)(\cos 100°)(\cos 140°) = \frac{1}{8}$$
$$(\cos 20°)(-\cos 80°)(-\cos 40°) = \frac{1}{8}$$
$$(\cos 20°)(\cos 40°)(\cos 80°) = \frac{1}{8}$$

(e) A calculator verifies that $(\cos 20°)(\cos 40°)(\cos 80°) = 0.125 = \frac{1}{8}$.

39. (a) Suppose $A \neq 0$ or $B \neq 0$. Then $f(x) = Ax^2 + Bx + C = 0$ can have at most two distinct roots by the Linear Factor Theorem. So $A = B = 0$. But this implies $C = 0$, and thus $f(x) = 0$ for all values of x.

(b)   Using the hint, let:

$$f(x) = \frac{a^2 - x^2}{(a-b)(a-c)} + \frac{b^2 - x^2}{(b-c)(b-a)} + \frac{c^2 - x^2}{(c-a)(c-b)} - 1$$

$$f(a) = 0 + \frac{b^2 - a^2}{(b-c)(b-a)} + \frac{c^2 - a^2}{(c-a)(c-b)} - 1$$

$$= \frac{b+a}{b-c} + \frac{c+a}{c-b} - 1$$

$$= \frac{-b-a+c+a}{c-b} - 1$$

$$= \frac{c-b}{c-b} - 1$$

$$= 0$$

$$f(b) = \frac{a^2 - b^2}{(a-b)(a-c)} + 0 + \frac{c^2 - b^2}{(c-a)(c-b)} - 1$$

$$= \frac{a+b}{a-c} + \frac{c+b}{c-a} - 1$$

$$= \frac{-a-b+c+b}{c-a} - 1$$

$$= \frac{c-a}{c-a} - 1$$

$$= 0$$

$$f(c) = \frac{a^2 - c^2}{(a-b)(a-c)} + \frac{b^2 - c^2}{(b-c)(b-a)} + 0 - 1$$

$$= \frac{a+c}{a-b} + \frac{b+c}{b-a} - 1$$

$$= \frac{-a-c+b+c}{b-a} - 1$$

$$= \frac{b-a}{b-a} - 1$$

$$= 0$$

Since a, b and c are distinct (for denominators to be nonzero), and all three are roots, by part (a) $f(x) = 0$ for all values of x. Thus the equation is an identity.

Graphing Utility Exercises for Section 12.4

1.   The equation does not have any real-number roots:

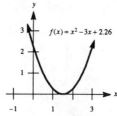

Computing the discriminant:
$$b^2 - 4ac = (-3)^2 - 4(1)(2.26) = 9 - 9.04 = -0.04 < 0$$

3.   The equation has three real-number roots:

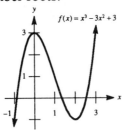

5.   The equation does not have any real-number roots:

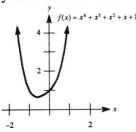

7.   The equation has one real-number root:

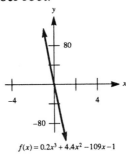

9.   (a)   Using $a = 3$ and $b = 76$:

$$x = \sqrt[3]{\frac{76}{2} + \sqrt{\frac{76^2}{4} + \frac{3^3}{27}}} - \sqrt[3]{-\frac{76}{2} + \sqrt{\frac{76^2}{4} + \frac{3^3}{27}}}$$

$$= \sqrt[3]{38 + \sqrt{1444 + 1}} - \sqrt[3]{-38 + \sqrt{1444 + 1}}$$

$$= \sqrt[3]{38 + \sqrt{1445}} - \sqrt[3]{-38 + \sqrt{1445}}$$

$$= 4$$

   (b)   Graphing the function in the suggested viewing rectangle:

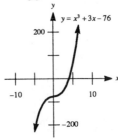

Note the x-intercept is 4. Substituting $x = 4$:
$$y = (4)^3 + 3(4) - 76 = 64 + 12 - 76 = 0$$

12.5 Rational and Irrational Roots

1. p-factors: $\pm 1, \pm 3$
   q-factors: $\pm 1, \pm 2, \pm 4$
   possible rational roots $\frac{p}{q}$: $\pm 1, \pm \frac{1}{2}, \pm \frac{1}{4}, \pm 3, \pm \frac{3}{2}, \pm \frac{3}{4}$

3. p-factors: $\pm 1, \pm 3, \pm 9$
   q-factors: $\pm 1, \pm 2, \pm 4, \pm 8$
   possible rational roots $\frac{p}{q}$: $\pm 1, \pm \frac{1}{2}, \pm \frac{1}{4}, \pm \frac{1}{8}, \pm 3, \pm \frac{3}{2}, \pm \frac{3}{4}, \pm \frac{3}{8}, \pm 9, \pm \frac{9}{2}, \pm \frac{9}{4}, \pm \frac{9}{8}$

5. First multiply by 3 (we need integer coefficients) to get:
   $$2x^3 - 3x^2 - 15x + 6 = 0$$
   p-factors: $\pm 1, \pm 2, \pm 3, \pm 6$
   q-factors: $\pm 1, \pm 2$
   possible rational roots $\frac{p}{q}$: $\pm 1, \pm \frac{1}{2}, \pm 2, \pm 3, \pm \frac{3}{2}, \pm 6$

7. The possible rational roots are ± 1. Using synthetic division:

   $$\underline{1|} \quad \begin{array}{cccc} 1 & 0 & -3 & 1 \\ & 1 & 1 & -2 \\ \hline 1 & 1 & -2 & -1 \end{array}$$

   So $x = 1$ is not a root.

   $$\underline{-1|} \quad \begin{array}{cccc} 1 & 0 & -3 & 1 \\ & -1 & 1 & 2 \\ \hline 1 & -1 & -2 & 3 \end{array}$$

   So $x = -1$ is not a root. There are no rational roots.

9. The possible rational roots are ± 1. Using synthetic division:

   $$\underline{1|} \quad \begin{array}{cccc} 1 & 1 & -1 & 1 \\ & 1 & 2 & 1 \\ \hline 1 & 2 & 1 & 2 \end{array}$$

   So $x = 1$ is not a root.

   $$\underline{-1|} \quad \begin{array}{cccc} 1 & 1 & -1 & 1 \\ & -1 & 0 & 1 \\ \hline 1 & 0 & -1 & 2 \end{array}$$

   So $x = -1$ is not a root. There are no rational roots.

11. The possible rational roots are $\pm 1, \pm \frac{1}{2}, \pm \frac{1}{3}, \pm \frac{1}{4}, \pm \frac{1}{6}, \pm \frac{1}{12}, \pm 2, \pm \frac{2}{3}, \pm 3, \pm \frac{3}{2}, \pm \frac{3}{4}, \pm 6$.
Using synthetic division:

$$\begin{array}{r|rrrrr}
1 & 12 & 0 & -1 & 0 & -6 \\
  &   & 12 & 12 & 11 & 11 \\
\hline
  & 12 & 12 & 11 & 11 & 5
\end{array}$$

So $x = 1$ is not a root. Since this row is all positive, we can exclude $x = 2, x = 3, \; x = \frac{3}{2}$
and $x = 6$.

$$\begin{array}{r|rrrrr}
-1 & 12 & 0 & -1 & 0 & -6 \\
  &   & -12 & 12 & -11 & 11 \\
\hline
  & 12 & -12 & 11 & -11 & 5
\end{array}$$

So $x = -1$ is not a root. Since this row alternates signs, we can exclude $x = -2, x = -3$,
$x = -\frac{3}{2}$ and $x = -6$.

$$\begin{array}{r|rrrrr}
1/4 & 12 & 0 & -1 & 0 & -6 \\
  &   & 3 & 3/4 & -1/16 & -1/64 \\
\hline
  & 12 & 3 & -1/4 & -1/16 & -385/64
\end{array}$$

So $x = \frac{1}{4}$ is not a root.

$$\begin{array}{r|rrrrr}
1/3 & 12 & 0 & -1 & 0 & -6 \\
  &   & 4 & 4/3 & 1/9 & 1/27 \\
\hline
  & 12 & 4 & 1/3 & 1/9 & -161/27
\end{array}$$

Proceeding in a similar fashion, we find that none of the candidates are rational roots.

13. The possible rational roots are $\pm 1, \pm 3$. Using synthetic division:

$$\begin{array}{r|rrrr}
1 & 1 & 3 & -1 & -3 \\
  &   & 1 & 4 & 3 \\
\hline
  & 1 & 4 & 3 & 0
\end{array}$$

So $x^3 + 3x^2 - x - 3 = (x - 1)(x^2 + 4x + 3) = (x - 1)(x + 1)(x + 3)$.
So the roots are 1, –1, and –3.

15. The possible rational roots are $\pm 1, \pm \frac{1}{2}, \pm \frac{1}{4}, \pm 5, \pm \frac{5}{2}, \pm \frac{5}{4}$. Using synthetic division:

$$\begin{array}{r|rrrr}
-1/4 & 4 & 1 & -20 & -5 \\
  &   & -1 & 0 & 5 \\
\hline
  & 4 & 0 & -20 & 0
\end{array}$$

Therefore:
$$\begin{aligned}
4x^3 + x^2 - 20x - 5 &= \left(x + \tfrac{1}{4}\right)\left(4x^2 - 20\right) \\
&= 4\left(x + \tfrac{1}{4}\right)\left(x^2 - 5\right) \\
&= 4\left(x + \tfrac{1}{4}\right)\left(x + \sqrt{5}\right)\left(x - \sqrt{5}\right)
\end{aligned}$$

So the roots are $-\frac{1}{4}, -\sqrt{5}$ and $\sqrt{5}$.

17. The possible rational roots are $\pm 1, \pm \frac{1}{3}, \pm \frac{1}{9}, \pm 2, \pm \frac{2}{3}, \pm \frac{2}{9}$. Using synthetic division:

$$\begin{array}{r|rrrr}
1 & 9 & 18 & 11 & 2 \\
  &   & 9 & 27 & 38 \\
\hline
  & 9 & 27 & 38 & 40
\end{array}$$

So $x = 2$ is excluded also (the row is positive).

$$
\begin{array}{r|rrrr}
-1 & 9 & 18 & 11 & 2 \\
   &   & -9 & -9 & -2 \\
\hline
   & 9 & 9 & 2 & 0
\end{array}
$$

So $9x^3 + 18x^2 + 11x + 2 = (x+1)(9x^2 + 9x + 2) = (x+1)(3x+2)(3x+1)$.

So the roots are $-1, -\frac{2}{3}$ and $-\frac{1}{3}$.

19. The possible rational roots are $\pm 1, \pm 2, \pm 3, \pm 4, \pm 6, \pm 8, \pm 12, \pm 24$. Using synthetic division:

$$
\begin{array}{r|rrrrr}
1 & 1 & 1 & -25 & -1 & 24 \\
  &   & 1 & 2 & -23 & -24 \\
\hline
  & 1 & 2 & -23 & -24 & 0
\end{array}
$$

So $x = 1$ is a root. Using synthetic division again:

$$
\begin{array}{r|rrrr}
-1 & 1 & 2 & -23 & -24 \\
   &   & -1 & -1 & 24 \\
\hline
   & 1 & 1 & -24 & 0
\end{array}
$$

So $x = -1$ is also a root. So $x^4 + x^3 - 25x^2 - x + 24 = (x-1)(x+1)(x^2 + x - 24)$. Use the quadratic formula:

$$
x = \frac{-1 \pm \sqrt{1+96}}{2} = \frac{-1 \pm \sqrt{97}}{2}
$$

So the roots are $1, -1, \frac{-1+\sqrt{97}}{2}$ and $\frac{-1-\sqrt{97}}{2}$.

21. The possible rational roots are ± 1. Using synthetic division:

$$
\begin{array}{r|rrrrr}
1 & 1 & -4 & 6 & -4 & 1 \\
  &   & 1 & -3 & 3 & -1 \\
\hline
  & 1 & -3 & 3 & -1 & 0
\end{array}
$$

Using synthetic division again:

$$
\begin{array}{r|rrrr}
1 & 1 & -3 & 3 & -1 \\
  &   & 1 & -2 & 1 \\
\hline
  & 1 & -2 & 1 & 0
\end{array}
$$

So $x^4 - 4x^3 + 6x^2 - 4x + 1 = (x-1)^2(x^2 - 2x + 1) = (x-1)^4$.

So the only root is 1 (with multiplicity 4).

23. First multiply by 2 to get integer coefficients:

$$2x^3 - 5x^2 - 46x + 24 = 0$$

The possible rational roots are $\pm 1, \pm \frac{1}{2}, \pm 2, \pm 3, \pm \frac{3}{2}, \pm 4, \pm 6, \pm 8, \pm 12, \pm 24$. Using synthetic division:

$$
\begin{array}{r|rrrr}
1/2 & 2 & -5 & -46 & 24 \\
    &   & 1 & -2 & -24 \\
\hline
    & 2 & -4 & -48 & 0
\end{array}
$$

Therefore:
$$2x^3 - 5x^2 - 46x + 24 = \left(x - \tfrac{1}{2}\right)\left(2x^2 - 4x - 48\right)$$
$$= 2\left(x - \tfrac{1}{2}\right)\left(x^2 - 2x - 24\right)$$
$$= 2\left(x - \tfrac{1}{2}\right)\left(x - 6\right)\left(x + 4\right)$$

So the roots are $\tfrac{1}{2}, 6$ and -4.

25. First multiply by 20 to get integer coefficients:
$$40x^4 - 18x^3 - 58x^2 + 27x - 3 = 0$$

The possible rational roots are $\pm 1, \pm\tfrac{1}{2}, \pm\tfrac{1}{4}, \pm\tfrac{1}{5}, \pm\tfrac{1}{8}, \pm\tfrac{1}{10}, \pm\tfrac{1}{20}, \pm\tfrac{1}{40}, \pm 3, \pm\tfrac{3}{2}, \pm\tfrac{3}{4}, \pm\tfrac{3}{5},$
$\pm\tfrac{3}{8}, \pm\tfrac{3}{10}, \pm\tfrac{3}{20}, \pm\tfrac{3}{40}$. Using synthetic division:

```
1/4│  40   -18   -58    27    -3
            10    -2   -15     3
      40    -8   -60    12     0
```

Using synthetic division again:

```
1/5│  40    -8   -60    12
             8     0   -12
      40     0   -60     0
```

Therefore:
$$40x^4 - 18x^3 - 58x^2 + 27x - 3 = \left(x - \tfrac{1}{4}\right)\left(x - \tfrac{1}{5}\right)\left(40x^2 - 60\right)$$
$$= 10\left(x - \tfrac{1}{4}\right)\left(x - \tfrac{1}{5}\right)\left(4x^2 - 6\right)$$
$$= 10\left(x - \tfrac{1}{4}\right)\left(x - \tfrac{1}{5}\right)\left(2x + \sqrt{6}\right)\left(2x - \sqrt{6}\right)$$

So the roots are $\tfrac{1}{4}, \tfrac{1}{5}, -\tfrac{\sqrt{6}}{2}$ and $\tfrac{\sqrt{6}}{2}$.

27. (a) The possible rational roots are $\pm 1, \pm\tfrac{1}{5}, \pm 2, \pm\tfrac{2}{5}, \pm 3, \pm\tfrac{3}{5}, \pm 4, \pm\tfrac{4}{5}, \pm 6, \pm\tfrac{6}{5}, \pm 12,$
$\pm\tfrac{12}{5}$. Using synthetic division:

```
2│   5     0     0   -10   -12
          10    20    40    60
     5    10    20    30    48
```

Since this row is all positive, $x = 2$ is an upper bound.

```
-1│   5     0     0   -10   -12
           -5     5    -5    15
      5    -5     5   -15     3
```

Since this row alternates signs, $x = -1$ is a lower bound.

(b) The possible rational roots are $\pm 1, \pm\tfrac{1}{3}, \pm 2, \pm\tfrac{2}{3}, \pm 4, \pm\tfrac{4}{3}$. Using synthetic division:

```
4/3│  3    -4     5    -2     -4
            4     0   20/3   56/9
      3     0     5   14/3   20/9
```

So $x = \tfrac{4}{3}$ is the upper bound.

$$
\begin{array}{r|rrrrr}
-2/3 & 3 & -4 & 5 & -2 & -4 \\
 & & -2 & 4 & -6 & 16/3 \\
\hline
 & 3 & -6 & 9 & -8 & 4/3
\end{array}
$$

So $x = -\frac{2}{3}$ is the lower bound. Since the question asked for the integral upper and lower bounds, the upper bound is 2 and the lower bound is −1.

(c) The possible rational roots are $\pm 1, \pm \frac{1}{2}, \pm 2, \pm 3, \pm \frac{3}{2}, \pm 4, \pm 6, \pm 12$. Using synthetic division:

$$
\begin{array}{r|rrrrr}
6 & 2 & -7 & -5 & 28 & -12 \\
 & & 12 & 30 & 150 & 1068 \\
\hline
 & 2 & 5 & 25 & 178 & 1056
\end{array}
$$

So $x = 6$ is an upper bound.

$$
\begin{array}{r|rrrrr}
-2 & 2 & -7 & -5 & 28 & -12 \\
 & & -4 & 22 & -34 & 12 \\
\hline
 & 2 & -11 & 17 & -6 & 0
\end{array}
$$

Actually −2 is a root, but also a lower bound.

29. Let $f(x) = x^3 + x - 1$.

$f(0) = -1$

$f(1) = 1$

So the root lies between 0 and 1.

$f(0.5) = -0.375$

$f(0.7) = 0.043$

$f(0.6) = -0.373$

So the root lies between 0.6 and 0.7.

$f(0.65) = -0.075$

$f(0.66) = -0.052$

$f(0.67) = -0.029$

$f(0.68) = -0.005$

$f(0.69) = 0.018$

So the root lies between 0.68 and 0.69.

31. Let $f(x) = x^5 - 200$.

$f(2) = -168$

$f(3) = 43$

So the root lies between 2 and 3.

$f(2.5) = -102$

$f(2.8) = -27.9$

$f(2.9) = 5.11$

So the root lies between 2.8 and 2.9.

$f(2.87) = -5.28$

$f(2.88) = -1.86$

$f(2.89) = 1.60$

So the root lies between 2.88 and 2.89.

33. Let $f(x) = x^3 - 8x^2 + 21x - 22$.

$$f(4) = -2$$
$$f(5) = 8$$

So the root lies between 4 and 5.

$$f(4.5) = 1.63$$
$$f(4.4) = 0.704$$
$$f(4.3) = -0.113$$

So the root lies between 4.3 and 4.4.

$$f(4.33) = 0.12$$
$$f(4.32) = 0.04$$
$$f(4.31) = -0.03$$

So the root lies between 4.31 and 4.32.

35. Let $f(x) = x^3 + x^2 - 2x + 1$.

$$f(-2) = 1$$
$$f(-3) = -11$$

So the root lies between -3 and -2.

$$f(-2.1) = 0.35$$
$$f(-2.2) = -0.41$$

So the root lies between -2.2 and -2.1.

$$f(-2.15) = -0.02$$
$$f(-2.14) = 0.05$$

So the root lies between -2.15 and -2.14.

37. Let $f(x) = x^3 + 2x^2 + 2x + 101$.

$$f(-5) = 16$$
$$f(-6) = -55$$

So the root lies between -6 and -5.

$$f(-5.4) = -8.94$$
$$f(-5.3) = -2.30$$
$$f(-5.2) = 4.07$$

So the root lies between -5.3 and -5.2.

$$f(-5.26) = 0.28$$
$$f(-5.27) = -0.36$$

So the root lies between -5.27 and -5.26.

39. (a) Since 2 is a factor of $8 \cdot 5 = 40$, and 2 is not a factor of 5, 2 must be a factor of 8, which is true.

(b) The result guarantees only that A is a factor of B in the case where A and C have no factor in common. Here A and C have a common factor of 5, so the result does not apply.

(c) Since $x = \frac{p}{q}$ is a root of the equation, this statement must be true.

(d)   Subtract a_0 and multiply by q^n to get:

$$a_n p^n + a_{n-1} q p^{n-1} + a_{n-2} q^2 p^{n-2} + \ldots + a_1 q^{n-1} p = -a_0 q^n$$

Therefore:

$$p\left(a_n p^{n-1} + a_{n-1} q p^{n-2} + a_{n-2} q^2 p^{n-3} + \ldots + a_1 q^{n-1}\right) = -a_0 q^n$$

41.   Sketch the graph:

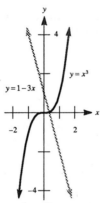

Find where $x^3 = 1 - 3x$, so $x^3 + 3x - 1 = 0$.

Let $f(x) = x^3 + 3x - 1$.

$\quad f(0) = -1$

$\quad f(1) = 3$

So the root lies between 0 and 1.

$\quad f(0.3) = -0.07$

$\quad f(0.4) = 0.264$

So the root lies between 0.3 and 0.4.

$\quad f(0.32) = -0.007$

$\quad f(0.33) = 0.02$

So the root lies between 0.32 and 0.33.

$\quad f(0.325) = 0.009$

So the root lies between 0.32 and 0.325, thus the x-coordinate is $x = 0.32$, accurate to two decimal places.

43.   Find where:

$$\sqrt{x} = x^2 - 1$$

$$x = x^4 - 2x^2 + 1$$

$$x^4 - 2x^2 - x + 1 = 0$$

Let $f(x) = x^4 - 2x^2 - x + 1$.

$\quad f(1) = -1$

$\quad f(2) = 7$

So the root lies between 1 and 2.

$\quad f(1.4) = -0.48$

$\quad f(1.5) = 0.06$

So the root lies between 1.4 and 1.5

$$f(1.48) = -0.06$$
$$f(1.49) = -0.001$$

So the root lies between 1.49 and 1.50.

$$f(1.495) = 0.03$$

So the root lies between 1.49 and 1.495, thus the x-coordinate is $x = 1.49$, accurate to two decimal places.

45. (a) Here $a_n = 9$, $a_0 = 27$ and $f(1) = 29$. Since these three numbers are odd, the equation has no rational roots.

   (b) Here $a_n = 5$, $a_0 = -25$ and $f(1) = -29$. Since these three numbers are odd, the equation has no rational roots.

47. (a) Substituting $x = -\tan 27° = \tan(-27°)$ into the equation:

   $$\tan^5(-27°) - 5\tan^4(-27°) - 10\tan^3(-27°) + 10\tan^2(-27°) + 5\tan(-27°) - 1$$
   $$= \left[\tan^5(-27°) - 10\tan^3(-27°) + 5\tan(-27°)\right] - \left[5\tan^4(-27°) - 10\tan^2(-27°) + 1\right]$$
   $$= \left[\tan(-135°)\right]\left[5\tan^4(-27°) - 10\tan^2(-27°) + 1\right] - \left[5\tan^4(-27°) - 10\tan^2(-27°) + 1\right]$$
   $$= \left[5\tan^4(-27°) - 10\tan^2(-27°) + 1\right]\left[\tan(-135°) - 1\right]$$
   $$= \left[5\tan^4(-27°) - 10\tan^2(-27°) + 1\right](1 - 1)$$
   $$= 0$$

   Following the exact same reasoning, since 1 is the only rational root of the equation, and $-\tan 27° = \tan(-27°) \neq 1$, then $-\tan 27° = \tan(-27°)$ must be an irrational root of the equation $x^4 - 4x^3 - 14x^2 - 4x + 1 = 0$.

   (b) Using the results of Exercise 32:

   $$\tan 9° \tan 27° \tan 63° \tan 81°$$
   $$= (\tan 9°)(-\tan 27°)(-\tan 63°)(\tan 81°) = r_1 r_2 r_3 r_4 = 1$$
   $$\tan 9° - \tan 27° - \tan 63° + \tan 81°$$
   $$= r_1 + r_2 + r_3 + r_4 = -(-4) = 4$$

   A calculator verifies these results.

49. The only possible rational roots are ± 1, $\pm p$ (since p is a prime number).
   Let $f(x) = x^2 + x - p$.

   $f(1) = 1 + 1 - p = 2 - p = 0$, so if $p = 2$ then 1 is a root
   $f(-1) = 1 - 1 - p = -p$

   Since $p \neq 0$, –1 is not a root.

   $$f(p) = p^2 + p - p = p^2$$

   Since $p^2 \neq 0$, p is not a root.

   $$f(-p) = p^2 - p - p = p^2 - 2p = p(p - 2)$$

   Here $p = 0$ and $p = 2$ are roots, but only $p = 2$ is prime. So $p = 2$ is the only prime number such that $f(x) = 0$ will have rational roots.

51. (a) If $x = 1$ is a root, then $1 + p - q = 0$, thus $q - p = 1$. But the only two prime numbers which are 1 unit apart would be $q = 3$ and $p = 2$. The remaining quadratic is then $x^2 + x + 3 = 0$, which we solve using the quadratic formula:

$$x = \frac{-1 \pm \sqrt{1-12}}{2} = \frac{-1 \pm i\sqrt{11}}{2}$$

So the remaining roots are $\frac{-1 \pm i\sqrt{11}}{2}$.

(b) Suppose -1 is a root. Then $-1 - p - q = 0$, so $p + q = -1$. Clearly this is impossible for prime numbers p and q. Suppose q is a root. If $x = q$, we have:

$$q^3 + pq - q = 0$$
$$q(q^2 + p - 1) = 0$$

Clearly $q \neq 0$, so $q^2 + p = 1$. But since $q > 1$ and $p > 1$, this is also impossible. Suppose $-q$ is a root. If $x = -q$, then we have:

$$-q^3 - pq - q = 0$$
$$-q(q^2 + p + 1) = 0$$

Clearly $q \neq 0$, so $q^2 + p = -1$. But since $q > 1$ and $p > 1$, this is also impossible.

53. Proceeding as in Exercise 49, if the equation had rational roots, they must be either ± 1, $\pm p$, $\pm q$, or $\pm pq$. Determine what happens in each case:

$x = 1$:
$$1 + p - pq = 0$$
$$1 = pq - p$$
$$1 = p(q - 1)$$
But this is impossible, since $p > 1$ and $q > 1$.

$x = -1$:
$$-1 - p - pq = 0$$
$$1 + p + pq = 0$$
This is clearly impossible, since $p > 1$ and $q > 1$.

$x = p$:
$$p^3 + p^2 - pq = 0$$
$$p(p^2 + p - q) = 0$$
Since $p \neq 0$, $p^2 + p = q$. But then $p(p + 1) = q$, which violates the fact that q is a prime number.

$x = -p$:
$$-p^3 - p^2 - pq = 0$$
$$-p(p^2 + p + q) = 0$$
Since $p \neq 0$, $p^2 + p = -q$. But then $p(p + 1) = -q$, which is impossible since $p > 1$ and $q > 1$.

$x = q$:
$$q^3 + pq - pq = 0$$
$$q^3 = 0, \text{ which is impossible.}$$

$x = -q$:
$$-q^3 - pq - pq = 0$$
$$-q(q^2 + 2p) = 0$$
Since $q \neq 0$, $q^2 = -2p$, which clearly is impossible.

$x = pq$:
$$p^3q^3 + p^2q - pq = 0$$
$$pq(p^2q^2 + p - 1) = 0$$

Since $pq \neq 0$, $p^2q^2 = 1 - p$. But since $p > 1$, this is clearly impossible.

$x = -pq$:
$$-p^3q^3 - p^2q - pq = 0$$
$$-pq(p^2q^2 + p + 1) = 0$$

Since $pq \neq 0$, $p^2q^2 = -(p + 1)$. But clearly this is impossible.

Since we have checked each possible rational root, the equation has no rational roots.

55. (a) Factor by grouping:
$$f(x) = x^3 + 3x^2 - x - 3$$
$$= x^2(x + 3) - 1(x + 3)$$
$$= (x + 3)(x^2 - 1)$$
$$= (x + 3)(x + 1)(x - 1)$$

(b) Graph $f(x)$, noting the x-intercepts are -3, -1, and 1:

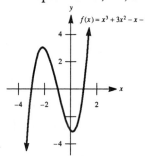

(c) Using synthetic division by -3:

$$
\begin{array}{r|rrrr}
-3 & 1 & 3 & -1 & -3 \\
   &   & -3 & 0 & 3 \\
\hline
   & 1 & 0 & -1 & 0
\end{array}
$$

Note that the numbers do not alternate signs, thus -3 fails the lower bound test.

Graphing Utility Exercises for Section 12.5

1. (a) Graphing the function:

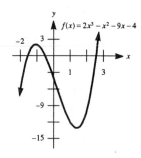

(b)   The x-intercepts are -0.5 (exact), -1.562 and 2.562 (approximate).

(c)   These intercepts are consistent with the roots determined in the text.

3.   (a)   Graphing the function:

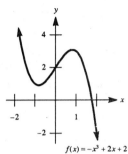

$f(x) = -x^3 + 2x + 2$

The root lies between 1 and 2.

(b)   Approximating the root between successive tenths:
$$f(1.7) = 0.487 \qquad\qquad f(1.8) = -0.232$$
The root lies between 1.7 and 1.8.
Approximating the root between successive hundredths:
$$f(1.76) \approx 0.068 \qquad\qquad f(1.77) \approx -0.005$$
The root lies between 1.76 and 1.77.
Approximating the root between successive thousandths:
$$f(1.769) \approx 0.0022 \qquad\qquad f(1.770) \approx -0.0052$$
The root lies between 1.769 and 1.770.

(c)   The x-intercept is approximately 1.7693, which matches our approximations from part (b).

5.   (a)   Graphing the function:

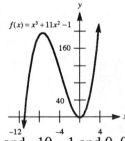

$f(x) = x^3 + 11x^2 - 1$

The roots lie between -11 and -10, -1 and 0, 0 and 1.

(b)   Approximating the roots between successive tenths:
$$f(-11.0) = -1 \qquad\qquad f(-10.9) = 10.881$$
$$f(-0.4) = 0.696 \qquad\qquad f(0.4) = -0.037$$
$$f(0.2) = -0.552 \qquad\qquad f(1.6) = 0.017$$
The roots lie between -11.0 and -10.9, -0.4 and -0.3, 0.2 and 0.3.

Approximating the roots between successive hundredths:

$$f(-11.00) = -1 \qquad\qquad f(-10.99) \approx 0.2078$$
$$f(-0.31) \approx 0.027 \qquad\qquad f(-0.30) = -0.037$$
$$f(0.29) \approx -0.0505 \qquad\qquad f(0.30) = 0.017$$

The roots lie between -11.00 and -10.99, -0.31 and -0.30, 0.29 and 0.30.

Approximating the roots between successive thousandths:

$$f(-10.992) \approx -0.0334 \qquad\qquad f(-10.991) \approx 0.0872$$
$$f(-0.306) \approx 0.0013 \qquad\qquad f(-0.305) \approx -0.0051$$
$$f(0.297) \approx -0.0035 \qquad\qquad f(0.298) \approx 0.0033$$

The roots lie between -10.992 and -10.991, -0.306 and -0.305, 0.297 and 0.298.

(c)   The x-intercepts are approximately -10.9917, -0.3058, and 0.2975, which match our approximations from part (b).

7.   (a)   Graphing the function:

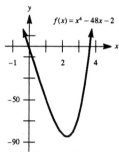

The roots lie between -1 and 0, 3 and 4.

(b)   Approximating the roots between successive tenths:

$$f(-0.1) = 2.8001 \qquad\qquad f(0.0) = -2$$
$$f(3.6) = -6.8384 \qquad\qquad f(3.7) = 7.8161$$

The roots lie between -0.1 and 0.0, 3.6 and 3.7.

Approximating the roots between successive hundredths:

$$f(-0.05) \approx 0.4000 \qquad\qquad f(-0.04) \approx -0.0800$$
$$f(3.64) \approx -1.1681 \qquad\qquad f(3.65) \approx 0.2890$$

The roots lie between -0.05 and -0.04, 3.64 and 3.65.

Approximating the roots between successive thousandths:

$$f(-0.042) \approx 0.0160 \qquad\qquad f(-0.041) \approx -0.0320$$
$$f(3.648) \approx -0.0037 \qquad\qquad f(3.649) \approx 0.1426$$

The roots lie between -0.042 and -0.041, 3.648 and 3.649.

(c)   The x-intercepts are approximately -0.0417 and 3.6480, which match our approximations from part (b).

9. Graphing the function:

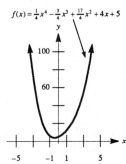

$$f(x) = \tfrac{1}{4}x^4 - \tfrac{3}{4}x^3 + \tfrac{17}{4}x^2 + 4x + 5$$

The graph has no x-intercepts, so the equation has no real roots.

11. Graphing the function:

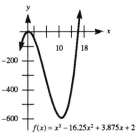

$$f(x) = x^3 - 16.25x^2 + 3.875x + 2$$

The roots are within the interval $[-1, 16]$.

12.6 Conjugate Roots and Descartes's Rule of Signs

1. The other root must be $7 + 2i$.

3. One other root must be $5 - 2i$. So $[x - (5 - 2i)][x - (5 + 2i)]$ are factors, or $x^2 - 10x + 29$. Using long division:

$$
\begin{array}{r}
x - 3 \\
x^2 - 10x + 29 \overline{\smash{\big)}\, x^3 - 13x^2 + 59x - 87} \\
\underline{x^3 - 10x^2 + 29x} \\
-3x^2 + 30x - 87 \\
\underline{-3x^2 + 30x - 87} \\
0
\end{array}
$$

Since $x - 3$ is the other factor, 3 is the other root. So the other two roots are $5 - 2i$ and 3.

5. One other root must be $-2 - i$, so $[x - (-2 + i)][x - (-2 - i)]$ are factors, or $x^2 + 4x + 5$. Using long division:

$$
\begin{array}{r}
x^2 + 6x + 9 \\
x^2 + 4x + 5 \overline{\smash{\big)}\ x^4 + 10x^3 + 38x^2 + 66x + 45} \\
\underline{x^4 + 4x^3 + 5x^2} \\
6x^3 + 33x^2 + 66x \\
\underline{6x^3 + 24x^2 + 30x} \\
9x^2 + 36x + 45 \\
\underline{9x^2 + 36x + 45} \\
0
\end{array}
$$

Since $x^2 + 6x + 9 = (x + 3)^2$, the other root is -3. So the remaining roots are $-2 - i$ and -3 (multiplicity 2).

7. One other root must be $6 + 5i$, so $[x - (6 - 5i)][x - (6 + 5i)]$ are factors, or $x^2 - 12x + 61$. Using long division:

$$
\begin{array}{r}
4x + 1 \\
x^2 - 12x + 61 \overline{\smash{\big)}\ 4x^3 - 47x^2 + 232x + 61} \\
\underline{4x^3 - 48x^2 + 244x} \\
x^2 - 12x + 61 \\
\underline{x^2 - 12x + 61} \\
0
\end{array}
$$

Since $4x + 1$ is the other factor, $-\frac{1}{4}$ is the other root. So the remaining roots are $6 + 5i$ and $-\frac{1}{4}$.

9. One other root must be $4 - \sqrt{2}i$, so $[x - (4 + \sqrt{2}i)][x - (4 - \sqrt{2}i)]$ are factors, or $x^2 - 8x + 18$. Using long division:

$$
\begin{array}{r}
4x^2 + 9 \\
x^2 - 8x + 18 \overline{\smash{\big)}\ 4x^4 - 32x^3 + 81x^2 - 72x + 162} \\
\underline{4x^4 - 32x^3 + 72x^2} \\
9x^2 - 72x + 162 \\
\underline{9x^2 - 72x + 162} \\
0
\end{array}
$$

Since $4x^2 + 9 = (2x + 3i)(2x - 3i)$, the other roots are $-\frac{3i}{2}$ and $\frac{3i}{2}$. So the remaining roots are $4 - \sqrt{2}i, -\frac{3i}{2}$ and $\frac{3i}{2}$.

11. One other root must be $10 - 2i$, so $[x - (10 + 2i)][x - (10 - 2i)]$ are factors, or $x^2 - 20x + 104$. Using long division:

$$
\begin{array}{r}
x^2 - 2x - 4 \\
x^2 - 20x + 104 \overline{\smash{\big)}\ x^4 - 22x^3 + 140x^2 - 128x - 416} \\
\underline{x^4 - 20x^3 + 104x^2} \\
-2x^3 + 36x^2 - 128x \\
\underline{-2x^3 + 40x^2 - 208x} \\
-4x^2 + 80x - 416 \\
\underline{-4x^2 + 80x - 416} \\
0
\end{array}
$$

So the other factor is $x^2 - 2x - 4$. We use the quadratic formula:

$$x = \frac{2 \pm \sqrt{4 + 16}}{2} = \frac{2 \pm 2\sqrt{5}}{2} = 1 \pm \sqrt{5}$$

So the remaining roots are $10 - 2i$, $1 + \sqrt{5}$ and $1 - \sqrt{5}$.

13. One other root must be $\frac{1 - \sqrt{2}i}{3}$ so $\left(x - \frac{1 + \sqrt{2}i}{3}\right)\left(x - \frac{1 - \sqrt{2}i}{3}\right)$ are factors, or $x^2 - \frac{2}{3}x + \frac{1}{3}$. Using long division:

$$
\begin{array}{r}
15x - 6 \\
x^2 - \frac{2}{3}x + \frac{1}{3} \overline{\smash{\big)}\ 15x^3 - 16x^2 + 9x - 2} \\
\underline{15x^3 - 10x^2 + 5x} \\
-6x^2 + 4x - 2 \\
\underline{-6x^2 + 4x - 2} \\
0
\end{array}
$$

So the other factor is $15x - 6$, so $x = \frac{2}{5}$ is a root. The remaining roots are $\frac{1 - i\sqrt{2}}{3}$ and $\frac{2}{5}$.

15. We know $3 + 2i$ is a root, so $[x - (3 - 2i)][x - (3 + 2i)]$ are factors, which is $x^2 - 6x + 13$. Using long division:

$$
\begin{array}{r}
x^5 + 3x^4 + x^3 - 3x^2 - 4x + 2 \\
x^2 - 6x + 13 \overline{\smash{\big)}\ x^7 - 3x^6 - 4x^5 + 30x^4 + 27x^3 - 13x^2 - 64x + 26} \\
\underline{x^7 - 6x^6 + 13x^5} \\
3x^6 - 17x^5 + 30x^4 \\
\underline{3x^6 - 18x^5 + 39x^4} \\
x^5 - 9x^4 + 27x^3 \\
\underline{x^5 - 6x^4 + 13x^3} \\
-3x^4 + 14x^3 - 13x^2 \\
\underline{-3x^4 + 18x^3 - 39x^2} \\
-4x^3 + 26x^2 - 64x \\
\underline{-4x^3 + 24x^2 - 52x} \\
2x^2 - 12x + 26 \\
\underline{2x^2 - 12x + 26} \\
0
\end{array}
$$

We know $x = -1 - i$ will be a root, so $[x - (-1 + i)][x - (-1 - i)]$ are factors, which is $x^2 + 2x + 2$. Using long division:

$$
\begin{array}{r}
x^3 + x^2 - 3x + 1 \\
x^2 + 2x + 2 \overline{\big)\ x^5 + 3x^4 + \ \ x^3 - 3x^2 - 4x + 2} \\
\underline{x^5 + 2x^4 + 2x^3} \\
x^4 - \ \ x^3 - 3x^2 \\
\underline{x^4 + 2x^3 + 2x^2} \\
-3x^3 - 5x^2 - 4x \\
\underline{-3x^3 - 6x^2 - 6x} \\
x^2 + 2x + 2 \\
\underline{x^2 + 2x + 2} \\
0
\end{array}
$$

Finally, we know $x = 1$ is a root, so we synthetically divide:

$$
\begin{array}{r|rrrr}
1 & 1 & 1 & -3 & 1 \\
  &   & 1 & 2 & -1 \\
\hline
  & 1 & 2 & -1 & 0
\end{array}
$$

So we are left with $x^2 + 2x - 1 = 0$. Using the quadratic formula:

$$
x = \frac{-2 \pm \sqrt{4 + 4}}{2} = \frac{-2 \pm 2\sqrt{2}}{2} = -1 \pm \sqrt{2}
$$

So the remaining roots are $3 + 2i$, $-1 - i$, $-1 + \sqrt{2}$, and $-1 - \sqrt{2}$.

17. Here $r_1 = 1 + \sqrt{6}$ and $r_2 = 1 - \sqrt{6}$, so $[x - (1 + \sqrt{6})][x - (1 - \sqrt{6})] = 0$. Multiplied out, we obtain $x^2 - 2x - 5 = 0$.

19. Here $r_1 = \frac{2 + \sqrt{10}}{3}$ and $r_2 = \frac{2 - \sqrt{10}}{3}$, so $\left(x - \frac{2 + \sqrt{10}}{3}\right)\left(x - \frac{2 - \sqrt{10}}{3}\right) = 0$. Multiplied out, we obtain $x^2 - \frac{4}{3}x - \frac{2}{3} = 0$.

21. Let $f(x) = x^3 + 5$. Since there are no sign changes, there will be no positive roots. Now $f(-x) = -x^3 + 5$. Since there is one sign change, there is 1 negative root. So the equation has 2 complex roots and 1 negative real root.

23. Let $f(x) = 2x^5 + 3x + 4$. Since there are no sign changes, there will be no positive roots. Now $f(-x) = -2x^5 - 3x + 4$. Since there is one sign change, there is 1 negative root. So the equation has 4 complex roots and 1 negative real root.

25. Let $f(x) = 5x^4 + 2x - 7$. Since there is one sign change, there will be 1 positive root. Now $f(-x) = 5x^4 - 2x - 7$. Since there is one sign change, there will be 1 negative root. So the equation has 2 complex roots, 1 positive real root and 1 negative real root.

27. Let $f(x) = x^3 - 4x^2 - x - 1$. Since there is one sign change, there will be 1 positive root. Now $f(-x) = -x^3 - 4x^2 + x - 1$. Since there are two sign changes, there will be 0 or 2 negative roots. So the equation has either 1 positive real root and 2 negative real roots, or 1 positive real root and 2 complex roots.

29. Let $f(x) = 3x^8 + x^6 - 2x^2 - 4$. Since there is one sign change, there is 1 positive root. Now $f(-x) = 3x^8 + x^6 - 2x^2 - 4$. Since there is one sign change, there is 1 negative root. So the equation has 1 positive real root, 1 negative real root and 6 complex roots.

31. Let $f(x) = x^9 - 2$. Since there is one sign change, there is 1 positive root. Now $f(-x) = -x^9 - 2$. Since there are no sign changes, there are no negative roots. So the equation has 1 positive real root and 8 complex roots.

33. Let $f(x) = x^8 - 2$. Since there is one sign change, there is 1 positive root. Now $f(-x) = x^8 - 2$. Since there is one sign change, there is 1 negative root. So the equation has 1 positive real root, 1 negative real root and 6 complex roots.

35. Let $f(x) = x^6 + x^2 - x - 1$. Since there is one sign change, there is 1 positive root. Now $f(-x) = x^6 + x^2 + x - 1$. Since there is one sign change, there is 1 negative root. So the equation has 1 positive real root, 1 negative real root and 4 complex roots.

37. Let $f(x) = x^4 + cx^2 + dx - e$. Since there is one sign change, there is 1 positive root. Now $f(-x) = x^4 + cx^2 - dx - e$. Since there is one sign change, there is 1 negative root. So the equation has 1 positive real root, 1 negative real root and 2 complex roots.

39. The two roots $\sqrt{3} + 2i$ and $\sqrt{3} - 2i$ must be included, so:
$$f(x) = \left[x - (\sqrt{3} + 2i)\right]\left[x - (\sqrt{3} - 2i)\right] = x^2 - 2\sqrt{3} + 7 = 0$$
Unfortunately, not all coefficients are rational:
$$x^2 + 7 = 2\sqrt{3}x$$
Squaring, we obtain:
$$x^4 + 14x^2 + 49 = 12x^2$$
$$x^4 + 2x^2 + 29 = 0$$
So $f(x) = x^4 + 2x^2 + 49$ is the desired polynomial.

41. (a) If $b = 0$, then $a + b\sqrt{c} = a = a - b\sqrt{c}$, so $a - b\sqrt{c}$ is also a root.

   (b) Compute $d(a + b\sqrt{c})$:
   $$d(a + b\sqrt{c}) = \left[a + b\sqrt{c} - (a + b\sqrt{c})\right]\left[a + b\sqrt{c} - (a - b\sqrt{c})\right] = 0(2b\sqrt{c}) = 0$$

   (c) Factor $d(x) = x^2 - 2ax + a^2 - b^2c = (x - a)^2 - b^2c$.

   (d) If $x = a + b\sqrt{c}$, we have:
   $$f(a + b\sqrt{c}) = d(a + b\sqrt{c})Q(a + b\sqrt{c}) + C(a + b\sqrt{c}) + D$$
   $$0 = 0 \cdot Q(a + b\sqrt{c}) + Ca + D + bC\sqrt{c}$$
   $$0 = Ca + D + bC\sqrt{c}$$

   But if C and D are rational, then $C = 0$ (otherwise $\sqrt{c}$ will not be "cancelled" out). If $C = 0$, then $D = 0$. So $C = D = 0$.

(e) Compute $f(a - b\sqrt{c})$:
$$f(a - b\sqrt{c}) = \left[a - b\sqrt{c} - (a + b\sqrt{c})\right]\left[a - b\sqrt{c} - (a - b\sqrt{c})\right]Q(x)$$
$$= (-2b\sqrt{c})(0)Q(x)$$
$$= 0$$

So $x = a - b\sqrt{c}$ is also a root of $f(x) = 0$.

43. Let $r_1 = a - bi$ and $r_2 = a + bi$ be the two roots of the equation $f(x) = x^2 + Bx + C = 0$.
Using Table 2:
$$r_1 + r_2 = -\frac{B}{1}$$
$$2a = -B$$
$$B = -2a$$

Also:
$$r_1 r_2 = \frac{C}{1}$$
$$a^2 + b^2 = C$$

Thus the function can be written as:
$$f(x) = x^2 - 2ax + \left(a^2 + b^2\right)$$
$$= \left(x^2 - 2ax + a^2\right) + b^2$$
$$= (x - a)^2 + b^2$$

So the vertex of the graph is $\left(a, b^2\right)$.

Graphing Utility Exercises for Section 12.6

1. (a) Computing $f(-x)$:
$$f(-x) = 2(-x)^4 - 3(-x)^3 + 12(-x)^2 + 22(-x) - 60$$
$$= 2x^4 + 3x^3 + 12x^2 - 22x - 60$$
Since the resulting function has only one sign change, there can only be one negative real root.

(b) Since $f(x) = 2x^4 - 3x^3 + 12x^2 + 22x - 60$ has three sign changes, there can be either one or three positive real roots.

(c) Graphing the function:

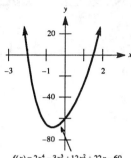

$$f(x) = 2x^4 - 3x^3 + 12x^2 + 22x - 60$$

From the graph it appears there is one positive real root.

(d) The roots are –2.000 and 1.500, which agrees with the roots $-2, \frac{1}{2}$ found in the text.

3.   (a) Since $f(x) = x^3 + 8x + 5$ has no sign changes, there are no positive real roots. Compute:
$$f(-x) = (-x)^3 + 8(-x) + 5 = -x^3 - 8x + 5$$
Since there is one sign change, there is exactly one negative real root.

(b) Graphing the function:

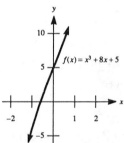

$$f(x) = x^3 + 8x + 5$$

Note that there is only one negative real root.

(c) The root is approximately –0.5982.

(d) Write the equation as $x^3 + 8x = -5$. Now use the formula with $a = 8$ and $b = -5$:
$$x = \sqrt[3]{\frac{-5}{2} + \sqrt{\frac{(-5)^2}{4} + \frac{8^3}{27}}} - \sqrt[3]{-\frac{-5}{2} + \sqrt{\frac{(-5)^2}{4} + \frac{8^3}{27}}}$$
$$= \sqrt[3]{-2.5 + \sqrt{\frac{25}{4} + \frac{512}{27}}} - \sqrt[3]{2.5 + \sqrt{\frac{25}{4} + \frac{512}{27}}}$$
$$= \sqrt[3]{-2.5 + \sqrt{\frac{2723}{108}}} - \sqrt[3]{2.5 + \sqrt{\frac{2723}{108}}}$$
$$\approx -0.5982$$
Note that our value agrees with that found in part (c).

12.7 Introduction to Partial Fractions

1.   (a) Clearing fractions, we have:
$$7x - 6 = A(x + 2) + B(x - 2) = (A + B)x + (2A - 2B)$$
Equating coefficients, we have:
$$A + B = 7$$
$$2A - 2B = -6$$
Multiplying the first equation by 2:
$$2A + 2B = 14$$
$$2A - 2B = -6$$
Adding yields $4A = 8$, so $A = 2$ and $B = 5$.

(b) Substituting $x = 2$ into $7x - 6 = A(x + 2) + B(x - 2)$:
$$14 - 6 = A(2 + 2) + B(2 - 2)$$
$$8 = 4A$$
$$2 = A$$
Substituting $x = -2$ into $7x - 6 = A(x + 2) + B(x - 2)$:
$$-14 - 6 = A(-2 + 2) + B(-2 - 2)$$
$$-20 = -4B$$
$$5 = B$$
So $A = 2$ and $B = 5$.

3.   (a) Clearing fractions, we have:
$$6x - 25 = A(2x - 5) + B(2x + 5) = (2A + 2B)x + (-5A + 5B)$$
Equating coefficients, we have:
$$2A + 2B = 6$$
$$-5A + 5B = -25$$
Dividing the first equation by 2 and the second equation by –5:
$$A + B = 3$$
$$A - B = 5$$
Adding yields $2A = 8$, so $A = 4$ and $B = -1$.

(b) Substituting $x = \frac{5}{2}$ into $6x - 25 = A(2x - 5) + B(2x + 5)$:
$$15 - 25 = A(5 - 5) + B(5 + 5)$$
$$-10 = 10B$$
$$-1 = B$$
Substituting $x = -\frac{5}{2}$ into $6x - 25 = A(2x - 5) + B(2x + 5)$:
$$-15 - 25 = A(-5 - 5) + B(-5 + 5)$$
$$-40 = -10A$$
$$4 = A$$
So $A = 4$ and $B = -1$.

5.   **(a)** Clearing fractions, we have:
$$1 = A(3x-1) + B(x+1) = (3A+B)x + (-A+B)$$
Equating coefficients, we have:
$$3A + B = 0$$
$$-A + B = 1$$
Multiplying the second equation by -1:
$$3A + B = 0$$
$$A - B = -1$$
Adding yields $4A = -1$, so $A = -\frac{1}{4}$. Substituting into $-A + B = 1$:
$$\tfrac{1}{4} + B = 1$$
$$B = \tfrac{3}{4}$$
So $A = -\frac{1}{4}$ and $B = \frac{3}{4}$.

(b) Substituting $x = \frac{1}{3}$ into $1 = A(3x-1) + B(x+1)$:
$$1 = A(1-1) + B\left(\tfrac{1}{3} + 1\right)$$
$$1 = \tfrac{4}{3}B$$
$$\tfrac{3}{4} = B$$
Substituting $x = -1$ into $1 = A(3x-1) + B(x+1)$:
$$1 = A(-3-1) + B(-1+1)$$
$$1 = -4A$$
$$-\tfrac{1}{4} = A$$
So $A = -\frac{1}{4}$ and $B = \frac{3}{4}$.

7.   Clearing fractions, we have:
$$8x + 3 = A(x+3) + B = Ax + (3A+B)$$
Equating coefficients, we have:
$$A = 8$$
$$3A + B = 3$$
Substituting $A = 8$ into $3A + B = 3$:
$$24 + B = 3$$
$$B = -21$$
So $A = 8$ and $B = -21$.

9.   Clearing fractions, we have:
$$6 - x = A(5x+4) + B = 5Ax + (4A+B)$$
Equating coefficients, we have:
$$5A = -1$$
$$4A + B = 6$$

So $A = -\frac{1}{5}$. Substituting $A = -\frac{1}{5}$ into $4A + B = 6$:

$$-\frac{4}{5} + B = 6$$
$$B = \frac{34}{5}$$

So $A = -\frac{1}{5}$ and $B = \frac{34}{5}$.

11. Clearing fractions, we have:

$$3x^2 + 7x - 2 = A\left(x^2 + 1\right) + (Bx + C)(x - 1)$$
$$= Ax^2 + A + Bx^2 + Cx - Bx - C$$
$$= (A + B)x^2 + (-B + C)x + (A - C)$$

Equating coefficients, we have:

$$A + B = 3$$
$$-B + C = 7$$
$$A - C = -2$$

Adding all three equations yields $2A = 8$, so $A = 4$, $B = -1$, and $C = 6$.

13. Clearing fractions, we have:

$$x^2 + 1 = A\left(x^2 + 4\right) + (Bx + C)(x + 1)$$
$$= Ax^2 + 4A + Bx^2 + Cx + Bx + C$$
$$= (A + B)x^2 + (B + C)x + (4A + C)$$

Equating coefficients, we have:

$$A + B = 1$$
$$B + C = 0$$
$$4A + C = 1$$

Form the augmented matrix:

$$\begin{pmatrix} 1 & 1 & 0 & 1 \\ 0 & 1 & 1 & 0 \\ 4 & 0 & 1 & 1 \end{pmatrix}$$

Adding -4 times row 1 to row 3:

$$\begin{pmatrix} 1 & 1 & 0 & 1 \\ 0 & 1 & 1 & 0 \\ 0 & -4 & 1 & -3 \end{pmatrix}$$

Adding 4 times row 2 to row 3:

$$\begin{pmatrix} 1 & 1 & 0 & 1 \\ 0 & 1 & 1 & 0 \\ 0 & 0 & 5 & -3 \end{pmatrix}$$

Dividing row 3 by 5, and adding −1 times row 3 to row 2:

$$\begin{pmatrix} 1 & 1 & 0 & 1 \\ 0 & 1 & 0 & \frac{3}{5} \\ 0 & 0 & 1 & -\frac{3}{5} \end{pmatrix}$$

Adding −1 times row 2 to row 1:

$$\begin{pmatrix} 1 & 0 & 0 & \frac{2}{5} \\ 0 & 1 & 0 & \frac{3}{5} \\ 0 & 0 & 1 & -\frac{3}{5} \end{pmatrix}$$

So $A = \frac{2}{5}$, $B = \frac{3}{5}$, and $C = -\frac{3}{5}$.

15. Clearing fractions, we have:
$$1 = A\left(x^2 - x + 1\right) + (Bx + C)x$$
$$= Ax^2 - Ax + A + Bx^2 + Cx$$
$$= (A + B)x^2 + (-A + C)x + A$$
Equating coefficients, we have:
$$A + B = 0$$
$$-A + C = 0$$
$$A = 1$$
So $A = 1$, $B = -1$, and $C = 1$.

17. Clearing fractions, we have:
$$3x^2 - 2 = A(x+1)(x-1) + B(x-2)(x-1) + C(x-2)(x+1)$$
Substituting $x = -1$:
$$3 - 2 = A(0)(-2) + B(-3)(-2) + C(-3)(0)$$
$$1 = 6B$$
$$B = \frac{1}{6}$$
Substituting $x = 1$:
$$3 - 2 = A(2)(0) + B(-1)(0) + C(-1)(-2)$$
$$1 = -2C$$
$$C = -\frac{1}{2}$$
Substituting $x = 2$:
$$12 - 2 = A(3)(1) + B(0)(1) + C(0)(3)$$
$$10 = 3A$$
$$A = \frac{10}{3}$$
So $A = \frac{10}{3}$, $B = \frac{1}{6}$, and $C = -\frac{1}{2}$.

19.   Clearing fractions, we have:

$$4x^2 - 47x + 133 = A(x-6)^2 + B(x-6) + C$$
$$= Ax^2 - 12Ax + 36A + Bx - 6B + C$$
$$= Ax^2 + (-12A+B)x + (36A - 6B + C)$$

Equating coefficients, we have:

$$A = 4$$
$$-12A + B = -47$$
$$36A - 6B + C = 133$$

Substituting $A = 4$ into $-12A + B = -47$:

$$-48 + B = -47$$
$$B = 1$$

Substituting into $36A - 6B + C = 133$:

$$144 - 6 + C = 133$$
$$138 + C = 133$$
$$C = -5$$

So $A = 4$, $B = 1$, and $C = -5$.

21.   Clearing fractions, we have:

$$x^2 - 2 = (Ax + B)\left(x^2 + 2\right) + (Cx + D)$$
$$= Ax^3 + Bx^2 + 2Ax + 2B + Cx + D$$
$$= Ax^3 + Bx^2 + (2A+C)x + (2B+D)$$

Equating coefficients, we have:

$$A = 0$$
$$B = 1$$
$$2A + C = 0$$
$$2B + D = -2$$

Substituting $A = 0$ into $2A + C = 0$ results in $C = 0$. Substituting $B = 1$ into $2B + D = -2$:

$$2 + D = -2$$
$$C = -4$$

So $A = 0$, $B = 1$, $C = 0$, and $D = -4$.

23.   Form the augmented matrix:

$$\begin{pmatrix} 1 & 1 & 0 & 7 \\ 0 & -2 & 1 & -9 \\ 9 & 0 & -2 & 29 \end{pmatrix}$$

Adding -9 times row 1 to row 3:

$$\begin{pmatrix} 1 & 1 & 0 & 7 \\ 0 & -2 & 1 & -9 \\ 0 & -9 & -2 & -34 \end{pmatrix}$$

Dividing row 2 by –2, and adding 9 times row 2 to row 3:

$$\begin{pmatrix} 1 & 1 & 0 & 7 \\ 0 & 1 & -\frac{1}{2} & \frac{9}{2} \\ 0 & 0 & -\frac{13}{2} & \frac{13}{2} \end{pmatrix}$$

Dividing row 3 by $-\frac{13}{2}$, and adding $\frac{1}{2}$ times row 3 to row 2:

$$\begin{pmatrix} 1 & 1 & 0 & 7 \\ 0 & 1 & 0 & 4 \\ 0 & 0 & 1 & -1 \end{pmatrix}$$

Adding –1 times row 2 to row 1:

$$\begin{pmatrix} 1 & 0 & 0 & 3 \\ 0 & 1 & 0 & 4 \\ 0 & 0 & 1 & -1 \end{pmatrix}$$

So $A = 3$, $B = 4$, and $C = -1$.

25. (a) Substituting $x = 0$:
$$0 + 1 = A(0 + 4) + (0 + C)(0)$$
$$1 = 4A$$
$$A = \frac{1}{4}$$

(b) Substituting $x = 2i$:
$$2i + 1 = A\left(4i^2 + 4\right) + (2iB + C)(2i)$$
$$2i + 1 = A(-4 + 4) + \left(4i^2 B + 2iC\right)$$
$$2i + 1 = -4B + 2iC$$

(c) Since $2i = 2iC$, $C = 1$. Since $-4B = 1$, $B = -\frac{1}{4}$.

27. Multiplying out the factors:
$$\left(x^2 + bx + 1\right)\left(x^2 + cx + 1\right) = x^4 + cx^3 + x^2 + bx^3 + bcx^2 + bx + x^2 + cx + 1$$
$$= x^4 + (b + c)x^3 + (2 + bc)x^2 + (b + c)x + 1$$

Equating coefficients, we have:
$$b + c = -1$$
$$2 + bc = 1$$

Since $b + c = -1$, $c = -1 - b$. Substituting:
$$2 + b(-1 - b) = 1$$
$$2 - b - b^2 = 1$$
$$0 = b^2 + b - 1$$

Using the quadratic formula:
$$b = \frac{-1 \pm \sqrt{1 - 4(1)(-1)}}{2(1)} = \frac{-1 \pm \sqrt{1+4}}{2} = \frac{-1 \pm \sqrt{5}}{2}$$

If $b = \frac{-1 + \sqrt{5}}{2}$:
$$c = -1 - \frac{-1 + \sqrt{5}}{2} = \frac{-2 + 1 - \sqrt{5}}{2} = \frac{-1 - \sqrt{5}}{2}$$

If $b = \frac{-1 - \sqrt{5}}{2}$:
$$c = -1 - \frac{-1 - \sqrt{5}}{2} = \frac{-2 + 1 + \sqrt{5}}{2} = \frac{-1 + \sqrt{5}}{2}$$

Thus one factorization is:
$$\left(x^2 + \frac{-1 + \sqrt{5}}{2} x + 1 \right)\left(x^2 + \frac{-1 - \sqrt{5}}{2} x + 1 \right)$$

29.   Multiplying out the factors:
$$\left(x^3 + ax^2 + bx + 1 \right)\left(x^3 + cx^2 + dx + 1 \right)$$
$$= x^6 + (a+c)x^5 + (ac+b+d)x^4 + (bc+ad+2)x^3$$
$$+ (c+bd+a)x^2 + (b+d)x + 1$$

Equating coefficients, we have:
$$a + c = -1$$
$$ac + b + d = 1$$
$$bc + ad + 2 = -1$$
$$c + bd + a = 1$$
$$b + d = -1$$

So $c = -1 - a$ and $d = -1 - b$. Substituting into $ac + b + d = 1$:
$$a(-1 - a) + b + (-1 - b) = 1$$
$$-a - a^2 + b - 1 - b = 1$$
$$-a - a^2 - 1 = 1$$
$$a^2 + a + 2 = 0$$

Using the quadratic formula:
$$a = \frac{-1 \pm \sqrt{1 - 4(1)(2)}}{2(1)} = \frac{-1 \pm \sqrt{1-8}}{2} = \frac{-1 \pm \sqrt{-7}}{2}$$

Since a is not a real number, the factorization (over the real numbers) is impossible.

12.8 More About Partial Fractions

1. (a) No, this is reducible, since:
$$x^2 - 16 = 0$$
$$(x+4)(x-4) = 0$$
$$x = -4, 4$$
Thus the polynomial equation $f(x) = 0$ has two real roots.

(b) Yes, this is irreducible, since $x^2 + 16 = 0$ has no real roots.

3. (a) No, this is reducible, since:
$$x^2 + 3x - 4 = 0$$
$$(x+4)(x-1) = 0$$
$$x = -4, 1$$
Thus the polynomial equation $f(x) = 0$ has two real roots.

(b) Yes, this is irreducible, since $x^2 + 3x + 4 = 0$ has no real roots.

5. (a) The denominator factors as $x^2 - 100 = (x+10)(x-10)$.

(b) The form of the partial fraction decomposition is:
$$\frac{11x + 30}{x^2 - 100} = \frac{A}{x+10} + \frac{B}{x-10}$$

(c) Clearing fractions yields:
$$11x + 30 = A(x-10) + B(x+10)$$
$$= Ax - 10A + Bx + 10B$$
$$= (A+B)x + (-10A + 10B)$$
Equating coefficients, we have:
$$A + B = 11$$
$$-10A + 10B = 30$$
Dividing the second equation by 10:
$$A + B = 11$$
$$-A + B = 3$$
Adding yields $2B = 14$, so $B = 7$ and $A = 4$. The decomposition is:
$$\frac{11x + 30}{x^2 - 100} = \frac{4}{x+10} + \frac{7}{x-10}$$

7. (a) The denominator factors as $x^2 - 5 = \left(x + \sqrt{5}\right)\left(x - \sqrt{5}\right)$.

(b) The form of the partial fraction decomposition is:
$$\frac{8x - 2\sqrt{5}}{x^2 - 5} = \frac{A}{x + \sqrt{5}} + \frac{B}{x - \sqrt{5}}$$

(c)   Clearing fractions yields:
$$8x - 2\sqrt{5} = A\left(x - \sqrt{5}\right) + B\left(x + \sqrt{5}\right)$$
$$= Ax - A\sqrt{5} + Bx + B\sqrt{5}$$
$$= (A + B)x + (-A + B)\sqrt{5}$$
Equating coefficients, we have:
$$A + B = 8$$
$$-A + B = -2$$
Adding yields $2B = 6$, so $B = 3$ and $A = 5$. The decomposition is:
$$\frac{8x - 2\sqrt{5}}{x^2 - 5} = \frac{5}{x + \sqrt{5}} + \frac{3}{x - \sqrt{5}}$$

9.   (a)   The denominator factors as $x^2 - x - 6 = (x - 3)(x + 2)$.

(b)   The form of the partial fraction decomposition is:
$$\frac{7x + 39}{x^2 - x - 6} = \frac{A}{x + 2} + \frac{B}{x - 3}$$

(c)   Clearing fractions yields:
$$7x + 39 = A(x - 3) + B(x + 2)$$
$$= Ax - 3A + Bx + 2B$$
$$= (A + B)x + (-3A + 2B)$$
Equating coefficients, we have:
$$A + B = 7$$
$$-3A + 2B = 39$$
Multiplying the first equation by 3:
$$3A + 3B = 21$$
$$-3A + 2B = 39$$
Adding yields $5B = 60$, so $B = 12$ and $A = -5$. The decomposition is:
$$\frac{7x + 39}{x^2 - x - 6} = \frac{-5}{x + 2} + \frac{12}{x - 3}$$

11.   (a)   The denominator factors as:
$$x^3 - 3x^2 - 4x + 12 = x^2(x - 3) - 4(x - 3)$$
$$= (x - 3)\left(x^2 - 4\right)$$
$$= (x - 3)(x - 2)(x + 2)$$

(b)   The form of the partial fraction decomposition is:
$$\frac{3x^2 + 17x - 38}{x^3 - 3x^2 - 4x + 12} = \frac{A}{x - 3} + \frac{B}{x - 2} + \frac{C}{x + 2}$$

(c) Clearing fractions yields:

$$3x^2 + 17x - 38 = A(x-2)(x+2) + B(x-3)(x+2) + C(x-3)(x-2)$$
$$= A(x^2-4) + B(x^2-x-6) + C(x^2-5x+6)$$
$$= Ax^2 - 4A + Bx^2 - Bx - 6B + Cx^2 - 5Cx + 6C$$
$$= (A+B+C)x^2 + (-B-5C)x + (-4A-6B+6C)$$

Equating coefficients, we have:

$$A + B + C = 3$$
$$-B - 5C = 17$$
$$-4A - 6B + 6C = -38$$

Multiplying the first equation by 4:

$$4A + 4B + 4C = 12$$
$$-4A - 6B + 6C = -38$$

Adding yields $-2B + 10C = -26$, so $B - 5C = 13$. We have:

$$B - 5C = 13$$
$$-B - 5C = 17$$

Adding yields $-10C = 30$, so $C = -3$. Substituting:

$$B - 5(-3) = 13$$
$$B + 15 = 13$$
$$B = -2$$

Substituting:

$$A - 2 - 3 = 3$$
$$A = 8$$

So $A = 8$, $B = -2$, and $C = -3$. The decomposition is:

$$\frac{3x^2 + 17x - 38}{x^3 - 3x^2 - 4x + 12} = \frac{8}{x-3} + \frac{-2}{x-2} + \frac{-3}{x+2}$$

13. (a) The denominator factors as:

$$x^3 + x^2 + x = x(x^2 + x + 1)$$

(b) The form of the partial fraction decomposition is:

$$\frac{5x^2 + 2x + 5}{x^3 + x^2 + x} = \frac{A}{x} + \frac{Bx + C}{x^2 + x + 1}$$

(c) Clearing fractions yields:

$$5x^2 + 2x + 5 = A(x^2 + x + 1) + (Bx + C)x$$
$$= Ax^2 + Ax + A + Bx^2 + Cx$$
$$= (A+B)x^2 + (A+C)x + A$$

Equating coefficients, we have:

$$A + B = 5$$
$$A + C = 2$$
$$A = 5$$

So $A = 5, B = 0$, and $C = -3$. The decomposition is:

$$\frac{5x^2 + 2x + 5}{x^3 + x^2 + x} = \frac{5}{x} + \frac{-3}{x^2 + x + 1}$$

15. (a) The denominator factors as $x^4 + 2x^2 + 1 = \left(x^2 + 1\right)^2$.

(b) The form of the partial fraction decomposition is:

$$\frac{2x^3 + 5x - 4}{x^4 + 2x^2 + 1} = \frac{Ax + B}{x^2 + 1} + \frac{Cx + D}{\left(x^2 + 1\right)^2}$$

(c) Clearing fractions yields:

$$\begin{aligned}
2x^3 + 5x - 4 &= (Ax + B)\left(x^2 + 1\right) + Cx + D \\
&= Ax^3 + Bx^2 + Ax + B + Cx + D \\
&= Ax^3 + Bx^2 + (A + C)x + (B + D)
\end{aligned}$$

Equating coefficients, we have:

$$\begin{aligned}
A &= 2 \\
B &= 0 \\
A + C &= 5 \\
B + D &= -4
\end{aligned}$$

So $A = 2, B = 0, C = 3$, and $D = -4$. The decomposition is:

$$\frac{2x^3 + 5x - 4}{x^4 + 2x^2 + 1} = \frac{2x}{x^2 + 1} + \frac{3x - 4}{\left(x^2 + 1\right)^2}$$

17. Since $x^3 - 3x^2 - 16x - 12 = (x - 6)(x + 1)(x + 2)$, the form of the decomposition is:

$$\frac{x^2 + 2}{x^3 - 3x^2 - 16x - 12} = \frac{A}{x - 6} + \frac{B}{x + 1} + \frac{C}{x + 2}$$

Clearing fractions yields:

$$x^2 + 2 = A(x + 1)(x + 2) + B(x - 6)(x + 2) + C(x - 6)(x + 1)$$

Substituting $x = -1$:

$$\begin{aligned}
1 + 2 &= A(0) + B(-7)(1) + C(0) \\
3 &= -7B \\
B &= -\frac{3}{7}
\end{aligned}$$

Substituting $x = -2$:

$$\begin{aligned}
4 + 2 &= A(0) + B(0) + C(-8)(-1) \\
6 &= 8C \\
C &= \frac{3}{4}
\end{aligned}$$

Substituting $x = 6$:

$$36 + 2 = A(7)(8) + B(0) + C(0)$$
$$38 = 56A$$
$$A = \frac{19}{28}$$

The partial fraction decomposition is:

$$\frac{x^2 + 2}{x^3 - 3x^2 - 16x - 12} = \frac{\frac{19}{28}}{x - 6} + \frac{-\frac{3}{7}}{x + 1} + \frac{\frac{3}{4}}{x + 2}$$

19. Since $6x^2 - 19x + 15 = (3x - 5)(2x - 3)$, the form of the decomposition is:

$$\frac{5 - x}{6x^2 - 19x + 15} = \frac{A}{3x - 5} + \frac{B}{2x - 3}$$

Clearing fractions yields:

$$5 - x = A(2x - 3) + B(3x - 5)$$
$$= 2Ax - 3A + 3Bx - 5B$$
$$= (2A + 3B)x + (-3A - 5B)$$

Equating coefficients, we have:

$$2A + 3B = -1$$
$$-3A - 5B = 5$$

Multiplying the first equation by 3 and the second equation by 2:

$$6A + 9B = -3$$
$$-6A - 10B = 10$$

Adding yields $-B = 7$, so $B = -7$ and $A = 10$. The partial fraction decomposition is:

$$\frac{5 - x}{6x^2 - 19x + 15} = \frac{10}{3x - 5} + \frac{-7}{2x - 3}$$

21. Since $x^3 - 5x = x(x^2 - 5) = x(x + \sqrt{5})(x - \sqrt{5})$, the form of the decomposition is:

$$\frac{2x + 1}{x^3 - 5x} = \frac{A}{x} + \frac{B}{x + \sqrt{5}} + \frac{C}{x - \sqrt{5}}$$

Clearing fractions yields:

$$2x + 1 = A(x + \sqrt{5})(x - \sqrt{5}) + Bx(x - \sqrt{5}) + Cx(x + \sqrt{5})$$
$$= Ax^2 - 5A + Bx^2 - B\sqrt{5}x + Cx^2 + C\sqrt{5}x$$
$$= (A + B + C)x^2 + (-\sqrt{5}B + \sqrt{5}C)x - 5A$$

Equating coefficients, we have:

$$A + B + C = 0$$
$$-\sqrt{5}B + \sqrt{5}C = 2$$
$$-5A = 1$$

Solving the third equation yields $A = -\frac{1}{5}$. Substituting into the first equation yields:

$$-\frac{1}{5} + B + C = 0$$
$$-\sqrt{5}B + \sqrt{5}C = 2$$

Multiplying the first equation by 5 and the second equation by $\sqrt{5}$:
$$5B + 5C = 1$$
$$-5B + 5C = 2\sqrt{5}$$

Adding yields $10C = 1 + 2\sqrt{5}$, so $C = \dfrac{1 + 2\sqrt{5}}{10}$. Subtracting yields $10B = 1 - 2\sqrt{5}$, so

$B = \dfrac{1 - 2\sqrt{5}}{10}$. The partial fraction decomposition is:

$$\frac{2x+1}{x^3 - 5x} = \frac{-1/5}{x} + \frac{\left(1 - 2\sqrt{5}\right)/10}{x + \sqrt{5}} + \frac{\left(1 + 2\sqrt{5}\right)/10}{x - \sqrt{5}}$$

23. Since $x^4 + 8x^2 + 16 = \left(x^2 + 4\right)^2$, the form of the decomposition is:

$$\frac{x^3 + 2}{x^4 + 8x^2 + 16} = \frac{Ax + B}{x^2 + 4} + \frac{Cx + D}{\left(x^2 + 4\right)^2}$$

Clearing fractions yields:
$$x^3 + 2 = (Ax + B)\left(x^2 + 4\right) + (Cx + D)$$
$$= Ax^3 + Bx^2 + 4Ax + 4B + Cx + D$$
$$= Ax^3 + Bx^2 + (4A + C)x + (4B + D)$$

Equating coefficients, we have:
$$A = 1$$
$$B = 0$$
$$4A + C = 0$$
$$4B + D = 2$$

Since $4(1) + C = 0$, $C = -4$, and since $4(0) + D = 2$, $D = 2$. The partial fraction decomposition is:

$$\frac{x^3 + 2}{x^4 + 8x^2 + 16} = \frac{x}{x^2 + 4} + \frac{-4x + 2}{\left(x^2 + 4\right)^2}$$

25. Factoring the denominator:
$$x^4 - 15x^3 + 75x^2 - 125x = x\left(x^3 - 15x^2 + 75x - 125\right) = x(x - 5)^3$$

The form of the partial fraction decomposition is:

$$\frac{x^3 + x - 3}{x(x - 5)^3} = \frac{A}{x} + \frac{B}{x - 5} + \frac{C}{(x - 5)^2} + \frac{D}{(x - 5)^3}$$

Clearing fractions yields:
$$x^3 + x - 3 = A(x - 5)^3 + Bx(x - 5)^2 + Cx(x - 5) + Dx$$

Substituting $x = 0$:

$$-3 = A(-5)^3$$
$$-3 = -125A$$
$$A = \frac{3}{125}$$

Substituting $x = 5$:

$$127 = 5D$$
$$D = \frac{127}{5}$$

Substituting $x = 4$:

$$64 + 4 - 3 = -A + 4B - 4C + 4D$$
$$65 = -\frac{3}{125} + 4B - 4C + \frac{508}{5}$$
$$65 = \frac{12697}{125} + 4B - 4C$$
$$-\frac{4572}{125} = 4B - 4C$$
$$-\frac{1143}{125} = B - C$$

Substituting $x = 6$:

$$216 + 6 - 3 = A + 6B + 6C + 6D$$
$$219 = \frac{3}{125} + 6B + 6C + \frac{762}{5}$$
$$219 = \frac{19053}{125} + 6B + 6C$$
$$\frac{8322}{125} = 6B + 6C$$
$$\frac{1387}{125} = B + C$$

So we have the system of equations:

$$B + C = \frac{1387}{125}$$
$$B - C = -\frac{1143}{125}$$

Adding yields $2B = \frac{244}{125}$, so $B = \frac{122}{125}$. Subtracting yields $2C = \frac{2530}{125}$, so $C = \frac{1265}{125} = \frac{253}{25}$.
The partial fraction decomposition is:

$$\frac{x^3 + x - 3}{x^4 - 15x^3 + 75x^2 - 125x} = \frac{\frac{3}{125}}{x} + \frac{\frac{122}{125}}{x - 5} + \frac{\frac{253}{25}}{(x-5)^2} + \frac{\frac{127}{5}}{(x-5)^3}$$

27. Since $x^3 - 1 = (x - 1)(x^2 + x + 1)$, the form of the decomposition is:

$$\frac{1}{x^3 - 1} = \frac{A}{x - 1} + \frac{Bx + C}{x^2 + x + 1}$$

Clearing fractions yields:

$$1 = A(x^2 + x + 1) + (Bx + C)(x - 1)$$
$$= Ax^2 + Ax + A + Bx^2 + Cx - Bx - C$$
$$= (A + B)x^2 + (A - B + C)x + (A - C)$$

Equating coefficients, we have:
$$A + B = 0$$
$$A - B + C = 0$$
$$A - C = 1$$
Adding the first two equations results in the system:
$$2A + C = 0$$
$$A - C = 1$$
Adding yields $3A = 1$, so $A = \frac{1}{3}$. Thus $B = -\frac{1}{3}$ and $C = -\frac{2}{3}$. The partial fraction decomposition is:
$$\frac{1}{x^3 - 1} = \frac{\frac{1}{3}}{x - 1} + \frac{-\frac{1}{3}x - \frac{2}{3}}{x^2 + x + 1}$$

29.   Factoring the denominator:
$$x^4 + 2x^3 + x^2 = x^2\left(x^2 + 2x + 1\right) = x^2(x+1)^2$$
The form of the partial fraction decomposition is:
$$\frac{7x^3 + 11x^2 - x - 2}{x^2(x+1)^2} = \frac{A}{x} + \frac{B}{x^2} + \frac{C}{x+1} + \frac{D}{(x+1)^2}$$
Clearing fractions yields:
$$7x^3 + 11x^2 - x - 2 = Ax(x+1)^2 + B(x+1)^2 + Cx^2(x+1) + Dx^2$$
Substituting $x = 0$:
$$-2 = B$$
Substituting $x = -1$:
$$-7 + 11 + 1 - 2 = D$$
$$3 = D$$
Substituting $x = 1$:
$$7 + 11 - 1 - 2 = 4A + 4B + 2C + D$$
$$15 = 4A - 8 + 2C + 3$$
$$20 = 4A + 2C$$
$$10 = 2A + C$$
Substituting $x = -2$:
$$-56 + 44 + 2 - 2 = -2A + B - 4C + 4D$$
$$-12 = -2A - 2 - 4C + 12$$
$$-22 = -2A - 4C$$
$$11 = A + 2C$$
So we have the system of equations:
$$2A + C = 10$$
$$A + 2C = 11$$
Multiplying the first equation by -2:
$$-4A - 2C = -20$$
$$A + 2C = 11$$

Adding yields $-3A = -9$, so $A = 3$. Substituting:

$$3 + 2C = 11$$
$$2C = 8$$
$$C = 4$$

The partial fraction decomposition is:

$$\frac{7x^3 + 11x^2 - x - 2}{x^4 + 2x^3 + x^2} = \frac{3}{x} + \frac{-2}{x^2} + \frac{4}{x+1} + \frac{3}{(x+1)^2}$$

31. Factoring the denominator:

$$x^4 - 81 = \left(x^2 + 9\right)\left(x^2 - 9\right) = \left(x^2 + 9\right)(x+3)(x-3)$$

The form of the partial fraction decomposition is:

$$\frac{x^3 - 5}{x^4 - 81} = \frac{A}{x+3} + \frac{B}{x-3} + \frac{Cx + D}{x^2 + 9}$$

Clearing fractions yields:

$$x^3 - 5 = A(x-3)\left(x^2 + 9\right) + B(x+3)\left(x^2 + 9\right) + (Cx + D)(x+3)(x-3)$$

Substituting $x = 3$:

$$27 - 5 = 6(9+9)B$$
$$22 = 108B$$
$$B = \tfrac{11}{54}$$

Substituting $x = -3$:

$$-27 - 5 = (-6)(9+9)A$$
$$-32 = -108A$$
$$A = \tfrac{8}{27}$$

Substituting $x = 0$:

$$-5 = -27A + 27B - 9D$$
$$-5 = -8 + \tfrac{11}{2} - 9D$$
$$-\tfrac{5}{2} = -9D$$
$$D = \tfrac{5}{18}$$

Substituting $x = 1$:

$$1 - 5 = -20A + 40B - 8C - 8D$$
$$1 = 5A - 10B + 2C + 2D$$
$$1 = \tfrac{40}{27} - \tfrac{55}{27} + 2C + \tfrac{5}{9}$$
$$1 = 2C$$
$$C = \tfrac{1}{2}$$

The partial fraction decomposition is:

$$\frac{x^3 - 5}{x^4 - 81} = \frac{\tfrac{8}{27}}{x+3} + \frac{\tfrac{11}{54}}{x-3} + \frac{\tfrac{1}{2}x + \tfrac{5}{18}}{x^2 + 9}$$

33. Following the hint, factor the denominator:

$$x^4 + x^3 + 2x^2 + x + 1 = (x^4 + x^3 + x^2) + (x^2 + x + 1)$$
$$= x^2(x^2 + x + 1) + (x^2 + x + 1)$$
$$= (x^2 + x + 1)(x^2 + 1)$$

The form of the partial fraction decomposition is:

$$\frac{1}{(x^2 + x + 1)(x^2 + 1)} = \frac{Ax + B}{x^2 + 1} + \frac{Cx + D}{x^2 + x + 1}$$

Clearing fractions yields:

$$1 = (Ax + B)(x^2 + x + 1) + (Cx + D)(x^2 + 1)$$
$$= Ax^3 + Ax^2 + Ax + Bx^2 + Bx + B + Cx^3 + Dx^2 + Cx + D$$
$$= (A + C)x^3 + (A + B + D)x^2 + (A + B + C)x + (B + D)$$

Equating coefficients:

$$A + C = 0$$
$$A + B + D = 0$$
$$A + B + C = 0$$
$$B + D = 1$$

Subtracting the first equation from the third equation yields $B = 0$, so $D = 1$, $A = -1$, and $C = 1$. The partial fraction decomposition is:

$$\frac{1}{x^4 + x^3 + 2x^2 + x + 1} = \frac{-x}{x^2 + 1} + \frac{x + 1}{x^2 + x + 1}$$

35. Using synthetic division with $x = 2$:

$$\begin{array}{r|rrrr}
2 & 1 & 2 & -5 & -6 \\
  &   & 2 & 8 & 6 \\
\hline
  & 1 & 4 & 3 & 0
\end{array}$$

Thus the equation factors as:

$$x^3 + 2x^2 - 5x - 6 = 0$$
$$(x - 2)(x^2 + 4x + 3) = 0$$
$$(x - 2)(x + 1)(x + 3) = 0$$
$$x = -3, -1, 2$$

The roots of the equation are $x = -3, -1, 2$.

37. (a) Simplifying identity (6):

$$3x^3 - x^2 + 7x - 3 = (Ax + B)(x^2 + 3) + (Cx + D)$$
$$= Ax^3 + Bx^2 + 3Ax + 3B + Cx + D$$
$$= Ax^3 + Bx^2 + (3A + C)x + (3B + D)$$

Equating coefficients:
$$A = 3$$
$$B = -1$$
$$3A + C = 7$$
$$3B + D = -3$$

(b) Substituting $A = 3$:
$$3(3) + C = 7$$
$$9 + C = 7$$
$$C = -2$$
Substituting $B = -1$:
$$3(-1) + D = -3$$
$$-3 + D = -3$$
$$D = 0$$
These agree with the values given in the text.

39. Using long division:

$$
\begin{array}{r}
6x + 8 \\
x^2 - 4x + 3 \overline{\smash{)}6x^3 - 16x^2 - 13x + 25} \\
\underline{6x^3 - 24x^2 + 18x} \\
8x^2 - 31x + 25 \\
\underline{8x^2 - 32x + 24} \\
x + 1
\end{array}
$$

The expression can be written as $6x + 8 + \dfrac{x+1}{x^2 - 4x + 3}$.

Since $x^2 - 4x + 3 = (x-3)(x-1)$, the form of the decomposition is :
$$\frac{x+1}{x^2 - 4x + 3} = \frac{A}{x-3} + \frac{B}{x-1}$$
Clearing fractions yields:
$$x + 1 = A(x-1) + B(x-3)$$
$$= Ax - A + Bx - 3B$$
$$= (A+B)x + (-A - 3B)$$
Equating coefficients:
$$A + B = 1$$
$$-A - 3B = 1$$
Adding yields $-2B = 2$, so $B = -1$ and $A = 2$. Thus we can write:
$$\frac{6x^3 - 16x^2 - 13x + 25}{x^2 - 4x + 3} = 6x + 8 + \frac{2}{x-3} + \frac{-1}{x-1}$$

41. Using long division:

$$\begin{array}{r}
x-4 \\
x^4-6x^3+12x^2-8x\overline{\smash{\big)}\,x^5-10x^4+36x^3-55x^2+32x+1} \\
\underline{x^5-\;6x^4+12x^3-\;8x^2} \\
-4x^4+24x^3-47x^2+32x \\
\underline{-4x^4+24x^3-48x^2+32x} \\
x^2\qquad\quad+1
\end{array}$$

The expression can be written as $x-4+\dfrac{x^2+1}{x^4-6x^3+12x^2-8x}$. Now factor the denominator:

$$x^4-6x^3+12x^2-8x = x\left(x^3-6x^2+12x-8\right) = x(x-2)^3$$

Thus the form of the decomposition is:

$$\frac{x^2+1}{x(x-2)^3} = \frac{A}{x}+\frac{B}{x-2}+\frac{C}{(x-2)^2}+\frac{D}{(x-2)^3}$$

Clearing fractions yields:

$$x^2+1 = A(x-2)^3 + Bx(x-2)^2 + Cx(x-2) + Dx$$

Substituting $x=0$:

$$1 = -8A$$
$$A = -\tfrac{1}{8}$$

Substituting $x=2$:

$$5 = 2D$$
$$D = \tfrac{5}{2}$$

Substituting $x=3$:

$$9+1 = A+3B+3C+3D$$
$$10 = -\tfrac{1}{8}+3B+3C+\tfrac{15}{2}$$
$$10 = \tfrac{59}{8}+3B+3C$$
$$\tfrac{21}{8} = 3B+3C$$
$$\tfrac{7}{8} = B+C$$

Substituting $x=1$:

$$1+1 = -A+B-C+D$$
$$2 = \tfrac{1}{8}+B-C+\tfrac{5}{2}$$
$$2 = \tfrac{21}{8}+B-C$$
$$-\tfrac{5}{8} = B-C$$

Thus the system of equations is:

$$B+C = \tfrac{7}{8}$$
$$B-C = -\tfrac{5}{8}$$

Adding yields $2B = \frac{1}{4}$, so $B = \frac{1}{8}$ and $C = \frac{3}{4}$. Thus we can write:

$$\frac{x^5 - 10x^4 + 36x^3 - 55x^2 + 32x + 1}{x^4 - 6x^3 + 12x^2 - 8x} = x - 4 + \frac{-\frac{1}{8}}{x} + \frac{\frac{1}{8}}{x-2} + \frac{\frac{3}{4}}{(x-2)^2} + \frac{\frac{5}{2}}{(x-2)^3}$$

43. Using long division:

$$
\begin{array}{r}
x^2 + 2x + 5 \\
x^4 - 1 \overline{\smash{\big)}\ x^6 + 2x^5 + 5x^4 - x^2 - 2x - 4} \\
\end{array}
$$

$$
\begin{array}{l}
\underline{x^6 \qquad\qquad\quad - x^2} \\
\quad 2x^5 + 5x^4 \qquad - 2x - 4 \\
\quad \underline{2x^5 \qquad\qquad - 2x} \\
\qquad\quad 5x^4 \qquad\qquad - 4 \\
\qquad\quad \underline{5x^4 \qquad\qquad - 5} \\
\qquad\qquad\qquad\qquad\quad 1
\end{array}
$$

The expression can be written as $x^2 + 2x + 5 + \dfrac{1}{x^4 - 1}$. Now factor the denominator:

$$x^4 - 1 = \left(x^2 + 1\right)\left(x^2 - 1\right) = \left(x^2 + 1\right)(x+1)(x-1)$$

Thus the form of the decomposition is:

$$\frac{1}{x^4 - 1} = \frac{A}{x+1} + \frac{B}{x-1} + \frac{Cx + D}{x^2 + 1}$$

Clearing fractions yields:

$$1 = A(x-1)\left(x^2 + 1\right) + B(x+1)\left(x^2 + 1\right) + (Cx + D)(x+1)(x-1)$$

Substituting $x = 1$:

$$1 = 4B$$

$$B = \tfrac{1}{4}$$

Substituting $x = -1$:

$$1 = -4A$$

$$A = -\tfrac{1}{4}$$

Substituting $x = 0$:

$$1 = -A + B - D$$

$$1 = \tfrac{1}{4} + \tfrac{1}{4} - D$$

$$\tfrac{1}{2} = -D$$

$$-\tfrac{1}{2} = D$$

Substituting $x = 2$:

$$1 = 5A + 15B + 6C + 3D$$

$$1 = -\tfrac{5}{4} + \tfrac{15}{4} + 6C - \tfrac{3}{2}$$

$$0 = 6C$$

$$0 = C$$

Thus we can write:

$$\frac{x^6 + 2x^5 + 5x^4 - x^2 - 2x - 4}{x^4 - 1} = x^2 + 2x + 5 + \frac{-\frac{1}{4}}{x+1} + \frac{\frac{1}{4}}{x-1} + \frac{-\frac{1}{2}}{x^2+1}$$

45. Rewriting the quadratic polynomial:

$$x^2 - 2ax + a^2 + b^2 = (x-a)^2 + b^2$$

Since this is a sum of two squares, it is irreducible.

47. The form of the partial fraction decomposition is:

$$\frac{px+q}{(x-a)(x-b)} = \frac{A}{x-a} + \frac{B}{x-b}$$

Clearing fractions yields:

$$px + q = A(x-b) + B(x-a)$$
$$= Ax - Ab + Bx - Ba$$
$$= (A+B)x + (-Ab - Ba)$$

Equating coefficients:

$$A + B = p$$
$$-Ab - Ba = q$$

Multiplying the first equation by a:

$$Aa + Ba = pa$$
$$-Ab - Ba = q$$

Adding yields $A(a-b) = pa + q$, so $A = \dfrac{pa+q}{a-b}$. Multiplying the first equation by b:

$$Ab + Bb = pb$$
$$-Ab - Ba = q$$

Adding yields $B(b-a) = pb + q$, so $B = \dfrac{pb+q}{b-a}$. The partial fraction decomposition is:

$$\frac{px+q}{(x-a)(x-b)} = \frac{(pa+q)/(a-b)}{x-a} + \frac{(pb+8)/(b-a)}{x-b}$$

49. The form of the partial fraction decomposition is:

$$\frac{px+q}{(x-a)(x+a)} = \frac{A}{x-a} + \frac{B}{x+a}$$

Clearing fractions yields:

$$px + q = A(x+a) + B(x-a)$$
$$= Ax + Aa + Bx - Ba$$
$$= (A+B)x + (Aa - Ba)$$

Equating coefficients:

$$A + B = p$$
$$Aa - Ba = q$$

Multiplying the first equation by a:
$$Aa + Ba = pa$$
$$Aa - Ba = q$$

Adding yields $2Aa = pa + q$, so $A = \dfrac{pa + q}{2a}$. Multiplying the first equation by $-a$:
$$-Aa - Ba = -pa$$
$$Aa - Ba = q$$

Adding yields $-2Ba = -pa + q$, so $B = \dfrac{pa - q}{2a}$. The partial fraction decomposition is:
$$\frac{px + q}{(x - a)(x + a)} = \frac{(pa + q)/2a}{x - a} + \frac{(pa - q)/2a}{x + a}$$

51. The form of the partial fraction decomposition is:
$$\frac{1}{(1 - ax)(1 - bx)(1 - cx)} = \frac{A}{1 - ax} + \frac{B}{1 - bx} + \frac{C}{1 - cx}$$
Clearing fractions yields:
$$1 = A(1 - bx)(1 - cx) + B(1 - ax)(1 - cx) + C(1 - ax)(1 - bx)$$
Substituting $x = \dfrac{1}{a}$:
$$1 = A\left(1 - \frac{b}{a}\right)\left(1 - \frac{c}{a}\right)$$
$$1 = A \cdot \frac{(a - b)(a - c)}{a^2}$$
$$A = \frac{a^2}{(a - b)(a - c)}$$
Substituting $x = \dfrac{1}{b}$:
$$1 = B\left(1 - \frac{a}{b}\right)\left(1 - \frac{c}{b}\right)$$
$$1 = B \cdot \frac{(b - a)(b - c)}{b^2}$$
$$B = \frac{b^2}{(b - a)(b - c)}$$

Substituting $x = \dfrac{1}{c}$:

$$1 = C\left(1 - \frac{a}{c}\right)\left(1 - \frac{b}{c}\right)$$

$$1 = C \cdot \frac{(c-a)(c-b)}{c^2}$$

$$C = \frac{c^2}{(c-a)(c-b)}$$

The partial fraction decomposition is:

$$\frac{1}{(1-ax)(1-bx)(1-cx)} = \frac{\frac{a^2}{(a-b)(a-c)}}{1-ax} + \frac{\frac{b^2}{(b-a)(b-c)}}{1-bx} + \frac{\frac{c^2}{(c-a)(c-b)}}{1-cx}$$

53. Factor the denominator by adding and subtracting $2x^2$:

$$x^4 + 1 = x^4 + 2x^2 + 1 - 2x^2 = \left(x^2 + 1\right)^2 - 2x^2 = \left(x^2 + \sqrt{2}x + 1\right)\left(x^2 - \sqrt{2}x + 1\right)$$

The form of the partial fraction decomposition is:

$$\frac{1}{x^4 + 1} = \frac{Ax + B}{x^2 + \sqrt{2}x + 1} + \frac{Cx + D}{x^2 - \sqrt{2}x + 1}$$

Clearing fractions yields:

$$1 = (Ax + B)\left(x^2 - \sqrt{2}x + 1\right) + (Cx + D)\left(x^2 + \sqrt{2}x + 1\right)$$

$$= Ax^3 - \sqrt{2}Ax^2 + Ax + Bx^2 - \sqrt{2}Bx + B + Cx^3 + \sqrt{2}Cx^2 + Cx + Dx^2 + \sqrt{2}Dx + D$$

$$= (A+C)x^3 + \left(-\sqrt{2}A + B + \sqrt{2}C + D\right)x^2 + \left(A - \sqrt{2}B + C + \sqrt{2}D\right)x + (B+D)$$

Equating coefficients:

$$A + C = 0$$
$$-\sqrt{2}A + B + \sqrt{2}C + D = 0$$
$$A - \sqrt{2}B + C + \sqrt{2}D = 0$$
$$B + D = 1$$

Substitute $C = -A$ and $D = 1 - B$ into the second equation:

$$-\sqrt{2}A + B + \sqrt{2}(-A) + 1 - B = 0$$

$$-2\sqrt{2}A + 1 = 0$$

$$1 = 2\sqrt{2}A$$

$$A = \frac{1}{2\sqrt{2}} = \frac{\sqrt{2}}{4}$$

$$C = -\frac{\sqrt{2}}{4}$$

Substituting into the third equation:

$$\frac{\sqrt{2}}{4} - \sqrt{2}B - \frac{\sqrt{2}}{4} + \sqrt{2}(1-B) = 0$$
$$-\sqrt{2}B + \sqrt{2} - \sqrt{2}B = 0$$
$$-2\sqrt{2}B = -\sqrt{2}$$
$$B = \tfrac{1}{2}$$
$$D = 1 - \tfrac{1}{2} = \tfrac{1}{2}$$

The partial fraction decomposition is:

$$\frac{1}{x^4+1} = \frac{\frac{\sqrt{2}}{4}x + \frac{1}{2}}{x^2 + \sqrt{2}x + 1} + \frac{-\frac{\sqrt{2}}{4}x + \frac{1}{2}}{x^2 - \sqrt{2}x + 1}$$

Chapter Twelve Review Exercises

1. Using synthetic division:

$$\begin{array}{r|rrrrr} -2 & 1 & 3 & -1 & -5 & 1 \\ & & -2 & -2 & 6 & -2 \\ \hline & 1 & 1 & -3 & 1 & -1 \end{array}$$

So $q(x) = x^3 + x^2 - 3x + 1$ and $R(x) = -1$.

3. Using synthetic division:

$$\begin{array}{r|rrrrr} 3 & 1 & 0 & -2 & 0 & 8 \\ & & 3 & 9 & 21 & 63 \\ \hline & 1 & 3 & 7 & 21 & 71 \end{array}$$

So the quotient is $x^3 + 3x^2 + 7x + 21$ and the remainder is 71.

5. Using synthetic division:

$$\begin{array}{r|rrrr} -4 & 2 & -5 & -6 & -3 \\ & & -8 & 52 & -184 \\ \hline & 2 & -13 & 46 & -187 \end{array}$$

So the quotient is $2x^2 - 13x + 46$ and the remainder is -187.

7. Using synthetic division:

$$\begin{array}{r|rrr} -0.2 & 5 & -19 & -4 \\ & & -1 & 4 \\ \hline & 5 & -20 & 0 \end{array}$$

So the quotient is $5x - 20$ and the remainder is zero.

9. Using synthetic division:

$$\underline{10} \begin{array}{|cccccc} & 1 & 0 & 0 & 0 & -10 & 4 \\ & & 10 & 100 & 1000 & 10000 & 99900 \\ \hline & 1 & 10 & 100 & 1000 & 9990 & 99904 \end{array}$$

So $f(10) = 99,904$.

11. Using synthetic division:

$$\underline{1/10} \begin{array}{|cccc} & 1 & -10 & 1 & -1 \\ & & 1/10 & -99/100 & 1/1000 \\ \hline & 1 & -99/10 & 1/100 & -999/1000 \end{array}$$

So $f\left(\tfrac{1}{10}\right) = -\tfrac{999}{1000}$.

13. Using synthetic division:

$$\underline{a-1} \begin{array}{|cccc} & 1 & 3 & 3 & 1 \\ & & a-1 & a^2+a-2 & a^3-1 \\ \hline & 1 & a+2 & a^2+a+1 & a^3 \end{array}$$

So $f(a-1) = a^3$.

15. (a) Using synthetic division:

$$\underline{-0.3} \begin{array}{|ccccc} & 1 & 4 & -6 & -8 & -2 \\ & & -0.3 & -1.11 & 2.133 & 1.7601 \\ \hline & 1 & 3.7 & -7.11 & -5.867 & -0.24 \end{array}$$

So $f(-0.3) \approx -0.24$.

(b) Using synthetic division:

$$\underline{-0.39} \begin{array}{|ccccc} & 1 & 4 & -6 & -8 & -2 \\ & & -0.39 & -1.4079 & 2.8891 & 1.9933 \\ \hline & 1 & 3.61 & -7.4079 & -5.1109 & -0.007 \end{array}$$

So $f(-0.39) \approx -0.007$.

(c) Using synthetic division:

$$\underline{-0.394} \begin{array}{|ccccc} & 1 & 4 & -6 & -8 & -2 \\ & & -0.394 & -1.420764 & 2.92378 & 2.00003 \\ \hline & 1 & 3.606 & -7.420764 & -5.07622 & 0.00003 \end{array}$$

So $f(-0.394) \approx 0.00003$

17. Using synthetic division:

$$\underline{3} \begin{array}{|cccc} & 1 & -4 & -a & -6 \\ & & 3 & -3 & -3a-9 \\ \hline & 1 & -1 & -a-3 & -3a-15 \end{array}$$

So if 3 is a root, then:

$$-3a - 15 = 0$$
$$-3a = 15$$
$$a = -5$$

19. Using synthetic division:

$$\begin{array}{r|rrrr} 1 & a^2 & 3a & 0 & 2 \\ & & a^2 & a^2+3a & a^2+3a \\ \hline & a^2 & a^2+3a & a^2+3a & a^2+3a+2 \end{array}$$

So if $x - 1$ is a factor, then:

$$a^2 + 3a + 2 = 0$$
$$(a+1)(a+2) = 0$$
$$a = -1, -2$$

So $a = -1$ or $a = -2$.

21. (a) Let $f(x + h) = 0$, and let $x = r - h$. Then $f(r - h + h) = f(r) = 0$, since r is a root of $f(x) = 0$.

   (b) Let $f(-x) = 0$, and let $x = -r$. Then $f(-(-r)) = f(r) = 0$, since r is a root of $f(x) = 0$.

   (c) Let $f\left(\frac{x}{k}\right) = 0$, and let $x = kr$. Then $f\left(\frac{kr}{k}\right) = f(r) = 0$, since r is a root of $f(x) = 0$.

23. p-factors: $\pm 1, \pm 2, \pm 3, \pm 6, \pm 9, \pm 18$
   q-factors: ± 1
   possible rational roots: $\pm 1, \pm 2, \pm 3, \pm 6, \pm 9, \pm 18$

25. p-factors: $\pm 1, \pm 2, \pm 4, \pm 8$
   q-factors: $\pm 1, \pm 2$
   possible rational roots: $\pm 1, \pm \frac{1}{2}, \pm 2, \pm 4, \pm 8$

27. p-factors: $\pm p, \pm 1$
   q-factors: ± 1
   possible rational roots: $\pm p, \pm 1$

29. The possible rational roots are $\pm 1, \pm \frac{1}{2}, \pm 2, \pm 3, \pm \frac{3}{2}, \pm 6$. Using synthetic division:

$$\begin{array}{r|rrrr} 2 & 2 & 1 & -7 & -6 \\ & & 4 & 10 & 6 \\ \hline & 2 & 5 & 3 & 0 \end{array}$$

So $2x^3 + x^2 - 7x - 6 = (x - 2)(2x^2 + 5x + 3) = (x - 2)(2x + 3)(x + 1)$.

So the roots are $2, -\frac{3}{2}$ and -1.

31. The possible rational roots are $\pm 1, \pm \frac{1}{2}, \pm 2, \pm 5, \pm \frac{5}{2}, \pm 10$. Using synthetic division:

$$\begin{array}{r|rrrr} 5/2 & 2 & -1 & -14 & 10 \\ & & 5 & 10 & -10 \\ \hline & 2 & 4 & -4 & 0 \end{array}$$

So $2x^3 - x^2 - 14x + 10 = \left(x - \frac{5}{2}\right)(2x^2 + 4x - 4) = 2\left(x - \frac{5}{2}\right)(x^2 + 2x - 2)$.

Using the quadratic formula:

$$x = \frac{-2 \pm \sqrt{4+8}}{2} = \frac{-2 \pm 2\sqrt{3}}{2} = -1 \pm \sqrt{3}$$

So the roots are $\frac{5}{2}, -1+\sqrt{3}$ and $-1-\sqrt{3}$.

33. First multiply by 2 to obtain $3x^3 + x^2 + x - 2 = 0$. The possible rational roots are $\pm 1, \pm \frac{1}{3}, \pm 2, \pm \frac{2}{3}$. Using synthetic division:

$$
\begin{array}{r|rrrr}
2/3 & 3 & 1 & 1 & -2 \\
    &   & 2 & 2 & 2 \\
\hline
    & 3 & 3 & 3 & 0
\end{array}
$$

So $3x^3 + x^2 + x - 2 = \left(x - \frac{2}{3}\right)\left(3x^2 + 3x + 3\right) = 3\left(x - \frac{2}{3}\right)\left(x^2 + x + 1\right)$.
Using the quadratic formula:

$$x = \frac{-1 \pm \sqrt{1-4}}{2} = \frac{-1 \pm i\sqrt{3}}{2}$$

So the roots are $\frac{2}{3}, \frac{-1+i\sqrt{3}}{2}$, and $\frac{-1-i\sqrt{3}}{2}$.

35. The possible rational roots are $\pm 1, \pm 7, \pm 49$. Using synthetic division:

$$
\begin{array}{r|rrrrrr}
-1 & 1 & 1 & -14 & -14 & 49 & 49 \\
   &   & -1 & 0 & 14 & 0 & -49 \\
\hline
   & 1 & 0 & -14 & 0 & 49 & 0
\end{array}
$$

Therefore:

$$
\begin{aligned}
x^5 + x^4 - 14x^3 - 14x^2 + 49x + 49 &= (x+1)\left(x^4 - 14x^2 + 49\right) \\
&= (x+1)\left(x^2 - 7\right)^2 \\
&= (x+1)\left(x + \sqrt{7}\right)^2\left(x - \sqrt{7}\right)^2
\end{aligned}
$$

So the roots are $-1, -\sqrt{7}$ (multiplicity 2), and $\sqrt{7}$ (multiplicity 2).

37. Since $x^3 - 9x^2 + 24x - 20 = 0$ has a root r with multiplicity 2, we can use synthetic division by r twice, each time the remainder must be 0:

$$
\begin{array}{r|rrrr}
r & 1 & -9 & 24 & -20 \\
  &   & r & r^2 - 9r & r^3 - 9r^2 + 24r \\
\hline
  & 1 & r-9 & r^2 - 9r + 24 & r^3 - 9r^2 + 24r - 20
\end{array}
$$

We continue the division:

$$
\begin{array}{r|rrr}
r & 1 & r-9 & r^2 - 9r + 24 \\
  &   & r & 2r^2 - 9r \\
\hline
  & 1 & 2r-9 & 3r^2 - 18r + 24
\end{array}
$$

So $3r^2 - 18r + 24 = 3(r^2 - 6r + 8) = 3(r-4)(r-2) = 0$, thus $r = 4$ or $r = 2$. We check these in $r^3 - 9r^2 + 24r - 20 = 0$:

$r = 4$: $(4)^3 - 9(4)^2 + 24(4) - 20 = 64 - 144 + 96 - 20 = -4 \neq 0$

So $r = 4$ cannot be a root.

$r = 2$: $(2)^3 - 9(2)^2 + 24(2) - 20 = 8 - 36 + 48 - 20 = 0$

So $r = 2$ is the root with multiplicity 2. Since the synthetic division resulted in $x + (2r - 9) = 0$, $x + (-5) = 0$ and thus $x = 5$. So the roots are 2 (multiplicity 2) and 5.

Note: Actually, an easier (and more direct) approach is to find the roots directly. The possible rational roots are $\pm 1, \pm 2, \pm 4, \pm 5, \pm 10, \pm 20$. Note that there are 3 sign changes, so there can be 1 or 3 positive real roots. Since $f(-x)$ has no sign changes, all 3 roots must be positive real numbers for one of them to have multiplicity of 2 (note that it cannot have a radical or be complex–why?). We use synthetic division:

$$\begin{array}{r|rrrr} 2 & 1 & -9 & 24 & -20 \\ & & 2 & -14 & 20 \\ \hline & 1 & -7 & 10 & 0 \end{array}$$

So $x^3 - 9x^2 + 24x - 20 = (x-2)(x^2 - 7x + 10) = (x-2)^2(x-5)$.

So the roots are 2 (with multiplicity 2) and 5.

39. (a) Let $p(x)$ and $d(x)$ be the polynomials where $d(x) \neq 0$. Then there are unique polynomials $q(x)$ and $R(x)$ such that
$$p(x) = d(x) \cdot q(x) + R(x)$$
where either $R(x) = 0$ or the degree of $R(x)$ is less than the degree of $d(x)$.

   (b) When a polynomial $f(x)$ is divided by $x - r$, the remainder is $f(r)$.

   (c) Let $f(x)$ be a polynomial. If $f(r) = 0$, then $x - r$ is a factor of $f(x)$. Conversely, if $x - r$ is a factor of $f(x)$, then $f(r) = 0$.

   (d) Every polynomial equation of the form
$$a_n x^n + a_{n-1} x^{n-1} + \ldots + a_1 x + a_0 = 0 \quad (n \geq 1, a_n \neq 0)$$
has at least one root among the complex numbers. (This root may be a real number.)

41. Factor:
$$6x^2 + 7x - 20 = (3x - 4)(2x + 5) = 6\left(x - \tfrac{4}{3}\right)\left[x - \left(-\tfrac{5}{2}\right)\right]$$

43. Factor:
$$\begin{aligned} x^4 - 4x^3 + 5x - 20 &= x^3(x - 4) + 5(x - 4) \\ &= (x - 4)(x^3 + 5) \\ &= (x - 4)\left(x + \sqrt[3]{5}\right)\left(x^2 - \sqrt[3]{5}x + \sqrt[3]{25}\right) \end{aligned}$$

Solve $x^2 - \sqrt[3]{5}x + \sqrt[3]{25} = 0$ using the quadratic formula, which yields:
$$x = \frac{\sqrt[3]{5} \pm i\sqrt{3\sqrt[3]{25}}}{2}$$

So $x^4 - 4x^3 + 3x - 20 = (x - 4)\left(x - \left(-\sqrt[3]{5}\right)\right)\left(x - \frac{\sqrt[3]{5} + i\sqrt{3\sqrt[3]{25}}}{2}\right)\left(x - \frac{\sqrt[3]{5} - i\sqrt{3\sqrt[3]{25}}}{2}\right)$.

45. One other root is $2 + 3i$, so $\left[x - (2 - 3i)\right]\left[x - (2 + 3i)\right]$ are factors, which is $x^2 - 4x + 13$. Using long division:

$$
\begin{array}{r}
x - 3 \\
x^2 - 4x + 13 \overline{\smash{\big)}\ x^3 - 7x^2 + 25x - 39} \\
\underline{x^3 - 4x^2 + 13x} \\
-3x^2 + 12x - 39 \\
\underline{-3x^2 + 12x - 39} \\
0
\end{array}
$$

So $x - 3$ is the other factor. So the roots are $2 - 3i$, $2 + 3i$, and 3.

47. One other root is $1 - i\sqrt{2}$, so $\left[x - (1 + i\sqrt{2})\right]\left[x - (1 - i\sqrt{2})\right]$ are factors, which is $x^2 - 2x + 3$. Using long division:

$$
\begin{array}{r}
x^2 - 4 \\
x^2 - 2x + 3 \overline{\smash{\big)}\ x^4 - 2x^3 - 4x^2 + 14x - 21} \\
\underline{x^4 - 2x^3 + 3x^2} \\
-7x^2 + 14x - 21 \\
\underline{-7x^2 + 14x - 21} \\
0
\end{array}
$$

So $x^2 - 7 = (x + \sqrt{7})(x - \sqrt{7})$ is the other factor. So the roots are $1 + i\sqrt{2}$, $1 - i\sqrt{2}$, $\sqrt{7}$, and $-\sqrt{7}$.

49. Let $f(x) = x^3 + 8x - 7$. There is one sign change, so there is 1 positive root. Now $f(-x) = -x^3 - 8x - 7$. Since there are no sign changes, there are no negative roots. So the equation has 1 positive real root and 2 complex roots.

51. Let $f(x) = x^3 + 3x + 1$. There are no sign changes, so there are no positive roots. Now $f(-x) = -x^3 - 3x + 1$. Since there is one sign change, there is 1 negative root. So the equation has 1 negative real root and 2 complex roots.

53. Let $f(x) = x^4 - 10$. There is one sign change, so there is 1 positive root. Since $f(-x) = f(x)$, there is also 1 negative root. So the equation has 1 positive real root, 1 negative real root and 2 complex roots.

55. (a) Let $f(x) = x^3 + x^2 + x + 1$, so $f(-x) = -x^3 + x^2 - x + 1$. There are three sign changes, so there are either 1 or 3 negative roots.

(b) Using the hint, we have $(x - 1)(x^3 + x^2 + x + 1) = x^4 - 1$. Let $g(x) = x^4 - 1$. Then $g(-x) = x^4 - 1$, so this can only have 1 negative root. Since $x = 1$ is a positive root, $x^3 + x^2 + x + 1 = 0$ can only have 1 negative root also.

(c)   Factoring, we have $x^3 + x^2 + x + 1 = x^2(x+1) + (x+1) = (x+1)(x^2+1)$.

So $x = -1$ is a root. Also:
$$x^2 + 1 = 0$$
$$x^2 = -1$$
$$x = \pm i$$
So the roots are -1, i, and $-i$.

57.   Substituting $y = x^3$, we have $x^2 + (x^3)^2 = 1$, so $x^6 + x^2 - 1 = 0$.

Let $f(x) = x^6 + x^2 - 1$.
$$f(0) = -1$$
$$f(1) = 1$$
So there is a root between 0 and 1.
$$f(0.8) = -0.09$$
$$f(0.9) = 0.34$$
So there is a root between 0.8 and 0.9.
$$f(0.82) = -0.02$$
$$f(0.83) = 0.01$$
So there is a root between 0.82 and 0.83.
So the x-coordinate lies between 0.82 and 0.83.

59.   (a)   Let $f(x) = x^3 - 36x - 84$. Since there is only one sign change, there is 1 positive real root.

(b)   Using synthetic division:

$$\begin{array}{r|rrrr} 7 & 1 & 0 & -36 & -84 \\ & & 7 & 49 & 91 \\ \hline & 1 & 7 & 13 & 7 \end{array}$$

Since the row is all positive, there are no real roots to the equation greater than 7.

(c)   Here $f(6) = -84$ and $f(7) = 7$, so there is a root between 6 and 7.
$f(6.8) = -14.4$ and $f(6.9) = -3.89$, so there is a root between 6.9 and 7.0.
$f(6.93) = -0.67$ and $f(6.94) = 0.42$, so there is a root between 6.93 and 6.94.

61.   Another root will be $4 + \sqrt{5}$, so:
$$\left[x - (4 - \sqrt{5}) \right]\left[x - (4 + \sqrt{5}) \right] = 0$$
$$x^2 - 8x + 11 = 0$$

63.   Other roots will be $6 + 2i$ and $-\sqrt{5}$:
$$(x - \sqrt{5})\left[x - (-\sqrt{5}) \right]\left[x - (6 - 2i) \right]\left[x - (6 + 2i) \right] = 0$$
$$(x^2 - 5)(x^2 - 12x + 40) = 0$$
$$x^4 - 12x^3 + 35x^2 + 60x - 200 = 0$$

65. Write $x - 1 = \sqrt{2} + \sqrt{3}$. Squaring both sides, we obtain:
$$x^2 - 2x + 1 = 5 + 2\sqrt{6}$$
$$x^2 - 2x - 4 = 2\sqrt{6}$$
Squaring both sides again, we obtain:
$$x^4 - 4x^3 - 4x^2 + 16x + 16 = 24$$
$$x^4 - 4x^3 - 4x^2 + 16x - 8 = 0$$

67. Factoring, we have $y = x^3 - 2x^2 - 3x = x(x^2 - 2x - 3) = x(x - 3)(x + 1)$. The zeros are 0, 3, and –1.

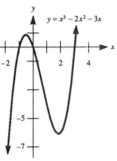

69. Factoring, we have $y = x^4 - 4x^2 = x^2(x^2 - 4) = x^2(x + 2)(x - 2)$. The zeros are 0, –2, and 2.

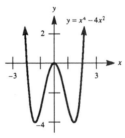

71. Performing the indicated operations yields:
$$(3 - 2i)(3 + 2i) + (1 + 3i)^2 = 9 - 4i^2 + 1 + 6i + 9i^2 = 9 + 4 + 1 + 6i - 9 = 5 + 6i$$

73. Performing the indicated operations yields:
$$(1 + i\sqrt{2})(1 - i\sqrt{2}) + (\sqrt{2} + i)(\sqrt{2} - i) = 1 - 2i^2 + 2 - i^2 = 1 + 2 + 2 + 1 = 6$$

75. Performing the indicated operations yields:
$$\frac{3 - i\sqrt{3}}{3 + i\sqrt{3}} \cdot \frac{3 - i\sqrt{3}}{3 - i\sqrt{3}} = \frac{9 - 6i\sqrt{3} - 3}{9 + 3} = \frac{6 - 6i\sqrt{3}}{12} = \frac{1 - i\sqrt{3}}{2} = \tfrac{1}{2} - \tfrac{\sqrt{3}}{2}i$$

77. Writing in complex form, we have:
$$-\sqrt{-2}\sqrt{-9} + \sqrt{-8} - \sqrt{-72} = -(i\sqrt{2})(3i) + 2i\sqrt{2} - 6i\sqrt{2} = 3\sqrt{2} - 4i\sqrt{2} = 3\sqrt{2} - 4\sqrt{2}i$$

79. Let $z = a + bi$, so $\bar{z} = a - bi$. Therefore:

$$\frac{z + \bar{z}}{2} = \frac{a + bi + a - bi}{2} = \frac{2a}{2} = a = Re(z)$$

81. (a) Compute the absolute values:
$$|6 + 2i| = \sqrt{6^2 + 2^2} = \sqrt{36 + 4} = \sqrt{40} = 2\sqrt{10}$$
$$|6 - 2i| = \sqrt{6^2 + (-2)^2} = \sqrt{36 + 4} = \sqrt{40} = 2\sqrt{10}$$

(b) Using the new definition:
$$|-3| = |-3 + 0i| = \sqrt{(-3)^2 + 0^2} = \sqrt{9} = 3$$

(c) Let $z = a + bi$, so $\bar{z} = a - bi$. We now compute:
$$z\bar{z} = (a + bi)(a - bi) = a^2 - b^2 i^2 = a^2 + b^2$$
$$|z|^2 = \left(\sqrt{a^2 + b^2}\right)^2 = a^2 + b^2$$
Thus $z\bar{z} = |z|^2$.

83. Combining the fractions, we have:
$$\frac{1}{a - bi} - \frac{1}{a + bi} = \frac{(a + bi) - (a - bi)}{(a - bi)(a + bi)} = \frac{2bi}{a^2 - b^2 i^2} = \frac{2bi}{a^2 + b^2}$$

85. Combining the fractions, we have:
$$\frac{a + bi}{a - bi} - \frac{a - bi}{a + bi} = \frac{(a + bi)^2 - (a - bi)^2}{(a - bi)(a + bi)}$$
$$= \frac{a^2 + 2abi - b^2 - a^2 + 2abi + b^2}{a^2 + b^2}$$
$$= \frac{4abi}{a^2 + b^2}$$

87. Since $100 - x^2 = (10 + x)(10 - x)$, the form of the decomposition is:
$$\frac{2x - 1}{100 - x^2} = \frac{A}{10 + x} + \frac{B}{10 - x}$$
Clearing fractions yields:
$$2x - 1 = A(10 - x) + B(10 + x) = (-A + B)x + (10A + 10B)$$
Equating coefficients, we have:
$$-A + B = 2$$
$$10A + 10B = -1$$
Multiplying the first equation by 10:
$$-10A + 10B = 20$$
$$10A + 10B = -1$$

Adding yields $20B = 19$, so $B = \frac{19}{20}$. Substituting into $-A + B = 2$:

$$-A + \frac{19}{20} = 2$$
$$-A = \frac{21}{20}$$
$$A = -\frac{21}{20}$$

The partial fraction decomposition is:

$$\frac{2x - 1}{100 - x^2} = \frac{-\frac{21}{20}}{10 + x} + \frac{\frac{19}{20}}{10 - x}$$

89. Since $x^3 + 2x^2 + x = x(x^2 + 2x + 1) = x(x + 1)^2$, the form of the decomposition is:

$$\frac{1}{x^3 + 2x^2 + x} = \frac{A}{x} + \frac{B}{x + 1} + \frac{C}{(x + 1)^2}$$

Clearing fractions yields:

$$1 = A(x + 1)^2 + Bx(x + 1) + Cx$$
$$= Ax^2 + 2Ax + A + Bx^2 + Bx + Cx$$
$$= (A + B)x^2 + (2A + B + C)x + A$$

Equating coefficients, we have:

$$A + B = 0$$
$$2A + B + C = 0$$
$$A = 1$$

So $B = -1$. Substituting into $2A + B + C = 0$:

$$2 - 1 + C = 0$$
$$1 + C = 0$$
$$C = -1$$

The partial fraction decomposition is:

$$\frac{1}{x^3 + 2x^2 + x} = \frac{1}{x} + \frac{-1}{x + 1} + \frac{-1}{(x + 1)^2}$$

91. Since $x^4 + 6x^2 + 9 = (x^2 + 3)^2$, the form of the decomposition is:

$$\frac{x^3 + 2}{x^4 + 6x^2 + 9} = \frac{Ax + B}{x^2 + 3} + \frac{Cx + D}{(x^2 + 3)^2}$$

Clearing fractions yields:

$$x^3 + 2 = (Ax + B)(x^2 + 3) + (Cx + D)$$
$$= Ax^3 + Bx^2 + 3Ax + 3B + Cx + D$$
$$= Ax^3 + Bx^2 + (3A + C)x + (3B + D)$$

Equating coefficients, we have:

$$A = 1$$
$$B = 0$$
$$3A + C = 0$$
$$3B + D = 2$$

So $C = -3$ and $D = 2$. The partial fraction decomposition is:

$$\frac{x^3 + 2}{x^4 + 6x^2 + 9} = \frac{x}{x^2 + 3} + \frac{-3x + 2}{\left(x^2 + 3\right)^2}$$

93. (a) Multiplying out factors:

$$\left(x^2 + bx + 1\right)\left(x^2 + cx - 1\right) = x^4 + (b+c)x^3 + bcx^2 + (-b+c)x - 1$$

Equating coefficients, we have:

$$b + c = -2$$
$$bc = 1$$
$$-b + c = 0$$

Adding the first and third equations yields $2c = -2$, so $c = -1$ and $b = -1$. Thus the factorization is given by:

$$x^4 - 2x^3 + x^2 - 1 = \left(x^2 - x + 1\right)\left(x^2 - x - 1\right)$$

(b) The form of the decomposition is:

$$\frac{x^3}{x^4 - 2x^3 + x^2 - 1} = \frac{Ax + B}{x^2 - x + 1} + \frac{Cx + D}{x^2 - x - 1}$$

Clearing fractions yields:

$$x^3 = (Ax + B)\left(x^2 - x - 1\right) + (Cx + D)\left(x^2 - x + 1\right)$$
$$= Ax^3 - Ax^2 - Ax + Bx^2 - Bx - B + Cx^3 - Cx^2 + Cx + Dx^2 - Dx + D$$
$$= (A + C)x^3 + (-A + B - C + D)x^2 + (-A - B + C - D)x + (-B + D)$$

Equating coefficients, we have:

$$A + C = 1$$
$$-A + B - C + D = 0$$
$$-A - B + C - D = 0$$
$$-B + D = 0$$

Adding the second and third equations yields $-2A = 0$, so $A = 0$ and $C = 1$. Using the second and fourth equations, we have:

$$B - 1 + D = 0$$
$$-B + D = 0$$

Adding yields $-1 + 2D = 0$, so $D = \frac{1}{2}$ and $B = \frac{1}{2}$. The partial fraction decomposition is:

$$\frac{x^3}{x^4 - 2x^3 + x^2 - 1} = \frac{\frac{1}{2}}{x^2 - x + 1} + \frac{x + \frac{1}{2}}{x^2 - x - 1}$$

Chapter Twelve Test

1.   Since $f\left(\frac{1}{2}\right)$ will be the remainder after synthetic division, we synthetically divide by $\frac{1}{2}$:

$$
\begin{array}{r|rrrrr}
1/2 & 6 & -5 & 7 & -2 & -2 \\
    &   & 3 & -1 & 3 & 1/2 \\
\hline
    & 6 & -2 & 6 & 1 & -3/2
\end{array}
$$

So $f\left(\frac{1}{2}\right) = -\frac{3}{2}$.

2.   Using synthetic division:

$$
\begin{array}{r|rrrr}
-3 & 1 & 1 & -11 & -15 \\
   &   & -3 & 6 & 15 \\
\hline
   & 1 & -2 & -5 & 0
\end{array}
$$

So $x^3 + x^2 - 11x - 15 = (x+3)(x^2 - 2x - 5)$. Using the quadratic formula, we have:

$$ x = \frac{2 \pm \sqrt{4+20}}{2} = \frac{2 \pm 2\sqrt{6}}{2} = 1 \pm \sqrt{6} $$

So the roots of the equation are -3, $1 + \sqrt{6}$, and $1 - \sqrt{6}$.

3. $\pm 1, \pm \frac{1}{2}, \pm 2, \pm 3, \pm \frac{3}{2}, \pm 6$

4.   Such a function would be $y = a(x - 1)(x + 8)$. Now $y = -24$ when $x = 0$, so $-24 = a(-1)(8)$, so $a = 3$. So the function is $y = 3(x - 1)(x + 8)$, or $f(x) = 3x^2 + 21x - 24$.

5.   Using synthetic division:

$$
\begin{array}{r|rrrr}
-1 & 4 & 1 & -8 & 3 \\
   &   & -4 & 3 & 5 \\
\hline
   & 4 & -3 & -5 & 8
\end{array}
$$

The quotient is $4x^2 - 3x - 5$ and the remainder is 8.

6.   (a)   Let $f(x)$ be a polynomial. If $f(r) = 0$, then $x - r$ is a factor of $f(x)$. Conversely, if $x - r$ is a factor of $f(x)$, then $f(r) = 0$.

(b)   Every polynomial of the form
$$ a_n x^n + a_{n-1} x^{n-1} + \ldots + a_1 x + a_0 = 0 \quad (n \geq 1, a_n \neq 0) $$
has at least one root among the complex numbers. (This root may be a real number.)

(c)   Let $f(x)$ be a polynomial, all of whose coefficients are real numbers. Then $f(x)$ can be factored (over the real numbers) into a product of linear and/or irreducible quadratic factors.

7. (a) Using synthetic division with 1:

$$\begin{array}{r|rrrr}
1 & 1 & -2 & 0 & -1 \\
  &   & 1 & -1 & -1 \\
\hline
  & 1 & -1 & -1 & -2
\end{array}$$

Now we use 2:

$$\begin{array}{r|rrrr}
2 & 1 & -2 & 0 & -1 \\
  &   & 2 & 0 & 0 \\
\hline
  & 1 & 0 & 0 & -1
\end{array}$$

Now use 3:

$$\begin{array}{r|rrrr}
3 & 1 & -2 & 0 & -1 \\
  &   & 3 & 3 & 9 \\
\hline
  & 1 & 1 & 3 & 8
\end{array}$$

So 3 is an upper bound for the roots.

(b) Since $f(2.2) = -0.032$ and $f(2.3) = 0.587$, the root lies between 2.2 and 2.3.

8. Two other roots are $1 - i$ and $3 + 2i$, so $x^2 - 2x + 2$ and $x^2 - 6x + 13$ are factors. Using long division:

$$
\begin{array}{r}
x^3 - 4x^2 + x + 26 \\
x^2 - 2x + 2 \overline{\smash{\big)}\, x^5 - 6x^4 + 11x^3 + 16x^2 - 50x + 52} \\
\underline{x^5 - 2x^4 + 2x^3} \\
-4x^4 + 9x^3 + 16x^2 \\
\underline{-4x^4 + 8x^3 - 8x^2} \\
x^3 + 24x^2 - 50x \\
\underline{x^3 - 2x^2 + 2x} \\
26x^2 - 52x + 52 \\
\underline{26x^2 - 52x + 52} \\
0
\end{array}
$$

Using long division again:

$$
\begin{array}{r}
x + 2 \\
x^2 - 6x + 13 \overline{\smash{\big)}\, x^3 - 4x^2 + x + 26} \\
\underline{x^3 - 6x^2 + 13x} \\
2x^2 - 12x + 26 \\
\underline{2x^2 - 12x + 26} \\
0
\end{array}
$$

So the roots are $1 \pm i$, $3 \pm 2i$, and -2.

9. Using long division:

$$
\begin{array}{r}
x^2 + 2x - 1 \\
x^2 + 1 \overline{\smash{\big)}\, x^4 + 2x^3 + 0x^2 - x + 6} \\
\underline{x^4 \qquad\quad + x^2} \\
2x^3 - x^2 - x \\
\underline{2x^3 \qquad + 2x} \\
-x^2 - 3x + 6 \\
\underline{-x^2 \qquad - 1} \\
-3x + 7
\end{array}
$$

So $q(x) = x^2 + 2x - 1$ and $R(x) = -3x + 7$.

10. First find the roots by the quadratic formula:
$$x = \frac{6 \pm \sqrt{36-40}}{4} = \frac{6 \pm 2i}{4} = \frac{3}{2} \pm \frac{1}{2}i$$
So $2x^2 - 6x + 5 = 2\left(x^2 - 3x + \frac{5}{2}\right) = 2\left[x - \left(\frac{3}{2} + \frac{1}{2}i\right)\right]\left[x - \left(\frac{3}{2} - \frac{1}{2}i\right)\right]$.

11. (a) The possible rational roots are ± 1, ± 2, ± 3, ± 4, ± 6, ± 8, ± 12, and ± 24.

  (b) Using synthetic division:

$$\begin{array}{r|rrrrr}
2 & 1 & -1 & 0 & 0 & 24 \\
  &   & 2 & 2 & 4 & 8 \\
\hline
  & 1 & 1 & 2 & 4 & 32
\end{array}$$

   Since this last row consists of all positive numbers, we know that 2 is an upper bound for the roots of this equation.

  (c) Only $x = 1$, since $x = 2$ is an upper bound and not a root.

  (d) Using synthetic division:

$$\begin{array}{r|rrrrr}
1 & 1 & -1 & 0 & 0 & 24 \\
  &   & 1 & 0 & 0 & 1 \\
\hline
  & 1 & 0 & 0 & 0 & 25
\end{array}$$

   Since $x = 1$ was the only possibility, there are no positive rational roots.

12. (a) The possible rational roots are ± 1, $\pm \frac{1}{2}$, ± 3, $\pm \frac{3}{2}$.
    Using synthetic division:

$$\begin{array}{r|rrrr}
3/2 & 2 & -1 & -1 & -3 \\
    &   & 3 & 3 & 3 \\
\hline
    & 2 & 2 & 2 & 0
\end{array}$$

   So $2x^3 - x^2 - x - 3 = \left(x - \frac{3}{2}\right)\left(2x^2 + 2x + 2\right) = 2\left(x - \frac{3}{2}\right)\left(x^2 + x + 1\right)$. Using the
   quadratic equation, we have:
$$x = \frac{-1 \pm \sqrt{1-4}}{2} = \frac{-1 \pm i\sqrt{3}}{2}$$
   So the only rational root is $\frac{3}{2}$.

  (b) All solutions are $\frac{3}{2}$, $\frac{-1+i\sqrt{3}}{2}$, and $\frac{-1-i\sqrt{3}}{2}$.

13. Let $f(x) = 3x^4 + x^2 - 5x - 1$. Since there is one sign change, there is 1 positive root.
   $f(-x) = 3x^4 + x^2 + 5x - 1$. Since there is one sign change, there is 1 negative root. So the
   equation has 1 positive real root, 1 negative real root, and 2 complex roots.

14. The factors are $x + 2$, $x - (1 - 3i)$, and $x - (1 + 3i)$, so:
$$(x+2)\left[x - (1-3i)\right]\left[x - (1+3i)\right] = 0$$
$$(x+2)\left(x^2 - 2x + 10\right) = 0$$
$$x^3 + 6x + 20 = 0$$

15. Here $f(x) = (x-2)(x-3i)^3 \left[x - (1+\sqrt{2})\right]^2$. Note that had a restriction of rational coefficients been added, we would also have $-3i$ (multiplicity 3) and $1 - \sqrt{2}$ (multiplicity 2) as roots.

16. Simplify the expression:

$$(4+2i)(4-2i) + \frac{3+i}{1+2i} = 16 - 4i^2 + \frac{3+i}{1+2i} \cdot \frac{1-2i}{1-2i}$$

$$= 16 + 4 + \frac{3 - 5i - 2i^2}{1 - 4i^2}$$

$$= 20 + \frac{5 - 5i}{5}$$

$$= 20 + 1 - i$$

$$= 21 - i$$

17. (a) Computing the discriminant:

$$b^2 - 4ac = (-4)^2 - 4(2)(3) = 16 - 24 = -8$$

Since the discriminant is negative, the equation will have two complex conjugate roots.

(b) Using the quadratic formula:

$$x = \frac{-(-4) \pm \sqrt{-8}}{2(2)} = \frac{4 \pm 2i\sqrt{2}}{4} = 1 \pm \frac{\sqrt{2}}{2}i$$

18. Since $x^3 - 16x = x(x^2 - 16) = x(x+4)(x-4)$, the decomposition will have the form:

$$\frac{3x-1}{x^3 - 16x} = \frac{A}{x} + \frac{B}{x+4} + \frac{C}{x-4}$$

Clearing fractions yields:

$$3x - 1 = A(x+4)(x-4) + Bx(x-4) + Cx(x+4)$$

Substituting $x = -4$:

$$-12 - 1 = A(0)(-8) + B(-4)(-8) + C(-4)(0)$$

$$-13 = 32B$$

$$B = -\tfrac{13}{32}$$

Substituting $x = 4$:

$$12 - 1 = A(8)(0) + B(4)(0) + C(4)(8)$$

$$11 = 32C$$

$$C = \tfrac{11}{32}$$

Substituting $x = 0$:

$$0 - 1 = A(4)(-4) + B(0)(-4) + C(0)(4)$$

$$-1 = -16A$$

$$A = \tfrac{1}{16}$$

So $A = \tfrac{1}{16}$, $B = -\tfrac{13}{32}$, and $C = \tfrac{11}{32}$. The partial fraction decomposition is:

$$\frac{3x - 1}{x^3 - 16x} = \frac{\tfrac{1}{16}}{x} + \frac{-\tfrac{13}{32}}{x + 4} + \frac{\tfrac{11}{32}}{x - 4}$$

19.   Factoring the denominator by grouping:

$$x^3 - x^2 + 3x - 3 = x^2(x - 1) + 3(x - 1) = (x - 1)\left(x^2 + 3\right)$$

The decomposition will have the form:

$$\frac{1}{x^3 - x^2 + 3x - 3} = \frac{A}{x - 1} + \frac{Bx + C}{x^2 + 3}$$

Clearing fractions yields:

$$1 = A\left(x^2 + 3\right) + (Bx + C)(x - 1)$$

$$= Ax^2 + 3A + Bx^2 + Cx - Bx - C$$

$$= (A + B)x^2 + (-B + C)x + (3A - C)$$

Equating coefficients, we have:

$$A + B = 0$$

$$-B + C = 0$$

$$3A - C = 1$$

Form the augmented matrix:

$$\begin{pmatrix} 1 & 1 & 0 & 0 \\ 0 & -1 & 1 & 0 \\ 3 & 0 & -1 & 1 \end{pmatrix}$$

Adding -3 times row 1 to row 3 and multiplying row 2 by -1:

$$\begin{pmatrix} 1 & 1 & 0 & 0 \\ 0 & 1 & -1 & 0 \\ 0 & -3 & -1 & 1 \end{pmatrix}$$

Adding 3 times row 2 to row 3:

$$\begin{pmatrix} 1 & 1 & 0 & 0 \\ 0 & 1 & -1 & 0 \\ 0 & 0 & -4 & 1 \end{pmatrix}$$

Dividing row 3 by –4, and adding it to row 2:

$$\begin{pmatrix} 1 & 1 & 0 & 0 \\ 0 & 1 & 0 & -\frac{1}{4} \\ 0 & 0 & 1 & -\frac{1}{4} \end{pmatrix}$$

Adding –1 times row 2 to row 1:

$$\begin{pmatrix} 1 & 0 & 0 & \frac{1}{4} \\ 0 & 1 & 0 & -\frac{1}{4} \\ 0 & 0 & 1 & -\frac{1}{4} \end{pmatrix}$$

So $A = \frac{1}{4}$, $B = -\frac{1}{4}$, and $C = -\frac{1}{4}$. The partial fraction decomposition is:

$$\frac{1}{x^3 - x^2 + 3x - 3} = \frac{\frac{1}{4}}{x - 1} + \frac{-\frac{1}{4}x - \frac{1}{4}}{x^2 + 3}$$

20. Since $x^3 - 4x^2 + 4x = x(x^2 - 4x + 4) = x(x - 2)^2$, the decomposition will have the form:

$$\frac{4x^2 - 15x + 20}{x^3 - 4x^2 + 4x} = \frac{A}{x} + \frac{B}{x - 2} + \frac{C}{(x - 2)^2}$$

Clearing fractions yields:

$$\begin{aligned} 4x^2 - 15x + 20 &= A(x - 2)^2 + Bx(x - 2) + Cx \\ &= Ax^2 - 4Ax + 4A + Bx^2 - 2Bx + Cx \\ &= (A + B)x^2 + (-4A - 2B + C)x + 4A \end{aligned}$$

Equating coefficients, we have:

$$A + B = 4$$
$$-4A - 2B + C = -15$$
$$4A = 20$$

So $A = 5$ and $B = -1$. Substituting into $-4A - 2B + C = -15$:

$$-4(5) - 2(-1) + C = -15$$
$$-20 + 2 + C = -15$$
$$C = 3$$

The partial fraction decomposition is:

$$\frac{4x^2 - 15x + 20}{x^3 - 4x^2 + 4x} = \frac{5}{x} + \frac{-1}{x - 2} + \frac{3}{(x - 2)^2}$$

Chapter Thirteen
Analytic Geometry

13.1 The Basic Equations

1. Using the distance formula:
$$d = \sqrt{(-5-3)^2 + (-6+1)^2} = \sqrt{64+25} = \sqrt{89}$$

3. Find the slope of the line:
$$4x - 5y - 20 = 0$$
$$-5y = -4x + 20$$
$$y = \tfrac{4}{5}x - 4$$

So the perpendicular slope is $m = -\tfrac{5}{4}$. Now find the y-intercept:
$$x - y + 1 = 0$$
$$-y = -x - 1$$
$$y = x + 1$$

So $b = 1$, and the equation is $y = -\tfrac{5}{4}x + 1$. Multiplying by 4:
$$4y = -5x + 4, \text{ or } 5x + 4y - 4 = 0$$

5. Find the slope of the line segment:
$$m = \tfrac{7-1}{6-2} = \tfrac{6}{4} = \tfrac{3}{2}$$

So the perpendicular slope is $m = -\tfrac{2}{3}$. Find the midpoint:
$$M = \left(\tfrac{2+6}{2}, \tfrac{1+7}{2}\right) = \left(\tfrac{8}{2}, \tfrac{8}{2}\right) = (4,4)$$

Now use the point-slope formula:
$$y - 4 = -\tfrac{2}{3}(x - 4)$$
$$y - 4 = -\tfrac{2}{3}x + \tfrac{8}{3}$$
$$y = -\tfrac{2}{3}x + \tfrac{20}{3}$$

Multiplying by 3, the equation is $3y = -2x + 20$, or $2x + 3y - 20 = 0$.

7. Since the center is $(1, 0)$ and the radius is 5, the equation must be $(x-1)^2 + y^2 = 25$. To find the x-intercepts, let $y = 0$ and solve the resulting equation for x:

$$y = 0$$
$$(x-1)^2 = 25$$
$$x - 1 = \pm 5$$
$$x = 6, -4$$

To find the y-intercepts, let $x = 0$ and solve the resulting equation for y:

$$x = 0$$
$$(-1)^2 + y^2 = 25$$
$$y^2 = 24$$
$$y = \pm\sqrt{24} = \pm 2\sqrt{6}$$

The x-intercepts are 6 and -4, and the y-intercepts are $\pm 2\sqrt{6}$.

9. First find the midpoint of $\overline{AB}$:

$$M = \left(\tfrac{1+6}{2}, \tfrac{2+1}{2}\right) = \left(\tfrac{7}{2}, \tfrac{3}{2}\right)$$

Now find the slope of the line:

$$m = \frac{8 - \frac{3}{2}}{7 - \frac{7}{2}} = \frac{\frac{13}{2}}{\frac{7}{2}} = \frac{13}{7}$$

Using the point $(7, 8)$ in the point-slope formula:

$$y - 8 = \tfrac{13}{7}(x - 7)$$
$$y - 8 = \tfrac{13}{7}x - 13$$
$$y = \tfrac{13}{7}x - 5$$

Multiply by 7 to get $7y = 13x - 35$, or $13x - 7y - 35 = 0$.

11. Since C is the x-intercept, let $y = 0$ to obtain $\frac{x}{7} = 1$, or $x = 7$. So the coordinates of C are $(7, 0)$. Similarly, since B is the y-intercept, let $x = 0$ to obtain $\frac{y}{5} = 1$, or $y = 5$. So the coordinates of B are $(0, 5)$. Find BC by the distance formula:

$$BC = \sqrt{(7-0)^2 + (0-5)^2} = \sqrt{49 + 25} = \sqrt{74}$$

The perimeter is given by:

$$P = AB + BC + AC = 5 + \sqrt{74} + 7 = 12 + \sqrt{74}$$

13. Since $\tan\theta = \sqrt{3}$, then $\theta = \frac{\pi}{3}$ or $60°$.

15. (a) Since $\tan\theta = 5$, then $\theta = 1.37$ or $78.69°$.

(b) Since $\tan\theta = -5$, then $\theta = 1.77$ or $101.31°$.

17. (a) Here $(x_0, y_0) = (1, 4)$, $m = 1$, $b = -2$, so:

$$d = \frac{|1 - 2 - 4|}{\sqrt{1+1}} = \frac{5}{\sqrt{2}} = \frac{5\sqrt{2}}{2}$$

(b) Using $x - y - 2 = 0$, $A = 1$, $B = -1$, $C = -2$:
$$d = \frac{|1 - 4 - 2|}{\sqrt{1+1}} = \frac{5}{\sqrt{2}} = \frac{5\sqrt{2}}{2}$$

19. (a) Here $(x_0, y_0) = (-3, 5)$. Convert to slope-intercept form:
$$4x + 5y + 6 = 0$$
$$5y = -4x - 6$$
$$y = -\tfrac{4}{5}x - \tfrac{6}{5}$$
So $m = -\tfrac{4}{5}$ and $b = -\tfrac{6}{5}$:
$$d = \frac{\left|\frac{12}{5} - \frac{6}{5} - 5\right|}{\sqrt{1 + \frac{16}{25}}} = \frac{\frac{19}{5}}{\frac{\sqrt{41}}{5}} = \frac{19}{\sqrt{41}} = \frac{19\sqrt{41}}{41}$$

(b) Given $A = 4$, $B = 5$, $C = 6$:
$$d = \frac{|-12 + 25 + 6|}{\sqrt{16 + 25}} = \frac{19}{\sqrt{41}} = \frac{19\sqrt{41}}{41}$$

21. (a) The radius of the circle is the distance from the point $(-2, -3)$ to the line $2x + 3y - 6 = 0$. Using $(x_0, y_0) = (-2, -3)$, $A = 2$, $B = 3$, $C = -6$:
$$r = \frac{|-4 - 9 - 6|}{\sqrt{4 + 9}} = \frac{19}{\sqrt{13}}$$
So the equation of the circle is $(x + 2)^2 + (y + 3)^2 = \frac{361}{13}$.

(b) Since the radius is the distance from the point $(1, 3)$ to the line $y = \tfrac{1}{2}x + 5$,
$(x_0, y_0) = (1, 3)$, $m = \tfrac{1}{2}$, $b = 5$:
$$r = \frac{\left|\frac{1}{2} + 5 - 3\right|}{\sqrt{1 + \frac{1}{4}}} = \frac{\frac{5}{2}}{\frac{\sqrt{5}}{2}} = \sqrt{5}$$

23. Using the suggestion, work with $\triangle ABC$ and $\triangle CDA$. For $\triangle ABC$, find the base using the distance formula:
$$AB = \sqrt{(8 - 0)^2 + (2 - 0)^2} = \sqrt{64 + 4} = \sqrt{68} = 2\sqrt{17}$$
Now find the equation of the line through A and B. Find the slope:
$$m = \tfrac{2-0}{8-0} = \tfrac{1}{4}$$
So the equation is $y = \tfrac{1}{4}x$. Find the distance from $C(4, 7)$ to this line:
$$h = \frac{|1 + 0 - 7|}{\sqrt{1 + \frac{1}{16}}} = \frac{6}{\frac{\sqrt{17}}{4}} = \frac{24}{\sqrt{17}}$$
So $\triangle ABC$ will have an area of:
$$\text{Area}_{\triangle ABC} = \tfrac{1}{2}(\text{base})(\text{height}) = \tfrac{1}{2}\left(2\sqrt{17}\right)\left(\tfrac{24}{\sqrt{17}}\right) = 24$$

For $\triangle CDA$, find the base using the distance formula:
$$CD = \sqrt{(1-4)^2 + (6-7)^2} = \sqrt{9+1} = \sqrt{10}$$
Now find the equation of the line through C and D. Find the slope:
$$m = \frac{6-7}{1-4} = \frac{-1}{-3} = \frac{1}{3}$$
Using $C(4, 7)$, in the point-slope formula:
$$y - 7 = \tfrac{1}{3}(x - 4)$$
$$y - 7 = \tfrac{1}{3}x - \tfrac{4}{3}$$
$$y = \tfrac{1}{3}x + \tfrac{17}{3}$$
$$3y = x + 17 \text{ or } x - 3y + 17 = 0$$
Find the distance from A (0, 0) to this line:
$$h = \frac{|0 + 0 + 17|}{\sqrt{1+9}} = \frac{17}{\sqrt{10}}$$

So $\triangle CDA$ will have an area of :
$$\text{Area}_{\triangle CDA} = \tfrac{1}{2}(\text{base})(\text{height}) = \tfrac{1}{2}\left(\sqrt{10}\right)\left(\tfrac{17}{\sqrt{10}}\right) = \tfrac{17}{2}$$
Thus, the total combined area of quadrilateral $ABCD$ is:
$$24 + \tfrac{17}{2} = \tfrac{65}{2}$$

25. Using the point (0, −5) in the point-slope formula:
$$y + 5 = m(x - 0)$$
$$y = mx - 5$$
Now, since the distance from the center (3, 0) to this line is 2:
$$2 = \frac{|3m - 5 - 0|}{\sqrt{1 + m^2}} = \frac{|3m - 5|}{\sqrt{1 + m^2}}$$
Squaring each side:
$$4 = \frac{9m^2 - 30m + 25}{1 + m^2}$$
$$4 + 4m^2 = 9m^2 - 30m + 25$$
$$0 = 5m^2 - 30m + 21$$
Using the quadratic formula:
$$m = \frac{30 \pm \sqrt{900 - 420}}{10} = \frac{30 \pm \sqrt{480}}{10} = \frac{30 \pm 4\sqrt{30}}{10} = \frac{15 \pm 2\sqrt{30}}{5}$$

27. Use the same approach as in the preceeding exercise. Let d_1 and d_2 represent the distances from (0, 0) to the lines $3x + 4y - 12 = 0$ and $3x + 4y - 24 = 0$, respectively. For d_1, $(x_0, y_0) = (0, 0)$, $A = 3$, $B = 4$, $C = -12$:
$$d_1 = \frac{|0 + 0 - 12|}{\sqrt{9 + 16}} = \frac{12}{\sqrt{25}} = \tfrac{12}{5}$$

For d_2, $(x_0, y_0) = (0, 0)$, $A = 3$, $B = 4$, $C = -24$:

$$d_2 = \frac{|0 + 0 - 24|}{\sqrt{9 + 16}} = \frac{24}{\sqrt{25}} = \frac{24}{5}$$

So the distance between the lines is:

$$d_2 - d_1 = \frac{24}{5} - \frac{12}{5} = \frac{12}{5}$$

29. For the described line segment to be bisected by the point $(2, 6)$, then $(2, 6)$ must be the midpoint of the x- and y-intercepts (as points) of the line. Using the point-slope formula:

$$y - 6 = m(x - 2)$$
$$y - 6 = mx - 2m$$
$$y = mx - 2m + 6$$

To find the x-intercept, let $y = 0$ and solve the resulting equation for x:

$$y = 0$$
$$mx - 2m + 6 = 0$$
$$mx = 2m - 6$$
$$x = \frac{2m - 6}{m}$$

To find the y-intercept, let $x = 0$ and solve the resulting equation for y:

$$x = 0$$
$$y = -2m + 6$$

Since $(2, 6)$ must be the midpoint of $\left(\frac{2m-6}{m}, 0\right)$ and $(0, -2m + 6)$, we have the two sets of equations:

$$\frac{2m - 6}{2m} = 2 \qquad\qquad \frac{-2m + 6}{2} = 6$$
$$2m - 6 = 4m \qquad\qquad -2m + 6 = 12$$
$$-6 = 2m \qquad\qquad -2m = 6$$
$$-3 = m \qquad\qquad m = -3$$

Now use the point-slope formula with the point $(2, 6)$:

$$y - 6 = -3(x - 2)$$
$$y - 6 = -3x + 6$$
$$y = -3x + 12$$

Here is an alternate solution. Let the intercepts be $(a, 0)$ and $(0, b)$. Since $(2, 6)$ is their midpoint:

$$2 = \frac{a + 0}{2}, \text{ so } a = 4$$
$$6 = \frac{b + 0}{2}, \text{ so } b = 12$$

Thus the slope is given by:

$$m = \frac{12 - 0}{0 - 4} = -3$$

Thus the equation is $y = -3x + 12$.

Special thanks to Johy Fay of Imperial Valley College for this solution.

31. Using the hint, we let d_1 and d_2 represent the distances from (x, y) to $x - y + 1 = 0$ and $x + 7y - 49 = 0$, respectively. Then:

$$d_1 = \frac{|x - y + 1|}{\sqrt{1 + 1}} = \frac{|x - y + 1|}{\sqrt{2}}$$

$$d_2 = \frac{|x + 7y - 49|}{\sqrt{1 + 49}} = \frac{|x + 7y - 49|}{\sqrt{50}}$$

Since $d_1 = d_2$:

$$\frac{|x - y + 1|}{\sqrt{2}} = \frac{|x + 7y - 49|}{\sqrt{50}}$$

Multiplying each side by $\sqrt{2}$:

$$|x - y + 1| = \frac{|x + 7y - 49|}{5}$$

$$5|x - y + 1| = |x + 7y - 49|$$

Rather than squaring each side, note that one of the two equations must hold:

$$5(x - y + 1) = x + 7y - 49 \qquad\qquad 5(x - y + 1) = -x - 7y + 49$$
$$5x - 5y + 5 = x + 7y - 49 \qquad\qquad 5x - 5y + 5 = -x - 7y + 49$$
$$4x - 12y + 54 = 0 \qquad\qquad\qquad 6x + 2y - 44 = 0$$
$$2x - 6y + 27 = 0 \qquad\qquad\qquad 3x + y - 22 = 0$$
$$y = \tfrac{1}{3}x + \tfrac{9}{2} \qquad\qquad\qquad\qquad y = -3x + 22$$

Note that the first of these lines, $y = \tfrac{1}{3}x + \tfrac{9}{2}$, is the solution as indicated in the figure. The second line we found is the second angle bisector passing through the same intersection point but bisecting the larger vertical angles.

33. (a) The standard form for a circle is $(x - h)^2 + (y - k)^2 = r^2$. Substitute each point for (x, y):

$$(-12 - h)^2 + (1 - k)^2 = r^2$$
$$(2 - h)^2 + (1 - k)^2 = r^2$$
$$(0 - h)^2 + (7 - k)^2 = r^2$$

Setting the first two equations equal:

$$(-12 - h)^2 + (1 - k)^2 = (2 - h)^2 + (1 - k)^2$$
$$144 + 24h + h^2 = 4 - 4h + h^2$$
$$28h = -140$$
$$h = -5$$

Setting the second two equations equal:

$$(2 - h)^2 + (1 - k)^2 = (0 - h)^2 + (7 - k)^2$$
$$4 - 4h + h^2 + 1 - 2k + k^2 = h^2 + 49 - 14k + k^2$$
$$-4h + 12k = 44$$
$$h - 3k = -11$$

Substituting $h = -5$:
$$-5 - 3k = -11$$
$$-3k = -6$$
$$k = 2$$
Now substitute into the second equation to find r:
$$(7)^2 + (-1)^2 = r^2$$
$$50 = r^2$$
$$5\sqrt{2} = r$$
So the center is $(-5, 2)$ and the radius is $5\sqrt{2}$.

(b) First draw the sketch:

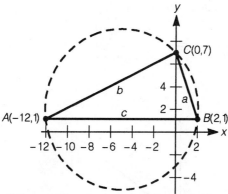

Find the lengths of three sides (using the distance formula):
$$a = \sqrt{(2-0)^2 + (1-7)^2} = \sqrt{4+36} = \sqrt{40} = 2\sqrt{10}$$
$$b = \sqrt{(-12-0)^2 + (1-7)^2} = \sqrt{144+36} = \sqrt{180} = 6\sqrt{5}$$
$$c = \sqrt{(-12-2)^2 + (1-1)^2} = \sqrt{196+0} = \sqrt{196} = 14$$
Using $R = 5\sqrt{2}$ from part (a):
$$\frac{abc}{4R} = \frac{(2\sqrt{10})(6\sqrt{5})(14)}{4(5\sqrt{2})} = \frac{42\sqrt{50}}{5\sqrt{2}} = 42$$
For the area of the triangle, note that the base = 14 and the height = 6, so:
$$\text{Area} = \tfrac{1}{2}(\text{base})(\text{height}) = \tfrac{1}{2}(14)(6) = 42$$
This verifies the result.

35. First find points P and Q. Solving for x, we have $x = 7y - 44$. Now substitute:
$$(7y - 44)^2 - 4(7y - 44) + y^2 - 6y = 12$$
$$49y^2 - 616y + 1936 - 28y + 176 + y^2 - 6y = 12$$
$$50y^2 - 650y + 2100 = 0$$
$$50(y^2 - 13y + 42) = 0$$
$$50(y - 7)(y - 6) = 0$$
$$y = 7, 6$$

When $y = 7$, $x = 49 - 44 = 5$. When $y = 6$, $x = 42 - 44 = -2$. So the two points are $P(5, 7)$ and $Q(-2, 6)$. Now use the distance formula:

$$PQ = \sqrt{(-2-5)^2 + (6-7)^2} = \sqrt{49+1} = \sqrt{50} = 5\sqrt{2}$$

37. Let d_1 be the distance from $(0, c)$ to $ax + y = 0$:

$$d_1 = \frac{|0+c+0|}{\sqrt{a^2+1}} = \frac{|c|}{\sqrt{a^2+1}}$$

Let d_2 be the distance from $(0, c)$ to $x + by = 0$:

$$d_1 = \frac{|0+bc+0|}{\sqrt{1+b^2}} = \frac{|bc|}{\sqrt{1+b^2}}$$

So the product of these distances is given by:

$$d_1 d_2 = \frac{|c|}{\sqrt{a^2+1}} \cdot \frac{|bc|}{\sqrt{1+b^2}} = \frac{|bc^2|}{\sqrt{a^2+a^2b^2+b^2+1}}$$

39. (a) Draw the triangle:

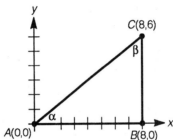

Since the triangle is a right triangle, note that $\sin\alpha = \frac{6}{10} = \frac{3}{5}$, $\cos\alpha = \frac{8}{10} = \frac{4}{5}$, $\sin\beta = \frac{8}{10} = \frac{4}{5}$ and $\cos\beta = \frac{6}{10} = \frac{3}{5}$. To find the bisector for A, note that:

$$m = \tan\frac{\alpha}{2} = \frac{\sin\alpha}{1+\cos\alpha} = \frac{\frac{3}{5}}{1+\frac{4}{5}} = \frac{3}{9} = \frac{1}{3}$$

Since this line passes through the point $(0, 0)$, its equation is $y = \frac{1}{3}x$. To find the bisector for B, note that its slope is -1 and that it passes through $(8, 0)$, so:

$$y - 0 = -1(x - 8)$$
$$y = -x + 8$$

To find the bisector for C, first draw the figure:

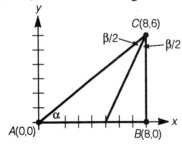

Now note that:

$$m = \tan\left(90° - \frac{\beta}{2}\right) = \cot\frac{\beta}{2} = \frac{1+\cos\beta}{\sin\beta} = \frac{1+\frac{3}{5}}{\frac{4}{5}} = 2$$

Using the point (8, 6), use the point-slope formula:

$$y - 6 = 2(x - 8)$$
$$y - 6 = 2x - 16$$
$$y = 2x - 10$$

So the bisectors at each vertex are:

A: $\quad y = \frac{1}{3}x$

B: $\quad y = -x + 8$

C: $\quad y = 2x - 10$

(b) Setting A and B equal:

$$\tfrac{1}{3}x = -x + 8$$
$$\tfrac{4}{3}x = 8$$
$$x = 6$$
$$y = 2$$

Setting B and C equal:

$$-x + 8 = 2x - 10$$
$$18 = 3x$$
$$6 = x$$
$$2 = y$$

Setting A and C equal:

$$\tfrac{1}{3}x = 2x - 10$$
$$-\tfrac{5}{3}x = -10$$
$$x = 6$$
$$y = 2$$

So the angle bisectors are concurrent at the point (6, 2).

41. The distance from the point (x, y) to $\left(0, \frac{1}{4}\right)$ is:

$$d = \sqrt{(x-0)^2 + \left(y - \tfrac{1}{4}\right)^2} = \sqrt{x^2 + y^2 - \tfrac{1}{2}y + \tfrac{1}{16}}$$

The distance from the point (x, y) to $y = -\frac{1}{4}$ is:

$$d = \frac{\left|0 - \tfrac{1}{4} - y\right|}{\sqrt{1+0}} = \left|y + \tfrac{1}{4}\right|$$

Setting these distances equal:

$$\left|y + \tfrac{1}{4}\right| = \sqrt{x^2 + y^2 - \tfrac{1}{2}y + \tfrac{1}{16}}$$

Squaring:

$$y^2 + \tfrac{1}{2}y + \tfrac{1}{16} = x^2 + y^2 - \tfrac{1}{2}y + \tfrac{1}{16}$$
$$y = x^2$$

Thus, the points satisfy the equation $y = x^2$.

43. (a) Write the line in slope-intercept form:

$$Ax + By + C = 0$$
$$By = -Ax - C$$
$$y = -\frac{A}{B}x - \frac{C}{B}$$

So the slope is $-\frac{A}{B}$ and the y-intercept is $-\frac{C}{B}$.

(b) Using the given formula:

$$d = \frac{|mx_0 + b - y_0|}{\sqrt{1 + m^2}} = \frac{\left|-\frac{Ax_0 + By_0 + C}{B}\right|}{\sqrt{1 + \frac{A^2}{B^2}}} = \frac{\frac{|Ax_0 + By_0 + C|}{|B|}}{\frac{\sqrt{A^2 + B^2}}{|B|}} = \frac{|Ax_0 + By_0 + C|}{\sqrt{A^2 + B^2}}$$

45. First find the slope of the line passing through (x_2, y_2) and (x_3, y_3):

$$m = \frac{y_2 - y_3}{x_2 - x_3}$$

Using the point-slope formula:

$$y - y_2 = \frac{y_2 - y_3}{x_2 - x_3}(x - x_2)$$
$$(x_2 - x_3)y - x_2 y_2 + x_3 y_2 = (y_2 - y_3)x - x_2 y_2 + x_2 y_3$$
$$(y_3 - y_2)x + (x_2 - x_3)y + (x_3 y_2 - x_2 y_3) = 0$$

Now find the distance from (x_1, y_1) to the line:

$$d = \frac{|x_1(y_3 - y_2) + y_1(x_2 - x_3) + x_3 y_2 - x_2 y_3|}{\sqrt{(y_3 - y_2)^2 + (x_2 - x_3)^2}}$$

$$= \frac{|x_1 y_3 - x_1 y_2 + x_2 y_1 - x_3 y_1 + x_3 y_2 - x_2 y_3|}{\sqrt{(x_2 - x_3)^2 + (y_2 - y_3)^2}}$$

Expanding D along the third column, we find that its terms exactly match the numerator of d.

47. Since the center can be written as $(x, 2x - 7)$, find the distance to each point. Let d_1 represent the distance to (6, 3) and d_2 represent the distance to (–4, –3) , then:

$$d_1 = \sqrt{(x-6)^2 + (2x-7-3)^2}$$
$$= \sqrt{x^2 - 12x + 36 + 4x^2 - 40x + 100}$$
$$= \sqrt{5x^2 - 52x + 136}$$
$$d_2 = \sqrt{(x+4)^2 + (2x-7+3)^2}$$
$$= \sqrt{x^2 + 8x + 16 + 4x^2 - 16x + 16}$$
$$= \sqrt{5x^2 - 8x + 32}$$

Since $d_1 = d_2$ (they both represent the radius of the circle):

$$\sqrt{5x^2 - 52x + 136} = \sqrt{5x^2 - 8x + 32}$$
$$5x^2 - 52x + 136 = 5x^2 - 8x + 32$$
$$-44x = -104$$
$$x = \tfrac{26}{11}$$
$$y = 2\left(\tfrac{26}{11}\right) - 7 = \tfrac{52}{11} - \tfrac{77}{11} = -\tfrac{25}{11}$$

So the center is $\left(\tfrac{26}{11}, -\tfrac{25}{11}\right)$. Find the radius using the point (6, 3) in the distance formula:

$$r = \sqrt{\left(6 - \tfrac{26}{11}\right)^2 + \left(3 + \tfrac{25}{11}\right)^2} = \sqrt{\tfrac{1600}{121} + \tfrac{3364}{121}} = \frac{\sqrt{4964}}{11}$$

So the equation of the circle is $\left(x - \tfrac{26}{11}\right)^2 + \left(y + \tfrac{25}{11}\right)^2 = \tfrac{4964}{121}$.

49. Since the line $2x + 3y = 26$ has a slope of $-\tfrac{2}{3}$, the radial line will have a slope of $\tfrac{3}{2}$. Call (h, k) the center of the circle, so:

$$\frac{k-6}{h-4} = \tfrac{3}{2}$$
$$2k - 12 = 3h - 12$$
$$2k = 3h$$
$$k = \tfrac{3}{2}h$$

The circle has an equation of $(x - h)^2 + (y - k)^2 = 25$, so using the point (4, 6):

$$(4 - h)^2 + (6 - k)^2 = 25$$
$$16 - 8h + h^2 + 36 - 12k + k^2 = 25$$
$$h^2 - 8h + k^2 - 12k = -27$$

Substituting $k = \frac{3}{2}h$:

$$h^2 - 8h + \frac{9}{4}h^2 - 18h = -27$$
$$4h^2 - 32h + 9h^2 - 72h = -108$$
$$13h^2 - 104h + 108 = 0$$
$$h = \frac{104 \pm \sqrt{5200}}{26} = 4 \pm \frac{10\sqrt{13}}{13}$$
$$k = 6 \pm \frac{15\sqrt{13}}{13}$$

So the two circles are:

$$\left(x - 4 - \frac{10\sqrt{13}}{13}\right)^2 + \left(y - 6 - \frac{15\sqrt{13}}{13}\right)^2 = 25$$
$$\left(x - 4 + \frac{10\sqrt{13}}{13}\right)^2 + \left(y - 6 + \frac{15\sqrt{13}}{13}\right)^2 = 25$$

51. (a) Replacing y with tx in the equation of the folium yields:

$$x^3 + t^3 x^3 = 6x(tx)$$
$$x^3 + t^3 x^3 = 6x^2 t \qquad \text{(since } x \neq 0 \text{ at point } Q\text{)}$$
$$x + t^3 x = 6t$$
$$x(1 + t^3) = 6t$$
$$x = \frac{6t}{1 + t^3}$$
$$y = tx = \frac{6t^2}{1 + t^3}$$

So the coordinates of point Q are $x = \dfrac{6t}{1 + t^3}$ and $y = \dfrac{6t^2}{1 + t^3}$.

(b) The slope of $\overline{PQ}$ is given by:

$$\frac{3 - \dfrac{6t^2}{1 + t^3}}{3 - \dfrac{6t}{1 + t^3}} = \frac{1 - \dfrac{2t^2}{1 + t^3}}{1 - \dfrac{2t}{1 + t^3}} \cdot \frac{1 + t^3}{1 + t^3} = \frac{1 + t^3 - 2t^2}{1 + t^3 - 2t} = \frac{t^3 - 2t^2 + 1}{t^3 - 2t + 1}$$

We can factor both the numerator and denominator of this expression. For the numerator, add and subtract the term t^2 as follows:

$$t^3 - 2t^2 + 1 = t^3 - 2t^2 + t^2 - t^2 + 1$$
$$= (t^3 - t^2) - (t^2 - 1)$$
$$= t^2(t - 1) - (t + 1)(t - 1)$$
$$= (t - 1)(t^2 - t - 1)$$

For the denominator, add and subtract the term t^2 as follows:
$$t^3 - 2t + 1 = t^3 - t^2 + t^2 - 2t + 1$$
$$= \left(t^3 - t^2\right) + \left(t^2 - 2t + 1\right)$$
$$= t^2(t-1) + (t-1)^2$$
$$= (t-1)\left(t^2 + t - 1\right)$$

So the slope of $\overline{PQ}$ is given by:
$$\frac{t^3 - 2t^2 + 1}{t^3 - 2t + 1} = \frac{(t-1)\left(t^2 - t - 1\right)}{(t-1)\left(t^2 + t - 1\right)} = \frac{t^2 - t - 1}{t^2 + t - 1}$$

(c) Cross-multiplying and combining like terms:
$$\left(t^2 - t - 1\right)\left(u^2 + u - 1\right) = \left(t^2 + t - 1\right)\left(u^2 - u - 1\right)$$
$$t^2 u^2 + t^2 u - t^2 - tu^2 - tu + t - u^2 - u + 1 = t^2 u^2 - t^2 u - t^2 + tu^2 - tu - t - u^2 + u + 1$$
$$2t^2 u - 2tu^2 + 2t - 2u = 0$$
$$t^2 u - tu^2 + t - u = 0$$
$$tu(t-u) + 1(t-u) = 0$$
$$(t-u)(tu+1) = 0$$
$$t = u \ \text{ or } \ tu = -1$$

Since points R, O and Q are not collinear, $t \neq u$, and thus $tu = -1$. Therefore $\overline{OR} \perp \overline{OQ}$, as required.

13.2 The Parabola

1. Note that $4p = 4$, so $p = 1$. So the focus is $(0, 1)$, the directrix is $y = -1$, and the focal width is 4.

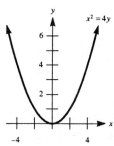

3.   Note that $4p = 8$, so $p = 2$. So the focus is $(-2, 0)$, the directrix is $x = 2$, and the focal width is 8.

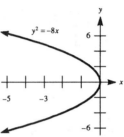

5.   Note that $4p = 20$, so $p = 5$. So the focus is $(0, -5)$, the directrix is $y = 5$, and the focal width is 20.

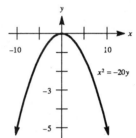

7.   Since $y^2 = -28x$, then $4p = 28$ so $p = 7$. So the focus is $(-7, 0)$, the directrix is $x = 7$, and the focal width is 28.

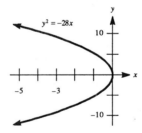

9.   Note that $4p = 6$, so $p = \frac{3}{2}$. So the focus is $\left(0, \frac{3}{2}\right)$, the directrix is $y = -\frac{3}{2}$, and the focal width is 6.

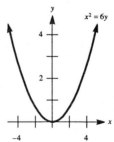

11. Since $x^2 = \frac{7}{4}y$, then $4p = \frac{7}{4}$ so $p = \frac{7}{16}$. So the focus is $\left(0, \frac{7}{16}\right)$, the directrix is $y = -\frac{7}{16}$, and the focal width is $\frac{7}{4}$.

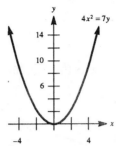

13. First complete the square:
$$y^2 - 6y - 4x + 17 = 0$$
$$y^2 - 6y = 4x - 17$$
$$y^2 - 6y + 9 = 4x - 17 + 9$$
$$(y - 3)^2 = 4x - 8 = 4(x - 2)$$
Note that $4p = 4$, so $p = 1$. The vertex is (2, 3), the focus is (3, 3), the directrix is $x = 1$, and the focal width is 4.

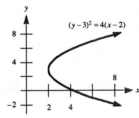

15. First complete the square:
$$x^2 - 8x - y + 18 = 0$$
$$x^2 - 8x = y - 18$$
$$x^2 - 8x + 16 = y - 18 + 16$$
$$(x - 4)^2 = y - 2$$
Note that $4p = 1$, so $p = \frac{1}{4}$. The vertex is (4, 2), the focus is $\left(4, \frac{9}{4}\right)$, the directrix is $y = \frac{7}{4}$, and the focal width is 1.

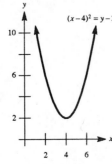

17. First complete the square:
$$y^2 + 2y - x + 1 = 0$$
$$y^2 + 2y = x - 1$$
$$y^2 + 2y + 1 = x - 1 + 1$$
$$(y+1)^2 = x$$

Note that $4p = 1$, so $p = \frac{1}{4}$. The vertex is $(0, -1)$, the focus is $\left(\frac{1}{4}, -1\right)$, the directrix is $x = -\frac{1}{4}$, and the focal width is 1.

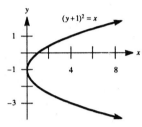

19. First complete the square:
$$2x^2 - 12x - y + 18 = 0$$
$$2x^2 - 12x = y - 18$$
$$x^2 - 6x = \frac{1}{2}y - 9$$
$$x^2 - 6x + 9 = \frac{1}{2}y - 9 + 9$$
$$(x-3)^2 = \frac{1}{2}y$$

Note that $4p = \frac{1}{2}$, so $p = \frac{1}{8}$. The vertex is $(3, 0)$, the focus is $\left(3, \frac{1}{8}\right)$, the directrix is $y = -\frac{1}{8}$, and the focal width is $\frac{1}{2}$.

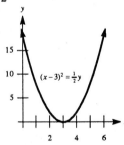

21. First complete the square:
$$2x^2 - 16x - y + 33 = 0$$
$$2x^2 - 16x = y - 33$$
$$2(x-4)^2 = y - 1$$
$$(x-4)^2 = \frac{1}{2}(y-1)$$

Note that $4p = \frac{1}{2}$, so $p = \frac{1}{8}$. The vertex is (4, 1), the focus is $\left(4, \frac{9}{8}\right)$, the directrix is $y = \frac{7}{8}$, and the focal width is $\frac{1}{2}$.

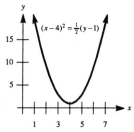

23. The line of symmetry has equation $y = 1$. The figure is graphed as:

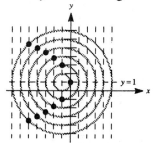

25. Given that the focus is at (0, 3), we see that $p = 3$ and thus $4p = 12$. Since the parabola opens up, it has the $x^2 = 4py$ form, and its equation must be $x^2 = 12y$.

27. The directrix is $x = -32$, so (32, 0) is the focus and the parabola opens to the right. So the parabola is of the form $y^2 = 4px$, where $p = 32$. Thus the equation is $y^2 = 128x$.

29. Since the parabola is symmetric about the x-axis, then $y^2 = \pm 4px$ is the form of the parabola. Since the x-coordinate of the focus is negative, then $y^2 = -4px$. Finally we are given $4p = 9$, so the equation is $y^2 = -9x$.

31. (a) Since $4p = 8$, then $p = 2$, and thus the focus is (0, 2). So the slope of the line is $\frac{8-2}{8-0} = \frac{3}{4}$, and thus the equation is $y = \frac{3}{4}x + 2$.

     (b) Solve the system:
     $$y = \frac{3}{4}x + 2$$
     $$x^2 = 8y$$

Substituting:
$$x^2 = 8\left(\tfrac{3}{4}x + 2\right)$$
$$x^2 = 6x + 16$$
$$x^2 - 6x - 16 = 0$$
$$(x - 8)(x + 2) = 0$$
$$x = 8, -2$$
$$y = 8, \tfrac{1}{2}$$

Since $(8, 8)$ repeats our original point, then Q must be $\left(-2, \tfrac{1}{2}\right)$.

(c)   Use the distance formula:
$$PQ = \sqrt{(8 + 2)^2 + \left(8 - \tfrac{1}{2}\right)^2} = \sqrt{100 + \tfrac{225}{4}} = \sqrt{\tfrac{625}{4}} = \tfrac{25}{2}$$

(d)   We know the radius is $\tfrac{25}{4}$, and the center of the circle will be the midpoint of the line segment $\overline{PQ}$:
$$\text{center} = \left(\frac{8 - 2}{2}, \frac{8 + \tfrac{1}{2}}{2}\right) = \left(3, \tfrac{17}{4}\right)$$

So the equation is $(x - 3)^2 + \left(y - \tfrac{17}{4}\right)^2 = \tfrac{625}{16}$.

(e)   The directrix is $y = -2$. The vertical distance of the center of the circle to the directrix is:
$$\left|\tfrac{17}{4} - (-2)\right| = \left|\tfrac{17}{4} + 2\right| = \tfrac{25}{4}$$

But the radius of the circle is also $\tfrac{25}{4}$, which implies that the directrix must be tangent to the circle.

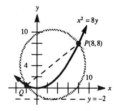

33.   Call $(0, 0)$ the vertex, thus the ends of the arch are at $(20, -15)$ and $(-20, -15)$. The parabola is of the form $x^2 = -4py$, so substitute:
$$400 = -4p(-15)$$
$$-4p = -\tfrac{80}{3}$$

So the equation is $x^2 = -\tfrac{80}{3}y$. We wish to find the y-coordinate when the base is 20 ft, so we substitute $x = \pm 10$ into the equation:
$$100 = -\tfrac{80}{3}y$$
$$y = -\tfrac{15}{4}$$

So the height above the base is $15 - \tfrac{15}{4} = \tfrac{45}{4} = 11.25$ ft.

35. (a) We have $4p = 2$, so $p = \frac{1}{2}$, and thus the focus is $\left(0, \frac{1}{2}\right)$. To find A', find the slope of the line containing A and the focus:

$$m = \frac{8 - \frac{1}{2}}{4 - 0} = \frac{\frac{15}{2}}{4} = \frac{15}{8}$$

So $y = \frac{15}{8}x + \frac{1}{2}$ is its equation. Now substitute into the parabola:

$$x^2 = 2\left(\frac{15}{8}x + \frac{1}{2}\right)$$
$$x^2 = \frac{15}{4}x + 1$$
$$4x^2 = 15x + 4$$
$$4x^2 - 15x - 4 = 0$$
$$x = -\frac{1}{4}, 4$$
$$y = \frac{1}{32}, 8$$

So $A' = \left(-\frac{1}{4}, \frac{1}{32}\right)$. To find B', find the slope of the line containing B and the focus:

$$m = \frac{2 - \frac{1}{2}}{-2 - 0} = \frac{\frac{3}{2}}{-2} = -\frac{3}{4}$$

So $y = -\frac{3}{4}x + \frac{1}{2}$ is its equation. Now substitute into the parabola:

$$x^2 = 2\left(-\frac{3}{4}x + \frac{1}{2}\right)$$
$$x^2 = -\frac{3}{2}x + 1$$
$$2x^2 + 3x - 2 = 0$$
$$x = \frac{1}{2}, -2$$
$$y = \frac{1}{8}, 2$$

So $B' = \left(\frac{1}{2}, \frac{1}{8}\right)$. Now find the slope through A and B':

$$m = \frac{8 - \frac{1}{8}}{4 - \frac{1}{2}} = \frac{\frac{63}{8}}{\frac{7}{2}} = \frac{9}{4}$$

Using $(4, 8)$ in the point-slope formula:

$$y - 8 = \frac{9}{4}(x - 4)$$
$$y - 8 = \frac{9}{4}x - 9$$
$$y = \frac{9}{4}x - 1$$

(b) Find the slope:

$$m = \frac{2 - \frac{1}{32}}{-2 + \frac{1}{4}} = \frac{\frac{63}{32}}{-\frac{7}{4}} = -\frac{9}{8}$$

Using (–2, 2) in the point-slope formula:
$$y - 2 = -\tfrac{9}{8}(x + 2)$$
$$y - 2 = -\tfrac{9}{8}x - \tfrac{9}{4}$$
$$y = -\tfrac{9}{8}x - \tfrac{1}{4}$$

(c) Setting the y-coordinates equal:
$$\tfrac{9}{4}x - 1 = -\tfrac{9}{8}x - \tfrac{1}{4}$$
$$18x - 8 = -9x - 2$$
$$27x = 6$$
$$x = \tfrac{2}{9}$$
$$y = \tfrac{1}{2} - 1 = -\tfrac{1}{2}$$

So the lines intersect at the point $\left(\tfrac{2}{9}, -\tfrac{1}{2}\right)$. Since the directrix is $y = -\tfrac{1}{2}$, then this point lies on the directrix.

37. (a) Since $4p = 1$, $p = \tfrac{1}{4}$, thus the focus is $\left(0, \tfrac{1}{4}\right)$. Find the slope of the focal chord from P:
$$m = \frac{4 - \tfrac{1}{4}}{2 - 0} = \frac{\tfrac{15}{4}}{2} = \tfrac{15}{8}$$

So its equation is $y = \tfrac{15}{8}x + \tfrac{1}{4}$. Now find the intersection of this line with the parabola $y = x^2$:
$$x^2 = \tfrac{15}{8}x + \tfrac{1}{4}$$
$$8x^2 = 15x + 2$$
$$8x^2 - 15x - 2 = 0$$
$$(8x + 1)(x - 2) = 0$$
$$x = -\tfrac{1}{8}, 2$$
$$y = \tfrac{1}{64}, 4$$

So the coordinates of Q are $\left(-\tfrac{1}{8}, \tfrac{1}{64}\right)$.

(b) Find the midpoint of $\overline{PQ}$:
$$M = \left(\frac{2 - \tfrac{1}{8}}{2}, \frac{4 + \tfrac{1}{64}}{2}\right) = \left(\tfrac{15}{16}, \tfrac{257}{128}\right)$$

(c)   The coordinates of S are $\left(0, \frac{257}{128}\right)$. We can find T, since the slope of this line (which is perpendicular to $\overline{PQ}$) is $-\frac{8}{15}$, and thus by the point-slope formula:

$$y - \frac{257}{128} = -\frac{8}{15}\left(x - \frac{15}{16}\right)$$

$$y - \frac{257}{128} = -\frac{8}{15}x + \frac{1}{2}$$

$$y = -\frac{8}{15}x + \frac{321}{128}$$

When $x = 0$, $y = \frac{321}{128}$, so the coordinates of T are $\left(0, \frac{321}{128}\right)$. Thus the length of $\overline{ST}$ is $\frac{321}{128} - \frac{257}{128} = \frac{64}{128} = \frac{1}{2}$. Since the focal width is $4p = 1$, this verifies that ST is one-half the focal width.

39.   The sketch in the book will have the following coordinates:

$$O(0,0), \; B\left(x, \frac{x^2}{4p}\right), \; A\left(-x, \frac{x^2}{4p}\right)$$

We require that $AB = OB$. So:

$$2x = \sqrt{(x-0)^2 + \left(\frac{x^2}{4p} - 0\right)^2} = \sqrt{x^2 + \frac{x^4}{16p^2}}$$

Squaring, we have:

$$4x^2 = x^2 + \frac{x^4}{16p^2}$$

$$0 = -3x^2 + \frac{x^4}{16p^2}$$

$$0 = -48x^2p^2 + x^4$$

$$0 = x^2\left(-48p^2 + x^2\right)$$

So $x^2 = 0$ or $x = 0$ is one root and $x^2 = 48p^2$ or $x = \pm 4\sqrt{3}p$ is the other. The distance from A to B is therefore:

$$2\left(4\sqrt{3}p\right) = 8\sqrt{3}p \text{ units}$$

To find the area we note that this is an equilateral triangle. Half of it is a $30°$- $60°$- $90°$ triangle whose height will be $\sqrt{3}$ times its base. The height is therefore $\sqrt{3} \cdot 4\sqrt{3}p = 12p$. The area is:

$$\tfrac{1}{2}\left(8\sqrt{3}p\right)(12p) = 48\sqrt{3}p^2 \text{ sq. units}$$

41.   Since the focus is $(p, 0)$, the slope of the focal chord is:

$$m = \frac{y_0 - 0}{\frac{y_0^2}{4p} - p} = \frac{4py_0}{y_0^2 - 4p^2}$$

Using the point-slope formula, the equation of this focal chord is:

$$y - y_0 = m\left(x - \frac{y_0^2}{4p}\right)$$

This line intersects the parabola when $x = \frac{y^2}{4p}$, so:

$$y - y_0 = m\left(\frac{y^2}{4p} - \frac{y_0^2}{4p}\right)$$

$$4p(y - y_0) = m(y + y_0)(y - y_0)$$

$$4p = m(y + y_0)$$

Substitute $m = \dfrac{4py_0}{y_0^2 - 4p^2}$:

$$4p = \frac{4py_0}{y_0^2 - 4p^2}(y + y_0)$$

$$\frac{y_0^2 - 4p^2}{y_0} = y + y_0$$

$$y = -\frac{4p^2}{y_0}$$

Therefore:

$$x = \frac{1}{4p}\left(-\frac{4p^2}{y_0}\right)^2 = \frac{1}{4p}\left(\frac{16p^4}{y_0^2}\right) = \frac{4p^3}{y_0^2}$$

But $y_0^2 = 4px_0$, so:

$$x = \frac{4p^3}{4px_0} = \frac{p^2}{x_0}$$

Thus the coordinates of Q are $\left(\dfrac{p^2}{x_0}, -\dfrac{4p^2}{y_0}\right)$.

43. Let P have coordinates (x_0, y_0). Then, from Exercise 41, Q has coordinates $\left(\dfrac{p^2}{x_0}, -\dfrac{4p^2}{y_0}\right)$. The following results summarize the calculations required for this problem. Note: At each step we have replaced the quantity y_0^2 by $4px_0$. This simplifies matters a great deal.

Coordinates of M: $\left(\dfrac{x_0^2 + p^2}{2x_0}, \dfrac{2p(x_0 - p)}{y_0}\right)$

Coordinates of S: $\left(\dfrac{p^2 + x_0^2}{2x_0}, 0\right)$

Slope of $\overline{PQ}$: $\dfrac{4px_0}{y_0(x_0 - p)}$

Slope of perpendicular: $\dfrac{y_0(p - x_0)}{4px_0}$

Since $-\frac{ST}{MS}$ is the slope, m, of the line through M and T, we have:

$$ST = -(MS)m = -\frac{2p(x_0 - p)}{y_0} \cdot \frac{4px_0}{y_0(p - x_0)} = \frac{8p^2 x_0}{y_0^2} = 2p$$

Thus ST is one-half of the focal width.

45. (a) Since the focus is $(0, p)$, then the slope of the focal chord is:

$$m = \frac{\frac{x_0^2}{4p} - p}{x_0 - 0} = \frac{x_0^2 - 4p^2}{4px_0}$$

The equation of the focal chord must be $y - p = mx$. This line intersects the parabola when $y = \frac{x^2}{4p}$, so we have:

$$\frac{x^2}{4p} - p = mx$$

$$\frac{x^2}{4p} - p = \frac{x_0^2 - 4p^2}{4px_0} \cdot x$$

$$x^2 x_0 - 4p^2 x_0 = xx_0^2 - 4p^2 x$$

$$xx_0(x - x_0) = -4p^2(x - x_0)$$

$$xx_0 = -4p^2$$

$$x = -\frac{4p^2}{x_0}$$

Therefore:

$$y = \frac{1}{4p}\left(-\frac{4p^2}{x_0}\right)^2 = \frac{1}{4p}\left(\frac{16p^4}{x_0^2}\right) = \frac{4p^3}{x_0^2}$$

But $x_0^2 = 4py_0$, so:

$$y = \frac{4p^3}{4py_0} = \frac{p^2}{y_0}$$

Thus the coordinates of Q are $\left(-\frac{4p^2}{x_0}, \frac{p^2}{y_0}\right)$.

(b) By the midpoint formula:

$$\text{Midpoint of } PQ = \left(\frac{x_0 - \frac{4p^2}{x_0}}{2}, \frac{y_0 + \frac{p^2}{y_0}}{2}\right) = \left(\frac{x_0^2 - 4p^2}{2x_0}, \frac{y_0^2 + p^2}{2y_0}\right)$$

(c) The distance from P to the directrix is $y_0 + p$. Therefore $PF = y_0 + p$. The distance from Q to the directrix is $\frac{p^2}{y_0} + p$. So we have $QF = \frac{p^2}{y_0} + p$ and consequently:

$$PQ = QF + PF = \frac{p^2}{y_0} + p + y_0 + p = \frac{p^2 + 2py_0 + y_0^2}{y_0} = \frac{(p + y_0)^2}{y_0}$$

(d) The distance from the center of the circle to the directrix is found by adding p to the y-coordinate of the midpoint of $\overline{PQ}$. This yields:

$$\frac{y_0^2 + p^2}{2y_0} + p = \frac{y_0^2 + p^2 + 2py_0}{2y_0} = \frac{(y_0 + p)^2}{2y_0}$$

Comparing this result with the expression for PQ determined in part (c), we conclude that the distance from the center of the circle to the directrix equals the radius of the circle. This implies that the circle is tangent to the directrix, as we wished to show.

Graphing Utility Exercises for Section 13.2

1. Entering $x^2 = 8y$ as $y = \frac{1}{8}x^2$ we have the parabola:

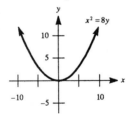

3. Entering $y^2 = 8x$ as the two functions $y = \sqrt{8x}$ and $y = -\sqrt{8x}$, we have the parabola:

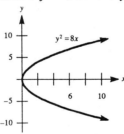

5. (a) Completing the square, we have:
$$4x + y^2 + 2y - 7 = 0$$
$$y^2 + 2y + 1 = -4x + 8$$
$$(y + 1)^2 = -4(x - 2)$$

(b) Entering $(y + 1)^2 = - 4(x - 2)$ as the two functions $y = -1 - \sqrt{-4(x-2)}$ and $y = -1 + \sqrt{-4(x-2)}$, we have the parabola:

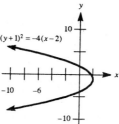

The graph is consistent with Figure 9.

7.   Since the tangent line passes through the points $(4, 2)$ and $(0, -2)$, we can find the slope:
$$m = \frac{-2 - 2}{0 - 4} = \frac{-4}{-4} = 1$$
Since the y-intercept is -2, the equation of the tangent line is $y = x - 2$. Graphing the parabola and tangent line:

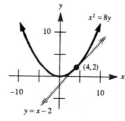

9.   Since the tangent line passes through the points $(-3, 9)$ and $(0, -9)$, we can find the slope:
$$m = \frac{-9 - 9}{0 - (-3)} = \frac{-18}{3} = -6$$
Since the y-intercept is -9, the equation of the tangent line is $y = - 6x - 9$. Graphing the parabola and tangent line:

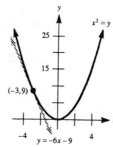

11.   Sketch the two curves:

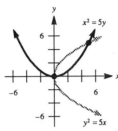

Solving $x^2 = 5y$ for y yields $y = \dfrac{x^2}{5}$, now substitute:

$$\left(\frac{x^2}{5}\right)^2 = 5x$$

$$\frac{x^4}{25} = 5x$$

$$x^4 = 125x$$

$$x^4 - 125x = 0$$

$$x\left(x^3 - 125\right) = 0$$

$$x = 0 \ \text{ or } \ x = \sqrt[3]{125} = 5$$
$$y = 0 \ \text{ or } \ y = 5$$

The intersection points are $(0, 0)$ and $(5, 5)$. These agree with the coordinates found using the graphing utility.

13.   Sketch the two curves:

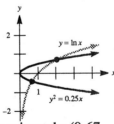

The intersections points are approximately $(0.67, -0.41)$ and $(2.04, 0.71)$.

13.3 Tangents to Parabolas (Optional)

1.   Call m the slope of the tangent line, so:
$$y - 4 = m(x - 2)$$
Now substitute $y = x^2$:
$$x^2 - 4 = m(x - 2)$$
$$(x + 2)(x - 2) = m(x - 2)$$
$$(x - 2)(x + 2 - m) = 0$$

So the solutions are $x = 2$ and $x = m - 2$. But since the line is tangent to the parabola, these two x-values must be equal:

$$m - 2 = 2$$
$$m = 4$$

Thus the tangent line is given by:

$$y - 4 = 4(x - 2)$$
$$y - 4 = 4x - 8$$
$$y = 4x - 4$$

Sketch the parabola and tangent line:

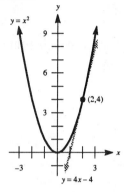

3.   Call m the slope of the tangent line, so:

$$y - 2 = m(x - 4)$$

Solving $x^2 = 8y$ for y yields $y = \dfrac{x^2}{8}$, now substitute:

$$\frac{x^2}{8} - 2 = m(x - 4)$$
$$x^2 - 16 = 8m(x - 4)$$
$$(x + 4)(x - 4) = 8m(x - 4)$$
$$(x - 4)(x + 4 - 8m) = 0$$

So the solutions are $x = 4$ and $x = 8m - 4$. But since the line is tangent to the parabola, these two x-values must be equal:

$$8m - 4 = 4$$
$$8m = 8$$
$$m = 1$$

Thus the tangent line is given by:

$$y - 2 = 1(x - 4)$$
$$y - 2 = x - 4$$
$$y = x - 2$$

Sketch the parabola and tangent line:

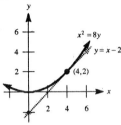

5.   Call m the slope of the tangent line, so:

$$y + 9 = m(x + 3)$$

Solving $x^2 = -y$ for y yields $y = -x^2$, now substitute:

$$-x^2 + 9 = m(x + 3)$$
$$x^2 - 9 = -m(x + 3)$$
$$(x + 3)(x - 3) = -m(x + 3)$$
$$(x + 3)(x - 3 + m) = 0$$

So the solutions are $x = -3$ and $x = 3 - m$. But since the line is tangent to the parabola, these two x-values must be equal:

$$3 - m = -3$$
$$-m = -6$$
$$m = 6$$

Thus the tangent line is given by:

$$y + 9 = 6(x + 3)$$
$$y + 9 = 6x + 18$$
$$y = 6x + 9$$

Sketch the parabola and tangent line:

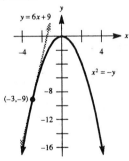

7. Call m the slope of the tangent line, so:

$$y - 2 = m(x - 1)$$

Solving $y^2 = 4x$ for x yields $x = \dfrac{y^2}{4}$, now substitute:

$$y - 2 = m\left(\frac{y^2}{4} - 1\right)$$

$$4(y - 2) = m\left(y^2 - 4\right)$$

$$4(y - 2) = m(y + 2)(y - 2)$$

$$(y - 2)[4 - m(y + 2)] = 0$$

So the solutions are $y = 2$ and $y = \dfrac{4}{m} - 2$. But since the line is tangent to the parabola, these two x-values must be equal:

$$\frac{4}{m} - 2 = 2$$

$$\frac{4}{m} = 4$$

$$4 = 4m$$

$$1 = m$$

Thus the tangent line is given by:

$$y - 2 = 1(x - 1)$$

$$y - 2 = x - 1$$

$$y = x + 1$$

Sketch the parabola and tangent line:

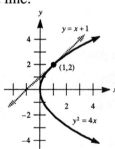

9. Let m be the slope, so $y - 2 = m(x - 4)$. Since $y = \sqrt{x}$, then $x = y^2$:

$$y - 2 = m\left(y^2 - 4\right)$$

$$y - 2 = m(y + 2)(y - 2)$$

$$0 = (y - 2)[m(y + 2) - 1]$$

So the solutions are $y = 2$ and $y = \frac{1}{m} - 2$. But since the line is tangent to the parabola, these two x-values must be equal:

$$\frac{1}{m} - 2 = 2$$

$$\frac{1}{m} = 4$$

$$m = \tfrac{1}{4}$$

11.   Let m be the slope, so $y - 8 = m(x - 2)$. Substitute $y = x^3$:

$$x^3 - 8 = m(x - 2)$$
$$(x - 2)(x^2 + 2x + 4) = m(x - 2)$$
$$(x - 2)(x^2 + 2x + 4 - m) = 0$$

So the solutions are $x = 2$ and $m = x^2 + 2x + 4$. But since the line is tangent to the parabola, we can substitute $x = 2$ to obtain:

$$m = 2^2 + 4 + 4 = 12$$

13.   Let the equation of the tangent line be $y - y_0 = m(x - x_0)$. Replacing y and y_0

by $\dfrac{x^2}{4p}$ and $\dfrac{x_0^2}{4p}$, respectively, we obtain:

$$\frac{x^2}{4p} - \frac{x_0^2}{4p} = m(x - x_0)$$

$$x^2 - x_0^2 - 4pm(x - x_0) = 0$$

$$(x - x_0)\left[(x + x_0) - 4pm\right] = 0$$

$$(x - x_0)(x + x_0 - 4pm) = 0$$

Setting the second factor equal to zero yields $x = 4pm - x_0$. But since $x = x_0$ is known

to be the unique solution, we have $x_0 = 4pm - x_0$. Thus $m = \dfrac{2x_0}{4p} = \dfrac{x_0}{2p}$. The equation

of the tangent line is therefore:

$$y - y_0 = \frac{x_0}{2p}(x - x_0)$$

$$y = \frac{x_0}{2p}x - \frac{x_0^2}{2p} + y_0$$

Now if in this last equation we replace x_0^2 by $4py_0$, we obtain $y = \dfrac{x_0}{2p}x - y_0$, as

required.

15. **(a)** Let the coordinates of P be (a, b), so that $a^2 = 4pb$. Then according to Exercise 40 of Section 9.1, the coordinates of Q are $\left(-\dfrac{4p^2}{a}, \dfrac{p^2}{b} \right)$. The slope of the tangent at P is $\dfrac{a}{2p}$. The slope of the tangent at Q is $\dfrac{-\dfrac{4p^2}{a}}{2p} = -\dfrac{2p}{a}$. Thus the two slopes are negative reciprocals, from which it follows that the tangents are perpendicular to one another.

(b) The equation of the tangent at P is $y = \dfrac{a}{2p} x - b$. The equation of the tangent at Q is found to be $y = -\dfrac{2p}{a} x - \dfrac{p^2}{b}$. By solving this system of two equations, we find that the coordinates of the point of intersection are $x = \dfrac{a(b-p)}{2b}$ and $y = -p$. Since the y-value here is $-p$, we conclude that the two tangents intersect at a point on the directrix.

(c) The coordinates of the intersection point D are given in the solution to part (b). Using those coordinates, the slope of the line from D to the focus is found to be $\dfrac{4bp}{a(p-b)}$. On the other hand, the slope of $\overline{PQ}$ is found to be $\dfrac{a(b-p)}{4bp}$. (**Note:** In computing the slope of $\overline{PQ}$, the relationship $a^2 = 4pb$ helps in simplifying the expressions that arise.) Now by inspection we see that the two expressions for slope are negative reciprocals, thus the lines are perpendicular.

17. Summarizing our calculations, we have:

Equation of normal: $y - y_0 = -\dfrac{y_0}{2p}\left(x - x_0 \right)$

Coordinates of A: $\left(2p + x_0, 0 \right)$

Equation of line $\overline{AZ}$: $y = \dfrac{p - x_0}{y_0}\left(x - 2p - x_0 \right)$

Equation of line $\overline{PF}$: $y = \dfrac{y_0}{x_0 - p}(x - p)$

Coordinates of Z: $x = \dfrac{2p^3 + p^2 x_0 + x_0^3}{\left(x_0 + p \right)^2} = \dfrac{2p^2 - px_0 + x_0^2}{x_0 + p}, \; y = \dfrac{y_0\left(x_0 - p \right)}{x_0 + p}$

With these coordinates for Z, the distance ZP can be determined by using the distance formula. The result is $ZP = 2p$, as indicated in the problem.

19. (a)   As indicated in Exercise 14, the x-intercept of the tangent line is $-x_0$. Therefore $FA = x_0 + p$. Next, according to the definition of a parabola, FP is equal to the distance from P to the directrix. But the distance from P to the directrix is $x_0 + p$. Consequently $FP = FA$ because both equal $x_0 + p$.

(b)   Since $AF = FP$, $\triangle AFP$ is isosceles, and consequently $\alpha = \gamma$. Also, since HP is parallel to the x-axis, we have $\beta = \gamma$.

(c)   We have $\alpha = \gamma = \beta$ and therefore $\alpha = \beta$, as required.

21.   The equation of the tangent line is $y = 2x - 4$. With $x = 3$ and $y = 2$, the equation becomes $2 = 2(3) - 4$, or $2 = 2$. Thus the point $(3, 2)$ lies on the line. The equation through $F(0, 1)$ and $P(4, 4)$ is $y = \frac{3}{4}x + 1$. The equation of the line containing AB is $y = -\frac{4}{3}x + 6$. By solving these last two equations for x and y, we find that the coordinates of B are $\left(\frac{12}{5}, \frac{14}{5}\right)$. The distance formula can now be used to find FB as well as FC. In both cases the result is 3. Thus $FB = FC$.

13.4 The Ellipse

1.   Dividing by 36, the standard form is $\dfrac{x^2}{9} + \dfrac{y^2}{4} = 1$. The length of the major axis is 6 and the length of the minor axis is 4. Using $c^2 = a^2 - b^2$, we find:
$$c = \sqrt{a^2 - b^2} = \sqrt{9 - 4} = \sqrt{5}$$
So the foci are $\left(\pm\sqrt{5}, 0\right)$ and the eccentricity is $\dfrac{\sqrt{5}}{3}$.

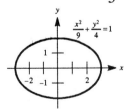

3. Dividing by 16, the standard form is $\dfrac{x^2}{16}+\dfrac{y^2}{1}=1$. The length of the major axis is 8 and the length of the minor axis is 2. Using $c^2 = a^2 - b^2$, we find:

$$c = \sqrt{a^2 - b^2} = \sqrt{16-1} = \sqrt{15}$$

So the foci are $\left(\pm\sqrt{15},0\right)$ and the eccentricity is $\dfrac{\sqrt{15}}{4}$.

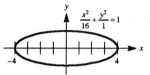

5. Dividing by 2, the standard form is $\dfrac{x^2}{2}+\dfrac{y^2}{1}=1$. The length of the major axis is $2\sqrt{2}$ and the length of the minor axis is 2. Using $c^2 = a^2 - b^2$, we find:

$$c = \sqrt{a^2 - b^2} = \sqrt{2-1} = 1$$

So the foci are $(\pm 1, 0)$ and the eccentricity is $\dfrac{\sqrt{2}}{2}$.

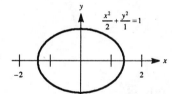

7. Dividing by 144, the standard form is $\dfrac{x^2}{9}+\dfrac{y^2}{16}=1$. The length of the major axis is 8 and the length of the minor axis is 6. Using $c^2 = a^2 - b^2$, we find:

$$c = \sqrt{a^2 - b^2} = \sqrt{16-9} = \sqrt{7}$$

So the foci are $\left(0,\pm\sqrt{7}\right)$ and the eccentricity is $\dfrac{\sqrt{7}}{4}$.

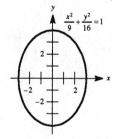

9.   Dividing by 5, the standard form is $\dfrac{x^2}{1/3}+\dfrac{y^2}{5/3}=1$. The lengths of the major and minor axes are given by:

major axis: $2\sqrt{\tfrac{5}{3}}=\dfrac{2\sqrt{5}}{\sqrt{3}}=\dfrac{2\sqrt{15}}{3}$

minor axis: $2\sqrt{\tfrac{1}{3}}=\dfrac{2}{\sqrt{3}}=\dfrac{2\sqrt{3}}{3}$

Using $c^2=a^2-b^2$, we find:

$$c=\sqrt{\tfrac{5}{3}-\tfrac{1}{3}}=\sqrt{\tfrac{4}{3}}=\dfrac{2}{\sqrt{3}}=\dfrac{2\sqrt{3}}{3}$$

So the foci are $\left(0,\pm\dfrac{2\sqrt{3}}{3}\right)$. The eccentricity is given by:

$$e=\dfrac{\frac{2\sqrt{3}}{3}}{\frac{\sqrt{15}}{3}}=\dfrac{2\sqrt{3}}{\sqrt{15}}=\dfrac{2}{\sqrt{5}}=\dfrac{2\sqrt{5}}{5}$$

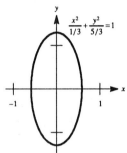

11.   Dividing by 4, the standard form is $\dfrac{x^2}{2}+\dfrac{y^2}{4}=1$. The length of the major axis is 4 and the length of the minor axis is $2\sqrt{2}$. Using $c^2=a^2-b^2$, we find:

$$c=\sqrt{a^2-b^2}=\sqrt{4-2}=\sqrt{2}$$

So the foci are $\left(0,\pm\sqrt{2}\right)$ and the eccentricity is $\dfrac{\sqrt{2}}{2}$.

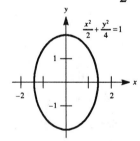

13. The equation is already in standard form with a center of $(5, -1)$. The length of the major axis is 10 and the length of the minor axis is 6. Using $c^2 = a^2 - b^2$, we find:

$$c = \sqrt{a^2 - b^2} = \sqrt{25 - 9} = \sqrt{16} = 4$$

So the foci are $(5 + 4, -1) = (9, -1)$ and $(5 - 4, -1) = (1, -1)$, and the eccentricity is $\frac{4}{5}$.

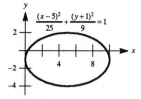

15. The equation is already in standard form with a center of $(1, 2)$. The length of the major axis is 4 and the length of the minor axis is 2. Using $c^2 = a^2 - b^2$, we find:

$$c = \sqrt{a^2 - b^2} = \sqrt{4 - 1} = \sqrt{3}$$

So the foci are $\left(1, 2 \pm \sqrt{3}\right)$ and the eccentricity is $\dfrac{\sqrt{3}}{2}$.

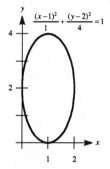

17. The equation is already in standard form with a center of $(-3, 0)$. The length of the major axis is 6 and the length of the minor axis is 2. Using $c^2 = a^2 - b^2$, we find:

$$c = \sqrt{a^2 - b^2} = \sqrt{9 - 1} = \sqrt{8} = 2\sqrt{2}$$

So the foci are $\left(-3 \pm 2\sqrt{2}, 0\right)$ and the eccentricity is $\dfrac{2\sqrt{2}}{3}$.

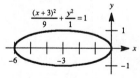

19. Complete the square to convert the equation to standard form:

$$3x^2 + 4y^2 - 6x + 16y + 7 = 0$$

$$3(x^2 - 2x) + 4(y^2 + 4y) = -7$$

$$3(x^2 - 2x + 1) + 4(y^2 + 4y + 4) = -7 + 3 + 16$$

$$3(x-1)^2 + 4(y+2)^2 = 12$$

$$\frac{(x-1)^2}{4} + \frac{(y+2)^2}{3} = 1$$

The center is (1, –2), the length of the major axis is 4, and the length of the minor axis is $2\sqrt{3}$. Using $c^2 = a^2 - b^2$, we find:

$$c = \sqrt{a^2 - b^2} = \sqrt{4-3} = 1$$

So the foci are (1 + 1, –2) = (2, –2) and (1 – 1, –2) = (0, –2), and the eccentricity is $\frac{1}{2}$.

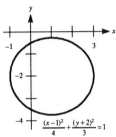

21. Complete the square to convert the equation to standard form:

$$5x^2 + 3y^2 - 40x - 36y + 188 = 0$$

$$5(x^2 - 8x) + 3(y^2 - 12y) = -188$$

$$5(x^2 - 8x + 16) + 3(y^2 - 12y + 36) = -188 + 80 + 108$$

$$5(x-4)^2 + 3(y-6)^2 = 0$$

Notice that the only solution to this is the center (4, 6). This is called a degenerate ellipse, or, more commonly, a point!

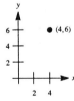

23. Complete the square to convert the equation to standard form:

$$16x^2 + 25y^2 - 64x - 100y + 564 = 0$$

$$16(x^2 - 4x) + 25(y^2 - 4y) = -564$$

$$16(x^2 - 4x + 4) + 25(y^2 - 4y + 4) = -564 + 64 + 100$$

$$16(x-2)^2 + 25(y-2)^2 = -400$$

Notice that there is no solution to this equation, since the left-hand side is non-negative. So there is no graph.

25. We are given $c = 3$ and $a = 5$, so we have the equation in the form:
$$\frac{x^2}{5^2} + \frac{y^2}{b^2} = 1$$
Since $c^2 = a^2 - b^2$, we find b:
$$9 = 25 - b^2$$
$$b^2 = 16$$
$$b = 4$$
So the equation is $\dfrac{x^2}{25} + \dfrac{y^2}{16} = 1$, or $16x^2 + 25y^2 = 400$.

27. We are given $a = 4$, so the equation has a form of:
$$\frac{x^2}{16} + \frac{y^2}{b^2} = 1$$
Now $\frac{c}{a} = \frac{1}{4}$, so $\frac{c}{4} = \frac{1}{4}$ and thus $c = 1$.
We find b:
$$c^2 = a^2 - b^2$$
$$1 = 16 - b^2$$
$$b^2 = 15$$
So the equation is $\dfrac{x^2}{16} + \dfrac{y^2}{15} = 1$, or $15x^2 + 16y^2 = 240$.

29. We have $c = 2$ and $a = 5$, so the equation has a form of:
$$\frac{x^2}{b^2} + \frac{y^2}{25} = 1$$
Find b:
$$c^2 = a^2 - b^2$$
$$4 = 25 - b^2$$
$$b^2 = 21$$
So the equation is $\dfrac{x^2}{21} + \dfrac{y^2}{25} = 1$, or $25x^2 + 21y^2 = 525$.

31. We know $a = 2b$ and that the equation has a form of:
$$\frac{x^2}{a^2} + \frac{y^2}{b^2} = 1$$
Using the point $\left(1, \sqrt{2}\right)$ and $a = 2b$, we have:
$$\frac{(1)^2}{(2b)^2} + \frac{\left(\sqrt{2}\right)^2}{b^2} = 1$$
$$\frac{1}{4b^2} + \frac{2}{b^2} = 1$$

Multiply by $4b^2$:
$$1 + 8 = 4b^2$$
$$b^2 = \tfrac{9}{4}$$
$$b = \tfrac{3}{2}$$

Since $a = 2b$, then $a = 3$. So the equation is $\dfrac{x^2}{9} + \dfrac{y^2}{9/4} = 1$, or $x^2 + 4y^2 = 9$.

33. (a) Using the equation $\dfrac{x_1 x}{a^2} + \dfrac{y_1 y}{b^2} = 1$, where $(x_1, y_1) = (8,2)$, $a^2 = 76$, $b^2 = \tfrac{76}{3}$, we have:

$$\frac{8x}{76} + \frac{2y}{76/3} = 1$$
$$8x + 6y = 76$$
$$6y = -8x + 76$$
$$y = -\tfrac{4}{3}x + \tfrac{38}{3}$$

(b) Here $(x_1, y_1) = (-7,3)$, $a^2 = 76$, $b^2 = \tfrac{76}{3}$:

$$-\frac{7x}{76} + \frac{3y}{76/3} = 1$$
$$-7x + 9y = 76$$
$$9y = 7x + 76$$
$$y = \tfrac{7}{9}x + \tfrac{76}{9}$$

(c) Here $(x_1, y_1) = (1,-5)$, $a^2 = 76$, $b^2 = \tfrac{76}{3}$:

$$\frac{x}{76} - \frac{5y}{76/3} = 1$$
$$x - 15y = 76$$
$$-15y = -x + 76$$
$$y = \tfrac{1}{15}x - \tfrac{76}{15}$$

35. (a) Using the tangent formula where $(x_1, y_1) = (4,2)$, $a^2 = \tfrac{52}{3}$, $b^2 = 52$:

$$\frac{4x}{52/3} + \frac{2y}{52} = 1$$
$$12x + 2y = 52$$
$$2y = -12x + 52$$
$$y = -6x + 26$$

(b)   The y-intercept is 26 and the x-intercept is $\frac{13}{3}$, so the area is:

$$\tfrac{1}{2}(\text{base})(\text{height}) = \tfrac{1}{2} \cdot \tfrac{13}{3} \cdot 26 = \tfrac{169}{3}$$

37.   (a)   Substituting the point, we have:

$$9x^2 + 25y^2 = 9(1)^2 + 25\left(\tfrac{6\sqrt{6}}{5}\right)^2 = 9 + 216 = 225$$

(b)   Here $\left(x_1, y\right) = \left(1, \tfrac{6\sqrt{6}}{5}\right)$, $a^2 = 25$, $b^2 = 9$:

$$\frac{x}{25} + \frac{\dfrac{6\sqrt{6}}{5}y}{9} = 1$$
$$9x + 30\sqrt{6}y = 225$$
$$3x + 10\sqrt{6}y - 75 = 0$$

(c)   We have $c^2 = a^2 - b^2 = 25 - 9 = 16$, so $c = 4$. Thus the coordinates of F_1 and F_2 are $F_1(-4,0)$ and $F_2(4,0)$. Now find each distance:

$$d_1 = \frac{|-12 + 0 - 75|}{\sqrt{9 + 600}} = \frac{87}{\sqrt{609}}$$
$$d_2 = \frac{|12 + 0 - 75|}{\sqrt{9 + 600}} = \frac{63}{\sqrt{609}}$$

(d)   Compute $d_1 d_2 = \dfrac{87}{\sqrt{609}} \cdot \dfrac{63}{\sqrt{609}} = \dfrac{5481}{609} = 9$, which is b^2.

39.   Begin by solving for y:

$$\frac{x^2}{1} + \frac{y^2}{16} = 1$$
$$\frac{y^2}{16} = 1 - x^2$$
$$y^2 = 16\left(1 - x^2\right)$$
$$y = \pm\sqrt{16\left(1 - x^2\right)} = \pm 4\sqrt{1 - x^2}$$

Use a calculator to complete the table:

x	0	0.1	0.2	0.3	0.4	0.5	0.6	0.7	0.8	0.9	1.0
y	±4	±3.98	±3.92	±3.82	±3.67	±3.46	±3.2	±2.86	±2.4	±1.74	0

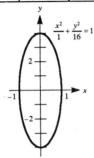

41. From Figure 4, we have $F_1P + F_2P > F_1F_2$. But by definition, $F_1P + F_2P = 2a$, while $F_1F_2 = 2c$. Thus $2a > 2c$, or $a > c$, so $a^2 > c^2$ (since a and c are positive), and consequently $a^2 - c^2 > 0$.

43. The tangent at (x_1, y_1) has equation $\dfrac{x_1 x}{a^2} + \dfrac{y_1 y}{b^2} = 1$, which has slope $\dfrac{-x_1 b^2}{y_1 a^2}$. Thus the

normal has slope $\dfrac{y_1 a^2}{x_1 b^2}$ and equation:

$$y - y_1 = \frac{y_1 a^2}{x_1 b^2}(x - x_1)$$
$$b^2 x_1 y - b^2 x_1 y_1 = y_1 a^2 x - a^2 x_1 y_1$$
$$a^2 y_1 x - b^2 x_1 y = (a^2 - b^2) x_1 y_1$$

45. If we first multiply each equation by $a^2 b^2$, we have:
$$b^2 x^2 + a^2 y^2 = a^2 b^2$$
$$a^2 x^2 + b^2 y^2 = a^2 b^2$$
Multiplying the first equation by $-a^2$ and the second equation by b^2 yields:
$$-a^2 b^2 x^2 - a^4 y^2 = -a^4 b^2$$
$$a^2 b^2 x^2 + b^4 y^2 = a^2 b^4$$
Adding, we have:
$$(b^4 - a^4) y^2 = a^2 b^2 (b^2 - a^2)$$
$$y^2 = \frac{a^2 b^2}{b^2 + a^2}$$
$$y = \pm \frac{ab}{\sqrt{a^2 + b^2}}$$

Substituting into the first equation:

$$b^2x^2 + \frac{a^4b^2}{a^2+b^2} = a^2b^2$$

$$b^2x^2 = \frac{a^2b^4}{a^2+b^2}$$

$$x^2 = \frac{a^2b^2}{a^2+b^2}$$

$$x = \pm\frac{ab}{\sqrt{a^2+b^2}}$$

So there are four intersection points:

$$\left(\frac{ab}{A},\frac{ab}{A}\right), \left(\frac{ab}{A},-\frac{ab}{A}\right), \left(-\frac{ab}{A},\frac{ab}{A}\right), \left(-\frac{ab}{A},-\frac{ab}{A}\right), \text{ where } A = \sqrt{a^2+b^2}$$

Sketch the intersection:

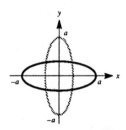

47. Using Exercise 43, we find that N is the point $\left(\frac{(a^2-b^2)x_1}{a^2},0\right)$. Thus:

$$FN = \frac{(a^2-b^2)x_1}{a^2}+c = \frac{c^2}{a^2}x_1+c = e^2x_1+c$$

It can be shown using $\frac{x_1^2}{a^2}+\frac{y_1^2}{b^2}=1$ that $y_1 = a^2-c^2-x_1^2+e^2x_1^2$. Now:

$$FP = \sqrt{(x_1+c)^2+y_1^2}$$
$$= \sqrt{x_1^2+2x_1c+c^2+a^2-c^2-x_1^2+e^2x_1^2}$$
$$= \sqrt{e^2x_1^2+2aex_1+a^2}$$
$$= ex_1+a$$

Thus:

$$\frac{FN}{FP} = \frac{e^2x_1+c}{ex_1+a} = \frac{e(ex_1+a)}{ex_1+a} = e$$

49. (a) Substitute the points:
$$5^2+3(1)^2 = 25+3 = 28$$
$$4^2+3(-2)^2 = 16+12 = 28$$
$$(-1)^2+3(3)^2 = 1+27 = 28$$

(b) $\overline{AB}$ has slope $\dfrac{-2-1}{4-5} = 3$. The line containing C parallel to $\overline{AB}$ has equation:

$$y - 3 = 3(x + 1)$$
$$y = 3x + 6$$

The intersections of this line with the ellipse have x-coordinates such that:

$$x^2 + 3(3x + 6)^2 = 28$$
$$x^2 + 27x^2 + 108x + 108 = 28$$
$$7x^2 + 27x + 20 = 0$$
$$(7x + 20)(x + 1) = 0$$
$$x = -\tfrac{20}{7}, -1$$

Thus the point D we want is $\left(-\tfrac{20}{7}, -\tfrac{18}{7}\right)$.

(c) The point O is $(0, 0)$. Use the suggested formula:

Area of $\triangle OAC = \tfrac{1}{2}\left|0 - 0 + 5 \bullet 3 - (-1)(1) + 0 - 0\right| = 8$

Area of $\triangle OBD = \tfrac{1}{2}\left|0 - 0 + 4\left(-\tfrac{18}{7}\right) - (-2)\left(-\tfrac{20}{7}\right) + 0 - 0\right| = 8$

Thus the areas are equal.

51. (a) We have $a\sqrt{(x + c)^2 + y^2} = a^2 + xc$. Dividing by a and noting that $e = \tfrac{c}{a}$:

$$\sqrt{(x + c)^2 + y^2} = a + xe$$

But the radical is $F_1 P$ by the distance formula, so:

$$F_1 P = a + xe$$

(b) We have:

$$F_1 P + F_2 P = 2a$$
$$(a + xe) + F_2 P = 2a$$
$$F_2 P = a - xe$$

53. (a) Use the tangent formula, and multiply by $a^2 b^2$:

$$\dfrac{x_1 x}{a^2} + \dfrac{y_1 y}{b^2} = 1$$
$$b^2 x_1 x + a^2 y_1 y = a^2 b^2$$

(b) Since (h, k) lies on this line, replacing (x, y) with (h, k) results in:

$$b^2 x_1 h + a^2 y_1 k = a^2 b^2$$

(c) Repeating part (a), we have $b^2 x_2 x + a^2 y_2 y = a^2 b^2$. Now substitute (h, k) to obtain

$$b^2 x_2 h + a^2 y_2 k = a^2 b^2.$$

(d) Replacing (x, y) with (x_1, y_1) and (x_2, y_2), respectively, results in the equations we have proved in (b) and (c). Thus this line must pass through the points (x_1, y_1) and (x_2, y_2).

55. By Exercise 51(b), $F_2 P = a - ex$ and $PR = \frac{a}{e} - x$. Thus:

$$\frac{F_2 P}{PR} = \frac{a - ex}{\frac{a}{e} - x} \cdot \frac{e}{e} = \frac{e(a - ex)}{a - ex} = e$$

57. Since $(0, 0)$ is the center, then $D = \frac{a}{e}$ and $d = \frac{a}{e} - c$, thus:

$$\frac{D}{d} = \frac{\frac{a}{e}}{\frac{a}{e} - c} = \frac{a}{a - ce} = \frac{a}{a - c \cdot \left(\frac{c}{a}\right)} = \frac{a^2}{a^2 - c^2} = \frac{a^2}{b^2}$$

59. (a) Substituting the point, we see that:

$$\frac{c^2}{a^2} + \frac{\left(b^2/a\right)^2}{b^2} = \frac{a^2 - b^2}{a^2} + \frac{b^2}{a^2} = 1$$

So $\left(c, \frac{b^2}{2a}\right)$ lies on the ellipse.

(b) The equation of the tangent is:

$$\frac{cx}{a^2} + \frac{\frac{b^2}{a} \bullet y}{b^2} = 1$$

$$\frac{cx}{a^2} + \frac{y}{a} = 1$$

The x-coordinate of the point of intersection is $\frac{a}{e}$, and the y-coordinate is the solution of:

$$\frac{c(a/e)}{a^2} + \frac{y}{a} = 1$$

$$\frac{a^2}{a^2} + \frac{y}{a} = 1$$

$$y = 0$$

Thus, the required point is $\left(\frac{a}{e}, 0\right)$.

61. Let P have coordinates (x_1, y_1) and Q have coordinates (x_1, y_2). The center of the circle is $(0, 0)$, and the slope of the normal through P is thus $\dfrac{y_1 - 0}{x_1 - 0} = \dfrac{y_1}{x_1}$. Thus the slope of the tangent through P is $-\dfrac{x_1}{y_1}$, and its equation is:

$$y - y_1 = -\frac{x_1}{y_1}(x - x_1)$$
$$yy_1 - y_1^2 = -x_1 x + x_1^2$$
$$x_1 x = -y_1 y + x_1^2 + y_1^2$$
$$x_1 x = -y_1 y + a^2$$

The tangent of the ellipse through Q has equation $\dfrac{x_1 x}{a^2} + \dfrac{y_2 y}{b^2} = 1$. Thus the y-coordinate of the point of intersection is the solution of:

$$\frac{-y_1 y + a^2}{a^2} + \frac{y_2 y}{b^2} = 1$$
$$\left(\frac{y_2}{b^2} - \frac{y_1}{a^2}\right) y = 0$$

If $\dfrac{y_2}{b^2} - \dfrac{y_1}{a^2} = 0$ then any y-value corresponds to a point of intersection, so the lines are the same. So the points are the same, contradicting our hypothesis that the points are distinct. Otherwise, $y = 0$, as required. Note: We implicitly assumed $x_1 \neq 0$ earlier, when assuming $x_1 x \neq 0$. Note that the lines do not intersect if $x_1 = 0$.

63. Let P have coordinates (x_1, y_1). Then by Exercise 62(a), Q has coordinates $(-x_1, -y_1)$. The tangent through P has equation:

$$\frac{x_1 x}{a^2} + \frac{y_1 y}{b^2} = 1$$
$$y = \frac{b^2}{y_1}\left(\frac{-x_1 x}{a^2} + 1\right)$$

So its slope is $-\dfrac{b^2 x_1}{y_1 a^2}$. The tangent through Q has equation:

$$-\frac{x_1 x}{a^2} + \frac{-y_1 y}{b^2} = 1$$
$$y = \frac{b^2}{y_1}\left(\frac{x_1 x}{a^2} + 1\right)$$

So its slope is $\dfrac{b^2 x_1}{y_1 a^2}$, the same as that of the other tangent. Thus the two tangents are parallel, as required.

65. **(a)** Using the formula $d = \dfrac{\left| Ax_0 + By_0 + C \right|}{\sqrt{A^2 + B^2}}$, we have:

$$F_1 A = \frac{\left| -c\left(\dfrac{x_1}{a^2}\right) + 0 - 1 \right|}{\sqrt{A^2 + B^2}} = \frac{\left| \dfrac{x_1 c}{a^2} + 1 \right|}{\sqrt{A^2 + B^2}},$$

where $A = \dfrac{x_1}{a^2}$ and $B = \dfrac{y_1}{b^2}$.

(b) Making the substitution $c = ae$ and $a + ex_1 = F_1 P$, we have:

$$F_1 A = \frac{\left| \dfrac{x_1 e}{a} + 1 \right|}{\sqrt{A^2 + B^2}} = \frac{\left| x_1 e + a \right|}{a\sqrt{A^2 + B^2}} = \frac{F_1 P}{a\sqrt{A^2 + B^2}}$$

(c) $F_2 B$ is the distance from the point $(c, 0)$ to $\dfrac{x_1 x}{a^2} + \dfrac{y_1 y}{b^2} = 1$, so:

$$F_2 B = \frac{\left| c\left(\dfrac{x_1}{a^2}\right) + 0 - 1 \right|}{\sqrt{A^2 + B^2}} = \frac{\left| \dfrac{x_1 c}{a^2} - 1 \right|}{\sqrt{A^2 + B^2}}$$

Now making the substitutions $c = ae$ and $a - ex_1 = F_2 P$, we have:

$$F_2 B = \frac{\left| \dfrac{x_1 e}{a} - 1 \right|}{\sqrt{A^2 + B^2}} = \frac{\left| x_1 e - a \right|}{a\sqrt{A^2 + B^2}} = \frac{\left| a - x_1 e \right|}{a\sqrt{A^2 + B^2}} = \frac{F_2 P}{a\sqrt{A^2 + B^2}}$$

(d) We evaluate each quotient:

$$\frac{F_1 A}{F_1 P} = \frac{1}{a\sqrt{A^2 + B^2}} \qquad \text{and} \qquad \frac{F_2 B}{F_2 P} = \frac{1}{a\sqrt{A^2 + B^2}}$$

Thus they are equal.

Graphing Utility Exercises for Section 13.4

1. **(a)** Solving for y, we obtain:

$$\frac{x^2}{9} + \frac{y^2}{4} = 1$$
$$4x^2 + 9y^2 = 36$$
$$9y^2 = 36 - 4x^2$$
$$y^2 = \frac{36 - 4x^2}{9}$$
$$y = \pm\tfrac{1}{3}\sqrt{36 - 4x^2}$$

Graphing the two functions:

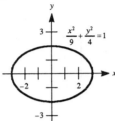

The graph appears to have symmetry with respect to the x-axis, the y-axis, and the origin.

(b) Solving for y, we obtain:

$$\frac{x^2}{1} + \frac{y^2}{16} = 1$$
$$16x^2 + y^2 = 16$$
$$y^2 = 16 - 16x^2$$
$$y = \pm 4\sqrt{1 - x^2}$$

Graphing the two functions:

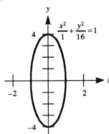

Again, the graph appears to have symmetry with respect to the x-axis, the y-axis, and the origin.

3. (a) Dividing by $16b^2$, we have:

$$\frac{x^2}{16} + \frac{y^2}{b^2} = 1$$

So the horizontal axis has length $2(4) = 8$.

(b) When $b^2 = 16$, $c^2 = 16 - 16 = 0$, so $c = 0$ and thus:

$$e = \frac{c}{a} = \frac{0}{4} = 0$$

When $b^2 = 13.44$, $c^2 = 16 - 13.44 = 2.56$, so $c = 1.6$ and thus:

$$e = \frac{c}{a} = \frac{1.6}{4} = 0.4$$

When $b^2 = 10.24$, $c^2 = 16 - 10.24 = 5.76$, so $c = 2.4$ and thus:

$$e = \frac{c}{a} = \frac{2.4}{4} = 0.6$$

When $b^2 = 5.76$, $c^2 = 16 - 5.76 = 10.24$, so $c = 3.2$ and thus:
$$e = \frac{c}{a} = \frac{3.2}{4} = 0.8$$

(c) Solving for y, we have:
$$b^2x^2 + 16y^2 = 16b^2$$
$$16y^2 = 16b^2 - b^2x^2$$
$$y^2 = \frac{b^2(16 - x^2)}{16}$$
$$y = \pm 0.25\sqrt{b^2(16 - x^2)}$$

Using the indicated settings, the graphs of all four ellipses appear as:

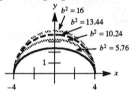

5. Dividing the ellipse equation by 12, we have:
$$\frac{x^2}{4} + \frac{y^2}{12} = 1$$

So $a^2 = 4$ and $b^2 = 12$. Using $(x_0, y_0) = (1, -3)$ in the tangent line formula yields:
$$\frac{1x}{4} + \frac{-3y}{12} = 1$$
$$3x - 3y = 12$$
$$x - y = 4$$
$$y = x - 4$$

Now graph the ellipse $\dfrac{x^2}{4} + \dfrac{y^2}{12} = 1$ and tangent line $y = x - 4$:

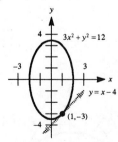

7. (a) Dividing by 12, graph the ellipse $\dfrac{x^2}{12} + \dfrac{y^2}{4} = 1$:

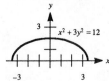

(b) Since $a^2 = 12$ and $b^2 = 4$, $a = 2\sqrt{3}$ and $b = 2$. Therefore:
$$c^2 = a^2 - b^2 = 12 - 4 = 8, \text{ so } c = \sqrt{8} = 2\sqrt{2}$$

(c) The auxiliary circle has an equation $x^2 + y^2 = 12$. Graphing the ellipse and auxiliary circle:

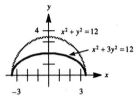

(d) Verify that the point $P(3, 1)$ lies on the ellipse:
$$(3)^2 + 3(1)^2 = 9 + 3 = 12$$
Using $(x_0, y_0) = (3, 1)$, find the equation of the tangent line to the ellipse at P:
$$\frac{3x}{12} + \frac{1y}{4} = 1$$
$$\frac{x}{4} + \frac{y}{4} = 1$$
$$x + y = 4$$
Now graph the upper halves of the ellipse and circle, and the line $y = -x + 4$:

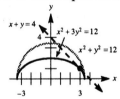

(e) The perpendicular to the tangent at $P(3, 1)$ has slope $m = 1$. Using the point $(-2\sqrt{2}, 0)$ and the point-slope formula:
$$y - 0 = 1(x + 2\sqrt{2})$$
$$y = x + 2\sqrt{2}$$
Now graph the upper halves of the ellipse and circle, as well as the lines $y = -x + 4$ and $y = x + 2\sqrt{2}$:

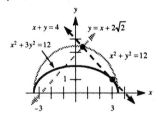

13.5 The Hyperbola

1. Dividing by 4, the standard form is $\dfrac{x^2}{4} - \dfrac{y^2}{1} = 1$. The vertices are $(\pm 2, 0)$, the length of the transverse axis is 4, the length of the conjugate axis is 2, and the asymptotes are $y = \pm\frac{1}{2}x$. Using $c^2 = a^2 + b^2$, we find:

$$c = \sqrt{a^2 + b^2} = \sqrt{4 + 1} = \sqrt{5}$$

The foci are $\left(\pm\sqrt{5}, 0\right)$ and the eccentricity is $\dfrac{\sqrt{5}}{2}$.

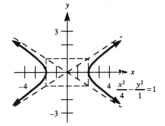

3. Dividing by 4, the standard form is $\dfrac{y^2}{4} - \dfrac{x^2}{1} = 1$. The vertices are $(0, \pm 2)$, the length of the transverse axis is 4, the length of the conjugate axis is 2, and the asymptotes are $y = \pm 2x$. Using $c^2 = a^2 + b^2$, we find:

$$c = \sqrt{a^2 + b^2} = \sqrt{4 + 1} = \sqrt{5}$$

The foci are $\left(0, \pm\sqrt{5}\right)$ and the eccentricity is $\dfrac{\sqrt{5}}{2}$.

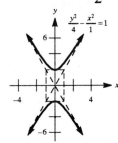

5.   Dividing by 400, the standard form is $\dfrac{x^2}{25} - \dfrac{y^2}{16} = 1$. The vertices are $(\pm 5, 0)$, the length of the transverse axis is 10, the length of the conjugate axis is 8, and the asymptotes are $y = \pm \frac{4}{5}x$. Using $c^2 = a^2 + b^2$, we find:

$$c = \sqrt{a^2 + b^2} = \sqrt{25 + 16} = \sqrt{41}$$

The foci are $\left(\pm\sqrt{41}, 0\right)$ and the eccentricity is $\dfrac{\sqrt{41}}{5}$.

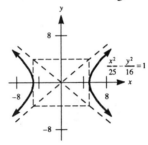

7.   Rewriting the equation, the standard form is $\dfrac{y^2}{1/2} - \dfrac{x^2}{1/3} = 1$. The vertices are $\left(0, \pm\sqrt{\tfrac{1}{2}}\right) = \left(0, \pm\tfrac{\sqrt{2}}{2}\right)$, the length of the transverse axis is $\sqrt{2}$, the length of the conjugate axis is $2\sqrt{\tfrac{1}{3}} = \tfrac{2\sqrt{3}}{3}$, and the asymptotes are $y = \pm\sqrt{\tfrac{3}{2}}x = \pm\tfrac{\sqrt{6}}{2}x$. Using $c^2 = a^2 + b^2$, we find:

$$c = \sqrt{a^2 + b^2} = \sqrt{\tfrac{1}{2} + \tfrac{1}{3}} = \sqrt{\tfrac{5}{6}} = \tfrac{\sqrt{5}}{\sqrt{6}} = \tfrac{\sqrt{30}}{6}$$

The foci are $\left(0, \pm\tfrac{\sqrt{30}}{6}\right)$ and the eccentricity is:

$$e = \dfrac{\frac{\sqrt{30}}{6}}{\frac{\sqrt{2}}{2}} = \dfrac{\sqrt{30}}{3\sqrt{2}} = \dfrac{\sqrt{15}}{3}$$

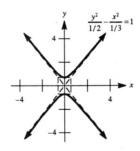

9. Dividing by 100, the standard form is $\dfrac{y^2}{25} - \dfrac{x^2}{4} = 1$. The vertices are $(0, \pm 5)$, the length of the transverse axis is 10, the length of the conjugate axis is 4, and the asymptotes are $y = \pm \frac{5}{2}x$. Using $c^2 = a^2 + b^2$, we have:

$$c = \sqrt{a^2 + b^2} = \sqrt{25 + 4} = \sqrt{29}$$

The foci are $\left(0, \pm \sqrt{29}\right)$ and the eccentricity is $\dfrac{\sqrt{29}}{5}$.

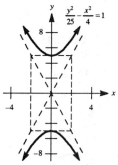

11. The equation is already in standard form with a center of $(5, -1)$. The vertices are $(5 + 5, -1) = (10, -1)$ and $(5 - 5, -1) = (0, -1)$, the length of the transverse axis is 10, and the length of the conjugate axis is 6. The asymptotes have slopes of $\pm \frac{3}{5}$, so using the point-slope formula:

$$y + 1 = \tfrac{3}{5}(x - 5) \qquad\qquad y + 1 = -\tfrac{3}{5}(x - 5)$$
$$y + 1 = \tfrac{3}{5}x - 3 \qquad\qquad y + 1 = -\tfrac{3}{5}x + 3$$
$$y = \tfrac{3}{5}x - 4 \qquad\qquad y = -\tfrac{3}{5}x + 2$$

Using $c^2 = a^2 + b^2$, we have:

$$c = \sqrt{a^2 + b^2} = \sqrt{25 + 9} = \sqrt{34}$$

The foci are $\left(5 \pm \sqrt{34}, -1\right)$ and the eccentricity is $\dfrac{\sqrt{34}}{5}$.

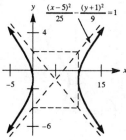

13.   The equation is already in standard form with a center of (1, 2). The vertices are
(1, 2 + 2) = (1, 4) and (1, 2 – 2) = (1, 0), the length of the transverse axis is 4, and the
length of the conjugate axis is 2. The asymptotes have slopes of ±2, so using the
point-slope formula:

$$y - 2 = 2(x - 1) \qquad\qquad y - 2 = -2(x - 1)$$
$$y - 2 = 2x - 2 \qquad\qquad y - 2 = -2x + 2$$
$$y = 2x \qquad\qquad\qquad y = -2x + 4$$

Using $c^2 = a^2 + b^2$, we have:

$$c = \sqrt{a^2 + b^2} = \sqrt{4 + 1} = \sqrt{5}$$

The foci are $\left(1, 2 \pm \sqrt{5}\right)$ and the eccentricity is $\dfrac{\sqrt{5}}{2}$.

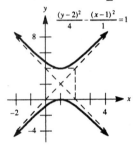

15.   The equation is already in standard form with a center of (–3, 4). The vertices are
(–3 + 4, 4) = (1, 4) and (–3 – 4, 4) = (–7, 4), the length of the transverse axis is 8, and the
length of the conjugate axis is 8. The asymptotes have slopes of ±1, so using the
point-slope formula:

$$y - 4 = 1(x + 3) \qquad\qquad y - 4 = -1(x + 3)$$
$$y - 4 = x + 3 \qquad\qquad y - 4 = -x - 3$$
$$y = x + 7 \qquad\qquad\qquad y = -x + 1$$

Using $c^2 = a^2 + b^2$, we have:

$$c = \sqrt{a^2 + b^2} = \sqrt{16 + 16} = \sqrt{32} = 4\sqrt{2}$$

The foci are $\left(-3 \pm 4\sqrt{2}, 4\right)$ and the eccentricity is $\dfrac{4\sqrt{2}}{4} = \sqrt{2}$.

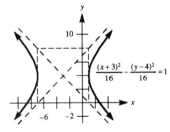

17. Complete the square to convert the equation to standard form:

$$x^2 - y^2 + 2y - 5 = 0$$
$$x^2 - \left(y^2 - 2y\right) = 5$$
$$x^2 - \left(y^2 - 2y + 1\right) = 5 - 1$$
$$x^2 - (y-1)^2 = 4$$
$$\frac{x^2}{4} - \frac{(y-1)^2}{4} = 1$$

The center is (0, 1), the vertices are (0 + 2, 1) = (2, 1) and (0 − 2, 1) = (−2, 1), and the lengths of both the transverse and conjugate axes are 4. The asymptotes have slopes of ±1, so using the point-slope formula:

$y - 1 = 1(x - 0)$	$y - 1 = -1(x - 0)$
$y - 1 = x$	$y - 1 = -x$
$y = x + 1$	$y = -x + 1$

Using $c^2 = a^2 + b^2$, we have:

$$c = \sqrt{a^2 + b^2} = \sqrt{4 + 4} = \sqrt{8} = 2\sqrt{2}$$

The foci are $\left(\pm 2\sqrt{2}, 1\right)$ and the eccentricity is $\dfrac{2\sqrt{2}}{2} = \sqrt{2}$.

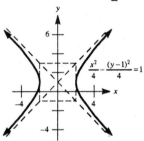

19. Complete the square to convert the equation to standard form:

$$x^2 - y^2 - 4x + 2y - 6 = 0$$
$$\left(x^2 - 4x\right) - \left(y^2 - 2y\right) = 6$$
$$\left(x^2 - 4x + 4\right) - \left(y^2 - 2y + 1\right) = 6 + 4 - 1$$
$$(x-2)^2 - (y-1)^2 = 9$$
$$\frac{(x-2)^2}{9} - \frac{(y-1)^2}{9} = 1$$

The center is (2, 1), the vertices are (2 + 3, 1) = (5, 1) and (2 − 3, 1) = (−1, 1), and the lengths of both the transverse and conjugate axes are 6. The asymptotes have slopes of ±1, so using the point-slope formula:

$y - 1 = 1(x - 2)$	$y - 1 = -1(x - 2)$
$y - 1 = x - 2$	$y - 1 = -x + 2$
$y = x - 1$	$y = -x + 3$

Using $c^2 = a^2 + b^2$, we have:
$$c = \sqrt{a^2 + b^2} = \sqrt{9+9} = \sqrt{18} = 3\sqrt{2}$$

The foci are $\left(2 \pm 3\sqrt{2}, 1\right)$ and the eccentricity is $\dfrac{3\sqrt{2}}{3} = \sqrt{2}$.

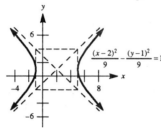

21.   Complete the square to convert the equation to standard form:
$$y^2 - 25x^2 + 8y - 9 = 0$$
$$\left(y^2 + 8y\right) - 25x^2 = 9$$
$$\left(y^2 + 8y + 16\right) - 25x^2 = 9 + 16$$
$$\left(y+4\right)^2 - 25x^2 = 25$$
$$\frac{\left(y+4\right)^2}{25} - \frac{x^2}{1} = 1$$

The center is $(0, -4)$, the vertices are $(0, -4+5) = (0, 1)$ and $(0, -4-5) = (0, -9)$, the length of the transverse axis is 10, and the length of the conjugate axis is 2. The asymptotes have slopes of ± 5, so using the point-slope formula:

$$y + 4 = 5(x - 0) \qquad\qquad y + 4 = -5(x - 0)$$
$$y + 4 = 5x \qquad\qquad\qquad y + 4 = -5x$$
$$y = 5x - 4 \qquad\qquad\qquad y = -5x - 4$$

Using $c^2 = a^2 + b^2$, we have:
$$c = \sqrt{a^2 + b^2} = \sqrt{25+1} = \sqrt{26}$$

The foci are $\left(0, -4 \pm \sqrt{26}\right)$ and the eccentricity is $\dfrac{\sqrt{26}}{5}$.

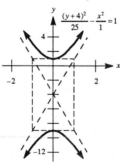

23. Complete the square to convert the equation to standard form:
$$x^2 + 7x - y^2 - y + 12 = 0$$
$$(x^2 + 7x) - (y^2 + y) = -12$$
$$\left(x^2 + 7x + \tfrac{49}{4}\right) - \left(y^2 + y + \tfrac{1}{4}\right) = -12 + \tfrac{49}{4} - \tfrac{1}{4}$$
$$\left(x + \tfrac{7}{2}\right)^2 - \left(y + \tfrac{1}{2}\right)^2 = 0$$

Notice that this is a degenerate hyperbola, and the graph consists of the "would-be" asymptotes with slopes ±1:

$$y + \tfrac{1}{2} = 1\left(x + \tfrac{7}{2}\right) \qquad\qquad y + \tfrac{1}{2} = -1\left(x + \tfrac{7}{2}\right)$$
$$y + \tfrac{1}{2} = x + \tfrac{7}{2} \qquad\qquad\quad y + \tfrac{1}{2} = -x - \tfrac{7}{2}$$
$$y = x + 3 \qquad\qquad\qquad\quad y = -x - 4$$

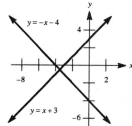

25. Since $P(x, y)$ lies on $\dfrac{x^2}{4} - \dfrac{y^2}{1} = 1$, we can find y in terms of x:

$$y^2 = \frac{x^2}{4} - 1 = \frac{x^2 - 4}{4}$$

Taking roots, we have $y = \dfrac{\sqrt{x^2 - 4}}{2}$, since $P(x, y)$ lies in the first quadrant. So the

coordinates of P are $\left(x, \dfrac{\sqrt{x^2 - 4}}{2}\right)$. Since $Q(x, y)$ lies in the first quadrant on the

asymptote, find the equation of the asymptote:
$$y - 0 = \tfrac{1}{2}(x - 0)$$
$$y = \tfrac{1}{2}x$$

So the coordinates of Q are $\left(x, \tfrac{1}{2}x\right)$. Since P and Q have the same x-coordinate PQ is the difference between their y-coordinates:

$$PQ = \frac{x}{2} - \frac{\sqrt{x^2 - 4}}{2} = \frac{x - \sqrt{x^2 - 4}}{2}$$

The order of subtraction is because the asymptote lies above the hyperbola in the first

quadrant, and thus $\dfrac{x}{2}$ is larger than $\dfrac{\sqrt{x^2 - 4}}{2}$. This proves the desired result.

27. Since the foci are $(\pm 4, 0)$ and the vertices are $(\pm 1, 0)$, then $c = 4$, $a = 1$, and the hyperbola has the form:

$$\frac{x^2}{1} - \frac{y^2}{b^2} = 1$$

Find b:

$$c^2 = a^2 + b^2$$
$$16 = 1 + b^2$$
$$b^2 = 15$$

So the equation is $\dfrac{x^2}{1} - \dfrac{y^2}{25} = 1$, or $15x^2 - y^2 = 15$.

29. The slope of the asymptotes is $\pm\frac{1}{2}$, which tells us that the ratio $\frac{b}{a} = \frac{1}{2}$ in this hyperbola. Also, since the vertices are $(\pm 2, 0)$ then $a = 2$. The required ratio is therefore:

$$\frac{b}{2} = \frac{1}{2} \text{ and } b = 1$$

The equation is $\dfrac{x^2}{4} - \dfrac{y^2}{1} = 1$, or $x^2 - 4y^2 = 4$.

31. Since the asymptotes are $y = \pm\frac{\sqrt{10}}{5}x$, then $\frac{b}{a} = \frac{\sqrt{10}}{5}$, so $b = \frac{\sqrt{10}}{5}a$. Now, since the foci are $\left(\pm\sqrt{7}, 0\right)$, then $c = \sqrt{7}$ and the hyperbola has the form:

$$\frac{x^2}{a^2} - \frac{y^2}{b^2} = 1$$

Since $b = \frac{\sqrt{10}}{5}a$ and $c = \sqrt{7}$, we have:

$$c^2 = a^2 + b^2$$
$$7 = a^2 + \left(\frac{\sqrt{10}}{5}a\right)^2$$
$$7 = a^2 + \frac{2}{5}a^2$$
$$7a^2 = 35$$
$$a^2 = 5$$
$$b^2 = \frac{2}{5}a^2 = \frac{2}{5}(5) = 2$$

So the equation is $\dfrac{x^2}{5} - \dfrac{y^2}{2} = 1$, or $2x^2 - 5y^2 = 10$.

33. The vertices are at $(0, \pm 7)$ so we know it is a "vertical" hyperbola. Its equation will be $\dfrac{y^2}{49} - \dfrac{x^2}{b^2} = 1$, but we also know that $(1, 9)$ is a point satisfying the equation. Use it to find b:

$$\frac{81}{49} - \frac{1}{b^2} = 1$$
$$81b^2 - 49 = 49b^2$$
$$32b^2 = 49$$
$$b^2 = \frac{49}{32}$$

So the equation is $\dfrac{y^2}{49} - \dfrac{x^2}{49/32} = 1$, or $y^2 - 32x^2 = 49$.

35. We have $2a = 6$, so $a = 3$. Also $2b = 2$, so $b = 1$. Since the foci are on the y-axis, the hyperbola will have the form:

$$\frac{y^2}{a^2} - \frac{x^2}{b^2} = 1$$

So the equation is $\dfrac{y^2}{9} - \dfrac{x^2}{1} = 1$, or $y^2 - 9x^2 = 9$.

37. Writing the equation as $\dfrac{x^2}{16} - \dfrac{y^2}{16} = 1$, we have $a = b = 4$. So the slopes of the asymptotes are $\pm \dfrac{b}{a} = \pm \dfrac{4}{4} = \pm 1$. But these are negative reciprocals of each other, so the asymptotes are perpendicular to each other.

39. (a) Substituting $P(5, 6)$ into $5y^2 - 4x^2 = 80$, we have:
$$5(6)^2 - 4(5)^2 = 5(36) - 4(25) = 180 - 100 = 80$$
So $P(5, 6)$ lies on the hyperbola.

(b) Dividing by 80 yields $\dfrac{y^2}{16} - \dfrac{x^2}{20} = 1$, so $a = 4$ and $b = 2\sqrt{5}$. Using $c^2 = a^2 + b^2$, we have:
$$c = \sqrt{a^2 + b^2} = \sqrt{16 + 20} = \sqrt{36} = 6$$
So $c = 6$ and the foci are $(0, \pm 6)$.

(c) Compute the distances:
$$F_1 P = \sqrt{(5-0)^2 + (6-6)^2} = 5$$
$$F_2 P = \sqrt{(5-0)^2 + (6-(-6))^2} = \sqrt{25 + 144} = 13$$

(d) Verify the result:
$$\left| F_1 P - F_2 P \right| = |5 - 13| = |-8| = 8 = 2(4) = 2a$$

41. (a) Since the asymptotes must have slopes of $\pm\frac{b}{a}$, then:

$$-\frac{b}{a} = \frac{-1}{\frac{b}{a}}$$

$$\frac{b^2}{a^2} = 1$$

$$b^2 = a^2$$

$$b = a$$

Now, since $c^2 = a^2 + b^2 = 2a^2$, then the eccentricity is:

$$\frac{c}{a} = \frac{\sqrt{2a^2}}{a} = \frac{\sqrt{2}a}{a} = \sqrt{2}$$

(b) The slopes of the asymptotes are $\pm\frac{a}{a} = \pm1$, so the hyperbola will have perpendicular asymptotes. The eccentricity is $\dfrac{\sqrt{2}a}{a} = \sqrt{2}$.

43. (a) This equation is just $F_1P - F_2P = 2a$, the defining relation of a hyperbola. Since P is on the right-hand branch, $F_1P > F_2P$.

(b) Squaring each side of the equation, we obtain:

$$\sqrt{(x+c)^2 + y^2} = 2a + \sqrt{(x-c)^2 + y^2}$$

$$(x+c)^2 + y^2 = 4a^2 + 4a\sqrt{(x-c)^2 + y^2} + (x-c)^2 + y^2$$

$$4xc = 4a^2 + 4a\sqrt{(x-c)^2 + y^2}$$

$$xc - a^2 = a\sqrt{(x-c)^2 + y^2}$$

$$xc - a^2 = a(F_2P)$$

(c) Dividing by a, we have:

$$\frac{xc}{a} - a = F_2P$$

$$xe - a = F_2P$$

45. For this hyperbola we find $a^2 = b^2 = k^2$, and $e = \sqrt{2}$.
Also $d^2 = x^2 + y^2 = x^2 + (x^2 - k^2) = 2x^2 - k^2$. Thus we want to show that
$F_1P \cdot F_2P = 2x^2 - k^2$. Using the formulas for F_1P and F_2P developed in Exercises 43 and 44, we have:

$$F_1P \cdot F_2P = (xe+a)(xe-a) = x^2e^2 - a^2 = x^2(2) - k^2 = 2x^2 - k^2 = d^2$$

47. The coordinates of D are $\left(\frac{a}{e}, \frac{b}{e}\right)$ and those of F are $(c, 0)$. Let O denote the center $(0, 0)$.

We show that $\angle ODF$ is a right angle by showing $OD^2 + DF^2 = OF^2$:

$$OF^2 = c^2$$

$$OD^2 = \left(\frac{a}{e}\right)^2 + \left(\frac{b}{e}\right)^2 = \frac{a^2 + b^2}{e^2} = \frac{c^2}{e^2} = a^2$$

$$DF^2 = \left(\frac{a}{e} - c\right)^2 + \left(\frac{b}{e}\right)^2$$

$$= \frac{(a - ec)^2 + b^2}{e^2}$$

$$= \frac{a^2 + b^2 - 2aec + e^2c^2}{e^2}$$

$$= \frac{c^2 - 2c^2}{e^2} + c^2$$

$$= -\frac{c^2}{e^2} + c^2$$

$$= -a^2 + c^2$$

Thus $OD^2 + DF^2 = OF^2$, as required.

Note: An alternate approach here is to show that the slope of $\overline{FD}$ is $-\frac{a}{b}$:

$$\text{slope } = \frac{\frac{b}{e}}{\frac{a}{e} - c} = \frac{b}{a - ce} = \frac{b}{a - \frac{c^2}{a}} = \frac{ab}{a^2 - c^2} = \frac{ab}{-b^2} = -\frac{a}{b}$$

Since the asymptote has a slope of $\frac{b}{a}$, the lines are perpendicular.

49. Draw the sketch:

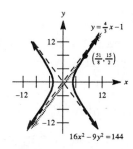

$y = \frac{4}{3}x - 1$

$\left(\frac{51}{8}, \frac{15}{2}\right)$

$16x^2 - 9y^2 = 144$

By substitution, we have:

$$16x^2 - 9\left(\tfrac{4}{3}x - 1\right)^2 = 144$$

$$16x^2 - 9\left(\tfrac{16}{9}x^2 - \tfrac{8}{3}x + 1\right) = 144$$

$$16x^2 - 16x^2 + 24x - 9 = 144$$

$$24x = 153$$

$$x = \tfrac{51}{8}$$

$$y = \tfrac{4}{3} \cdot \tfrac{51}{8} - 1 = \tfrac{17}{2} - 1 = \tfrac{15}{2}$$

Therefore $\left(\tfrac{51}{8}, \tfrac{15}{2}\right)$ is the intersection point.

51.   We have $(x_1, y_1) = (4,6)$, $a^2 = 4$, $b^2 = 12$:

$$\frac{4x}{4} - \frac{6y}{12} = 1$$

$$x - \frac{y}{2} = 1$$

$$2x - y = 2$$

$$-y = -2x + 2$$

$$y = 2x - 2$$

53.   The equation $\dfrac{x^2}{a^2} - \dfrac{y^2}{b^2} = 1$ is equivalent to $b^2x^2 - a^2y^2 = a^2b^2$ (1). So (x_1, y_1) is on the hyperbola if and only if:

$$b^2x_1^2 - a^2y_1^2 = a^2b^2$$

Thus if (x_1, y_1) is on the hyperbola then the equation of the hyperbola can be written as:

$$b^2\left(x^2 - x_1^2\right) - a^2\left(y^2 - y_1^2\right) = 0$$

Let $y - y_1 = m(x - x_1)$ be the equation of the tangent to the hyperbola through (x_1, y_1). Then as in Exercise 54 of Section 9.3 the system

$$b^2\left(x^2 - x_1^2\right) - a^2\left(y^2 - y_1^2\right) = 0$$

$$y - y_1 = m(x - x_1)$$

has exactly one solution, (x_1, y_1). Substutituting $y = m(x - x_1) + y_1$ into the first equation, we have:

$$b^2\left(x^2 - x_1^2\right) - a^2\left[\left(m(x - x_1) + y_1\right)^2 - y_1^2\right] = 0$$

$$b^2\left(x^2 - x_1^2\right) - a^2m^2\left(x - x_1\right)^2 - 2a^2my_1\left(x - x_1\right) = 0$$

$$\left(x - x_1\right)\left[b^2\left(x + x_1\right) - a^2m^2\left(x - x_1\right) - 2a^2my_1\right] = 0$$

Since x_1 is the only solution, it is a double root, so:

$$b^2(x_1 + x_1) - a^2 m^2(x_1 - x_1) - 2a^2 m y_1 = 0$$
$$2a^2 m y_1 = 2b^2 x_1$$
$$m = \frac{b^2 x_1}{a^2 y_1}$$

Thus the tangent line has equation:

$$y - y_1 = \frac{b^2 x_1}{a^2 y_1}(x - x_1)$$
$$a^2 y_1 y - a^2 y_1^2 = b^2 x_1 x - b^2 x_1^2$$
$$b^2 x_1 x - a^2 y_1 y = b^2 x_1 - a^2 y_1^2$$
$$b^2 x_1 x - a^2 y_1 y = a^2 b^2$$
$$\frac{x_1 x}{a^2} - \frac{y_1 y}{b^2} = 1$$

Note that $m = \dfrac{b^2 x_1}{a^2 y_1}$ cannot be the slope $\pm\dfrac{b}{a}$ of the asymtote if (x_1, y_1) lies on the hyperbola, since if so then:

$$\pm\frac{b}{a} = \frac{b^2 x_1}{a^2 y_1}$$

$$\pm\frac{a}{b} y_1 = x_1$$

Therefore:

$$\frac{\dfrac{a^2}{b^2} \bullet y_1^2}{a^2} - \frac{y_1^2}{b^2} = 0 \neq 1$$

55. By the law of sines:

$$\frac{F_2 P}{\sin\theta} = \frac{F_2 A}{\sin\alpha}$$
$$\frac{F_2 P}{F_2 A} = \frac{\sin\theta}{\sin\alpha}$$

Also by the law of sines:

$$\frac{F_1 P}{\sin(180° - \theta)} = \frac{F_1 A}{\sin\beta}$$
$$\frac{F_1 P}{F_1 A} = \frac{\sin\theta}{\sin\beta}$$

Thus if $\dfrac{F_2 P}{F_2 A} = \dfrac{F_1 P}{F_1 A}$, then $\dfrac{\sin\theta}{\sin\alpha} = \dfrac{\sin\theta}{\sin\beta}$, so $\sin\alpha = \sin\beta$. By construction α and β are acute, so $\alpha = \beta$.

57. We know the slope of the tangent line at P is $\dfrac{b^2 x_1}{a^2 y_1}$, so the slope of the normal

is $\dfrac{-a^2 y_1}{b^2 x_1}$ and its equation is:

$$y - y_1 = \frac{-a^2 y_1}{b^2 x_1}(x - x_1)$$

$$b^2 x_1 y - b^2 x_1 y_1 = -a^2 y_1 x + a^2 y_1 x_1$$

$$a^2 y_1 x + b^2 x_1 y = x_1 y_1 (a^2 + b^2)$$

59. The equation of the normal line is $a^2 y_1 x + b^2 x_1 y = x_1 y_1(a^2 + b^2)$. Letting $y = 0$ we see

that the x-intercept of the line is $x = \dfrac{x_1(a^2 + b^2)}{a^2}$. Similarily, the y-intercept is found to

be $y = \dfrac{y_1(a^2 + b^2)}{b^2}$. Thus the coordinates of Q and R are $\left(\dfrac{x_1(a^2 + b^2)}{a^2}, 0 \right)$ and

$\left(0, \dfrac{y_1(a^2 + b^2)}{b^2} \right)$. The area of $\triangle OQR$ is :

$$\tfrac{1}{2} OQ \cdot OR = \frac{x_1 y_1 (a^2 + b^2)^2}{2 a^2 b^2}$$

61. The equation of the tangent line at the point P with coordinates (x_1, y_1) is

$\dfrac{x_1 x}{a^2} - \dfrac{y_1 y}{b^2} = 1$. The asymtotes are $y = \pm \dfrac{b}{a} x$. Suppose the tangent intersects $y = \dfrac{b}{a} x$ at A

and $y = -\dfrac{b}{a} x$ at B. The x-coordinate of A is found by solving:

$$\frac{x_1 x}{a^2} - \frac{y_1 \cdot \dfrac{b}{a} x}{b^2} = 1$$

$$b x_1 x - a y_1 x = b a^2$$

$$x = \frac{b a^2}{b x_1 - a y_1}$$

So $y = \dfrac{b}{a} x = \dfrac{b^2 a}{b x_1 - a y_1}$. So A is the point $\left(\dfrac{b a^2}{b x_1 - a y_1}, \dfrac{b^2 a}{b x_1 - a y_1} \right)$. The point B is

similarly found to be $\left(\dfrac{b a^2}{b x_1 + a y_1}, \dfrac{b^2 a}{b x_1 + a y_1} \right)$. The midpoint of $\overline{AB}$ has x-coordinate:

$$\frac{1}{2}\left(\frac{b a^2}{b x_1 - a y_1} + \frac{b a^2}{b x_1 - a y_1} \right) = \frac{1}{2}\left(\frac{b a^2 (2 b x_1)}{b^2 x_1^2 - a^2 y_1^2} \right) = \frac{b^2 a^2 x_1}{b^2 x_1^2 - a^2 y_1^2}$$

Since $b^2 x_1^2 - a^2 y_1^2 = a^2 b^2$, this is equal to:

$$\frac{b^2 a^2 x_1}{a^2 b^2} = x_1$$

Similarly the y-coordinate of the midpoint is found to be y_1, so P is the midpoint of $\overline{AB}$ as required.

Graphing Utility Exercises for Section 13.5

1.   (a)   Solving for y, we have:

$$16 x^2 - 9 y^2 = 144$$
$$16 x^2 - 144 = 9 y^2$$
$$y^2 = \frac{16 x^2 - 144}{9}$$
$$y = \pm \tfrac{1}{3} \sqrt{16 x^2 - 144}$$

Graphing the two functions we obtain the hyperbola:

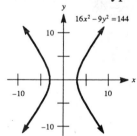

(b)   Dividing by 144, the standard form of the hyperbola is $\dfrac{x^2}{9} - \dfrac{y^2}{16} = 1$, so $a = 3$ and $b = 4$. Thus the asymptotes are $y = \pm\frac{4}{3}x$. Now graph the hyperbola and two asymptotes:

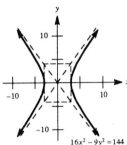

(c)   Using a larger viewing rectangle, we obtain the graphs:

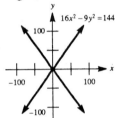

3.   (a)   Using the quadratic formula to solve $-2y^2 + 4xy + \left(x^2 - 6\right) = 0$:

$$
\begin{aligned}
y &= \frac{-4x \pm \sqrt{(4x)^2 - 4(-2)\left(x^2 - 6\right)}}{2(-2)} \\
&= \frac{-4x \pm \sqrt{16x^2 + 8x^2 - 48}}{-4} \\
&= \frac{-4x \pm \sqrt{24x^2 - 48}}{-4} \\
&= \frac{-4x \pm 2\sqrt{6x^2 - 12}}{-4} \\
&= x \pm \tfrac{1}{2}\sqrt{6x^2 - 12}
\end{aligned}
$$

(b)   Graphing these two functions in the standard viewing rectangle:

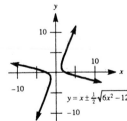

(c)   Adding the graphs of the asymptotes in the standard viewing rectangle:

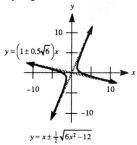

$$y = \left(1 \pm 0.5\sqrt{6}\right)x$$

$$y = x \pm \tfrac{1}{2}\sqrt{6x^2 - 12}$$

(d)   Expanding the viewing rectangle:

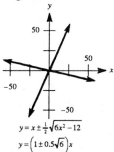

$$y = x \pm \tfrac{1}{2}\sqrt{6x^2 - 12}$$

$$y = \left(1 \pm 0.5\sqrt{6}\right)x$$

Note that the curve is virtually identical to the asymptotes.

13.6 The Focus-Directrix Property of Conics

1.   In order to use the formulas for the focal radii, we must find a and e. Dividing by 76, the

standard form is $\dfrac{x^2}{76} + \dfrac{y^2}{76/3} = 1$. So $a = \sqrt{76} = 2\sqrt{19}$. To find the eccentricity, first

find c:

$$c = \sqrt{a^2 - b^2} = \sqrt{76 - \tfrac{76}{3}} = \sqrt{\tfrac{152}{3}} = \frac{2\sqrt{38}}{\sqrt{3}} = \frac{2\sqrt{114}}{3}$$

The eccentricity is:

$$e = \frac{c}{a} = \frac{\frac{2\sqrt{114}}{3}}{2\sqrt{19}} = \frac{\sqrt{6}}{3}$$

The focal radii are given by:

$$F_1 P = a + ex = 2\sqrt{19} + \frac{\sqrt{6}}{3}(-8) = 2\sqrt{19} - \frac{8\sqrt{6}}{3} = \frac{6\sqrt{19} - 8\sqrt{6}}{3}$$

$$F_2 P = a - ex = 2\sqrt{19} - \frac{\sqrt{6}}{3}(-8) = 2\sqrt{19} + \frac{8\sqrt{6}}{3} = \frac{6\sqrt{19} + 8\sqrt{6}}{3}$$

3.   The equation is already in standard form with $a = 15$ and $b = 5$. Finding c:
$$c = \sqrt{a^2 - b^2} = \sqrt{225 - 25} = \sqrt{200} = 10\sqrt{2}$$
The eccentricity is:
$$e = \frac{c}{a} = \frac{10\sqrt{2}}{15} = \frac{2\sqrt{2}}{3}$$
The focal radii are given by:
$$F_1P = a + ex = 15 + \frac{2\sqrt{2}}{3}(9) = 15 + 6\sqrt{2}$$
$$F_2P = a - ex = 15 - \frac{2\sqrt{2}}{3}(9) = 15 - 6\sqrt{2}$$

5.   (a)   The ellipse is already in standard form with $a = 4$ and $b = 3$. Finding c:
$$c = \sqrt{a^2 - b^2} = \sqrt{16 - 9} = \sqrt{7}$$
The foci are $(\pm\sqrt{7}, 0)$ and the eccentricity is $\frac{\sqrt{7}}{4}$. The directrices are given by:
$$x = \pm\frac{a}{e} = \pm\frac{4}{\frac{\sqrt{7}}{4}} = \pm\frac{16}{\sqrt{7}} = \pm\frac{16\sqrt{7}}{7}$$

   (b)   The hyperbola is already in standard form with $a = 4$ and $b = 3$. Finding c:
$$c = \sqrt{a^2 + b^2} = \sqrt{16 + 9} = \sqrt{25} = 5$$
The foci are $(\pm 5, 0)$ and the eccentricity is $\frac{5}{4}$. The directrices are given by:
$$x = \pm\frac{a}{e} = \pm\frac{4}{\frac{5}{4}} = \pm\frac{16}{5}$$

7.   (a)   Dividing by 156, the standard form for the ellipse is $\dfrac{x^2}{13} + \dfrac{y^2}{12} = 1$, and so $a = \sqrt{13}$
   and $b = \sqrt{12} = 2\sqrt{3}$. Finding c:
$$c = \sqrt{a^2 - b^2} = \sqrt{13 - 12} = \sqrt{1} = 1$$
The foci are $(\pm 1, 0)$ and the eccentricity is $\frac{1}{\sqrt{13}} = \frac{\sqrt{13}}{13}$. The directrices are given by:
$$x = \pm\frac{a}{e} = \pm\frac{\sqrt{13}}{\frac{\sqrt{13}}{13}} = \pm 13$$

   (b)   Dividing by 156, the standard form for the hyperbola is $\dfrac{x^2}{13} - \dfrac{y^2}{12} = 1$, and so $a = \sqrt{13}$
   and $b = \sqrt{12} = 2\sqrt{3}$. Finding c:
$$c = \sqrt{a^2 + b^2} = \sqrt{13 + 12} = \sqrt{25} = 5$$
The foci are $(\pm 5, 0)$ and the eccentricity is $\frac{5}{\sqrt{13}} = \frac{5\sqrt{13}}{13}$. The directrices are given by:
$$x = \pm\frac{a}{e} = \pm\frac{\sqrt{13}}{\frac{5\sqrt{13}}{13}} = \pm\frac{13}{5}$$

9. (a) Dividing by 900, the standard form for the ellipse is $\dfrac{x^2}{36} + \dfrac{y^2}{25} = 1$, and so $a = 6$ and $b = 5$. Finding c:
$$c = \sqrt{a^2 - b^2} = \sqrt{36 - 25} = \sqrt{11}$$
The foci are $(\pm\sqrt{11}, 0)$ and the eccentricity is $\dfrac{\sqrt{11}}{6}$. The directrices are given by:
$$x = \pm\frac{a}{e} = \pm\frac{6}{\frac{\sqrt{11}}{6}} = \pm\frac{36}{\sqrt{11}} = \pm\frac{36\sqrt{11}}{11}$$

(b) Dividing by 900, the standard form for the hyperbola is $\dfrac{x^2}{36} - \dfrac{y^2}{25} = 1$, and so $a = 6$ and $b = 5$. Finding c:
$$c = \sqrt{a^2 + b^2} = \sqrt{36 + 25} = \sqrt{61}$$
The foci are $(\pm\sqrt{61}, 0)$ and the eccentricity is $\dfrac{\sqrt{61}}{6}$. The directrices are given by:
$$x = \pm\frac{a}{e} = \pm\frac{6}{\frac{\sqrt{61}}{6}} = \pm\frac{36}{\sqrt{61}} = \pm\frac{36\sqrt{61}}{61}$$

11. Since the foci are $(\pm 1, 0)$, $c = 1$. Since the directrices are $x = \pm 4$, $\frac{a}{e} = 4$, so $a = 4e$. But since $e = \frac{c}{a} = \frac{1}{a}$, we have:
$$a = 4 \cdot \frac{1}{a}$$
$$a^2 = 4$$
Since $a^2 - b^2 = c^2$, we can find b^2:
$$2^2 - b^2 = 1^2$$
$$-b^2 = -3$$
$$b^2 = 3$$
The equation of the ellipse is $\dfrac{x^2}{4} + \dfrac{y^2}{3} = 1$, or $3x^2 + 4y^2 = 12$.

13. Since the foci are $(\pm 2, 0)$, $c = 2$. Since the directrices are $x = \pm 1$, $\frac{a}{e} = 1$, so $a = e$. But since $e = \frac{c}{a} = \frac{2}{a}$, we have:
$$a = \frac{2}{a}$$
$$a^2 = 2$$
Since $a^2 + b^2 = c^2$, we can find b^2:
$$2 + b^2 = 2^2$$
$$b^2 = 2$$
So the equation of the hyperbola is $\dfrac{x^2}{2} - \dfrac{y^2}{2} = 1$, or $x^2 - y^2 = 2$.

15. (a) By the distance formula:
$$d_1 = \sqrt{(x+c)^2 + (y-0)^2} = \sqrt{(x+c)^2 + y^2}$$
$$d_2 = \sqrt{(x-c)^2 + (y-0)^2} = \sqrt{(x-c)^2 + y^2}$$
Squaring:
$$d_1^2 = (x+c)^2 + y^2$$
$$d_2^2 = (x-c)^2 + y^2$$

(b) Working from the left-hand side:
$$d_1^2 - d_2^2 = (x+c)^2 - (x-c)^2 = x^2 + 2cx + c^2 - x^2 + 2cx - c^2 = 4cx$$

(c) Since d_1 and d_2 represent the distances from the foci to a point on the ellipse, $d_1 + d_2 = 2a$ by the definition of an ellipse.

(d) Factoring:
$$d_1^2 - d_2^2 = 4cx$$
$$\left(d_1 + d_2\right)\left(d_1 - d_2\right) = 4cx$$
$$2a\left(d_1 - d_2\right) = 4cx$$
$$d_1 - d_2 = \frac{2cx}{a}$$

(e) Adding the two equations:
$$2d_1 = 2a + \frac{2cx}{a}$$
$$d_1 = a + \frac{c}{a}x = a + ex$$

(f) Substituting the result from (e):
$$a + ex + d_2 = 2a$$
$$d_2 = a - ex$$

13.7 The Conics in Polar Coordinates

1. (a) Comparing the given equation with the four basic types, it appears this is the type associated with Figure 2. Divide both numerator and denominator by 3 to obtain:
$$r = \frac{2}{1 + \frac{2}{3}\cos\theta} = \frac{\frac{2}{3} \cdot 3}{1 + \frac{2}{3}\cos\theta}$$
Therefore $e = \frac{2}{3}$ and $d = 3$. Since $e < 1$, this confirms the given conic is an ellipse.

The eccentricity is $\frac{2}{3}$ and the directrix is $x = 3$. Computing the values of r when $\theta = 0, \frac{\pi}{2}, \pi$ and $\frac{3\pi}{2}$:

θ	0	$\frac{\pi}{2}$	π	$\frac{3\pi}{2}$
r	$\frac{6}{5}$	2	6	2

Since the major axis of this ellipse lies along the x-axis, the length of the major axis is:

$$2a = \frac{6}{5} + 6 = \frac{36}{5}, \text{ so } a = \frac{18}{5}$$

The endpoints of the major axis are at $\left(\frac{6}{5}, 0\right)$ and $(-6, 0)$, so the x-coordinate of the center is:

$$\frac{1}{2}\left(-6 + \frac{6}{5}\right) = \frac{1}{2}\left(-\frac{24}{5}\right) = -\frac{12}{5}$$

So the center is $\left(-\frac{12}{5}, 0\right)$. Finally, we calculate b:

$$b = a\sqrt{1 - e^2} = \frac{18}{5}\sqrt{1 - \frac{4}{9}} = \frac{18}{5}\sqrt{\frac{5}{9}} = \frac{18\sqrt{5}}{15} = \frac{6\sqrt{5}}{5}$$

So the endpoints of the minor axis are $\left(-\frac{12}{5}, \pm\frac{6\sqrt{5}}{5}\right)$. Graph the ellipse:

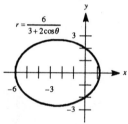

(b) Comparing the given equation with the four basic types, it appears this is the type associated with Figure 3. Divide both numerator and denominator by 3 to obtain:

$$r = \frac{2}{1 - \frac{2}{3}\cos\theta} = \frac{\frac{2}{3} \cdot 3}{1 - \frac{2}{3}\cos\theta}$$

Therefore $e = \frac{2}{3}$ and $d = 3$. Since $e < 1$, this confirms the given conic is an ellipse.

The eccentricity is $\frac{2}{3}$ and the directrix is $x = -3$. Computing the values of r when $\theta = 0, \frac{\pi}{2}, \pi$ and $\frac{3\pi}{2}$:

θ	0	$\frac{\pi}{2}$	π	$\frac{3\pi}{2}$
r	6	2	$\frac{6}{5}$	2

Since the major axis of this ellipse lies along the x-axis, the length of the major axis is:

$$2a = \frac{6}{5} + 6 = \frac{36}{5}, \text{ so } a = \frac{18}{5}$$

The endpoints of the major axis are at $(6, 0)$ and $\left(-\frac{6}{5}, 0\right)$, so the x-coordinate of the center is:

$$\tfrac{1}{2}\left(6 - \tfrac{6}{5}\right) = \tfrac{1}{2}\left(\tfrac{24}{5}\right) = \tfrac{12}{5}$$

So the center is $\left(\frac{12}{5}, 0\right)$. Finally, we calculate b:

$$b = a\sqrt{1 - e^2} = \tfrac{18}{5}\sqrt{1 - \tfrac{4}{9}} = \tfrac{18}{5}\sqrt{\tfrac{5}{9}} = \tfrac{18\sqrt{5}}{15} = \tfrac{6\sqrt{5}}{5}$$

So the endpoints of the minor axis are $\left(\frac{12}{5}, \pm\frac{6\sqrt{5}}{5}\right)$. Graph the ellipse:

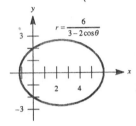

3.   (a)   Comparing the given equation with the four basic types, it appears this is the type associated with Figure 2. Divide both numerator and denominator by 2 to obtain:

$$r = \frac{\frac{5}{2}}{1 + \cos\theta} = \frac{1 \cdot \frac{5}{2}}{1 + \cos\theta}$$

Therefore $e = 1$ and $d = \frac{5}{2}$. Since $e = 1$, this confirms the given conic is a parabola. The directrix is $x = \frac{5}{2}$. Computing the value of r when $\theta = 0$ yields $r = \frac{5}{4}$, so the vertex is $\left(\frac{5}{4}, 0\right)$. Graph the parabola:

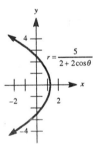

(b)   Comparing the given equation with the four basic types, it appears this is the type associated with Figure 3. Divide both numerator and denominator by 2 to obtain:

$$r = \frac{\frac{5}{2}}{1 - \cos\theta} = \frac{1 \cdot \frac{5}{2}}{1 - \cos\theta}$$

Therefore $e = 1$ and $d = \frac{5}{2}$. Since $e = 1$, this confirms the given conic is a parabola.

The directrix is $x = -\frac{5}{2}$. Computing the value of r when $\theta = \pi$ yields $r = \frac{5}{4}$, so the vertex is $\left(-\frac{5}{4}, 0\right)$. Graph the parabola:

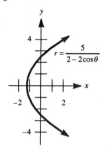

5.   **(a)**   Comparing the given equation with the four basic types, it appears this is the type associated with Figure 2. Divide both numerator and denominator by 2 to obtain:

$$r = \frac{3}{2 + 4\cos\theta} = \frac{\frac{3}{2}}{1 + 2\cos\theta} = \frac{2 \cdot \frac{3}{4}}{1 + 2\cos\theta}$$

Therefore $e = 2$ and $d = \frac{3}{4}$. Since $e > 1$, this confirms the given conic is a hyperbola.

The eccentricity is 2 and the directrix is $x = \frac{3}{4}$. Computing the values of r when $\theta = 0, \frac{\pi}{2}, \pi$ and $\frac{3\pi}{2}$:

θ	0	$\frac{\pi}{2}$	π	$\frac{3\pi}{2}$
r	$\frac{1}{2}$	$\frac{3}{2}$	$-\frac{3}{2}$	$\frac{3}{2}$

Since the two vertices $\left(\frac{1}{2}, 0\right)$ and $\left(\frac{3}{2}, 0\right)$ lie on the transverse axis, then:

$$2a = \frac{3}{2} - \frac{1}{2} = 1, \text{ so } a = \frac{1}{2}$$

The center of the hyperbola is the midpoint of these two vertices, which is $(1, 0)$. Since a focus is $(0, 0)$, then $c = 1$. Finally, we find b:

$$b = \sqrt{c^2 - a^2} = \sqrt{1 - \frac{1}{4}} = \sqrt{\frac{3}{4}} = \frac{1}{2}\sqrt{3}$$

Notice that we could also find b from the eccentricity:

$$b = a\sqrt{e^2 - 1} = \frac{1}{2}\sqrt{4 - 1} = \frac{1}{2}\sqrt{3}$$

Graph the hyperbola:

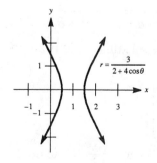

(b)   Comparing the given equation with the four basic types, it appears this is the type associated with Figure 3. Divide both numerator and denominator by 2 to obtain:

$$r = \frac{3}{2 - 4\cos\theta} = \frac{\frac{3}{2}}{1 - 2\cos\theta} = \frac{2 \cdot \frac{3}{4}}{1 - 2\cos\theta}$$

Therefore $e = 2$ and $d = \frac{3}{4}$. Since $e > 1$, this confirms the given conic is a hyperbola.

The eccentricity is 2 and the directrix is $x = -\frac{3}{4}$. Computing the values of r when

$\theta = 0, \frac{\pi}{2}, \pi$ and $\frac{3\pi}{2}$:

θ	0	$\frac{\pi}{2}$	π	$\frac{3\pi}{2}$
r	$-\frac{3}{2}$	$\frac{3}{2}$	$\frac{1}{2}$	$\frac{3}{2}$

Since the two vertices $\left(-\frac{3}{2}, 0\right)$ and $\left(-\frac{1}{2}, 0\right)$ lie on the transverse axis, then:

$$2a = -\frac{1}{2} + \frac{3}{2} = 1, \text{ so } a = \frac{1}{2}$$

The center of the hyperbola is the midpoint of these two vertices, which is $(-1, 0)$.
Since a focus is $(0, 0)$, then $c = 1$. Finally, we find b:

$$b = \sqrt{c^2 - a^2} = \sqrt{1 - \frac{1}{4}} = \sqrt{\frac{3}{4}} = \frac{1}{2}\sqrt{3}$$

Notice that we could also find b from the eccentricity:

$$b = a\sqrt{e^2 - 1} = \frac{1}{2}\sqrt{4 - 1} = \frac{1}{2}\sqrt{3}$$

Graph the hyperbola:

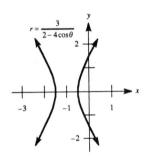

7.   Comparing the given equation with the four basic types, it appears this is the type associated with Figure 3. Divide both numerator and denominator by 2 to obtain:

$$r = \frac{24}{2 - 3\cos\theta} = \frac{12}{1 - \frac{3}{2}\cos\theta} = \frac{\frac{3}{2} \cdot 8}{1 - \frac{3}{2}\cos\theta}$$

Since the eccentricity is $e = \frac{3}{2} > 1$, this conic is a hyperbola. Computing the values of

r when $\theta = 0, \frac{\pi}{2}, \pi$ and $\frac{3\pi}{2}$:

θ	0	$\frac{\pi}{2}$	π	$\frac{3\pi}{2}$
r	-24	12	$\frac{24}{5}$	12

Since the two vertices $(-24, 0)$ and $\left(-\frac{24}{5}, 0\right)$ lie on the transverse axis, its length must be:

$$2a = -\frac{24}{5} + 24 = \frac{96}{5}, \text{ so } a = \frac{48}{5}$$

The center of the hyperbola is the midpoint of these two vertices, which is $\left(-\frac{72}{5}, 0\right)$.

Since a focus is $(0, 0)$, $c = \frac{72}{5}$ and thus:

$$b = \sqrt{c^2 - a^2} = \sqrt{\frac{5184}{25} - \frac{2304}{25}} = \sqrt{\frac{2880}{25}} = \frac{24\sqrt{5}}{5}$$

So the length of the conjugate axis is $2b = \frac{48\sqrt{5}}{5}$. Graph the hyperbola:

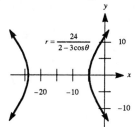

9. Comparing the given equation with the four basic types, it appears this is the type associated with Figure 4. Divide both numerator and denominator by 5 to obtain:

$$r = \frac{8}{5 + 3\sin\theta} = \frac{\frac{8}{5}}{1 + \frac{3}{5}\sin\theta} = \frac{\frac{3}{5} \cdot \frac{8}{3}}{1 + \frac{3}{5}\sin\theta}$$

Since the eccentricity is $e = \frac{3}{5} < 1$, this conic is an ellipse. Computing the values of r when $\theta = 0, \frac{\pi}{2}, \pi$ and $\frac{3\pi}{2}$:

θ	0	$\frac{\pi}{2}$	π	$\frac{3\pi}{2}$
r	$\frac{8}{5}$	1	$\frac{8}{5}$	4

Since the two vertices $(0, 1)$ and $(0, -4)$ lie on the major axis, its length must be:

$$2a = 1 + 4 = 5, \text{ so } a = \frac{5}{2}$$

The center of the ellipse is the midpoint of these two vertices, which is $\left(0, -\frac{3}{2}\right)$. Since a focus is $(0, 0)$, then $c = \frac{3}{2}$ and thus:

$$b = \sqrt{a^2 - c^2} = \sqrt{\frac{25}{4} - \frac{9}{4}} = \sqrt{\frac{16}{4}} = 2$$

So the length of the minor axis is $2b = 4$. Graph the ellipse:

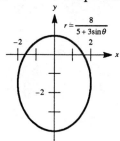

11. Comparing the given equation with the four basic types, it appears this is the type associated with Figure 5. Divide both numerator and denominator by 5 to obtain:

$$r = \frac{12}{5 - 5\sin\theta} = \frac{\frac{12}{5}}{1 - \sin\theta} = \frac{1 \cdot \frac{12}{5}}{1 - \sin\theta}$$

Since the eccentricity is $e = 1$, this conic is a parabola with directrix $y = -\frac{12}{5}$. Since the focus is $(0, 0)$, the vertex must be the midpoint of $(0, 0)$ and $\left(0, -\frac{12}{5}\right)$, which is $\left(0, -\frac{6}{5}\right)$. Graph the parabola:

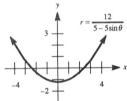

13. Comparing the given equation with the four basic types, it appears this is the type associated with Figure 2. Divide both numerator and denominator by 7 to obtain:

$$r = \frac{12}{7 + 5\cos\theta} = \frac{\frac{12}{7}}{1 + \frac{5}{7}\cos\theta} = \frac{\frac{5}{7} \cdot \frac{12}{5}}{1 + \frac{5}{7}\cos\theta}$$

Since the eccentricity is $e = \frac{5}{7} < 1$, this conic is an ellipse. Computing the values of r when $\theta = 0, \frac{\pi}{2}, \pi$ and $\frac{3\pi}{2}$:

θ	0	$\frac{\pi}{2}$	π	$\frac{3\pi}{2}$
r	1	$\frac{12}{7}$	6	$\frac{12}{7}$

Since the two vertices $(1, 0)$ and $(-6, 0)$ lie on the major axis, its length must be:

$$2a = 1 + 6 = 7, \text{ so } a = \frac{7}{2}$$

The center of the ellipse is the midpoint of these two vertices, which is $\left(-\frac{5}{2}, 0\right)$. Since a focus is $(0, 0)$, then $c = \frac{5}{2}$ and thus:

$$b = \sqrt{a^2 - c^2} = \sqrt{\frac{49}{4} - \frac{25}{4}} = \sqrt{6}$$

So the length of the minor axis is $2b = 2\sqrt{6}$. Graph the ellipse:

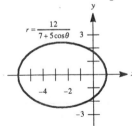

15. Comparing the given equation with the four basic types, it appears this is the type associated with Figure 4. Divide both numerator and denominator by 5 to obtain:

$$r = \frac{4}{5 + 5\sin\theta} = \frac{\frac{4}{5}}{1 + \sin\theta} = \frac{1 \cdot \frac{4}{5}}{1 + \sin\theta}$$

Since the eccentricity is $e = 1$, this conic is a parabola with directrix $y = \frac{4}{5}$. Since the focus is $(0, 0)$, the vertex must be the midpoint of $(0, 0)$ and $\left(0, \frac{4}{5}\right)$, which is $\left(0, \frac{2}{5}\right)$. Graph the parabola:

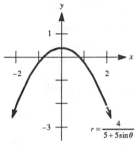

17. Comparing the given equation with the four basic types, it appears this is the type associated with Figure 3. The equation is already in standard form with eccentricity $e = 2 > 1$, so this conic is a hyperbola. Computing the values of r when $\theta = 0, \frac{\pi}{2}, \pi$ and $\frac{3\pi}{2}$:

θ	0	$\frac{\pi}{2}$	π	$\frac{3\pi}{2}$
r	-9	9	3	9

Since the two vertices $(-9, 0)$ and $(-3, 0)$ lie on the transverse axis, its length must be:

$$2a = -3 + 9 = 6, \text{ so } a = 3$$

The center of the hyperbola is the midpoint of these two vertices, which is $(-6, 0)$. Since a focus is $(0, 0)$, then $c = 6$ and thus:

$$b = \sqrt{c^2 - a^2} = \sqrt{36 - 9} = \sqrt{27} = 3\sqrt{3}$$

So the length of the conjugate axis is $2b = 6\sqrt{3}$. Graph the hyperbola:

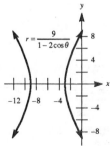

19. Since the coordinates of P are (r, θ), then the coordinates of Q are $(r, \theta + \pi)$. Now find FP and FQ, noting that $\cos(\theta + \pi) = -\cos\theta$.

$$FP = r = \frac{ed}{1 - e\cos\theta}$$

$$FQ = r = \frac{ed}{1 - e\cos(\theta + \pi)} = \frac{ed}{1 + e\cos\theta}$$

Therefore:

$$\frac{1}{FP} + \frac{1}{FQ} = \frac{1 - e\cos\theta}{ed} + \frac{1 + e\cos\theta}{ed} = \frac{2}{ed}$$

This is remarkable in that $\frac{2}{ed}$ is a constant, even though P is a variable point.

21. Draw a focal chord $\overline{AB}$, with A representing the endpoint on the "left". Since $\overline{AB}$ is a 90° rotation from $\overline{PQ}$, then A corresponds to the polar coordinates $A\left(r, \theta + \frac{\pi}{2}\right)$ and B corresponds to the coordinates $B\left(r, \theta + \frac{3\pi}{2}\right)$. Using the identities $\cos\left(\theta + \frac{\pi}{2}\right) = -\sin\theta$ and $\cos\left(\theta + \frac{3\pi}{2}\right) = \sin\theta$:

$$AF = r = \frac{ed}{1 - e\cos\left(\theta + \frac{\pi}{2}\right)} = \frac{ed}{1 + e\sin\theta}$$

$$FB = r = \frac{ed}{1 - e\cos\left(\theta + \frac{3\pi}{2}\right)} = \frac{ed}{1 - e\sin\theta}$$

Since $AB = AF + FB$:

$$AB = \frac{ed}{1 + e\sin\theta} + \frac{ed}{1 - e\sin\theta}$$

$$= \frac{ed(1 - e\sin\theta) + ed(1 + e\sin\theta)}{(1 + e\sin\theta)(1 - e\sin\theta)}$$

$$= \frac{ed - e^2 d\sin\theta + ed + e^2 d\sin\theta}{1 - e^2\sin^2\theta}$$

$$= \frac{2ed}{1 - e^2\sin^2\theta}$$

Using the result from Exercise 10, we show the required sum is constant:

$$\frac{1}{PQ} + \frac{1}{AB} = \frac{1 - e^2\cos^2\theta}{2ed} + \frac{1 - e^2\sin^2\theta}{2ed} = \frac{2 - e^2(\sin^2\theta + \cos^2\theta)}{2ed} = \frac{2 - e^2}{2ed}$$

But since e and d are constants, we have proven the desired result.

13.8 Rotation of Axes

1. For x, we have:
$$x = x'\cos\theta - y'\sin\theta$$
$$= \sqrt{3}\cos 30° - 2\sin 30°$$
$$= \sqrt{3}\cdot\frac{\sqrt{3}}{2} - 2\cdot\frac{1}{2}$$
$$= \frac{3}{2} - 1$$
$$= \frac{1}{2}$$

For y, we have:
$$y = x'\sin\theta + y'\cos\theta$$
$$= \sqrt{3}\sin 30° + 2\cos 30°$$
$$= \sqrt{3}\cdot\frac{1}{2} + 2\cdot\frac{\sqrt{3}}{2}$$
$$= \frac{\sqrt{3}}{2} + \sqrt{3}$$
$$= \frac{3\sqrt{3}}{2}$$

So the coordinates in the x-y system are $\left(\frac{1}{2}, \frac{3\sqrt{3}}{2}\right)$.

3. For x, we have:
$$x = x'\cos\theta - y'\sin\theta$$
$$= \sqrt{2}\cos 45° + \sqrt{2}\sin 45°$$
$$= \sqrt{2}\cdot\frac{1}{\sqrt{2}} + \sqrt{2}\cdot\frac{1}{\sqrt{2}}$$
$$= 1 + 1$$
$$= 2$$

For y, we have:
$$y = x'\sin\theta + y'\cos\theta$$
$$= \sqrt{2}\sin 45° - \sqrt{2}\cos 45°$$
$$= \sqrt{2}\cdot\frac{1}{\sqrt{2}} - \sqrt{2}\cdot\frac{1}{\sqrt{2}}$$
$$= 1 - 1$$
$$= 0$$

So the coordinates in the x-y system are $(2, 0)$.

5. For x', we have:
$$x' = x\cos\theta + y\sin\theta$$
$$= -3\cos\left[\sin^{-1}\left(\frac{5}{13}\right)\right] + 1\sin\left[\sin^{-1}\left(\frac{5}{13}\right)\right]$$
$$= -3\cdot\frac{12}{13} + 1\cdot\frac{5}{13}$$
$$= -\frac{31}{13}$$

For y', we have:

$$y' = -x\sin\theta + y\cos\theta$$

$$= 3\sin\left[\sin^{-1}\left(\tfrac{5}{13}\right)\right] + 1\cos\left[\sin^{-1}\left(\tfrac{5}{13}\right)\right]$$

$$= 3\cdot\tfrac{5}{13} + 1\cdot\tfrac{12}{13}$$

$$= \tfrac{27}{13}$$

So the coordinates in the x'-y' system are $\left(-\tfrac{31}{13},\tfrac{27}{13}\right)$.

7.   We have:

$$\cot 2\theta = \frac{A-C}{B} = \frac{25-18}{-24} = -\tfrac{7}{24}, \text{ so } \tan 2\theta = -\tfrac{24}{7}$$

$$\sec^2 2\theta = 1+\tan^2 2\theta = 1+\left(-\tfrac{24}{7}\right)^2 = \tfrac{625}{49}$$

So $\sec 2\theta = -\tfrac{25}{7}$ (second quadrant, since $\cot 2\theta < 0$), and thus $\cos 2\theta = -\tfrac{7}{25}$. Since θ is in the first quadrant:

$$\sin\theta = \sqrt{\frac{1-\cos 2\theta}{2}} = \sqrt{\frac{1+\tfrac{7}{25}}{2}} = \sqrt{\tfrac{16}{25}} = \tfrac{4}{5}$$

$$\cos\theta = \sqrt{\frac{1+\cos 2\theta}{2}} = \sqrt{\frac{1-\tfrac{7}{25}}{2}} = \sqrt{\tfrac{9}{25}} = \tfrac{3}{5}$$

9.   We have:

$$\cot 2\theta = \frac{1-8}{-24} = \frac{-7}{-24} = \tfrac{7}{24}, \text{ so } \tan 2\theta = \tfrac{24}{7}$$

$$\sec^2 2\theta = 1+\tan^2 2\theta = 1+\left(\tfrac{24}{7}\right)^2 = \tfrac{625}{49}$$

So $\sec 2\theta = \tfrac{25}{7}$ (first quadrant, since $\cot 2\theta > 0$), and thus $\cos 2\theta = \tfrac{7}{25}$. Since θ is in the first quadrant:

$$\sin\theta = \sqrt{\frac{1-\cos 2\theta}{2}} = \sqrt{\frac{1-\tfrac{7}{25}}{2}} = \sqrt{\tfrac{9}{25}} = \tfrac{3}{5}$$

$$\cos\theta = \sqrt{\frac{1+\cos 2\theta}{2}} = \sqrt{\frac{1+\tfrac{7}{25}}{2}} = \sqrt{\tfrac{16}{25}} = \tfrac{4}{5}$$

11.   We have:

$$\cot 2\theta = \frac{A-C}{B} = \frac{1-(-1)}{-2\sqrt{3}} = -\frac{1}{\sqrt{3}}, \text{ so } \tan 2\theta = -\sqrt{3}$$

Therefore $2\theta = 120°$, and thus $\theta = 60°$. So:

$$\sin\theta = \sin 60° = \tfrac{\sqrt{3}}{2}$$

$$\cos\theta = \cos 60° = \tfrac{1}{2}$$

13. We have:

$$\cot 2\theta = \frac{A-C}{B} = \frac{0-(-240)}{161} = \frac{240}{161}, \text{ so } \tan 2\theta = \frac{161}{240}$$

$$\sec^2 2\theta = 1 + \tan^2 2\theta = 1 + \left(\frac{161}{240}\right)^2 = \frac{83521}{57600}$$

So $\sec 2\theta = \frac{289}{240}$ (first quadrant, since $\cot 2\theta > 0$), and thus $\cos 2\theta = \frac{240}{289}$. Since θ is in the first quadrant:

$$\sin\theta = \sqrt{\frac{1-\cos 2\theta}{2}} = \sqrt{\frac{1-\frac{240}{289}}{2}} = \sqrt{\frac{49}{578}} = \frac{7\sqrt{2}}{34}$$

$$\cos\theta = \sqrt{\frac{1+\cos 2\theta}{2}} = \sqrt{\frac{1+\frac{240}{289}}{2}} = \sqrt{\frac{529}{578}} = \frac{23\sqrt{2}}{34}$$

15. Using the rotation equations:

$$x = x'\cos\theta - y'\sin\theta = x'\cos 45° - y'\sin 45° = \frac{\sqrt{2}}{2}x' - \frac{\sqrt{2}}{2}y'$$

$$y = x'\sin\theta + y'\cos\theta = x'\sin 45° + y'\cos 45° = \frac{\sqrt{2}}{2}x' + \frac{\sqrt{2}}{2}y'$$

So the equation $2xy = 9$ becomes:

$$2\left(\frac{\sqrt{2}}{2}x' - \frac{\sqrt{2}}{2}y'\right)\left(\frac{\sqrt{2}}{2}x' + \frac{\sqrt{2}}{2}y'\right) = 9$$

$$2\left(\tfrac{1}{2}x'^2 - \tfrac{1}{2}y'^2\right) = 9$$

$$x'^2 - y'^2 = 9$$

Graph the equation:

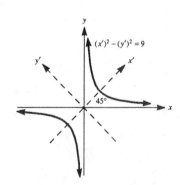

17. We find:

$$\cot 2\theta = \frac{7-1}{8} = \frac{3}{4}, \text{ so } \tan 2\theta = \frac{4}{3}$$

$$\sec^2 2\theta = 1 + \tan^2 2\theta = 1 + \left(\frac{4}{3}\right)^2 = \frac{25}{9}, \text{ so } \sec 2\theta = \frac{5}{3} \quad \text{(since } 2\theta < 90°\text{)}$$

Thus $\cos 2\theta = \frac{3}{5}$, and therefore:

$$\sin\theta = \sqrt{\frac{1-\cos 2\theta}{2}} = \sqrt{\frac{1-\frac{3}{5}}{2}} = \sqrt{\frac{1}{5}} = \frac{\sqrt{5}}{5}$$

$$\cos\theta = \sqrt{\frac{1+\cos 2\theta}{2}} = \sqrt{\frac{1+\frac{3}{5}}{2}} = \sqrt{\frac{4}{5}} = \frac{2\sqrt{5}}{5}$$

Thus $\theta = \sin^{-1}\left(\frac{\sqrt{5}}{5}\right) \approx 26.6°$. Now:

$$x = x'\cos\theta - y'\sin\theta = \frac{2\sqrt{5}}{5}x' - \frac{\sqrt{5}}{5}y'$$

$$y = x'\sin\theta + y'\cos\theta = \frac{\sqrt{5}}{5}x' + \frac{2\sqrt{5}}{5}y'$$

Making the substitutions into $7x^2 + 8xy + y^2 - 1 = 0$:

$$7\left(\frac{4}{5}x'^2 - \frac{4}{5}x'y' + \frac{1}{5}y'^2\right) + 8\left(\frac{2}{5}x'^2 + \frac{3}{5}x'y' - \frac{2}{5}y'^2\right) + \left(\frac{1}{5}x'^2 + \frac{4}{5}x'y' + \frac{4}{5}y'^2\right) - 1 = 0$$

$$\frac{28}{5}x'^2 - \frac{28}{5}x'y' + \frac{7}{5}y'^2 + \frac{16}{5}x'^2 + \frac{24}{5}x'y' - \frac{16}{5}y'^2 + \frac{1}{5}x'^2 + \frac{4}{5}x'y' + \frac{4}{5}y'^2 - 1 = 0$$

$$9x'^2 - y'^2 = 1$$

$$\frac{x'^2}{\frac{1}{9}} - \frac{y'^2}{1} = 1$$

Rotating $\theta = 26.6°$, sketch the hyperbola:

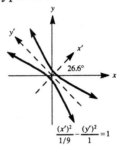

19. Use an alternate approach here:

$$x^2 + 4xy + 4y^2 = 1$$

$$(x + 2y)^2 = 1$$

$$x + 2y = 1 \qquad \text{or} \qquad x + 2y = -1$$

$$y = -\tfrac{1}{2}x + \tfrac{1}{2} \qquad\qquad\qquad y = -\tfrac{1}{2}x - \tfrac{1}{2}$$

The graph consists of two lines:

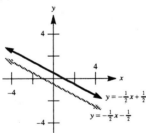

21. We find:

$$\cot 2\theta = \frac{9 - 16}{-24} = \frac{7}{24}, \text{ so } \tan 2\theta = \frac{24}{7}$$

$$\sec^2 2\theta = 1 + \tan^2 2\theta = 1 + \left(\frac{24}{7}\right)^2 = \frac{625}{49}, \text{ so } \sec 2\theta = \frac{25}{7} \quad (\text{since } 2\theta < 90°)$$

Thus $\cos 2\theta = \frac{7}{25}$, and therefore:

$$\sin \theta = \sqrt{\frac{1 - \cos 2\theta}{2}} = \sqrt{\frac{1 - \frac{7}{25}}{2}} = \sqrt{\frac{9}{25}} = \frac{3}{5}$$

$$\cos \theta = \sqrt{\frac{1 + \cos 2\theta}{2}} = \sqrt{\frac{1 + \frac{7}{25}}{2}} = \sqrt{\frac{16}{25}} = \frac{4}{5}$$

Thus $\theta = \sin^{-1}\left(\frac{3}{5}\right) \approx 36.9°$. Now:

$$x = x'\cos\theta - y'\sin\theta = \frac{4}{5}x' - \frac{3}{5}y'$$

$$y = x'\sin\theta + y'\cos\theta = \frac{3}{5}x' + \frac{4}{5}y'$$

Making the substitutions into $9x^2 - 24xy + 16y^2 - 400x - 300y = 0$ and collecting like terms:

$$25y'^2 - 500x' = 0$$

$$y'^2 = 20x'$$

Rotating $36.9°$, sketch the parabola:

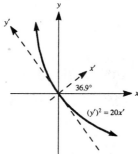

23.   We find:

$$\cot 2\theta = \frac{0 - 3}{4} = -\frac{3}{4}, \text{ so } \tan 2\theta = -\frac{4}{3}$$

$$\sec^2 2\theta = 1 + \tan^2 2\theta = 1 + \left(-\frac{4}{3}\right)^2 = \frac{25}{9}$$

$$\sec 2\theta = -\frac{5}{3} \quad (\text{since } 2\theta > 90°)$$

Thus $\cos 2\theta = -\frac{3}{5}$, and therefore:

$$\sin \theta = \sqrt{\frac{1 - \cos 2\theta}{2}} = \sqrt{\frac{1 + \frac{3}{5}}{2}} = \sqrt{\frac{4}{5}} = \frac{2\sqrt{5}}{5}$$

$$\cos \theta = \sqrt{\frac{1 + \cos 2\theta}{2}} = \sqrt{\frac{1 - \frac{3}{5}}{2}} = \sqrt{\frac{1}{5}} = \frac{\sqrt{5}}{5}$$

Thus $\theta = \cos^{-1}\left(\frac{\sqrt{5}}{5}\right) \approx 63.4°$. Now:

$$x = x'\cos\theta - y'\sin\theta = \frac{\sqrt{5}}{5}x' - \frac{2\sqrt{5}}{5}y'$$

$$y = x'\sin\theta + y'\cos\theta = \frac{2\sqrt{5}}{5}x' + \frac{\sqrt{5}}{5}y'$$

Making the substitutions into $4xy + 3y^2 + 4x + 6y = 1$ and completing the square on x' and y' terms:

$$\frac{\left(x' + \frac{2\sqrt{5}}{5}\right)^2}{1} - \frac{\left(y' + \frac{\sqrt{5}}{5}\right)^2}{4} = 1$$

Rotating $63.4°$, sketch the hyperbola:

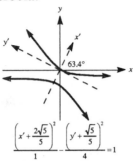

25.   We find:

$$\cot 2\theta = \frac{3-3}{-2} = 0, \text{ so } 2\theta = 90° \text{ and thus } \theta = 45°$$

Therefore:

$$x = x'\cos\theta - y'\sin\theta = x'\cos 45° - y'\sin 45° = \frac{\sqrt{2}}{2}x' - \frac{\sqrt{2}}{2}y'$$

$$y = x'\sin\theta + y'\cos\theta = x'\sin 45° + y'\cos 45° = \frac{\sqrt{2}}{2}x' + \frac{\sqrt{2}}{2}y'$$

Making the substitutions into $3x^2 - 2xy + 3y^2 - 6\sqrt{2}x + 2\sqrt{2}y + 4 = 0$ and completing the square on x' and y' terms:

$$\frac{(x'-1)^2}{1} + \frac{(y'+1)^2}{\frac{1}{2}} = 1$$

Rotating $45°$, sketch the ellipse:

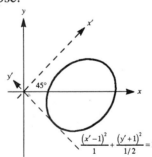

27.   First multiply out to get:

$$x^2 - 2xy + y^2 = 8y - 48$$

$$x^2 - 2xy + y^2 - 8y + 48 = 0$$

Now:

$$\cot 2\theta = \frac{1-1}{-2} = 0, \text{ so } 2\theta = 90° \text{ and thus } \theta = 45°$$

Therefore:

$$x = x'\cos\theta - y'\sin\theta = \tfrac{\sqrt{2}}{2}x' - \tfrac{\sqrt{2}}{2}y'$$

$$y = x'\sin\theta + y'\cos\theta = \tfrac{\sqrt{2}}{2}x' + \tfrac{\sqrt{2}}{2}y'$$

Making the substitutions and completing the square on x' and y' terms yields:

$$\left(y' - \sqrt{2}\right)^2 = 2\sqrt{2}\left(x' - \tfrac{11\sqrt{2}}{2}\right)$$

Rotating $45°$, sketch the parabola:

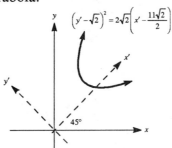

29. We find:

$$\cot 2\theta = \frac{3-6}{4} = -\tfrac{3}{4}, \text{ so } \tan 2\theta = -\tfrac{4}{3}$$

$$\sec^2 2\theta = 1 + \tan^2 2\theta = 1 + \left(-\tfrac{4}{3}\right)^2 = \tfrac{25}{9}$$

So $\sec 2\theta = -\tfrac{5}{3}$ (since $2\theta > 90°$), and thus $\cos 2\theta = -\tfrac{3}{5}$. Therefore:

$$\sin\theta = \sqrt{\frac{1-\cos 2\theta}{2}} = \sqrt{\frac{1+\tfrac{3}{5}}{2}} = \frac{2\sqrt{5}}{5}$$

$$\cos\theta = \sqrt{\frac{1+\cos 2\theta}{2}} = \sqrt{\frac{1-\tfrac{3}{5}}{2}} = \frac{\sqrt{5}}{5}$$

Thus $\theta = \cos^{-1}\left(\tfrac{\sqrt{5}}{5}\right) \approx 63.4°$. Now:

$$x = x'\cos\theta - y'\sin\theta = \tfrac{\sqrt{5}}{5}x' - \tfrac{2\sqrt{5}}{5}y'$$

$$y = x'\sin\theta + y'\cos\theta = \tfrac{2\sqrt{5}}{5}x' + \tfrac{\sqrt{5}}{5}y'$$

Making the substitutions into $3x^2 + 4xy + 6y^2 = 7$:

$$\frac{x'^2}{1} + \frac{y'^2}{\tfrac{7}{2}} = 1$$

Rotating $63.4°$, sketch the ellipse:

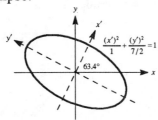

31. We find:

$$\cot 2\theta = \frac{17-8}{-12} = -\frac{3}{4}, \text{ so } \tan 2\theta = -\frac{4}{3}$$

As with Exercise 29 we find $\sin\theta = \frac{2\sqrt{5}}{5}$ and $\cos\theta = \frac{\sqrt{5}}{5}$, so:

$$x = \frac{\sqrt{5}}{5}x' - \frac{2\sqrt{5}}{5}y'$$
$$y = \frac{2\sqrt{5}}{5}x' + \frac{\sqrt{5}}{5}y'$$

Substituting into $17x^2 - 12xy + 8y^2 - 80 = 0$ yields:

$$\frac{x'^2}{16} + \frac{y'^2}{4} = 1$$

Rotating $63.4°$, sketch the ellipse:

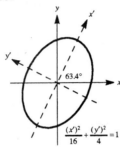

$$\frac{(x')^2}{16} + \frac{(y')^2}{4} = 1$$

33. We find:

$$\cot 2\theta = \frac{0+4}{3} = \frac{4}{3}, \text{ so } \tan 2\theta = \frac{3}{4}$$

$$\sec^2 2\theta = 1 + \tan^2 2\theta = 1 + \frac{9}{16} = \frac{25}{16}, \text{ so } \sec 2\theta = \frac{5}{4}$$

Then $\cos 2\theta = \frac{4}{5}$, and thus:

$$\sin\theta = \sqrt{\frac{1-\cos 2\theta}{2}} = \sqrt{\frac{1-\frac{4}{5}}{2}} = \frac{\sqrt{10}}{10}$$

$$\cos\theta = \sqrt{\frac{1+\cos 2\theta}{2}} = \sqrt{\frac{1+\frac{4}{5}}{2}} = \frac{3\sqrt{10}}{10}$$

Then $\theta = \sin^{-1}\left(\frac{\sqrt{10}}{10}\right) \approx 18.4°$, and:

$$x = \frac{3\sqrt{10}}{10}x' - \frac{\sqrt{10}}{10}y'$$
$$y = \frac{\sqrt{10}}{10}x' + \frac{3\sqrt{10}}{10}y'$$

Substituting into $3xy - 4y^2 + 18 = 0$ results in:

$$\frac{y'^2}{4} - \frac{x'^2}{36} = 1$$

Rotating 18.4°, sketch the hyperbola:

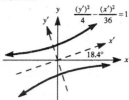

35. First multiply out terms to obtain:
$$x^2 + 2xy + y^2 + 4\sqrt{2}x - 4\sqrt{2}y = 0$$
We find:
$$\cot 2\theta = \frac{1-1}{2} = 0, \text{ so } 2\theta = 90° \text{ and thus } \theta = 45°$$
Now:
$$x = \frac{\sqrt{2}}{2}x' - \frac{\sqrt{2}}{2}y'$$
$$y = \frac{\sqrt{2}}{2}x' + \frac{\sqrt{2}}{2}y'$$
Substituting into $x^2 + 2xy + y^2 + 4\sqrt{2}x - 4\sqrt{2}y = 0$ results in:
$$x'^2 = 4y'$$
Rotating 45°, sketch the parabola:

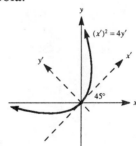

37. We find:
$$\cot 2\theta = \frac{3-2}{-\sqrt{15}} = -\frac{1}{\sqrt{15}}, \text{ so } \tan 2\theta = -\sqrt{15}$$
$$\sec^2 2\theta = 1 + \tan^2 2\theta = 1 + 15 = 16, \text{ so } \sec 2\theta = -4 \quad (\text{since } 2\theta > 90°)$$
Thus $\cos 2\theta = -\frac{1}{4}$ and we find:
$$\sin \theta = \sqrt{\frac{1-\cos 2\theta}{2}} = \sqrt{\frac{1+\frac{1}{4}}{2}} = \sqrt{\frac{5}{8}} = \frac{\sqrt{10}}{4}$$
$$\cos \theta = \sqrt{\frac{1+\cos 2\theta}{2}} = \sqrt{\frac{1-\frac{1}{4}}{2}} = \sqrt{\frac{3}{8}} = \frac{\sqrt{6}}{4}$$
Then $\theta = \cos^{-1}\left(\frac{\sqrt{6}}{4}\right) \approx 52.2°$, and:
$$x = \frac{\sqrt{6}}{4}x' - \frac{\sqrt{10}}{4}y'$$
$$y = \frac{\sqrt{10}}{4}x' + \frac{\sqrt{6}}{4}y'$$

Substituting into $3x^2 - \sqrt{15}xy + 2y^2 = 3$ results in:
$$\frac{x'^2}{6} + \frac{y'^2}{\frac{2}{3}} = 1$$
Rotating 52.2°, sketch the ellipse:

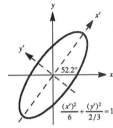

39.   We find:
$$\cot 2\theta = \frac{3-3}{-2} = 0, \text{ so } 2\theta = 90° \text{ and thus } \theta = 45°$$
Now:
$$x = \tfrac{\sqrt{2}}{2}x' - \tfrac{\sqrt{2}}{2}y'$$
$$y = \tfrac{\sqrt{2}}{2}x' + \tfrac{\sqrt{2}}{2}y'$$
Substituting into $3x^2 - 2xy + 3y^2 + 2 = 0$ results in:
$$x'^2 + 2y'^2 = -1$$
But clearly this is impossible, so there is no graph.

41.   Multiplying the first equation by $\sin\theta$ and the second equation by $\cos\theta$ yields:
$$(\sin\theta\cos\theta)x + (\sin^2\theta)y = x'\sin\theta$$
$$(-\sin\theta\cos\theta)x + (\cos^2\theta)y = y'\cos\theta$$
Adding:
$$(\sin^2\theta + \cos^2\theta)y = x'\sin\theta + y'\cos\theta$$
$$y = x'\sin\theta + y'\cos\theta$$
Multiplying the first equation by $\cos\theta$ and the second equation by $-\sin\theta$ yields:
$$(\cos^2\theta)x + (\sin\theta\cos\theta)y = x'\cos\theta$$
$$(\sin^2\theta)x - (\sin\theta\cos\theta)y = -y'\sin\theta$$
Adding:
$$(\sin^2\theta + \cos^2\theta)x = x'\cos\theta - y'\sin\theta$$
$$x = x'\cos\theta - y'\sin\theta$$

43.   (a)   Since $\cot 2\theta = -\frac{7}{24}$, $\tan 2\theta = -\frac{24}{7}$ so:
$$\sec^2 2\theta = 1 + \left(\frac{24}{7}\right)^2 = \frac{49}{49} + \frac{576}{49} = \frac{625}{49}$$
Thus $\sec 2\theta = \pm\frac{25}{7}$. Since 2θ is in either the first or second quadrant, and $\cot 2\theta < 0$, then 2θ is in the second quadrant, so $\sec 2\theta = -\frac{25}{7}$ and thus $\cos 2\theta = -\frac{7}{25}$.

(b)   Compute $\sin\theta$ and $\cos\theta$:

$$\sin\theta = \sqrt{\frac{1-\cos 2\theta}{2}} = \sqrt{\frac{1+\frac{7}{25}}{2}} = \sqrt{\frac{32}{50}} = \sqrt{\frac{16}{25}} = \frac{4}{5}$$

$$\cos\theta = \sqrt{\frac{1+\cos 2\theta}{2}} = \sqrt{\frac{1-\frac{7}{25}}{2}} = \sqrt{\frac{18}{50}} = \sqrt{\frac{9}{25}} = \frac{3}{5}$$

(c)   Making the substitutions and simplifying:

$$16\left(\tfrac{1}{5}(3x'-4y')\right)^2 - 24\left(\tfrac{1}{5}(3x'-4y')\right)\left(\tfrac{1}{5}(4x'+3y')\right)$$

$$+9\left(\tfrac{1}{5}(4x'+3y')\right)^2 + 110\left(\tfrac{1}{5}(3x'-4y')\right) - 20\bullet\tfrac{1}{5}(4x'+3y')+100 = 0$$

$$\tfrac{16}{25}\left(9x'^2 - 24x'y' + 16y'^2\right) - \tfrac{24}{25}\left(12x'^2 - 7x'y' - 12y'^2\right)$$

$$+\tfrac{9}{25}\left(16x'^2 + 24x'y' + 9y'^2\right) + 22(3x'-4y') - 4(4x'+3y') + 100 = 0$$

$$\frac{144-288+144}{25}x'^2 + \frac{-384+168+216}{25}x'y' + \frac{256+288+81}{25}y'^2$$

$$+(66-16)x' + (-88-12)y' + 100 = 0$$

$$25y'^2 + 50x' - 100y' + 100 = 0$$

$$y'^2 + 2x' - 4y' + 4 = 0$$

45.   Working from the right-hand side:

$$A' + C' = A\cos^2\theta + B\sin\theta\cos\theta + C\sin^2\theta + A\sin^2\theta - B\sin\theta\cos\theta + C\cos^2\theta$$

$$= (A+C)\left(\cos^2\theta + \sin^2\theta\right)$$

$$= A+C$$

Chapter Thirteen Review Exercises

1.   The x-axis contains $\overline{AB}$ and has equation $y=0$. The slope of the line containing $\overline{BC}$ is $-\frac{1}{b}$ and the equation is:

$$y = -\tfrac{1}{b}(x-6b)$$
$$by = 6b - x$$
$$x + by = 6b$$

The slope of the line containing $\overline{AC}$ is $-\frac{1}{a}$ and the equation is:

$$y = -\tfrac{1}{a}(x-6a)$$
$$x + ay = 6a$$

3.   Using the equations from Exercise 2, we show $G(2a+2b, 2)$ lies on all three medians:

$$2(2a+2b) + (a+b)(2) = 4a+4b+2a+2b = 6(a+b)$$
$$2a+2b - (b-2a)(2) = 2a+2b-2b+4a = 6a$$
$$2a+2b - (a-2b)(2) = 2a+2b-2a+4b = 6b$$

5. Using the equations from Exercise 4, we show $H(0, -6ab)$ lies on each altitude:
$$x = 0$$
$$-6ab = b \cdot 0 - 6ab = -6ab$$
$$-6ab = a \cdot 0 - 6ab = -6ab$$

7. Using the equations from Exercise 6, we show $O(3a + 3b, 3ab + 3)$ lies on each perpendicular bisector:
$$x = 3a + 3b$$
$$b(3a + 3b) - (3ab + 3) = 3ab + 3b^2 - 3ab - 3 = 3b^2 - 3$$
$$a(3a + 3b) - (3ab + 3) = 3a^2 + 3ab - 3ab - 3 = 3a^2 - 3$$

9. Find each distance and use the result from Exercise 8:
$$p = \sqrt{(6b)^2 + 6^2} = 6\sqrt{b^2 + 1}$$
$$q = \sqrt{(6a)^2 + 6^2} = 6\sqrt{a^2 + 1}$$
$$r = 6(a - b)$$
$$R = 3\sqrt{(a^2 + 1)(b^2 + 1)}$$
Thus:
$$\frac{pqr}{4R} = \frac{\left(6\sqrt{b^2 + 1}\right)\left(6\sqrt{a^2 + 1}\right)(6(a - b))}{4 \cdot 3\sqrt{(a^2 + 1)(b^2 + 1)}} = 18(a - b)$$
$$\text{Area of } \triangle ABC = \tfrac{1}{2}(6(a - b))(6) = 18(a - b)$$

So the area is $\frac{pqr}{4R}$, as required.

11. Compute each side of the identity:
$$OH^2 = (3a + 3b - 0)^2 + (3ab + 3 + 6ab)^2$$
$$= 9(a^2 + 2ab + b^2) + 9(3ab + 1)^2$$
$$= 9(a^2 + 2ab + b^2 + 9a^2b^2 + 6ab + 1)$$
$$= 81a^2b^2 + 9a^2 + 9b^2 + 72ab + 9$$
$$9R^2 - (p^2 + q^2 + r^2)$$
$$= 9\left[9(a^2 + 1)(b^2 + 1)\right] - 36(b^2 + 1) - 36(a^2 + 1) - 36(a^2 - 2ab + b^2)$$
$$= 81a^2b^2 + 81a^2 + 81b^2 + 81 - 36b^2 - 36 - 36a^2 - 36 - 36a^2 + 72ab - 36b^2$$
$$= 81a^2b^2 + 9a^2 + 9b^2 + 72ab + 9$$
Thus $OH^2 = 9R^2 - (p^2 + q^2 + r^2)$.

13. First compute the squares:
$$HA^2 = (6a)^2 + (-6ab)^2 = 36a^2 + 36a^2b^2$$
$$HB^2 = (6b)^2 + (-6ab)^2 = 36b^2 + 36a^2b^2$$
$$HC^2 = 0^2 + (6 + 6ab)^2 = 36 + 72ab + 36a^2b^2$$
So $HA^2 + HB^2 + HC^2 = 108a^2b^2 + 36a^2 + 36b^2 + 72ab + 36$.

Now compute the right-hand side:

$$12R^2 - \left(p^2 + q^2 + r^2\right)$$

$$= 12\left[9\left(a^2+1\right)\left(b^2+1\right)\right] - \left[36\left(b^2+1\right) + 36\left(a^2+1\right) + 36\left(a^2 - 2ab + b^2\right)\right]$$

$$= 108a^2b^2 + 108a^2 + 108b^2 + 108 - 36b^2 - 36 - 36a^2 - 36 - 36a^2 + 72ab - 36b^2$$

$$= 108a^2b^2 + 36a^2 + 36b^2 + 72ab + 36$$

Thus $HA^2 + HB^2 + HC^2 = 12R^2 - (p^2 + q^2 + r^2)$.

15. In Exercise 12 we saw that:

$$GH^2 = 4\left(9a^2b^2 + a^2 + b^2 + 8ab + 1\right)$$

Therefore:

$$GH = 2\sqrt{9a^2b^2 + a^2 + b^2 + 8ab + 1}$$

Now:

$$2GO = 2\sqrt{(2a + 2b - 3a - 3b)^2 + (2 - 3ab - 3)^2}$$

$$= 2\sqrt{a^2 + 2ab + b^2 + 1 + 6ab + 9a^2b^2}$$

$$= 2\sqrt{9a^2b^2 + a^2 + b^2 + 8ab + 1}$$

Thus $GH = 2GO$.

17. Since $\tan \theta = -\frac{2}{3}$, then $\theta \approx 146.3°$.

19. Given $(x_0, y_0) = (-1, -3)$, $A = 5$, $B = 6$ and $C = -30$, so using the distance formula from a point to a line yields:

$$d = \frac{|5(-1) + 6(-3) - 30|}{\sqrt{5^2 + 6^2}} = \frac{53}{\sqrt{61}} = \frac{53\sqrt{61}}{61}$$

21. Label $A(-6, 0)$, $B(6, 0)$ and $C\left(0, 6\sqrt{3}\right)$. The height of the triangle is $6\sqrt{3}$. Since the line $\overline{AB}$ is the x-axis, the distance from $(1, 2)$ to $\overline{AB}$ is 2. The line containing $\overline{AC}$ has slope $\sqrt{3}$ and equation:

$$y - 0 = \sqrt{3}(x + 6)$$
$$y = \sqrt{3}x + 6\sqrt{3}$$

The distance from $(1, 2)$ to $\overline{AC}$ is thus:

$$\frac{\left|\sqrt{3}(1) + 6\sqrt{3} - 2\right|}{\sqrt{1^2 + \left(-\sqrt{3}\right)^2}} = \frac{7\sqrt{3} - 2}{2}$$

The line containing $\overline{BC}$ has slope $-\sqrt{3}$ and equation:

$$y - 0 = -\sqrt{3}(x - 6)$$
$$y = -\sqrt{3}x + 6\sqrt{3}$$

The distance from $(1, 2)$ to $\overline{BC}$ is thus:

$$\frac{\left|-\sqrt{3}(1) + 6\sqrt{3} - 2\right|}{\sqrt{1^2 + \left(\sqrt{3}\right)^2}} = \frac{5\sqrt{3} - 2}{2}$$

The sum of these distances is $6\sqrt{3}$, which is also the height.

23. (a) The form is $y^2 = 4px$, where $p = 4$. Thus the equation is $y^2 = 16x$.
    (b) The form is $x^2 = 4py$, where $p = 4$. Thus the equation is $x^2 = 16y$.

25. Since the parabola is symmetric about the positive y-axis, its equation must be of the form $x^2 = 4py$, where $p > 0$. Now the focal width is 12, so $4p = 12$. Thus the equation is $x^2 = 12y$.

27. We have $c = 2$ and $a = 8$, and the ellipse must have the form:
$$\frac{x^2}{8^2} + \frac{y^2}{b^2} = 1$$
Since $c^2 = a^2 - b^2$, we can find b:
$$4 = 64 - b^2$$
$$b^2 = 60$$
The equation is $\dfrac{x^2}{64} + \dfrac{y^2}{60} = 1$, or $15x^2 + 16y^2 = 960$.

29. Since one end of the minor axis is $(-6, 0)$, $b = 6$ and the ellipse has a form of:
$$\frac{x^2}{36} + \frac{y^2}{a^2} = 1$$
Now $\frac{c}{a} = \frac{4}{5}$, so $c = \frac{4}{5}a$. Finding a:
$$c^2 = a^2 - b^2$$
$$\left(\tfrac{4}{5}a\right)^2 = a^2 - 36$$
$$\tfrac{16}{25}a^2 = a^2 - 36$$
$$36 = \tfrac{9}{25}a^2$$
$$100 = a^2$$
The equation is $\dfrac{x^2}{36} + \dfrac{y^2}{100} = 1$, or $25x^2 + 9y^2 = 900$.

31. Since the foci are $(\pm 6, 0)$ and the vertices are $(\pm 2, 0)$, $c = 6$, $a = 2$, and the equation has the form:
$$\frac{x^2}{4} - \frac{y^2}{b^2} = 1$$
Finding b:
$$c^2 = a^2 + b^2$$
$$36 = 4 + b^2$$
$$32 = b^2$$
The equation is $\dfrac{x^2}{4} - \dfrac{y^2}{32} = 1$, or $8x^2 - y^2 = 32$.

33. Since the foci are $(\pm 3, 0)$, $c = 3$ and the equation has the form:

$$\frac{x^2}{a^2} - \frac{y^2}{b^2} = 1$$

Now $\frac{c}{a} = 4$, so $\frac{3}{a} = 4$, thus $a = \frac{3}{4}$. Substitute to find b:

$$c^2 = a^2 + b^2$$
$$9 = \tfrac{9}{16} + b^2$$
$$144 = 9 + 16b^2$$
$$135 = 16b^2$$
$$\tfrac{135}{16} = b^2$$

The equation is $\dfrac{x^2}{\frac{9}{16}} - \dfrac{y^2}{\frac{135}{16}} = 1$, or $240x^2 - 16y^2 = 135$.

35. Note that $4p = 10$, so $p = \frac{5}{2}$. The vertex is $(0, 0)$, the focus is $\left(0, \frac{5}{2}\right)$, the directrix is $y = -\frac{5}{2}$, and the focal width is 10.

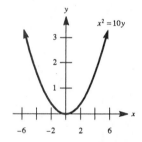

37. Note that $4p = 12$, so $p = 3$. The vertex is $(0, 3)$, the focus is $(0, 0)$, the directrix is $y = 6$, and the focal width is 12.

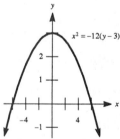

39. Note that $4p = 4$, so $p = 1$. The vertex is $(1, 1)$, the focus is $(0, 1)$, the directrix is $x = 2$, and the focal width is 4.

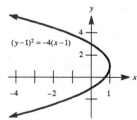

41.   Dividing by 144, the standard form is $\frac{x^2}{36} + \frac{y^2}{16} = 1$. The center is (0, 0), the length of the major axis is 12, and the length of the minor axis is 8. Using $c^2 = a^2 - b^2$, we find:

$$c = \sqrt{a^2 - b^2} = \sqrt{36 - 16} = \sqrt{20} = 2\sqrt{5}$$

The foci are $(\pm 2\sqrt{5}, 0)$ and the eccentricity is $\dfrac{2\sqrt{5}}{6} = \dfrac{\sqrt{5}}{3}$.

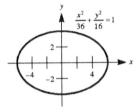

43.   Dividing by 9, the standard form is $\frac{x^2}{1} + \frac{y^2}{9} = 1$. The center is (0, 0), the length of the major axis is 6, and the length of the minor axis is 2. Using $c^2 = a^2 - b^2$, we find:

$$c = \sqrt{a^2 - b^2} = \sqrt{9 - 1} = \sqrt{8} = 2\sqrt{2}$$

The foci are $(0, \pm 2\sqrt{2})$ and the eccentricity is $\dfrac{2\sqrt{2}}{3}$.

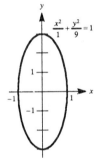

45.   The equation is already in standard form where the center is (–3, 0), and the lengths of the major and minor axes are 6. Since the lengths of the major and minor axes are equal this ellipse is actually a circle. Using $c^2 = a^2 - b^2$, we find:

$$c = \sqrt{a^2 - b^2} = \sqrt{9 - 9} = 0$$

The only focus is at (–3, 0), which is the center of the circle, and the eccentricity is $\frac{0}{3} = 0$.

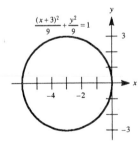

47.   Dividing by 144, the standard form is $\frac{x^2}{36} - \frac{y^2}{16} = 1$. The center is $(0, 0)$, the vertices are $(\pm 6, 0)$, and the asymptotes are $y = \pm \frac{4}{6}x = \pm \frac{2}{3}x$. Using $c^2 = a^2 + b^2$, we have:
$$c = \sqrt{a^2 + b^2} = \sqrt{36 + 16} = \sqrt{52} = 2\sqrt{13}$$
The foci are $\left(\pm 2\sqrt{13}, 0\right)$ and the eccentricity is $\dfrac{2\sqrt{13}}{6} = \dfrac{\sqrt{13}}{3}$.

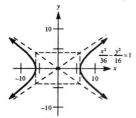

49.   Dividing by 9, the standard form is $\frac{y^2}{1} - \frac{x^2}{9} = 1$. The center is $(0, 0)$, the vertices are $(0, \pm 1)$, and the asymptotes are $y = \pm \frac{1}{3}x$. Using $c^2 = a^2 + b^2$, we have:
$$c = \sqrt{a^2 + b^2} = \sqrt{1 + 9} = \sqrt{10}$$
The foci are $\left(0, \pm \sqrt{10}\right)$ and the eccentricity is $\dfrac{\sqrt{10}}{1} = \sqrt{10}$.

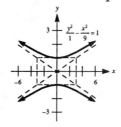

51.   The equation is already in standard form where the center is $(0, -3)$ and the vertices are $(0, -3 + 3) = (0, 0)$ and $(0, -3 - 3) = (0, -6)$. The asymptotes have slopes of $\pm \frac{3}{3} = \pm 1$, so using the point-slope formula:

$$y - (-3) = 1(x - 0) \qquad\qquad y - (-3) = -1(x - 0)$$
$$y + 3 = x \qquad\qquad\qquad\quad y + 3 = -x$$
$$y = x - 3 \qquad\qquad\qquad\quad y = -x - 3$$

Using $c^2 = a^2 + b^2$, we have:
$$c = \sqrt{a^2 + b^2} = \sqrt{9 + 9} = \sqrt{18} = 3\sqrt{2}$$
The foci are $\left(0, -3 \pm 3\sqrt{2}\right)$ and the eccentricity is $\dfrac{3\sqrt{2}}{3} = \sqrt{2}$.

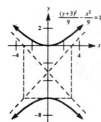

53. Complete the square to convert the equation to standard form:

$$y^2 - 8y = 16x - 80$$
$$y^2 - 8y + 16 = 16x - 64$$
$$(y - 4)^2 = 16(x - 4)$$

This is the equation of a parabola. Since $4p = 16$, $p = 4$. The vertex is $(4, 4)$, the axis of symmetry is $y = 4$, the focus is $(8, 4)$, and the directrix is $x = 0$.

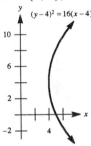

55. Complete the square to convert the equation to standard form:

$$16x^2 + 64x + 9y^2 - 54y = -1$$
$$16(x^2 + 4x) + 9(y^2 - 6y) = -1$$
$$16(x^2 + 4x + 4) + 9(y^2 - 6y + 9) = -1 + 64 + 81$$
$$16(x + 2)^2 + 9(y - 3)^2 = 144$$
$$\frac{(x + 2)^2}{9} + \frac{(y - 3)^2}{16} = 1$$

This is the equation of an ellipse. Its center is $(-2, 3)$, the length of the major axis is 8, and the length of the minor axis is 6. Using $c^2 = a^2 - b^2$, we find:

$$c = \sqrt{a^2 - b^2} = \sqrt{16 - 9} = \sqrt{7}$$

The foci are $\left(-2, 3 \pm \sqrt{7}\right)$.

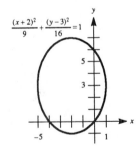

57. Complete the square to convert the equation to standard form:
$$x^2 + 6x = 12y - 33$$
$$(x+3)^2 = 12y - 24$$
$$(x+3)^2 = 12(y-2)$$
This is the equation of a parabola. Since $4p = 12$, $p = 3$. The vertex is $(-3, 2)$, the axis of symmetry is $x = -3$, the focus is $(-3, 5)$, and the directrix is $y = -1$.

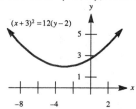

59. Complete the square to convert the equation to standard form:
$$x^2 - 4x - y^2 + 2y = 6$$
$$\left(x^2 - 4x + 4\right) - \left(y^2 - 2y + 1\right) = 6 + 4 - 1$$
$$(x-2)^2 - (y-1)^2 = 9$$
$$\frac{(x-2)^2}{9} - \frac{(y-1)^2}{9} = 1$$
This is the equation of a hyperbola. Its center is $(2, 1)$ and its vertices are $(2 + 3, 1) = (5, 1)$ and $(2 - 3, 1) = (-1, 1)$. The asymptotes have slopes of $\pm\frac{3}{3} = \pm1$, so using the point-slope formula:

$$y - 1 = 1(x - 2)$$ $$y - 1 = -1(x - 2)$$
$$y - 1 = x - 2$$ $$y - 1 = -x + 2$$
$$y = x - 1$$ $$y = -x + 3$$

Using $c^2 = a^2 + b^2$, we have:
$$c = \sqrt{a^2 + b^2} = \sqrt{9+9} = \sqrt{18} = 3\sqrt{2}$$
The foci are $\left(2 \pm 3\sqrt{2}, 1\right)$.

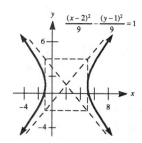

61. Complete the square to convert the equation to standard form:

$$x^2 + 2y - 12 = 0$$
$$x^2 = -2y + 12$$
$$x^2 = -2(y - 6)$$

This is the equation of a parabola. Since $4p = -2$, $p = -\frac{1}{2}$. The vertex is $(0, 6)$, the axis of symmetry is $x = 0$, the focus is $\left(0, \frac{11}{2}\right)$, and the directrix is $y = \frac{13}{2}$.

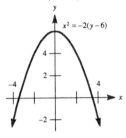

63. Complete the square to convert the equation to standard form:

$$x^2 + 16y^2 - 160y = -384$$
$$x^2 + 16(y^2 - 10y) = -384$$
$$x^2 + 16(y - 5)^2 = -384 + 400$$
$$x^2 + 16(y - 5)^2 = 16$$
$$\frac{x^2}{16} + \frac{(y - 5)^2}{1} = 1$$

This is the equation of an ellipse. Its center is $(0, 5)$, the length of the major axis is 8, and the length of the minor axis is 2. Using $c^2 = a^2 - b^2$, we find:

$$c = \sqrt{a^2 - b^2} = \sqrt{16 - 1} = \sqrt{15}$$

The foci are $\left(\pm\sqrt{15}, 5\right)$.

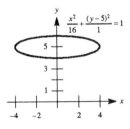

65. Complete the square to convert the equation to standard form:

$$16x^2 - 64x - 25y^2 + 100y = 36$$
$$16(x^2 - 4x) - 25(y^2 - 4y) = 36$$
$$16(x - 2)^2 - 25(y - 2)^2 = 36 + 64 - 100$$
$$16(x - 2)^2 - 25(y - 2)^2 = 0$$
$$16(x - 2)^2 = 25(y - 2)^2$$
$$\pm 4(x - 2) = 5(y - 2)$$

The graph of this equation consists of two lines:

$$4(x-2) = 5(y-2) \qquad\qquad -4(x-2) = 5(y-2)$$
$$4x - 8 = 5y - 10 \qquad\qquad -4x + 8 = 5y - 10$$
$$4x - 5y = -2 \qquad\qquad 4x + 5y = 18$$

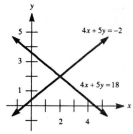

67. Complete the square:

$$Ax^2 + Dx + Ey + F = 0$$

$$Ey = -A\left(x^2 + \frac{D}{A}x\right) - F$$

$$y = -\frac{A}{E}\left(x^2 + \frac{D}{A}x + \frac{D^2}{4A^2}\right) + \frac{D^2}{4AE} - \frac{F}{E}$$

$$y = -\frac{A}{E}\left(x + \frac{D}{2A}\right)^2 + \frac{D^2 - 4AF}{4AE}$$

Thus the vertex has x-coordinate $-\dfrac{D}{2A}$ and y-coordinate $\dfrac{D^2 - 4AF}{4AE}$.

69. Since the distance VF is equal to p, we wish to show the intersection point has a y-coordinate of $\frac{p}{2}$. The parabola has equation $x^2 = 4py$ and the circle has equation $x^2 + y^2 = \left(\frac{3p}{2}\right)^2$. Substitute:

$$4py + y^2 = \frac{9p^2}{4}$$
$$16py + 4y^2 = 9p^2$$
$$4y^2 + 16py - 9p^2 = 0$$
$$(2y - p)(2y + 9p) = 0$$
$$2y = p \qquad \text{or} \qquad 2y = -9p$$
$$y = \frac{p}{2} \qquad\qquad\quad y = -\frac{9p}{2}$$

Clearly $y = -\frac{9p}{2}$ indicates a parabola opening downward, so $y = \frac{p}{2}$. Thus $\overline{VF}$ is bisected by the indicated chord.

71. The given line will be tangent to the circle at (a, b) if and only if (a, b) lies on the line and the perpendicular distance from the center of the circle (h, k) to the line is the radius r. Since (a, b) is on the given circle:

$$(a - h)(a - h) + (b - k)(b - k) = (a - h)^2 + (b - k)^2 = r^2$$

So (a, b) lies on the line. Using the formula $d = \dfrac{|Ax_0 + By_0 + C|}{\sqrt{A^2 + B^2}}$ we obtain:

$$d = \frac{\left|(a - h)h + (b - k)k - ah + h^2 - bk + k^2 - r^2\right|}{\sqrt{(a - h)^2 + (b - k)^2}} = \frac{\left|-r^2\right|}{\sqrt{r^2}} = r$$

The perpendicular distance from (h, k) to the line is r. Thus the given line is the tangent line at (a, b).

73. Use $a = 5$ and $b = 3$ in each formula:

	Approximation Obtained	Percentage Error
C_1	25.531776	0.019
C_2	25.526986	0.000049
C_3	25.519489	0.029

Since C_2 has the smallest percentage error, it would be the best approximation of the circumference of the three.

Chapter Thirteen Test

1. Since $4p = 12$, then $p = 3$. The focus is $(-3, 0)$ and the directrix is $x = 3$.

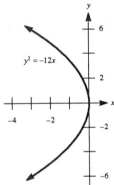

2. Dividing by 4, the standard form is $\dfrac{x^2}{4} - \dfrac{y^2}{1} = 1$. The asymptotes are $y = \pm \tfrac{1}{2}x$. Using $c^2 = a^2 + b^2$, we have:
$$c = \sqrt{a^2 + b^2} = \sqrt{4+1} = \sqrt{5}$$
The foci are $\left(\pm\sqrt{5}, 0\right)$.

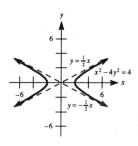

3. (a) We have:
$$\cot 2\theta = \frac{A-C}{B} = \frac{1-3}{2\sqrt{3}} = -\frac{1}{\sqrt{3}}, \text{ so } \tan 2\theta = -\sqrt{3}$$
Thus $2\theta = 120°$, or $\theta = 60°$.

(b) Applying the rotation formulas:
$$x = x'\cos 60° - y'\sin 60° = \tfrac{1}{2}x' - \tfrac{\sqrt{3}}{2}y'$$
$$y = x'\sin 60° + y'\cos 60° = \tfrac{\sqrt{3}}{2}x' + \tfrac{1}{2}y'$$
Substituting into $x^2 + 2\sqrt{3}xy + 3y^2 - 12\sqrt{3}x + 12y = 0$ yields:
$$x'^2 = -6y'$$
Rotating $60°$, graph the parabola:

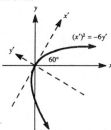

4. Since $\tan\theta = \tfrac{1}{\sqrt{3}}$, then $\theta = 30°$.

5. Since $e = \tfrac{c}{a}$, then $\tfrac{c}{a} = \tfrac{1}{2}$, so $a = 2c$. Since the foci are $(0, \pm 2)$, $c = 2$ and thus $a = 4$.
Now find b^2:
$$c^2 = a^2 - b^2$$
$$4 = 16 - b^2$$
$$b^2 = 12$$
The equation of the ellipse is $\dfrac{x^2}{12} + \dfrac{y^2}{16} = 1$.

6.　Call m the slope of the tangents, so:
$$y - 0 = m(x + 4)$$
$$y = mx + 4m$$
We now find the distance from the center of the circle $(0, 0)$ to this line:
$$r = \frac{|0 + 0 - 4m|}{\sqrt{1 + m^2}} = \frac{|4m|}{\sqrt{1 + m^2}}$$
Since this radius is 1, we have:
$$1 = \frac{|4m|}{\sqrt{1 + m^2}}$$
Squaring, we have:
$$1 = \frac{16m^2}{1 + m^2}$$
$$1 + m^2 = 16m^2$$
$$1 = 15m^2$$
$$m^2 = \tfrac{1}{15}$$
$$m = \pm\frac{\sqrt{15}}{15}$$

7.　The slope is given by $m = \tan 60° = \sqrt{3}$. Using the point $(2, 0)$ in the point-slope formula:
$$y - 0 = \sqrt{3}(x - 2)$$
$$y = \sqrt{3}x - 2\sqrt{3}$$
$$\sqrt{3}x - y - 2\sqrt{3} = 0$$

8.　Since the foci are $(\pm 2, 0)$, $c = 2$. Also $\frac{b}{a} = \frac{1}{\sqrt{3}}$, so $a = b\sqrt{3}$. So:
$$c^2 = a^2 + b^2$$
$$4 = \left(b\sqrt{3}\right)^2 + b^2$$
$$4 = 4b^2$$
$$1 = b^2$$
$$1 = b$$
So $a = 1\sqrt{3} = \sqrt{3}$, thus the equation is $\dfrac{x^2}{3} - \dfrac{y^2}{1} = 1$.

9.　(a)　Substituting $P(6, 5)$ into the equation $5x^2 - 4y^2 = 80$, we have:
$$5(6)^2 - 4(5)^2 = 5(36) - 4(25) = 180 - 100 = 80$$
So $P(6, 5)$ lies on the hyperbola.

(b) The quantity $(F_1P - F_2P)^2$ can be computed without determining the coordinates of F_1 and F_2. By definition we have $|F_1P - F_2P| = 2a$ for any point P on the hyperbola. Squaring both sides here yields $(F_1P - F_2P)^2 = 4a^2$. Now to compute a, we convert the equation $5x^2 - 4y^2 = 80$ to standard form. The result is $\dfrac{x^2}{16} - \dfrac{y^2}{20} = 1$. Therefore $a = 4$ and we obtain $(F_1P - F_2P)^2 = 4a^2 = 4(16) = 64$.

10. Dividing by 100 yields the standard form $\frac{x^2}{25} + \frac{y^2}{4} = 1$. The length of the major axis is 10 and the length of the minor axis is 4. Using $c^2 = a^2 - b^2$, we find:
$$c = \sqrt{a^2 - b^2} = \sqrt{25 - 4} = \sqrt{21}$$
The foci are $(\pm\sqrt{21}, 0)$.

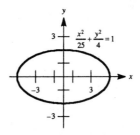

11. Here $(x_0, y_0) = (-1, 0)$, $A = 2$, $B = -1$ and $C = -1$, so using the distance formula from a point to a line yields:
$$d = \frac{|Ax_0 + By_0 + C|}{\sqrt{A^2 + B^2}} = \frac{|-2 + 0 - 1|}{\sqrt{4 + 1}} = \frac{3}{\sqrt{5}} = \frac{3\sqrt{5}}{5}$$

12. Complete the square to convert the equation to standard form:
$$16x^2 + y^2 - 64x + 2y + 65 = 0$$
$$16(x^2 - 4x) + (y^2 + 2y) = -65$$
$$16(x^2 - 4x + 4) + (y^2 + 2y + 1) = -65 + 64 + 1$$
$$16(x - 2)^2 + (y + 1)^2 = 0$$
Since the only solution to this equation is the point $(2, -1)$, the graph consists of a single point.

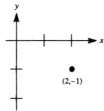

13.   The equation represents a hyperbola with center $(-4, 4)$. Graphing the hyperbola:

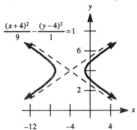

14.   Comparing the given equation with the four basic types, it appears this is the type associated with Figure 3 (Section 11.7). Divide both numerator and denominator by 5 to obtain:

$$r = \frac{9}{5 - 4\cos\theta} = \frac{\frac{9}{5}}{1 - \frac{4}{5}\cos\theta}$$

Since the eccentricity is $e = \frac{4}{5} < 1$, this conic is an ellipse. Computing the values of r when $\theta = 0, \frac{\pi}{2}, \pi$ and $\frac{3\pi}{2}$:

θ	0	$\frac{\pi}{2}$	π	$\frac{3\pi}{2}$
r	9	$\frac{9}{5}$	1	$\frac{9}{5}$

So the vertices on the major axis are $(9, 0)$ and $(-1, 0)$, and thus the center is $(4, 0)$ and the length of the major axis is $2a = 10$, so $a = 5$. Since a focus is $(0, 0)$, then $c = 4$ and thus:

$$b = \sqrt{a^2 - c^2} = \sqrt{25 - 16} = \sqrt{9} = 3$$

Graph the ellipse:

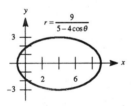

15.   Since $4p = 8$, then $p = 2$. The focal width is 8 and the vertex is $(1, 2)$.

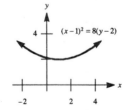

16. (a) Since $a = 6$ and $b = 5$, we find c:
$$c = \sqrt{a^2 - b^2} = \sqrt{36 - 25} = \sqrt{11}$$
So the eccentricity is $e = \frac{c}{a} = \frac{\sqrt{11}}{6}$. Thus the directrices are given by:
$$x = \pm\frac{a}{e} = \pm\frac{6}{\frac{\sqrt{11}}{6}} = \pm\frac{36}{\sqrt{11}} = \pm\frac{36\sqrt{11}}{11}$$

(b) The focal radii are given by:
$$F_1P = a + ex = 6 + \frac{\sqrt{11}}{6}(3) = 6 + \frac{\sqrt{11}}{2} = \frac{12 + \sqrt{11}}{2}$$
$$F_2P = a - ex = 6 - \frac{\sqrt{11}}{6}(3) = 6 - \frac{\sqrt{11}}{2} = \frac{12 - \sqrt{11}}{2}$$

17. The equation of the tangent line is $\frac{x_1 x}{a^2} + \frac{y_1 y}{b^2} = 1$. With $x_1 = -2$ and $y_1 = 4$, this

becomes $\frac{-2x}{a^2} + \frac{4y}{b^2} = 1$. To determine a^2 and b^2, we divide both sides of the

equation $x^2 + 3y^2 = 52$ by 52. This yields $\frac{x^2}{52} + \frac{y^2}{52/3} = 1$. Thus $a^2 = 52$ and $b^2 = \frac{52}{3}$,

and the equation of the tangent becomes $\frac{-2x}{52} + \frac{4y}{52/3} = 1$. When we simplify and solve

for y, the result is $y = \frac{1}{6}x + \frac{13}{3}$. This is the equation of the tangent line, as required.

18. Let m denote the slope for the required tangent line. Then the equation of the line is
$y - 8 = m(x - 4)$. Because this is the tangent line, the system
$$y - 8 = m(x - 4)$$
$$x^2 = 2y$$
must have exactly one solution, namely $(4, 8)$. Solving the second equation yields
$y = \frac{x^2}{2}$. Substitute into the first equation:

$$\frac{x^2}{2} - 8 = m(x - 4)$$
$$x^2 - 16 = 2m(x - 4)$$
$$(x - 4)(x + 4) = 2m(x - 4)$$
$$(x - 4)(x + 4 - 2m) = 0$$
The solutions are $x = 4$ and $x = 2m - 4$. But since the line is tangent to the parabola,
these two x-values must be equal:
$$2m - 4 = 4$$
$$2m = 8$$
$$m = 4$$

With this value for m, the equation $y - 8 = m(x - 4)$ becomes:

$$y - 8 = 4(x - 4)$$
$$y - 8 = 4x - 16$$
$$y = 4x - 8$$

This is the required tangent line.

Chapter Fourteen
Additional Topics in Algebra

14.1 Mathematical Induction

1. Here P_n denotes the statement:

$$1 + 2 + 3 + \ldots + n = \frac{n(n+1)}{2}$$

Since $1 = \frac{1(1+1)}{2} = 1$, P_1 is true. It remains to show that P_k implies P_{k+1}. Assume P_k is true:

$$1 + 2 + \ldots + k = \frac{k(k+1)}{2}$$

Thus:

$$1 + 2 + \ldots + k + k + 1 = \frac{k(k+1)}{2} + k + 1$$
$$= (k+1)\left(\frac{k}{2} + 1\right)$$
$$= \frac{(k+1)(k+2)}{2}$$
$$= \frac{(k+1)[(k+1)+1]}{2}$$

So P_{k+1} is true and the induction is complete.

3. Here P_n denotes the statement:
$$1 + 4 + 7 + \ldots + (3n - 2) = \frac{n(3n - 1)}{2}$$

Since $1 = \frac{1(3(1)-1)}{2} = 1$, P_1 is true. Assume P_k is true:
$$1 + 4 + \ldots + (3k - 2) = \frac{k(3k - 1)}{2}$$

Thus:
$$1 + 4 + \ldots + (3k - 2) + [3(k + 1) - 2] = \frac{k(3k - 1)}{2} + 3(k + 1) - 2$$
$$= \tfrac{1}{2}\left[k(3k - 1) + 6(k + 1) - 4\right]$$
$$= \tfrac{1}{2}\left(3k^2 - k + 6k + 2\right)$$
$$= \tfrac{1}{2}\left(3k^2 + 5k + 2\right)$$
$$= \frac{(k + 1)(3k + 2)}{2}$$
$$= \frac{(k + 1)[3(k + 1) - 1]}{2}$$

So P_{k+1} is true and the induction is complete.

5. Here P_n denotes:
$$1^2 + 2^2 + 3^2 + \ldots + n^2 = \frac{n(n + 1)(2n + 1)}{6}$$

Since $1^2 = 1 = \frac{1(1+1)[2(1)+1]}{6} = \frac{1(2)(3)}{6} = 1$, P_1 is true. Assume P_k is true:
$$1^2 + 2^2 + \ldots + k^2 = \frac{k(k + 1)(2k + 1)}{6}$$

Thus:
$$1^2 + 2^2 + \ldots + k^2 + (k + 1)^2 = \frac{k(k + 1)(2k + 1)}{6} + (k + 1)^2$$
$$= \frac{(k + 1)[k(2k + 1) + 6(k + 1)^2]}{6}$$
$$= \frac{(k + 1)\left(2k^2 + k + 6k + 6\right)}{6}$$
$$= \frac{(k + 1)\left(2k^2 + 7k + 6\right)}{6}$$
$$= \frac{(k + 1)(k + 2)(2k + 3)}{6}$$
$$= \frac{(k + 1)[(k + 1) + 1][2(k + 1) + 1]}{6}$$

So P_{k+1} is true and the induction is complete.

7.   Here P_n denotes:

$$1^2 + 3^2 + 5^2 + ... + (2n-1)^2 = \frac{n(2n-1)(2n+1)}{3}$$

As $1^2 = 1 = \dfrac{1[2(1)-1][2(1)+1]}{3} = 1$, P_1 is true. Assume P_k is true:

$$1^2 + 3^2 + ... + (2k-1)^2 = \frac{k(2k-1)(2k+1)}{3}$$

Thus:

$$1^2 + 3^2 + ... + (2k-1)^2 + [2(k+1)-1]^2 = \frac{k(2k-1)(2k+1)}{3} + [2(k+1)-1]^2$$

$$= \frac{k(2k-1)(2k+1)}{3} + \frac{3(2k+1)^2}{3}$$

$$= \frac{2k+1}{3}[k(2k-1) + 3(2k+1)]$$

$$= \frac{[2(k+1)-1]}{3}(2k^2 + 5k + 3)$$

$$= \frac{(k+1)[2(k+1)-1](2k+3)}{3}$$

$$= \frac{(k+1)[2(k+1)-1][2(k+1)+1]}{3}$$

So P_{k+1} is true and the induction is complete.

9.   Here P_n denotes:

$$3 + 3^2 + 3^3 + ... + 3^n = \tfrac{1}{2}(3^{n+1} - 3)$$

Since $3 = \tfrac{1}{2}(3^2 - 3) = \tfrac{1}{2}(6) = 3$, P_1 is true. Assume P_k is true:

$$3 + 3^2 + ... + 3^k = \tfrac{1}{2}(3^{k+1} - 3)$$

Thus:

$$3 + 3^2 + ... + 3^k + 3^{k+1} = \tfrac{1}{2}(3^{k+1} - 3) + 3^{k+1}$$

$$= \tfrac{3}{2}(3^{k+1}) - \tfrac{3}{2}$$

$$= \tfrac{1}{2}[3(3^{k+1}) - 3]$$

$$= \tfrac{1}{2}(3^{k+2} - 3)$$

$$= \tfrac{1}{2}[3^{(k+1)+1} - 3]$$

So P_{k+1} is true and the induction is complete.

11. Here P_n denotes:

$$1^3 + 2^3 + 3^3 + \ldots + n^3 = \left[\frac{n(n+1)}{2}\right]^2$$

Since $1^3 = 1 = \left[\frac{1(2)}{2}\right]^2 = 1$, P_1 is true. Assume P_k is true:

$$1^3 + 2^3 + \ldots + k^3 = \left[\frac{k(k+1)}{2}\right]^2$$

Thus:

$$1^3 + 2^3 + \ldots + k^3 + (k+1)^3 = \left[\frac{k(k+1)}{2}\right]^2 + (k+1)^3$$

$$= (k+1)^2\left[\frac{k^2}{4} + (k+1)\right]$$

$$= \frac{(k+1)^2}{4}\left[k^2 + 4k + 4\right]$$

$$= \frac{(k+1)^2(k+2)^2}{4}$$

$$= \frac{(k+1)^2[(k+1)+1]^2}{2^2}$$

$$= \left(\frac{(k+1)[(k+1)+1]}{2}\right)^2$$

So P_{k+1} is true and the induction is complete.

13. Here P_n denotes:

$$1^3 + 3^3 + 5^3 + \ldots + (2n-1)^3 = n^2(2n^2 - 1)$$

Since $1^3 = 1(1)^2[2(1)^2 - 1] = (1)(1) = 1$, P_1 is true. Assume P_k is true:

$$1^3 + 3^3 + \ldots + (2k-1)^3 = k^2(2k^2 - 1)$$

Thus:

$$1^3 + 3^3 + \ldots + (2k-1)^3 + [2(k+1)-1]^3 = k^2(2k^2 - 1) + (2k+1)^3$$

$$= 2k^4 - k^2 + 8k^3 + 12k^2 + 6k + 1$$

$$= 2k^4 + 8k^3 + 11k^2 + 6k + 1$$

$$= (k+1)^2(2k^2 + 4k + 1)$$

$$= (k+1)^2[2(k^2 + 2k + 1) - 1]$$

$$= (k+1)^2[2(k+1)^2 - 1]$$

So P_{k+1} is true and the induction is complete.

15. Here P_n denotes:

$$1 \times 3 + 3 \times 5 + 5 \times 7 + \ldots + (2n-1)(2n+1) = \frac{n(4n^2 + 6n - 1)}{3}$$

Since $1 \times 3 = 3 = \dfrac{(1)\left[4(1)^2 + 6(1) - 1\right]}{3} = \frac{9}{3} = 3$, P_1 is true. Assume P_k is true:

$$1 \times 3 + 3 \times 5 + \ldots + (2k-1)(2k+1) = \frac{k(4k^2 + 6k - 1)}{3}$$

Thus:

$$1 \times 3 + 3 \times 5 + \ldots + (2k-1)(2k+1) + \left[2(k+1) - 1\right]\left[2(k+1) + 1\right]$$

$$= \frac{k(4k^2 + 6k - 1)}{3} + (2k+1)(2k+3)$$

$$= \frac{4k^3 + 6k^2 - k + 12k^2 + 24k + 9}{3}$$

$$= \frac{4k^3 + 18k^2 + 23k + 9}{3}$$

Noting that -1 is a root:

$$= \frac{(k+1)(4k^2 + 14k + 9)}{3}$$

$$= \frac{(k+1)\left[4(k+1)^2 + 6(k+1) - 1\right]}{3}$$

So P_{k+1} is true and the induction is complete.

17. Here P_n denotes:

$$1 + \frac{3}{2} + \frac{5}{2^2} + \ldots + \frac{2n-1}{2^{n-1}} = 6 - \frac{2n+3}{2^{n-1}}$$

Since $1 = 6 - \dfrac{[2(1) + 3]}{2^0} = 6 - 5 = 1$, P_1 is true. Assume P_k is true:

$$1 + \frac{3}{2} + \ldots + \frac{2k-1}{2^{k-1}} = 6 - \frac{2k+3}{2^{k-1}}$$

Thus:

$$1 + \frac{3}{2} + \ldots + \frac{2k-1}{2^{k-1}} + \frac{2(k+1) - 1}{2^k} = 6 - \frac{2k+3}{2^{k-1}} + \frac{2k+1}{2^k}$$

$$= 6 - \frac{4k + 6 - 2k - 1}{2^k}$$

$$= 6 - \frac{2k+5}{2^k}$$

$$= 6 - \frac{2(k+1) + 3}{2^{(k+1)-1}}$$

So P_{k+1} is true and the induction is complete.

19. Let P_n denote the statement:

$$n \leq 2^{n-1}$$

Since $2^{1-1} = 2^0 = 1 \geq 1$, P_1 is true. Assume P_k is true:

$$k \leq 2^{k-1}$$

Hence:

$$k + 1 \leq 2^{k-1} + 1$$

Since $1 \leq 2^{k-1}$ (use induction to prove this), we have:

$$k + 1 \leq 2^{k-1} + 1 \leq 2^{k-1} + 2^{k-1} = 2\left(2^{k-1}\right) = 2^k = 2^{(k+1)-1}$$

So P_{k+1} is true and the induction is complete.

21. For $n = 2$:

$$2^2 + 4 = 8 < (2+1)^2$$

Assuming:

$$k^2 + 4 < (k+1)^2$$

Then:

$$k^2 + 2k + 4 < (k+1)^2 + 2k$$
$$(k+1)^2 + 3 < k^2 + 4k + 1$$
$$(k+1)^2 + 4 < k^2 + 4k + 2$$
$$(k+1)^2 + 4 < k^2 + 4k + 4$$
$$(k+1)^2 + 4 < (k+2)^2$$

This completes the induction for $n \geq 2$.

23. For $n = 7$, we have:

$$(1.5)^7 \approx 17.09 > 2(7), \text{ so } P_1 \text{ is true}$$

Assume P_k is true, so:

$$(1.5)^k > 2k$$

Then for $k \geq 7$:

$$\begin{aligned} (1.5)^{k+1} &= 1.5(1.5)^k \\ &> 1.5(2k) \\ &= 3k \\ &= 2k + k \\ &> 2k + 2 \quad (\text{since } k > 2) \\ &= 2(k+1) \end{aligned}$$

So P_{k+1} is true and the induction is complete.

25. For $n = 39$, we have:

$$(1.1)^{39} \approx 41.1 > 39, \text{ so } P_{39} \text{ is true}$$

Assume P_k is true for some $k \geq 39$, so:

$$(1.1)^k > k$$

Then:

$$(1.1)^{k+1} = 1.1(1.1)^k > 1.1k = k + 0.1k > k + 1 \quad (\text{since } k > 10)$$

So P_{k+1} is true and the induction is complete.

27. (a) Complete the table:

n	1	2	3	4	5
$f(n)$	$\frac{1}{2}$	$\frac{2}{3}$	$\frac{3}{4}$	$\frac{4}{5}$	$\frac{5}{6}$

(b) $\frac{6}{7}$, $f(6) = f(5) + \frac{1}{42} = \frac{5}{6} + \frac{1}{42} = \frac{36}{42} = \frac{6}{7}$

(c) $f(n) = \dfrac{1}{1 \times 2} + \dfrac{1}{2 \times 3} + \ldots + \dfrac{1}{n(n+1)} = \dfrac{n}{n+1}$

$f(1) = \frac{1}{2} = \dfrac{1}{1+1} = \frac{1}{2}$

Assuming $f(k) = \dfrac{k}{k+1}$, then:

$$f(k+1) = f(k) + \frac{1}{(k+1)(k+2)}$$
$$= \frac{k}{k+1} + \frac{1}{(k+1)(k+2)}$$
$$= \frac{k(k+2)+1}{(k+1)(k+2)}$$
$$= \frac{k^2 + 2k + 1}{(k+1)(k+2)}$$
$$= \frac{(k+1)^2}{(k+1)(k+2)}$$
$$= \frac{k+1}{k+2}$$

This completes the induction.

29. (a) Complete the table:

n	1	2	3	4	5
$f(n)$	1	4	9	16	25

(b) 36, $f(6) = f(5) + 2\sqrt{f(5)} + 1 = 25 + 2\sqrt{25} + 1 = 36$

(c) $f(n) = n^2$
$f(1) = 1 = 1^2$
Assume that for some $k \geq 1$ that $f(k) = k^2$. Then we have:
$$f(k+1) = f(k) + 2\sqrt{f(k)} + 1 = k^2 + 2\sqrt{k^2} + 1 = k^2 + 2k + 1 = (k+1)^2$$
The induction is complete.

31. For $n = 1$: 13 is prime and P_1 is true
For $n = 2$: 17 is prime and P_2 is true
For $n = 3$: 23 is prime and P_3 is true
For $n = 4$: 31 is prime and P_4 is true
For $n = 5$: 41 is prime and P_5 is true
For $n = 6$: 53 is prime and P_6 is true
For $n = 7$: 67 is prime and P_7 is true
For $n = 8$: 83 is prime and P_8 is true
For $n = 9$: 101 is prime and P_9 is true
However, for $n = 10$, we have $(10)^2 + 10 + 11 = 121 = (11)(11)$ and so P_{10} is false.

33. Let P_n denote:

$$1 + r + r^2 + \ldots + r^{n-1} = \frac{r^n - 1}{r - 1} \text{ when } r \neq 1$$

Then since $1 = \dfrac{r - 1}{r - 1} = 1$, P_1 is true. Assuming P_k is true, then:

$$1 + r + \ldots + r^{k-1} = \frac{r^k - 1}{r - 1}$$

Thus:

$$1 + r + \ldots + r^{k-1} + r^k = \frac{r^k - 1}{r - 1} + r^k = \frac{r^k - 1 + r^k(r - 1)}{r - 1} = \frac{r^{k+1} - 1}{r - 1}$$

So P_{k+1} is true and the induction is complete.

35. Let P_n denote:
$$n^5 - n = 5R \text{ for some natural number } R$$
Since $2^5 - 2 = 32 - 2 = 30 = 5(6)$, P_2 is true. Since:

$$(k + 1)^5 - (k + 1) = k^5 + 5k^4 + 10k^3 + 10k^2 + 4k = k^5 - k + 5\left(k^4 + 2k^3 + 2k^2 + k\right)$$

Assuming P_k is true results in:
$$(k + 1)^5 - (k + 1) = 5(R + k^4 + 2k^3 + 2k^2 + k)$$
So P_{k+1} is true and the induction is complete.

37. For $n = 0$, we have:

$$2^1 + 3^1 = 5 = 5(1), \text{ so } P_0 \text{ is true}$$

Assume P_k is true, so:

$$2^{2k+1} + 3^{2k+1} = 5A, \text{ for some natural number } A$$

Then:

$$2^{2k+3} + 3^{2k+3} = 4 \cdot 2^{2k+1} + 9 \cdot 3^{2k+1}$$
$$= 4 \cdot 2^{2k+1} + 9\left(5A - 2^{2k+1}\right)$$
$$= 4 \cdot 2^{2k+1} + 45A - 9 \cdot 2^{2k+1}$$
$$= 45A - 5 \cdot 2^{2k+1}$$
$$= 5\left(9A - 2^{2k+1}\right)$$

Since 5 is a factor of this expression, P_{k+1} is true and the induction is complete.

39. For $n = 0$, we have:

$$2^1 + (-1)^0 = 2 + 1 = 3 = 3(1), \text{ so } P_0 \text{ is true}$$

Assume P_k is true, so:

$$2^{k+1} + (-1)^k = 3A, \text{ for some natural number } A$$

Then:

$$2^{k+2} + (-1)^{k+1} = 2 \cdot 2^{k+1} - (-1)^k$$
$$= 2\left[3A - (-1)^k\right] - (-1)^k$$
$$= 6A - 2(-1)^k - (-1)^k$$
$$= 6A - 3(-1)^k$$
$$= 3\left[2A - (-1)^k\right]$$

Since 3 is a factor of this expression, P_{k+1} is true and the induction is complete.

41. Let P_n denote:

$$x^n - y^n = (x - y)(Q(x, y)) \text{ where } Q(x, y) \text{ is a polynomial in } x \text{ and } y.$$

Then:

$$x - y = (x - y)(1), \text{ so } P_1 \text{ is true}$$

While using the suggested identity:

$$x^{k+1} - y^{k+1} = x^k(x - y) + (x^k - y^k)y$$

So assuming P_k, we have $x^k - y^k = (x - y)Q(x, y)$:

$$x^{k+1} - y^{k+1} = \left(x^k + yQ(x, y)\right)(x - y)$$

So P_{k+1} is true and the induction is complete.

43. For $n = 1$:

$$(1 + p) \geq (1 + p), \text{ so } P_1 \text{ is true}$$

When $p > -1$ then $p + 1 > 0$. Assuming $(1 + p)^k \geq 1 + kp$ then:

$$(1 + p)^{k+1} \geq (1 + kp)(1 + p) \geq 1 + p + kp + kp^2 = 1 + (k + 1)p + kp^2 \geq 1 + (k + 1)p$$

The induction is complete.

14.2 The Binomial Theorem

1.   (a) Compute:
$$(a+b)^2 = (a+b)(a+b) = a^2 + ab + ba + b^2 = a^2 + 2ab + b^2$$

   (b) Compute:
$$\begin{aligned}(a+b)^3 &= (a+b)(a+b)^2\\ &= (a+b)(a^2 + 2ab + b^2)\\ &= a^3 + 2a^2b + ab^2 + ba^2 + 2ab^2 + b^3\\ &= a^3 + 3a^2b + 3ab^2 + b^3\end{aligned}$$

3.   Expand using the binomial theorem:
$$\begin{aligned}(a+b)^9 &= \binom{9}{0}a^9 + \binom{9}{1}a^8b + \binom{9}{2}a^7b^2 + \binom{9}{3}a^6b^3 + \binom{9}{4}a^5b^4 + \binom{9}{5}a^4b^5\\ &\quad + \binom{9}{6}a^3b^6 + \binom{9}{7}a^2b^7 + \binom{9}{8}ab^8 + \binom{9}{9}b^9\\ &= a^9 + 9a^8b + 36a^7b^2 + 84a^6b^3 + 126a^5b^4 + 126a^4b^5 + 84a^3b^6\\ &\quad + 36a^2b^7 + 9ab^8 + b^9\end{aligned}$$

5.   Expand using the binomial theorem:
$$\begin{aligned}(2A+B)^3 &= \binom{3}{0}(2A)^3 + \binom{3}{1}(2A)^2B + \binom{3}{2}(2A)B^2 + \binom{3}{3}B^3\\ &= 8A^3 + 12A^2B + 6AB^2 + B^3\end{aligned}$$

7.   Expand using the binomial theorem:
$$\begin{aligned}(1-2x)^6 &= \binom{6}{0}1^6 - \binom{6}{1}1^5(2x) + \binom{6}{2}1^4(2x)^2 - \binom{6}{3}1^3(2x)^3\\ &\quad + \binom{6}{4}1^2(2x)^4 - \binom{6}{5}1(2x)^5 + \binom{6}{6}(2x)^6\\ &= 1 - 12x + 60x^2 - 160x^3 + 240x^4 - 192x^5 + 64x^6\end{aligned}$$

9.   Expand using the binomial theorem:
$$\left(\sqrt{x} + \sqrt{y}\right)^4$$
$$\begin{aligned} &= \binom{4}{0}(\sqrt{x})^4 + \binom{4}{1}(\sqrt{x})^3(\sqrt{y}) + \binom{4}{2}(\sqrt{x})^2(\sqrt{y})^2 + \binom{4}{3}(\sqrt{x})(\sqrt{y})^3 + \binom{4}{4}(\sqrt{y})^4\\ &= x^2 + 4x\sqrt{xy} + 6xy + 4y\sqrt{xy} + y^2\end{aligned}$$

11. Expand using the binomial theorem:

$$(x^2 + y^2)^5 = \binom{5}{0}(x^2)^5 + \binom{5}{1}(x^2)^4 y^2 + \binom{5}{2}(x^2)^3(y^2)^2 + \binom{5}{3}(x^2)^2(y^2)^3$$
$$+ \binom{5}{4}(x^2)(y^2)^4 + \binom{5}{5}(y^2)^5$$
$$= x^{10} + 5x^8 y^2 + 10x^6 y^4 + 10x^4 y^6 + 5x^2 y^8 + y^{10}$$

13. Expand using the binomial theorem:

$$\left(1 - \frac{1}{x}\right)^6 = \binom{6}{0} - \binom{6}{1}\frac{1}{x} + \binom{6}{2}\left(\frac{1}{x}\right)^2 - \binom{6}{3}\left(\frac{1}{x}\right)^3 + \binom{6}{4}\left(\frac{1}{x}\right)^4 - \binom{6}{5}\left(\frac{1}{x}\right)^5 + \binom{6}{6}\left(\frac{1}{x}\right)^6$$
$$= 1 - \frac{6}{x} + \frac{15}{x^2} - \frac{20}{x^3} + \frac{15}{x^4} - \frac{6}{x^5} + \frac{1}{x^6}$$

15. Expand using the binomial theorem:

$$\left(\frac{x}{2} - \frac{y}{3}\right)^3 = \binom{3}{0}\left(\frac{x}{2}\right)^3 - \binom{3}{1}\left(\frac{x}{2}\right)^2\left(\frac{y}{3}\right) + \binom{3}{2}\left(\frac{x}{2}\right)\left(\frac{y}{3}\right)^2 - \binom{3}{3}\left(\frac{y}{3}\right)^3$$
$$= \frac{x^3}{8} - \frac{x^2 y}{4} + \frac{xy^2}{6} - \frac{y^3}{27}$$

17. Expand using the binomial theorem:

$$(ab^2 + c)^7 = \binom{7}{0}(ab^2)^7 + \binom{7}{1}(ab^2)^6 c + \binom{7}{2}(ab^2)^5 c^2 + \binom{7}{3}(ab^2)^4 c^3 + \binom{7}{4}(ab^2)^3 c^4$$
$$+ \binom{7}{5}(ab^2)^2 c^5 + \binom{7}{6}(ab^2)c^6 + \binom{7}{7}c^7$$
$$= a^7 b^{14} + 7a^6 b^{12} c + 21a^5 b^{10} c^2 + 35a^4 b^8 c^3 + 35a^3 b^6 c^4 + 21a^2 b^4 c^5$$
$$+ 7ab^2 c^6 + c^7$$

19. Expand using the binomial theorem:

$$(x + \sqrt{2})^8 = \binom{8}{0}x^8 + \binom{8}{1}x^7\sqrt{2} + \binom{8}{2}x^6(\sqrt{2})^2 + \binom{8}{3}x^5(\sqrt{2})^3 + \binom{8}{4}x^4(\sqrt{2})^4$$
$$+ \binom{8}{5}x^3(\sqrt{2})^5 + \binom{8}{6}x^2(\sqrt{2})^6 + \binom{8}{7}x(\sqrt{2})^7 + \binom{8}{8}(\sqrt{2})^8$$
$$= x^8 + 8\sqrt{2}x^7 + 56x^6 + 112\sqrt{2}x^5 + 280x^4 + 224\sqrt{2}x^3 + 224x^2$$
$$+ 64\sqrt{2}x + 16$$

21. Expand using the binomial theorem:

$$(\sqrt{2} - 1)^3 = \binom{3}{0}(\sqrt{2})^3 - \binom{3}{1}(\sqrt{2})^2 + \binom{3}{2}(\sqrt{2}) - \binom{3}{3} = 2\sqrt{2} - 6 + 3\sqrt{2} - 1 = 5\sqrt{2} - 7$$

23. Expand using the binomial theorem:

$$(\sqrt{2}+\sqrt{3})^5 = \binom{5}{0}(\sqrt{2})^5 + \binom{5}{1}(\sqrt{2})^4\sqrt{3} + \binom{5}{2}(\sqrt{2})^3(\sqrt{3})^2 + \binom{5}{3}(\sqrt{2})^2(\sqrt{3})^3$$

$$+ \binom{5}{4}(\sqrt{2})(\sqrt{3})^4 + \binom{5}{5}(\sqrt{3})^5$$

$$= 4\sqrt{2} + 20\sqrt{3} + 60\sqrt{2} + 60\sqrt{3} + 45\sqrt{2} + 9\sqrt{3}$$

$$= 89\sqrt{3} + 109\sqrt{2}$$

25. Expand using the binomial theorem:

$$(2\sqrt[3]{2}-\sqrt[3]{4})^3 = \binom{3}{0}(2\sqrt[3]{2})^3 - \binom{3}{1}(2\sqrt[3]{2})^2(\sqrt[3]{4}) + \binom{3}{2}(2\sqrt[3]{2})(\sqrt[3]{4})^2 - \binom{3}{3}(\sqrt[3]{4})^3$$

$$= 16 - 12\sqrt[3]{16} + 6\sqrt[3]{32} - 4$$

$$= 12 - 24\sqrt[3]{2} + 12\sqrt[3]{4}$$

27. Expand using the binomial theorem:

$$\left[x^2 - (2x+1)\right]^5$$

$$= \binom{5}{0}(x^2)^5 - \binom{5}{1}(x^2)^4(2x+1) + \binom{5}{2}(x^2)^3(2x+1)^2 - \binom{5}{3}(x^2)^2(2x+1)^3$$

$$+ \binom{5}{4}(x^2)(2x+1)^4 - \binom{5}{5}(2x+1)^5$$

$$= x^{10} - \left[5x^8(2x+1)\right] + \left[10x^6(4x^2+4x+1)\right] - \left[10x^4(8x^3+12x^2+6x+1)\right]$$

$$+ \left[5x^2(16x^4+32x^3+24x^2+8x+1)\right]$$

$$- (32x^5+80x^4+80x^3+40x^2+10x+1)$$

$$= x^{10} - (10x^9+5x^8) + (40x^8+40x^7+10x^6) - (80x^7+120x^6+60x^5+10x^4)$$

$$+ (80x^6+160x^5+120x^4+40x^3+5x^2)$$

$$- (32x^5+80x^4+80x^3+40x^2+10x+1)$$

$$= x^{10} - 10x^9 + 35x^8 - 40x^7 - 30x^6 + 68x^5 + 30x^4 - 40x^3 - 35x^2 - 10x - 1$$

29. $5! = (5)(4)(3)(2)(1) = 120$

31. Compute:

$$\binom{7}{3}\binom{3}{2} = \frac{7!}{3!(4!)} \cdot \frac{3!}{2!(1!)} = \frac{7\cdot 6\cdot 5}{3\cdot 2} \cdot 3 = 105$$

33. (a) Compute:

$$\binom{5}{3} = \frac{5!}{3!(5-3)!} = \frac{5(4)}{2} = 10$$

(b) Compute:

$$\binom{5}{4} = \frac{5!}{4!(5-4)!} = 5$$

35. Simplify:

$$\frac{(n+2)!}{n!} = \frac{(n+2)(n+1)n!}{n!} = n^2 + 3n + 2$$

37. Compute:

$$\binom{6}{4} + \binom{6}{3} - \binom{7}{4} = \frac{6!}{4!2!} + \frac{6!}{3!3!} - \frac{7!}{4!3!}$$
$$= \frac{(6)(5)}{2} + \frac{(6)(5)(4)}{6} - \frac{(7)(6)(5)}{6}$$
$$= 15 + 20 - 35$$
$$= 0$$

39. The fifteenth term will be given by:

$$\binom{16}{14} a^2 b^{14} = 120 a^2 b^{14}$$

41. The one-hundredth term will be given by:

$$\binom{100}{99} x^{99} = 100 x^{99}$$

43. Here $n - r + 1 = 8$, so $10 - r + 1 = 8$, so $r = 3$. Hence the coefficient is:

$$\binom{10}{2} = 45$$

45. Here $r - 1 = 8$, thus $r = 9$. Hence the coefficient is:

$$\frac{1}{2}(-4)^8 \binom{9}{8} = 9 \cdot \frac{1}{2}(65536) = 294912$$

47. Here $r - 1 = 6$. Hence $r = 7$, and the coefficient is:

$$\binom{8}{6} = \frac{8(7)}{2} = 28$$

49. Here $(12 - r + 1)(-1) + 2(r - 1) = 0$, so $-12 + r - 1 + 2r - 2 = 0$, then $3r = 15$, so $r = 5$ and the coefficient is:

$$(3)^4\binom{12}{4} = \frac{(12)(11)(10)(9)}{(4)(3)(2)}(3)^4 = 495(81) = 40095$$

51. Here $r - 1 = n$, so $r = n + 1$, so the coefficient is:

$$\binom{2n}{n} = \frac{(2n)!}{n!n!} = \frac{(2n)!}{(n!)^2}$$

53. (a) Complete the table:

k	0	1	2	3	4	5	6	7	8
$\binom{8}{k}$	1	8	28	56	70	56	28	8	1

   (b) We have $1 + 8 + 28 + 56 + 70 + 56 + 28 + 8 + 1 = 256$, while $2^8 = 256$.

   (c) Since a^L and b^L equal 1 for any L because $a = b = 1$, from the binomial theorem:

$$2^n = (1+1)^n = \binom{n}{0} + \binom{n}{1} + \ldots + \binom{n}{n}$$

55. (a) Simplify:

$$(1+x)^n\left(1+\frac{1}{x}\right)^n = (1+x)^n\left(\frac{x+1}{x}\right)^n = (1+x)^n \cdot \frac{(1+x)^n}{x^n} = \frac{(1+x)^{2n}}{x^n}$$

   (b) The nth term of expansion on the right is:

$$\frac{\binom{2n}{n}(1)^n(x)^n}{x^n} = \binom{2n}{n}$$

   This verifies the result.

   (c) Using the binomial theorem:

$$(1+x)^n = \binom{n}{0} + \binom{n}{1}x + \binom{n}{2}x^2 + \binom{n}{3}x^3 + \ldots + \binom{n}{n}x^n$$

$$\left(1+\frac{1}{x}\right)^n = \binom{n}{0} + \binom{n}{1}x^{-1} + \binom{n}{2}x^{-2} + \binom{n}{3}x^{-3} + \ldots + \binom{n}{n}x^{-n}$$

When multiplied, the terms that do not contain x come from terms with corresponding positive and negative exponents:

$$\binom{n}{1}x \cdot \binom{n}{1}x^{-1} + \binom{n}{2}x^2 \cdot \binom{n}{2}x^{-2} + \dots + \binom{n}{n}x^n \cdot \binom{n}{n}x^{-n}$$

$$= \binom{n}{1}^2 + \binom{n}{2}^2 + \dots + \binom{n}{n}^2$$

This verifies the identity.

14.3 Introduction to Sequences and Series

1. Since $a_n = \dfrac{n}{n+1}$, we can write out the first four terms by setting n equal to the natural numbers 1 through 4:

$$a_1 = \frac{1}{1+1} = \tfrac{1}{2}, \quad a_2 = \frac{2}{2+1} = \tfrac{2}{3}, \quad a_3 = \frac{3}{3+1} = \tfrac{3}{4}, \quad a_4 = \frac{4}{4+1} = \tfrac{4}{5}$$

So the first four terms are $\tfrac{1}{2}, \tfrac{2}{3}, \tfrac{3}{4}, \tfrac{4}{5}$.

3. Here $b_n = (-1)^n$, so $b_1 = -1$, $b_2 = 1$, $b_3 = -1$, and $b_4 = 1$. The first four terms are $-1, 1, -1, 1$.

5. Here $c_n = 2^{-n}$, so:

$$c_1 = 2^{-1} = \tfrac{1}{2}, \quad c_2 = 2^{-2} = \tfrac{1}{4}, \quad c_3 = 2^{-3} = \tfrac{1}{8}, \quad c_4 = 2^{-4} = \tfrac{1}{16}$$

The first four terms are $\tfrac{1}{2}, \tfrac{1}{4}, \tfrac{1}{8}, \tfrac{1}{16}$.

7. Here $x_n = 3n$, so:

$$x_1 = 3(1) = 3, \quad x_2 = 3(2) = 6, \quad x_3 = 3(3) = 9, \quad x_4 = 3(4) = 12$$

The first four terms are $3, 6, 9, 12$.

9. Here $b_n = \left(1 + \tfrac{1}{n}\right)^n$, so:

$$b_1 = \left(1 + \tfrac{1}{1}\right)^1 = 2, \quad b_2 = \left(1 + \tfrac{1}{2}\right)^2 = \tfrac{9}{4}, \quad b_3 = \left(1 + \tfrac{1}{3}\right)^3 = \tfrac{64}{27}, \quad b_4 = \left(1 + \tfrac{1}{4}\right)^4 = \tfrac{625}{256}$$

The first four terms are $2, \tfrac{9}{4}, \tfrac{64}{27}, \tfrac{625}{256}$.

11. Here $a_n = \dfrac{n-1}{n+1}$, so:

$$a_0 = \frac{0-1}{0+1} = -1, \quad a_1 = \frac{1-1}{1+1} = 0, \quad a_2 = \frac{2-1}{2+1} = \tfrac{1}{3}, \quad a_3 = \frac{3-1}{3+1} = \tfrac{1}{2}$$

The first four terms are $-1, 0, \tfrac{1}{3}, \tfrac{1}{2}$.

13. Here $b_n = \dfrac{(-2)^{n+1}}{(n+1)^2}$, so:

$$b_0 = \frac{(-2)^{0+1}}{(0+1)^2} = -2, \quad b_1 = \frac{(-2)^{1+1}}{(1+1)^2} = 1, \quad b_2 = \frac{(-2)^{2+1}}{(2+1)^2} = -\frac{8}{9}, \quad b_3 = \frac{(-2)^{3+1}}{(3+1)^2} = 1$$

The first four terms are $-2, 1, -\frac{8}{9}, 1$.

15. Here $a_1 = 1$ and $a_n = \left(1 + a_{n-1}\right)^2$ for $n \geq 2$, so:

$$a_1 = 1$$
$$a_2 = \left(1 + a_1\right)^2 = 2^2 = 4$$
$$a_3 = \left(1 + a_2\right)^2 = 5^2 = 25$$
$$a_4 = \left(1 + a_3\right)^2 = 26^2 = 676$$
$$a_5 = \left(1 + a_4\right)^2 = 677^2 = 458329$$

The first five terms are $1, 4, 25, 676, 458329$.

17. Here $a_1 = 2$, $a_2 = 2$, and $a_n = a_{n-1}a_{n-2}$ for $n \geq 3$, so $a_1 = 2$, $a_2 = 2$, $a_3 = 4$, $a_4 = 8$, and $a_5 = 32$. The first five terms are $2, 2, 4, 8, 32$.

19. Here $a_1 = 1$, $a_{n+1} = na_n$ for $n \geq 1$, so $a_1 = 1$, $a_2 = 1$, $a_3 = 2$, $a_4 = 6$, and $a_5 = 24$. The first five terms are $1, 1, 2, 6, 24$.

21. Here $a_1 = 0$, $a_n = 2^{a_{n-1}}$ for $n \geq 2$, so $a_1 = 0$, $a_2 = 1$, $a_3 = 2$, $a_4 = 4$, and $a_5 = 16$. The first five terms are $0, 1, 2, 4, 16$.

23. Since $a_n = 2^n$, the sum is $2 + 4 + 8 + 16 + 32 = 62$.

25. Since $a_n = n^2 - n$, the sum is $0 + 2 + 6 + 12 + 20 = 40$.

27. Since $a_n = \dfrac{(-1)^n}{n!}$, the sum is $-1 + \frac{1}{2} - \frac{1}{6} + \frac{1}{24} - \frac{1}{120} = -\frac{19}{30}$.

29. Here $a_1 = 1$, $a_2 = 2$, and $a_n = a_{n-1}^2 + a_{n-2}^2$ for $n \geq 3$, so the sum is $1 + 2 + 5 + 29 + 866 = 903$.

31. Here $a_1 = 2$ and $a_n = \left(a_{n-1}\right)^2$ for $n \geq 2$, so the sum is $2 + 4 + 16 + 256 = 278$.

33. First compute the required values of a_n as shown in the table:

n	1	2	3	4
a_n	-1	1	-1	1

Now graph these values:

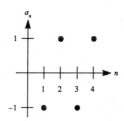

35. First compute the required values of a_n as shown in the table:

n	1	2	3	4
a_n	1	$\frac{1}{2}$	$\frac{1}{3}$	$\frac{1}{4}$

Now graph these values:

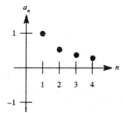

37. First compute the required values of a_n as shown in the table:

n	1	2	3
a_n	0	$\frac{1}{3}$	$\frac{1}{2}$

Now graph these values:

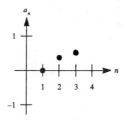

39. Expand the sum:

$$\sum_{k=1}^{3} (k-1) = 0 + 1 + 2 = 3$$

41. Expand the sum:

$$\sum_{k=4}^{5} k^2 = 16 + 25 = 41$$

43. Expand the sum:

$$\sum_{n=1}^{3} x^n = x + x^2 + x^3$$

45.   Expand the sum:

$$\sum_{n=1}^{4}\frac{1}{n} = 1 + \tfrac{1}{2} + \tfrac{1}{3} + \tfrac{1}{4} = \tfrac{25}{12}$$

47.   Expand the sum:

$$\sum_{j=1}^{9}\log_{10}\frac{j}{j+1} = \log_{10}\tfrac{1}{2} + \log_{10}\tfrac{2}{3} + \log_{10}\tfrac{3}{4} + \log_{10}\tfrac{4}{5} + \log_{10}\tfrac{5}{6} + \log_{10}\tfrac{6}{7}$$
$$+ \log_{10}\tfrac{7}{8} + \log_{10}\tfrac{8}{9} + \log_{10}\tfrac{9}{10}$$
$$= \log_{10}\left(\tfrac{1}{2}\cdot\tfrac{2}{3}\cdot\tfrac{3}{4}\cdot\tfrac{4}{5}\cdot\tfrac{5}{6}\cdot\tfrac{6}{7}\cdot\tfrac{7}{8}\cdot\tfrac{8}{9}\cdot\tfrac{9}{10}\right)$$
$$= \log_{10}\tfrac{1}{10}$$
$$= -1$$

49.   Expand the sum:

$$\sum_{j=1}^{6}\left(\frac{1}{j} - \frac{1}{j+1}\right) = \left(1 - \tfrac{1}{2}\right) + \left(\tfrac{1}{2} - \tfrac{1}{3}\right) + \left(\tfrac{1}{3} - \tfrac{1}{4}\right) + \left(\tfrac{1}{4} - \tfrac{1}{5}\right) + \left(\tfrac{1}{5} - \tfrac{1}{6}\right) + \left(\tfrac{1}{6} - \tfrac{1}{7}\right) = 1 - \tfrac{1}{7} = \tfrac{6}{7}$$

51.   Write the sum in sigma notation:

$$5 + 5^2 + 5^3 + 5^4 = \sum_{j=1}^{4} 5^j$$

53.   Write the sum in sigma notation:

$$x + x^2 + x^3 + x^4 + x^5 + x^6 = \sum_{j=1}^{6} x^j$$

55.   Write the sum in sigma notation:

$$1 + \tfrac{1}{2} + \tfrac{1}{3} + \ldots + \tfrac{1}{12} = \sum_{k=1}^{12}\frac{1}{k}$$

57.   Write the sum in sigma notation:

$$2 - 2^2 + 2^3 - 2^4 + 2^5 = \sum_{j=1}^{5}(-1)^{j+1}2^j$$

59.   Write the sum in sigma notation:

$$1 - 2 + 3 - 4 + 5 = \sum_{j=1}^{5}(-1)^{j+1}j$$

61. (a) The first eight terms are given by:

$$s_1 = 5$$

$$s_2 = \sqrt{s_1} = \sqrt{5} \approx 2.236$$

$$s_3 = \sqrt{s_2} \approx \sqrt{2.236} \approx 1.495$$

$$s_4 = \sqrt{s_3} \approx \sqrt{1.495} \approx 1.223$$

$$s_5 = \sqrt{s_4} \approx \sqrt{1.223} \approx 1.106$$

$$s_6 = \sqrt{s_5} \approx \sqrt{1.106} \approx 1.052$$

$$s_7 = \sqrt{s_6} \approx \sqrt{1.052} \approx 1.025$$

$$s_8 = \sqrt{s_7} \approx \sqrt{1.025} \approx 1.013$$

The answers appear to be approaching 1.

(b) $s_{15} \approx 1.000098$

(c) It is reasonable to expect s_{100} to be very close to 1.

63. (a) The first six terms are given by:

$$s_1 = 0.7$$

$$s_2 = \left(s_1\right)^2 = (0.7)^2 = 0.49$$

$$s_3 = \left(s_2\right)^2 = (0.49)^2 = 0.2401$$

$$s_4 = \left(s_3\right)^2 = (0.2401)^2 \approx 0.05765$$

$$s_5 = \left(s_4\right)^2 \approx (0.05765)^2 \approx 0.00332$$

$$s_6 = \left(s_5\right)^2 \approx (0.00332)^2 \approx 0.00001$$

(b) $s_{10} \approx 4.90 \times 10^{-80}$

(c) The value of s_{100} would be virtually identical to 0.

65. (a) Completing the table:

F_1	F_2	F_3	F_4	F_5	F_6	F_7	F_8	F_9	F_{10}
1	1	2	3	5	8	13	21	34	55

(b) Since $F_{22} = F_{20} + F_{21}$, $F_{22} = 6,765 + 10,946 = 17,711$. Also:

$$F_{19} + F_{20} = F_{21}$$
$$F_{19} + 6765 = 10946$$
$$F_{19} = 4181$$

(c)   We have:
$$F_{29} + F_{30} = F_{31}$$
$$514,229 + F_{30} = 1,346,269$$
$$F_{30} = 832,040$$

67. (a)   For $n = 1$, we have:
$$F_1 = 1$$
$$F_3 - 1 = 2 - 1 = 1$$
For $n = 2$, we have:
$$F_1 + F_2 = 1 + 1 = 2$$
$$F_4 - 1 = 3 - 1 = 2$$
For $n = 3$, we have:
$$F_1 + F_2 + F_3 = 1 + 1 + 2 = 4$$
$$F_5 - 1 = 5 - 1 = 4$$

(b)   The values completed in (a) show the hypothesis to be true for $n = 1$. Assuming:
$$F_1 + F_2 + \ldots + F_k = F_{k+2} - 1$$
Therefore:
$$F_1 + F_2 + \ldots + F_k + F_{k+1} = F_{k+2} - 1 + F_{k+1} = F_{k+3} - 1 = F_{(k+1)+2} - 1$$
This completes the induction.

69. (a)   For $n = 1$, we have:
$$F_2^2 = 1^2 = 1$$
$$F_1 F_3 + (-1)^1 = 1 \cdot 2 - 1 = 1$$
For $n = 2$, we have:
$$F_3^2 = 2^2 = 4$$
$$F_2 F_4 + (-1)^2 = 1 \cdot 3 + 1 = 4$$
For $n = 3$, we have:
$$F_4^2 = 3^2 = 9$$
$$F_3 F_5 + (-1)^3 = 2 \cdot 5 - 1 = 9$$

(b)   Since $F_2^2 = 1^2 = 1 = F_1 F_3 + (-1)^1 = 2 - 1 = 1$, the hypothesis is true for $n = 1$.
Assume:
$$F_{k+1}^2 = F_k F_{k+2} + (-1)^k$$
Then:
$$F_{k+1} F_{k+2} + F_{k+1}^2 = F_k F_{k+2} + F_{k+1} F_{k+2} + (-1)^k$$
$$F_{k+1}(F_{k+2} + F_{k+1}) = F_{k+2}(F_k + F_{k+1}) + (-1)^k$$
$$F_{k+1} F_{k+3} = F_{k+2}^2 + (-1)^k$$
$$F_{k+2}^2 = F_{k+1} F_{k+3} + (-1)^{k+1}$$
This completes the induction.

71. (a) For $n = 1$, we have:

$$F_1 = \frac{\left(1+\sqrt{5}\right)^1 - \left(1-\sqrt{5}\right)^1}{2^1\sqrt{5}} = \frac{1+\sqrt{5}-1+\sqrt{5}}{2\sqrt{5}} = \frac{2\sqrt{5}}{2\sqrt{5}} = 1$$

For $n = 2$, we have:

$$F_2 = \frac{\left(1+\sqrt{5}\right)^2 - \left(1-\sqrt{5}\right)^2}{2^2\sqrt{5}} = \frac{\left(6+2\sqrt{5}\right)-\left(6-2\sqrt{5}\right)}{4\sqrt{5}} = \frac{4\sqrt{5}}{4\sqrt{5}} = 1$$

(b) Substituting $n = 24$ and $n = 25$:

$$F_{24} = \frac{\left(1+\sqrt{5}\right)^{24} - \left(1-\sqrt{5}\right)^{24}}{2^{24}\sqrt{5}} = 46,368$$

$$F_{25} = \frac{\left(1+\sqrt{5}\right)^{25} - \left(1-\sqrt{5}\right)^{25}}{2^{25}\sqrt{5}} = 75,025$$

(c) Using the formula:

$$F_{26} = \frac{\left(1+\sqrt{5}\right)^{26} - \left(1-\sqrt{5}\right)^{26}}{2^{26}\sqrt{5}} = 121,393$$

Checking using the results from part (b):

$$F_{24} + F_{25} = 46,368 + 75,025 = 121,393$$

73. (a) Subtracting the two equations:

$$\alpha^n - \beta^n = \left(F_n\alpha + F_{n-1}\right) - \left(F_n\beta + F_{n-1}\right)$$

$$\alpha^n - \beta^n = F_n\alpha - F_n\beta$$

$$\alpha^n - \beta^n = F_n\left(\alpha - \beta\right)$$

$$F_n = \frac{\alpha^n - \beta^n}{\alpha - \beta}$$

(b) Using the quadratic formula to solve $x^2 - x - 1 = 0$:

$$x = \frac{-(-1) \pm \sqrt{(-1)^2 - 4(1)(-1)}}{2(1)} = \frac{1 \pm \sqrt{1+4}}{2} = \frac{1 \pm \sqrt{5}}{2}$$

Choosing α to be the larger root yields $\alpha = \frac{1+\sqrt{5}}{2}$ and $\beta = \frac{1-\sqrt{5}}{2}$.

(c)   Substituting these values for α and β:

$$F_n = \frac{\alpha^n - \beta^n}{\alpha - \beta}$$

$$= \frac{\left(\frac{1+\sqrt{5}}{2}\right)^n - \left(\frac{1-\sqrt{5}}{2}\right)^n}{\frac{1+\sqrt{5}}{2} - \frac{1-\sqrt{5}}{2}}$$

$$= \frac{\frac{\left(1+\sqrt{5}\right)^n}{2^n} - \frac{\left(1-\sqrt{5}\right)^n}{2^n}}{\frac{1+\sqrt{5}-1+\sqrt{5}}{2}}$$

$$= \frac{\left(1+\sqrt{5}\right)^n - \left(1-\sqrt{5}\right)^n}{2^n\sqrt{5}}$$

(d)   For $n = 1$, the equation states:

$$F_1 = \frac{\left(1+\sqrt{5}\right)^1 - \left(1-\sqrt{5}\right)^1}{2^1\sqrt{5}} = \frac{1+\sqrt{5}-1+\sqrt{5}}{2\sqrt{5}} = \frac{2\sqrt{5}}{2\sqrt{5}} = 1$$

Thus the equation holds for $n \geq 1$.

14.4 Arithmetic Sequences and Series

1.   (a)   To find the common difference d, subtract any term from the succeeding term.
Here that can be:
$$3 - 1 = 2$$
$$5 - 3 = 2$$
$$7 - 5 = 2$$
So 2 is the required common difference.

(b)   Again subtract terms:
$$6 - 10 = -4$$
$$2 - 6 = -4$$
$$-2 - 2 = -4$$
The common difference is -4. It is not necessary to try all three pairs, but this helps to verify that we have an arithmetic sequence.

(c)   Subtract:
$$1 - \tfrac{2}{3} = \tfrac{4}{3} - 1 = \tfrac{5}{3} - \tfrac{4}{3} = \tfrac{1}{3}$$
So $\tfrac{1}{3}$ is the common difference.

(d)   Subtract:
$$1 + \sqrt{2} - 1 = \sqrt{2} \text{ or } \left(1 + 2\sqrt{2}\right) - \left(1 + \sqrt{2}\right) = \sqrt{2}$$
Here $\sqrt{2}$ is the common difference.

3.   Since $a = 10$ and $d = 11$, using $a_n = a + (n-1)d$ results in:
$$a_{12} = 10 + (12 - 1)11 = 131$$

5. $a_{100} = 6 + (100 - 1)(5) = 6 + 495 = 501$

7. $a_{1000} = -1 + (1000 - 1)(1) = 998$

9.   Here $a_4 = -6$ and $a_{10} = 5$, thus:
$$-6 = a + 3d$$
$$5 = a + 9d$$
Subtracting the second equation from the first yields:
$$-6d = -11$$
$$d = \tfrac{11}{6}$$
Therefore:
$$a = 5 - 9\left(\tfrac{11}{6}\right) = 5 - \tfrac{33}{2} = -\tfrac{23}{2}$$

11.   Here $a_{60} = 105$ and $d = 5$, so:
$$105 = a + 59(5)$$
$$105 = a + 295$$
$$a = -190$$

13.   Since $a_{15} = a + 14d$ and $a_7 = a + 6d$, we have:
$$a_{15} - a_7 = 8d$$
$$-1 = 8d$$
$$d = -\tfrac{1}{8}$$

15.   Since $a = 1$ and $d = 1$, we have:
$$S_{1000} = \tfrac{1000}{2}(2 + 999) = 500500$$

17.   For $\tfrac{\pi}{3} + \tfrac{2\pi}{3} + \pi + \tfrac{4\pi}{3} + \ldots + \tfrac{13\pi}{3}$, $a = \tfrac{\pi}{3}$, $d = \tfrac{\pi}{3}$, so the sum is:
$$S_{13} = \frac{13}{2}\left(\frac{2\pi}{3} + \frac{12\pi}{3}\right) = \frac{14(13)(\pi)}{6} = \frac{91\pi}{3}$$

19.   Since $d = 5$ and $S_{38} = 3534$, we have:
$$3534 = \tfrac{38}{2}[2a + 37(5)]$$
$$3534 = 38a + 3515$$
$$19 = 38a$$
$$a = \tfrac{1}{2}$$

21. Here $a = 4$, $a_{16} = -100$, therefore:
$$S_{16} = 16\left(\tfrac{4-100}{2}\right) = -768$$
So we have:
$$-768 = \tfrac{16}{2}(2(4) + 15d)$$
$$-768 = 64 + 120d$$
$$-832 = 120d$$
$$d = -\tfrac{104}{15}$$

23. Since $a_8 = 5$ and $S_{10} = 20$, we have:
$$5 = a + 7d$$
$$20 = 5(2a + 9d)$$
$$4 = 2a + 9d$$
Solve the system:
$$a + 7d = 5$$
$$2a + 9d = 4$$
Multiplying the first equation by -2 and adding to the second equation:
$$-2a - 14d = -10$$
$$2a + 9d = 4$$
$$-5d = -6$$
$$d = \tfrac{6}{5}$$
Therefore:
$$a = 5 - 7\left(\tfrac{6}{5}\right) = -\tfrac{17}{5}$$

25. $S = S_{20} = \displaystyle\sum_{k=1}^{20}(4k + 3)$ so $a = 7$ and $a_{20} = 83$, thus:
$$S = 20\left(\tfrac{7+83}{2}\right) = 900$$

27. Let x denote the middle term, so the three terms are $x - d, x, x + d$:
$$x - d + x + x + d = 30$$
$$3x = 30$$
$$x = 10$$
Therefore:
$$x(x - d)(x + d) = 360$$
$$10(100 - d^2) = 360$$
$$100 - d^2 = 36$$
$$-d^2 = -64$$
$$d = \pm 8$$
The terms are 2, 10, 18 or 18, 10, 2.

29. Using equations as in Exercise 27:

$3x = 6$, so $x = 2$

$$(x - d)^3 + x^3 + (x + d)^3 = 132$$
$$(2 - d)^3 + 8 + (2 + d)^3 = 132$$
$$(2 - d)^3 + (2 + d)^3 = 124$$
$$16 + 12d^2 = 124$$
$$d^2 = 9$$
$$d = \pm 3$$

The terms are $-1, 2, 5$ or $5, 2, -1$.

31. (a) Subtracting yields:

$$a_2 - a_1 = -1 - \frac{1}{1 + \sqrt{2}} = \frac{-2 - \sqrt{2}}{1 + \sqrt{2}} = \frac{(1 - \sqrt{2})(-2 - \sqrt{2})}{-1} = -\sqrt{2}$$

$$a_3 - a_2 = \frac{1}{1 - \sqrt{2}} + 1 = \frac{2 - \sqrt{2}}{1 - \sqrt{2}} = \frac{(2 - \sqrt{2})(1 + \sqrt{2})}{-1} = -\sqrt{2}$$

So $a_2 - a_1 = a_3 - a_2 = -\sqrt{2}$.

(b) Using $a = \frac{1}{1 + \sqrt{2}}$ and $d = -\sqrt{2}$ results in:

$$S_6 = 3\left[2\left(\frac{1}{1 + \sqrt{2}}\right) + 5(-\sqrt{2})\right]$$
$$= 3\left(\frac{2}{1 + \sqrt{2}} - \frac{10 + 5\sqrt{2}}{1 + \sqrt{2}}\right)$$
$$= 3\left(\frac{-8 - 5\sqrt{2}}{1 + \sqrt{2}}\right)$$
$$= -\frac{24 + 15\sqrt{2}}{1 + \sqrt{2}}$$
$$= (24 + 15\sqrt{2})(1 - \sqrt{2})$$
$$= 24 - 30 - 9\sqrt{2}$$
$$= -6 - 9\sqrt{2}$$

33. Using $a = \frac{1}{1 + \sqrt{b}}$ results in:

$$d = \frac{1}{2}\left(\frac{1}{1 - \sqrt{b}} - \frac{1}{1 + \sqrt{b}}\right) = \frac{1}{2}\left(\frac{2\sqrt{b}}{1 - b}\right) = \frac{\sqrt{b}}{1 - b}$$

Therefore:

$$S_n = \frac{n}{2}\left[\frac{2}{1 + \sqrt{b}} + (n - 1)\frac{\sqrt{b}}{1 - b}\right]$$
$$= \frac{n}{2}\left[\frac{(2 - 2\sqrt{b}) + (n - 1)\sqrt{b}}{1 - b}\right]$$
$$= \frac{n}{2(1 - b)}\left[2 + (n - 3)\sqrt{b}\right]$$

35.   Using the given ratio, we have:

$$\frac{n^2}{m^2} = \frac{\frac{n}{2}[2a+(n-1)d]}{\frac{m}{2}[2a+(m-1)d]}$$

$$\frac{2a+(n-1)d}{2a+(m-1)d} = \frac{n}{m}$$

$$m[2a+(n-1)d] = n[2a+(m-1)d]$$

$$2am+m(n-1)d = 2an+n(m-1)d$$

$$2a(m-n) = [n(m-1)-m(n-1)]d$$

$$d = \frac{2a(m-n)}{m-n}$$

$$d = 2a \quad \text{(assuming } m \neq n\text{)}$$

Consequently:

$$\frac{a_n}{a_m} = \frac{a+(n-1)(2a)}{a+(m-1)(2a)} = \frac{1+2n-2}{1+2m-2} = \frac{2n-1}{2m-1}$$

37.   Let the three consecutive terms be $x-d$, x, and $x+d$, then using the Pythagorean theorem:

$$(x-d)^2 + x^2 = (x+d)^2$$

$$x^2 - 2xd + d^2 + x^2 = x^2 + 2xd + d^2$$

$$x^2 - 4xd = 0$$

$$x = 0 \quad \text{or} \quad x = 4d$$

Since the side of the triangle cannot have zero length, $x = 4d$ and the three terms are $3d$, $4d$, and $5d$, which is clearly similar to a 3-4-5 right triangle.

39.   We have $\frac{1}{b} - \frac{1}{a} = \frac{1}{c} - \frac{1}{b}$, from which it follows that $b = \dfrac{2ac}{a+c}$. Therefore:

$$\ln(a+c) + \ln(a-2b+c) = \ln(a+c) + \ln\left(a - \frac{4ac}{a+c} + c\right)$$

$$= \ln(a+c) + \ln\left(\frac{a^2 + ac - 4ac + ac + c^2}{a+c}\right)$$

$$= \ln(a+c) + \ln\frac{(a-c)^2}{a+c}$$

$$= \ln(a-c)^2$$

$$= 2\ln(a-c)$$

14.5 Geometric Sequences and Series

1.   The first three terms have a common ratio, hence:

$$\frac{x}{9} = \frac{4}{x}, \text{ therefore } x^2 = 36 \text{ and } x = \pm 6$$

Since the ratio is positive the second term is 6.

3. Letting the common ratio be r, we have:

$\quad\quad$ 4, 4r, and $4r^2$ so $64r^3 = 8000$

Therefore, $r^3 = 125$, so $r = 5$, and the second and third terms are 20 and 100, respectively.

5. Here $a_1 = -1$ and $r = -1$, so $a_{100} = -1(-1)^{99} = 1$.

7. Here $a_1 = \frac{2}{3}$ and $r = \frac{2}{3}$, so $a_8 = \frac{2}{3}\left(\frac{2}{3}\right)^7 = \frac{256}{6561}$.

9. Here $4096 = a_7 = r^6$, so $r = \pm\sqrt[6]{4096} = \pm 4$.

11. Compute the sum:

$$S_{10} = \frac{7\left(1 - 2^{10}\right)}{1 - 2} = -7(1 - 1024) = 7161$$

13. Compute the sum:

$$1 + \sqrt{2} + 2 + \dots + 32 = S_{11} = \frac{1\left[1 - \left(\sqrt{2}\right)^{11}\right]}{1 - \sqrt{2}} = \frac{1 - 32\sqrt{2}}{1 - \sqrt{2}} = 63 + 31\sqrt{2}$$

15. Compute the sum:

$$\sum_{k=1}^{6}\left(\frac{3}{2}\right)^k = S_6 = \frac{\frac{3}{2}\left[1 - \left(\frac{3}{2}\right)^6\right]}{1 - \frac{3}{2}} = -3\left(1 - \frac{729}{64}\right) = \frac{1995}{64}$$

17. Compute the sum:

$$\sum_{k=2}^{6}\left(\frac{1}{10}\right)^k = S_6 - a_1$$

$$= \frac{\frac{1}{10}\left[1 - \left(\frac{1}{10}\right)^6\right]}{1 - \frac{1}{10}} - \frac{1}{10}$$

$$= \frac{1}{9}\left(1 - \frac{1}{1000000}\right) - \frac{1}{10}$$

$$= \frac{1}{9} \cdot \frac{999999}{1000000} - \frac{1}{10}$$

$$= \frac{111111}{1000000} - \frac{1}{10}$$

$$= \frac{11111}{1000000}$$

This is equivalent is 0.011111.

19. Compute the sum:

$$\frac{2}{3} - \frac{4}{9} + \frac{8}{27} - \dots = \frac{\frac{2}{3}}{1 + \frac{2}{3}} = \frac{2}{5}$$

21.   Compute the sum:

$$1 + \frac{1}{1.01} + \frac{1}{(1.01)^2} + \ldots = \frac{1}{1 - \frac{1}{1.01}} = 101$$

23.   Writing as a geometric sum, we have:

$$0.555\ldots = \frac{5}{10} + \frac{5}{100} + \ldots = \frac{\frac{5}{10}}{1 - \frac{1}{10}} = \frac{5}{9}$$

25.   Writing as a geometric sum, we have:

$$0.12323\ldots = \frac{1}{10} + \frac{23}{1000} + \frac{23}{100000} + \ldots = \frac{1}{10} + \frac{\frac{23}{1000}}{1 - \frac{1}{100}} = \frac{1}{10} + \frac{23}{990} = \frac{122}{990} = \frac{61}{495}$$

27.   Writing as a geometric sum, we have:

$$0.432\ldots = \frac{432}{1000} + \frac{432}{1000000} + \ldots = \frac{\frac{432}{1000}}{1 - \frac{1}{1000}} = \frac{432}{999} = \frac{16}{37}$$

29.   For $\frac{a}{r}$, a, ar we have:

$$\frac{a}{r}(a)(ar) = -1000$$
$$a^3 = -1000$$
$$a = -10$$

So we have:

$$\frac{a}{r} + a + ar = 15$$
$$\frac{-10}{r} - 10 - 10r = 15$$
$$2r^2 + 5r + 2 = 0$$
$$(2r + 1)(r + 2) = 0$$
$$r = -\tfrac{1}{2}, -2$$

31.   Compute the sum:

$$\frac{\sqrt{3}}{\sqrt{3}+1} + \frac{\sqrt{3}}{\sqrt{3}+3} + \ldots = \frac{\frac{\sqrt{3}}{\sqrt{3}+1}}{1 - \frac{\sqrt{3}+1}{\sqrt{3}+3}} = \frac{\frac{\sqrt{3}}{\sqrt{3}+1}}{\frac{2}{\sqrt{3}+3}} = \frac{\sqrt{3}(\sqrt{3}+3)}{2(\sqrt{3}+1)} = \frac{3 + 3\sqrt{3}}{2(1+\sqrt{3})} = \frac{3}{2}$$

33. For $a_1, a_2, \ldots$ a geometric sequence, with ratio r:

$$S = a_1 + a_2 + \ldots + a_n = \frac{a_1(1 - r^n)}{1 - r}$$

Also:

$$T = \frac{1}{a_1} + \frac{1}{a_2} + \ldots + \frac{1}{a_n} = \frac{\frac{1}{a_1}\left(1 - \frac{1}{r^n}\right)}{1 - \frac{1}{r}}$$

Therefore:

$$\frac{S}{T} = \frac{\frac{a_1(1-r^n)}{1-r}}{\frac{\frac{1}{a_1}\left(1 - \frac{1}{r^n}\right)}{\frac{r-1}{r}}} = \frac{a_1(1-r^n)a_1(r-1)r^n}{(1-r)r(r^n - 1)} = a_1^2 r^{n-1} = a_1\left(a_1 r^{n-1}\right) = a_1 a_n$$

35. The total distance traveled by the ball consists of two infinite geometric series:

down: $6 + \frac{1}{3}(6) + \frac{1}{9}(6) + \ldots$

up: $2 + \frac{1}{3}(2) + \frac{1}{9}(2) + \ldots$

Now use the formula $S = \frac{a}{1-r}$ to find each sum:

down: $a = 6, r = \frac{1}{3}$

$$S = \frac{6}{1 - \frac{1}{3}} = \frac{6}{\frac{2}{3}} = 9 \text{ ft.}$$

up: $a = 1, r = \frac{1}{3}$

$$S = \frac{2}{1 - \frac{1}{3}} = \frac{2}{\frac{2}{3}} = 3 \text{ ft.}$$

So the total distance traveled is 12 ft.

14.6 Permutations and Combinations

1. There are 6 routes: $3(2) = 6$

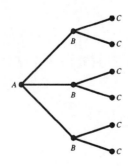

3. There are 12 routes: $3(2)(2) = 12$

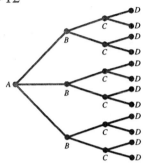

5. This is $P(4,2) = 4 \cdot 3 = 12$.

7. This is $P(5,3) = 5 \cdot 4 \cdot 3 = 60$.

9. Using the multiplication principle, the amount of numbers is:
$(3)(6)(6) = 108$

11. Using the multiplication principle, the number of license plates is:
$(10)(9)(8)(26)(25)(24) = 11,232,000$

13. This is $P(10, 10) = 10! = 3,628,800$

15. Since $P(n, n) = 24$, we have $n! = 24$ and thus $n = 4$ letters.

17. This is $P(8, 1) = 8$.

19. This is $P(6,3) = 6 \cdot 5 \cdot 4 = 120$.

21. We have:
$$P(n,n-1) = n(n-1)(n-2)\ldots[n-(n-1)+1] = n(n-1)(n-2)\ldots(2) = n!$$

23. Expanding the products:
$$P(100,50) = (100)(99)(98)\cdots(51)$$
$$P(101,51) = (101)(100)(99)\cdots(51) = 101 \cdot P(100,50)$$
So $P(101, 51)$ is larger.

25. Following the solution of Example 9:
$2P(5, 5) = 2(5!) = 240$

27. As in Example 9:
$2P(3, 3) = 2(3!) = 12$

29. (a) $9(10)(10) = 900$
(b) $9(9)(8) = 648$

31. (a) $9(9)(9) = 729$
(b) $9(8)(7) = 504$
(c) $1(8)(7) = 56$
(d) $5(7)(4) = 140$

33. (a) Compute:
$$C(12,11) = \frac{12!}{11!1!} = 12$$
(b) Compute:
$$C(12,1) = \frac{12!}{1!11!} = 12$$

35. (a) Compute:
$$C(8,3) = \frac{8!}{3!5!} = 56$$
(b) Compute:
$$C(8,5) = \frac{8!}{3!5!} = 56$$

37. (a) Compute:
$$C(12,5) = \frac{12!}{5!7!} = 792$$
(b) Compute:
$$C(12,7) = 792$$

39. Compute:
$$C(n,n-2) = \frac{n!}{(n-2)!2!} = \frac{n(n-1)}{2}$$

41. Compute:
$$C(8,3) = \frac{8!}{3!5!} = 56$$

43. (a) Compute:
$$C(10,6) = \frac{10!}{6!4!} = 210$$
(b) Compute:
$$C(4,3) \cdot C(6,3) = \frac{4!}{3!1!} \cdot \frac{6!}{3!3!} = 80$$

45. The number of diagonals will be $C(6, 2) - 6$, which is:
$$C(6,2) - 6 = \frac{6!}{2!4!} - 6 = 15 - 6 = 9$$

47.   Compute:

$$C(8,3) = \frac{8!}{3!5!} = 56 \text{ triangles}$$

49.   First notice that no two different combinations of these coins can add up to the same sum. Then we add the number of ways of combining the four coins, one at a time, two at a time, three and four at a time, thus:

$$C(4,1) + C(4,2) + C(4,3) + C(4,4) = 4 + 6 + 4 + 1 = 15 \text{ sums}$$

51.   (a)   Choose 6 of 9 Republicans and 2 of 8 Democrats, so the number of committees is:

$$C(9,6) \cdot C(8,2) = \frac{9!}{6!3!} \cdot \frac{8!}{2!6!} = 2352 \text{ committees}$$

   (b)   There are $C(9,6) \cdot C(8,2)$ ways to pick the committee with 6 Republicans, $C(9,7) \cdot C(8,1)$ ways to pick it with 7 Republicans, and $C(9,8) \cdot C(8,0)$ ways to pick it with 8 Republicans, so the total is:

$$C(9,6) \cdot C(8,2) + C(9,7) \cdot C(8,1) + C(9,8) \cdot C(8,0)$$
$$= \frac{9!}{6!3!} \cdot \frac{8!}{2!6!} + \frac{9!}{7!2!} \cdot \frac{8!}{1!7!} + \frac{9!}{8!1!} \cdot \frac{8!}{0!8!}$$
$$= 2352 + 288 + 9$$
$$= 2649 \text{ committees}$$

53.   In a polygon with 10 vertices, first choose a particular vertex (there are 9 other vertices to which a line could be drawn.) If lines are drawn to the two adjacent vertices, the lines would become sides, not diagonals; thus, there are 7 diagonals from each vertex. Not wanting to count a diagonal from one vertex to another twice, we must divide by 2. Thus there are $\frac{7 \times 10}{2} = 35$ diagonals.

14.7 Introduction to Probability

1.   (a)   {2, 4, 6}
   (b)   {4, 5, 6}
   (c)   {2, 4, 5, 6}
   (d)   {4, 6}

3.   (a)   {1, 3, 5}
   (b)   {1, 2, 3, 4, 5, 6}
   (c)   {1, 2, 3, 4, 5, 6}
   (d) $\varnothing$

5.   No, since $A \cap B = \{4,6\} \neq \varnothing$.

7.   Yes, since $C \cap D = \varnothing$.

9. (a) $n(A) = 3$
   (b) $n(B) = 3$
   (c) $n(A \cup B) = 4$

11. (a) $n(E) = 3$
   (b) $n(F) = 6$
   (c) $n(G) = 0$

13. (a) $P(A) = \frac{3}{6} = \frac{1}{2}$
   (b) $P(B) = \frac{3}{6} = \frac{1}{2}$

15. (a) $P(E) = \frac{3}{6} = \frac{1}{2}$
   (b) $P(F) = \frac{6}{6} = 1$

17. (a) $P(A \cup B) = \frac{4}{6} = \frac{2}{3}$
   (b) $P(A) + P(B) = \frac{3}{6} + \frac{3}{6} = 1$

19. (a) $P(B \cup E) = \frac{5}{6}$
   (b) $P(B) + P(E) = \frac{3}{6} + \frac{3}{6} = 1$

21. $P(A) = \frac{3}{8}$

23. $P(C) = \frac{4}{8} = \frac{1}{2}$

25. $P(E) = \frac{2}{8} = \frac{1}{4}$

27. Compute:
$$P(A \cup B) = P(A) + P(B) - P(A \cap B) = \frac{3}{8} + \frac{1}{4} - 0 = \frac{5}{8}$$

29. Compute:
$$P(A \cup C) = P(A) + P(C) - P(A \cap C) = \frac{3}{8} + \frac{1}{2} - \frac{1}{8} = \frac{6}{8} = \frac{3}{4}$$

31. Compute:
$$P(C \cup F) = P(C) + P(F) - P(C \cap F) = \frac{1}{2} + \frac{1}{2} - 0 = 1$$

33. Compute:
$$P(E \cup F) = P(E) + P(F) - P(E \cap F) = \frac{1}{4} + \frac{1}{2} - \frac{1}{8} = \frac{5}{8}$$

35. Compute:
$$P(B \cup C) = P(B) + P(C) - P(B \cap C) = \frac{1}{4} + \frac{1}{2} - \frac{1}{4} = \frac{1}{2}$$

37. $P(D \cap A) = 0$

39. Compute:

$$P(D \cup E) = P(D) + P(E) - P(D \cap E) = \tfrac{1}{8} + \tfrac{1}{4} - \tfrac{1}{8} = \tfrac{1}{4}$$

41. We can compute this as:

$$P(1\underline{\text{st}} \text{ is } d) \cdot P(2\underline{\text{nd}} \text{ is not } d) + P(1\underline{\text{st}} \text{ is not } d) \cdot P(2\underline{\text{nd}} \text{ is } d)$$
$$= \tfrac{1}{6} \cdot \tfrac{5}{5} + \tfrac{5}{6} \cdot \tfrac{1}{5} = \tfrac{1}{6} + \tfrac{1}{6} = \tfrac{1}{3}$$

43. (a) The probability is:
$$\tfrac{12}{14} = \tfrac{6}{7}$$

(b) Using the multiplication principle, the probability is:
$$\tfrac{12}{14} \cdot \tfrac{11}{13} \cdot \tfrac{10}{12} \cdot \tfrac{9}{11} \cdot \tfrac{8}{10} = \tfrac{72}{182} = \tfrac{36}{91}$$

45. Compute:

$$P(A \cup B) = P(A) + P(B) - P(A \cap B) = \tfrac{1}{3} + \tfrac{1}{4} - 0 = \tfrac{7}{12}$$

47. Solve:

$$P(A \cup B) = P(A) + P(B) - P(A \cap B)$$
$$\tfrac{7}{9} = \tfrac{5}{9} + \tfrac{4}{9} - P(A \cap B)$$
$$\tfrac{7}{9} = 1 - P(A \cap B)$$
$$P(A \cap B) = \tfrac{2}{9}$$

49. (a) Compute:
$$\tfrac{6}{15} \cdot \tfrac{5}{15} = \tfrac{1}{8}$$
(b) Compute:
$$\tfrac{10}{16} \cdot \tfrac{9}{15} = \tfrac{3}{8}$$
(c) Compute:
$$\tfrac{1}{8} + \tfrac{3}{8} = \tfrac{1}{2}$$

51. Compute:

$$P\left(A\ 1\underline{\text{st}}\ \text{or}\ A\ 2\underline{nd}\right) = P\left(A\ 1\underline{\text{st}}\right) + P\left(A\ 2\underline{nd}\right) - P(A\ \text{both}) = \tfrac{1}{2} + \tfrac{3}{10} - \tfrac{1}{10} = \tfrac{7}{10}$$

53. Using the multiplication principle, we have:

$$\tfrac{13}{52} \cdot \tfrac{12}{51} \cdot \tfrac{11}{50} \cdot \tfrac{10}{49} \cdot \tfrac{9}{48} = \tfrac{33}{66640}$$

14.8 DeMoivre's Theorem

1. The complex number $4 + 2i$ is identified with the point $(4, 2)$:

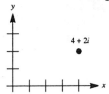

3. The complex number $-5 + i$ is identified with the point $(-5, 1)$:

5. The complex number $1 - 4i$ is identified with the point $(1, -4)$:

7. The complex number $-i$, or $0 - 1i$, is identified with the point $(0, -1)$:

9. We convert the complex number to rectangular form:
$$2\left[\cos\tfrac{\pi}{4} + i\sin\tfrac{\pi}{4}\right] = 2\left[\tfrac{\sqrt{2}}{2} + \tfrac{\sqrt{2}}{2}i\right] = \sqrt{2} + \sqrt{2}i$$

11. We convert the complex number to rectangular form:
$$4\left[\cos\tfrac{5\pi}{6} + i\sin\tfrac{5\pi}{6}\right] = 4\left[-\tfrac{\sqrt{3}}{2} + \tfrac{1}{2}i\right] = -2\sqrt{3} + 2i$$

13. We convert the complex number to rectangular form:
$$\sqrt{2}\left[\cos 225° + i\sin 225°\right] = \sqrt{2}\left[-\tfrac{\sqrt{2}}{2} - \tfrac{\sqrt{2}}{2}i\right] = -1 - i$$

15. We convert the complex number to rectangular form:
$$\sqrt{3}\left[\cos\tfrac{\pi}{2} + i\sin\tfrac{\pi}{2}\right] = \sqrt{3}(0 + 1i) = \sqrt{3}i$$

17.   We use the hint to find $\cos 75°$ and $\sin 75°$:

$$\cos 75° = \cos(30° + 45°) = \cos 30° \cos 45° - \sin 30° \sin 45° = \frac{\sqrt{3}}{2} \cdot \frac{\sqrt{2}}{2} - \frac{1}{2} \cdot \frac{\sqrt{2}}{2} = \frac{\sqrt{6} - \sqrt{2}}{4}$$

$$\sin 75° = \sin(30° + 45°) = \sin 30° \cos 45° + \cos 30° \sin 45° = \frac{1}{2} \cdot \frac{\sqrt{2}}{2} + \frac{\sqrt{3}}{2} \cdot \frac{\sqrt{2}}{2} = \frac{\sqrt{6} + \sqrt{2}}{4}$$

Therefore:

$$4[\cos 75° + i \sin 75°] = 4\left[\frac{\sqrt{6} - \sqrt{2}}{4} + \frac{\sqrt{6} + \sqrt{2}}{4} i\right] = (\sqrt{6} - \sqrt{2}) + (\sqrt{6} + \sqrt{2})i$$

19.   Here $a = \frac{\sqrt{3}}{2}$ and $b = \frac{1}{2}$, so:

$$r = \sqrt{a^2 + b^2} = \sqrt{\frac{3}{4} + \frac{1}{4}} = 1$$

We now find θ such that $\cos \theta = \frac{a}{r} = \frac{\sqrt{3}}{2}$ and $\sin \theta = \frac{b}{r} = \frac{1}{2}$. Such a θ is $\theta = \frac{\pi}{6}$. Thus:

$$\frac{\sqrt{3}}{2} + \frac{1}{2}i = \cos \frac{\pi}{6} + i \sin \frac{\pi}{6}$$

21.   Here $a = -1$ and $b = \sqrt{3}$, so:

$$r = \sqrt{a^2 + b^2} = \sqrt{1 + 3} = \sqrt{4} = 2$$

We now find θ such that $\cos \theta = \frac{a}{r} = -\frac{1}{2}$ and $\sin \theta = \frac{b}{r} = \frac{\sqrt{3}}{2}$. Such a θ is $\theta = \frac{2\pi}{3}$. Thus:

$$-1 + \sqrt{3}i = 2\left[\cos \frac{2\pi}{3} + i \sin \frac{2\pi}{3}\right]$$

23.   Here $a = -2\sqrt{3}$ and $b = -2$, so:

$$r = \sqrt{a^2 + b^2} = \sqrt{12 + 4} = \sqrt{16} = 4$$

We now find θ such that $\cos \theta = \frac{a}{r} = -\frac{2\sqrt{3}}{4} = -\frac{\sqrt{3}}{2}$ and $\sin \theta = \frac{b}{r} = -\frac{2}{4} = -\frac{1}{2}$. Such a θ is $\theta = \frac{7\pi}{6}$. Thus:

$$-2\sqrt{3} - 2i = 4\left[\cos \frac{7\pi}{6} + i \sin \frac{7\pi}{6}\right]$$

25.   Here $a = 0$ and $b = -6$, so:

$$r = \sqrt{a^2 + b^2} = \sqrt{0 + 36} = 6$$

We now find θ such that $\cos \theta = \frac{a}{r} = \frac{0}{6} = 0$ and $\sin \theta = \frac{b}{r} = -\frac{6}{6} = -1$. Such a θ is $\theta = \frac{3\pi}{2}$. Thus:

$$-6i = 6\left[\cos \frac{3\pi}{2} + i \sin \frac{3\pi}{2}\right]$$

27.   Here $a = \frac{\sqrt{3}}{4}$ and $b = -\frac{1}{4}$, so:

$$r = \sqrt{a^2 + b^2} = \sqrt{\frac{3}{16} + \frac{1}{16}} = \sqrt{\frac{1}{4}} = \frac{1}{2}$$

We now find θ such that $\cos \theta = \frac{a}{r} = \frac{\sqrt{3}/4}{1/2} = \frac{\sqrt{3}}{2}$ and $\sin \theta = \frac{b}{r} = \frac{-1/4}{1/2} = -\frac{1}{2}$. Such a θ is $\theta = \frac{11\pi}{6}$. Thus:

$$\frac{\sqrt{3}}{4} - \frac{1}{4}i = \frac{1}{2}\left[\cos \frac{11\pi}{6} + i \sin \frac{11\pi}{6}\right]$$

29. Performing the multiplication and adding angles, we have:
$$2[\cos 22° + i\sin 22°] \cdot 3[\cos 38° + i\sin 38°] = 6[\cos 60° + i\sin 60°]$$
$$= 6\left(\tfrac{1}{2} + \tfrac{\sqrt{3}}{2}i\right)$$
$$= 3 + 3\sqrt{3}i$$

31. Performing the multiplication and adding angles, we have:
$$\sqrt{2}\left[\cos\tfrac{\pi}{3} + i\sin\tfrac{\pi}{3}\right] \cdot \sqrt{2}\left[\cos\tfrac{4\pi}{3} + i\sin\tfrac{4\pi}{3}\right] = 2\left[\cos\tfrac{5\pi}{3} + i\sin\tfrac{5\pi}{3}\right] = 2\left(\tfrac{1}{2} - \tfrac{\sqrt{3}}{2}i\right) = 1 - \sqrt{3}i$$

33. Performing the multiplication and adding angles, we have:
$$3\left[\cos\tfrac{\pi}{7} + i\sin\tfrac{\pi}{7}\right] \cdot \sqrt{2}\left[\cos\tfrac{\pi}{7} + i\sin\tfrac{\pi}{7}\right] = 3\sqrt{2}\left[\cos\tfrac{2\pi}{7} + i\sin\tfrac{2\pi}{7}\right]$$
$$= 3\sqrt{2}\cos\tfrac{2\pi}{7} + \left(3\sqrt{2}\sin\tfrac{2\pi}{7}\right)i$$

35. Performing the division and subtracting angles, we have:
$$6[\cos 50° + i\sin 50°] \div 2[\cos 5° + i\sin 5°] = 3[\cos 45° + i\sin 45°]$$
$$= 3\left(\tfrac{\sqrt{2}}{2} + \tfrac{\sqrt{2}}{2}i\right)$$
$$= \tfrac{3\sqrt{2}}{2} + \tfrac{3\sqrt{2}}{2}i$$

37. Performing the division and subtracting angles, we have:
$$2^{4/3}\left[\cos\tfrac{5\pi}{12} + i\sin\tfrac{5\pi}{12}\right] \div 2^{1/3}\left[\cos\tfrac{\pi}{4} + i\sin\tfrac{\pi}{4}\right] = 2\left[\cos\tfrac{\pi}{6} + i\sin\tfrac{\pi}{6}\right] = 2\left(\tfrac{\sqrt{3}}{2} + \tfrac{1}{2}i\right) = \sqrt{3} + i$$

39. Performing the division and subtracting angles, we have:
$$\left[\cos\tfrac{2\pi}{5} + i\sin\tfrac{2\pi}{5}\right] \div \left[\cos\tfrac{2\pi}{5} + i\sin\tfrac{2\pi}{5}\right] = \cos 0 + i\sin 0 = 1 + 0i = 1$$

41. Using DeMoivre's theorem, we have:
$$\left[3\left(\cos\tfrac{\pi}{3} + i\sin\tfrac{\pi}{3}\right)\right]^5 = 3^5\left(\cos\tfrac{5\pi}{3} + i\sin\tfrac{5\pi}{3}\right) = 243\left(\tfrac{1}{2} - \tfrac{\sqrt{3}}{2}i\right) = \tfrac{243}{2} - \tfrac{243\sqrt{3}}{2}i$$

43. Using DeMoivre's theorem, we have:
$$\left[\tfrac{1}{2}\left(\cos\tfrac{\pi}{24} + i\sin\tfrac{\pi}{24}\right)\right]^6 = \left(\tfrac{1}{2}\right)^6\left(\cos\tfrac{\pi}{4} + i\sin\tfrac{\pi}{4}\right) = \tfrac{1}{64}\left(\tfrac{\sqrt{2}}{2} + \tfrac{\sqrt{2}}{2}i\right) = \tfrac{\sqrt{2}}{128} + \tfrac{\sqrt{2}}{128}i$$

45. Using DeMoivre's theorem, we have:
$$\left[2^{1/5}(\cos 63° + i\sin 63°)\right]^{10} = \left(2^{1/5}\right)^{10}(\cos 630° + i\sin 630°) = 4(0 - 1i) = -4i$$

47. Performing the multiplication and adding angles, we have:
$$2(\cos 100° + i\sin 200°) \times \sqrt{2}(\cos 20° + i\sin 20°) \times \tfrac{1}{2}(\cos 5° + i\sin 5°)$$
$$= \sqrt{2}(\cos 225° + i\sin 225°)$$
$$= \sqrt{2}\left(-\tfrac{\sqrt{2}}{2} - \tfrac{\sqrt{2}}{2}i\right)$$
$$= -1 - i$$

49. We first write $\frac{1}{2} - \frac{\sqrt{3}}{2}i$ in trigonometric form:

$$\frac{1}{2} - \frac{\sqrt{3}}{2}i = \cos\frac{5\pi}{3} + i\sin\frac{5\pi}{3}$$

Now using DeMoivre's theorem, we have:

$$\left(\frac{1}{2} - \frac{\sqrt{3}}{2}i\right)^5 = \cos\frac{25\pi}{3} + i\sin\frac{25\pi}{3} = \frac{1}{2} + \frac{\sqrt{3}}{2}i$$

51. We first write $-2 - 2i$ in trigonometric form:

$$-2 - 2i = 2\sqrt{2}\left(-\frac{\sqrt{2}}{2} - \frac{\sqrt{2}}{2}i\right) = 2\sqrt{2}\left(\cos\frac{5\pi}{4} + i\sin\frac{5\pi}{4}\right)$$

Now using DeMoivre's theorem, we have:

$$(-2 - 2i)^5 = (2\sqrt{2})^5\left(\cos\frac{25\pi}{4} + i\sin\frac{25\pi}{4}\right) = 128\sqrt{2}\left(\frac{\sqrt{2}}{2} + \frac{\sqrt{2}}{2}i\right) = 128 + 128i$$

53. We first write $-2\sqrt{3} - 2i$ in trigonometric form:

$$-2\sqrt{3} - 2i = 4\left(-\frac{\sqrt{3}}{2} - \frac{1}{2}i\right) = 4\left(\cos\frac{7\pi}{6} + i\sin\frac{7\pi}{6}\right)$$

Now using DeMoivre's theorem, we have:

$$(-2\sqrt{3} - 2i)^4 = 4^4\left[\cos\frac{14\pi}{3} + i\sin\frac{14\pi}{3}\right] = 256\left(-\frac{1}{2} + \frac{\sqrt{3}}{2}i\right) = -128 + 128\sqrt{3}i$$

55. We begin by writing $-27i$ in trigonometric form:

$$-27i = 27(0 - 1i) = 27\left[\cos\frac{3\pi}{2} + i\sin\frac{3\pi}{2}\right]$$

Now let $z = r(\cos\theta + i\sin\theta)$ denote a cube root of $-27i$. Then:

$$z^3 = r^3(\cos3\theta + i\sin3\theta) = 27\left[\cos\frac{3\pi}{2} + i\sin\frac{3\pi}{2}\right]$$

Then $r^3 = 27$ so $r = 3$, and $3\theta = \frac{3\pi}{2} + 2\pi k$, so $\theta = \frac{\pi}{2} + \frac{2\pi}{3}k$.

When $k = 0$, we have:

$$z_1 = 3\left[\cos\frac{\pi}{2} + i\sin\frac{\pi}{2}\right] = 3(0 + i) = 3i$$

When $k = 1$, we have:

$$z_2 = 3\left[\cos\frac{7\pi}{6} + i\sin\frac{7\pi}{6}\right] = 3\left(-\frac{\sqrt{3}}{2} - \frac{1}{2}i\right) = -\frac{3\sqrt{3}}{2} - \frac{3}{2}i$$

When $k = 2$, we have:

$$z_3 = 3\left[\cos\frac{11\pi}{6} + i\sin\frac{11\pi}{6}\right] = 3\left(\frac{\sqrt{3}}{2} - \frac{1}{2}i\right) = \frac{3\sqrt{3}}{2} - \frac{3}{2}i$$

So the cube roots of $-27i$ are $3i$, $-\frac{3\sqrt{3}}{2} - \frac{3}{2}i$ and $\frac{3\sqrt{3}}{2} - \frac{3}{2}i$.

57. We begin by writing 1 in trigonometric form:

$$1 = 1(1 + 0i) = 1[\cos0 + i\sin0]$$

Now let $z = r(\cos\theta + i\sin\theta)$ denote an eighth root of 1. Then:

$$z^8 = r^8(\cos8\theta + i\sin8\theta) = 1[\cos0 + i\sin0]$$

Then $r^8 = 1$ so $r = 1$, and $8\theta = 0 + 2\pi k$, so $\theta = \frac{\pi}{4}k$.

When $k = 0$, we have:

$$z_1 = 1[\cos0 + i\sin0] = 1 + 0i = 1$$

When $k = 1$, we have:

$$z_2 = 1\left[\cos\tfrac{\pi}{4} + i\sin\tfrac{\pi}{4}\right] = \tfrac{\sqrt{2}}{2} + \tfrac{\sqrt{2}}{2}i$$

When $k = 2$, we have:

$$z_3 = 1\left[\cos\tfrac{\pi}{2} + i\sin\tfrac{\pi}{2}\right] = 0 + 1i = i$$

When $k = 3$, we have:

$$z_4 = 1\left[\cos\tfrac{3\pi}{4} + i\sin\tfrac{3\pi}{4}\right] = -\tfrac{\sqrt{2}}{2} + \tfrac{\sqrt{2}}{2}i$$

When $k = 4$, we have:

$$z_5 = 1[\cos\pi + i\sin\pi] = -1 + 0i = -1$$

When $k = 5$, we have:

$$z_6 = 1\left[\cos\tfrac{5\pi}{4} + i\sin\tfrac{5\pi}{4}\right] = -\tfrac{\sqrt{2}}{2} - \tfrac{\sqrt{2}}{2}i$$

When $k = 6$, we have:

$$z_7 = 1\left[\cos\tfrac{3\pi}{2} + i\sin\tfrac{3\pi}{2}\right] = 0 - 1i = -i$$

When $k = 7$, we have:

$$z_8 = 1\left[\cos\tfrac{7\pi}{4} + i\sin\tfrac{7\pi}{4}\right] = \tfrac{\sqrt{2}}{2} - \tfrac{\sqrt{2}}{2}i$$

So the eighth roots of 1 are 1, $\tfrac{\sqrt{2}}{2} + \tfrac{\sqrt{2}}{2}i$, i, $-\tfrac{\sqrt{2}}{2} + \tfrac{\sqrt{2}}{2}i$, -1, $-\tfrac{\sqrt{2}}{2} - \tfrac{\sqrt{2}}{2}i$, $-i$, and

$\tfrac{\sqrt{2}}{2} - \tfrac{\sqrt{2}}{2}i$. These final results were obtained using the half-angle formulas for sine and cosine.

59. We begin by writing 64 in trigonometric form:

$$64 = 64(1 + 0i) = 64[\cos 0 + i\sin 0]$$

Now let $z = r(\cos\theta + i\sin\theta)$ denote a cube root of 64. Then:

$$z^3 = r^3(\cos 3\theta + i\sin 3\theta) = 64[\cos 0 + i\sin 0]$$

Then $r^3 = 64$ so $r = 4$, and $3\theta = 0 + 2\pi k$, so $\theta = 0 + \tfrac{2\pi}{3}k$.

When $k = 0$, we have:

$$z_1 = 4[\cos 0 + i\sin 0] = 4(1 + 0i) = 4$$

When $k = 1$, we have:

$$z_2 = 4\left[\cos\tfrac{2\pi}{3} + i\sin\tfrac{2\pi}{3}\right] = 4\left(-\tfrac{1}{2} + \tfrac{\sqrt{3}}{2}i\right) = -2 + 2\sqrt{3}i$$

When $k = 2$, we have:

$$z_3 = 4\left[\cos\tfrac{4\pi}{3} + i\sin\tfrac{4\pi}{3}\right] = 4\left(-\tfrac{1}{2} - \tfrac{\sqrt{3}}{2}i\right) = -2 - 2\sqrt{3}i$$

So the cube roots of 64 are 4, $-2 + 2\sqrt{3}i$, and $-2 - 2\sqrt{3}i$.

61. We first write 729 in trigonometric form:

$$729 = 729(1 + 0i) = 729[\cos 0 + i\sin 0]$$

Now let $z = r(\cos\theta + i\sin\theta)$ denote a sixth root of 729. Then:

$$z^6 = r^6[\cos 6\theta + i\sin 6\theta] = 729[\cos 0 + i\sin 0]$$

Then $r^6 = 729$ so $r = 3$, and $6\theta = 0 + 2\pi k$, so $\theta = 0 + \tfrac{\pi}{3}k$.

When $k = 0$, we have:
$$z_1 = 3[\cos 0 + i \sin 0] = 3(1 + 0i) = 3$$
When $k = 1$, we have:
$$z_2 = 3\left[\cos \tfrac{\pi}{3} + i \sin \tfrac{\pi}{3}\right] = 3\left(\tfrac{1}{2} + \tfrac{\sqrt{3}}{2}i\right) = \tfrac{3}{2} + \tfrac{3\sqrt{3}}{2}i$$
When $k = 2$, we have:
$$z_3 = 3\left[\cos \tfrac{2\pi}{3} + i \sin \tfrac{2\pi}{3}\right] = 3\left(-\tfrac{1}{2} + \tfrac{\sqrt{3}}{2}i\right) = -\tfrac{3}{2} + \tfrac{3\sqrt{3}}{2}i$$
When $k = 3$, we have:
$$z_4 = 3[\cos \pi + i \sin \pi] = 3(-1 + 0i) = -3$$
When $k = 4$, we have:
$$z_5 = 3\left[\cos \tfrac{4\pi}{3} + i \sin \tfrac{4\pi}{3}\right] = 3\left(-\tfrac{1}{2} - \tfrac{\sqrt{3}}{2}i\right) = -\tfrac{3}{2} - \tfrac{3\sqrt{3}}{2}i$$
When $k = 5$, we have:
$$z_6 = 3\left[\cos \tfrac{5\pi}{3} + i \sin \tfrac{5\pi}{3}\right] = 3\left(\tfrac{1}{2} - \tfrac{\sqrt{3}}{2}i\right) = \tfrac{3}{2} - \tfrac{3\sqrt{3}}{2}i$$
So the sixth roots of 729 are $3, \tfrac{3}{2} + \tfrac{3\sqrt{3}}{2}i, -\tfrac{3}{2} + \tfrac{3\sqrt{3}}{2}i, -3, -\tfrac{3}{2} - \tfrac{3\sqrt{3}}{2}i$ and $\tfrac{3}{2} - \tfrac{3\sqrt{3}}{2}i$.

63. We first write $7 - 7i$ in trigonometric form:
$$7 - 7i = 7\sqrt{2}\left(\tfrac{1}{\sqrt{2}} - \tfrac{1}{\sqrt{2}}i\right) = 7\sqrt{2}\left[\cos \tfrac{7\pi}{4} + i \sin \tfrac{7\pi}{4}\right]$$
Using DeMoivre's theorem we have:
$$(7 - 7i)^8 = \left(7\sqrt{2}\right)^8\left[\cos 14\pi + i \sin 14\pi\right] = \left(7^8\right)\left(2^4\right)(1 + 0i) = 92,236,816$$

65. We begin by writing i in trigonometric form:
$$i = 1(0 + 1i) = 1\left[\cos \tfrac{\pi}{2} + i \sin \tfrac{\pi}{2}\right]$$
Now let z be a fifth root of i, where $z = r(\cos \theta + i \sin \theta)$. Then:
$$z^5 = r^5[\cos 5\theta + i \sin 5\theta] = 1\left[\cos \tfrac{\pi}{2} + i \sin \tfrac{\pi}{2}\right]$$
So $r^5 = 1$ thus $r = 1$, and $5\theta = \tfrac{\pi}{2} + 2\pi k = 90° + 360°k$, so $\theta = 18° + 72°k$.

When $k = 0$: $z_1 = 1\left[\cos 18° + i \sin 18°\right] = 0.95 + 0.31i$

When $k = 1$: $z_2 = 1\left[\cos 90° + i \sin 90°\right] = i$

When $k = 2$: $z_3 = 1\left[\cos 162° + i \sin 162°\right] = -0.95 + 0.31i$

When $k = 3$: $z_4 = 1\left[\cos 234° + i \sin 234°\right] = -0.59 - 0.81i$

When $k = 4$: $z_5 = 1\left[\cos 306° + i \sin 306°\right] = 0.59 - 0.81i$

So the fifth roots of i are $0.95 + 0.31i, i, -0.95 + 0.31i, -0.59 - 0.81i$ and $0.59 - 0.81i$. Each of the real and imaginary parts here has been rounded off to two decimal places.

67. We begin by writing $8 - 8\sqrt{3}i$ in trigonometric form:
$$8 - 8\sqrt{3}i = 16\left(\tfrac{1}{2} - \tfrac{\sqrt{3}}{2}i\right) = 16\left[\cos \tfrac{5\pi}{3} + i \sin \tfrac{5\pi}{3}\right]$$
Now let $z = r(\cos \theta + i \sin \theta)$ denote a fourth root of $8 - 8\sqrt{3}i$. Then:
$$z^4 = r^4[\cos 4\theta + i \sin 4\theta] = 16\left[\cos \tfrac{5\pi}{3} + i \sin \tfrac{5\pi}{3}\right]$$

Then $r^4 = 16$ so $r = 2$, and $4\theta = \frac{5\pi}{3} + 2\pi k$, so $\theta = \frac{5\pi}{12} + \frac{\pi}{2}k$.

When $k = 0$, we have:

$$z_1 = 2\left[\cos\frac{5\pi}{12} + i\sin\frac{5\pi}{12}\right] = 2\left[\frac{\sqrt{6}-\sqrt{2}}{4} + \frac{\sqrt{6}+\sqrt{2}}{4}i\right] = \frac{\sqrt{6}-\sqrt{2}}{2} + \frac{\sqrt{6}+\sqrt{2}}{2}i$$

When $k = 1$, we have:

$$z_2 = 2\left[\cos\frac{11\pi}{12} + i\sin\frac{11\pi}{12}\right] = 2\left[\frac{-\sqrt{2}-\sqrt{6}}{4} + \frac{\sqrt{6}-\sqrt{2}}{4}i\right] = \frac{-\sqrt{2}-\sqrt{6}}{2} + \frac{\sqrt{6}-\sqrt{2}}{2}i$$

When $k = 2$, we have:

$$z_3 = 2\left[\cos\frac{17\pi}{12} + i\sin\frac{17\pi}{12}\right] = 2\left[\frac{\sqrt{2}-\sqrt{6}}{4} + \frac{-\sqrt{2}-\sqrt{6}}{4}i\right] = \frac{\sqrt{2}-\sqrt{6}}{2} - \frac{\sqrt{2}+\sqrt{6}}{2}i$$

When $k = 3$, we have:

$$z_4 = 2\left[\cos\frac{23\pi}{12} + i\sin\frac{23\pi}{12}\right] = 2\left[\frac{\sqrt{2}+\sqrt{6}}{4} + \frac{\sqrt{2}-\sqrt{6}}{4}i\right] = \frac{\sqrt{2}+\sqrt{6}}{2} + \frac{\sqrt{2}-\sqrt{6}}{2}i$$

So the fourth roots of $8 - 8\sqrt{3}i$ are $\frac{\sqrt{6}-\sqrt{2}}{2} + \frac{\sqrt{6}+\sqrt{2}}{2}i$, $\frac{-\sqrt{2}-\sqrt{6}}{2} + \frac{\sqrt{6}-\sqrt{2}}{2}i$,

$\frac{\sqrt{2}-\sqrt{6}}{2} + \frac{-\sqrt{2}-\sqrt{6}}{2}i$ and $\frac{\sqrt{2}+\sqrt{6}}{2} + \frac{\sqrt{2}-\sqrt{6}}{2}i$. These final results were obtained using the addition formulas for sine and cosine.

Note: If you used half-angle formulas, your answers, though identical in value, may "look" vastly different. Those answers (which are correct) are $\sqrt{2-\sqrt{3}} + \sqrt{2+\sqrt{3}}i$, $-\sqrt{2+\sqrt{3}} + \sqrt{2-\sqrt{3}}i$, $-\sqrt{2-\sqrt{3}} - \sqrt{2+\sqrt{3}}i$ and $\sqrt{2+\sqrt{3}} - \sqrt{2-\sqrt{3}}i$.

69. (a) Let $z = r(\cos\theta + i\sin\theta)$, so $z^3 = r^3(\cos 3\theta + i\sin 3\theta)$. Since $1 = 1(\cos 0 + i\sin 0)$, we have $r^3 = 1$ so $r = 1$, and $3\theta = 0 + 2\pi k$, so $\theta = 0 + \frac{2\pi}{3}k$.

When $k = 0$, we have:

$$z_1 = 1[\cos 0 + i\sin 0] = 1(1 + 0i) = 1$$

When $k = 1$, we have:

$$z_2 = 1\left[\cos\frac{2\pi}{3} + i\sin\frac{2\pi}{3}\right] = 1\left(-\frac{1}{2} + \frac{\sqrt{3}}{2}i\right) = -\frac{1}{2} + \frac{\sqrt{3}}{2}i$$

When $k = 2$, we have:

$$z_3 = 1\left[\cos\frac{4\pi}{3} + i\sin\frac{4\pi}{3}\right] = 1\left(-\frac{1}{2} - \frac{\sqrt{3}}{2}i\right) = -\frac{1}{2} - \frac{\sqrt{3}}{2}i$$

So the cube roots of 1 are 1, $-\frac{1}{2} + \frac{\sqrt{3}}{2}i$ and $-\frac{1}{2} - \frac{\sqrt{3}}{2}i$.

(b) We compute the sum:

$$z_1 + z_2 + z_3 = 1 + \left(-\frac{1}{2} + \frac{\sqrt{3}}{2}i\right) + \left(-\frac{1}{2} - \frac{\sqrt{3}}{2}i\right) = 0 + 0i = 0$$

Now compute the products:

$$z_1 z_2 = 1\left[\cos\frac{2\pi}{3} + i\sin\frac{2\pi}{3}\right] = -\frac{1}{2} + \frac{\sqrt{3}}{2}i \qquad \text{(Note this is } z_2)$$

$$z_2 z_3 = 1[\cos 2\pi + i\sin 2\pi] = 1 \qquad \text{(Note this is } z_1)$$

$$z_3 z_1 = 1\left[\cos\frac{4\pi}{3} + i\sin\frac{4\pi}{3}\right] = -\frac{1}{2} - \frac{\sqrt{3}}{2}i \qquad \text{(Note this is } z_3)$$

So $z_1 z_2 + z_2 z_3 + z_3 z_1 = z_1 + z_2 + z_3 = 0$.

71. Using trigonometric forms and DeMoivre's theorem, we compute the powers:

$$\left[\frac{-1+i\sqrt{3}}{2}\right]^5 = \left[-\frac{1}{2}+\frac{\sqrt{3}}{2}i\right]^5 = \left[\cos\frac{2\pi}{3}+i\sin\frac{2\pi}{3}\right]^5 = \cos\frac{10\pi}{3}+i\sin\frac{10\pi}{3} = -\frac{1}{2}-\frac{\sqrt{3}}{2}i$$

$$\left[\frac{-1-i\sqrt{3}}{2}\right]^5 = \left[-\frac{1}{2}-\frac{\sqrt{3}}{2}i\right]^5 = \left[\cos\frac{4\pi}{3}+i\sin\frac{4\pi}{3}\right]^5 = \cos\frac{20\pi}{3}+i\sin\frac{20\pi}{3} = -\frac{1}{2}+\frac{\sqrt{3}}{2}i$$

Now compute the sum:

$$\left[\frac{-1+i\sqrt{3}}{2}\right]^5 + \left[\frac{-1-i\sqrt{3}}{2}\right]^5 = \left(-\frac{1}{2}-\frac{\sqrt{3}}{2}i\right)+\left(-\frac{1}{2}+\frac{\sqrt{3}}{2}i\right) = -1$$

73. Since the multiplication results in a difference of squares, we have:

$$(\cos\theta+i\sin\theta)(\cos\theta-i\sin\theta) = \cos^2\theta - i^2\sin^2\theta$$
$$= \cos^2\theta+\sin^2\theta \quad (\text{since } i^2=-1)$$
$$= 1$$

75. Using the hint, we have:

$$\frac{r(\cos\alpha+i\sin\alpha)}{R(\cos\beta+i\sin\beta)} \cdot \frac{(\cos\beta-i\sin\beta)}{(\cos\beta-i\sin\beta)}$$

$$= \frac{r(\cos\alpha\cos\beta+i\sin\alpha\cos\beta-i\cos\alpha\sin\beta-i^2\sin\alpha\sin\beta)}{R(\cos^2\beta-i^2\sin^2\beta)}$$

$$= \frac{r[(\cos\alpha\cos\beta+\sin\alpha\sin\beta)+i(\sin\alpha\cos\beta-\cos\alpha\sin\beta)]}{R(\cos^2\beta+\sin^2\beta)}$$

Using the difference identities for sine and cosine:

$$\frac{r[\cos(\alpha-\beta)+i\sin(\alpha-\beta)]}{R} = \frac{r}{R}[\cos(\alpha-\beta)+i\sin(\alpha-\beta)]$$

77. Using the hint and the identities $\cos(-\theta) = \cos\theta$ and $\sin(-\theta) = -\sin\theta$, we have:

$$\frac{1}{z} = \frac{1(\cos 0+i\sin 0)}{r(\cos\theta+i\sin\theta)}$$

$$= \frac{1}{r}[\cos(0-\theta)+i\sin(0-\theta)] \quad (\text{by Exercise 75})$$

$$= \frac{1}{r}[\cos(-\theta)+i\sin(-\theta)]$$

$$= \frac{1}{r}(\cos\theta-i\sin\theta)$$

79. First convert the equation to a quadratic form:

$$w + \frac{1}{w} = 2\cos\theta$$

$$w^2 + 1 = (2\cos\theta)w$$

$$w^2 - (2\cos\theta)w + 1 = 0$$

Using the quadratic formula:

$$w = \frac{2\cos\theta \pm \sqrt{4\cos^2\theta - 4}}{2}$$

$$= \frac{2\cos\theta \pm \sqrt{-4(1 - \cos^2\theta)}}{2}$$

$$= \frac{2\cos\theta \pm i\sqrt{4\sin^2\theta}}{2}$$

$$= \cos\theta \pm i\sin\theta$$

Chapter Fourteen Review Exercises

1. When $n = 1$:

$$5 = \tfrac{5}{2}(1)(1 + 1) = 5$$

Assuming:

$$5 + 10 + \ldots + 5k = \tfrac{5}{2}k(k + 1)$$

Then:

$$5 + 10 + \ldots + 5k + 5(k + 1) = \tfrac{5}{2}k(k + 1) + 5(k + 1) = (k + 1)\left(\tfrac{5}{2}k + 5\right) = \tfrac{5}{2}(k + 1)(k + 2)$$

This completes the induction.

3. When $n = 1$:

$$1 \cdot 2 = 2 = \tfrac{1}{3}(1)(1 + 1)(1 + 2) = 2$$

Assuming:

$$1 \cdot 2 + 2 \cdot 3 + \ldots + k(k + 1) = \tfrac{1}{3}k(k + 1)(k + 2)$$

Then:

$$1 \cdot 2 + 2 \cdot 3 + \ldots + k(k + 1) + (k + 1)(k + 2)$$

$$= \tfrac{1}{3}k(k + 1)(k + 2) + (k + 1)(k + 2)$$

$$= \left(\tfrac{1}{3}k + 1\right)(k + 1)(k + 2)$$

$$= \tfrac{1}{3}(k + 1)(k + 2)(k + 3)$$

This completes the induction.

5. For $n = 1$:
$$1 = 3 + (2-3)2 = 3 - 2 = 1$$
Assuming:
$$1 + 3 \cdot 2 + 5 \cdot 2^2 + \ldots + (2k-1) \cdot 2^{k-1} = 3 + (2k-3) \cdot 2^k$$
Then:
$$1 + 3 \cdot 2 + \ldots + (2k-1)2^{k-1} + (2k+1)2^k$$
$$= 3 + (2k-3)2^k + (2k+1)2^k$$
$$= 3 + (4k-2)2^k$$
$$= 3 + (2k-1)2^{k+1}$$
$$= 3 + (2(k+1)-3) \cdot 2^{k+1}$$
The induction is complete.

7. For $n = 1$:
$$1 = \left(1^2 - 2 + 3\right)2 - 3 = 1$$
Assuming:
$$1 + 2^2 \cdot 2 + 3^2 \cdot 2^2 + 4^2 \cdot 2^3 + \ldots + k^2 \cdot 2^{k-1} = \left(k^2 - 2k + 3\right)2^k - 3$$
Then:
$$1 + 2^2 \cdot 2 + \ldots + (k+1)^2 2^k = \left(k^2 - 2k + 3\right)2^k - 3 + (k+1)^2 2^k$$
$$= 2^k\left[k^2 - 2k + 3 + (k+1)^2\right] - 3$$
$$= 2^k\left(2k^2 + 4\right) - 3$$
$$= 2^{k+1}\left(k^2 + 2\right) - 3$$
$$= 2^{k+1}\left[(k+1)^2 - 2k + 1\right] - 3$$
$$= 2^{k+1}\left[(k+1)^2 - 2(k+1) + 3\right] - 3$$
The induction is complete.

9. For $n = 1$:
$$7^1 - 1 = 6 = 3(2)$$
Assuming:
$$7^k - 1 = 3L$$
Then:
$$7^{k+1} - 1 = 3L + 7^{k+1} - 7^k = 3L + 7^k(7-1) = 3L + 6\left(7^k\right) = 3\left(L + 2 \cdot 7^k\right)$$
The induction is complete.

11. Expand using the binomial theorem:
$$\left(3a + b^2\right)^4 = \binom{4}{0}(3a)^4 + \binom{4}{1}(3a)^3(b^2) + \binom{4}{2}(3a)^2(b^2)^2 + \binom{4}{3}(3a)(b^2)^3 + \binom{4}{4}(b^2)^4$$
$$= 81a^4 + 108a^3b^2 + 54a^2b^4 + 12ab^6 + b^8$$

13. Expand using the binomial theorem:

$$(x+\sqrt{x})^4 = \binom{4}{0}x^4 + \binom{4}{1}x^3(\sqrt{x}) + \binom{4}{2}x^2(\sqrt{x})^2 + \binom{4}{3}x(\sqrt{x})^3 + \binom{4}{4}(\sqrt{x})^4$$
$$= x^4 + 4x^3\sqrt{x} + 6x^3 + 4x^2\sqrt{x} + x^2$$

15. Expand using the binomial theorem:

$$(x^2 - 2y^2)^5 = \binom{5}{0}(x^2)^5 - \binom{5}{1}(x^2)^4(2y^2) + \binom{5}{2}(x^2)^3(2y^2)^2 - \binom{5}{3}(x^2)^2(2y^2)^3$$
$$+ \binom{5}{4}(x^2)(2y^2)^4 - \binom{5}{5}(2y^2)^5$$
$$= x^{10} - 10x^8y^2 + 40x^6y^4 - 80x^4y^6 + 80x^2y^8 - 32y^{10}$$

17. Expand using the binomial theorem:

$$\left(1+\frac{1}{x}\right)^5 = \binom{5}{0} + \binom{5}{1}\frac{1}{x} + \binom{5}{2}\left(\frac{1}{x}\right)^2 + \binom{5}{3}\left(\frac{1}{x}\right)^3 + \binom{5}{4}\left(\frac{1}{x}\right)^4 + \binom{5}{5}\left(\frac{1}{x}\right)^5$$
$$= 1 + \frac{5}{x} + \frac{10}{x^2} + \frac{10}{x^3} + \frac{5}{x^4} + \frac{1}{x^5}$$

19. Expand using the binomial theorem:

$$(a\sqrt{b} - b\sqrt{a})^4 = a^2b^2(\sqrt{a} - \sqrt{b})^4$$
$$= a^2b^2\left[(\sqrt{a})^4 - 4(\sqrt{a})^3\sqrt{b} + 6(\sqrt{a})^2(\sqrt{b})^2 - 4(\sqrt{a})(\sqrt{b})^3 + (\sqrt{b})^4\right]$$
$$= a^2b^2(a^2 - 4a\sqrt{ab} + 6ab - 4b\sqrt{ab} + b^2)$$
$$= a^4b^2 - 4a^3b^2\sqrt{ab} + 6a^3b^3 - 4a^2b^3\sqrt{ab} + a^2b^4$$

21. The fifth term is given by:

$$\binom{5}{5-1}(3x)(y^2)^4 = 5(3x)y^8 = 15xy^8$$

23. Here $7 - r + 1 = 5$, so $r = 3$, and the coefficient of the third term is:

$$\binom{7}{2}(2)^2 = (4)\frac{7!}{5!2!} = 84$$

25. Here $r - 1 = 6$, so $r = 7$, and the coefficient of the seventh term is:

$$\binom{8}{6} = \frac{8!}{6!2!} = 28$$

27. Computing each side:

$$\binom{2}{0}^2 + \binom{2}{1}^2 + \binom{2}{2}^2 = (1)^2 + (2)^2 + (1)^2 = 6$$

$$\binom{4}{2} = \frac{4!}{2!2!} = 6$$

29. Computing each side:

$$\binom{4}{0}^2 + \binom{4}{1}^2 + \binom{4}{2}^2 + \binom{4}{3}^2 + \binom{4}{4}^2 = 1^2 + 4^2 + 6^2 + 4^2 + 1^2 = 70$$

$$\binom{8}{4} = \frac{8!}{4!4!} = 70$$

31. Computing values:

$$\binom{3}{0} + \binom{3}{1} + \binom{3}{2} + \binom{3}{3} = 1 + 3 + 3 + 1 = 8 = 2^3$$

33. Computing values, we have:

$$a_1 = \frac{2}{1+1} = 1, \quad a_2 = \frac{4}{2+1} = \frac{4}{3}, \quad a_3 = \frac{6}{3+1} = \frac{3}{2}, \quad a_4 = \frac{8}{4+1} = \frac{8}{5}$$

The first four terms are $1, \frac{4}{3}, \frac{3}{2}, \frac{8}{5}$.
Graph these points:

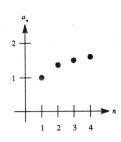

35. Computing values, we have:

$$a_1 = (-1)\left(1 - \tfrac{1}{2}\right) = -\tfrac{1}{2}$$

$$a_2 = (1)\left(1 - \tfrac{1}{3}\right) = \tfrac{2}{3}$$

$$a_3 = (-1)\left(1 - \tfrac{1}{4}\right) = -\tfrac{3}{4}$$

$$a_4 = (1)\left(1 - \tfrac{1}{5}\right) = \tfrac{4}{5}$$

The first four terms are $-\frac{1}{2}, \frac{2}{3}, -\frac{3}{4}, \frac{4}{5}$.

Graph these points:

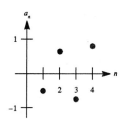

37. Computing values, we have:
$$a_0 = -3, \quad a_1 = -12, \quad a_2 = -48, \quad a_3 = -192$$
The first four terms are $-3, -12, -48, -192$.

39. (a) Expanding the sum, we have:
$$\sum_{k=1}^{3} (-1)^k (2k+1) = -3 + 5 - 7 = -5$$

  (b) Expanding the sum, we have:
$$\sum_{k=0}^{8} \left(\frac{1}{k+1} - \frac{1}{k+2} \right) = \left(1 - \tfrac{1}{2}\right) + \left(\tfrac{1}{2} - \tfrac{1}{3}\right) + \left(\tfrac{1}{3} - \tfrac{1}{4}\right) + \left(\tfrac{1}{4} - \tfrac{1}{5}\right) + \left(\tfrac{1}{5} - \tfrac{1}{6}\right)$$
$$+ \left(\tfrac{1}{6} - \tfrac{1}{7}\right) + \left(\tfrac{1}{7} - \tfrac{1}{8}\right) + \left(\tfrac{1}{8} - \tfrac{1}{9}\right) + \left(\tfrac{1}{9} - \tfrac{1}{10}\right)$$
$$= 1 - \tfrac{1}{10}$$
$$= \tfrac{9}{10}$$

41. The sum can be written as:
$$\frac{5}{3} + \frac{5}{3^2} + \frac{5}{3^3} + \frac{5}{3^4} + \frac{5}{3^5} = \sum_{k=1}^{5} \frac{5}{3k}$$

43. Here $a_n = 1 + 4n$, so $a_{18} = 1 + 4(18) = 73$.

45. Here $a_n = \dfrac{10}{2^{n-1}}$, so $a_{12} = \dfrac{10}{2^{11}} = \dfrac{5}{1024}$.

47. The sum is given by:
$$S_{12} = (12)\left(\frac{8 + \frac{43}{2}}{2} \right) = \tfrac{59}{2}(6) = 177$$

49. By inspection $S_1 = 7$, $S_2 = 77$, ... so $S_{10} = 7{,}777{,}777{,}777$.

51. Compute:
$$a_3 = 4$$
$$a_5 = 10$$
$$\frac{a_5}{a_3} = \frac{10}{4} = \frac{5}{2} = r^2 \text{ and } r = \pm\sqrt{\frac{5}{2}}$$

Since r is known to be negative, $r = -\sqrt{\frac{5}{2}}$ and $a_6 = -\sqrt{\frac{5}{2}}(10) = -5\sqrt{10}$.

53. The sum is given by:
$$S = \frac{\frac{3}{5}}{1 - \frac{1}{5}} = \frac{3}{4}$$

55. The sum is given by:
$$S = \frac{\frac{1}{9}}{1 + \frac{1}{9}} = \frac{1}{10}$$

57. We have $0.\overline{45} = 0.45 + 0.0045 + 0.000045 + \ldots$. This is a geometric sequence with $a = 0.45$, $r = \frac{1}{100}$, so:
$$0.\overline{45} = \frac{0.45}{1 - \frac{1}{100}} = \frac{45}{99} = \frac{15}{33} = \frac{5}{11}$$

59. The sum is given by:
$$S_n = \frac{n}{2}\left[2(1) + (n-1)^1\right]$$
$$= \frac{n}{2}(2 + n - 1)$$
$$= \frac{n}{2}(1 + n)$$
$$= \frac{n^2}{2} + \frac{n}{2}$$
$$= n + \frac{n^2}{2} - \frac{n}{2}$$
$$= n + \frac{n(n-1)}{2}$$

61. The sum is given by:

$$S_n = \frac{n}{2}[2(1) + (n-1)3]$$

$$= \frac{n}{2}(2 + 3n - 3)$$

$$= \frac{n}{2}(3n - 1)$$

$$= \frac{3n^2}{2} - \frac{n}{2}$$

$$= n + \frac{3n^2}{2} - \frac{3n}{2}$$

$$= n + \frac{3n(n-1)}{2}$$

63. (a) Using formula (1), we have:

$$1^2 + 2^2 + \ldots + 50^2 \approx \frac{\left(50 + \frac{1}{2}\right)^{2+1}}{2+1} = \frac{\left(\frac{101}{2}\right)^3}{3} = \frac{1030301}{24} \approx 42929$$

(b) The exact sum is given by:

$$\frac{50(51)(101)}{6} = 42925$$

The percent error is:

$$100 \cdot \frac{4}{42925} \approx 0.00932\%$$

(c) Using formula (1), we have:

$$1^4 + 2^4 + \ldots + 200^4 \approx \frac{(200.5)^5}{5} \approx 6.48040 \times 10^{10}$$

(d) Compute the sum:

$$1^4 + 2^4 + \ldots + 200^4 \approx \frac{200(201)(401)(120599)}{30} \approx 6.48027 \times 10^{10}$$

The percent error is:

$$100 \cdot \frac{0.00013 \times 10^{10}}{6.48027 \times 10^{10}} = 2 \times 10^{-3}\%$$

65. There are $4 \times 4 = 16$ different numbers (assumes repetition).

67. The permutations are:
{$abc, acb, bac, bca, cab, cba, abd, adb, bad, bda, dab, dba, acd, adc, cad, cda,$
$dac, dca, bcd, bdc, cbd, cdb, dbc, dcb$}
So $P(4, 3) = 4(3)(2) = 24$.

69. Compute $P(12, 8) = (12)(11)(10)(9)(8)(7)(6)(5) = 19,958,400$.

71. There are a total of 6! different arrangements, while $2(5!)$ have e, f next to one another, hence:
$$6! - 2(5!) = 4(5!) = 4(120) = 480$$

73. Compute:
$$C(9,2) \cdot C(10,2) = \frac{9!}{7!2!} \cdot \frac{10!}{8!2!} = (36)(45) = 1620 \text{ subcommittees}$$

75. Compute:
$$\tfrac{9}{12} \cdot \tfrac{8}{11} = \tfrac{72}{132} = \tfrac{6}{11}$$

77. (a) Using the multiplication principle, the probability is:
$$\tfrac{5}{12} \cdot \tfrac{4}{11} = \tfrac{5}{33}$$

   (b) Using the multiplication principle, the probability is:
$$\tfrac{7}{12} \cdot \tfrac{6}{11} = \tfrac{7}{22}$$

   (c) Since the two events are mutually exclusive, the probability is:
$$\tfrac{5}{33} + \tfrac{7}{22} - 0 = \tfrac{31}{66}$$

79. The possible die combinations are $(1, 2)$ and $(2, 1)$, so the probability is $\tfrac{2}{36} = \tfrac{1}{18}$.

81. The possible die combinations are $(1, 6)$, $(6, 1)$, $(2, 5)$, $(5, 2)$, $(3, 4)$ and $(4, 3)$, so the probability is $\tfrac{6}{36} = \tfrac{1}{6}$.

83. The possible die combinations are those from Exercise 81 together with $(5, 6)$ and $(6, 5)$, so the probability is $\tfrac{8}{36} = \tfrac{2}{9}$.

85. The possible die combinations are $(4, 6)$, $(6, 4)$, $(5, 5)$, $(5, 6)$, $(6, 5)$ and $(6, 6)$, so the probability is $\tfrac{6}{36} = \tfrac{1}{6}$.

87. There are 11 die combinations resulting in a sum of either 7 or 8, so there are $36 - 11 = 25$ die combinations resulting in a sum which is not 7 or 8. So the probability is $\tfrac{25}{36}$.

89. (a) The probability is $\tfrac{7}{15}$.

   (b) The probability is $\tfrac{8}{15}$.

   (c) The probability is $\tfrac{10}{15} = \tfrac{2}{3}$.

   (d) The probability is $\tfrac{10}{15} = \tfrac{2}{3}$.

91. (a) There are $9 \cdot 10 \cdot 10 = 900$ three-digit numbers (remember, the first digit cannot be 0), and there are $9 \cdot 9 \cdot 8 = 648$ three-digit numbers with no repetitions, so there are $900 - 648 = 252$ three-digit numbers containing one or more repeated digits.

    (b) The probability is $\frac{252}{900} = \frac{7}{25}$.

93. Call $b = ra$ and $c = r^2a$, so $a + ar + ar^2 = 70$. Since $4a$, $5b$, $4c$ are consecutive terms in an arithmetic sequence, their common difference is the same. So:
$$5b - 4a = 4c - 5b$$
$$10b = 4a + 4c$$
$$5b = 2a + 2c$$
$$5(ra) = 2a + 2(r^2a)$$
$$5ra = 2a + 2r^2a$$
Since a is nonzero, we divide by a:
$$5r = 2 + 2r^2$$
$$2r^2 - 5r + 2 = 0$$
$$(2r - 1)(r - 2) = 0$$
$$r = \tfrac{1}{2}, 2$$
If $r = \frac{1}{2}$, we find a:
$$a + a \bullet \tfrac{1}{2} + a \bullet \tfrac{1}{4} = 70$$
$$7a = 280$$
$$a = 40$$
So the terms are 40, 20, 10.
If $r = 2$, we find a:
$$a + 2a + 4a = 70$$
$$7a = 70$$
$$a = 10$$
So the terms are 10, 20, 40.

95. Assume:
$$\frac{1}{c+a} - \frac{1}{b+c} = \frac{1}{a+b} - \frac{1}{c+a}$$
Our goal is to prove that:
$$b^2 - a^2 = c^2 - b^2$$
Multiply the first equation by $(c + a)(b + c)(a + b)$:
$$(b+c)(a+b) - (c+a)(a+b) = (c+a)(b+c) - (b+c)(a+b)$$
$$\left(ab + ac + bc + b^2\right) - \left(ac + bc + ab + a^2\right) = \left(bc + ab + ac + c^2\right) - \left(ab + ac + bc + b^2\right)$$
$$b^2 - a^2 = c^2 - b^2$$
This proves the desired result.

97. (a) We must have $4 + x = r(3 + x)$ and $5 + x = r^2(3 + x)$, so $r = \frac{4+x}{3+x}$. Substituting into the second equality:

$$5 + x = \left(\frac{4+x}{3+x}\right)^2 (3 + x)$$

$$5 + x = \frac{(4+x)^2}{3+x}$$

Multiplying each side by $3 + x$:

$$15 + 8x + x^2 = 16 + 8x + x^2$$
$$15 = 16$$

So no such value for x exists.

(b) Here $b + x = r(a + x)$ and $c + x = r^2(a + x)$, so $r = \frac{b+x}{a+x}$. Substituting:

$$c + x = \left(\frac{b+x}{a+x}\right)^2 (a + x)$$

$$(c + x)(a + x) = (b + x)^2$$

$$ac + (a + c)x + x^2 = b^2 + 2bx + x^2$$

$$(a + c - 2b)x = b^2 - ac$$

$$x = \frac{b^2 - ac}{a + c - 2b}$$

(Note that $a + c - 2b = 0$ for (a), explaining why no x could be found.)

99. (a) Let $S = a_1 + ra_2 + r^2 a_3 + \ldots + r^{n-1}a_n$, so:

$$rS = ra_1 + r^2 a_2 + r^3 a_3 + \ldots + r^n a_n$$

Then:

$$S - rS = a_1 + r(a_2 - a_1) + r^2(a_3 - a_2) + \ldots + r^{n-1}(a_n - a_{n-1}) - r^n a_n$$

$$= a_1 + rd + r^2 d + \ldots + r^{n-1}d - r^n a_n$$

$$= a_1 + \frac{d(r - r^n)}{1 - r} - r^n a_n$$

(b) Continuing from (a), we have:

$$S(1 - r) = a_1 - r^n a_n + \frac{d(r - r^n)}{1 - r}$$

$$S = \frac{a_1 - r^n a_n}{1 - r} + \frac{d(r - r^n)}{(1 - r)^2}$$

101. Convert to rectangular form:

$$3\left[\cos\tfrac{\pi}{3} + i\sin\tfrac{\pi}{3}\right] = 3\left(\tfrac{1}{2} + \tfrac{\sqrt{3}}{2}i\right) = \tfrac{3}{2} + \tfrac{3\sqrt{3}}{2}i$$

103. Convert to rectangular form:

$$2^{1/4}\left[\cos\tfrac{7\pi}{4} + i\sin\tfrac{7\pi}{4}\right] = 2^{1/4}\left(\tfrac{\sqrt{2}}{2} - \tfrac{\sqrt{2}}{2}i\right) = 2^{-1/4} - 2^{-1/4}i$$

105. Convert to trigonometric form:

$$\tfrac{1}{2} + \tfrac{\sqrt{3}}{2}i = 1\left[\cos\tfrac{\pi}{3} + i\sin\tfrac{\pi}{3}\right]$$

107. Convert to trigonometric form:

$$-3\sqrt{2} - 3\sqrt{2}i = 6\left(-\tfrac{\sqrt{2}}{2} - \tfrac{\sqrt{2}}{2}i\right) = 6\left[\cos\tfrac{5\pi}{4} + i\sin\tfrac{5\pi}{4}\right]$$

109. Compute the product:

$$5\left[\cos\tfrac{\pi}{7} + i\sin\tfrac{\pi}{7}\right] \cdot 2\left[\cos\tfrac{3\pi}{28} + i\sin\tfrac{3\pi}{28}\right] = 10\left[\cos\tfrac{7\pi}{28} + i\sin\tfrac{7\pi}{28}\right]$$
$$= 10\left[\cos\tfrac{\pi}{4} + i\sin\tfrac{\pi}{4}\right]$$
$$= 10\left(\tfrac{\sqrt{2}}{2} + \tfrac{\sqrt{2}}{2}i\right)$$
$$= 5\sqrt{2} + 5\sqrt{2}i$$

111. Compute the quotient:

$$8\left[\cos\tfrac{\pi}{12} + i\sin\tfrac{\pi}{12}\right] \div 4\left[\cos\tfrac{\pi}{3} + i\sin\tfrac{\pi}{3}\right] = 2\left[\cos\left(-\tfrac{3\pi}{12}\right) + i\sin\left(-\tfrac{3\pi}{12}\right)\right]$$
$$= 2\left[\cos\left(-\tfrac{\pi}{4}\right) + i\sin\left(-\tfrac{\pi}{4}\right)\right]$$
$$= 2\left(\tfrac{\sqrt{2}}{2} - \tfrac{\sqrt{2}}{2}i\right)$$
$$= \sqrt{2} - \sqrt{2}i$$

113. Compute the power using DeMoivre's theorem:

$$\left[3^{1/4}\left(\cos\tfrac{\pi}{36} + i\sin\tfrac{\pi}{36}\right)\right]^{12} = 3^3\left[\cos\tfrac{12\pi}{36} + i\sin\tfrac{12\pi}{36}\right]$$
$$= 27\left[\cos\tfrac{\pi}{3} + i\sin\tfrac{\pi}{3}\right]$$
$$= 27\left(\tfrac{1}{2} + \tfrac{\sqrt{3}}{2}i\right)$$
$$= \tfrac{27}{2} + \tfrac{27\sqrt{3}}{2}i$$

115. First write $\sqrt{3} + i$ in trigonometric form:

$$\sqrt{3} + i = 2\left(\tfrac{\sqrt{3}}{2} + \tfrac{1}{2}i\right) = 2\left[\cos\tfrac{\pi}{6} + i\sin\tfrac{\pi}{6}\right]$$

Now compute the power using DeMoivre's theorem:

$$\left(\sqrt{3} + i\right)^{10} = \left[2\left(\cos\tfrac{\pi}{6} + i\sin\tfrac{\pi}{6}\right)\right]^{10}$$
$$= 2^{10}\left[\cos\tfrac{5\pi}{3} + i\sin\tfrac{5\pi}{3}\right]$$
$$= 2^{10}\left(\tfrac{1}{2} - \tfrac{\sqrt{3}}{2}i\right)$$
$$= 2^9 - 2^9\sqrt{3}i$$
$$= 512 - 512\sqrt{3}i$$

117. Let $z = r(\cos\theta + i\sin\theta)$ and $\quad 1 = 1[\cos 0 + i\sin 0]$, so:
$$z^6 = r^6(\cos 6\theta + i\sin 6\theta) = 1(\cos 0 + i\sin 0)$$

Thus $r = 1$ and $6\theta = 0 + 2\pi k$, so $\theta = 0 + \frac{\pi}{3}k$.

When $k = 0$, we have:
$$z_1 = 1[\cos 0 + i\sin 0] = 1(1 + 0i) = 1$$

When $k = 1$, we have:
$$z_2 = 1\left[\cos\frac{\pi}{3} + i\sin\frac{\pi}{3}\right] = \frac{1}{2} + \frac{\sqrt{3}}{2}i$$

When $k = 2$, we have:
$$z_3 = 1\left[\cos\frac{2\pi}{3} + i\sin\frac{2\pi}{3}\right] = -\frac{1}{2} + \frac{\sqrt{3}}{2}i$$

When $k = 3$, we have:
$$z_4 = 1[\cos\pi + i\sin\pi] = -1 + 0i = -1$$

When $k = 4$, we have:
$$z_5 = 1\left[\cos\frac{4\pi}{3} + i\sin\frac{4\pi}{3}\right] = -\frac{1}{2} - \frac{\sqrt{3}}{2}i$$

When $k = 5$, we have:
$$z_6 = 1\left[\cos\frac{5\pi}{3} + i\sin\frac{5\pi}{3}\right] = \frac{1}{2} - \frac{\sqrt{3}}{2}i$$

So the sixth roots of 1 are $1, \frac{1}{2} + \frac{\sqrt{3}}{2}i, -\frac{1}{2} + \frac{\sqrt{3}}{2}i, -1, -\frac{1}{2} - \frac{\sqrt{3}}{2}i$ and $\frac{1}{2} - \frac{\sqrt{3}}{2}i$.

119. Let $z = r(\cos\theta + i\sin\theta)$ and write $\sqrt{2} - \sqrt{2}i$ in trigonometric form:
$$\sqrt{2} - \sqrt{2}i = 2\left(\frac{\sqrt{2}}{2} - \frac{\sqrt{2}}{2}i\right) = 2\left[\cos\frac{7\pi}{4} + i\sin\frac{7\pi}{4}\right]$$

Then $z^2 = r^2(\cos 2\theta + i\sin 2\theta) = 2\left[\cos\frac{7\pi}{4} + i\sin\frac{7\pi}{4}\right]$.

So $r^2 = 2$ and $r = \sqrt{2}$, and also $2\theta = \frac{7\pi}{4} + 2\pi k$, so $\theta = \frac{7\pi}{8} + \pi k$.

When $k = 0$, we have:
$$z_1 = \sqrt{2}\left[\cos\frac{7\pi}{8} + i\sin\frac{7\pi}{8}\right]$$
$$= \sqrt{2}\left[-\frac{\sqrt{2+\sqrt{2}}}{2} + i\frac{\sqrt{2-\sqrt{2}}}{2}\right]$$
(by the half - angle identities)
$$= -\frac{\sqrt{4+2\sqrt{2}}}{2} + \frac{\sqrt{4-2\sqrt{2}}}{2}i$$

When $k = 1$, we have:
$$z_2 = \sqrt{2}\left[\cos\frac{15\pi}{8} + i\sin\frac{15\pi}{8}\right] = \sqrt{2}\left[\frac{\sqrt{2+\sqrt{2}}}{2} - i\frac{\sqrt{2-\sqrt{2}}}{2}\right] = \frac{\sqrt{4+2\sqrt{2}}}{2} - \frac{\sqrt{4-2\sqrt{2}}}{2}i$$

So the square roots of $\sqrt{2} - \sqrt{2}i$ are $-\frac{\sqrt{4+2\sqrt{2}}}{2} + \frac{\sqrt{4-2\sqrt{2}}}{2}i$ and $\frac{\sqrt{4+2\sqrt{2}}}{2} - \frac{\sqrt{4-2\sqrt{2}}}{2}i$.

121. Let $z = r(\cos\theta + i\sin\theta)$, and write $1 + i$ in trigonometric form:

$$1 + i = \sqrt{2}\left(\frac{\sqrt{2}}{2} + \frac{\sqrt{2}}{2}i\right) = \sqrt{2}\left[\cos\frac{\pi}{4} + i\sin\frac{\pi}{4}\right]$$

Then $z^5 = r^5[\cos 5\theta + i\sin 5\theta] = \sqrt{2}[\cos 45° + i\sin 45°]$.

So $r^5 = \sqrt{2}$ and $r = 2^{1/10} \approx 1.0718$, and $5\theta = 45° + 360°k$, so $\theta = 9° + 72°k$.

When $k = 0$: $z_1 = 1.0718[\cos 9° + i\sin 9°] \approx 1.06 + 0.17i$

When $k = 1$: $z_2 = 1.0718[\cos 81° + i\sin 81°] \approx 0.17 + 1.06i$

When $k = 2$: $z_3 = 1.0718[\cos 153° + i\sin 153°] \approx -0.95 + 0.49i$

When $k = 3$: $z_4 = 1.0718[\cos 225° + i\sin 225°] \approx -0.76 - 0.76i$

When $k = 4$: $z_5 = 1.0718[\cos 297° + i\sin 297°] \approx 0.49 - 0.95i$

So the fifth roots of $1 + i$ are $1.06 + 0.17i$, $0.17 + 1.06i$, $-0.95 + 0.49i$, $-0.76 - 0.76i$ and $0.49 - 0.95i$.

Chapter Fourteen Test

1. For $n = 1$, we have:

$$1^2 = \frac{1(1+1)(2+1)}{6} = \frac{6}{6} = 1$$

Assume P_k is true, so:

$$1^2 + 2^2 + 3^2 + \ldots + k^2 = \frac{k(k+1)(2k+1)}{6}$$

Then:

$$1^2 + 2^2 + \ldots + k^2 + (k+1)^2 = \frac{k(k+1)(2k+1)}{6} + (k+1)^2$$

$$= (k+1)\left[\frac{k(2k+1)}{6} + k + 1\right]$$

$$= (k+1)\left(\frac{2k^2 + k + 6k + 6}{6}\right)$$

$$= \frac{(k+1)(2k^2 + 7k + 6)}{6}$$

$$= \frac{(k+1)(k+2)(2k+3)}{6}$$

$$= \frac{(k+1)(k+2)[2(k+1)+1]}{6}$$

So P_{k+1} is true and the induction is complete.

2. (a) Expanding the sum:

$$\sum_{k=0}^{2}(10k-1) = -1+9+19 = 27$$

(b) Expanding the sum:

$$\sum_{k=1}^{3}(-1)^k k^2 = -1+4-9 = -6$$

3. (a) $S_n = \dfrac{a(1-r^n)}{1-r}$

(b) Here $a = \frac{3}{2}$, $r = \frac{3}{2}$, and $n = 10$, so:

$$S_{10} = \frac{\frac{3}{2}\left[1-\left(\frac{3}{2}\right)^{10}\right]}{1-\frac{3}{2}} = \frac{\frac{3}{2}\left(1-\frac{59049}{1024}\right)}{-\frac{1}{2}} = -3\left(-\frac{58025}{1024}\right) = \frac{174075}{1024}$$

4. (a) Here $11 - r + 1 = 3$, so $r = 9$, and the coefficient is:

$$\binom{11}{8}(-2)^8 = \frac{(11)(10)(9)}{(3)(2)}(256) = 42240$$

(b) The fifth term is:

$$\binom{11}{4}(a)^7(-2b^3)^4 = \frac{(11)(10)(9)(8)}{(4)(3)(2)}a^7(16b^{12}) = 5280a^7b^{12}$$

5. Expand using the binomial theorem:

$$(3x^2+y^3)^5 = \binom{5}{0}(3x^2)^5 + \binom{5}{1}(3x^2)^4(y^3) + \binom{5}{2}(3x^2)^3(y^3)^2 + \binom{5}{3}(3x^2)^2(y^3)^3$$
$$+ \binom{5}{4}(3x^2)(y^3)^4 + \binom{5}{5}(y^3)^5$$
$$= 243x^{10} + 405x^8y^3 + 270x^6y^6 + 90x^4y^9 + 15x^2y^{12} + y^{15}$$

6. The sum is:

$$S_{12} = 12\left(\frac{8+\frac{43}{2}}{2}\right) = \frac{59}{2}(6) = 177$$

7. The sum is:

$$S = \frac{a}{1-r} = \frac{\frac{7}{10}}{1-\frac{1}{10}} = \frac{7}{9}$$

8.   Here $a_1 = 1$ and $a_2 = 1$, so:
$$a_3 = a_2^2 + a_1 = 1 + 1 = 2$$
$$a_4 = a_3^2 + a_2 = 2^2 + 1 = 5$$
$$a_5 = a_4^2 + a_3 = 5^2 + 2 = 27$$
The fourth and fifth terms are 5 and 27, respectively.

9.   Since $a_3 = 4$ and $a_5 = 10$, $a_5 = a_3 r^2$, so:
$$10 = 4r^2$$
$$\tfrac{5}{2} = r^2$$
$$r = -\sqrt{\tfrac{5}{2}} = -\tfrac{\sqrt{10}}{2}$$
Then $a_6 = ra_5 = -\tfrac{\sqrt{10}}{2}(10) = -5\sqrt{10}$.

10.   Here $a_n = -61 + 15(n-1)$, so:
$$a_{20} = -61 + 15(19) = 224$$

11.   Convert to rectangular form:
$$z = 2\left(\cos\tfrac{2\pi}{3} + i\sin\tfrac{2\pi}{3}\right) = 2\left(-\tfrac{1}{2} + \tfrac{\sqrt{3}}{2}i\right) = -1 + \sqrt{3}i$$

12.   Convert to trigonometric form:
$$\sqrt{2} - \sqrt{2}i = 2\left(\tfrac{\sqrt{2}}{2} - \tfrac{\sqrt{2}}{2}i\right) = 2\left(\cos\tfrac{7\pi}{4} + i\sin\tfrac{7\pi}{4}\right)$$

13.   Compute the product:
$$zw = 3\left(\cos\tfrac{2\pi}{9} + i\sin\tfrac{2\pi}{9}\right) \bullet 5\left(\cos\tfrac{\pi}{9} + i\sin\tfrac{\pi}{9}\right)$$
$$= 15\left(\cos\tfrac{\pi}{3} + i\sin\tfrac{\pi}{3}\right)$$
$$= 15\left(\tfrac{1}{2} + \tfrac{\sqrt{3}}{2}i\right)$$
$$= \tfrac{15}{2} + \tfrac{15\sqrt{3}}{2}i$$

14.   Let $z = r(\cos\theta + i\sin\theta)$ be a cube root of $64i$. Write $64i$ in trigonometric form:
$$64i = 64(0 + 1i) = 64\left(\cos\tfrac{\pi}{2} + i\sin\tfrac{\pi}{2}\right)$$
Thus:
$$z^3 = r^3(\cos 3\theta + i\sin 3\theta) = 64\left(\cos\tfrac{\pi}{2} + i\sin\tfrac{\pi}{2}\right)$$
So $r^3 = 64$ and thus $r = 4$, and $3\theta = \tfrac{\pi}{2} + 2\pi k$ thus $\theta = \tfrac{\pi}{6} + \tfrac{2\pi}{3}k$.
When $k = 0$, we have:
$$z_1 = 4\left(\cos\tfrac{\pi}{6} + i\sin\tfrac{\pi}{6}\right) = 4\left(\tfrac{\sqrt{3}}{2} + \tfrac{1}{2}i\right) = 2\sqrt{3} + 2i$$
When $k = 1$, we have:
$$z_2 = 4\left(\cos\tfrac{5\pi}{6} + i\sin\tfrac{5\pi}{6}\right) = 4\left(-\tfrac{\sqrt{3}}{2} + \tfrac{1}{2}i\right) = -2\sqrt{3} + 2i$$

When $k = 2$, we have:
$$z_3 = 4\left(\cos\tfrac{3\pi}{2} + i\sin\tfrac{3\pi}{2}\right) = 4(0 - i) = -4i$$
So the cube roots of $64i$ are $2\sqrt{3} + 2i$, $-2\sqrt{3} + 2i$, and $-4i$.

15. Using the multiplication principle, the total number of area codes is $8 \cdot 2 \cdot 9 = 144$. Since there were 5 area codes not yet assigned, the amount of area codes in use in 1991 was 139.

16. There will be $P(14,8) = 14 \cdot 13 \cdot 12 \cdot 11 \cdot 10 \cdot 9 \cdot 8 \cdot 7 = 121{,}080{,}960$ possible arrangements.

17. Using the multiplication principle, the probability is:
$$\tfrac{12}{15} \cdot \tfrac{11}{14} = \tfrac{22}{35}$$

18. (a) Using the multiplication principle, the probability is:
$$\tfrac{3}{10} \cdot \tfrac{2}{9} = \tfrac{1}{15}$$

(b) Using the multiplication principle, the probability is:
$$\tfrac{7}{10} \cdot \tfrac{6}{9} = \tfrac{7}{15}$$

(c) Since the two events are mutually exclusive, the probability is:
$$\tfrac{1}{15} + \tfrac{7}{15} - 0 = \tfrac{8}{15}$$

Appendix

A.1 Using a Graphing Utility

1.   Sketch the graph in the standard viewing rectangle:

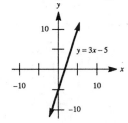

3.   Sketch the graph in the standard viewing rectangle:

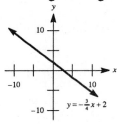

5.   Sketch the graph in the standard viewing rectangle:

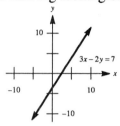

$$3x - 2y = 7$$

7.   Sketch the graph in the standard viewing rectangle:

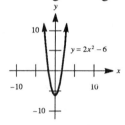

$$y = 2x^2 - 6$$

9.   Sketch the graph in the standard viewing rectangle:

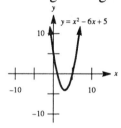

$$y = x^2 - 6x + 5$$

11.   Sketch the graph in the standard viewing rectangle:

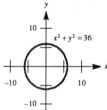

$$x^2 + y^2 = 36$$

13.   Sketch the graph in the standard viewing rectangle:

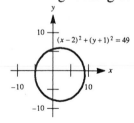

$$(x - 2)^2 + (y + 1)^2 = 49$$

15. Graphing the equation using modified range settings:

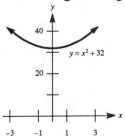

17. Graphing the equation using modified range settings:

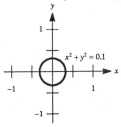

19. Graphing the equation using modified range settings:

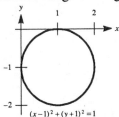

21. Graphing the equation using modified range settings:

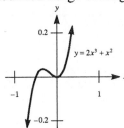

23. Graphing the curve $y = x^2 + 2x - 4$:

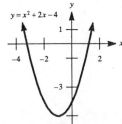

The x-intercepts are $x \approx -3.236$ and 1.236, which are the solutions to the equation.

25. Graphing the curve $y = x^3 - 4x^2 - 5x$:

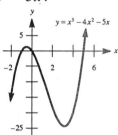

The x-intercepts are $x = -1, 0$, and 5, which are the solutions to the equation.

27. Graphing the curve $y = x^3 + 5x^2 + 5x + 1$:

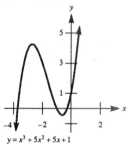

The x-intercepts are $x \approx -3.732$, -1 (exact), and -0.268, which are the solutions to the equation.

29. Graphing the curve $y = x^4 + 3x^3 - 5x^2 - 9x + 10$:

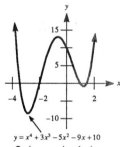

The x-intercepts are $x \approx -3.449$, -2 (exact), 1 (exact), and 1.449, which are the solutions to the equation.

31. Graphing the curve:

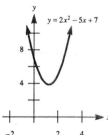

The vertex of the parabola is approximately $(1.25, 3.88)$.

33. Graphing the curve:

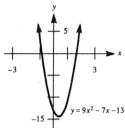

The vertex of the parabola is approximately (0.39, –14.36).

35. Graphing the curve:

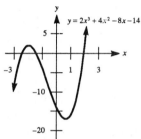

The high point is (–2, 2) and the low point is approximately (0.7, –17.0).

37. Graphing the curve:

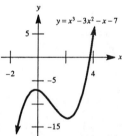

The high point is approximately (–0.2, –6.9) and the low point is approximately (2.2, –13.1).

39. Graphing the curve:

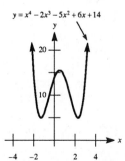

The high point is approximately (0.5, 15.6) and the low points are approximately (–1.3, 5.0) and (2.3, 5.0).

A.2 Notation and Language

1.   The expression is equivalent to $x^5 y^2$.

3.   The expression is equivalent to $(x+1)^3$.

5.   (a)   The expression is equivalent to $4x$.

    (b)   The expression is equivalent to $3(x^2+1)$, or $3x^2+3$.

7.   (a)   The expression is equivalent to $3a+2b$.

    (b)   The expression is equivalent to $3a^2+2b^2$.

9.   (a)   The expression is equivalent to $(2a+1)^3(2b+1)^2$.

    (b)   Since $1+2a=2a+1$ and $1+2b=2b+1$, the expression is equivalent to
$(2a+1)^3(2b+1)^2$.

11.   Substituting $x=4$ and $y=6$, we have:
$$x+3y=4+3(6)=4+18=22$$

13.   Substituting $a=3$ and $b=4$, we have:
$$a^2+b^2=(3)^2+(4)^2=9+16=25$$

15.   Substituting $x=10$ and $y=4$, we have:
$$x^2-4y^2=(10)^2-4(4)^2=100-64=36$$

17.   Substituting $x=5$, we have:
$$x^2-x+1=(5)^2-5+1=25-5+1=21$$

19.   Substituting $x=3$ and $y=2$, we have:
$$2x^3-3y^2=2(3)^3-3(2)^2=54-12=42$$

21.   Substituting $a=1$ and $b=1$, we have:
$$1\div a^2+b^2=1\div 1^2+1^2=1\div 1+1=1+1=2$$

23.   Substituting $a=\tfrac{1}{2}$ and $b=\tfrac{1}{2}$, we have:
$$\frac{1}{a^2}+\frac{1}{b^2}-\frac{1}{a^2+b^2}=\frac{1}{\left(\tfrac{1}{2}\right)^2}+\frac{1}{\left(\tfrac{1}{2}\right)^2}-\frac{1}{\left(\tfrac{1}{2}\right)^2+\left(\tfrac{1}{2}\right)^2}=\frac{1}{\tfrac{1}{4}}+\frac{1}{\tfrac{1}{4}}-\frac{1}{\tfrac{1}{4}+\tfrac{1}{4}}=4+4-2=6$$

25.   (a)   Substituting $A=2$, $B=2$ and $C=3$, we have:
$$A^{B^C}=2^{2^3}=2^8=256$$

    (b)   Substituting $A=2$, $B=3$ and $C=2$, we have:
$$A^{B^C}=2^{3^2}=2^9=512$$

(c) Substituting $A = 3$, $B = 2$ and $C = 2$, we have:
$$A^{B^C} = 3^{2^2} = 3^4 = 81$$

27. Compute each expression:
$$\left(2^3\right)^4 = 8^4 = 4096$$
$$2^{3^4} = 2^{81} \approx 2.4 \times 10^{24}$$
So 2^{3^4} is much larger than $\left(2^3\right)^4$.

29. (a) Substituting $A = 5$, $B = 2$ and $C = 3$, we have:
$$A^{B^C} = 5^{2^3} = 5^8 = 390,625$$

(b) Substituting $A = 5$, $B = 3$ and $C = 2$, we have:
$$A^{B^C} = 5^{3^2} = 5^9 = 1,953,125$$

31. (a) Since $2^5 = 32$ and $2^2 = 4$, the first quantity 2^{2^5} should be larger since it has a much larger exponent.

(b) Compute each quantity:
$$2^{2^5} = 2^{32} = 4,294,967,296$$
$$5^{2^2} = 5^4 = 625$$
This confirms our conjecture from part (a).

33. "Two more than x" would translate to $x + 2$.

35. "Four times the sum of a and b^2" would translate to $4\left(a + b^2\right)$.

37. "The sum of x^2 and y^2" would translate to $x^2 + y^2$.

39. "The sum of x and twice the square of x" would translate to $x + 2x^2$.

41. "The average of x, y, and z" would translate to $\dfrac{x + y + z}{3}$.

43. "The average of the squares of x, y, and z" would translate to $\dfrac{x^2 + y^2 + z^2}{3}$.

45. "One less than twice the product of x and y" would translate to $2xy - 1$.

47. (a) (i) Compute each quantity:
$$6^7 = 279,936 \text{ while } 7^6 = 117,649$$
So 6^7 is larger.

(ii)   Compute each quantity:
$$7^8 = 5{,}764{,}801 \text{ while } 8^7 = 2{,}097{,}152$$
So 7^8 is larger.

(iii)   Compute each quantity:
$$9^8 = 43{,}046{,}721 \text{ while } 8^9 = 134{,}217{,}728$$
So 8^9 is larger.

(b)   There are two such numbers. If $a = 1$, we have:
$$(a+1)^a = (1+1)^1 = 2^1 = 2$$
$$a^{a+1} = 1^{1+1} = 1^2 = 1$$
So $(a+1)^a > a^{a+1}$. If $a = 2$, we have:
$$(a+1)^a = (2+1)^2 = 3^2 = 9$$
$$a^{a+1} = 2^{2+1} = 2^3 = 8$$
So $(a+1)^a > a^{a+1}$. The two positive whole numbers are $a = 1$ and $a = 2$.

A.3 Properties of Real Numbers

1. $8(100) + 8(4) = 800 + 32 = 832$

3. $45(100) + 45(-2) = 4500 - 90 = 4410$

5.   (a)   Using the distributive property, we have:
$$(A+B)(A+B) = A(A+B) + B(A+B) = A^2 + AB + BA + B^2 = A^2 + 2AB + B^2$$

(b)   Using the distributive property, we have:
$$(A+B)(A-B) = A(A-B) + B(A-B) = A^2 - AB + BA - B^2 = A^2 - B^2$$

(c)   Using the distributive property, we have:
$$(A-B)(A-B) = A(A-B) - B(A-B) = A^2 - AB - BA + B^2 = A^2 - 2AB + B^2$$

7.   (a)   Using the distributive property, we have:
$$\begin{aligned}(x+2y)(x+2y) &= x(x+2y) + 2y(x+2y) \\ &= x^2 + 2xy + 2yx + 4y^2 \\ &= x^2 + 4xy + 4y^2\end{aligned}$$

(b)   Using the distributive property, we have:
$$(x+2y)(x-2y) = x(x-2y) + 2y(x-2y) = x^2 - 2xy + 2yx - 4y^2 = x^2 - 4y^2$$

(c)   Using the distributive property, we have:
$$\begin{aligned}(x-2y)(x-2y) &= x(x-2y) - 2y(x-2y) \\ &= x^2 - 2xy - 2yx + 4y^2 \\ &= x^2 - 4xy + 4y^2\end{aligned}$$

9. (a) Using the distributive property, we have:
$$A(x^2 - x + 1) = Ax^2 - Ax + A$$

(b) Using the distributive property, we have:
$$(x+1)(x^2 - x + 1) = x(x^2 - x + 1) + 1(x^2 - x + 1)$$
$$= x^3 - x^2 + x + x^2 - x + 1$$
$$= x^3 + 1$$

11. commutative property of addition

13. additive inverse property

15. associative property of addition

17. associative property of multiplication

19. identity property of addition

21. closure property with respect to addition

23. distributive property

25. If $b = c = 1$, then:
$$\frac{1}{b} + \frac{1}{c} = \tfrac{1}{1} + \tfrac{1}{1} = 1 + 1 = 2$$
$$\frac{1}{b+c} = \frac{1}{1+1} = \tfrac{1}{2}$$
So $\dfrac{1}{b} + \dfrac{1}{c} \neq \dfrac{1}{b+c}$.

27. If $x = 1$ and $y = 2$, then:
$$3(x + y) = 3(1 + 2) = 3(3) = 9$$
$$3x + y = 3(1) + 2 = 3 + 2 = 5$$
So $3(x + y) \neq 3x + y$.

29. If $x = 2$ and $y = 1$, then:
$$x^2(x + y) = 2^2(2 + 1) = 4(3) = 12$$
$$x^3 + y = 2^3 + 1 = 8 + 1 = 9$$
So $x^2(x + y) \neq x^3 + y$.

31. If $x = 2$ and $y = 3$, then:
$$3(xy) = 3(2 \cdot 3) = 3(6) = 18$$
$$(3x)(3y) = (3 \cdot 2)(3 \cdot 3) = (6)(9) = 54$$
So $3(xy) \neq (3x)(3y)$.

33.   If $a = 5$, $b = 3$ and $c = 1$, then:
$$a - (b - c) = 5 - (3 - 1) = 5 - 2 = 3$$
$$a - b - c = 5 - 3 - 1 = 1$$
So $a - (b - c) \neq a - b - c$.

35. $1 + 2(-3)(4) = 1 - 24 = -23$

37. $4 - (-3 + 5) = 4 - (2) = 2$

39. $-\left[(-1)^4\right] = -\left[(-1)(-1)(-1)(-1)\right] = -(1) = -1$

41. $1 - \left\{1 - \left[-(-1 - 1)\right]\right\} = 1 - \left\{1 - \left[-(-2)\right]\right\} = 1 - \left\{1 - 2\right\} = 1 - (-1) = 2$

43.   Obtaining common denominators, we have:
$$\frac{7}{5} + \frac{2}{3} = \frac{21 + 10}{15} = \frac{31}{15}$$

45.   Subtracting the two fractions (they already have a common denominator) yields:
$$\frac{7}{x} - \frac{12}{x} = \frac{7 - 12}{x} = -\frac{5}{x}$$

47.   Subtracting the two fractions yields:
$$\frac{x^2 + y^2}{x + y} - \frac{x^2 - y^2}{x + y} = \frac{\left(x^2 + y^2\right) - \left(x^2 - y^2\right)}{x + y} = \frac{x^2 + y^2 - x^2 + y^2}{x + y} = \frac{2y^2}{x + y}$$

49.   Multiplying the two fractions yields:
$$\frac{7}{12} \cdot \left(\frac{-7}{3}\right) = \frac{-49}{36} = -\frac{49}{36}$$

51.   Dividing the two fractions yields:
$$\frac{2}{5} \div \frac{7}{3} = \frac{2}{5} \cdot \frac{3}{7} = \frac{6}{35}$$

53.   By first simplifying the parentheses, we have:
$$1 \div \left(\frac{2}{3} + 1\right) = 1 \div \left(\frac{2}{3} + \frac{3}{3}\right) = 1 \div \frac{5}{3} = 1 \cdot \frac{3}{5} = \frac{3}{5}$$

55.   (a)   Multiplying both numerator and denominator by 12 (the least common multiple) yields:
$$\frac{\frac{4}{3} + \frac{1}{2}}{\frac{11}{6} + \frac{5}{4}} = \frac{12\left(\frac{4}{3} + \frac{1}{2}\right)}{12\left(\frac{11}{6} + \frac{5}{4}\right)} = \frac{16 + 6}{22 + 15} = \frac{22}{37}$$

(b) Since $\dfrac{a-b}{b-a} = -1$, we have:

$$\dfrac{\frac{4}{3}-\frac{1}{2}}{\frac{1}{2}-\frac{4}{3}} = -1$$

57. Since $\dfrac{a-b}{b-a} = -1$, we have:

$$\dfrac{1-(x+y)}{x+y-1} = \dfrac{1-(x+y)}{(x+y)-1} = -1$$

59. Multiplying both numerator and denominator by 12 (the least common multiple) yields:

$$\dfrac{\frac{1}{2}-\frac{1}{3}+\frac{1}{4}}{\frac{1}{2}+\frac{1}{3}-\frac{1}{4}} = \dfrac{12\left(\frac{1}{2}-\frac{1}{3}+\frac{1}{4}\right)}{12\left(\frac{1}{2}+\frac{1}{3}-\frac{1}{4}\right)} = \dfrac{6-4+3}{6+4-3} = \frac{5}{7}$$

61. Substituting $x = -2$, we have:

$$(-2)^2 - 8(-2) - 4 = 4 + 16 - 4 = 16$$

63. Substituting $x = -\frac{1}{2}$, we have:

$$8\left(-\tfrac{1}{2}\right)^3 - 4\left(-\tfrac{1}{2}\right)^2 - 4\left(-\tfrac{1}{2}\right) = 8\left(-\tfrac{1}{8}\right) - 4\left(\tfrac{1}{4}\right) + 2 = -1 - 1 + 2 = 0$$

65. Substituting $x = -1$, we have:

$$1 - (-1) - (-1)^2 - (-1)^3 - (-1)^4 = 1 + 1 - 1 + 1 - 1 = 1$$

67. Substituting $x_2 = -1$, $x_1 = -2$, $y_2 = 6$ and $y_1 = -6$, we have:

$$\begin{aligned}
d^2 &= \left(x_2 - x_1\right)^2 + \left(y_2 - y_1\right)^2 \\
&= \left[-1 - (-2)\right]^2 + \left[6 - (-6)\right]^2 \\
&= (-1 + 2)^2 + (6 + 6)^2 \\
&= 1 + 144 \\
&= 145
\end{aligned}$$